THE OXFORD HAND:

ZOOARCHAEOLOGY

Umberto Albarella is Professor in Zooarchaeology at the University of Sheffield. He obtained his PhD from the University of Durham, having first become interested in anthropology and then archaeology as an undergraduate student, and worked at the Universities of Lecce, Birmingham, and Durham before moving to the University of Sheffield in 2004. Specializing in the study of animal bones from archaeological sites, his main areas of research are wide-ranging and include animal domestication and husbandry intensification, ethnoarchaeology, the ritual use of animals, husbandry evidence of Romanization, animals and medieval life, integration in archaeology, and archaeology and politics. He is widely published in these fields and has previously served as Secretary of the International Council of Archaeozoology (ICAZ) from 2006 until 2012.

Mauro Rizzetto is a PhD student at the University of Sheffield whose research concerns the development of animal husbandry during the late Roman to early medieval transition in Britain and the lower Rhine region, with particular regard to biometrical changes. He has also been working at a number of archaeological sites in Italy, Britain, France, Greece, and Spain, dating from the Neolithic to the post-medieval period. He previously obtained an undergraduate degree in Archaeological Science in 2013 and a Master's degree in Osteoarchaeology in 2015, both at the University of Sheffield.

Hannah Russ is a freelance zooarchaeologist working with archaeology.biz, based in Barnard Castle, and an Honorary Research fellow at the University of Wales, Trinity Saint David. She is a zooarchaeologist specializing in the study of aquatic animals, including fish, molluscs, and crustaceans, and has worked on remains from five UNESCO World Heritage sites, as well as other sites in Western Europe and the Middle East dating from the Upper Palaeolithic through to the post-medieval period. Hannah completed her PhD in Archaeological Sciences in 2011 at the University of Bradford and subsequently held positions at the University of Sheffield and Oxford Brookes University. She was appointed a Fellow of the Society of Antiquaries of London (FSA) in 2019.

Kim Vickers completed her PhD on the palaeoentomology of the North Atlantic islands in 2007. Her research has focused on the environmental impact of medieval human settlement and activity in Iceland, Greenland, and the Faroe islands, and on the nature of resource use and contact between Norse and Inuit cultures in Greenland, while her other research interests include the Iron Age to Roman transition in Britain and the effects of the Roman invasion of Britain on farming practices and animal husbandry in the early first millennium AD.

Sarah Viner-Daniels completed her PhD at the University of Sheffield and was subsequently appointed as a Research Associate to the Feeding Stonehenge project. Her main areas of interest include animal exploitation in Mesolithic and Neolithic Britain and the application of isotopic analysis (using strontium and oxygen) to the understanding of prehistoric livestock mobility.

Oxford Handbooks offer authoritative and up-to-date surveys of original research in a particular subject area. Specially commissioned essays from leading figures in the discipline give critical examinations of the progress and direction of debates, as well as a foundation for future research. *Oxford Handbooks* provide scholars and graduate students with compelling new perspectives upon a wide range of subjects in the humanities, social sciences, and sciences.

Also published by
OXFORD UNIVERSITY PRESS

The Oxford Handbook of Archaeology
Edited by Barry Cunliffe, Chris Gosden, and Rosemary A. Joyce

The Oxford Handbook of the Archaeology and Anthropology of Hunter-Gatherers
Edited by Vicki Cummings, Peter Jordan, and Marek Zvelebil

The Oxford Handbook of Maritime Archaeology
Edited by Alexis Catsambis, Ben Ford, and Donny L. Hamilton

The Oxford Handbook of the Archaeology of Death and Burial
Edited by Sarah Tarlow and Liv Nilsson Stutz

The Oxford Handbook of Caribbean Archaeology
Edited by William F. Keegan, Corinne L. Hofman, and Reniel Rodríguez Ramos

THE OXFORD HANDBOOK OF

ZOOARCHAEOLOGY

Edited by

UMBERTO ALBARELLA

with

MAURO RIZZETTO, HANNAH RUSS,
KIM VICKERS,

and

SARAH VINER-DANIELS

OXFORD

UNIVERSITY PRESS

OXFORD
UNIVERSITY PRESS

Great Clarendon Street, Oxford, OX2 6DP,
United Kingdom

Oxford University Press is a department of the University of Oxford.
It furthers the University's objective of excellence in research, scholarship,
and education by publishing worldwide. Oxford is a registered trade mark of
Oxford University Press in the UK and in certain other countries

Published in the United States of America by Oxford University Press
198 Madison Avenue, New York, NY 10016, United States of America

British Library Cataloguing in Publication Data
Data available

Library of Congress Cataloging in Publication Data
Data available

ISBN 978–0–19–968647–6 (Hbk.)
ISBN 978–0–19–885443–2 (Pbk.)

Printed and bound by
CPI Group (UK) Ltd, Croydon, CRO 4YY

Animals have played a fundamental role in shaping human history, and the study of their remains from archaeological sites—zooarchaeology—has gradually been emerging as a powerful discipline and crucible for forging an understanding of our past. *The Oxford Handbook of Zooarchaeology* offers a cutting edge compendium of zooarchaeology the world over that transcends environmental, economic, and social approaches, seeking instead to provide a holistic view of the roles played by animals in past human cultures.

Incisive chapters written by leading scholars in the field incorporate case studies from across five continents, providing a sense of the dynamism of the discipline, the many approaches and methods adopted by different schools and traditions, and an idea of the huge range of interactions that have occurred between people and animals throughout the world and its history. With an introduction that clearly contextualizes the current practice of zooarchaeology in relation to both its history and the challenges and opportunities that can be expected for the future, and a methodological glossary illuminating the way in which zooarchaeologists approach the study of their material, this *Handbook* will be invaluable not only for specialists in the field, but for anybody who has an interest in our past and the role that animals have played in forging it.

FOREWORD

FEW categories of archaeological remains are as informative as nice, large, well-preserved samples of animal bones. With apologies to colleagues who find pottery and stone tools more fascinating, I have always found boxes of bones, tabular counts of anatomical elements by species, and survivorship charts to be the sort of data that will open up a narrative of what people did and how a society lived. If archaeology can be called a jigsaw puzzle in which most of the pieces are missing (and no picture on the box), zooarchaeological data provide some of the edges from which interior spaces can be filled in with other information. Since animal bones *should* be a part of the archaeological record, and were either discarded thoughtlessly or buried in structured deposits, their recovery, identification, tabulation, measurement, and quantification have genuine meaning for interpreting patterns of human behaviour.

I am biased in my enthusiasm for zooarchaeology. As a graduate student, I learned its principles from Richard Meadow and made the study of faunal remains from Neolithic sites in Poland the subject of my dissertation. I fully intended to become a committed zooarchaeologist until life events intervened and took my career in another direction. Even though I became mainly a consumer rather than a producer of zooarchaeological data, the practices that people constructed around animals have always been central to my archaeological interests. Perhaps I could be classified as an archaeologist with faunal leanings, not a card-carrying member of ICAZ but a fellow-traveller nonetheless.

It was my privilege to read this entire volume in manuscript. I am pleased to say that I learned as much, if not more, about the archaeology of the various times and places discussed in the case studies as I did about the faunal remains. In fact, one might say that this book can be read as an overview of world archaeology through the lens of animal bones and their interpretation. About halfway through, I began to make a list of the major archaeological themes that I was encountering repeatedly: human-environment interactions/historical ecology; land use/landscapes; sedentism/mobility; migration; storage/surplus; inequality/status; specialization/crafting; ritual/feasting; agency/choices; contact/interaction/trade; culture contact/creolization. Relative abundance of species is only the starting point of these studies, not their goal. Collectively, they draw a portrait of an evolved discipline in which fundamental methods are extended by continuous innovation in analytical techniques and imaginative interpretations.

This book is thus an outstanding illustration of zooarchaeology as *archaeology*, as practised by a globally integrated community of scholars. Whether they come from the archaeological or the zoological tradition, the authors describe research motivated by broad questions in which animal bones are interrogated to explain human activities and practices. I hope that the reader finds the case studies in this volume as illuminating as I have, for they demonstrate how the modern study of animal bones from archaeological sites can be the source of empirically grounded and conceptually sophisticated narratives of the human experience.

Peter Bogucki
Princeton, New Jersey, USA

PREFACE

....

TOWARDS the end of 2011 I received an email from Hilary O'Shea, then Senior Editor of Classics, Ancient History, and Archaeology at Oxford University Press, asking whether I would be interested in editing an Oxford Handbook of Zooarchaeology. The Oxford Handbook series had been in existence for a number of years, producing important reference books for archaeology and beyond, and there was no question that a volume dedicated to zooarchaeology was desirable. Yet my initial reaction was 'no way!'. I was already far too over-committed to embark myself in another gigantic project; yet to pass the 'hot potato' to somebody else sounded unfair and, while I was pondering whether to accept the proposal or not, I was already getting excited about how the volume could be organized. In other words, torn between a wise and a crazy choice, I opted for the latter, and the consequences of this decision have ended up dominating large parts of my life over these past four years. The fact that, somewhat miraculously and despite circumstances that would prove to be far less than favourable, the volume is now complete arguably vindicates my initial choice. It will, however, be up to the readers to decide whether it was worth the huge time investment.

What is, however, unquestionably pleasing for me is that the volume is what I wanted it to be—a showcase of 'world zooarchaeology', illustrating the many facets of the discipline and its huge potential for our understanding of the human past. At its onset we considered three main possible avenues for the structure of the volume: methods, themes, and geography. It became soon clear that thematic investigations would naturally arise from a geographic approach and, as far as methods were concerned, we did not want to replicate the many excellent zooarchaeology textbooks that were in existence. We preferred 'methods' to emerge in terms of their application to specific research questions, and the Methodological Glossary that we have included at the end will provide some more information for the uninitiated. The glossary shows that most methodological applications are on display in one chapter or another.

By giving priority to a geographic approach we could guarantee that zooarchaeology would emerge as the truly international discipline that it is. We were keen to cover all major cultural areas of the world, most key stages in human history, and also to display the richness of approaches provided by the many research schools across the globe. Needless to say, coverage could never be truly complete and the gaps that unquestionably exist still make my heart bleed; but one must accept that no single project can be fully comprehensive. As is inevitable with a project of this scale, some contributions were lost on the way and the three missing chapters from western Asia may explain why this area does not enjoy the kind of coverage that its astonishing archaeological richness would have deserved. Of the fifty-three chapters that we commissioned, forty-six are published in this volume, which represents a reasonably good strike rate. This has allowed us to have good representation from all continents and no less than twenty-five countries (in terms of authors' addresses). The number of countries whose zooarchaeology is discussed as part of this volume is much greater. Coverage is not entirely even, as this also reflects intensity of research in different areas. Contributors were given complete freedom in terms of how they approached their chapters' structures and themes, as we wanted regional research traditions to emerge naturally from the volume. The widespread and diverse backgrounds of the authors has meant that the full range of zooarchaeological approaches is on display.

A project of such magnitude would have not been possible without the help and support of many colleagues and friends. First of all I would like to acknowledge my co-editors, who were instrumental to its success. Hannah, Kim, and Sarah were involved in the earlier phase of editing and Mauro in the latter. I am particularly grateful to Mauro, who helped me to get through the difficult final stages of the project, working almost full time on it. All contributors, without exception, were wonderfully supportive and sympathetic, and I am ever so grateful for their understanding, even when the pace of progress became frustratingly slow. They were also unfailingly patient in dealing with our many editorial comments. We are very lucky to have developed such a great level of solidarity, helpfulness, and friendship in our research community. Special thanks to Peter Bogucki, who kindly agreed to write a foreword for the volume, and waited very patiently for its completion. I am honoured that the Oxford University Press has had faith in my ability to complete such a complex task, and I am very grateful to the many staff who have helped us with suggestions and answered our many queries: Hilary O'Shea, Charlotte Loveridge, Annie Rose, Alex Johnson, Céline Louasli, Sarah Kain, Blaine Alleluia, Michael Dela Cruz, and Lauren Konopko. My deepest debt of gratitude is to the British Academy, which funded my research leave for the academic year 2014–2015, making it possible for me to complete this project.

During the production of this volume, in Papua New Guinea a little angel flew to the sky. This book is dedicated to her.

Umberto Albarella
10 September 2015

Contents

PART III ASIA

PART IV AFRICA

PART V NORTH AMERICA

PART VI   SOUTH AMERICA

PART VII   OCEANIA

FIGURES

Tables

ONLINE SUPPLEMENTARY MATERIAL

The following supplementary material is available in the endmatter of the book where it has been published on Oxford Handbooks Online: https://www. oxfordhandbooks.com/view/10.1093/oxfordhb/9780199686476.001.0001/ oxfordhb-9780199686476-miscMatter-10.

PART I

INTRODUCTION

CHAPTER 1

ZOOARCHAEOLOGY IN THE TWENTY-FIRST CENTURY

where we come from, where we are now, and where we are going

UMBERTO ALBARELLA

INTRODUCTION

ZOOARCHAEOLOGY is today a thriving area of archaeological research, well recognized for the vitality of its community of researchers and the depth and breadth of its approaches. A long and bumpy road had, however, to be negotiated to get to that point and there is still some way to go. This chapter introduces a volume that is intended to present an overview of world zooarchaeology, covering a multitude of geographic areas, cultural periods, approaches, and themes. In this introduction I will present my personal view on the current state of play in zooarchaeology, with some considerations regarding the nature of the discipline, its roots, and its potential.

WHAT IS ZOOARCHAEOLOGY?

Definitions never work entirely, but it is probably not outrageously wrong to consider zooarchaeology as 'the study of animal remains from archaeological sites'. Exceptions can of course exist—e.g. animal footprints should be considered within the remit of zooarchaeology, some non-anthropogenic sites may also produce relevant finds—but the bulk of zooarchaeology is probably covered by that definition. What is more interesting of the ultimate definition is, however, a consideration of what that implies. 'Animal remains' are hugely varied in their nature, size, and composition, yet a lot of zooarchaeology deals with just bones and teeth—the remains of vertebrates. It would, however, be wrong to confine the discipline to such finds, as invertebrates are also animals, and have an important potential in archaeology. Nevertheless, traditions and thematic investigations inevitably contribute to shape a discipline, so you will find that entomological studies only feature marginally in this volume, as they tend to be more commonly associated with more strictly palaeoenvironmental investigations. The same is the case for the study of land snails, but marine molluscs—part of the same phylum— are more commonly studied in conjunction with vertebrate remains. Consequently, they contribute significantly to this volume, and not just as a potential source of food (see for instance Daniela Klokler's chapter).

As important as the consideration regarding what actual material zooarchaeologists study is the concept that the zooarchaeological evidence derives from 'archaeological sites'. Zooarchaeology is no more (and no less) than one aspect of archaeology (O'Connor, 1998). Archaeology deals with the physical remains of our past, and zooarchaeology analyses the remains of animals that contributed to characterize human life. As such the distinction between the sister disciplines of palaeontology and zooarchaeology is obvious—while palaeontologists will focus on the animals themselves, zooarchaeologists investigate their relationships with humans.

It is for this reason that we have chosen, for the title of the book, to use the term 'zooarchaeology', rather than 'archaeozoology', as the former places its emphasis on the archaeological side of the discipline, and therefore more properly defines it. The issue of which is the better term has been lingering long in the literature (e.g. Legge, 1978) but, in reality, both expressions are widely accepted and used, and their adoption is mainly the result of different scholarly traditions (cf. Bartosiewicz, 2001; Steele, 2015). It would have been churlish to be strict about the adoption of a single term, and it is therefore appropriate, for a volume that intends to promote diversity, that contributors were given the freedom to use 'zooarchaeology' or 'archaeozoology' as they saw fit.

It is also useful, and far more than a merely semantic exercise, to reflect on the position of zooarchaeology within archaeology. There is no question that animal remains represent what is left of what once were living organisms, and it is therefore appropriate to consider zooarchaeology within the realm of 'bioarchaeology'. The categorization is, however, not particularly useful, and is made more problematic by the common, and unfortunate, use in American literature of the term 'bioarchaeology' to indicate the study of human bones from archaeological sites (e.g. Spencer Larsen, 1999; Martin et al., 2013). Humans are of course animals and the study of their remains has much to share with zooarchaeology, but in terms of approaches and nature of the evidence, zooarchaeology and 'human osteoarchaeology' (a better term than 'bioarchaeology') tend to represent separate, although related, sub-disciplines.

More problematic is the frequent categorization of zooarchaeology as part of 'environmental archaeology', particularly, but not only, in British literature. The understanding and reconstruction of palaeoenvironments where people lived is well within the remit of zooarchaeology, but there is much more to zooarchaeology than environmental analysis, and therefore the classification of zooarchaeology as part of environmental archaeology is misleading; it is rather the product of a common misconception regarding what environmental archaeology is (cf. Albarella, 2001; Thomas, 2001; Wilkinson and Stevens, 2003). As already mentioned, aspects of zooarchaeology that are more strictly palaeoenvironmental have had a tendency to develop into independent research strands.

If we accept the definition found in the Encyclopaedia Britannica of science as 'knowledge of the world of nature' then it necessarily follows that zooarchaeology is a scientific discipline. The investigation of the natural world, of which human communities are part, is central to the concern of zooarchaeologists, but they also investigate the cultural attitude and behaviour of human societies towards animals—and therefore they operate within the realm of 'humanities'. Zooarchaeology represents a primary example of the inter-disciplinarity of archaeology, a discipline that constantly operates at the intersection between nature and culture. Some of the chapters of this book may lean more towards a scientific or a humanities approach, but never exclusively so.

Animals are a ubiquitous and important presence in all aspects of human life and, consequently, zooarchaeology can contribute to almost any strand of archaeological investigation (Steele, 2015). Although this should be obvious, this is a concept that has proven strangely difficult to put across, with zooarchaeology often ghettoized to rather limited (and limiting) research themes. Although animal remains had already caught the interest of archaeologists by the nineteenth century, the persistence of an antiquarian tradition in archaeology meant that, for many decades, there was limited interest in zooarchaeology, particularly as concerned the historical periods. Zooarchaeology, like other bioarchaeological disciplines, was relegated to the notorious 'appendix'. The emergence of the so-called processual, or new, archaeology in the late 1960s and '70s, with its focus on human behaviour and an anthropological approach to archaeology, led to an enhanced attention to the role that animals played in human societies. The downside was the frequent excessive focus on taphonomic (e.g. Binford, 1981) and/or economic and ecological (e.g. Higgs, 1972; 1975) aspects, to the detriment of other lines of investigation (and interpretation). Other research schools that started emerging in the 1980s (broadly defined as 'post-processualism') pointed out this fault of the New Archaeology, and emphasized the need to pay greater attention to the social and ideological components of human society. Paradoxically, and unnecessarily, this became a battle between 'nature' and 'culture' approaches to archaeology, with zooarchaeology becoming sidelined, or even ending up being considered some kind of backwards burden to the development of this new conceptual approach in archaeology (cf. Thomas, 1990). Zooarchaeologists reacted rather slowly to this new challenge, but eventually the concern caught up with them and the last decade has seen a new strand of zooarchaeology focusing more on social aspects (e.g. Marciniak, 2005; Russell, 2012; Overton and Hamilakis, 2013). Commendable as such an attempt is, it also carries the risk of re-emphasizing once again the old and false dichotomy between 'nature' and 'economy' on the one hand and 'culture' and 'society' on the other—merely seen from the opposite viewpoint.

The warning of the post-processualists to avoid purely mechanistic interpretations in zooarchaeology was welcome, but the reality is that the portrayal of the zooarchaeologist as an environmental determinist is largely caricatural. Examples of zooarchaeological approaches—even from decades ago—which, in addition to the ecological and economic elements, deal with issues related to the structure of a society, as well as its cultural preferences, religion, and ideology, abound (e.g. Reitz, 1987; Grant, 1988; Ijzereef, 1989; Meniel, 1989). A mere browse of the chapters included in this book will demonstrate that ecological, economic, social, and ritual elements cannot be neatly separated in archaeological interpretations, as they all play a role in the shaping of human societies. Animals contribute to all of them and zooarchaeology today is at the forefront of a new integrated approach to archaeological interpretation, which will hopefully once and forever overcome the artificially constructed divisions of the past.

The reason why zooarchaeology can be in this prime position is due to its interdisciplinary nature. Zooarchaeologists have familiarity with both biological and cultural phenomena and, as such, feel at ease in communicating with scholars across different disciplines. It is not uncommon for zooarchaeological interpretations to consider evidence from disciplines as disparate as ethnography, history, architecture, arts, genetics, bio- and geochemistry, and many others (for a range of interesting examples see Maltby, 2006; more evidence can be found in this volume).

What has also been emerging more and more powerfully in the past few years is the potential of zooarchaeology to inform on issues of relevance to the contemporary world. The volume by Lauwerier and Plug (2003) is a prime example of how nature conservation and heritage management issues can be productively informed by zooarchaeological evidence, but there are other cases in point (e.g. Lyman, 1996; Lyman and Cannon, 2004). Additionally, ethnographic work, aimed at addressing zooarchaeological questions, is highlighting the cultural and ecological merits of various forms of traditional husbandry, as well as the value of traditional domestic breeds, many of them today rare or on the verge of extinction (Albarella et al., 2007; Albarella et al., 2011; Hadjikoumis, 2012).

ZOOARCHAEOLOGY: METHODS AND APPROACHES

In order to address broader research enquiries in archaeology, zooarchaeologists have had to develop methods and research strategies to answer questions relevant to the more specific evidence they analyse. Although these questions have been refined and developed over the years, the core lines of investigations have remained the same. Like in the earlier days of the discipline, zooarchaeologists will want to know which animals are represented in a certain assemblage, which parts of their carcasses, the age and sex make-up of the populations, the size and shape of the animals, the occurrence of any pathological conditions, and the evidence of human-induced modifications of the bones, such as butchery or burning. A more recently developed technique, tooth microwear analysis (e.g. Mainland, 1998), has also provided the opportunity to collect some evidence regarding the nature of animal diet, which can be very useful for a better understanding of the forms of animal management, as well as the range of habitats used by both domestic and wild species. Examples of all, most, or at least some of these investigations can be found in all chapters in this book. The diversity of emphasis that is placed on different strands of evidence in each chapter typifies geographic regions, chronological and cultural periods, research themes, and/or the interests and expertise of different contributors.

Although the issue was neglected in the early days of zooarchaeology (and can still be, in some unfortunate situations) zooarchaeologists have for quite some time been aware that their interpretations must rely on an understanding of the processes that led to the formation of the assemblages they study (Schiffer, 1987). Therefore, modifications of the animal remains, such as those caused by scavengers and various natural agents before and after burial, are also important to observe and record systematically. Evaluations of issues such as preservation and fragmentation can be important for a reconstruction of the history of an assemblage, as well as for an understanding of the biases that will affect the evidence. Since the pioneering work carried out by Payne (1975) there has also been increasing awareness of the effect that recovery bias can have on the frequency of species, body parts, age, and sex categories (e.g. Gamble and Bailey, 1994). Potentially this is the greatest bias that can affect an assemblage. In theory it can be controlled during archaeological excavation, for instance through a carefully considered sieving programme, but, in practice, it only occasionally is. Zooarchaeologists, however, have developed various systems that allow them to assess the degree of recovery bias. They cannot retrieve information that has been lost but they can at least assess the magnitude of the error. There are still unfortunate cases in which the issue is entirely ignored—inevitably leading to spurious interpretations—but it is heart-warming to see how strongly a discussion of recovery bias features in many chapters of this book.

The maturity of the discipline cannot be better demonstrated than by the availability of not one, but a plethora of textbooks outlining the key principles and methods of zooarchaeology (Cornwall, 1956; Ryder, 1968; Chaplin, 1971; Klein and Cruz-Uribe, 1984; Hesse and Wapnish, 1985; Davis, 1987; Rackham, 1994; Reitz and Wing, 1999; O'Connor, 2000; De Grossi Mazzorin, 2008; Matsui, 2008; Beisaw, 2013). Some of these have also been translated in other languages—for instance Davis' book, originally in English, has long been available in a Spanish version and recently has been translated into Korean. In addition, zooarchaeology has benefitted from a number of papers highlighting the main potential pitfalls in the interpretation of animal remains (e.g. Payne, 1972; Uerpmann, 1973; Meadow, 1980), which have been instrumental for the appropriate methodological development of the discipline. Identification atlases have long been available—e.g. Schmid (1972) and Hillson (1992) for European mammals, Miles Gilbert (1993) for American mammals, Walker (1985) for African mammals, Cohen and Serjeantson (1996) for European birds, Miles Gilbert et al. (1985) for American birds—and have now been supplemented with new productions (e.g. Yamazaki and Uyeno, 2008; Bocheński and Tomek, 2009; Plug, 2014a) as well as web-based online sources (e.g. ArchéoZooThèque http://archeozoo.org/archeozootheque for mammals, Aves 3D http://aves3d.org/ for birds, and the Archaeological Fish Resource http://fishbone.nottingham.ac.uk/). None of these can of course replace skeletal reference collections, whose importance is widely acknowledged (e.g. Coy, 1978; Henry, 1991), and which represent key magnets of activity for some institutions, despite the challenge of ever-shrinking research budgets.

In other areas of investigation zooarchaeology has also well-established methodological procedures. For tooth ageing the works of Payne (1973) and Grant (1982) are widely used and new methodological developments have also been put forward (e.g. Jones and Sadler, 2012; Lemoine et al., 2014; Wright et al., 2014). Biometrical analysis has hugely benefitted from the standardization of measurements proposed by von den Driesch (1976) for mammals and birds, and Morales and Rosenlud (1979) for fishes. Both are almost universally used, without, however, stifling further considerations regarding which measurements should be taken and why (e.g. Payne and Bull, 1988; Wheeler and Jones, 1989; Davis, 1996; 2000; Albarella and Payne, 2005; Popkin et al., 2012). In terms of biometrical data analysis, the current easy access to statistical and graphics computer packages has immensely facilitated the work of the zooarchaeologist, and the ever-growing application of scaling index techniques (Ducos, 1968; Uerpmann, 1979; Meadow, 1999; Albarella, 2002) is contributing to address the common problem of small sample size. Shape analysis in the form of the so-called 'geometric morphometrics', a technique long used by biologists and palaeontologists (e.g. Bookstein, 1991), has made some inroads in zooarchaeology (e.g. Bignon and Eisenmann, 2006). Useful as it is, this method is time-consuming and requires expensive equipment. Most importantly it needs to be built on a solid understanding of the potential of linear measurements, something that current scholarship has often shied away from—see Rowley-Conwy and Zeder (2014) for an effective critical analysis of the risks of a superficial application of the technique, combined with palaeogenetics, with insufficient understanding of basic biometry.

In palaeopathology, the classic work of Baker and Brothwell (1980), which has been intensively used by generations of zooarchaeologists, has now finally been complemented by a new textbook on the subject (Bartosiewicz with Gál, 2013), which undoubtedly will prove to be equally useful. Miles and Grigson's (1990) survey of tooth conditions represents a very useful reference for the identification of dental pathologies.

In summary, the literature on zooarchaeology methods that exists today is vast, and students and new trainees are spoilt for choice in terms of accessible resources. This is all made easier by the availability of much information through 'open access' and, in general, on the web. In fact, the young zooarchaeologist has today the opposite problem to that faced by my generation—rather than a scarcity of information, over-abundance. It is therefore necessary to skilfully plough your way through an extensive literature, applying critical thinking in the distinction of what is useful from what is redundant.

Further indication that the discipline of zooarchaeology has now reached its full maturity is demonstrated by the fact that debate on the adoption of 'minimum standards', which still raged in the late 1980s, now appears to be a thing of the past. Data comparability is very important, but this cannot be achieved through the imposition of standard methods of recording and analysis that would stifle creativity and reduce the work of the zooarchaeologist to that of a mere technician. Rather, we need to carry on refining our methods, making them accessible and affordable to as many practitioners as possible. Comparability can also be greatly enhanced by the constant encouragement to fully explain the adopted methods, to improve both accountability and the opportunity to compare datasets appropriately. Assemblages of animal remains are hugely varied in their composition and may require substantially different approaches, which are also dependent on logistic conditions such as available time and money. It should also be right for a zooarchaeologist to approach the study of an assemblage in an original and personal way, driven by specific research interests and questions. The study of a zooarchaeological assemblage represents an intellectual undertaking rather than a mechanical collection of data. It is for this reason that it is essential that assemblages are preserved for future use—they can be read in a number of different ways, emphasizing either one aspect or the other. The notion that an assemblage, once studied, can be preserved by record is not only wrong, but supremely arrogant.

It is with such awareness that the International Council of Archaeozoology (ICAZ) drafted a 'professional protocol', which is represented by a set of useful recommendations rather than prescriptive or detailed procedures (Reitz, 2009 http://alexandriaarchive.org/icaz/pdf/protocols2009.pdf). New systems for the recording and analysis of animal remains keep being published (e.g. Schibler, 1998), and this is to be welcomed, as it provides opportunities for new researchers to get a starting point, and for experienced ones to reconsider their systems and priorities. What remains essential is that diversity of approaches is not sacrificed on the altar of data comparability.

One methodological area in which more reflection is required in zooarchaeology regards the only apparently simple task of counting and recording. I am carefully using my words here as I do not mean 'quantification', which has, conversely, been amply debated (e.g. Grayson, 1984; Lyman, 2008). All quantification systems, however, rely on what is recorded and counted and in that area we still have a great level of ambiguity in zooarchaeology. It may be useful to debate about the virtues and problems of systems such as the Number of Identified Specimens (NISP) and the Minimum Number of Individuals (MNI), but if we are not sure about what a 'specimen' is, the whole quantification edifice collapses. Many years ago Watson (1979) tried to circumvent this problem by proposing the recording of 'diagnostic zones', a system that, with substantial differences and modifications, has been adopted by many zooarchaeologists (e.g. Serjeantson, 1991; Davis, 1992; Albarella and Davis, 2010), who still, however, probably represent a minority. This is not the place to go into a detailed discussion of this issue, but I remain disconcerted by the fact that a discipline that has made such huge progress in the critical evaluation of how it operates, is prepared to leave the definition of what it records and counts to the vagaries of variables such as the skill of the researcher, time pressure, light conditions, tiredness, completeness of a reference collection, the identifiability and preservation of the material, and many others.

Today zooarchaeologists can also benefit from a level of analysis that goes beyond the macroscopic level. The study of amino acid peptides has proved its usefulness in taxonomic identifications (Buckley et al., 2010) and it is developing as a valuable technique to use in conjunction with macroscopic identifications. Isotopic studies are helping in clarifying issues associated with animal diet (e.g. Pearson et al., 2007), seasonality (e.g. Balasse, 2003), palaeoclimates (e.g. Stevens and Hedges, 2004), and mobility (e.g. Towers et al., 2010; Viner et al., 2010; Minniti et al., 2014), all areas in which traditional zooarchaeological approaches can helpfully be integrated by other lines of evidence.

Studies of the DNA of modern animals have contributed to our understanding of the variability of animal species and populations (Luikart et al., 2001; Larson et al., 2005; Bruford and Townsend, 2006), therefore throwing some light also on their evolution and past history. Palaeogenetic applications are more problematic due to potential issues of preservation and contamination (Geigl, 2008; Pruvost et al., 2008), but can be very effective as they will offer direct evidence of the genetic make-up of past animals. The volume by Zeder et al. (2006) provides a good summary of the interplay between zooarchaeologists, geneticists, and palaeogeneticists in tackling the study of animal domestication. This is an area of research that has seen rapid development, but has its downside too. The broad scale approach that is often characteristic of palaeogenetics may lead to the risk of over-simplifications, and much genetic work has been insufficiently or inappropriately integrated with archaeological analysis. We must also be careful not to rush to conclusions that may be a consequence of erratic sampling. For instance, the issue of the nature of the introduction of cattle domestication into Europe, which appeared to have been solved through palaeogenetic analysis, has proved to be far more complex than originally thought, once the sample size was increased (cf. Troy et al., 2001; Beja-Pereira et al., 2006; Edwards et al., 2007; Mona et al., 2010).

Most of the palaeogenetic work carried out so far in zooarchaeology deals with mitochondrial DNA, which is present in greater abundance in a cell, and has therefore better chances of survival. Improvements in extraction and replication techniques have, however, meant that palaeogeneticists have also, in some cases, managed to access nuclear DNA. In addition to further information on the characteristics of an animal genotype (which, unlike mitochondrial DNA, is not transmitted exclusively matrilinearly) the nuclear DNA can also help in sexing specimens, a highly valuable type of information, when coupled with morphometric analysis (for applications see Svensson et al., 2008; Davis et al., 2012).

In order to adhere to the principle of integration, for this volume we did not commission any chapter to deal specifically with biochemical evidence. The evidence from DNA and isotopes is, however, discussed in many contributions, in conjunction with the rest of zooarchaeology.

The Internationality
of Zooarchaeology

One of the most impressive achievements of zooarchaeology has been its ability to develop as a worldwide discipline, with a high level of exchange and communication between researchers from all corners of the world. This internationality has been promoted, to a substantial extent, by the work of the International Council of Archaeozoology (ICAZ), which is an important reference organization for zooarchaeologists. With its quadrennial international conferences, the meeting of its Working Groups, and a plethora of other activities, ICAZ has for many years guaranteed that zooarchaeologists from across the world had a common house, which would support the exchange of data and ideas, as well as diversity and inclusiveness. It is such internationality that this book wants to celebrate.

ICAZ has a very interesting history and, by following it, we can gain a sense of the overall development of zooarchaeology as a discipline. The first ICAZ international meeting was held in Budapest in 1971 (Grigson, 2014), which may mean little to researchers of the latest generations, but it is very significant when one thinks that this was the time of the 'Iron Curtain' when communication between the East and West of Europe (and, to some extent, the world) could be strained. Hungary was of course under Soviet influence but this did not prevent western researchers from attending and contributing to the take-off of the organization. Thus, from its early days, zooarchaeologists demonstrated their determination to join forces despite the many economic, cultural, and political barriers that existed between them. The following years would see many more examples of such an attitude.

For several decades ICAZ kept to a relatively small scale but the London 1982 conference organized by Juliet Clutton-Brock and Caroline Grigson was attended by more than a hundred delegates (Grigson, 2014). By the time of the 1994 conference in Constance (Germany), it was clear that the organization had grown to the point that the informality of its early days had become insufficient to guarantee transparency and efficiency. A more formal structure, with proper membership, elected officers, and committee members, had to be set up. Over the years, the composition of the committees has invariably been highly international, with representation from all continents. The main conferences have also moved around across the continents, with two of the last three being held in Latin American countries, and the next one (2018) planned to take place, for the first time, in Asia.

Another important milestone of the 1994 conference was the move away from the adoption of ICAZ 'official languages'. It became clear that the concept was impractical and unsuited to the ethos of inclusiveness that ICAZ was increasingly keen to promote. This idea took a further step forward when Keith Dobney, Peter Rowley-Conwy, and myself organized the 9th ICAZ conference in Durham (UK) in 2002. It was decided that the conference itself would not have official languages, with contributors free to speak in whatever language they preferred, ranging from Swahili to Urdu (as long as the paper abstract was in the same language, to warn the audience of what to expect). Eventually, several hundred presentations were delivered—all in English—a triumph for freedom of expression, respect of other cultures and ... common sense! The 2002 conference also saw the introduction of the concept that conference sessions would be centred on research themes, rather than chronological periods or geographic areas, in order to promote greater exchange between researchers from different parts of the world. The idea was so successful that it has become a constant feature of all successive ICAZ conferences.

The ICAZ 2010 Paris conference organized by Jean-Denis Vigne, Christine Lefèvre, and Marilène Patou became the largest aggregation of zooarchaeologists ever known, with more than 700 delegates from 56 countries (Vigne and Lefèvre, 2010). Equally impressive was the achievement of the 2014 ICAZ conference in San Rafael (Argentina), which, despite being held in what by many would be regarded as a remote place (on the verge of Patagonia), still attracted a large international crowd. Personally, I regarded the conference in Argentina as a triumph. In the occasion of the Durham conference I had become extremely impressed by the very good number of Argentinian colleagues who had attended, in the very year—2002—the country had experienced a serious economic crash. Once again, zooarchaeologists had shown great resilience in the face of adversity. That experience convinced me that sooner or later a conference in Argentina was due.

ICAZ alone cannot sustain full responsibility for the internationalization of zoo-archaeology and it is, fortunately, well supported by other initiatives, which facilitate exchange and communication. Prominent among these is the role carried out by BoneCommons (http://www.alexandriaarchive.org/bonecommons/). Part of the Alexandria Archive Institute and managed by Sarah Whitcher Kansa, BoneCommons, as specified in its heading, is 'an online community, building and sharing resources for archaeozoology'. Working in close collaboration with ICAZ, this resource has, for many years, proven its worth, once again encouraging participation and promoting a sense of mutual aid in zooarchaeology.

Complementary to BoneCommons is the email discussion list Zooarch (https://www.jiscmail.ac.uk/cgi-bin/webadmin?A0=ZOOARCH). Founded in 2000 by Jacqui Mulville and myself, and counting almost 1,200 subscribers, Zooarch was regarded to be the most valuable communication tool in zooarchaeology in a survey undertaken by Jim Morris (Morris, 2010). In addition, Morris himself has created the zooarchaeology version of a social network (http://zooarchaeology.ning.com/), which has also proven to be most helpful, and it is widely used, especially by the younger generation of zooar-chaeologists. What is heartening is that all these resources operate in an excellent spirit of collaboration, helping and supporting each other, and joining forces in promoting zooarchaeology worldwide.

Although there is much to be cheered regarding the huge forays that zooarchaeology has made in guaranteeing participation from all areas of the world, there are still considerable challenges ahead. However widespread zooarchaeology is, the bulk of its practitioners are still concentrated in the wealthiest areas of the world, with the north–south divide being particularly striking. Progress has been made, particularly in South America, but large parts of Africa (cf. Plug, 2014b) and Asia still lag behind, inevitably as a consequence of the inequality of wealth distribution in the world. Although we may have come to accept this as normality, there is something disturbingly wrong with the notion that countries such as France or the UK have many dozens of active researchers in zooarchaeology and Nigeria and Bangladesh, which are about three time as populous, have none, or at least very few (no ICAZ members). We are moderately satisfied with the fact that two of our eight chapters dealing with Asia and three of eight dealing with Africa are written by researchers based in those continents, but, sadly, our book also reflects the imbalance in the distribution of research and researchers across the globe.

An additional, and increasingly serious, obstacle to international participation is represented by limitations that may occur in crossing borders. In the age of free circulation of goods ('free trade'), it is ironic that more and more barriers exist in the movement of people between countries. The Middle East, a traditional area of prime zooarchaeological research, is ravaged by wars, which generate constant misery in the local populations and prevent them from becoming engaged in academic activities. Several countries in that part of the world (and others) are today no-go areas, preventing therefore the promotion of cultural activities, with the consequent risk that their future is jeopardized too. The zooarchaeological community has come of age also in dealing with these issues. Conferences have in some cases provided restricted or no access to delegates of certain nationalities—which is very much against the spirit of free circulation of ideas that our research community endorses. Once again the zooarchaeological community has responded to these challenges with maturity and, rather that burying its head in the sand, has been prepared to discuss these issues openly, trying to find reasonable solutions to intractable problems. A robust discussion was held on Zooarch (see archives at https://www.jiscmail.ac.uk/cgi-bin/webadmin?Ao=zooarch) regarding the organization of the meeting of the Archaeozoology of Southwest Asia (ASWA) ICAZ Working Group, first in Abu Dhabi and then Israel. The debate continued for several years and was also featured in ICAZ newsletters, which can be downloaded from the ICAZ webpage at http://www.alexandriaarchive.org/icaz/publications-newsletter (Bartosiewicz, 2011; Kolska Horwitz, 2011; Albarella, 2012).

THE FUTURE

Zooarchaeology has a rich history, a bright and exciting present, and an unpredictable future. Zooarchaeologists have come a long way from the days when their research was just regarded as an addendum to the core of archaeological investigations, as can be attested, in a diversity of styles and approaches, by contributions to this book. It would be unwise, however, to rest on our laurels, as there are many important challenges that still need to be tackled. Below is a very personal excursus of some of the areas in which, for the better or worse, I think the future of zooarchaeology will be decided:

- The excitement associated with the opportunities offered by new, lab-based, sophisticated techniques should not make zooarchaeologists neglect the roots of their original work, and the constant methodological advances that it requires.
- Zooarchaeologists have been excellent at providing—as much as the context allowed them—equal opportunities to their practitioners. The overwhelming majority of zooarchaeological work is today still undertaken on very limited budgets. To develop zooarchaeology in directions that are unaffordable to most would mean to create a fracture in the research community between the elected few and a majority left behind—ironically replicating the current ills of world society. This would be regrettable and inconsistent with the aims of a discipline that claims to be inclusive.
- The current world economic creed is unsympathetic to research that does not have direct application to industrial production or other money-making enterprises. The expectation is that the years ahead will be lean, with many academic departments, museums, and commercial units likely to close their business. Solidarity and reciprocal support, rather than competition, can help us in getting through such difficult times. Zooarchaeologists have done it before.
- Training in zooarchaeology will remain a challenge, with some countries imposing enormous tuition fees, which are increasingly unaffordable for many. It will be important to fight this trend and provide opportunities outside the more traditional academic courses. Community-based learning has great potential and may develop well beyond the training of amateurs.
- Large-scale skeletal reference collections are essential for good quality zooarchaeology work, but they require such a huge investment in time and money that is impracticable to think that there can be very many of them. Those institutions holding reference collections have the opportunity to promote them as regional centres of research, where zooarchaeologists can congregate and contribute to their development in exchange for freedom of access. Charging for the use of reference collections goes against this spirit, and should be resisted.
- Zooarchaeologists should continue championing inter-disciplinarity by maintaining a good level of communication with other archaeologists, as well as scholars from other disciplines. For this to be sustained it is also important that the community of zooarchaeology will preserve its diversity in terms of both backgrounds and interests. Zooarchaeology has now rightly affirmed its position at the core of archaeological enquiry, but this should not occur at the expenses of a loss of biological knowledge. Zooarchaeologists with a biological background remain an important asset in zooarchaeology and the risk for animal remains to be interpreted devoid of the living creatures they once belonged to should definitely be avoided.
- Zooarchaeologists should continue exploring the impact they can make on our understanding of contemporary society, making clearer that they possess

unique and essential information on the history and composition of the world in which we live.

- Zooarchaeology needs the intellectual and cultural input that comes from the developing world, as well as from the least privileged members of society. Much of the future of the discipline will depend on its ability to fight the tyranny of the direct proportionality currently existing between monetary wealth and intensity of research. As a generous, inclusive, and supportive community of researchers, zooarchaeologists are in a prime position to achieve that objective.

ACKNOWLEDGEMENTS

I am very grateful to Mauro Rizzetto and Lizzie Wright for comments on an earlier draft. Generous financial support from the British Academy gave me the opportunity to be on research leave for the academic year 2014–15 and therefore find the time to write—among other things—this chapter.

REFERENCES

Albarella, U. (2001) 'Exploring the real nature of environmental archaeology', in Albarella, U. (ed.) *Environmental Archaeology: Meaning and Purpose*, pp. 3–13. Dordrecht: Kluwer Academic Publishers.

Albarella, U. (2002) "Size matters': how and why biometry is still important in zooarchaeology', in Dobney, K. and O'Connor, T. (eds) *Bones and the Man: Studies in Honour of Don Brothwell*, pp. 51–62. Oxford: Oxbow Books.

Albarella, U. (2012) 'ICAZ, ASWA and the issue of academic freedom of movement', *ICAZ Newsletter*, 13(1), 3–4.

Albarella, U. and Davis, S. (2010) 'The animal bones', in Chapman, A. (ed.) *West Cotton, Raunds: A Study of Medieval Settlement Dynamics ad 450–1450*, pp. 516–37. Oxford: Oxbow Books.

Albarella, U., Manconi, F., and Trentacoste, A. (2011) 'A week on the plateau: pig husbandry, mobility and resource exploitation in central Sardinia', in Albarella, U. and Trentacoste, A. (eds) *Ethnozooarchaeology: The Present and Past of Human-Animal Relationships*, pp. 143–59. Oxford: Oxbow Books.

Albarella, U., Manconi, F., Vigne, J.-D., and Rowley-Conwy, P. (2007) 'The ethnoarchaeology of traditional pig husbandry in Sardinia and Corsica', in Albarella, U., Dobney, K., Ervynck, A., and Rowley-Conwy, P. (eds) *Pigs and Humans: 10,000 Years of Interaction*, pp. 285–307. Oxford: Oxford University Press.

Albarella, U. and Payne, S. (2005) 'Neolithic pigs from Durrington Walls, Wiltshire, England: a biometrical database', *Journal of Archaeological Science*, 32(4), 589–99.

Baker, J. and Brothwell, D. (1980) *Animal Diseases in Archaeology*, London: Academic Press.

Balasse, M. (2003) 'Determining sheep birth seasonality by analysis of tooth enamel oxygen isotope ratios: the Late Stone Age site of Kasteelberg (South Africa)', *Journal of Archaeological Science*, 30, 305–15.

Bartosiewicz, L. (2001) 'Archaeozoology or zooarchaeology?: a problem from the last century', *Archaeologia Polona*, 39, 75–86.

Bartosiewicz, L. (2011) 'Letter from the president', *ICAZ Newsletter*, 12(2), 2–3.

Bartosiewicz, L. with Gál, E. (2013) *Shuffling Nags, Lame Ducks: The Archaeology of Animal Disease*, Oxford: Oxbow Books.

Beisaw, A. M. (2013) *Identifying and Interpreting Animal Bones*, College Station, TX: Texas A&M University Press.

Beja-Pereira, A., Caramelli, D., Lalueza-Fox, C., Vernesi, C., Ferrand, N., Casoli, A., Goyache, F., Royo, L. J., Conti, S., Lari, M., Martini, A., Ouragh, L., Magid, A., Atash, A., Zsolnai, A., Boscato, P., Triantaphylidis, C., Ploumi, K., Sineo, L., Mallegni, F., Taberlet, P., Erhardt, G., Sampietro, L., Bertranpetit, J., Barbujani, G., Luikart, G., and Bertorelle, G. (2006) 'The origin of European cattle: evidence from modern and ancient DNA', *Proceedings of the National Academy of Sciences of the United States of America*, 103, 8113–18.

Bignon, O. and Eisenmann, V. (2006) 'Western European Late Glacial horse diversity and its ecological implications', in Mashkour, M. (ed.) *Equids in Time and Space: Papers in Honour of Véra Eisenmann*, pp. 161–71. Oxford: Oxbow Books.

Binford, L. R. (1981) *Bones: Ancient Men and Modern Myths*, Orlando: Academic Press.

Bocheński, Z. M. and Tomek, T. (2009) *A Key for the Identification of Domestic Bird Bones in Europe: Preliminary Determination*, Cracow: Institute of Systematics and Evolution of Animals, Polish Academy of Sciences.

Bookstein, F. L. (1991) *Morphometric Tools for Landmark Data: Geometry and Biology*, Cambridge: Cambridge University Press.

Bruford, M. W. and Townsend, S. J. (2006) 'Mitochondrial DNA diversity in sheep: implications for domestication', in Zeder, M. A., Bradley, D. G., Emschwiller, E., and Smith, B. D. (eds) *Documenting Domestication: New Genetic and Archaeological Paradigms*, pp. 306–16. London: University of California Press.

Buckley, M., Whitcher Kansa, S., Howard, S., Campbell, S., Thomas-Oates, J., and Collins, M. (2010) 'Distinguishing between archaeological sheep and goat bones using a single collagen peptide', *Journal of Archaeological Science*, 37, 13–20.

Chaplin, R. E. (1971) *The Study of Animal Bones from Archaeological Sites*, New York: Seminar.

Cohen, A. and Serjeantson, D. (1996) *A Manual for the Identification of Bird Bones from Archaeological Sites*, London: Archetype.

Cornwall, I. W. (1956) *Bones for the Archaeologist*, London: Phoenix House.

Coy, J. (1978) 'Comparative collections for zooarchaeology', in Brothwell, D. R., Thomas, K. D., and Clutton-Brock, J. (eds) *Research Problems in Zooarchaeology*. Institute of Archaeology Occasional Publications 3, pp. 143–4. London: University of London, Institute of Archaeology.

Davis, S. J. M. (1987) *The Archaeology of Animals*, London: Batsford.

Davis, S. J. M. (1992) 'A Rapid Method for Recording Information about Mammal Bones from Archaeological Sites. English Heritage, Ancient Monuments Laboratory Report 19/92'.

Davis, S. J. M. (1996) 'Measurements of a group of adult female Shetland sheep skeletons from a single flock: a baseline for zooarchaeologists', *Journal of Archaeological Science*, 23, 593–612.

Davis, S. J. M. (2000) 'The effect of castration and age on the development of the Shetland sheep skeleton and a metric comparison between bones of males, females and castrates', *Journal of Archaeological Science*, 27, 373–90.

Davis, S., Svensson, E., Albarella, U., Detry, C., Götherström, A., Pires, A. E., and Ginja, C. (2012) 'Molecular and osteometric sexing of cattle metacarpals: a case study from 15th century AD Beja, Portugal', *Journal of Archaeological Science*, 39, 1445–54.

De Grossi Mazzorin, J. (2008) *Archeozoologia: lo studio dei resti animali in archeologia*, Bari: Laterza.

Driesch, von den, A. (1976) *A Guide to the Measurement of Animal Bones from Archaeological Sites*, Harvard: Peabody Museum.

Ducos, P. (1968) *L'origine des animaux domestiques en Palestine*, Bordeux: Delmas.

Edwards, C. J., Bollongino, R., Scheu, A., Chamberlain, A., Tresset, A., Vigne, J. D., Baird, J. F., Larson, G., Ho, S. Y. W., Heupink, T. H., Shapiro, B., Freeman, A. R., Thomas, M. G., Arbogast, R. M., Arndt, B., Bartosiewicz, L., Benecke, N., Budja, M., Chaix, L., Choyke, A. M., Coqueugniot, E., Dohle, H. J., Goldner, H., Hartz, S., Helmer, D., Herzig, B., Hongo, H., Mashkour, M., Ozdogan, M., Pucher, E., Roth, G., Schade-Lindig, S., Schmolcke, U., Schulting, R. J., Stephan, E., Uerpmann, H. P., Voros, I., Voytek, B., Bradley, D. G., and Burger, J. (2007) 'Mitochondrial DNA analysis shows a Near Eastern Neolithic origin for domestic cattle and no indication of domestication of European aurochs', *Proceedings of the Royal Society B-Biological Sciences*, 274, 1377–85.

Gamble, C. and Bailey, G. (1994) 'The faunal specialist as excavator: the impact of recovery techniques on faunal interpretation at Klithi', in Luff, R. and Rowley-Conwy, P. (eds) *Whither Environmental Archaeology?*, pp. 81–9. Oxford: Oxbow Books.

Geigl, E.-M. (2008) 'Palaeogenetics of cattle domestication: methodological challenges for the study of fossil bones preserved in the domestication centre in southwest Asia', *Comptes Rendus Palevol*, 7(2–3), 99–112.

Grant, A. (1982) 'The use of tooth wear as a guide to the age of domestic ungulates', in Wilson, B., Grigson, C., and Payne, S. (eds) *Ageing and Sexing Animal Bones from Archaeological Sites*. BAR British Series 109, pp. 91–108. Oxford: Archaeopress.

Grant, A. (1988) 'Food, status and religion in England in the Middle Ages: an archaeozoological perspective', in Bodson, L. (ed.) *L'animal dans l'alimentation humaine: les critères de choix*. Anthropozoologica, second numéro special, pp. 139–46. Paris: Bulletin de l'Association L'Homme et l'Animal.

Grayson, D. K. (1984) *Quantitative Zooarchaeology: Topics in the Analysis of Archaeological Faunas*, Orlando: Academic Press.

Grigson, C. (2014) 'My ICAZ history', *ICAZ Newsletter*, 15(1), 4–6.

Hadjikoumis, A. (2012) 'Traditional pig herding practices in southwest Iberia: questions of scale and zooarchaeological implications', *Journal of Anthropological Archaeology*, 31, 353–64.

Henry, E. (ed.) (1991) *Guide to the Curation of Archaeozoological Collections: Proceedings of the Curation Workshop Held at the Smithsonian Institution Washington D.C.*, Gainesville: Florida Museum of Natural History.

Hesse, B. and Wapnish, P. (1985) *Animal Bone Archaeology: From Objectives to Analysis*, Washington: Taraxacum.

Higgs, E. S. (ed.) (1972) *Papers in Economic Prehistory*, Cambridge: Cambridge University Press.

Higgs, E. S. (ed.) (1975) *Palaeoeconomy*, Cambridge: Cambridge University Press.

Hillson, S. (1992) *Mammal Bones and Teeth: An Introductory Guide to Methods of Identification*, London: UCL.

Ijzereef, F. G. (1989) 'Social differentiation from animal bone studies', in Serjeantson, D. and Waldron, T. (eds) *Diet and Crafts in Towns: The Evidence of Animal Remains from the Roman to Post-Medieval Periods*. BAR British Series 199, pp. 41–54. Oxford: Archaeopress.

Jones, G. G. and Sadler, P. (2012) 'Age at death in cattle: methods, older cattle and known-age reference material', *Environmental Archaeology*, 17, 11–28.

Klein, R. G. and Cruz-Uribe, K. (1984) *The Analysis of Animal Bones from Archaeological Sites*, Chicago: University of Chicago Press.

Kolska Horwitz, L. (2011) 'Where is the ASWA heading?', *ICAZ Newsletter*, 12(2), 3.

Larson, G., Dobney, K., Albarella, U., Fang, M., Matisoo-Smith, E., Robins, J., Lowden, S., Finlayson, H., Brand, T., Willerslev, E., Rowley-Conwy, P., Andersson, L., and Cooper, A. (2005) 'Worldwide phylogeography of wild boar reveals multiple centres of pig domestication', *Science*, 307, 1618–21.

Lauwerier, R. C. G. M. and Plug, I. (eds) (2003) *The Future from the Past: Archaeozoology in Wildlife Conservation and Heritage Management*, Oxford: Oxbow Books.

Legge, A. J. (1978) 'Archaeozoology—or zooarchaeology?', in Brothwell, D. R., Thomas, K. D., and Clutton-Brock, J. (eds) *Research Problems in Zooarchaeology*. Institute of Archaeology Occasional Publications 3, pp. 129–32. London: University of London, Institute of Archaeology.

Lemoine, X., Zeder, M. A., Bishop, K. J., and Rufolo, S. J. A. (2014) 'New system for computing dentition-based age profiles in *Sus scrofa*', *Journal of Archaeological Science*, 47, 179–93.

Luikart, G., Gielly, L., Excoffier, L., Vigne, J. D., Bouvet, J., and Taberlet, P. (2001) 'Multiple maternal origins and weak phylogeographic structure in domestic goats', *Proceedings of the National Academy of Sciences of the United States of America*, 98(10), 5927–32.

Lyman, R. L. (1996) 'Applied zooarchaeology: the relevance of faunal analysis to wildlife management', *World Archaeology*, 28(1), 110–25.

Lyman, R. L. (2008) *Quantitative Palaeozoology*, Cambridge: Cambridge University Press.

Lyman, R. L. and Cannon, K. P. (eds) (2004) *Zooarchaeology and Conservation Biology*, Salt Lake City: University of Utah Press.

Mainland, I. L. (1998) 'Dental microwear and diet in domestic sheep (*Ovis aries*) and goats (*Capra hircus*): distinguishing grazing and fodder-fed ovicaprids using a quantitative analytical approach', *Journal of Archaeological Science*, 25(12), 1259–71.

Maltby, M. (ed.) (2006) *Integrating Zooarchaeology*, Oxford: Oxbow Books.

Marciniak, A. (2005) *Placing Animals in the Neolithic: Social Zooarchaeology of Prehistoric Farming Communities*, London: UCL Press.

Martin, D. L., Harrod, L. P., and Perez, V. R. (2013) *Bioarchaeology: An Integrated Approach to Working with Human Remains*, New York: Springer.

Matsui, A. (2008) *Fundamentals of Zooarchaeology in Japan*, Kyoto: Kyoto University Press.

Meadow, R. H. (1980) 'Animal bones: problems for the archaeologist together with some possible solutions', *Paléorient*, 6, 65–77.

Meadow, R. H. (1999) 'The use of size index scaling techniques for research on archaeozoological collections from the Middle East', in Becker, C., Manhart, H., Peters, J., and Schibler, J. (eds) *Historia Animalium ex Ossibus: Festschrift für Angela von den Driesch*, pp. 285–300. Rahden: Verlag Marie Leidorf GmbH.

Meniel, P. (ed.) (1989) *Animal et pratique religieuses: les manifestations materielles*. Anthropozoologica, second numéro special. Paris: Bulletin de l'Association L'Homme et l'Animal.

Miles, A. E. W. and Grigson, C. (1990) *Colyer's Variations and Diseases of the Teeth of Animals*, Cambridge: Cambridge University Press.

Miles Gilbert, B. (1993) *Mammalian Osteology*, Columbia: Missouri Archaeological Society.

Miles Gilbert, B., Martin, L. D., and Savage, H. G. (1985) *Avian Osteology*, Flagstaff: B. Miles Gilbert Publisher.

Minniti, C., Valenzuela-Lamas, S., Evans, J., and Albarella, U. (2014) 'Widening the market: strontium isotope analysis on cattle teeth from Owslebury (Hampshire, UK) highlights changes in livestock supply between the Iron Age and the Roman period', *Journal of Archaeological Science*, 42, 305–14.

Mona, S., Catalano, G., Lari, M., Larson, G., Boscato, P., Casoli, A., Sineo, L., Di Patti, C., Pecchioli, E., Caramelli, D., and Bertorelle, G. (2010) 'Population dynamic of the extinct European aurochs: genetic evidence of a north-south differentiation pattern and no evidence of post-glacial expansion', *Bmc Evolutionary Biology*, 10, 83.

Morales, A. and Rosenlund, K. (1979) *Fish Bone Measurements: An Attempt to Standardize the Measuring of Fish Bones from Archaeological Sites*, Copenhagen: Steenstrupia.

Morris, J. (2010) 'Commercial zooarchaeology in the United Kingdom', *Environmental Archaeology*, 15(1), 81–91.

O'Connor, T. (1998) 'A critical overview of archaeological animal bone studies', *World Archaeology*, 28(1), 5–19.

O'Connor, T. (2000) *The Archaeology of Animal Bones*, Stroud: Sutton.

Overton, N. J. and Hamilakis, Y. (2013) 'A manifesto for a social zooarchaeology. Swans and other beings in the Mesolithic', *Archaeological Dialogues*, 20(2), 111–36.

Payne, S. (1972) 'On the interpretation of bone samples from archaeological sites', in Higgs, E. S. (ed.) *Papers in Economic Prehistory*, pp. 65–81. Cambridge: Cambridge University Press.

Payne, S. (1973) 'Kill-off patterns in sheep and goats: the mandibles from Aşvan Kale', *Anatolian Studies*, 23, 281–303.

Payne, S. (1975) 'Partial recovery and sample bias', in Clason, A. T. (ed.) *Archaeozoological Studies*, pp. 7–17. Amsterdam: North Holland.

Payne, S. and Bull, G. (1988) 'Components of variation in measurements of pig bones and teeth, and the use of measurements to distinguish wild from domestic pig remains', *Archaeozoologia*, 2, 27–65.

Pearson, J. A., Buitenhuis, H., Hedges, R. E. M., Martin, L., Russell, N., and Twiss, K. C. (2007) 'New light on early caprine herding strategies from isotope analysis: a case study from Neolithic Anatolia', *Journal of Archaeological Science*, 34, 2170–9.

Plug, I. (2014a) *What Bone is That? A Guide to the Identification of Southern African Mammal Bones*, Pretoria: Rosslyn Press.

Plug, I. (2014b) 'My ICAZ history', *ICAZ Newsletter*, 15(2), 11.

Popkin, P. R. W., Baker, P., Worley, F., Payne, S., and Hammon, A. (2012) 'The Sheep Project (1): determining skeletal growth, timing of epiphyseal fusion and morphometric variation in unimproved Shetland sheep of known age, sex, castration status and nutrition', *Journal of Archaeological Science*, 39, 1775–92.

Pruvost, M., Schwarz, R., Bessa Correia, V., Champlot, S., Grange, T., and Geigl, E.-M. (2008) 'DNA diagenesis and palaeogenetic analysis: critical assessment and methodological progress', *Palaeogeography, Palaeoclimatology, Palaeoecology*, 266(3–4), 211–19.

Rackham, J. (1994) *Animal Bones (Interpreting the Past)*, Berkeley: University of California Press.

Reitz, E. J. (1987) 'Vertebrate fauna and socioeconomic status', in Spencer-Wood, S. M. (ed.) *Consumer Choice in Historical Archaeology*, pp. 101–19. New York: Plenum Press.

Reitz, E. J. (2009) 'International Council for Archaeozoology (Icaz) Professional Protocols for Archaeozoology'. http://alexandriaarchive.org/icaz/pdf/protocols2009.pdf.

Reitz, E. J. and Wing, E. S. (1999) *Zooarchaeology*, Cambridge: Cambridge University Press.

Rowley-Conwy, P. and Zeder, M. (2014) 'Mesolithic domestic pigs at Rosenhof—or wild boar? A critical re-appraisal of ancient DNA and geometric morphometrics', *World Archaeology*, 46(5), 813–24.

Russell, N. (2012) *Social Zooarchaeology: Humans and Animals in Prehistory*, New York: Cambridge University Press.

Ryder, M. L. (1968) *Animal Bones in Archaeology: A Book of Notes and Drawings for Beginners*, Oxford: Blackwell Scientific.

Schibler, J. (1998) 'OSSOBOOK, a database system for archaeozoology', in Anreiter, P., Bartosiewicz, L., Jerem, E., and Meid, W. (eds) *Man and the Animal World: Studies in Archaeozoology, Archaeology, Anthropology and Palaeolinguistics in Memoriam of Sándor Bökönyi*, pp. 491–510. Budapest: Archaeolingua.

Schiffer, M. B. (1987) *Formation Processes of the Archaeological Record*, Albuquerque: University of New Mexico Press.

Schmid, E. (1972) *Atlas of Animal Bones for Prehistorians, Archaeologists and Quaternary Geologists*, Amsterdam: Elsevier.

Serjeantson, D. (1991) "Rid grasse of bones': a taphonomic study of the bones from midden deposits at the Neolithic and Bronze Age site of Runnymede, Surrey, England', *International Journal of Osteoarchaeology*, 1(2), 73–89.

Spencer Larsen, C. (1999) *Bioarchaeology: Interpreting Behavior from the Human Skeleton*, Cambridge: Cambridge University Press.

Steele, T. E. (2015) 'The contribution of animal bones from archaeological sites: the past and future of zooarchaeology', *Journal of Archaeological Science*, 56, 168–76.

Stevens, R. E. and Hedges, R. E. M. (2004) 'Carbon and nitrogen stable isotope analysis of northwest European horse bone and tooth collagen, 40,000 BP–present: palaeoclimatic interpretations', *Quaternary Science Reviews*, 23, 977–91.

Svensson, E. M., Gotherstrom, A., and Vretemark, M. (2008) 'A DNA test for sex identification in cattle confirms osteometric results', *Journal of Archaeological Science*, 35, 942–6.

Thomas, J. (1990) 'Silent running: the ills of environmental archaeology', *Scottish Archaeological Review*, 7, 2–7.

Thomas, K. (2001) 'Environmental archaeology is dead: long live bioarchaeology, geoarchaeology and human palaeoecology. A comment on "Environmental archaeology is not human palaeoecology"', in Albarella, U. (ed.) *Environmental Archaeology: Meaning and Purpose*, pp. 55–8. Dordrecht: Kluwer Academic Publishers.

Towers, J., Mongomery, J., Evans, J., Jay, M., and Parker Pearson, M. (2010) 'An investigation of the origins of cattle and aurochs deposited in the Early Bronze Age barrows at Gayhurst and Irthlingborough', *Journal of Archaeological Science*, 37, 508–15.

Troy, C. S., Machugh, D. E., Bailey, J. F., Magee, D. A., Loftus, R. T., Cunningham, P., Chamberlain, A., Sykes, B. C., and Bradley, D. G. (2001) 'Genetic evidence for Near-Eastern origins of European cattle', *Nature*, 410, 1088–91.

Uerpmann, H.-P. (1973) 'Animal bone finds and economic archaeology: a critical study of 'osteo-archaeological' method', *World Archaeology*, 4(3), 307–22.

Uerpmann, H.-P. (1979) *Probleme der Neolithisierung des Mittelmeerraums*, Wiesbaden: Dr Ludwig Reicher Verlag.

Vigne, J.-D. and Lefèvre, C. (2010) 'ICAZ 2010 International Conference hugely successful', *ICAZ Newsletter*, 11(2), 1–3.

Viner, S., Evans, J., Albarella, U., and Parker Pearson, M. (2010) 'Cattle mobility in prehistoric Britain: strontium isotope analysis of cattle teeth from Durrington Walls (Wiltshire, Britain)', *Journal of Archaeological Science*, 37, 2812–20.

Walker, R. (1985) *A Guide to Post-Cranial Bones of East African Animals: Mrs. Walker's Bone Book*, Palo Alto: Hylochoerus Press.

Watson, J. P. N. (1979) 'The estimation of the relative frequencies of mammalian species: Khirokitia 1972', *Journal of Archaeological Science*, 6, 127–37.

Wheeler, A. and Jones, A. (1989) *Fishes*, Cambridge: Cambridge University Press.

Wilkinson, K. and Stevens, C. (2003) *Environmental Archaeology: Approaches, Techniques, and Applications*, Stroud: Tempus.

Wright, E., Viner-Daniels, S., Parker Pearson, M., and Albarella, U. (2014) 'Age and season of pig slaughter at late Neolithic Durrington Walls (Wiltshire, UK) as detected through a new system for recording tooth wear', *Journal of Archaeological Science*, 52, 497–514.

Yamazaki, K. and Uyeno, T. (2008) *Jaws of Bony Fishes*, Japan: Art & Science KoBo TAKAI.

Zeder, M. A., Bradley, D. G., Emschwiller, E., and Smith, B. D. (eds) (2006) *Documenting Domestication: New Genetic and Archaeological Paradigms*, London: University of California Press.

PART II

EUROPE

CHAPTER 2

··

HUMANS AND MAMMALS IN THE UPPER PALAEOLITHIC OF RUSSIA

··

MIETJE GERMONPRÉ AND MIKHAIL V. SABLIN

INTRODUCTION

THIS chapter reviews faunal assemblages of Upper Palaeolithic sites in Russia. It considers the human and mammal relationships during the Upper Palaeolithic and focuses on mammoths and large canids—wolf (*Canis lupus*) and/or dog (*Canis familiaris*)—and their association with prehistoric humans. We are especially interested in the topics of mammoth hunting and dog domestication. Both are highly controversial. Whether or not mammoth hunting was practised during the Upper Palaeolithic is still a matter of debate and, despite considerable research, no consensus has been reached about the timing of the domestication of the wolf. We wish to widen and enrich these debates by reviewing the mammoth and canid bone material from the Upper Palaeolithic sites in Russia. Therefore, this chapter pursues the following questions: (1) Is the mammoth ubiquitously found in the Upper Palaeolithic sites of Russia? (2) Are large canids as often present at Siberian sites as they are in the Russian Plain? (3) Could the high frequency of the mammoth remains in several Upper Palaeolithic assemblages be due to hunting by prehistoric humans?

Full geographical coverage is not provided, but some key regions and sites in the Russian Plain and Siberia are selected. For the Early, Middle, and Late Upper Palaeolithic 47 sites from the Russian Plain and 180 Siberian sites with mammal assemblages are listed (Table S2.1). These non-exhaustive lists are compiled from data provided by Vereshchagin and Kuz'mina (1977), Vasil'ev (2003), Goebel (2004), Amirkhanov et al. (2009), Hoffecker et al. (2010), and our own data. A summary is given in Table 2.1.

The environment in Russia during the Upper Palaeolithic has been described as an open landscape with a cold and dry climate. The fauna was composed of

Table 2.1 Summary of the frequency of mammoths and large canids in Upper Palaeolithic sites from the Russian Plain and Siberia

	number	mammoth		large canids	
	sites/assemblages	present	%	present	%
Russian Plain					
Early Upper Palaeolithic	13	12	92.3	10	76.9
Middle Upper Palaeolithic	30	25	83.3	23	76.7
Late Upper Palaeolithic	4	4	100.0	3	75.0
Siberia					
Early Upper Palaeolithic	24	10	41.7	6	25.0
Middle Upper Palaeolithic	31	17	54.8	6	19.4
Late Upper Palaeolithic	125	33	26.4	40	32.0

cold-tolerant species such as woolly mammoth (*Mammuthus primigenius*), woolly rhinoceros (*Coelodonta antiquitatis*), steppe bison (*Bison priscus*), wild horse (*Equus* sp.), reindeer (*Rangifer tarandus*), and arctic hare (*Lepus timidus*). Southern Siberia was also inhabited by large herbivores such as yak (*Poephagus grunniens*), Mongolian gazelle (*Procapra gutturosa*), and argali sheep (*Ovis ammon*). Carnivores included species such as cave lion (*Panthera spelaea*), cave hyena (*Crocuta crocuta spelaea*), wolf (*Canis lupus*), and red (*Vulpes vulpes*) and arctic foxes (*Alopex lagopus*) (Ukraintseva, 1993).

The radiocarbon and AMS dates from the sites discussed in the text are given in Table S2.2. The dates in the text are calibrated in calendar years before 1950 (cal BP), are derived from the radiocarbon dates given in Table S2.2, and have been calculated using the CalPal software (Weninger et al., 2013). The presence of mammoth, mammoth mass accumulation, and large canids is noted. The measurements on skeletal elements of the large canids are compared with those of recent northern wolves from Belgium, Sweden, and Russia (Germonpré et al., 2012; 2015a), Pleistocene European wolves from the Jaurens and Maldier caves in France (Boudadi-Maligne, 2010), Belgium, and the Czech Republic (Germonpré et al., 2012; this study), Palaeolithic dogs (Germonpré et al., 2012), and recent Archaic dogs from Siberia, Greenland, and Northern Canada (Germonpré et al., 2015b) (Tables S2.3, S2.4, and S2.5). In most cases, the fossil canid material is too fragmented to identify these remains as from Palaeolithic dog or Pleistocene wolf; for a description of these two morphotypes, the reader is referred to Germonpré et al. (2009; 2012; 2013; 2015a; 2015b). The canid fragments are termed 'dog-like in size' when their size corresponds best to the size ranges of recent and Palaeolithic dogs and falls outside the ranges of the recent and Pleistocene wolves. Their size suggests in that case that these fragments could be from domestic Palaeolithic dogs. The canid specimens are termed 'wolf-like in size' when their size corresponds best to the size ranges of the wolves.

Is the Mammoth Ubiquitously Found in the Upper Palaeolithic Sites of Russia?

The Russian Plain

Sites/assemblages from the Russian Plain dating from the Early Upper Palaeolithic period amount to at least thirteen. Most assemblages contain mammoth bones (Table 2.1), but in low quantities (Table S2.1). The Kostenki-Borshevo region, near the city of Voronezh, witnessed intensive habitation periods starting from the Early Upper Palaeolithic. Here, twenty-six sites are concentrated in an area of about 20 km². The sites are found primarily on the first and second terrace of the Don River (Klein, 1969; Sinitsyn and Hoffecker, 2006). Kostenki-17 is located on the second terrace (Fig. 2.1). Level II occurs below a layer of volcanic ash, which is dated by $^{40}Ar/^{39}Ar$ to 38,000 years BP (Holliday et al., 2007). In this level, which has a mean calibrated age of 39,000 years BP, an early Upper Palaeolithic artefact assemblage was found. Mammoth is only represented by worked ivory (Praslov and Rogachev, 1982). Cultural level III of Kostenki-14 yielded an early Upper Palaeolithic blade industry including bone tools and beads made from bird long bones (Praslov and Rogachev, 1982). Its mean calibrated age is 34,300 years BP. The three most common mammals in the assemblage are

FIGURE 2.1 Map of northern Eurasia with the most important sites mentioned in the text. 1: Transbaikal sites, 2: Kostenki-Borshevo sites, 3: Belgian sites, 4: Yana and Nikita sites, 5: Altai, 6: Předmostí (Czech Republic), 7: Yudinovo and Eliseevichi sites, 8: Afontova Gora, 9: Verholenskaya and Ust-Khaita sites, 10: Ushki sites.

in decreasing order arctic hare, horse, and large canids; mammoth is represented by only three bones (Vereshchagin and Kuz'mina, 1977).

The Russian Plain assemblages dating from the Middle Upper Palaeolithic period number at least thirty and include, beside the Kostenki sites, other well-known sites such as Avdeevo and Zaraisk. The mammoth occurs in 83.3% of the assemblages (Table 2.1; Table S2.1). Kostenki-12 is located on the second terrace of the Don River (Fig. 2.1). The upper horizon (K-12/I) has a mean calibrated age of 32,900 years BP. A peculiar find from this horizon is a paddle-shaped shovel made from mammoth bone (Klein, 1969). The horse is the most abundant mammal in the assemblage with 95.5%; mammoth remains account for 1% (Vereshchagin and Kuz'mina, 1977). Kostenki-8 is also located on the second terrace (Fig. 2.1). The artifact assemblage from the second layer (K-8/II) is attributed to the Early Gravettian and has a mean calibrated age of 29,400 years BP. Remains of five dwelling structures have been documented here (Sinitsyn and Hoffecker, 2006). The hare is, with 33%, the best represented species in the mammal assemblage; the mammoth has a frequency of 7%. The upper horizon of Kostenki-8 (K-8/I) has a mean calibrated age of 27,000 years BP. Remains of a dwelling with a central hearth and several storage pits were here discovered. Mammoth and hare occur with respectively 23% and 7% in the mammal assemblage (Vereshchagin and Kuz'mina, 1977).

Kostenki-1 is located on the second terrace (Fig. 2.1). The upper layer (K-1/I) is well known for its dwelling complexes with aligned hearths. The cultural industry of K-1/I is assigned to the Eastern Gravettian (Praslov and Rogachev, 1982) and has a mean calibrated age of 26,600 years BP. Mammoth predominates in the mammal assemblage (Vereshchagin and Kuz'mina, 1977). Interestingly, at Kostenki-1/I direct evidence for mammoth hunting exists: a fragment of a flint point that is embedded in a mammoth rib testifies to the direct contact between the animal and a prehistoric hunter (Praslov, 2000). A subset from the mammoth assemblage from K-1/I, from the southern edge of dwelling structure 2, consists of 5 tusks, 1 mandible, and 142 postcranial bones. The mammoth bones of this subset are only slightly weathered and merely 7% has been gnawed. Almost half of this subset is composed of subadult bones, with one or both unfused epiphyses present with the diaphysis. Remains from at least thirteen mammoths were found. Seven animals are juveniles on the basis of the unfused diaphyses or epiphyses of their humeri. Furthermore, a portion of the postcranial material was found here in an articulated state: a partial front foot with the carpals and metacarpals, two series of articulated thoracic vertebrae and several diaphyses with their unfused epiphyses. This suggests that body parts from freshly killed mammoths were deposited at the site. The reconstructed shoulder heights of the adult animals range from 2.2 m to 2.7 m, with a mean of 2.5 m (Germonpré, unpublished data) and are much smaller than the shoulder heights (2.8 m–3.6 m, mean 3.2 m) from the Pleniglacial and Early Glacial mammoths from Belgium (Germonpré et al., 2008). The features of this subset of the mammoth assemblage from Kostenki-1/I compare well with several taphonomic and palaeobiological characteristics of the mammoth complexes of the Epigravettian Yudinovo site (Germonpré et al., 2008) (see below). The relative small size of the Kostenki mammoths suggests that they could have been easier to hunt than the larger-sized western European mammoths. The upper horizon of Kostenki-11 (K-11/Ia) is famous for the remains of a dwelling complex that has a diameter of about 8 m and is composed of mammoth tusks, mandibles, scapulae, pelves, and long bones (Klein, 1969; Praslov and Rogachev, 1982). The mean age of this horizon is 23,100 years BP. The mammal assemblage is almost exclusively composed of mammoth bones (98%) (Vereshchagin and Kuz'mina, 1977).

The non-exhaustive list from the Late Upper Palaeolithic of the Russian Plain counts four sites dating from this period. The mammoth occurs at all of them (Table 2.1, Table S2.1). The Epigravettian Yudinovo site is situated on the right bank of the Sudost' River, a tributary of the Desna (Fig. 2.1). The mean calibrated age of the site is 18,140 years BP. Four mammoth-bone complexes were discovered on the site. In addition to the mammoth bones, huge quantities of ivory hunting tools and ivory ornaments were recovered. Germonpré et al. (2008) studied in detail the taphonomic and palaeobiological characteristics of two of these complexes that comprise mainly mammoth skulls, scapulae, ilia, and limb bones. Many of the scapulae, three ilia, and even one humerus have artificial holes. The human-made holes are typical of cultural mammoth sites of the region. Such perforations were found at Gontsy, Berdyzh, and Mezhirich. The phenomenon has been interpreted as an architectural feature that is involved in the use of these mammoth bones for construction purposes (Soffer, 1985). Most of these perforations at Yudinovo have fresh rims, but the borders of some perforations are smoothed, which suggests that these holes were abraded by use; possibly posts were inserted in these holes (cf. Klein, 1973). Germonpré et al. (2008) demonstrated that the mammoth bones from these complexes originated from freshly killed mammoths. The age and size distribution of the mammoth remains indicate that the bulk of this material is from young animals and prime-aged cows. The adult mammoths had shoulder heights ranging from 2.1 m to 2.9 m (mean 2.3 m), similar to the shoulder heights of the hunted mammoths at Kostenki-1/I (see p. 28) and Yana (see p. 30). The Yudinovo mammoths were killed to obtain large quantities of meat, fat, organs, brain, and fatty bones. The skeletal parts from the adult mammoths were used to build the complexes.

The Epigravettian Eliseevichi site is located about 65 km northeast of Yudinovo (Fig. 2.1) and is slightly younger (calibrated age: 17,150 years BP). Here, remains of at least eight complexes made from mammoth skulls and bones were recovered. In addition, the presence of numerous mammoth milk molars suggest that mammoth calves were hunted for food or for their fur, since calf bones are in general not used in mammoth-bone complexes (cf. Germonpré et al., 2008). In contrast to the other Epigravettian mammoth sites from the Russian Plain and the Ukraine, at Eliseevichi no human-made perforations occur in mammoth scapulae or ilia (Khlopachev, pers. comm.).

Siberia

At least twenty-four sites with remains of large mammals dating from the Early Upper Palaeolithic are known from Siberia. The mammoth occurs at 41.7% of these sites (Table 2.1, Table S2.1). Kamenka, a site located in southern Siberia, lies in the valley of the Brianka River, which belongs to the Selenga River Basin (Fig. 2.1). The two main stratigraphic units are Complex A and Complex B (Germonpré and Lbova, 1996). The early age of Complex A is confirmed by the calibrated age of 42,030 years BP of a cut-marked tibia from a Mongolian gazelle. The gazelle predominates in the faunal list, while the second most common mammal is the horse (Germonpré and Lbova, 1996). Mammoth bones are missing in Kamenka Complex A, but this species is represented by an ivory bracelet (Goebel, 2004). The younger assemblage of Kamenka Complex B (mean calibrated age 33,670 years BP) is not associated with artefacts. The mammoth is represented with one bone, a proximal femur fragment. Varvarina Gora is also located in the Brianka valley (Fig. 2.1). It has a mean calibrated age of 39,970 years BP and is thus somewhat younger than Kamenka Complex A. Bone tools include one ivory implement (Goebel, 2004). The faunal assemblage is, like Kamenka Complex A, dominated by the Mongolian gazelle and the horse, but the woolly rhinoceros is here better represented (Germonpré and Lbova, 1996). Tolbaga, another Transbaikal site, is located in the valley of the Khilok River, a tributary of the Selenga River (Fig. 2.1). Layer 4 contains remains of several dwellings (Vasiliev and Rybin, 2009). The wide range of the radiocarbon dates suggests repeated phases of human occupation (Table S2.2). Layer 4 has a mean calibrated age of 35,400 years BP. The artefact assemblage includes a mammoth rib fragment with traces of polishing. The woolly rhinoceros is the most common animal; other well-represented prey species are the horse and the argali sheep (Vasiliev and Rybin, 2009).

There are at least thirty-one Siberian sites dated to the Middle Upper Palaeolithic. The mammoth occurs in 54.8% of these sites (Table 2.1, Table S2.1). Yana, a remarkable site in the Yana-Indigirka lowland (northern Yakutia), with a mean calibrated age of 32,500 years BP, documents probably the initial stage of the dispersal of Anatomically Modern Humans (AMH) in the Arctic (Fig. 2.1) (Pitulko et al., 2004). Mass accumulations of mammoth bones are here encountered. The site consists of separate habitation areas along a 2.5 km stretch of the Yana River. The Yana site is especially notable since it is one of the few sites with direct evidence of mammoth hunting: two right mammoth scapulae display embedded stone tools. The Yana people hunted mainly young adults (probably females). The shoulder heights (1.8 m–3.2 m, mean 2.3 m) of the adult mammoths (Nikolskiy and Pitulko, 2013) are similar to the sizes of the mammoths from the Gravettian Kostenki-1/I site (see p. 28) and the Epigravettian Yudinovo site (see p. 29).

The number of Siberian sites dating from the Late Upper Palaeolithic adds up to 125; they are thus much more numerous than in the previous Upper Palaeolithic periods. The mammoth is limited to 26.4% of these sites. There is a clear decline in the amount of Siberian sites containing mammoth bones, compared to the earlier periods of the Upper Palaeolithic (Table 2.1, Table S2.1). Furthermore, the range of the mammoth seems to move northward during the Late Glacial (Vasil'ev, 2003). Afontova Gora-1 is situated on the western bank of the Enisei River (Fig. 2.1). The culture horizon yielded numerous remains from mammoth, reindeer, steppe bison, horse, and large canids (Pavlow, 1930). A canid tibia has a calibrated age of 16,900 years BP. The Nikita site, located in the Yana–Indigirka lowland, is slightly younger and dates from 14,000 years BP. Here, direct evidence of mammoth hunting is present: several mammoth ribs have embedded fragments from lithic tools (Pitulko et al., 2014).

The mammoth is almost ubiquitous in Upper Palaeolithic sites of the Russian Plain and is clearly less regularly encountered in the Siberian sites of this period (Table 2.1, Table S2.1). Mammoth mass accumulations, however, are rare. They date from the Middle and Late Upper Palaeolithic and can be found both in Siberia and the Russian Plain (Table S2.2).

ARE LARGE CANIDS AS OFTEN PRESENT AT SIBERIAN SITES AS AT SITES FROM THE RUSSIAN PLAIN?

The Russian Plain

Large canids are present in 76.9% of the assemblages dating from the Early Upper Palaeolithic (Table 2.1, Table S2.1).The faunal assemblage of Kostenki-17/II is dominated by remains of large canids and the second most common species is the horse (Sablin, 2007). A complete skull and two mandibles from Pleistocene wolves have been identified (Germonpré et al., 2012; 2015a). The postcranial bones are all 'wolf-like in size' (Table S2.3). A fragment of a canid humerus from Cultural level III of Kostenki-14 is 'dog-like in size' (Table S2.3).

Large canids occur in 76.7% of the assemblages dating from the Middle Upper Palaeolithic (Table 2.1). Two distal fragments of canid humeri from two individuals found at Kostenki-12/I are 'dog-like in size' (Table S2.3). In the faunal assemblage from the second layer of Kostenki-8 (K-8/II), 'large canid' is the second best represented group in the mammal assemblage with 22% of the remains (Vereshchagin and Kuz'mina, 1977). A canid mandible was identified as a dog (Germonpré et al., 2015a). A maxilla fragment is probably from this same individual (Fig. 2.2). The maxilla shows tooth crowding, a feature regularly encountered in recent and prehistoric dogs (e.g. Germonpré et al., 2012): its third premolar is rotated (Fig. 2.3).

In the upper horizon of Kostenki-8 (K-8/I), 'large canid' is the most common group in the mammal assemblage (50.5%); they are mainly represented by elements of the feet, often in anatomical connection. This suggests the use of pelts at this site (Vereshchagin and Kuz'mina, 1977). A canid femur is 'dog-like in size' (Table S2.3). Large canids are the second best represented group in the Kostenki-1/I assemblage (Vereshchagin and Kuz'mina, 1977). An isolated carnassial (P^4) has a crown length of 22.8 mm. It is smaller than the carnassials from Pleistocene wolves and Palaeolithic dogs, falls inside the size range of the recent Archaic dogs, and is barely larger than the minimum length in recent Northern wolves (Table S2.4). Its size therefore suggests that the animal was 'dog-like in size', and the same applies to the distal fragment of a radius (Table S2.3). In the mammal

FIGURE 2.2 Skull fragment and mandible from the Palaeolithic dog found at Kostenki-8/II.

FIGURE 2.3 Occlusal view of the maxilla of the Palaeolithic dog from Kostenki-8/II, showing the rotated position of the third premolar.

assemblage from Kostenki-11/Ia only a few remains are from large canids (Vereshchagin and Kuz'mina, 1977). Two long bone fragments are 'dog-like in size' (Table S2.3).

Our list of sites dating from the Late Upper Palaeolithic contains four sites; large canids occur at three of them (Table 2.1; Table S2.1). Very few large canid remains are recorded at the Epigravettian Yudinovo site. A distal fragment of a humerus compares well in size with those of recent Archaic dogs (Table S2.3). Therefore, it is described here as 'dog-like in size'. The low frequency of mammoth bones gnawed by carnivores (3%) (Germonpré et al., 2008) suggests that the mammoth-bone complexes were somehow protected or that the dogs were not allowed to roam freely at the site.

At the Epigravettian Eliseevichi site, two canids skulls and one mandible were identified as being from Palaeolithic dogs (Sablin and Khlopachev, 2002; Germonpré et al., 2015a). The most complete dog skull was found in a hearth deposit, near a concentration of mammoth skulls (Polikarpovich, 1968). Its braincase has been perforated on the left and right side. Cut marks are present on the zygomatic and frontal bones. Both carnassial teeth were extracted by the Epigravettian people; for this they damaged the alveoli (Sablin and Khlopachev, 2002; Germonpré et al., 2009). The human modifications and the location of the skull suggest that this dog could have been part of an Epigravettian ritual. Apart from the skulls and the lower jaw, fragments from a humerus, a radius, and a femur are 'dog-like in size' (Table S2.3).

Siberia

Twenty-five percent of the Siberian sites dating from the Early Upper Palaeolithic contain remains from large canids (Table 2.1, Table S2.1). Two sites located in the Transbaikal, Varvarina Gora and Tolbaga, yielded remains of large canids. At Tolbaga, a canid calcaneum is 'wolf-like in size' (Ovodov, 1987) (Table S2.3). Large canids are present at 19.4% of the Siberian sites dating from the Middle Upper Palaeolithic (Table S2.1). In the Razboinichya cave in the Altai (Fig. 2.1), a large canid was described by Ovodov et al. (2011) as an 'incipient dog'. This cave is not a prehistoric site but a cave hyena den (Ovodov et al., 2011). Currently, it is not clear if and how much this specimen differs from local Pleistocene wolves (see Germonpré et al., 2013; 2015a). The arctic Yana site has yielded several remains of large canids. A metatarsal was made into an awl; the size of the bone falls inside the range of modern wolves (Pitulko et al., 2004). Measurements on other canid bones are not available, so it is currently not possible to identify these animals.

Large canids occur in 32% of the Siberian sites dating from the Late Upper Palaeolithic (Table 2.1, Table S2.1). At Afontova Gora-1, a skull from a prehistoric dog was described by Pavlow (1930). This specimen is now lost, but an illustration is available in Pavlow (1930: Plate 1: Fig. 2). Its length (23 cm) compares well with that from recent and Palaeolithic dogs and falls outside the range of Pleistocene and recent Northern wolves (Table S2.5). Therefore, the identification by Pavlow (1930) is here accepted. The Zoological Institute of the Russian Academy of Science holds about ten canid specimens from Afontova Gora-1. The dated tibia (Table S2.2) is incomplete; its distal epiphysis is smaller than those of recent and Pleistocene wolves and is barely larger than that of recent Archaic dogs (Table S2.3). This suggests that this animal is 'dog-like in size'. Verholenskaya Gora is situated on the third terrace of the Angara River (Fig. 2.1). It has a calibrated age of 14,900 years BP. Birulja (1929) indicates the presence of prehistoric dog remains from the site. A multivariate analysis of a lower jaw confirms this (Germonpré et al., 2015a). The canid assemblage from Verholenskaya also includes an incomplete tibia that can be described as 'dog-like in size' (Table S2.3).

Large canids (Pleistocene wolves and/or Palaeolithic dogs) are noticeably less numerous in Siberian sites compared to contemporaneous sites from the Russian Plain, during the whole Upper Palaeolithic (Table S2.1). Palaeolithic dogs and/or canids that are 'dog-like in size' are found in the Russian Plain from the end of the Early Upper Palaeolithic on. In Siberia, clear evidence for the presence of Palaeolithic dogs is so far only recorded in Late Upper Palaeolithic sites. 'Dog-like in size' canids are in general present at sites with mammoth mass accumulations, but occur also in sites dominated by other herbivore species (Tables S2.1, S2.2).

COULD THE HIGH FREQUENCY OF THE MAMMOTH REMAINS IN SEVERAL UPPER PALAEOLITHIC ASSEMBLAGES BE DUE TO HUNTING BY PREHISTORIC HUMANS?

In several of the Russian sites that date from the Middle and Late Upper Palaeolithic both direct (Kostenki-1/I, Yana, Nikita) and indirect evidence (Yudonovo, Eliseevichi) for mammoth hunting by prehistoric humans has been demonstrated (see pp. 28–30). Also at Předmostí (the Czech Republic), a Gravettian site where Palaeolithic dogs are present (Germonpré et al., 2012; 2015a), mammoth hunting has been put forward to explain the extremely rich mammoth assemblage (Oliva, 1997). We have postulated that Palaeolithic dogs could have been used for helping in transporting mammoth body parts from the kill site to the camp site (Germonpré et al., 2012). Shipman (2015) proposes that a tradition of mammoth hunting with Palaeolithic dogs, in combination with the use of complex projectile weaponry, could have originated in the European Gravettian. This tradition would then have spread later in the Gravettian to the east, to European Russia, and continued to exist there until the Epigravettian. Indeed, Palaeolithic dogs and/or 'dog-like in size' canids occur at Předmostí, Kostenki-1/I, Kostenki-11/Ia, Eliseevichi, and Yudinovo, sites that are characterized by mammoth mass accumulations (Table S2.2) and for most of which direct or indirect evidence of mammoth hunting is present. This could imply that prehistoric people hunted the mammoth during the Middle and Late Upper Palaeolithic at these locations. This hypothesis could be further confirmed if, as has been predicted by Shipman (2015: 44), Palaeolithic 'dog-like in size' canids will be found at other mammoth sites, such as those of Yana or Nikita in the Siberian arctic. This essay therefore highlights the need for further detailed analyses (morphometric, biogeochemical and genetic studies) of the large canids that are present at sites with high frequencies of mammoth remains.

CONCLUSION

During the Early Upper Palaeolithic, mammoth remains are ubiquitous at sites from the Russian Plain. In Siberian sites from this period the mammoth is less regularly encountered (Table 2.1, Table S2.1). In both regions, mammoth remains are usually present in small numbers. It is only during the Middle and the Late Upper Palaeolithic that mammoth mass accumulations occur in several sites from the Russian Plain and Siberia. Furthermore, Palaeolithic dog-like canids are also present in many of these sites (Tables S2.1 and S2.2). Interestingly, a recent genetic study has shown that the beginning of the domestication process of the wolf can be situated in Europe before or during the Last Glacial Maximum (Thalmann et al., 2013). The presence of 'dog-like in size' canids in the Russian Plain dating from the Early and Middle Upper Palaeolithic corroborates these results. In Siberia, large canids occur less frequently than at contemporaneous sites of the Russian Plain. And it is only in Late Upper Palaeolithic sites that the first unequivocal traces of early dogs appear in Siberia (Table S2.2). From that time on, sites yielding material from Palaeolithic dogs are more numerous and widespread, from Eliseevichi in the west, Ust'Khaita in the south (Losey et al., 2013), and to Uskhi in the east (Dikov, 1996) (Fig. 2.1). The number of prehistoric sites in Siberia with mammoth bones decreases during the Late Upper Palaeolithic (Table 2.1, Table S2.1). Genetic research has revealed a severe and sudden decline in the size of mammoth populations starting at 20,000–15,000 years BP (Palkopoulou et al., 2013). It is possible that the environmental fluctuations at the end of the Pleistocene, which had a significant role in the demographic changes that the mammoth experienced during this time, combined with the additional stresses caused by prehistoric humans and their Palaeolithic dogs, finally led to the extinction of the woolly mammoth on the continent.

ACKNOWLEDGMENTS

We thank Umberto Albarella and Hannah Russ for inviting us to contribute to this volume. Mogens Andersen (Natural History Museum of Denmark) and Daniela Kalthoff (Swedish Museum of Natural History) permitted us to study material held in their care. Financial support to Mikhail V. Sablin was provided by the Russian Foundation for Basic Research (Grant 13-04-00203). This paper is dedicated to the memory of our friend and colleague Nikulai Praslov (1937–2009).

REFERENCES

Amirkhanov, H., Akhmetgaleeva, N., Buzhilova, A., Burova, N., Lev, S., and Maschenko, E. (2009) *Palaeolithic Studies in Zaraysk 1999–2005*, Moscow: Paleograph Press.

Birulya, A. A. (1929) *Rapport préliminaire sur les mammifères 'des débris de cuisine' d'une station de l'homme préhistorique de l'âge de pierre sur le mont Verholensk près Irkoutsk*. Doklady Akademii Nauk SSSR: 91–3.

Boudadi-Maligne, M. (2010) 'Les Canis pléistocènes du sud de la France: approche biosystématique, evolutive et biochronologique'. Thèse pour obtenir le grade de Docteur, Université Bordeaux 1, Spécialité : Préhistoire et Géologie du Quaternaire.

Dikov, N. N. (1996) 'The Ushki sites, Kamchatka Peninsula', in West, F. H. (ed.) *American Beginnings, the Prehistory and Palaeoecology of Beringia*, pp. 244–50. Chicago: University of Chicago Press.

Germonpré, M., Lázničková-Galetová, M., Losey, R. J., Jannikke Räikkönen, J., and Sablin, M. V. (2015a) 'Large canids at the Gravettian Předmostí site, the Czech Republic: the mandible', *Quaternary International*, 359–60, 261–79.

Germonpré, M., Lázničková-Galetová, M., and Sablin, M. (2012) 'Palaeolithic dog skulls at the Gravettian Předmostí site, the Czech Republic', *Journal of Archaeological Science*, 39, 184–202.

Germonpré, M. and Lbova, L. (1996) 'Mammalian remains from the Upper Palaeolithic site of Kamenka, Buryatia (Siberia)', *Journal of Archaeological Science*, 23, 35–57.

Germonpré, M., Sablin, M. V., Després, V., Hofreiter, M., Lázničková-Galetová, M., Stevens, R. E., and Stiller, M. (2013) 'Palaeolithic dogs and the early domestication of the wolf: a reply to the comments of Crockford and Kuzmin', *Journal of Archaeological Science*, 40, 786–92.

Germonpré, M., Sablin, M. V., Khlopachev, G. A., and Grigoreiva, G. V. (2008) 'Possible evidence of mammoth hunting during the Epigravettian at Yudinovo, Russian Plain', *Journal of Anthropological Archaeology*, 27, 475–92.

Germonpré, M., Sablin, M. V., Lázničková-Galetová, M., Després, V., Stevens, R. E., Stiller, M., and Hofreiter, M. (2015b) 'Palaeolithic dogs and Pleistocene wolves revisited: a reply to Morey (2014)', *Journal of Archaeological Science*, 54, 210–16.

Germonpré, M., Sablin, M. V., Stevens, R. E., Hedges, R. E. M., Hofreiter, M., and Després, V. (2009) 'Fossil dogs and wolves from Palaeolithic sites in Belgium, the Ukraine and Russia: osteometry, ancient DNA and stable isotopes', *Journal of Archaeological Science*, 36, 473–90.

Goebel, T. (2004) 'The Early Upper Palaeolithic of Siberia', in Brantingham, P. J., Kuh, S. L., and Derry, K. W. (eds) *The Early Upper Palaeolithic beyond Western Europe*, pp. 162–95. Berkeley: University of California Press.

Hoffecker, J. F., Kuz'mina, I. E., Syromyatnikova, E. V., Anikovich, M. V., Sinitsyn, A. A., Popov, V. V., and Holliday, V. T. (2010) 'Evidence for kill-butchery events of Early Upper Paleolithic age at Kostenki, Russia', *Journal of Archaeological Science*, 37, 1073–89.

Holliday, V. T., Hoffecker, J. F., Goldberg, P., Macphail, R. I., Forman, S. L., Anikovich, M., and Sinitsyn, A. (2007) 'Geoarchaeology of the Kostenki-Borshchevo sites, Don River valley, Russia', *Geoarchaeology*, 22, 181–228.

Klein, R. G. (1969) *Man and Culture in the Late Pleistocene: A Case Study*, San Francisco: Chandler Publishing Company.

Klein, R. G. (1973) *Ice-Age Hunters of the Ukraine*, Chicago: The University Press.

Losey, R. J., Garvie-Lok, S., Leonard, J. A, Katzenberg, M. A., Germonpré, M., Nomokonova, T., Sablin, M. V., Goriunova, O. I., Berdnikova, N. E., and Savel'ev, N. A. (2013) 'Burying dogs in ancient Cis-Baikal, Siberia: temporal trends and relationships with human diet and subsistence practices', *PLoS ONE*, 8(5), DOI: 10.1371/journal.pone.0063740.

Nikolskiy, P. and Pitulko, V. (2013) 'Evidence from the Yana Palaeolithic site, Arctic Siberia, yields clues to the riddle of mammoth hunting', *Journal of Archaeological Science*, 40, 4189–97.

Oliva, M. (1997) 'Les sites pavloviens près de Předmostí: a propos de la chasse au mammouth au Paléolithique Supérieur', *Acta Musei Moraviae, Scientiae Sociales*, 82, 3–64.

Ovodov, N. D. (1987) 'Fauna paleoliticheskih poselenii Tobalga i Varvarina Gora v zapadnom Zabaikal'e', in Rezanova, V. V., Bazarova, L. D., and Hamzina, E. A. (ed.) *Prirodnaya Sreda i Drevnie Ludi Pozdnego Antropogena*, pp. 122–40. Ulan-Ude: Akademiya nauk SSSR.

Ovodov, N. D., Crockford, S. J., Kuzmin, Y. V., Higham, T. F. G., Hodgins, G. W. L., and van der Plicht, J. (2011) 'A 33,000-year-old incipient dog from the Altai Mountains of Siberia: evidence of the earliest domestication disrupted by the Last Glacial Maximum', *PLoS One*, 6(7), DOI: 10.1371/journal.pone.0022821.

Palkopoulou, E., Dalén, L., Lister, L. M., Vartanyan, S., Sablin, M., Sher, A., Nyström Edmark, V., Brandström, M. D., Germonpré, M., Barnes, I., and Thomas, J. A. (2013) 'Holarctic genetic structure and range dynamics in the woolly mammoth', *Proceedings of the Royal Society, B*, 280, DOI: 10.1098/rspb.2013.1910.

Pavlow, M. (1930) 'Mammifères post-tertiaires trouvés aux bords du Volga près de Senguiley et quelques formes provenant d'autres localités', *Annuaire de la Société Paléontologique de Russie*, 9, 1–42.

Pitulko, V. V., Nikolsky, P. A., Girya, E. U., Basilyan, A. A., Tumskoy, V. E., Koulakov, S. A., Astakhov, S. N., Pavlova, E. Y., and Anisimov, M. A. (2004) 'The Yana RHS site: humans in the Arctic before the Last Glacial Maximum', *Science*, 303, 52–6.

Pitulko, V. V., Pavlova, E., and Basilyan, A. A. (2014) 'Mammoth 'graveyards' of the northern Yana-Indighirka lowland, Arctic Siberia', in Kostopoulos, D. S., Vlachos, E., and Tsoukala, E. (eds) *Abstract Book of the VIth International Conference on Mammoths and Their Relatives*. Scientific Annals, School of Geology, Special Volume 102, p. 155. Thessaloniki: Aristotle University of Thessaloniki.

Polikarpovich, K. M. (1968) *Paleolit Verhnego Podneprov'ya*, Minsk: Nauka i Technika.

Praslov, N. D. (2000) 'Outils de chasse du Paleolithique de Kostenki', *Anthropologie et Préhistoire*, 111, 37.

Praslov, N. D. and Rogachev, A. N. (1982) *Palaeolithic of the Kostenki-Borshchevo Area on the River Don 1879–1979*, Leningrad: Nauka.

Sablin, M. V. (2007) 'New researches of the Kostenki faunal complexes of the deposits below the Campanian Ignimbrite (CI) Y5 tephra (K 6, K 12, K 17 sites)', Uglešić, A., Maršić, D., and Fabijanić, T. (eds) *Abstract Book, European Association of Archaeologist (EAA) 13th Annual Meeting September 18th–23rd, 2007, Zadar, Croatia*, pp. 286–7. Zadar: University of Zadar.

Sablin, M. V. and Khlopachev, G. A. (2002) 'The earliest Ice Age dogs: evidence from Eliseevichi', *Current Anthropology*, 43, 795–9.

Shipman, P. (2015) 'How do you kill 86 mammoths? Taphonomic investigations of mammoth megasites', *Quaternary International*, 359–60, 38–46.

Sinitsyn, A. A. and Hoffecker, J. F. (2006) 'Radiocarbon dating and chronology of the Early Upper Paleolithic at Kostenki', *Quaternary International*, 152–3, 164–74.

Soffer, O. (1985) *The Upper Paleolithic of the Central Russian Plain*, San Diego: Academic Press.

Thalmann, O., Shapiro, B., Cui, P., Schuenemann, V. J., Sawyer, S. K., Greenfield, D. L., Germonpré, M. B., Sablin, M. V., López-Giráldez, F., Domingo-Roura, X., Napierala, H., Uerpmann, H.-P., Loponte, D. M., Acosta, A. A., Giemsch, L., Schmitz, R. W., Worthington, B., Buikstra, J. E., Druzhkova, A., Graphodatsky, A. S., Ovodov, N. D., Wahlberg, N., Freedman, A. H., Schweizer, R. M., Koepfli, K.-P., Leonard, J. A., Meyer, M., Krause, J., Pääbo, S., Green, R. E., and Wayne, R. K. (2013) 'Complete mitochondrial genomes of ancient canids suggest a European origin of domestic dogs', *Science*, 342(6160), 871–4.

Ukraintseva, V. V. (1993) *Vegetation Cover and Environment of the 'Mammoth Epoch' in Siberia*, The Mammoth Site of Hot Springs: Larry Agenbroad.

Vasil'ev, S. A. (2003) 'Faunal exploitation, subsistence practices and Pleistocene extinctions in Paleolithic Siberia', *Deinsea*, 9, 513–56.

Vasiliev, S. G. and Rybin, E. P. (2009) 'Tolbaga: Upper Paleolithic settlement patterns in the Trans-Baikal region', *Archaeology, Ethnology and Anthropology of Eurasia*, 37, 13–34.

Vereshchagin, N. K. and Kuz'mina, I. E. (1977) 'Remains of mammals from Palaeolithic sites on the Don and Desna Rivers', *Proceedings of the Zoological Institute, Academy of Sciences of the USSR*, 72, 77–110.

Weninger, B., Jöris, O., and Danzeglocke, U. (2013) *The Monrepos Radiocarbon Calibration and Paleoclimate Research Package (CALPAL)*, Monrepos: Monrepos Archäologisches Forschungszentrum und Museum für menschliche Verhaltensevolution.

CHAPTER 3

THE ZOOARCHAEOLOGY OF COMPLEXITY AND SPECIALIZATION DURING THE UPPER PALAEOLITHIC IN WESTERN EUROPE

changing diversity and evenness

KATHERINE BOYLE

INTRODUCTION

THE Western European Upper Palaeolithic is a broadly defined period dated very approximately to 40,000 to 11,000 BP and has long been a focus of attention in archaeo-zoological studies (see Serangeli, 2006 and references therein). As long ago as the mid-nineteenth century, Lartet and Christy (1865–1875) wrote of the 'Reindeer Age' in the Périgord region of southwest France. Since then the archetypal picture of regional faunas of much of Western Europe has been one of herds of reindeer (*Rangifer tarandus*) crossing open landscapes much as caribou do today in North America (Spiess, 1979). Across open tundra small groups of hunters eked out an existence by dispatching reindeer to the exclusion of almost everything else while simultaneously painting images on cave walls which reflect ability likened by the Abbé Henri Breuil to that of Michelangelo in the Sistine Chapel (Breuil, 1952; Smith, 2004; Fagan, 2010). Thus the image of the specialized reindeer hunter emerged, with faunal assemblages frequently dominated by reindeer to a degree unlikely to reflect local or regional natural faunal environments. We know that other animals, including those of comparable size, were regularly hunted: bones of all the large herbivores encountered in faunal assemblages bear cut-marks. What remains uncertain is the role of the various species in human diet and subsistence. Ranging from mammoth (*Mammuthus primigenius*) to hare (*Lepus* sp.), birds (Laroulandie, 2000) and fish (Le Gall, 1984; 2009; 2010), all types of faunal resources were exploited somewhere at some point during the Upper Palaeolithic, in particular wild horse (*Equus ferus*), red deer (*Cervus elaphus*), steppe bison (*Bison priscus*), and aurochs (*Bos primigenius*).

Over the last twenty-five years several attempts have been made to determine the degree to which Upper Palaeolithic subsistence was characterized by a strategy based on some degree of specialization, especially in southwest France (Enloe, 1993; Mellars, 1996; 2004; Fontana, 2000; Grayson et al., 2001; Grayson and Delpech, 2002; Costamagno, 2004). These studies have focused on the two chronological extremes of the Upper Palaeolithic, with greatest attention devoted to the Early Upper Palaeolithic and comparatively little given to the period between the two (see Boyle, 1990). This chapter also focuses on southwest France but it begins to address this chronological gap, tracing assemblage structure through a 35,000-year period up to and including the final Upper Palaeolithic on which it provides some focus. We move away from the assumption that assemblages which include 90% (NISP or MNI) or more of the same species are specialized (see Mellars 1973; 1989; 2004) and seek to determine whether such assemblages are in fact much more diverse than commonly assumed (assemblages of only one species are excluded, as examples of either human hunting of just one resource, or occurrences of natural or carnivore-caused single-species death sites). Most faunal assemblages are time-averaged palimpsest deposits which can be dated only approximately and result from several behavioural episodes exploiting more than one species. We also move away from traditional site-based approaches to Palaeolithic archaeozoology. Whereas most studies focus on one site, this chapter advocates a macroscale approach, comparing assemblages from multiple sites. Although we see distinct variation in assemblage formation and preservation, excavation, recording, and analytical techniques and standards (e.g. not all publications record raw data, some only list species present), it does not mean that data are unusable, only that a common denominator must be found and that the information available must be treated with care. The data used here are derived from reports published during the last fifty years so many issues involved in working with 'old' excavations do not apply. Assemblages included are those with at least some rodent, bird, or fish, as these assemblages are likely to be relatively complete despite frequent lack of sieving and flotation during the 1960s and '70s. All the reports consulted provide counts of bones per species (NISP). Minimum Number of Individual values are not used. When only these values are available, sites are omitted, other than in a record of number of large herbivore species recorded (NTAXA).

The faunal data considered here derive from an as yet unpublished Upper Palaeolithic database compiled by the author. Although doubtless far from complete, it covers most of Western Europe and is derived primarily from published sources. Discussion here focuses on NISP data from more than three hundred occurrences in southwest France, south of the Charente and north of the Garonne rivers. Each occurrence is a level defined primarily according to material culture content or sediment characteristic. Where available C14 dates are noted and location is recorded as latitude and longitude, plus altitude. In this chapter geographic and topographic variation is not considered.

That dimension remains to be considered with a fully macroecological approach (*sensu* Brown, 1995; Franklin, 2009; Boyle, work in progress).

THE UPPER PALAEOLITHIC FAUNA
OF SOUTHWEST FRANCE

Throughout most of the Upper Palaeolithic of southwest France, the most regularly observed species are the reindeer and horse (Mellars, 1973; Delpech, 1983; Boyle, 1990; Costamagno, 1999; Fontana, 2000), although mean species frequencies vary considerably from period to period as the values presented in Table 3.1 show.

Châtelperronian through Proto- and Lower Aurignacian assemblages from sites in southwest France dated to between *c.*45,000–40,000, 42,000–39,500, and 40,000–38,000 cal BP respectively, are often dominated by reindeer, although horse runs a close second and actually occurs at a greater proportion of sites than does reindeer (Boyle, 2010). However, in isolated occurrences the reindeer is the only large herbivore species recorded (e.g. Abri Pataud levels 13/14, Bouchud, 1975), something which cannot be said of the horse. Such values help to account for much of the heated debate regarding (reindeer) specialization as a marker of the development of the true Early Upper Palaeolithic and the demise of the Neanderthals (Grayson and Delpech, 2002; Mellars, 2004). However, there are a further twelve large herbivore species regularly recorded and their presence, with cut bones demonstrating undeniable human activity, shows that hunting was not entirely focused on one or even two resources.

During the 'Classic' or Later Aurignacian (*c.*38,000–34,000 cal BP) reindeer values decline noticeably, falling to a mean value of 36% NISP, accompanied by higher frequencies of a wider range of species observed at more sites than was the case earlier on, e.g. red deer, steppe bison, aurochs, Alpine ibex (*Capra ibex*), and chamois (*Rupicapra rupicapra*). This decline, however, is relatively short-lived. In fact, average reindeer figures peak during a period roughly equivalent to *c.*34,000–31,000 cal BP (97% NISP) or the 'Gravettian' in its strictly French sense (i.e. Perigordian IV). Average reindeer values then decline to *c.*64% NISP during the Upper Perigordian (Perigordian Va, b, and c, *c.*31,000–29,000 cal BP: Middle Gravettian *sensu lato*) but rise again, this time to a mean of 91% NISP, during the Protomagdalenian (*c.*29,000–25,000 cal BP; Later Gravettian). This period and the Solutrean (25,000–22,500 cal BP) see Late Glacial Maximum (LGM) conditions, and an associated narrowing of the available resource base. The Later Solutrean, however, also sees the beginning of a slow but inexorable decline in average reindeer values.

The period traditionally and most frequently associated with the Upper Palaeolithic reindeer hunter is the Magdalenian (22,500–14,000 cal BP)—see Kuntz and Costamagno (2011) for a useful general study of the human–reindeer relationship in southwest France. However, average relative frequencies of the reindeer are often lower than

Table 3.1 Average % NISP frequencies of the major species regularly observed at Upper Palaeolithic sites in southwest France and across the rest of Western Europe

	Southwest France								Western Europe							
	Reindeer	Red deer	Bovids	Horse	Chamois	Ibex	Roe deer	Wild Boar	Reindeer	Red deer	Bovids	Horse	Chamois	Ibex	Roe deer	Wild Boar
Châtelperronian	36.2	3.83	19.53	33.4	0.26	0.29	1.16	0.39	32.02	5.30	18.25	29.44	1.22	1.13	1.28	0.88
Lower Aurignacian	78.3	0.59	6.42	12.3	0.29	1.25	0	0.14	58.50	6.81	7.22	13.74	1.47	4.55	0.48	0.42
Later Aurignacian	36.6	12.65	19.7	13.2	6.67	3.67	3.23	1.25	31.99	13.65	8.29	13.44	5.19	5.82	2.61	1.23
Gravettian	97.3	0.48	0.33	1.73	0.04	0.09	0.04	0	10.25	28.02	4.07	16.26	5.85	14.37	3.43	2.32
Upper Perigordian	63.5	15.68	4.85	7.7	2.19	4.79	0.48	0.05	59.17	19.11	5.53	8.37	2.04	4.26	0.42	0.11
Protomagdalenian	91.3	1.02	2.43	2.96	1.29	0.52	0.36	0	-	-	-	-	-	-	-	-
Solutrean	87.6	0.28	0.83	7.86	0.48	0.67	0.02	0.02	26.79	29.91	2.87	10.54	4.97	18.91	2.18	0.90
Lower Magdalenian	72.5	0.44	0.56	18.4	1.31	6.53	0	0.06	41.06	21.75	3.46	15.82	4.10	14.27	0.45	0.07
Middle Magdalenian	86.8	1.27	1.64	2.87	3.11	2.98	0.06	0.22	63.51	9.11	3.35	5.40	2.71	5.57	0.27	0.18
Upper Magdalenian	57.5	5.46	4.59	18	0.81	9.41	0.7	0.78	28.71	16.06	3.25	18.38	2.71	12.98	1.32	4.33
Azilian	0.09	42.94	4.4	5.72	0.79	10.9	8.95	26.2	4.82	44.45	4.84	6.31	1.71	12.22	5.30	8.63

expected (see Table 3.1) and many assemblages include a wider range of species than the Upper Palaeolithic mean. In fact, the Magdalenian sees reindeer falling to 73% NISP in phases I/II, with horse and ibex a distant second and third respectively and the range of species similarly declining. The Middle Magdalenian (*c*.21,000–19,000 cal BP) sees a mean of 87% NISP, a value which hides considerable variation as values in some places exceed 90% but in others are considerably lower (e.g. St Germain-la-Rivière *c*.6% NISP; Lachaud 71% NISP; Combe Cullier 4'2 > 92% NISP).

The Upper Magdalenian through Azilian (*c*.19,000–11,000 cal BP) see reindeer gradually begin to disappear as early post-glacial conditions begin to develop. Horse population levels also appear to begin to decline, although, unlike the reindeer, they do not disappear (Boyle, 2006). The mammoth, woolly rhinoceros (*Coelodonta antiquitatis*), and saiga (*Saiga tatarica*), to name just three species, have already disappeared. Red deer, roe deer (*Capreolus capreolus*), and wild boar (*Sus scrofa*) increase (Delpech, 1983; Boyle, 1990: 151–3), and large bovids are now usually represented by aurochs, rather than bison. This new faunal spectrum is associated with increasing humidity, temperature, and woodland vegetation, frequently assumed to result in a period of increased resource variability/richness—a 'Broad Spectrum' (*sensu* Stiner, 2001). However, period-mean numbers of large herbivore prey species, here indicated as NTAXA, actually remain below the Upper Palaeolithic average. Evidence for the broadening of the resource base and its exploitation which is frequently associated with early post-glacial conditions (Binford, 1968; Stiner, 2001; Stutz et al., 2009) is not seen. It is the distribution of species frequencies across NTAXA which changes, a trend which can be explored in greater detail through the use of quantitative diversity and related measures.

Specialization vs Diversity

Today archaeozoologists regularly measure faunal assemblage structure using standard indices of ecological biodiversity (Magurran, 1988) and evenness (Smith and Wilson, 1996; Heip et al., 1998), measuring the distribution of abundance amongst species rather than the abundance itself. They provide an alternative means of determining whether faunal assemblage structure is 'specialized' (here defined as being dominated by one species in a ratio of at least 5:1), 'even' (showing how similar species are in their abundance (Magurran, 2004: 18)), or 'diverse' (a measure which assesses evenness against richness or number of species, NTAXA). The latter are also known as measures of 'heterogeneity'. Some interest in assemblage 'diversity' has been paid to later phases of the Upper Palaeolithic at individual sites in southwest France (Grayson et al., 2001), but most has focused on the Mousterian through Aurignacian (Grayson and Delpech, 2002; Hockett and Haws, 2009), tackling issues relating to the origins of modern human behaviour and subsistence 'specialization' (see Mellars, 2004). Only limited attention has been given to patterns over the period between these two chronological extremes (see Boyle, 1990).

The Simpson and Shannon indices are two of many such measures but it is these which have received greatest attention in archaeozoology and their use here, by means of PAST (http://folk.uio.no/ohammer/past/) and MVSP (http://www.kovcomp.com/) software, ensures broad comparability with existing results, such as those presented and discussed by Grayson and Delpech (2002). Simpson diversity values reflect the dominance of an assemblage by a species and can be argued to be most appropriate here, measuring the degree of species concentration in an assemblage (see e.g. Grayson, 1984; Grayson and Delpech, 2002). The Shannon Index shown below enables us to measure evenness (the reader is referred to Magurran (1988) for a detailed discussion of individual indices).

Simpson	Shannon Index (Evenness)
$B=(\sum p_i^2)^{-1}$	$-\sum p_i \ln p_i / \ln S$
where p_i is the proportion rather than percentage of species i in an assemblage, and S is the number of species	

Table 3.2 presents mean herbivore (prey) NTAXA, diversity, and evenness for each period in the Dordogne/Vézère area of the Périgord region, from the Châtelperronian to Azilian and equivalent values for Western Europe as a whole. Fig. 3.1 shows the chronological trend in evenness values plotted against absolute age (dates cal BP).

Upper Palaeolithic values of both diversity and evenness are at their highest during the Transition (Châtelperronian) and Protoaurignacian. They are periods of particularly diverse species representation (Boyle, 2010; see also Grayson and Delpech, 2002), which contrast with the subsequent Lower Aurignacian which sees slightly more assemblages dominated by reindeer (Mellars, 1973; Grayson and Delpech, 2002; Boyle, 2010) and reduced average diversity and evenness values. Despite the fact that assemblages such as Abri Pataud level 14 yield reindeer frequencies in excess of 99% NISP, with four other herbivore prey species making up the remaining 1%, most assemblages are much more diverse than is commonly asserted, an observation which supports Grayson and Delpech's assertion of little difference between the Late Middle Palaeolithic (Mousterian, MIS5-3), Châtelperronian, and Lower Aurignacian (Grayson and Delpech, 2002: 1445). Moving on several thousand years, average evenness and diversity values associated with the Later Aurignacian are almost identical to those of the earlier Transition period (see Table 3.2) and are thereby significantly higher than those observed during the Lower Aurignacian. Mean reindeer values are also very similar during these periods, both under 37%, whereas other species do not show this degree of similarity. These high diversity and evenness values are reflected by a relatively high mean number of species (6.5) during the Later Aurignacian. In summary, the Aurignacian is therefore a period of significant variation in southwest France.

As we move on into the Middle Upper Palaeolithic (MUP) things change. Assemblage diversity values are at their lowest at this point. Corresponding broadly with the cold, glacial conditions and discontinuous permafrost of the LGM. The Gravettian

Table 3.2 Table listing mean NTAXA$_{herbivore}$, Evenness and Diversity values according to individual periods

	Southwest France			Western Europe (excluding southwest France)		
	NTAXA$_{herb}$	Evenness	Diversity	NTAXA$_{herb}$	Evenness	Diversity
Châtelperronian	5.37	0.66	0.62	5.88	0.65	0.63
Protoaurignacian	5.5	0.67	0.65	5.5	0.67	0.65
Lower Aurignacian	4.88	0.43	0.29	4.97	0.54	0.41
Later Aurignacian	6.5	0.64	0.64	5.8	0.61	0.55
Gravettian	6.6	0.31	0.19	6.23	0.53	0.53
Upper Perigordian	6.27	0.38	0.34	6.23	0.4	0.37
Protomagdalenian	4.8	0.37	0.16	-	-	-
Solutrean	3.89	0.42	0.2	4.71	0.57	0.45
Badegoulian	4.27	0.52	0.37	4.4	0.49	0.39
Lower Magdalenian	5	0.37	0.26	5.02	0.5	0.39
Middle Magdalenian	5	0.43	0.33	5.41	0.39	0.3
Upper Magdalenian	4.57	0.5	0.3	5.36	0.55	0.46
Azilian	4.75	0.74	0.65	5.38	0.57	0.49

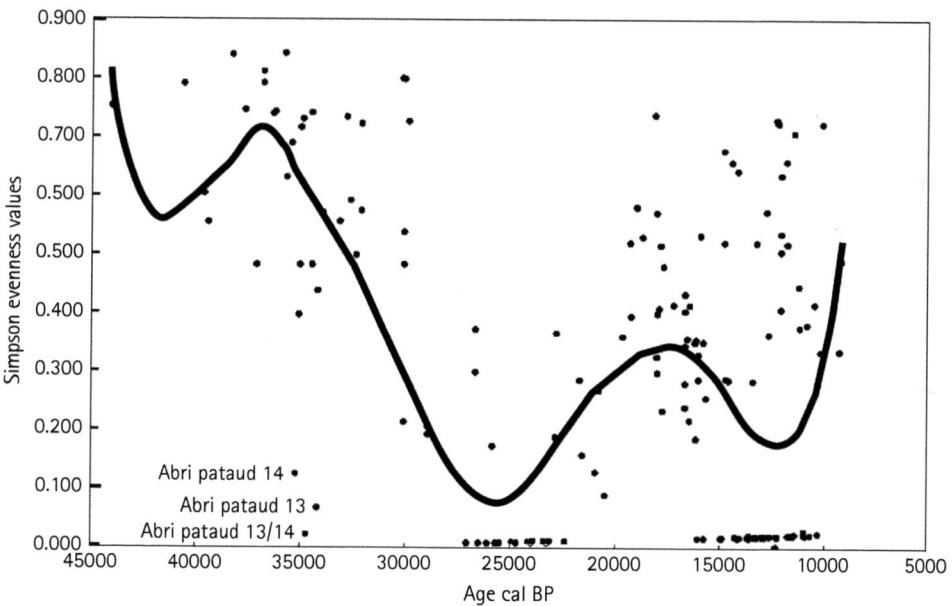

FIGURE 3.1 Chronological distribution of Simpson diversity index values at individual sites in southwest France, identifying a small group of Early Aurignacian specialized assemblages. Graph by author.

sensu lato (including Gravettian or Perigordian IV, the Upper Perigordian, and Protomagdalenian in southwest France) sees greatest assemblage specialization or narrowness, with average diversity values at their lowest (Table 3.2). Assemblages are frequently dominated by a single species—usually, but not invariably, reindeer. After this point values rise again. During the Solutrean the range of values is smaller and focuses on low-medium figures (> 0.200, < 0.600), but extreme low values of less than 0.2 are not observed.

As the Magdalenian takes over we see a greater degree of variation. Indeed, the primary difference between the Early and Late Upper Palaeolithic is the fact that during the later period there is a clear mixture of high and low diversity. Some sites yield narrow or 'specialized' assemblages while others display clear evidence of broad, evenly or relatively evenly distributed species frequencies per site, with several taxa observed. The Late Upper Palaeolithic is not therefore, strictly speaking, characterized by single-species hunting. Some assemblages can be described as specialized and others as diverse. The diverse are significantly less diverse than those of the earliest periods of the Upper Palaeolithic. One interesting observation of the Magdalenian is the fact that at sites where there are several levels yielding NISP data there is relatively little variation in diversity and evenness values between these levels, especially where reindeer is numerically important. At La Madeleine, for example, calculations employing data presented by Delpech (1983), show that at no point do specialized assemblages immediately follow diverse ones and vice versa. Trends in inter-site assemblage structure are gradual and relatively constant within archaeological periods.

Meanwhile, viewing assemblage evenness and diversity in relation to individual species frequency suggests that the role of reindeer is indeed of some importance, for only reindeer yields a Pearson correlation coefficient value in excess of 0.500 when correlated with evenness values for all regional Upper Palaeolithic assemblages (-0.541). Although not a strong value, it suggests that the species is in some way determinant of assemblage structure, implying greater system-specialization focused on reindeer than any other species. This is reinforced by the relationship between reindeer frequency and assemblage diversity, with a value of Pearson's Correlation Coefficient $r = -0.795$ ($r^2 = 0.632$) for the entire Upper Palaeolithic, $r = -0.856$ ($r^2 = 0.731$) for the Gravettian and Upper Perigordian, and $r = -0.895$ $r^2 = 0.801$ for the Upper Magdalenian, values not seen for any other species.

The recent work of Kuntz (2011: 315–18), which suggests that the reindeer was exploited throughout much of the year is of particular interest here as it indicates that the species was probably available for exploitation on an almost permanent basis so that its role as a determinant of assemblage structure may not be entirely surprising. Her results run contrary to the traditional picture (see Gordon, 1988) of season-specific exploitation, but, if the reindeer was taken throughout the year, its apparent numerically controlling role in assemblage structure can be explained, as it was taken at times of both high and low carcass/meat quality. It may have been taken in abundance when meat was of relatively good quality. Alternatively, it may have been processed intensively or taken for non-food use when on poorer quality. Either way, the archaeozoological record will display abundant reindeer, perhaps due to discard of body parts excess to requirement or intensive processing and excessive bone fracturing. Of course, low fat animals, such as is the reindeer, are frequently regarded as poor value among hunters (Speth, 1983; 1991; Kelly, 2013). However, protein, provided it is not consumed in excess, provides essential amino-acids, vitamins, and minerals and is crucial for metabolic functioning, and low fat intake in the presence of high protein can be counterbalanced by additional sugars. Sugars can be obtained from the carbohydrate in many plants of course, but horse meat also provides a small quantity of sugar (Ducommun, 1982; Migaud, 1993: 78). Perhaps this combination of protein and carbohydrate goes some way towards explaining the diversity in and between assemblages, and the close association of reindeer and horse throughout the Upper Palaeolithic. Dispatch of a single horse would have yielded significantly more than a single reindeer, especially at a time of poor reindeer quality. Therefore, when nutritional value is added to the equation, the importance of reindeer falls still further, which again reduces its apparent importance to the Palaeolithic hunter.

DISCUSSION: THE FINAL PALAEOLITHIC IN WESTERN EUROPE

Although this account has focused particularly on southwest France, much of Western Europe sees a similar structure in the archaeozoological record: reindeer and horse are the major species in many areas, as shown by Serangeli's (2006) useful pan-European survey. However, as Grayson and Delpech (2002: 1445) make clear, apparent concentration, in the archaeozoological record, on just one or two species does not warrant the terms 'specialization' or 'specialized', especially with regard to hunting and hunters. Instead, such patterns reflect just one aspect of a strategy which, in all probability, saw variety and diversity in target species cull and exploitation alike.

As previously demonstrated (p. 000), faunal assemblage diversity values vary through the Upper Palaeolithic in both southwest France and a broadly similar pattern can be seen across Europe as a whole (see Table 3.2). Both diversity and evenness values are at their highest during the Initial and Final Upper Palaeolithic and at their lowest during the period broadly corresponding to the cold, glacial conditions and discontinuous permafrost of the LGM. The Gravettian of Western Europe, like the more precisely defined Protomagdalenian and Solutrean in southwest France, sees 'narrow' assemblage structure, with assemblages frequently dominated by just one or two species (Serangeli, 2006). The subsequent Magdalenian sees greater evidence of significant regional variability in resource base as the range of species begins to broaden slightly. Not only do different large prey (reindeer, bison, red deer, saiga antelope, etc.) occupy the primary position at different sites, but there is also greater evidence of increasing use of small game, birds, and fish than we see during earlier periods (Le Gall, 1984; 2009; Costamagno, 2001; Jones, 2004). Therefore, uncertainty regarding the reality and nature of specialization during the early Upper Palaeolithic (Mellars, 1973; Grayson and Delpech, 2002) extends to the Magdalenian, long viewed as the endpoint of growing dominance by reindeer (see Grayson et al., 2001).

CONCLUSION

In this chapter we have focused on the classic region associated with the so-called Upper Palaeolithic specialized reindeer-hunter (southwest France). Despite the fact that some faunal assemblages are more than nine-tenths reindeer, the overall picture is one of a rich strategy (i.e. many prey targets) and an apparent mixture of biotopes—mosaic environments yielding a variety of resources, which may reflect seasonal variation in subsistence strategy or perhaps chronologically undetectable temporal changes in local environment.

A subsistence strategy focused on just one or two species is not a necessary requirement to account for the structure of much of the Upper Palaeolithic faunal record of Western Europe. Assemblages may result to reflect one or any combination of many natural, human, and carnivore-related variables: environmental conditions and mosaic-availability of resources, single or multiple 'events', different seasons of activity by human or carnivore hunters, etc. Although faunal assemblages may be 'specialized' in the sense of being numerically dominated by a single taxon in a proportion unlikely to have been seen in the natural environment, it does not necessarily mean that the diet and subsistence-related activity were equally focused on this species. Indeed the reverse may have been the case: a resource taken in large numbers potentially provides more than can be processed and used at any one time. If an animal is killed when meat quality is not high then much of the carcass may be discarded as unusable, thereby leaving an artificially high frequency of unprocessed bones that are more likely to survive than are the intensively processed equivalents of other species. It is also possible that processing of the major taxa was selective for purposes unrelated to subsistence/diet and that only a part of the carcass was therefore used. A 'specialized' faunal assemblage does not therefore necessarily mean a specialized dietary strategy, nor does it need to mean a specialist hunter. Similarly, a record of diversity may be the sum of several separate specialized or purpose-specific activities; a specialized record may reflect just one aspect of a diverse and broad-based strategy. Such is the nature of the archaeozoological record of the Upper Palaeolithic, and its true complexity is something which only a regional or multi-site and multi-species approach will enable us to understand.

REFERENCES

Binford, L. R. (1968) 'Post-Pleistocene adaptations', in Binford, S. R. and Binford, L. R. (eds) *New Perspectives in Archaeology*, pp. 313–41. Chicago: Aldine.

Bouchud, J. (1975) 'Étude de la faune de l'Abri Pataud', in Movius, L. H. (ed.) *Excavation of the Abri Pataud, Les Eyzies (Dordogne)*, pp. 69–153. Harvard: Peabody Museum Press.

Boyle, K. V. (1990) *Upper Palaeolithic Faunas from South-West France: A Zoogeographic Perspective.* BAR International Series 557. Oxford: Archaeopress.

Boyle, K. V. (2006) 'Neolithic wild animals in Western Europe: the question of hunting', in Serjeantson, D. and Field, D. (eds) *Animals in the Neolithic of Britain and Europe*, pp. 10–25. Oxford: Oxbow.

Boyle, K. V. (2010) 'Rethinking the "ecological basis of social complexity"', in Boyle, K., Gamble, C., and Bar-Yosef, O. (eds) *The Upper Palaeolithic Revolution in Global Perspective: Essays in Honour of Paul Mellars*, pp. 137–51. Cambridge: McDonald Institute for Archaeological Research.

Breuil, H. (1952) *Quatre cents siècles d'art pariétal: les cavernes ornées de l'âge du renne*, Montignac: Centre d'études et de documentation préhistorique.

Brown, J. H. (1995) *Macroecology*, Chicago: The University of Chicago Press.

Costamagno, S. (1999) 'Stratégies de chasse et fonction des sites au Magdalénien dans le sud de la France'. Unpublished PhD dissertation, University of Bordeaux I (Talence).

Costamagno, S. (2001) 'Exploitation de l'antilope saiga au Magdalénien en Aquitaine', *Paléo*, 13, 111–28.

Costamagno, S. (2004) 'Sites magdaléniens du sud de la France n'étaient pas des chasseurs spécialisés, qu'étaient-ils?', in Bodu, P. and Costamagno, S. (eds) *Approches fonctionenelles en préhistoire*, pp. 361–9. Paris: Société Préhistorique Française.

Delpech, F. (1983) *Les Faunes du Paléolithique du sud-ouest de la France*, Paris: CNRS.

Ducommun, J.-F. (1982) *Gastronomie de la viande de cheval*, Paris: Fédération de la Boucherie Hippophagique de France.

Enloe, J. (1993) 'Subsistence organization in the Early Upper Paleolithic: reindeer hunters of the Abri du Flageolet, Couche V', in Knecht, H., Pike-Tay, A., and White, R. (eds) *Before Lascaux: The Complex Record of the Early Upper Paleolithic*, pp. 101–15. Boca Raton: CRC Press.

Fagan, B. (2010) *Cro-Magnon: How the Ice Age Gave Birth to the First Modern Humans*, London: Bloomsbury Press.

Fontana, L. (2000) 'La chasse au renne au Paléolithique supérieur dans le sud-ouest de la France: nouvelles hypothèses de travail', *Paléo*, 12, 141–64.

Franklin, J. (2009) *Mapping Species Distributions: Spatial Inference and Prediction*, Cambridge: Cambridge University Press.

Gordon, B. (1988) *Of Men and Reindeer Herds in French Magdalenian Prehistory.* BAR International Series 390. Oxford: Archaeopress.

Grayson, D. (1984) *Quantitative Zooarchaeology: Topics in the Analysis of Archaeological Faunas*, Orlando: Academic Press.

Grayson, D. K. and Delpech, F. (2002) 'Specialized Early Upper Palaeolithic hunters in south-western France?', *Journal of Archaeological Science*, 29, 1439–49.

Grayson, D. K., Delpech, F., Rigaud, J.-Ph., and Simek, J. F. (2001) 'Explaining the development of dietary dominance by a single ungulate taxon at Grotte XVI, Dordogne, France', *Journal of Archaeological Science*, 28, 115–25.

Heip, C. H. R., Herman, P. M. J., and Soetaert, K. (1998) 'Indices of diversity and evenness', *Océanis*, 24, 61–87.

Hockett, B. and Haws, J. (2009) 'Continuity in animal resource diversity in the Late Pleistocene human diet of Central Portugal', *Before Farming*, 2, DOI: 10.3828/bfarm.2009.2.2.

Jones, E. L. (2004) 'Broad Spectrum Diets and the European Rabbit (*Oryctolagus cuniculus*): Dietary Change during the Pleistocene-Holocene Transition in the Dordogne, Southwestern France'. Unpublished PhD dissertation, University of Washington (Seattle).

Kelly, R. L. (2013) *The Lifeways of Hunter-Gatherers: The Foraging Spectrum*, Cambridge: Cambridge University Press.

Kuntz, D. (2011) 'Ostéométrie et migration(s) du renne (*Rangifer tarandus*) dans le sud-ouest de la France au cours du dernier Pléniglaciaire et du Tardiglaciaire (21 500–13 000 Cal. BP)'. Unpublished PhD dissertation, University of Toulouse II—Le Mirail (Toulouse).

Kuntz, D. and Costamagno, S. (2011) 'Relationships between reindeer and man in southwestern France during the Magdalenian', *Quaternary International*, 238, 12–24.

Laroulandie, V. (2000) 'Taphonomie et archéozoologie des oiseaux en Grotte: applications aux sites Paléolithiques du Bois-Ragot (Vienne), de Combe Saunière (Dordogne) et de La Vache (Ariège)'. Unpublished PhD dissertation, University of Bordeaux I (Talence).

Lartet, E. and Christy, H. (1865–75) *Reliquiae Aquitanicae: Being Contributions to the Archaeology and Palaeontology of Périgord and the Adjoining Provinces of Southern France*, London: Williams & Norgate.

Le Gall, O. (1984) *L'ichtyofaune d'eau douce dans les sites préhistoriques: ostéologie-paléoecologie-palethnologie*. Les cahiers du quaternaire 8. Paris: CNRS.

Le Gall, O. (2009) 'Archéo-ichtyologie et pêches préhistoriques: résultats et perspectives', *Archéopages*, 26, 52–5.

Le Gall, O. (2010) 'Influences des glaciaires/interglaciaires sur les ichtyofaunes des eaux douces européennes', *Quaternaire*, 21, 203–14.

Magurran, A. E. (1988) *Ecological Diversity and Its Measurement*, London: Croom Helm.

Magurran, A. E. (2004) *Measuring Biological Diversity*, Oxford: Blackwell Publishing.

Mellars, P. (1973) 'The character of the Middle-Upper Palaeolithic transition in south-west France', in Renfrew, C. (ed.) *The Explanation of Culture Change: Models in Prehistory*, pp. 255–76. London: Duckworth.

Mellars, P. (1989) 'Major issues in the emergence of modern humans', *Current Anthropology*, 30, 349–85.

Mellars, P. (1996) *The Neanderthal Legacy: An Archaeological Perspective of Western Europe*, Princeton: Princeton University Press.

Mellars, P. (2004) 'Reindeer specialization in the Early Upper Palaeolithic: the evidence from south west France', *Journal of Archaeological Science*, 31, 613–17.

Migaud, P. (1993) 'Le cheval dans le haut Moyen Age d'après occidental'. Unpublished PhD dissertation, Ecole Nationale Vétérinaire de Nantes (Nantes).

Serangeli, J. (2006) *Verbreitung der großen Jagdfauna in Mittel- und Westeuropa im oberen Jungpleistozän: ein kritischer Beitrag*. Tübinger Arbeiten zur Urgeschichte 3. Rahden: Marie Leidorf Verlag.

Smith, D. (2004) 'Beyond the cave: Lascaux and the prehistoric in post-war French culture', *French Studies*, 58, 219–32.

Smith, B. and Wilson, J. B. (1996) 'A consumer's guide to evenness indices', *Oikos*, 76, 70–82.

Speth, J. D. (1983) *Bison Kills and Bone Counts*, Chicago: University of Chicago Press.

Speth, J. D. (1991) 'Protein selection and avoidance strategies of contemporary and ancestral foragers: unresolved issues', *Philosophical Transactions of the Royal Society of London* B, 334, 265–70.

Spiess, A. E. (1979) *Reindeer and Caribou Hunters: An Archaeological Study*, New York: Academic Press.

Stiner, M. (2001) 'Thirty years on the 'Broad Spectrum Revolution' and paleolithic demography', *Proceedings of the National Academy of Science*, 98, 6993–6.

Stutz, A. J., Munro, N. D., and Bar-Oz, G. (2009) 'Increasing the resolution of the Broad Spectrum Revolution in the southern Levantine Epipaleolithic (19–12 KA)', *Journal of Human Evolution*, 56, 294–306.

CHAPTER 4

MESOLITHIC HUNTING AND FISHING IN THE COASTAL AND TERRESTRIAL ENVIRONMENTS OF THE EASTERN BALTIC

LEMBI LÕUGAS

INTRODUCTION

THIS chapter focuses on hunting and fishing in the Mesolithic of the eastern Baltic region. Therefore, it should be important to clearly define the period and area under investigation. A clear definition of the Mesolithic is a very complex matter and depends on different, regionally varied, ecological, techno-stylistic, economic, social, and/or chronological criteria (e.g. Kozłowski, 2003; 2009). In this chapter the Mesolithic is defined, rather simply, as the time span and cultural background which marks the post-glacial period from *c.*9000 cal BC until the first appearance of ceramics *c.*5500 cal BC or the first indications of agricultural activities in the eastern Baltic region. The archaeological culture most commonly used to describe the Mesolithic in this region is the Kunda Culture (Indreko, 1948), but the use of such a Mesolithic mono-cultural picture to time the end of the Mesolithic has recently been reconsidered by some researchers (e.g. Ostrauskas, 2000; Antainaitis-Jacobs and Girininkas, 2002; Kriiska, 2009). An important question is whether the Late Mesolithic should comprise the period when ceramics were taken into use, but before agricultural activities become traceable in the material culture (i.e. 'ceramic Mesolithic'), or if it should follow the previously agreed periodization in which the Mesolithic ends with the appearance of ceramics into the eastern Baltic archaeological material. The time gap between these two events is about eight hundred years in the area of modern Estonia, but somewhat shorter in southern areas such as Latvia and Lithuania. This time gap is occupied by people of the so-called Narva Culture, when rather basic types of ceramics were taken into use, whereas the hunting and fishing economy remained similar to that of the Late Mesolithic period. Because of the appearance of ceramics and a noticeable increase in biodiversity of fauna during this period, the time gap has come to the attention of researchers, regardless of differing opinions about the beginning of Neolithization in the area. Thus, the end of the Mesolithic in this study is referred to as 'Late Mesolithic/Early Neolithic', which includes also the Narva Culture.

The eastern Baltic region can be defined geographically as the area today occupied by the Baltic countries including, from north to south, Estonia, Latvia, and Lithuania. It is obvious though that we should not consider modern political borders too strictly as they had no meaning for Mesolithic people. Therefore, the geographical coordinates in this study, even when using the names of today's states, are not bounded by a physical line, but rather mark zones with a complexity of environmental conditions, where people lived and obtained their subsistence as well as established their cultural identities.

Colonization of new land by humans depends very much on the potential resources that can support human settlement and, thereafter, the availability of technology by which to exploit these resources. Studies of Mesolithic hunting and fishing should not be based only on the finds of material culture, but also take into account different palaeo-environmental records. The eastern Baltic region has been influenced by Ice Age glaciations and since the last glaciation the Baltic Sea coastline has been subjected to many changes. Data on the late and post-glacial formation of natural resources have been a valuable source of information on this subject. In addition to the colonization of terrestrial environments, the coast and islands, which have much narrower ecological niches, were also settled during the Mesolithic period. Colonization of these areas would have required different technology and knowledge to that required for hunting in terrestrial surroundings.

Basic information about Mesolithic hunting and fishing in the eastern Baltic region is generally well known. It focused on the hunting of European elk (*Alces alces*), beaver (*Castor fiber*), and fishing for pike (*Esox lucius*) and perch (*Perca fluviatilis*). However, the nuances which come from environmental, geographical, and chronological specificity are still not clear. The overview presented in this chapter is based mainly on faunal remains from Mesolithic sites combined with other lines of archaeological evidence as well as palaeo-environmental information. An attempt will also be made to separate the types of fauna used during the different parts of the Mesolithic.

MATERIAL AND METHODS

The archaeological dataset used here comes from the study of Mesolithic settlement sites in the study area. Geographically this area is situated in the western part of the East European Plain, on the east coast of the Baltic Sea, and provides a good opportunity to study both hunting and fishing in terrestrial and coastal environments.

So far, more than twelve Mesolithic settlement sites of this region (Fig. 4.1) yielded a considerable amount of animal bone, which allows conclusions to be drawn about hunting and fishing, as well as the faunal history of the period. The faunal assemblages of these sites consist primarily of bones, teeth, and antlers, whereas other faunal remains, such as shells, hair, keratin, and cartilage, are much more rarely found. The taxonomic composition of assemblages from the Mesolithic period is often similar to the modern composition, i.e. it is represented by species of the boreal forest zone. The beginning and end of the period, when climatic conditions varied and Mesolithic people had to adapt to changing environmental conditions, deserve, however, special attention and will be discussed in greater detail below. Hunting and fishing activities are discussed in

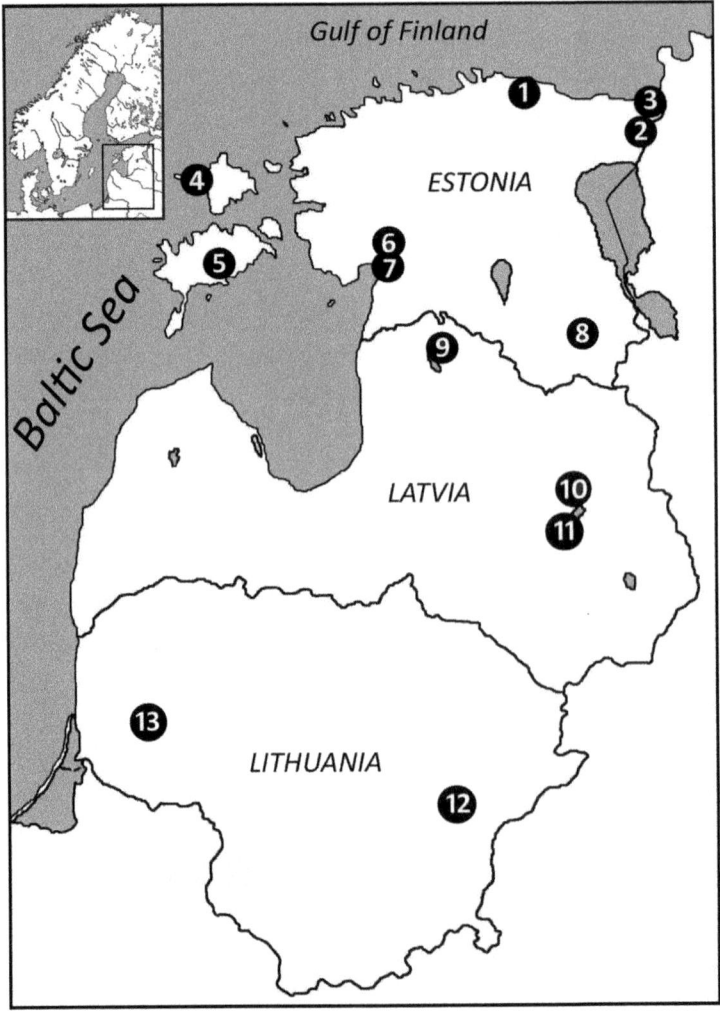

FIGURE 4.1 Geographical locations of the sites mentioned in text. 1: Lammasmägi at Kunda, 2: Joaoru at Narva, 3: Riigiküla, 4: Kõpu, 5: Kõnnu, 6: Pulli, 7: Sindi-Lodja, 8: Kääpa, 9: Zvejnieki, 10: Zvidze, 11: Sūļagals, 12: Žemaitiškės, 13: Donkalnis.

this overview mainly by reference to archaeological animal remains, but environmental evidence will also be used.

Animal bones from twelve Mesolithic settlement sites and one human grave were chosen to investigate the faunal composition of the periods in question (Fig. 4.1). The selection of sites was based on their chrono-stratigraphy and abundance of animal bones. The ideal material comes from a short, well-dated time span and is abundant. A high level of taxonomic variation was not considered an important characteristic for site selection, as hunting or fishing activities may be focused on one or a small selection of animals. Most of the selected sites are in the northern part of the area (Estonia and Latvia), while Lithuanian sites have generally produced smaller faunal assemblages and are therefore less well represented.

In this study, the Early Mesolithic site of Pulli, southwestern Estonia, produced such 'ideal material'. The excavations there took place between 1969 and 1976 (Jaanits and Jaanits, 1978; Lõugas, 1997; 2008) and produced the earliest traces of human activity in Estonia (no Late Palaeolithic sites are known from the country). Stone Age hunters and fishermen occupied the site around 9230–8230 cal BC. Archaeological layers occur between well-dated Baltic Sea sediments, i.e. on different Baltic Sea developmental stages known as the Yoldia Sea and under the Ancylus Lake sediments. The Ancylus Lake first inundated the area at Sindi-Lodja, reached the Paikuse area around 8400–8600 cal BC, and finally Pulli and other higher elevation sites around 8200 cal BC (Veski et al., 2005). As a result of the rise of the water level, Pulli was inundated and levels containing evidence of human activity were buried under the Ancylus sediments.

The Early Mesolithic sites of Sūļagals, eastern Latvia, and the Zvejnieki II lower layer, northern Latvia, were excavated in the 1970s. The sites are considered contemporaneous with Pulli, although Zvejnieki II lower layer is perhaps a somewhat later habitation period (Zagorska, 1992). The sedimentation of chronologically different archaeological layers at these sites is not as well distinguishable as at Pulli, where the archaeological layers lay between water sediments (see above).

The important Middle Mesolithic site of Lammasmägi ('sheep hill') is located at Kunda, northern Estonia. It is better known by the name of Kunda, which also provided the name for a Mesolithic culture—the Kunda Culture. Lammasmägi and its surroundings (for example, the deposits of the ancient lake) were already known to the archaeological community at the end of the nineteenth century (Grewingk, 1882). Its investigation in the 1930s is one of the earliest examples of interdisciplinary cooperation between archaeologists and natural scientists in Estonia (Indreko, 1948). Johannes Lepiksaar, later the curator of the Gothenburg Museum of Natural History, Sweden, analysed animal bones excavated from Kunda in the 1930s and gave a first broad impression of the fauna exploited during the Mesolithic in Estonia (published by Indreko, 1936; 1948). More recent studies at Kunda indicate different occupation intervals at the site: 8590–8030 cal BC, 7530–6820 cal BC, 2490–1780 cal BC (Åkerlund et al., 1996) and also 6230–6340 cal BC as demonstrated by the date of a wild horse bone (Sommer et al., 2011). The stratigraphy of the site is quite complicated, but nevertheless the material is most probably from the period c.7500 to 6800 cal BC.

The other Middle Mesolithic sites which yielded a considerable amount of animal remains and thus provided good data are Joaoru at Narva (sites II and III), northeastern Estonia, and the Zvejnieki II settlement site, upper layer (Paaver, 1965; Zagorska, 1992; Lõugas, 1996b; 2006; Rosentau et al., 2013). These sites are situated on the banks of lakes or rivers and are considered to be inland sites. However, the Middle Mesolithic is known also as a coastal transition period, and more settlement sites from the ancient coastal formations and islands are being discovered.

In Estonia, Sindi-Lodja I and II sites, and Kõpu IV/V and VII/VIII sites on Hiiumaa (Dagö) Island seem to represent a shift to more specialized hunting strategies of aquatic fauna, including seals. Here, the Narva Joaoru and Sindi-Lodja sites are dated to the period of c.7200–6500 cal BC (Kriiska, 1996; Kriiska and Lõugas, 2009) and Kõpu sites c.5600–5000 cal BC (Kriiska and Lõugas, 1999).

Late Mesolithic/Early Neolithic sites for which archaeological evidence suggests a Mesolithic economy, but also contain ceramic finds, are included in this analysis. Kääpa, southeastern Estonia, and Žemaitiškės 3B, northeastern Lithuania, are situated inland, on the bank of a river (Paaver, 1965; Daugnora and Girininkas, 2004). Narva Joaoru I, and Riigiküla III, northeastern Estonia, are located near to the sea coast (Paaver, 1965; Kriiska, 1996), but Kõpu I, Hiiumaa Island, and Kõnnu, Saaremaa Island, are both sites representing an island habitation (Lõugas, 1997; Kriiska and Lõugas, 1999). Zvidze, eastern Latvia, is located on the bank of the Lubana Lake (Loze, 1988).

Animal bones were identified by morphological criteria using reference collections and bone atlases. Methods used during the archaeological excavations substantially affected the results of bone analyses. More precisely, the excavated soil was not sieved in many cases and therefore we do not have many bones of small animals, especially small fish, in the material. Also, in some cases, the collection of bones was selective, which means that many, probably non-determinable bones, were discarded already during the excavations. Regardless of these limitations, the bone assemblages preserved so far allow some detailed study for addressing questions about the Mesolithic period. In contrast, excavations from the 1990s onwards were focused on the collection of all kinds of small finds, including fish bones (e.g. Kriiska, 1996; Kriiska and Lõugas, 2009) and can therefore provide a more detailed account of the breadth of resource exploitation during this period.

ANIMALS HUNTED IN THE MESOLITHIC
OF THE EASTERN BALTIC

Results of the zooarchaeological analysis of all the sites are summarized in Table 4.1. Here, the number of identified specimens (NISP) recorded for mammals, reptiles, and fish are presented. Mesolithic fowling has received little attention in previous works of zooarchaeology in the study area. There are few identifications available, and they are

Table 4.1 Animal bones from Mesolithic sites of the eastern Baltic region (NISP values)

	Early Mesolithic		Middle Mesolithic					Late Mesolithic/Early Neolithic							
	Pulli	Zvejnieki II lower	Sūlagals	Kunda	Zvejnieki II upper	Narva Joaoru II+III	Sindi-Lodja	Kõpu IV/VIII	Narva Joaoru I	Riiģikula III	Kääpa	Zvidze	Žemaitiškes	Kõpu I	Kõnnu
MAMMALS															
Castor fiber	390	51	20	1,893	804	152			137	24	3,920	157	5		8
Alces alces	374	1,035	343	4,756	5,023	186	5		124	226	7,166	2,095	26		5
Cervus elaphus	10	10		7	62	35			12		143	1,048	23		3
Sus scrofa			23	59	341	71	8		140	83	3,214	2,103	4		4
Capreolus capreolus	2	3	3	4	21	10			4	6	24	163	1		
Bos primigenius				65	21	6			20	155	1,065	91			
Equus ferus				30	29	29					282	39			
Ursus arctos	41	6	2	77	37	9			25	13	613	191	5		
Martes martes	6		2	5	17	4			5	4	460	133			4
Mustela putorius											2				
Mustela lutreola											3				
Lutra lutra	2	2	1	36	33	2			2	1	139	14			
Meles meles				3						2	167	6			
Vulpes vulpes		2	2			1					3	18	1		
Canis lupus	5	1	2	23	4	1	1				23	25			3

(Continued)

Table 4.1 Continued

	Early Mesolithic		Middle Mesolithic				Late Mesolithic/Early Neolithic									
	Pulli	Zvejnieki II lower	Sulagals	Kunda	Zvejnieki II upper	Narva Joaoru II+III	Sindi-Lodja	Kõpu IV/V/III	Narva Joaoru I	Riigiküla III	Kääpa	Zvidze	Žemaitiškės	Kõpu I	Kõnnu
Canis lupus f. familiaris	1	8		43	112	27	1		27	32	311	16	1		1
Lynx lynx				1							5				
Felis silvestris					7						2				
Phoca hispida				4	1		34	24						54	76
Halichoerus grypus				78		12		6	8					45	25
Phocidae				1						178				433	141
Phocoena phocoena										1					1
Lepus timidus			2	4	8	1			1		26	57	1		
Sciurus vulgaris											3				
Erinaceus europaeus										1	1			4	
REPTILES															
Emys orbicularis					1	3					19				

FISH

Esox lucius	21	304	66	437	5	277	95	1	12
Salmo sp.				1					
Coregonus sp.					1				
Perca fluviatilis		31	6	1	12	4	5		
Sander lucioperca	164			256	47	75	4		
Abramis brama		208				4	4		
Rutilus rutilus									7
Tinca tinca		5	1			1			
Cyprinidae	1	1				18			
Silurus glanis				98		6	41		
Gadus morhua								4	
Scophthalmus maximus								1	

mostly in the form of reports and a few separate articles. A thorough overview of the role of avian fauna in the Baltic regions is presented by Mannermaa (2008), but includes only few data on the Mesolithic and Early Neolithic material from the Baltic countries. Therefore, a precise overview about the topic is not yet available. According to the limited data, it is clear that the most commonly hunted avian group in the Mesolithic period was that of the Anatids; long-tailed duck (*Clangula hyemalis*), common eider (*Somateria mollissima*), mallard (*Anas platyrhynchos*), tufted duck (*Aythya fuligula*), velvet scoter (*Melanitta fusca*), goosander (*Mergus merganser*), greylag goose (*Anser anser*), and whooper swan (*Cygnus cygnus*) were recorded in the Kunda Lammasmägi material (Lõugas, 1996). There are exceptions according to the location of sites, for example the divers (Gaviidae) dominate in the Early Mesolithic material from Pulli. Here, the red-throated diver (*Gavia stellata*) and black-throated diver (*Gavia arctica*) were both recorded (Lõugas, 2008). The great cormorant (*Phalacrocorax carbo*) seems to be quite common in the Mesolithic as well as in the Neolithic, where it is found in small numbers on almost all of the sites studied (Lõugas, 2008; Mannermaa, 2008). The black grouse (*Tetrao tetrix*) and white-tailed sea eagle (*Haliaëtus albicilla*) were also found on most of the Mesolithic sites (Lõugas, 1997; 2008; Daugnora and Girininkas, 2004).

Based on bone records, the most numerous hunted animals during the Mesolithic period were the (European) elk and beaver. This pattern is found at all of the sites studied with the exception of some coastal ones, for which the zooarchaeological record indicates a greater reliance on aquatic resources. It seems that elk immigrated to the eastern Baltic areas just after the late glacial, during the cooler Younger Dryas period. It is conceivable that reindeer (*Rangifer tarandus*) and elk could have inhabited the area at the same time, but the zooarchaeological evidence does not support it. The latest radiocarbon date comes from a reindeer recovered from the marl lake close to Kunda (9970±85 years BP (Hela-598)) (Ukkonen et al., 2006), while the earliest elk finds come from Pulli and are few hundred years younger (according to charcoal from archaeological layer, 9385±105 (Ua-13351) or, according to one dated elk bone, 9095±90 years BP (Ua-13352)) (Poska and Veski, 1999). The Late Palaeolithic sites from the banks of the Daugava River, Latvia (Zagorska, 1995), do not contain any elk bone and do not support the co-existence of these species here either. Elk was heavily exploited by Mesolithic humans as evidenced by the amount of bones of this species found on archaeological sites. Besides lithic material, elk bones and antlers were the main raw materials for tool-making. A large number of spear-heads, arrows, points, etc. is recorded from Mesolithic contexts (Fig. 4.2).

Some other species of Artiodactyla were found in the Early Mesolithic sites in Estonia and Latvia, such as wild boar (*Sus scrofa*) and roe deer (*Capreolus capreolus*), whereas aurochs (*Bos primigenius*) and red deer (*Cervus elaphus*) appear only in the Middle Mesolithic. Unfortunately, there is no such representative Early Mesolithic settlement site discovered in Lithuania so far which could illustrate the fauna hunted there. Some bone assemblages, for example that containing tooth pendants from the Donkalnis burials, may indicate that the red deer had become an important resource in southern Baltic areas earlier than in the northern part of the study area. Throughout the Holocene

FIGURE 4.2 Some of the raw materials for tool-making included elk bone. Spear-heads from the Lammasmägi site at Kunda (Archaeological Collections of the Institute of History, Tallinn University). Author's own image.

the northern limit of the red deer distribution was somewhere in the eastern Baltic and depended on the prevailing climatic conditions. It reached as far north as Estonia during the Atlantic period (Late Mesolithic/Early Neolithic).

The distribution of the aurochs reaches also its most northern border in the eastern Baltic area in the Boreal and Atlantic periods (Middle Mesolithic until Early Neolithic), when it spread to almost every part of the Estonian mainland. The aurochs is not known from the islands, with the exception of some tooth pendants made of aurochs' incisors found from Kõnnu burials, Saaremaa Island. However, this cannot be seen as evidence that the aurochs was the object of hunting activity on Saaremaa Island because pendants, as well as humans buried in these graves, could have come from the mainland or had direct contacts there. The same concerns the pendants made of wild horse (*Equus ferus*) incisors found in Kõnnu (Lõugas, 1997).

The ecology of the wild horse is usually linked to a high-silicate diet (e.g. silicate rich components such as sedges, grain, millets, etc.) and an open landscape, so its occurrence in mainly wooded environments is not common. Regardless of this, the wild horse is represented in Mesolithic/Early Neolithic material of northeastern Europe, and its distribution here is interpreted as a result of sporadic invasions from the steppe or forest-steppe zones of eastern Europe (e.g. Lepiksaar, 1986). Alternatively, wild horses of this period could have belonged to the fauna of the temperate forest zone, inhabiting the more open areas at the riverbanks and the margins of other water bodies (e.g. Paaver, 1965). The wild horse is not known from the Early Mesolithic (at Pulli, Zvejnieki, and Sūļagals), but does occur from the Middle Mesolithic to the Early Neolithic in quite high numbers (see Table 4.1). Thus, the horse inhabited the eastern Baltic at the time when Neolithization took place in most of Europe, making the opening of forested landscapes suitable for horse (Sommer et al., 2011). However, such opening of the landscape did not occur in the eastern Baltic and the adaptation of the wild horse to broad leaf forests should therefore be considered.

Although elk dominates in the bone assemblages of Mesolithic Estonia, this species is replaced by red deer and aurochs in southern areas of the eastern Baltic. The Middle and Late Mesolithic materials from Latvia and Lithuania prove the decline of the elk in those areas and the parallel increase of red deer and aurochs. The pattern observed clearly suggests that the number of red deer, aurochs, and also wild boar increased in the Atlantic period, when the forests of broad leaf trees dominated even in the most northern parts of Estonia.

CARNIVORES

Carnivores of the temperate zone are not as sensitive to climatic fluctuations as the taxa discussed above, but are more dependent on their prey, which are usually herbivores. When the environment is suitable for herbivores, carnivores can easily expand their distribution into new areas. Brown bear (*Ursus arctos*) was the most hunted carnivore in the northern areas. This species was probably not hunted only for dietary reasons, but it had an important role in the art and beliefs of northern people throughout their history (e.g. Zachrisson and Iregren, 1974; Asplund, 2005; Helskog, 2012). Thus, the brown bear is a dominant species among the Mesolithic carnivores hunted by people. Other species, such as fox (*Vulpes vulpes*), pine marten (*Martes martes*), otter (*Lutra lutra*), and badger (*Meles meles*), were more probably hunted for practical reason—providing fur for clothing and overwintering. Wolf (*Canis lupus*) and dog (*Canis lupus* f. *familiaris*), both recorded in the Mesolithic material, hardly represent any food remains, otherwise more fragmented and hashed bones would be expected in the refuse material. Wolf could also have been a cult animal, similar in status to the brown bear, and dogs likely had a practical function in the lives of people, i.e. they had working roles in hunting and/ or hauling sledges. However, some evidence of the consumption of dogs in the study area does exist. Some dog remains from Riigiküla and Kääpa appear to represent food refuse based on the large number of bones found from these settlement sites, the pattern of fragmentation of the bones, and the fact they are commingled with the bones of other butchered animals. However, many complete dog bones are likely to derive from buried animals, though some of them (especially limb bones) have cut marks probably caused by skinning (personal observation).

AQUATIC RESOURCES

Some coastal and island sites were clearly inhabited by hunters who focused on seal hunting. Even Lammasmägi (Kunda) yielded a few bones of the ringed seal (*Pusa hispida*), a seal whose ecology is connected with the littoral zone of water bodies and rivers where it can swim upstream to feed on migratory fish. During the Early Mesolithic period the Baltic Sea was first a brackish body of water (during the Yoldia Sea stage), followed by a fresh-water stage (the Ancylus Lake), which means that marine influence was almost non-existent on the eastern side of the Baltic in the early part of the Mesolithic. The only sea mammal recorded from this time is the ringed seal, which was getting into the Baltic basin probably during the Baltic Ice Lake stage (Lepiksaar, 1986) or, according to bone evidence, during the Yoldia Sea stage (Ukkonen et al., 2014).

Middle Mesolithic Sindi-Lodja sites I and II were situated on the bank of Pärnu River, the estuary of which was much closer to the sites at the time of occupation than at present, and are considered as specialized ringed seal hunting sites (Kriiska and Lõugas, 2009). The same can be assumed for many other Mesolithic sites on the Saaremaa and Hiiumaa Islands, though unfortunately only Kõpu yielded a sufficiently large sample of identifiable bones (Moora and Lõugas, 1995; Kriiska and Lõugas, 1999). Common to these Mesolithic seal hunting sites is the absence of bones from marine fish (e.g. cod, flounder, herring) although the bones of fresh-water fish do occur. This can be partly explained by the development of the Baltic Sea, which in the early post-glacial stages was mainly composed of fresh water. Even the beginning of the more saline Littorina Sea stage probably did not affect the eastern coast much, with the result that human groups did not undertake marine fishing at that time. The peak of salinity in the Baltic is connected with the Littorina transgression at about 8000–7000 cal BP (e.g. Veski et al., 2005), though marine fish do not appear in human settlements of this part of the coast until about 1,000–1,500 years later. Only Kõpu I site, which is one of the most recent sites in this review, yielded few bones of marine fish. Unfortunately, the preservation of bones in Kõpu and other island sites is very poor and the number of fish bones is correspondingly low. This poor preservation precludes an estimation of the role of fishing in the island environments, where terrestrial and fresh-water resources were not at hand. The adoption of marine fishery technology in the Mesolithic eastern Baltic was nevertheless inhibited, probably due to the fact that there existed a comparatively rich fresh-water fauna, which was utilized by hunting and gathering populations for many millennia. Such a delay, which is reflected in the archaeological material, does not indicate that marine fish were absent on the eastern coast of the Baltic, but can be better explained by the time-consuming nature of the activities required to discover, learn, and/or find a use for this new resource.

Fishing in the Early Mesolithic was focused on the nearby lakes, rivers, and streams. Fish was clearly an important food resource for Mesolithic people, but only a few species have been identified in the bone refuse of the Early Mesolithic, and these come mainly from pike, pike-perch (*Sander lucioperca*), and perch, with some cyprinids also found. Wels catfish (*Silurus glanis*), which was also among the early post-glacial immigrants of the eastern Baltic water systems, is represented by more bone finds in the later settlement sites. A low number of species is not necessarily a consequence of the bad preservation of bones or recovery biases, but more likely of the utilization of local water bodies during certain seasons, when one species was more abundant and easier to catch in that location. Early Mesolithic sites are often interpreted as seasonal camp sites, and the short-term utilization of aquatic resources would provide a good explanation for their temporary, seasonal exploitation. Thus, as is the case today, pike-perch was likely to be more common in the estuary of the Pärnu River at the time when Pulli was occupied, and pike was more common in the Burtnieks Lake, where fishermen from Zvejnieki could easily catch them.

From the Middle Mesolithic onwards fishing seems to be more wide-ranging and fishermen are likely to have maintained barriers and traps (though there is no actual evidence), especially at lake inlets and outlets, in order to catch more varied fish species during different seasons. This does not necessarily mean that settlement sites during this period were sedentary, but they were probably more frequently settled than before. Stable isotope (^{13}C and ^{15}N) studies undertaken on the Zvejnieki material (human, animal, and fish bones) indicate the importance of fresh-water fish in the diet of Middle Mesolithic people. It showed that these people were almost totally dependent on fish for their diet as their protein intake was most similar to otters' (Eriksson et al., 2003; Eriksson, 2006). Although there is evidence that later, from the Middle Neolithic onwards, the contribution of terrestrial mammals and/or birds to the diet increased at Zvejnieki, fish still remained an important component of the diet, and a marine signature appeared in the remains of some people from that period (Eriksson et al., 2003; Eriksson, 2006).

CHANGING ANIMAL USE THROUGH THE MESOLITHIC

Despite the limitations of material availability discussed in the first part of this chapter, the archaeofaunal record of the Mesolithic period can provide a great deal of useful information about the exploitation strategies and environment of the period. On the basis of the archaeofaunal record of the eastern Baltic, the Mesolithic can be divided into three phases according to the information provided by archaeological as well as environmental data:

(1) Early Mesolithic: the taxonomic variability is not high and it reflects the Pre-Boreal/Boreal conditions of the environment, i.e. cool to moderate climate, sparse birch and pine forests, with leaf groves on the banks of rivers and lakes. The dominant species is elk followed by beaver. Other species have marginal importance. Pike, pike-perch, and perch are characteristic fish species. An assemblage typical of the period is that from Pulli.

(2) Middle Mesolithic: the taxonomic variability is higher and it reflects the Boreal/ (Early) Atlantic conditions of the environment, i.e. somewhat warmer climate, mixed forests, higher biodiversity. Even though there are many more taxa exploited, the dominant species are still elk and beaver. Red deer is more important in the southern part of the study area. Coastal sites provide evidence for the exploitation of the ringed seal. Wels catfish seems to increase in importance in the ichthyofauna. Assemblages typical of the period come from Kunda Lammasmägi (terrestrial) and Sindi-Lodja I/II (coastal).

(3) Late Mesolithic/Early Neolithic: the taxonomic variability is at its highest for the period, for both herbivores and carnivores, and refers to the Atlantic conditions of the environment, i.e. the climate optimum—warm and humid climate, with a vegetation dominated by broad leaf forests. Such conditions (warmer and humid) are evidenced also by the occurrence of the marsh tortoise (*Emys orbicularis*) as far north as northern Estonia. Although the elk remains dominate the northern part of the region, other herbivores such as red deer and aurochs were available in large numbers, especially in the southern part of the study area. Fish remains still consist entirely of fresh-water species. Typical assemblages are represented by Kääpa (terrestrial), and Kõnnu and Kõpu I (coastal).

References

Åkerlund, A., Regnell, M., and Possnert, G. (1996) 'Stratigraphy and chronology of the Lammasmägi site at Kunda', in Hackens, T., Hicks, S., Lang, V., Miller, U., and Saarse, L. (eds) *Coastal Estonia: Recent Advances in Environmental and Cultural History*. PACT vol. 51, pp. 253–72. Rixensart: Pact Belgium.

Antainaitis-Jacobs, I. and Girininkas, A. (2002) 'Periodization and chronology of the Neolithic in Lithuania', *Archaeologia Baltica*, 5, 9–39.

Asplund, H. (2005) 'The bear and the female. Bear-tooth pendants in Late Iron Age Finland', in Mäntyla, S. (ed.) *Rituals and Relations: Studies on the Society and Material Culture of the Baltic Finns*. Humaniora vol. 336, pp. 13–30. Saarijärvi: Suomalaisen Tiedeakatemian Toimituksia.

Daugnora, L. and Girininkas, A. (2004) *Rytų Pabaltijo Bendruomenių Gyvensena XI–II Tūkst. Pr. Kr.*, Kaunas: Lietuvos veterinarijos akademija ir Lietuvos istorijos institutas.

Eriksson, G. (2006) 'Stable isotope analyses of human and faunal remains from Zvejnieki', in Larsson, L. and Zagorska, I. (eds) *Back to the Origin. New Research in the Mesolithic-Neolithic Zvejnieki Cemetery and Environment, Northern Latvia.* Acta Archaeologica Lundensia vol. 52, pp. 183–215. Stockholm: Almqvist and Wiksell International.

Eriksson, G., Lõugas, L., and Zagorska, I. (2003) 'Stone Age hunter-fisher-gatherers at Zvejnieki, northern Latvia: radiocarbon, stable isotope and archaeozoology data', *Before Farming*, 1(2), 1–25.

Grewingk, C. (1882) 'Geologie und Archäologie des Mergellagers von Kunda in Estland', *Archiv für die Naturkunde Liv-, Est- und Kurlands*, 1(9), 1.

Helskog, K. (2012) 'Bears and meanings among hunter-fisher-gatherers in northern Fennoscandia 9000–2500 BC', *Cambridge Archaeological Journal*, 22, 209–36.

Indreko, R. (1936) 'Vorläufige bemerkungen über die Kunda-funde', *Õpetatud Eesti Seltsi Aastaraamat*, 1934, 225–98.

Indreko, R. (1948) *Die Mittlere Steinzeit in Estland.* Kungliga Vitterhets Historie och Antikvitets Akademiens Handlingar vol. 66. Uppsala: Almqvist and Wiksell.

Jaanits, L. and Jaanits, K. (1978) 'Ausgrabungen der Frühmesolithischen Siedlung von Pulli', *Eesti NSV Teaduste Akadeemia Toimetised. Ühiskonnateadused*, 27, 56–63.

Kozłowski, S. K. (2003) 'The Mesolithic: What do we know and what do we believe?', in Larsson, L., Kindgren, H., Knutsson, K., Loeffer, D., and Åkerlund, A. (eds) *Mesolithic on the Move*, pp. xvii–xxi. Oxford: Oxbow Books.

Kozłowski, S. K. (2009) 'Mapping the European Mesolithic', in McCartan, S., Schulting, R., Warren, G., and Woodman, P. (eds) *Mesolithic Horizons*, Vol. 1, pp. xx–xxvi. Oxford: Oxbow Books.

Kriiska, A. (1996) 'Stone Age settlements in the lower reaches of the Narva River, north-eastern Estonia', in Hackens, T., Hicks, S., Lang, V., Miller, U., and Saarse, L. (eds) *Coastal Estonia: Recent Advances in Environmental and Cultural History.* PACT vol. 51, pp. 359–69. Rixensart: Pact Belgium.

Kriiska, A. (2009) 'The beginning of farming in the eastern Baltic', in Dolukhanov, P. M., Sarson, G. R., and Shukurov, A. M. (eds) *The East European Plain on the Eve of Agriculture.* BAR International Series 1964, pp. 159–79. Oxford: Archaeopress.

Kriiska, A. and Lõugas, L. (1999) 'Late Mesolithic and Early Neolithic seasonal settlement at Kõpu, Hiiumaa Island, Estonia', in Miller, U., Hackens, T., Lang, V., Raukas, A., and Hicks, S. (eds) *Environmental and Cultural History of the Eastern Baltic Region.* PACT vol. 57, pp. 157–72. Rixensart: Pact Belgium.

Kriiska, A. and Lõugas, L. (2009) 'Stone Age settlement sites on an environmentally sensitive coastal area along the lower reaches of the River Pärnu (south-western Estonia), as indicators of changing settlement patterns, technologies and economies', in McCartan, S., Schulting, R., Warren, G., and Woodman, P. (eds) *Mesolithic Horizons*, Vol. 1, pp. 167–75. Oxford: Oxbow Books.

Lepiksaar, J. (1986) 'The Holocene history of theriofauna in Fennoscandia and Baltic countries', *Striae*, 51–70.

Loze, I. A. (1988) *Poselenya Kamennogo veka Lybanskoy Nizini: Mesolit, Rannij i Srednij Neolit*, Riga: Zinatne.

Lõugas, L. (1996a) 'Analyses of animal remains from the excavations at the Lammasmägi site, Kunda, north-east Estonia', in Hackens, T., Hicks, S., Lang, V., Miller, U., and Saarse, L. (eds)

Coastal Estonia: Recent Advances in Environmental and Cultural History. PACT vol. 51, pp. 273–91. Rixensart: Pact Belgium.

Lõugas, L. (1996b) 'A subfossil fauna complex from the Narva Region', in Hackens, T., Hicks, S., Lang, V., Miller, U., and Saarse, L. (eds) *Coastal Estonia: Recent Advances in Environmental and Cultural History*. PACT vol. 51, pp. 369–72. Rixensart: Pact Belgium.

Lõugas, L. (1997) 'Post-Glacial Development of Vertebrate Fauna in Estonian Water Bodies: A Palaeozoological Study'. Dissertationes Biologicae Universitatis Tartuensis 32. Tartu: Tartu University Press.

Lõugas, L. (2006) 'Animals as subsistence and bones as raw material for settlers of Prehistoric Zvejnieki', in Larsson, L. and Zagorska, I. (eds) *Back to the Origin: New Research in the Mesolithic-Neolithic Zvejnieki Cemetery and Environment, Northern Latvia*. Acta Archaeologica Lundensia vol. 52, pp. 75–89. Stockholm: Almqvist and Wiksell International.

Lõugas, L. (2008) 'Mõnedest mesoliitilistest faunakompleksidest Läänemere idarannikult', in Jaanits, L., Lang, V. and Peets, J. (eds) *Loodus, Inimene ja Tehnoloogia 2*. Muinasaja teadus vol. 17, pp. 253–62. Tallin: Ajaloo Instituut.

Lõugas, L., Kriiska, A. & Moora, H. (1996) 'Coastal adaptation and marine exploitation of the Island Hiiumaa, during the Stone Age with special emphasis on the Kõpu I site', in Robertsson, A.-M., Hicks, S., Åkerlund, A., Risberg, J., and Hackens, T. (eds) *Landscapes and Life*. PACT vol. 50, pp. 197–211. Rixensart: Pact Belgium.

Mannermaa, K. (2008) 'The Archaeology of Wings. Birds and People in the Baltic Sea Region during the Stone Age'. Academic dissertation. Printed by Gummerus Kirjapaino OY. University of Helsinki (Helsinki).

Moora, H. and Lõugas, L. (1995) 'Natural conditions at the time of primary habitation of the Hiiumaa Island', *Proceedings of the Estonian Academy of Sciences. Humanities and Social Sciences*, 44/4, 472–81.

Ostrauskas, T. (2000) 'Mesolithic Kunda Culture: A Glimpse from Lithuania', in Lang, V. and Kriiska, A. (eds) *De Temporibus Antiquissimis ad Honorem Lembit Jaanits*. Muinasaja teadus vol. 8, pp. 167–80. Tallin: Ajaloo Instituut.

Paaver, K. (1965) *Formirovanie Teriofauny i Izmencivost Mlekopitajuschich Pribaltiki v Holocene*, Tartu: Akademija Nauk Estonskoj SSR.

Poska, A. and Veski, S. (1999) 'Man and environment at 9500 BP: a palynological study of an Early-Mesolithic settlement site in south-west Estonia', *Acta Palaeobotanica*, Supplement 2, 603–7.

Rosentau, A., Muru, M., Kriiska, A., Subetto, D., Vassiljev, J., Hang, T., Gerasimov, D., Nordqvist, K., Ludikova, A., Lõugas, L., Raig, H., Kihno, K., Aunap, R., and Letyka, N. (2013) 'Stone Age settlement and Holocene shore displacement in the Narva-Luga Klint Bay area, eastern Gulf of Finland', *Boreas*, 42(4), 912–31.

Sommer, R., Benecke, N., Lõugas, L., Nelle, O., and Schmölcke, U. (2011) 'Holocene survival of the wild horse in Europe: a matter of open landscape?', *Journal of Quaternary Science*, 26(8), 805–12.

Ukkonen, P., Lõugas, L., Zagorska, I., Lukševica, L., Lukševics, E., Daugnora, L., and Jungner, H. (2006) 'History of the reindeer (*Rangifer tarandus*) in the eastern Baltic region and its implications for the origin and immigration routes of the recent northern European wild reindeer populations', *Boreas* 35(2), 222–30.

Ukkonen, P., Aaris-Sørensen, K., Arppe, L., Daugnora, L., Halkka, A., Lõugas, L., Oinonen, M. J., Pilot, M., and Storå, J. (2014) 'An Arctic seal in temperate waters: history of the ringed seal (*Pusa hispida*) in the Baltic Sea and its adaptation to the changing environment', *The Holocene*, 24(12), 1694–1706.

Veski, S., Heinsalu, A., Klassen, V., Kriiska, A., Lõugas, L., and Saluäär, U. (2005) 'Early Holocene coastal palaeoenvironment and ancient man on the shore of the Baltic Sea at Pärnu, southwestern Estonia', *Journal of Quaternary International*, 130(1), 75–85.

Zachrisson, I. and Iregren, E. (1974) *Lappish Bear Graves in Northern Sweden: An Archaeological and Osteological Study*, Stockholm: Kungl. Vitterhets historie och antikvitetsakademien.

Zagorska, I. (1992) 'The Mesolithic in Latvia', *Acta Archaeologica*, 63, 97–117.

Zagorska, I. (1995) 'Late Glacial and Early Postglacial finds in the Latvian coastal area', in Fisher, A. (ed.) *Man and Sea in the Mesolithic*, pp. 251–8. Oxford: Oxbow Books.

CHAPTER 5

ARCHAEOZOOLOGICAL TECHNIQUES AND PROTOCOLS FOR ELABORATING SCENARIOS OF EARLY COLONIZATION AND NEOLITHIZATION OF CYPRUS

JEAN-DENIS VIGNE

INTRODUCTION: ARCHAEOZOOLOGY, TECHNICAL SYSTEMS, PRACTICES, AND CULTURAL HISTORY

EVERY archaeological animal remain is potentially bearing macroscopic, microscopic, and molecular information. Some of these characteristics developed during the life of the animal (intrinsic characters; Poplin, 1973), while others affected the skeleton after the animal died or even after burial (extrinsic characters). Both intrinsic and extrinsic characters reflect the relationships between humans and animals, the reconstruction of which is the main aim of archaeozoology. Quantitative approaches of large datasets of archaeological animal remains increase the quality of this information. Providing appropriate statistical techniques were applied to the study of bone collections, they allow distinguishing the anecdotic facts from the actual ones, only the latter truly reflecting historical events.

Animals play a major role in techno-economic and food-related activities, as well as in the social relationships and in the beliefs and religions of all human societies (Poplin, 1989; 2008). In archaeology, animal remains deriving from food refuses reflect different *chaînes opératoires* (Cresswell, 1983) of management of animal resources, and of acquisition, transformation, and utilization (including consumption) of animal products. In the structuralist's conceptual framework (Lévi-Stauss, 1958), they are techniques (*sensu* Mauss, 1967) and, altogether, they constitute the technical animal subsystem of the society (Vigne, 1998; Horard-Herbin et al., 2005). This sub-system is tightly connected with the other technical subsystems of the society, but also with its physiological, social, and symbolic systems and with the surrounding ecosystems. Therefore, animal remains from domestic contexts are not only associated with techno-economic activities, but also with the socio-symbolic sphere, together with the faunal remains found in funeral and religious deposits (Vigne et al., 2005). In other words, they not only speak of the *chaînes opératoires*, but also of the practices, that is, for a human group, the 'proper way-to-do which leads to a succession of operations depending of the technical, social and cognitive characteristics of the group, and of its traditions, beliefs and mental representations' (Horard-Herbin and Vigne, 2005: 189). Consequently, the animal subsystem can contribute to the characterization of past societies no less than flint technology, pottery, or metal productions, providing its study is not restricted to a pure description of (series of) objects, but aims to contribute to more general anthropological and/or environmental questions.

In this chapter, I will illustrate this conception of archaeozoology through the example of the study of the Cypriot Pre-Pottery Neolithic of Shillourokambos. I will emphasize the particular techniques and protocols developed with the aim to secure the quality of the archaeozoological information as well as a correct approach to the study of historical trajectories.

The Cypriot Pre-Pottery Neolithic and the Site of Shillourokambos

The island of Cyprus was always separated from the mainland by at least 70 km of open sea (Held, 1989; Vigne et al., 2014). Due to this insularity, only four endemic mammal species lived there during the Late Pleistocene, a dwarf hippo (*Phanourios minutus*), a dwarf elephant (*Elephas cypriotes*), a genet (*Genetta plesictoides*), and a mouse (*Mus cypriacus*) (Simmons, 1999). Sometimes at the end of the Late Glacial, hippos and elephants became extinct and people introduced wild boar (*Sus scrofa*), which rapidly proliferated and decreased in size (Vigne et al., 2009). The chronology of these events is still debated (Vigne et al., 2014; Zazzo et al., 2015), but we can date the earliest unquestionable evidence of the presence of humans in Cyprus to 10,500 cal BC, at the small Aetokremnos rockshelter (Simmons, 1999). There is then a gap of information about the presence of humans in Cyprus in the tenth millennium BC, but two Pre-Pottery Neolithic villages have recently been discovered, Asprokremnos (Manning et al., 2010) and Klimonas (Vigne et al., 2012), dating to the beginning of the ninth millennium. They are characterized by a typical Levantine Late PPNA way of life, with cultivation of pre-domestic cereals and hunting, the only available large game being the Cypriot wild boar. Domestic dogs (*Canis familiaris*) and commensal cats (*Felis silvestris lybica*) were introduced to the island at that time or shortly before (Vigne et al., 2012). The endemic Cypriot mouse (*Mus cypriacus*) was living in the villages, and still survives today, besides the commensal mouse (*Mus musculus domesticus*) (Cucchi et al., 2006).

Several sites indicate the development of the Cypro-PPNB culture starting from 8400–8300 cal BC, in parallel to similar events on the mainland. The most important one is Shillourokambos, a large open air site excavated on 5,000 m² from 1992 to 2004 (Guilaine et al., 2011). The site was divided into two main sectors, 1 and 3. In sector 1 it is mostly composed of more than 470 archaeological pits (wells, silos, trenches) and of some thick layers of demolition. The less eroded sector 3 provided the remains of about fifteen buildings. The excavation revealed the existence of four chronological phases, based on a small number of real superimpositions of stratigraphic units (SU), a series of more than forty radiocarbon dates, the flint technology and the rates of the different categories of lithic rough material, and the animal species. The phases are defined as follows:

- Early A (8400/8300–8200 cal BC)
- Early B (8000–7500)
- Middle (7500–7400)
- Late (7400–7100/7000).

The two main challenges for archaeozoology consisted in establishing the chronology of the immigration/introduction of a series of new mammal species coming from the mainland, and to characterize the emergence of animal husbandry over the 1,400 years of the site occupation. The sectors 1 and 3 provided respectively 32,500 and 80,000 animal remains, but only the archaeozoological results of sector 1 have been exhaustively published (Guilaine et al., 2011; Vigne et al., 2011b). Animal remains only occasionally come from funeral contexts (Vigne et al., 2004; Le Mort et al., 2008), most of them being from domestic refuses. Apart from the presence of sometimes thick and persistent calcareous crusts, the preservation of the animal bones was generally good, due to the absence of any carnivore gnawing: dogs were very rare, except in the earliest phase, and there were no wild large carnivores in Cyprus.

All the classical osteo-archaeological techniques have been adopted: taxonomic, age and sex identification, morphometrics, fragmentation and butchery mark analysis, frequencies of the skeletal parts, and spatial distributions. In this chapter, we will focus on more original techniques and protocols aiming to improve the chrono-stratigraphic reliability of the samples and to address as accurately as possible the question of the status of each of the ungulate species. I took part in all the excavation seasons since 1995 and I excavated myself most of the contexts rich in animal bones.

FIELD ARCHAEOZOOLOGY: SAMPLING
AND QUESTIONING THE SAMPLES
HOMOGENEITY

The reliability of the final interpretation depends on the intrinsic qualities of the archaeo(zoo)logical series, which can be divided into two components: their significance and their validity (Poplin, 1977; Vigne, 1988). Concerning the significance of the faunal assemblages, we only took into consideration domestic refuses for the techno-economic reconstruction, the burial deposits being studied separately. The validity of each of the faunal assemblage was determined by: (1) the chrono-stratigraphic reliability and representativeness of the archaeological context of origin, which have been systematically assessed during excavation (often by myself) and through the composition and the homogeneity of the material included in it; all the questionable contexts were excluded from the study; (2) the number of faunal specimens, which determines the statistical reliability of the faunal dataset, all the faunal assemblages below 100 NISP being excluded from this study; (3) the degree of degradation, which generates differential effects on animal assemblages and may drastically bias the original proportions; the best way to assess the degradation is the relative proportion of the skeletal parts (Binford, 1981).

In order to investigate the question of the distance between the places of slaughtering/butchering and consumption of caprines and deer, which is an important argument for discussing their status, we relied on samples large enough to provide more than 15 Minimum Number of Elements (MNE) for most of the skeletal parts. This was the case of SU-1000 (section 3), a 30 m² wide dense accumulation of animal remains (15–40 kg/m²). In order to make sure of the homogeneity of a subsample representing nearly half of the area of SU-1000 (c.3,200 NISP), we compared the proportions of the three main taxa (deer, caprines, and pigs) between the 16 m² delimited during the excavation, on the basis of Chi² tests. We found significant differences (Chi² = 117, df = 28, p < .000), indicating an heterogeneity which had not been detected during excavation. But we also noticed that the main contributions to the Chi² were coming from five adjoining square metres, on the southwestern part of the SU. After excluding them from the study, we got a non-significant Chi² for the eleven remaining squares (Chi² = 37, df = 20, p > .01), which could therefore been considered as a non-heterogeneous set. We applied this protocol to all the large bone accumulations and to the successive layers accumulated in the wells (Guilaine et al., 2011).

For Shillourokambos, the critical analyses of the significance and validity of the various hundreds of faunal subsets led to a selection of a priority group of twenty-eight (total NISP = 5,535), considered to be excellent for the analysis of the technico-economic aspects, and a second group of twenty-two, which were good but slightly less reliable (NISP = 1,383).

INVESTIGATING THE EVOLUTION OF
THE SUBSYSTEM OF EXPLOITATION
OF ANIMALS

As I mentioned above, the attribution of the different SU to the four phases of site occupation was determined on the basis of a small number of criteria, including some faunal markers. Deducing the evolution of the system of exploitation of animals based on such chronological attributions was therefore at least partly marred by a circular reasoning. In order to escape this bias, we decided to ignore these chronological attributions, at least for the first step of the data processing. We preferred to reconstruct the proper chronological organization of the faunal assemblages according to the following protocol (Vigne and Carrère, 2011):

1. Investigating the organization of the diversity of the faunal spectra (NISP of 10 taxa) based on the Correspondence Analysis (CA) of the twenty-eight more valid assemblages, a calculation which takes into account the differences in sample sizes and which estimates multivariate distances between them; the twenty-two slightly less reliable assemblages were used as supplementary data, a procedure which excludes them from the calculation but provides information about their degree of similarity with each of the best assemblages; CA evidenced four principal groups of spectra (grey polygons on Fig. 5.1), and two intermediate groups of lesser importance (light-grey dotted lines);

2. Reconstructing the chronological relationships between these groups on the basis of the thirteen most reliable stratigraphic relationships (*sensu* Harris, 1979) observed during excavation; this revealed that the six groups of spectra corresponded to successive chronological phases (Fig. 5.1), the variation within each of them being likely due to their respective functional origin or taphonomic evolution;

3. Comparing this faunal chronology with the one based on chipped stones industries; we observed only eleven discrepancies (21% of the 28 + 22 assemblages), all of them concerning the phase just before or just after; in three of these cases, the faunal chronological attribution was the same as the one based on the lithic macro-tools, which suggests that the rhythm of evolution of the different technical subsystems could have been slightly different.

This protocol allowed us to secure the attribution of the most important faunal assemblages to the Early A (group 1), Early B (groups 2 and 3), Middle (groups 4 and 5), and Late (group 6) phases. It also provided a refined chronological framework for the whole site, subdividing the Early B into Early B and Early C (the latter having subsequently been characterized by a different lithic industry and dated to 7700–7500 cal BC) and the Middle phase into two sub-phases (Middle A and B).

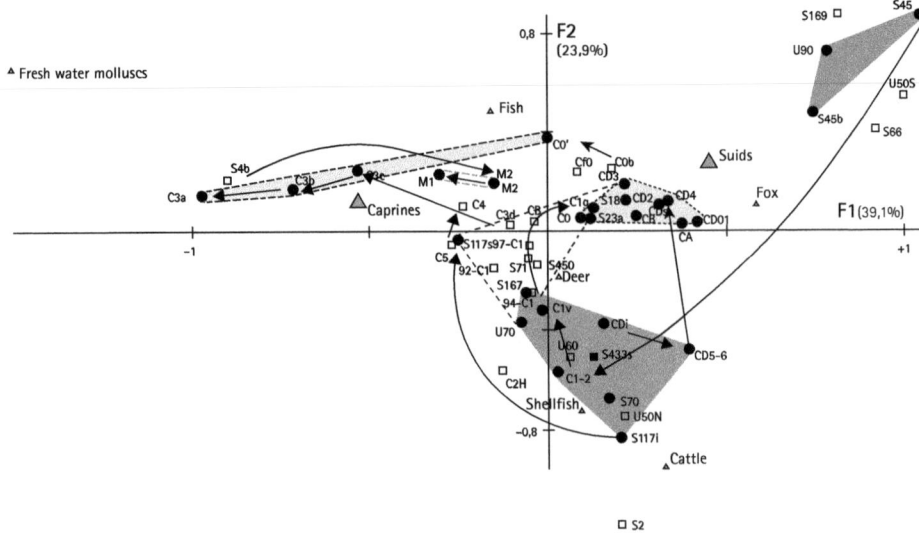

FIGURE 5.1 F1xF2 scatter diagram of the Correspondence Analysis of the 28 most reliable faunal spectra (solid circles) and of the 22 slightly less reliable ones (empty squares) from Shillourokambos. The arrows point to the 13 stratigraphic relationships. The size of the triangles corresponding to the taxa is proportional to their contribution to the total variance of the analysis. Author's own image.

A posteriori Validation: Analytic Results and Chronological Scenarios

The last step of the protocol consisted in testing the coherence of the evolution of the faunal composition and the morphometry of each of the ungulate species from 8400 to 7000 cal BC, based on this consolidated chronological framework. I will only give some examples of this investigation (for more details, see Guilaine et al., 2011 and Vigne et al., 2011b).

First of all, we observed that the list of the species present on site evolved through time. Only suids (*Sus scrofa*), goat (*Capra hircus*), cattle (*Bos taurus*), dog, and cat are attested during the Early A phase, with an overwhelming dominance of suids (87%). With reference to the animal species attested five hundred years earlier at Klimonas, located at less than 10 km from Shillourokambos (Vigne et al., 2012), this suggested that goat and cattle were introduced to the island between 8800 and 8400 cal BC. Fox (*Vulpes vulpes*), Persian fallow deer (*Dama mesopotamica*), and sheep (*Ovis aries*) appeared at the beginning of the Early phase B, the two latter representing 41% and 15% of the NISP (n = 1,586), respectively. I concluded that these three new taxa were introduced between

8300 and 8000 cal BC, and that the period between 8800 and 8000 cal BC was characterized by an intense introduction of new species to Shillourokambos, and likely to the whole island. The end of the sequence (Middle and Late phases) was marked by a drastic decrease of the importance of dog and cattle, two species absent from Cypriot sites dating to the following millennium (Davis, 2003).

For investigating the evolution of each ungulate species, I relied mostly on adult bone/tooth measurements (n = 10,689 from 3,547 items) and on postcranial and dental age at death estimates (respectively: n = 4,278 and 1,578, representing a minimum number of 394 individuals). Since deer, goats, and cattle, and to a lesser extent sheep are sexually dimorphic, the mean of their measurements reflects both the average morphology of the species and the sex proportions. Consequently, size increase/decrease from one phase to another may also only be a consequence of variations in the proportions of the sexes. In order to bypass this bias, I used Gaussian mixture analyses (Monchot and Léchelle, 2002). Starting from the bimodal distribution of the frequencies of the measurements, this calculation proposes a separate Gaussian distribution for each of the two groups, which can be assimilated to the two sexes (Fig. 5.2A). When it is supported by a good Akaike Information Criterion, it provides estimates of the means and variance of each of the measurements for males and females, and allows comparing males with males and females with females without any bias due to the sex ratio fluctuations. Additionally, it provides estimates of the sex ratio and of the sexual dimorphism, the latter being a good marker of domestication (Helmer et al., 2005). When the sample sizes are too small, this technique can also be applied to the Log Size Index (LSI, Fig. 5.2B), keeping in mind the recommendations of Meadow (1999).

For goat, the combination of all these data produced an interesting scenario (Vigne, 2011). Despite possessing very strong horncores of the wild type, it seems that the earliest goats at Shillourokambos were already smaller in size than their wild ancestor, the bezoar goat (*Capra aegagrus*), due to an incipient domestication on the mainland before their introduction to Cyprus. However, between the Early A and Middle B phases (one millennium), there was no significant change in size in either males or females (Fig. 5.2B). In that period there is also no evidence of specialization as illustrated by the slaughtering profiles, the sex ratios of adult being kept well balanced (Fig. 5.2C). The horncores were still of a wild type and the parts of the skeleton of high meat return were those mainly brought back to the site. We have even direct evidence of hunting thanks to a flint micro-flake embedded into a goat bone. It appears that goats were released into the wild shortly after their introduction, and exploited by hunting. However, during the Middle A phase, the slaughtering profiles became more specialized and the sex ratio tends to decrease, sometimes down to values which are significantly lower than 50% (dark-grey in Fig. 5.2C). This suggests that the process of (re-)domestication of the feral goats was initiated then. Consecutive morphological modifications only become visible in the course of the Middle B or Late phases, with the appearance of horncores of the domestic type and with a tendency to size decrease for both males and females. This decrease becomes statistically significant only during the next millennium (Khirokitia), together with a visible sexual dimorphism diminution, at a time when the sex ratio was

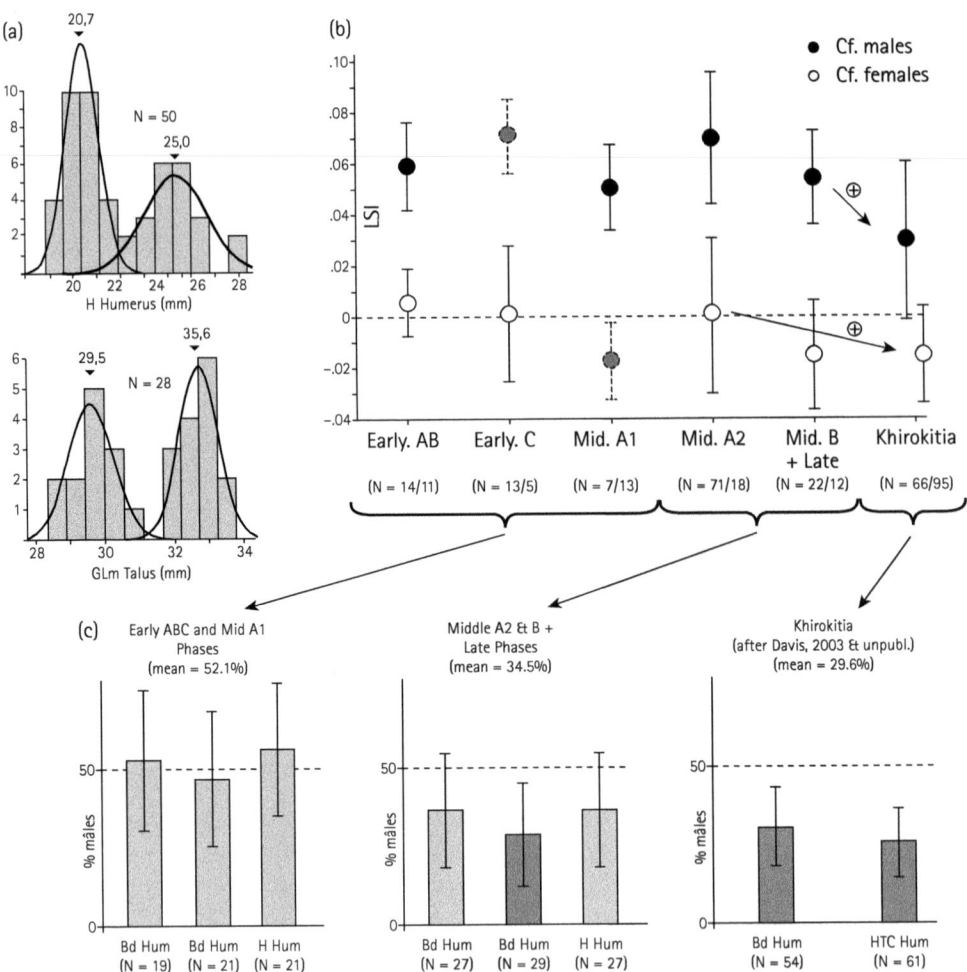

FIGURE 5.2 The process of re-domestication of the feral goat at Shillourokambos, after Vigne (2011):

A. Histograms of distribution of the densities and mixture analyses for two measurements (according to the standards of von den Driesch, 1976), one for the humerus and one for the talus, as examples of the protocol which has been used for estimating the sex ratio and the separate mean and standard deviation for males and females;

B. Evolution of the average size (expressed by the Log Size Index, LSI) of males and females (separated by mixture analyses) during the different phases of occupation of the village of Shillourokambos (Early AB to Mid B + Late) and at Khirokitia; grey dots correspond to small samples; arrows and (+) indicate statistically significant differences; significant size decreases due to domestication appear at the end of the sequences, *c.*7500–6500 cal BC;

C. Evolution of the goat sex ratio over three successive phases of the evolution of the PPN of Shillourokambos and Khirokitia, according to the mixture analyses of different measurements of the humerus distal part (Bd, BT, H, HTC according to von den Driesch, 1976, and Davis, unpublished) of adult goats; the vertical segments delimit the standard deviation of the mean; light grey columns indicate no statistically significant differences between the sex ratio estimate and 50%; conversely, dark grey columns indicate significant differences; the sex ratio is well balanced during the early step; it tends to decrease below 50% (i.e. less adult males than females) during the Middle phase; it is significantly biased towards adult females during the recent phase; together with the size decrease, this unbalanced sex ratio argues for local domestication of the goat. Author's own image.

definitely unbalanced. This scenario indicates that the Cypriot PPNB villagers were experiencing local domestication, exactly as their contemporaneous continental counterparts, except that they did it starting from feral goats. This is important information about the mechanism and rhythm of the domestication processes.

The same methods revealed the different scenarios of evolution of the other ungulates at Shillourokambos (Vigne et al., 2011b). Cattle and sheep were also introduced to Shillourokambos as early domesticates. However, unlike the goat they were not released to the wild but herded throughout the period of the site occupation. During the Early B phase, the evidence suggests a sophisticated mixed herding of sheep for milk and meat. As illustrated by the age profiles, suids were subject to a constant intensification of their exploitation, but it is still not clear if they were locally domesticated and/or introduced from elsewhere as early domesticates. Throughout the occupation of the site, deer were not subjected to any decrease in size or in sexual dimorphism; the age profiles continued to indicate a lack of specialized exploitation and the sex ratio was permanently well balanced. Domestication was initiated with the introduction to the island, which implied some form of control, but it never went further (Vigne et al., 2016).

The Shillourokambos villagers appear to have combined or even alternated hunting and husbandry. We must wonder, however, whether this strategy resulted in a stable level of food production or rather in a tendency similar to the one which led to the final emergence of farming on the nearby continent, at the end of the eighth millennium (Vigne, 2008). I investigated the importance of production *vs.* predation in the meat supply for each of the Shillourokambos phases, using the bone weight and taking into account the sometimes large uncertainty about the status of the species (Fig. 5.3). Unsurprisingly, this shows that the contribution of husbandry slowly increased through time. This tendency was, however, not linear, but subject to at least one recurrence, in connection with the failure/collapse of sheep herding and with several important modifications in the lithic raw material and knapping techniques (Guilaine et al., 2011).

CONCLUSION: ISLAND COLONIZATION AND CULTURAL HISTORY

Although it is still necessary to test our interpretations on other Cypriot sites, this scenario illustrates a series of components of the processes affecting the Near East Neolithic transition.

Feralization was frequently mentioned as an epiphenomenon of the neolithization process, especially in Europe, North America, and Australia (T. Mac Knight, 1964, cited by Digard, 1990). It has, however, rarely been demonstrated through archaeological evidence. In Cyprus, it is not only apparent for several species, but it also represented an alternative strategy to husbandry for people whose tradition was still embedded in their hunting past. It is interesting to see that the early Cypriot villagers were able to combine

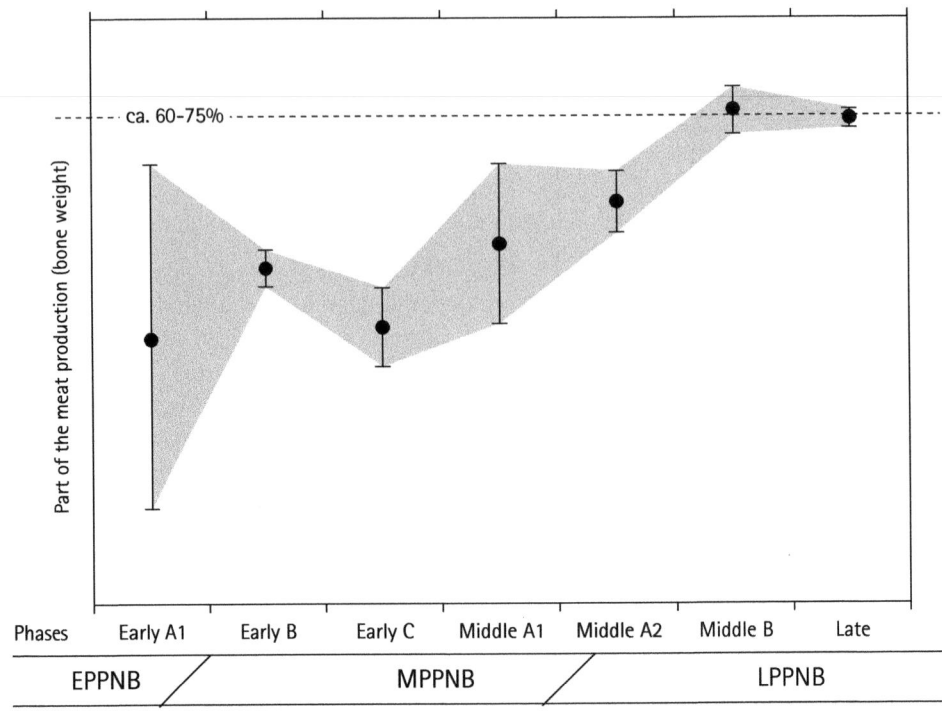

FIGURE 5.3 Evolution of meat production through the different chronological phases of Shillourokambos, after Vigne et al. (2011a). The part of the meat which came from husbandry with reference to the whole meat intake was estimated on the basis of the bone weight. The vertical segment delimitates the domain of uncertainty due to the questionable status of some species such as suids for the Early A1 phase. Correlations between the bone weight and the Meat and Offal Weight estimate (Vigne, 1992) could be done for the Middle phase only, and allowed the estimate that the maximal proportion of bone weight corresponds to 60% to 75% of the total meat consumption (Vigne et al., 2011a: 1178–80). Author's own image.

a large array of strategies, from hunting to herding, which endowed them with high adaptability. The non-linear tendency to herding at Shillourokambos, however, suggests that these societies were in constant search of stability, torn as they were between their increasing demography and social complexity, and their changing environment and resources.

The processes of domestication as observed at Shillourokambos also indicate that Cyprus, although an island, was part of the vast Near East area where people were experimenting with domestication. This strengthens the conception of a large area of diffuse origin of the Neolithic, *contra* a core area from which the innovations would have diffused (Fuller et al., 2011). This model is tightly connected with intense and long distance exchanges, a component of the Neolithic transition often under-evaluated but well highlighted by our observations (Vigne, 2013).

Another unexpected contribution of the archaeozoological results from Shillourokambos concerns early navigation (Vigne et al., 2014). In the absence of any boat wreck earlier than the Early Bronze Age in the Mediterranean, all we knew was that people transported obsidian in the Aegean Sea during the Epipalaeolithic, an operation which could be done with a simple canoe boat. Voyaging with large ruminants required much more sophisticated boats. For different biological reasons (Vigne, 2014), the ruminants must have been transported standing and already weaned (that is to say more than 120 kg for calves), as fast as possible over more than 70 km of open sea. The absence of morphological divergence in the domestic mouse between Cyprus and the continent across the PPN indicates that the evolutionary drift that should have occurred due to isolation was offset by an important and constant gene flow coming from the continent to Cyprus; based on quantitative modern observations, it was even possible to estimate that this gene flow was the equivalent of at least two successful immigrations per year (Vigne and Cucchi, 2005, p. 191). There is no doubt that the PPN boats were more similar to the fourth millennium Naqadian 'sewn' boats from Egypt (wooden boards assembled with tongue and mortise and tied up with plant fibres) than to the Mesolithic dugouts (Vigne et al., 2014).

Such important and sometimes amazing contributions of archaeozoology to larger questions results from several factors. First of all, islands are excellent sources of information, because they provide simplified systems, easier to analyse. The absence of any native ancestor of domesticates on Cyprus helped to demonstrate animal transportations, which would have been much more difficult to detect on the mainland. Secondly, we benefited from a large site, covering a long time span, with extensive and careful excavations and well-preserved animal remains. The participation of the archaeozoologist in the excavation and his involvement in the critical approach of the archaeological contexts were decisive at Shillourokambos. Also important was the quest for high chronological resolution, at least at the scale of the Neolithic. Refining the chronology as much as possible provided access to the complexity of the processes. The quantitative and integrated approaches of as many archaeozoological aspects as possible also contributed to clarify PPN practices further.

Last but not least, this would not have been possible without the ambition to contribute not only to the reconstruction of the animal exploitation strategies, but also to a series of broader questions, including the critical approach to the site itself, the dynamics of the Early Neolithic societies at the regional scale, and their navigation history. Such an ambition is the only way for archaeozoology to play a significant role in the concert of the historical, social, and environmental sciences.

Acknowledgements

I am grateful to Simon Davis, who provided unpublished osteometric data from Khirokitia, to Isabelle Carrère, who assisted me with the cleaning and sorting of the Shillourokambos faunal remains for many years, to Jean Guilaine, who trusted me for the faunal study of Shillourokambos, and to Umberto Albarella who edited a draft of this paper and made useful suggestions for improving it. This study was granted by the French Schoool at Athens, by the French Ministry of Foreign Affairs, and by the Site d'Étude en Écologie Globale 'Limassol' (CNRS, INEE), with constant support from the Cyprus National Department of Antiquities.

REFERENCES

Binford, L. S. (1981) *Bones: Ancient Men and Modern Myths*, London: Academic Press.

Cresswell, R. (1983) 'Transfert de techniques et chaînes opératoires', *Techniques et Cultures*, 2, 143–59.

Cucchi, T., Orth, A., Auffray, J.-C., Renaud, S., Fabre, L., Catalan, J., Hadjisterkotis, E., Bonhomme, F., and Vigne, J.-D. (2006) 'A new endemic species of the subgenus *Mus* (Rodentia, Mammalia) on the Island of Cyprus', *Zootaxa*, 261, 1–36.

Davis, S. (2003) 'The zooarchaeology of Khirokitia (Neolithic Cyprus), including a view from the mainland', in Guilaine, J. and Le Brun, A. (eds) *Le Néolithique de Chypre, Bulletin de Correspondance Hellénique, Suppl. 43*, pp. 253–78. Athens: École Française d'Athènes.

Digard, J.-P. (1990) *L'Homme et les animaux domestiques: anthropologie d'une passion*, Paris: Fayard.

Driesch, von den, A. (1976) *A Guide to the Measurement of Animal Bones from Archaeological Sites*, Harvard: Peabody Museum of Archaeology and Ethnology.

Fuller, D. Q., Willcox, G., and Allaby, R. G. (2011) 'Cultivation and domestication had multiple origins: arguments against the core area hypothesis for the origins of agriculture in the Near-East', *World Archaeology*, 43(4), 628–52.

Guilaine, J., Briois, F., and Vigne, J.-D. (eds) (2011) *Shillourokambos: un établissement néolithique pré-céramique à Chypre. Les fouilles du secteur 1*, Paris: Errance/École française d'Athènes.

Harris, E. (1979) *Principles of Archaeological Stratigraphy*, London: Academic Press.

Held, S. O. (1989) 'Colonization cycles on Cyprus. 1: The biogeographic and palaeontological foundation of early prehistoric settlement', *Report of the Department of Antiquities, Cyprus*, pp. 1–28.

Helmer, D., Gourichon, L., Monchot, H., Peters, J., and Saña Segui, M. (2005) 'Identifying early domestic cattle from Pre-Pottery Neolithic sites on the Middle Euphrates using sexual dimorphism', in Vigne, J.-D., Peters, J., and Helmer, D. (eds) *The First Steps of Animal Domestication: New Archaeobiological Approaches*, pp. 86–95. London: Oxbow.

Horard-Herbin, M.-P., Lefèvre, C., and Vigne, J.-D. (2005) 'L'alimentation carnée et les produits alimentaires', in Horard-Herbin, M.-P. and Vigne, J.-D. (eds) *Animaux, environnements et sociétés*, pp. 63–92. Paris: Errance.

Horard-Herbin, M.-P. and Vigne, J.-D. (eds) (2005) *Animaux, environnements et sociétés*, Paris: Errance.

Le Mort, F., Vigne, J.-D., Davis, S., Guilaine, J., and Le Brun, A. (2008) 'Man-animal relationships in the Pre-Pottery burials at Shillourokambos and Khirokitia (Cyprus, 8th and 7th millennia cal. BC)', in Vila, E., Gourichon, L., Choike, A. and Buitenhuis, H. (eds) *Proceedings of the Eighth International Symposium on the Archaeozoology of Southwestern Asia and Adjacent Areas*, pp. 219–41. Lyon: Archéorient, Maison de l'Orient et de la Méditerranée.

Lévi-Strauss, C. (1958) *Anthropologie Structurale*, Paris: Plon.

Manning, S. W., McCartney, C., Kromer, B., and Stewart, S. T. (2010) 'The earlier Neolithic in Cyprus: recognition and dating of a Pre-Pottery Neolithic A occupation', *Antiquity*, 84, 693–706.

Mauss, M. (1967) *Manuel d'ethnographie*, Paris: Payot.

Meadow, R. H. (1999) 'The use of size index scaling techniques for research on archaeozoological collections from the Middle East', *Internationale Archäologie*, 8, 285–300.

Monchot, H. and Léchelle, J. (2002) 'Statistical nonparametrics methods for the study of fossil populations', *Paleobiology*, 28(1), 55–69.

Poplin, F. (1973) 'Interprétation ethnologique des vestiges animaux', in *L'homme, hier et aujourd'hui: recueil d'études en hommage à André Leroi-Gourhan*, pp. 345–54. Paris: Cujas-CNRS.

Poplin, F. (1977) 'Problèmes d'ostéologie quantitative relatifs à l'étude de l'écologie des hommes fossiles', *Bulletin de l'Association Française pour l'Étude du Quaternaire*, 47, 63–8.

Poplin, F. (1989) 'Matière, animal, homme, esprit: introduction à l'animal dans les pratiques religieuses', in Méniel, P. (ed.) *Animal et pratiques religieuses: les manifestations matérielles*. *Anthropozoologica*, 3rd Special Issue, 13–21.

Poplin, F. (2008) 'True animal, sacrifice and the domestication of dairy animals', in Vila, E., Gourichon, L., Choyke, A. M., and Buitenhuis, H. (eds) *Archaeozoology of the Near East VIII: Proceedings of the Eighth International Symposium on the Archaeozoology of Southwestern Asia and Adjacent Areas*, pp. 33–44. Lyon: Archéorient, Maison de l'Orient et de la Méditerranée.

Simmons, A. (ed.) (1999) *Faunal Extinction in an Island Society. Pygmy Hippopotamus Hunters of Cyprus*, New York: Kluwer Academy—Plenum Publisher.

Vigne, J.-D. (1988) *Les mammifères post-glaciaires de Corse*, Paris: CNRS (Gallia Préhistoire).

Vigne, J.-D. (1992) 'The meat and offal weight (MOW) method and the relative proportion of ovicaprines in some ancient meat diets of the north-western Mediterranean', *Rivista di Studi Liguri, A*, 57(2), 21–47.

Vigne, J.-D. (1998) 'Faciès culturels et sous-système technique de l'acquisition des ressources animales: application au Néolithique ancien méditerranéen', in D'Anna, A. and Binder, D. (eds) *Production et identité culturelle*, pp. 27–45. Antibes: APDCA.

Vigne, J.-D. (2008) 'Zooarchaeological aspects of the Neolithic diet transition in the Near-East and Europe, and their putative relationships with the Neolithic Demographic Transition', in Bocquet-Appel, J.-P. and Bar-Yosef, O. (eds) *The Neolithic Demographic Transition and Its Consequences*, pp. 179–205. New York: Springer.

Vigne, J.-D. (2011) 'La chèvre (*Capra aegagrus/Capra hircus*)', in Guilaine, J., Briois, F. and Vigne, J.-D. (eds) *Shillourokambos: un établissement Néolithique pré-céramique à Chypre. Les fouilles du secteur 1*, pp. 1001–20. Paris: Errance / École française d' Athènes.

Vigne, J.-D. (2013) 'Domestication process and domestic ungulates: new observations from Cyprus', in College, S., Conolly, J., Dobney, K., Manning, K., and Shennan, S. (eds) *The Origins and Spread of Domestic Animals in Southwest Asia and Europe*, pp. 115–28. Walnut Creek: Left Coast Press.

Vigne, J.-D. (2014) 'The origins of mammals on the Mediterranean islands as an indicator of early voyaging', *Eurasian Prehistory*, 10(1–2), 45–56.

Vigne, J.-D., Arbogast, R.-M., Horard-Herbin, M.-P., Méniel, P., and Lepetz, S. (2005) 'Animaux, sociétés et cultures', in Horard-Herbin, M.-P. and Vigne, J.-D. (eds) *Animaux, environnements et sociétés*, pp. 151–82. Paris: Errance.

Vigne, J.-D., Bailon, S., Carrère, I., Cucchi, T., Desse-Berset, N., Desse, J., le Dosseur, G., Serrand, N., and Stordeur, D. (2011a) 'Apports des faunes de Shillourokambos à la caractérisation historique, techno-économique et symbolique des sociétés du Néolithique pré-céramique chypriote et proche-oriental', in Guilaine, J., Briois, F., and Vigne, J.-D. (eds) *Shillourokambos: un établissement Néolithique pré-céramique à Chypre. Les fouilles du secteur 1*, pp. 1161–94. Paris: Errance / École française d' Athènes.

Vigne, J.-D., Briois, F., Zazzo, A., Willcox, G., Cucchi, T., Thiébault, S., Carrère, I., Franel, Y., Touquet, R., Martin, C., Moreau, C., Comby, C., and Guilaine, J. (2012) 'The first wave of cultivators spread to Cyprus earlier than 10,600 years ago', *Proceedings of the National Academy of Sciences of the United States of America*, 109(22), 8445–9.

Vigne, J.-D. and Carrère, I. (2011) 'Les ossements de vertébrés: méthodologie, spectres de faune et périodisation', in Guilaine, J., Briois, F., and Vigne, J.-D. (eds) *Shillourokambos: un établissement Néolithique pré-céramique à Chypre. Les fouilles du secteur 1*, pp. 539–78. Paris: Errance / École française d' Athènes.

Vigne, J.-D., Carrère, I., Briois, F., and Guilaine, J. (2011b) 'The early process of the mammal domestication in the Near-East: new evidence from the Pre-Neolithic and Pre-Pottery Neolithic in Cyprus', *Current Anthropology*, S52(4), S255–S271.

Vigne, J.-D. and Cucchi, T. (2005) 'Premières navigations au Proche-Orient: les informations indirectes de Chypre', *Paléorient*, 31(1), 186–94.

Vigne, J.-D., Cucchi, T., Zazzo, A., Carrère, I., Briois, F., and Guilaine, J. (2014) 'Transportation of mammals to Cyprus sheds light on early seafaring and boats in the Mediterranean', *Eurasian Prehistory*, 10(1–2), 157–78.

Vigne, J.-D., Daujat J., and Monchot H. (2016) 'First introduction and early exploitation of the Persian fallow deer on Cyprus (8,000–6,000 cal. BC)', *International Journal of Osteoarchaeology*, 26, 853–866.

Vigne, J.-D., Guilaine, J., Debue, K., Haye, L., and Gérard, P. (2004) 'Early taming of the cat in Cyprus', *Science*, 304, 259.

Vigne, J.-D., Zazzo, A., Saliège, J.-F., Poplin, F., Guilaine, J., and Simmons, A. (2009) 'Pre-Neolithic wild boar management and introduction to Cyprus more than 11,400 years ago', *Proceedings of the National Academy of Sciences of the United States of America*, 106(38), 16131–8.

Zazzo, A., Lebon, M., Quilès, A., Reiche, I., and Vigne, J.-D. (2015) 'Direct dating and physico-chemical analyses cast doubts on the coexistence of humans and dwarf hippos in Cyprus', *PlosONE*, DOI:10.1371/journal.pone.0134429.

···

CHAPTER 6

ZOOARCHAEOLOGICAL RESULTS FROM NEOLITHIC AND BRONZE AGE WETLAND AND DRYLAND SITES IN THE CENTRAL ALPINE FORELAND

economic, ecologic, and taphonomic relevance

···

JÖRG SCHIBLER

INTRODUCTION

···

SWITZERLAND is a small central European country with an area of about 41,285 km², of which the presently habitable area is about 15,628 km² (38%) (http://www.partinational-istesuisse.ch/documents). The landscape can be characterized into three different topographic units: the Alpine region in the south (highest elevation about 4,500 m); the Jura mountain region in the north (highest elevation about 1,700 m) and the flatter and hilly area in between—the midlands (Fig. 6.1). Each of these units is oriented from northeast to southwest. This orientation is responsible for the fact that most of the archaeologically visible cultural influences come from either the north and east (Danubian influence) or south and west (Mediterranean influence), with a meeting point somewhere in the midlands (Stöckli et al., 1995). The large Alpine valleys, such as the Rhine valley in the east and the Rhone valley (Valais) in the southwestern part of Switzerland, mainly show, respectively, strong Danubian or Mediterranean cultural influences.

In addition to their topographical differentiation, Swiss Neolithic and Bronze Age sites can be divided by their pedological and geomorphological characteristics: dryland (mineral soil) or wetland sites. Wetland sites are primarily situated on the shores of smaller and larger lakes in the midlands, whereas dryland sites can be found across

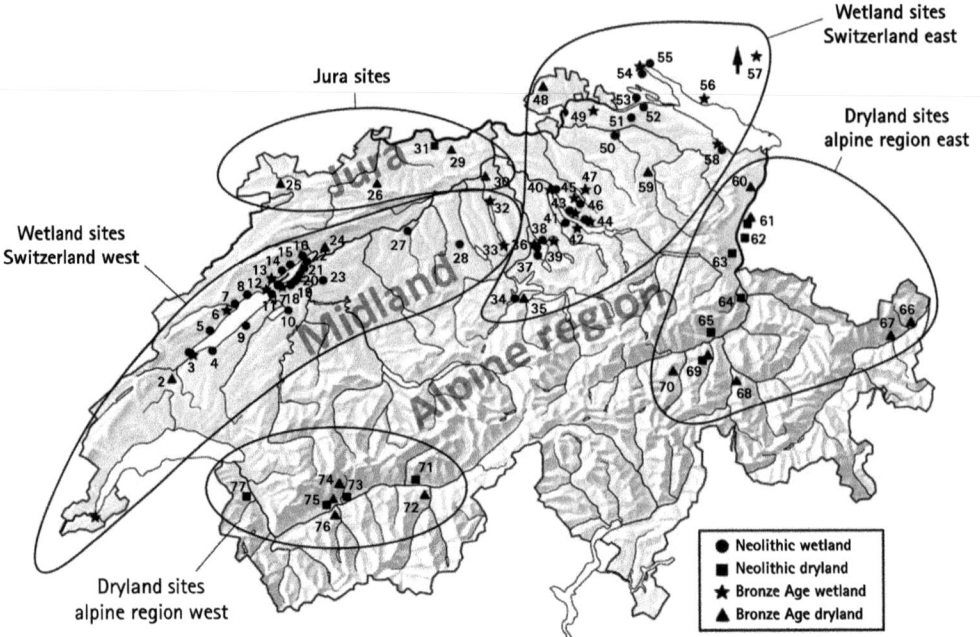

FIGURE 6.1 Large topographic units of Switzerland and location of Neolithic and Bronze Age wetland and dryland sites with archaeozoological data. For numbers see Table S61–3.

all three topographic units. Due to the deposition of finds below the groundwater level, the preservation quality of all archaeological finds is much higher in wetland sites than in their dryland counterparts. Anaerobic conditions below groundwater are responsible for a limited—or even non-existent—decomposition of organic material. Therefore, botanical remains such as seeds, fruits, and wood are preserved in a non-carbonized state, and animal bones are perfectly preserved. These excellent preservation conditions allow significant economic and ecological insights into the daily life of Neolithic and Bronze Age farmers. All known houses from wetland sites were made of timber, which has resulted in the survival of many larger wooden structural elements, such as posts or planks, which can be used to accurately date settlements, and even single houses, through dendrochronology. Therefore, the precise dating of settlement layers and archaeological finds, e.g. animal bones, is possible in wetland sites.

The combination of these site characteristics with a detailed archaeological and archaeozoological survey results in a unique opportunity to gain a comprehensive overview of the Swiss Neolithic and Bronze Age.

MATERIALS

The following overview is based on archaeozoological reports from 227 different settlements, providing 383,936 identified animal bone fragments (Table S6.1–3). The 169

Neolithic sites (303,865 fragments) are far more numerous than the 58 Bronze Age settlements with archaeozoological data (80,071 fragments). The number of wetland sites (175) outweighs those from dryland locations (52), largely due to the Neolithic emphasis on wetland compared to dryland sites (148 and 21 respectively). Archaeozoological data from Bronze Age wetland and dryland sites are more equally represented (27 and 31 respectively). The highest number of identifiable animal bones from a single site (32,863 fragments) is recorded from Arbon Bleiche 3, a Neolithic wetland site. The smallest quantity of identifiable bones from a single site, included in this overview, is sixty-two fragments. Only eight included sites produced less than one hundred identified animal bones, and most of the results are based on assemblages of several hundred up to two thousand identified bones.

Excluded from this dataset are animal bones recovered from sieved soil samples. This is because only in a few recently (last ten years) excavated sites have soil samples been taken, and investigations of bones from smaller animal species undertaken. Therefore, to ensure data comparability, only hand-picked animal bones are included in this study.

DATING

Most of the Neolithic wetland sites can be dated by dendrochronology to a precision of only a few years. Archaeozoological results of these sites can, therefore, be grouped chronologically, with high precision, between 4,300 and 2,500 BC. The same quality of dating is not yet possible for the Bronze Age wetland sites; results from these sites are therefore grouped into the three relative chronology phases: Early Bronze Age, Middle Bronze Age, and Late Bronze Age. Neolithic and Bronze Age dryland sites are mostly dated through C14 analysis or typological information. Therefore no precise absolute dating—of the kind available for Neolithic wetland sites—is possible for these dryland sites.

IMPORTANCE OF HUNTING

Comparing the proportion of bones of wild animals between all four categories of sites—Neolithic / Bronze Age, wetland / dryland—it becomes obvious that hunting was very important in Neolithic wetland settlements. As the median values indicate, in Neolithic lakeshore settlements generally around one third of the bones are from hunted animals (Fig. 6.2).

The comparison of summarized data for the importance of wild animal bones shows two other facts. First, wild animal bones are, on average, more frequently found in Neolithic than Bronze Age sites (Fig. 6.2). This can be seen when wetland and dryland sites are compared separately. Secondly, if looking separately at Neolithic and Bronze Age sites, wild animal bones are clearly more frequent in wetland sites. To explain these

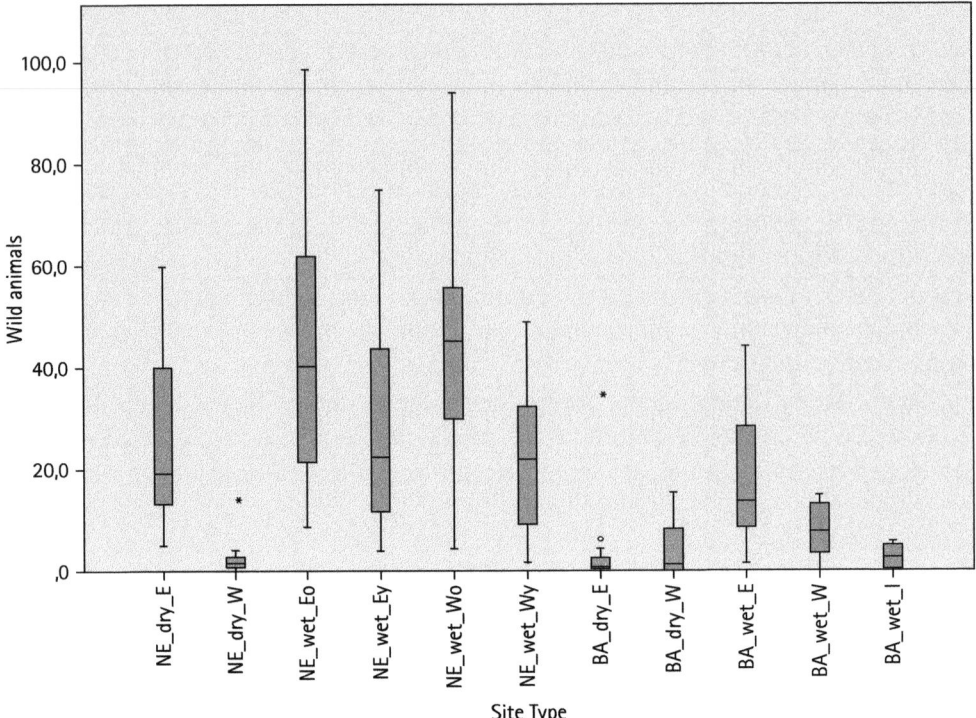

FIGURE 6.2 Boxplot of the importance (% of fragment numbers) of wild animal remains in Neolithic and Bronze Age wetland and dryland sites of Switzerland. NE_dry_E: Neolithic dryland sites in the eastern part of Switzerland (10 sites); NE_dry_W: Neolithic dryland sites in the western part of Switzerland (10 sites); NE_wet_Eo: older Neolithic (4300–3400 BC) wetland sites in the eastern part of Switzerland (42 sites); NE_wet_Ey: younger Neolithic (3400 BC–2500 BC) wetland sites in the eastern part of Switzerland (38 sites); NE_wet_Wo: older Neolithic (4300–3400 BC) wetland sites in the western part of Switzerland (34 sites); NE_wet_y: younger Neolithic (3400 BC–2500 BC) wetland sites in the western part of Switzerland (34 sites); BA_dry_E: Bronze Age (2200–800 BC) dryland sites in the eastern part of Switzerland (19 sites); BA_dry_W: Bronze Age dryland sites in the western part of Switzerland (8 sites); BA_wet_E: Bronze Age wetland sites in the eastern part of Switzerland (21 sites); BA_wet_W: Bronze Age wetland sites in the western part of Switzerland (6 sites); BA_dry_J: Bronze Age dryland sites in the Jura mountains (4 sites); one Neolithic dryland site in northwest Switzerland not included. For names of sites, their dating, and archaeozoological references see Table S6.1–3. Graph by author.

observations the precisely dated Neolithic wetland sites need to be examined in more detail. All 148 Neolithic wetland sites can be divided into two geographically separated regions: western midlands and eastern part of Switzerland (Fig. S6.1). The high precision of dendrochronological dating makes it possible to arrange the data (importance of wild animal bones) for each of these regions precisely along the timespan from 4,300 BC to 2,500 BC. As a consequence, it is possible to detect periods of a few years to a few decades where no settlements and no archaeozoological information is available (Fig. S6.1). Typological or even C14 dating rarely permit such possibilities.

The importance of wild animal bones at the 148 Neolithic wetland sites fluctuates considerably: the lowest value is at 1.8% and the highest value at 98.6% (Fig. S6.1). Even though there is no gradual chronological decrease of the importance of wild animal bones, in the group of younger settlements dated from 3,400 BC to 2,500 BC the importance of wild animal bones, on average, is reduced (Fig. 6.2). This trend is clear for the Bronze Age wetland sites, even though the importance of wild animals fluctuates somewhat (Fig. S6.2). But why was the importance of wild animals so variable within only a few decades?

Former investigations have indicated that the importance of hunting can most probably be correlated with short-term climatic fluctuations (Schibler et al., 1997; Arbogast et al., 2006; Schibler, 2006; Schibler and Jacomet, 2010). The relation between hunting and short climatic fluctuation can be explained by success or loss of subsistence production. During periods with more years of bad weather conditions (lower temperature and increased precipitation) one of the most important calorie providers—cereals—could not be produced in the required quantities. Therefore, wild resources, such as meat from hunted animals, were more intensively exploited. A direct confirmation of this hypothesis is not possible due to the difficulty in estimating cereal frequencies because of taphonomic reasons. The prevalence of cereal remains is strongly dependant on the periods when settlements had to be abandoned due to conflagrations, or which parts of the settlements were excavated or sampled. During periods of reduced cereal production, to increase the production of meat, milk, or milk products by using domestic animals is not easily possible due to the environmental conditions limiting the amount of fodder, which is needed to feed the domestic animals (Schibler et al., 1997a). Therefore, the easiest way to provide enough calories during periods of limited food production was to hunt wild animals and collect nuts and other wild vegetable resources more intensively. Bones from wild animals are easier to quantify than plant remains, but it is still worth pointing out that wild plants are extremely frequent in all sites (Hosch and Jacomet, 2004; Colledge and Conolly, 2014). The very obvious correlation between climatic deterioration illustrated by climatic proxies, e.g. C14-residuals, and the importance of wild animal bones supports this hypothesis (Schibler and Jacomet, 2010). It is also clear that the importance of hunting is not culturally determined. This is shown by a simple comparison of the chronological definition of Neolithic cultures and the importance of wild animal bones; there is no correlation between these factors (Schibler, 2006). Short climatic fluctuations had a stronger impact on early farming economies in the Alpine foreland than in other regions with much more balanced climatic conditions. Therefore, Neolithic farmers in the Alpine foreland were more frequently approaching the limit of their ability to produce a sufficient food supply.

The gradually decreasing average values of importance of hunting from the oldest Neolithic to the Bronze Age lakeshore settlements (Figs. 6.2, S6.1–2) can be explained by the increasing diversity of carbohydrate (cereals) and protein (pulses) rich cultivated plants (Jacomet and Brombacher, 2009). Some of these plants are more resistant to bad weather conditions (e.g. spelt). Therefore, in younger Neolithic and especially Bronze Age settlements economic problems caused by bad weather conditions could be overcome because the risk of a complete loss of carbohydrate and protein rich plants was minimized through spreading it across a greater variety of crop types. The higher proportion of wild animal bones in the eastern settlements indicates that climatic deterioration had a higher impact on the communities of eastern Switzerland (Fig. 6.2).

Another argument for an essentially economic driver of hunting during periods of bad weather conditions is represented by the observed species diversity. Normally in bone samples the number of animal species, including wild, increases proportionally to sample size as the number of bone fragments analysed increases (Casteel, 1979). However, in settlements that have a high frequency of wild animal bones the number of species is significantly lower than in settlements where hunting played a lesser role, even when the sample size is large (Schibler et al., 1997a; Schibler and Jacomet, 2010). This means that when Neolithic farmers were forced to rely more intensively on hunting they concentrated their efforts on hunting fewer and larger species to maximize their meat supply. The most common large wild animal inhabiting the Swiss Neolithic and Bronze Age forests was the red deer (*Cervus elaphus*), which, therefore, became the main target species. Periodic intensive red deer hunting over 3,500 years (4,300–800 BC) had an impact on the animal populations themselves. A reduction in size is documented for red deer during periods of intensive hunting (Schibler and Steppan, 1999), due to the selection of large and mature, mostly male, animals to produce the highest possible amount of meat. This focus on red deer hunting also had a negative impact on technological developments. Neolithic farmers used sockets made of red deer antler to fit stone axes into wooden handles, and because the axe was one of the technologies making the Neolithic economy possible, antler became a very important type of raw material (Schibler, 2013). After several decades of intensive red deer hunting, in the thirty-seventh century BC the importance of antler artefacts clearly decreased (Schibler, 2001). It is possible that a shortage of antler was responsible for changing axe technology, with antler sockets no longer produced and stone axes directly hafted to wooden handles as it is well recorded in settlements of the thirty-fourth and thirty-third centuries BC.

Turning to the question of the higher frequency of wild animal bones at wetland sites (Fig. 6.2), can we suggest that the subsistence economy of dryland settlements differed from that of the wetland settlements? Are we dealing with an economy that was less susceptible to episodes of climatic instability, and was therefore less dependent on hunting and gathering?

The regional distribution of dryland settlements indicates that this is likely to be the case for sites located in Valais. The Valais, an inner-alpine dry region, is at a climatic advantage when compared to other areas. Neolithic and Bronze Age settlements in this area would definitely have been less susceptible to crop failure as a result of climatic worsening. On the other hand, this line of argument does not hold true for the dryland sites in the Swiss plateau or Grison region, where any climatic deterioration experienced by the lakeshore settlements would have had an equally intensive impact on the dryland subsistence economy. Nevertheless, there are very few Neolithic and Bronze Age sites found in the plateau or Grison, with the increased wild animal presence indicative of a more intensive use of the available wild resources. The majority of the known sites show some exceptional taphonomic conditions that were responsible for protecting the archaeological layer from any erosional disturbance. These include the covering of the Neolithic layer by a Bronze Age rampart (Schellenberg Borscht), the preservation of archaeological material in rock fissures (Cazis Petrushügel), and covering cultural layers by a rockslide (Late Bronze Age to Early Iron Age site 'Haldenstein auf dem Stein' GR; Schibler and Jacomet, 2005). These observations indicate that it is most likely that in dryland sites cultural layers dated to periods of short climatic deterioration are only preserved when special taphonomic features protect them from erosion. In wetland sites these phases have evidently been preserved more often. Therefore, taphonomic differences between wetland and dryland sites could be one of the driving forces influencing our knowledge of the economic behaviour of Neolithic farmers. Without knowing and considering all kinds of taphonomic factors influencing the quality of preservation of archaeological remains, the economic interpretation of archaeozoological data remains fragmentary.

HUSBANDRY AND THE IMPORTANCE OF DOMESTIC SPECIES

In Neolithic settlements the five domestic species—cattle (*Bos taurus*), sheep (*Ovis aries*), goat (*Capra hircus*), pig (*Sus domesticus*), and dog (*Canis familiaris*)—are all present. Only a few horse bones, from very small animals, have been uncovered (Hüster Plogmann and Schibler, 1997). Before 3,400 BC horse bones are only present in settlements where hunting dominates. Therefore, it is most likely that these bones relate to small wild horses (*Equus ferus silvaticus*) that lived in very small populations in the catchment area of the larger rivers, such as the Rhine and Aare. Nineteen bones from the site of Auvernier la Saunerie, on Lake Neuchâtel and dating to around 2,600 BC (Desse, 1976), may represent the first remains of domestic horses in Switzerland. However, their status—domestic or wild—is not yet defined. It is only during the Bronze Age that horse bones are more regularly found in settlement contexts (Fig. S6.3). All of these bones are identified as belonging to domestic forms. This identification is not proven by osteometric or genetic results, but is suggested on the basis of their more frequent occurrence. Only in Late Bronze Age settlements are horse bones encountered in higher percentages, i.e. up to 10% of the total number of domestic animal bones (Fig. S6.4). This is true for both wetland and dryland sites. It is also evident that horse bones are very rarely present in Bronze Age settlements in the smaller eastern Alpine valleys (Grison region). It appears that domestic horses first appear in Bronze Age or perhaps Late Neolithic settlements of the midlands and entered smaller Alpine valleys only during the Later Bronze Age (Fig. S6.3).

The frequency of cattle bones varies strongly in Neolithic and Bronze Age sites. Chronological and regional patterns are noticeable, but, in contrast to those noted for the bones of wild animals, no taphonomically induced differences are observable. Chronologically, it is possible to detect a tendency towards lower proportions of cattle bones during the sequence of Neolithic and Bronze Age settlements (Fig. 6.3).

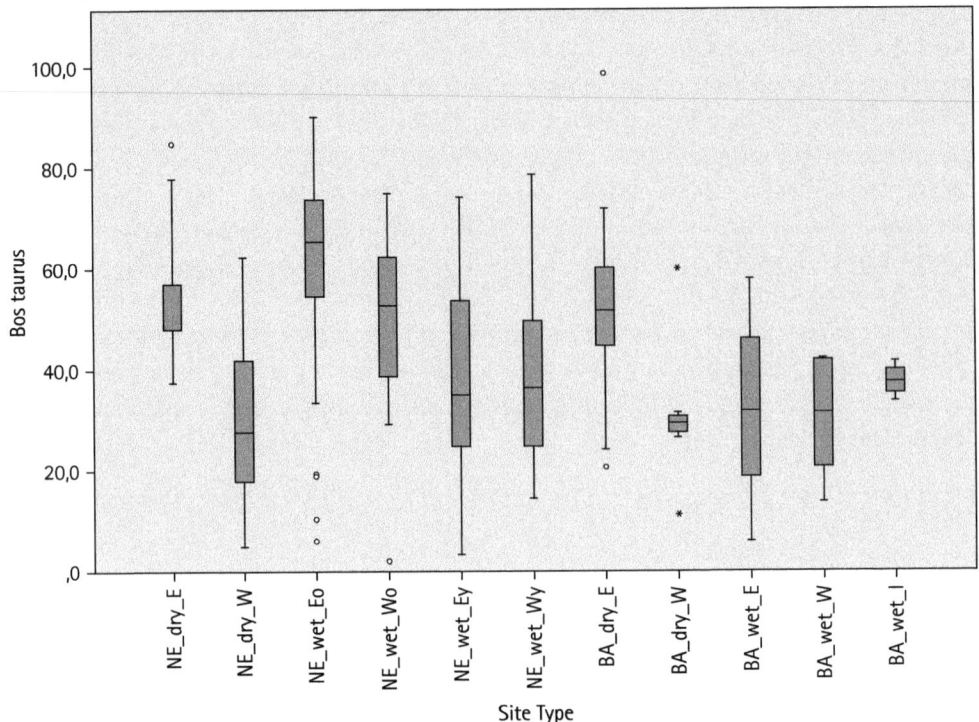

FIGURE 6.3 Box plots of the importance (% of fragment numbers) of cattle (*Bos taurus*) remains in Neolithic and Bronze Age wetland and dryland sites of Switzerland. For label definitions see Fig. 6.2. Graph by author.

Geographically, cattle bones are, on average, more frequent in settlements of the eastern part of Switzerland. This is particularly evident for the dryland sites in the Alpine region, with Neolithic and Bronze Age sites in the western part (Valais) showing, on average, lower frequencies of cattle bones (Fig. 6.3). The same pattern is only present in wetland sites of the early phase of the Neolithic period (4,300–3,400 BC). Archaeological finds indicate that during this period of wetland occupation the communities received strong cultural influences from eastern (so-called 'Danubian') and western or southern (so-called 'Mediterranean') regions (Stöckli et al., 1995). Therefore, the observed variations can be related to different culturally driven attitudes towards the economic importance of cattle. It is, however, also possible that topographic and climatic factors could be responsible. In the western parts of the Swiss Midlands and also the Alpine region—especially Valais—warmer and dryer climatic conditions predominate. Furthermore, a steep and hilly hinterland is typical for most of the sites in the western midlands. A combination of cultural, topographic, and climatic conditions are, therefore, mainly responsible for the observed differences in the importance of cattle bones.

The precisely dated, through dendrochronology, Neolithic wetland sites provide an opportunity to gain a more detailed insight to the exploitation of cattle. In these sites—especially the sites of the Zurich region—it is possible to include another method of quantification: the calculation of the number of bone fragments per square metre and settlement phase (Schibler et al., 1997a; Schibler and Jacomet, 2010). The values thus obtained, however, are not entirely accurate because the dendrochronologically dated settlement phases vary from fifteen to twenty-five years in duration. Small differences among these values should, therefore, not be considered significant. The real advantage of this quantification method lies in the fact that, in contrast to percentages based on the number of bone fragments, the densities are absolute rather than relative measures of abundance. Neolithic settlements of the Zurich region, therefore, provide the unique opportunity to directly compare the two methods of quantification: percentages based on fragment numbers vs bone densities. On the basis of these data four chronological phases of cattle economy can be distinguished. These phases correspond neatly with archaeological identified cultures (Schibler, 2006), and can also be used to describe patterns of exploitation of all other domestic species:

- The oldest sites, dated between 4,300 and 4,000 BC, show very low percentages (Fig. S6.4) and very low find density values (Fig. S6.5) of cattle bones. Evidently, during this period only small cattle herds were kept, and it is even questionable if each of these early wetland sites kept an independent cattle herd. This low intensity of cattle herding could be due to the very dense forest cover that is indicated by palynological investigations (Rösch, 1996; Van der Knaap and Ammann, 1997; Wehrli et al., 2007). Fodder production for early cattle herds must have been the limiting factor (Schibler et al., 1997a). It is suggested that during this early phase of wetland occupation cattle were only exploited for meat as indicated by the high frequency of immature individuals (Hüster Plogmann and Schibler, 1997).
- During the following phase, between 4,000 and 3,400 BC, very high percentages (40–80%) of cattle bones can be observed (Fig. S6.4). Find densities of cattle bones are significantly higher than before, but not remarkably so (Fig. S6.5). In the settlement of Arbon Bleiche 3 (Lake of Constance) dating to around 3,400 BC, milk exploitation has been demonstrated by the chemical analysis of fatty acids in potsherds (Spangenberg et al., 2006). This represents the only chemical analysis of Neolithic potsherds in Switzerland undertaken so far, meaning that the beginning of milk exploitation is not yet known. It is, therefore, only assumed that during the second phase of the Neolithic wetland sites (4,000–3,400 BC) slightly larger cattle herds were kept for both milk and meat exploitation.
- Proportions of cattle bones during the third phase, between 3,400 and 2,800 BC, seem to be lower than in the previous period (Fig. S6.4). However, if the find density values from settlements in the Zurich region are considered, no decrease can be seen (Fig. S6.5). This means that other domestic animal species—especially pigs—became more important and therefore concurrently reduced the percentages of cattle bones. Quantification on the basis of find densities leads to the conclusion that during the period 3,400 to 2,800 BC, Neolithic farmers practised cattle husbandry with the same intensity as before, but also that pigs were kept in greater numbers to produce meat. From the osteometrical and pathological point of view,

the first evidence for the use of cattle as draught animals appears in the settlements between 3,400 and 2,800 BC (Hüster-Plogmann and Schibler, 1997; Deschler-Erb and Marti-Grädel, 2004). Cattle appear to have been smaller but more robust than in preceding and subsequent periods, and are contemporaneous to the occurrence of wooden wheels and small carts (Jacomet and Schibler, 2006). Therefore, the third period of wetland occupation is characterized by the exploitation of cattle for meat, milk, and as draught animals.

- The final phase (2,800–2,500 BC) is characterized by high proportions (Fig. S6.4) and high find densities (Fig. S6.5) of cattle bones, with the conclusion that larger cattle herds were maintained. In addition, the fist evidence of castrates (Hüster-Plogmann and Schibler, 1997) clearly supports the exploitation of cattle as draught animals.

The importance of the domesticated small ruminants—sheep (*Ovis aries*) and goat (*Capra hircus*)—has to be evaluated in combination because only a small number of their bone fragments have been differentiated. In cases in which identification has been possible, sheep invariably results to be more frequent—particularly in the western part of Switzerland. The summary of the data indicate the sheep and goat bones are most common in Neolithic and Bronze Age dryland sites of the western Alpine region. These are sites in the Valais, where Neolithic and Bronze Age settlements show median values of around 60% sheep and goat bones (Fig. 6.4). The Rhone valley (Valais) with its steep slopes and very high Alpine elevations has drier and warmer climatic conditions than the other Swiss regions, and archaeological data clearly show that the Rhone valley was culturally influenced by the Mediterranean region, where sheep and goats were the most important domestic animals. Both climatic and cultural factors are responsible for the high economic importance of the small ruminants in this region until the Roman period. Slightly higher proportions of sheep and goat bones can also be recognized in the Neolithic and Bronze Age wetland sites of western Switzerland (Fig. 6.4). This could also be due to different topographic, climatic, and cultural conditions. Overall, a clear tendency for the higher importance of sheep and goat bones in Bronze Age wetland sites than in their Neolithic predecessors can be recognized.

The precisely dated Neolithic wetland sites can, however, provide more detailed insights. During the first phase of sites—between 4,300 and 4,000 BC—the higher importance of sheep and goat bones can be seen (Fig. S6.6). This is possibly due to a stronger Mediterranean influence during the Egolzwil and the early Cortaillod cultures. Perhaps this Mediterranean influence, together with a more hilly topography, could also be responsible for the higher percentages of these small ruminants at most sites in the western part of Switzerland (Fig. 6.4 and S6.6). Between 4,000 and 2,800 BC a slightly variable smaller proportion—between 10% and 30%—of sheep and goat bones can be observed for the wetland sites. Therefore, the economic relevance of meat from small ruminants could not have been significant. However, in the aforementioned lakeshore settlement of Arbon Bleiche 3, chemical analysis of potsherds prove the exploitation of milk and milk products from sheep and goat (Spangenberg et al., 2006). During the

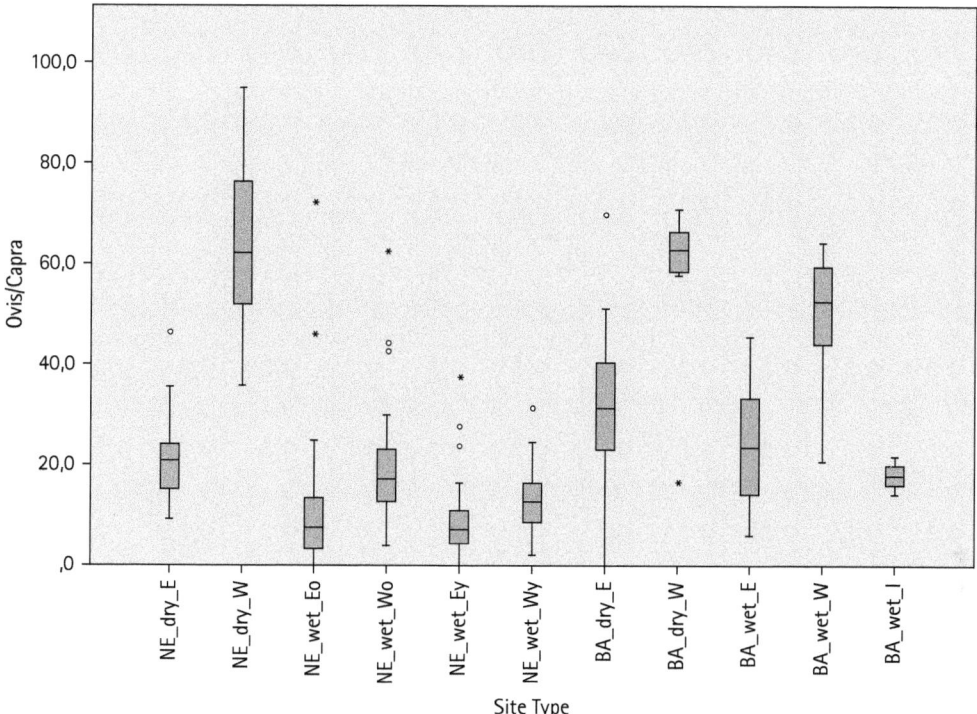

FIGURE 6.4 Box plots of the importance (% of fragment numbers) of caprine (*Ovis aries/Capra hircus*) remains in Neolithic and Bronze Age wetland and dryland sites of Switzerland. For label definitions see Fig. 6.2. Graph by author.

youngest phase of Neolithic wetland sites (2,800–2,500 BC) greater densities of bones (Fig. S6.5), higher slaughter ages, and increased size (Hüster-Plogmann and Schibler, 1997) are possible indicators for the keeping of a different breed of sheep for wool production. This idea is supported by the earliest finds of bone and antler needles and buttons in the Corded Ware sites around Zurich (Schibler, 1997). The higher importance of small ruminants—especially sheep—together with the decreasing importance and quantities of the remains of fibrous plant linen in Bronze Age wetland sites (Jacomet et al., 1998) strongly supports the concept of the beginning of sheep wool exploitation at the end of the Neolithic period and the intensification of such practices in the Bronze Age.

The highest importance of pig bones can be observed in the younger phase of Neolithic wetland sites (3,400–2,500 BC), as well as Bronze Age dryland sites in the Jura region (Fig. 6.5). Comparing the data geographically, a weak tendency towards a higher importance of pig bones in the eastern sites is visible. The examination of the precisely dated Neolithic wetland sites reveals increasing proportions of pig bones in the eastern lakeshore sites dated to between 3,900 and 3,400 BC (Fig. S6.7). After this trend of increasingly important pig economy in both regions a high proportion—over 40–70%—of pig bones can be seen between 3,400 and 2,800 BC. This is the period of the

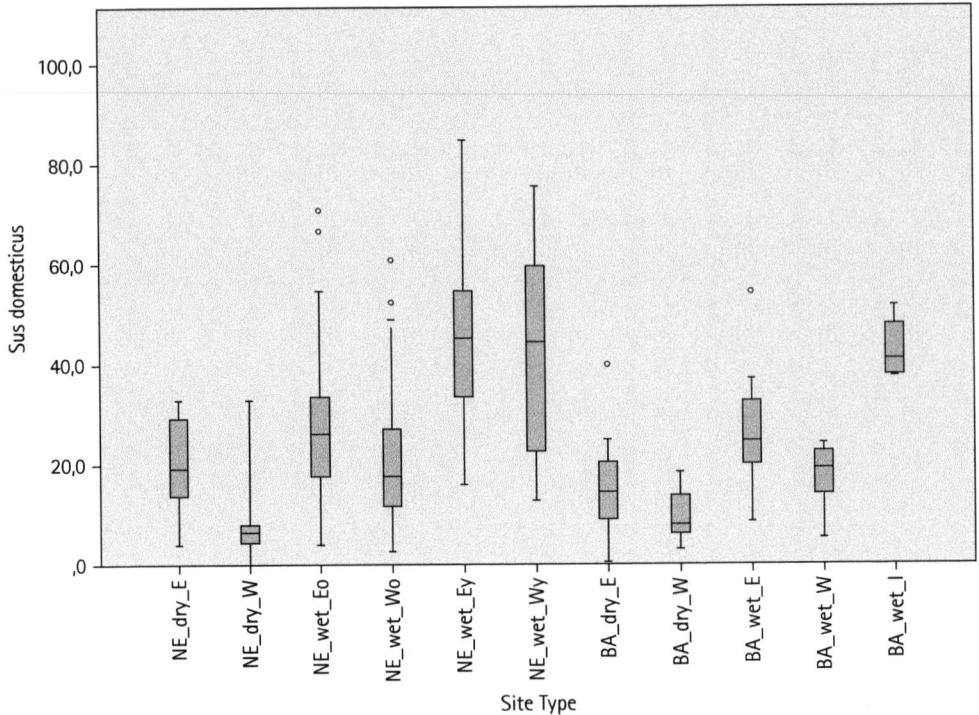

FIGURE 6.5 Box plots of the importance (% of fragment numbers) of pig (*Sus domesticus*) remains in Neolithic and Bronze Age wetland and dryland sites of Switzerland. For label definitions see Fig. 6.2. Graph by author.

Horgen culture—the first archaeological culture to influence both geographical regions of the midlands. Initial influences of Horgen culture can be detected in the eastern part of the Swiss midlands, with subsequent spread over the western part (Stöckli et al., 1995). It appears that intensive pig herding emerged in the lakeshore settlements of the eastern midlands and became a typical attribute of the Horgen culture. Only with the development of the Horgen culture in the western midlands (around 3,400 BC) can high proportions of pig bones be recognized in lakeshore settlements of this region (Fig. S6.7). Higher pig bone density values between 3,400 and 2,500 BC illustrate more intensive pig meat exploitation (Fig. S6.5). After 2,800 BC lower percentages of pig bones are typical of all Neolithic wetland sites (Fig. S6.7). This is only because all other domestic species like cattle, sheep, and goat show higher find densities (Fig. S6.5), and therefore were kept in larger herds. The very high proportions of pig bones of the Bronze Age settlements in the Jura region (Fig. 6.5) is possibly due to the special type of settlement. All these sites are located on the very top of hills, and thus naturally protected against all kind of aggressors. During periods of incertitude people had the possibility to occupy a secure location for a certain time span. At such places people had to store sufficient food stocks, and the easiest way to stock meat was to take living pigs to these refuges and slaughter them as required.

Across all types of sites dog bones are rarely found; mean and median values normally range around 1–2% for all bones of domestic species. Two exceptions are known from Neolithic lakeshore settlements. Higher frequencies of dog bones are only found at sites in the western part of the midlands between 3,700 and 3,500 BC, and at sites from eastern Switzerland during the second half of the 4th millennium BC. From a chronological point of view, keeping more dogs first of all became important in the late Cortaillod sites in the western midlands, and then reached the eastern sites of the Horgen culture around 3,400 BC. In both phases dog bones frequently show cut marks, and in both regions, during the aforementioned periods, pendants made from dog metapodials are regularly found (Schibler, 1981: 104–6; Deschler-Erb et al., 2002: 313). This fashion was exported eastwards from the western late Cortaillod culture sites, and both observations suggest a special relationship between man and dog during these periods. The reason for this changing relationship is as yet unknown.

CONCLUSIONS

Very dense archaeological and archaeozoological survey and information in a small but topographically strongly structured landscape provides a great opportunity for a detailed analysis of Swiss archaeozoological data. The quality of knowledge especially benefits from the fact that numerous Neolithic and Bronze Age sites, in both water-logged and dryland conditions, are known. The comparison of data and results from both kinds of sites provides information about the influence of taphonomic factors. Waterlogged conditions not only lead to a better quality of preservation of animal bones, but also permit the combination of archaeozoological data with precise dating. Only this possibility allows the arrangement of archaeozoological data in a proper chronological sequence permitting the identification of changes and variation in animal exploitation by time, and serves to highlight periods during which no sites—and therefore no data—are known. Waterlogged conditions also allow the preservation of significantly more plant remains than are encountered in dryland sites and a large part of these occur in non-charred conditions. Therefore, archaeobotanical investigations provide far more detailed information about not only economy, nutrition, and environment, but also taphonomy. Together with the archaeological data, it is possible to reconstruct the relationship and form of interaction between prehistoric communities, animals, plants, and environment to a high chronological resolution.

The Swiss zooarchaeological data primarily illustrates a strong influence of weather conditions—short climatic fluctuations—on food production. This relationship is ultimately responsible for the strongly fluctuating importance of hunting in Neolithic and Bronze Age settlements. Such climatic influence becomes especially important in Switzerland because of the highly structured topography and landscape. In contrast, cultural influences appear to have little correlation with the varying importance of hunting. The diverse topographic and climatic conditions are also responsible for the differing importance of domestic animal species. However, the importance of domestic animal species is also strongly dependent on cultural traditions and influences (i.e. Danubian and Mediterranean). For the region of the Alpine foreland it is evident that hunting (and gathering) was primarily an activity to survive climatically induced food crises, whereas the importance of the domestic animal species was influenced by several factors such as cultural tradition and prevailing local climatic conditions and topography. To define which of these factors are responsible for the importance of the different domestic species is not always easy. In certain situations all of these factors favour one domestic species; for instance in the Valais, where from the beginning of the Neolithic until the end of Iron Age sheep (and goat) was the most important species.

ACKNOWLEDGEMENTS

I thank Benjamin Jennings for improving the English version of the text.

REFERENCES

Arbogast, R.-M., Jacomet, S., Magny, M., and Schibler, J. (2006) 'The significance of climate fluctuations for lake level changes and shifts in subsistence economy during the Late Neolithic (4,300–2,400 BC) in central Europe', *Vegetation History and Archaeobotany*, 15(4), 403–18.

Casteel, R. W. (1979) 'Taxonomic abundance and sample size in archaeological faunal assemblages' in Kubasiewicz, M. (ed.), *Archaeozoology. Proceedings of the 3rd International Archaeozoological Conference held 23–26th April 1978 at the Agricultural Academy, Szcecin—Poland*, pp. 129–36. Szcecin: Agricultural Academy.

Colledge, S. and Conolly, J. (2014) 'Wild plant use in European Neolithic subsistence economies: a formal assessment of preservation bias in archaeobotanical assemblages and the implications for understanding changes in plant diet breadth', *Quaternary Science Reviews*, 101, 193–206.

Deschler-Erb, S. and Marti-Grädel, E. (2004) 'Viehhaltung und Jagd. Ergebnisse der Untersuchung der handaufgelesenen Tierknochen', in Jacomet, S., Leuzinger U., and Schibler, J. (eds) *Die jungsteinzeitliche Seeufersiedlung Arbon Bleiche 3. Umwelt und Wirtschaft*. Archäologie im Thurgau 12, pp. 158–251. Frauenfeld: Amt für Archäologie.

Deschler-Erb, S., Marti-Grädel, E., and Schibler, J. (2002) 'Die Knochen-, Zahn- und Geweihartefakte', in de Capitani, A., Deschler-Erb, S., Leuzinger, U., Marti-Grädel, E., and Schibler, J. (eds) *Die jungsteinzeitliche Seeufersiedlung Arbon Bleiche 3. Funde*. Archäologie im Thurgau 11, pp. 277–366. Frauenfeld: Amt für Archäologie.

Desse, J. (1976) 'La faune du site archéologique Auvernier-Brise Lames, canton de Neuchâtel (Suisse)'. Unpublished PhD dissertation, University of Poitiers (Poitiers).

Hosch, S. and Jacomet, S. (2004) 'Ackerbau und Sammelwirtschaft. Ergebnisse der Untersuchung von Samen und Früchten', in Jacomet, S., Leuzinger, U., and Schibler, J. (eds) *Die jungsteinzeitliche Seeufersiedlung Arbon Bleiche 3. Umwelt und Wirtschaft*. Archäologie im Thurgau 12, pp. 112–57. Frauenfeld: Amt für Archäologie.

Hüster-Plogmann, H. and Schibler, J. (1997) 'Archäozoologie', in Schibler, J., Hüster-Plogmann, H., Jacomet, S., Brombacher, C., Gross-Klee, E., and Rast-Eicher, A. (eds) *Ökonomie und Ökologie neolithischer und bronzezeitlicher Ufersiedlungen am Zürichsee*. Monographien der Kantonsarchäologie Zürich 20, pp. 40–121. Zürich and Egg: Fotorotar.

Jacomet, S. and Brombacher, Ch. (2009) 'Geschichte der Flora in der Regio Basiliensis seit 7,500 Jahren: Ergebnisse von Untersuchungen pflanzlicher Makroreste aus archäologischen Ausgrabungen', *Mitteilungen der Naturforschenden Gesellschaften beider Basel*, 11, 27–106.

Jacomet, S., Rachoud-Schneider, A.-M., and Zoller, H. (1998) 'Vegetationsentwicklung, Vegetationsveränderung durch menschlichen Einfluss, Ackerbau und Sammelwirtschaft', in Hochuli, S., Niffeler, U., and Rychner, V. (eds) *Bronzezeit. Die Schweiz vom Paläolithikum bis zum frühen Mittelalter 3*, pp. 141–70. Basel: Schweizerische Gesellschaft für Ur- und Frühgeschichte.

Jacomet, S. and Schibler, J. (2006) 'Traction animale et données paléoenvironnementales au Néolithique dans le nord des Alpes', in Pétrequin, P., Arbogast, R., Pétrequin, A.-M., van Willigen, S., and Bailly, M. (eds) *Premiers chariots, premiers araires. La diffusion de la traction animale en Europe pendant les IVe et IIIe millénaires avant notre ère*. CRA Monographies 29, pp. 141–55. Paris: CNRS éditions.

Knaap, Van der, O.W. and Ammann, B. (1997) 'Depth-age relationships of 25 well-dated Swiss Holocene pollen sequences archived in the Alpine Palynological Data-Base', *Revue de Paléobiologie*, 16 (2), 433–80.

Rösch, M. (1996) 'New approaches to prehistoric land-use reconstruction in south-western Germany', *Vegetation History and Archaeobotany*, 5, 65–79.

Schibler, J. (1981) *Typologische Untersuchungen der cortaillodzeitlichen Knochenartefakte. Die neolithischen Ufersiedlungen von Twann 17*. Bern: Staatlicher Lehrmittelverlag.

Schibler, J. (1997) 'Knochen- und Geweihartefakte', in Schibler, J., Hüster-Plogmann, H., Jacomet, S., Brombacher, C., Gross-Klee, E., and Rast-Eicher, A. (eds) *Ökonomie und Ökologie neolithischer und bronzezeitlicher Ufersiedlungen am Zürichsee*. Monographien der Kantonsarchäologie Zürich 20, pp. 122–219. Zürich and Egg: Fotorotar.

Schibler, J. (2001) 'Red deer antler: exploitation and raw material management in Neolithic lake dwelling sites from Zürich, Switzerland', in Buitenhuis, H. and Prummel, W. (eds) *Animals and Man in the Past. Essays in Honour of Dr. A. T. Clason Emeritus Professor of Archaeozoology Rijksuniversiteit Groningen, the Netherlands*. ARC-Publicatie 41, pp. 82–94. Groningen: ARC.

Schibler, J. (2006) 'The economy and environment of the 4th and 3rd millennia BC in the northern Alpine foreland based on studies of animal bones', *Environmental Archaeology*, 11(1), 49–64.

Schibler, J. (2013) 'Bone and antler artefacts in wetland sites', in Menotti, F. and O'Sullivan, A. (eds) *The Oxford Handbook of Wetland Archaeology*, pp. 329–45. Oxford: Oxford University Press.

Schibler, J., Hüster-Plogmann, H., Jacomet, S., Brombacher, Ch., Gross-Klee, E., and Rast-Eicher, A. (eds) (1997a) *Ökonomie und Ökologie neolithischer und bronzezeitlicher Ufersiedlungen am Zürichsee. Ergebnisse der Ausgrabungen Mozartstrasse, Kanalisationssanierung Seefeld, AKAD/Pressehaus und Mythenschloss in Zürich*. Monographien der Kantonsarchäologie Zürich 20. Zürich and Egg: Fotorotar.

Schibler, J. and Jacomet, S. (2005) 'Fair-weather archaeology? A possible relationship between climate and the quality of archaeological sources', in Gronenborn, D. (ed.), *Klimaveränderung und Kulturwandel in neolithischen Gesellschaften Mitteleuropas, 6,700–2,200 v. Chr.* RGZM-Tagungen 1, pp. 27–39. Mainz: RGZM.

Schibler, J. and Jacomet, S. (2010) 'Short climatic fluctuations and their impact on human economies and societies: the potential of the Neolithic lake shore settlements in the Alpine foreland', *Environmental Archaeology*, 15(2), 173–82.

Schibler, J., Jacomet, S., Hüster-Plogmann, H., and Brombacher, C. (1997) 'Economic crash in the 37th and 36th centuries cal. BC in Neolithic lake shore sites in Switzerland', *Anthropozoologica*, 25/26, 553–70.

Schibler, J. and Steppan, K. (1999) 'Human impact on the habitat of large herbivores in eastern Switzerland and southwest Germany in the Neolithic', *Archeofauna*, 8, 87–99.

Spangenberg, J. E., Jacomet, S., and Schibler, J. (2006) 'Chemical analyses of organic residues in archaeological pottery from Arbon Bleiche 3, Switzerland—evidence for dairying in the late Neolithic', *Journal of Archaeological Science*, 33, 1–13.

Stöckli, W. E., Niffeler, U., and Gross-Klee, E. (eds) (1995) *Die Schweiz vom Paläolithikum bis zum frühen Mittelalter. Vom Neandertaler bis zu Karl dem Großen. Band II Neolithikum.* Basel: Verlag Schweizerische Gesellschaft für Ur- und Frühgeschichte.

Wehrli, M., Tinner, W., and Ammann, B. (2007) '16,000 years of vegetation and settlement history from Egelsee (Menzingen, central Switzerland)', *The Holocene*, 17, 747–61.

......

CHAPTER 7

ZOOARCHAEOLOGY IN THE CARPATHIAN BASIN AND ADJACENT AREAS

......

LÁSZLÓ BARTOSIEWICZ

GEOGRAPHICAL DEFINITION

......

THE Carpathian Basin occupies about 200,000 km² between the Alps, the Carpathians, and the Dinaric Alps. The mountain ranges mark clear topographic boundaries. It is environmentally distinct from the Balkans since its climate is more humid and continental and the basin is completely land-locked. Divided by the Danube and its lowland habitats (the westernmost extension of the east European steppe) it mostly comprises only moderately high hills. The Danube represented a corridor along which peoples, goods, and information spread. It also forms a natural barrier.

Most data originate from modern-day Hungary because the centre of the Carpathian Basin was more intensively inhabited and has been more thoroughly researched than surrounding mountain areas. The plains, however, extend into neighbouring countries. The regions discussed are as follows (Fig. 7.1):

1. *Transdanubia* (Fig. 7.1/1) occupies the right bank of the Danube in Hungary. It is characterized by rolling hills, barely exceeding 700 m asl. In the North, the Danube floodplain continues toward Slovakia. The central portion of this region played a key role in the emergence of the Linearband Culture during the Neolithic. The Danube served as the *limes*, the actual political border for the Roman province of Pannonia, from the 1st to the 4th cent. AD.

2. The *Great Plain* (Fig. 7.1/2) is divided by the Tisza, the largest tributary of the Danube. The section in Hungary is called the Great Hungarian Plain and it extends into Romania (Banat), Serbia (Vojvodina), and Croatia (Baranja/Syrmia). In order to avoid imposing political terminology on this contiguous

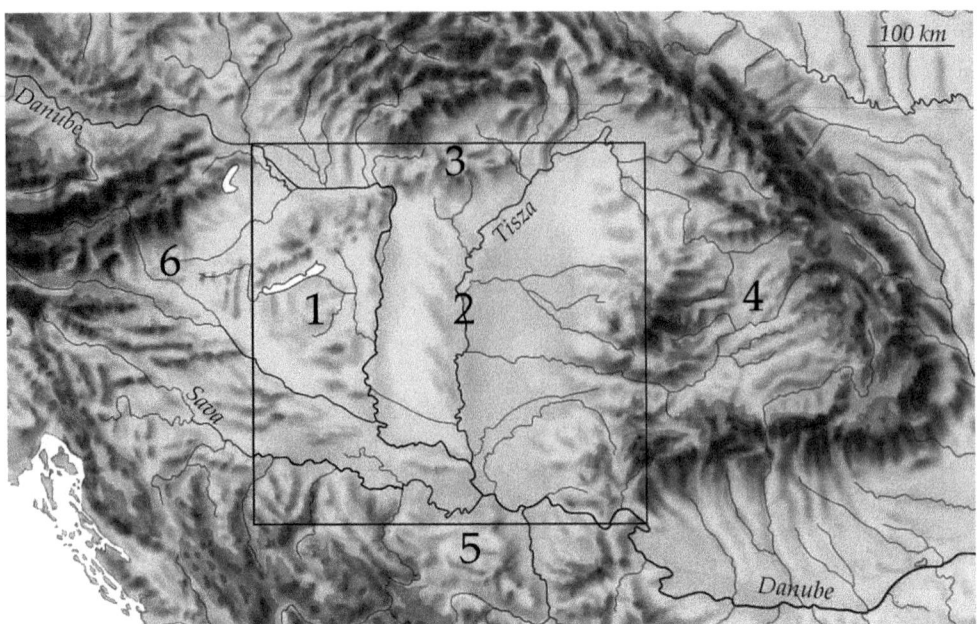

FIGURE 7.1 Location of regions discussed. 1: Transdanubia, 2: Great Plain, 3: Northern Hills, 4: Transylvania, 5: Balkans, 6: Alpine Foreland. Framing marks the Middle Danube Valley between Devín, Slovakia (northwest) and Turnu Severin, Romania (southeast).

area, it is often referred to as the 'Pannonian Plain'. However, Roman Pannonia occupied only Transdanubia. The Great Plain at that time corresponded to the *Barbaricum*, wedged between the Danubian *limes* and the Roman province of Dacia in the East (Bartosiewicz et al., 2011: footnote 1). Open toward the South, this region is the 'soft underbelly' of the Carpathian Basin (Tóth et al., 2010: 246), prone to influences from the South and East: Neolithization entered from here. Subsequently, the Great Plain marked the northernmost distributions of pre-historic tell cultures in Europe, as well as of the Ottoman Empire during the 16th–17th cent.

3. The *Northern Hills* (Fig.7.1/3) are located within the Carpathian range. The high-est peak reaches 1,000 m asl. From the 4th cent. BC, Celtic groups penetrated the Carpathians and then, partly from this region, pushed towards the Balkans result-ing in an interesting admixture of Iron Age Celtic and Scythian influences.

4. *Transylvania* (Fig. 7.1/4) is a hilly region in western Romania that played a role in mediating early Neolithization. Much of it became the Roman province of Dacia. During the Ottoman Period, Transylvania was a Turkish protectorate at a time when the Great Plain and southern Transdanubia fell under Turkish occupation.

5. The northern border of the *Balkans* (Fig. 7.1/5) is marked by the Sava and Danube rivers flowing east. This is an artificial line since most archaeological cultures occupied both banks of these rivers.

6. The *Alpine Foreland* (Fig. 7.1/6) is contiguous with Transdanubia. This region was least affected by eastern influences, but marked the border zone between the Avar and Frankish Empires in the 8th–9th cent. and formed the frontier of Hungarian conquest during the 10th cent. This region never fell under Ottoman occupation.

Brief Research History

Since the mid-nineteenth century, prehistoric archaeology has been a driving force in the study of animal remains: the earlier the site, the more archaeologists have enlisted the help of natural scientists trained in biological disciplines (hence the term 'archaeo-zoologist'; Bartosiewicz, 2001: 80). For subsequent periods (Classical, Medieval, etc.) archaeologists usually rely on documentary sources and tend to disregard animals as material culture. Thus, in spite of its long research tradition in the region, relatively little success has been achieved in organically integrating archaeozoology into the mainstream of historically inspired archaeological narrative.

The narrow field of archaeozoology in the discussed area developed under different circumstances. Sándor Bökönyi (1926–1994) established the Archaeozoological Collections in the Hungarian National Museum in 1951 and began analysing animal bones full time. In his ground-breaking publication on the regional history of domestic mammals (Bökönyi, 1974) he listed material from 294 sites. Bökönyi's lifelong interest lay with early domestication. János Matolcsi (1923–1983) published animal remains from thirty-seven sites, many of them representing historical periods. István Vörös, a zooarchaeologist interested in the earliest, Palaeolithic periods has recently turned to trying to reconcile Medieval osteological evidence with data in the contemporaneous written record (Vörös, 2009). Major trends in twentieth-century Hungarian research were analysed by Bartosiewicz and Choyke (2002).

In Romania, research by Olga Necrasov (1910–2000) and Sergiu Haimovici (1929–2009) gained momentum in the late 1950s at the University of Iaşi. Initiatives by Alexandra Bolomey (1932–1993) in Bucharest resulted in the emergence of another important centre of zooarchaeological research. While both Iaşi and Bucharest fall well outside the Carpathian Basin, work by Georgeta El Susi (1996) in the Banat as well as by Alexandru Gudea (2007) and Diana Bindea (2008) in Transylvania represent local research. A catalogue of Romanian archaeozoological investigations nationwide contained 420 reports (Bălăşescu et al., 2003), including sites outside the Carpathian Basin.

Developments differed in former Yugoslavia. Excavations were financed within the framework "of an interstate convention between Yugoslavia and the United States" (Benac, 1973: 7). With no local archaeozoologists available, Sándor Bökönyi was employed in these projects. This was a historic occasion as German-style morphology-inspired archaeozoology was synergetically combined with the Anglo-American view of anthropologically oriented archaeology (Bartosiewicz et al., 2011). Joint excavations also employed western zooarchaeologists (e.g. Clason, 1979; Greenfield, 1986), as well as Bökönyi (1981). By the 1980s a generation of local experts emerged in Serbia. Svetlana Blažić, a biologist, began publishing animal remains from Vojvodina (Blažić, 1978). In addition to numerous papers on Pleistocene faunas, Vesna Dimitrijević expanded her research into the much discussed Mesolithic/Neolithic transition of the Iron Gates gorge (Dimitrijević, 2004). Approximately half of the forty-three sites studied in Serbia to date (Stojanović and Bulatović, 2013: 16–17, Table 1) fall within the southern reaches of the Carpathian Basin. In Croatia most archaeozoological research relevant to the Carpathian Basin is focused on the Vučedol culture (Trbojević-Vukicević et al., 2003).

Since 1990, international cooperative projects in the entire area have introduced an increasingly multidisciplinary approach and have facilitated access to costly laboratory techniques such as stable isotope and DNA analyses (e.g. Whittle, 2007; Bonsall et al., 2008).

DIACHRONIC REVIEW

The Neolithic is indubitably the most investigated period in the study area. The first domesticates arrived in the Carpathian Basin from the Balkans during the time of the Körös/Criş cultures (*c.*5,800–5,300 cal BC), a period of remarkable mobility. Sheep (*Ovis aries*) and goat (*Capra hircus*), caprines domesticated in southwest Asia, provided the bulk of the meat consumed in small herding communities. Hunting was of negligible importance, even though sheep and goat were ill-adapted to the marshy floodplain environments (Bartosiewicz and Gál, 2013: 119). The Linearband Culture (LBK), named after its style of pottery decoration, dominated the early Middle Neolithic period (5,400–5,000 cal BC). Animal keeping became in harmony with the floodplain environments: the previous near-monocultural reliance on caprines shifted toward pigs (*Sus domesticus*) and cattle (*Bos taurus*). Sheep remained important, but their bones do not markedly exceed those of pigs in most assemblages. During the Middle Neolithic, multi-layered tell sites emerged on the Great Plain and lasted through the Late Neolithic (5,000–4,500 cal BC). Cattle and pig grew in importance while caprines further declined. Although domestication was fully developed, hunting resumed at tells in the Tisza valley with game contributing as much as 23–48% to the meat diet (Kovács and Gál, 2009: 151–2). Bökönyi (1974: 112) hypothesized that excessive aurochs (*Bos primigenius*) hunting noted at the multi-layered tell settlements of Berettyóújfau-Herpály and Polgár-Csőszhalom was motivated by a 'domestication fever': he considered bovine bones of intermediate sizes as originating from crosses between large aurochs and small domestic cattle. This interpretation was refuted decades later by DNA evidence suggesting that the primary domestication of Neolithic cattle from Hungary was unlikely to have taken place in Europe (Edwards et al., 2007). While interbreeding between domestic cattle and its wild ancestor, the aurochs, was possible, it does not necessarily imply local domestication on a large scale. Bökönyi's hypothesis was based on the unicausal interpretation of intraspecies variability.

An increase in hunting is clear at contemporaneous tells along the Lower Danube such as Borduşani-Popină, Pietrele-Măgura Gorgana, Hârşova, etc. (Bréhard and Bălăşescu, 2012; Benecke et al., 2013). Tells increasingly appear to have been focal features within complex horizontal settlement networks. While the latter yielded evidence of ordinary domesticates, activity on mounds left behind concentrations of bones from aurochs (Polgár-Csőszhalom; Bartosiewicz, 2005: 56, Fig. 6.5) or wild boar (*Sus scrofa*) (Pietrele; Benecke et al., 2013: 182, Fig. 7). The increase in hunting at Late Neolithic/Copper Age tells may have been triggered by both social and environmental factors. Mounds in combination with horizontal settlements represented complex communities on floodplains. In addition to serving as an alternative source of meat, hunting must have represented a way of asserting high status in a competitive social environment.

Neolithic animal exploitation is summarized using a principal component analysis of 154 faunal lists (Bartosiewicz, 2005; Bréhard and Bălăşescu, 2012; Orton, 2012; Benecke et al., 2013). Although this synthesis of metadata lacks detail (taphonomy, spatial distributions, age profiles), the large sample numbers make diachronic trends quite evident. The two axes in Fig. 7.2 explain 45% and 33% of the total variance represented by the data. The horizontal axis shows the importance of caprines vs cattle, between Early Neolithic

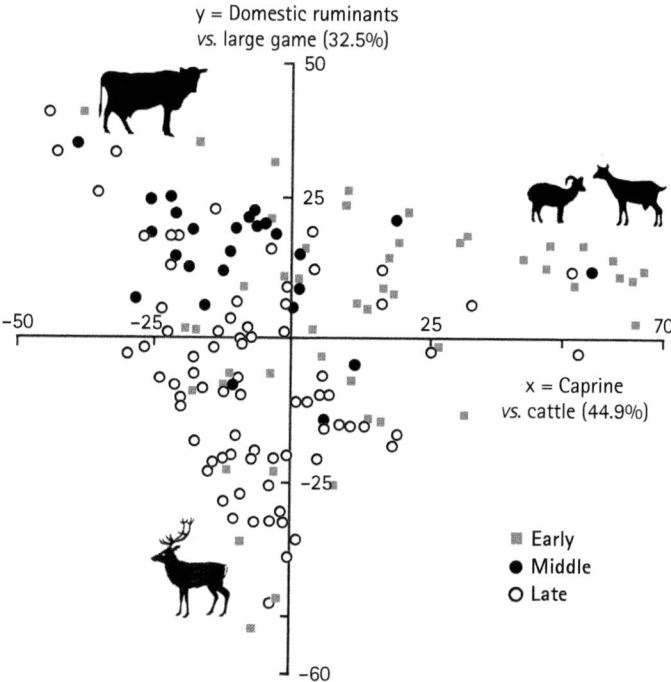

FIGURE 7.2 The distribution of Neolithic settlements by species composition in the Carpathian Basin and adjacent areas. Axes (principal components) represent typical combinations of animals. The main diachronic trend may be followed counter-clockwise in the graph (see explanation in the text). Graph by author.

reliance on mutton and Middle and Late Neolithic beef (and pork) consumption. The vertical axis indicates complementarity between domestic ruminants (cattle/caprines) and large game such as red deer (*Cervus elaphus*), aurochs, and wild pig. Two Late and Middle Neolithic outliers mixed among Early Neolithic settlements shown in the right-hand side of the graph represent sites geographically peripheral to the Carpathian Basin: Late Neolithic assemblages in the warmer and drier northern Balkans, far more favourable habitats for caprines than the Carpathian Basin (Bartosiewicz, 2005). On the other hand, Early Neolithic assemblages mixed among late sites (high degree of hunting) represent the Early Neolithic of the Iron Gates gorge (Bökönyi, 1970) and other Starčevo culture settlements.

Domestic pig and dog (*Canis familiaris*) co-occur on a third principal component (only 9% explanatory value): their bone numbers did not correlate strongly with those of distinctive ruminants. Bones of large predators such as bear (*Ursus arctos*), lynx (*Lynx lynx*), and even lions (*Panthera leo*) (Vörös, 1983; Bartosiewicz, 2009) occur in Late Neolithic/Copper Age assemblages. Fowling at such sites concentrated on aquatic/wetland species: ducks, geese (Anatidae), herons (Ardeidae), and birds of prey (Gál, 2007).

Animals from the metal ages have received less attention in research in comparison with the Neolithic. Some Early Bronze Age (*c.*2,800–1,800 BC) sites are characterized by large-scale horse (*Equus caballus*) eating in the Carpathian Basin with up to 40% horse bone present in certain assemblages (Bökönyi, 1974; Choyke, 1984). In contrast to much contested sporadic Copper Age equid finds (Vörös, 1981: 54), these horses are considered to be domestic.

The Middle and Late Bronze Ages brought about a second wave of tell settlements in the Great Plain (first half of the second millennium BC) and deeply stratified hill-fort settlements in Transdanubia, taking advantage of the natural topography. Animal remains from fourteen sites have been studied as sources of raw materials for manufacturing by Choyke (1984). Domesticates dominated, with cattle bones being most numerous at many sites. The ratio between sheep/goat and pig remains alternated depending on local habitats. Importantly, horses gained a new role: together with high-status bronze weaponry many were tokens of self-representation. The contribution of game to the diet was less than 5% although red deer antler became an important raw material for tools and ornaments. While spectacular bronze finds from a few tells and hill-forts have been studied since the nineteenth century, little is known of the network of smaller satellite settlements and their modes of animal exploitation.

Iron Age Hallstatt and La Tène cultures influenced the Carpathian Basin from the west, while Scythians occupied the Great Plain from the east. Domestic hen (*Gallus domesticus*) of Asian extraction first arrived to Central Europe during the Hallstatt C–D period (Bökönyi, 1974: 371; Kyselý, 2010: 10–11) probably through Scythian or Greek mediation. Domestic ass (*Equus asinus*) encountered at La Tène settlements (Vörös, 1982) is considered a Mediterranean element. Celts are the best studied of Iron Age populations in the region, in part for being the ethnographically documented 'others' in Classical written sources. They formed a network of rural settlements by the time of

Roman occupation at the beginning of the first century AD. Their cultic use of red deer and the interment of pigs in mortuary rituals are characteristic features of Celtic animal exploitation in the Carpathian Basin.

Roman colonization meant the intensification of military as well as trade movements. Romans dismantled the 'indigenous' worlds in their provinces, integrating local populations into the Empire. There was little evidence of hunting at settlement sites in the province of Pannonia. However, the proportion of red deer still exceeded 25% in the Early Roman strata of the Gellérthegy–Tabán oppidum and Celto-Roman layers of Budapest-Corvin Square (Lyublyanovics, 2010). Hunting may have been a form of demonstrating social identity by Celtic elites seeking to distinguish themselves from Romans. Among the Romans, evidence of hunting, perhaps as an upper-class 'sport', was more common at forts and other military establishments (Choyke, 2003).

Both in the Roman provinces of Pannonia and Dacia, the contribution of beef to the diet increased in comparison with pre-Roman Celtic and Dacian times due to military provisioning and urban development (Bökönyi, 1984; Gudea, 2007). A wide variability in size became apparent in cattle and horse in both provinces. Increasing stature in the latter species may be explained by the influence of highly developed Roman horse breeds, while smaller equid bones may originate from imported mules/hinnies, widely documented in Roman sources. The size of sheep seems less effected, but changes in kill-off patterns may indicate different strategies of secondary exploitation under Roman influence. Pigs, which were important in traditional indigenous subsistence, seem to have changed least.

The fauna of the Carpathian Basin was enriched by Roman colonization. Domestic cats (*Felis catus*) and camels appeared for the first time. In Hungary, so far eleven Roman period assemblages yielded camel bones, and four such sites are known in Serbia (Vuković and Bogdanović, 2013). While the majority of finds originate from dromedaries (*Camelus dromedarius*) (Bartosiewicz and Dirjec, 2001), some robust bones seem to represent large Bactrian camels (*Camelus bactrianus*). In contrast to dromedaries commonly used in the Eastern Mediterranean at the time, Bactrian camels originate from Central Asia thus being indicative of long-distance contacts (Daróczi-Szabó et al., 2014). Commensal black rats (*Rattus rattus*) dependent on human habitats were also brought to the Carpathian Basin by the Romans at a time when a gradation of rats seems to have peaked across Europe (Kovács, 2012: 383).

The Roman period (1st–4th cent. AD) saw waves of pastoralists from the east and the west infiltrate the Great Plain between Pannonia and Dacia. They consisted of competing Germanic tribes, Asiatic Sarmatians, and Huns, followed by Avars and Hungarians after the fall of the Roman Empire (Bartosiewicz, 2003). These groups typically relied on keeping caprines, and horses were of outstanding significance in their mobile military culture. While a slight shift from mutton to pork consumption may be detected along with increasing sedentism among Sarmatian and Avar communities, the central role played by horses is a source of striking functional similarities between ethnically diverse groups, fuelling analogies and as yet unsubstantiated theories of continuity. Uncertain cranio-morphological differences between Avar period and Early Hungarian horses

(Bartosiewicz, 2006) have recently been reconfirmed using DNA analyses (Priskin et al., 2010), showing that the two mobile pastoralist cultures used genetically different stocks.

Fifth- to ninth-century AD Migration period horse burials have yielded numerous complete skeletons in the Carpathian Basin. Best represented is the Avar Khanate (AD c.568–822), the first political entity to unite the Carpathian Basin in written history (Bökönyi, 1974; Müller and Ambros, 1994). These horse burials offer data on individuals in their proper biological context (age, sex, morbidity) rarely available for ordinary domesticates whose bones are mostly found as disarticulated, butchered, and re-deposited refuse. Horse burials are rich in palaeopathological information, as illustrated by the example of vertebral disorders in Fig. 7.3. An admixture of minor pathological symptoms is concentrated between the 13th thoracic and 2nd lumbar vertebrae in these horses, mainly attributable to riding (Fig. 7.3: light columns). Extreme cases of vertebral fusion from Keszthely and Tiszafüred represent chronic illness: horses useless as mounts, but kept alive for years (Bartosiewicz and Bartosiewicz, 2002: 827). Bökönyi's (1974: 290–2) romantic interpretation of such animals as 'shaman horses' directs attention to the special relationship between horses and humans in these cultures, whose traditional way of life depended on horses.

The AD 1000 establishment of a Christian kingdom (occupying the Carpathian Basin) is considered the beginning of the Middle Ages in Hungary. The Early Medieval

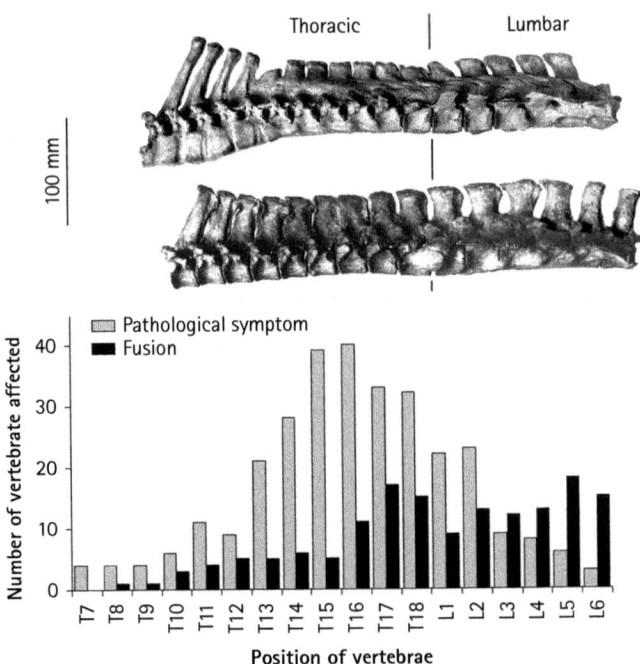

FIGURE 7.3 Vertebral pathologies in horse. Top: left lateral aspect of fused vertebral columns from Hungary (Keszthely-Általános iskola, Germanic; Tiszafüred-Majoros, Avar). Bottom: cumulative histogram showing the distribution of disorders in 83 Migration period horses in Hungary and Slovakia. Graph by author.

period of the Árpád Dynasty (AD 1000–1301), is dominated by issues of pastoralists (including the Hungarians themselves) adapting to sedentism in a centralized state system. A number of rural settlements show that the eastern pastoral tradition of eating horse flesh survived well into the eleventh to thirteenth centuries. Following the mid-thirteenth-century Mongol Tartar invasion, horse consumption was largely abandoned and pork became staple as western settlers of largely Germanic origins replaced locals who perished during Mongol attacks. This change in diet, however, remains invisible in written sources. Another undocumented Árpád period phenomenon is the surviving use of chickens and dogs in animist rituals among Christian populations. Upside-down pots, covering the skeletons of hens and eggs have been found across Hungary. The Árpád period village of Kána yielded several individual puppies and a pike buried covered by pots the same way (Daróczi-Szabó, 2010).

While Early Medieval research in the Carpathian Basin is best represented by domesticates from rural settlements, many post-fourteenth century, Late Medieval materials originate from emerging urban centres, manorial houses, and high status royal/clerical settlements. Until the Ottoman occupation, the popularity of pork steadily increased, although beef dominated in the meat diet of the aristocracy and of the emerging merchant class. Many of the latter represent local livestock owners who for centuries profited from massive cattle drives towards urbanized areas in southern Germany and northern Italy. Cattle bones dominate at many settlements along international herding routes (Bartosiewicz, 1995). In forested, mountainous areas such as the eastern Carpathians in Transylvania and Moldova as well as the northern hills of Hungary, hunting was an important source of high-status game including red deer, bison (*Bison bonasus*), and in some cases elk (*Alces alces*) (Bejenaru, 2003; 2006: 75; Bartosiewicz and Gál, 2010: 118). Falconry was pursued by aristocracy, even though ordinary raptors, including goshawk and sparrowhawk (*Accipiter* sp.), may have been used by commoners. Bones of a lanner falcon (*Falco biarmicus*) of Mediterranean origins were recovered in Buda, the royal centre (14th cent.). Tawny eagle (*Aquila rapax*), identified at the Ottoman period fort at Bajcsavár, inhabited by western mercenaries, may have been an eastern import (Gál, 2012).

The Middle Ages ended with the Ottoman Turkish occupation of central Hungary in 1526. Lucrative cattle drives generating tax revenues for the Sublime Porte, the imperial centre in Istanbul, continued. Although pork consumption declined somewhat, pigs remained important for the non-Muslim population and Christian mercenaries in the service of the Ottoman army. Foreign occupation brought new species and breeds (Bartosiewicz and Gál, 2003). Thousands of dromedaries were deployed by the Ottoman army, chiefly transporting artillery supplies. Early finds of water buffalo (*Bubalus bubalis*) also date to this period (Bökönyi, 1974: 154). Turkeys (*Meleagris gallopavo*) are known from five sites in Hungary, with three specimens coming from settlements in northern areas out of Ottoman control. It is difficult to tell whether they represent western or Turkish imports. Skulls of another luxury bird, the crested hen (a breed characterized by benign brain hernia and called *Gallina turcica* at the time in Italy), are known from several sites near the Pasha's palace in Buda (Gál et al., 2010).

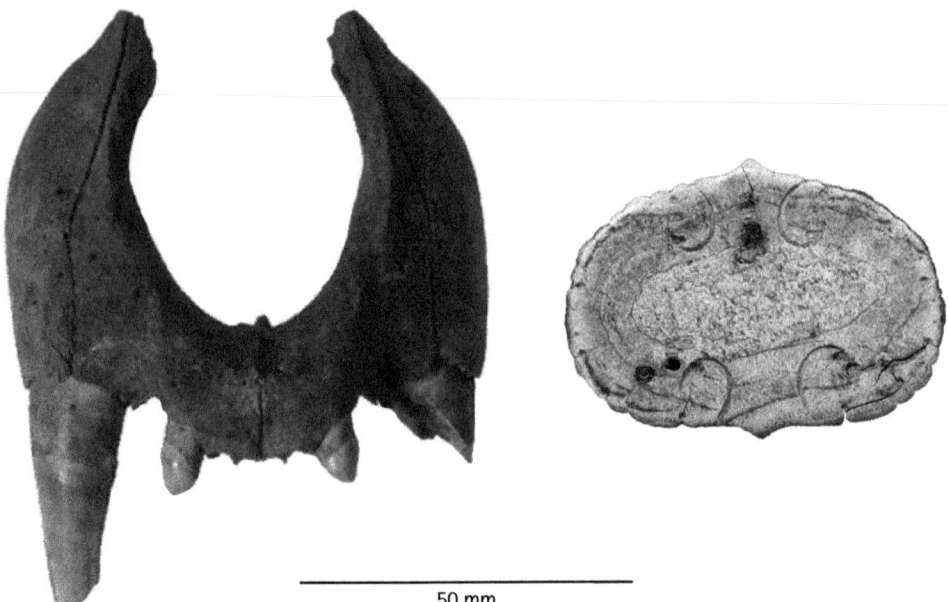

50 mm

FIGURE 7.4 Exotic finds from Hungary. Decorative leopard skull rostral segment with teeth (Segesd-Pékóföld, 14th C) and ornate walrus ivory fitting (Barcs, 16th C).

Worked skeletal parts of exotic animals such as objects of elephant (Elephantidae) ivory typically end up as 'small finds' in archaeological collections when zoological identification is neglected. A fortunate exception is the rostral segment of a large leopard (*Panthera pardus*) skull sporting canine teeth, found at the fourteenth-century queen's town at Segesd-Pékóföld, Hungary (Fig. 7.4). This bone may have been attached to a skin as suggested by high polishing on its reverse (aboral) side. Leopard skins were an attribute, among others, of Teutonic knights, as their order was established in the Levant. Given the rarity of walrus (*Odobenus rosmarus*) ivory artefacts in Hungary, the belt plaque found at a Turkish fortress along the Drava (Fig. 7.4) is a surprising find. It may have come from Russia either as a personal accessory or a trade item with Tartar and Ottoman Turkish mediation (Gál and Kovács, 2011).

CONCLUSIONS

The representation of zooarchaeology in the area and periods discussed here are far from homogeneous. The degree, resolution, and especially the foci of research vary strongly. Therefore, only characteristic zooarchaeological problems could be singled out for illustrating a diachronic sketch of animal exploitation. Due to the absence of water-sieving at most excavations, fish exploitation in the past is little known in the area. Carp (Cyprinidae), pike (*Esox lucius*), and catfish (*Silurus glanis*) have sporadically been found in all periods while large, anadromous Danubian sturgeons (Acipenseridae) were equally important from prehistoric times through the twentieth century (Bartosiewicz et al., 2008).

At the other extreme, studying manufactured skeletal materials has developed into a distinct sub-discipline (Choyke and Bartosiewicz, 2001), far beyond the scope of this summary.

Inevitably, this broad-stroke description does not reveal many subtleties of human–animal relationships in the Carpathian Basin over the last 8,000 years. New data contradicting old hypotheses are constantly appearing and it is hoped that the gaps, particularly in the understanding of animal exploitation in later prehistory and historical periods, will be narrowed by future research. New laboratory techniques will become increasingly available to help fine-tuning the picture, but the hands-on identification of large assemblages will remain indispensable in appraising patterns of consumption and increasing the probability of recovering special finds.

REFERENCES

Bălăşescu, A., Udrescu, U., Radu, V., and Popovici, D. (2003) *Archéozoologie en Roumanie. Corpus de Données*, Târgovişte: Editura Cetatea de Scaun.

Bartosiewicz, L. (1995) *Animals in the Urban Landscape in the Wake of the Middle Ages*. BAR International Series 609. Oxford: Archaeopress.

Bartosiewicz, L. (2001) 'Archaeozoology or zooarchaeology? A problem from the last century', *Archaeologia Polona*, 39, 75–86.

Bartosiewicz, L. (2003) 'A millennium of migrations: protohistoric mobile pastoralism in Hungary', *Bulletin of the Florida Museum of Natural History*, 44, 101–30.

Bartosiewicz, L. (2005) 'Plain talk: animals, environment and culture in the Neolithic of the Carpathian Basin and adjacent areas', in Bailey, D. and Whittle, A. (eds) *(Un)settling the Neolithic*, pp. 51–63. Oxford: Oxbow.

Bartosiewicz, L. (2006) 'Phenotype and age in protohistoric horses', in Ruscillo, D. (ed.) *Recent Advances in Ageing and Sexing Animal Bones*, pp. 204–15. Oxford: Oxbow.

Bartosiewicz, L. (2009) 'Lion's share of attention', *Acta Archaeologica Academiae Scientiarum Hungariae*, 59, 759–73.

Bartosiewicz, L. and Bartosiewicz, G. (2002) ' "Bamboo spine" in a Migration Period horse from Hungary', *Journal of Archaeological Science*, 29(8), 819–30.

Bartosiewicz, L., Bonsall, C., and Şişu, V. (2008) 'Sturgeon fishing along the middle and lower Danube', in Bonsall, C., Boroneanţ, V., and Radovanović, I. (eds) *The Iron Gates in Prehistory: New Perspectives*. BAR International Series 1893, pp. 39–54. Oxford: Archaeopress.

Bartosiewicz, L. and Choyke, A. M. (2002) 'Archaeozoology in Hungary', *Archaeofauna*, 11, 117–29.

Bartosiewicz, L. and Dirjec, J. (2001) 'Camels in antiquity', *Antiquity*, 75, 279–85.

Bartosiewicz, L. and Gál, E. (2003) 'Animal exploitation in Hungary during the Ottoman Era', in Gerelyes, I. and Kovács, Gy. (eds) *Archaeology of the Ottoman Period in Hungary*, Vol. III, pp. 365–76. Budapest: Opuscula Hungarica.

Bartosiewicz, L. and Gál, E. (2010) 'Living on the edge: "Scythian" and "Celtic" meat consumption in Iron Age Hungary', in Campana, D., Choyke, A. M., Crabtree, P., deFrance, S., and

Lev-Tov, J. (eds) *Anthropological Approaches to Zooarchaeology: Colonialism, Complexity and Animal Transformations*, pp. 115–27. Oxford: Oxbow Books.

Bartosiewicz, L. and Gál, E. (2013) *Shuffling Nags, Lame Ducks*, Oxford: Oxbow.

Bartosiewicz, L., Mérai, D., and Csippán, P. (2011) 'Dig up-Dig in', in Lozny, L. R. (ed.) *Comparative Archaeologies*, pp. 273–337. New York: Springer.

Bejenaru, L. (2003) *Archaeozoologia Spaţiului Românesc Medieval*, Iaşi: Editura Universităţii 'Alexandru Ioan Cuza'.

Bejenaru, L. (2006) *Archaeozoologia Moldovei Medievale*, Iaşi: Editura Universităţii 'Alexandru Ioan Cuza'.

Benac, A. (1973) 'A neolithic settlement of the Butmir group at Gornje polje', in Benac, A. and Gimbutas, M. (eds) *Obre I and II. Wissenschaftliche Mitteilungen des Bosnisch-Herzegowinischen Landesmuseums, Band III, Heft A*, pp. 5–191. Sarajevo: Zemalski Museum.

Benecke, N., Hansen, S., Nowacki, D., Reingruber, A., Ritchie, K., and Wunderlich, J. (2013) 'Pietrele in the lower Danube region', *Documenta Praehistorica*, 40, 175–93.

Bindea, D. (2008) *Arheozoologia Transilvaniei în Pre- şi Protoistorie*, Cluj-Napoca: Editura Teognost.

Blažić, S. (1978) 'Ostaci faune u keltskom oppidumu kod Backe Palanke (Faunal remains from the Celtic *oppidum* in Bačka Palanka)', *Грађа за проучавање споменика културе Војводине*, 6/7, 14–16.

Bökönyi, S. (1970) 'Animal remains from Lepenski Vir', *Science*, 167, 1702–4.

Bökönyi, S. (1974) *History of Domestic Mammals in Central and Eastern Europe*, Budapest: Akadémiai Kiadó.

Bökönyi, S. (1981) 'Eisenzeitliche Tierhaltung und Jagd im Jugoslawischen Donaugebiet', *Materijali*, 19, 105–19.

Bökönyi, S. (1984) *Animal Husbandry and Hunting in Tác-Gorsium*, Budapest: Akadémiai Kiadó.

Bonsall, C., Boroneanţ, V., and Radovanović, I. (eds) (2008) *The Iron Gates in Prehistory*. BAR International Series 1893. Oxford: Archaeopress.

Bréhard, S. and Bălăşescu, A. (2012) 'What's behind the tell phenomenon? An archaeozoological approach of Eneolithic sites in Romania', *Journal of Archaeological Science*, 39(10), 3167–83.

Choyke, A. M. (1984) 'An analysis of bone, antler and tooth tools from Bronze Age Hungary', *Mitteilungen des Archäologischen Instituts der Ungarischen Akademie der Wissenschaften*, 12/13, 1–20.

Choyke, A. M. (2003) 'Animals and Roman lifeways in Aquincum', in Zsidi, P. (ed.) *Forschungen in Aquincum 1969–2002*, pp. 210–32. Budapest: Aquincum Nostrum II/2.

Choyke, A. M. and Bartosiewicz, L. (eds) (2001) *Crafting Bone-Skeletal Technologies through Time and Space*. BAR International Series 937. Oxford: Archaeopress.

Clason, A. (1979) 'The farmers of Gomolava in the Vinča and La Tène period', *Rad Vojvođanskih Muzeja*, 25, 60–114.

Daróczi-Szabó, L., Daróczi-Szabó, M., Kovács, Zs. E., Kőrösi, A., and Tugya, B. (2014) 'Recent camel finds from Hungary', in Mashkour, M. and Beech, M. J. (eds) Old World Ancient Camelids between Arabia and Europe. Proceedings of the 11th ICAZ conference in Paris, pp. 265–80. *Anthropozoologica* 49(2).

Daróczi-Szabó, M. (2010) 'Pets in pots: superstitious belief in a Medieval Christian (12th–14th century) village in Hungary', in Campana, D., Crabtree, P., deFrance, S. D., Lev Tov, J., and Choyke, A. M. (eds) *Anthropological Approaches to Zooarchaeology: Complexity, Colonialism, and Animal Transformations*, pp. 211–15. Oxford: Oxbow.

Dimitrijević, V. (2004) 'Pleistocene survivors in the Iron Gates Mesolithic/Neolithic', *Antaeus*, 27, 293–302.

Edwards, C. J., Bollongino, R., Scheu, A., Chamberlain, A., Tresset, A., Vigne, J-D., Baird, J. F., Larson, G., Ho, S. Y. W., Heuping, T. W., Shapiro, B., Freeman, A. R., Thomas, M. G., Arbogast, R-M., Arndt, B., Bartosiewicz, L., Benecke, N., Budja, M., Chaix, L., Choyke, A. M., Conqueugniot, E., Döhle, H.-J., Göldner, H., Hartz, S., Helmer, D., Herzig, B., Hongo, H., Mashkour, M., Özdogan, M., Pucher, E., Roth, G., Schade-Lindig, S., Schmölke, U., Schulting, R. J., Stephan, E., Uerpmann, H.-P., Vörös, I., Voytek, B., Bradley, D. G., and Burger, J. (2007) 'Mitochondrial DNA shows a Near Eastern Neolithic origin of domestic cattle and no indication of domestication of European aurochs', *Proceedings of the Royal Society B.*, 274, 1377–85.

El Susi, G. (1996) *Vânători, Pescari si Crescători de Animale în Banatul Mileniilor VI î. ch— i d. ch.: Studiu Arheozoologic (Jäger, Fischer und Viehzüchter im Banat im 6. Jahrtausend v. Chr.—1. Jahrtausend n. Chr.)*. Bibliotheca historica et archaeologica banatica 3. Timişoara: Editura Mirton.

Gál, E. (2007) *Fowling in Lowlands. Neolithic and Copper Age Bird Bone Remains from the Great Hungarian Plain and South-East Romania*, Budapest: Archaeolingua, Series Minor.

Gál, E. (2012) 'Possible evidence for hawking from a 16th century Styrian Castle (Bajcsa, Hungary)', in Raemaekers, D. C. M., Esser, E., Lauwerier, R. C. G. M., and Zeiler, J. T. (eds) *A Bouquet of Archaeozoological Studies: Essays in Honour of Wietske Prummel*, pp. 173–9. Groningen: Bakhuis & University of Groningen Library.

Gál, E., Csippán, P., Daróczi-Szabó, L., and Daróczi-Szabó, M. (2010) 'Evidence of the crested form of domestic hen (*Gallus gallus* f. *domestica*) from three Post-Medieval sites in Hungary', *Journal of Archaeological Science*, 37, 1065–72.

Gál, E. and Kovács, Gy. (2011) 'A walrus-tusk belt plaque from an Ottoman-Turkish castle at Barcs, Hungary', *Antiquity*, 85(329), Project Gallery. http://antiquity.ac.uk/projgall/gal329/

Greenfield, H. J. (1986) *The Paleoeconomy of the Central Balkans (Serbia): a Zooarchaeological Perspective on the Late Neolithic and Bronze Age (ca 4,500–1,000 BC)*. BAR International Series 304. Oxford: Archaeopress.

Gudea, A. (2007) *Contribuţii la Istoria Economică a Daciei Romane. Studiu Arheozoologic*, Cluj Napoca: Editura Mega.

Kovács, Zs. E. (2012) 'Dispersal history of an invasive rodent in Hungary—subfossil finds of *Rattus rattus*', *Acta Zoologica Academiae Scientiarum Hungaricae*, 58(4), 379–94.

Kovács, Zs. E. and Gál, E. (2009) 'Animal remains from the site of Öcsöd-Kováshalom', in Draşovean, F., Ciobotaru, D. L., and Maddison, M. (eds) *Ten Years After: The Neolithic of the Balkans, as Uncovered by the Last Decade of Research*, pp. 152–7. Timişoara: Editura Marineasa.

Kyselý, R. (2010) 'Review of the oldest evidence of domestic fowl (*Gallus gallus* f. *domestica*) from the Czech Republic in its European context', *Acta Zoologica Cracoviensia*, 53(1–2), 9–34.

Lyublyanovics, K. (2010) 'Animal keeping and Roman colonization in the province of Pannonia Inferior, western Hungary', in Campana, D., Crabtree, P., deFrance, S. D., Lev Tov, J., and Choyke, A. M. (eds) *Anthropological Approaches to Zooarchaeology: Complexity, Colonialism, and Animal Transformations*, pp. 178–89. Oxford: Oxbow.

Müller, H.-H. and Ambros, C. (1994) 'Neue frühgeschichtliche Pferdeskelettfunde aus dem Gebiet der Slowakei', *Študijné Zvesti* 30, 117–75.

Orton, D. (2012) 'Herding, settlement, and chronology in the Balkan Neolithic', *European Journal of Archaeology*, 15(1), 5–40.

Priskin, K., Szabó, K., Tömöry, G., Bogácsi-Szabó, E., Csányi, B., Eördögh, R., Downes, C. S., and Raskó, I. (2010) 'Mitochondrial sequence variation in ancient horses from the Carpathian Basin and possible modern relatives', *Genetica*, 138, 211–18.

Stojanović, I. and Bulatović, J. (2013) 'Arhaozoooloshka istrazhivanjie mladse praistorije na territoriji Srbie (Archaeozoological research of Late Prehistory in Serbia)', in Miladinović-Radmilović, N. and Vitezović, S. (eds) *Bioarchaeology in the Balkans: Balance and Perspectives*, pp. 13–24. Beograd: Srpsko Arheološko Društvo.

Tóth, A. J., Daróczi-Szabó, L., Kovács, Zs. E., Gál, E., and Bartosiewicz, L. (2010) 'In the light of the crescent moon', in VanDerwarker, A. M. and Peres, T. M. (eds) *Integrating Zooarchaeology and Paleoethnobotany: A Consideration of Issues, Methods, and Cases*, pp. 245–86. New York: Springer.

Trbojević-Vukicević, T., Kužir, S., Babić, K., Mihelić, D., and Radionov, D. (2003) 'An archaeozoological analysis of teeth of lower jaw of pigs coming from the Kostolac culture (3,250–3,000 BC), originating from Vučedol, including comparison with recent', *Collegium Antropologicum*, 27(Suppl. 2), 31–7.

Vörös, I. (1981) 'Wild equids from the Early Holocene in the Carpathian Basin', *Folia Archaelogica*, 32, 37–68.

Vörös, I. (1982) 'The animal bones from the La Tène and Roman settlement of Szakály–Rétiföldek', in Gabler, D., Patek, E., and Vörös, I. (eds) *Studies in the Iron Age of Hungary*. BAR International Series 144, pp. 129–79. Oxford: Archaeopress.

Vörös, I. (1983) 'Lion remains from the Late Neolithic and Copper Age of the Carpathian Basin', *Folia Archaeologica*, 36, 33–50.

Vörös, I. (2009) 'Adatok a dunakanyar régió Árpád-kori állattartásához (Data on animal keeping in the Danube Bend region (Hungary) during the 11th–13th century period of the Árpád Dynasty)', in Bartosiewicz, L., Gál, E., and Kováts, I. (eds) *Csontvázak a Szekrényből (Skeletons from the cupboard)*, pp. 105–23. Budapest: Martin Opitz Kiadó.

Vuković, S. and Bogdanović, I. (2013) 'A camel skeleton from the Viminacium amphitheatre', *Starinar*, 58, 251–67.

Whittle, A. (ed.) (2007) *The Early Neolithic on the Great Hungarian Plain: Investigations of the Körös Culture Site of Ecsegfalva 23, County Békés I*. Varia Archaeologica Hungarica 21. Budapest: Archaeological Institute of the Hungarian Academy of Sciences.

CHAPTER 8

SHEEP, SACRIFICES, AND SYMBOLS

animals in Later Bronze Age Greece

PAUL HALSTEAD AND VALASIA ISAAKIDOU

INTRODUCTION

THE Minoan and Mycenaean 'palatial' civilizations of later Bronze Age southern Greece (Table 8.1) have excited popular and scholarly interest for their rich material culture and Europe's earliest known scripts. Animals are prominent in Minoan and Mycenaean iconography and a major focus of Mycenaean clay documents in the deciphered Linear B script. Together, images, texts, and skeletal remains offer rare insight into the ideological, political, and economic importance of animals to these societies. This chapter focuses on animals in later Bronze Age (second millennium cal BC) southern Greece, with brief comparative consideration of the preceding Neolithic and Early Bronze Age (seventh–third millennia cal BC) and non-palatial north of Greece to highlight any changes associated with palatial society. We consider first the osteological evidence for which species people exploited and how; then textual evidence for their role in Mycenaean political economy; and finally iconography as a guide to their place in elite ideology and cosmology. We focus on the terrestrial fauna that dominates the osteological and textual data.

HUSBANDRY, HUNTING, HOSPITALITY, AND SYMBOLISM: THE BARE BONES

A handful of large, multi-period faunal assemblages, each studied to internally consistent protocols, provides an overview of long-term change in animal management in

Table 8.1 Outline chronology for Neolithic and Bronze Age Greece

cal BC	Period	Social and faunal changes
7,000		
6,000	Early Neolithic (EN)	domestic sheep, goat, cow, pig, and dog appear; meat mortality for sheep, goats, and cattle; draught cows at Knossos; intensive bone breakage
	Middle Neolithic (MN)	
5,000	Late Neolithic (LN)	milk residues on ceramic; less intensive bone breakage; 'dressed' carcasses
4,000	Final Neolithic (FN)	more widespread structured depositions of whole/partial carcasses
3,000		
	Early Bronze Age (EB)	more adult males = more emphasis on traction (cattle), wool (sheep), and hair (goat)?
2,000		
	Middle Bronze Age (MB)	
	Late Bronze Age (LB)	Minoan 'palaces'
1,000		Mycenaean 'palaces'; palace draught oxen and wool flocks
		post-palatial

Greece that reduces the difficulty of comparing sites analysed with incompatible zooarchaeological methods. Knossos on Crete (see Fig. 8.1 for location) offers the most continuous record, from the early seventh millennium cal BC foundation of Europe's oldest known farming settlement to the second millennium cal BC construction and abandonment of the 'palace' complex.

Husbandry

As elsewhere in Greece, domestic sheep (*Ovis aries*), goats (*Capra hircus*), cattle (*Bos taurus*), pigs (*Sus domesticus*), and dogs (*Canis familiaris*) were present at Knossos throughout the Neolithic and Bronze Age. On the mainland, domestic cattle and pigs grew smaller from the Early Neolithic (EN—for chronological abbreviations, see Table 8.1) onwards (e.g. von den Driesch, 1987), suggesting sufficiently close control

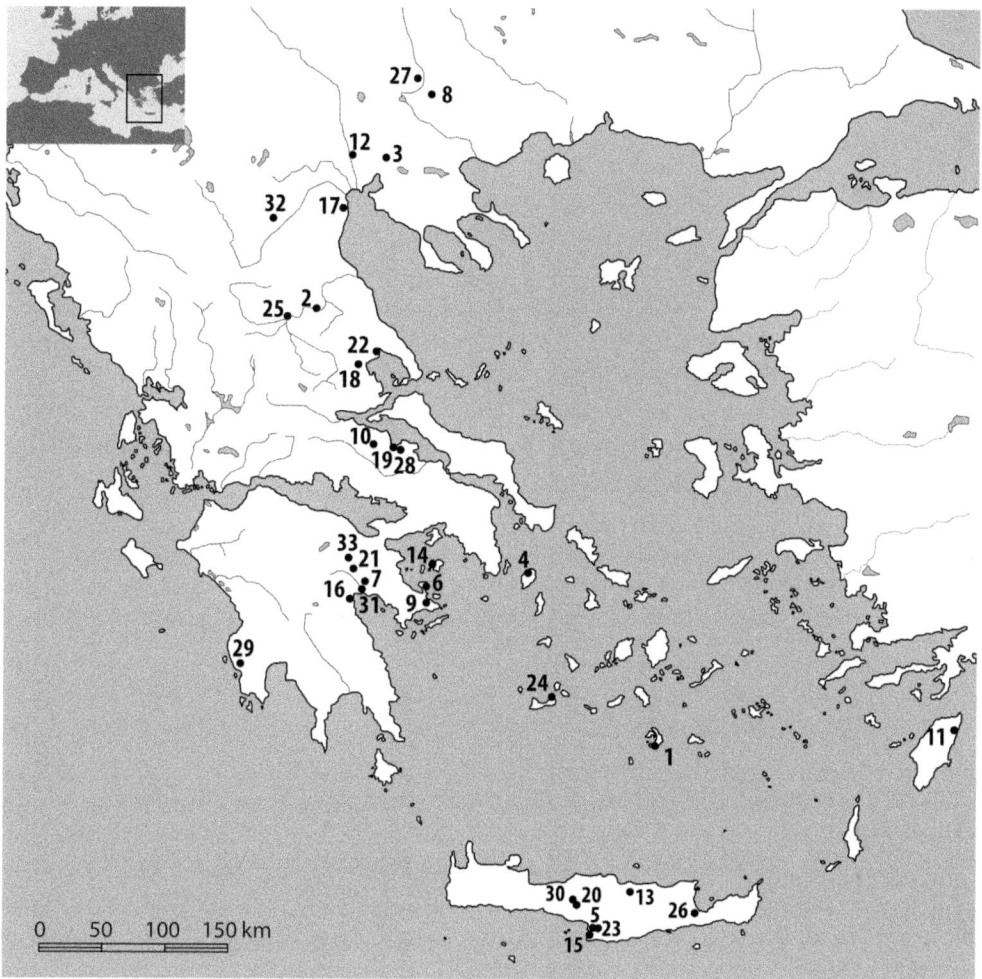

FIGURE 8.1 Map of Greece showing the location of regions and sites mentioned in the text. 1: Akrotiri (Gamble, 1978; Trantalidou, 2013), 2: Argissa (Boessneck, 1962), 3. Assiros Toumba (Halstead, unpublished data), 4: Ayia Irini (Coy, 1986), 5: Ayia Triada, 6: Ayios Konstantinos (Hamilakis and Konsolaki, 2004), 7: Dendra (Pappi and Isaakidou, 2015), 8: Faia Petra (Valla et al., 2013), 9: Galatas (Hamilakis, 1996), 10: Kalapodi (Stanzel, 1991), 11: Kalythies Cave (Halstead and Jones, 1987), 12: Kastanas (Becker, 1986), 13: Knossos (Isaakidou, 2006; 2011), 14: Kolonna (Forstenpointner et al., 2010), 15: Kommos (Ruscillo, 2012), 16: Lerna (Gejvall, 1969), 17: Makriyalos (Pappa et al., 2004), 18: Mikrothives (Isaakidou and Halstead, 2013), 19: Mitrou (Isaakidou and Halstead, 2013), 20: Monastiraki (Mylona, 2012), 21: Mycenae, 22: Pevkakia (Jordan, 1975; Amberger, 1979; Hinz, 1979), 23: Phaistos, 24: Phylakopi (Gamble, 1982), 25: Platia Magoula Zarkou (Becker, 1991), 26: Priniatikos Pyrgos (Molloy et al., 2014), 27: Promachon (Trantalidou, 2010), 28: Proskunas (Isaakidou and Halstead, 2013), 29: Pylos (Halstead and Isaakidou, 2004), 30: Thronos-Kephala (D. Mylona, pers. comm.), 31: Tiryns (Von den Driesch and Boessneck, 1990), 32: Toumba Kremastis-Koiladas (Tzevelekidi et al., 2014), 33: Tsoungiza (Dabney et al., 2004).

for reproductive isolation from wild aurochs (*Bos primigenius*) and boar (*Sus scrofa*). Conversely, at Knossos, on an island lacking large indigenous wild animals, persistently large cattle and pigs suggest free-range management or the development of hunted feral populations.

More or less intact skeletons of horse (*Equus caballus*) and donkey (*Equus asinus*) are known from southern mainland mortuary contexts of LB date (see discussion p. 119), when fragments of horse, donkey, and perhaps mule also occur from southerly Tiryns to northerly Kastanas in sufficient numbers to preclude interpretation as later intrusions. A few specimens in earlier Bronze Age and later Neolithic levels must be treated cautiously without direct dating. Either way, the founder domesticates dominated Bronze Age assemblages (e.g. Halstead, 1996).

At open-air settlements on the mainland, a more or less balanced mixture of cattle, pigs, sheep, and to a lesser extent goats is found at EB-LB Kastanas in the north, EB-MB Argissa and EB-LB Pevkakia in central Greece, and EB-MB Lerna and EB-LB Tiryns in the south. In some cases, LB increases in sheep *might* reflect growing importance of wool (see discussion p. 120) or extension of cultivation and fallow/stubble pasture suitable for sheep. On the southern Aegean islands, however, as in the Neolithic, cattle and pigs are relatively scarce, while sheep and goats (where these species are distinguished) are more or less equally abundant at MB Ayia Irini on Kea, at EB and early LB Akrotiri on Thira, at EB-LB Phylakopi on Melos, and at EB Knossos, LB Kommos, LB Monastiraki, and EB Priniatikos Pyrgos on Crete. The persistent insular abundance of goats as well as sheep suggests that environmental constraints were as influential as any demand for wool in shaping species frequencies. Other potential factors include settlement nucleation or emerging inter- and intra-site hierarchy and the more balanced composition of the assemblage from MB-LB Knossos may reflect its overwhelming derivation from the site's 'public-elite' core and thus privileged access to local and regional animal resources.

Ruminant milk residues have been identified in LN ceramics from northern Greece (Evershed et al., 2008), but mortality patterns for Neolithic sheep, goats, and cattle at Knossos and elsewhere, exhibiting selective slaughter of juvenile-subadult males, suggest prioritization of meat production. The scarcity of infant deaths implies limited availability of milk for humans and so limited dairy exploitation. Throughout the Neolithic and perhaps Bronze Age at Knossos, stress 'pathologies' in hip and 'remodelling' of foot joints suggest traction use of paired cattle including females (Isaakidou, 2006), but similar evidence is otherwise sparse in Neolithic-Bronze Age Greece. Sheep, goats, and (less markedly) cattle exhibit older mortality and higher adult male survivorship at Bronze Age than Neolithic Knossos. The Bronze Age sample, largely from the 'public-elite' area, is perhaps biased towards impressively large, adult male carcasses, but a similar trend occurs elsewhere. Thus there are more adult male sheep (including castrates) at Tiryns and (with improved adult survivorship) at Pevkakia and Kastanas. There is also a more balanced adult sex ratio in cattle, coupled with fairly high adult survivorship at Tiryns, metrical evidence for castrated males at Tiryns and Pevkakia, and traction 'pathologies' at Tiryns and Kastanas. Secondary products (sheep wool, goat hair, and cattle traction), therefore, were probably higher priorities in the Bronze Age than in the Neolithic. Animals made a minor direct contribution to Bronze Age diet, however, given apparently rising human population densities and the lack of intensive dairying (which offers far more calories than meat per head of live-stock or hectare of pasture).

Hunting

At Knossos there is no hint that early farmers encountered the endemic Pleistocene deer. Rare Neolithic specimens of exotic red (*Cervus elaphus*) and fallow deer (*Dama dama*) may be intrusive, but LB fallow deer at Knossos and elsewhere on Crete are frequent enough to indicate successful introduction. Fallow deer was introduced to several other Aegean islands in the Neolithic and apparently hunted, at least at LN-FN Kalythies cave on Rhodes. Here the common domesticates are well represented by all carcass parts, consistent with nearby slaughter of herded animals, but deer are over-whelmingly represented by meat-rich upper limbs, suggesting discard of unwanted parts at a distant kill-site. On the mainland, wild animals contribute on average less than 5% of mammal bones at Neolithic open-air sites, but substantially more at EB-MB Argissa, EB-LB Pevkakia, and EB Platia Magoula Zarkou in central Greece and EB-LB Kastanas in the north. The commonest game was wild boar and red deer, the latter giving way to fallow deer in the later Bronze Age, while increasingly widespread finds of isolated lion (*Panthera leo*) bones (Yannouli, 2003—to which several findspots must now be added) have attracted most interest.

Hospitality

Excluding dogs and equids (see discussion p. 119) and intrusive material from animal burrows, the overwhelming majority of bones found at Knossos and other Neolithic and Bronze Age sites bears traces of human carcass processing: butchery marks, 'bar-becue' burning, and fragmentation patterns incompatible with natural or accidental breakage. Carcasses were normally skinned and dismembered and some meat was fil-leted from the bone with small knives—chipped-stone in the Neolithic and metal in the Bronze Age, judging from cut-mark morphology. Chopping with larger metal cleavers is frequent at some later Bronze Age sites, but rare at others. Cut marks are concentrated in fewer parts of the body in Neolithic than Bronze Age deposits. Neolithic carcasses were thus cooked in large pieces, perhaps on open fires, with meat subsequently shared quite widely. Smaller Bronze Age joints could have been cooked in ceramic or metal pots, enabling more varied cuisine, and were probably shared among fewer people (Isaakidou, 2007).

Throughout the Neolithic and Bronze Age, bones were fractured for marrow and perhaps grease more intensively in cattle than sheep, goats, and pigs, and more so in adults than juveniles, consistent with the higher returns on labour from larger bones.

Over time, however, carcass processing became less intensive. In earlier Neolithic levels, even small sheep and goat phalanges were often splintered, but from the later Neolithic onwards bones of the smaller domesticates were merely broken open and even the large phalanges of cattle were mostly left intact (Halstead and Isaakidou, 2011). Furthermore, in skinning cattle at Knossos, more of the foot was discarded in the palatial later Bronze Age than in earlier periods (Isaakidou, 2007). Carcasses were also prepared for consumption more formally over time. In the earlier Neolithic, all parts were apparently processed and discarded together. From the later Neolithic onwards, however, heads and/or feet were often removed at a place or time separate from consumption and are absent from articulated skeletons at LN Toumba Kremastis-Koiladas in northern Greece or underrepresented in commingled material elsewhere (Tzevelekidi et al., 2014). Such carcass dressing drew a distinction of context and perhaps participants between slaughter / primary butchery and consumption (Halstead and Isaakidou, 2011), while clusters of young pig femurs from a house at Pevkakia and a serving vessel at Mitrou in LB central Greece imply that consumption of particular anatomical parts signalled further commensal distinctions (Isaakidou and Halstead, 2013).

Discard practices too underlined the significance of some commensal occasions. While most animal bone on Neolithic and Bronze Age settlements comprises commingled body parts and species, from the later Neolithic onwards whole or partial carcasses were widely deposited in pits, sometimes *after* butchery and presumably consumption (e.g. LN Toumba Kremastis-Koiladas; FN Mikrothives and EB Proskunas in central Greece; LB Knossos and Thronos-Kephala on Crete). Such formal discard also distinguished a third stage after carcass dressing and consumption (Isaakidou and Halstead, 2013). Around the Mycenaean palace of Pylos, five groups of burnt cattle and deer bones were deposited that comprised selected body parts (mandible, humerus, and femur) not normally discarded together during routine butchery. Before burning, the bones had been filleted but—unusually—not broken to remove marrow. The repeated wastage of marrow, non-practical selection of body parts and careful deposition identify these as ritual deposits, while similarity with rites in later sanctuaries and with descriptions in Homer's *Odyssey* invites identification of sacrifices.

The 'burnt bone sacrifices' at LB Pylos apparently involved more or less simultaneous slaughter of several cattle, at least nineteen (probably male) and one deer in the largest deposit, and thus enough meat (a few hundred kilograms to a few tonnes) for hundreds or even thousands of guests. The palace pantries held enough cooking pots and tableware to host large-scale commensality, while access routes and available space suggest segregated reception of higher- and lower-status guests. Sacrifice signalled divine participation in, and hence approval of, these inegalitarian arrangements (Halstead and Isaakidou, 2004).

Rural sanctuaries provide roughly contemporary evidence for sacrifice with feasting. At Ayios Konstantinos on the southern mainland, a cult room contained libation vessels, human and animal figurines interpreted as votives, and a hearth in which piglets (some with bones stripped of meat) had been burnt. In adjacent rooms, unburnt bones from dressed carcasses principally of sheep or goat suggest that small-scale feasting accompanied the burnt sacrifice of piglets (Hamilakis and Konsolaki, 2004). In the same region at Tsoungiza, 10 km from the palatial centre of Mycenae and perhaps sub-ordinate to it, a series of pits yielded tableware indicating large-scale commensality, for which a female figure of a type known from sanctuaries suggests a religious rather than purely secular context. The associated faunal material provides no evidence for sacri-fice, but includes the heads and feet of several cattle, the missing meaty upper limbs of which were—consistently over several decades—discarded elsewhere at Tsoungiza or removed by visitors from other settlements. Conceivably, the rulers of Mycenae hosted gatherings here to promote their regional interests. At Kalapodi in central Greece, where superimposed altars and temples span the Late Bronze and Early Iron Ages, post-palatial LB faunal remains are predominantly of sheep and goats (in similar numbers), followed by cattle, then pigs and red deer, and an array of other wild animals. As in classical antiquity, the animals offered to deities, or consumed in their honour, varied greatly.

Despite the importance of feasting in Mycenaean society, comparison with LN northern Greece is instructive. Cattle skulls on house facades at Promachon per-haps advertised past hospitality, but remains of several hundred thoroughly butch-ered domestic animals in a pit at Makriyalos, representing tens of tonnes of meat and suggesting a regional catchment for consumers and provisions, were associated with highly standardized cooking and serving vessels, emphasizing equality and cohe-sion. Conversely, Bronze Age tableware, ingredients, cuisine, and physical setting distinguished increasingly overtly between commensal participants. Palatial feast-ing differed from earlier commensality in distinction of status rather than scale of consumption.

Symbolism

Osteological hints of the symbolic meanings of animals include hunting trophies, such as deer 'skull plates' with attached antlers and helmets of cut boars' tusks (Morris, 1990), found in high-status LB graves in southern Greece. In the LB settlement at Assiros Toumba in northern Greece, the selection of deer bones for carefully manufactured and intensively used awls, while domesticate bones were used expediently, perhaps had similar connotations. Among the domesticates, only dog (throughout the Neolithic and Bronze Age) and equids (from the later Bronze Age) occur frequently as more or less intact skeletons, as well as commingled butchered fragments. Osteologically elusive dif-ferences of appearance (e.g. coat colour) might explain why these animals were some-times eaten and sometimes not. Intact skeletons may not represent burials but carcasses discarded as unfit for consumption or, as perhaps with two skinned dogs in a LB tomb at Galatas, deposited as grave goods. More or less complete skeletons of other domes-ticates are also found in tombs, but several of these had been dressed and, in three LB mortuary enclosures at northerly Faia Petra, dismembering and filleting cut-marks identified a calf and two yearling sheep as remains of funerary meals. At the Dendra LB cemetery in the southern mainland, however, equids arguably display mortuary treat-ment in their own right. Here, paired horse 'burials' were not associated directly with human graves and, in at least one case, were placed on the ground and covered with a mound, while a group of at least four donkeys had been buried, disinterred, and re-buried, recalling secondary burial treatment of humans.

Symbolic significance of particular body parts, implied in the LB pig femur groups at Mitrou and Pevkakia and the 'burnt bone sacrifices' at Pylos (see discussion p. 118), is especially evident in EB funerary assemblages. On the Cycladic islands, palettes and decorated pigment tubes were made from cattle and sheep/goat femurs, respectively; and on Crete, ring-seals made from cattle metatarsals inspired skeuomorphs in exotic ivory.

PALATIAL POLITICAL ECONOMY: TEXTS
WITH BONES

The Linear B texts from Mycenaean palaces were temporary records that overwhelmingly monitored the fulfilment of obligations—primarily to the palace. Most texts citing animals document herd composition, lamb or wool returns, or allocation of livestock to named persons and places. A minority of texts lists animals destined for slaughter. The latter are effectively 'deadstock' records and, with allowance for contrasting formation processes and the shorter time-depth of texts, can be compared relatively directly with faunal data, whereas the former 'livestock' records can only be compared indirectly, allowing for each species' average lifespan. Even with such allowance, 'livestock' and 'deadstock' texts are mutually inconsistent, while consideration of demographic parameters (e.g. potential lifespan of male sheep, lambing rate of females) shows that breeding ewes, young replacement lambs and older sheep ready for culling must be heavily under-recorded in both groups of texts. Both 'livestock' and 'deadstock' texts, therefore, deal selectively with animals and aspects of their management that interested the palace administration (Halstead, 2001a).

The 'deadstock' records confirm faunal evidence that various animals were slaughtered for feasts and sacrifices, add that such animals were often fattened beforehand and sometimes selected for their colour (Killen, 1994), and highlight the unusually large scale of the event at Pylos for which faunal evidence indicates slaughter of at least nineteen cattle (see discussion p. 118). They also confirm that palaces sponsored outlying festivals (as perhaps at Tsoungiza—see discussion pp. 118–19), identify some of the occasions celebrated (religious festivals, a royal initiation), and suggest that the palace contributed little of what was consumed.

The majority of 'livestock' texts at Knossos record 80,000–100,000 sheep distributed across central Crete, from which the palace annually procured 30–50 tonnes of wool for fine textiles, which were probably dispensed in local and distant gift exchange / trade (Killen, 1993). The palace entrusted wool flocks to named individuals on a yearly basis, expecting return of the original number of sheep and a proportionate payment of wool, but was not concerned with the age and sex of the returned sheep. These 'herders' apparently culled elderly animals from, provided replacements to, and otherwise added to and subtracted from the supposedly 'palatial' flocks, and so must also have run their own sheep (Halstead, 2001a). Other texts record location at major sub-centres of pairs of working oxen, with detailed descriptions implying palatial concern to retain specific animals. These oxen probably enabled the large-scale grain production, apparently on the land of local communities, which provided rations for palace dependents, including hundreds of textile workers. Recorded interests in draught oxen and wool sheep thus secured two vital and linked palatial resources—grain and prestige craft goods. Both interests involved the palace in 'sharecropping' arrangements with non-dependent farmers/herders (Halstead, 2001b).

Among relatively few texts listing equids, a set from Knossos lists sixty-eight horses alongside armour and chariots. Although livestock records offer no hint of milking or milk products, 'deadstock' records list cheese among banquet provisions or offerings. Deadstock products other than meat also appear in records of 'taxation' in hides and goat horns and of the supply of hide, horn (or perhaps antler), and ivory (imported!) to workshops.

The palace of Pylos has yielded good samples of both texts and bones from the period preceding its final destruction and comparison reinforces the selectivity of the former. Faunal analysis has revealed several species that are absent in the texts (dogs, several wild species in addition to the textual 'deer') and far higher proportions of *female* sheep and cattle than 'deadstock' or 'livestock' texts record. At Tiryns, a palatial centre that has yielded few texts, butchered bones of dog, horse, and donkey contrast with the absence of these species from 'deadstock' (consumption) texts elsewhere. Furthermore, substantial pre-palatial (EB, early LB), palatial (late LB), and post-palatial (final LB) faunal assemblages from Tiryns exhibit little change in the relative proportions of different domesticates. Likewise, assemblages spanning the Bronze Age from Pevkakia, at the margins of Mycenaean palatial society, and Kastanas, in the non-palatial north, resemble southerly Knossos, Tiryns, and Pylos in their more or less balanced mixture of cattle, pigs, sheep, and goats and in mortality patterns implying greater emphasis on secondary products than local Neolithic assemblages. Despite the textually evident importance of livestock and deadstock to palatial political economy, therefore, the rise and subsequent collapse of the palaces apparently transformed rights to animal produce rather than patterns of animal management.

ULTIMATE TRUTHS: GRIFFINS, LIONS, AND BULLS

Minoan and Mycenaean animal imagery ranges from the fantastic to the realistic, the boundaries between which are sometimes unclear. Winged griffins are plainly fantastic, but does the crocodile on an inlaid dagger from Mycenae depict an actual adventure of a traveller to Egypt? Lions are depicted fighting wild and domestic animals and human hunters, but did not roam Crete (unless temporarily introduced). Even on the mainland, sparse lion bones may represent occasional imported skins or dangerous 'pets', as inevitably at Kolonna on the small island of Aegina. Either way, the lion as symbol of power—most famously in the heraldic pair over the gateway at Mycenae—may be one of many borrowings from elite culture further east. On the other hand, Cretan depictions of humans hunting cattle and goats could represent encounters with real feral populations (see discussion p. 114). Further zooarchaeological research may resolve these biogeographical uncertainties, but imaginary beasts and 'real' wild or feral animals may have shared similar cultural significance as representing prestigious encounters with the unfamiliar (Hamilakis, 2003). Images of paired horses in harness or pulling chariots match indications of elite associations from paired horse burials and texts. Conversely, a LB figurine from Phaistos on Crete depicts a laden donkey in its traditional role as pack animal, but a LB fresco from Mycenae, showing donkey- or mule-headed 'genii' in procession, supports osteological indications from Dendra and Tiryns that horses may not have enjoyed higher status than donkeys (Pappi and Isaakidou, 2015).

The most frequently depicted of the founder domesticates (e.g. Vanschoonwinkel, 1996) are cattle. Goats too are recurrent subjects, but sheep infrequent. The emphasis on cattle, paralleled in Neolithic zoomorphic figurines, underscores the preeminent commensal and sacrificial value of this large domesticate. Cattle, usually bulls, are more often seen under attack from hunters or predators or in ceremonial scenes such as 'bull leaping' and sacrifice, than in mundane poses (e.g. cows suckling calves). Minoan depictions of sacrifice, such as the trussed bull on the LB Ayia Triada sarcophagus, suggest blood libations, in contrast with emerging osteological evidence for burnt offerings on the mainland. Intriguingly, and in marked contrast with Bronze Age Egypt, Aegean iconography offers no hint, other than the probable winnowing gang on the so-called 'Harvesters Vase' from LB Crete, of the role of oxen and extensive plough-agriculture in elite production of grain staples.

CONCLUSION

Iconography highlights the cultural importance to Bronze Age elites in Greece of hunting large game (native, exotic, and mythical), riding horse-drawn chariots, and sacrificing or offering bulls. Linear B texts largely overlook wild animals, availability of which was inevitably unpredictable. They confirm elite use of horse-drawn chariots and reveal palatial control of draught oxen. They confirm palatial involvement in sacrificial and commensal slaughter, while contradicting the iconographic predominance of cattle and showing that palatial subjects provided many of these animals. The largest group of extant texts, however, records sheep—largely ignored in iconography—and provision of wool to palace-sponsored textile-workers. Specialist weaving and ox-powered grain production were two key elements in palatial resource mobilization.

Bronze Age animal bones offer less idealized, socially more inclusive, and chronologically and regionally broader coverage than texts or iconography. They confirm textual indications that LB cattle, sheep, goats, and pigs were all sacrificed and show that horses and donkeys were variously deposited intact in ceremonial contexts or butchered for consumption. Improved male survivorship of cattle, sheep, and goats suggests more emphasis on draught-power, wool, and perhaps hair than in the Neolithic, but does not replicate the textual specialization in wool sheep and draught oxen.

Iconography and texts reveal the importance of animals to Aegean Bronze Age society as aids to staple grain production, sources of raw materials for prestige goods, key ingredients in ritual and commensal politics, and inspirations for symbols of power, while comparison with faunal data highlights the selective nature of elite interests in regional livestock and rights to animal resources. Commensality involving meat was a central concern of palatial, but also of preceding Neolithic and EB, society. Bronze Age and especially LB palatial feasting contrasted with Neolithic commensality not in increased scale of meat consumption, but in greater ceremonial elaboration and overt emphasis on inequality between participants.

References

Amberger, K.-P. (1979) 'Neue Tierknochenfunde aus der Magula Pevkakia in Thessalien, 2: die Wiederkäuer'. Unpublished dissertation, University of Munich (Munich).

Becker, C. (1986) *Kastanas: die Tierknochenfunde*, Berlin: Volker Spiess.

Becker, C. (1991) 'Die Tierknochenfunde von der Platia Magoula Zarkou—neue Untersuchungen zu Haustierhaltung, Jagd und Rohstoffverwendung im neolithisch-bronzezeitlichen Thessalien', *Prähistorische Zeitschrift*, 66, 14–78.

Boessneck, J. (1962) 'Die Tierreste aus der Argissa-Magula vom präkeramischen Neolithikum bis zur mittleren Bronzezeit', in Milojcic, V., Boessneck, J., and Hopf, M. *Argissa-Magula 1: Das präkeramische Neolithikum sowie die Tier- und Pflanzenreste*, pp. 27–99. Bonn: R. Habelt.

Coy, J. (1986) 'Appendix 2: the faunal remains from Period V', in Davis, J. L. (ed.) *Keos 5: Ayia Irini. Period V*, pp. 109–11. Mainz: von Zabern.

Dabney, M., Halstead, P., and Thomas, P. (2004) 'Late Mycenaean feasting on Tsoungiza at ancient Nemea', *Hesperia*, 73, 197–215.

Driesch, von den, A. (1987) 'Haus- und Jagdtiere im vorgeschichtlichen Thessalien', *Prähistorische Zeitschrift*, 62, 1–21.

Driesch, von den, A. and Boessneck, J. (1990) 'Die Tierreste von der mykenischen Burg Tiryns bei Nafplion/Peloponnes', in Weisshaar, H.-J., Weber-Hiden, I., von den Driesch, A., Boessneck, J., Rieger, A., and Böser, W. *Tiryns Forschungen und Berichte 11*, pp. 87–164. Mainz: von Zabern.

Evershed, R. P., Payne, S., Sherratt, A. G., Copley, M. S., Coolidge, J., Urem-Kotsu, D., Kotsakis, K., Özdogan, M., Özdogan, A. E., Nieuwenhuyse, O., Akkermans, P. M. M. G., Bailey, D., Andeescu, R.-R., Campbell, S., Farid, S., Hodder, I., Yalman, N., Özbasaran, M., Bıçakcı, E.,

Garfinkel, Y., Levy, T., and Burton, M. M. (2008) 'Earliest date for milk use in the Near East and southeastern Europe linked to cattle herding', *Nature*, 455, 528–31.

Forstenpointner, G., Galik, A., Weissengruber, G. E., Zohmann, S., Thanheiser, U., and Gauss, W. (2010) 'Subsistence and more in Middle Bronze Age Aegina Kolonna: patterns of husbandry, hunting and agriculture', in Philippa-Touchais, A., Touchais, G., Voutsaki, S., and Wright, J. (eds) *Mesohelladika: The Greek Mainland in the Middle Bronze Age*, pp. 733–42. Paris: École française d'Athènes.

Gamble, C. (1978) 'The Bronze Age animal economy from Akrotiri: a preliminary report', in Doumas, C. (ed.) *Thera and the Aegean World I*, pp. 745–53. London: Thera Foundation.

Gamble, C. (1982) 'Animal husbandry, population and urbanisation', in Renfrew, C. and Wagstaff, M. (eds) *An Island Polity: The Archaeology of Exploitation in Melos*, pp. 161–71. Cambridge: Cambridge University Press.

Gejvall, N.-G. (1969) *Lerna 1, the Fauna*, Princeton: American School of Classical Studies at Athens.

Halstead, P. (1996) 'Pastoralism or household herding? Problems of scale and specialisation in early Greek animal husbandry', *World Archaeology*, 28, 20–42.

Halstead, P. (2001a) 'Texts, bones and herders: approaches to animal husbandry in Late Bronze Age Greece', *Minos*, 33–4, 149–89.

Halstead, P. (2001b) 'Mycenaean wheat, flax and sheep: palatial intervention in farming and its implications for rural society', in Voutsaki, S. and Killen, J. (eds) *Economy and Politics in the Mycenaean Palace States*, pp. 38–50. Cambridge: Cambridge Philological Society.

Halstead, P. and Isaakidou, V. (2004) 'Faunal evidence for feasting: burnt offerings from the Palace of Nestor at Pylos', in Halstead, P. and Barrett, J. C. (eds) *Food, Cuisine and Society in Prehistoric Greece*, pp. 136–54. Oxford: Oxbow.

Halstead, P. and Isaakidou, V. (2011) 'Political cuisine: rituals of commensality in the Neolithic and Bronze Age Aegean', in Aranda Jiménez, G., Montón-Subías, S., and Romero, S. (eds) *Guess Who's Coming to Dinner: Feasting Rituals in the Prehistoric Societies of Europe and the Near East*, pp. 91–108. Oxford: Oxbow.

Halstead, P. and Jones, G. (1987) 'Bioarchaeological remains from Kalythies cave, Rhodes', in Sampson, A. *Η Νεολιθική Περίοδος στα Δωδεκάνησα* (The Neolithic Period in the Dodecanese), pp. 135–52. Athens: Ministry of Culture.

Hamilakis, Y. (1996) 'A footnote on the archaeology of power: animal bones from a Mycenaean chamber tomb at Galatas, NE Peloponnese', *Annual of the British School at Athens*, 91, 153–66.

Hamilakis, Y. (2003) 'The sacred geography of hunting: wild animals, social power and gender in early farming societies', in Kotjabopoulou, E., Hamilakis, Y., Halstead, P., Gamble, C., and Elefanti, P. (eds) *Zooarchaeology in Greece: Recent Advances*, pp. 239–48. London: British School at Athens.

Hamilakis, Y. and Konsolaki, E. (2004) 'Pigs for the gods: burnt animal sacrifices as embodied rituals at a Mycenaean sanctuary', *Oxford Journal of Archaeology*, 23, 135–51.

Hinz, G. (1979) 'Neue Tierknochenfunde aus der Magula Pevkakia in Thessalien, 1: die Nichtwiederkaüer'. Unpublished dissertation, University of Munich (Munich).

Isaakidou, V. (2006) 'Ploughing with cows: Knossos and the 'secondary products revolution', in Serjeantson, D. and Field, D. (eds) *Animals in the Neolithic of Britain and Europe*, pp. 95–112. Oxford: Oxbow.

Isaakidou, V. (2007) 'Cooking in the labyrinth: exploring "cuisine" at Bronze Age Knossos', in Mee, C. and Renard, J. (eds) *Cooking up the Past: Food and Culinary Practices in the Neolithic and Bronze Age Aegean*, pp. 5–24. Oxford: Oxbow.

Isaakidou, V. (2011) 'Early Minoan I, the Palace Well: faunal remains and taphonomy', 'Early Minoan II-III, Area A. Royal Road North: faunal remains', 'Early Minoan II-III, Area B. The early houses: faunal remains', in Hood, S. and Cadogan, G. (eds) *Knossos Excavations 1957–1961: Early Minoan*. British School at Athens Suppl. 46, pp. 63–7, 229–33, 237. London: British School at Athens.

Isaakidou, V. and Halstead, P. (2013) 'Bones and the body politic? A diachronic analysis of structured deposition in the Neolithic-Early Iron Age Aegean', in Ekroth, G. and Wallensten, J. (eds) *Bones, Behaviour and Belief: The Zooarchaeological Evidence as a Source for Ritual Practice in Ancient Greece and Beyond*, pp. 87–99. Stockholm: Swedish Institute at Athens.

Jordan, B. (1975) 'Tierknochenfunde aus der Magula Pevkakia in Thessalien'. Unpublished dissertation, University of Munich (Munich).

Killen, J. T. (1993) 'Records of sheep and goats at Mycenaean Knossos and Pylos', *Bulletin on Sumerian Agriculture*, 7, 209–18.

Killen, J. T. (1994) 'Thebes sealings, Knossos tablets and Mycenaean state banquets', *Bulletin of the Institute of Classical Studies*, 39, 67–84.

Molloy, B., Day, J., Bridgford, S., Isaakidou, V., Nodarou, E., Kotzamani, G., Milic, M., Carter, T., Westlake, P., Klontza-Jaklova, V., Larsson, E., and Hayden, B. J. (2014) 'Life and death of a Bronze Age house: excavation of Early Minoan I levels at Priniatikos Pyrgos', *American Journal of Archaeology*, 118, 307–58.

Morris, C. E. (1990) 'In pursuit of the white tusked boar: aspects of hunting in Mycenaean society', in Hägg, R. and Nordquist, G. C. (eds) *Celebrations of Death and Divinity in the Bronze Age Argolid*, pp. 149–55. Stockholm: Swedish Institute at Athens.

Mylona, D. (2012) 'Chapter VI. The animal bones from the Archive Building area', in Kanta, A., Tzigounaki, A., Godart, L., Pecoraro, G., Mylona, D., and Speliotopoulou, A. (eds) *Monastiraki IIA: The Archive Building and Associated Finds*, pp. 199–206. Herakleio: A. Kanta.

Pappa, M., Halstead, P., Kotsakis, K., and Urem-Kotsou, D. (2004) 'Evidence for large-scale feasting at Late Neolithic Makriyalos, N Greece', in Halstead, P. and Barrett, J. C. (eds) *Food, Cuisine and Society in Prehistoric Greece*, pp. 16–44. Oxford: Oxbow.

Pappi, E. and Isaakidou, V. (2015) 'On the significance of equids in the Late Bronze Age Aegean: new and old finds from the cemetery of Dendra in context', in Schallin, A.-L. and Tournavitou, I. (eds) *Mycenaeans up to Date: The Archaeology of the NE Peloponnese, Current Concepts and New Directions*, pp. 469–81. Stockholm: Swedish Institute in Athens.

Ruscillo, D. (2012) 'The faunal remains', in Shaw, M. C. and Shaw, J. W. (eds) *House X at Kommos, a Minoan Mansion near the Sea*, part 1: *Architecture, Stratigraphy, and Selected Finds*, pp. 93–116. Philadelphia: INSTAP Academic Press.

Stanzel, M. (1991) 'Die Tierreste aus dem Artemis-/Apollon-Heiligtum bei Kalapodi in Böotien/Griechenland'. Unpublished dissertation, University of Munich (Munich).

Trantalidou, K. (2010) 'Bovid skulls in southeastern European Neolithic dwellings: the case of the subterranean circular room at Promachon-Topolniča in the Strymon Valley, Greece', in Campana, D., Crabtree, P., deFrance, S. D., Lev-Tov, J., and Choyke, A. (eds) *Anthropological Approaches to Zooarchaeology: Complexity, Colonialism, and Animal Transformations*, pp. 213–19. Oxford: Oxbow.

Trantalidou, K. (2013) 'The animal bones: the exploitation of livestock', in Renfrew, A. C., Philaniotou, O., Brodie, N., Gavalas, G., and Boyd, M. J. (eds) *The Settlement at Dhaskalio*, pp. 429–41. Cambridge: McDonald Institute.

Tzevelekidi, V., Halstead, P. and Isaakidou, V. (2014) 'Invitation to dinner: practices of animal consumption and bone deposition at Makriyalos I (Pieria) and Toumba Kremastis-Koiladas (Kozani)', in Stefani, E., Merousis, N., and Dimoula, A. (eds) *100 Χρόνια Έρευνας στην Προϊστορική Μακεδονία* (A Century of Research in Prehistoric Macedonia), pp. 425–36. Thessaloniki: Archaeological Museum of Thessaloniki.

Valla, M., Triantaphyllou, S., Halstead, P., and Isaakidou, V. (2013) 'Manipulating death at the end of the Late Bronze Age: the case of Faia Petra, 13th c. BC, eastern Macedonia, Greece', in Lochner, M. and Ruppenstein, F. (eds) *Cremation Burials in the Region between the Middle Danube and the Aegean 1300–750 BC*, pp. 231–48. Vienna: Austrian Academy of Sciences.

Vanschoonwinkel, J. (1996) 'Les animaux dans l'art Minoen', in Reese, D. S. (ed.) *Pleistocene and Holocene Fauna of Crete and its First Settlers*, pp. 351–412. Madison: Prehistory Press.

Yannouli, E. (2003) 'Non-domestic carnivores in Greek prehistory: a review', in Kotjabopoulou, E., Hamilakis, Y., Halstead, P., Gamble, C., and Elefanti, P. (eds) *Zooarchaeology in Greece: Recent Advances*, pp. 175–92. London: British School at Athens.

CHAPTER 9

CHANGES IN LIFESTYLE IN ANCIENT ROME (ITALY) ACROSS THE IRON AGE/ ROMAN TRANSITION

the evidence from animal remains

JACOPO DE GROSSI MAZZORIN
AND CLAUDIA MINNITI

INTRODUCTION

ROME currently provides a most promising area of investigation, as the modern city has been the scene of extensive archaeological activity. A large number of assemblages of animal bones have come to light and have been studied providing a large amount of information on aspects of animal consumption and exploitation in Roman times, which were only hinted at by written sources.

Archaeozoological studies carried out in many provinces conquered by the Romans have demonstrated that their arrival caused significant changes in the economies of local communities and led to an extension of market transactions in comparison with the preceding Iron Age (Albarella et al., 2008 and references therein; Minniti et al., 2014). In some areas the change was fairly abrupt and led to a significant change in diet and significant improvements in livestock husbandry as a result of the grafting of Roman elements and the introduction of new genetic types of animals. In other areas innovations occurred more gradually.

The zooarchaeology of Rome offers the opportunity to see in detail that changes in animal exploitation at the Iron Age-Roman transition and throughout the Roman period also occurred at the core of the Roman Empire and the centre of *Romanitas*.

The urban nature of the contexts allows us to gather information on diet and aspects of animal exploitation that were closely linked to the market and the lifestyle of the city, bearing in mind that the place of consumption does not coincide with the place of production.

Our study involves the analysis of over thirty thousand identified specimens coming from fifty-two different archaeological contexts located in Rome and in the neighbouring areas. Table 9.1 lists all assemblages that are included in this study, along with the reference for the original publication in each case. The data pertaining to the different chronological phases are not distributed evenly throughout the assemblages and there are voids that need to be filled in. Most of the assemblages span the whole period of the Roman Empire (second half of first century BC–fifth century AD), but the samples dating to the Iron Age (second half of tenth–fifth century BC) and the Republican period (fourth–first half of first century BC) are less common due to stratigraphic overlap and preservation bias. Samples from fourth–third century BC are represented, but very few data currently concern the second century and the first half of the first century BC. More numerous are the Iron Age and Republican period samples from neighbouring areas which include the modern regions of Latium, Tuscany, and Abruzzi (Fig. 9.1). These are included in Table 9.1, and are compared with urban samples with the aim of investigating changes in animal husbandry in Italy after the introduction and spread of Roman lifestyle, politics, and culture. We also must bear in mind that the samples vary in size and that statistically significant results were not always obtained. Small samples were only taken into account when combined, and only used to draw more general conclusions regarding the dataset as a whole. The synthesis presented here results from the work mainly carried out by the two authors, who have contributed equally to the paper.

MAIN DOMESTIC SPECIES FREQUENCY

The Roman period saw a substantial change in the overall relative proportion of the main domesticates in comparison with the preceding Iron Age. The Iron Age economy of central Italy was generally dominated by caprines (sheep, *Ovis aries*, and goat, *Capra hircus*) and cattle (*Bos taurus*) husbandry, following the trend identified in previous periods of the Middle (seventeenth–fourteenth century BC), Recent (thirteenth–first half of twelfth century BC), and Final Bronze Age (second half of twelfth–first half of tenth century BC), with pigs (*Sus domesticus*) being the third most common domesticate, although this species sees an initial increase in importance in some sites of this period (De Grossi Mazzorin and Minniti, 2009b; Minniti, 2012a). Whereas in Abruzzi the importance of pig is already attested at sites dated to the Final Bronze Age or the Early Iron Age, in Latium and in Tuscany its importance only increases in the eighth century BC (up to 47%, with an average of 31%, of the domestic animal remains according to the NISP account). The increase in pig remains in Rome is especially striking as frequencies rise to as high as 60% (with an average of 48%) during the Orientalizing (eighth–seventh

Table 9.1 Italian sites from Tuscany, Abruzzi, and Latium with indication of NISP and NISD (i.e. Number of Identified Specimens of the three main Domesticates) and percentages of cattle, caprine, and pig within the NISD. The NISP count refers to the total number of identified specimens per site. Periods: EIA = Early Iron Age; Or/Arch = Orientalizing/Archaic; Rep = Republican; I = Imperial; LA = Late Antiquity. For details on the references listed, see King et al., 1985; De Grossi Mazzorin, 1989; 2006; Cerilli, 2005; Minniti, 2005; 2012a; De Grossi Mazzorin and Coppola, 2008; De Grossi Mazzorin and Minniti, 2009b; Alhaique and Fortunato, 2010; Fiore and Tagliacozzo, 2011; Perrone, 2012

Site	Chronology	Period	Reference	NISP/NISD	% cattle	% caprines	% pig
Tarquinia—Cretoncini	9th cent. BC	EIA	De Grossi Mazzorin, 1995a	226/197	30.9	50.7	18.3
Fidene—hut	9th cent. BC	EIA	De Grossi Mazzorin, 1989a; Minniti, 2012a	37/35	22.9	54.3	22.9
Tarquinia—ph. 1	9th–7th cent. BC	Or/Arch	Bedini, 1997	916/627	23.1	34.0	42.9
Fidene—area A	9th–7th cent. BC	Or/Arch	De Grossi Mazzorin, 1989a	240/229	29.3	39.7	31.0
Teramo	9th–6th cent. BC	Or/Arch	De Grossi Mazzorin and Minniti, 2003	551/526	39.9	27.9	32.1
Punta D'Erce	9th–6th cent. BC	Or/Arch	De Grossi Mazzorin and Minniti, 2003	149/123	29.3	49.6	21.1
Tortoreto	9th–8th cent. BC	Or/Arch	De Grossi Mazzorin et al., 2008	1,281/950	35.1	36.5	28.4
Madonna degli Angeli	9th–6th cent. BC	Or/Arch	De Grossi Mazzorin and Minniti, 2003	484/442	32.6	33.0	34.4
Fidene—hut	9th–6th cent. BC	Or/Arch	Minniti, 2012a	302/291	28.9	41.2	29.9
Fidene—U.P.F.	8th cent. BC	Or/Arch	De Grossi Mazzorin, 1989a; Minniti, 2012a	303/239	44.4	38.1	17.6
Campassini—ph. I	8th cent. BC	Or/Arch	Bartoloni et al., 1997	54/43	18.6	34.8	46.5
Ficana—area 3b–c (II)	8th–7th cent. BC	Or/Arch	De Grossi Mazzorin, 1997a; Minniti, 2012a	658/598	26.6	38.0	35.5
Rome—Domus Regia	8th–7th cent. BC	Or/Arch	Minniti, 2007; 2012a	65/61	19.7	44.3	36.1
San Giovenale—Spring Building	8th–7th cent. BC?	Or/Arch	Sorrentino, 1981a	743/280	62.1	15.7	22.1
San Giovenale—acropolis	8th–6th cent. BC	Or/Arch	Sorrentino, 1981b	73/64	18.7	40.6	40.6
San Giovenale—cistern I	7th–6th cent. BC	Or/Arch	Sorrentino, 1981a	61/43	62.8	11.6	25.6
Ficana—area 5a	7th cent. BC	Or/Arch	De Grossi Mazzorin, 1989a	418/366	18.0	52.5	29.5
Campassini—ph. II	7th cent. BC	Or/Arch	Bartoloni et al., 1997	279/250	13.6	49.6	36.8
Ficana—area 3b–c (III)	7th–6th cent. BC	Or/Arch	De Grossi Mazzorin, 1989a; Minniti, 2012a	523/492	33.3	33.9	32.7
Ficana—area 3b–c (II + III)	8th–6th cent. BC	Or/Arch	De Grossi Mazzorin, 1989a; Minniti, 2012a	227/205	48.3	29.3	22.4

(Continued)

Table 9.1 Continued

Site	Chronology	Period	Reference	NISP/NISD	% cattle	% caprines	% pig
Acquarossa—area A	7th–6th cent. BC	Or/Arch	Gejvall, 1982	403/372	82.7	12.9	4.3
Acquarossa—trenches	7th–6th cent. BC	Or/Arch	Tagliacozzo, 1994	178/167	52.7	26.3	21.0
Roselle (Donati's excavation)	6th cent. BC	Or/Arch	Corridi, 1989	158/140	30.0	28.5	41.4
Roselle (Cygielman's excavation)	6th cent. BC	Or/Arch	Corridi, 1989	58/54	35.1	20.3	44.4
Tarquinia—ph. 2	6th–5th cent. BC	Or/Arch	Bedini, 1997	416/392	17.1	33.7	49.2
Cerveteri	6th–5th cent. BC	Or/Arch	Clark, 1989	505/472	37.1	34.3	28.6
Rome—Velia	7th–6th cent. BC	Or/Arch	Minniti, 2012a	233/222	6.8	40.5	52.7
Rome—Forum of Caesar (well B)	6th–5th cent. BC	Or/Arch	De Grossi Mazzorin, 2014	107/103	1.0	38.8	60.2
La Castellina	6th–5th cent. BC	Or/Arch	Fiore and Tagliacozzo, 2011	49/46	23.9	34.8	41.3
Case Veldon	6th–4th cent. BC	Or/Arch	Perrone, 2012	630/523	48.2	22.8	29.1
Rome—Palatine (Boni's excavation)	8th–5th cent. BC	Or/Arch	De Grossi Mazzorin and Minniti, 2009b	234/217	24.0	38.7	37.3
Rome—Palatine (Puglisi's excavation)	5th–4th cent. BC	Rep	De Grossi Mazzorin, in progress	37/35	42.9	20.0	37.1
Rome—Palatine (temenos area)	4th–3rd cent. BC	Rep	Minniti, in progress	746/659	5.3	32.0	62.7
La Castellina	4th–3rd cent. BC	Rep	Fiore and Tagliacozzo, 2011	31/31	48.4	22.6	29.0
Rome—Palatine (Puglisi's excavation)	3rd cent. BC	Rep	De Grossi Mazzorin, in progress	36/28	14.3	32.1	53.6
Populonia	3rd cent. BC	Rep	De Grossi Mazzorin, 1985	2,114/1,988	10.3	43	46.7
Tarquinia—ph. 3	3rd–2nd cent. BC	Rep	Bedini, 1997	88/85	27.1	31.7	41.2
Settefinestre—ph. I	1st cent. BC	—	King et al., 1985	233/176	10.8	42.6	46.6
Ferento	1st cent. BC–1st AD	—	Alhaique and Fortunato, 2010	128/95	7.4	41.1	51.6
Rome—Forum of Caesar (taberna 11)	1st cent. BC–1st AD	—	Minniti, 2012b	32/32	–	3.1	96.9
Rome—Forum of Caesar (tholos cistern)	1st cent. BC–1st AD	—	Minniti, 2014	40/35	17.1	17.1	65.8
Rome—Aqua Marcia	1st cent. BC–1st AD	—	De Grossi Mazzorin, 1996b	178/152	28.3	5.3	66.4

Site	Date		Reference				
Rome—Meta Sudans (US 3399)	1st cent. BC–1st AD	—	De Grossi Mazzorin and Minniti, 1995	566/382	6.5	18.8	74.6
Rome—Aqua Marcia	1st cent. AD	—	De Grossi Mazzorin, 1996b	149/139	7.9	14.4	77.7
Rome—Arch of Constantine	1st cent. AD	—	De Grossi Mazzorin, in progress	51/49	12.2	10.2	77.6
Rome—Forum of Nerva	1st cent. AD	—	De Grossi Mazzorin, 1989b	82/72	12.5	9.7	77.8
Rome—Quirinal	1st cent. AD	—	De Grossi Mazzorin, 1998b	1,445/1,201	4	15.5	80.5
Rome—Caput Africae	1st cent. AD	—	Tagliacozzo, 1993	87/40	5	10	85
Rome—Via Sacchi	1st cent. AD	—	De Grossi Mazzorin and Coppola, 2008	200/199	71.4	6	22.6
Rome—Villa dei Quintili	1st–2nd cent AD	—	De Grossi Mazzorin, in progress	262/132	–	13.6	86.4
Rome—Colosseum (underground corridors)	1st–2nd cent. AD	—	Minniti, in progress	81/60	13.3	16.7	70
Ad Vacanas (rooms 126 + 53 alfa)	1st–2nd cent. AD	—	Cerilli, 2005	117/117	41.9	13.7	44.4
Laurentina—Tenuta di Vallerano	1st–2nd cent. AD	—	Minniti, 2005	471/231	50.2	35.9	13.9
Settefinestre–ph. II	2nd cent. AD	—	King et al., 1985	1,773/1,524	13.0	17.3	69.7
Rome—Caput Africae	2nd cent. AD	—	Tagliacozzo, 1993	207/163	0.6	25.8	73.6
Rome—Baths of Trajan	2nd cent. AD	—	De Grossi Mazzorin et al., in progress	253/144	6.9	14.6	78.5
Rome—Via Sacchi	2nd cent. AD	—	De Grossi Mazzorin and Coppola, 2008	929/826	9.2	46.6	44.2
Rome—Arch of Constantine	2nd cent. AD	—	De Grossi Mazzorin, in progress	293/265	6	20.4	73.6
Rome—Colosseum (sewer east)	2nd–3rd cent. AD	—	Minniti, in progress	726/531	5.3	16.4	78.3
Rome—Aqua Marcia	2nd–3rd cent. AD	—	De Grossi Mazzorin, 1996b	86/79	–	25.3	74.7
Rome—Colosseum (aisle 68)	3rd cent. AD	—	Minniti, in progress	253/206	1.9	6.8	91.3
Rome—Colosseum (wedge 33)	3rd cent. AD	—	Delfino and Minniti, 2005	610/438	0.9	11.2	87.9
Rome—Cryptoporticum of Commodus	3rd cent. AD	—	Minniti, in progress	291/195	2.6	14.9	82.6

(Continued)

Table 9.1 Continued

Site	Chronology	Period	Reference	NISP/NISD	% cattle	% caprines	% pig
Rome—Cryptoporticum of Commodus	4th cent. AD	LA	Minniti, in progress	71/66	3	9.1	87.9
Settefinestre—ph. III	3rd–5th cent AD	LA	King et al., 1985	886/713	6.0	15.4	78.5
Rome—Colosseum (sewer east)	3rd–5th cent. AD	LA	Minniti, in progress	672/375	4.5	16.3	79.2
Rome—Colosseum (sewer east–west)	2nd–5th cent. AD	LA	Minniti, in progress	984/817	3.1	19	78
Rome—Colosseum (sewer west)	4th–5th cent AD	LA	Minniti, in progress	1,097/635	9	23.6	67.4
Rome—Colosseum (undergorund corridors)	5th–6th cent. AD	LA	Minniti, in progress	21/8	25	12.5	62.5
Rome—Meta Sudans (SU 3180)	5th–6th cent. AD	LA	De Grossi Mazzorin, 1995c	305/42	26.2	23.8	50
Rome—Meta Sudans (SU 3641)	5th–6th cent. AD	LA	De Grossi Mazzorin, 1995c	2,521/1,852	18.8	26.3	54.8
Rome—Baths of Trajan	6th cent. AD	LA	De Grossi Mazzorin et al., in progress	646/523	10.3	21.4	68.3

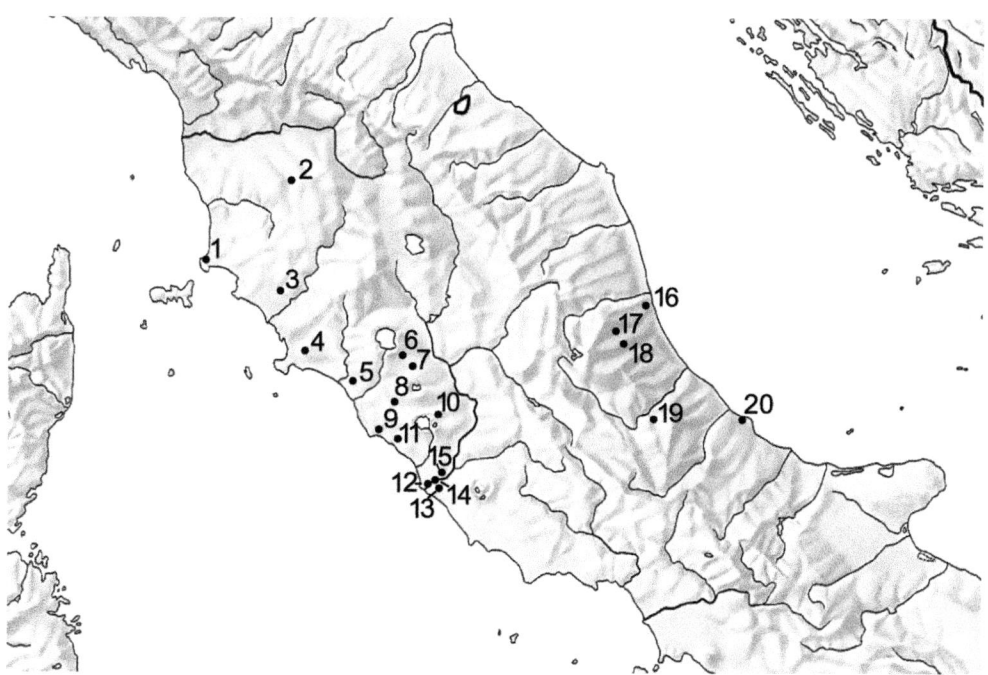

FIGURE 9.1 Location of the Italian sites mentioned in the text: 1. Populonia, 2. Campassini, 3. Roselle, 4. Settefinestre, 5. Tarquinia, 6. Ferento, 7. Acquarossa, 8. San Giovenale, 9. La Castellina, 10. Ad Vacanas, 11. Cerveteri, 12. Ficana, 13. Laurentina, 14. Roma, 15. Fidene, 16. Tortoreto, 17. Case Veldon, 18. Teramo, 19. Madonna degli Angeli, 20. Punta D'Erce. Authors' own image.

century BC) and the Archaic periods (sixth–fifth century BC) and slightly further during the Republican period (fourth–first half of first century BC). This differs considerably from other important contemporary settlements of the region, such as Ficana and Fidene, where pig remains do not increase so dramatically, probably due to Rome's greater population size and the process of transformation from a village to a city (De Grossi Mazzorin, 2001). Changing demographic conditions could cause pork to become an important part of the diet of urban people. The surge in the consumption of pork could be explained with an increased demand for meat, which is likely to be fulfilled with the supply of meat of the most prolific and exclusive meat producer domesticate.

In the following centuries of the Roman Empire, pork consumption increased still further. The increase (up to 78% of the domesticate assemblage) is observed in first century AD deposits in Rome, and is further enhanced, up to 84%, in the third and fourth century AD. The urban contexts of Via Sacchi, dated to the first and second century AD, seem to be an exception to this trend, as only 23% and 44% of pig remains are reported. The particular nature of the feature, a bone refuse midden accumulated by nearby workshops, may have been the determining factor for the relatively low percentages of pig remains, compared to the high percentage of cattle, whose bones are ideal for craft activities (De Grossi Mazzorin and Coppola, 2008).

Ancient texts testify that since the first century AD Roman inhabitants received free pork from the government during the games held at the Colosseum (Suetonius, *Domitian*, 4.12, 7.1; Delfino and Minniti, 2005). Pork should have been then the type of meat mostly eaten in Rome during the Roman period. Neighbouring areas and other distant provinces supplied the meat market of ancient Rome, although differently. In the Roman Republic and Imperial period, regions of northern Italy were the largest meat producers and exporters, whereas from the second century AD these were replaced by provinces of southern Italy (Belli Pasqua, 1995).

Only after the end of the Roman Empire does the frequency of pig remains decrease, accounting for less than 55% of the domestic species. This change corresponds with a period of crisis and partial dismantling of the urban buildings, according to a demographic decrease, which occurred in Rome in the second half of the fifth century after several barbarian sacks of the city (Santangeli Valenzani, 2012). The generous gifts of pork to the populace that had been established by Aurelian (270–275 AD) decreased greatly between the end of the fourth and the end of the fifth century AD, proceeding in parallel to the progressive depopulation of the city. According to Mazzarino (1951), *c.*320,000 people enjoyed free distribution of meat during the reign of Valentinian I (364–375 AD), but this fell to *c.*120,000 during the reign of Honorius (395–423 AD) and *c.*140,000 during the reign of Valentinian III (425–475 AD) (Fig. 9.2).

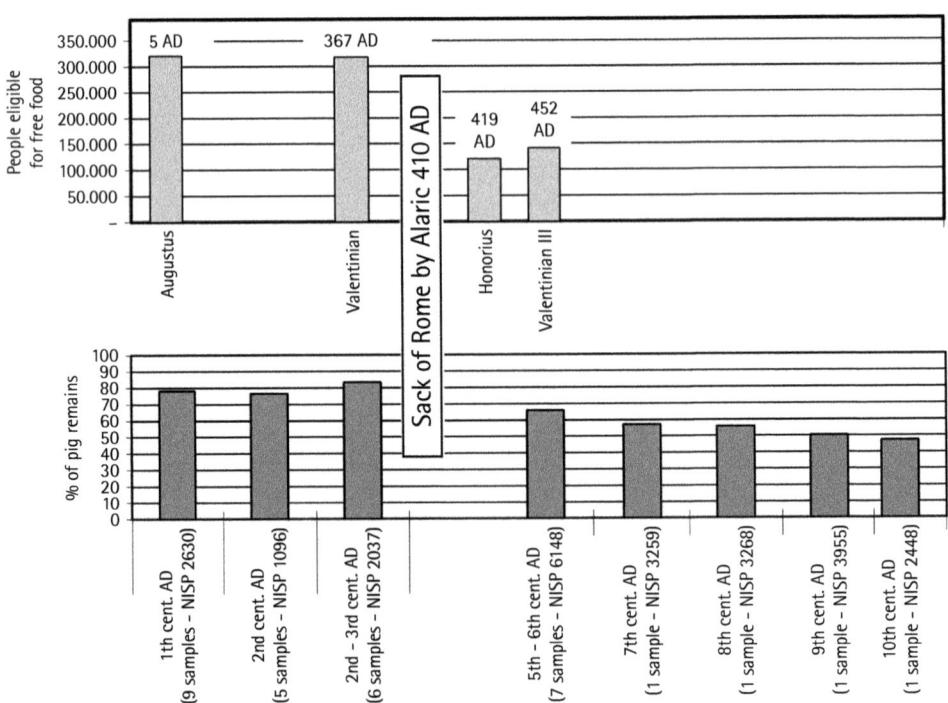

FIGURE 9.2 Percentages of pig remains from a number of urban contexts in Rome (below) compared to the number of people eligible for free grain (above, light grey: donations of grain; dark grey: donations of meat). Authors' own image.

Not all scholars agree with the demographic fluctuations mentioned above, though there is consensus regarding a considerable decrease in population throughout the fifth–sixth century AD and the following Medieval period. At the time of Theodoric (493–526 AD) Rome had no more than 200,000 inhabitants and this figure decreased further in the following centuries. The food supply (*annona*) to the town, gifts of pork included, definitely ceased in the Byzantine period.

Sheep, goats, and cattle are less well represented than pigs in all urban samples from Rome, but this mainly reflects consumption rather than production. Caprine mortality patterns show an unsurprising emphasis on the slaughter of immature animals, killed between the second and the third year of age, with very few remains of juvenile animals, suggesting a low consumption of lambs and goat kids, probably confined to the diet of the wealthy sector of the urban population. The low level of beef consumption by the Romans is consistent with the prevailing use of adult cattle as traction animals, as described by ancient sources (Aelian, *De natura animalium* 2.57; Pliny, *Historia naturalis* 8.180).

An important document (*De re coquinaria*, a collection of Roman cookery recipes, usually thought to have been compiled in the late fourth or early fifth century AD and ascribed to *Marcus Gavius Apicius*) provides further evidence of Roman diet and food taste and mirrors the evidence from the animal remains. In addition to the seventeen recipes exclusively devoted to the preparation of pork, the first ten chapters of *Liber VII* are dedicated to the preparation of particular body parts of the pig considered as delicacies (vulva, teats, liver, and ham). Only eleven and four recipes are instead dedicated to the cooking of beef and veal, respectively.

The chicken was probably introduced in Italy during the Early Iron Age, but chicken remains only become well documented in archaeozoological samples from much later chronological phases (third century BC onwards). Therefore, we can argue that, until then, the chicken was known but not widely distributed or a common component of the diet (De Grossi Mazzorin, 2005). Chicken remains are commonly found in Roman contexts, but not in such numbers to suggest that the bird represented a very important aspect of the Roman diet. A small increase in the occurrence of chicken remains is documented archaeologically from the sixth century AD onwards.

LIVESTOCK TYPES

Biometric data on pig remains are not particularly numerous, mainly due to the fact that most bones belong to immature individuals and cannot provide information on size; however, a limited dataset is still available. These data provide valuable information on domestic types and changes across the Iron Age / Roman transition and throughout the Roman period.

The analysis of individual measurements that are taken following von den Driesch (1976), particularly the width of the distal articulation of the tibia (Bd) and the greatest length of the astragalus (GLl) indicate that there was an increase in pig size from the Imperial period onwards (Table 9.2). The occurrence of two main size ranges in pig throughout the Roman period has been linked to different swine husbandries practised in Italy at that time, with smaller pigs seen as the product of sty keeping, and larger animals reflecting free-range husbandry (MacKinnon, 2001).

The increase in pig size is also apparent in the length and width dimensions, kept separate according to Davis (1996), when the log ratio technique of Simpson et al. (1960) was used. This method allowed us to use a larger biometric dataset combining different measurements and comparing them with a standard. The standards here used for all three domesticate log ratios follow Albarella et al. (2008). The Wilcoxon–Whitney–Mann test indicates that the increase in length is statistically significant between the Orientalizing-Archaic, the Republican period, and the Imperial period (Table 9.2; Fig. 9.3). For width, a statistically significant increase seems to occur later, during Late Antiquity (from the fourth century AD).

Measurements on caprine remains are more numerous than those for pig bones (Table 9.3). The individual measurements of the width of the distal articulation of the humerus (BT) and the tibia (Bd) have been considered. These data indicate an increase in size from the Republican period onwards, suggesting livestock improvement. The measurements of the astragalus (GLl and Bd) from two samples relating to two tombs respectively located near the Roman city of Populonia in Tuscany and at Poggio Picenze in Abruzzi, both dated to the fourth–third century BC (De Grossi Mazzorin and Minniti, 2013), also confirm that an increase in size likely occurred since the Republican period. The log ratio analysis also confirms that there was an increase in length and width measurements from the Republican period and further in Late Antiquity (Fig. 9.4). The increase in length measurements is statistically significant (Table 9.3).

Cattle bones provided no large body of biometric data. The most frequent measurements were those of the astragalus, which show an increase in cattle size during the Roman period (Table 9.4). Not enough data are currently available to verify if the increase occurred through time.

The effect of Roman culture therefore included an increase in the size of domestic species, but to various degrees. The observed changes in pig and caprine size suggest that Romans may have improved animals by introducing and crossbreeding different stocks, a practice encouraged by a wider range of available technologies and trades.

A similar process can be seen in other domestic animals, such as dogs. Changes in the morphology and size of dog bones from Roman contexts clearly reflect an emergence of different canine breeds and varieties (De Grossi Mazzorin and Tagliacozzo, 1997). Substantial morphological variability started occurring in the Iron Age, but became more evident during the Roman period. The process appears to be complete by the late Roman period, when dogs with strong differences in the size and general shape of the skull are well defined. Small dogs characterized by very short but slightly twisted, or brachymelic, limbs as well as distinct skull shapes also occurred in the Roman period. The spread of different dog breeds in Roman times is well documented by ancient sources (for instance see Columella, *De re rustica* 7.12; Oppian, *Cynegetica* 1.370–5; Aelian, *De natura animalium* 7.38).

Table 9.2 Pig: biometric data. Summary table of measurements for the sites in Tuscany, Abruzzi, and Latium mentioned in Table 9.1. Results of statistical test: N, not significant; * significant at the 95% confidence interval; ** significant at the 99% confidence interval; *** significant at the 99.9% confidence interval. Data and statistical test only refer to samples larger than 10; period II: Orientalizing/Archaic; period III: Republican; period IV: Imperial; period V: Late Antiquity

Period	Measurement		Summary						Statistical difference between periods			
			n	Min.	Max.	Mean	SD	V	II	III	IV	V
Orientalizing/Archaic	humerus	Bd	31	26.3	42.0	36.3	2.8	7.6		N	N	
	radius	Bp	32	23.5	36.0	27.9	2.2	8.0		N	N	
	III metacarpal	GL	12	59.2	73.0	67.0	3.3	4.9		N		
	IV metacarpal	GL	13	57.3	73.4	67.8	4.5	6.7		N		
	Log ratio lengths		41	-0.12	0.02	-0.042	0.03	-71.4		N	***	**
	Log ratio widths		66	-0.19	0.08	-0.040	0.04	-100.0		N	N	N
Republican	humerus	Bd	20	31.9	42.0	36.9	2.5	6.8			N	
	radius	Bp	17	24.0	29.9	27.2	1.6	5.9			N	
	III metacarpal	GL	13	61.0	75.0	69.0	4.04	5.9				
	IV metacarpal	GL	10	63.8	72.0	68.8	2.6	3.8				
	astragalus	GLI	21	31.7	42.5	36.8	2.9	7.7			***	
	Log ratio lengths		53	-0.11	0.02	-0.04	0.03	-76.3			***	**
	Log ratio widths		42	-0.11	0.03	-0.04	0.029	-72.5			N	*
Imperial	humerus	Bd	23	27.0	40.0	35.9	2.8	7.7				
	radius	Bp	37	24.0	31.0	27.6	1.8	6.6				
	tibia	Bd	30	25.0	32.5	28.5	1.9	6.7				***
	astragalus	GLI	44	34.4	45.4	39.5	2.5	6.3				
	Log ratio lengths		67	-0.09	0.05	-0.02	0.03	-172.2				N
	Log ratio widths		94	-0.18	0.07	-0.04	0.04	-97.2				*
Late Antiquity	tibia	Bd	10	25.0	27.0	26.2	0.7	2.8				
	astragalus	GLI	23	33.4	45.8	39.7	2.8	6.9				
	Log ratio lengths		40	-0.08	0.05	-0.02	0.03	-145.5				
	Log ratio widths		28	-0.09	0.047	-0.05	0.03	-63.0				

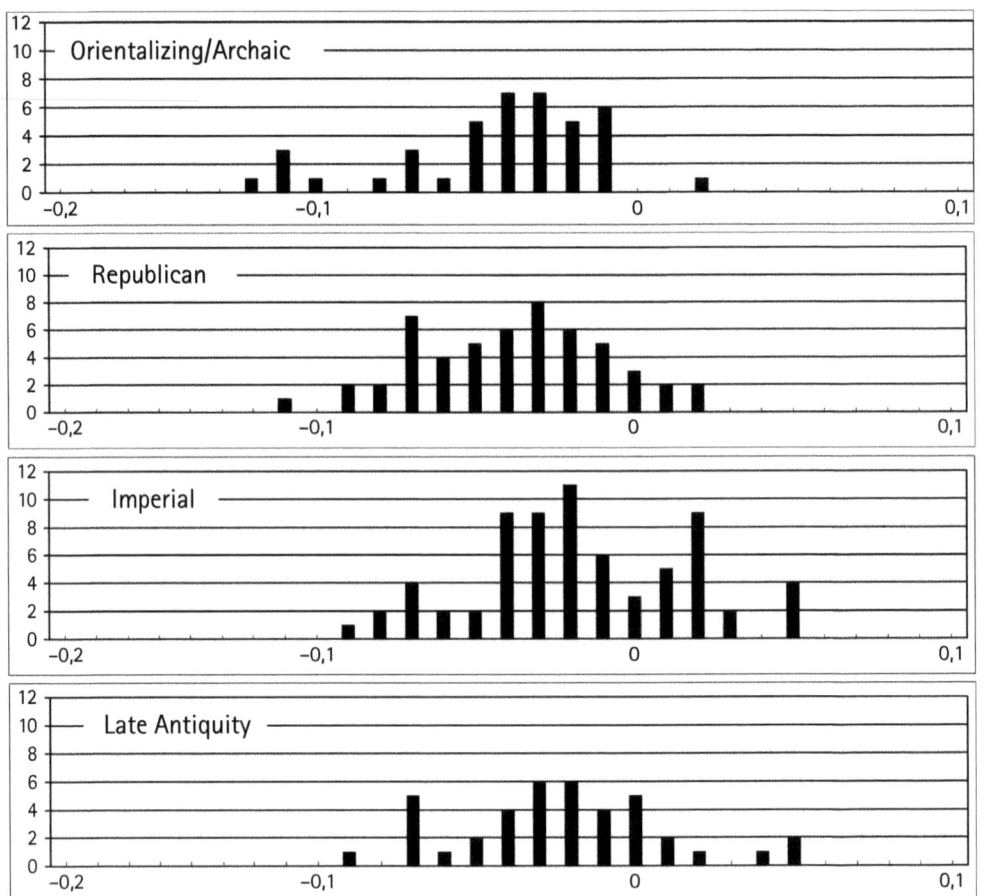

FIGURE 9.3 Pig: Log ratio diagrams for length measurements by period from a number of archaeological sites located in Tuscany, Abruzzi, and Latium. Authors' own image.

Roman times include the appearance of exotic (non-native) animals, even if most of them, including lion (*Panthera leo*), leopard (*Panthera pardus*), and ostrich (*Struthio camelus*), had mainly been imported for the exhibitions and the games held in the Colosseum, rather than being of dietary importance (De Grossi Mazzorin et al., 2005).

Ancient texts and epigraphic and iconographic documents, mostly dated between the second and the fifth century AD, inform us of the introduction of exotic animals by the Romans (Toynbee, 1973: 21–3, 74, 116; Sarà, 1995; MacKinnon, 2006). Since the beginning of the second century BC the taste for hunting game (*venatio*) spread to Italy as a consequence of Greek influence (Polybius, *Historiae* 31.29.22). The exhibitions held in the Colosseum constituted the highest expression of Roman passion for hunting, lasting about seven hundred years and reaching its peak in the first three centuries of the Roman Empire.

For their shows the Romans used both exotic animals and native species, such as the bear (*Ursus arctos*), which was widely used in the exhibitions of the Colosseum held

Table 9.3 Caprines: biometric data. Summary table of measurements for the sites in Tuscany, Abruzzi, and Latium mentioned in Table 9.1. Results of statistical test: N, not significant; * significant at the 95% confidence interval; ** significant at the 99% confidence interval; *** significant at the 99.9% confidence interval. Data and statistical test only refer to samples larger than 10

Period	Measurement		Summary						Statistical difference between periods			
			n	Min.	Max.	Mean	SD	V	II	III	IV	V
Early Iron Age	Log ratio widths		14	−0.04	0.06	0.006	0.038	633.3	N	***	**	N
Orientalizing/Archaic	humerus	BT	25	23.0	31.0	26.6	2.2	8.1		**	*	
	radius	Bp	22	25.5	33.3	29.8	2.1	7.1		**	**	
	metacarpal	GL	10	107.5	128.1	118.2	6.8	5.8				
	metacarpal	Bp	23	18.0	28.5	22.2	2.1	9.5		***		
	metacarpal	Bd	21	21.2	27.5	24.2	1.6	6.6		***	***	
	tibia	Bd	50	22.4	31.4	25.6	1.8	7.0		***	***	***
	astragalus	GLl	16	24.8	32.0	27.9	2	7.2		N		***
	astragalus	Bd	15	15.8	19.5	18	1.2	6.7		N	***	***
	Log ratio lengths		46	−0.03	0.08	0.022	0.028	127.3		N	***	N
	Log ratio widths		164	−0.07	0.13	0.005	0.033	660.0		N	N	N
Republican	humerus	BT	16	26.0	33.0	28.6	1.8	6.3				
	radius	Bp	22	26.8	35.0	31.4	2.2	7.0			N	
	metacarpal	Bp	23	19.5	27.0	24	2.3	9.6			N	
	metacarpal	Bd	23	22.0	32.0	26.5	2.7	10.1				
	tibia	Bd	57	21.2	30.5	26.8	1.8	6.6			N	
	metatarsal	Bd	18	21.0	28.0	25.2	1.9	7.7			N	
	calcaneum	GL	19	53.0	66.0	59.3	3.3	5.6			N	
	astragalus	GLl	13	26.5	34.1	28.9	2	6.9				**
	astragalus	Bd	13	16.5	20.8	18.6	1.1	6.0				*
	Log ratio lengths		43	−0.01	0.11	0.05	0.029	60.4			N	**
	Log ratio widths		172	−0.04	0.13	0.035	0.036	102.9			N	**

(Continued)

Table 9.3 Continued

Period	Measurement	Summary						Statistical difference between periods			
		n	Min.	Max.	Mean	SD	V	II	III	IV	V
Imperial											
	humerus BT	10	27.0	30.7	28.3	1.0	3.5				
	radius Bp	13	29.4	38.8	32.3	2.45	7.6				
	metacarpal Bd	14	23.9	31.0	27	2.2	8.1				
	tibia Bd	13	23.5	32.1	27.6	2.1	7.6				
	metatarsal Bd	13	22.5	28.0	25	1.53	6.1				
	Log ratio lengths	24	−0.02	0.11	0.047	0.029	61.7				**
	Log ratio widths	78	−0.03	0.12	0.042	0.032	76.2				N
Late Antiquity											
	astragalus GLI	15	27.5	35.3	31.2	2.4	7.7				
	astragalus Bd	15	17.5	23.3	20	2.8	14				
	Log ratio lengths	22	0.01	0.12	0.07	0.03	42.9				
	Log ratio widths	35	−0.02	0.12	0.053	0.03	56.6				

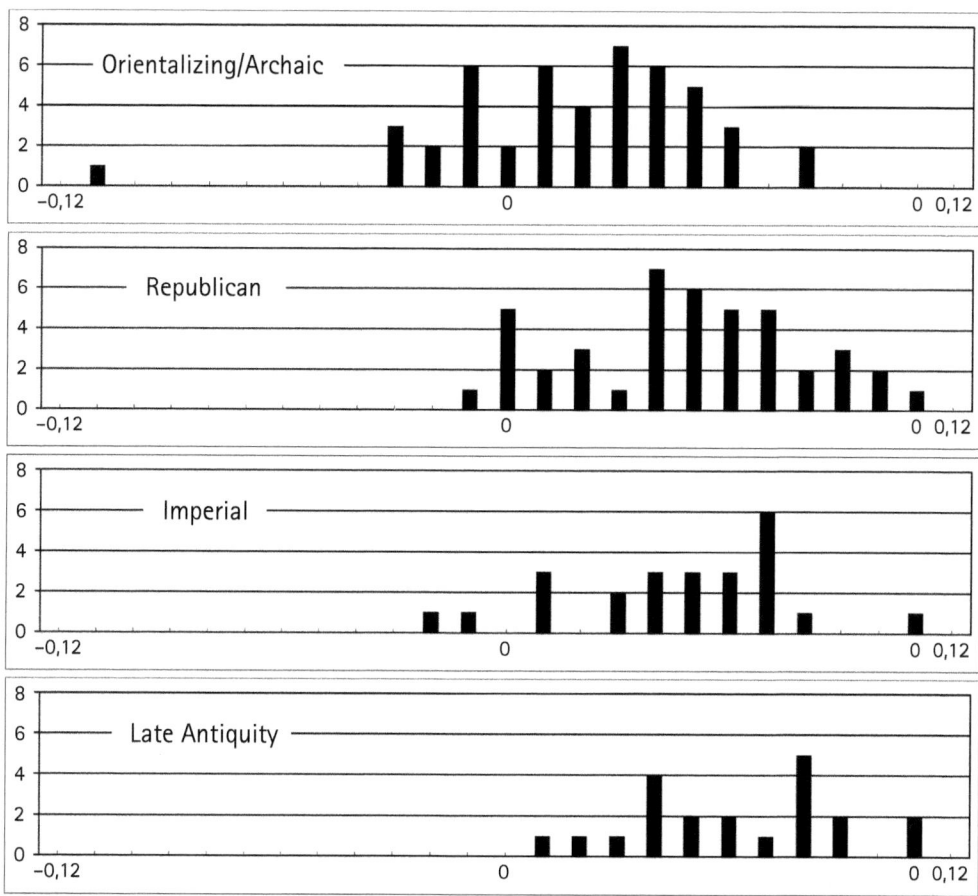

FIGURE 9.4 Sheep/goat: Log ratio diagrams for length measurements by period from a number of archaeological sites located in Tuscany, Abruzzi, and Latium. Authors' own image.

76; 7.122; Ovid, *Fasti* 5.371), were imported from northern Africa but also Asia. A small number of bones belonging to these species were recovered from areas of the Colosseum (aisle 68, east and west sewers and the area of Meta Sudans, located close to the amphitheatre), dating to between the second and the fifth century AD (De Grossi Mazzorin et al., 2005). The distal articulation of an ostrich tarsometatarsus also came from the area of Meta Sudans (De Grossi Mazzorin et al., 2005).

Other exotic species documented in urban contexts seem to be unrelated to theatre games or other forms of social entertainment. This is the case of the bactrian camel (*Camelus bactrianus*) and/or the Arabian camel (or dromedary) (*Camelus dromedarius*). A camel (*Camelus* sp.) first phalanx was recovered from the cemetery located close to the Via Latina (Tomb of Valerii) and dated to the second–third century AD, and a distal articulation of a metapodial came from the area of Templum Pacis, dated to the fifth–sixth century AD (De Grossi Mazzorin et al., 2005; De Grossi Mazzorin, 2010). These animals were occasionally used in Roman shows, but are more likely to have been kept

Table 9.4 Cattle: biometric data. Summary table of measurements for the sites in Tuscany, Abruzzi, and Latium mentioned in Table 9.1. Results of statistical test: N, not significant; * significant at the 95% confidence interval; ** significant at the 99% confidence interval; *** significant at the 99.9% confidence interval. Data and statistical test only refer to samples larger than 10

Period	Measurement		Summary						Statistical difference between periods			
			n	Min.	Max.	Mean	SD	V	II	III	IV	V
Orientalizing/Archaic	metacarpal	Bd	26	44.8	71.3	54.2	6.1	11.3				N
	metatarsal	Bd	21	42.4	60.8	50.5	5.1	10.1			N	N
	astragalus	GLl	17	54.0	69.1	60.6	4.3	7.1				
	astragalus	Bd	16	34.5	46.2	38.4	3	7.8				
	Log ratio lengths		25	−0.04	0.07	0.009	0.030	333.3		N	N	N
	Log ratio widths		77	−0.09	0.12	0.003	0.043	1433.3		N	N	N
Republican	Log ratio widths		12	−0.02	0.11	0.08	0.037	46.3			N	N
Imperial	metacarpal	Bd	19	55.2	72.2	64.2	4.7	7.3				
	metatarsal	Bd	19	53.5	71.0	60.8	4.5	7.4				
	Log ratio lengths		22	0.03	0.13	0.064	0.022	34.4				N
	Log ratio widths		45	0.00	0.15	0.079	0.033	41.8				N
Late Antiquity	Log ratio widths		20	−0.02	0.12	0.063	0.041	65.1				

as beasts of burden and for military use, and possibly also as meat and milk providers (Procopius, *Anecdota* 30.15–16; Pliny, *Historia naturalis* 11.237; 28.123).

Fragments of the pectoral fin of a Nilotic North African catfish (*Clarias* cf. *gariepinus*) have been recovered from the Cryptoporticus of Commodus, in the area of the Colosseum, and also at Tenuta di Vallerano, a rural settlement near Rome (De Grossi Mazzorin, 2000; Minniti, 2005). They are of particular interest as an exotic introduction from Africa or the northern region located between Syria and Turkey. Pliny the Elder refers to the Nilotic catfish as *silurus*. Several ancient authors mention the use of the catfish pectoral fin in medical treatments, for example to release points/tips stuck in the body as a consequence of an injury (Pliny, *Historia naturalis* 32.90, 93, 94, 104, 111, 119, 125, 131; Dioscorides, *De materia medica*, 2.27). This particular use may explain the recovery of catfish pectoral fins in Rome, as better tasting fishes than the catfish live in the Mediterranean Sea, and it is therefore unlikely that the introduction of this species was as food.

Another exotic species whose remains have been found in the Colosseum is the roach (*Rutilus pigus*). This fish lives only in the northern regions of Italy and may have been imported to Rome from those provinces probably preserved in salt.

CONCLUSION

The evidence presented in this paper shows that studies of animal remains from ancient Rome and the neighbouring geographic areas have provided important information on changes in diet and animal exploitation that occurred in the city. These are likely to be linked to a process of radical change of the city in political, cultural, and urban planning.

A change in the meat diet of the population in favour of pork occurred from the Late Iron Age onwards. The increase in pork consumption could have been in response to a rapid population growth that followed the transition from village to city. A size improvement of domestic species, although in variable degrees, occurred from the Republican period onwards. Several animal species were also moved outside the area of their natural distribution, as proven by the appearance of exotic animals in Rome. The increase in morphological variability of domestic species, such as dogs, in Roman times, is consistent with the word of the ancient sources that report that the Romans developed an intentional and selective form of breeding. In general, these phenomena reflect the increase in mobility of people and animals, which was the consequence of the gradual expansion in international trade.

REFERENCES

Albarella, U., Johnstone, C., and Vickers, K. (2008) 'The development of animal husbandry from the Late Iron Age to the end of the Roman period: a case study from south-east Britain', *Journal of Archaeological Science*, 35, 1828–48.

Alhaique, F. and Fortunato, M. T. (2010) 'Il campione faunistico del pozzo 593 dal sito di Ferento (Viterbo): tra alimentazione ed artigianato', in Tagliacozzo, A., Fiore, I., Marconi, S., and Tecchiati, U. (eds) *Atti del 5° convegno nazionale di archeozoologia*, pp. 261–4. Rovereto: Edizioni Osiride.

Belli Pasqua, R. (1995) 'Il rifornimento alimentare di carne a Roma nel I–V secolo d.C.', in Quilici, L. and Quilici Gigli, S. (eds) *Agricoltura e commerci nell'Italia antica*, pp. 257–72. Roma: L'Erma di Bretschneider Editore.

Cerilli, E. (2005) 'Consumi alimentari in una mansio romana: il caso della mansio ad Vacanas (Valle di Baccano, Capagnano di Roma, Lazio)', in Fiore, I., Malerba, G., and Chilardi, S. (eds) *Atti del 3° convegno nazionale di archeozoologia*. Studi di Paletnologia II, pp. 433–42. Roma: Istituto Poligrafico e Zecca dello Stato.

Davis, S. (1996) 'Measurements of a group of adult female Shetland sheep skeletons from a single flock: a baseline for zoo-archaeologists', *Journal of Archaeological Science*, 23, 593–612.

De Grossi Mazzorin, J. (1989) 'Testimonianze di allevamento e caccia nel Lazio antico tra l'VIII e il VII secolo a.C.', *Dialoghi d'Archeologia*, Serie III, 7(1), 125–42.

De Grossi Mazzorin, J. (2000) 'État de nos connaissances concernant le traitement et la consommation du poisson dans l'Antiquité, à la lumière de l'archéologie: l'exemple de Rome', *Mélanges de l'École Française de Rome*, 112, 155–67.

De Grossi Mazzorin, J. (2001) 'Archaeozoology and habitation models: from a subsistence to a productive economy in central Italy', in Brandt, J. R. and Karlsson, L. (eds) *From Huts to Houses: Transformations of Ancient Societies*. Acta Instituti Romani Regni Sueciae IV 56, pp. 323–30. Stockholm: Paul Åströms Förlag.

De Grossi Mazzorin, J. (2005) 'Introduzione e diffusione del pollame in Italia ed evoluzione delle sue forme di allevamento fino al Medioevo', in Fiore, I., Malerba, G., and Chilardi, S. (eds) *Atti del 3° convegno nazionale di archeozoologia*. Studi di Paletnologia II, pp. 351–64. Roma: Istituto Poligrafico e Zecca dello Stato.

De Grossi Mazzorin, J. (2010) 'Presenze di cammelli nell'Antichità in Italia e in Europa: aggiornamenti', in Volpe, G., Buglione, A., and De Venuto, G. (eds) *Vie degli animali, vie degli uomini: Transumanza e altri spostamenti di animali nell'Europa tardoantica e medievale*, pp. 91–106. Bari: Edipuglia.

De Grossi Mazzorin, J. (2014) 'Analisi faunistica dei resti osteologici provenienti dal pozzo B', in Delfino, A. (ed.) *Forum Iulium: l'Area del Foro di Cesare alla luce delle campagne di scavo 2005–2008*. BAR International Series 2607, pp. 83–7. Oxford: Archaeopress.

De Grossi Mazzorin, J. and Coppola, F. (2008) 'L'analisi dei resti faunistici nel quadro delle strategie di allevamento e alimentazione nella Roma imperiale', in Filippi, F. (ed.) *Horti et sordes*, pp. 410–19. Roma: Quasar Editore.

De Grossi Mazzorin, J. and Minniti, C. (2009a) 'Appendice', in Magagnini, A. and Van Kampen, I. (eds) I pozzi della Velia: la lettura di un contesto, in Rendeli, M. (ed.) *Ceramica, abitati, territori nella bassa valle del Tevere e Latium Vetus*. Collection de l'Ecole Francaise de Rome 425, pp. 85–91. Rome: École Française de Rome.

De Grossi Mazzorin, J. and Minniti, C. (2009b) 'L'utilizzazione degli animali nella documentazione archeozoologica a Roma e nel Lazio dalla preistoria recente all'età classica', in Drago Troccoli, L. (ed.) *Il Lazio dai Colli Albani ai Monti Lepini tra preistoria ed età moderna*, pp. 39–68. Roma: Edizioni Quasar.

De Grossi Mazzorin, J. and Minniti, C. (2013) 'Ancient use of the knuckle-bone for rituals and gaming piece', *Anthropozoologica*, 48(2), 371–80.

De Grossi Mazzorin, J., Minniti, C., and Rea, R. (2005) 'De ossibus in anphitheatro Flavio effossis: 110 anni dopo i rinvenimenti di Francesco Luzj', in Malerba, G. and Vicentini, P. (eds) Atti del 4° convegno nazionale di archeozoologia. Quaderni del Museo Archeologico del Friuli Occidentale 6, pp. 337–48. Pordenone: Comune di Pordenone, Museo Archaeologico.

De Grossi Mazzorin, J. and Tagliacozzo, A. (1997) 'Dog remains in Italy from the Neolithic to the Roman period', Anthropozoologica, 25–6, 429–40.

Delfino, A. and Minniti, C. (2005) 'Oggetti in osso, avorio e pasta vitrea dal Cuneo XXXIII dell'Anfiteatro Flavio. I resti ossei animali dal Cuneo XXXIII dell'Anfiteatro Flavio', Bullettino della Commissione Archeologica Comunale di Roma, 106, 287–93.

Driesch, von den, A. (1976) A Guide to the Measurement of Animal Bones from Archaeological Sites, Harvard: Peabody Museum.

Fiore, I. and Tagliacozzo, A. (2011) 'I resti ossei faunistici provenienti dal sito de La Castellina', in Grand-Aymerich, J. and Dominguez-Arranz, A. (eds) La Castellina a sud di Civitavecchia: origini ed eredità, pp. 1080–96. Roma: L'Erma di Bretschneider.

King, A. C., Rhodes, P. A., Rielley, K., and Thomas, K. D. (1985) 'I resti animali', in Carandini, A. and Ricci, A. (eds) Settefinestre, una villa schiavistica nell'Etruria Romana, Vol. 2, pp. 278–300. Modena: Panini Editore.

MacKinnon, M. (2001) 'High on the hog: linking zooarchaeological, literary, and artistic data for pig breeds in Roman Italy', American Journal of Archaeology, 105, 649–73.

MacKinnon, M. (2006) 'Supplying exotic animals for the Roman amphitheatre games: new reconstructions combining archaeological, ancient textual, historical and ethnographic data', Mouseion, Series III, 6, 137–61.

Mazzarino, S. (1951) Aspetti sociali del quarto secolo: ricerche di storia tardo-romana, Roma: L'Erma di Bretschneider.

Minniti, C. (2005) 'Analisi dei resti faunistici provenienti da tre pozzi (nn. 6, 7 e 11) della Tenuta di Vallerano (Roma, I–II secolo d.C.)', in Fiore, I., Malerba, G., and Chilardi, S. (eds) Atti del 3° convegno nazionale di archeozoologia. Studi di Paletnologia II, pp. 419–32. Roma: Istituto Poligrafico e Zecca dello Stato.

Minniti, C. (2012a) Ambiente, sussistenza e articolazione sociale nell'Italia centrale tra Bronzo medio e primo Ferro. BAR International Series 2394. Oxford: Archaeopress.

Minniti, C. (2012b) 'I resti ossei animali', in Delfino, A., de Luca, I., Minniti, C., Munzi, M., and Zampini, S. (eds) Lo scavo di una fornace metallurgica nella taberna XI del Foro di Cesare (con appendice di Andrea Pernella, Ulderico Santamaria, Fabio Morresi), in Ceci, M. (ed.) Contesti ceramici dai Fori Imperiali. BAR International Series 2455, pp. 118–21. Oxford: Archaeopress.

Minniti, C. (2014) 'I resti ossei dalla cisterna a tholos: i resti ossei animali dai livelli augustei della taberna XI', in Delfino, A. (ed.) Forum Iulium. L'area del Foro di Cesare alla luce delle campagne di scavo 2005-2008. Le fasi arcaica, repubblicana e cesariano-augustea. BAR International Series 2607, pp. 207–10. Oxford: Archaeopress.

Minniti, C., Valenzuela-Lamas, S., Evans, J., and Albarella, U. (2014) 'Widening the market: strontium isotope analysis on cattle teeth from Owslebury (Hampshire, UK) highlights changes in livestock supply between the Iron Age and the Roman period', Journal of Archaeological Science, 42C, 305–14.

Perrone, N. (2012) 'L'economia produttiva animale di 'Case Veldon' (Sant'Egidio alla Vibrata, TE) durante l'età del Ferro', in De Grossi Mazzorin, J., Saccà, D., and Tozzi, C. (eds) Atti del 6° convegno nazionale di archeologia, pp. 315–18. Lucca: AIAZ.

Santangeli Valenzani, R. (2012) 'I quartieri residenziali: deprezzamento, crisi e mutamenti proprietari delle domus aristocratiche', in Di Berardino, A., Pilara, G., and Spera, L. (eds) *Roma e il sacco del 410: realtà, interpretazione, mito. Atti della Giornata di studio (Roma, 6 Dicembre 2010)*. Studia Ephemeridis Augustinianum 131, pp. 219–28. Roma: Institutum Patristicum Augustinianum.

Sarà, M. (1995) 'Animali e ambiente negli apparati musivi', in Sposito, A. (ed.) *Natura e arteficio nell'iconografia ennese: architettura, arte e ambiente nelle fonti letterarie-artistiche dal sec. V a.C. al sec. VII*, pp. 83–9. Palermo: Arti Grafiche S. Pezzino & F. Di Pezzino Salvatore.

Simpson, G. G., Roe, A., and Lewontin, R. C. (1960) *Quantitative Zoology*, New York: Harcourt Brace.

Toynbee, J. M. C. (1973) *Animals in Roman Life and Art*, New York: Cornell University Press.

CHAPTER 10

ZOOARCHAEOLOGY OF THE SCANDINAVIAN SETTLEMENTS IN ICELAND AND GREENLAND

diverging pathways

KONRAD SMIAROWSKI, RAMONA HARRISON,
SETH BREWINGTON, MEGAN HICKS,
FRANK J. FEELEY, CÉLINE DUPONT-HÉBERT,
BRENDA PREHAL, GEORGE HAMBRECHT,
JAMES WOOLLETT, AND THOMAS H. MCGOVERN

COMMON ORIGINS, DIFFERENT ENDS

BOTH Iceland and Greenland were settled in the wave of sea-borne colonization that took European agricultural settlements to far offshore North Atlantic islands, reaching Iceland *c.* AD 875, Greenland *c.* AD 985, and briefly to Newfoundland by AD 1000 (Fig. 10.1). Both modern and ancient DNA analyses (Helgason et al., 2000a; 2000b; 2001) confirm the strong British Isles genetic component of Icelanders, and recent aDNA from Greenlandic cemeteries (Lynnerup and Nørby, 2004) provide confirmation of the traditional accounts of Greenlandic settlement from Iceland. The long-term fate of these communities presents a stark contrast: despite challenges of climate cooling, soil erosion, volcanic eruption, famine, smallpox, and plague, the Icelanders survived to become a fully developed twenty-first-century Scandinavian society. The Greenlanders were not as successful, and while the end of their community around AD 1450 has become a classic case of 'collapse' (Diamond, 2005), their dramatic fate remains an active subject for international, interdisciplinary research (Dugmore et al., 2012; 2013).

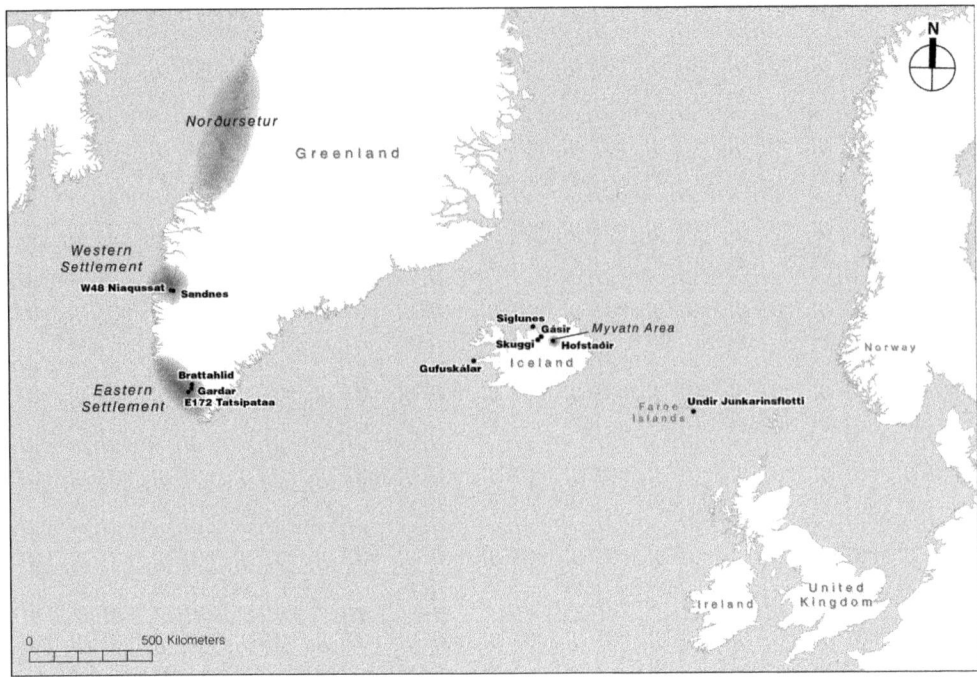

FIGURE 10.1 Map of the North Atlantic with regions and sites mentioned in the text. Authors' own image.

Zooarchaeology came early to Iceland and Greenland. The Danish Captain Daniel Bruun regularly collected unmodified animal bone remains from his very professionally conducted excavations in both islands just over a century ago (Bruun, 1895; 1896; 1899; 1903; 1917; Bruun and Jónsson, 1911), with pioneering zooarchaeological reports produced by Herluf Winge of the University of Copenhagen Zoological Museum (Winge in Bruun, 1895; 1917). Magnus Degerbøl and Ulrik Møhl continued the Zoological Museum tradition with a series of now-classic reports based on major Danish projects in Greenland and Iceland (Degerbøl, 1929; 1934; 1936; 1939). Post–World War II zooarchaeology in Greenland added radiocarbon-dated, stratified collections and sieved recovery (McGovern, 1985a; Buckland et al., 1996; McGovern et al., 1996; Enghoff, 2003). Since 1975 there has been an explosion of new zooarchaeological work in both Iceland and Greenland, as in the rest of the North Atlantic, much of which is now available through the North Atlantic Biocultural Organization (NABO) website (www.nabohome.org). During the 2007–11 International Polar Year and under the 2012–15 Comparative Island Ecodynamics Project, NABO research has focused on the diverging pathways and differing outcomes of 'long-term human ecodynamics' in Iceland and Greenland. This paper draws both upon this new work and upon a zooarchaeological research tradition extending over a century. The pace of research and large new archaeofauna still under analysis from both Iceland and Greenland will inevitably make this overview something of an interim report of work in progress, but the rich zooarchaeological record now in hand allows for some productive broad scale comparison.

Diverging Pathways at Landnám

The initial settlement (*Landnám*) of Iceland and Greenland was also a moment for initial branching of economic pathways. In Iceland, Norse settlement became widespread around the time of the volcanic ash fall datable to 871±2 AD with full-scale settlement spreading surprisingly rapidly to inland areas such as the well-studied Lake Mývatn basin up to 70 km from the coast (McGovern et al., 2007; Vésteinsson and McGovern, 2012). While the traditional written accounts (composed centuries after the event) emphasize chiefly land-taking in agricultural regions, there is both place name and zoo-archaeological evidence for the major role of wild species in the Viking Age economy (McGovern et al., 2006; 2009). Since 1944, investigations in the Aðalstraeti area under modern downtown Reykjavik have produced walrus (*Odobenus rosmarus*) bone and substantial amounts of sea-bird including great auk (*Pinguinus impennis*) and marine fish bone datable to first settlement (summary in Harrison et al., in press). Walrus fragments include both tusks and post-cranial bone (some from very young individuals), suggesting nearby kill sites and local breeding populations. A segment of walrus vertebra and a scapula embedded in the exterior turf wall of the early long hall at Aðalstraeti, as apparent trophies, and walrus place names down the nearby Reykjanes peninsula suggest walrus hunting may have been an initial motive for the Icelandic *Landnám* (Pierce, 2009; McGovern, 2011). As in Faroese Viking Age archaeofauna (Brewington, 2006; 2010b; 2011; 2014; Brewington and McGovern, 2008; Church et al., 2005), sea-bird bones initially outnumber those of imported sheep (*Ovis aries*), goat (*Capra hircus*), pig (*Sus domesticus*), cattle (*Bos taurus*), and horse (*Equus caballus*) in the early southern Icelandic archaeofauna, indicating that both sea mammals and bird colonies provided a source of natural capital that was drawn down to support the early phases of colonization in southern Iceland.

In the north of Iceland multiple projects have demonstrated early use of marine fish and marine mammals, especially from the cod family (Gadidae) on both coastal and inland sites (McGovern et al., 2006; Gísladóttir et al., 2013). Volcanic tephra horizons have allowed secure dating of multiple contemporary sites in the inland Lake Mývatn basin prior to 940 AD that are rich in headless gadid fish, with post-cranial element distributions suggesting large scale consumption both as flat-dried and round-dried ('stockfish') products, along with smaller numbers of marine mammal, bird, and molluscan remains. These well-dated inland Mývatn archaeofauna provided critical evidence that the widespread 'Fish Event Horizon' observed in British and Continental archaeofauna (Barrett et al., 2004) has a Scandinavian origin, and may well represent one of the most lasting heritages of the Viking Age in Europe (Perdikaris and McGovern, 2008a; 2008b). Recent work at the eroding coastal site of Siglunes, an early chieftain's farm at the mouth of Siglufjord in northern Iceland, has provided radiocarbon and tephra dates extending well into the Viking Age (Harrison, 2014b). It has produced large, stratified archaeofauna dominated by cod-family fish and demonstrating the relative surplus of cranial *vs* post-cranial bones characterizing later 'producer sites' (Krivogorskaya et al., 2005).

Domestic mammals in the Icelandic Viking Age archaeofauna are dominated by cattle and caprines (in most cases mainly sheep but with a significant proportion of goats), with substantial numbers of pigs in some collections. Ancient DNA analysis of house mice (*Mus musculus*) accidentally imported to Iceland show connections to populations in both Continental Europe and Norse Greenland (Jones et al., 2012).

Horse bones are comparatively rare in all archaeofauna, but in pre-Christian contexts it is clear that they were occasionally butchered and consumed on most sites. Both horses and dogs were regularly included in pagan burials, and the horse bones (unaffected by partly marine or freshwater fish diet, Ascough et al., 2010) are now regularly used for radiocarbon dating of the rapidly growing corpus of pre-Christian Icelandic graves (Friðriksson, 2013). Stable nitrogen (N), carbon (C), and strontium (Sr) isotope analyses in the Mývatn area have all documented a significant freshwater reservoir effect (FRE) in local arctic char (*Salvelinus alpinus*) and trout (*Salmo trutta*), and allowed for the identification of a special freshwater-fish consumption signature in a few of the Viking Age pigs (Ascough et al., 2007; Sayle et al., 2013).

Recent work on pre-Christian cemetery complexes in northern Iceland (Roberts and Hreiðarsdóttir, 2012) and at the temple farm complex at Hofstaðir, near Mývatn (Lucas, 2009), suggests that cats (*Felis domesticus*) as well as dogs (*Canis familiaris*) and horses (*Equus caballus*) may have played a role in rituals (Maher, 2009; Prehal, 2011). Cats are found in rare and unusual circumstances, such as at the pre-Christian grave field at Ingiríðarstaðir, where one was found in a pit amongst human skull fragments (Brewington, 2010a). The modern large-scale excavations at Hofstaðir conducted by the Archaeological Institute of Iceland and NABO, followed the initial work by Daniel Bruun in 1908 (Bruun and Jónsson, 1911). The project produced both a substantial archaeofauna, dated *c.*940–1000 AD and evidence for skinning cats for fur as well as a recurring ritual beheading of bulls, with their skulls displayed along the exterior of the great hall (Lucas and McGovern, 2008; Lucas, 2009; McGovern et al., 2009).

The Mývatn archaeofauna also document a millennial-scale case of successful, community-level management of migratory waterfowl, beginning at Landnám and continuing down to the present (McGovern et al., 2006; Hicks et al., 2013; 2015). The Mývatn lake basin annually hosts up to 30,000 pairs of migratory waterfowl coming from both sides of the Atlantic, and modern lakeside farmers regularly collect 10,000 eggs annually without adversely impacting these species (Guðmundsson, 1979). Modern farmers carefully monitor nesting birds, take only a few eggs per nest, and only rarely consume the adults, while protecting the nesting grounds against predators (Beck-Guðmundsdóttir, 2013). This pattern can be documented back to the mid-nineteenth century AD, and current archaeological excavations around Mývatn have generated archaeofauna rich in eggshells but with only a small number of waterfowl bones (McGovern et al., 2006). Ongoing collaborative work making use of modern comparative specimens and SEM imagery is combining wildlife management, ethnography, and zooarchaeology to both document this case of long-term traditional ecological knowledge and apply lessons learned to future management for long-term sustainability (Hicks et al., 2015).

The zooarchaeology of Viking Age Iceland is thus producing an increasingly rich record of a North Atlantic community similar in many respects to contemporary communities in northwestern Europe. Wild species supplemented domestic stock (and initially widespread barley cultivation; Trigg et al., 2008), animals played varied roles in pre-Christian rituals, and cases of both rapid draw-down and long term sustainable management of animals as natural capital can be documented dating back to the first years of settlement. By the time Iceland was Christianized, in c.1000 AD, it had become a well-populated island community integrating farming with hunting and fishing, and producing modest surpluses of wool and dried fish mainly for internal exchange. By the end of the eleventh century AD, Iceland's population had probably neared its pre-modern maximum of 50,000–60,000 and supported two bishoprics and many large estates.

Greenland was always different. When Icelandic settlers crossed the Denmark Strait around AD 985–1000 to found two communities on the western coast (Eastern Settlement in modern Kujalleq district; Western Settlement further north in Sermersooq district, near Nuuk) they crossed significant climatic and biological frontiers, though these may not have all been immediately apparent (Dugmore et al., 2013). In Greenland they encountered caribou (*Rangifer tarandus*), polar bear (*Ursus maritimus*), and huge populations of walrus as well as both familiar North Atlantic and unfamiliar Arctic seals and whales. Greenland was probably always beyond the reach of cereal agriculture, and the two pockets of farmland in the inner fjords of the southwest were isolated by thousands of kilometres of barren coast and the interior ice sheet.

Strontium (Sr) isotope calibration samples unexpectedly identified two early cattle in Greenland who had been born in Iceland (Price, in press). Initially, the full Icelandic range of cattle, sheep, goats, dogs, horses, and pigs appear in early Greenlandic collections (Smiarowski, 2012; 2013; 2014).

Recent comparative research has increasingly underlined the character of the Greenlandic settlements as always something of a specialized arctic resource extraction community, with export-orientated hunting for ivory and furs being supported by subsistence hunting and farming, rather than a farming community supplementing agriculture with subsistence hunting and fishing as in Iceland (Dugmore et al., 2007b; Keller, 2010). The historic concentration of walrus and walrus hunting has centred on Disko Bay on the central western coast, in an area the Norse called the 'Norðursetur' or northern hunting grounds (Gad, 1970; McGovern, 1985b). Written sources indicate that annual hunting trips were launched from both settlement areas to the Norðursetur, up to 800 km one way from the farming districts. The zooarchaeological evidence for this remarkable long-range hunt has been found in virtually every archaeofauna from the home farms in the form of fragments of walrus maxilla from around the tusk root, left behind by careful extraction of the ivory from the maxilla. These walrus maxillary fragments are found on inland as well as coastal farms in both settlement areas and throughout the stratigraphic sequences, indicating the active participation of most of the community in the Norðursetur hunt and ivory preparation (McGovern et al., 1996). Tusk ivory or finished ivory pieces are rare on the home farms (though walrus penis-bone trophies and post-canines used for craftwork are not unusual). There are no concentrations of walrus post-cranial elements as found in the Aðalstraeti deposits in Iceland, as the Greenlandic walrus kill sites were regularly hundreds of kilometres from the home farm processing area. This long-range Norðursetur walrus hunt thus seems to have been of a very different character and intensity from the sort of exploitation of nearby local walrus pods that we can now document from early Iceland. Processing of the furs and hides mentioned in written sources is harder to document through zooarchaeology, but new collections from the Greenlandic Bishop's manor at Gardar/Igaliku in the Eastern Settlement have produced multiple polar bear third phalanges with cut marks suggesting on-site final finishing of bear skins (Smiarowski, 2013; Frei et al., 2016).

On the Greenlandic home farms, shorter growing seasons and lower overall pasture productivity levels constrained stock production. Dairy cattle probably spent nearly nine months a year indoors being hand-fed fodder harvested in autumn (McGovern, 1992), and evidence of preserved dung concentrations suggests that at least some goats and sheep were also regularly stabled indoors in winter (Enghoff, 2003). The spectre of late winter shortfall in stored fodder and human provisions was a recurring threat to North Atlantic farmers (McGovern et al., 1988; Amorosi et al., 1998). Nevertheless, cattle were still maintained on all farms and there are no archaeofaunas indicating specialized caprine herding, even on the smallest farms with poor pastures. Among the caprines, goats were often more numerous than sheep in the Greenlandic archaeofauna from first settlement onwards (McGovern et al., 2014; Smiarowski, 2014).

Greenlandic settlers' encounter with the immense populations of migratory harp (*Pagophilus groenlandicus*) and hooded seals (*Cystophora cristata*) (rare or absent in Iceland and the eastern North Atlantic) had immediate and lasting impact on their subsistence economy. Current zooarchaeological evidence from both the Western Settlement (McGovern, 1985a) and the Eastern Settlement (Smiarowski et al., 2007; Smiarowski, 2012; 2013; 2014) dating to the early settlement period, indicates a rapid and radical shift in use of wild species by the original colonists. Marine fishing and dried fish production seems to have been immediately supplanted by large scale (probably communal) hunting of the newly encountered migratory seals, supplemented by sea-bird and caribou hunting. Seals were regularly taken throughout the North Atlantic from prehistoric times, but the harbour (*Phoca vitulina*) and grey seals (*Halichoerus grypus*) found in most of the eastern North Atlantic form comparatively small non-migratory pods and are very vulnerable to over-hunting. In Iceland, law codes regulated sealing beaches and harbour and grey seal populations seem to have generally been harvested sustainably at a fairly low level, with seal bones appearing as trace species in most archaeofauna in the Viking and early Middle Ages. In Greenland, harbour seal colonies were present, and the bones of this species appear regularly in Greenlandic archaeofauna (McGovern, 1985a; Ogilvie et al., 2009; Smiarowski, 2013), but they are greatly outnumbered by the bones of migratory harp seals (both settlements) and hooded seal (Eastern Settlement only). The bones of non-migratory arctic ringed (*Phoca hispida*) and bearded seals (*Erignathus barbatus*), which make breathing holes in winter ice but are not so readily taken with boat drives and other communal hunting strategies, are rare in Norse collections from Greenland. The Norse Greenlanders apparently adapted communal seal hunting techniques to the newly encountered migratory species and did not make extensive use of the sea ice sealing practices of either the Dorset or Thule peoples or of the Nordic ringed-seal hunters of the contemporary northern Baltic (Storå and Lõugas, 2005).

Seal bones vary from about 25% of major identified taxa to nearly 80% on small farms with limited pasture, and are common on far inland Greenlandic farms. While a few marine and freshwater fish bones have now been identified in Greenlandic archaeofauna, they represent a trace element (less than 1%, Smiarowski, 2013; 2014), far less than the staple represented by marine fish (25 to over 80%) in Viking and Early Medieval Icelandic archaeofauna. Seals appear to have replaced marine fish almost entirely in the Greenlandic subsistence strategy, and this seems to have happened in the very first years of settlement. There has been extensive debate about the cause of this clear pattern, which seems strongly counter-intuitive given both the role of marine fisheries in modern Greenland and the now well-documented late ninth-century AD Icelandic fishing record. Scheduling issues, rather than ritual prohibitions (Diamond, 2005), are likely at the core of this unexpected divergence. In Iceland, marine fishing was regularly practised in winter, and the air drying of stockfish requires prolonged temperatures hovering around the freezing point for curing. Winter was also the agricultural slack season, and in later time periods Icelandic farm hands were regularly put to sea in winter as fishermen. In Greenland, winter sea conditions, even during a warmer climate, are far more affected by sea ice, and winter temperature ranges, for most of even the southwest, tend to be too cold for effective stockfish curing. Greenlandic seagoing boats and labour were needed for most of the summer for the weeks-long voyages to the Norðursetur and the walrus hunt, which thus would compete directly with a summer fishing effort.

The migratory seals would arrive in the outer fjords of the two Greenlandic settlements in late May and early June; before the probable start of the Norðursetur voyages and during the worst of any recurring late-winter household provisioning gap. Despite centuries of hunting, harp seals still number in the millions, and unlike the non-migratory harbour seals they could support a large-scale harvest sustainably. The Greenlandic choice of intensifying migratory seal hunting and de-emphasizing marine fishing thus appears rational, given the environmental conditions and the scheduling limitations imposed by the long-range Norðursetur hunt. As in more recent cases, the demands of production for export may have limited the options for viable local subsistence strategies.

The Gardar bishopric was established in AD 1126, and current evidence suggests that this became by far the largest manor and elite centre in Greenland, with cattle byres capable of housing nearly a hundred cattle (in contrast to the 3–5 stalls usually encountered on smaller farms). This site also contains the largest concentration of caribou bone in the Eastern Settlement (Smiarowski, 2013), adding to the pattern of elite caribou consumption suggested for the Western Settlement (McGovern et al., 1996).

Comparative investigations of church and settlement patterns between Iceland and Greenland during the period c. AD 1000–1200 suggest a pattern of consolidation by higher-ranking elites who in both communities seem to have gathered power at the expense of middle-ranking chieftains and farmers (Arneborg et al., 2008). The Greenlandic settlements remained much smaller than the Icelandic, with maximum population probably well under 4,000 at peak.

By the thirteenth century AD both Greenland and Iceland were well-established Medieval communities, with ecclesiastical and secular hierarchies in place. Though by AD 1264 both were part of a trans-Atlantic Norwegian realm, they had become very different places, with a similar mix of imported northwestern European domestic live-stock masking major contrasts in the use of wild species and the role of surplus extrac-tion. While Icelandic fisheries and marine-mammal hunting seem to have been initially focused on supplying local subsistence demand and could be readily integrated into an annual agricultural cycle, the Greenlandic Norðursetur hunt generated inedible trans-Atlantic trade goods while creating significant conflict with the subsistence round (McGovern, 1985b).

HIGH MEDIEVAL COURSE CHANGES

In the mid-thirteenth to early fourteenth century AD a conjuncture of local, regional, and extra-regional social, economic, and environmental changes placed both stresses and opportunities before these two westernmost Scandinavian communities. Growing links between East Asia, the Mediterranean, and northern Europe during the *Pax Mongolica* of the mid-thirteenth to the mid-fourteenth century AD developed into a Medieval proto-world-system, with distant echoes in the Scandinavian North Atlantic (Abu-Lughod, 1981). In Iceland, this period saw the establishment of a number of seasonal trading centres distributed around the coastal fjords, with Gásir in Eyjafjord currently the best archaeologically documented (Hermannsdóttir, 1987; Roberts, 2002; Roberts et al., 2009; 2010; Harrison et al., 2004; 2008; Harrison, 2005; 2006; 2009; Pálsdóttir and Roberts, 2006; 2007; Vésteinsson et al., 2008; 2011 Vésteinsson, 2009; 2011). The inves-tigations at Gásir have developed into a multi-site investigation of the impact of this seasonal trading centre on a broader hinterland (Harrison, 2009; 2010a; 2010b; 2010c; 2011a; 2011b; 2013; 2014a). Among the findings of this ongoing research is that farms in this hinterland altered the traditional dairying economy in order to provision Gásir with prime-beef aged cattle. In return, some of these farms were consuming imported barley, and even had access to fashionable continental lap dogs; this area of rural Iceland was clearly connected to the larger world on multiple levels. While the Gásir excavations have confirmed documentary references to Medieval trade in Icelandic falcons and sul-fur, the major exports seem to have been woolen cloth and dried fish (Harrison et al., 2008; Harrison, 2014a, 2014b).

Coastal fishing sites increase in numbers and distribution in northern Iceland and the West Fjords after *c.* AD 1250, with the small seasonal site of Akurvík producing two large, fish-dominated archaeofauna, the first datable to the thirteenth century AD and the second to the fifteenth (Amundsen et al., 2005). These both show a clear 'producer signature' of surplus fish heads, as well as indications of the production of both flat-dried and round-dried cod and haddock in the thirteenth century AD, switching in the fifteenth century AD to a concentration on round-dried 'stockfish' cod. The nearby farm at Gjögur shows a dramatic increase in fish bone after *c.* AD 1250 and in later times was known as a major fishing farm (Krivogorskaya et al., 2005). By the late thirteenth century AD Icelandic magnate families were switching their core holdings from the main agricultural areas to the prime fishing regions, and it seems clear that the local-level artisanal subsistence fisheries of Viking Age Iceland were undergoing intensification for wider export (Vésteinsson, 2016).

In the thirteenth century AD some Icelandic archaeofaunas show a dramatic change in cattle to caprine bone ratios, shifting from the 1: 3 to 1: 5 ratios common in the Viking Age and Early Medieval periods to the 1: 20 ratios characteristic of eighteenth-century AD stock records. Goats become very rare, and the zooarchaeological data suggest higher proportions of older (and larger) sheep likely representing wethers or older ewes maintained for wool production (McGovern et al., 2007; Harrison, 2013). In Eyjafjord, the Gásir hinterlands were spatially re-organized, with small subsistence farms like Skuggi replaced by specialized sheep-herding structures on valley floors (Harrison, 2010a; 2013). Woolen cloth fragments show standardization into the legally defined *vaðmal*, suitable for exchange and valuation as a commodity (Hayeur-Smith, 2011). In Iceland by the mid-thirteenth century AD it appears that wool production and marine fishing were both being intensified, and that both woolen goods and dried-fish products were undergoing standardization and commoditization for a new export market as well as domestic consumption.

In Greenland, there is no indication of similar alterations in the relation of subsistence and surplus production for trade. The amount of walrus maxillary bone tusk-extraction debris remains constant or increases in the stratified Western and Eastern Settlement archaeofauna (McGovern et al., 1996; Smiarowski, 2013; 2014). Documentary records indicate that while hundreds of kilos of Greenlandic walrus ivory were still being collected by church factors in the mid-fourteenth century AD, this product was increasingly difficult to market profitably (Keller, 2010). Cattle-to-caprine ratios remain fairly stable on larger manor farms, and where caprine bones increase on smaller farms many of these are goats rather than sheep (Smiarowski 2014; McGovern et al., 2014). No evidence for standardization of woolen cloth production has yet been identified in the Greenlandic collections (Hayeur-Smith, 2014). While initially probably far more engaged in cash hunting for low-bulk, high-value exchange products in the Viking Age, the Greenlandic community proved less able than the Icelanders to shift to high-bulk, low-value commoditized trade in the thirteenth century AD.

Climate change as well as early globalization impacted both Medieval Iceland and Greenland. In AD 1257–1258 a massive volcanic eruption on Lombok (Indonesia) triggered an immediate cooling across the North Atlantic, and between 1275–1300 AD a threshold-crossing increase in summer sea ice impacted both northern Iceland and southwest Greenland (Miller et al., 2012). Pasture productivity in both communities was adversely affected, and the summer drift ice impacted trans-Atlantic voyages to Greenland, local travel, and the viability of harbour seal colonies in the Eastern Settlement area (Ogilvie et al., 2009). In Iceland, a *c.* AD 1300 archaeofauna from Hofstaðir exhibits both intensive bone processing for collagen extraction and the sudden appearance of substantial numbers of harp seal bones on this inland farm; both patterns indicating not only 'hard times' but also the resilient use of a newly available wild resource (McGovern et al., 2014). In Greenland, later archaeofauna show a marked intensification of the existing harp seal hunt, a pattern mirrored by the human stable isotope data that indicate Norse Greenlanders moving decisively into the marine food web after *c.* AD 1250 (Arneborg et al., 2012). The combined zooarchaeological and bioarchaeological record indicates that the Norse Greenlanders successfully survived the climate shocks of AD 1275–1300 by intensifying their existing communal seal hunting strategies to compensate for stress on the farming economy. Around AD 1425, a second climate shock impacted the whole region, with a dramatic increase in storminess (Dugmore et al., 2007a). The successful Greenlandic response to the initial climate impact may have rendered this small community tragically vulnerable to loss of life at sea in a radically stormier North Atlantic, and by around AD 1450 Norse Greenland was extinct.

While the Icelanders were impacted by both increased storminess and the appearance of the Black Death in AD 1402, their larger population and more effective combination of subsistence and exchange economies may have provided critical buffering (Streeter et al., 2012). European demand for stockfish continued to expand. Recent work on the Snæfellsnes peninsula has revealed nearly a kilometre of exposed dense fishbone midden 50–75 cm thick with radiocarbon dates indicating an accumulation within a few decades in the mid- to late fifteenth century AD (Pálsdóttir, 2011; 2013). While field and laboratory work at Gufuskálar continues, mammal bones suggest a pattern of provisioning with cuts of high quality lamb and beef more similar to the consumption patterns at the earlier Gásir trading site than the contemporary but much smaller Akurvík fishing station (Feeley, 2012; 2013). Finds of amber and pewter rosary beads, a bronze finger ring, fragments of chain mail armour, and a concentration of imported glazed red wares within a substantial stone structure suggests that this 'near industrial scale' fishing station may have been occupied and run by English or other Europeans as well as Icelanders. While Norse Greenland was slipping into final obscurity by the mid-fifteenth century AD, Iceland remained vital, with an economy that now appears more complex and diverse than previously thought (Boulhosa, 2010).

REFERENCES

Abu-Lughod, J. (1981) *Before European Hegemony: The World System* AD *1250–1350*, New York: Oxford University Press.

Amorosi, T., Buckland, P. C., Edwards, K., Mainland, I., McGovern, T. H., Sadler J., and Skidmore, P. (1998) 'They did not live by grass alone: the politics and palaeoecology of animal fodder in the North Atlantic region', *Environmental Archaeology*, 1, 41–55.

Amundsen, C., Perdikaris, S., McGovern, T. H., Krivogorskaya, Y., Brown, M., Smiarowski, K., Storm, S., Modugno, S., Frik, M., and Koczela, M. (2005) 'Fishing booths and fishing strategies in Medieval Iceland: an archaeofauna from the site of Akurvík, north-west Iceland', *Environmental Archaeology*, 10(2), 141–98.

Arneborg, J., Heinemeier, J., and Lynnerup, N. (eds) (2012) 'Greenland isotope project: diet in Norse Greenland AD 1000–AD 1450'. *Journal of the North Atlantic*, Special Volume 3.

Arneborg, J., Nyegaard, G., and Vésteinsson, O. (eds) (2008) 'Selected papers from the Hvalsey Conference 2008'. *Journal of the North Atlantic*, Special Volume 2.

Ascough, P. L., Cook, G. T., Church, A., Dugmore, A., McGovern, T. H., Dunbar, E., Einarsson, A., Friðriksson, A., and Gestsdóttir, H. (2007) 'Reservoirs and radiocarbon: 14C dating problems in Mývatnssveit, northern Iceland', *Radiocarbon*, 49(2), 1–15.

Ascough, P. L., Cook, G. T., Church, M. J., Dunbar, E., Einarsson, Á., McGovern, T. H., Dugmore, A. J., Perdikaris, S., Hastie, H., Friðriksson, A., and Gestsdóttir, H. (2010) 'Temporal and spatial variations in freshwater 14C reservoir effects: Lake Myvatn, northern Iceland', *Radiocarbon*, 86(3), 211–15.

Barrett, J., Locker, A. M., and Roberts, C. M. (2004) 'Dark Age economics revisited: the English fish bone evidence AD 600–1600', *Antiquity*, 78(301), 618–36.

Beck-Guðmundsdóttir, S. (2013) 'Exploitation of wild birds in Iceland from the Settlement period to the 19th century and its reflection in archaeology', *Archaeologia Islandica*, 10, 28–52.

Boulhosa, P. P. (2010) 'Of fish and ships in Medieval Iceland', in Imsen, S. (ed.) *The Norwegian Domination and the Norse World c.1100–c.1400*. Trondheim Studies in History 3, pp. 175–94. Trondheim: Tapir Academic Press.

Brewington, S. D. (2006) 'Interim Report on Archaeofauna from Undir Junkarinsfløtti, Sandoy, Faroe Islands'. NORSEC Zooarchaeology Laboratory Report No. 32.

Brewington, S. D. (2010a) 'Report: Analysis of Animal Bones Recovered during 2010 Excavations at Ingiríðarstaðir (ING), N Iceland'. NORSEC Zooarchaeology Laboratory Report No. 51.

Brewington, S. D. (2010b) 'Third Interim Report on Analysis of Archaeofauna from Undir Junkarinsfløtti, Sandoy, Faroe Islands'. NORSEC Zooarchaeology Laboratory Report No. 46.

Brewington, S. D. (2011) 'Fourth Interim Report on Analysis of Archaeofauna from Undir Junkarinsfløtti, Sandoy, Faroe Islands'. NORSEC Zooarchaeology Laboratory Report No. 56.

Brewington, S. D. (2014) 'The key role of wild resources in the Viking-Age to Late-Norse palaeo-economy of the Faroe Islands: the zooarchaeological evidence from Undir Junkarinsfløtti, Sandoy', in Kulyk, S., Tremain, C. G., and Sawyer, M. (eds) *Climates of Change: The Shifting Environments of Archaeology. Proceedings of the 44th Annual Chacmool Conference*, pp. 297–306. Calgary: Department of Archaeology, The University of Calgary.

Brewington, S. D. and McGovern, T. H. (2008) 'Plentiful puffins: zooarchaeological evidence for early seabird exploitation in the Faroe Islands', in Michelsen, H. and Paulsen, C. (eds) *Símunarbók: Heiðursrit til Símun V. Arge á 60 ára Degnum*, pp. 23–30. Torshavn: Faroe University Press.

Bruun, D. (1895) 'Arkaeologiske Undersøgelser i Julianehaabs Distrikt', *Meddelelser om Grønland*, 16, 171–462.

Bruun, D. (1896) 'Arkæologisk Undersøgelsesrejse til Færøerne og Island 1896', *Geografisk Tidskrift*, 13, 175–7.

Bruun, D. (1899) 'Arkæologiske Undersølgelser paa Island', *Geografiske Tidsskrift*, 15, 71–87.

Bruun, D. (1903) 'Arkæologiske Undersøgelser i Godthaabs og Frederikshaabs Distrikter i Grønland foretagne i Aaret 1903', *Geografisk Tidskrift*, 17, 187–206.

Bruun, D. (1917) 'Oversigt over Nordboruniner i Godthaabs og Frederikhaabs Distrikter', *Meddelelser om Grønland*, 56, 55–148.

Bruun, D. and Jónsson, F. (1911) 'Finds and excavations of Heathen Temples in Iceland', *Saga Book of the Viking Club*, 7VII, 25–37.

Buckland, P. C., Amorosi, T., Barlow, L. K., Dugmore, A. J., Mayewski, P. A., McGovern, T. H., Ogilvie, A. E. J., Sadler, J. P., and Skidmore, P. (1996) 'Bioarchaeological and climatological evidence for the fate of the Norse farmers in Medieval Greenland', *Antiquity*, 70(1), 88–96.

Church, M., Arge, S., Brewington, S., McGovern, T. H., Woollett, J., Perdikaris, S., Lawson, I. T., Cook, G. C., Amundsen, C., Harrison, R., and Krivogorskaya, Y. (2005) 'Puffins, pigs, cod and barley: palaeoeconomy at Undir Junkarinsfløtti, Sandoy, Faroe Islands', *Environmental Archaeology*, 10(2), 198–221.

Degerbøl, M. (1929) 'Animal bones from the Norse ruins at Gardar, Greenland', *Meddelelser om Grønland*, 76, 183–92.

Degerbøl, M. (1934) 'Animal bones from the Norse ruins at Brattahlið', *Meddelelser om Grønland*, 88, 149–55.

Degerbøl, M. (1936) 'Animal remains from the West Settlement in Greenland with special reference to livestock', *Meddelelser om Grønland*, 88(3), 1–54.

Degerbøl, M. (1939) 'Nogle bemærkninger om husdyrene pa Island i Middelalderen', in Stenberger, M. K. H. and Roussell, A. (eds) *Forntida Gårdar i Island: Meddelanden från den Nordiska Arkeologiska Undersökningen i Island Sommaren 1939*, pp. 261–8. Copenhagen: E. Munksgaard.

Diamond, J. (2005) *Collapse: How Societies Choose to Fail or Survive*, London: Allen Lane.

Dugmore, A. J., Borthwick, D. M., Church, M. J., Dawson, A., Edwards, K., Keller, C., Mayewski, P., McGovern, T. H., Mairs, K., and Sveinbjarnardóttir, G. (2007a) 'The role of climate in settlement and landscape change in the North Atlantic islands: an assessment of cumulative deviations in high-resolution proxy climate records', *Human Ecology*, 35, 169–78.

Dugmore, A. J., Keller, C., and McGovern, T. H. (2007b) 'Reflections on climate change, trade, and the contrasting fates of human settlements in the North Atlantic islands', *Arctic Anthropology*, 44(1), 12–37.

Dugmore, A. J., McGovern, T. H., Streeter, R., Madsen, C. K., Smiarowski, K., and Keller, C. (2013) ' "Clumsy solutions" and "elegant failures": lessons on climate change adaptation from the settlement of the North Atlantic islands, chapter 38', in Sygna, L., O'Brien, K., and Wolf, J. (eds) *A Changing Environment for Human Security: Transformative Approaches to Research, Policy and Action*, pp. 435–50. London: Routledge.

Dugmore, A. J., McGovern, T. H., Vésteinsson, O., Arneborg, J., Streeter, R., and Keller, C. (2012) 'Cultural adaptation, compounding vulnerabilities, and conjunctures in Norse Greenland', *Proceedings of the National Academy of Sciences of the United States of America*, 109(10), 3011–16.

Enghoff, I. B. (2003) 'Hunting, Fishing, and Animal Husbandry at the Farm Beneath the Sand, Western Greenland: An Archaeozoological Analysis of a Norse Farm in the Western Settlement'. Meddelelser om Grønland/Man and Society 28. Copenhagen: Danish Polar Center.

Feeley, F. J. (2012) 'Mammal Consumption at the Medieval Fishing Station at Gufuskálar'. NORSEC Zooarchaeology Laboratory Report No. 62.

Feeley, F. J. (2013) 'Medieval commercial fishing at Gufuskálar, Snæfellsnes, Western Iceland', Paper presented at the 2013 NABO Stefansson Arctic Institute Conference, Akureyri, Iceland, 12 July 2013.

Frei, Karin M., Coutu, Ashley N., Smiarowski, Konrad, Harrison, Ramona, Madsen, Christian K., Arneborg, Jette, Frei, Robert, Guðmundsson, Gardar, Sindbæk, Søren, M., Woollett, James, Hartman, Steven, Hicks, Megan, and McGovern, Thomas, H. (2015). 'Was it for walrus? Viking Age settlement and medieval walrus ivory trade in Iceland and Greenland', *World Archaeology*, DOI: 10.1080/00438243.2015.1025912.

Friðriksson, A. (2013) 'La place du mort: les tombes Vikings dans le Paysage Culturel Islandais'. Unpublished PhD dissertation, University of Paris-Sorbonne (Paris).

Gad, F. (1970) *The History of Greenland: Earliest Times to 1700*, Vol. 1, London: Hurst and Co.

Gísladóttir, G. A., Woollett, J. M., Ævarsson, U., Dupont-Hébert, C., Newton, A., and Vésteinsson, O. (2013) 'The Svalbard project', *Archaeologia Islandica*, 10, 69–103.

Guðmundsson, F. (1979) 'The past status and exploitation of the Mývatn waterfowl populations', *Oikos*, 32, 232.

Harrison, R. (2005) 'Faunal analysis results from the 2004 excavations at Gásir, Eyjafjörður, N Iceland', in Roberts, H. (ed.) *Excavations at Gásir 2004: An Interim Report*. CUNY Northern Science and Education Center; FSÍ (Fornleifastofnun Íslands/Icelandic Archaeological Institute), FS280-01076, June 2005, Reykjavík, Iceland.

Harrison, R. (2006) 'Faunal analysis results from the 2005 excavations at Gásir, Eyjafjörður, N Iceland', in Pálsdóttir, L. B. and Roberts, H. (eds) *Excavations at Gásir 2005: An Interim Report*. CUNY Northern Science and Education Center; FSÍ (Fornleifastofnun Íslands/ Icelandic Archaeological Institute), FS312-01078, May 2006, Reykjavík, Iceland.

Harrison, R. (2009) 'The Gásir Area A Archaeofauna: An Update of the Results from the Faunal Analysis of the High Medieval Trading Site in Eyjafjörður, N Iceland'. NORSEC Zooarchaeology Laboratory Report No. 44.

Harrison, R. (2010a) 'Small holder farming in Early Medieval Iceland: Skuggi in Hörgárdalur', *Archaeologia Islandica*, 8, 51–76.

Harrison, R. (2010b) 'Skuggi in Hörgárdalur, N Iceland: Preliminary Report of the 2008/2009 Archaeofauna'. NORSEC Zooarchaeology Laboratory Report No. 50.

Harrison, R. (2010c) 'Gásir Hinterlands Project 2009: Midden Prospection and Excavation'. FSÍ (Fornleifastofnun Íslands/Icelandic Archaeological Institute), Reykjavík and NORSEC, New York, FS440-06384, February 2010.

Harrison, R. (2011a) 'Myrkárdalur in Hörgárdalur, N Iceland: Brief Summary of the 2008/2009 Archaeofauna'. NORSEC/HERC Zooarchaeology Laboratory Report No. 57.

Harrison, R. (2011b) 'Möðruvellir in Hörgárdalur, N Iceland: General Overview of the Faunal Remains Analyzed from the 2006–08 Midden Mound Excavations'. NORSEC/HERC Zooarchaeology Laboratory Report No. 59.

Harrison, R. (2013) 'World Systems and Human Ecodynamics in Medieval Eyjafjörður, North Iceland: Gásir and Its Hinterlands'. Unpublished PhD dissertation, City University of New York (New York).

Harrison, R. (2014a) 'Connecting the land to the sea at Gásir: international exchange and long-term Eyjafjörður ecodynamics in Medieval Iceland', in Harrison, R. and Maher, R. (eds) *Human Ecodynamics in the North Atlantic: A Collaborative Model of Humans and Nature through Space and Time*, pp. 117–36. Lanham: Lexington Publishers.

Harrison, R. (2014b). 'The Siglunes 2011/12 Archaeofauna. Interim Report on the Fishing Station's Sampled Faunal Remains'. NORSEC/HERC Zooarchaeology Laboratory Report No. 62.

Harrison, R., Brewington, S., Woollett, J., and McGovern, T. H. (2004) 'Interim Report of Animal Bones from the 2003 Excavations at Gásir, Eyjafjörður, N Iceland'. NORSEC Zooarchaeology Laboratory Report No. 16.

Harrison, R., Roberts, H. M., and Adderley, W. P. (2008) 'Gásir in Eyjafjörður: international exchange and local economy in Medieval Iceland', *Journal of the North Atlantic*, 1(1), 99–119.

Harrison, R., McGovern, T. H., and Tinsley, C. (in press) 'The Zooarchaeology of Aðalstræti 14–18: Revised Report on the Aðalstræti Viking Age Archaeofauna', in Vésteinsson, O. (ed.) *Excavations at Aðalstræti Reykjavik Iceland*. Reykjavik: City Museum of Reykjavik.

Hayeur Smith, M. (2011) 'Preliminary Textile Report: Möðruvellir, Iceland, 2011'. Research Report of the Circumpolar Laboratoy No. 3, Haffenreffer Museum of Anthropology, Brown University.

Hayeur Smith, M. (2014) 'Dress, cloth, and the farmer's wife: textiles from Ø 172 Tatsipataa, Greenland, with comparative data from Iceland', in Arneborg, J., McGovern, T. H., and Nyegaard, G. (eds) *In the Footsteps of Vebæk: Vatnahverfi Studies 2005–2011*, pp. 64–81. *Journal of the North Atlantic*, Special Volume 6.

Helgason, A., Hickey, E., Goodacre, S., Bosnes, V., Stefansson, K., Ward, R., and Sykes, B. (2001) 'mtDNA and the islands of the North Atlantic: estimating the proportions of Norse and Gaelic ancestry', *American Journal of Human Genetics*, 68, 723–37.

Helgason, A., Siguroardottir, S., Gulcher, J. R., Ward, R., and Stefansson, K. (2000a) 'mtDNA and the origin of the Icelanders: deciphering signals of recent population history', *American Journal of Human Genetics*, 66, 999–1016.

Helgason, A., Siguroardottir, S., Nicholson, J., Sykes, B., Hill, E. W., Bradley, D. G., Bosnes, V., Gulcher, J. R., Ward, R., and Stefansson, K. (2000b) 'Estimating Scandinavian and Gaelic ancestry in the male settlers of Iceland', *American Journal of Human Genetics*, 67, 697–717.

Hermannsdóttir, M. (1987) 'Fornleifarannsóknir að Gásum og víðar í Eyjafirði árið 1986', *Tímaritið Súlur*, 1987, 3–39.

Hicks, M., Edwald, A., Einarsson, A., Anamthatwat Jónsson, K., Þór Þórsson, Æ., Friðriksson, A., Hambrecht, G. and McGovern, T. H. (2015) 'Local sustainable management on the millennial scale', in Isendahl, C., and Stump, D. (eds) *Oxford Handbook of Historical Ecology and Applied Archaeology*. Oxford: Oxford University Press.

Hicks, M., Einarsson, A., Anamthatwat Jónsson, K., and Þór Þórsson, Æ. (2013) 'A Preliminary Report of the Mývatn Bird Egg Archaeofaunal Identification Project'. NORSEC/HERC Zooarchaeology Laboratory Report No. 65.

Jones, E. P., Skinisson, K., McGovern, T. H., Gilbert, M. Y. P., Willerslev, E., and Searle, J. B. (2012) 'Fellow travelers: a concordance of colonization patterns between mice and men in the North Atlantic region', *BMC Evolutionary Biology*, 12, 35–43.

Keller, C. (2010) 'Furs, fish, and ivory: Medieval Norsemen at the Arctic fringe', *Journal of the North Atlantic*, 3, 1–23.

Krivogorskaya, Y., Perdikaris, S., and McGovern, T. H. (2005) 'Fish bones and fishermen: the potential of zooarchaeology in the Westfjords', *Archaeologia Islandica*, 4, 31–51.

Lucas, G. (ed.) (2009) *Hofstaðir: Excavations of a Viking Age Feasting Hall in North Eastern Iceland*, Reykjavik: Institute of Archaeology.

Lucas, G. and McGovern, T. H. (2008) 'Bloody slaughter: ritual decapitation and display at Viking age Hofstaðir N Iceland', *Journal of European Archaeology*, 10(1), 7–30.

Lynnerup, N. and Nørby, S. (2004) 'The Greenland Norse: bones, graves, computers, and DNA', *Polar Record*, 40(2), 107–11.

Maher, R. A. (2009) 'Landscapes of Life and Death: Social Dimensions of a Perceived Landscape in Viking Age Iceland'. Unpublished PhD dissertation, City University of New York (New York).

McGovern, T. H. (1985a) 'Contributions to the paleoeconomy of Norse Greenland', *Acta Archaeologica*, 54, 73–122.

McGovern, T. H. (1985b) 'The arctic frontier of Norse Greenland', in Green, S. and Perlman, S. (eds) *The Archaeology of Frontiers and Boundaries*, pp. 275–323. New York: Academic Press.

McGovern, T. H. (1992) 'Bones, buildings, and boundaries: palaeoeconomic approaches to Norse Greenland', in Morris, C. D. and Rackham, D. J. (eds) *Norse and Later Settlement and Subsistence in the North Atlantic*, pp. 193–230. Glasgow: Department of Archaeology, University of Glasgow.

McGovern, T. H. (2011) 'Walrus Tusks and Bone from Aðalstræti 14–18, Reykjavík, Iceland'. NORSEC Zooarchaeology Laboratory Report No. 55.

McGovern, T. H., Amorosi, T., Perdikaris, S., and Woollett, J. W. (1996) 'Zooarchaeology of Sandnes V51: economic change at a chieftain's farm in West Greenland', *Arctic Anthropology*, 33(2), 94–122.

McGovern, T. H., Bigelow, G. F., Amorosi, T., and Russell, D. (1988) 'Northern islands, human error, and environmental degradation: a preliminary model for social and ecological change in the Medieval North Atlantic', *Human Ecology*, 16(3), 45–105.

McGovern, T. H., Gestsdóttir, H., Brewington, S., Harrison, R., Hicks, M., Smiarowski, K., and Woollett, J. (in press) 'Medieval climate impact and human response: an archaeofauna circa 1300 AD from Hofstaðir in Mývatnssveit, N Iceland', *Journal of the North Atlantic*.

McGovern, T. H., Harrison, R., and Smiarowski, K. (2014) 'Sorting sheep and goats in Medieval Iceland and Greenland: local subsistence, climate change, or world system impacts?', in Harrison, R. and Maher, R. (eds) *Human Ecodynamics in the North Atlantic: A Collaborative Model of Humans and Nature through Space and Time*, pp. 153–76. Lanham: Lexington Publishers.

McGovern, T. H., Perdikaris, S., Einarsson, A., and Sidell, J. (2006) 'Coastal connections, local fishing, and sustainable egg harvesting: patterns of Viking age inland wild resource use in Mývatn District, northern Iceland', *Environmental Archaeology*, 11(1), 102–28.

McGovern, T. H., Perdikaris, S., Mainland, I., Ascough, P., Ewens, V., Einarsson, A., Sidell, J., Hambrecht, G., and Harrison, R. (2009) 'Chapter 4: the archaeofauna', in Lucas, G. (ed.) *Hofstaðir: Excavations of a Viking Age Feasting Hall in North-Eastern Iceland*. Institute of Archaeology Monograph Series 1, pp. 168–252. Reykjavik: Institute of Archaeology.

McGovern, T. H., Vésteinsson, O., Friðriksson, A., Church, M. J., Lawson, I. T., Simpson, I. A., Einarsson, A., Dugmore, I. A., Cook, A. J., Perdikaris, S., Edwards, K., Thomson, A. M., Adderley, P. W., Newton, A. J., Lucas, G., Edvardsson, R., Aldred, O., and Dunbar, E. (2007) 'Landscapes of settlement in northern Iceland: historical ecology of human impact and climate fluctuation on the millennial scale', *American Anthropologist*, 109(1), 27–51.

Miller, G. H., Geirsdóttir, Á., Zhong, Y., Larsen, D. J., Otto-Bliesner, B. L., Holland, M. M., Bailey, D. A., Refsnider, K. A., Lehman, S. J., Southon, J. R., Anderson, C., Björnsson, H., and Thordarson, T. (2012) 'Abrupt onset of the Little Ice Age triggered by volcanism and sustained by sea-ice/ocean feedbacks', *Geophysical Research Letters*, 39, DOI:10.1029/2011GL050168.

Ogilvie, A. E. J., Woollett, J. M., Smiarowski, K., Arneborg, J., Troelstra, S., Kuijpers, A., Pálsdóttir, A., and McGovern, T. H. (2009) 'Seals and sea ice in Medieval Greenland', *Journal of the North Atlantic*, 2, 60–80.

Pálsdóttir, L. B. (2011) 'Archaeological Investigations on the Fishing Station at Gufuskálar, Snæfellsnes: Preliminary Report'. FSÍ (Fornleifastofnun Íslands/Icelandic Archaeological Institute), FS407-08232.

Pálsdóttir, L. B. (2013) 'Archaeological Investigations on the Fishing Station at Gufuskálar, Snæfellsnes: Preliminary Report'. FSÍ (Fornleifastofnun Íslands/Icelandic Archaeological Institute), FS407-08233.

Pálsdóttir, L. B. and Roberts, H. M. (eds) (2006) 'Excavations at Gásir 2005 An Interim Report/ Framvinduskýrsla. Fornleifastofnun Islands FS312-01078'. Reykjavík: Fornleifastofnun Íslands.

Pálsdóttir, L. B. and Roberts, H. M. (eds) (2007) 'Excavations at Gásir 2006: an Interim Report/ Framvinduskýrsla. Fornleifastofnun Íslands FS355-010710'. Reykjavík: Fornleifastofnun Íslands.

Perdikaris, S. and McGovern, T. H. (2008a) 'Codfish and kings, seals and subsistence: Norse marine resource use in the North Atlantic', in Rick, T. and Erlandson, J. (eds) Human Impacts on Ancient Marine Ecosystems: A Global Perspective, pp. 157–90. Berkeley: University of California Press.

Perdikaris, S. and McGovern, T. H. (2008b) 'Viking Age economics and the origins of commercial cod fisheries in the North Atlantic', in Sickling, L. and Abreu-Ferreira, D. (eds) The North Atlantic Fisheries in the Middle Ages and Early Modern Period: Interdisciplinary Approaches in History, Archaeology, and Biology, pp. 61–90. Leiden: Brill Publishers.

Pierce, E. (2009) 'Walrus hunting and the ivory trade in early Iceland', Archaeologia Islandica, 7, 55–63.

Prehal, B. (2011) 'Freyja's Cats: Perspectives on Recent Viking Age Finds in Thegjandadalur, North Iceland'. Unpublished MA dissertation, Hunter College of the City University of New York (New York).

Price, T. D. (in press) 'Viking settlers of the North Atlantic: an isotopic approach', Journal of the North Atlantic.

Roberts, H. M. (2002) 'Archaeological Investigations at Gásir 2002: A Preliminary Report. Fornleifastofnun Norðurlands, FS180-01072'. Reykjavík: Fornleifastofnun Íslands.

Roberts, H. M. and Hreiðarsdóttir, E. O. (2012) 'The Litlu-Núpar burials', Archaeologia Islandica, 10, 1–40.

Roberts, H. M., Vésteinsson, O., Brorsson, T., Konráðsdóttir, H., Harrison, R., Ólafsson, S., Gílsadóttir, G. A., and Snæsdóttir, M. (2009) 'Gásir Post Excavation Reports: Volume 1. 2009, FS423-010712'. Reykjavík: Fornleifastofnun Íslands.

Roberts, H. M., Vésteinsson, O., Guðmundsdóttir Beck, S., Mould, Q., Konrádsdóttir, H., and Hansen, S. C. J. (2010) 'Gasir Post Excavation Reports: Volume 2. 2010, FS450-010713'. Reykjavík: Fornleifastofnun Íslands.

Sayle, K. L., Cook, G. T., Ascough, P. L., McGovern, T. H., and Hicks, M. (2013) 'Application of ^{34}S analysis for elucidating terrestrial, marine and freshwater ecosystems: evidence of animal movement/husbandry practices in an early Viking community around Lake Mývatn, Iceland', Geochimica et Cosmochimica Acta, 120, 531–44.

Smiarowski, K. (2012) 'E172 Tatsip Ataa Midden Excavation: 2009 & 2010 Preliminary Excavation Report'. NABO Field Report Series.

Smiarowski, K. (2013) 'Preliminary Report on the 2012 Archaeofauna from E47 Gardar in the Eastern Settlement, Greenland'. HERC-NORSEC Zooarchaeology Laboratory Report No. 61.

Smiarowski, K. (2014) 'Climate related farm-to-shieling transition at E74 Qorlortorsuaq in Norse Greenland', in Harrison, R. and Maher, R. (eds) Human Ecodynamics in the North Atlantic: A Collaborative Model of Humans and Nature through Space and Time, pp. 177–94. Lanham: Lexington Publishers.

Smiarowski, K., Pálsdóttir, A., and McGovern, T. H. (2007) 'Preliminary Assessment Report of the Archaeofauna from KNK 203 (E 74), a Norse Farm in the Eastern Settlement, Greenland'. NORSEC Zooarchaeology Laboratory Report No. 39.

Storå, J. and Lõugas, L. (2005) 'Human exploitation and history of seals in the Baltic during the Late Holocene', in Monks, G. G. (ed.) *The Exploitation and Cultural Importance of Sea Mammals*, pp. 95–106. Oxford: Oxbow Books.

Streeter, R. T., Dugmore, A. J., and Vésteinsson, O. (2012) 'Plague and landscape resilience in premodern Iceland', *Proceedings of the National Academy of Sciences of the United States of America*, 109(10), 3664–9.

Trigg, H. B., Bolender, D. J., Johnson, K. M., Patalano, M. D., and Steinberg, J. M. (2008) 'Note on barley found in dung in the lowest levels of the farm mound midden at Reynistaður, Skagafjörður, Iceland', *Archaeologia Islandica*, 7, 64–72.

Vésteinsson, O. (2009) 'A Medieval merchants' church in Gásir, North Iceland', *Hikuin*, 36, 159–70.

Vésteinsson, O. (ed.) (2011) 'Gásir Post Excavation Reports—Volume 3'. FSÍ (Fornleifastofnun Íslands/Icelandic Archaeological Institute), FS466-010714.

Vésteinsson, O. (2016) 'Commercial fishing and the political economy of Medieval Iceland', in Barrett, H. H. and Ortion, D. C. (eds.) *Cod and Herring: The Archaeology and History of Medieval Sea Fishing*, pp. 71–80. Oxford: Oxbow Books.

Vésteinsson, O., Gísladóttir, G. A., and Harrison, R. (2008) 'The Church in Gásir: Interim Report on Excavations in 2004 and 2006. Fornleifastofnun Íslands, FS385-010711'. Reykjavík: Fornleifastofnun Íslands.

Vésteinsson, O. and McGovern, T. H. (2012) 'The peopling of Iceland', *Norwegian Archaeological Review*, 45(2), 206–18.

Vésteinsson, O., Þorgeirsdóttir, S., and Roberts, H. M. (2011) 'Efniviður Íslandssögunnar: vitnisburður fornleifa um einokun og neyslu', in Vésteinsson, O., Lucas, G., Þórsdóttir, K., and Gylfadóttir, R. G. (eds) *Upp á Yfirbórðið: Nýjar Rannsóknir í Íslenskri Fornleifafræði*, pp. 71–93. Reykjavik: Fornleifastofnun Islands.

....

CHAPTER 11

FISHING, WILDFOWLING, AND MARINE MAMMAL EXPLOITATION IN NORTHERN SCOTLAND FROM PREHISTORY TO EARLY MODERN TIMES

....

DALE SERJEANTSON

INTRODUCTION

....

THE resources of the shore and the sea have made a contribution to the food supply of foragers and farmers living around the coast of Scotland since the end of the last Ice Age. In the Mesolithic period marine foods were an important part of the hunter-gatherer diet. From Neolithic times onwards people gave most of their energy and attention to growing crops and raising animals. The eating of fish and birds came to be avoided on inland sites, but marine foods continued to be exploited around the north of Scotland, especially in the islands, though in greatly reduced quantities. Bone preservation is good on many sites and sieving has routinely been used, which has been crucial for understanding the consumption of small fish and seabirds. Fig. 11.1 shows the location of sites referred to in the text and Table 11.1 sets out the time periods discussed.

In Scotland historical accounts compiled between the sixteenth and eighteenth centuries describe long-standing subsistence traditions. The most valuable accounts are by two ministers who visited the Hebrides and the Northern Isles respectively in the seventeenth century (Brand, 1701; Martin, 1716). The somewhat later *Statistical Account of Scotland* (Sinclair, 1791–1799) describes modes of agriculture and fishing that recent archaeological research has shown to go back for millennia (Serjeantson, 2013). Scotland

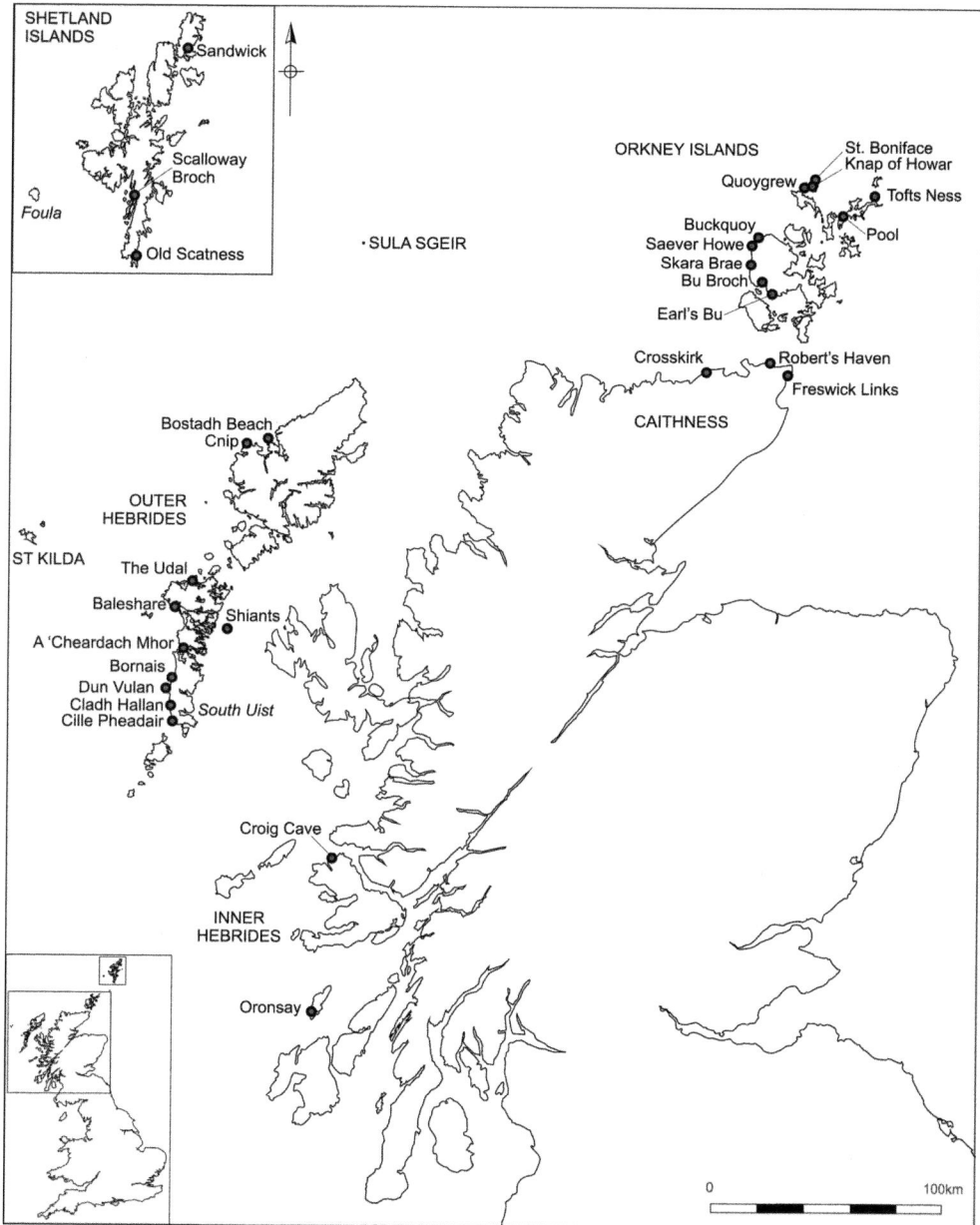

FIGURE 11.1 Map of Scotland showing sites referred to in the text. Illustration by P. Copeland.

also has a rich ethnographic tradition (Fenton, 1978; Baldwin, 2005; Ceron-Carrasco, 2005). Both sources of evidence help to interpret marine exploitation in the region.

This chapter focuses on fish, seabirds, and cetaceans, but other marine resources were also exploited. It is clear from both ethnographic sources and archaeological research that people collected crustaceans and shellfish for food in all periods and used shellfish

Table 11.1 Archaeological time periods (with abbreviations for Table 11.2) and dates for the north and west of Scotland

Period	Abbreviation	Approximate date range
Mesolithic	Meso	9th–5th millennium bc
Neolithic	Neo	late 5th–mid-3rd millennium bc
Bronze Age	BA	mid-3rd millennium–8th century bc
Early Iron Age	EIA	8th–2nd century bc
Middle Iron Age	MIA	1st century bc–4th century ad
Late Iron Age and Pictish	LIA	5th–9th century ad
Viking and Norse	Nor	10th–mid-15th century ad
Medieval	Med	mid-15th–17th century ad

as bait in early historic times. Some seaweeds were collected for food. Kelp (*Laminaria* spp.) that had been washed up on the shore was used as fodder for domestic animals and as manure on the cultivated fields. From at least the mid-1st millennium seaweed ash was used as lye for scouring wool as well as to preserve fish and seabirds.

FISHING AND FISH CONSUMPTION

The fisheries fall into three main types. The first focused on small saithe (*Pollachius virens*) caught close to the shore, the second on cod (*Gadus morhua*) and related large fish that were usually caught in deeper waters, and the third, in the Hebrides but not in the Northern Isles, was for herring (*Clupea harengus*) (Barrett et al., 1999; Ceron-Carrasco, 2005; Ingrem, 2005). Only a few sites do not fit this general pattern.

Inshore Fishing for Saithe

Immature saithe remains are found on nearly all prehistoric sites. As well as being one of the mainstays of subsistence at the Mesolithic site of Oronsay, they predominated at Tofts Ness in the Bronze Age and many Iron Age sites. At Cnip, thousands of saithe bones, all of small fish, were recovered in 1 mm mesh sieves, compared with fewer than one hundred for any other species. Most of the saithe there were below 30 cm in length. Other small pelagic fish were caught incidentally, including wrasse (Labridae) and whiting (*Merlangius merlangus*). Inshore fishing could be carried out by those living in small farming communities as it carried a low risk (Colley, 1983) and could be combined with farming and husbanding animals. The fishing techniques included the use of simple dip nets and hooks and line which were used from the shore or from boats close to the shore (Ceron-Carrasco, 2005).

The small saithe were caught for home consumption. In the Northern Isles, in the eighteenth to twentieth centuries, they were split and hung on a line above the hearth to be preserved by smoke from the fire (Fenton, 1978). They were also hung outdoors to dry in the wind and sun. In the Hebrides fish were preserved in the (salty) ash of burnt sea-weed (Martin, 1716). Excavation has revealed a Late Iron Age building at Old Scatness that is thought to have been used for this purpose (Bond et al., 2010). Saithe were valued for the oil from their livers as well as for food. Domestic middens containing a lot of ash as well as saithe bones suggest that extracting oil from saithe livers took place at least as early as the Iron Age (Nicholson, 1998). The fishery for small saithe continued even after the cod fishery began. At Quoygrew, for instance, there was a fishery for saithe 20–50 cm in length in the Norse period at the same time as offshore fishing was taking place (Harland and Barrett, 2013).

Deep-Water Fishery for Cod and Related Species

Offshore fishing for cod and other large fish took place from the ninth century AD onwards. The practice was brought by the Norse, having started in Norway in the Late Iron Age (Barrett et al., 2001). The focus on cod and saithe was striking: at Quoygrew cod and saithe make up more than 90% of all fish. In the Hebrides the concentration on cod and saithe was less pronounced because herrings were also caught. Large fish other than cod were also sometimes caught, including large saithe, ling (*Molva molva*), and hake (*Merluccius merluccius*) (especially in the Hebrides). At the site of Earl's Bu, haddock (*Melanogrammus aeglefinus*) was the most frequent species. This anomalous fishery is thought to be because the settlement was the seat of the Earl of Orkney who had access to restricted haddock fishing grounds (Barrett et al., 1999).

The cod caught were 60 cm or more in length and many reached lengths greater than 100 cm, the maximum size reached by cod. Even in the early Norse period, the cod at Pool were this size, as were those from the later sites of Quoygrew and Bornais. The ling were also large, 100 cm or more (Barrett et al., 1999: Fig. 4). These large fish normally live offshore at 100–400 metres depth. They were caught by men in small boats with four or six oars who used long lines with multiple hooks (Wheeler, 1983).

Cod fishing intensified from the eleventh century onwards. Several middens have been excavated in the Northern Isles and Caithness from this time which comprise almost exclusively fish remains together with ash and shells of shellfish that had been used as bait (Barrett et al., 1999; Harland and Barrett, 2013). Those at Freswick Links, Quoygrew, Robert's Haven, St Boniface, and Sandwick contained mainly cod with a few large saithe and ling. It was originally thought that the cod at Freswick Links were for domestic consumption but the scale of middens such as that at Robert's Haven make it much more likely that the fish were processed for export. From this time onwards the cod exported from the Northern Isles formed part of a trade that reached all of Medieval Europe. The trade coincided with, and must have been driven by, reforms in the Christian church which permitted the consumption of fish on days when the faithful were required to refrain from eating flesh meat (Barrett et al., 2011).

In Orkney from about the fourteenth century there was a trend towards catching more large saithe and fewer cod. Provisional findings from the Udal suggest that this was also the case in the Hebrides (Serjeantson, 2013: 71–9). The reasons have yet to be fully explored but may be associated with declining cod stocks, changes in the market for cod, or worsening weather which restricted offshore fishing.

Large-scale preservation of cod was done by air drying, sometimes combined with salting, and sometimes by smoking (Brand, 1701: 121). A smoke house for fish has tentatively been identified in a Norse period building at Bornais. One method of butchering large cod for preservation was to split the fish, cut off the head, and remove the central portion of the vertebral column (Fig. 11.2). The cleithrum, a post-cranial bone from the 'shoulder', was left on some large fish to keep them rigid. It was under-represented in some fish middens but over-represented at the consumption site of Earls Bu.

Remains of large cod and other gadid fish have been found on Neolithic as well as Norse sites, for instance at Tofts Ness and Knap of Howar. Their presence raises the question of whether the cod were obtained by offshore fishing at that time. Wheeler (1983) concluded that some of the cod were so large that they must have been caught at least two miles (3.2 km) out at sea. As deep-sea fishing involves greater risk than inshore fishing—according to Brand (1701: 128) it was "toilsome and dangerous"—it is thought that it would only have been undertaken if surplus production were required (Colley, 1983; Nicholson, 1998). This may indeed have been the case in the Neolithic period. However, another possibility is that before the days of intensive fishing, large fish occurred inshore more often than today (Wheeler, 1983; Nicholson, 1998).

There is very little evidence for herring bones on prehistoric sites, but from the Norse period onwards herring bones are found in notable quantities in several sites in the Hebrides. They were abundant at Bostadh Beach (Ceron-Carrasco, 2005), the Udal, and Bornais. The size and age class of the herring at Bornais show that adult herring were caught at the edge of the continental shelf at 200 m depth in what must have been an early example of drift netting (Ingrem, 2005: 157–8). The absence of herring from archaeological deposits in Orkney and Shetland is intriguing in view of their importance there a few centuries later. This may be because there was no market for herring from the Northern Isles at the time, whereas the Hebrides supplied Dublin and the boroughs of Lowland Scotland (Harland and Barrett, 2013).

Other Fisheries

Trout or small salmon (Salmonidae) were caught at some sites, including at Neolithic Skara Brae. They were also fairly common at Bornais in the Late Iron Age, though rarer than saithe. Very small pelagic fish were eaten in the Mesolithic (Ingrem, 2012) but when their remains are found on later sites they are usually interpreted as the stomach contents of large fish or as coming from otter spraint (Colley, 1983; Barrett et al., 1999).

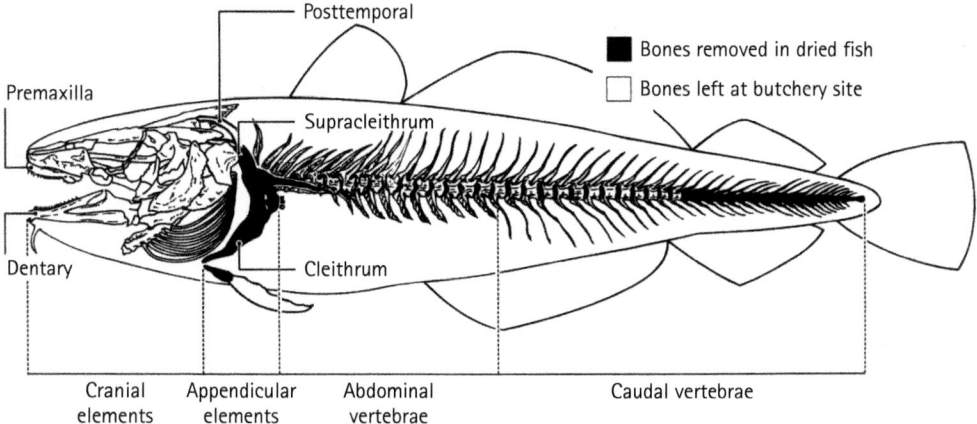

FIGURE 11.2 Skeleton of a fish of the cod family showing elements remaining on the fish and elements left at the butchery site in the production of stock-fish. Reproduced with permission, © McDonald Institute, Cambridge.

SEABIRDS, EGGS, AND WILDFOWLING

In the north and west of Scotland fowling for seabirds has taken place for as long as fishing. In the Mesolithic, people caught the now extinct great auk and other seabirds (McCormick and Buckland, 1997) and even today some people still harvest young gannets from the offshore island of Sula Sgeir each year (Murray, 2008). However, since the Neolithic, with the exception of the remote islands of St Kilda and Foula, birds have been a less important source of food than farm produce and fish. In a sample of eleven assemblages, bird bones made up only 1–9% of identified bones of birds and mammals together. It might be expected that the highest percentage of birds would have been from prehistoric settlements, but in fact it is from Norse and later Medieval sites (Serjeantson, 2014).

People also collected seabird eggs (Fenton, 1978). Eggshell fragments, until recently impossible to identify, have been found on many later prehistoric and early historic sites in the region. Research by John Stewart on eggshell from the Norse period has identified eggs of both wild birds and domestic chickens (Best, 2014).

Seabird fowling was focused on fewer than a dozen species (Table 11.2). They are all seabirds and, with the exception of gulls, all feed principally on fish. They range in size from the great auk, which weighed 5 kg, to the puffin, which weighs only 500 g. All breed in large colonies, often on offshore islands where they are free from disturbance and ground predators. The colonies of guillemots and razorbills, which usually nest together, are very difficult to access, as these birds breed on steep cliff ledges. Gannets and puffins breed on cliff tops, but on offshore islands that are often hard to reach. Unlike the other seabirds, which come to land only in the breeding season, shags and cormorants (the former

Table 11.2 Seabird species that have been the main target of wildfowling, with sites of origin of the assemblages and their time period

Species	Target at
Great auk (*Pinguinus impennis*)	Cnoc Coig (Meso); Knap of Howar (Neo); Dun Vulan (EIA); Baleshare (EIA); Cnip (MIA)
Large gulls (*Larus marinus; L. argentatus; L. fuscus*)	Tofts Ness (Neo); Bornais Mound 1 (LIA); Bornais Mound 1 (Nor); Bornais Mound 3 (Nor); Cille Pheadair (Nor)
Gannet (*Morus bassanus*)	Skara Brae (Neo); Clad Hallan (BA); Crosskirk (IA); Bu Broch (IA); Udal (LIA); Buckquoy (LIA); Saevar Howe (LIA); Udal (Nor); Buckquoy (Nor); Pool 8 (Nor); Udal (Med)
Puffin (*Fratercula arctica*)	Shiant (IA); Scalloway (LIA)
Shag and cormorant (*Phalacrocorax aristotelis; P. carbo*)	Dun Vulan (EIA); Crosskirk (IA); Bornais Md 1 (LIA); Pool 6 (LIA); Pool 7 (LIA/Nor); Quoygrew (Nor)
Guillemot and razorbill (*Uria aalge; Alca torda*)	Croig Cave (Meso); Knap of Howar (Neo); Dun Vulan (EIA); Quoygrew (Nor)

more common in Scotland than the latter) are present all year round. They breed on rocky shores and islands. The three large gulls breed on sand dunes and other flat areas, sometimes sharing sites.

While some of the methods used for harvesting seabirds and eggs from the breeding colonies were fairly simple, all involved great guile and some involved danger, strength and courage (Baldwin, 2005). Nests on dunes and cliff tops could be robbed of their eggs without risk, but the eggs of auks and other cliff-nesting birds could only be reached using ropes. The birds themselves—provided they could be approached—were clubbed to death or caught in snares or nets. Only the great auk, which was flightless, was easy prey.

The reasons why the principal species differed between sites and over time at the same site must have depended on the presence of a colony that was close by or that could be accessed conveniently. In historic times, local colonies were not necessarily owned by the local community but were claimed or controlled by the local earls or chiefs who exacted birds as tribute (Serjeantson, 2001).

The great auk was the most frequent bird species at many prehistoric sites and its remains have also been found at later sites. After the mid-1st millennium AD bones are only found occasionally, no doubt because predation had wiped out most of its breeding sites in Scotland (Serjeantson, 2001). Cormorants and shags were favoured at Iron Age sites, perhaps because they were the only birds that could be caught in winter. The gannet was the main species in many assemblages, especially in later periods. The puffin is the main species at a few sites only; a colony on the Shiants survived heavy predation when the islands were occupied in the Iron Age (Best, 2014). Gulls have been among the main species from all periods and from both Orkney and the Hebrides. The colony that was harvested from the South Uist sites persisted for centuries, so it must have been protected from ground predators, perhaps on an offshore island. Though guillemots and razorbills were the main species in only a few assemblages, their remains are found on most sites. They continued to be caught, probably because their cliff breeding colonies were resilient to a certain level of predation.

Though most communities focused on one or a few species, remains from a diverse range of other birds are usually present (Best, 2014; Serjeantson, 2014). In particular, auks (guillemot, razorbill, and puffin) are found on most sites. This diversity makes it unlikely that all were caught from breeding colonies. As all the main species except the gulls catch fish by diving, they would also have been caught using methods used for fishing, i.e. on fish hooks or in fishing nets, either accidentally or even deliberately (Serjeantson, 2014).

Like fish, seabirds were eaten both fresh and preserved. The methods of preservation were the same as those used for fish: salting, drying, and smoking. At Old Scatness charring on gull bones suggests that some were cooked over an open fire but some bird bones were also found in the building which is thought to have been used to smoke fish, showing that they were treated like fish (Bond et al., 2010). Seabirds were traded in historical times, but never on the scale of fish.

Though seabirds were only a minor source of food, they continued to be caught until the twentieth century in the north and west of Scotland (Baldwin, 2005). The Norse introduced domestic chickens, but even after that time seabirds were preferred at many sites, no doubt due to their high fat content and the oil that some yielded. At sites where the main target species changed over time (Table 11.2) it suggests that local colonies were wiped out over time, presumably by predation and disturbance from domestic and commensal animals. Once colonies were restricted to remote islands, they became accessible only to people in boats. Throughout prehistoric and indeed historic times seabirds filled the role of an occasional and fallback foodstuff that, like fish, could be stored and eaten when crops and dairy products were in short supply.

Marine Mammals

More than twenty species of cetaceans are regularly found in the seas around Scotland. The smaller whales come close to the shore but most of the larger ones remain in deep water. Seals are very common around Scotland and their remains are found on most sites. Bones of dolphin (Delphinidae) and porpoise (*Phocoena phocoena*) have also occasionally been found. The otter (*Lutra lutra*) is a marine species around the Scottish coast: its remains are present on Mesolithic and Neolithic sites, but are surprisingly rare later (McCormick and Buckland, 1997).

Identifying sea mammals, especially the whales, to species is difficult because the remains are fragmented, and because there are large size variations within each species. Even the two seal species found in Scotland, the grey seal (*Halichoerus grypus*) and the common seal (*Phoca vitulina*), cannot always be separated (Buckley et al., 2013). As a result, some seal bones and the great majority of pieces of other cetacean bone have not been identified to species.

The remains of marine mammals found up to 2002 were the subject of a comprehensive survey by Mulville (2002). Other research has extended the range of species and elements which could be identified and widened our understanding of cetacean exploitation (Szabo, 1997; 2008). The reports on finds published subsequently have not extended the range of species identified, but continue to enhance our understanding of whalebone as a raw material. Cetaceans provide not just meat and bone but also blubber for food and oil, skins, and, depending on species, baleen and ambergris (Szabo, 1997). The scarcity of wood in the north of Scotland from the Bronze Age onwards (Clark, 1952: 71; Edwards and Whittington, 1997) would have meant that whalebone was an important raw material for artefacts and building materials.

Consumption

The meat and blubber of whales and other cetaceans is very nutritious. Blubber, though today distasteful to most European palates, was valued as food for people living outdoor lives in high latitudes and an exposed environment. As sea mammals were butchered on the shore, it is difficult to establish from settlement finds how much food they would have provided. The presence of seal bone suggests that seals were eaten, but a bone from a whale, even of one from which the meat had been used, does not imply that the whole carcass was available. In historic times—and possibly earlier—stranded whales had to be shared by several communities, or between a community and its overlord (Fenton, 1978; Szabo, 2008).

The consumption of whale meat rather than just the collection of bones is suggested by two things. Vertebrae carry more meat than the ribs, so finds of these suggest that they were carried back for the attached meat. The flipper also carries meat and the bones are less useful for tools, so the presence of flipper bones also indicates consumption. Immature rather than mature cetaceans were hunted for food as they were easier to capture. Most of the bones from Pool in the Pictish period, which are from immature dolphins and whales, are thought to have this origin (Szabo, 2008: 175).

Raw Materials

Blubber and the cancellous tissue of whalebone yield a light oil, known as train oil. This was another source of oil for lighting. Nearly half the cetacean bones from Mound 3 at Bornais were burnt, suggesting that they had been processed to extract oil. The skins of seals and other small cetaceans were used for clothing and made into ropes. The skins of seal pups were particularly valued for the colour and softness, so remains that come from immature rather than adult seals are likely to be from animals killed for their skins.

As discussed, much whalebone from excavated sites comes from building materials and the manufacture of artefacts. At Scalloway Broch almost one third of the worked bone objects were made from cetacean bone (Sharples, 1998). Whalebone was used as a building material. Mulville's survey lists several examples. Two vertebrae from a large whale were used as sockets for door posts at the Iron Age site of A' Cheardach Mhor; the skull of a sperm whale (*Physeter macrocephalus*) was used as drain cover at Iron Age Dun Vulan; the humerus of a blue whale (*Balaenoptera musculus*) was used in the wall of a building at Viking Cille Pheadair and the skull of a whale was built into a wall at Norse Freswick Links. While some of these uses were purely functional, some are thought also to have have had a symbolic or ritual role (Mulville, 2002).

The bones most valuable for tools and artefacts were the mandible and rib, which are dense and flat. Some bladelike implements made from ribs may have been blubber mattocks and use-wear suggests that some were used as ards in the Iron Age (Mulville, 2002). From the Norse period onwards whalebone was used for implements such as flax beaters, scutching knives, and smoothers (Fig. 11.3), all used in linen manufacture. Whale vertebrae were used as chopping boards and were hollowed out for various uses. The unfused vertebral epiphyses of small cetaceans which have been found on some prehistoric sites are thought to have been used as pot lids (Mulville, 2002). Combs were sometimes made of whalebone as were many small tools including stakes and pegs, all common on sites of the 1st and early 2nd millennium AD.

Were Cetaceans Hunted or Stranded?

Historical as well as archaeological evidence makes it clear that seals were hunted. Hunting took place when the seals came ashore to have their pups, in June and July for common seals and between September and November for grey seals. Porpoises, the smallest of the cetaceans, and the smaller dolphins were probably incidental catches while fishing (Gardiner, 1997). Whether whales were hunted in prehistoric times as well as being exploited after stranding has been discussed for many years (Clark, 1952: 63–72). The earliest references to active whaling in northern Europe date from the ninth century AD (Clark, 1952; Gardiner, 1997; Szabo, 1997). There is little mention of hunting large whales but pods of small whales and dolphins that swam into bays and inlets were chased by boats and driven onto the shore.

Strandings occurred regularly if unpredictably (Fenton, 1978). The Norse *Saga of Grettir the Strong* describes a stranded whale—and the fight that took place to claim ownership of it (Clark, 1952: 65). It is likely that the remains of the large whales such as the blue whale at Cille Pheadair are from beached carcasses. Whales become stranded while still alive as well as after death, so meat and blubber as well as the bones could be scavenged from whales washed up on the shore.

FIGURE 11.3 Norse whalebone paddle or flax beater from Mound 3 at Bornais. Photograph by N. Sharples.

DISCUSSION

The resources of the sea were important in coastal Scotland and on the islands for two reasons: marine productivity was high at all trophic levels in the region so the resources of the sea are particularly rich. In addition, the climate and soils limited the potential for growing crops, which meant other food sources were needed. The climate, however, did constrain what could be harvested from the sea and when. The region is exceptionally stormy, with gales occurring not only in winter but at all times of year. Exploiting stranded whales, fishing close to the shore, and catching shags and cormorants was possible in bad weather, but deep-sea fishing and harvesting birds from offshore islands were possible only in calm weather. Changes in storminess in the region between 3,500 BC and AD 1500 are poorly understood, but they may help to explain variations in the exploitation of large cod, seabirds (especially gannets), and small cetaceans at different times. An additional factor, as discussed, was that regular exploitation quickly wiped out bird colonies, and eventually had an effect on fish stocks and the abundance of small cetaceans.

From the Neolithic until the middle of the 1st millennium AD exploitation of marine resources was a minor but constant part of subsistence, but from the ninth century AD there was a distinct change in the the Northern Isles to an economy that was based on marine exploitation combined with farming (Barrett, 1997; Nicholson, 1998; Barrett et al., 1999). The Norse introduced a tradition of offshore fishing together with the boats and techniques which allowed it to be carried out. The bone evidence for the increased consumption of marine foods in Orkney is confirmed by isotopes in human skeletons: whereas prehistoric skeletons show little or no effect of fish consumption, up to 30% of the diet in the Norse period was made up of fish (Barrett et al., 2001). In the Hebrides, a similar increase in fish consumption took place from the ninth century, with herring as well as cod exploited. There, however, fishing declined in the Late Middle Ages as a result of deteriorating weather, poor government, and the lack of markets (Serjeantson, 2013: 98–201). By contrast, from the beginning of the 2nd millennium AD onwards, stockfish from the Northern Isles entered the European market economy.

References

Baldwin, J. R. (2005) 'Seabirds, subsistence and coastal communities: an overview of cultural traditions in the British Isles', in Randall, J. (ed.) *Traditions of Seabird Fowling in the North Atlantic Region*, pp. 12–36. Stornaway: Islands Book Trust.

Barrett, J. H. (1997) 'Fish trade in Norse Orkney and Caithness: a zooarchaeological approach', *Antiquity*, 71, 616–38.

Barrett, J. H., Beukens, R. P., and Nicholson, R. A. (2001) 'Diet and ethnicity during the Viking colonization of northern Scotland: evidence from fish bones and stable carbon isotopes', *Antiquity*, 75, 145–54.

Barrett, J. H., Nicholson, R. A., and Ceron-Carrasco, R. (1999) 'Archaeo-ichthyological evidence for long-term socioeconomic trends in northern Scotland: 3500 BC to AD 1500', *Journal of Archaeological Science*, 26, 353–88.

Barrett, J. H., Orton, D., Johnstone, C., Harland, J., Van Neer, W., Ervynck, A., Roberts, C., Locker, A., Amundsen, C., Bødker Enghoff, I., Hamilton-Dyer, S., Heinrich, D., Hufthammer, A. K., Jones, A. K. G., Jonsson, L., Makowiecki, D., Pope, P., O'Connell, T. C., de Roo, T., and Richards, M. (2011) 'Interpreting the expansion of sea fishing in Medieval Europe using stable isotope analysis of archaeological cod bones', *Journal of Archaeological Science*, 38, 1516–24.

Best, J. (2014) 'Living in Liminality: An Osteoarchaeological Investigation into the Use of Avian Resources in North Atlantic Island Environments'. Unpublished PhD dissertation, Cardiff University (Cardiff).

Bond, J., Nicholson, R. A., and Cussans, J. (2010) 'Biological evidence', in Dockrill, S. J., Bond, J. M., Turner, V. E., Brown, L. D., Bashford, D. J., Cussans, J. E., and Nicholson, R. A. (eds) *Excavations at Old Scatness, Shetland*, pp. 131–205. Lerwick: Shetland Heritage.

Brand, J. (1701) *A Brief Description of Orkney, Zetland, Pightland-Firth, and Caithness*, Edinburgh.

Buckley, M., Fraser, S., Herman, J., Melton, N. D., Mulville, J., and Palsdottir, A. (2013) 'Species identification of archaeological marine mammals using collagen fingerprinting', *Journal of Archaeological Science*, 41(1), 631–41.

Ceron-Carrasco, R. (2005) *'Of Fish and Men' ('De Iasg Agus Dhaione'): A Study of the Utilization of Marine Resources as Recovered from Selected Hebridean Archaeological Sites*. BAR British Series 4000. Oxford: Archaeopress.

Clark, J. G. D. (1952) *Prehistoric Europe: The Economic Basis*, Cambridge: Cambridge University Press.

Colley, S. (1983) 'Interpreting prehistoric fishing strategies: an Orkney case study', in Grigson, C. and Clutton-Brock, J. (eds) *Animals in Archaeology 2: Shell Middens, Fishes and Birds*. BAR International Series 183, pp. 157–71. Oxford: Archaeopress.

Edwards, K. and Whittington, G. (1997) 'Vegetation change', in Edwards, K. J. and Ralston, I. B. M. (eds) *Scotland: Environment and Archaeology, 8000 BC–AD 1000*, pp. 63–82. London: Wiley.

Fenton, A. (1978) *The Northern Isles*, Edinburgh: John Donald.

Gardiner, M. (1997) 'The exploitation of sea mammals in Medieval England: bones and their social context', *Archaeological Journal*, 154, 173–95.

Harland, J. F. and Barrett, J. H. (2013) 'The maritime economy: fish bone', in Barrett, J. H. (ed.) *Being an Islander: Production and Identity at Quoygrew, Orkney, AD 900–1600*, pp. 115–54. Cambridge: McDonald Institute.

Ingrem, C. (2005) 'Fish ecology and fishing techniques', in Sharples, N. M. (ed.) *A Norse Farmstead in the Outer Hebrides: Excavations at Mound 3, Bornais, South Uist*, pp. 157–8. Oxford: Oxbow.

Ingrem, C. (2012) 'Fish, mammal, bird and amphibian bones', in Mithen, S. and Wicks, K. (eds) Croig Cave: A Late Bronze Age Ornament Deposit and Three Millennia of Fishing and Foraging on the North-West Coast of Mull, Scotland, pp. 95–109. *Proceedings of the Society of Antiquaries of Scotland*, 142, 63–132.

Martin, M. (1716) *A Description of the Western Islands of Scotland*, London: Bell.

McCormick, F. and Buckland, P. C. (1997) 'The vertebrate fauna', in Edwards, K. and Ralston, I. B. M. (eds) *Scotland: Environment and Archaeology, 8000 bc–ad 1000*, pp. 83–103. London: Wiley.

Mulville, J. (2002) 'The role of cetacea in prehistoric and historic Atlantic Scotland', *International Journal of Osteoarchaeology*, 12, 34–48.

Murray, D. S. (2008) *The Guga Hunters*, Edinburgh: Birlinn.

Nicholson, R. A. (1998) 'Fishing in the Northern Isles: a case-study based on fish bone assemblages from two multi-period sites on Sanday, Orkney', *Environmental Archaeology*, 2, 15–28.

Sharples, N. (1998) *Scalloway Broch*, Oxford: Oxbow.

Serjeantson, D. (2001) 'The great auk and the gannet: a prehistoric perspective on the extinction of the great auk', *International Journal of Osteoarchaeology*, 11, 43–55.

Serjeantson, D. (2013) *Farming and Fishing in the Outer Hebrides AD 600 to 1700: The Udal, North Uist*, Southampton: Highfield Southampton Press.

Serjeantson, D. (2014) 'The diverse origins of bird bones from Scottish coastal sites', *International Journal of Osteoarchaeology*, 24(3), DOI: 10.1002/oa.2387.

Sinclair, J. (ed.) (1791–1799) *The Statistical Account of Scotland*, London: Creech.

Szabo, V. E. (1997) 'The use of whales in early Medieval Britain', *Haskins Society Journal*, 9, 137–57.

Szabo, V. E. (2008) *Monstrous Fishes and the Mead-Dark Sea: Whaling in the Medieval North Atlantic*, Leiden: Brill.

Wheeler, A. (1983) 'Fish remains from Knap of Howar, Orkney', in Ritchie, A. (ed.) *Excavation of a Neolithic farmstead at Knap of Howar, Papa Westray, Orkney. Proceedings of the Society of Antiquaries of Scotland*, 113, 40–121.

CHAPTER 12

ZOOARCHAEOLOGICAL EVIDENCE FOR MUSLIM IMPROVEMENT OF SHEEP (*OVIS ARIES*) IN PORTUGAL

SIMON J. M. DAVIS

Old Fernando, who told me the Moors were the best thing that ever happened to Spain, had at the same time the common Andaluz prejudice against eating lamb on the grounds that it was 'Moors' food' and therefore not worthy of Christians

From *Andalucia: A Portrait of Southern Spain* by Nicholas Luard, published by Century (Luard, 1984: 117). Reproduced by permission of The Random House Group Ltd.

INTRODUCTION

ON 28 April AD 711, just ninety-two years after Mohammed's flight from Mecca, Tariq bin Ziyad and some 12,000 men, mostly Berbers, landed near Gibraltar and invaded the Iberian Peninsula. This region was to become an important centre of culture and learning. Not only were song, literature, and mathematics encouraged, but so too was agronomy. Agriculture flourished: the Muslims introduced new irrigation techniques and new plants like sugar cane, rice, cotton, spinach, bananas, pomegranates, and citrus trees, to name just a few. This is referred to as the 'Arab Green Revolution' (Watson, 1974; 1983; Glick, 1979; Araújo, 1983; Guichard, 2000). In his introduction to the *Kitâb al-Filâha* (Book of Agriculture), written towards the end of the twelfth century AD by Abū Zakariyā Yaḥyā ibn Muḥammad ibn Aḥmad ibn al-'Awwām, better known as Ibn al-'Awwām al-Ishbīlī, 'the Sevillian', El Faïz (2000: 23–49) refers to the eleventh and twelfth centuries as 'le moment andalou' in Hispano-Arab history. Seville had become a Mecca for agronomists, and its hinterland, or *Aljarafe*, their laboratory. But while the literature speaks much of oranges and lemons, and apart from the famous Arab horses, we know little about the rest of the livestock sector in both the Muslim period and following the subsequent Christian conquest—generally referred to as the *reconquista*. For over five centuries Muslims ruled the southern part of what later became the Kingdom of Portugal. With the aid of the Crusaders, many of whom hailed from northern Europe, the Christians gradually advanced south and brought about the demise of Muslim rule (see Fig. 12.1). Both *Shantarin* (Santarém) and *al-Ušbuna* (Lisbon) were captured in AD 1147, and by AD 1250, *al-gharb al-Andalus* (Algarve), the last bastion of Islam in the south, fell to forces of the cross under Dom Afonso III. An important question is to what extent these historical changes affected domestic animals in Portugal? This chapter considers the osteometric variation of sheep (*Ovis aries*), essentially their size, in southern Portugal (that part of the country once under Muslim rule) and aims to determine if and when sheep were improved. Is it possible to link osteometry with what we know about the Muslims who lived in the Iberian Peninsula?

This study is therefore one of a number that deal with the size variation of domesticated animal remains from archaeological sites dating to the last three millennia (see, for example, Matolsci, 1970; Teichert, 1984; Albarella and Davis, 1996; Albarella, 1997; Audoin-Rouzeau, 1997; Clavel et al., 1997; Peters, 1998; Breuer et al., 2001; Johnstone and Albarella, 2002; Albarella, 2003; Schlumbaum et al., 2003; Johnstone, 2004; Thomas et al., 2013; Colominas et al., 2014).

MATERIAL

There are abundant collections of animal remains from archaeological sites in Portuguese museums. Most come from excavations undertaken in recent times. It is mainly the larger assemblages which form the basis for this study and these include Neolithic Lameiras, Alcáçova de Santarém with its Iron Age, Roman, and Muslim period levels, Chalcolithic Leceia and Zambujal, Iron Age and Roman Castro Marim, Roman Torre de Palma, Almohad Muslim Silves, and the fifteenth-century AD silos in Beja. These and the smaller assemblages of animal bones considered here are listed in Table 12.1. The sizes of each sample of measurements are shown on the figures and also in Davis (2008) and range from 1 to 77.

An attempt was made to have each of the major periods (Neolithic, Chalcolithic, Iron Age, Roman, Muslim, and post-Muslim) represented by at least two reasonably large assemblages; some, like the Bronze Age, remain unrepresented and the only sizeable Neolithic and post-Muslim samples come from Lameiras and Beja respectively.

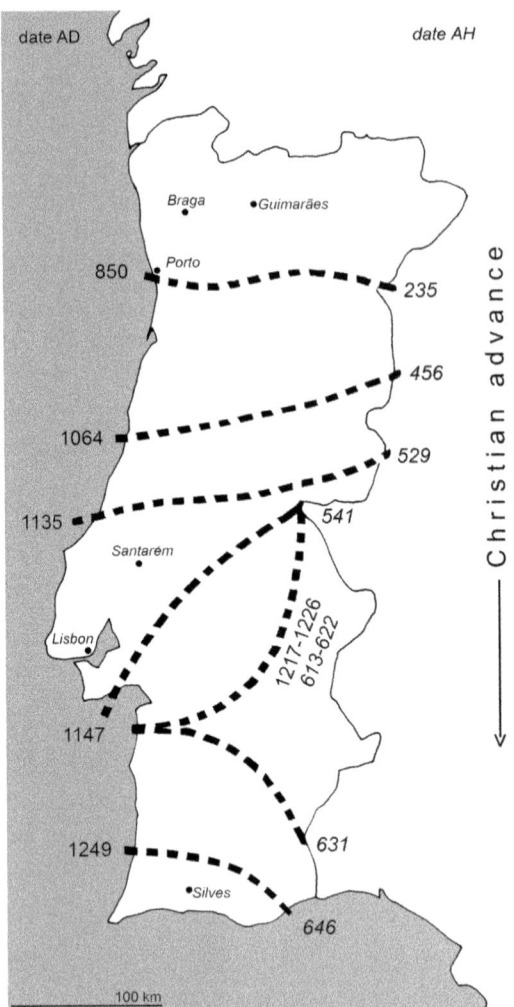

FIGURE 12.1 Map showing the gradual seizure of Muslim Portugal by Christian forces, each thick line being the approximate frontier at a given date (*Anno Domini* on the left and *Anno Hegirae* on the right). While the far north was never under permanent Muslim rule, the northern-most frontier between Islam and Christianity was approximately along the Douro River. Algarve in the south was the last bastion of Islam that finally succumbed to the forces of the cross in 1249 when Afonso III captured Faro (modified from Saraiva, 1983 and Mattoso et al., 1994). Authors' own image.

The DGPC zooarchaeology laboratory (formerly part of the Portuguese Institute of Archaeology—IPA) reference collections of modern Portuguese sheep breeds (Merino [Preto and Branco] and Churra da Terra Quente) serve as a metrical baseline for the sheep.

Table 12.1 List of sites with their locations, cultural assignation, and date from which *Ovis* bones were studied in this survey

Site	District	Period	Date	Storage location	Reference
Avenida Miguel Fernandes, Beja (silos)	Baixa Alentejo	Christian	15th/16th C ad	Crivarque, Torres Novas	Martins et al. (2010)
Silves biblioteca (*lixeira*)	Algarve	Moslem (Almohad)	12th/13th C ad	Museu de Silves	Davis et al. (2008)
Alcáçova de Santarém	Ribatejo	Moslem	11th–12th C ad	CM de Santarém	Davis (2006)
Alcáçova de Santarém	Ribatejo	Roman	2nd C bc–5th C ad	CM de Santarém	Davis (2006)
Torre de Palma	Alto Alentejo	Roman	mainly ad 320–400	MNA Lisbon	McKinnon (unp.)
São Pedro Fronteira	Alto Alentejo	Roman	3rd–5th C ad	CM de Fronteira	Davis (2005)
Castro Marim	Algarve	Roman	1st C ad	CM de Castro Marim	Davis (2007)
Castro Marim	Algarve	Iron Age	8th–3rd C bc	CM de Castro Marim	Davis (2007)
Alcáçova de Santarém	Ribatejo	Iron Age	8th–3rd C bc (mainly 3rd C bc)	CM de Santarém	Davis (2006)
Leceia	Estremadura	Chalcolithic	2600–1800 bc	Centro de Estudos Arqueológicos, Oeiras	Cardoso and Detry (2002)
Zambujal	Estremadura	Chalcolithic	2600–1800 bc	Museu de Torres Vedras	Driesch and Boessneck (1976)
Mercador	Baixa Alentejo	Chalcolithic		Era, Dafundo, Lisbon	Moreno-Garcia (2003)
Lameiras	Estremadura	Neolithic	5500–3000 cal bc	Museu de Odrinhas, Sintra	Davis and Simões (2006)

METHODS AND THE CONTROL OF VARIABLES AFFECTING SIZE

Many of the samples of bones are sufficiently large. Moreover, the correlation between many measurements is not always high (see Davis, 1996). Therefore metrical data for each bone have been considered separately. The sizes of sheep bones from different sites are compared by plotting histograms and stacking these in chronological order, the oldest at the bottom and the youngest at the top, as in a geological or archaeological sequence. Data from the various sites within each period are pooled and mean values for each period are compared by a series of pair-wise Student's t-tests (see Davis, 2008).

Since the principal aim here is to determine human influence upon the size of a species of domesticated animal over the course of time, it is important to rule out or 'control' other complicating factors that may also affect bone size. These include observer variation, the presence of closely related taxa and wild forms, age, and sex.

Observer Error

One important aspect of taking measurements is that each investigator may take her/his measurements in a different way. The importance of this problem was emphasized by several biometricians. One of the first was Francis Sumner (1927) who referred to it as the 'personal equation'. Subsequently others have written about this problem such as Jewell and Fullagar (1966) and Yablokov (1974: 14–17). Zooarchaeologists for example may hold their callipers at a different angle and the points across which the jaws of the callipers are laid may not always be the same. This problem was avoided here since all measurements were taken by the author with vernier callipers to the nearest tenth of a millimetre in the manner recommended by Driesch (1976) and Davis (1996).

The Presence of Other Closely Related and Osteologically Similar Species

For sheep in Portugal there is no possibility of confusion with its wild relatives as these were absent from western Europe. However, more serious is the problem, well known to zooarchaeologists, of confusing sheep (*Ovis*) and goat (*Capra*) bones. These two animals are closely related (current thinking is that goat and sheep lineages separated some seven million years ago during the Late Miocene; Randi et al., 1991; Bibi, 2013). They are, along with the tahr (*Hemitragus*), Barbary sheep (*Ammotragus*), chamois

(*Rupicapra*), mountain goat (*Oreamnos*), and musk ox (*Ovibos*), both members of the same sub-family *Caprinae*, and for most bones that comprise their skeleton it is difficult or impossible to identify them to the species level (i.e. definite sheep or definite goat). The morphological criteria of Boessneck (1969) and Boessneck et al. (1964) and in addition the metric method of Payne (1969) for metacarpals give, with some confidence, a method by which it is possible to separate sheep from goat distal humeri, distal metacarpals, calcanea, astragali, and distal metatarsals. A metrical method for separating sheep from goat astragali that takes into account their slightly different shape has also been proposed (Davis, in press). Other caprine bones and teeth such as the lower third molar and the tibia remain in the well-known zooarchaeological taxon 'sheep/goat' and their measurements are not considered.

Age

Unfused epiphyses and astragali with spongy, incompletely ossified, surfaces (i.e. from juvenile animals) were excluded from the study. Some parts of some bones do continue growing perhaps for some time into adulthood and so mean size change of samples of these measurements such as scapula SLC, humerus BT, and astragalus Bd could become affected by age-at-death composition of the animals. However, many measurements such as humerus HTC and astragalus GLl show little increase in size with age as can be seen in Table 5 of Davis (2000). Moreover, and as discussed below, the Muslim period sheep which show an increase in size were on average slaughtered younger than sheep from the earlier Roman and Iron Age periods and so if age-at-death were a complicating factor this should have lead to an average size *reduction* and not an *increase*.

Sex

In most mammals males are larger than females. This means that the average size of a sample consisting of more males will be greater than that of a sample from the same population consisting of more females. The amount of this sexual size difference may vary, not only according to species, but also according to which measurements and which bones are considered. In order to discern a real size change of a species in the course of time it is therefore important to consider measurements that show little or no inter-sex difference. Examples are the humerus HTC and astragalus GLl in sheep (Davis, 2000). The third molar width has proven useful for cattle, but this measurement could not be used for the sheep due to my inability to distinguish sheep from goat molars and both taxa are common on post-Neolithic sites in southern Portugal. The presence of castrates, whose limb bones tend to be longer and slenderer, could be a complicating factor.

RESULTS

The stacked histograms (Figs 12.2, 12.3) show size variation of sheep bones. Most of the measurements taken of the different bones appear to tell a similar story. In most cases there is little evidence for any substantial change between Neolithic and Roman times, but by the Muslim period sheep were clearly larger. Note the increase in size between Roman and Muslim periods in these figures. While the evidence for a Roman-to-Muslim size increase is clearer in some bone measurements such as those cited above than in others such as metacarpal BFd and metatarsal BFd (see Davis, 2008), they do all show the same general trend of increasing size over time.

A series of t-tests (Davis, 2008) indicate that the mean differences are significant when most measurements of sheep bones from the Muslim period are compared to measurements of sheep bones from earlier periods. Just to give a few examples: the values of 't' for a comparison between Muslim and Roman periods of humerus BT and HTC, astragalus GLl, Bd, and Dl are 3.5; 4.2; 5.7; 5.4; and 5.8 respectively. All these are significantly different at the 1% level. Following the Muslim period there was a further increase in size, though this is less apparent in the case of the astragalus. The amount of increase seems to be less and although most are significantly different at the 1% level, not all are. The modern Churra da Terra Quente ewes are large by Roman standards, and the Merino ewes are similar in terms of size to the sheep from fifteenth-century Beja.

DISCUSSION

The general scarcity of data and fine-tuning of the chronology makes it difficult to determine just when size changes occurred during the Muslim period, so that at this stage of zooarchaeological studies in Portugal we are unable to say whether improvements occurred under Umayyad, Taifa, Almoravid, Almohad, or even, for that matter, Visigoth rule.

That the increase of humerus HTC between Roman and Muslim periods is far greater than the 1% difference observed between rams and ewes in Shetland sheep (i.e. these are measurements that show almost no sexual dimorphism; Davis, 2000) suggests that the Roman–Muslim size increase is a real one and not one due to a change in the sex ratio.

One other, admittedly unlikely possibility, is a change of climate in the Iberian Peninsula at the times of this size increase. Many mammals and birds show an inverse correlation between the size of their bodies and the temperature of the environment, an observation first made by Bergmann (1847; see also Mayr, 1956). However, the magnitudes of the size increase of sheep between Roman and Muslim times would require a huge drop in temperature for which there is little evidence compared, for example, with the changes at the end of the Pleistocene in the Near East (Davis, 1981). Besides, if

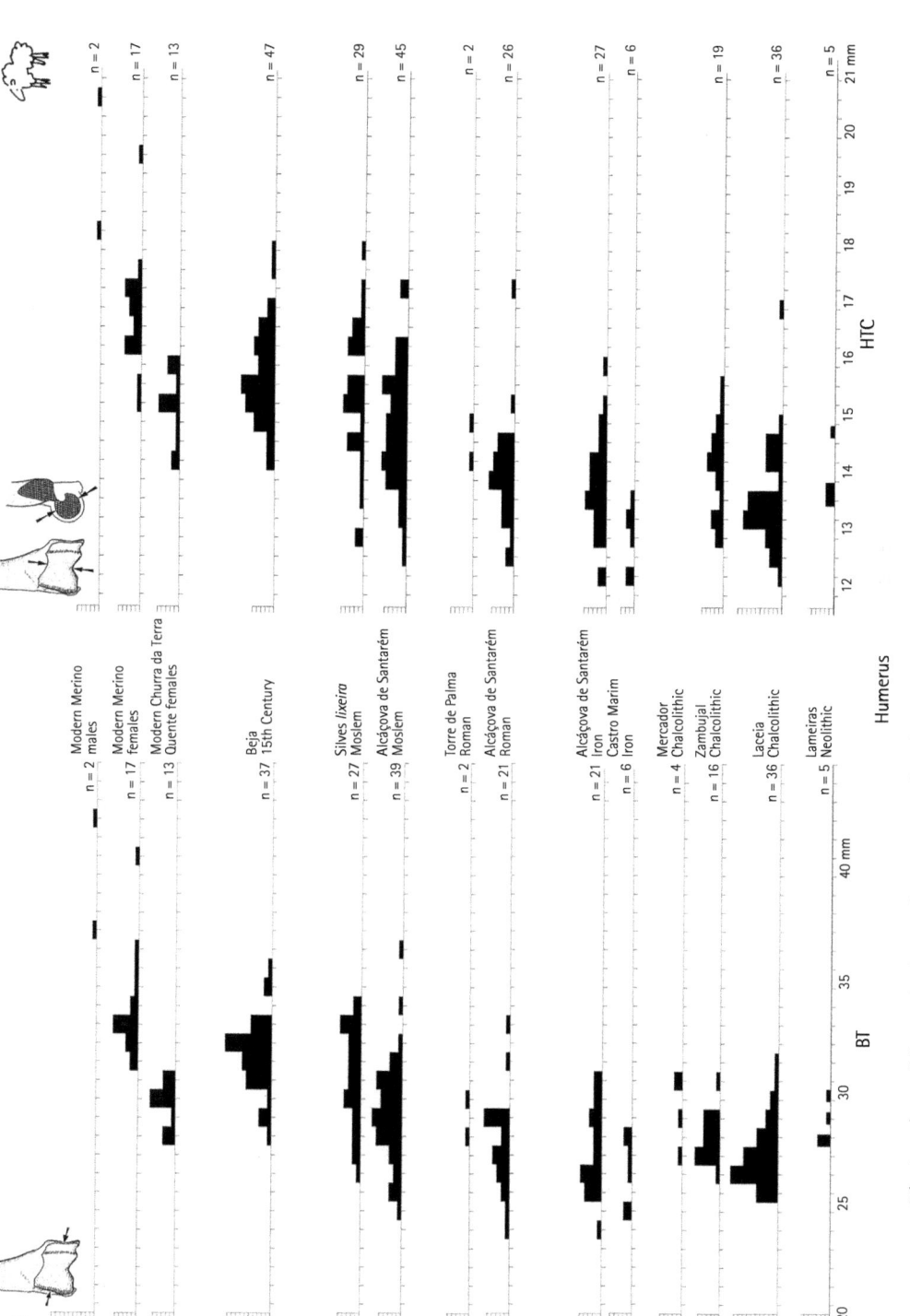

FIGURE 12.2 The increase of sheep size in southern Portugal in the course of time. Stacked histograms of measurements of the trochlea width (BT; left) and the minimum trochlea diameter (HTC; right) of sheep humeri from Neolithic, Chalcolithic, Iron Age, Roman, Moslem, 15th CE Beja and modern Churra da Terra Quente ewes, Merino ewes and two Merino males above. 'n' refers to sample size. Author's own image.

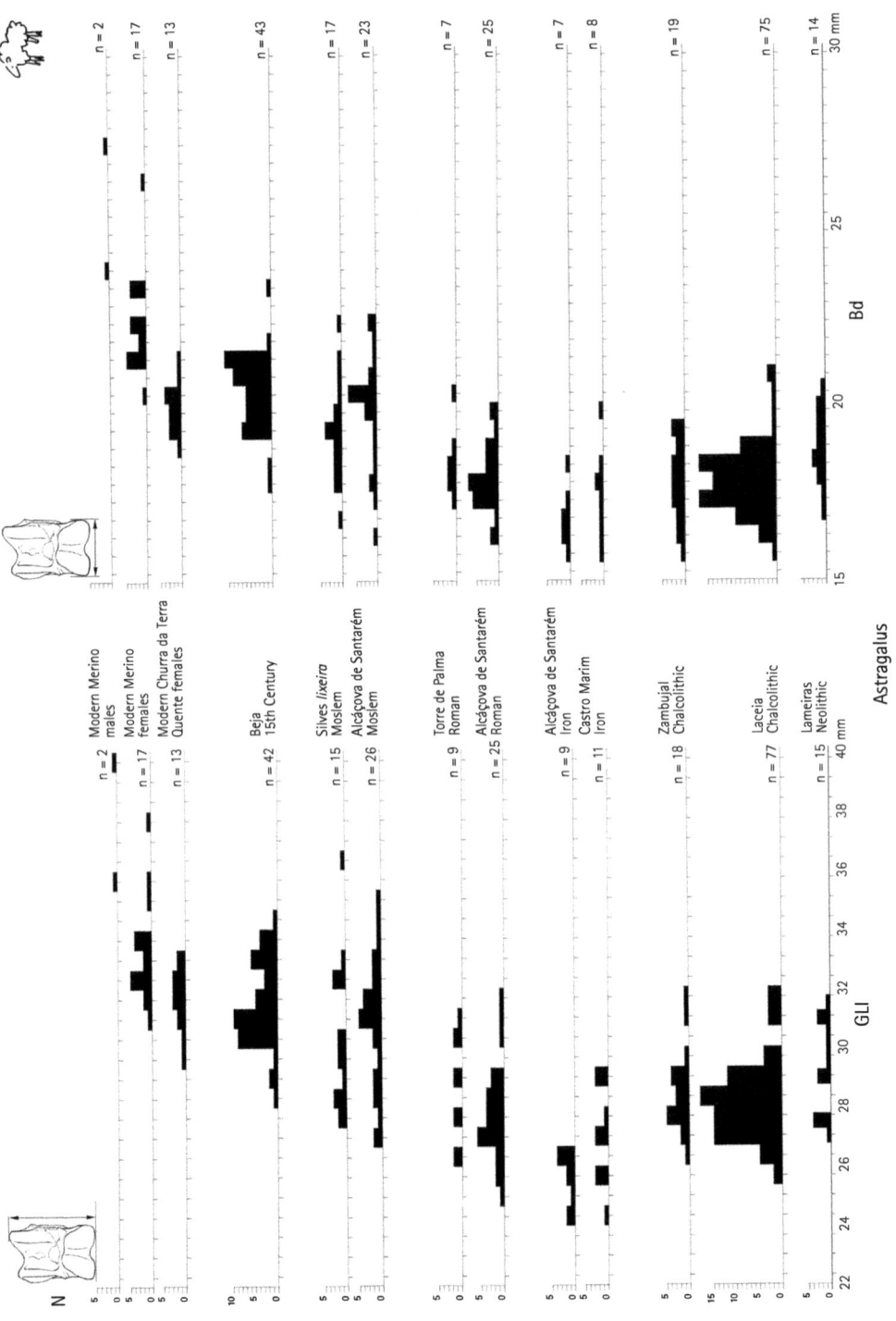

FIGURE 12.3 The increase of sheep size in southern Portugal in the course of time. Stacked histograms of measurements of the greatest lateral length (GLl; left) and the distal width (Bd; right) of sheep astragali from Neolithic, Chalcolithic, Iron Age, Roman, Muslim, 15th C AD Beja and modern Churra da Terra Quente ewes, Merino ewes and two Merino males above. 'n' refers to sample size. Note the increase in size between Roman and Moslem periods. Author's own image.

temperature was a factor then we would expect both sheep as well as other taxa like cattle to have increased in size simultaneously, and they did not. Cattle only increased in size following the Christian *reconquista* (Davis, 2008).

In brief, sheep increased in size over the course of time in southern Portugal in Muslim times. Why? If we accept the assumption that a size increase in a lineage of domesticated animals signifies their improvement (i.e. that their carcasses for example carried greater amounts of meat or in some way there was a promotion of desired qualities), then we need to understand why Portuguese sheep were improved in Muslim times. Can we link improvement to what we know about the Muslims of the Iberian Peninsula and to what we know about Arab dietary preferences and farm animal exploitation? Such an improvement of sheep by the Moslems is hardly a great surprise given their well-known improvements to Iberian agriculture and the esteem with which they held, and still hold, mutton.

Perhaps in part because Islam forbids the consumption of pork, the Arabs have a well-known preference for lamb and mutton—'the favourite meat of the people' (Khayat and Keatinge, 1959: x). In his review of early Arab cuisine, Rosenberger (1999) writes that in the Arab world beef was not much liked and cows and oxen gave milk or laboured in the fields. Most meat came from the vast flocks of sheep. The Arabs liked the taste of mutton and the abundant fat that it provided, and Arab physicians regarded the meat of the yearling lamb as being close to perfection. Glick (1979: 66) notes that in 400 years the pattern of agriculture that emerged in al-Andalus included an increase, over Roman times, in the economic significance of sheepherding. Glick's interesting remarks concerning Muslim *vs* Christian attitudes are relevant here. He writes (1979: 103): 'To a society of town-dwellers and agriculturalists the sheep was an animal primarily raised for meat; its wool was a by-product. The Christians of the later Middle Ages turned the equation around: they cared only for wool and ascribed a low value to the meat.'

This corroborates what 'Old Fernando' had to say about mutton in the quotation beginning this chapter. Thus the Christians, obsessed with wool production, may have preferred to eat pork and beef rather than mutton, and this is certainly the case in many parts of Portugal today, the Alentejo being an exception. It is probably an exaggeration to suggest that mutton/lamb was avoided, but more likely there occurred a shift of emphasis from mutton to beef/pork. No doubt mutton continued to play a significant dietary role. The zooarchaeology of post-Muslim Portugal is still in its infancy, but among the large mammal bones at the eighteenth/nineteenth-century site of Palácio Centeno in Lisbon (Davis, 2009), the percentage of caprines is a mere 9%, while at Moslem Silves (Davis et al., 2008) and Muslim Alcáçova de Santarém (Davis, 2006) it is 77% and 59% respectively. Admittedly recovery at all these sites was by hand and most probably of varying care.

Age-at-death data can also shed some light upon the nature of the animal economy. A high cull of young animals (say up to 2 years old) suggests an emphasis upon meat as, in terms of food input and meat gained, it makes little sense to maintain sheep or cattle much beyond their second or third year. In contrast, an economy geared towards the so-called secondary products such as milk, wool, and power will maintain cattle and sheep until they are quite old. The age-at-death data for the Iron Age, Roman, and Muslim assemblages at Alcáçova de Santarém (Davis, 2006: 49–52) appear to corroborate the thesis here. Thus the caprines were slaughtered at a somewhat younger age in the Muslim period than in the Roman and Iron Ages indicating a shift by the Muslims towards meat. Thus the Payne (1973) sheep and goat mandible tooth eruption and wear stages in the Iron Age and Roman periods show peaks in stages 'E' and 'F' (2–3 and 3–4 years respectively), in the Muslim period there is a peak in stage 'D' (1–2 years). In the Muslim period fewer calves were slaughtered suggesting that cattle were now kept more for power and/or milk. The Almohad (Muslim; twelfth/thirteenth century) period bones from Silves (Davis et al., 2008) also indicate that the sheep and goats were kept primarily for their meat. Thus their mandibles show a distinct peak of slaughter at 2–3 years (Payne, 1973 wear stage 'E') while the majority of the cattle derive from much older animals. For example there are thirteen cattle P_4s (adult fourth premolars), and no dP_4s (deciduous fourth premolars), at Silves.

Since higher meat yield in sheep is correlated with larger bones (Hammond, 1960: 131), it is logical to link the increased size of Muslim period sheep with their improved meat yield. This leads us to query how this may have happened. Did the Muslims improve the local sheep or did they import new stock from, say, the Maghreb or the Middle East? Evidence from the Cairo Genizeh indicates quite clearly that the Mediterranean of the eleventh and twelfth century was a kind of medieval common market with the Islamic world forming a free trade area (Goitein, 1967). (The Cairo Genizeh is the famous store of fragments of manuscripts found in the Ben Ezra synagogue in Fustat, old Cairo; they are an important source of social and economic information from the tenth to thirteenth centuries AD.) This communication network, shared by Muslims, Jews, and Christians, expressed the notion (Glick, 1979: 27) that in mobility there is blessing as the Arab maxim states "fi'l-haraka baraka". Moreover, Klein (1920: 4–6) suggested that it was the Beni Merin Berbers who introduced the Merinos from northern Morocco during the Almohad expansion into southern Iberia. Not only was the Mediterranean important, but the Atlantic maritime trade between Spain, Portugal, and the Maghreb at this time is also well documented (Picard, 1997). Klein (1920: 4–6) also noted that many of the pastoral terms used to this day in Spain are of Arabic origin. Some examples that Klein gives include *zagal* and *rabadan* (shepherd's assistants), *rafala* (a pen for strays), *morrueco* (breeding ram), *ganado* (domestic animal), *cabaña* (herd, sheepfold, shepherd's cabin), and *mechta* (winter sheep encampment). He also notes that the methods used in medieval Spain to select breeding rams, to castrate and prepare sheep for slaughter, and to clip and wash wool, were like those of the North African tribes, and were commonly believed by Spanish herdsmen to be of Berber origin. There are indeed several likely etymologies of the word *merino* and possible origins of this most important breed of sheep (see for example Laguna Sanz, 1986; Sanchez Belda and Sanchez Trujillano, 1986) although Riu (1983) suggests that the Merinos resulted from cross-breeding of coarse-woolled ewes with North African fine-woolled rams in the mid-fourteenth century. Even today Merinos tend to be reared in the southern part of Spain and Portugal and they are genetically somewhat distinct from other breeds kept in central and northern

Spain (Arranz et al., 1998). A genetic (mitochondrial DNA) study of seven modern breeds of Portuguese sheep (Pereira et al., 2006) reveals the presence of maternal lineages that until now were only found in the Middle East and Asia. A broad north–south pattern indicates a trend with southern Portuguese sheep clearly distinct from most other breeds. This is interpreted in terms of an influx of new genetic diversity, via a maritime route, although it is impossible at the moment to know when this happened. Clearly further studies, both osteological and genetic, of sheep remains dating back over the last two or three millennia in Portugal are needed, but it is tempting to presume that at least some live sheep accompanied the oranges and lemons brought into the Iberian Peninsula.

In many parts of Europe there is now substantial zooarchaeological evidence that livestock and even fowl were improved in later Medieval and post-Medieval times (Matolsci, 1970; Albarella and Davis, 1996; Audoin-Rouzeau, 1997; Clavel et al., 1997; Davis and Beckett, 1999; Thomas et al., 2013). A pre-fifteenth-century AD date for improved cattle in Portugal (Davis, 2008) indeed seemed somewhat early in comparison and may indicate an advanced state of farming here in Portugal at that time. Furthermore, the even earlier size increase of the sheep comes as a greater surprise. However, more recent zooarchaeological investigations by Thomas (2005) and Thomas et al. (2013) are revealing evidence for agricultural changes as early as the fourteenth century in England as Dyer (1981) had found in his studies of the documentary evidence. Thomas, like Dyer, links these fourteenth-century improvements with the Black Death (1348–1350) and the resulting demographic decline, and suggests that the demand to feed an expanding population had dissipated and the market in grain crashed. Animal husbandry became a viable alternative, being less labour intensive but requiring more land—which was plentiful following the effect of the Black Death. A possible chain of explanations for these fourteenth-century changes in England, which these authors propose, include a downward social distribution of access to land and the tendency for peasants to become landowners. Peasants who were in more 'intimate contact' with animals were better able to take 'technological initiatives'. Disease and demographic factors may explain changes in agriculture and livestock in fourteenth- to sixteenth-century Christian Portugal. However, these factors seem to be less likely to have affected the Muslims' influence upon sheep. It is probable that the Moslems improved sheep for economic reasons—to produce heavier carcasses with more meat for their tajines. This must surely reflect the general state of affairs and advancement in the Iberian Peninsula under Muslim rule.

Conclusions

This metrical study of sheep bones from archaeological sites in southern Portugal indicates that sheep became larger in Muslim times. A size increase of the sheep may have been 'meat-driven'; reflecting selection for heavier-boned animals with greater meat yield. This seems quite logical since mutton was and still is much favoured in the Muslim world. Whether new stock such as the Merino was imported from abroad or whether local animals were 'improved' must remain within the realm of speculation. While an absence of any observable shape change of the sheep bones between Roman and Muslim periods (see Davis, 2008) tends to point in favour of a local improvement rather than import of stock, the genetic evidence based on modern sheep in the Iberian Peninsula does indicate some input from overseas, though just when this occurred is unknown (Pereira et al., 2006). Following the Christian conquest of southern Portugal the sheep took on a new role as provider of wool—a prime source of wealth for Medieval Portugal and Castile. However, as a source of meat, this animal was relegated to a subsidiary role, with beef and pork becoming the favoured meats. It is hoped that continuing archaeological investigations in southern Portugal will produce more data with refined dates which should in turn improve our understanding of the relations between people and their domesticated animals during the last two millennia.

ACKNOWLEDGEMENTS

It is a pleasure to acknowledge the many archaeologists who have invited me to study the animal bones from their sites. They include Maria Antónia Amaral, Ana Arruda, André Carneiro, Maria José Gonçalves, Maia Langley, Michael Kunst, Rui Mataloto, Teresa Simões, António Valera, Catarina Viegas, and Jõao Zilhão. João Luis Cardoso, Cláudia Costa, Cleia Detry, Andréa Martins, Michael MacKinnon, Marta Moreno García, Adelaide Pinto (and 'Crivarque' staff), staff of the Museu Nacional de Arqueologia in Lisbon, Museu de Torres Vedras and the Serviços Geológicos, Lisbon all very kindly gave me access to collections in their care. I have benefited greatly from an exchange of ideas with Umberto Albarella, Albano Beja-Pereira, and John Watson. With their knowledge of history, Jane Bridgeman, Jacinta Bugalhão, and Carlos Pimenta have also been a source of help. Umberto Albarella, Joris Peters, and Jörg Schibler told me about Roman improvements in several other parts of Europe. Cathy Douzil and José-Paulo Ruas helped with the figures and Umberto Albarella, Cleia Detry, Cathy Douzil, and John Watson offered many invaluable comments on previous drafts of this script— part of an article on both sheep and cattle improvement, published in the *Journal of Archaeological Science*. António Abel, Alfredo Sendim, and Miguel Madeira have, over the years, generously supplied me with merino sheep carcasses. Jorge Azevedo, Ana Luísa Lourenço, and staff and students of UTAD, Vila Real, generously donated and helped prepare the Churra skeletons.

REFERENCES

Albarella, U. (1997) 'Size, power, wool and veal: zooarchaeological evidence for late Medieval innovations', in De Boe, G. and Verhaeghe, F. (eds) *Environment and Subsistence in Medieval Europe: Papers of the 'Medieval Europe Brugge 1997' Conference, Vol. 9*, pp. 19–30. Zellik: Institut vor het Archeologisch Patrimonium.

Albarella, U. (2003) 'Animal bone', in Germany, M. (ed.) *Excavations at Great Holts Farm, Boreham, Essex, 1992–94.* East Anglian Archaeology Monographs 105, pp. 193–200. Oxford: Oxbow Books.

Albarella, U. and Davis, S. J. M. (1996) 'Mammals and birds from Launceston castle, Cornwall: decline in status and the rise of agriculture', *Circaea*, 12, 1–156.

Araújo, de, L. M. (1983) 'Os muçulmanos no Ocidente peninsular', in Saraiva, J. H. (ed.) *História de Portugal, 1*, pp. 245–89. Lisboa: Publicações Alfa.

Arranz, J. J., Bayón, Y., and Primitivo, F. S. (1998) 'Genetic relationships among Spanish sheep using microsatellites', *Animal Genetics*, 29(6), 435–40.

Audoin-Rouzeau, F. (1997) 'Les éléments nouveaux de l'élevage aux temps modernes', *Cahiers d'Histoire*, 42, 481–509.

Bergmann, C. (1847) 'Über die Verhältnisse der Wärmeökonomie der Thiere zu ihrer Grosse', *Göttingen Studien*, 3, 595–708.

Bibi, F. (2013) 'A multi-calibrated mitochondrial phylogeny of extant Bovidae (Artiodactyla, Ruminantia) and the importance of the fossil record to systematics', *BMC Evolutionary Biology*, 13, 166.

Boessneck, J. (1969) 'Osteological differences between sheep (*Ovis aries* Linne) and goat (*Capra hircus* Linne)', in Brothwell, D. and Higgs, E. S. (eds) *Science in Archaeology*, 2nd edn, pp. 331–58. London: Thames and Hudson.

Boessneck, J., Müller, H.-H., and Teichert, M. (1964) 'Osteologische Unterscheidungsmerkmale zwischen Schaf (*Ovis aries* Linné) und Ziege (*Capra hircus* Linné)', *Kühn-Archiv*, 78, 1–129.

Breuer, G., Rehazek, A., and Stopp, B. (2001) 'Veränderung der Körpergrösse von Haustieren aus Fundstellen der Nordschweiz von der Spätlatènezeit bis ins Frühmittelalter', *Jahresberichte aus Augst und Kaiseraugst*, 22, 161–78.

Cardoso, J. L. and Detry, C. (2002) 'Estudo arqueozoológico dos restos de ungulados do povoado pré-histórico de Leceia (Oeiras)', *Estudos Arqueológicos de Oeiras*, 10, 131–82.

Clavel, B., Marinval-Vigne, M. C., Lepetz, S., and Yvinec, J.-H. (1997) 'Évolution de la taille et de la morphologie du coq au cours de périodes historiques en France du Nord', *LE COQ Ethnozootechnie*, 58, 3–12.

Colominas, L., Schlumbaum, A., and Saña, M. (2014) 'The impact of the Roman Empire on animal husbandry practices: study of the changes in cattle morphology in the north-east of the Iberian Peninsula through osteometric and ancient DNA analyses', *Archaeological and Anthropological Sciences*, 6, 1–16.

Davis, S. J. M. (1981) 'The effects of temperature change and domestication on the body size of Late Pleistocene to Holocene mammals of Israel', *Paleobiology*, 7, 101–14.

Davis, S. J. M. (1996) 'Measurements of a group of adult female Shetland sheep skeletons from a single flock: a baseline for zooarchaeologists', *Journal of Archaeological Science*, 23, 593–612.

Davis, S. J. M. (2000) 'The effect of castration and age on the development of the Shetland sheep skeleton and a metric comparison between bones of males, females and castrates', *Journal of Archaeological Science*, 27, 373–90.

Davis, S. J. M. (2005) *Animal Bones from Roman São Pedro, Fronteira, Alentejo*. Trabalhos do CIPA 88. Lisbon: Instituto Português de Arqueologia.

Davis, S. J. M. (2006) *Faunal Remains from Alcáçova de Santarém, Portugal*. Trabalhos de Arqueologia 43. Lisbon: Instituto Português de Arqueologia.

Davis, S. J. M. (2007) *The Mammals and Birds from the Iron Age and Roman Periods of Castro Marim, Algarve, Portugal*. Trabalhos do CIPA 107. Lisbon: Instituto Português de Arqueologia.

Davis, S. J. M. (2008) 'Zooarchaeological evidence for Moslem and Christian improvements of sheep and cattle in Portugal', *Journal of Archaeological Science*, 35(4), 991–1010.

Davis, S. J. M. (2009) 'Animal remains from an 18th–19th century AD pit in the Palácio Centeno, Lisbon', *Revista Portuguesa de Arqueologia*, 12(2), 271–82.

Davis, S. J. M. (in press) 'A metrical distinction between sheep and goat astragali', in, Rowley-Conwy, P., Serjeantson, D., and Halstead, P. (eds) *Economic Zooarchaeology: Studies in Hunting, Herding and Early Agriculture*. Oxford, Oxbow.

Davis, S. J. M. and Beckett, J. (1999) 'Animal husbandry and agricultural improvement: the archaeological evidence from animal bones and teeth', *Rural History: Economy, Society, Culture*, 10, 1–17.

Davis, S. J. M., Gonçalves, M.-J., and Gabriel, S. (2008) 'Animal remains from a Moslem period (12th/13th C AD) *lixeira* (garbage dump) in Silves, Algarve, Portugal', *Revista Portuguêsa de Arqueologia*, 11(1), 183–258

Davis, S. J. M. and Simões, T. (2016). 'The velocity of *Ovis* in prehistoric times: the sheep bones from Early Neolithic Lameiras, Sintra, Portugal', in: Diniz, M., Neves, C., and Martins, A. (eds), *O Neolítico em Portugal antes do horizonte 2020: perspectivas em debate*, pp. 51–66. Associação dos Arqueólogos Portugueses, Monografias 2.

Driesch, von den, A. (1976) *A Guide to the Measurement of Animal Bones from Archaeological Sites*, Harvard: Peabody Museum.

Driesch, von den, A. and Boessneck, J. (1976) 'Die Fauna vom Castro do Zambujal (Fundmaterial der Grabungen von 1966 bis 1973 mit Ausnahme der Zwingerfunde)', in Driesch, von den, A. and Boessneck, J. (eds) *Studien über frühe Tierknochenfunde von der Iberischen Halbinsel 5: Institut für Palaeoanatomie, Domestikationsforschung und Geschichte der Tiermedizin der Universität München*, pp. 4–129. München: Deutsches Archäologisches Institut Abteilung Madrid.

Dyer, C. (1981) 'Warwickshire Farming 1349–c.1520: Preparations for Agricultural Revolution'. Dugdale Society Occasional Papers 27. Oxford: Dugdale Society.

El Faïz, M. (2000) 'Introduction à l'œuvre agronomique d'Ibn al-'Awwâm', in Clément-Mullet, J.-J. and El Faïz, M. (eds) *Ibn al-'Awwâm, le livre de l'agriculture, Kitâb al-Filâha*, pp. 9–40. Arles: Actes Sud.

Glick, T. F. (1979) *Islamic and Christian Spain in the Early Middle Ages*, New Jersey: Princeton University Press.

Goitein, S. D. (1967) *A Mediterranean Society*, Vol. 1: *Economic Foundations*, Los Angeles: University of California Press.

Guichard, P. (2000) *Al-Andalus 711–1492: une histoire de l'Espagne musulmane*, Paris: Hachette Littératures.

Hammond, J. (1960) *Farm Animals: Their Breeding, Growth, and Inheritance*, 3rd edn, London: Edward Arnold.

Jewell, P. A. and Fullagar, P. J. (1966) 'Body measurements of small mammals: sources of error and anatomical changes', *Journal of Zoology, London*, 150, 501–9.

Johnstone, C. J. (2004) 'A Biometric Study of Equids in the Roman World'. Unpublished PhD dissertation, University of York (York).

Johnstone, C. J. and Albarella, U. (2002) 'The Late Iron Age and Romano-British Mammal and Bird Bone Assemblage from Elms Farm, Heybridge, Essex (Site code: HYEF93-95)'. Portsmouth, English Heritage, Centre for Archaeology Report 45/2002.

Khayat, M. K. and Keatinge, M. C. (1959) *Food from the Arab World*, Beirut: Khayats.

Klein, J. (1920) *The Mesta: A Study in Spanish Economic History 1273–1836*, Cambridge (MA): Harvard University Press.

Laguna Sanz, E. (1986) 'Historia del Merino', Madrid: Ministerio de Agricultura, Pesca y Alimentacion and Direccion General de la Produccion Agraria.

Luard, N. (1984) *Andalucia: A Portrait of Southern Spain*, London: Century.

Martins, A., Neves, C., Costa, C., and Lopes, G. (2010) 'Sobre um conjunto de silos em Beja: a Avenida Miguel Fernandes', *Revista Portuguesa de Arqueologia*, 13(1), 145–65.

Matolsci, J. (1970) 'Historische Erforschung der Körpergrösse des Rindes auf Grund von ungarischem Knochenmaterial', *Zeitschrift für Tierzüctung und Züchtungsbiologie*, 87, 89–137.

Mattoso, J. Magalhães, A. M., and Alçada, I. (1994) *No reino de Portugal; História de Portugal*, Vol. 2. Lisbon: Caminho.

Mayr, E. (1956) 'Geographical character gradients and climatic adaptation', *Evolution*, 10, 105–8.

Moreno García, M. (2003) *Estudo arqueozoológico dos restos faunísticos do Povoado Calcolítico do Mercador, Mourão*. Trabalhos do CIPA 56. Lisbon: Instituto Português de Arqueologia.

Payne, S. (1969) 'A metrical distinction between sheep and goat metacarpals', in Ucko, P. J. and Dimbleby, G. W. (eds) *The Domestication and Exploitation of Plants and Animals*, pp. 295–305. London: Duckworth.

Payne, S. (1973) 'Kill-off patterns in sheep and goats: the mandibles from Aşvan Kale', *Anatolian Studies*, 23, 281–303.

Pereira, F., Davis, S. J. M., Pereira, L., McEvoy, B., Bradley, D. G., and Amorim, A. (2006) 'Genetic signatures of a Mediterranean influence in Iberian Peninsula sheep husbandry', *Molecular Biology and Evolution*, 23(7), 140–6.

Peters, J. (1998) *Römische Tierhaltung und Tierzucht. Eine Synthese aus archäozoologischer Untersuchung und schriftlich-bildlicher Überlieferung*. Passauer Universitätsschriften zur Archäologie 5. Rahden: Leidorf.

Picard, C. (1997) *L'Océan Atlantique musulman: de la conquête arabe à l'époque almohade. navigation et mise en valeur des côtes d'al-Andalus et du Maghreb occidental (Portugal-Espagne-Maroc)*, Paris: Maisonneuve et Larose.

Randi, E., Fusco, G., Lorenzini, R., Toso, S., and Tosi, G. (1991) 'Allozyme divergence and phylogenetic relationships among *Capra*, *Ovis* and *Rupicapra* (Artiodactyla, Bovidae)', *Heredity (Edinburgh)*, 67(3), 281–6.

Riu, M. (1983) 'The woollen industry in Catalonia in the later Middle Ages', in Harte, N. B. and Ponting, K. G. (eds) *Cloth and Clothing in Medieval Europe: Essays in Memory of Prof. E. M. Carus-Wilson*, pp. 205–29. London: Heinemann Educational Books and Pasold Research Fund.

Rosenberger, B. (1999) 'Arab cuisine and its contribution to European culture', in Flandrian, J.-L., Montanari, M. (eds) *Food: A Culinary History from Antiquity to the Present*, pp. 207–23. New York: Columbia University Press.

Sanchez Belda, A. and Sanchez Trujillano, M. C. (1986) 'Razas ovinas españolas', Madrid: Ministerio de Agricultura, Pesca y Alimentación.

Saraiva, J. H. (1983) *História de Portugal*, Vol. 2. Lisbon: Alfa.

Schlumbaum, A., Stopp, B., Breuer, G., Rehazek, A., Blatter, R., Turgay, M., and Schibler, J. (2003) 'Combining archaeozoology and molecular genetics: the reason behind the changes in cattle size between 150 BC and 700 AD in northern Switzerland', *Antiquity*, 77(298), http://antiquity.ac.uk/ProjGall/schlumbaum/index.html.

Sumner, F. B. (1927) 'Linear and colorimetric measurements of small mammals', *Journal of Mammalogy*, 8, 177–206.

Teichert, M. (1984) 'Size variation in cattle from Germania Romana and Germania Libera', in Grigson, C. and Clutton-Brock, J. (eds) *Animals and Archaeology 4: Husbandry in Europe.* BAR International Series 227, pp. 93–103. Oxford: Archaeopress.

Thomas, R. (2005) 'Zooarchaeology, improvement and the British agricultural revolution', *International Journal of Historical Archaeology*, 9(2), 71–88.

Thomas, R., Holmes, M., and Morris, J. (2013) ' "So bigge as bigge may be": tracking size and shape change in domestic livestock in London (AD 1220–1900)', *Journal of Archaeological Science*, 40(8), 3309–25.

Watson, A. M. (1974) 'The Arab agricultural revolution and its diffusion, 700–1110', *Journal of Economic History*, 34(1), 8–35.

Watson, A. M. (1983) *Agricultural Innovation in the Early Islamic World: The Diffusion of Crops and Farming Techniques, 700–1100.* Cambridge Studies in Islamic Civilisation. Cambridge: Cambridge University Press.

Yablokov, A. V. (1974) *Variability of Mammals [Izmenchivost' Mlekopitayushchikh]* (translated from Russian), Washington: Smithsonian Institution and National Science Foundation.

CHAPTER 13

THE ZOOARCHAEOLOGY OF MEDIEVAL IRELAND

FINBAR MCCORMICK AND EMILY MURRAY

INTRODUCTION

THE Medieval period in Ireland traditionally starts with the arrival of St Patrick in the latter half of the fifth century AD and closes at the end of the sixteenth century. It can be subdivided into early and late phases, separated by the arrival of the Anglo-Normans in AD 1169. The Anglo-Normans had a significant impact on the Irish economy and farming practices (McCormick, 1991a) with the expansion of towns leading to the widespread commercialization of agriculture. The Normans also introduced a number of new species to Ireland. In this study the main trends in livestock agricultural change in Ireland will be examined across these eleven centuries using samples from sites (Fig. 13.1) where the MNI (minimum numbers of individuals) value totals for the three main domesticates in an assemblage, i.e. cattle (*Bos taurus*), sheep (*Ovis aries*), and pig (*Sus domesticus*), is greater than forty.

EARLY MEDIEVAL PERIOD

During the Early Medieval period Ireland was politically organized into a large number of small kingdoms ('*tuatha*'), estimated to have been about 150 in number (Byrne, 2001: 7), each with a projected population of around three thousand (Kelly, 1988: 4). Ireland was never conquered by the Romans and therefore their centralized political structure that was rolled out across the Empire did not reach Ireland. As a consequence, settlement remained exclusively rural in character until the arrival of the Vikings in the ninth century. The settlement type-site of the period is the ringfort, a small circular enclosure which is regarded as a family farmstead, and of which nearly 50,000 have been identified (Stout, 1997: 53). With the exception of the larger monastic complexes, it was thought that settlement was generally isolated and scattered, with no equivalent to the villages of Anglo-Saxon England. More recently, however, large multiple-enclosure settlements have been discovered that were clearly occupied by more than one family (Corlett and Potterton, 2010; 2011). Some of the sites show species distribution similar to those noted on ringforts and crannogs (artificial islands used as dwellings), but a small number produced unusual distributions, the most atypical being the very high incidence of pig at Castlefarm, Co. Meath (see Fig. 13.1).

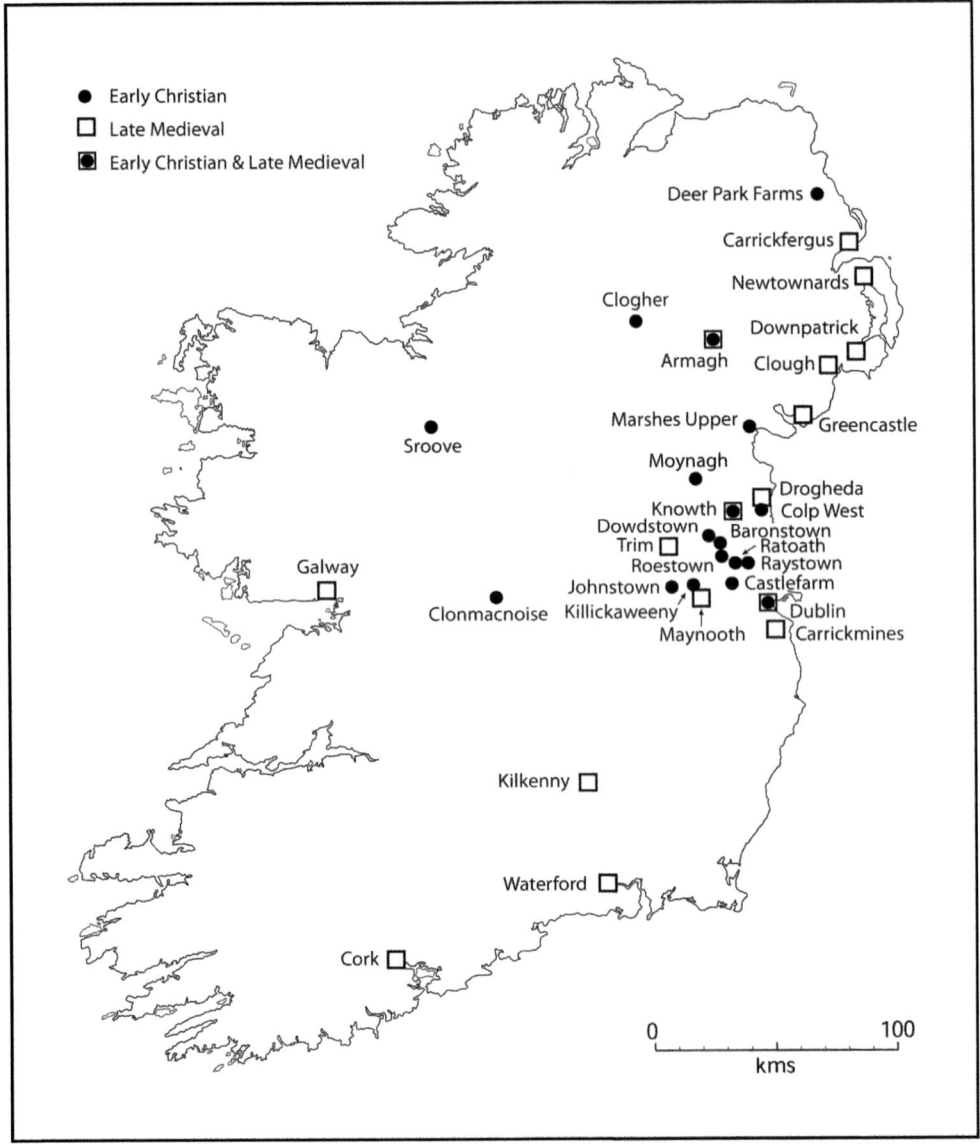

FIGURE 13.1 Map showing the distribution of sites mentioned in the text that have produced Early Christian and/or Late Medieval faunal assemblages where the total MNI for domesticates is > 40. Authors' own image.

The zooarchaeological dataset for Early Medieval Ireland is complemented by a large body of documentary sources, in particular legal texts from the seventh and eighth centuries and an annalistic record that spans the entire era. The pollen evidence for Early Medieval Ireland indicates a landscape of scrubby woodland and mixed farming, dominated by pastureland. Palaeoclimatic proxies for Ireland suggest that, in general terms, the Early Medieval climate was warm and wet and recent palaeoenvironmental studies suggest that climatic conditions took a downturn during the latter part of the Early Medieval period (Kerr et al., 2009). Given the long growing season, the Irish did not save hay until relatively recently and it was possible to leave livestock outdoors all year around (Kelly, 1997: 44).

Species Frequency

Cattle are generally the most numerous species present and account for 40% to 50% of the main domesticates on most sites (Table 13.1). The almost universal predominance of cattle in the Early Medieval Irish economy differs from contemporary England where sheep tend to be the dominant species (O'Connor, 2011: 367). Pig played a primary role in urban Viking Dublin, presumably reflecting an attempt by the town's inhabitants to minimize their dependence on outside producers. Sheep are generally of lesser importance, with small coastal assemblages sometimes recording a relatively high level, reflecting their marginal status (Murray et al., 2004).

Cattle

The early law tracts indicate that cattle, and more specifically the cow, was the basic unit of wealth and an individual's social status in this hierarchical society was to a large extent dependent on the number of cows that one had (Lucas, 1989). The sources make it clear that cows were kept primarily for the production of milk; the laws note that a 'dry cow' has only half the value of a milk cow (Kelly, 1997: 65). In zooarchaeology attempts to differentiate the sexes of cattle have been undertaken through the metrical analysis of their skeletal remains, in particular metacarpals which display a high degree of sexual dimorphism (Higham, 1969; Grigson, 1982; Thomas, 1988). Bulls generally have shorter and more robust metacarpals than cows, while castrates fall between the two with a tendency to have longer bones than either of the other two sexes (Albarella, 1997: 38). This biometrical analysis when applied to material from Ireland (see Fig. 13.2) tends to confirm the importance of cows as the majority of adult cattle present in Early Medieval assemblages are female (McCormick, 1992a). The implication of this is that the preponderance of sub-adult cattle (whose bones cannot be sexed), i.e. those killed-off for their meat, were male and this is confirmed by the legal sources. Male cattle, unless they were kept for traction, were not ascribed any monetary value after they were two years of age (McCormick and Murray, 2007: 54–5). A commentary on animal values indicates that a bull held only half

Table 13.1 MNI percentages for the main domesticates from Early Medieval Irish sites

Site and phase/feature	Date (centuries ad)	Site type	Cattle %	Sheep/Goat %	Pig %	MNI Total	Source
Moynagh, D	7th/8th	crannog–royal	40	22	37	258	McCormick, 1987
Moynagh, A1	8th–9th	crannog–royal	19	53	28	57	McCormick, 1987
Sroove, Ph. 3–4	7th–9th	crannog	53	22	25	68	Lofquist, 2002
Armagh, Cathedral Hill	5th–8th	monastic	45	19	36	58	Higgins, 1984a
Clonmacnoise, Ph. 1	7th–8th	monastic	48	21	32	244	Soderberg, 2003
Clonmacnoise, Ph. 2	10th	monastic	44	22	34	336	Soderberg, 2003
Clonmacnoise, Ph. 3	11th–13th	monastic	40	20	41	384	Soderberg, 2003
Marshes Upper, ditch 3	7th/8th	ringfort	45	20	35	40	McCormick, 1992b
Clogher, ring ditch	5th–7th	ringfort–royal	55	10	34	116	Bonar, 2001
Clogher, ringfort	7th–9th	ringfort–royal	57	14	29	76	Bonar, 2001
Deer Park Farms, I–V	7th–8th	ringfort (raised)	46	19	34	119	McCormick and Murray, 2011
Baronstown 1, Ph.1	6th–7th	rural enclosure complex	53	25	23	208	Sloane, 2009a
Baronstown 1, Ph. 2	7th–9th	rural enclosure complex	47	28	26	43	Sloane, 2009a
Castlefarm, II	5th–8th	rural enclosure complex	40	15	45	110	Foster, 2009
Castlefarm, III–IV	8th–10th	rural enclosure complex	34	18	48	73	Foster, 2009
Colp West, Ph. 2	7th–8th	rural enclosure complex	56	27	17	59	McQuade, 2001
Dowdstown 2, Ph. 2	5th–7th	rural enclosure complex	49	28	22	81	Coles, 2009
Dowdstown 2, Ph. 3	7th–9th	rural enclosure complex	55	22	22	67	Coles, 2009
Killickaweeney	7th–9th	rural enclosure complex	46	31	24	59	Lofquist, 2008
Ratoath, early	5th–7th	rural enclosure complex	48	20	31	64	Beglane, undated
Raystown, II	6th–7th	rural enclosure complex	47	29	24	107	Murray, in press
Raystown, III	7th–8th	rural enclosure complex	38	41	22	64	Murray, in press
Roestown 2, 1a	M6th	rural enclosure complex	53	28	19	43	Sloane, 2009b
Roestown 2, 2a	M7th	rural enclosure complex	41	26	33	70	Sloane, 2009b
Roestown 2, 2b	8th	rural enclosure complex	44	31	25	81	Sloane, 2009b
Knowth, ST 8	6th–7th	rural enclosure–royal	57	22	22	83	McCormick and Murray, 2007
Knowth, ST 9	10th–11th	rural enclosure–royal	41	27	32	41	McCormick and Murray, 2007
Johnstown, Ph. 2	5th–7th	rural settlement cemetery	55	11	34	47	Bonar, 2003
Dublin, Fishamble St, Plots 1 and 2	10th–11th	urban Viking	35	11	56	1,053	McCormick, 1987
Dublin, Fishamble St, banks and wall	E12th	urban Viking	59	11	30	97	McCormick, 1987

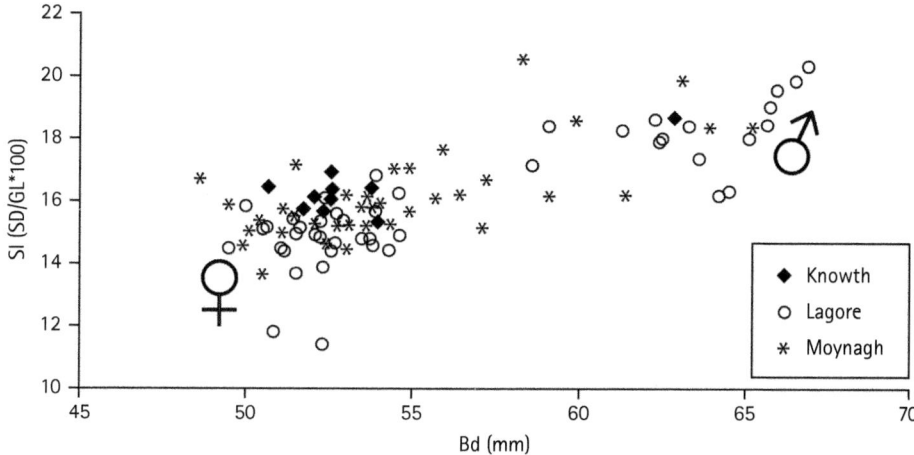

FIGURE 13.2 Cattle metacarpals from the Early Medieval sites of Knowth, Co. Meath and the contemporary crannog sites of Lagore and Moynagh, both in Co. Meath. The distal breadths (Bd) are compared with the slenderness index (SI = SD/GL x 100), showing a cluster to the left-hand side of the plot that can be interpreted as representing cows, while the more diffuse scatter on the right represents bulls. Authors' own image.

of the value of a cow of the same age which could be subdivided according to the value of its flesh (and hide), for its work and for its potential (Kelly, 1997: 66).

Despite the fact that we are certain that dairying was the basis of the cattle economy, the age-slaughter pattern does not conform to Payne's (1973) model for dairying. In his model dairying is indicated by the deliberate slaughter of large numbers of young calves. In Early Medieval Ireland young calves are rarely present with the peaks in slaughter instead being generally for cattle in their second year and in older animals (McCormick, 2014). This age-slaughter distribution would seem to be indicative of self-sufficient producer/consumer sites, and rural sites almost invariably conform to this pattern. In 'consumption' sites, where meat is mainly sourced from outside producers, the pattern is dominated by the presence of old animals. This is particularly notable in tenth- to eleventh-century Viking Dublin but also in a small number of rural, large enclosure sites which implies a more complex rural economy than previously recognized (McCormick, 2014). A similar age-slaughter pattern has been noted at the major monastery at Clonmacnoise (Soderberg, 2003) and this and other contemporary large monasteries have been identified as 'proto-urban complexes' (Edwards, 1996: 100).

Pig

On most sites the age-slaughter patterns for pigs indicates a peak in killing occurring when they were between 1.5 and 2 years old. It can be assumed that this represents the time when the animals were reaching full size and at the optimal stage for slaughter.

In urban Dublin there is an additional peak with the slaughter of animals aged 6–12 months, reflecting an alternative rearing strategy in the urban environment. This killing-off of considerable numbers of pigs during their first year may be an adaption to the confined environment in which they were raised. Where the presence of the canines allows tooth rows to be sexed, there is a clear tendency for the killing-off of male pigs at an earlier age on most sites (McCormick and Murray, 2007: 63–4).

The early documentary sources complement the zooarchaeological evidence for pigs. Farrowing was limited to once a year, in spring, as in the case of wild pig. Litters of up to nine are recorded with the weakest often being hand-reared on milk (Kelly, 1997: 81). Pigs were fed on acorns in the woods but were frequently fattened on cereals or milk before slaughter (Kelly, 1997: 81). Documentary sources record the killing of sows after they have produced two or three litters and sows who failed to become pregnant were also killed. Much pork was consumed in salted form, as indicated, for example, in the descriptions given in the twelfth-century satirical poem Aislinge Meic Con Glinne (Meyer, 1892) but it has not been possible to demonstrate this in the zoo-archaeological record.

Sheep/Goat

The majority of caprine bones present on Early Medieval sites are of sheep. Goat (*Capra hircus*) tends to be absent from high status sites: none were noted at the royal site of Lagore with only a few remains being noted at the royal site of Knowth, though these included two juvenile goat mandibles possibly representing the consumption of kid flesh (McCormick and Murray, 2007: 60). In contrast, relatively large numbers are present in urban contexts with up to 20% of the identifiable caprine bones in Viking Dublin being of goat (McCormick, 1987: 144). This would suggest that the keeping of goats was an urban phenomenon and it is a trend that continues into the later period. Goats may have been kept in town as a source of milk and the goat horncores included a high proportion of females which would support this suggestion. Mandibular evidence indicates that the majority of sheep were killed before the age of 28 months implying that they were primarily exploited for their meat rather than for other secondary uses (i.e. wool or milk).

The documentary sources again provide valuable information to complement the faunal evidence. Except for the few rams kept for breeding all males were castrated after they were weaned and it is specifically noted that wethers were considered to be summer and autumn food (Kelly, 1997: 69). Only a small number of sheep were white in colour with the majority being dun-coloured or black. As a consequence the law tracts indicate that white sheep, presumably because of their more desirable wool, were considered much more valuable than the others (Kelly, 1997: 70). Female lambs were considered to be of more value than males, likely because of their breeding abilities and potential to produce milk. That said, sheep milk was regarded as inferior to that of goat (Kelly, 1997: 75–6).

LATE MEDIEVAL PERIOD

The onset of the Late Medieval period in Ireland is traditionally identified with the arrival of the Anglo-Normans in AD 1169. This also coincides with a period of increased trade and urbanization, the reform of the Irish Church, and the introduction of continental religious orders (O'Conor, 1998: xi). The political and social structure of the era can therefore be divided broadly into those lands and populations that were subject to the crown, those that remained under Gaelic control and lands held by the church. The conquest was incomplete and concentrated along the east coast of the island, centred on Dublin, and coinciding with many of the urban trading centres previously established by the Vikings (i.e. Dublin, Wexford, and Waterford). The introduction of cash rents as opposed to the older system of clientship and tributes paid in 'foodstuffs', also brought about significant changes in the livestock economy of the period.

Most of the Late Medieval faunal assemblages are from urban sites or castles and, geographically, there is a bias towards the eastern part of the country with the majority of the large assemblages coming from coastal towns. Very little material has been found on Late Medieval rural sites. Most of the assemblages date from between the late twelfth and the fourteenth century and, as yet, few fifteenth- to sixteenth-century sites have produced useful faunal material.

Species Frequency

There are no clear trends in the animal economy in terms of site type or date (Table 13.2). The small collection of eleventh- and twelfth-century sites mirrors the Early Medieval trend with the predominance of cattle followed by pig and then sheep/goat, though the percentage values are more variable. This would suggest a degree of conservatism in agricultural practices until the English trading networks and urban centres became more established in the thirteenth and fourteenth centuries, when the distribution of the main species becomes more changeable.

Cattle still dominate the livestock economy in Late Medieval Ireland and they are the principal species exploited for the majority (77%) of sites considered. The average value for the relative proportion of cattle is 45% of the MNI, which is comparable with Early Medieval sites (see Table 13.1) although this calculation belies the wide variability demonstrated with values ranging from 15% to 70%. Such extremes in the exploitation of cattle are not evident in the earlier period.

The two sites with the highest value for pigs, 61% and 53%, are castle sites. As pigs are not multipurpose animals and are only exploited for their meat, the consumption of pork carried a status value (see below). The occurrence of pig in higher proportions on Late Medieval elite sites, such as castles, is therefore unsurprising and is a pattern recorded across Europe (Thomas, 2007: 138). An alternative explanation for the increased exploi-tation of pigs is that they were imported for the provisioning of billeted soldiers. Irish State Papers indicate that pigs, both live and as carcasses, were sent to Ireland in the late twelfth and early thirteenth centuries to supply the Anglo-Norman army (McCormick, 1991a: 48). In contrast to Viking Dublin, pigs are never the dominant species in urban settlements of this period as, unlike Viking Dublin, Anglo-Norman towns were located within friendly hinterlands and the need to be self-sufficient was less of a concern.

Historical records for the period would suggest that we should see a significant rise in the exploitation of sheep given the importance of the wool trade (O'Neill, 1987: 58–67). In Ireland, this is certainly noted at a number of sites in the east and southeast (samples too small for inclusion in the tables here) which, it has been argued, reflect the influence of Cistercian foundations in these regions (McCormick, 1991a: 46). Relatively high pro-portions of sheep are also noted in urban sites including Cornmarket in Dublin which may be a reflection of the wool trade and the supply of sheep to the meat market after shearing. Over time, when the proportions of pig and sheep/goat are compared, a grad-ual increase in the latter at the expense of the former is observed, suggesting a decline in the role of pork and increase in mutton in the diet. This is a trend also recorded in Medieval England (Thomas, 2007: 138).

Cattle

The ageing evidence for cattle indicates that older animals (36+ months), and predom-inately female, dominated the urban assemblages with an increase in the percentage of older individuals over time (Denham, 2007: 204–5, 215). Older cattle, again mostly female, also dominated the contemporary castle assemblages (Denham, 2007: 207–8). This suggests that the Medieval urban markets were being supplied by cattle where meat was a secondary use of the animals and that the suppliers, not the consumers, were dictating the meat supplied which evidently derived from former dairy herds. Females were not used for traction at this time. There is some evidence that urban dwellers sometimes raised animals. In Dublin, for instance, cattle kept by freeman of the city were grazed within the city and its suburbs though regulations were introduced to try and manage these practices (Cantwell, 2001: 75–6). The fact that young calves are rarely encountered in the zooarchaeological record might imply that this refers to the fattening of animals for the meat market rather than rearing the animals from birth. In the later Medieval period, cow hides and calfskins, along with fish, were Ireland's most important exports (O'Neill, 1987: 77). There is some limited contemporary zooarchae-ological evidence for the processing of calf skins, and hornworking, from a number of the Late Medieval Dublin sites (Denham, 2007: 226–7). In particular, it is suggested that large collections of horns from very young cattle represent debris from vellum-making (McCormick, 2004).

Table 13.2 MNI percentages for the main domesticates from Late Medieval Irish sites

Site and phase/feature	Date (centuries ad)	Site type	Cattle %	Sheep/Goat %	Pig %	MNI Total	Source
Trim Castle, S1	L13th–E14th	castle	41	31	28	68	Murray and McCormick, 2011
Clough Castle	E13th	castle	27	12	61	92	Jope, 1954
Greencastle	M13th	castle	56	21	23	57	McCormick, undated (a)
Greencastle	14th–15th	castle	43	36	21	107	McCormick, undated (a)
Carrickmines castle	medieval	castle	23	39	39	109	Denham, 2007
Maynooth castle	14th–15th	castle	18	29	53	55	Murray, undated
Galway, Courthouse Lane A2	High medieval	castle/urban	50	29	21	48	Murray, 2004
Newtownards, Movilla Abbey	13th–14th	ecclesiastical	43	40	17	58	Higgins, 1984b
Knowth, ST9	10th–11th	rural	43	24	33	240	McCormick and Murray, 2007
Knowth, ST10	L12th–16th	rural	41	27	32	41	McCormick and Murray, 2007
High St, Trim	13th	rural	53	23	23	43	McCormick, 1991b
Dublin, Essex St. West, T1&T2	13th	urban	39	43	18	51	Bermingham, 1995
Dublin, Arran Quay	L14th–L15th	urban	34	42	24	144	McCormick, 2004
Dublin, Patrick St, Sites B&C	L13th–14th	urban	33	36	31	87	McCormick and Murphy, 1997
Dublin, Wood Quay	13th	urban	40	36	24	917	Butler, 1984
Dublin, Back Lane	E12th–13th	urban	43	14	43	44	McCormick and Murphy, undated
Dublin, Fishamble St.	L11th–12th	urban	59	11	30	97	McCormick, 1987
Carrickfergus	13th–14th	urban	49	25	25	63	Murphy, 1999a
Carrickfergus	14th–16th	urban	70	26	4	46	Murphy, 1999a

(Continued)

Table 13.2 Continued

Site and phase/feature	Date (centuries ad)	Site type	Cattle %	Sheep/Goat %	Pig %	MNI Total	Source
Armagh, Scotch Street	medieval	urban	47	22	31	45	McCormick, undated (b)
Cork, Barrack St.	L12th–E14th	urban	53	24	22	58	McCarthy, 1993
Cork, French's Quay	L12th–E14th	urban	48	25	27	75	McCarthy, 1993
Dublin, Bridge St. Upper	14th–15th	urban	52	24	24	42	MacManus, 1995
Dublin, Christchurch place	10th–12th	urban	63	13	24	437	Cremin, 1996
Dublin, Cornmarket St.	L13th	urban	37	47	16	57	MacManus, 1995
Dublin, Cornmarket St.	14th–15th	urban	37	54	10	82	MacManus, 1995
Dublin, High St.	L12th–E13th	urban	41	17	41	46	McCormick, undated (c)
Dublin, Wood Quay	13th	urban	40	23	36	94	Butler, 1984
Kilkenny, Patrick St./ Pudding Lane	E13th–M14th	urban	61	21	18	99	Murphy, 1999b
Drogheda, James St.	E16th	urban	44	36	20	86	McCormick, undated (d)
Waterford, Bakehouse lane	M12th	urban	42	38	21	173	McCormick, 1997
Waterford, High St.	L13th–E14th	urban	15	47	38	95	McCormick, 1997
Waterford, Peter St.	M11th–E12th	urban	43	21	36	91	McCormick, 1997
Waterford, Peter St	E/M12th	urban	49	22	30	105	McCormick, 1997
Waterford, Peter St	M12th–E13th	urban	47	29	24	154	McCormick, 1997
Waterford, Peter St	13th	urban	41	30	29	66	McCormick, 1997
Downpatrick	medieval	urban	61	20	20	92	McCormick, undated (e)

Pig

In the Late Medieval period the majority of pigs were killed in their second year. However, in urban assemblages, like Viking Dublin, there is a secondary peak in the 7–12 month age-bracket with younger and older individuals also present (Denham, 2007: 254) indicating a continuation of urban pig rearing strategies. The keeping of pigs in Dublin is demonstrated by the numerous regulations from the fifteenth and sixteenth centuries forbidding their presence, along with surviving accounts of their often destructive interactions (Cantwell, 2001: 76–7). On castle sites there is a single peak with the majority of animals killed towards the end of their second year, mirroring the pattern of Early Medieval rural sites (Denham, 2007: 256). A trend observed on castle sites is a higher frequency of cranial elements, and to a lesser extent limb extremities, compared to pig bone assemblages from urban sites. All sites indicate on-site butchery but the pattern noted at castles would suggest that there was a greater emphasis on maximizing the meat resource available (Denham, 2007: 260–2). The flesh of piglets appears to have been considered a high status delicacy as an assemblage of very young pig bones were found in association with high status imported German pottery in thirteenth-century Waterford (McCormick, 1997: 831).

Sheep/Goat

The age-slaughter data for sheep for this period indicates two peaks—sheep killed in their second year and older individuals. This suggests that they were exploited for both their meat (the younger age group) and for their wool but that there is not a single economic strategy. Perhaps the fact that the main breeds were hairy and dark in colour made the fleeces less desirable and profitable at market. Some sites in Dublin and Limerick show a decrease over time in the exploitation of older animals, mirroring the decline in the wool trade by the fifteenth and sixteenth centuries (Denham, 2007: 232–3). This, however, is not universal. The pattern from castle sites is more mixed with an apparent emphasis on younger animals, and presumably meat, at Trim and older animals at Carrickmines (Denham, 2007: 235–6). The biometrical data indicates that the size and stature of sheep does not change post-conquest (Denham, 2007: 239).

In Late Medieval urban assemblages goats frequently constitute a much higher percentage of the sheep-to-goat ratio than noted in earlier periods and in many cases this is accounted for largely, but not exclusively, by horncores. The presence of the latter is generally interpreted as evidence for horn working while postcranial elements are seen as an indicator of urban goat rearing and both practices appear to have been employed in Medieval towns. The sex ratio of horncores shows a predominance of females which supports the idea of urban goat rearing for dairying. The disparity between the frequency of horncores and postcranial elements, however, would suggest that horncores were being deliberately imported, perhaps along with their skins, a phenomenon also recorded in Medieval animal bones assemblages from England (Serjeantson, 1989). Unfortunately useful comparative data from rural sites from the period is lacking, which might otherwise have allowed a determination of whether they were being supplied from the rural hinterlands or further afield. Castle sites do not follow this urban trend (Denham, 2007: 248).

Other Domesticates

The other typical domesticates—horse (*Equus caballus*), dog (*Canis familiaris*), cat (*Felis catus*), and domestic fowl (*Gallus gallus*)—are widely recorded on sites from both periods, though in small numbers, typically accounting for less than 10% of the total MNI in the larger assemblages (Denham, 2007; McCormick and Murray, 2007).

An urban–rural dichotomy in how dogs and cats were treated has been noted with feline skeletal remains on urban sites, both Viking and Anglo-Norman, often being on average smaller and younger than those from rural assemblages suggesting that they were not treated with the same care in towns (McCormick, 1988). Zooarchaeological evidence for skinning, for both dogs and cats, is also predominately an urban trait (McCormick and Murray, 2007: 49–50) and surviving accounts record the trading in cat skins and export of dog pelts from Youghal in the fourteenth and sixteenth centuries (McCormick, 1991a: 49). The biometrical data for dogs shows a bimodal distribution in the early period indicating two distinct size groups, but by the later period these merge (Fig. 13.3). This would suggest strict supervision of breeding in Early Medieval Ireland which is not sustained over time, implying a freely interbreeding 'mongrel' population in the towns, at least, in later centuries. The evidence would suggest that the role of both cats and dogs changed over the centuries from that of pets to one of an economic commodity (McCormick, 1991a: 44–5). The early literature refers to specific cats kept as pets, the most famous being *Pangur bán*, the pet cat of a monk, while they were also valued for their ability to purr (Kelly, 1997: 121–2). This is also supported by the zooarchaeological data which indicate that cats on rural Early Medieval sites tended to be larger and have a longer life expectancy than those from either Scandinavian Dublin or Anglo-Norman towns (McCormick, 1988).

Horse bones are found in small numbers on the majority of sites right across the period. Their bones are recovered with the food refuse of other domesticates, are most often bones of adult animals, and frequently with signs of butchery, indicating that they were exploited to their full capacity (McCormick, 2007: 92). The eating of horse meat may have been confined to times of stress or food shortages although it was a religious taboo according to an old Irish penitential (Kelly, 1997: 352). It can be noted that the highest incidence of horse remains found on an Early Medieval site was at a site of low status (McCormick, 2007: 92) implying that horsemeat was consumed by persons who could less afford other meats. An alternative possibility is that on low status sites, due to lower meat consumption, non-meat species tend to be proportionally better represented. Horses were associated with the aristocracy and people of high rank (Kelly,

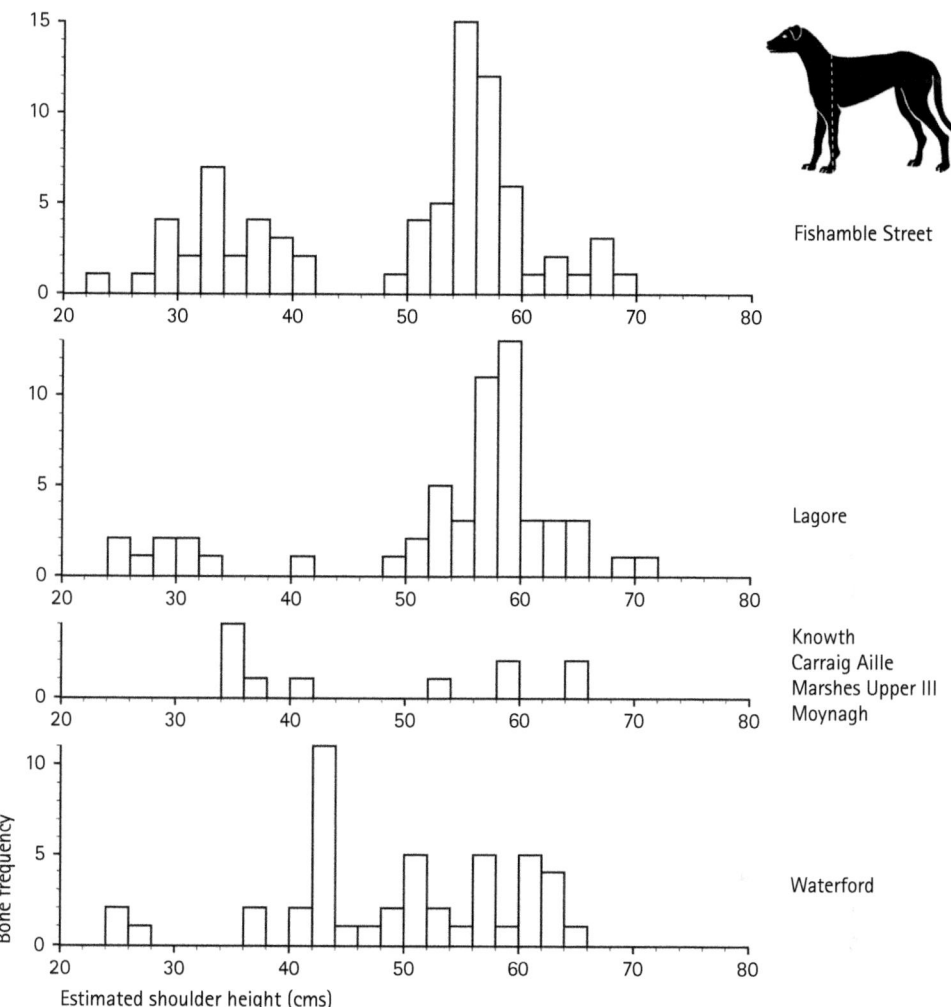

FIGURE 13.3 Estimated shoulder heights of Irish dogs from Medieval sites. With the exception of the Waterford material, which dates from the early 12th C to early 14th C, all of the sites date to the Early Medieval period (from McCormick, 1991a: 45).

1997: 89) and it is of note that the largest horses recorded in faunal assemblages from the early period are from known royal sites (McCormick, 2007: 95). Horses were used for riding and agricultural work though much of the contemporary references are concerned with horse racing and their use in warfare (Kelly, 1997: 89). This is also the case in the Late Medieval period (Denham, 2007: 104). The earliest evidence for the use of horses for ploughing in Ireland is the late thirteenth century, though the replacement of oxen with horses for ploughing was a gradual shift, and not fully realized until the fifteenth century (McCormick, 2007: 96). This contrasts with England, where plough teams comprised only of horse were in use by the mid-twelfth century alongside teams of oxen, as well as mixed teams of horses and oxen (Langdon, 1986: 51).

According to early written sources, hens, and their eggs, were highly valued goods (Kelly, 1997: 102–4). This contrasts with the zooarchaeological record where bones of domestic fowl are commonly present but they are infrequent with the exception of Viking Dublin (Hamilton-Dyer, 2007: 108; McCormick and Murray, 2007: 75). In later centuries the Anglo-Normans were much keener consumers of fowl, both domestic hens and geese (*Anser* sp.) and these dominate the bird bone assemblages from urban and castle sites (Hamilton-Dyer, 2007: 108). The English introduced dovecotes, domestic geese, and pheasants (*Phasianus colchicus*), with the earliest evidence for the latter recovered from Anglo-Norman deposits at Trim Castle (Hamilton-Dyer, 2007: 109). The limited zooarchaeological data from contemporary Gaelic sites suggests that they remained conservative in their meat-eating practices and the frequency of fowl bones stays low (McCormick, 1991a: 49; Hamilton-Dyer, 2007: 109).

Wild Animals

The Anglo-Normans introduced both fallow deer (*Dama dama*) and rabbits (*Oryctolagus cuniculus*) to Ireland, and possibly also the hedgehog (*Erinaceus europaeus*) (McCormick, 1999) with the former imported from the thirteenth century onwards to stock deer parks, another English introduction (Beglane, 2010: 81). Red deer (*Cervus elaphus*) remains, antler and postcranial elements, are present on most Early Medieval sites (McCormick and Murray, 2007: 192–5). The frequency of deer elements during both periods, however, is low, indicating that venison played a very minor role in the diet. Where specialized antler workshops have been identified, the material used is mainly derived from shed antler (McCormick, 1997: 836). In written sources deer-hunting is typically associated with royalty and a king would have had at least one hunter in his employment (Kelly, 1997: 273–4). On Gaelic Late Medieval sites red deer remains have also been recovered though none, to date, have yielded any fallow deer bones which have been found exclusively on castle and urban sites from the eastern half of the country (Beglane, 2010: 81). Hunting would have been important to the nobility of both cultures, but how this was approached and executed was evidently very different.

There is limited evidence for the exploitation of wildfowl in Medieval Ireland. In the early period, on sites where bird bones have been recovered, the range of species represented suggests opportunistic exploitation of the local environment such as wetlands or coastal cliffs, probably seasonal, with wildfowl making an insignificant contribution to the diet (McCormick and Murray, 2007: 74–5). The zooarchaeological data suggest that the ducks and geese exploited in the early period were not domesticated and this is corroborated by the documentary sources (McCormick and Murray, 2007: 74–5). Bones of raptors are relatively common on rural and urban Medieval sites (Hamilton-Dyer, 2007: 109), reflecting the fact that they were then more widespread in the countryside and, as the laws indicate, would have been considered scavengers and vermin (Kelly, 1997: 303). The use of raptors for hawking and falconry was an Anglo-Norman introduction with many references to the practice in contemporary documents (Kelly,

1997: 303); however, direct zooarchaeological evidence for this activity has yet to be found (Hamilton-Dyer, 2007: 110).

As with birds, fish and shell-fish were exploited opportunistically in the Early Medieval period, almost exclusively at coastal sites although the presence of cod (*Gadus morhua*) and hake (*Merluccius merluccius*), along with salmon (*Salmo salar*), at Knowth (*c.*12 miles inland) implies some form of trade or exchange in fish at this time (McCormick and Murray, 2007: 75). The monastic sites of Illaunloughan and Clonmacnoise, coastal and riverine respectively, also had a relatively higher frequency of fish (Hamilton-Dyer, 2007: 113), which may be a reflection of their restricted diets. In the early written sources fish and specifically salmon are frequently referred to (Kelly, 1997: 285) and this discrepancy with the zooarchaeological record may in part be taphonomic, though attitudes and culture must have played a role. This all changed with the arrival of the Anglo-Normans who developed both an internal and external trade in fish and shell-fish and introduced new fresh-water species. Ireland's fresh-water fish population was depleted during the last Ice Age and, post-glaciation, was limited to migrating species: salmon, trout (*Salmo trutta*), eel (*Anguilla anguilla*) and lamprey (*Lampetra* sp.), and possibly a few landlocked species, e.g. pollan (*Coregonus pollan*). The majority of fresh-water fish are therefore post-Norman introductions (McCormick, 1999: 367). Coastal urban sites have produced the largest fish-bone assemblages which are dominated by large offshore species, principally Gadidae, which fits in with trends, observed across contemporary Europe (Hamilton-Dyer, 2007: 113–15). The one contemporary assemblage from a castle, Trim, was also dominated by Gadidae and produced the earliest evidence for pike (*Esox lucius*) in Ireland dating to late thirteenth-and early fourteenth-century contexts (Hamilton-Dyer, 2007: 115). Recent genetic studies, however, suggest that pike may have colonized Ireland independently much earlier, with a second 'wave' in the Late Medieval period (Pedreschi et al., 2013). The fishing and trade in herring (*Clupea harengus*) was a hugely important industry in the Medieval period and the principal fisheries in Ireland were off the western and southwestern coasts and in the Irish Sea, with their trade conducted in respective regional port towns (O'Neill, 1987: 30–6). This is not reflected in the zooarchaeological data with relatively few herring bones represented in contemporary assemblages (Hamilton-Dyer, 2007: 115).

Conclusion

The study of faunal remains from Medieval Irish sites has a long history. William Wilde's study of the animal bones from the crannog at Dunshaughlin, Co. Meath published in 1840 must be one of the earliest studies of such material from an archaeological site. During the 1950s Margaret Jope was amongst the first to include estimations of minimum numbers of individual values in her reports (e.g. Waterman and Jope, 1954). A complete overview of the Early Medieval period was published in 2007 (McCormick and Murray, 2007). Since then there has been a significant amount of archaeological excavations undertaken in advance of the massive motorway building programmes in recent years which has generated a large body of zooarchaeological data (McCormick et al., 2014). Hopefully, the full publication of these sites will fill in many of the gaps in our knowledge of this period. In general the zooarchaeological evidence from the early period displays considerable uniformity, which may reflect the former role of cows as a currency. However, during the later period livestock regimes become more variable reflecting both particular geographical adaptations and the growing market economy of animal produce.

REFERENCES

Albarella, U. (1997) 'Shape variation of cattle metapodials: age, sex or breed? Some examples from Medieval and Postmedieval sites', *Anthropozoologica*, 25–6, 37–47.

Beglane, F. (undated) 'Report on Faunal Material from Ratoath, Co. Meath. Lic no. 03E1781'. Unpublished report prepared for Arch-Tech.

Beglane, F. (2010) 'Deer and identity in Medieval Ireland', in Kucera, M. and Kunst, G. K. (eds) *Bestial Mirrors: Animals as Material Culture in the Middle Ages—3. Using Animals to Construct Human Identities in Medieval Europe*, pp. 77–84. Vienna: Vienna Institute for Archaeological Science.

Bermingham, N. (1995) 'Animal remains', in Simpson, L. (ed.) *Excavations at Essex Street West, Dublin*. Temple Bar Archaeological Report 2, pp. 101–3. Dublin: Temple Bar Properties.

Bonar, C. (2001) 'The Faunal Remains from Clogher'. Unpublished MSc dissertation, Queen's University Belfast (Belfast).

Bonar, C. (2003) 'Analysis of mammal bones from Johnstown 1, Co. Meath (Appendix 7)', in Carlin, N., Clarke, L., and Walsh, F. (eds) *The Archaeology of Life and Death in the Boyne Floodplain: The Linear Landscape of the M4*. National Roads Authority Scheme Monographs 2, pp. 1–57. Dublin: National Roads Authority/Wordwell.

Butler, V. (1984) 'Cattle in Thirteenth Century Dublin: An Osteological Examination of Its Remains'. Unpublished BA dissertation, The National University of Ireland (Dublin).

Byrne, F. J. (2001) *Irish Kings and High Kings*, Dublin: Four Courts Press.

Cantwell, I. (2001) 'Anthropozoological relationships in Late Medieval Dublin', *Dublin Historical Records*, 54(1), 73–80.

Coles, C. (2009) 'Animal bone report', in Cagney, L., O'Hara, R., Kelleher, G., and Morkan, R. (eds) *Report on the Archaeological Excavation of Dowdstown 2, Co. Meath*. Unpublished report prepared for Archaeological Consultancy Services Ltd.

Corlett, C. and Potterton, M. (eds) (2010) *Death and Burial in Early Medieval Ireland in the Light of Recent Archaeological Excavations*, Bray: Wordwell.

Corlett, C. and Potterton, M. (eds) (2011) *Settlement in Early Medieval Ireland in the Light of Recent Archaeological Excavations*, Bray: Wordwell.

Cremin, A. (1996) 'Animal Bones from Christ Church Place, Dublin, 10th–12th Century'. Unpublished MA dissertation, Queen's University Belfast (Belfast).

Denham, S. (2007) 'Animal Exploitation in Medieval Ireland'. Unpublished PhD dissertation, Queen's University Belfast (Belfast).

Edwards, N. (1996) *The Archaeology of Medieval Ireland*, London: B.T. Batsford Ltd.

Foster, H. (2009) 'Animal bone report from Castlefarm I, Co. Meath', in O'Connell, A. and Clark, A. (eds) *Report on the Archaeological Excavation of Castlefarm 1, Co. Meath.* Unpublished report prepared for Archaeological Consultancy Services Ltd.

Grigson, C. (1982) 'Sex and age determination of some bones and teeth of domestic cattle: a review of the literature', in Wilson, B., Grigson, C., and Payne, S. (eds) *Ageing and Sexing Animal Bones from Archaeological Sites.* BAR British Series 109, pp. 7–23. Oxford: Archaeopress.

Hamilton-Dyer, S. (2007) 'Exploitation of bird and fish in historic Ireland: a brief review of the evidence', in Murphy, E. M. and Whitehouse, N. J. (eds) *Environmental Archaeology in Ireland*, pp. 102–18. Oxford: Oxbow.

Higgins, V. (1984a) 'The animal bone', in Gaskell Brown, C. and Harper, A. E. T. (eds) Excavations on Cathedral Hill, Armagh, 1968, *Ulster Journal of Archaeology*, 47, 109–61.

Higgins, V. (1984b) 'The animal remains', in Ivens, R. (ed.) Movilla Abbey Newtownards, Co. Down: excavations 1981, *Ulster Journal of Archaeology*, 47, 71–108.

Higham, C. F. W. (1969) 'The metrical attributes of two samples of bovine limb bones', *Journal of Zoology: Proceedings of the Zoological Society of London*, 157, 63–74.

Jope, M. (1954) 'Animal remains from Clough Castle', in Waterman, D. M. (ed.) Excavations at Clough Castle, *Ulster Journal of Archaeology*, 17, 103–63.

Kelly, F. (1988) *A Guide to Early Irish Law*, Dublin: Dublin Institute for Advanced Studies.

Kelly, F. (1997) *Early Irish Farming*, Dublin: Dublin Institute for Advanced Studies.

Kerr, T. R., Swindles, G. T., and Plunkett, G. (2009) 'Making hay while the sun shines? Socio-economic change, cereal production and climatic deterioration in Early Medieval Ireland', *Journal of Archaeological Science*, 36(12), 2868–74.

Langdon, J. (1986) *Horses, Oxen and Technological Innovation*, Cambridge: Cambridge University Press.

Lofquist, C. (2002) 'Animal bones', in Fredengren, C. (ed.) *Crannogs: A Study of People's Interaction with Lakes, with Particular References to Lough Gara in the North-West of Ireland*, pp. 142–84. Bray: Wordwell.

Lofquist, C. (2008) 'Osteological report from Killickaweeny: Appendix 1', in Carlin, N., Clarke, L., and Walsh, F. (eds) *The Archaeology of Life and Death in the Boyne Floodplain: The Linear Landscape of the M4.* National Roads Authority Scheme Monographs 2, pp. 27–54. Dublin: National Roads Authority/Wordwell.

Lucas, A. T. (1989) *Cattle in Ancient Ireland*, Kilkenny: Boethius.

MacManus, C. (1995) 'A Study of Excavated Bones from Thirteenth and Eighteenth Century Dublin'. Unpublished BA dissertation, Queen's University Belfast (Belfast).

McCarthy, M. (1993) 'Medieval remains', in O'Brien, M. (ed.) *Excavations at Barrack Street-French's Quay, Cork*, pp. 43–5. *Journal of Cork Historical and Archaeological Society*, 98, 27–49.

McCormick, F. (undated a) 'The Animal Bones from Greencastle, Co. Down'. Unpublished report.

McCormick, F. (undated b) 'The Animal Bones from Scotch Street Armagh'. Unpublished report.

McCormick, F. (undated c) 'The Animal Bones from High Street, Dublin'. Unpublished report.

McCormick, F. (undated d) 'The Animal Bones from James Street, Drogheda, Co. Louth'. Unpublished report.

McCormick, F. (undated e) 'The Animal Bones from Downpatrick, Co. Down'. Unpublished report.

McCormick, F. (1987) 'Stockrearing in Early Christian Ireland'. Unpublished PhD dissertation, Queen's University Belfast (Belfast).

McCormick, F. (1988) 'The domesticated cat in early Christian and Medieval Ireland', in MacNiocaill, G. and Wallace, P. F. (eds) *Keimelia: Studies in Medieval Archaeology and History in Memory of Tom Delaney*, pp. 218–28. Galway: Galway University Press.

McCormick, F. (1991a) 'The effect of the Anglo-Norman settlement in Ireland's wild and domesticated fauna', in Crabtree, P. J. and Ryan, K. (eds) *Animal Use and Culture Change*. MASCA Research Papers in Science and Archaeology 8, pp. 40–52. Philadelphia: University of Pennsylvania.

McCormick, F. (1991b) 'The animal bones from High Street, Trim', in Walsh, C. (ed.) An excavation at the library site, High Street, Trim, *Riocht na Mide*, 8, 53–7.

McCormick, F. (1992a) 'Early faunal evidence for dairying', *Oxford Journal of Archaeology*, 11(2), 201–9.

McCormick, F. (1992b) 'The animal bones', in Gowen, M. (ed.) Excavation of two souterrain complexes at Marshes Upper, Dundalk, Co. Louth, *Proceedings from the Royal Irish Academy*, 92C, 113–19.

McCormick, F. (1997) 'The animal bones', in Hurley, M. F., Scully, O. M. B, and McCutcheon, S. (eds) *Late Viking Age and Medieval Waterford: Excavations 1986–92*, pp. 819–53. Waterford: Waterford Corporation.

McCormick, F. (1999) 'Early evidence for wild animals in Ireland', in Benecke, N. (ed.) *The Holocene History of European Vertebrate Fauna: Modern Aspects of Research*. Archaeologie in Eurasien 6, pp. 355–71. Rahden: Verlag Marie Leidorf.

McCormick, F. (2004) 'The mammal bone', in Hayden, A. (ed.) Excavation of the Medieval river frontage at Arran Quay, Dublin, *Medieval Dublin*, 5, 221–31.

McCormick, F. (2007) 'The horse in early Ireland', *Anthropozoologica*, 42(1), 85–104.

McCormick, F. (2014) 'Agriculture, settlement and society', *Quaternary International*, 346, 119–30.

McCormick, F., Kerr, T., McClatchie, M., and O'Sullivan, A. (2014) *Early Medieval Agriculture, Livestock and Cereal Production in Ireland, ad 400–1100*. BAR International Series 2647. Oxford: Archaeopress.

McCormick, F. and Murphy, E. (undated) 'The Animal Bone from Back Lane, Dublin'. Unpublished report.

McCormick, F. and Murphy, E. M. (1997) 'Mammal bones', in Walsh, C. (ed.) *Archaeological Excavations at Patrick, Nicholas and Winetavern Streets*, pp. 199–218. Dingle: Brandon Book Publishers.

McCormick, F. and Murray, E. (2007) *Knowth and the Zooarchaeology of Early Christian Ireland*, Dublin: Royal Irish Academy.

McCormick, F. and Murray, E. (2011) 'The animal bones from Deer Park Farms', in Lynn, C. J. and McDowell, J. A. (eds) *Deer Park Farms: The Excavation of a Raised Rath in the Glenarm Valley, Co. Antrim*, pp. 469–88. Belfast: TSO.

McQuade, M. (2001) 'Analysis of the faunal remains: Colp West, Co. Meath', in Clarke, L. and Murphy, D. (eds) Report on the Archaeological Resolution of a Multiperiod Settlement Site at Colp West, Co. Meath. Unpublished report prepared for Archaeological Consultancy Services Ltd.

Meyer, K. (ed.) (1892) *Aislinge Meic Conglinne, The Vision of MacConglinne, a Middle-Irish Wonder Tale, with a Translation*, London: D. Nutt.

Murphy, E. M. (1999a) 'Osteological Report on the Mammal Bones from Carrickfergus Castle, Co. Antrim'. Unpublished report.

Murphy, E. M. (1999b) 'Osteological Report on the Mammal Bones from Patrick Street/ Pudding Lane, Kilkenny'. Unpublished report for Judith Carroll Archaeological Consultancy.

Murray, E. (undated) 'The Faunal Remains from Maynooth Castle, Co. Meath'. Unpublished report.

Murray, E. (2004) 'Animal bone', in Fitzpatrick, E., O'Brien, M. and Walsh, P. (eds) *Archaeological Excavation in Galway City, 1987–1999*, pp. 562–601. Bray: Wordwell.

Murray, E. (in press) 'Hogget, beef and corncrake: the animal bone assemblage', in Seaver, M. (ed.), *Meitheal: The Archaeology of Lives, Labours and Beliefs at Raystown, Co. Meath*, pp. 104–21. Dublin: Transport Infrastructure Ireland.

Murray, E. and McCormick, F. (2011) 'The animal bones', in Hayden, A. R. (ed.) *Trim Castle, Co. Meath: Excavations 1995–8*. Archaeological Monograph Series 6, pp. 419–31. Dublin: The Stationery Office.

Murray, E., McCormick, F., and Plunkett, G. (2004) 'The food economies of Atlantic Irish monasteries', *Environmental Archaeology*, 9, 179–89.

O'Connor, T. (2011) 'Animal husbandry', in Hinton, D. and Hamerow, H. (eds) *Handbook of Anglo-Saxon Archaeology*, pp. 363–78. Oxford: Oxford University Press.

O'Conor, K. (1998) *The Archaeology of Medieval Rural Settlement in Ireland*, Dublin: Discovery Programme.

O'Neill, T. (1987) *Merchants and Mariners in Medieval Ireland*, Dublin: Irish Academic Press.

Payne, S. (1973) 'Kill-off patterns in sheep and goats: the mandibles from Aşvan Kale', *Anatolian Studies*, 23, 281–305.

Pedreschi, D., Kelly-Quinn, M., Caffrey, J., O'Grady, M., and Mariani, S. (2013) 'Genetic structure of pike (*Esox Lucius*) reveals a complex and previously unrecognized colonization history of Ireland', *Journal of Biogeography*, 40, 1–13.

Serjeantson, D. (1989) 'Animal remains in the tanning trade', in Serjeantson, D. and Waldron, T. (eds) *Diet and Crafts in Towns*. BAR British Series 199, pp. 129–46. Oxford: Archaeopress.

Sloane, R. (2009a) 'Appendix 8: faunal remains report', in Linnane, S. and Kinsella, J. (eds) Report on the Archaeological Excavations of Baronstown 1, Co. Meath. Unpublished report.

Sloane, R. (2009b) 'Animal Bone Analysis: Roestown 2, Co. Meath'. Unpublished report.

Soderberg, J. (2003) 'Feeding the Community: Urbanization, Religion, and Zooarchaeology at Clonmacnoise, an Early Medieval Irish Monastery'. Unpublished PhD dissertation, University of Minnesota (Minneapolis).

Stout, M. (1997) *The Irish Ringfort*, Dublin: Four Courts Press.

Thomas, R. M. (2007) 'Food and the maintenance of social boundaries in Medieval England', in Twiss, K. C. (ed.) *The Archaeology of Food and Identity*. Occasional Paper 34, pp. 130–51. Illinois: Centre for Archaeological Investigations.

Thomas, R. N. (1988) 'A statistical evaluation of criteria used in sexing cattle metapodials', *Archaeolzoologia*, 2, 83–92.

Waterman, D. and Jope, M. (1954) 'Excavations at Clough Castle', *Ulster Journal of Archaeology*, 17, 103–63.

Wilde, W. (1840) 'Animal remains and antiquities found at Dunshaughlin, County Meath', *Proceedings of the Royal Irish Academy*, 1, 420–8.

CHAPTER 14

ANIMALS IN URBAN LIFE IN MEDIEVAL TO EARLY MODERN ENGLAND

TERRY O'CONNOR

INTRODUCTION AND GENERAL CONTEXT

ANIMALS are omnipresent in human lives, and animal bones are ubiquitous on excavations in English historic towns. Distinctive urban geochemistry often allows the survival of large quantities of bone fragments even in regions where the prevailing geology might not seem favourable (Fig. 14.1). The great majority of those bones derive from human activities within the town, such as acquiring, distributing, and consuming meat and other carcass products, or working horn and bone into artefacts. From the rebirth of English towns in the eighth and ninth centuries to the recent past, animals have come into towns, and their remains are an abundant and informative part of the archaeological record.

This paper reviews the place of animals in the lives of the people who populated towns in England from Saxon times to the nineteenth century. This is not a précis of the zoo-archaeological record, but an attempt to understand how animals, live and dead, featured in urban life through that millennium. To do so, it has been necessary to consider the animals in expedient categories that have porous boundaries. In most assemblages, the predominant remains are those of animals that were of economic value, raised as domestic animals often some distance from where their remains are excavated. The term *livestock* covers these animals satisfactorily. Other remains will include animals kept in urban households, serving a variety of functions but principally acting as *companions*. A third group are animals that adopted the urban environment for the resources and opportunities it offered: these we can term *commensal* animals (O'Connor, 2013). Between them, livestock, companion, and commensal animals constitute the great majority of all animal remains excavated from English towns and, crucially, the great

FIGURE 14.1 Excavations in English historic towns often produce copious quantities of well-preserved animal bones. Author's own image.

majority of the interactions that the people of those towns had with animals. Some species feature in more than one category: a ewe may have been livestock but her lamb adopted as a companion, or a tame jackdaw may have made the transition from commensal to companion. The object of this essay is to understand how different species, populations of species, and individuals within those populations may have featured in urban human lives, taking the zooarchaeological record as the main source of evidence.

LIVESTOCK

Zooarchaeological studies commonly use the evidence from urban contexts to infer husbandry decisions and strategies carried out in the pastoral hinterland, e.g. Landon (1997) for Colonial America and O'Connor (2010) for Medieval northern Europe. Mortality profiles are key to such investigations, focusing on the production of specific primary and secondary resources. Historical zooarchaeology shares this with the zooarchaeology of prehistoric sites, and the methodologies involved are seldom specifically adapted to the historical context (Landon, 2005). Less often addressed is the place of livestock in everyday urban lives. The animal bone record shows that (mostly) cattle (*Bos taurus*), sheep (*Ovis aries*), pigs (*Sus domesticus*), and occasionally goats (*Capra hircus*) were brought into Medieval and later towns from their hinterlands, and from considerable distances from Late Medieval times onwards (Finberg, 1954; Keene, 2012). The frequent presence of cranial and foot bones in urban refuse indicates that these livestock generally entered towns as live animals, and some historic towns retain areas of common pasture where incoming livestock could be held (Bowden and Smith, 2013). The noise and smell of livestock must have been familiar, reducing the contrast between 'town' and 'country'. Although household slaughtering of the occasional sheep or pig may have continued into relatively recent times, animals were increasingly slaughtered by specialized butchers as the Medieval period wore on, shown by the documented emergence of butchers' guilds (Rixson, 2000). Townsfolk would therefore have been familiar from childhood with the full supply chain from noisy and noisome living beasts to skins, bones, and joints of meat. Medieval urban ordnances bristle with restrictions on the dumping of entrails and other foul wastes, showing both the scale of the problem and the ineffectiveness of fines and other threats. As Carr (2008: 461) says: 'Try as the towns might and did, butchers were a tough group to deodorize.'

How general was access to fresh meat in different towns at different times? The abundance of animal bone refuse seems consistent with quite general availability. Furthermore, dietary stable isotope analysis of Medieval human skeletons shows that most people had at least some animal protein in their diet (Müldner and Richards, 2005). The quality of the meat available to different people is another matter. Intensity of carcass utilization may indicate socioeconomic 'status', showing that some households needed to extract food value from elements of low utility, for example by splitting cattle metapodials and other limb elements to extract marrow (e.g. Crabtree, 2014), while others could be both selective and relatively wasteful. In fact, households that had no need of marrow and other bone products may not have acquired beef 'on the bone' and so may yield relatively little cattle bone in their refuse. The rise in relative abundance of sheep bones in Late to post-Medieval assemblages from English towns, coinciding with the rise in wealth largely based on wool, may reflect changing attitudes to carcass utility. Such interpretations are complicated by the emergence of greater control over waste disposal (Sabine, 1933; Carr, 2008), potentially leading to an over-representation of cattle in 'town dumps' and of sheep and other smaller-boned taxa in household refuse. Very broadly, cattle bones predominate in urban bone assemblages from England. In the more easterly parts of the country, and particularly in post-Medieval assemblages, sheep bones sometimes outnumber those of cattle, though cattle would still have predominated in terms of meat yield and other carcass products.

Cattle featured in urban lives in many other ways. Each beef carcass would have yielded hide, hoof, horn, and bone to be worked into artefacts familiar to everyone. Where water-logging and careful excavation have yielded Medieval leather artefacts and off-cuts, cattle and calf leathers predominate (e.g. Cameron, 1998; Mould et al., 2003). Coppergate, York, presented an unusual opportunity to compare bones and leather from the same phases of one neighbourhood. Cattle predominated in both sources of evidence, but the age profile of the leathers showed an appreciably higher proportion of young animals than the bone debris indicated, perhaps showing that the source of leather was not simply whatever the butchers produced but included some selectivity. Other keratinous materials such as horn and hoof seldom survive even in water-logged sediments. However, urban excavations often encounter concentrations of cattle and goat horn cores (Armitage, 1990; Huntley and Stallibrass, 1995: 187–9; Serjeantson and Rees, 2009: 176–7), and the regular identification of horn in mineral-preserved organic remains on, for example, the handles of iron knives, reminds us that horn was an important and versatile everyday material. Like the leather-workers, horners seem to have acquired raw material selectively, not only from the regular butchering of cattle for meat. Goats are disproportionately represented in urban deposits by horn cores. Differentiation of sheep and goat horn cores is simpler than for many other parts of the skeleton. However, little other goat is found even in assemblages recorded by analysts familiar with the identification of, and actively looking for, goats (e.g. Bond and O'Connor, 1999: 410–11), showing that goat horn was a valued commodity.

The other significant carcass product is the bones themselves, a source of robust raw material for artefacts (Choyke and O'Connor, 2013). Here there is little evidence of raw material importation, other than of antler. Deer bones are recovered from most urban assemblages, including introduced fallow deer (*Dama dama*) as well as native roe (*Capreolus capreolus*) and red deer (*Cervus elaphus*). However, as with goats, antler greatly outnumbers the post-cranial remains, especially from the eighth to twelfth centuries. Antler-working was a significant craft right across northern Europe during those centuries, with stylistic and analytical evidence for long-distance exchange (Ashby, 2005; von Holstein et al., 2014). Objects in bone and antler are commonly found (Fig. 14.2). Even allowing for the durability of the material in most burial environments,

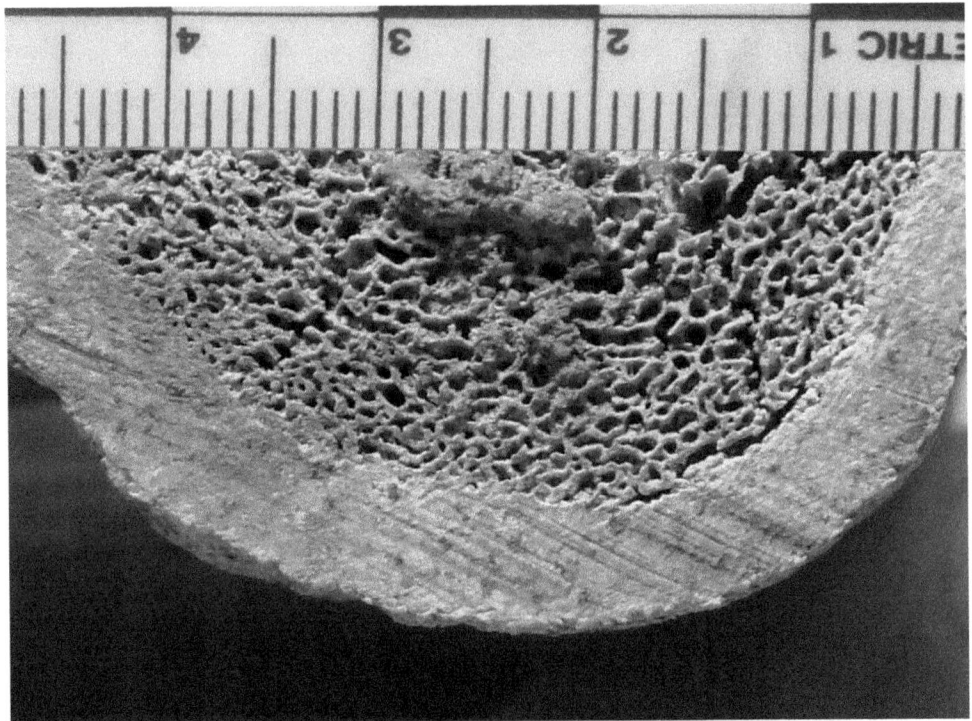

FIGURE 14.2 Red deer antler from Medieval York, sawn up as artefact raw material. Author's own image.

these objects (as hair-pins, knife-handles, gaming-pieces, clothes-fasteners) must have been as familiar to Medieval people as plastic objects are today. It is unlikely that the identity of the animal mattered. Simple objects such as bone pins are often unidentifiable beyond 'mammalian cortical bone', so it seems unlikely that one made of horse, for example, would have out-ranked one made of cattle bone. The abundance of antler in Early Medieval contexts, in contrast with its later decline, may simply be because the emparkment of the English landscape under Norman and Angevin ruling elites cut off the ready supply of antler to urban craftsmen.

Identifying dairy produce in the archaeological record is notoriously difficult, though some analytical progress is being made (Craig et al., 2005; 2011). Mortality profiles with the bimodal age distribution of young (presumably male) calves and old cows expected of a specialized dairy herd are seldom encountered before Tudor times (for example Wilson, 1994; for an exception, Bond and O'Connor, 1999: 384–6). However, Medieval texts and illustrations make it quite clear that cows were milked, even if not kept specifically for dairying, and that cheese and butter were made in most village and estate households (Woolgar, 2006), so these products of live cattle were familiar to urban populations.

Horses must have become an increasingly common sight in towns as they replaced cattle as the main source of traction power in later Medieval times, though ridden horses would have been familiar from Saxon times onwards (Langdon, 2002). Urban bone assemblages commonly include a few horse bones, often butchered and disposed of in the same way and places as cattle bones (e.g. Bond and O'Connor, 1999) and more substantial deposits of horse remains sometimes occur (e.g. Baxter, 1996). Occasional finds of butchered horse bones suggest that horse meat was sometimes eaten, despite papal interdiction, though the 'passing off' of horse as mature beef or venison cannot be ruled out. Horses seem to be under-represented in the zooarchaeological record given their likely importance in towns, at least from High Medieval times onwards and there are few published excavation records of stables and blacksmiths' premises. The evidence suggests that horses were kept and maintained, and their carcasses mostly disposed of, outside urban areas. The zooarchaeological record reflects the place of *dead* horses in towns rather than *live* ones and therefore understates their importance. Indeed, Gunn and Gromelski (2012) estimate that 10% of accidental deaths in Tudor England occurred whilst working with horses.

The place of pigs in towns may have been quite different from that of cattle and sheep. Pigs lend themselves more readily to being kept in a backyard area, leading O'Connor to suggest that pigs were kept within the tenements and yards of Viking Age York, though subsequent work has cast doubt on this interpretation (Hammond and O'Connor, 2013). Pigs are a ubiquitous but seldom abundant component of English urban assemblages, not attaining the high relative abundance seen in eastern Europe (O'Connor, 2010) and their urban status remains debatable.

Poultry occur quite regularly, mostly as bones of chickens (*Gallus gallus*) and geese (*Anser* and *Branta* spp.). As chickens are not an endemic species, there is little doubt they are present as domestic livestock. Reliable separation of wild and domestic forms of greylag goose (*Anser anser*) is rarely possible, so context is often the only indication of domestic status. Ducks, too, are commonly present, though separating wild and domestic forms of *Anas platyrhynchos* is problematic. Chicken bones in Saxon and Medieval assemblages are predominantly of adult birds, even where preservation and recovery have been good enough for sub-adult bones to have survived. Although this is supposition, the predominance of adults suggests chickens were kept for their eggs at least as much as for their meat. Fragmented eggshell can be abundant where preservation and recovery allow and new biomolecular procedures enable the species identification of eggshell fragments, giving a more nuanced assessment of the role of chicken, duck, and goose eggs (Stewart et al., 2013). Immature chicken bones are encountered rather more often in post-Medieval assemblages, perhaps showing the rise of capons as a table bird. An important consideration regarding chickens, and perhaps geese, is their value as currency, allowing small-scale exchange to go on between households. Poultry and eggs often feature in documentary records of commodities traded specifically by women (Hilton, 1984; Whittle, 2005).

Poultry can therefore be regarded as 'household livestock', a potential source of food and trade for urban residents with little or no land, and the same applies to rabbits

(*Oryctolagus cuniculus*). In Medieval England, rabbits were animals of the landed estates, economically important where poor soils made grain uneconomic (Bailey, 1988) and their remains are relatively scarce on urban sites. On post-Medieval and Early Modern sites, we encounter rabbits more frequently, reflecting their wider availability as a feral animal and perhaps the 'backyard' keeping of rabbits for the pot.

A final point about livestock is the social importance of regional types or 'breeds'. To the occupants of a Saxon or Medieval town, the livestock of their immediate rural hinterland probably had particular characteristics of conformation, colour, horns, fleece, and so on that marked those animals out as 'local' (Trow Smith, 1957; Armitage, 1982). When livestock were brought in from further afield, how were those more unfamiliar beasts received? Morphometric studies hint at the presence of different 'types' amongst samples of cattle and sheep bones (Armitage, 1990; Davis and Beckett, 1999) and advances in genomics make it possible to ask more specific questions about livestock populations and demes (Edwards et al., 2003; Speller et al., 2013). In interpreting those results, the conflicting influences of conservatism and novelty must be kept in mind. The importance of long-distance cattle droving from Medieval times onwards is well known, but detailed sources such as Skeel (1926) show that much of this consisted of the Crown or nobility acquiring cattle at a distance. Those beasts may not have found their way into the general urban food supply, nor their bones into general urban refuse. There is a pressing need for some astute integration of stable isotope and genomic studies with urban zooarchaeology to investigate the population diversity and geographical origins of the cattle that featured in the lives of Medieval and later towns in England.

COMPANIONS

Companion animals present something of a challenge. First, an individual of almost any species could be a 'pet' or companion animal in particular circumstances: consider poet Gérard de Nerval's pet lobster (Cavanaugh, 2011). Second, we have no archaeological template for recognizing the bones of a companion animal. Instead, we naively assume that species that are usually companion animals now, such as dogs and cats, were companion animals in the past. Despite the thorough research of scholars such as Walker-Meikle (2012), Medieval texts are mostly uninformative, as they reflect the literate groups of society. Serpell (1996: 47–8) points out that the Church frowned upon the keeping of animals for other than utilitarian purposes, excluding the bonds of affection usually associated with pet-keeping. We might question whether God-fearing English households would have kept companion animals at all.

Cats (*Felis catus*) feature frequently in assemblages from Medieval and later English towns. Occasional examples of apparent mass-felicide (McCormick, 1988; Luff and Moreno García, 1995) and more frequent finds of cat bones with cut-marks consistent with skinning show that not all were companions (Fig. 14.3). Mortality profiles of urban Medieval cats are seldom published, but this author's impression is that sub-adult cats

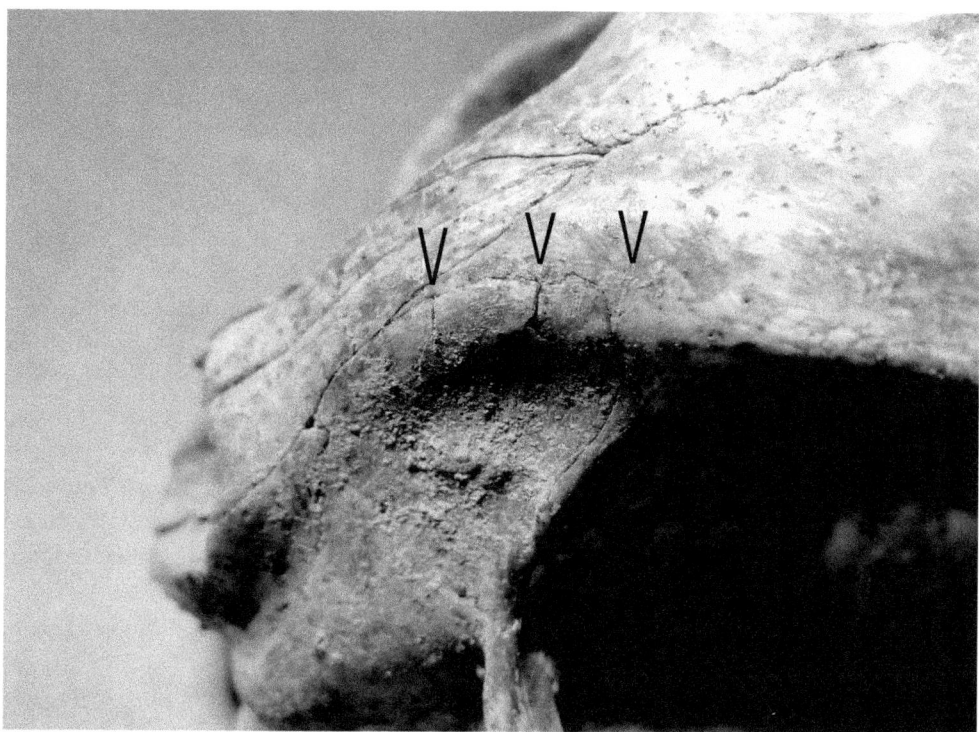

FIGURE 14.3 Post-Medieval cat skull from York showing cut-marks consistent with skinning. Author's own image.

make up a substantial proportion in most towns. Cat bones were deposited into general urban refuse and only rarely in discrete interments, more consistent with feral populations of commensal cats than with cared-for companions. In a rare systematic taphonomic study of cat and dog remains in a post-Medieval town, Clare Rainsford's currently unpublished analysis for Hungate, York, shows little evidence of deliberate burial prior to the eighteenth century, with mainly casual disposal of feral animals prior to that date.

Dogs (*Canis familiaris*) were potentially more useful to a Medieval household than cats and so a companion dog could more readily be justified. Pathologies consistent with rough treatment of dogs are occasionally reported from English towns, though there is to date no full published review to match those for the Classical world (MacKinnon, 2010), France (Binois et al., 2013), Scotland (Smith, 1998), Russia (Zinoviev, 2012), or Germany (Teegen, 2005). Biometric studies show the majority of Medieval and later dogs to have been of medium size, with a few larger individuals, consistent with working and guard dogs rather than small 'lap' dogs. In East Anglia, for example, Crabtree (2013) notes an increase in the morphological variation of dogs from the mainly rural evidence of the fifth to eighth centuries to the more urban ninth to eleventh centuries, suggesting that urban dogs may have served multiple roles. The wealthier Medieval households kept companion, as well as hunting, dogs. In 1440, the Prioress of Langley complained that her lodger Eleanor, Lady Audley, kept too many dogs, up to a dozen of which would follow her into church (Power, 1922). Presumably the town houses of those same families accommodated more than a few dogs, whose remains we encounter amongst the urban refuse. As with cats, individual dog burials are rare in urban Medieval and later contexts, perhaps in part a reflection of the status of dogs in urban society and in part a consequence of the inevitable re-deposition in urban stratigraphy.

The scarcity of guinea pigs (*Cavia porcellus*) and parrots (Psittacoidea), in post-Medieval English towns is surprising. Guinea pigs originate in South America and were depicted in European art by the mid-sixteenth century. One account from eighteenth-century France indicates that they were eaten in Europe as well as kept as curiosities (Van Dijk and Silkens, 2013). There are few English records of these endearing animals, the earliest being a late sixteenth-century specimen from a rural manor house (Hamilton-Dyer, 2009). Guinea pigs are not even found where sieving has recovered appreciable numbers of other rodents, indicating genuine rarity not poor recovery. Parrots and other colourful, exotic birds feature in Late Medieval and post-Medieval texts and illustrations (Yapp, 1982; Albarella, 2007), though not in the archaeological record. A mid- to late seventeenth-century pit at the Castle Mall site in Norwich yielded what may be the only example, identifiable only to the sub-family Psittacinae (Albarella et al., 1997: 51–2).

Categorization of urban bird taxa can be problematic. Corvid birds, particularly jackdaws (*Corvus monedula*), commonly occur in urban assemblages, and it is a reasonable assumption that most were free-living urban residents. However, corvids in general are highly social and intelligent birds that readily adapt to life as tame companions (e.g. Marzluff and Angell, 2007; Woolfson, 2010) and it is possible that a few of the corvids that occur in urban bone assemblages were tame birds kept as companions. Documentary sources are unhelpful on this point: Yapp (1979) notes only a few clear examples of corvids on Medieval documents and texts offer little in support or contradiction.

COMMENSALS

A more likely role for corvid birds, and certainly for rodents, is that of urban commensal, the synanthropic animals that lived in towns of their own volition, exploiting refuse and stored foodstuffs (O'Connor, 2013). Corvids were listed in the Vermin Acts of 1532 and 1566, designated as animals that were injurious to human health and wealth and therefore to be exterminated (Lovegrove, 2007: 79–85). Corvids were condemned particularly for eating cereal crops, so urban populations of crows and jackdaws may have been less reviled than their rural conspecifics. The remains of corvids occur dispersed through occupation and refuse deposits, not in the concentrations that might result from a cull of urban populations. Specimens of raven (*Corvus corax*) are frequently recorded from Medieval towns throughout England, though seldom in abundance and a few post-Medieval records show that this species persisted at least into Tudor times (Yalden and Albarella, 2009: 127). White-tailed eagle (*Haliaeetus albicilla*), on the other hand, is only sparsely represented in the Medieval urban record and not thereafter (Yalden and Albarella, 2009: 148–9). Commensal 'street' pigeons (*Columba livia*) are recorded in small numbers from many Medieval to Early Modern urban sites, though it is problematic to differentiate free-living commensal populations from pigeons that were fed and housed as a source of meat and eggs. Most pigeon records consist of bones dispersed in the general urban refuse, which is more consistent with commensal birds than with maintained domestic pigeons.

Apart from noisy jackdaws and opportunistic pigeons, urban people would have encountered rodents such as house mouse (*Mus domesticus*) and ship rat (*Rattus rattus*). Both species have a continuous record in English towns from Late Saxon times onwards, apparent absences mostly being attributable to inappropriate sampling and recovery. House mice, of course, are still with us though ship rats have ceased to be part of the urban scene. Arguments continue for and against ship rats as a vector for plague and hence the Black Death: Antoine (2008) gives a useful overview and Benedictow (2010) and Hufthammer and Walløe (2012) demonstrate the differences of opinion. Urban zooarchaeology could record the displacement of ship rats by common rat (*Rattus norvegicus*), probably during the eighteenth century. Regrettably, survival and sampling of Early Modern deposits is a rarity and we have few good records for common rat. Recent work in York demonstrated the presence of both species in an early nineteenth-century property (Fig. 14.4). Twigg (1992) noted the close association of ship rat colonies with active ports and coastal locations in England in the latter half of the twentieth century and O'Connor (2013: 90–2) has suggested that ship rat populations were most persistent where they were regularly 'topped up' by new introductions.

People in Medieval and Early Modern towns experienced a range of commensal animals similar to those that we see in England today (corvids, pigeons, rats, and mice), indicating that urban living generated much the same opportunities. It is surprising, therefore, that records of red fox (*Vulpes vulpes*) are infrequent, given that this species has become such a successful urban commensal in recent decades. Medieval and later towns certainly had refuse accumulations, yet foxes seem not to have taken advantage. Perhaps foxes were subject to competitive exclusion by other scavengers, such as feral dogs and cats. The scarcity of foxes in the urban zooarchaeological record may support the inference that many of the cats whose remains we encounter were living as feral animals rather than household 'pets'.

These commensal animals must have been a significant part of the ecology of Medieval and later towns, occupying an important niche as scavengers of the organic debris of human lives. More than that, they would have been a familiar part of people's lives, an intrusion of the 'wild' into the constructed environment of the town. Whether they were regarded as 'vermin' may have been locally contingent, with attitudes perhaps becoming as polarized as modern attitudes to, for example, badgers (Cassidy, 2012).

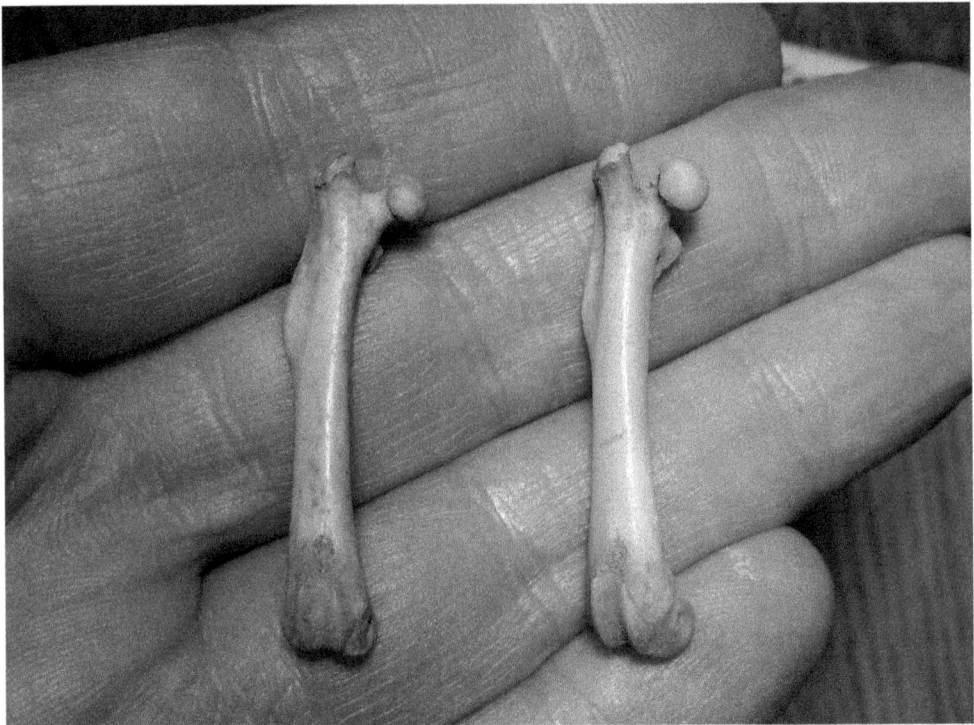

FIGURE 14.4 Femora of ship rat (*Rattus rattus*, left) and common rat (*Rattus norvegicus*, right) from York, found in the same early 19th C context which produced the cat skull in Fig. 14.3. Author's own image.

AND THE WILD THINGS

Apart from the commensal animals of the town, people's experience of 'wild' animals was largely limited to fish and wildfowl. Other than antler, deer remains are infrequent and venison may have been only a rare luxury or the occasional product of poaching. Where bone preservation is good and sieving has been deployed, fish bones may be abundant. Eels (*Anguilla anguilla*) and freshwater fish such as cyprinids tend to be the majority in Saxon towns, broadening to include quantities of herring (*Clupea harengus*) by late tenth to eleventh centuries and increasingly marine fish, especially gadids, as the Medieval period went on (Barrett et al., 2004). Post-Medieval and Early Modern fish assemblages show that 'modern' patterns of marine fish exploitation, focused on large gadids and flatfish, were established by Late Medieval times and changes in shipping and capture technology seem to have made little difference to the species composition of catches. To the people of Medieval towns, fish represented two rather different realms: the locally available eels, pike (*Esox lucius*), and cyprinids that any enterprising angler could procure and the generally larger marine species that reflected an environment and way of life beyond the experience of nearly all urban residents.

Apart from poultry and commensal urban birds, Medieval and later towns regularly yield bones of wildfowl such as waders and the typical 'game' birds of the Medieval literature (Yapp, 1983). Documentary records show that some birds, such as crane (*Grus grus*) and bittern (*Botaurus stellaris*) were favoured for high-status feasting (Albarella and Thomas, 2002) and their remains are infrequent in the general urban refuse compared with those of wild ducks, plovers (*Pluvialis* spp.), and wood pigeon (*Columba palumbus*). As with the fish, urban people may have regarded wildfowl in distinct categories: the local and mundane birds of the surrounding woods and fields and the higher-status birds from further afield or from exclusive estates.

Summing Up

Urban zooarchaeology in England has delivered a substantial volume of data on the occurrence of bones of different species at different times and places. Although those data are open to analysis in terms of mortality, biometry, pathology, and much more, what they ultimately represent are the roles that living animals and their dead remains played in the economic life of historic towns and the individual lives of their people. The bones show the predominance of cattle on the streets but chickens in the home and the place of dogs as working animals, pets, and vermin. Medieval to Early Modern towns offer the unusual opportunity to trace those roles over a millennium, noting changes in response to larger historical or climatic events and in response to the developing ecology of the town itself. Somewhere between the sixteenth and twentieth centuries, Medieval towns became modern and we can explore the place of animals in that transition. There are fewer large-scale urban excavations in England today than in the 1970s and '80s: today's challenge is to understand assemblage formation processes and to frame the higher-level research questions that link people and animals, now and in the past, in the constructed urban environment.

References

Albarella, U. (2007) 'Companions of our travel: the archaeological evidence of animals in exile', in Hartmann, S. (ed.) *Fauna and Flora in the Middle Ages*, pp. 133–53. Frankfurt: Verlag Peter Lang.

Albarella, U., Beech, M., and Mulville, J. (1997) 'The Saxon, Medieval and Post-Medieval Mammal and Bird Bones Excavated 1989–91 from Castle Mall, Norwich, Norfolk. London, English Heritage, Ancient Monuments Laboratory Report 72/97'.

Albarella, U. and Thomas, R. (2002) 'They dined on crane: bird consumption, wild fowling and status in Medieval England', *Acta Zoologica Cracoviensia*, 45(Special issue), 23–8.

Antoine, D. (2008) '5: The archaeology of "plague"', *Medical History. Supplement*, 27, 101.

Armitage, P. L. (1982) 'Developments in British cattle husbandry', *The Ark*, 9, 50–4.

Armitage, P. L. (1990) 'Post-Medieval cattle horn cores from the Greyfriars site, Chichester, West Sussex, England', *Circaea*, 7(2), 81–90.

Ashby, S. P. (2005) 'Bone and antler combs: towards a methodology for the understanding of trade and identity in Viking Age England and Scotland', in Luik, H. (ed.) *From Hooves to Horns, from Mollusc to Mammoth: Manufacture and Use of Bone Artefacts from Prehistoric Times to the Present. Proceedings of the 4th Meeting of the Worked Bone Research Group, Tallinn, Estonia, August 2003*. Muinasaja Teadus 15, pp. 255–62. Tallinn: University of Tartu.

Bailey, M. (1988) 'The rabbit and the Medieval East Anglian economy', *The Agricultural History Review*, 36(1), 1–20.

Barrett, J. H., Locker, A. M., and Roberts, C. M. (2004) 'The origins of intensive marine fishing in Medieval Europe: the English evidence', *Proceedings of the Royal Society of London. Series B: Biological Sciences*, 271(1556), 2417–21.

Baxter, I. L. (1996) 'Medieval and early Post-Medieval horse bones from Market Harborough, Leicestershire, England, UK', *Circaea*, 11(2), 65–79.

Benedictow, O. J. (2010) *What Disease Was Plague? On the Controversy over the Microbiological Identity of Plague Epidemics of the Past*, Leiden: Brill.

Binois, A., Wardius, C., Rio, P., Bridault, A., and Petit, C. (2013) 'A dog's life: multiple trauma and potential abuse in a Medieval dog from Guimps (Charente, France)', *International Journal of Paleopathology*, 3(1), 39–47.

Bond, J. M. and O'Connor, T. P. (1999) *Bones from Medieval Deposits at 16–22 Coppergate and Other Sites in York*, York: Council for British Archaeology.

Bowden, M. and Smith, N. (2013) '"A very fair field indeed …": an archaeology of the common lands of English towns', in Rotherham, I. D. (ed.) *Cultural Severance and the Environment*, pp. 217–27. Dordrecht: Springer Netherlands.

Cameron, E. (1998) *Leather and Fur: Aspects of Early Medieval Trade and Technology*, London: Archetype Publications.

Carr, D. R. (2008) 'Controlling the butchers in Late Medieval English towns', *Historian*, 70(3), 450–61.

Cassidy, A. (2012) 'Vermin, victims and disease: UK framings of badgers in and beyond the bovine TB controversy', *Sociologia Ruralis*, 52(2), 192–214.

Cavanaugh, R. (2011) 'Gérard de Nerval—the man who walked lobsters—fallen stars', *The British Journal of Psychiatry*, 198(2), 108.

Choyke, A. and O'Connor, S. A. (eds) (2013) *From These Bare Bones: Raw Materials and the Study of Worked Osseous Objects*, Oxford: Oxbow.

Crabtree, P. J. (2013) 'A note on the role of dogs in Anglo-Saxon society: evidence from East Anglia', *International Journal of Osteoarchaeology*, 25, 976–80. DOI:10.1002/oa.2358.

Crabtree, P. J. (2014) 'Animal husbandry and farming in East Anglia from the 5th to the 10th centuries CE', *Quaternary International*, 346, 102–8.

Craig, O. E., Chapman, J., Heron, C., Willis, L. H., Bartosiewicz, L., Taylor, G., and Collins, M. (2005) 'Did the first farmers of central and eastern Europe produce dairy foods?', *Antiquity*, 79(306), 882.

Craig, O. E., Steele, V. J., Fischer, A., Hartz, S., Andersen, S. H., Donohoe, P., and Heron, C. P. (2011) 'Ancient lipids reveal continuity in culinary practices across the transition to agriculture in northern Europe', *Proceedings of the National Academy of Sciences of the United States of America*, 108(44), 17910–15.

Davis, S. J. and Beckett, J. V. (1999) 'Animal husbandry and agricultural improvement: the archaeological evidence from animal bones and teeth', *Rural History*, 10(1), 1–17.

Dijk, van, J. and Silkens, B. (2013) 18. The first archaeological find of a guinea pig in the Netherlands', in Raemaekers, D. C. M., Esser, E., Lauwerier, R. C. G. M., and Zeiler, J. T. (eds) *Bouquet of Archaeozoological Studies: Essays in Honour of Wietske Prummel*, pp. 188–94. Eelde: Barkhuis.

Edwards, C. J., Connellan, J., Wallace, P. F., Park, S. D. E., McCormick, F. M., Olsaker, I., Eythorsdottir, E., MacHugh, D. E., Bailey, J. F., and Bradley, D. G. (2003) 'Feasibility and utility of microsatellite markers in archaeological cattle remains from a Viking Age settlement in Dublin', *Animal Genetics*, 34(6), 410–16.

Finberg, H. P. R. (1954) 'An early reference to the Welsh cattle trade', *Agricultural History Review*, 2, 12–14.

Gunn, S. and Gromelski, T. (2012) 'For whom the bell tolls: accidental deaths in Tudor England', *The Lancet*, 380(9849), 1222–3.

Hamilton-Dyer, S. (2009) 'Animal bones', in Dury, P. and Simpson, R. (eds) *Hill Hall: A Singular House Devised by a Tudor Intellectual*, pp. 345–51. London: Society of Antiquaries/English Heritage.

Hammond, C. and O'Connor, T. (2013) 'Pig diet in Medieval York: carbon and nitrogen stable isotopes', *Archaeological and Anthropological Sciences*, 5(2), 123–7.

Hilton, R. H. (1984) 'Women traders in Medieval England', *Women's Studies: An Interdisciplinary Journal*, 11(1–2), 139–55.

Holstein, von, I. C., Ashby, S. P., van Doorn, N. L., Sachs, S. M., Buckley, M., Meiri, M., and Collins, M. J. (2014) 'Searching for Scandinavians in pre-Viking Scotland: molecular fingerprinting of Early Medieval combs', *Journal of Archaeological Science*, 41, 1–6.

Hufthammer, A. K. and Walløe, L. (2012) 'Rats cannot have been intermediate hosts for *Yersinia pestis* during Medieval plague epidemics in northern Europe', *Journal of Archaeological Science*, 40, 1752–9.

Huntley, J. R. and Stalllibrass, S. (1995) *Plant and Vertebrate Remains from Archaeological Sites in Northern England: Data Reviews and Future Directions*, Durham: Architectural and Archaeological Society of Durham and Newcastle.

Keene, D. (2012) 'Medieval London and its supply hinterlands', *Regional Environmental Change*, 12(2), 263–81.

Landon, D. (1997) 'Interpreting urban food supply and distribution systems from faunal assemblages: an example from colonial Massachusetts', *International Journal of Osteoarchaeology*, 7, 51–64.

Landon, D. (2005) 'Zooarchaeology and historical archaeology: progress and prospects', *Journal of Archaeological Method and Theory*, 12(1), 1–36.

Langdon, J. (2002) *Horses, Oxen and Technological Innovation: The Use of Draught Animals in English Farming from 1066–1500*, Cambridge: Cambridge University Press.

Lovegrove, R. (2007) *Silent Fields: The Long Decline of a Nation's Wildlife*, Oxford: Oxford University Press.

Luff, R. M. and Moreno García, M. (1995) 'Killing cats in the Medieval period: an unusual episode in the history of Cambridge, England', *Archaeofauna*, 4, 93–114.

MacKinnon, M. (2010) ' "Sick as a dog": zooarchaeological evidence for pet dog health and welfare in the Roman world', *World Archaeology*, 42(2), 290–309.

Marzluff, J. M. and Angell, T. (2007) *In the Company of Crows and Ravens*, New Haven: Yale University Press.

McCormick, F. (1988) 'The domesticated cat in early Christian and Medieval Ireland', in Wallace, T., MacNiocaill, G., and Wallace, P. F. (eds) *Keimelia: Studies in Medieval Archaeology and History in Memory of Tom Delaney*, pp. 218–28. Gaillimh: Galway University Press.

Mould, Q., Carlisle, I. and Cameron, E. A. (2003) *Craft, Industry and Everyday Life: Leather and Leatherworking in Anglo-Scandinavian and Medieval York*, York: Council for British Archaeology.

Müldner, G. and Richards, M. P. (2005) 'Fast or feast: reconstructing diet in later Medieval England by stable isotope analysis', *Journal of Archaeological Science*, 32(1), 39–48.

O'Connor, T. (2010) 'Livestock and deadstock in Early Medieval Europe from the North Sea to the Baltic', *Environmental Archaeology*, 15(1), 1–15.

O'Connor, T. (2013) *Animals as Neighbors: The Past and Present of Commensal Animals*, East Lansing: Michigan State University Press.

Power, E. (1922) *Mediaeval English Nunneries*, Cambridge: Cambridge University Press.

Rixson, D. E. (2000) *The History of Meat Trading*, Nottingham: Nottingham University Press.

Sabine, E. L. (1933) 'Butchering in mediaeval London', *Speculum: A Journal of Mediaeval Studies*, 8(3), 335–53.

Serjeantson, D. and Rees, H. (2009) *Food, Craft and Status in Medieval Winchester*. Winchester: Winchester Museums.

Serpell, J. (1996) *In the Company of Animals: A Study of Human–Animal Relationships*, Cambridge: Cambridge University Press.

Skeel, C. (1926) 'The cattle trade between Wales and England from the fifteenth to the nineteenth centuries', *Transactions of the Royal Historical Society*, 9(2), 135–58.

Smith, C. (1998) 'Dogs, cats and horses in the Scottish Medieval town', *Proceedings of the Society of Antiquaries of Scotland*, 128, 859–85.

Speller, C. F., Burley, D. V., Woodward, R. P., and Yang, D. Y. (2013) 'Ancient mtDNA analysis of early 16th century Caribbean cattle provides insight into founding populations of New World creole cattle breeds', *PloS ONE*, 8(7), DOI:10.1371/journal.pone.0069584.

Stewart, J. R. M., Allen, R. B., Jones, A. K. G., Kendall, T., Penkman, K. E. H., DeMarchi, B., O'Connor, T., and Collins, M. J. (2013) 'Walking on eggshells: a study of egg use in Anglo-Scandinavian York based on eggshell identification using ZooMS', *International Journal of Osteoarchaeology*, 24(3), 247–55.

Teegen, W. (2005) 'Rib and vertebral fractures in Medieval dogs from Haithabu, Starigrad and Schleswig', in Davies, J., Fabiš, M., Mainland, I., Richards, M., and Thomas, R. (eds) *Diet and Health in Past Animal Populations: Current Research and Future Directions*, pp. 34–8. Oxford:Oxbow.

Trow-Smith, R. (1957) *A History of British Livestock Husbandry to 1700*, London: Routledge.

Twigg, G. (1992) 'The black rat *Rattus rattus* in the United Kingdom in 1989', *Mammal Review*, 22(1), 33–42.

Walker-Meikle, K. (2012) *Medieval Pets*, Woodbridge: Boydell Press.

Whittle, J. (2005) 'Housewives and servants in rural England, 1440–1650: evidence of women's work from probate documents', *Transactions of the Royal Historical Society (Sixth Series)*, 15, 51–74.

Wilson, B. (1994) 'Mortality patterns, animal husbandry and marketing in and around Medieval and Post-Medieval Oxford', in Hall, A. R. and Kenward, H. K. (eds) *Urban–Rural Connexions: Perspectives from Environmental Archaeology*, pp. 103–15. Oxford: Oxbow.

Woolfson, E. (2010) *Corvus: A Life with Birds*, London: Granta Books.

Woolgar, C. M. (2006) 'Meat and dairy products in Late Medieval England', in Woolgar, C. M., Serjeantson, D., and Waldron, T. (eds) *Food in Medieval England: Diet and Nutrition*, pp. 88–101. Oxford: Oxford University Press.

Yalden, D. W. and Albarella, U. (2009) *The History of British Birds*, Oxford: Oxford University Press.

Yapp, W. B. (1979) 'The birds of English Medieval manuscripts', *Journal of Medieval History*, 5(4), 315–48.

Yapp, W. B. (1982) 'Birds in captivity in the Middle Ages', *Archives of Natural History*, 10(3), 479–500.

Yapp, W. B. (1983) 'Game-birds in Medieval England', *Ibis*, 125(2), 218–21.

Zinoviev, A. V. (2012) 'Study of the Medieval dogs from Novgorod, Russia (X–XIV century)', *International Journal of Osteoarchaeology*, 22(2), 145–57.

CHAPTER 15

··

FROM BOVID TO BEAVER

mammal exploitation in Medieval northwest Russia

··

MARK MALTBY

INTRODUCTION

··

THIS chapter reviews the current state of knowledge regarding the exploitation of mammals in northwest Russia from AD *c*.900 to 1500. The discussion draws largely upon evidence obtained from the town of Novgorod and its hinterland with reference to other sites in the region. Although based principally on zooarchaeological data, this summary will also refer to other types of archaeological and documentary evidence. This survey will be concerned only with mammals. For information on birds and fish, readers are referred to Maltby (2012).

Novgorod (Fig. 15.1) was, according to some chronicles, founded in the mid-ninth century AD, although the earliest archaeological features date to the early tenth century. It is situated on the River Volkhov, to the north of Lake Ilmen, in an area that began to be heavily populated from the seventh century AD. Average temperatures now range from -8°C in January to 18°C in July with 120–135 days of snow cover and annual precipitation of 650–700 mm (Spiridonova and Aleshinskaya, 2012). There are large numbers of small ninth- to tenth-century settlements on the more fertile lands in this area, the most important of which was the fortified centre of Ryurik Gorodishche, from where a substantial faunal assemblage has been collected mainly from ninth- to tenth-century deposits (Maltby, 2012). The area was originally heavily forested but clearances took place throughout the Medieval period and by the late thirteenth century much of the forest had been replaced by arable and meadows.

The emergence of Novgorod was due principally to its role as an international trading centre. Along with Staraya Ladoga to the north, Novgorod acted as a gateway for access to the region's forest resources and other commodities and had extensive trading links to the Baltic in particular. It acquired direct or indirect control of huge territories, which at times stretched from the Baltic to the Urals. Wealthy merchants (boyars)

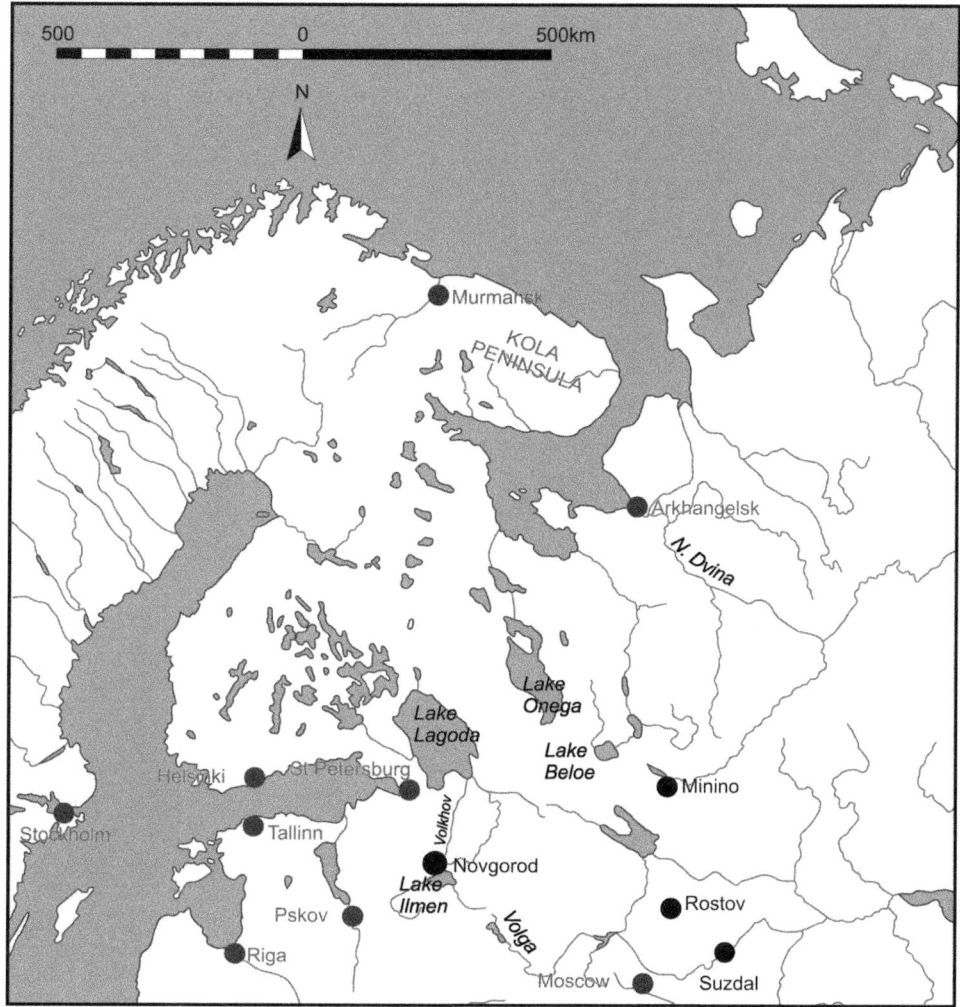

FIGURE 15.1 Location of sites mentioned in the text.

built properties within the town and they had a complex political relationship with the Novgorodian royal dynasty. Overseas traders, particularly from Germany, also took up residence in the town (Brisbane et al., 2012: 4–9).

Extensive excavations since the 1930s in various parts of Novgorod have revealed very deep anaerobic layers that have provided superb preservation conditions for organic materials including wood, plants, textiles, and leather. Sequences of wooden roadways and associated buildings have been investigated. Nearly one thousand birch-bark documents have been discovered, affording insights into the lives of the literate classes and providing records of items brought to the town as tribute. Animal bones also survive well and summaries of data from some excavations have been published (Tsalkin, 1956). Unfortunately, until quite recently, animal bones have subsequently not been retained.

Most of the evidence discussed here comes from the Troitsky sites situated in the south-west of the town (Maltby, 2012).

Evidence for animal exploitation in the forest zone comes mainly from around the area of Beloozero, where several sites at Minino on Lake Kubenskoye have been exca-vated (Makarov, 2012). This region lay on the edge of the Rostov-Suzdahl territory that bounded Novgorodian lands about 500 km to the east of Novgorod.

The Exploitation of Cattle

Novgorod faunal assemblages have consistently been dominated by cattle, which usually provide over 70% of the NISP counts of cattle, sheep/goat, and pig (Table 15.1). Although this may partly reflect bias towards the recovery of large bones in hand-collected assemblages, it is clear that beef was the dominant meat product consumed in the town throughout the Medieval period. Slightly earlier ninth- to tenth-century assemblages from sites in Novgorod's hinterland have included relatively fewer cattle bones, and they are outnumbered by pig at Georgii and in the northern area of Ryurik Gorodishche. This may reflect more efficient recovery of pig bones in assemblages partly collected by sieving. However, the percentage of cattle at Ryurik Gorodishche also increased to levels comparable to Novogord in the eleventh- to twelfth-century deposits (Sablin, in press), which suggests that more cattle were being kept and consumed in the region after Novgorod was founded. Local pollen profiles show significant increases in meadowland during this period (Spiridonova and Aleshinskaya, 2012). Novgorod's hinterland is well suited for cattle with abundant pasture available in the Volkhov floodplain and along the shores of Lake Ilmen.

Assemblages from other western Russian Medieval towns also show that cattle are consistently very well represented usually forming over 50% of the total cattle, pig, and sheep/goat NISP counts. Comparing other urban and castle assemblages from northeast Europe, only the post-Medieval assemblage from Tornio in Finland (Puputti, 2008) has produced over 60% cattle (Table 15.1). It would appear that the urban populations of northwest Russia were more reliant on cattle than in areas to the west.

Tooth ageing analysis from Ryurik Gorodishche and Novgorod shows that most cattle were killed between 3 and 8 years old (Table 15.2: stage 6). Many of these cat-tle would have provided calves, milk, and/or traction power prior to slaughter. However, quite a small percentage of mandibles belonged to old animals (Table 15.2: stage 7). In Novgorod there was also a significant kill-off of second- and third-year cattle (Table 15.2: stages 4–5). Animals slaughtered in their second year (mainly stage 4) were much less common in the earlier Ryurik Gorodishche assemblage, which suggests that the focus on beef production became more pronounced in the later Medieval period.

At both Ryurik Gorodishche and Novgorod, substantial numbers of calves aged 2–6 months old were also represented (Table 15.2: stage 2). This suggests that veal was quite

Table 15.1 NISP percentages and ratios of cattle, pig, and sheep/goat from Novgorod and its hinterland and from some other northeastern European towns and castles

Town/Site	Date	% NISP				C:P	S/G:P	NISP	Source
		Cattle	Pig	S/G					
Novgorod Troitsky IX	10–E12	70	22	8		3.19	0.34	3,423	Maltby, 2012
Novgorod Troitsky X	10–E12	74	20	6		3.60	0.29	6,464	Maltby, 2012
Novgorod Troitsky XI S22-14	10–E12	66	23	11		2.89	0.49	12,376	Maltby, 2012
Novgorod Troitsky XI S13-8	M12–E13	71	22	8		3.22	0.34	4,971	Maltby, 2012
Novgorod Troitsky XI S7-1	M13–E15	73	18	9		4.11	0.49	4,718	Maltby, 2012
Novgorod, Desyatinny-1	10–16	75	21	4		3.57	0.20	5,503	Zinoviev, in press
Novgorod, Nerevsky	11–15	80	15	6		5.45	0.39	10,416	Tsalkin, 1956
Novgorod, Slavensky	13–17	80	13	7		5.98	0.51	10,056	Tsalkin, 1956
Gorodishche 1979–1996 (all)*	9–10	53	40	6		1.31	0.16	3,906	Maltby, 2012
Gorodishche 1979–1996 (North)	9–10	35	58	7		0.60	0.13	835	Maltby, 2012
Gorodishche 1979–1996 (South)	9–10	72	27	2		2.66	0.06	1,322	Maltby, 2012
Gorodishche 2000–2004 (North)	9–10	43	48	9		0.90	0.19	2,572	Sablin, in press
Gorodishche 2000–2004 (North)	11–12	70	21	8		3.27	0.39	862	Sablin, in press
Georgii	9–10	42	52	6		0.82	0.12	483	Maltby, 2012
Prost	9–10	47	45	8		1.02	0.18	187	Maltby, 2012
Minino	11–13	40	16	44		2.41	2.64	794	Savinetsky, in press
Staraya Ladoga	8–10	56	36	8		1.55	0.22	12,216	Tsalkin, 1956
Pskov	9–12	70	22	8		3.14	0.35	8,001	Tsalkin, 1956
Moscow, Zaryadrey	10–17	65	25	10		2.62	0.39	22,468	Tsalkin, 1956
Moscow, Romanov Dvor	12–19	44	26	30		1.49	1.13	11,092	Maltby, in press

(Continued)

Table 15.1 Continued

Town/Site	Date	% NISP			C:P	S/G:P	NISP	Source
		Cattle	Pig	S/G				
Staraya Ryazan	11–13	60	28	12	2.18	0.44	7,234	Tsalkin, 1956
Tver, Kremlin	13–E15	61	27	12	2.23	0.45	31,485	Lantseva and Lapshin, 2001
Grodno, Belarus	11–15	40	38	22	1.06	0.57	6,546	Tsalkin, 1956
Vilanji, Estonia	13–17	52	14	34	3.77	2.48	23,926	Rannamäe, 2010
Turku, Finland	13–19	49	13	38	3.95	3.08	37,667	Tourunen, 2008
Tornio, Finland	17	61	10	29	6.01	2.84	1,939	Puputti, 2008
Wolin, Poland	9–12	17	71	12	0.24	0.17	28,717	O'Connor, 2010
Szczecin, Poland	9–11	12	78	9	0.16	0.12	22,088	O'Connor, 2010
Poznan, Poland	10–12	35	24	41	1.51	1.74	10,081	O'Connor, 2010
Legnica, Poland	10–12	43	50	6	0.86	0.12	4,727	O'Connor, 2010
Opole, Poland	10–12	30	56	14	0.52	0.25	11,561	O'Connor, 2010
Gdansk, Poland	10–12	30	53	17	0.56	0.31	18,320	O'Connor, 2010
Malbork Castle, Poland	14–18	45	34	21	1.34	0.62	1,611	Maltby et al., 2009
Jersika hillfort, Latvia	9–13	51	33	16	0.65	0.31	2,257	Maltby et al., in press
Cēsis Castle, Latvia	13–18	53	12	35	1.53	2.96	42,385	Maltby et al., in press
Vecdole Castle, Latvia	13	49	36	15	1.34	0.43	973	Maltby et al., in press
Ventspils town, Latvia	14–15	53	20	28	1.91	1.41	459	Maltby et al., in press
Riga Castle, Latvia	13–16	54	22	24	2.40	1.09	368	Maltby et al., in press

Table 15.2 Cattle mandible ageing stage percentages from Gorodishche and Novgorod Troitsky sites IX–XI

Stage	Gorodishche (N = 50)	Troitsky IX (N = 178)	Troitsky X (N = 240)	Troitsky XI (N = 364)
1	-	1	1	1
2	16	6	7	11
2–3	-	–	1	0
2–4	2	3	3	3
3	-	4	4	6
3–4	-	2	5	3
4	2	14	14	12
4–5	-	3	2	5
5	6	5	3	8
5–6	12	7	8	9
5–7	2	4	9	8
6	50	36	35	27
6–7	8	9	7	5
7	2	6	3	4

frequently eaten and that milk production was an important consideration in cattle husbandry. Metrical analysis of the metacarpals has indicated that around three-quarters of the adult cattle from Ryurik Gorodishche and from the tenth- to twelfth-century deposits from Novgorod were females. Even fewer adult males were represented in the later Medieval deposits from Novgorod, concomitant with an increase in the percentage of mandibles of young (presumably mainly male) calves (Maltby, 2012). This implies that dairy production had become more important.

Butchery marks, mostly made by heavy blades and cleavers, were observed on 30% of the cattle bones from the Troitsky sites in Novgorod. There was a lot of variability in the location and nature of these butchery marks. This inconsistency, combined with the presence of bones from body areas of poor meat quality alongside those with high protein value, strongly suggests that complete carcasses were commonly processed within all of the properties investigated. In contrast, the ninth- to tenth-century layers from the southern area of Ryurik Gorodishche produced a much higher percentage of cattle elements than the northern area (Table 15.1), and a significantly higher proportion of foot and cranial elements (Maltby, 2012). This could indicate that the southern area, on the periphery of the settlement, was where much of the primary processing of cattle was carried out. Similar dumps have as yet not been recognized in Novgorod.

Leather artefacts and offcuts have been found on many northern Russian Medieval settlements including Ryurik Gorodishche and Novgorod (Kurbatov, 2012). Many of those from Novgorod were from cattle hides and calfskins (Solovyov, 2012). Some insects found on the Troitsky sites are types that infest hides stored for processing (Reilly, 2012). Fine incisions, indicative of initial skinning, were observed on many cattle first phalanges (Maltby, in press). This suggests that hide preparation and leather production took place within a number of different properties. Footwear was a particularly common product and the eleventh- to twelfth-century deposits at Ryurik Gorodishche have produced evidence for the presence of craftspeople making high-quality footwear on this high-status site, whereas contemporary craftspeople in Novgorod were producing less elaborate products perhaps for a wider market (Kurbatov, 2012: 406).

Recent metrical analysis of cattle bones from Novgorod and Ryurik Gorodishche has confirmed previous observations that cattle in this region were of small stature (Tsalkin, 1956). Measurements of 155 complete limb bones from the Troitsky sites produced withers heights estimates ranging between 95.6 cm and 127.5 cm with a mean of 109.8 cm. Twenty-nine specimens from ninth- to tenth-century levels at Ryurik Gorodishche ranged between 100.5 cm and 123.2 cm, with a slightly higher mean of 112.0 cm. It seems that the average size of cattle decreased during the Medieval period. Mean values of length, breadth, and depth measurements of limb bones were 3–10% higher in the ninth- to tenth-century levels at Ryurik Gorodishche than in the thirteenth- to fifteenth-century levels from Novgorod (Maltby, in press). Although the low numbers of bones of bulls and oxen in the later deposits could account for some of this decrease, measurements of loose teeth, which are less susceptible to sexual dimorphism, also indicate a slight decrease in size. Travellers to Russia in the sixteenth and seventeenth century commented on the small size of cattle in the region (Kurbatov, 2012: 393).

The Exploitation of Pigs

Pigs have been the second most common species recorded from the various sites in Novgorod, forming 13–23% of cattle, pig, and sheep/goat counts. Pigs are generally better represented in the earlier assemblages at Ryurik Gorodishche and Georgii, where they sometimes outnumbered cattle (Table 15.1). The relative frequencies of pig and sheep/goat have been compared using the sheep/goat/pig index (after O'Connor, 2010). This ratio ranges between 0.20 and 0.49 in the hand-collected assemblages from Novgorod, indicating that pigs were consistently better represented than sheep/goat. The ratio of sheep/goat is even lower in the ninth- to tenth-century assemblages from Georgii, Prost, and Ryurik Gorodishche (0.06–0.19), probably reflecting that more forest was available for foraging pigs prior to the main phases of agricultural clearances during the later Medieval period. In contrast, pigs contributed only 27% of the total pig and sheep/goat assemblage from Minino, despite high percentages of woodland wild mammal bones in that sample. It is possible that climatic conditions were less favourable for pig keeping. The area has lower mean winter temperatures and longer and deeper coverings of snow (Spiridonova and Aleshinskaya, 2012) that would be less favourable for winter foraging by free-range pigs.

Comparisons with other Russian and eastern European urban and castle assemblages show substantial variations in the abundance of pigs. Pig percentages from

Novgorod are broadly similar to those from Pskov, the Zaryadrey site in Moscow, Staraya Ryazan, and Tver. Pigs are substantially better represented at Staraya Ladoga, although still heavily outnumbered by cattle (Table 15.1). None of the Russian assemblages are dominated by pigs in contrast to most urban sites in northern Poland (O'Connor, 2010).

Tooth ageing analyses from Ryurik Gorodishche and earlier layers from Novgorod have shown that substantial numbers of pigs were killed in their first year (*c.*30% of the mandibles recovered). The presence of neonatal mortalities from Ryurik Gorodishche indicates that pigs were bred there. Neonatal pigs have not been recorded at Novgorod, which perhaps indicates they were not commonly kept within the town, although further sieving is needed to confirm this. The proportion of juvenile pigs decreased significantly (to *c.*5%) in the thirteenth- to fifteenth-century deposits from Novgorod, suggesting that pigs became less intensively exploited in the town. Most of the other pigs recovered from Novgorod and its hinterland were culled in their second and third years with jaws of third-year animals becoming more common in the later Medieval assemblages (Maltby, in press).

Over 45% of the pig elements from the Troitsky sites in Novgorod are cranial elements (Maltby, in press), probably indicating that butchery often took place within the properties. However, percentages of upper limb bones gradually increased in the later Medieval deposits, possibly indicating that joints of fresh and preserved meat were more commonly brought to the urban properties. The low numbers of foot bones from all the Novgorod sites may also indicate that trotters were often previously removed, although further sieving is again required to establish whether the paucity of these relatively small bones is not simply a factor of retrieval bias.

Metrical analysis of bones and teeth from Novgorod and its hinterland sites showed that the vast majority of the porcine specimens fell comfortably within the size range of domestic pigs rather than the larger wild boar. The pigs from the ninth- to tenth-century assemblages from Ryurik Gorodishche, Georgii, and Prost tended to be slightly larger than those from Novgorod. More bones of smaller pigs were found in later Medieval deposits from the Troitsky sites. The decrease may reflect the presence of a larger proportion of sows in the later assemblages (although this needs to be confirmed by analysis of sexing data). Another possibility is that pigs generally became slightly smaller in the later Medieval period, perhaps reflecting the decrease in prime woodland habitats in the region.

THE EXPLOITATION OF GOAT AND SHEEP

In the ninth- to tenth-century assemblages from Ryurik Gorodishche, 43% of the thirty-five diagnostic sheep/goat bones belonged to goats (Maltby, 2012). About one-third of the forty-six measured metapodials from the Troitsky sites in Novgorod belonged to goats. Goats are reported to outnumber sheep on the neighbouring Desyatinny site (Zinoviev, in press). It seems that goats formed a much higher proportion of the ovicaprine assemblage in Novgorod than in most central and eastern European sites. For example, in the seventeenth-century sample from Tornio, Finland, only 1% of the diagnostic elements were identified as goat (Puputti, 2008). Some goats were probably kept in Novgorod itself both for meat and milk. Goat horns (particularly large ones of males) were also frequently imported for working in Novgorod and other towns. At Viljandi, Estonia, 90% of the horn cores and skulls belonged to goat, whereas 66% of the measured metacarpals were classified as sheep (Rannamäe, 2010).

Sheep/goat elements are poorly represented on most Medieval sites in northwest Russia, often forming < 10% of cattle, pig and sheep/goat NISP counts (Table 15.1). This is largely due to environmental factors. Much of the land around Novgorod, for example, was unsuitably wet or wooded for sheep grazing. In contrast, at Minino, sheep/goat formed 44% of the cattle, pig, and sheep/goat counts, reflecting the development of agricultural land and the increase in dry pasture more suitable for sheep.

Not all the sheep/goat mandibles used in ageing analyses from these sites have been identified specifically as sheep or goat, which limits interpretations of slaughter patterns, as the two species may have had different mortality profiles. However, over half the mandibles from the Troitsky sites belonged to sheep and goats under 2 years old. Such a high percentage of immature animals indicates that sheep and goat exploitation was mainly focused on meat production. The presence of significant numbers of mandibles from lambs and kids younger than three 3 months old may also indicate exploitation for milk. On Troitsky XI, the percentage of immature sheep and goats decreased in the thirteenth- to fifteenth-century deposits, indicating that meat production became less intensive. The corresponding increase in the proportion of mandibles from mature adults indicates that wool production had increased in importance. Most of the textiles from Novgorod dating from the tenth century onwards were made of woollen cloth. Wool and cloth were very important commodities traded in vast quantities throughout Medieval Europe. Some of the textiles found in Novgorod were clearly imported (Kublo, 2012). However, local production was more common. Racks for cleaning wool have been discovered in one Troitsky property and other wooden objects associated with cloth production are common finds in Novgorod. Goat skins were common components of footwear (Solovyov, 2012) and skins of both sheep and goats are mentioned in birch bark documents (Rybina, 2001).

Measurements of twenty-one complete sheep limb bones from the Troitsky excavations provided withers height estimates ranging between 55.0 and 65 cm with a mean of 59.7 cm. There were no clear chronological variations.

THE EXPLOITATION OF HORSE

Horse bones formed 9% of the domestic mammal remains of the ninth- to tenth-century Ryurik Gorodishche assemblage, although these include associated groups, the presence of which implies that many horse carcasses were not fully processed for food (Maltby, 2012). However, excluding the skulls discussed below, 5% of the horse bones from Ryurik Gorodishche bear processing marks, including those made during dismemberment and filleting, indicating some consumption of horsemeat. In addition, at least twenty-two complete horse skulls (some associated with mandibles) were deposited in the defensive ditch when it was infilled in the late ninth century (Sablin, in press). The skulls were spread across at least twenty square metres. This appears to have been a ritual deposition. All these horses were male and ranged between 5 and 15 years of age. Bit wear damage to the second premolars indicated that many of them had been used for riding. These may have been horses acquired by the Prince of Novgorod's armed forces.

Horse bones have a small but consistent presence in the Novgorod assemblages. They formed 5% of the domestic mammal counts on the Troitsky sites and the ratio of horse to cattle ranged between 0.05:1 to 0.15:1 on Novgorod sites. Although horses were not a regular source of food, 17% of their bones from the Troitsky sites bore processing marks. Some marks, particularly those on the phalanges, metapodials, and radii, were made during skinning and bone-working but many others, particularly those located on the scapula and upper limb bones, indicate dismemberment and filleting (Maltby, 2012). It seems that some inhabitants of Novgorod consumed horseflesh throughout the Medieval period. Elsewhere in the region, unusually high ratios of horse to cattle were recorded at Grodno, Belarus (0.35:1), and Staraya Ryazan (0.28:1) (Tsalkin, 1956), although it is unclear whether these counts were biased by the inclusion of bones from horse skeletons. These results lie in stark contrast to those from many towns in the northern Baltic, where horse bones form very low percentages of the faunal assemblages.

Although horses were sometimes eaten, their hides made into leather, and some of their bones made into artefacts, particularly skates, they were mainly valued for riding and as pack animals and could expect a long life. Nearly all the horse bones found in Novgorod are from adults. Plant macrofossil and insect evidence indicates that horses were stabled in various Novgorod properties (Monk and Johnston, 2012; Reilly, 2012). Birch-bark documents allude to horses much more frequently than to any other species (Rybina, 2001), reflecting their high status.

Withers heights of horses at Ryurik Gorodishche averaged around 142–144 cm, the size of large ponies, but these are c.10 cm larger than the average size of those recorded on later Medieval sites in Russia (Tsalkin, 1956). Estimates of horse withers heights from ninty-three complete limb bones from the Troitsky sites ranged between 117 cm and 153 cm, with a mean of 133.4 cm (Maltby, in press).

THE EXPLOITATION OF WILD MAMMALS

There is little evidence that wild mammals played a significant role in the diet of the inhabitants of Medieval Novgorod or in other major centres in the region. Bones of wild mammals formed 1% or less of the total mammal assemblage from the Troitsky excavations. Similarly low percentages have been obtained from other Novgorod sites (Table 15.3). The high percentage from the Desyatinny site is the result of a much higher proportion of the porcine bones being designated as wild boar (Zinoviev, in press). Excluding wild boar, the percentages of wild species from this site are as low as those from other Novgorod sites.

Hare has been the most common wild species recovered from the Troitsky sites. Many hare bones, however, may have been overlooked during excavation. Elk was the species which probably provided most meat. Bones from all parts of their skeletons have been found in Novgorod, indicating that their carcasses were sometimes brought to the town for final processing. In addition, large numbers of elk antlers were imported for specialist manufacture of combs and other artefacts. Many antler offcuts have been found within some properties.

The pivotal role of Novgorod in the international fur trade is well attested by documentary sources (Martin, 1986). Thousands of pelts were collected annually from the forest zones of northern Russia, with beavers and squirrels being particularly important. There are frequent references to squirrel pelts in Novgorodian birch-bark documents and seals of cylinders containing furs brought as tribute have also been discovered (Makarov, 2012). However, the great importance of fur-bearing species is not reflected in the zooarchaeological record from Novgorod and settlements in its immediate hinterland.

Only small numbers of beaver bones have been recorded from Novgorod sites (Table 15.3). Foot bones are under-represented, probably previously being removed with the skins, although again retrieval bias may be a problem. Butchery marks were observed on 35% of their bones, including skinning marks on crania. However, most of the marks were associated with meat processing (Maltby, in press).

Even fewer squirrel bones have been recovered from Novgood itself. Further sieving could produce more evidence for these and other fur-bearing species (marten, otter, polecat, stoat, fox, lynx, bear, wolf, sable) which have also only rarely been found in excavations of the town (Table 15.3). Third phalanges were the only bear bones identified in the Troitsky assemblage. Some of these could have been amulets but it is probable that others were attached to bearskins brought to the town.

Nor is there much evidence for the exploitation of wild mammals in the immediate hinterland of Novgorod. Despite extensive sieving, bones of wild species provided < 3% of the mammal assemblage from Ryurik Gorodishche, although smaller species such as hare and squirrel were slightly better represented, and wolf, fox, stoat, and marten have also been identified. Wild mammal bones were slightly more common in the Georgii assemblage. Most of these belonged to elk, although several fur-bearing species (including bear and lynx) were present. At Prost, foot bones from the processed skins of at least four pine martens were found in one deposit (Table 15.3).

Evidence for large-scale fur procurement has mainly been found in more remote parts of the forest zone, for example, at Minino, where wild species made up 65% of the

Table 15.3 Wild mammal species counts (NISP) from recent excavations in Novgorod and its territories

Site	Date	Bos	Elk	Rein.	Roe	Hare	Boar*	Bear	Squir.	Beav.	Otter	Mart.	Pole.	Stoat	Fox	Lynx	Wolf	Badg.	Mac.	Total wild	Total mamm.	% Wild	Wild ex. boar
Troitsky IX	10–E12		9	2		14		1		18										44	3,701	1.2	1.2
Troitsky X	10–E12		5			5			1	16										27	6,876	0.4	0.4
Troitsky XI S22-14	10–E12		16			26	1	2	2	64		1								112	13,333	0.8	0.8
Troitsky XI S13-8	M12–E13		34			20	1	4		2										61	5,302	1.2	1.1
Troitsky XI S7-1	M13–E15		16	1		49	2	2							1					71	5,084	1.4	1.3
Troitsky IX–XI	*Total*		*80*	*3*		*114*	*4*	*9*	*3*	*100*		*1*			*1*					*315*	*34,296*	*0.9*	*0.9*
*Desyatinny-1**	*10–16*		*1*		*4*	*3*	*177*			*2*		*1*			*3*		*1*			*192*	*5,503*	*3.4*	*0.3*
Nerevsky	11–15		37			9		5		29		5			6		8	5		104	9,850	1.0	1.0
Slavensky	13–17		13				6			3							1			23	8,424	0.3	0.2
Gorodishche 79-96	9–10		7			44		1	11	23		1		1			4			92	4,351	2.1	2.1
Gorodishche 2000-4	9–10		3			24				12		2			10					51	3,232	1.6	1.6
Gorodishche 2000-4	11–12		27			3	4					1			1					36	1,085	3.3	2.9
Georgii	9–10		18			1			1	3						1			1	25	558	4.5	4.5
Prost	9–10	1								1		54								56	260	21.5	21.5
Minino	11–E12		31	7	2	6	2		214	423	4	45	2		5					741	1,021	72.6	72.5
Minino	L12–13		56	4	2	13	6	2	153	183	7	40	4		5			1		476	819	58.1	57.8
Minino	11–13		20	2	2	5	7		56	252	1	19	1		1					366	611	59.9	59.4
Minino	*Total*		*107*	*13*	*6*	*24*	*15*	*2*	*423*	*858*	*12*	*104*	*7*		*11*			*1*		*1,583*	*2,451*	*64.6*	*64.4*

identified mammals. Beaver, which also provided significant amounts of meat for the local community, formed 35% of the assemblage. In addition, large numbers of squirrel and marten bones were recovered and bones of otter, polecat, fox, and bear were also found (Savinetsky, in press). Several other settlements in the Beloozero and Vologda regions have also produced assemblages containing substantial percentages of bones of fur-bearing species (Makarov, 2012).

The percentage of beaver halved in the thirteenth-century assemblage from Minino and percentages of squirrel and marten also decreased (Table 15.3). Over-exploitation and woodland clearance are both likely to have been factors in their decline. This is also reflected in the Novgorod assemblages. Beaver bones were largely absent from thirteenth-century and later deposits on the Troitsky sites. References to beavers on birch-bark documents disappear after the early thirteenth century (Rybina, 2001). Most of the beaver population in European Russia had disappeared by the nineteenth century because of over-exploitation (Zinoviev, in press). Therefore, evidence for Novgorod's role in the international fur trade is best reflected in the composition of the animal bone assemblages on supply sites like Minino rather than in the town itself. This is not surprising as the trade was in furs and pelts, not meat and bones.

Variations in the presence and relative abundance of different wild mammal species depends upon a variety of factors including local climatic conditions and vegetation, date, and the social status of the settlement. Thus, it is no surprise to see that bones of seals occur only on the most northerly sites such as Tornio and Staraya Ladoga (Tsalkin, 1956; Puputti, 2008). The discovery of a macaque skull from a late twelfth-century deposit in Ryurik Gorodishche indicates the import of an exotic species (native to North Africa) to this high-status site. The macaque may have been a gift to the Prince or part of a menagerie (Brisbane et al., 2007).

Conclusions

This brief review of mammal exploitation in Medieval northwest Russia relies heavily on evidence from Novgorod and its hinterland. Although many of the same species as elsewhere in northeastern Europe were exploited, there were some distinctive variations. Novgorodians ate relatively more beef than the inhabitants of many contemporary towns. Lamb and mutton were much less important and goats may have been as important as sheep as a food resource. Dairy produce may have become more prevalent in the later Medieval period. Meat from horses and beavers supplemented their diet, at least until the latter became a rare commodity because of over-exploitation for its fur. Wild mammals generally provided only a small portion of the diet. The vast importance of the fur trade is not reflected in the bone assemblages from Novgorod itself but is evident in areas where fur-bearing species were hunted.

Zooarchaeological studies can provide many insights into the lives of residents of Medieval Novgorod where the superb preservation conditions for organic materials, allied with documentary evidence, provides exciting opportunities for multidisciplinary research. Future intra-site comparisons have the potential to study fine-grained chronological variations in animal exploitation between different neighbourhoods. Detailed examination of assemblages from within individual properties can provide insights into how space was utilized within them. However, to achieve these and other aims, there needs to be a sampling policy that is fit for this purpose. As yet, the importance of birds and especially fish in the Novgorodian diet cannot be fully assessed because very few sieved samples have been analysed with which to make comparisons to assemblages from other sites in the region where such sampling has been carried out.

REFERENCES

Brisbane, M., Hambleton, E., Maltby, M., and Nosov, E. (2007) 'A monkey's tale: the skull of a macaque found at Ryurik Gorodishche during excavations in 2003', *Medieval Archaeology*, 51, 185–91.

Brisbane, M., Makarov, N., and Nosov, E. (eds) (2012) *The Archaeology of Medieval Novgorod in Context*, Oxford: Oxbow.

Grant, A. (1982) 'The use of toothwear as a guide to the age of domestic ungulates', in Wilson, R., Grigson, C., and Payne, S. (eds) *Ageing and Sexing Animal Bones from Archaeological Sites*. BAR British Series 109, pp. 91–108. Oxford: Archaeopress.

Kublo, E. (2012) 'The production of textiles in Novgorod from the 10th to the 14th centuries', in Brisbane, M., Makarov, N., and Nosov, E. (eds) *The Archaeology of Medieval Novgorod in Context*, pp. 224–58. Oxford: Oxbow.

Kurbatov, A. (2012) 'Leather working in north-west Russia', in Brisbane, M., Makarov, N., and Nosov, E. (eds) *The Archaeology of Medieval Novgorod in Context*, pp. 391–415. Oxford: Oxbow.

Lantseva, M. and Lapshin, V. (2001) 'Results of identification of osseous mammalian remains from excavations in 1994–1997 at Tver Kremlin', in *Tver Kremlin Complex Archaeological Source Study*, pp. 171–80. Saint Petersburg: Evropeysky Dom.

Makarov, N. (2012) 'The fur trade in the economy of the northern borderlands of Medieval Russia', in Brisbane, M., Makarov, N., and Nosov, E. (eds) *The Archaeology of Medieval Novgorod in Context*, pp. 381–90. Oxford: Oxbow.

Maltby, M. (2012) 'From *Alces* to *Zander*: a summary of the zooarchaeological evidence from Novgorod, Gorodishche and Minino', in Brisbane, M., Makarov, N., and Nosov, E. (eds) *The Archaeology of Medieval Novgorod in Context*, pp. 351–80. Oxford: Oxbow.

Maltby, M. (ed.) (in press) *Animals and Archaeology in Northern Medieval Russia: Zooarchaeological Studies in Novgorod and Its Region*, Oxford: Oxbow.

Maltby, M., Pluskowski, A. and Rannamäe, E. (in press) 'Farming, hunting and fishing in Medieval Livonia: zooarchaeological data', in A. Pluskowski (ed.) *Terra Sacra: The Ecology of Crusading, Colonisation and Religious Conversion in the Medieval Eastern Baltic Volume 1*. Turnhout: Brepols.

Maltby, M., Pluskowski, A., and Seetah, K. (2009) 'Animal bones from the industrial quarter at Malbork, Poland: towards an ecology of a castle built in Prussia by the Teutonic Order', *Crusades*, 8, 191–212.

Martin, J. (1986) *Treasure of the Land of Darkness: The Fur Trade and Its Significance for Medieval Russia*, Cambridge: Cambridge University Press.

Monk, M. and Johnston, P. (2012) 'Perspectives on non-wood plants in the sampled assemblage from the Troitsky excavations of Medieval Novgorod', in Brisbane, M., Makarov, N., and Nosov, E. (eds) *The Archaeology of Medieval Novgorod in Context*, pp. 283–320. Oxford: Oxbow.

O'Connor, T. (2010) 'Livestock and deadstock in Early Medieval Europe from the North Sea to the Baltic', *Environmental Archaeology*, 15, 1–15.

Puputti, A. (2008) 'A zooarchaeogy of modernizing human–animal relationships in Tornio, northern Finland, 1620–1800', *Post-Medieval Archaeology*, 42, 304–16.

Rannamäe, E. (2010) *A Zooarchaeological Study of Animal Consumption in Medieval Viljandi*, Lund: University of Tartu.

Reilly, E. (2012) 'Fair and foul: analysis of sub-fossil insect remains from Troitsky XI–XIII, Novgorod (1996–2002)', in Brisbane, M., Makarov, N., and Nosov, E. (eds) *The Archaeology of Medieval Novgorod in Context*, pp. 265–82. Oxford: Oxbow.

Rybina, E. (2001) 'The birch-bark letters: the domestic economy of Medieval Novgorod', in Brisbane, M. and Gaimster, D. (eds) *Novgorod: The Archaeology of a Russian Medieval City and Its Hinterland*. British Museum Occasional Papers 141, pp. 127–31. London: British Museum Press.

Sablin, M. (in press) 'Fauna from Ryurik Gorodishche (2000–2004)', in Maltby, M. (ed.) *Animals and Archaeology in Northern Medieval Russia: Zooarchaeological Studies in Novgorod and Its Region*. Oxford: Oxbow.

Savinetsky, A. (in press) 'Archaeozoological materials from Minino and changes in populations of utilized mammals from the north of European Russia from the Mesolithic to the Medieval period', in Maltby, M. (ed.) *Animals and Archaeology in Northern Medieval Russia: Zooarchaeological Studies in Novgorod and Its Region*. Oxford: Oxbow.

Solovyov, D. (2012) 'Leather objects from Troitsky XI, Novgorod', in Brisbane, M., Makarov, N., and Nosov, E. (eds) *The Archaeology of Medieval Novgorod in Context*, pp. 416–20. Oxford: Oxbow.

Spiridonova, E. and Aleshinskaya, A. (2012) 'Results of palynological investigations of the archaeological sites in the Lake Ilmen and Lake Kubenskoye study areas', in Brisbane, M., Makarov, N., and Nosov, E. (eds) *The Archaeology of Medieval Novgorod in Context*, pp. 10–39. Oxford: Oxbow.

Tourunen, A. (2008) *Animals in an Urban Context: A Zooarchaeological Study of the Medieval and Post-Medieval Town of Turku*, Turun Yliopisto: University of Turku.

Tsalkin, V. (1956) *Material Concerning the History of Animal Husbandry and Hunting in Medieval Russia*, Moscow: Academy of Sciences, Materials and Investigations of Archaeology, USSR 51.

Zinoviev, A. (in press) 'From pike to *Sus*: a summary of the zooarchaeological evidence from the Desyatinny-1 site, Novgorod', in Maltby, M. (ed.) *Animals and Archaeology in Northern Medieval Russia: Zooarchaeological Studies in Novgorod and Its Region*. Oxford: Oxbow.

PART III

ASIA

CHAPTER 16

THE EMERGENCE OF LIVESTOCK HUSBANDRY IN EARLY NEOLITHIC ANATOLIA

JORIS PETERS, NADJA PÖLLATH,
AND BENJAMIN S. ARBUCKLE

INTRODUCTION

THE transition from foraging to farming in Neolithic southwest Asia is a milestone in the cultural history of humankind. The earliest archaeological signature of this process is found in the microlithic cultures of the Levantine Epipalaeolithic, where semi-sedentary communities practising intensive plant exploitation arose. Further north in Anatolia, culturally related Epipalaeolithic foragers have been identified in the Antalya region at Karain B and Öküzini caves, at Direkli Cave in the central Taurus region, and at Pınarbaşı rock shelter in central Anatolia (Kartal 2009; Erek 2010; Baird, 2012) (Fig. 16.1). At present, however, it is in Upper Mesopotamia that evidence suggests livestock husbandry had its origins. In the Upper Euphrates basin sedentary communities with links to the Levantine Natufian tradition emerged towards the end of the Epipalaeolithic (10,700–9600 cal BC), e.g. at Tell Abu Hureyra and Tell Mureybet, while contemporary sites in the Upper Tigris basin, including Hallan Çemi, Körtik Tepe, and Hasankeyf Höyük exhibit lithic assemblages linking them to the northern Zagros region. Whereas the Upper Tigris region was largely abandoned in the ninth millennium cal BC (Miyake et al., 2012), settlement is continuous in the Upper Euphrates basin through the Pre-Pottery Neolithic (PPN), c.9600–7000 cal BC. It is here that food production technologies emerged comparably early (Wilcox and Stordeur, 2012). In central Anatolia, sedentary communities practising plant cultivation appear in the mid-ninth millennium cal BC, as seen at Boncuklu and Aşıklı Höyük (Baird, 2012; Özbaşaran, 2012), while further west, excavations in the Lakes and Izmir regions indicate a much later expansion of Neolithic lifeways in the late eighth–early seventh millennium cal BC (Çilingiroğlu and Çakırlar, 2013).

Paralleling the long phase of pre-domestic cereal cultivation in the PPNA (9600–8700 cal BC) and Early PPNB (EPPNB, 8700–8200 cal BC) of the northern Fertile Crescent (Tanno and Willcox, 2006), recent archaeozoological literature suggests an extended period of interaction between humans and ungulates prior to their domestication. This model often assumes long-term management of free-ranging herds of wild ungulates within their respective natural environments. For the moment, however, neither the practicality nor the duration of this early management phase is well understood. Herds of wild sheep (*Ovis orientalis*) and goat (*Capra aegagrus*), for instance, undertake large-scale seasonal migrations challenging the sustainability and productivity of managing free-ranging populations in the vast, hilly landscapes that characterize the northern Fertile Crescent. In contrast, wild boar (*Sus scrofa*) is behaviourally pre-adapted for developing a closer relationship with sedentary human populations due to its wide diet breadth and habitat preferences which parallel, in many

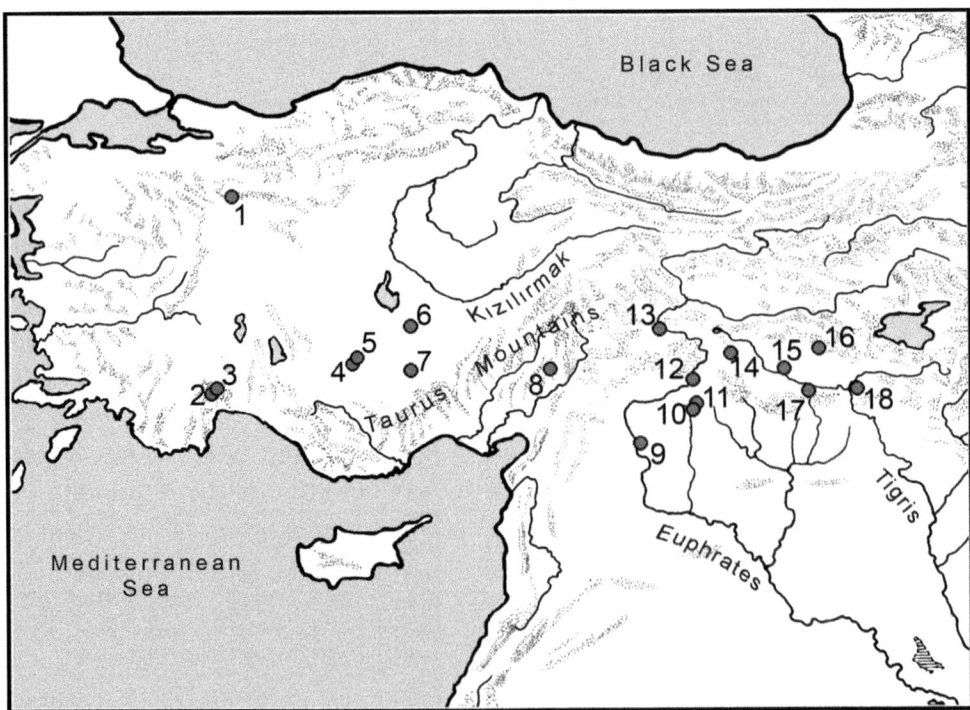

FIGURE 16.1 Map of Anatolia showing the sites mentioned in the text. 1 Demircihüyük, 2 Karain, 3 Öküzini, 4 Çatalhöyük, 5 Boncuklu, 6 Aşıklı Höyük, 7 Pınarbaşı, 8 Direkli Cave, 9 Mezraa-Teleilat, 10 Gürcütepe, 11 Göbekli Tepe, 12 Nevalı Çori, 13 Cafer Höyük, 14 Çayönü, 15 Körtik Tepe, 16 Hallan Çemi Tepesi, 17 Hasankeyf, 18 Gusir Höyük. Authors' own image.

ways, those of humans (Redding and Rosenberg, 1998). As a result, it is only in the wild boar that we find tangible evidence for a long-lasting commensal relationship with humans, developing over time into pig husbandry (Hongo and Meadow, 1998; Ervynck et al., 2001). In contrast to this situation, cultural control over herds of wild caprines necessitated restriction of movement (e.g. penning individuals in or near a settlement) in order to avoid valuable 'walking larders' wandering off or being killed by predators. Penning caprines and their reproduction in captivity would limit interbreeding with free-ranging relatives, explaining why domestic traits could appear more rapidly in ruminants compared to pigs, which likely experienced very little reproductive isolation during the early stages of interaction with sedentary communities.

Although experiments have shown that morphological changes can occur within thirty generations of breeding in strict isolation from the wild (Trut et al., 2009), because animal management within an anthropogenic environment proceeded as a process of learning-by-doing, early stock-keepers must have faced many set-backs, resulting in repeated restocking from wild herds. Consequently, it is likely that a pioneer stage of on-site management characterized by continuous gene flow and hence phenotypically wild animals continued well past the estimates produced from experimental domestication studies. However, due to the difficulties of identifying this early stage of cultural control over ungulates in the archaeological record, the length of time represented by this initial management stage has yet to be demonstrated. Arguably, spatial isolation from the wild worked as a catalyst for the active role of humans in the process leading to livestock domestication and husbandry, but even for confined ungulate populations, archaeozoologists still need to refine the osteological and biomolecular criteria that allow us to define domestic status.

As a result of differences in mammalian behaviour and human exploitation strategies in distinct environments, pathways to domestication must be seen as species- and perhaps even site-specific (Zeder, 2012; Peters et al., 2013). The domestication process involved both unconscious and intentional human intervention in the reproduction of livestock species, and of natural and artificial selection operating in anthropogenic environments (Zohary et al., 1998; Albarella et al., 2006). During this process the interplay of environmental, biological, and cultural variables gradually, or rapidly, altered the behaviour and genetics of animals kept under (different) forms of human control (Ervynck et al., 2001), which likely varied between settlements. Considering variables such as supplementing fodder or social stress and health problems due to penning, pioneer efforts to control ungulates presumably involved few individuals or small flocks. As a result, the archaeo(zoo)logical signatures of these are expected to be obscured by those of hunting, which remained the dominant mode of exploitation in the early stages of animal management and domestication. Clear evidence for husbandry is expected to emerge only with the appearance of a post-pioneer phase in animal management characterized by the development of larger, more stable herds and effective management practices visible in the archaeozoological (e.g. sex-biased culling) and archaeological (e.g. stabling areas, dung) records.

Although the osteological markers most directly relevant to the recognition of early husbandry are a matter of debate, traditional osteomorphological and metrical criteria combined with the assessment of demographic profiles and sex ratios have revealed a generalized and robust spatio-temporal framework for the beginnings and early spread of animal husbandry in the Fertile Crescent (Conolly et al., 2011). Spatio-temporal comparison across southwest Asia assigns a pioneer role to PPN Anatolia for ungulate domestication, with local efforts in managing ruminants and pigs being found in Upper Mesopotamia (Peters et al., 1999; 2013) and ruminants in central Anatolia (Peters et al., 2013; Stiner et al., 2014). Neolithization in Anatolia beyond these regions set in much later (i.e. after 7500 cal BC) and represents the spread of already domesticated livestock into southern, western, and northwestern regions (Arbuckle, 2013; Peters et al., 2013; 2014; Arbuckle et al., 2014).

DOMESTICATION HISTORIES

A broad methodological approach is essential to provide insight into animal exploitation and the emergence of livestock husbandry. We therefore consider a combination of biogeographic, morphological, demographic, and light stable isotopic evidence for documenting the domestication process of the four major livestock taxa (for details on methodology, see Dobney et al., 2013; Peters et al., 2013; 2014). However, presenting a coherent picture of the faunal developments in Anatolia is problematic since it is based on less than thirty (partly unpublished) faunal assemblages (Peters et al., 2013). Moreover, PPN archaeofaunas currently under study do not necessarily reflect primary domestication sites but rather settings where humans had deliberately introduced founder stock soon after the benefits of this innovative subsistence strategy became obvious. Such secondary contexts are equally important for understanding the process of ungulate domestication in Neolithic Anatolia.

Pig (Sus)

As a result of a shared preference for woodland habitats located near permanent water bodies, humans and wild boar have a long history of interaction. In PPN Anatolia the abundance of wild boar remains in Epipalaeolithic and PPNA sites including Hallan Çemi, Çayönü, and Boncuklu has been noted. At PPNA Hallan Çemi a high frequency of juvenile remains, combined with skeletal part representations indicating the presence of complete carcasses at the site, has been presented as evidence for management and possible pig husbandry (Rosenberg and Redding, 1998). However, given the absence of osteomorphological change, the proximity of wild boar habitats to the settlement, and the abundance of juveniles in hunted suid populations, recent work now favours an interpretation of pig exploitation in the tradition of Epipalaeolithic hunting, rather than a transitional Neolithic farming strategy (Starkovich and Stiner, 2009; Lemoine, 2012).

Clearer evidence of pig husbandry is found at Çayönü, where demographic and biometric data suggest a suid population living in an intermediary relationship with humans, gaining in intensity through time, and gradually resulting in characteristics that discriminate them from the original wild population (Ervynck et al., 2001). Subtle changes in the slaughter schedule implying slightly earlier kill-off correspond with the appearance of smaller-sized individuals. Hongo and Meadow (1998) therefore concluded that domesticated pigs were kept at the site by c.8300–8200 cal BC.

At EPPNB Cafer Höyük and Nevalı Çori, human control over pigs is suggested by the presence of individuals smaller in size than those found in Early Holocene Syrian and modern Near Eastern wild boar (Peters et al., 2005; Helmer, 2008). At Nevalı Çori cereals as well as legumes including lentil (*Lens* spp.), field pea (*Pisum* spp.), chickpea (*Cicer* sp.), grass pea (*Lathyrus* sp.), bitter vetch (*Vicia ervilia*), and horse bean (*Vicia faba*) were dietary staples consumed in great enough quantities (Pasternak, 1998) to result in a reduction in δ^{15}N values in the bone collagen of the Nevalı Çori inhabitants compared to other PPN human groups. Since some pigs, sheep, and goat also exhibited lowered δ^{15}N values, it has been postulated that these animals had access to the refuse of human consumption of pulses (Lösch et al., 2006; Grupe and Peters, 2011). Given their broad-spectrum diet and opportunistic feeding behaviour, the δ^{15}N signatures found in pigs do not exclude wild boar a priori (Rowley-Conwy et al., 2012), yet the combination of the small size and reduced nitrogen values suggest that some of the suids at Nevalı Çori were likely living, eating, and breeding within the community (for a similar argument see Barton et al., 2009).

The fact that the pigs at Çayönü exhibit only gradual changes in age profiles and phenotype over two millennia has been taken as proof that osteomorphological changes are a delayed component of the domestication process (Zeder, 2008). Pig management at Çayönü may, however, have been 'relaxed' at first, with no intention of keeping animals strictly isolated from the wild, since controlling the mobility, reproduction, and feeding in this commensal species would have been unnecessary for providing predictable access to suids (Redding and Rosenberg, 1998). Nevertheless, once plant cultivation became the focus of Early Neolithic economies, the threat posed by crop raiders such as pigs had to be eliminated and the mobility in livestock more closely controlled. It is perhaps not a coincidence that at Çayönü, the emergence of early pig husbandry goes along with other innovations in subsistence, e.g. the cultivation of wild emmer (*Triticum dicoccum*) and the onset of caprine and cattle (*Bos*) husbandry (van Zeist and de Roller, 1994; Hongo et al., 2009). However, even after its emergence in Upper Mesopotamia in the ninth millennium cal BC, pig husbandry was slow to spread into adjacent regions: it was introduced in southwest Anatolia with the initial movement of food-producing economies in the early seventh millennium cal BC, but did not appear in central Anatolia, the southern Levant, or the Zagros region until the Chalcolithic period (Arbuckle, 2013; Price and Arbuckle, 2015).

Sheep and Goat (Ovis and Capra)

Changes in the human–caprine relationship have traditionally been documented by diachronic comparison in bone width and depth (rather than length) dimensions (Uerpmann, 1979). In southeast Anatolia, sheep and goats significantly smaller in size than their wild relatives are evident in the late EPPNB, c.8400–8200 cal BC, indicating that human management had intensified by the mid-ninth millennium cal BC, if not earlier. Conversely, bone size in Persian Goitered gazelle (*Gazella subgutturosa*) did not vary significantly throughout the PPN (Peters et al., 2013: Fig. 5.8), ruling out variation in climate as a possible cause for size decrease in caprines (cf. Bergmann's rule).

Based on the analysis of goat remains from the Zagros region, Zeder and Hesse (2000) argued that the earliest signal for human control over ruminants would be the targeted culling of young males, rather than morphological change. Their study draws upon Payne's (1973) seminal work delineating the relationship between the goals of sheep/goat herd management and the age and sex characteristics of the animals chosen for slaughter. In taxa with pronounced sexual dimorphism like wild goats, sex-biased culling practices can be identified in the biometric data through the analysis of skewness in size-distribution histograms: young male kill-off will result in positive skewing, i.e. a tall peak on the left (skeletally mature females) and a tail extending to the right (mature males).

Although relevant to the domestication debate, young male kill-off has hardly been tested in the Fertile Crescent beyond the Zagros cultural zone. Size profiles for caprines from PPNA sites in southeast Anatolia clearly exhibit negative skewing or symmetrical distributions, pointing to hunting economies targeting adult males (Peters et al., 2014: Figs 5, 7). Sheep and goat remains from EPPNB Cafer Höyük and Nevalı Çori do not show selective kill-off either, despite the first appearance of osteometric and dietary changes thought to be associated with caprine management and husbandry. As such, a dominance of smaller (i.e. female) animals is observed first in contexts dating to the MPPNB, but with significant inter-site variability. At Çayönü, for instance, young male kill-off in goats clearly pre-dates the same strategy in sheep (Hongo et al., 2002), whilst at Mezraa-Teleilat, the situation is reversed. Based on their overview, Arbuckle and Atici (2014) conclude that husbandry in PPNB Anatolia was characterized by "initial diversity", suggesting that early herd management strategies did not conform to modes of herd management used by later pastoralists. Despite early variation in herd management strategies, the practice of young male culling became widespread in the mid-eighth millennium cal BC to become the dominant management strategy for herding caprines across southwest Asia.

With its dominance of goats aged 6–12 months, the caprine (primarily goat) assemblage from Cafer Höyük exhibits uniquely intense juvenile slaughter. Low temperatures and heavy snow certainly forced (wild) goats to pasture at lower altitudes in wintertime, providing opportunities to hunt yearlings. However, the highly selective culling of this age class (c.70% of ageable mandibles), makes it very unlikely that only wild animals contributed to the sample. Osteometric data also suggest human management of at least part of the goat population. The fact that in the course of the site's occupation, adult goats exhibit slightly later kill-off suggests intensification of breeding and perhaps even the use of milk (Helmer, 2008: Figs 15–17). In this respect, it is noteworthy that in the absence of their offspring, female goats let down milk easier than ewes and cows (McKusick et al., 2002).

At LPPNB Gürcütepe, the *Capra* bones represent domesticates, because in Neolithic times wild goat was alien to the southern Anti-Taurus piedmont. For sheep, both wild and domestic animals are present, but osteometrics and the much higher frequency of *Ovis* compared to the assemblage from PPNA Göbekli Tepe imply that domesticates predominated. Demographic profiling shows that in sheep, nearly two-thirds of the population did not reach two years of age (Fig. 16.2), while for goats almost half of the animals survived beyond this age. Because the size distribution in *Capra* shows a dominance of small (i.e. female) animals, the slaughtering of juveniles mainly concerned male goats, again suggesting that milk may have played a role in the human diet. We conclude that by the late eighth millennium cal BC, sheep and goat in southeast Anatolia already experienced distinct exploitation strategies and complementary production goals.

Moving to central Anatolia, both sheep and goat husbandry was already well developed at the onset of the occupation of Çatalhöyük *c.*7200 cal BC (Russell and Martin, 2005), whereas analysis of the Level 2 fauna from Aşıklı Höyük dating to the first half of the eighth millennium cal BC did not yield conclusive evidence on caprine husbandry (Buitenhuis, 1997). However, when compared to wild sheep from Göbekli Tepe and later domesticates, the *Ovis* from Aşıklı Höyük (Level 2) exhibit an intermediate body size

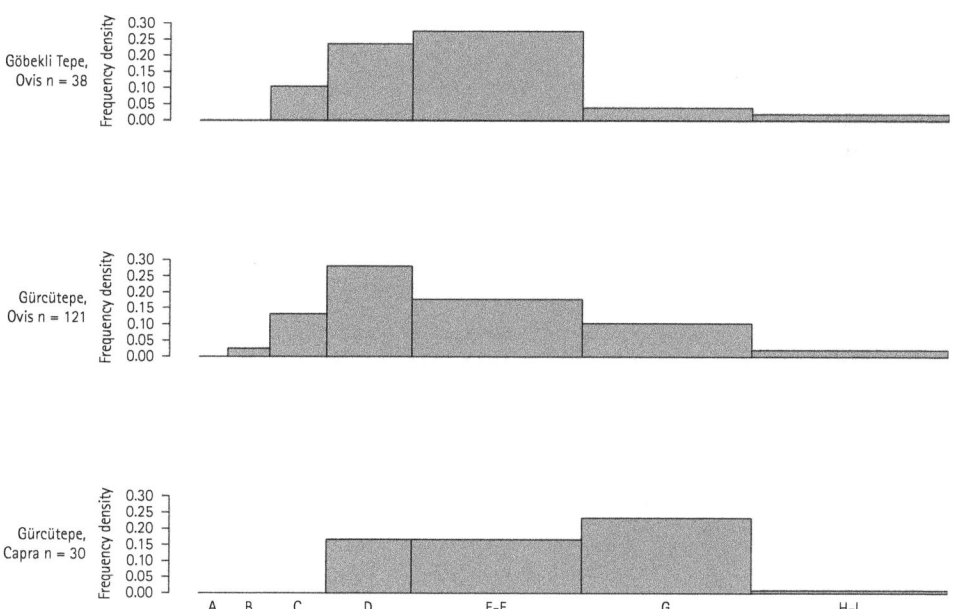

FIGURE 16.2 Age profiles for *Ovis* and *Capra* from Göbekli Tepe and Gürcütepe.
Age groups based on mandibles and mandibular teeth: A 0–2 months, B 3–6 months, C 7–12 months, D 1–2 years, EF 2–4 years, G 4–6 years, HI more than 6 years. Authors' own image.

(Peters et al., 2013). Since it is predicted, based on climate and latitude (Bergmann, 1847), that central Anatolian wild sheep likely equaled or even surpassed in size their relatives inhabiting the climatically milder southern Anti-Taurus piedmont, the relatively small size of the Aşıklı sheep points to their management in the eighth millennium cal BC. In addition, the size of Aşıklı Höyük goats (Level 2) is intermediate when compared to the domestic goats from Çatalhöyük and wild goats from PPNA Direkli Cave, again suggesting human manipulation (Peters et al., 2013). This biometric evidence for morphological change is further supported by limited variation in carbon and nitrogen stable isotope values in caprine teeth, implying that sheep and goats were pastured together or confined to restricted areas (Pearson et al., 2007). In addition, an unusually high frequency of foetal and neonate caprines and micromorphological evidence for ruminant dung within the settlement suggest that caprines were regularly present on site (Özbaşaran, 2012; Stiner et al., 2014). In sum, management of herds of sheep and goat seems already well established in central Anatolia by the early eighth and possibly even late ninth millennium cal BC. However, whether sheep management was adopted as a response to conflicts in resource scheduling once people became sedentary (Stiner et al., 2014) remains to be seen. Increased control over the movements of wild caprines frequenting the site catchment may have become essential with ongoing intensification of plant cultivation at Aşıklı. Moreover, the capacity of caprines to turn residues of crop harvesting and processing into valuable protein and fat whilst producing dung for manuring agricultural plots provides an additional explanation for why people started experimenting with the crop–caprine combination in this particular landscape setting.

Cattle (Bos)

Diachronic analysis of changes in the size of cattle remains from Early Neolithic settlements along the Syrian Euphrates has shown that the degree of sexual dimorphism diminished markedly through the PPN. A decrease in the size of male *Bos* in the EPPNB indicates that human manipulation of the cattle life cycle extends back to the mid-ninth millennium cal BC (Helmer et al., 2005). Considering its high profile role in the symbolic world of the PPN (Peters and Schmidt, 2004), it is possible that the aurochs (*Bos primigenius*) was initially brought under human control for reasons other than subsistence.

Cattle measurements from PPN sites located in the Upper Euphrates River valley and its tributaries confirm that the bimodal size distribution observed in aurochs from PPNA Göbekli Tepe becomes less clear over time. Size profiling illustrates a higher proportion of females in assemblages in the EPPNB and the appearance of less robust cattle in the MPPNB; all of which are interpreted as evidence for a developing relationship between humans and cattle (Peters et al., 2013: Fig. 5.11). These changes are also evidenced at PPN Çayönü, where, additionally, a subtle trend towards younger kill-off is observed. Diachronic changes in the diet of cattle—but not of red deer (*Cervus elaphus*)—are captured in shifting carbon isotope values, suggesting that the movement of cattle was increasingly restricted and/or herds were provided with supplementary fodder. Hongo et al. (2009) conclude that the relationship of humans and cattle changed slowly throughout the site's occupation, with the appearance of domesticates c.8300–8200 cal BC.

In central Anatolia, *Bos* finds from Aceramic Neolithic contexts represent morphologically wild animals. In the LPPNB levels at Çatalhöyük, for instance, the size distribution in cattle is almost identical to that observed in wild cattle from PPNA Göbekli Tepe, with few slightly larger and some smaller individuals. Although the Çatalhöyük samples exhibit a trend towards a higher proportion of females and younger kill-off in the early seventh millennium cal BC, the lack of change in cattle size and morphology between 7200 and 6500 cal BC seems to confirm their wild status (Russell et al., 2005). The rather sudden appearance of small-sized domestic cattle with reduced sexual dimorphism in central Anatolia after c.6500 cal BC, however, suggests that domesticates may have been imported from elsewhere (Arbuckle and Makarewicz, 2009; Russell et al., 2013).

Of course, the absence of wild cattle from tenth-millennium cal BC contexts in central Anatolia complicates the interpretation of osteometric evidence from the early levels in Çatalhöyük. Russell and Martin (2005) noted that the skeletal distribution in cattle does not show a 'schlepp' effect (Perkins and Daly, 1968), implying that all body parts were consistently transported from kill sites back to the settlement, independent of their utility value (Binford, 1979). Although it is possible that this pattern represents the hunting of aurochs close to the site, with an extent of 13.5 ha and 3,500–8,000 people exploiting the site catchment at any one time (Hodder, 2005), the local landscape around Çatalhöyük certainly would have suffered from resource depression, especially in regards to slow-reproducing large mammals such as aurochs. This implies that big game hunting would have required forays to more distant hunting territories. As an alternative to the interpretation that wild cattle were hunted for the first millennium of occupation at Çatalhöyük, is it possible that the site inhabitants kept (domestic) female cattle in enclosures and practised crossbreeding with (young) male aurochs? Such management could explain the persistence of large-sized, long-horned cattle at Çatalhöyük until the mid-seventh millennium cal BC (Twiss and Russell, 2010), when this practice must have ceased. Moreover, increasingly young kill-off and a higher proportion of females amongst adult cattle would also fit with cultural control over a herd rather than hunting. Provided that the Çatalhöyük inhabitants raised cattle in captivity, mobility in these valuable animals might also have been restricted. This scenario is supported by stable isotope evidence, which indicates that cattle were not grazing over as great a diversity of environments as were domesticated sheep and goat (Hodder, 2005; Pearson et al., 2007).

THE EMERGENCE OF HUSBANDRY

As stated above, early stock-keepers certainly faced a host of problems as they invented techniques necessary for the successful management of their herds. However, once techniques such as (semi-)mobile grazing and foddering were developed, which provided access to adequate year-round nutrition, and herd reproduction was mastered, herd size could increase and livestock husbandry become a central part of Neolithic economic life. Conceivably, Early Neolithic husbandry strategies also aimed at risk reduction as, even today, pastoralists prefer incorporating two or more species in their herds, provided they are complementary in terms of diet, breeding goals, and resistance to diseases. Although husbanding more than one species complicates matters considerably, except perhaps for sheep and goat which can be herded together, the long-term benefit of just such a commitment seems to be at the foundation of the emergence of the 'Neolithic package' during the PPNB.

In southeast Anatolia, risk reduction is suggested by the findings from EPPNB Nevalı Çori. Here sheep and goats were managed in small numbers, probably together with pigs, while gazelle hunting remained the focus of the animal economy. With the growing economic importance of stock-keeping along the Upper Euphrates and its tributaries, however, competition for bio-resources increased, affecting the richness and diversity of local wildlife. In the Upper Euphrates basin biodiversity was already on the decline in the late ninth millennium cal BC, a situation which continued during subsequent centuries (Peters et al., 2013: Fig. 5.5b). As such, the replacement of Persian Goitered gazelle by domestic sheep and goat as the mainstay of animal food economies shortly after 8000 cal BC marks the turning point in the process of Neolithization of southeast Anatolia. It is noteworthy that during Neolithic times, the intensification of husbandry practices amplified the rate and scale of ecological deterioration around settlements. For example, overgrazing is indicated by carbon isotope signatures in the bones of livestock pastured near LPPNB Gürcütepe, offering a potential explanation for the abandonment of the site at the end of the PPN (Grupe and Peters, 2011).

Further north, diachronic change is less spectacular, but nonetheless indicative of mixed husbandry practices, and hence risk reduction. At PPNA and EPPNB Çayönü, for example, (wild and/or domestic) *Sus* and caprines were both parts of the economy. At the transition of the Early to the Middle PPNB, however, the ratio of *Sus* to *Ovis/ Capra* shifts in favour of the latter (Peters et al., 2013: Fig. 5.4b), implying that the returns of caprine husbandry were increasingly valued. At contemporaneous Cafer Höyük, goats played a key role in subsistence, but throughout the site's occupation the proportion of pigs increased significantly (Peters et al., 2013: Fig. 5.4c). At Cafer, pig husbandry likely emerged as a complementary strategy to goat herding promoted by the location and permanence of the settlement and the presence of edible waste resulting from plant harvesting and the storage, preparation, and consumption of plant foods.

With domestic cattle appearing in southeast Anatolia at the transition from the EPPNB to the MPPNB, the 'Neolithic barnyard' was completed. Parallel to the growing importance of livestock husbandry, cultivated and/or domesticated food plants contributed increasingly to the human diet as well. The success of this newly developed agro-pastoralism in the Upper Euphrates and Tigris River basins was rooted in the nutritional and labour complementarity and productivity of the crop–livestock combination (Harris, 2002; Peters et al., 2005), in which the role of animal dung for manuring cultivated plots should not be underestimated. Conceivably, it was this combined system of animal and plant food production which fuelled the spread of Neolithic technologies within and outside of the Fertile Crescent region.

Documenting the emergence of animal husbandry in central Anatolia is complicated due to the lack of PPNA faunal assemblages that can be used to trace diachronic change. Nonetheless, over time, species composition in Neolithic assemblages in this region increasingly resembles that observed in local post-Neolithic pastoralist economies. We can quantify this transition from hunting to herding economies using Percentage Similarity analysis, a method applied in ecological studies (Peters et al., 2013). Thus, from EPPNB Pınarbaşı A to Pottery Neolithic Pınarbaşı B, Percentage Similarity with the reference assemblage (= 100% of Bronze Age Demircihüyük) ranges from 50% to 77%, whilst for Çatalhöyük a value of c.85% has been calculated. Livestock husbandry obviously played an important role at the latter sites, which correlates well with the archaeozoological findings (Russel and Martin, 2005; Baird et al., 2011).

Because cattle and pig husbandry were incorporated much later into central Anatolian subsistence systems, Early Neolithic risk reduction strategies there differed from those developed in Early Neolithic southeast Anatolia. Ethnographic case studies illustrate the risks inherent to economies building on (too) few taxa, and it remains to be seen how early herders in central Anatolia dealt with caprine feeding, stalling, breeding, and health issues. However, the presence of large numbers of foetal and neonate sheep at Aşıklı, indicating high infant mortality and spontaneous abortion rates (Stiner et al., 2014), is one sign that on the central Anatolian plateau, the learning-by-doing process of ungulate management and husbandry was not without its problems.

The Neolithization of Anatolia beyond the southeast and central regions occurred much later. The full package of domestic livestock appears in southwest and west Anatolia in the late eighth and/or early seventh millennium, while in the northwest domestic caprines and cattle appear in the mid-seventh millennium cal BC, pig husbandry being adopted in the early sixth millennium cal BC (Özdoğan, 2011; Çilingiroğlu and Cakırlar, 2013; Arbuckle et al., 2014).

THE NEOLITHIC MINDSET

Animals play important roles in human societies beyond subsistence (e.g. Paine, 1971; Ingold, 1988) and scholars are increasingly exploring the social and religious uses of animals as motivating factors in the process of livestock domestication (Russell, 2012). In the material culture of the Neolithic Near East, animals figure prominently in the PPN socio-cultural landscape. This is illustrated by growing evidence for feasting events as well as cult practices and animal representations with complex social and cosmological meaning. This and other aspects led Cauvin (1997) to argue that the emergence of religious concepts played a catalytic role in the transition from hunting to farming.

Evidence for feasting events (Twiss, 2008), and the social competition that is played out within them, is also found in the PPN of Anatolia. In the Upper Tigris River basin, for example, pig-centric feasting events were socially important, as seen at Çayönü and Hallan Çemi. In central Anatolia, bovine-centric feasting practices predominate (Buitenhuis, 1996; Baird, 2012). At Çatalhöyük, cattle were procured, shared, and consumed among community members and special events were commemorated through the curation, display, and deposition of selected elements, e.g. bucrania and scapulae (Twiss and Russell, 2010). Social obligations to participate in feasting events and the status gained through provisioning feasting events with animals may have been prominent motivational factors that drove the costly and failure-prone initial stages of ungulate management and husbandry.

In addition to feasting, the abundant and rich iconography including paintings, engravings, plastered relief figures, mounted bucrania, figurines, and sculptures

FIGURE 16.3 Göbekli Tepe, Enclosure D, Pillar 43. This external pillar is covered with depictions of birds, snakes, and arthropods among others. © GEO & DAI, Photo B. Steinhilber.

indicates a prominent symbolic role for animals in Neolithic cosmology and ritual practice. Although the precise relationships between animal symbolism, ritual practices, and the origins of animal husbandry are difficult to articulate, the early stages of livestock management were certainly situated within a cognitive world in which animals played a central and symbolic role. It is therefore likely that the symbolic associations of animals in the PPNA, richly expressed at sites such as Göbekli Tepe, and the ontologies that they represent provided a meaningful context which framed and affected the emergence of animal husbandry.

The presence of diverse animals including insects, arachnids, snakes, birds, and mammals on the anthropomorphic T-shaped stone pillars at PPNA Göbekli (Figs 16.3; 16.4)

FIGURE 16.4 Göbekli Tepe, Enclosure D, Pillar 18. Arms, hands, and a fox pelt serving as loincloth indicate the anthropomorphic character of this central pillar. © DAI, Photo, N. Becker.

suggests that these anthropomorphs exerted power over (dangerous) wild animals and thus—in the figurative sense—over the living world. If animal symbolism at Göbekli Tepe reflects 'mental' control over the living world, 'physical' and hence 'cultural' control over animals may have been logical next steps, resulting over time in the development of systems of management and eventually domestication. The introduction of several mammal species in Cyprus (Vigne, 2011) seems to illustrate this new mind-set. In this context it may be significant that it was the least symbolically potent animals—sheep and goats—which were the first to come under human control and spread throughout Neolithic Anatolia, while the management and domestication of more symbolically potent animals such as aurochs and wild boar was both later and slower to spread (Arbuckle, 2013).

CONCLUSION

In the course of the ninth and early eighth millennia cal BC, the domestication histories of the four candidates to become the earliest livestock animals started at different places with different paces and in different ways in Upper Mesopotamia and central Anatolia. From there, the husbandry of pigs, sheep, goats, and cattle began to spread in varying configurations into the regions south, west, and north of the core areas of early animal management and husbandry. Conceivably, megalithic ritual centres bringing together hunter-gatherers from a wide catchment area such as Göbekli Tepe facilitated the initial spread of innovative subsistence technologies including those related to ungulate management and husbandry. In this respect, large-scale sharing of know-how relative to the practicality, costs, and economic benefits of managing ungulates in captivity would also explain why in the Early Neolithic Near East, domestication was neither restricted to one ungulate taxon nor to a particular eco-geographic setting. Thus, PPNA ritual practices with their prominent animal themes, large-scale social gatherings, and feasting may have played a central role in promoting and facilitating the early stages of management and animal husbandry which then, along with plant cultivation, signalled the end of the hunting and gathering way of life in Anatolia.

ACKNOWLEDGEMENTS

This paper could not have been written without the results of our work on the faunal assemblages excavated at Aşıklı Höyük, Erbaba Höyük, Göbekli Tepe, Gürcütepe, Gusir Höyük, Körtik Tepe, Köşk Höyük, and Nevalı Çori. We are grateful to the directors of excavation H. Hauptmann, N. Karul, K. Schmidt (†), M. Özbaşaran, A. Öztan, and V. Özkaya and their teams as well as the General Directorate of Antiquities of Turkey,

Ankara, for allowing us to study these interesting collections. This paper benefitted also from discussions with them and many other colleagues.

The financial support of the German Research Foundation (DFG) to JP and NP (Grants PE 424/9, 1–2 and PE 424/10, 1–2) and of the National Science Foundation (BCS0530699) and the American Research Institute in Turkey to BA is gratefully acknowledged.

References

Albarella, U., Dobney, K., and Rowley-Conwy, P. A. (2006) 'The domestication of the pig (*Sus scrofa*): new challenges and approaches', in Zeder, M., Bradley, D. G., Emschwiller, E., and Smith, B. D. (eds) *Documenting Domestication: New Genetic and Archaeological Paradigms*, pp. 209–27. Berkeley: University of California Press.

Arbuckle, B. S. (2013) 'The late adoption of cattle and pig husbandry in Neolithic Central Turkey', *Journal of Archaeological Science*, 40(4), 1805–15.

Arbuckle, B. S. and Atici, L. (2014) 'Initial diversity in sheep and goat management in Neolithic southwestern Asia', *Levant*, 45(2), 219–35.

Arbuckle, B. S., Kansa, S., Kansa, E., Atici, L., Buitenhuis, H., Çakırlar, C., Carruthers, D., De Cupere, B., Frame, S., Galik, A., Gourichon, L., Helmer, D., Marciniak, A., Mulville, J., Orton, D., Peters, J., Pöllath, N., Powloska, K., Russell, N., Twiss, K., and Würtenberger, D. (2014) 'Data sharing reveals early evidence for westward spread of domestic livestock across Neolithic Turkey', *PLoS ONE*, 9(6), DOI:10.1371/journal.pone.0099845.

Arbuckle, B. S. and Makarewicz, C. A. (2009) 'The early management of cattle (*Bos taurus*) in Neolithic central Anatolia', *Antiquity*, 83, 669–86.

Baird, D. (2012) 'The Late Epipaleolithic, Neolithic and Chalcolithic of the Anatolian Plateau, 13,000–4000 BC', in Potts, D. T. (ed.) *A Companion to the Archaeology of the Ancient Near East*, Vol. 1, pp. 431–65. Malden: Wiley-Blackwell.

Baird, D., Carruthers, D., Fairbairn, A., and Pearson, J. (2011) 'Ritual in the landscape: evidence from Pınarbası in the seventh-millennium cal BC Konya Plain', *Antiquity*, 85, 380–94.

Barton, L., Newsome, S. D., Chen, F.-H., Wang, H., Guilderson, T. P., and Bettinger, R. L. (2009) 'Agricultural origins and the isotopic identity of domestication in northern China', *Proceedings of the National Academy of Sciences of the United States of America*, 106(14), 5523–8.

Bergmann, C. (1847) 'Über die Verhältnisse der Wärmeökonomie der Thiere zu ihrer Größe', *Göttinger Studien*, 3(1), 595–708.

Binford, L. R. (1979) *Nunamuit Ethnoarchaeology*, New York: Academic Press.

Buitenhuis, H. (1996) 'Archaeozoology of the Holocene in Anatolia: a review', in Demirci, S., Ozer, A. M., and Summer, G. D. (eds) *Archaeometry 94: Proceedings of the 29th International Symposium on Archaeometry, Ankara, 1994*, pp. 411–21. Ankara: Tübitak Publications.

Buitenhuis, H. (1997) 'Aşıklı Höyük: a "protodomestication" site', *Anthropozoologica*, 25–6, 655–62.

Cauvin, J. (1997) *Naissance des divinités. Naissance de l'agriculture: la révolution des symboles au Néolithique*, 2nd edn, Paris: C.N.R.S.

Çilingiroğlu, Ç. and Çakırlar, C. (2013) 'Towards configuring the neolithisation of Aegean Turkey', *Documenta Praehistorica*, 40, 21–9.

Conolly, J., Colledge, S., Dobney, K., Vigne, J.-D., Peters, J., Stopp, B., Manning, K., and Shennan, S. (2011) 'Meta-analysis of zooarchaeological data from SW Asia and SE Europe provides insight into the origins and spread of animal husbandry', *Journal of Achaeological Science*, 38, 538–45.

Dobney, K., Colledge, S., Conolly, J., Manning, K., Peters, J., and Shennan, S. (2013) 'The origins and spread of stock-keeping', in Colledge, S., Conolly, J., Dobney, K., Manning, K., and Shennan, S. (eds) *The Origins and Spread of Domestic Animals in Southwest Asia and Europe*. Publications of the Institute of Archaeology, University College, London 59, pp. 17–26. Walnut Creek: Left Coast Press.

Erek, C. M. (2010) 'A new Epi-paleolithic site in the Northeast Mediterranean region: Direkli Cave (Kahramanmaras, Turkey)'. *Adalya*, 13, 1–19.

Ervynck, A., Dobney, K., Hongo, H., and Meadow, R. (2001) 'Born free! New evidence for the status of pigs from Çayönü Tepesi, Eastern Anatolia', *Paléorient*, 27, 47–73.

Grupe, G. and Peters, J. (2011) 'Climate conditions, hunting activities and husbandry practices in the course of the Neolithic transition: the story told by stable isotope analysis of human and animal skeletal remains', in Pinhasi, R. and Stock, J. T. (eds) *Human Bioarchaeology of the Transition to Agriculture*, pp. 63–85. Hoboken: Wiley-Blackwell.

Harris, D. R. (2002) 'Development of the agro-pastoral economy in the Fertile Crescent during the Pre-Pottery Neolithic period', in Cappers, R. T. and Bottema, S. (eds) *The Dawn of Farming in the Near East*. Studies in Early Near Eastern Production, Subsistence, and Environment 6, pp. 67–83. Berlin: Ex Oriente.

Helmer, D. L. (2008) 'Révision de la faune de Cafer Höyük (Malatya, Turquie): apports des méthodes de l'analyse de mélanges et de l'analyse de Kernel à la mise en évidence de la domestication', in Vila, E., Gourichon, L., Choyke, A., and Buitenhuis, H. (eds) *Archaeozoology of the Near East*, Vol. 8. Travaux de la Maison de l'Orient et de la Méditerranée 49, pp. 169–96. Lyon: Maison de l'Orient et de la Méditerranée.

Helmer, D. L., Gourichon, L., Monchot, H., Peters, J., and Saña Segui, M. (2005) 'Identifying early domestic cattle from Pre-Pottery Neolithic sites on the middle Euphrates using sexual dimorphism', in Vigne J.-D., Peters, J., and Helmer, D. (eds) *The First Steps of Animal Domestication: New Archaeozoological Approaches*, pp. 86–95. Oxford: Oxbow Books.

Hodder, I. (2005) 'Memory', in Hodder, I. (ed.) *Çatalhöyük perspectives: Reports from the 1995-99 Seasons by Members of the Çatalhöyük Teams*. Çatalhöyük Research Project 6, British Institute at Ankara Monograph 40, pp. 183–95. Cambridge: McDonald Institute for Archaeological Research/British Institute at Ankara.

Hongo, H. and Meadow, R. H. (1998) 'Pig exploitation at Neolithic Çayönü Tepesi, southeastern Anatolia', in Nelson, S. M. (ed.) *Ancestors for the Pigs: Pigs in Prehistory*. MASCA Research Papers in Science and Archaeology 15, pp. 77–89. Philadelphia: University of Pennsylvania, Museum of Archaeology and Anthropology.

Hongo, H., Meadow, R. H., Öksuz, B., and İlgezdi, G. (2002) 'The process of ungulate domestication in Pre-Pottery Neolithic Çayönü, southeastern Turkey', in Al-Shiyab, A. H., Choyke, A. M., and Buitenhuis, H. (eds) *Archaeozoology of the Near East*, Vol. 5, pp. 153–65. Groningen: Archaeological Research & Consultancy.

Hongo, H., Pearson, J., Öksüz, B., and İlgezdi, G. (2009) 'The process of ungulate domestication at Çayönü, southeastern Turkey: a multidisciplinary approach focusing on *Bos* sp. and *Cervus elaphus*', *Anthropozoologica*, 44(1), 63–73.

Ingold, T. (1988) 'The animal in the study of humanity', in Ingold, T. (ed.) *What Is an Animal?*, pp. 84–99. London: Unwin.

Kartal, M. (2009) *Epi-Paleolitik Dönem Türkiye'de Son Avcı Toplayıcılar: Konar-Göçerlikten Yerleşik Yaşama Geçiş*, Istanbul: Arkeoloji ve Sanat Yayınları.

Lemoine, X. (2012) 'Pig (*Sus scrofa*) Exploitation at Hallan Çemi, Southeastern Anatolia: Proposing an Alternative Model'. Unpublished BA dissertation, Portland State University (Portland).

Lösch, S., Grupe, G., and Peters, J. (2006) 'Stable isotopes and dietary adaptations in humans and animals at Pre-Pottery Neolithic Nevalı Çori, southeast Anatolia', *American Journal of Physical Anthropology*, 131, 181–93.

McKusick, B. C., Thomas, D. L., Berger, Y. M., and Marnet, P. G. (2002) 'Effects of milking systems on alveolar and cistern milk accumulation and milk production and composition in dairy ewes', *Journal of Dairy Science*, 85, 2197–206.

Miyake, Y., Maeda, O., Tanno, K., Hongo, H., and Gündem, C. Y. (2012) 'New excavations at Hasankeyf Höyük: a 10th millennium cal BC site on the Upper Tigris, southeast Anatolia', *Neo-lithics*, 1(12), 3–7.

Özbaşaran, M. (2012) 'Aşıklı', in Özdoğan, M., Başgelen, N. and Kuniholm, P. (eds) *The Neolithic in Turkey, Central Turkey and Mediterranean*, Vol. 3, pp. 135–58. Istanbul: Arkeoloji ve Sanat Yayınları.

Özdoğan, M. (2011) 'Archaeological evidence on the westward expansion of farming communities from eastern Anatolia to the Aegean and the Balkans', *Current Anthropology*, 52, S415–S430.

Paine, R. (1971) 'Animals as capital: comparisons among northern nomadic herders and hunters', *Anthropological Quarterly*, 44, 157–72.

Pasternak, R. (1998) 'Investigations of botanical remains from Nevalı Çori PPNB, Turkey', in Damania, A., Valkoun, J., Willcox, G., and Qualset, C. (eds) *The Origins of Agriculture and Crop Domestication. Proceedings of the Harlan Symposium, 10–14 May 1997, Aleppo, Syria*, pp. 170–7. Aleppo: International Center for Agricultural Research in the Dry Areas.

Payne, S. (1973) 'Kill-off patterns in sheep and goats: the mandibles from Asvan Kale', *Anatolian Studies*, 23, 281–303.

Pearson, J. A., Buitenhuis, H., Hedges, R. E. M., Martin, L., Russell, N., and Twiss, K. (2007) 'New light on early caprine herding strategies from isotope analysis: a case study from Neolithic Anatolia', *Journal of Archaeological Science*, 34, 2170–9.

Perkins, D. P. and Daly, P. (1968) 'A hunters' village in Neolithic Turkey', *Scientific American*, 219, 96–106.

Peters, J., Arbuckle, B. S., and Pöllath, N. (2014) 'Subsistence and beyond: animals in Neolithic Anatolia', in Özdoğan, M., Başgelen, N., and Kuniholm, P. (eds) *The Neolithic in Turkey, 10,500–5200 BC: Environment, Settlement, Flora, Fauna, Dating, Symbols of Belief, with views from North, South, East and West*, Vol. 6, pp. 135–203. Istanbul: Arkeoloji ve Sanat Yayınları.

Peters, J., Buitenhuis, H., Grupe, G., Schmidt, K., and Pöllath, N. (2013) 'The long and winding road: ungulate exploitation and domestication in Early Neolithic Anatolia (10,000–7000 cal BC)', in Colledge, S., Conolly, J., Dobney, K., Manning, K., and Shennan, S. (eds) *The Origins and Spread of Domestic Animals in Southwest Asia and Europe*. Publications of the Institute of Archaeology, University College, London 59, pp. 83–114. Walnut Creek: Left Coast Press.

Peters, J., von den Driesch, A., and Helmer, D. (2005) 'The Upper Euphrates-Tigris basin: cradle of agro-pastoralism?', in Vigne, J.-D., Peters, J., and Helmer, D. (eds) *The First Steps of Animal Domestication: New Archaeozoological Approaches*, pp. 96–124. Oxford: Oxbow Books.

Peters, J., Helmer, D., von den Driesch, A., and Saña Segui, M. (1999) 'Early animal husbandry in the northern Levant', *Paléorient*, 25(2), 27–48.

Peters, J. and Schmidt, K. (2004) 'Animals in the symbolic world of Pre-Pottery Neolithic Göbekli Tepe, south-eastern Turkey: a preliminary assessment', *Anthropozoologica*, 39(1), 179–218.

Price, M. D. and Arbuckle, B. S. (2015) 'Early pig management in the Zagros flanks: re-analysis of the fauna from Neolithic Jarmo, northern Iraq', *International Journal of Osteoarchaeology*, 25, 441–53, DOI: 10.1002/oa.2312.

Redding, R. W. and Rosenberg, M. (1998) 'Ancestral pigs: a New Guinea model for pig domestication in the Middle East', in Nelson, S. M. (ed.) *Ancestors for the Pigs: Pigs in Prehistory*. MASCA Research Papers in Science and Archaeology 15, pp. 65–76. Philadelphia: University of Pennsylvania, Museum of Archaeology and Anthropology.

Rosenberg, M. and Redding, R. W. (1998) 'Early pig husbandry in southwestern Asia and its implications for modeling the origins of food production', in Nelson, S. M. (ed.) *Ancestors for the Pigs: Pigs in Prehistory*. MASCA Research Papers in Science and Archaeology 15, pp. 55–64. Philadelphia: University of Pennsylvania, Museum of Archaeology and Anthropology.

Rowley-Conwy, P. A., Albarella, U., and Dobney, K. (2012) 'Distinguishing wild boar and domestic pigs in prehistory: a review of approaches and recent results', *Journal of World Prehistory*, 25(1), 1–44.

Russell, N. (2012) *Social Zooarchaeology: Humans and Animals in Prehistory*, Cambridge: Cambridge University Press.

Russell, N. and Martin, L. (2005) 'The Çatalhöyük mammal remains', in Hodder, I. (ed.) *Inhabiting Çatalhöyük: Reports from the 1995–1999 Seasons*. McDonald Institute Monographs 4, pp. 33–98. Cambridge: McDonald Institute for Archaeological Research.

Russell, N., Martin, L., and Buitenhuis, H. (2005) 'Cattle domestication at Çatalhöyük revisited', *Current Anthropology*, 46, 101–8.

Russell, N., Twiss, K., Orton, D. C., and Demirergi, A. (2013) 'More on the Çatalhöyük mammal remains', in Hodder, I. (ed.) *Humans and Landscapes of Çatalhöyük: Reports from the 2000–2008 Seasons*. British Institute at Ankara Monograph 47, pp. 213–58. Los Angeles: Cotsen Institute of Archaeology Press.

Starkovich, B. M. and Stiner, M. C. (2009) 'Hallan Çemi Tepesi: high-ranked game exploitation alongside intensive seed processing at the Epipaleolithic-Neolithic transition in southeastern Turkey', *Anthropozoologica*, 44(1), 41–62.

Stiner, M. C., Buitenhuis, H., Duru, G., Kuhn, S. L., Mentzer, S. M., Munro, N. D., Pöllath, N., Quade, J., Tsartsidou, G., and Özbaşaran, M. (2014) 'The forager-herder trade off, from broad spectrum hunting to sheep management at Aşıklı Höyük, Turkey', *Proceedings of the National Academy of Sciences of the United States of America*, 111(23), 8404–9.

Tanno, K. and Willcox, G. (2006) 'How fast was wild wheat domesticated?', *Science*, 311, 1886.

Trut, L., Oskina, I., and Kharlamova, A. (2009) 'Animal evolution during domestication: the domesticated fox as a model', *BioEssays*, 31, 349–60.

Twiss, K. (2008) 'Transformations in an early agricultural society: feasting in the southern Levantine Pre-Pottery Neolithic', *Journal of Anthropological Archaeology*, 27, 418–42.

Twiss, K. C. and Russell, N. (2010) 'Taking the bull by the horns: ideology, masculinity, and cattle horns at Çatalhöyük', *Paléorient*, 35, 17–29.

Uerpmann, H.-P. (1979) *Probleme der Neolithisierung des Mittelmeerraumes*. Beihefte zum Tübinger Atlas des Vorderen Orients (TAVO), Reihe B 28. Wiesbaden: Dr Ludwig Reichert.

Vigne, J.-D. (2011) 'Le mouton (*Ovis aries*)', in Guilaine, J., Briois, F., and Vigne, J.-D. (eds) *Shillourokambos: Un établissement néolithique pré-céramique à Chypre. Les Fouilles du Secteur 1*, pp. 1021–38. Paris: Éditions Errance.

Willcox, G. and Stordeur, D. (2012) 'Large-scale cereal processing before domestication during the 10th millennium cal BC in northern Syria', *Antiquity*, 86, 99–114.

Zeder, M. A. (2008) 'Animal domestication in the Zagros: an update and directions for future research', in Vila, E., Gourichon, L., Choyke, A., and Buitenhuis, H. (eds) *Archaeozoology of the Near East*, Vol. 8, No. 1. Travaux de la Maison de l'Orient et de la Méditerranée 49, pp. 243–78. Lyon: Maison de l'Orient et de la Méditerranée.

Zeder, M. A. (2012) 'The broad spectrum revolution at 40: resource diversity, intensification, and an alternative to optimal foraging explanations', *Journal of Anthropological Archaeology*, 31, 241–64.

Zeder, M. A. and Hesse, B. (2000) 'The initial domestication of goats (*Capra hircus*) in the Zagros Mountains 10,000 years ago', *Science*, 287, 2254–7.

Zeist, van, W. and de Roller, G. J. (1994) 'The plant husbandry of Aceramic Çayönü, SE Turkey', *Palaeohistoria*, 33–4, 65–96.

Zohary, D., Tchernov, E., and Kolska Horwitz, L. R. (1998) 'The role of unconscious selection in the domestication of sheep and goats', *Journal of Zoology*, 245, 129–35.

CHAPTER 17

PATTERNS OF ANIMAL EXPLOITATION IN WESTERN TURKEY

from Palaeolithic molluscs to Byzantine elephants

CANAN ÇAKIRLAR AND LEVENT ATICI

THE REGION: HISTORY, GEOGRAPHY, AND BIOGEOGRAPHY

WESTERN Turkey (Fig. 17.1), the vast region covering an area of *c.*150,000 square kilometres west of the central Anatolian basin, is an area of great interest to zooarchaeological research because of its well recognized but hardly fulfilled potential to address pivotal questions about crucial transformations in the history of western cultures. The region's key role in western history is shaped to a certain extent by its distinct geographical position between arid southwest Asia and temperate Europe. The productive environments of the region contained habitats with rich and diverse resources that sustained Epipalaeolithic foragers for more than 10,000 years, even after the onset of farming economies in other regions of Anatolia (Atici, 2011a). Indeed, as is the case with the central Anatolian plateau, the archaeological record attests to the persistence of foraging economies in western Anatolia during the Early Holocene (Atici, 2011b). Early communities of western Anatolia then faced, for the first time, the challenges and opportunities of the temperate European climate and biogeography during the westward dispersals out of the semi-arid Neolithic core. Thanks to its geographical position, the region has since been home to continuous blending of genes, languages, religious ideologies, and patterns of socio-political organization. Due to this frontier character, it has also been subject to endless disputes between Balkan, Aegean, and Anatolian polities, oscillating between innovation centre and buffer zone. For these reasons, the region enjoys

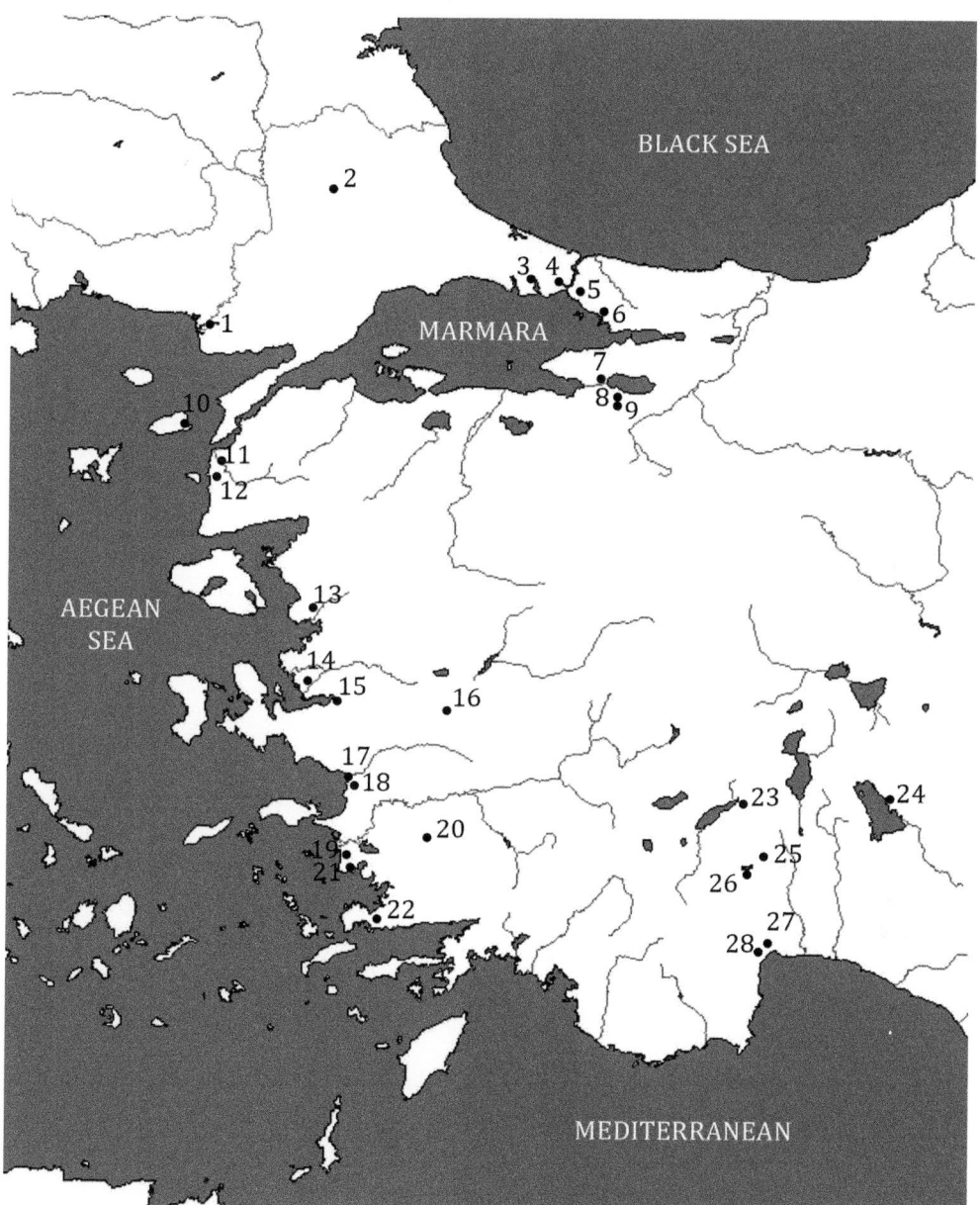

FIGURE 17.1 Map of sites used in the overview (1. Hocaçeşme, 2. Kanlıgeçit, 3. Yarımburgaz, 4. Yenikapı, 5. Fikirtepe, 6. Pendik, 7. Ilıpınar, 8. Menteşe, 9. Barcın, 10. Yenibademli, 11. Troy, 12. Kumtepe, 13. Pergamon, 14. Panaztepe, 15. Ulucak, 16. Sardis, 17. Çukuriçi, 18. Ephesus, 19. Miletus, 20. Aphrodisias, 21. Didyma, 22. Halikarnassos, 23. Sagalassos, 24. Erbaba, 25. Höyücek, 26. Bademağacı, 27. Karain, 28. Öküzini).

the longest running excavations in Turkey that encompass many major milestones in human history from the Palaeolithic to the Medieval period.

Whilst forming a distinct frontier as a whole, the region's geographic elements are diverse, including the straits of Bosporus and Dardanelles, the western Anatolian mountains that rise up to 2,500 m asl, coastal alluvial plains that fill the indented and shallow Aegean continental shelf, and two lake districts, one that lines the southeastern coast of the Marmara Sea, and the other located in the mountainous hinterland of the western Taurus.

The frontier nature and the diverse geography of western Turkey generate a unique zoogeography that serves as an ideal host for the Boreal elements of Eurasia, such as the red squirrel (*Sciurus vulgaris*) from the west, and that allows minimal permeation of steppic (e.g. Equids) and Afrotropic (e.g. *Gazella* sp.) elements from the east. It is the homeland of the European (or Anatolian) fallow deer (*Dama dama*), which subsequently spread throughout the world through human agency (Sykes et al., 2011).

General Character and Methods of Zooarchaeological Research

The influence of Childean research agendas (Düring, 2013) and classical traditions (Hamilakis, 1999), and a sandwiched position between the literate Hittites and the imagined border of Europe, have hampered zooarchaeological research in the region much longer than adjacent Greece and southwest Asia. Consequently, there is a notable patterning in the number of detailed zooarchaeological projects by chronological period. The fact that the zooarchaeological research has often focused on the Neolithic, despite the presence of a much greater number of sites of the Classical period, perhaps attests to a conspicuous divergence between research agendas of archaeologists and zooarchaeologists working in the region.

Nevertheless, the history of zooarchaeological research in western Turkey goes back to the nineteenth century with Virchow's (1881) report on the faunal remains from Schliemann's excavations at Troy. Since these early days, zooarchaeological studies in the region have been site-based, reflecting the dominant mode of archaeological investigations. This factor contributes to the fragmented and disarticulated nature of the data that are synthesized in this chapter.

Site-based reports typically present number of identified specimens (NISP) counts and sometimes also tallied bone weight for principal taxa. The majority of these reports also include information on skeletal elements, long bone fusion, tooth eruption and wear, and osteometric data. Molluscs, fish, and birds are also reported frequently, but yields are relatively smaller for fish and birds because of their inadequate and incomplete recovery, i.e. collection via hand picking. Since most of the original archaeofaunal assemblages have been lost in museum depots, in accidental fires, or re-buried due to lack of storage space, re-analysis is often impossible. Thus, these reports become invaluable sources for synthetic zooarchaeological research frameworks. The recent pioneering work of the *Central and Western Anatolian Neolithic Economies Working Group* provides us with a good model: this collaboration has published online a substantial amount of primary data from Epipalaeolithic and Neolithic zooarchaeological studies, allowing direct access to worksheets and primary or raw data of a fairly large number of zooarchaeologists working in the region (Russell et al., 2013).

Research questions that drive zooarchaeological work in the region include, but are not limited to, forager adaptations, patterns of aquatic foraging, emergence of pastoralism, beginnings of animal secondary products, urban economies, and symbolic roles of animals. Environmental issues such as effects of geomorphological and climatic changes have also been addressed using zooarchaeological methods.

We must emphasize from the onset that there are both spatial and temporal gaps in the zooarchaeological record of western Turkey reflecting the scanty, fragmentary, and disarticulated nature of past research. Inadequate field work and the small number of excavated sites with well-dated and rigorously established cultural sequences across western Turkey prevent us from developing a fine-resolution, regional, and diachronic picture of human–animal interactions. Clusters of sites in the sub-regions (e.g. western Taurus, the Marmara, and the Aegean) represent chronological periods unevenly, generating disrupted and incomparable cultural sequences. This, in turn, hampers our understanding of the developmental trajectories in general and the evolution of human–animal interactions in particular. For instance, most of what we know about the entire Epipalaeolithic period largely comes from western Taurus caves, whereas no zooarchaeological data are available for this period from the other sub-regions. Similarly, the relatively well-investigated zooarchaeological record of the seventh millennium BC comes largely from two disconnected sub-regions: the Marmara and the central Aegean. Along the same line, published assemblages from the two-and-a-half millennia between 5500 and 3000 BC stem from four sites only. The patchy nature of the record is inevitably reflected in the overview provided in this chapter, as we are currently not able to investigate long-term trends from the Palaeolithic to the historical times in one sub-region of western Turkey. Still, we combine the sequences currently available from different sub-regions to bridge the foresaid gaps in order to present a more complete zooarchaeological overview.

FIRST AND LAST FORAGERS

Luckily, the only two sites that shed light on early human evolution, hunter-gatherer lifeways, and the natural history of Turkey with artefactual and ecofactual assemblages from dated strata are located in western Turkey. The available data, however, are not sufficient to construct plausible palaeo-stories beyond indicating, in broad strokes, changes in environmental and climatic conditions.

Yarım Burgaz Cave near Istanbul in the Marmara yielded numerous bones of cave bear (*Ursus deningeri*) accompanied by herbivores including wild horse (*Equus* sp.), wild boar (*Sus scrofa*), fallow deer, roe deer (*Capreolus* sp.), giant elk (*Megaceros* sp.), *Bos/Bison*, gazelle (*Gazella* sp.), and wild goat (*Capra* sp.). Other carnivores identified, although in much smaller numbers, included wolf (?) (*Canis* sp.), fox (*Vulpes* sp.), leopard (*Panthera* sp.), cat (*Felis* sp.), and spotted hyena (*Crocuta* sp.) (Farrand and McMahon, 1997). Electronic spin resonance (ESR) dating of ten bear teeth generated a range from 390,000 to 270,000 BP, dating this early human occupation to the Lower Palaeolithic (Farrand and McMahon, 1997: 561). A large lithic assemblage recovered from the lower cave include many retouched and used cores and flakes, pebble tools, and choppers, providing sound evidence for the Lower Palaeolithic human occupation of the site (Kuhn et al., 1996: 34).

Karain Cave, a complex of seven interconnected chambers (A through G), near Antalya in the western Taurus mountains is by far the most exhaustively excavated and thoroughly investigated Palaeolithic site in Turkey. Excavations at Chamber E yielded a ten-metre deep sequence representing the Lower Palaeolithic and a long and rich Middle Palaeolithic, as evidenced by dates ranging from 350,000 to 60,000 BP (Yalçınkaya et al., 1997). A simple core reduction technology and extensively retouched tools on thick flakes represent a two-phase Lower Palaeolithic respectively (Kuhn, 2002: 201), whereas a long Middle Palaeolithic sequence features Levallois core reduction technology and a rich repertoire of tools at Chamber E (Yalçınkaya et al., 1993). Karain E has also yielded non-diagnostic hominid remains from the Lower Palaeolithic strata and 'Neandertaloid' cranial and postcranial remains from the Middle Palaeolithic strata, providing firm direct evidence for human occupation of the cave (Kuhn, 2002: 203). Among the principal taxa identified were elephants (*Elephas* sp.), hippos (*Hippopotamus amphibius*), rhinos (*Dicerorhinus* sp.), aurochs (*Bos primigenius*), wild horses, European ass (*Equus hydruntinus*), red deer (*Cervus elaphus*), Anatolian fallow deer, bezoar goats (*Capra aegagrus*), wild sheep (*Ovis orientalis*), wild boars, cave bears (*Ursus spelaeus*), leopards (*Panthera pardus*), cave hyenas (*Crocuta spelaea*), and red foxes (*Vulpes vulpes*) (Yalçınkaya et al., 2002 and references therein).

Spatial and temporal gaps also exist in Epipalaeolithic archaeology with two western Taurus sites, Karain B and Öküzini Caves, forming the basis for most of what we know about the Epipalaeolithic foragers (Atici, 2011b). The sites have finely stratified and rigorously dated (fifty-nine radiocarbon dates from Öküzini and twenty-nine radiocarbon dates from Karain B) sequences spanning the entire Epipalaeolithic. Atici's (2009a; 2009b; 2011a) extensive zooarchaeological work on time-series data from the two caves provide us a detailed picture of forager adaptations in western Turkey during a period of rapid environmental and cultural change. At Karain B, wild sheep and wild goat exclusively dominated the menu of foragers, while at Öküzini wild sheep and goat remained as the primary food animals throughout the Epipalaeolithic, but were supplemented by secondary fallow deer hunting and tertiary exploitation of roe deer and wild boar. Thus, western Taurus Mountains Epipalaeolithic subsistence can be characterized by specialized caprine hunting complemented by fallow deer hunting between 17,500 and 12,500 cal BC. A trend towards *relatively* broader dietary breadth is evident by 12,500 cal BC with the addition of high-yield tertiary taxa such as roe deer and wild boar, small and fast-moving taxa such as hare (*Lepus europaeus*) and partridge (*Alectoris chukar*), and small and slow-moving taxa such as tortoise (*Testudo graeca*). The observed shift in species trend at Karain B and Öküzini took place during the environmentally more favourable Bölling/Allerød climate optimum, most likely reflecting the availability, accessibility, predictability, and abundance of resources. This dietary expansion then may have led to changes in patterns of site use and seasonality of hunting, with a shift from restricted seasonal to multiseasonal site use, i.e. increased sedentism. The key here, however, is that the causal factor for this shift was not the environmental deterioration, but amelioration and resource availability. There is no evidence currently to support other scenaria, such as population increase.

EMERGENCE AND DEVELOPMENT OF ANIMAL HUSBANDRY

Recent Neolithic investigations in western Turkey have revealed ground-breaking results, pushing the appearance of herders in the Marmara and Izmir regions back to the early seventh millennium BC and revealing decidedly Aceramic Neolithic layers at Ulucak, now the type-site for the eastern Aegean Neolithic (Çilingiroğlu and Çakırlar, 2013).

Sedentary herders with no ceramic tradition appear in the Izmir region around 6800 cal BC, focusing heavily on four-tiered husbandry involving sheep, goats, cattle, and pigs (Çakırlar, 2012a). Evidence for hunting is thin. In contrast, earliest herding with ceramic tradition appears in the Lake District around 6500 cal BC with domestic sheep, goats, cattle, and pigs in juxtaposition with wild goat and boar hunting (De Cupere et al., 2008; Ottoni et al., 2013). Herders become visible in the zooarchaeological record of the Marmara region around 6600 cal BC with domestic sheep, goat, and cattle, but without domestic pigs (Çakırlar, 2013). In the Neolithic of the Marmara region, *Sus* exploitation remains infrequent until *c.*6000 cal BC, when domestic pigs are integrated into the herding system (Çakırlar, 2013).

Biomarker research coupled with zooarchaeological analysis in western Turkey has revealed compelling results to redefine the modes of Neolithic animal husbandry. Compound-specific stable carbon isotope analysis of lipid residues in potsherds documented dairy use in the Marmara and Izmir regions beginning around 6500 cal BC (Özbal et al., 2013). In the Marmara region, milk production is tied to cattle-rearing based on the high proportion of cattle NISP counts (Evershed et al., 2008), although mortality data are not conclusive about this link (Boessneck and von den Driesch, 1979). Culling patterns of domestic ruminants at Ulucak also point at dairy production starting at 6500 cal BC, confirming the results of lipid residue analysis (Çakırlar, 2012b).

There are fifteen sites and twenty distinct chronological assemblages to probe the development of post-Neolithic animal husbandry in western Turkey. The diversity of environments represented (e.g. island, mountains, plains, etc.) and the high inter-analyst variation in data recording, analysis, and presentation decrease the comparability of the zooarchaeological data. Nonetheless, some general patterns do emerge from their evaluations.

While sheep and goat seem to remain staple resources through the millennia, increased frequencies of cattle and pig remains from major Classical (Hellenistic to Late Roman) cities like Sagalassos, Pergamon, and Ephesus (Boessneck and von den Driesch, 1985; De Cupere, 2001; Galik et al., 2010a) indicate that beef and pork became major foodstuffs. Certainly, the importance of fleece/wool production that began in the Early Bronze Age (EBA) at the latest (Gündem, 2010) along with the pressing need to feed large urban populations were factors contributing to this shift. The use of cattle as a draught animal in agricultural production and transport has been inferred from culling patterns and body size since the EBA (von den Driesch, 1999; Gündem, 2010). However, unambiguous evidence for the systematic use of cattle as labour comes at a much later period with a clear demonstration of draught-related pathologies at Roman Sagalassos (De Cupere et al., 2000).

Small numbers of remains show that the donkey (*Equus asinus*) joined the animal work force in the EBA (Crabtree and Monge, 1986; von den Driesch, 1999; Gündem, 2010). The appearance of domestic horses (*Equus caballus*) in Aegean Turkey during the Late Bronze Age can be related to their spread during this period, with coinciding earliest occurrences in Mycenaean Greece and Hittite Anatolia (Uerpmann, 2003). The earlier emergence of horses in eastern Thrace during the EBA, directly dated to 2600–2300 cal BC, indicates a different trajectory for the adoption of the domestic horse in that area (Benecke, 2009).

EXPLOITATION OF TERRESTRIAL
WILD FAUNA

Hunting practices, represented most unambiguously by the relative proportion of deer remains—predominantly fallow deer, but also red deer and roe deer (*Capreolus capreolus*)—show a general increase around *c*.6000 cal BC at most prehistoric settlements (Fig. 17.2).

Assemblages dating to the first Dark Ages of western Turkey (Düring 2011), the Middle and Late Chalcolithic periods (*c*.5500–3000 cal BC), show high occurrences of fallow deer in both inland and coastal sites (Crabtree and Monge, 1986; Uerpmann, 2003), especially in the Aegean part of western Turkey. Although this pattern was previously attributed to environmental changes (Uerpmann, 2003; Riehl and Marinova, 2008), since the transportation of fallow deer to Aegean islands pre-dates this period (Sykes et al., 2011), it is more plausible to surmise that controlled management took place

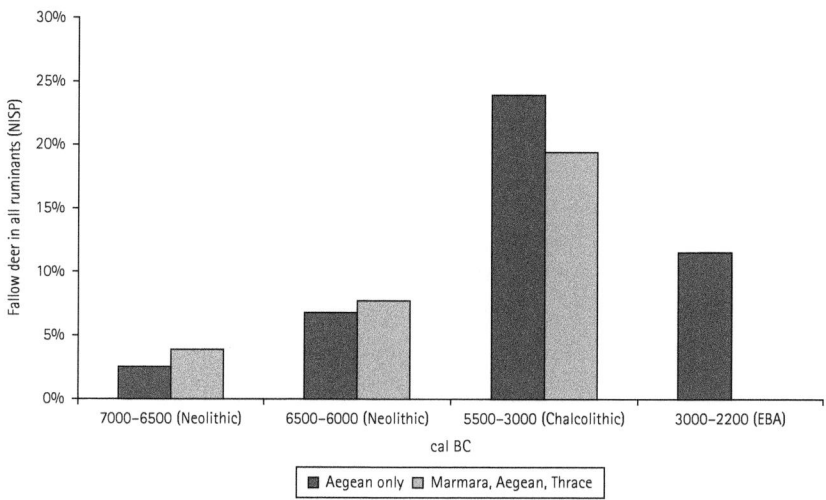

FIGURE 17.2 Proportion of fallow deer among all ruminants in 18 different archaeofaunal assemblages, each with > 500 NISP; total NISP = > 83,000. No EBA data are available for the Marmara region. (Boessneck and Von den Driesch, 1979; Crabtree and Monge, 1986; Buitenhuis, 1995; Uerpmann, 2003; Gerritsen et al., 2010; Gündem, 2010; Horejs et al., 2011; Çakırlar, 2013; Russell et al., 2013). Graph by author.

also on the mainland starting towards the end of the seventh millennium and increasing during the Chalcolithic.

That fallow deer exploitation continues to be significant during the EBA led some to consider their value as status symbols for ruling elites at urban sites such as Troy (Gündem, 2010). There is, however, no contextual evidence to sustain this view ubiquitously for the EBA. For the Classical periods, as argued by Fabiš (2003) based on high proportion of fallow deer (*c.*30% of the total NISP of ruminants) from the Hellenistic sanctuary of Troy, zooarchaeological evidence for the symbolic attributes of the species is more substantial. After the Hellenistic period, remains of deer and other terrestrial wild fauna almost disappear from the zooarchaeological record of western Turkey, remaining invariably below 1% of the NISP counts regardless of their context.

Remains of large carnivores are rare but ubiquitous, with a range which includes locally extinct species such as leopards. A recent study interprets the leopard remains from central western Turkey as indicators of ritual activities in the Neolithic period and as remnants of elite hunts during the Bronze Age (Galik et al., 2013).

EXPLOITATION OF AQUATIC RESOURCES

Understanding the role and scale of fishing in this peninsular region is greatly hampered by the lack of systematic sieving. Hand-picked assemblages indicate that site location dictates fisheries exploitation at least until the Classical period. While inshore species such as sea breams (Sparidae) and groupers (Epinephelinae) dominate the assemblages from sites located on estuaries and lagoons (Uerpmann and Van Neer, 2000; Çakırlar, 2013), pelagic fish such as bluefin tuna (*Thunnus thynnus*) are abundant at sites located directly on migratory routes (von den Driesch, 1999). Further inferences about fisheries utilization are based on comparisons of fish biometric data (Uerpmann and Van Neer, 2000). The involvement of Medieval ports in processed fish trade, already known from historical sources (Dalby, 2003: 67–8), has been recently attested by the discovery of a shipwreck containing amphorae filled with preserved fish in one of the Byzantine harbours in Istanbul (Kocabaş, 2008). The same site also produced evidence for bluefin tuna, dolphin (Delphinidae), and sea turtle (Cheloniidae) consumption (Onar et al., 2013a).

In contrast, the high visibility of marine shells has led to their more complete recovery, especially in coastal sites. Imported and worked marine shells appear already in the Epipalaeolithic deposits of Öküzini. Most of the molluscs at Öküzini belong to the small sea snails, namely ballaruga (*Columbella rustica*) and the swollen nassa (*Arcularia gibbosula*), and were most likely used as ornamental objects (Atici, 2011a). The abundance of oyster and mussel shells in Neolithic villages in Istanbul became a repeated clause in claims for Mesolithic influence in the Marmara region (Özdoğan, 2011). However, growing evidence shows that shellfish-gathering is not a marker of the Mesolithic, but a continuously important aspect of coastal economies in western Turkey from the beginnings of settlement until historical periods (von den Driesch, 1999; Reese, 2006; Çakırlar, 2013). Overexploitation of lagoonal and rocky shore molluscs such as the lagoon cockle (*Cerastoderma glaucum*) and the rayed Mediterranean limpet (*Patella caerulea*) is suggested by the negative correlation between shell size and relative proportion in the faunal assemblages of Bronze Age sites (Çakırlar, 2008; 2009). Evidence for large-scale production of murex dye, considered to be a Minoan innovation and a luxurious trade item, dates to the end of the second millennium BC in the western Aegean (Çakırlar and Becks, 2009).

ASPECTS OF ANIMAL EXPLOITATION
BEYOND SUBSISTENCE

Zooarchaeological research into later periods focuses more commonly on the symbolic role of animals, studying material from ritual contexts. Examples include puppy offerings in a funerary context in Iron Age Sardis (Greenewalt and Payne, 1976); concentrations of piglet remains, burnt hind legs, and amulets made of carnivore teeth and extremities at the Artemis temple in Ephesos (Forstenpointner, 1993); a Roman well with an unusually high proportion of cattle remains including the remains of ritual animal sacrifices performed at the nearby temple at Didyma (Boessneck and Schäffer, 1986; Tuchelt, 1992); and a fourteenth-century AD shamanistic ritual deposit with complete skeletons and purposefully placed skulls of equids at Ephesos (Galik et al., 2010b).

Long-distance trade in captured wild animals, livestock, and their body parts becomes evident during the Classical period. The occurrences of African catfish (*Clarias gariepinus*) from the Levant or Egypt and Nile perch (*Lates niloticus*) from Egypt in Archaic and Roman cities of western Turkey provide firm evidence for long-distance exchange in processed fish (Van Neer et al., 2004; Galik et al., 2010a). The palaeogenetic study of the *Clarias* remains indicates that the Sagalassos *Clarias* likely originated from Egypt (Arndt et al., 2003).

The zooarchaeological study of the recently discovered harbour of Theodosius in Istanbul demonstrates how long-distance trade in animals and their parts was a thriving venture in the capital of the Byzantine Empire. Exotic taxa found in the harbour include elephants (*Elephas* sp.), ostriches (*Struthio camelus*), gazelles (*Gazella* sp.), and macaques (*Macaca* sp.) (Onar et al., 2013b). Some of these remains, such as the postcranial bones of elephants and macaques, possibly represent refuse from the royal menagery of Byzantium, known from literary sources (Dalby, 2003: 35), whereas others signify the high demands of elite consumers.

CONCLUSIONS AND FUTURE DIRECTIONS

This chapter demonstrates the explanatory potential of zooarchaeological research in western Turkey through reviewing various aspects of human–animal interactions including Terminal Pleistocene forager adaptations, dispersal of livestock from the Fertile Crescent to Europe, beginnings of secondary product utilization, specialized pastoral economies, ritualized use of animals, and anthropogenic impact on marine resources.

At the moment, the potential of zooarchaeology in western Turkey remains largely underexplored. Materializing this potential depends on a number of factors. Firstly, adoption of full recovery techniques and systematic collection and storage of all data classes must become a default and basic method of every excavation to ensure consistent, comparable, and comprehensive analysis by the zooarchaeologist. Secondly, sites with finely stratified and rigorously dated strata must be used to establish chronologies for each of the sub-regions in western Turkey. Lastly, best practices in data recording, analysing, and publishing must be adopted and promoted to facilitate data sharing and comparisons in order to develop synthetic research paradigms (*sensu* Atici et al., 2013). This would then enable the zooarchaeologist to employ a fine-comb approach to local, regional, and supraregional zooarchaeological and archaeological events like environmental degradation and its effects, the rise of territorial polities, urbanization, and Romanization.

REFERENCES

Arndt, A., Van Neer, A., Hellemans, B., Robben, J., Volckaert, F., and Waelkens, M. (2003) 'Roman trade relationships at Sagalassos (Turkey) elucidated by ancient DNA of fish remains', *Journal of Archaeological Science*, 30(9), 1095–1105.

Atici, L. (2009a) 'Implications of age structures for Epipaleolithic hunting strategies in the western Taurus Mountains, southwest Turkey', *Anthropozoologica*, 44, 13–39.

Atici, L. (2009b) 'Specialization & diversification: animal exploitation strategies in the terminal Pleistocene, Mediterranean Turkey', *Before Farming*, 3, 136–52.

Atici, L. (2011a) *Before the Revolution: Epipaleolithic Subsistence in the Western Taurus Mountains, Turkey*. BAR International Series 2251. Oxford: Archaeopress.

Atici, L. (2011b) 'Epipaleolithic archaeology in Turkey', in Taşkıran, H., Kartal, M., Özçelik, K., Kösem, M. B., and Kartal, G. (eds) *Studies in Honour of Işın Yalçınkaya*, pp. 27–47. Ankara: Bilgin Kultur Sanat.

Atici, L., Kansa, S., Lev-Tov, J., and Kansa, E. (2013) 'Other people's data: a demonstration of the imperative of publishing primary data', *Journal of Archaeological Method and Theory*, 20(4), 663–81.

Benecke, N. (2009) 'On the beginning of horse husbandry in the southern Balkan Peninsula: the horse bones from Kirklareli-Kanligeçit (Turkish Thrace)', *Turkish Academy of Sciences Journal of Archaeology*, 12, 13–24.

Boessneck, J. and Schäffer, J. (1986) 'Tierknochenfunde aus Didyma II', *Archaeologischer Anzeiger*, 251–301.

Boessneck, J. and von den Driesch, A. (1979) *Die Tierknochenfunde aus der Neolithischen Siedlung auf dem Fikirtepe bei Kadıköy am Marmarameer*, München: Institut für Palaeoanatomie, Domestikationsforschung und Geschichte der Tiermedizin der Universität München.

Boessneck, J. and von den Driesch, A. (1985) *Knochenfunde aus Zisternen in Pergamon*, Muenchen: Institut für Palaeoanatomie, Domestikationsforschung und Geschichte der Tiermedizin der Universität München, Deutsches Archäologisches Insitut Abteilung Istanbul.

Buitenhuis, H. (1995) 'The faunal remains', in Roodenberg, J. (ed.) *The Ilıpınar Excavations I: Five Seasons of Fieldwork in NW Anatolia, 1987–91*, pp. 151–6. Istanbul: Nederlands Historisch-Archaeologisch Instituut.

Çakırlar, C. (2008) 'Investigations on archaeological *Cerastoderma glaucum* populations from Troia (Turkey) and their potential for palaeoeconomical reconstruction', *Archaeofauna*, 17, 91–102.

Çakırlar, C. (2009) *Mollusk Shells in Troia, Yenibademli and Ulucak: An Archaeomalacological Approach to Environment and Economy in the Aegean*. BAR S2051. Oxford: John and Erica Hedges.

Çakırlar, C. (2012a) 'The evolution of animal husbandry in Neolithic central-west Anatolia: the zooarchaeological record from Ulucak Höyük (*c.*7040–5660 cal. BC, Izmir, Turkey)', *Anatolian Studies*, 62(1), 1–33.

Çakırlar, C. (2012b) 'Neolithic dairy technology at the European-Anatolian frontier? Implications of the archaeozoological evidence from Ulucak Höyük, İzmir, ca. 7040–5660 BC cal.', *Anthropozoologica*, 47(1), 78–99.

Çakırlar, C. (2013) 'Rethinking Neolithic subsistence at the gateway to Europe with new archaeozoological evidence from Istanbul', in Groot, M., Lentjes, D., and Zeiler, J. (eds) *Barely Surviving or More than Enough? The Environmental Archaeology of Subsistence, Specialisation and Surplus Food Production*, pp. 59–79. Amsterdam: Sidestone Press.

Çakırlar, C. and Becks, R. (2009) 'Murex dye production at Troia: assessment of archaeomalacological data from old and new excavations', *Studia Troica*, 2, 87–103.

Çilingiroğlu, Ç. and Çakırlar, C. (2013) 'Towards configuring the Neolithization of Aegean Turkey', *Documenta Praehistorica*, 40, 21–9.

Crabtree, P. and Monge, J. (1986) 'Faunal analysis', in Joukowsky, M. S. (ed.) *Prehistoric Aphrodisias: An Account of the Excavations and Artifact Studies*. Archaeologica Transatlantica 3, pp. 180–90. Providence: Brown University, Center for Old World Archaeology and Art.

Tuchelt, K. (1992) 'Tieropfer in Didyma—Ein Nachtrag' *Archaeologischer Anzeiger*, 61–81.

Dalby, A. (2003) *Tastes of Byzantium: The Cuisine of a Legendary Empire*, London: I. B. Tauris.

De Cupere, B. (2001) *Animals at Ancient Sagalassos: Evidence from Animal Remains*. Studies in Eastern Mediterranean Archaeology 4. Turnhout: Brepols Publishers.

De Cupere, B., Duru, R., and Umurtak, G. (2008) 'Animal husbandry at the Early Neolithic to Early Bronze Age site of Bademağacı (Antalya province, SW Turkey): evidence from the faunal remains', in Vila, E., Gourichon, L., Choyke, A. M., and Buitenhuis, H. (eds) *Proceedings of the 8th International Symposium on the Archaeozoology of Southwestern Asia and Adjacent Areas*, pp. 367–405. Lyon: Archéorient, Maison de l'Orient et de la Méditerranée.

De Cupere, B., Lentacker, A., Van Neer, W., Waelkens, M., and Verslype, L. (2000) 'Osteological evidence for the draught exploitation of cattle: first applications of a new methodology', *International Journal of Osteoarchaeology*, 10(4), 254–67.

Driesch, von den, A. (1999) 'Archäozoologische Untersuchungen an Tierknochen aus dem dritten und ersten vorchristlichen Jahrtausend vom Beşik-Yassıtepe, Westtürkei', *Studia Troica*, 9, 439–74.

Düring, B. S. (2011) 'Millenia in the Middle? Reconsidering the Chalcolithic of Asia Minor', in Steadman, S. and McMahon, G. (eds) *Oxford Handbook of Anatolian Studies, 8000–323 bc*, pp. 796–812. Oxford: Oxford University Press.

Düring, B. S. (2013) 'Breaking the bond: investigating the Neolithic expansion in Asia Minor in the seventh millennium BC', *Journal of World Prehistory*, 26(2), 75–100.

Evershed, R. P., Payne, S., Sheratt, A. G., Coolidge, J., Urem-Kotsu, D., Kotsakis, K., Özdoğan, M., Özdoğan, A., Nieuwenhuyse, O., Akkermans, P. M. M. G., Bailey, D., Andeescu, R.-R., Campbell, S., Farid, S., Hodder, I., Yalman, N., Özbaşaran, M., Bıçakçı, E., Garfinkel, Y., Levy, T., and Burton, M. M. (2008) 'Earliest date for milk use in the Near East and southeastern Europe linked to cattle herding', *Nature*, 455, 528–31.

Fabiš, M. (2003) 'Troia and fallow deer', in Wagner, G. A., Pernicka, E., and Uerpmann, H. P. (eds) *Troia and the Troad: Scientific Approaches*, pp. 264–75. Berlin: Springer Verlag.

Farrand, W. R. and McMahon J. P. (1997) 'History of the sedimentary infilling of Yarimburgaz Cave, Turkey', *Geoarchaeology*, 12, 537–65.

Forstenpointner, G. (1993) 'Kurzbericht zur archäozoologischen Befundungstätigkeit am Tierknochenmaterial der Artemisiongrabung', *Jahreshefte des Österreichischen Archäologischen Instituts in Wien*, 62, 10–12.

Galik, A., Forstenpointner, G., and Weissengruber, G. (2010a) 'The expression of demand for particular fish food implied by aquatic facilities in living areas of noble households in Terrace House 2 in Ephesos', in Ladstätter, S. and Scheibelreiter, V. (eds) *Städtisches wohnen im östlichen Mittelmeerraum 4.Jh. v.Chr.—1. Jh. n.Chr. Akten des Internationalen Kolloquiums vom 24.–27. Oktober 2007*, pp. 667–74. Vienna: Österreichischen Akademie der Wissenschaften.

Galik, A., Forstenpointner, G., Zohmann, S., and Weisssengruber, G. (2010b) 'Die Tierreste aus dem Schachtbrunnen und der Nische des Präfurniums', in Pfeiffer-Tas, S. (ed.) *Funde und Befunde aus dem Schachtbrunnen im Hamam III in Aysuluk/Ephesos: Eine schamanistische Bestattung des 15. Jahrunderts*. Archaologische Forschungen 16, pp. 77–126. Vienna: Österreichischen Akademie der Wissenschaften.

Galik, A., Horejs, B., and Nessel, B. (2013) 'Der nächtliche Jäger als Beute. Studien zur prähistorischen Leopardenjagd', *Praehistorische Zeitschrift*, 87(2), 261–307.

Gerritsen, F., Özbal, R., Thissen, L., Özbal, H., and Galik, A. (2010) 'The Late Chalcolithic settlement of Barcin Höyük', *Anatolica*, 36, 197–255.

Greenewalt, C. H. and Payne, S. (1976) *Ritual Dinners in Early Historic Sardis*. Classical Studies 17. Berkeley: University of California Press.

Gündem, C. Y. (2010) 'Animal Based Economy in Troia and the Troas Düring the Maritime Troy Culture (c.3000–2200 BC) and a General Summary for West Anatolia'. Unpublished PhD dissertation, Universität Tübingen (Tübingen).

Hamilakis, Y. (1999) 'The anthropology of food and drink consumption and Aegean archaeology', in Coulson, W. D. E. and Vaughan, S. J. (eds) *Palaeodiet in the Aegean*, pp. 55–63. Oxford: Oxbow Books.

Horejs, B., Galik, A., Thanheiser, U., and Wiesinger, S. (2011) 'Aktivitäten und Subsistenz in den Siedlungen des Çukuriçi Höyük. Der Forschungsstand nach den Ausgrabungen 2006–2009', *Praehistorische Zeitschrift*, 86(1), 31–66.

Kocabaş, U. (2008) *Old Ships of the New Gate*, Istanbul: Zero Productions.

Kuhn, S. L. (2002) 'Paleolithic archaeology in Turkey', *Evolutionary Anthropology*, 11, 198–210.

Kuhn, S. L., Arsebuk, G., and Howell, C. (1996) The Middle Pleistocene lithic assemblage from Yarim Burgaz Cave, Turkey', *Paléorient*, 22, 31–49.

Neer, Van, W., Lernau, O., Friedman, R., Mumford, G., Poblóme, J., and Waelkens, M. (2004) 'Fish remains from archaeological sites as indicators of former trade connections in the Eastern Mediterranean', *Paléorient*, 30(1), 101–47.

Onar, V., Alpak, H., Pazvant, G., Armutak, A., Ince, N. G., and Kiziltan, Z. (2013a) A bridge from Byzantium to modern day Istanbul: an overview of animal skeleton remains found during Metro and Marmaray excavations', *Journal of the Faculty of Veterinary Medicine of Istanbul University*, 39(1), 1–8.

Onar, V., Pazvant, G., Alpak, H., Ince, N. G., Armutak, A., and Kiziltan, Z. (2013b) 'Animal skeletal remains of the Theodosius harbor: general overview', *Turkish Journal of Veterinary Animal Sciences*, 37, 81–5.

Ottoni, C., Girdland Fink, L., Evin, A., Geörg, C., De Cupere, B., Van Neer, W., Bartosiewicz, L., Linderholm, A., Barnett, R., Peters, J., Decorte, R., Waelkens, M., Vanderheyden, N., Ricaut, F. X., Çakırlar, C., Çevik, Ö., Hoelzel, A. R., Mashkour, M., Karimlu, A. F. M., Seno, S. S., Daujat, J., Brock, F., Pinhasi, R., Hongo, H., Perez-Enciso, M., Rasmussen, M., Frantz, L., Megens, H. J., Crooijmans, R., Groenen, M., Arbuckle, B., Benecke, N., Vidartottir, U. S., Burger, J., Cucchi, T., Dobney, K., and Larson, G. (2013) 'Pig domestication and human-mediated dispersal in western Eurasia revealed through ancient DNA and geometric morphometrics', *Molecular Biology and Evolution*, 30(4), 824–32.

Özbal, H., Thissen, L., Doğan, T., Gerritsen, F., Özbal, R., and Türkekul-Biyik, A. (2013) 'Neolitik Batı Anadolu ve Marmara yerleşimleri çanak çömleklerinde organik kalıntı analizleri' *Arkeometri Sonuçları Toplantısı*, 28, 105–14.

Özdoğan, M. (2011) 'An Anatolian perspective on the Neolithization process in the Balkans. New questions, new prospects', in Krauß, R. (ed.) *Beginnings … New Research in the Appearance of the Neolithic between Northwestern Anatolia and the Carpathian Basin. Papers of the International Workshop 8th–9th April 2009, Istanbul (Menschen-Kulturen-Traditionen 1)*, pp. 23–33. Rahden: Leidorf.

Reese, D. (2006) 'The exploitation of aquatic resources at Bakla Tepe, Liman Tepe and Panaztepe', in Avunç, B. (ed.), *Studies in Honor of Hayat Erkanal: Cultural Reflections*, pp. 626–30. Istanbul, Homer Kitabevi.

Riehl, S., and Marinova, E. (2008) 'Mid-Holocene vegetation change in the Troad (NW Anatolia): man-made or natural?', *Vegetation Hisotory and Archaeobotany*, 17(3), 297–312.

Russell, N., Twiss, K., Buitenhuis, H., Frame, S., Yeomans, L., Martin, L., Marciniak, A., Pawlowska, K., Christensen, C., Orton, D., Demirergi, A., Arbuckle, B. S., Meese, S., Atici, L., Çakırlar, C., Whitcher Kansa, S., Henton, L., Aydinuloglu, B., Galik, A., Watson, A., Carruthers, D., Cameron, I., Mayon-White, R., Erwin, A., Dimitrijevic, V., Daly, R., Symmons, N., Yeni, S., Hills, C., Buitenhuis, H., Leblanc, L., Boric, D., and Powell, A. (2013-08-18) 'EOL Computational Data Challenge: Primary Zooarchaeology Dataset', in Arbuckle, B. S., Orton, D., Buitenhuis, H., Marciniak, A., Atici, L., Çakırlar, C., Galik, A., Carruthers, D., and Whitcher Kansa, S. (eds) *Open Context*. <http://opencontext.org/tables/f07bce4fb-08cfe926505c9e534d89a09> DOI 10.6078/M75H7D6D

Sykes, N., Carden, R., and Harris, K. (2011) 'Changes in the size and shape of fallow deer—evidence for the movement and management of a species', *International Journal of Osteoarchaeology*, 23(1), 55–68.

Uerpmann, H. P. (2003) 'Environmental aspects of economic changes in Troia', in Wagner, G. A., Pernicka, E., and Uerpmann, H. P. (eds) *Troia and the Troad: Scientific Approaches*, pp. 251–62. Berlin: Springer Verlag.

Uerpmann, M. and Van Neer, W. (2000) 'Fischreste aus den neuen grabungen in Troia (1989–1999)', *Studia Troica*, 10, 145–79.

Virchow, R. (1881) 'Fauna der Troas', in Schliemann, H. (ed.) *Ilios: Stadt und Land der Trojaner*, pp. 130–5. Leipzig: F. A. Brockhaus.

Yalçınkaya, I., Otte, M., Bar-Yosef, O., Léotard, J.-M., and Taşkıran, H. (1993) 'The excavations in Karain Cave, southwestern Turkey', in Olszewski, D. and Dibble, H. (eds) *The Paleolithic Prehistory of the Zagros-Taurus*, pp. 101–17. Philadelphia: University of Pennsylvania Press.

Yalçınkaya, I., Otte, M., Taşkıran, H., Kösem, M. B., and Ceylan, K. (1997) '1985 ve 1995 Karain kazıları ışığında Anadolu Paleolitiğinin önemi', *Kazı Sonuçları Toplantısı*, 23(1), 1–9.

Yalçınkaya, I.,Taşkıran, H., Kösem, M. B., Özçelik, K., and Atici A. L. (2002) '2000 yılı Karain kazısı', *Kazı Sonuçları Toplantısı*, 23(1), 163–70.

CHAPTER 18

SOUTH ASIAN
CONTRIBUTIONS TO
ANIMAL DOMESTICATION
AND PASTORALISM

bones, genes, and archaeology

AJITA K. PATEL AND RICHARD H. MEADOW

INTRODUCTION

SOUTH Asia is an amalgam—a *masala*—of past and present cultural practices and traditions that has been sustained by the subcontinent's paradox of isolation and connection. As a peninsula to the south of the high Himalayas jutting out into the Indian Ocean, access to and from the subcontinent is challenging, with even traditional overland routes in the northwest and northeast presenting their share of obstacles with major river systems, deserts, and rainforests to be crossed. Yet traversed they have been in history and prehistory, with influences from outside reaching deep into the peninsula and those from within spreading west and east by land and eventually by sea and even north through high mountain passes. During the Holocene, the region has seen the episodic and uneven development of animal and plant management, domestication, and agro-pastoral practices, resulting in a mosaic of commitment to the use of varied domesticated resources.

In this essay, we focus on the northwest, a zone that has provided evidence for two major cultural phenomena, namely, the earliest development of plant and animal husbandry and the first manifestations of urbanism in the subcontinent. The latter phenomenon—the Indus or Harappan civilization—was spread across much of the northwest in the second half of the third millennium cal BC, with a subsistence base founded on agro-pastoral developments of the preceding four or five millennia.

Throughout that long period of socio-economic development, there was complex use of the landscape. From at least as early as the eighth millennium cal BC, in steppic and semi-arid areas of North Gujarat, southern Rajasthan, and eastern parts of Sindh, mobile hunter-gatherer communities with microlithic tool technologies often occupied stabilized sand dunes, where they took advantage of the fauna attracted to the water and vegetation in the interdunal depressions. By the early fourth millennium cal BC, mobile pastoralists using ceramics and herding primarily cattle began taking advantage of these same dunes and interdunal depressions, while other Chalcolithic populations began establishing sedentary agro-pastoral communities of various sizes throughout the region where arable land was available. Through this period and even into the Iron Age, there is evidence for groups continuing to use microlithic tool technologies of the kind traditionally associated with hunter-gatherers, while also incorporating domestic animals and plants as well as ceramics and eventually metals into their lifeways.

Farther north and west, settlements are found in areas of the semi-arid Balochistan highlands and along the arid western margins of the Indus plain dating as early as the late eighth millennium cal BC. While there are Upper Palaeolithic assemblages in the area, the immediate antecedents of the early settlements have yet to be found. To the east, available archaeological evidence indicates that significant settled occupation did not occur before the fourth millennium cal BC, by which time agro-pastoral practices that were developed in the highlands and along the alluvial margins had been transferred deep into the riverine zones of the Indus River and its tributaries. When this transfer began to take place, however, remains to be determined. There has been little research into the archaeology of the fifth and sixth millennia cal BC in this area, probably due as much to the focus on the later Indus civilization and its immediate antecedents as to the inherent difficulties of locating small settlements in active riverine zones that also have been further transformed by recent human activity (C. Jarrige et al., 1995; Possehl, 1999).

The transfer of agro-pastoralism into the alluvial zones included notably the crop domesticates wheat and barley and domestic sheep, goat, and cattle. Although these forms tended to dominate, wild plants and animals were widely exploited, with some becoming managed and eventually domesticated including local millets, pulses, sesame, cotton, and rice as well as water buffalo (Meadow, 1996; Patel and Meadow, 1998; Fuller and Madella, 2002).

In the following pages an effort is made to describe some of our current understandings of the domestication and exploitation of goat, sheep, cattle, and water buffalo in northwestern South Asia from the eighth through the third millennium cal BC and the nature of the evidence on which those understandings are based. That evidence derives primarily from archaeological, zooarchaeological, and genetic investigations,

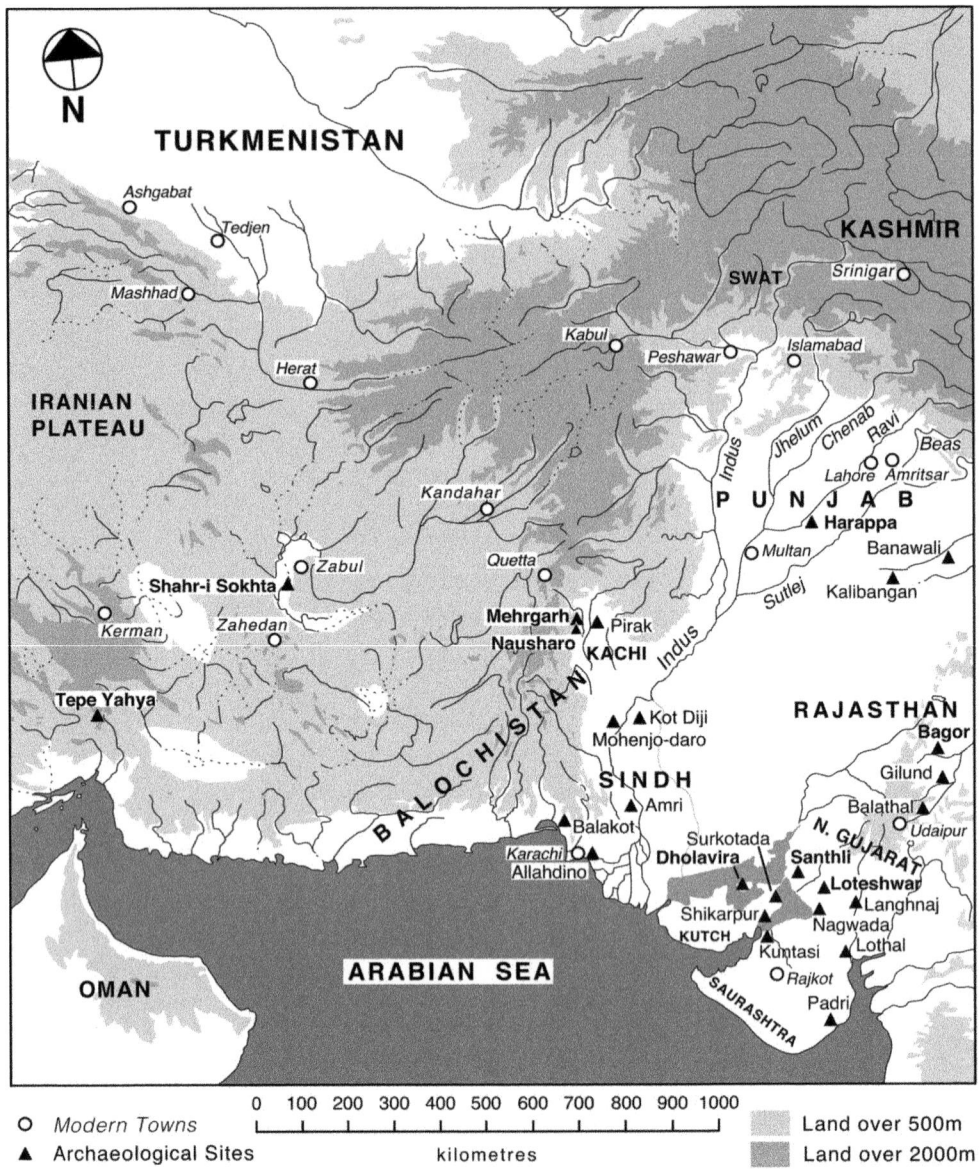

FIGURE 18.1 Map of northwestern South Asia and lands to the west with the approximate locations of key archaeological sites plotted. Names of sites featuring prominently in the text are displayed in bold. Authors' own image.

each of which has its own contributions to make but also limitations, which are discussed as well. A map is provided of northwestern South Asia and areas to the west with the locations of significant archaeological sites including those discussed in the text (Fig. 18.1).

WILD UNGULATES OF THE NORTHWEST

The indigenous wild fauna of the northwestern South Asia is largely Indo-Malayan in the lowlands of the Indus River system and continuing farther east and south; it is mostly Palearctic through the western and northern highlands and onto the Iranian plateau (Roberts, 1997; Menon, 2009). The taxa most frequently attested in the archaeological record from this region include the wild ungulates listed in Table 18.1 (with the addition of the elephant). Today, many large mammals other than domesticates are scarce and their ranges restricted in response to centuries of human predation and landscape modification. The small bovids blackbuck and chinkara, however, still occur through much of sub-Himalayan India. In Pakistan the blackbuck has been hunted to extinction as has most large game, but the chinkara is still found across Balochistan and onto the eastern Iranian plateau. Wild boar live throughout much of the sub-continent as do the nilgai, chowsingha, sambar, chital, and muntjack. The swamp deer and hog deer today have more restricted distributions than they did in the past, as is also the case for elephant and rhinoceros, which is reflected in the archaeofaunal record. Historically, the onager was widely distributed through the northwest, including in the western highlands, but today it is confined to preserves in Kutch and North Gujarat. Wild goats and sheep of different species are found throughout the western and northern highlands. Those of concern here, the bezoar, urial, and mouflon, are discussed below, as is the wild water buffalo. The wild cattle of South Asia are extinct, known only from the palaeontological and archaeological records.

GOAT AND SHEEP

The initial domestication of both goat (*Capra hircus*) and sheep (*Ovis aries*) is generally thought to have taken place in Southwest Asia. Yet the distributions of their likely wild ancestral forms *Capra aegagrus* (the bezoar) and *Ovis orientalis* (the mouflon) today extend across the Iranian plateau and into northwestern South Asia. Thus possibilities may have existed for local management leading to domestication and for cross-breeding between wild and domestic forms.

Goat

The first managed or domesticated bovid in northwestern South Asia was the goat, evidence for which comes from the earliest levels at the site of Mehrgarh located at the foot of the Bolan Pass in eastern Balochistan. An extensive Aceramic Neolithic occupation (Period I, late eighth into the sixth millennium cal BC) is represented by over six metres

Table 18.1 Wild ungulates (and elephant), the remains of which have been identified from northwestern South Asian archaeological sites prominently mentioned in the text

Taxonomic designation	Common name	From the site(s) of:
Elephas maximus indicus Cuvier, 1798	Indian elephant	Mehrgarh
Rhinoceros unicornis Linnaeus, 1758	Indian rhinoceros	Mehrgarh?
Bos namadicus Falconer, 1859	South Asian wild cattle	Mehrgarh, Loteshwar
Bubalus arnee Kerr, 1792	wild water buffalo	Mehrgarh, Loteshwar, Santhli, Bagor?, Dholavira
Boselaphus tragocamelus Pallas, 1766	nilgai	Mehrgarh, Loteshwar, Santhli, Dholavira?
Capra aegagrus Erxleben, 1777	bezoar	Mehrgarh
Ovis orientalis Gmelin, 1774	mouflon	Mehrgarh?, Nausharo?
Ovis vignei Blyth, 1841	urial	Mehrgarh?
Antilope cervicapra Linnaeus, 1758	blackbuck	Mehrgarh, Loteshwar, Santhli, Bagor, Dholavira
Gazella bennettii Sykes, 1831	Indian gazelle, chinkara	Mehrgarh, Loteshwar, Santhli, Bagor, Dholavira
Tetracerus quadricornis de Blainville, 1816	chowsingha	Reported from other sites in northwestern India
Cervus duvaucelii Cuvier, 1823	swamp deer, barasingha	Mehrgarh, Loteshwar?, Bagor, Dholavira?
Cervus unicolor Kerr, 1792	sambar	Bagor
Axis axis Erxleben, 1777	chital, spotted deer	Mehrgarh, Loteshwar, Bagor, Dholavira
Axis porcinus Zimmermann, 1780	hog deer	Loteshwar, Bagor
Muntiacus muntjak Zimmermann, 1780	muntjak, barking deer	Reported from other sites in northwestern India
Sus scrofa cristatus Wagner, 1839	Indian wild boar	Mehrgarh, Loteshwar, Santhli?, Bagor, Dholavira?
Equus hemionus Pallas, 1775	Indian wild ass, onager	Mehrgarh, Loteshwar, Dholavira

Bagor (Thomas, 1977); Dholavira (Patel, 1997; 2015); Loteshwar (Patel, 2009); Mehrgarh (Meadow, 1993); Nausharo (sheep only: Meadow, 1989); Santhli (Patel and Meadow, 1998). See Roberts (1997) and Menon (2009) for information on the mammals of India and Pakistan, respectively. For other sites in northwestern India, see Thomas (2002).

of superimposed deposits that include nine phases of habitation remains alternating with burials. Dug into the earliest habitation level were three elliptical pit-graves of women who had been buried with goats under three months old arranged in semicircles around their flexed legs. Two of the graves yielded five kids with the third having four kids (C. Jarrige et al., 1995; J.-F. Jarrige et al., 2013). The ability to obtain these numbers of young animals to place together in human interments indicates ready access to breeding stock, indicating that at least some goats were managed, in the process of domestication, or domesticated (*sensu* Zeder, 2015).

Breadth and depth dimensions of fused postcranial bones from habitation debris demonstrate that a wide range of goat sizes are represented in Period I, with the larger end of the size continuum disappearing by the end of the Aceramic Neolithic. One interpretation of this body-mass reduction through time is that, during the course of the Period I, kill-off in a single population of increasingly managed goats shifted from being non-targeted to being focused on young males (the unfused elements of which were not measured). Another is that two populations of goats are represented in the faunal assemblage, namely, a husbanded one and a hunted one. Domestic stock might have been brought to the site by its original settlers, who also hunted local wild goats along with the eleven other ungulates that have been identified from the site (Table 18.1). Over time, the contributions of hunted stock decreased while those of domestic stock increased. Indeed, a decrease in wild mammal bone percentages is clearly attested through the course of Period I and into later periods at Mehrgarh, with the assemblage becoming dominated by the remains of domestic cattle, sheep, and goat (Meadow, 1981; 1993).

Populations of wild goat or bezoar are still found in southern Balochistan including in the highlands west of Mehrgarh (Roberts, 1997). Analysis of the mitochondrial DNA (mtDNA) recovered from twenty-eight bezoars from that region has shown that some animals have 'wild' haplotypes not included in any modern domestic goat haplogroup, while others have haplotypes that group with today's domestic C and F haplogroups (Naderi et al., 2008). Based on mtDNA evidence from C haplogroup bezoars of southern and eastern Iran, Naderi and colleagues suggest that population expansion took place from that area during the Neolithic, which included animals being translocated northwest during a first phase of management or 'incipient domestication' on the Iranian plateau. Such an interpretation is predicated on movement of animals and accompanying human interaction across much of the Iranian plateau west to the Zagros Mountains and eastern Anatolia and east to the margins of the Indus valley, which is not improbable.

That said, the complexity of interpreting modern wild goat mtDNA distribution is exemplified by haplotype 134 within the F haplogroup. The five specimens reported with this haplotype each come from different highland localities between eastern Turkey and western Sindh, with the Pakistan specimen being the most geographically isolated (Naderi et al., 2008). Such a situation might result from the movement of captive wild or managed females over long distances and subsequent breeding with local wild males or through feralization of domesticates with that haplotype. In this regard the absence of evidence for the common domestic mtDNA haplogroups A and B introgressing into the wild populations of Balochistan is intriguing. Given the limitations of mtDNA for providing a comprehensive picture of phylogenetic history, however, approaches to the genetic study of goats in the Mehrgarh area would be well served to include both Y-chromosome and autosomal analyses of modern domestic and wild samples together with ancient DNA studies on archaeofaunal specimens.

Sheep

The phylogeography and evolutionary history of different sheep is more complicated than for goats. The wild sheep found today in the mountains of eastern Balochistan is the urial (*Ovis vignei*), a species with a chromosome number (2n = 58) that differs from that of the Asiatic mouflon (*Ovis orientalis* 2n = 54). In spite of such long recognized cytogenetic differences, only recently have debates about relationships between wild sheep been clarified by mtDNA, Y-chromosome, and autosomal studies of modern forms, with the urial and mouflon seen as having diverged deep in the past (e.g. Meadows and Kijas, 2009; Rezaei et al., 2010; Meadows et al., 2011). Although genetic relationships between wild and domestic forms remain less well understood than for goats, researchers generally agree that one or more mouflon population(s) within their range of distribution in western Asia provided the original ancestors for modern domestic sheep. Nevertheless, animals of another species of sheep might have been managed or even domesticated in another region.

The Asiatic mouflon today is an animal of highland and foothill zones of Southwest Asia, with its range extending into eastern Iran where there is a zone of hybridity with the urial. The urial continues to be found north from Balochistan into the highlands of the Punjab and then northwest into Afghanistan and southern Central Asia and northeast into Kashmir. Urial and mouflon likely remained reproductively isolated until the Holocene, during which the retreat and eventual destruction of forests removed an effective barrier between the species (Valdez et al., 1978). Over what period of time that may have occurred is not clear, nor is it known whether the current distribution of these forms today is close to the same that it was in the past.

One possibility is that past ranges of distribution were different, with the mouflon early in the Holocene extending farther to the east and the urial being a more northern form that later spread south, where the two came to interbreed and produce fertile offspring. There is weak zooarchaeological support for this scenario in the form of a single wild sheep horn-core from the third-millennium cal BC site of Nausharo, located near Mehrgarh. The specimen has an overall morphology closer to the mouflon than to the urial (Meadow, 1989). Unfortunately, well-preserved horn-cores from archaeological sites in the region are rare and, in addition, we have inadequate understanding of variability in both modern and ancient horn-core morphologies.

The earliest evidence for sheep exploitation comes from the basal deposits excavated at Mehrgarh. By the middle of the Aceramic Neolithic (Period I) and continuing through the Ceramic Neolithic (Period II), sheep bones are at least as common as those from goats. In contrast to goats, however, only large-bodied individuals are represented in the early Neolithic. Sheep remains were not placed in burials, which lends support to an interpretation that their relationship to humans was initially different than that of goats. Furthermore, specimens from small-bodied sheep only begin to occur after the middle of the Aceramic Neolithic and do not dominate sheep assemblages until the Chalcolithic (Period III: mid-fifth millennium cal BC). By that time there are very few specimens from large-bodied animals (Meadow, 1981; 1993).

The rather abrupt appearance of bones from smaller-bodied sheep in the later levels of Mehrgarh Period I may indicate a change in exploitation practices to include greater numbers of sub-adult or adult ewes. This occurred at the same time that kill-off of large sheep continued. Such a pattern of exploitation may be interpreted as initial hunting followed by management and eventual domestication of sheep populations from local stock. This might be considered a 'directed pathway' to domestication (*sensu* Zeder, 2015), in which experience with an already domesticated animal (goat) forms the foundation for the intentional domestication of a local wild species (Option 1). Another possibility is that the hunting of wild populations during the early Aceramic Neolithic was followed by continued hunting through the end of the Neolithic combined with the exploitation of stock already domesticated elsewhere and introduced to the Mehrgarh area (Option 2).

Option 1 flies in the face of a widely held belief in a single zone in western Asia for sheep domestication. While the earliest levels of Mehrgarh are later than the beginning of the sheep domestication process in the west, the site's inhabitants had experience with the domestic goat as well as with wild sheep and goats. In addition, genetic studies are making it increasingly clear that there have been varying degrees of introgression between wild and domestic forms in the past. Whether the wild sheep in the Mehrgarh area was the urial or a hybrid, just because there is no genetic signature of such a form in modern sheep is not sufficient reason to eliminate local domestication as a possibility. Original domestic stock in the region were likely replaced or genetically swamped when wool sheep were developed and became widely favoured across much of Eurasia. The rather abrupt appearance of bones from small sheep toward the end of the Aceramic Neolithic, however, might favor Option 2—importation of domestic stock from elsewhere, although Option 1—local domestication—could also lead to rapid size diminution if animals were isolated from wild stock and kept under conditions of nutritional stress.

Goat and Sheep beyond Balochistan

By the mid-fourth millennium cal BC, remains of domestic sheep and goats have been identified from ancient settlements across northwestern South Asia beyond the zones of distribution of their wild forms. Together with cattle and crop plants, they became part of the foundation of an agro-pastoral economy that developed and flourished in this area through the period of the Indus civilization and beyond (e.g. Meadow, 1991; Patel, 1997; Meadow and Patel, 2002; Thomas, 2002; Chase, 2010). Understandings of the antecedents to the fourth-millennium cal BC sites, as well as their founding, development, and chronology remain rudimentary. Also poorly understood is the nature of the interactions that inhabitants of fourth-millennium cal BC agro-pastoral settlements had with peoples represented at more ephemeral and episodically occupied contemporary sites termed 'Mesolithic'—or better 'Microlithic', some of which had begun to be occupied at least by the eighth millennium cal BC (Patel, 2008).

Microlithic sites are typically situated atop stabilized sand dunes in arid and semi-arid environments to the east and southeast of the Indus alluvium through southern Rajasthan and into North Gujarat. These sites are seen as representing populations distinct from those living in more substantial settlements. They are characterized by relatively shallow cultural deposits without significant structural remains and by sediments that appear quite homogeneous, making it difficult to define stratigraphic units. They often have multiple components, termed 'phases' or 'periods', which are differentiated on the basis of cultural material. All components share a microlithic tool industry, which in later phases may be supplemented by ceramics and by metal implements of copper/bronze or iron. It is often not clear whether the deposits within each phase represent a single continuous occupation or might reflect multiple short-term resettlements through time. Each component is often seen as leading directly into the next without consideration of whether there might be temporal gaps between them.

Further complicating understandings of microlithic sites are issues related to identification of the faunal remains. What has not been recognized until recently is the interpretative significance of the difficulties inherent in distinguishing not only bones of sheep from those of goats, but from comparably sized wild bovids such as blackbuck and chinkara and even from deer such as the chital and hog deer (Table 18.1). This methodological issue has been explored in the literature (e.g. Meadow, 1996; Meadow and Patel, 2003; Patel, 2009) and, especially when combined with unrecognized depositional mixing, has significant implications for understanding dune-top occupations throughout the northwest and for the timing and processes of introduction of sheep and goat into the region.

For example, at the site of Bagor in the Mewar region of southwestern Rajasthan (Misra, 1973), up to 60% of the faunal materials from Microlithic Phase I (dating from the sixth millennium cal BC) have been identified as coming from domestic animals (primarily goat and sheep but also cattle and pig: Thomas, 1977). These problematic identifications for goat and sheep have led to suggestions that such animals were obtained by the microlithic hunter-gatherers from as far away as contemporary settlements on the western margins of the Indus plain (Possehl, 1999) or even that they were domesticated locally within the context of an *in situ* transformation from hunting-gathering to use of domesticated forms (Shinde et al., 2004; and also Shinde, 2008 for the site of Gilund). Both scenarios seem unlikely.

In contrast, at the site of Loteshwar in North Gujarat careful consideration of site formation processes, dating, and faunal characterization has helped to understand some of the characteristics of dune-top Microlithic sites in the region. Loteshwar has two components—an Aceramic Microlithic (Period I) underlying a Chalcolithic (Period II) with no evident stratigraphic discontinuity between the two. The Period II deposits included pits that were dug into the Period I sediment. As a result, Aceramic Microlithic cultural materials, including lithics and bones, were incorporated into the later levels of the site by the Chalcolithic pit-diggers. Carefully selected animal bone samples from unmixed Period I deposits and from three Period II pits generated two series of radio-carbon dates that provide both the time spans for each component and a *c.*1,000 year gap between the Aceramic Microlithic (*c.*7300–4700 cal BC) and Chalcolithic (*c.*3700–2200 cal BC) (Patel, 2008). The duration of each period and the shallowness of the deposits, taken together, indicate a depositional history of multiple episodic occupations within each component. Faunal identifications based on comparative osteology show that the Period I assemblages are dominated by the remains of blackbuck along with specimens of nine other wild ungulates (Patel, 2009) (Table 18.1). More importantly, no sheep or goat bones were identified from these deposits (Patel, 2008). Indeed, no Microlithic component of any North Gujarat site for which the fauna has been studied has provided evidence for sheep or goat.

Details of the spread of domestic goat and sheep from Balochistan into the remainder of northwestern South Asia are limited. By the third millennium cal BC, however, it is possible to discern differences between populations in different areas that likely reflect local environmental conditions and/or pastoral practices. For example, at Nausharo (Balochistan) and Dholavira (Kutch, Gujarat), both situated in hot arid zones, the frequency of sheep remains is similar to those for goats. In contrast, at Harappa in better-watered Punjab, the remains of sheep greatly outnumber those for goats. Furthermore the sheep at both Nausharo and Dholavira were smaller-bodied than at Harappa, where the sheep included many heavily built animals. This could reflect the development of regional breeds (Meadow, 1991; Patel, 1997). The third millennium cal BC was also a time of increased interactions across the whole of West Asia, which could have included the introduction of new breeds from outside the region and hybridization with local forms.

CATTLE

While South Asia is at the eastern and southern margins of the distribution of wild *Ovis* and *Capra*, it forms the western part of the Indo-Malayan arc that includes all of the extant wild taxa of the genera *Bos* and *Bubalus*. In the past it was also home to now-extinct *Bos namadicus* (sometimes designated as *Bos primigenius namadicus*), which is likely to have been the ancestor of domestic zebu cattle (*Bos indicus*) (e.g. Grigson, 1985). Notable by its hump, drooping ears, and large dewlap, the domestic zebu is also characterized by a physiology that makes it able to tolerate heat stress better than taurine cattle (*Bos taurus*) (Hansen, 2004). Thus it is well adapted to tropical and subtropical regions from eastern Iran to southern China and throughout both mainland and island

Southeast Asia, where it continues to be the dominant form of domestic cattle.

A case can be made from the archaeofaunal record for the local domestication of zebu in northwestern South Asia by considering both changes in size and proportion and in taxonomic abundance. Cattle bone breadths and depths can be used to track body mass change and bone lengths to evaluate stature (Meadow, 1981; 1993). Through the Aceramic Neolithic (Period I) at Mehrgarh, diminution in stature is attested in both median and smallest values. In the earliest levels, the few length measurements available are all from tall animals. Bones from shorter individuals are first encountered in the middle levels, with their numbers increasing through the end of the Aceramic and into the Ceramic Neolithic (Period II). During this same time frame, specimens from the largest-stature animals become increasingly rare, although the infrequent tall individual is still represented through the Chalcolithic (Period III). A similar pattern can be seen in the breadth and depth measurements. Diminution in body mass is attested by the increasing numbers of smaller-sized bones in the last phases of Period I and continuing through Period II. Bones from the heaviest animals largely disappear by the beginning of the Ceramic Neolithic. This can be interpreted as cattle becoming more gracile, which could reflect changes in diet and activity level.

As for taxonomic abundance, cattle bones are not common in the earliest levels of Aceramic Neolithic Mehrgarh (< 5%). By the end of Period I, however, they make up between 40% and 60% of the ungulate remains. Similarly high proportions continue into later periods at Mehrgarh and are characteristic of many fourth-, third-, and second-millennium cal BC sites in northwestern South Asia as well as elsewhere in India, reflecting the importance of cattle for both primary and secondary products (Patel, 1997; Thomas, 2002; Meadow and Patel, 2003).

The duration of Period I at Mehrgarh was as long as 1,500 years, which would have provided ample time for increasingly intensive animal management to develop into captive breeding leading to body-size reduction. The parallel trends of increasing exploitation and diminution in both body mass and animal height through time are consistent with an interpretation of local cattle domestication. The process may have begun with management of females and later incorporated the use of male offspring for breeding. There are other possible scenarios as well, including capturing and raising wild infants, but a problem with cattle as opposed to other bovids is that there are no extant wild populations, the behaviours of which could be employed as models for the past.

Evidence for the cattle at Mehrgarh sharing a similar phenotype to modern *Bos indicus* is varied. Zygomatic bones from Periods I and II all have the flat orbital rim that is a 'fairly good indicator for *indicus* skulls' in older animals (Grigson, 1980: 23). In addition, from Late Period I come a partial skull with intact occipital region that has zebu morphology as well as a gracile bifid thoracic vertebra of the type characteristic of zebu. Finally, also from Late Period I, there is a clay cattle figurine with a pronounced hump, which is the earliest known in a long tradition of making zebu figurines in South Asia (Meadow, 1981; 1993).

The taxonomic relationship between taurine and zebu cattle was debated since at least 1758, when Linnaeus assigned them to the different species *Bos taurus* L. and *Bos indicus*

L. based on their morphological differences. Darwin (1868: 80; see also 1859: 28) supported the position of Linnaeus, writing: 'there can hardly be a doubt, notwithstanding the adverse opinion of some naturalists, that the humped and non-humped cattle must be ranked as specifically distinct'. The differences of opinion on zebu origins continued up to the 1990s, when genetic analyses of modern zebu and taurine cattle resolved the issue, showing that the genetic differences between the two forms are too great to have taken place within the time frame of the Holocene. In a remarkably comprehensive approach for the time, studies included analysis of mitochondrial DNA, Y-chromosomes, and autosomal microsatellites (e.g. Loftus et al., 1994; Bradley et al., 1996; MacHugh et al., 1997; Bradley et al., 1998), with subsequent work bringing out the complex genetic structures and histories of various cattle populations (for a review, see Magee et al., 2014).

Two major mtDNA haplogroups have been defined for modern zebu: I1 (= Z1) and I2 (= Z2). Analyses of diversity within each haplogroup suggest that northwestern South Asia is the best candidate for the origin of the dominant haplogroup I1 and thus for zebu domestication. I1 is also the dominant haplogroup in other parts of India as well as throughout Southeast and East Asia. Haplogroup I2 shows a more complex pattern of regionalized genetic diversity, which may have resulted from domestication(s) of individuals from one or more wild I2 populations or from introgression through recruitment of local wild I2 females into domestic stock in various parts of India (Baig et al., 2005; Magee et al., 2007; Chen et al., 2010).

In contrast, only one haplogroup (Y3) has been identified in zebu Y-chromosomes (Götherström et al., 2005). That haplogroup has very low haplotype diversity, which suggests high genetic uniformity among Asian zebu ancestral male breeding stock. This in turn supports a single domestication hypothesis for males and introgression through the crossing of domestic bulls with wild cows to account for the mtDNA diversity noted above (Pérez-Pardal et al., 2010).

Autosomal SNP-based demographic modeling has also presented a complex domestication history for zebu (Murray et al., 2010). This includes an initial domestication-related bottleneck in the Holocene, followed by population-size recovery and subsequent admixtures from the ancestral South Asian wild cattle population. Yet, the genetic complexity of zebu also does not exclude the possibility of multiple wild ox ancestries contributing to two or more separate lines of domestication. And in addition to recruitment of local wild females into domestic herds, insemination of domestic females by wild males is indicated as a likely introgression scenario.

Taking an overall view of the implications of the genetic data for *Bos indicus*, a number of scenarios for domestication and subsequent breeding practices can be envisioned. These include single or multiple localities of management and eventual domestication; cross-breeding of wild males with domestic females and of domestic males with wild females, both leading to genetic introgression; crossing of domestic breeds; and feralizations of males and/or females. The problem with such data from modern domestic animals, however, is that they provide information based on the offspring of continued reproductive successes far removed in time from the period when domestication processes took place. They do not record unrealized or failed efforts, swamped or extinct genetic lines, or domestic population replacements. To address such issues requires study of ancient DNA. Given the absence of such data for South Asian cattle, it is not possible to understand the true complexity of zebu genetic and domestication history. That said, there is archaeofaunal evidence from parts of India and eastern Iran that can help us frame some of those complexities archaeologically.

At the site of Loteshwar in North Gujarat (discussed above), there is evidence for both wild and domestic forms of cattle. The cattle remains from the Aceramic Microlithic (Period I) are all exclusively from large-bodied wild animals. Those from the Chalcolithic (Period II) have considerably smaller dimensions and include the remains of some young animals. Three Period II cattle bones were directly dated. Two produced results between *c.*3700 and 3500 cal BC and one dated to between *c.*2500 and 2000 cal BC, showing that cattle pastoralism was practised in North Gujarat by the fourth millennium cal BC (Patel, 2008; 2009).

This Loteshwar evidence highlights a number of important issues. The presence of a wild form of cattle in the Aceramic Microlithic establishes that humans would have been familiar with wild cattle behaviour, which is an underlying prerequisite for local domestication. Potentially they also would have had access to young animals that could be more readily captured and managed. By the Chalcolithic, when domestic cattle are attested at Loteshwar, introgressions from local wild forms would also have been a possibility if wild cattle were still at hand. Given the diversity of mtDNA haplotypes represented in modern *Bos indicus*, introgressions may have occurred rather widely until the time of final extinction of the wild form *Bos namadicus*. Localities beyond the northwest that have been highlighted for introgressions include the Gangetic region of northern India as well as central and southern India (Chen et al., 2010). In these areas, however, evidence for cattle pastoralism begins to occur only in the third millennium cal BC, considerably later than at Loteshwar, which is among the earliest-dated sites with evidence for cattle pastoralism beyond the Mehrgarh area.

Whether cattle pastoralism developed locally in North Gujarat or was the result of groups making their way over time south through Sindh remains unclear. It is intriguing, however, that the ceramics found with the domestic cattle bones at Loteshwar are a local variety, not currently known from Sindh (Ajithprasad and Sonawane, 2011). Other outstanding issues are the nature of the shift from hunting and gathering to agriculture and pastoralism in the region and whether that change was a case of local populations making the transition, whether such groups were replaced by newcomers or whether processes of acculturation took place.

By the beginning of the third millennium cal BC, domestic plant- and animal-producing economies were common across most areas of northwestern South Asia and by the middle of that millennium had given birth to the Indus civilization. Harappan iconography includes representations of humped and non-humped cattle, quite likely reflecting the important connections that peoples of northwestern South Asia had with contemporary agro-pastoral societies in Iran, Afghanistan, and beyond. Cross-breeding of zebu and taurine cattle could also have been taking place by that time, especially in the northwest. Such admixture is attested genetically in various modern breeds that have zebu or taurine phenotypes and is said to have considerable time depth (P. Kumar et al, 2003; Decker et al., 2014).

Two sites from southeastern Iran—Tepe Yahya and Shahr-i Sokhta—have provided evidence for zebu and for taurine cattle as well. From Tepe Yahya, twelve orbital rims have been illustrated and described (Meadow, 1986). Of these, one has a zebu-like morphology, one is taurine in its morphology and the other ten have morphologies that fall between the two. All date between between the mid-sixth and end of the fourth millennium cal BC, with the most taurine-like specimen being the latest of the group. In addition, from the nearby site of Tepe Gaz Tavila comes a single bifid vertebra spine, dating to the mid-sixth millennium cal BC. Northeast of Tepe Yahya is the site of Shahr-i Sokhta dating from the late fourth to the late third millennium cal BC. Some of the complete cattle horn-cores from the site have been identified as coming from *Bos namadicus* based on their size and cross-sectional morphology and some of the metapodials are particularly gracile, a feature common in zebu. In addition, bifid thoracic vertebrae and humped-cattle figurines have been recovered from the site (Bökönyi, 1997).

Epstein (1971) has argued that the barriers posed by the central deserts of Iran provided the western limits of *Bos namadicus* distribution and that eastern Iran was a prime location for zebu domestication. Bökönyi (1997), while acknowledging the Mehrgarh evidence, notes that zebu could have been domesticated anywhere the wild form was present, including in eastern Iran. Larson and Burger (2013) argue that domestic zebu resulted from early cross-breeding of domestic taurine cattle from the west with *Bos namadicus*. Our proposal is that cattle from one or more wild populations in northwestern South Asia were brought under increasing human control beginning as early as the late eighth millennium cal BC. That said, it is important to emphasize that *Bos primigenius* and *Bos namadicus* are extinct, *Bos taurus* and *Bos indicus* are difficult to differentiate based on most of their skeletal parts, and that, to date, there is no archaeogenetic evidence from eastern Iran or South Asia. Because the archaeology of northwestern South Asia and eastern Iran is so poorly known for the first half of the Holocene, we understand little about the antecedents to Mehrgarh, Loteshwar, and Tepe Yahya. As a result, more robust answers to questions about changing human–cattle relations in the region will come only when it is possible to investigate that archaeological record in greater detail and to successfully extract and sequence ancient cattle DNA from the region.

WATER BUFFALO

The domestic water buffalo (*Bubalus bubalis*) is another important large bovine of South and Southeast Asia that continues to have a major impact on the economies of the region and beyond. Two forms have been morphologically and genetically differentiated— the river buffalo of much of South Asia and the swamp buffalo in northeastern India,

Southeast Asia, and much of China. Unlike the case for cattle, the wild form (*Bubalus arnee*) is still extant although endangered, with isolated populations in parts of India, Nepal, Bhutan, Thailand, and Cambodia. There are also numerous wild–domestic crossbreeds (generally from domestic females breeding with wild males) and feral populations throughout the region (Barker, 2014; Choudhury, 2014).

Phenotypic differences between the river and swamp forms have been recognized in horn morphology. Swamp animals have outward- and backward-sweeping horns in the plane of the frontal, while river breeds have a variety of morphologies that share a downward and backward orientation with a more or less well-developed spiral (Mason, 1974). Other phenotypic differences, such as hair, skin, and body characteristics and development and behaviour, as well as geographic distribution, have also been observed (Mason, 1974; Cockrill, 1984). In addition, the two forms have been found to differ in their chromosome numbers with river buffalo being $2n = 50$ and swamp buffalo being $2n = 48$. Fertile hybrids occur routinely in areas where both occur. Both forms do well under hot conditions so long as they have access to sufficient standing or running water in which to wallow or swim or from which they can be doused to assist in thermoregulation.

Archaeologically, our understanding of water buffalo domestication and exploitation is limited. In northwestern South Asia, water buffalo postcranial specimens have been identified from the early levels of the Aceramic Neolithic at Mehrgarh (Period I), horn-core fragments from the Ceramic Neolithic (Period II) and a complete horn-core from the Chalcolithic (Period III) (Meadow, 1981; Patel and Meadow, 1998). These specimens are considerably larger than those from modern domestic forms and likely came from wild animals. The horn-core morphology is similar to that of modern swamp and wild water buffalo.

Farther south on a low stabilized sand dune in the alluvial plains of North Gujarat east of the Little Rann of Kutch, an impressive collection of water buffalo remains was uncovered at the site of Santhli (Patel and Meadow, 1998). As at Loteshwar, Santhli has both an earlier Aceramic component with microliths and a later Chalcolithic one with human burials and ceramics, each component constituting a shallow deposit. Although not directly dated, the Santhli buffalo remains are likely to have been contemporary with some part of the Aceramic Microlithic at Loteshwar (mid-eighth to mid-fifth millennium cal BC) because of similarities between the nature of the microlithic settlements and assemblages at the two sites. Indeed a few water buffalo postcranial specimens have been identified from the Aceramic Microlithic deposits at Loteshwar.

The Santhli water buffalo remains comprise clusters including both cranial and postcranial specimens from at least eight morphometrically wild animals (Patel and Meadow, 1998). Their horn-core morphologies are similar to those of domestic swamp buffalo and of extant wild water buffalo from India. Basal horn-core dimensions and postcranial measurements show that the Santhli animals were considerably larger than modern domestic forms and comparable in size to the wild water buffalo of Mehrgarh. Indeed, a particularly important feature of the Santhli assemblage is that it provides a unique mid-Holocene wild water buffalo standard with which other buffalo material can be compared.

Within the clusters, in addition to conjoining cervical vertebrae, many of the post-cranial bones are parts of articulated units from meat-poor zones of the appendicular skeleton. These joints may have been left behind by hunters who focused on the meat-rich parts of the carcasses. Based on their mandibular tooth eruption and wear, five of these animals ranged in age from an infant to a young adult, while epiphyseal fusion of the limb-bones indicates an age-range from early juvenile into adult. The broad span of ages represented in this deposit with many articulated skeletal segments from a number of different animals can be interpreted in two ways. One is that this part of Santhli was a disposal area for the remains that accumulated over time from multiple hunts. The other is that the deposit resulted from a single catastrophic kill-off event possibly related to an animal drive. In either case, the deposits demonstrate that microlithic groups had access to wild water buffalo of different ages and were familiar with the animal's behaviour within the landscape of North Gujarat. This availability and knowledge could have been translated into management and eventual domestication of water buffalo in this region, a possibility that is explored below.

Also in Gujarat, on an island in the Great Rann of Kutch, is the urban Harappan settlement of Dholavira, which dates from the beginning of the third to the middle of the second millennium cal BC (Bisht, 2015). Bovines contributed between 50 and 70% of the faunal remains from deposits of occupation Stages III through V (between c.2800 and 2000 cal BC) (Patel, 1997). In the subset of those bones sufficiently diagnostic to permit differentiation between the two bovine taxa, the proportions of cattle to water buffalo vary depending on time period and context but generally favour cattle (Patel, 1997; 2015). Measurements of postcranial specimens show that almost all water buffalo bones come from animals comparable in size to modern domestic forms, with only two elements from buffalo of a size similar to those represented at Santhli and Mehrgarh (Patel and Meadow, 1998). Small-bodied animals such as those represented at Dholavira would have resulted from processes of animal management that ultimately led to water buffalo domestication. These are likely to have taken place sometime between the sixth and fourth millennium cal BC, a timeframe for which the archaeofaunal record is inadequate to provide meaningful information on changing human–water buffalo interactions.

Although water buffalo remains have been identified from Harappan period sites across much of northwestern South Asia (summarized in Thomas, 2002), such high proportions as those documented at Dholavira have not been reported elsewhere. For example, only a very few specimens have been identified in the bovine-rich assemblages from Balakot near the Arabian Sea coast in eastern Balochistan (Meadow, 1979) and from Harappa in Pakistani Punjab (Meadow, 1991). Even so, at Harappa and other Indus sites north of Gujarat, water buffalo are well represented in the iconography of stamp seals and terracotta figurines, which indicates their familiarity and significance—whether practical or conceptual—to local populations. The frequency of their remains in the faunal assemblages at Dholavira reflects their practical importance to the local economy. The site was constructed with water harvesting and storage facilities to mitigate its location on an island without perennial rivers in an area that receives fluctuating amounts of rainfall on an irregular basis (Bisht, 2005). This water-centric focus of Dholavira would have been ideal for keeping water buffalo that require moisture for thermoregulation (Patel, 1997).

Given the evidence for wild buffalo in North Gujarat and for both wild and especially domestic populations at Dholavira, this region is important to explore for evidence of local water buffalo domestication. In considering that process, however, there are some outstanding phenotypic and genotypic issues that also need to be explored. The iconographic and archaeofaunal evidence both show horn morphologies like those of modern swamp and wild buffalo but not like those of river buffalo, even though it is the latter that dominate much of India today. Genetic studies have confirmed that phenotypic characteristics are not always reliable for identifying to which water buffalo form a modern individual, breed, or population belongs. For example, the domestic Chilika and Toda buffalo populations of eastern and southern India, respectively, were generally believed to be of swamp type based on phenotypic and behavioural characteristics. Cytogenetic, microsatellite, and mtDNA analyses, however, all show that both breeds group with riverine buffalo (e.g. Nair et al., 1986; S. Kumar et al., 2006; Mishra et al., 2009; Kathiravan et al., 2011). In another genetic study, a sample of ten Nepalese wild buffalo with swamp-type morphology including horn shape were found to share a genetic structure with river buffalo. Microsatellite analysis also determined that the animals tested included eight wild and two wild–domestic hybrids (Flamand et al., 2003; Zhang et al., 2011).

Genetic approaches have also been employed to investigate the phylogenetic relationships between modern river and swamp buffalo in order to illuminate their domestication histories. S. Kumar and colleagues (2007a; 2007b) approached these issues and determined that there is a sufficiently high degree of mtDNA sequence divergence between river and swamp buffalo to separate them into two clades. The divergence was determined to be comparable to that between *Bos indicus* and *Bos taurus*. A similar degree of divergence was also found to be represented in the Y-chromosome (Yindee et al., 2010). Time of divergence has been variously estimated, and all results indicate that the two wild ancestral populations from which modern domestic water buffalo descended had diverged well before domestication occurred. S. Kumar and colleagues (2007a; 2007b) also suggest independent domestication of river and swamp buffalo. They interpret the complexity of mtDNA genetic structure in modern river buffalo to indicate that there has been continuous introgression of wild animals into domestic stock in South Asia from early in the domestication process (also Nagarajan et al., 2015).

Unfortunately, there is limited archaeological evidence from eastern India through southern China to bring to bear on the question of swamp buffalo domestication. What information there is comes from southern China and Thailand and it indicates that domestic buffalo may not have become economically important or even present there until the late first millennium cal BC (Higham, 2012; Yue et al., 2013). If valid, this late date underscores a significant difference in the domestication histories of river and swamp buffalo, raises the issue of the impetus for domestication of the swamp form, and underlines the need for increased understandings of interactions between South and East Asia during the second half of the first millennium cal BC.

In addition to the genetic studies focusing on modern river buffalo cited previously, there have been a number of similar analyses carried out on swamp buffalo particularly in China (e.g. Lei et al., 2011; Zhang et al., 2011; Yue et al., 2013) but also in northeastern India (Mishra et al., 2015). Missing from these studies, however, with the exception of that on Nepalese animals (noted previously), is genetic evidence from modern wild buffalo populations. Such studies would provide insights into the genetic structure and variability within and between different wild buffalo populations as well as into their relationships to modern domestic forms. Genetic introgression from the domestic into the wild is a significant issue for conservation of the endangered wild buffalo today. It continues to occur with farmers turning out their females to breed with wild males, which if not retrieved, can give birth to F1 hybrids that join the wild herd (Barker, 2014). This and other pathways to introgression and admixture would have occurred in the past including between river and swamp forms, as has been demonstrated for some modern populations in northeastern India, Nepal, China, and Southeast Asia (Zhang et al., 2011; Mishra et al., 2015).

Understanding processes of water buffalo domestication are still in their infancy. They are hampered by a dearth of zooarchaeological studies that focus on the bovines of South and Southeast Asia and by the difficulties inherent in reliable taxonomic identification of their remains. In northwestern South Asia, there is evidence for third-millennium cal BC water buffalo domesticates and for the presence of contemporary and earlier wild forms. Yet the archaeological record for the sixth, fifth, and fourth millennia cal BC that would be key for investigating domestication processes remains sparse. While the modern genetic data and their interpretation are intriguing in the complexities that they reveal, their relevance for understanding the various facets of the ancient situation is limited, and no ancient DNA research on South Asian water buffalo remains has been successfully carried out to date.

Conclusions

Animal domestication was a complex process of developing human–animal relationships and behaviours that varied according to the animal form and the human population involved under specific circumstances at any given time and in any given place. In the preceding pages, we have highlighted some of these complexities by exploring aspects of the archaeological, zooarchaeological, and genetic evidence for the domestication of goat, sheep, zebu cattle, and water buffalo in northwestern South Asia.

Almost all the earliest evidence for the transition from exploiting wild fauna to the development of animal husbandry comes, to date, from one site—Mehrgarh—and that only for goat, sheep, and cattle. In other areas to the east and south, remains of shallow and episodic occupations from the eighth through the fifth millennium cal BC are those of hunter-gatherers who exploited the diverse wild resources of the region, as seen at the sites of Loteshwar and Santhli. The spread of goat and sheep outside the zone of their wild relatives likely would have taken place by the fourth millennium cal BC as the result of the movement of herders or agro-pastoralists or through trade, although there is no archaeological evidence for the nature of the processes involved. For cattle, since the distribution of the wild form as attested in the archaeofaunal record included most of South Asia, the possibility existed for local management leading to domestication. This could have been accompanied by periodic exchange of genetic material between domestic animals and their free-ranging relatives. While wild cattle are extinct, wild water buffalo are still found in South Asia. Remains of domestic or wild buffalo have been identified from a number of sites in the northwest, but the bulk of zooarchaeological evidence comes from Gujarat—for wild animals from Santhli and for domesticates from Dholavira. Management leading to domestication likely took place sometime from the sixth to fourth millennia, possibly in Gujarat. Although direct evidence for this process is lacking, it could have involved long-term genetic exchange due to cross-breeding between wild, managed, and domestic forms.

Thus, for these animals there is no common narrative, but there are some common challenges in the context of South Asia. These include lack of adequate archaeological evidence for key time periods especially that between the sixth and fourth millennium cal BC, difficulties in differentiating the remains of similar-sized bovid species, and poor preservation of ancient DNA. A final challenge is to not underestimate variability and complexity in past human behaviour in the region.

ACKNOWLEDGEMENTS

Our synthesis is a reflection of research carried out over many years that was encouraged and facilitated by many friends and colleagues from various institutions and supported by funding sources too numerous to individually acknowledge here. They are extensively credited in our various publications that are included in the bibliography. Here the authors wish to thank Umberto Albarella along with Hannah Russ, Mauro Rizzetto, Kim Vickers, and Sarah Viner-Daniels for inviting us to contribute our essay and for all the hard work that they have put in to getting this volume ready. In particular we would like to thank Umberto for his utmost patience and understanding while we were drafting a paper that came to take on a life of its own. We dedicate this contribution to the memory of Jean-François Jarrige, whose extensive work at archaeological sites in Balochistan, including Mehrgarh, provided the solid foundations needed to better understand the development of major cultural phenomena in northwestern South Asia.

REFERENCES

Ajithprasad, P. and Sonawane, V. H. (2011) 'The Harappa culture in North Gujarat: a regional paradigm', in Osada, T. and Endo, H. (eds) *Linguistics, Archaeology, and Human Past*, pp. 223–69. Kyoto: Research Institute for Humanity and Nature.

Baig, M., Beja-Pereira, A., Mohammad, R., Kulkarni, K., Farah, S., and Luikart, G. (2005) 'Phylogeography and origin of Indian domestic cattle', *Current Science*, 89(1), 38–40.

Barker, J. S. F. (2014) 'Genetics of wild water buffalo', in Melletti, M. and Burton, J. (eds) *Ecology, Evolution and Behavior of Wild Cattle*, Box 16.1. Cambridge: Cambridge University Press.

Bisht, R. S. (2005) 'The water structures and engineering of the Harappans at Dholavira (India)', in Jarrige, C. and Lefèvre, V. (eds) *South Asian Archaeology 2001*, Vol. 1, pp. 11–25. Paris: Éditions Recherche sur les Civilisations.

Bisht, R. S. (2015) *Excavations at Dholavira 1989-90 to 2004-2005*, New Delhi: Archaeological Survey of India.

Bökönyi, S. (1997) 'Zebus and Indian wild cattle', *Anthropozoologica*, 25–6, 647–54.

Bradley, D. G., MacHugh, D. E., Cunninghan, P., and Loftus, R. T. (1996) 'Mitochondrial diversity and origins of African and European cattle', *Proceedings of the National Academy of Sciences of the United States of America*, 93, 5131–5.

Bradley, D. G., Loftus, R. T., Cunningham, P., and MacHugh, D. E. (1998) 'Genetics and domestic cattle origins', *Evolutionary Anthropology*, 6(3), 79–86.

Chase, B. (2010) 'Social change at the Harappan settlement of Gola Dhoro: a reading from animal bones', *Antiquity*, 84, 528–43.

Chen, S., Lin, B.-Z., Baig, M., Mitra, B., Lopes, R. J., Santos, A. M., Magee, D. A., Azevedo, M., Tarroso, P., Sasazaki, S., Ostrowski, S., Mahqoub, O., Chaudhuri, T. K., Zhang, Y. P., Costa, V., Royo, L. J., Goyache, F., Luikart, G., Boivin, N., Fuller, D. Q., Mannen, H., Bradley, D. G., and Beja-Pereira, A. (2010) 'Zebu cattle are an exclusive legacy of the South Asia Neolithic', *Molecular Biology and Evolution*, 27(1), 1–6.

Choudhury, A. (2014) 'Wild water buffalo *Bubalus arnee* (Kerr, 1792)', in Melletti, M. and Burton, J. (eds) *Ecology, Evolution and Behavior of Wild Cattle*, pp. 255–301. Cambridge: Cambridge University Press.

Cockrill, W. R. (1984) 'Water buffalo', in Mason, I. L. (ed.) *Evolution of Domesticated Animals*, pp. 52–63. London: Longman.

Darwin, C. (1859) *On the Origin of Species by Means of Natural Selection*, London: John Murray.

Darwin, C. (1868) *The Variation of Animals and Plants under Domestication*, Vol. 1, London: John Murray.

Decker, J. E., McKay, S. D., Rolf, M. M., Kim, J.-W., Molina Alcalá, A., Sonstegard, T. S., Hanotte, O., Götherström, A., Seabury, C. M., Praharani, L., Ellahi Babar, M., Correia de Almeida Regitano, L., Ali Yilzid, M., Heaton, M. P., Liu, W.-S., Lei, C.-Z., Reecy, J. M., Saif-Ur-Rehman, M., Schnabel, R. D., and Taylor, J. F. (2014) 'Worldwide patterns of ancestry, divergence, and admixture in domesticated cattle', *PLOS Genetics*, 10(3), DOI: 10.1371/journal.pgen.1004254.

Epstein, H. (1971) *The Origins of Domestic Animals of Africa*, revised edn, New York: Africana Publishing.

Flamand, J. R. B., Vankan, D., Gairhe, K. P., Duong, H., and Barker, J. S. F. (2003) 'Genetic identification of wild Asian water buffalo in Nepal', *Animal Conservation*, 6, 1–10.

Fuller, D. Q. and Madella, M. (2002) 'Issues in Harappan archaeobotany: retrospect and prospect', in Settar, S. and Korisettar, R. (eds) *Indian Archaeology in Retrospect: Protohistory*, Vol. 2, pp. 317–90. New Delhi: Indian Council of Historic Research & Manohar.

Götherstörm, A., Anderung, C., Hellborg, L., Elburg, R., Smith, C., Bradley, D. G., and Ellegren, H. (2005) 'Cattle domestication in the Near East was followed by hybridization with aurochs bulls in Europe', *Proceedings of the Royal Society B*, 272, 2345–50.

Grigson, C. (1980) 'The craniology and relationships of four species of *Bos*: 5. *Bos indicus* L.', *Journal of Archaeological Science*, 7, 3–32.

Grigson, C. (1985) '*Bos indicus* and *Bos namadicus* and the problem of autochthonous domestication in India', in Misra, V. N. and Bellwood, P. (eds) *Recent Advances in Indo-Pacific Prehistory*, pp. 425–8. New Delhi: Oxford & IBH.

Hansen, P. J. (2004) 'Physiological and cellular adaptations of zebu cattle to thermal stress', *Animal Reproduction Science*, 82–3, 349–60.

Higham, C. (2012) 'The long and winding road that leads to Angkor', *Cambridge Archaeological Journal*, 22(2), 265–89.

Jarrige, C., Jarrige, J.-F., Meadow, R. H., and Quivron, G. (eds) (1995) *Mehrgarh Field Reports 1974–1985 from Neolithic Times to the Indus Civilization*, Karachi: Department of Culture and Tourism, Government of Sindh.

Jarrige, J.-F., Jarrige, C., and Quivron, G. (2013) *Mehrgarh Neolithic Period Seasons 1997–2000*, Paris: Éditions de Boccard.

Kathiravan, P., Kataria, R. S., Mishra, B. P., Dubey, P. K., Sadana, D. K., and Joshi, B. K. (2011) 'Population structure and phylogeography of Toda buffalo in Nilgiris throw light on possible origin of aboriginal Toda tribe of South India', *Journal of Animal Breeding and Genetics*, 128, 295–304.

Kumar, P., Freeman, A. R., Loftus, R. T., Gaillard, C., Fuller, D. Q., and Bradley, D. G. (2003) 'Admixture analysis of South Asian cattle', *Heredity*, 91, 43–50.

Kumar, S., Gupta, J., Kumar, N., Dikshit, K., Navani, N., Jain, P., and Nagarajan, M. (2006) 'Genetic variation and relationships among eight Indian riverine buffalo breeds', *Molecular Ecology*, 15, 593–600.

Kumar, S., Nagarajan, M., Sandhu, J. S., Kumar, N., Behl, V., and Nishanth, G. (2007a) 'Mitochondrial DNA analyses of Indian water buffalo support a distinct genetic origin of river and swamp buffalo', *Animal Genetics*, 38, 227–32.

Kumar, S., Nagarajan, M., Sandhu, J. S., Kumar, N., and Behl, V. (2007b) 'Phylogeography and domestication of Indian river buffalo', *BMC Evolutionary Biology*, 7, 186.

Larson, G. and Burger, J. (2013) 'A population genetics view of animal domestication', *Trends in Genetics*, 29(4), 197–205.

Lei, C. Z., Zhang, C. M., Weining, S., Campana, M. G., Bower, M. A., Zhang, X. M., Liu, L., Lan, X. Y., and Chen, H. (2011) 'Genetic diversity of mitochondrial cytochrome b gene in Chinese native buffalo', *Animal Genetics*, 42, 432–6.

Linnaeus, C. (1758) *Systema Naturae per Regna Tria Naturae*, 10th edn, Stockholm: Laurentii Salvii.

Loftus, R. T., MacHugh, D. E., Bradley, D. G., Sharp, R. M., and Cunningham, P. (1994) 'Evidence of two independent domestications of cattle', *Proceedings of the National Academy of Sciences of the United States of America*, 91, 2757–61.

MacHugh, D. E., Shriver, M. D., Loftus, R. T., Cunningham, P., and Bradley, D. G. (1997) 'Microsatellite DNA variation and the evolution, domestication and phylogeography of taurine and zebu cattle (*Bos taurus* and *Bos indicus*)', *Genetics*, 146, 1071–86.

Magee, D. A., Mannen, H., and Bradley, D. G. (2007) 'Duality in *Bos indicus* mtDNA diversity: support for geographical complexity in zebu domestication', in Petraglia, M. D. and Allchin, B. (eds) *The Evolution and History of Human Populations in South Asia: Inter-Disciplinary Studies in Archaeology, Biological Anthropology, Linguistics and Genetics*, pp. 385–91. Dordrecht: Springer.

Magee, D. A., MacHugh, D. E., and Edwards, C. J. (2014) 'Interrogation of modern and ancient genomes reveals the complex domestic history of cattle', *Animal Frontiers*, 4(3), 7–22.

Mason, I. L. (1974) 'Species, types and breeds', in Cockrill, W. R. (ed.) *The Husbandry and Health of Domestic Buffalo*, pp. 1–47. Rome: Food and Agriculture Organization of the United Nations.

Meadow, R. H. (1979) 'Prehistoric subsistence at Balakot: initial consideration of the faunal remains', in Taddei, M. (ed.) *South Asian Archaeology 1977*, pp. 275–315. Naples: Istituto Universitario Orientale, Seminario di Studi Asiatici.

Meadow, R. H. (1981) 'Early animal domestication in South Asia: a first report of the faunal remains from Mehrgarh, Pakistan', in Härtel, H. (ed.) *South Asian Archaeology 1979*, pp. 143–79. Berlin: Dietrich Reimer Verlag.

Meadow, R. H. (1986) 'Animal Exploitation in Prehistoric Southeastern Iran: Faunal Remains from Tepe Yahya and Tepe Gaz Tavila-R37 5500–3000 BC'. Unpublished PhD dissertation, Harvard University (Cambridge, Massachusetts).

Meadow, R. H. (1989) 'Prehistoric wild sheep and sheep domestication on the eastern margin of the Middle East', in Crabtree, P. J., Campana, D., and Ryan, K. (eds) *Early Animal Domestication in Its Cultural Context*. MASCA Research Papers in Science and Archaeology 6, Special Supplement, pp. 24–36. Philadelphia: University of Pennsylvania.

Meadow, R. H. (1991) 'Faunal remains and urbanism at Harappa', in Meadow, R. H. (ed.) *Harappa Excavations 1986–1990: A Multidisciplinary Approach to Third Millennium Urbanism*, pp. 89–106. Madison: Prehistory Press.

Meadow, R. H. (1993) 'Animal domestication in the Middle East: a revised view from the eastern margin', in Possehl, G. L. (ed.) *Harappan Civilization: A Recent Perspective*, 2nd revised edn, pp. 295–320. New Delhi: Oxford & IBH.

Meadow, R. H. (1996) 'The origins and spread of agriculture and pastoralism in northwestern South Asia', in Harris, D. R. (ed.) *The Origins and Spread of Agriculture and Pastoralism in Eurasia*, pp. 390–412. London: UCL Press.

Meadow, R. H. and Patel, A. K. (2002) 'From Mehrgarh to Harappa and Dholavira: prehistoric pastoralism in north-western South Asia through the Harappan period', in Settar, S. and Korisettar, R. (eds) *Indian Archaeology in Retrospect: Protohistory*, Vol. 2, pp. 391–408. New Delhi: Indian Council of Historic Research & Manohar.

Meadow, R. H. and Patel, A. K. (2003) 'Prehistoric pastoralism in northwestern South Asia from the Neolithic through the Harappan period', in Weber, S. A. and Belcher, W. R. (eds) *Indus Ethnobiology: New Perspectives from the Field*, pp. 65–93. Lanham: Lexington Books.

Meadows, J. R. S., Hiendleder, S., and Kijas, J. W. (2011) 'Haplogroup relationships between domestic and wild sheep resolved using a mitogenome panel', *Heredity*, 106, 700–6.

Meadows, J. R. S. and Kijas, J. W. (2009) 'Re-sequencing regions of ovine Y chromosome in domestic and wild sheep reveals novel paternal haplotypes', *Animal Genetics*, 40, 119–23.

Menon, V. (2009) *Mammals of India*, Princeton: Princeton University Press.

Mishra, B. P., Kataria, R. S., Bulandi, S. S., Prakash, B., Kathiravan, P., Mukesh, M., and Sadana, D. K. (2009) 'Riverine status and genetic structure of Chilika buffalo of eastern India as inferred from cytogenetic and molecular marker-based analysis', *Journal of Animal Breeding and Genetics*, 126, 69–79.

Mishra, B. P., Dubey, P. K., Prakash, B., Kathiravan, P., Goyal, S., Sadana, D. K., Das, G. C., Goswami, R. N., Bhasin, V., Joshi, B. K., and Kataria, R. S. (2015) 'Genetic analysis of river, swamp and hybrid buffaloes of north-east India throw new light on phylogeography of water buffalo (*Bubalus bubalis*)', *Journal of Animal Breeding and Genetics*, DOI: 10.1111/jbg.12141.

Misra, V. N. (1973) 'Bagor, a Late Mesolithic settlement in north-west India', *World Archaeology*, 5, 92–110.

Murray, C., Huerta-Sanchez, E., Casey, F., and Bradley, D. G. (2010) 'Cattle demographic history modeled from autosomal sequence variation', *Philosophical Transactions of Royal Society B*, 365, 2531–9.

Nagarajan, M., Nimisha, K., and Kumar, S. (2015) 'Mitochondrial DNA variability of domestic river buffalo (*Bubalus bubalis*) populations: genetic evidence for domestication of river buffalo in Indian subcontinent', *Genome Biology and Evolution*, 7(5), 1252–9.

Naderi, S., Rezaei, H.-R., Pompanon, F., Blum, M. G. B., Negrini, R., Naghash, H.-R., Balkız, Ö., Mashkour, M., Gaggiotti, O. E., Ajmone-Marsan, P., Kence, A., Vigne, J.-D., and Taberlet, P. (2008) 'The goat domestication process inferred from large-scale mitochondrial DNA analysis of wild and domestic individuals', *Proceedings of the National Academy of Sciences of the United States of America*, 105(46), 17659–64.

Nair, P. G., Balakrishnan, M., and Yadav, B. R. (1986) 'The Toda buffaloes of Nilgiri', *Buffalo Journal*, 2, 169–78.

Patel, A. K. (1997) 'The pastoral economy of Dholavira: a first look at animals and urban life in third millennium Kutch', in Allchin, R. and Allchin, B. (eds) *South Asian Archaeology 1995*, pp. 101–13. New Delhi: Oxford & IBH.

Patel, A. K. (2008) 'New radiocarbon determinations from Loteshwar and their implications for understanding Holocene settlement and subsistence in North Gujarat and adjoining areas', in Raven, E. M. (ed.) *South Asian Archaeology 1999*, pp. 123–34. Groningen: Egbert Forsten.

Patel, A. K. (2009) 'Occupational histories, settlements and subsistence in western India: what bones and genes can tell us about the origins and spread of pastoralism', *Anthropozoologica*, 44(1), 173–88.

Patel, A. K. (2015) 'Analysis of faunal remains from excavations at Dholavira (District Kutch, Gujarat)', in Bisht, R. S. (ed.) *Excavations at Dholavira 1989–90 to 2004–2005*, pp. 839–69. New Delhi: Archaeological Survey of India.

Patel, A. K. and Meadow, R. H. (1998) 'The exploitation of wild and domestic water buffalo in northwestern South Asia', in Buitenhuis, H., Bartosiewicz, L., and Choyke, A. M. (eds) *Archaeozoology of the Near East III*, pp. 180–99. Groningen: Centre for Archaeological Research and Consultancy, Rijksuniversiteit.

Pérez-Pardal, L., Royo, L. J., Beja-Pereira, A., Chen, S., Cantet, R. J., Traoré, A., Curik, I., Sölkner, J., Bozzi, R., Fernández, I., Alvarez, I., Gutiérrez, J. P., Gómez, E., Ponce de León, F. A., and Goyache, F. (2010) 'Multiple paternal origins of domestic cattle revealed by Y-specific interspersed multilocus microsatellites', *Heredity*, 105, 511–19.

Possehl, G. L. (1999) *Indus Age: The Beginnings*, New Delhi: Oxford & IBH.

Rezaei, H. R., Naderi, S., Chintauan-Marquier, I. C., Taberlet, P., Virk, A. T., Naghash, H. R., Rioux, D., Kaboli, M., and Pompanon, F. (2010) 'Evolution and taxonomy of the wild species of genus *Ovis* (Mammalia, Artiodactyla, Bovidae)', *Molecular Phylogenetics and Evolution*, 54, 315–26.

Roberts, T. J. (1997) *The Mammals of Pakistan*, revised edn, Karachi: Oxford University Press.

Shinde, V. (2008) 'Cultural development from Mesolithic to Chalcolithic in the Mewar region of Rajasthan, India', *Pragdhara*, 18, 201–13.

Shinde, V., Deshpande, S. S., and Yasuda, Y. (2004) 'Human response of Holocene climatic change: a case study of western India between the 5th and 3rd millennia BC', in Yasuda, Y. and

Shinde, V. (eds) *Monsoon and Civilization*, pp. 383–406. New Delhi: International Research Center for Japanese Studies & Roli Books.

Thomas, P. K. (1977) 'Archaeozoological Aspects of Prehistoric Cultures of Western India'. Unpublished PhD dissertation, Deccan College Post-Graduate and Research Institute (Pune).

Thomas, P. K. (2002) 'Investigations into the archaeofauna of Harappan sites in western India', in Settar, S. and Korisettar, R. (eds) *Indian Archaeology in Retrospect: Protohistory*, Vol. 2, pp. 409–20. New Delhi: Indian Council of Historic Research & Manohar.

Valdez, R., Nadier, C. F., and Bunch, T. D. (1978) 'Evolution of wild sheep in Iran', *Evolution*, 32, 56–72.

Yindee, M., Vlamings, B. H., Wajjwalku, W., Techakumphu, M., Lohachit, C., Sirivaidyapong, S., Thitaram, C., Amarasinghe, A. A., Alexander, P. A., Colenbrander, B., and Lenstra, J. A. (2010) 'Y-chromosomal variation confirms independent domestications of swamp and river buffalo', *Animal Genetics*, 41, 433–5.

Yue, X.-P., Li, R., Xie, W.-M., Xu, P., Chang, T.-C., Liu, L., Cheng, F., Zhang, R.-F., Lan, X.-Y., Chen, H., and Lei, C.-Z. (2013) 'Phylogeography and domestication of Chinese swamp buffalo', *PLOS One*, 8(2), DOI: 10.1371/journal.pone.0056552.

Zeder, M. A. (2015) 'Core questions in domestication research', *Proceedings of the National Academy of Sciences of the United States of America*, 112(11), 3191–8.

Zhang, Y., Vankan, D., Zhang, Y., and Barker, J. S. F. (2011) 'Genetic differentiation of water buffalo populations in China, Nepal and south-east Asia: inferences on the region of domestication of the swamp buffalo', *Animal Genetics*, 42, 366–77.

CHAPTER 19

THE ZOOARCHAEOLOGY OF NEOLITHIC CHINA

LI LIU AND XIAOLIN MA

INTRODUCTION

CHINA is one of the few primary loci of animal domestication and of emergent agriculture in the world; Chinese society has been predominantly agrarian for thousands of years, a process that began in the Neolithic (7000–2000 cal BC). As in other parts of the world, agriculture formed the economic foundation for the rise of civilization in China, and domesticated animals played a significant role in this development. The domestication of animals during the Neolithic was crucial, not only as a part of subsistence strategies for supporting growing populations, but also as an essential component in ritual practice, which facilitated communication between humans and the supernatural, assisted ambitious individuals' competition for social status, and helped to reinforce power structures. Animal domestication contributed, economically and ideologically, to the formation of some long-lasting cultural traditions in Chinese civilization.

Neolithic beginnings in China can be traced back to at least 9,000 years ago (Liu, 2004; Liu and Chen, 2012). This era is defined by the appearance of several new developments, including animal and plant domestication, sedentism, and the common use of pottery and ground stone tools. During the Neolithic period, the archaeological record shows great regional variations in terms of the level of socioeconomic development, adaptation strategies, and human–animal relationships.

Studies of animal remains in Chinese archaeology have changed dramatically, from an emphasis on taxonomic identification to the employment of multidisciplinary approaches to address diverse issues relating to human–animal relationships (Yuan, 2002; Liu and Chen, 2012). Since the 1990s this field has been characterized by a rapid growth of research and publications. Researchers now use systematic analytical methods to record and interpret faunal remains, employing scientific approaches, such as DNA and isotope analyses, to investigate the origins and dispersal of domesticated animals, and explore changing relationships between humans and animals in relation to environmental fluctuations and social formations.

In this chapter we discuss several major issues in the zooarchaeology of Neolithic China. Fig. 19.1 shows the location of sites mentioned.

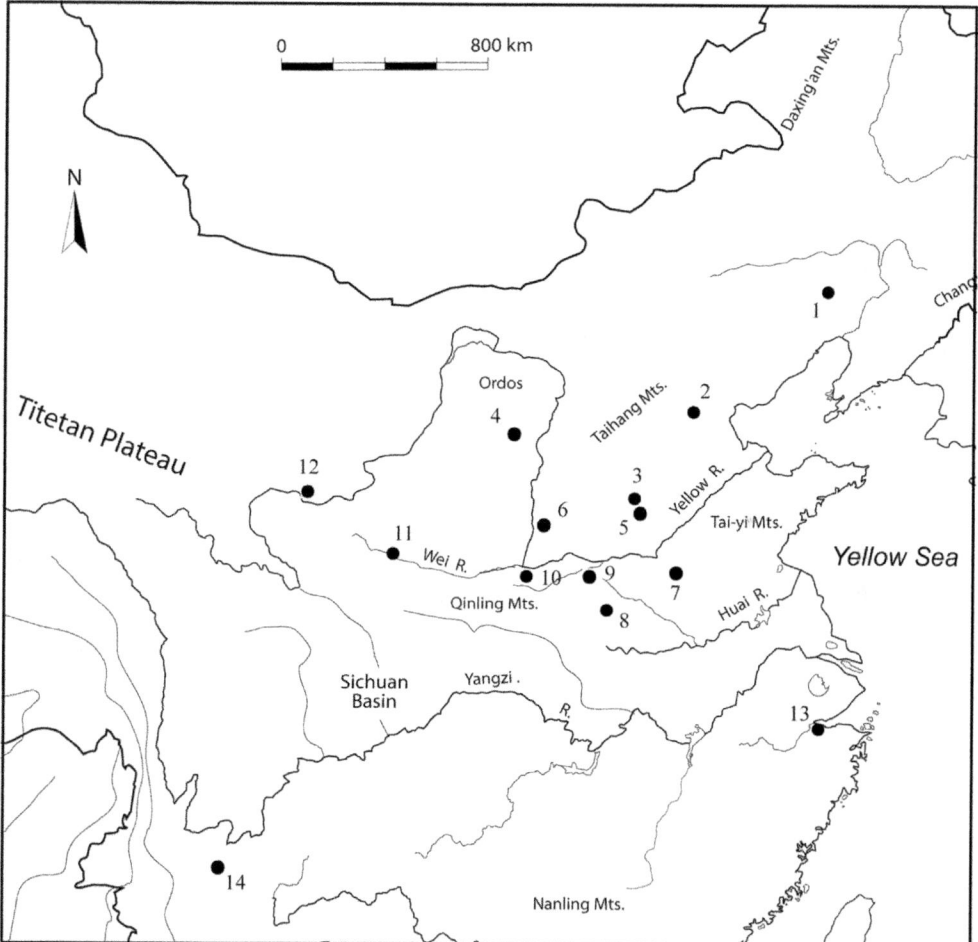

FIGURE 19.1 Location of archaeological sites with domesticated animals discussed in the text. 1: Xinglongwa, 2: Nanzhuangtou, 3: Cishan, 4: Huoshiliang, 5: Yin Dynasty ruins (Anyang), 6: Taosi, 7: Shantaisi, 8: Jiahu, 9: Yanshi Shang city, 10: Xipo, 11: Dadiwan, 12: Donghuishan, 13: Kuahuqiao, 14: Shizhaishan. Author's own image.

THE ORIGINS AND SPREAD
OF ANIMAL DOMESTICATION

Among the most important domestic animals known in Neolithic China, pig (*Sus domesticus*) and dog (*Canis familiaris*) were first domesticated indigenously. Sheep (*Ovis aries*), goat (*Capra hircus*), cattle (*Bos taurus*), and horse (*Equus caballus*) were introduced into northern China later from the Eurasian steppes. Zebu (*Bos indicus*) and buffalo (*Bubalus bubalis*) were probably introduced first into southwest China. The chicken (*Gallus domesticus*) is the least understood among these domesticates, and its origins remain unclear archaeologically.

Dog

The domestic dog, used for hunting, is the earliest domesticated animal in the world. It has been argued that the Chinese dog (*Canis familiaris*) originated from the Chinese wolf (*Canis lupus chanco*) (Olsen and Olsen, 1977). The earliest archaeological evidence for domestic dog in China dates to the transitional phase between the Upper Palaeolithic and the Neolithic, *c.*10,000 cal BP, from Nanzhuangtou in Hebei (site 2 in Fig. 19.1). The domestic status of the specimen is based on the shortened dental length in a mandible compared with that from known examples of wolf (Yuan and Li, 2010).

More archaeological evidence for the early domestic dog comes from the Early Neolithic (7000–5000 cal BC) in the Yellow River Valley. Complete dog skeletons were intentionally buried in cemeteries or near houses at Jiahu in Henan, site 8 in Fig. 19.1 (Henansheng Wenwu Kaogu Yanjiusuo, 1999), and one dog was buried in a pit at Cishan in Hebei, site 3 in Fig. 19.1 (Sun et al., 1981). At Dadiwan in Gansu (site 11 in Fig. 19.1), isotopic analyses on dog remains suggest provisioning by humans with millet (Barton et al., 2009). In these Early Neolithic villages, hunting was still an important component in the subsistence strategies, as indicated by finds of abundant hunting tools; ritual burial of dogs at these sites seems to suggest their close relationship with humans, perhaps as aids for hunting.

Pig

The origin of the domestic pig has been the major focus of multidisciplinary studies in China. Genetic studies indicate that pigs were domesticated independently in multiple areas across Eurasia, including probably twice in China (Larson et al., 2005; 2010). *Sus* bones, likely a mixture of domesticated and wild, have frequently been found in Early Neolithic sedentary settlements in both northern and southern China.

The earliest evidence of pig domestication in China has been revealed at Jiahu (site 8 in Fig. 19.1), dating to 7000 cal BC. This conclusion is based on several lines of evidence, including tooth morphology, which is similar to the domesticated forms, a kill-off pattern suggesting a culling strategy often used in domesticated populations (younger than three years old), and linear enamel hypoplasia (LEH) frequency higher than that of wild boar (*Sus scrofa*) (Luo and Zhang, 2008; Wei et al., 2014).

Most Dadiwan Phase I pigs (site 11 in Fig. 19.1, 5800–2800 cal BC) have also been identified as domestic, on the basis of kill-off patterns and lower third molar size (Qi et al., 2006). However, isotope analysis of pig bones from this assemblage suggests that all pigs were wild, eating predominantly C3 plants, which is presumably different from the millet-based human diet (C4 plant) (Barton et al., 2009). This conclusion contradicts the faunal analysis, and the inconsistency is probably due to the small sample size of the pig bones (n = 4) and the lack of human bones for comparison in the isotope study. It is also possible that humans fed pigs with C3 foods, such as tubers, or pigs were kept free-range.

The evidence for pig domestication at Cishan (site 3 in Fig. 19.1) is apparently mixed. Several complete skeletons of young pigs were found at the bottom of ash pits, probably used as ritual offerings, while the lower third molar size range (39.2–45 mm) is consistent with the wild form (Zhou, 1981; Yuan and Flad, 2002). A similar situation has been found in the Xinglongwa culture in northeast China (site 1 in Fig. 19.1, 6200–5200 cal BC). The pig skeletons from these Early Neolithic settlements represent large animals, but tooth measurements overlap with the range for domestic forms. Two complete adult pig skeletons, a male and a female, were found in a human burial at Xinglongwa (Fig. 19.2), a phenomenon leading some archaeologists to believe that pig domestication may have indeed been underway (Yang and Liu, 1997; Tang et al., 2004). However, it cannot rule out the possibility that wild animals, young or old, were used as sacrifice in burials.

The earliest evidence for pig domestication in the Yangzi River region is found at Kuahuqiao in Zhejiang (site 13 in Fig. 19.1, 6000–5000 cal BC). Three *Sus* mandibles

FIGURE 19.2 A human burial associated with two pigs at Xinglongwa (Yang and Liu, 1997: Plate 2.1). Courtesy of Guoxiang Liu.

show distorted alignment of teeth, and third molars show a general trend of reduction in size from early to late phases. The distribution of lower third molar sizes also falls into both wild and domesticated ranges (Yuan and Yang, 2004).

Archaeological data described above seem to support the theory of multi-centred pig domestication. At the least, Jiahu and Kuahuqiao pigs (sites 8 and 13 in Fig. 19.1) may represent two separate early domestication events in north and central China, respectively. Northeast China may have been another centre of pig domestication, as the large size of Xinglongwa pigs may reflect a local form, perhaps the result of adaptation to a colder environment in this region.

Sheep, Goat, and Cattle

These animals were first domesticated in the Fertile Crescent, western Asia, about 11,500 years ago (Zeder, 2011). Ancient DNA analysis suggests that sheep were introduced into northern China during the Late Neolithic (Cai et al., 2011; 2014), although exactly when and through which routes the diffusion took place remains unclear. The origins of the domesticated goat in China remain controversial among geneticists, and multiple centres have been suggested, including various regions in China (Larson et al., 2010; Cucchi et al., 2011). Archaeologically, the earliest remains of domestic sheep, dating to the fourth millennium cal BC, have been reportedly found in areas to the west and north of the Central Plain. Sheep and goat began to occur in the middle and lower Yellow River Valley during the Early Longshan period (3000–2500 cal BC), and became more widely dispersed during the Late Longshan period (2500–1900 cal BC) and the Bronze Age (c.1900–500 cal BC) (for a summary see Liu, 2004: 59; Flad et al., 2007).

Ancient DNA analysis suggests that domestic cattle were introduced to northern China between 3000 and 2000 cal BC (Cai et al., 2014). Archaeologically, about thirty sites have yielded cattle remains, dating to the third millennium BC, particularly from the Late Longshan and Qijia cultures (2500–1800 cal BC). The cattle assemblages show characteristics of domestication, such as reduced body size and evidence of being slaughtered at a young age. There is also a marked increase in the overall frequency of this species at several sites, such as Shantaisi in Zhecheng, Henan, site 7 in Fig. 19.1 (Lü, 2010). Domestic cattle, having once appeared in northern China, became distributed through the entire Yellow River region, and often coexisted with sheep/goats as part of pastoralist assemblages, especially in the Ordos region and the Upper Yellow River region, suggesting a likely derivation from the Eurasian steppe.

The introduction of sheep/goat and cattle appears to be associated with the post-Holocene climatic optimum after 3000 cal BC, when north China became generally colder and drier; a condition more favourable to pastoralism, especially in the north-west. Increased interaction between pastoralists in the Eurasian steppe and early Chinese agriculturalists may have also contributed to the appearance of new animals and plants in China, which added to the existing subsistence economy as supplementary food sources. This situation is clearly exemplified at Donghuishan, in Minle, Gansu (site 12 in Fig. 19.1, 2000–1500 cal BC), where the faunal assemblage is dominated by pig and deer bones, accompanied by remains of sheep and dog, in lesser proportions (Qi, 1998). Remains of various crops, including millets (*Setaria italica* and *Panicum miliaceum*), wheat (*Triticum aestivum*), barley (*Hordeum vulgare* var. nudum), and rye (cf. *Secale montanum*) were also recovered (Gansu Institute of Archaeology, 1998: 140; Li et al., 1989). Conversely, primary reliance on sheep and goat (62% of the total NISP of nineteen species) is evident in the faunal assemblage at Huoshiliang (site 4 in Fig. 19.1, 2150–1900 cal BC) in Yulin, northern Shaanxi. These domesticates were supplemented by pig, cattle, dog, and fourteen species of wild fauna, including gazelle (*Gazella* sp.) and three forms of deer (*Cervus nippon*, *Cervus elaphus*, and *Capreolus capreolus*) (Hu et al., 2008). These finds suggest that by the end of the third millennium BC a mature agropastoral economy, focused on grazing animals, became established in the Ordos region. By the early second millennium BC, at latest, an agro-pastoralist subsistence strategy had already emerged in northern China, producing both indigenous and introduced domestic crops and animals.

Zebu

Little is known about the first archaeological occurrence of domestic zebu in China. Bronze sculptures found at Shizhaishan in Yunnan (site 14 in Fig. 19.1, third century BC–first century AD) depict zebu in scenes of ritual, sacrifice, and husbandry, most of which seem to show domestic animals, as they appear to be herded by humans riding horses (Zhang, 1998). Based on representations in art, therefore, we may infer that zebu was introduced to southwest China no later than the third century cal BC.

Water Buffalo

It was originally thought that the swamp type of water buffalo (*Bubalus bubalis*) was first domesticated in the Yangzi River Valley around 5000 cal BC as a component of rice cultivation. New studies suggest, however, that indigenous buffalo (*Bubalus mephistopheles*) during the Holocene in China were wild and became extinct in antiquity (Liu et al., 2006a). An ancient DNA study of *B. mephistopheles* bones dating between 5000 and 1600 cal BC shows a clear separation between this ancient bovine and modern domestic buffalo (*B. bubalis*), suggesting that the indigenous Chinese buffalo is unlikely to have been involved in the process of buffalo domestication (Yang et al., 2008). Domesticated swamp buffalo may have originated in Southeast Asia. Modern buffalo populations in Yunnan, southwest China, show a high diversity in mtDNA (Yue et al., 2013), but whether or not Yunnan was part of the region in which swamp buffalo domestication took place is currently uncertain. A better understanding of the timing and the locations of this animal's domestication must rely on the discovery and zooarchaeological analysis of more buffalo remains in China, as well as on DNA tests of archaeological buffalo remains from southwest China and its bordering regions.

Horse

The origin of domestic horses has been a controversial issue. Genetic studies show no sequence diversity on the male-inherited Y chromosome, but abundant genetic diversity within female-inherited mtDNA (Lippold et al., 2011). These mtDNA results can be explained as follows: as the original domesticated population expanded, female horses from wild populations were occasionally introduced into the domestic herds (Levine, 2006), and some of these events may have taken place in China (Lei et al., 2009). Nevertheless, other genetic studies do not support the hypothesis that early Chinese domestic horses were derived from the indigenous horse (*Equus przewalskii*) (Cai et al., 2009).

Most archaeologists and historians believe that horses were introduced from the steppe region to China by the later part of the second millennium cal BC (Yuan and Flad, 2006; Flad et al., 2007). Isolated horse bones, likely wild species, have been unearthed from many Early and Middle Neolithic sites in north China. Beginning from the Late Longshan and Qijia periods, horse bones, in small proportions in the faunal assemblages, have often been unearthed together with those of sheep/goat and cattle. Some researchers argue for their being wild, based on their relative scarcity at any given site (Linduff, 2003). Others favour the possibility of domestication based on their frequency of incidence in the archaeological record (Yuan and Flad, 2006). Current data provide no hard evidence by which to determine their domesticity. The archaeological horse remains from the northwest region point to limited human utilization, where isolated horse bones were present in domestic refuse or human burials. This situation is rather different from the horse remains found in the Central Plain during the Shang dynasty, where complete horse skeletons and chariots were buried together as ritual sacrifice (Linduff, 2003).

Chicken

The most probable wild progenitor of the domestic chicken was the red jungle fowl (*Gallus gallus*) (Fumihito et al., 1994); however, the location of its domestication remains controversial. Based on genetic studies, chickens have multiple maternal origins with domestication occurring in South Asia, Southeast Asia, and southwest China, including Yunnan (Liu et al., 2006b; Miao et al., 2013). Archaeologically, we know little about the timing and location of the chicken's original domestication in China. Many Neolithic sites have revealed 'domestic' chicken remains, but these identifications are questionable. Recent examinations of chicken remains from China, using systematic identification methods, suggest that the most reliable evidence for domestic chicken is from

Anyang (site 5 in Fig. 19.1, Late Shang dynasty); but the first domestication of chicken could have occurred much earlier (Deng et al., 2013).

Summary

In summary, the earliest evidence for dog domestication in China can be traced back to 8000 cal BC and of pig to 7000 cal BC, but the first attempts to domesticate these animals may have been earlier. Already-domesticated sheep/goat and cattle were introduced to China by the second half of the third millennium cal BC. Horses became common in northwest China when sheep/goats and cattle were introduced, and these grazing animals tend to co-occur at some sites, suggesting that they were introduced together as a part of a pastoralist economy. By the later part of the second millennium cal BC, domestic horses were widespread in northern China, and also became a part of elite culture. How these animals were brought into China is a matter of debate.

Due to Xinjiang's geographical position, it has been seen as an area with potential for further understanding east–west interactions; however, as regards domesticated animals, to date, the earliest evidence for cattle, sheep/goat, and horse in Xinjiang are no earlier than those from the Yellow River Valley. Thus, these domesticated grazing animals were likely diffused in multiple events and through various routes from the Eurasian steppe to a very broad region of northern China. These routes may have included precursors of the historical lines of communication, such as the Silk Road in the west and many routes connecting north China and Mongolia in the north.

The domestication processes of buffalo and zebu are likely to have been related to interregional interactions between southwest China and its neighbouring areas. The development of the so-called Southwest Silk Road, a modern name for an ancient route that connected southwest China with Burma and India during the Han dynasty, very likely encouraged trade between different regions and promoted the dispersal of domestic bovines. Such interactions may have existed long before the Han dynasty, and future research may reveal earlier evidence for the origins of these bovines.

DYNAMICS OF DOMESTICATION

There are two competing theories regarding the underlying dynamics of food-producing economies. One takes a cultural-ecological view, which in general emphasizes the emergence of farming in environmentally marginal areas, where severe climatic change forced human populations to find new food sources (Watson, 1995). Animal domestication, therefore, is a solution for food shortage. This model, however, does not explain archaeological findings from Kuahuqiao, Jiahu, and Xinglongwa (sites 13, 8, and 1 in Fig. 19.1, respectively) that have yielded domestic pig remains. All these sites reveal conditions with abundant natural resources, without meat shortage;

evidently the people relied on a broad-spectrum subsistence economy, with farming as only a minor component.

A second theory offering explanation for the origins of domestication involves a social-political approach, which follows Bender's (1978) original argument, that competition between neighbouring groups, to achieve local dominance through community feasting, was the driving force behind food production. This motivation required increased subsistence resources and therefore intensified the process of food production. This view has gained increasing attention in more recent years. Hayden (1995) has argued that domestication occurs when status distinctions and socioeconomic inequalities first appear within societies in a resource-rich natural environment. Although this is by no means accepted by all archaeologists (see Smith, 2001), this approach, often referred to as the Socioeconomic Competition Model, or Food-Fight Theory, may explain pig domestication in China. If pigs were first domesticated in areas with abundant food resources, we cannot rule out the possibility that pork was favoured over other animal flesh, perhaps as a delicacy at competitive feasts, at least occasionally. This may have also reinforced the incentive for pig domestication, particularly as we have evidence that pigs were used as ritual offerings in mortuary practices. This is exemplified by pig mandibles entered as grave goods in three burials at the Early Neolithic Dadiwan site (site 11 in Fig. 19.1) (Gansu Institute of Archaeology, 2006).

Each of these two theoretical models is useful, in some degree, for explaining the human and natural dynamics underlying early food production, but each also displays biases influenced by a particular theoretical framework (Processual vs. Post-processual approaches). Indeed, some studies have revealed that agriculture was often initiated in areas of relatively abundant resources (e.g. Price and Gebauer, 1995; Smith, 2011). It is not difficult to see that all the sites in China associated with early domesticates were situated in rich environmental surroundings, and that these domesticates, however small their role in the overall subsistence economy, nevertheless formed a stable part of local human diet for a long period of time before intensified farming was undertaken. During this extended process, these domesticates were probably used both as routine staples and as luxuries for conspicuous consumption. Therefore, the dynamics of domestication were both ecological and social.

SECONDARY PRODUCTS IN ANIMAL DOMESTICATION

Domesticated livestock in China were likely first used for meat; the exploitation of wool, riding, traction, and pack transport probably developed later. The identification of dairy fat residues in archaeological potsherds suggests that the use of milk and the emergence of dairying across Europe and the Near East can be traced back as early as 8,500 years ago, indicating that dairying was an early feature of Neolithic subsistence in these region

(Gerbault et al., 2013). However, such early evidence is lacking in China, probably due to the late introduction of domestic bovines. Few studies based on zooarchaeological data from China have considered animal secondary products. Kill-off patterns for sheep from Taosi in Shanxi (site 6 in Fig. 19.1, 2500–1900 cal BC) suggest wool production (Li et al., 2014). In Xinjiang (northwestern China), where artefacts are well preserved, sheep–cattle pastoralism was an important part of the economy; dairying and wool production are evident in the second millennium cal BC (Xinjiang Institute, 2008; Yang et al., 2014).

In the large part of China where agriculture was predominant, secondary products were mainly associated with the muscle power of bovines. While zooarchaeological information is scarce, many discussions are based on artefacts. For example, stone plough-shaped artefacts have been recovered from several Late Neolithic sites in the lower Yangzi River Valley. However, use-wear analysis suggests that they were multi-functional tools, used for various tasks rather than ploughing. There is no conclusive archaeological evidence for bovine-plough technology during the Neolithic in China (Liu et al., 2012).

DOMESTICATED ANIMALS IN RITUAL PRACTICE

During the Early Neolithic, pigs and dogs were the first domesticates to be used as sacrificial offerings. Pig mandibles were placed in burials at Dadiwan (site 11 in Fig. 19.1), and two pigs, probably domesticated, were placed next to a male in an indoor burial at Xinglongwa (site 1 in Fig. 19.1; Fig. 19.2). Pig and dog skeletons were placed in pits, at Cishan (site 3 in Fig. 19.1), that appear to have then been filled with millet; likely related to ritual practices. The early use of animal sacrifice in mortuary contexts may have been closely associated with ancestor worship, which became a cultural tradition that lasted for thousands of years in China. Pigs and dogs continued to be used as sacrificial offerings throughout the Neolithic period, and were particularly common in the Lower Yellow River region. The occurrences of pig and dog sacrifice in mortuary contexts are closely correlated with the presence of elite grave goods, such as jade and fine pottery, strongly suggesting that animal sacrifice was an important strategy for power-building among Neolithic elites (Liu, 1996).

Cattle, sheep/goats, and horses also joined the inventory of sacrificial offerings after their introduction to China. At the Late Longshan culture site of Shantaisi in Henan (site 7 in Fig. 19.1), a sacrificial pit containing nine articulated cattle skeletons is considered to represent the earliest-known origin of cattle sacrifice, a ritual practice which later became common during the Shang period (Murowchick and Cohen, 2001). The earliest evidence for sheep sacrifice comes from Yanshi Shang city in Henan (site 9 in Fig. 19.1, 1600–1300 cal BC), a walled Bronze Age settlement. Horses, often together with chariots, became sacrificial victims in the Late Shang capital city, Anyang (site 5 in Fig. 19.1, 1250–1046 cal BC). Sacrifice of domestic animals—including cattle, sheep/goats, pig, and dog—also became more frequently and elaborately performed in the ancestor-worship rituals to demonstrate the royal power of the Shang elite at Anyang (Yuan and Flad, 2005).

ANIMAL-BASED SUBSISTENCE STRATEGY IN RELATION TO THE DEVELOPMENT OF COMPLEX SOCIETY

The development of complex society during the Neolithic period has received much attention in Chinese archaeology, and faunal remains have been used to illustrate this theme through investigation of animal-based subsistence strategies. For example, during the Pre-Yangshao (7000–5000 cal BC) and Early Yangshao (5000–4000 cal BC) periods, wild animals, particularly various deer species, comprised the major sources of human meat diet in the Yellow River Valley (Yuan, 1999). In contrast, domestic pigs became the dominant element in meat consumption during the Middle Yangshao period (4000–3500 cal BC), as seen at Xipo (site 10 in Fig. 19.1). This marked change in the major sources of meat coincided in time with a dramatic increase in both human population density and the level of social complexity in the region (Ma, 2005: 73). Meat-acquisition patterns varied regionally across China between 3500 and 1500 cal BC, which was the formative period of Chinese civilization. In the Yangzi River Valley, hunting was still the major mode of meat acquisition, supplemented by a small proportion of domestic animals. In contrast, domestic pigs, sheep, and cattle became dominant in meat consumption in the middle Yellow River Valley after 2500 cal BC. The latter pattern indicates the development of an advanced, diverse, and stable subsistence economy, which may have contributed to the economic foundation for the rise of early states in the Yellow River Valley (Yuan et al., 2008; Yuan and Luo, 2011).

CONCLUSIONS

Whether domestication in China occurred intentionally or unintentionally, the first domesticates, particularly pig, played rather minor roles for a few millennia before they became major food sources, during the Middle Neolithic or later. The introduction of cattle, sheep/goats, and horses during the third and second millennium cal BC is likely to have resulted from a combination of factors including changing environmental conditions and movement of populations from the Eurasian steppe to north China with economically valuable technologies. The lack of clear evidence for domestic buffalo and zebu in south China is probably affected by insufficient zooarchaeological investigation in this region.

Domestic animals and plants not only played crucial roles in the subsistence economy, but were also used as ritual offerings at various ceremonial events, facilitating social elites' negotiation for power. While indigenous domesticates formed the economic basis for the emergence of complex societies in the Neolithic, introduced domesticates contributed to accelerate socio-political changes in the Early Bronze Age, leading to the formation of early states / civilization in northern China.

Faunal remains are an important proxy in the archaeological record for the reconstruction of social processes, as well as the human–environment relationships in Neolithic China.

REFERENCES

Barton, L., Newsome, S. D., Chen, F.-H., Wang, H., Guilderson, T. P., and Bettinger, R.L. (2009) 'Agricultural origins and the isotopic identity of domestication in northern China', *Proceedings of the National Academy of Sciences of the United States of America*, 106(14), 5523–8.

Bender, B. (1978) 'Gatherer-hunter to farmer: a social perspective', *World Archaeology*, 10, 204–22.

Cai, D., Sun, Y., Tang, Z., Hu, S., Li, W., Zhao, X., Xiang, H., and Zhou, H. (2014) 'The origins of Chinese domestic cattle as revealed by ancient DNA analysis', *Journal of Archaeological Science*, 41, 423–34.

Cai, D., Tang, Z., Han, L., Speller, C. F., Yang, D. Y., Ma, X., Cao, J. E., Zhu, H., and Zhou, H. (2009) 'Ancient DNA provides new insights into the origin of the Chinese domestic horse', *Journal of Archaeological Science*, 36, 835–42.

Cai, D., Tang, Z., Yu, H., Han, L., Ren, X., Zhao, X., Zhu, H., and Zhou, H. (2011) 'Early history of Chinese domestic sheep indicated by ancient DNA analysis of Bronze Age individuals', *Journal of Archaeological Science*, 38, 896–902.

Cucchi, T., Hulme-Beaman, A., Yuan, J., and Dobney, K. (2011) 'Early Neolithic pig domestication at Jiahu, Henan Province, China: clues from molar shape analyses using geometric morphometric approaches', *Journal of Archaeological Science*, 38(1), 11–22.

Deng, H., Yuan, J., Song, G., Wang, C., and Masaki, E. (2013) 'Zhongguo gudai jiaji de zai tantao', *Kaogu*, 6, 83–96.

Flad, R., Yuan, J., and Li, S. (2007) 'Zooarcheological evidence of animal domestication in northwest China', in Madsen, D. B., Chen, F.-H., and Gao, X. (eds) *Late Quaternary Climate Change and Human Adaptation in Arid China*, pp. 167–203. Amsterdam: Elsevier.

Fumihito, A., Miyake, T., Sumi, S., Takada, M., Ohno, S., and Kondo, N. (1994) 'One subspecies of the red jungle fowl (*Gallus gallus gallus*) suffices as the matriarchic ancestor of all domestic breeds', *Proceedings of the National Academy of Sciences of the United States of America*, 91, 12505–9.

Gansu Institute of Archaeology (ed.) (1998) *Minle Donghuishan Kaogu*, Beijing: Kexue Press.

Gansu Institute of Archaeology (ed.) (2006) *Qin'an Dadiwan*, Beijing: Wenwu Press.

Gerbault, P., Roffet-Salque, M. Evershed, R. P., and Thomas, M. G. (2013) 'How long have adult humans been consuming milk?', *International Union of Biochemistry and Molecular Biology*, 65(12), 983–90.

Hayden, B. (1995) 'A new overview of domestication', in Price, T. D. and Gebauer, A. B. (eds) *Last Hunters-First Farmers*, pp. 273–99. Santa Fe: School of American Research Press.

Henansheng Wenwu Kaogu Yanjiusuo (ed.) (1999) *Wuyang Jiahu*, Beijing: Kexue Chubanshe.

Hu, S., Zhang, P., and Yuan, M. (2008) 'Yulin Huoshiliang yizhi dongwu yicun yanjiu', *Renleixue Xuebao*, 3, 232–48.

Larson, G., Dobney, K., Albarella, U., Fang, M., Matisoo-Smith, E., Robins, J., Lowden, S., Finlayson, H., Brand, T., Willerslev, E., Rowley-Conwy, P., Andersson, L., and Cooper, A. (2005) 'Worldwide phylogeography of wild boar reveals multiple centers of pig domestication', *Science*, 307(5715), 1618–22.

Larson, G., Liu, R., Zhao, X., Yuan, J., Fullere, D., Barton, L., Dobney, K., Fan, Q., Gu, Z., Liu, X.-H., Luo, Y., Lv, P., Andersson, L., and Li, N. (2010) 'Patterns of East Asian pig domestication, migration, and turnover revealed by modern and ancient DNA', *Proceeding of the National Academy of Sciences of the United States of America*, 107, 7686–91.

Lei, C. Z., Wang, X. B., Bower, M. A., Edwards, C. J., Su, R., Weining, S., Liu, L., Xie, W. M., Li, F., Liu, R. Y., Zhang, Y. S., Zhang, C. M., and Chen, H. (2009) 'Multiple maternal origins of Chinese modern horse and ancient horse', *Animal Genetics*, 4, 933–44.

Levine, M. (2006) 'MtDNA and horse domestication: the archaeologist's cut', in Mashkour, M. (ed.) *Equids in Time and Space*, pp. 192–201. Oxford: Oxbow Books.

Li, F., Li, J., Lu, Y., Bai, P., and Cheng, H. (1989) 'Gansusheng Minlexian Donghuishan xinshiqi yizhi gunongye yicun xinfaxian', *Nongye Kaogu*, 1, 56–69, 73–4.

Li, Z., Brunson, K. and Dai, L. (2014) 'Zhongyuan diqu xinshiqi shidai dao qingtong shidai zaoqi yangmao kaifa de dongwu kaoguxue yanjiu', *Quaternary Sciences*, 34(1), 149–57.

Linduff, K. M. (2003) 'A walk on the wild side: late Shang appropriation of horses in China', in Levine, M., Renfrew, C., and Boyle, K. (eds) *Prehistoric Steppe Adaptation and the Horse*, pp. 139–62. Cambridge: McDonald Institute Monographs.

Lippold, S., Knapp, M., Kuznetsova, T., Leonard, J. A., Benecke, N., Ludwig, A., Rasmussen, M., Cooper, A., Weinstock, J., Willerslev, E., Shapiro, B., and Hofreiter, M. (2011) 'Discovery of lost diversity of paternal horse lineages using ancient DNA', *Nature Communications*, 2, 1–6.

Liu, L. (1996) 'Mortuary ritual and social hierarchy in the Longshan culture', *Early China*, 21, 1–46.

Liu, L. (2004) *The Chinese Neolithic: Trajectories to Early States*, Cambridge: Cambridge University Press.

Liu, L. and Chen, X. (2012) *The Archaeology of China: From the Late Palaeolithic to the Early Bronze Age*, Cambridge: Cambridge University Press.

Liu, L., Chen, X., and Jiang, L. (2004) 'A study of Neolithic water buffalo remains from Zhejiang, China', *Bulletin of the Indo-Pacific Prehistory Association: The Taipei Papers*, 24(2), 113–20.

Liu, L., Chen, X., Pan, L., Min, Q., and Jiang, L. (2012) 'Were Neolithic rice paddies plowed? Usewear analysis of plow-shaped tools from Pishan in the Lower Yangzi River region, China', *Vestnik*, 11(10), 14–28.

Liu, L., Yang, D., and Chen, X. (2006a) 'Zhongguo jiayang shuiniu de qiyuan', *Kaogu Xuebao*, 2, 141–78.

Liu, Y.-P., Wu, G.-S., Yao, Y.-G., Miao, Y.-W., Luikart, G., Baig, M., Beja-Pereira, A., Ding, Z.-L., Palanichamy, M. G., and Zhang, Y.-P. (2006b) 'Multiple maternal origins of chickens: out of the Asian jungles', *Molecular Phylogenetics and Evolution*, 38, 12–19.

Lü, P. (2010) 'Shilun Zhongguo jiayang huangniu de qiyuan', in Henan Institute of Archaeology (ed.) *Dongwu Kaogu (Di 1 Ji)*, pp. 152–76. Beijing: Wenwu Press.

Luo, Y. and Zhang, J. (2008) 'Henan Wuyangxian Jiahu yizhi chutu zhugu de zaiyanjiu', *Kaogu*, 1, 90–6.

Ma, X. (2005) *Emergent Social Complexity in the Yangshao Culture: Analyses of Settlement Patterns and Faunal Remains from Lingbao, Western Henan, China (c.4900–3000 BC)*. BAR International Series 1453. Oxford: Archaeopress.

Miao, Y-W, Peng, M.-S., Wu, G.-S., Ouyang, Y.-N., Yang, Z.-Y., Yu, N., Liang, J.-P., Pianchou, G., Beja-Pereira, A., Mitra, B., Palanichamy, M., Baig, M., Chaudhuri, T., Shen, Y.-Y., Kong, Q.-P., Murphy, R., Yao, Y.-G., and Zhang, Y.-P. (2013) 'Chicken domestication: an updated perspective based on mitochondrial genomes', *Heredity*, 110, 277–82.

Murowchick, R. and Cohen, D. (2001) 'Searching for Shang's beginnings: Great City Shang, city Song, and collaborative archaeology in Shangqiu, Henan', *The Review of Archaeology*, 22(2), 47–60.

Olsen, S. J. and Olsen, J. W. (1977) 'The Chinese wolf, ancestor of New World dogs', *Science*, 197, 533–5.

Price, D. T. and Gebauer, A. B. (1995) *Last Hunters-First Farmers*, Santa Fe: School of American Research Press.

Qi, G. (1998) 'Donghuishan mudi shougu jianding baogao', in Gansu Institute of Archaeology and Jilin University (eds) *Minle Donghuishan Kaogu*, pp. 184–5. Beijing: Kexue Press.

Qi, G., Lin, Z., and An, J. (2006) 'Dadiwan yizhi dongwu yicun jianding baogao', in Gansu Institute of Archaeology (ed.) *Qin'an Dadiwan*, pp. 861–910. Beijing: Wenwu Press.

Smith, B. D. (2001) 'The transition to food production', in Feinman, G. M. and Price, T. D. (eds) *Archaeology at the Millennium: A Sourcebook*, pp. 199–230. New York: Kluwer Academic/ Plenum Publishers.

Smith, B. D. (2011) 'A cultural niche construction theory of initial domestication', *Biological Theory*, 6(3), 260–71.

Sun, D., Liu, Y., and Chen, G. (1981) 'Hebei Wu'an Cishan yizhi', *Kaogu Xuebao*, 3, 303–38.

Tang, Z., Guo, Z., and Suo, X. (2004) 'Baiyinchanghan yizhi chutu de dongwu yicun', in Inner Mongolia Institute of Archaeology (ed.) *Baiyinchanghan*, pp. 546–75. Beijing: Kexue Press.

Watson, P. J. (1995) 'Explaining the transition to agriculture', in Price, T. D. and Gebauer, A. B. (eds) *Last Hunters-First Farmers: New Perspectives on the Prehistoric Transition to Agriculture*, pp. 21–37. Santa Fe: School of American Research Press.

Wei, C., Lu, J., Xu, L., Liu, G., Wang, Z., Zhao, F., Zhang, L., Han, X., Du, L., and Liu, C. (2014) 'Genetic structure of Chinese indigenous goats and the special geographical structure in the southwest China as a geographic barrier driving the fragmentation of a large population', *PLOS ONE*, 9(4), DOI: 10.1371/journal.pone.0094435.

Xinjiang Institute of Cultural Relics and Archaeology (2008) 'The Xiaohe graveyard in Luobupo, Xinjiang', *Chinese Archaeology*, 8, 85–95.

Yang, D., Liu, L., Chen, X., and Speller, C. F. (2008) 'Wild or domesticated: ancient DNA examination of water buffalo remains from north China', *Journal of Archaeological Science*, 35, 2778–85.

Yang, H. and Liu, G. (1997) 'Neimenggu Aohanqi Xinglongwa juluo yizhi 1992 nian fajue jianbao', *Kaogu*, 1, 1–26.

Yang, Y., Shevchenko, A., Knaust, A., Abuduresule, I., Li, W., Hu, X., Wang, C., and Shevchenko, A. (2014) 'Proteomics evidence for kefir dairy in Early Bronze Age China', *Journal of Archaeological Science*, 45, 178–86.

Yuan, J. (1999) 'Lun Zhongguo xinshiqi shidai jumin huoqu roushi ziyuan de fangshi', *Kaogu Xuebao*, 1, 1–22.

Yuan, J. (2002) 'The formation and development of Chinese zooarchaeology: a preliminary review', *Archaeofauna*, 11, 205–12.

Yuan, J. and Flad, R. (2002) 'Pig domestication in ancient China', *Antiquity*, 76, 724–32.

Yuan, J. and Flad, R. (2005) 'New zooarchaeological evidence for changes in Shang dynasty animal sacrifice', *Journal of Anthropological Archaeology*, 24, 252–70.

Yuan, J. and Flad, R. (2006) 'Research on early horse domestication in China', in Mashkour, M. (ed.) *Equids in Time and Space*, pp. 124–31. Oxford: Oxbow Books.

Yuan, J., Flad, R., and Luo, Y. (2008) 'Meat-acquisition patterns in the Neolithic Yangzi River Valley', *Antiquity*, 82(316), 351–66.

Yuan, J. and Li, J. (2010) 'Hebei Xushui Nanzhuangtou yizhi chutu dongwu yicun yanjiu baogao', *Kaogu Xuebao*, 3, 385–91.

Yuan, J. and Luo, Y. (2011) 'Zhonghua wenming xingcheng shiqi de dongwu kaoguxue yanjiu', *Keji Kaogu*, 3, 80–99.

Yuan, J. and Yang, M. (2004) 'Dongwu yanjiu', in Zhejiang Institute of Archaeology and Xiaoshan Musuem (eds) *Kuahuqiao*, pp. 241–69. Beijing: Wenwu Press.

Yue, X.-P., Li, R., Xie, W.-M., Xu, P., Chang, T.-C., Liu, L., Cheng, F., Zhang, R.-F., Lan, X.-Y., Chen, H., and Lei, C.-Z. (2013) 'Phylogeography and domestication of Chinese swamp buffalo', *PLOS ONE*, 8(2), DOI: 10.1371/journal.pone.0056552.

Zeder, M. A. (2011) 'The origins of agriculture in the Near East', *Current Anthropology*, 52(Supplement 4), 221–35.

Zhang, Z. (1998) *Jinning Shizhaishan*, Kunming: Yunnan Meishu Press.

Zhou, B. (1981) 'Hebei Wu'an Cishan yizhi de dongwu guhai', *Kaogu Xuebao*, 3, 339–47.

CHAPTER 20

···

SUBSISTENCE ECONOMY, ANIMAL DOMESTICATION, AND HERD MANAGEMENT IN PREHISTORIC CENTRAL ASIA (NEOLITHIC–IRON AGE)

···

NORBERT BENECKE

INTRODUCTION

···

CENTRAL Asia is the core region of the Asian continent and stretches from the Caspian Sea in the west to China in the east and from Afghanistan in the south to Russia in the north. Politically, the area encompasses the five countries of Kazakhstan, Uzbekistan, Turkmenistan, Kyrgyzstan, and Tajikistan. Central Asia is a large region of varied geography, including mountain ranges in the south (Tian Shan), vast deserts and semi-deserts (Kara Kum, Kyzyl Kum) in its central part, and grassy steppes in the north. In contrast to Europe, research on animal remains from archaeological sites has no tradition in the countries of Central Asia. Aside from a few exceptional works, publications are quite rare, generally brief, and often consist of nothing more than a list of species. This situation has slightly improved in the last two decades due to greater international cooperation in the field of archaeology in this region. However, the available published faunal record of Central Asia is still fragmentary and permits only a restricted insight into the development of the exploitation of animal resources by humans from prehistory to the present.

NEOLITHIC

Sites of the Jeitun culture located in the Kopetdag piedmont (Turkmenistan) pre-sent the earliest evidence of settled village life in Central Asia, when people lived in houses and supported themselves mainly by cereal cultivation and the herding of animals. The Jeitun culture spanned some 1,400 years, from *c.*6100 to 4500 BC. The most intensively studied site of this culture is the type site of Jeitun (north of Ashgabat) which has been the subject of different excavations starting already in the 1950s (Harris, 2010) (Fig. 20.1). They produced animal bone assemblages of vary-ing sizes which have been studied by various specialists (last report by Dobney and Jaques, 2010). These studies show that caprine husbandry (with a bias toward goats) was the basis of stock-keeping at Jeitun. There was no evidence of the occurrence of cattle (*Bos taurus*) and pigs (*Sus domesticus*). Hunted wild animals that added protein and other nutrients to the diet included gazelles (*Gazella* sp.), probably wild boar (*Sus scrofa*), and other relatively abundant local prey such as foxes (*Vulpes* sp.), hares (*Lepus tolai*), and tortoises (*Agrionemys horsfieldii*). Despite extensive sieving in the last excavation campaigns (1989–1994), almost no bird or fish bones

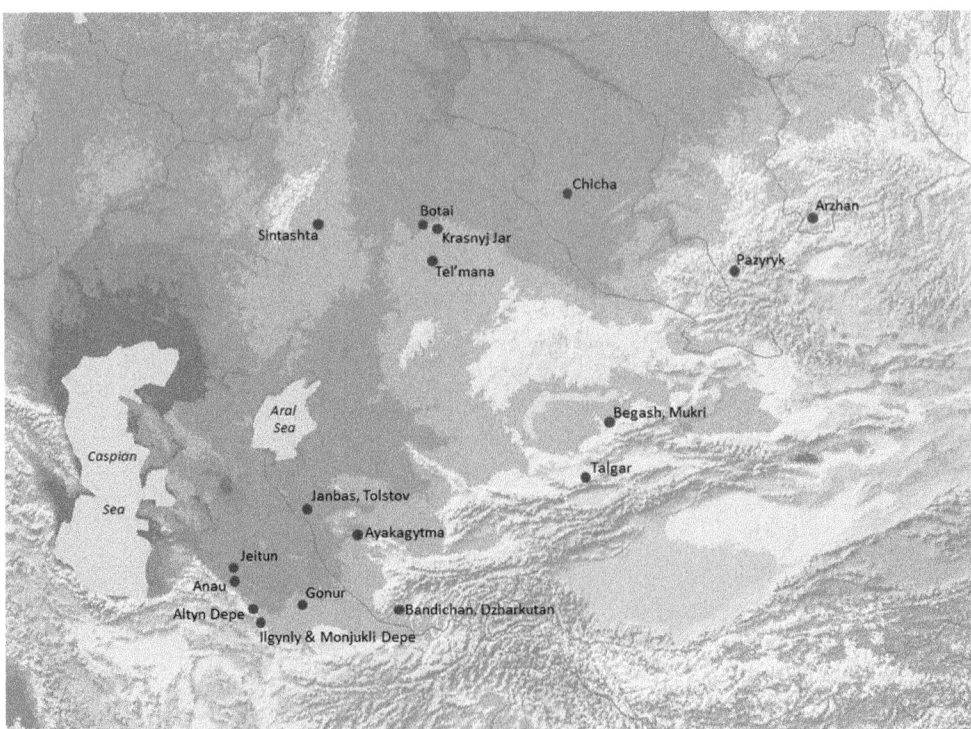

FIGURE 20.1 Location of prehistoric sites in Central Asia mentioned in the text. Author's own image.

were recovered, suggesting that these resources were of minor subsistence importance at the settlement. Bio-archaeological evidence from other sites like Chopan, Chagylly, and Monjukli Depe confirm the mainly agro-pastoral character of the Jeitun economy.

To the north of the sedentary Jeitun culture, in the areas of the Karakum and Kyzylkum deserts and the deltas of the Amudarya and Zerevshan rivers, there is evidence of a different, more mobile type of Neolithic occupation—the Keltiminar culture (c.5500–3500 BC). Faunal analyses on bone collections of this culture (e.g. at Janbas, Tolstov, Beshbulak, and Lyavlyakan in Uzbekistan) demonstrate that most of the sites had been occupied by seasonally mobile hunter-fisher-gatherers (Harris, 2010). The question of whether stock-breeding was part of the Keltiminar subsistence during the Neolithic is difficult to resolve because of the paucity of well-analysed assemblages from stratified sites. Recent excavations at the Keltiminar site of Ayakagytma in the southern Kyzylkum (Fig. 20.1) have recovered bone remains from two occupation phases (c.6000–5500 and c.4000–3000 BC). The specialists who undertook the analysis classified the cattle, sheep/goat, and pig as domesticated, concluding that in both the earlier and the later occupation phases the economy was nearly fully based on domestic animals, with hunting, fishing, and gathering playing a secondary role (Lasota-Moskalewska et al., 2006). Although there are some doubts regarding the accuracy of the identification of domestic animals at Ayakagytma (Harris, 2010), the Keltiminar appears to have depended on a combination of some form of livestock-herding supplemented by hunting, fishing, and gathering.

In the area of north and central Kazakhstan the development of the Neolithic was linked with the Atbasar culture (c.6000–4000 BC). The few faunal assemblages available so far exhibit a clear dominance of wild horses (*Equus ferus*) among the bone finds, pointing to a significant role for this species in the diet and as a source of raw materials (e.g. Tel'mana, Vinogradovka, Javlenka). As a general pattern these collections also include a few remains of cattle and sheep/goat (Akhinzhanov et al., 1992; Zaibert, 1992). Obviously, elements of animal husbandry became known and were adopted in these areas during the Atbasar culture. Due to the lack of radiocarbon dates for this period the exact timing of the first occurrence/introduction of cattle and sheep/goat remains unknown. According to the archaeozoological evidence from neighbouring territories in the west, i.e. the south Urals and the north Caspian region, cattle and sheep husbandry became part of the local subsistence economy in the forest-steppe and steppe to the east of the Volga some time in the sixth or fifth millennium BC (Wechler, 2001; Matjushin, 2003).

CHALCOLITHIC

In the forest-steppe and steppe zones of northern Central Asia the development of the Chalcolithic is connected with the closely related Botai and Tersek cultures (north Kazakhstan), dating to the period 3700–2900 BC. Concerning the pattern of animal exploitation, two groups of sites can be distinguished. The first group comprises settlements of the Botai culture located in the region of the upper Tobol and Ishim rivers, e.g. Botai, Krasnyj Jar, Livanovka, and Solenoe Ozero I (Fig. 20.1). The faunal materials of these places nearly exclusively consist of horse remains with only a few bones coming from other species (Akhinzhanov et al., 1992; Benecke and von den Driesch, 2003; Olsen, 2003). For example, among the faunal remains from Botai (sites 31–33) c.99% of the bones and teeth recovered belong to horses. The rest of the fauna consists of the remains of dogs (*Canis familiaris*) and several wild species, such as elk (*Alces alces*), aurochs (*Bos primigenius*), saiga antelope (*Saiga tatarica*), corsac fox (*Vulpes corsac*), beaver (*Castor fiber*), hare (*Lepus europaeus*), bean goose (*Anser fabalis*), whooper swan (*Cygnus cygnus*), willow grouse (*Lagopus lagopus*), and crane (*Grus* sp.). Bones of cattle and sheep/goat are missing at Botai. Fish remains from sieved soil samples as well as isotopic studies on human remains indicate that fishing was part of the economy at this site (O'Connell et al., 2003). The second group of Chalcolithic sites from north Kazakhstan encompasses Tersek culture settlements located in the Turgai plain. Unlike the sites of the first group, the faunal assemblages of these sites (e.g. Kozhaj I, Kumkeshu I, and Kaindy III) exhibit high diversity and variability in economically important species. Aside from horse, other species such as aurochs, saiga antelope, and onager (*Equus hemionus*) played a significant role in the local subsistence economy. There is also evidence for fowling and fishing (Kalieva and Logvin, 1997).

The key role of horses as a source of food and raw materials in the Chalcolithic of north Kazakhstan has raised the question whether the horses of these sites represent still-wild or already-domesticated animals and whether the steppes to the east of the Ural Mountains played a prominent role in the domestication of the horse at that time. For more than two decades this topic has been the subject of considerable research, debate, and controversy, largely because it is difficult to establish robust zoological criteria for distinguishing the bones of wild and domesticated horses (cf. critical reviews by Levine, 2004 and Olsen, 2006). Recently, a fresh approach to the problem based on three independent lines of evidence, i.e. metrical analysis of metacarpal bones, examination of damage to teeth caused by the use of bridles, and organic-residue analysis of potsherds indicating that horse milk was processed, has provided strong evidence that at least some of the horses from the site of Botai were closely managed, milked, and possibly ridden (Outram et al., 2009). These horses, probably domestic, date to the mid-fourth millennium BC. Analyses of ancient DNA, targeting nuclear genes responsible for coat-colour variation, have further indicated the domestication of horses in the Eurasian steppe region some time prior to 3000 BC (Ludwig et al., 2009). This study identified a rapid and substantial increase in the number of coat colorations around this time, which is best explained by selective breeding after domestication.

In the areas of the Karakum and Kyzylkum deserts, the Chalcolithic is represented by the late Keltiminar culture. Except for the site of Ayakagytma in the southern Kyzylkum (see above), little is known on the subsistence economy of this period.

In the piedmont zone of the Kopetdag, tell sites like Monjukli Depe, Ilgynly Depe, and Anau (Turkmenistan) (Fig. 20.1) have yielded data on the subsistence economy during the Chalcolithic. Large collections of animal bones are available from recent excavations of Chalcolithic layers (c.5100–4500 BC) at Monjukli Depe in the Meana-Čaača region of the eastern Kopetdag foothills (Pollock and Bernbeck, 2011: Table 13). Remains of domestic animals form the bulk of the collections from this site, constituting almost 97% of the bone finds. Four species have been identified, i.e. sheep, goat, cattle, and dog. Animal husbandry was mainly based on sheep and goats. Data on age distribution and sex ratio suggest that, in addition to meat products, both species were also exploited for their milk. Wild animals are represented by only a small number of bones from various species of mammals (e.g. gazelle, onager, camel, bear (*Ursus arctos*), and hare), birds (e.g. great bustard (*Otis tarda*), sand grouse (*Syrrhaptes paradoxus*)), and fish, indicating a limited use of natural resources for subsistence purposes at Monjukli Depe. The fauna of the nearby site of Ilgynly Depe, a large proto-urban centre of the Middle Chalcolithic period, also points to an intensive herding of sheep and goats as part of the pastoral component of the local subsistence economy (Kasparov, 1994: Table 1). But in contrast to Monjukli Depe, hunting was still of great significance for providing food and raw materials at this site. About 30% of the mammal bones represent wild animals such as gazelle, onager, and wild sheep (*Ovis vignei*). At the famous site of Anau, located in the central Kopetdag foothills, faunal studies indicate a mixed strategy of herding and hunting throughout its Chalcolithic occupation (c.4500–2900 BC). Sheep, goats, and cattle formed the basis of animal husbandry in all settlement phases (Moore et al., 2003: Table 12.1). Unlike Monjukli Depe and Ilgynly Depe, domestic pigs were kept in significant numbers at Anau from the earliest periods of occupation onwards. Besides animal-keeping, hunting was of some significance. There is a substantial proportion of hunted animals, mainly onagers, representing as much as 25% of the food remains.

BRONZE AND IRON AGE

In the Bronze Age, a new species, the two-humped or Bactrian camel, became a common livestock animal in Central Asia. Without any doubt, this animal promoted the cultural and economic development of prehistoric human civilizations in the Eurasian dry zones. The Bactrian camel continues to be used today by modern rural and nomadic societies as a source of transportation, labour, meat, milk, and wool. Despite the great economic impact of domestic camels for the inhabitants of Central

Asia, the time and place of their domestication is still uncertain. This is mainly due to the incomplete knowledge of the range and biological variability of the wild Bactrian camel population at the onset of the domestication process and the resulting difficulties in differentiating between the bones of wild and domestic camel (Peters and von den Driesch, 1997).

Archaeozoological records give evidence of a closer relationship between humans and the Bactrian camel at sites in the Kyzylkum desert and Kopetdag foothills beginning in the fourth millennium BC. Bulliet (1975) suggested that this species may have been domesticated in the borderland between Turkmenistan and Iran several centuries prior to 2500 BC. The earliest direct evidence of domesticated Bactrian camels comes from the site Shahr-i Sokhta in east central Iran, especially from layers dating to the period 2700–2400 BC, where remains of camel bones were found, as well as pieces of camel dung, and fibres of camel hair in fragments of woven cloth (Compagnoni and Tosi, 1978). In southern Turkmenistan, clay models of four-wheeled carts pulled by camels have been found at Early and Middle Bronze Age sites, for example at Altyn Depe (Masson and Sarianidi, 1972: Plate 36), clearly indicating that by c.2000 BC Bactrian camel had been incorporated as draft animals into systems of transport and agricultural production. Also, by the end of the third millennium BC, representations of domesticated Bactrian camels appear in the archaeological record of Margiana and Bactria in the form of terracotta figurines and amulets (Moore et al., 1994). Recent DNA-studies on modern and prehistoric Bactrian camels have favoured the idea of a single domestication process that took place in the southwestern parts of Central Asia rather than in Mongolia or even East Asia (Trinks et al., 2012).

In the Bronze Age, settlement sites in southern Central Asia developed an urban character both in the Kopetdag foothills and in the desert oases of Margiana, Geoksyur, Bactria, and Zaman Baba (Hiebert, 1994). Archaeologically, they are subsumed under the term Bactria-Margiana Archaeological Complex or Oxus civilization. According to faunal data from sites like Gonur (Figs. 20.1 and 20.2), Togolok, and Kelleli (Moore et al., 1994) herding formed an important component of the Bronze Age oasis economy. Sheep and goats are the most common species among the domestic animals. The presence of older individuals from these species points to the exploitation of secondary products (milk, wool, etc.). Cattle, traditionally not pastured in the desert, are only represented by low numbers in the faunal assemblages of the oasis settlements. As single-bone finds show, camel, horse, and probably also donkey (*Equus asinus*) were part of the livestock animals of the Oxus civilization (Ermolova, 1983; Moore et al., 1994; 2003). Wild animals like wild boar, gazelle, onager, hare, and birds were hunted as a small complement to the domestic animals. Their remains in the faunal assemblages range from 5% to 10%. Recent excavations of Bronze and Iron Age sites in southern Uzbekistan, such as Dzharkutan, Tilla Bulak, and sites in the Bandichan microregion, have yielded large collections of animal remains (Fig. 20.1). Their analysis (which is still in progress) suggests

FIGURE 20.2 Excavations at Gonur (sector 19) in 2012. Photo: Nikolaus Boroffka, German Archaeological Institute, Berlin.

a developed stock-breeding, mainly founded on sheep, goats, and cattle, as part of the subsistence economy at these sites.

In the forest-steppe and steppe zones of northern Central Asia the beginning of the Bronze Age is connected with the Sintashta and Petrovka cultures. Faunal remains from sites like Sintashta (region of Chelyabinsk), Petrovka II, and Novonikol'skoe I (northern Kazakhstan) show that the subsistence economy in these areas underwent great changes, i.e. from an economy nearly solely based on horse exploitation in earlier times (Chalcolithic) to one that was mainly founded on herding cattle, sheep, and horses (Benecke and von den Driesch, 2003: Fig. 6.2). This type of animal husbandry persisted through the centuries of the Bronze and Iron Ages, forming the basis of the pastoralist economy in the forest-steppe and steppe zones (Akhinzhanov et al., 1992).

The Late Bronze Age-Iron Age site of Chicha in the Baraba forest-steppe (Russia) (Fig. 20.1) provides a good example of how a pastoralist subsistence economy developed in the steppe region over a period of about 1,500 years (Privat et al., 2006). Thanks to extensive excavations and various analytical approaches (e.g. faunal studies, analysis of stable isotopes, vessel residues, and pollen) a reconstruction of the economy and diet of the inhabitants was possible for the three main phases of occupation at Chicha. The subsistence economy seems to have been dominated to varying degrees by animal husbandry, which was mainly based on cattle and horses. While the faunal and palaeodietary evidence show that hunting and fishing were clearly important activities during the first part of the Late Bronze Age phase of occupation (fourteenth to thirteenth centuries BC), the high percentage of cattle bones in the faunal assemblages of the two Iron Age periods (tenth to ninth centuries BC and first century BC to first century AD) highlights the increasing role of this animal at Chicha in later times. Residues extracted from vessels, mainly obtained from the Early Iron Age occupational phase, also reflect the intensive exploitation of ruminant animal products such as meat, fat, and milk. The use of natural animal resources by hunting (elk), fowling (ducks, geese, galliform birds), and fishing (Crucian carp (*Carassius carassius*), pike (*Esox lucius*), perch (*Perca fluviatilis*)) persisted in the younger periods of occupation but its significance for dietary purposes clearly decreased.

Sequences of faunal assemblages covering the centuries of the Bronze and Iron Age could also be studied at the sites Begash and Mukri in the piedmont zone of the Dzhungar Mountains, southeastern Kazakhstan (Frachetti and Benecke, 2009;

FIGURE 20.3 Geographic setting of excavations at Begash. Photo: Michael Frachetti, Washington University, St. Louis.

Frachetti et al., 2010) (Figs. 20.1 and 20.3). The geography of both sites, located at a height of c.900 m above sea level, indicates that Begash and Mukri were most favourable as winter-time settlements for regionally mobile pastoralists. According to the archaeozoological evidence from Begash, the settlement's population relied on animal keeping rather than hunting. Wild animals are generally represented by low numbers of various taxa of both mammals and birds. From the earliest phase of occupation at Begash (c.2500–1950 BC, phase 1a) sheep/goat and cattle dominate the fauna, while ibex (*Capra sibirica*) and red deer (*Cervus elaphus*) are represented among the wild animals. This assemblage is consistent with a pattern of wintering in midland elevations and summering in higher elevation pastures. The initial presence of goitered gazelle (*Gazella subgutturosa*) in phase 1b (c.1950–1690 BC) suggests that the pastoralists of Begash may have made sporadic hunting forays into more arid, low-elevation territories at the beginning of the second millennium BC. Starting in phase 2 (c.1625–1000 BC), increasing diversity of wild taxa indicates that limited hunting still augmented the established herding strategy and Begash's pastoralists widened their range of exploited territory. For example, the habitat preferences of wild boar and mountain sheep (argali; *Ovis ammon*), which appear in the faunal sequence for the first time, suggest that the pastoralists were ranging from the deltaic river fan of the Koksu River to the highland pastures in the Dzhungar Mountains by the mid-second millennium BC—a considerably wider range than in previous phases. Faunal data clearly show that this pattern of vertical transhumance continued in the settlement phases 3a and 3b of Begash (c.970 BC– AD 30).

Detailed archaeozoological research on the Iron Age economy of steppe pastoralists has been conducted on faunal assemblages from various sites in the Talgar region (Fig. 20.1), located in the northern foothills of the Zailiisky Alatau range of the Tian Shan Mountains, southeastern Kazakhstan (Benecke, 2003; Chang et al., 2003). The region at 550 to 1,100 metres in elevation is comprised of the river Talgar and other smaller streams that form a broad alluvial fan or delta. The bone collections, which date to the period 800 BC– AD 100, indicate that herd composition did not undergo great changes in the Talgar region during the Iron Age period. Sheep and cattle were the most frequent species, being of fundamental importance in providing meat and other products (e.g. milk, wool). Beside these species, horses played an important role in the settlements of the Talgar region. Age and sex data point to a mode of horse-keeping where only a small part of the herd was raised solely for providing meat, while most horses were used at first for other purposes (e.g. riding, traction) before being slaughtered at an advanced age. Noteworthy identifications are those of Bactrian camel and donkey. Both species were primarily exploited for purposes of transport, i.e. as pack animals and for riding. Additionally, camels could also have been used for their milk. According to ethnographic accounts, camel milk was one of the staple foods of the Adai Kazakhs in the summer months (Khazanov, 1994). Wild animals are represented by remains of red deer, roe deer (*Capreolus capreolus*), wild boar, fox, hare, and birds. Considering their low numbers, hunting was only of marginal significance for providing food and raw materials.

The archaeozoological evidence from the main sites (Taldy Bulak 2, Tuzusai 1, Tseganka 8), in conjunction with ethnographic accounts, implies pastoral systems characterized by a predominantly sedentary animal economy in the Talgar region during the Iron Age period. The data obtained are consistent with two forms of pastoralism, which can be described as sedentary animal husbandry or herdsman husbandry according to Khazanov's classification (Khazanov, 1994). In the first case, livestock is kept on pastures adjacent to the settlement (free grazing), whereby the laying-in of fodder and the maintenance of livestock in enclosures or stables is absent or limited. In the case of herdsman husbandry the majority of the population leads a sedentary life and is occupied for the most part with agriculture, while the livestock or, more often, some of it, is maintained all year round on pastures, sometimes quite far from the settlement, and tended by herdsmen especially assigned to this task. For part of the year the livestock is usually kept in enclosures, pens, or stalls, which entail the laying-in of fodder. Today in the Talgar area, the Kazakh herdsmen practice a kind of 'yaylag' pastoralism. They often graze their cattle, sheep, and horses in the upland summer pastures of the Tian Shan foothills, while some family members remain in the lowlands to cultivate fields and gardens. At the end of summer, livestock is driven to lower zones.

FIGURE 20.4 Arzhan 2, grave 16 (southern part). View of the unearthed horse skeletons. Photo: Michael Hochmuth, German Archaeological Institute, Berlin.

In the Iron Age, the Scythians, a nomadic horse-riding people who populated the Eurasian steppes from China in the east to the Hungarian Plain in the west, left their traces in the steppe zone of Central Asia. The most outstanding monuments of their presence are the huge burial mounds, the so-called kurgans. According to written sources (e.g. Herodotus, Strabo), artistic representations, and archaeological finds, the domestic horse played a pivotal role in Scythian life. Horses provided food and raw materials, were used as a mode of transportation for civilian daily life and at the time were employed as a 'weapon' in the realm of military conflict (Rolle, 1980). The unique status horses were bestowed with in Scythian life is embodied by a system of beliefs and burial rituals which included the sacrifice and offering of animals. The inclusion of horses during interment is seen as an essential part amongst the grave associations, in particular for the graves of the Scythian elite. The spectacular archaeological finds consisting of Altai-Scythian horses, preserved by freezing conditions, from the Pazyryk kurgans in the Altai Mountains (fifth to fourth centuries BC) (Fig. 20.1) are well known (Rudenko, 1970). Archaeozoological studies on horse skeletons from the recently excavated Scythian royal grave mound Arzhan 2 (Tuva, Russia) (Figs 20.1 and 20.4) dating to the late seventh century BC could prove that there was a selection of horses designated for burial in terms of age, sex, and body size (Benecke, 2007). The animals buried here were all male horses being in their prime (8–18 years of age), i.e. of optimal age as mounts, and they attained a relatively large size with withers heights ranging between 135 and 145 cm. The results of the molecular genetic analysis further suggest that the animals were chosen from various herds (Benecke et al., 2010).

From an archaeological standpoint it is thought that the horse burial at the royal grave mound of Arzhan 2 had a ritually motivated purpose, possibly as sacrifice to pay homage to the deceased and as a gesture in remembrance of the dead. Customs involving horses such as these have persisted in some cultures, for example the Ossets and Kazakhs, into the Early Modern Age (Čugunov et al., 2003). This includes a ritual horse race staged to honour a highly esteemed man on the first anniversary of his death.

REFERENCES

Akhinzhanov, S. M., Makarova, L. A., and Nurumov, T. N. (1992) *K Istorii Skotovodstva i Okhoty v Kazakhstane*, Alma-Ata: Gylym.

Benecke, N. (2003) 'Iron Age economy of the inner Asian steppe: a bioarchaeological perspective from the Talgar Region in the Ily River Valley (southeastern Kazakhstan)', *Eurasia Antiqua*, 9, 63–84.

Benecke, N. (2007) 'The horse skeletons from the Scythian royal grave mound at Aržan 2 (Tuva, W. Siberia)', in Grupe, G. and Peters, J. (eds) *Skeletal Series and Their Socio-Economic Context*. Documenta Archaeobiologiae 5, pp. 115–31. Rahden: Verlag Marie Leidorf.

Benecke, N. and von den Driesch, A. (2003) 'Horse exploitation in the Kazakh Steppes during the Eneolithic and Bronze Age', in Levine, M. A., Renfrew, A. C., and Boyle, K. (eds) *Prehistoric Steppe Adaptation and the Horse*, pp. 69–82. Cambridge: McDonald Institute for Archaeological Research.

Benecke, N., Pruvost, M., and Weber, C. (2010) 'Die Pferdeskelette—Archäozoologie und Molekulargenetik', in Čugunov, K. V., Parzinger, H., and Nagler, A. (eds) *Der skythenzeitliche Fürstenkurgan Aržan 2 in Tuva*. Archäologie in Eurasien 26, pp. 249–56. Mainz: Philipp von Zabern.

Bulliet, R. W. (1975) *The Camel and the Wheel*, Cambridge: Harvard University Press.

Chang, C., Benecke, N., Grigoriev, F. P., Rosen, A. M., and Tourtellotte, P. A. (2003) 'Iron Age society and chronology in south-east Kazakhstan', *Antiquity*, 77, 298–312.

Compagnoni, B. and Tosi, M. (1978) 'The camel: its distribution and state of domestication during the third millennium BC in light of finds from Shahr-i Soktha', in Meadow, R. and Zeder, M. A. (eds) *Approaches to Faunal Analysis in the Middle East*. Peabody Museum Bulletin 2, pp. 91–103. Cambridge (MA): Peabody Museum of Archaeology and Ethnology.

Čugunov, K. V., Parzinger, H., and Nagler, A. (2003) 'Der skythische Fürstengrabhügel Aržan 2 in Tuva', *Eurasia Antiqua*, 9, 113–62.

Dobney, K. and Jaques, D. (2010) 'The vertebrate assemblage from excavations at Jeitun, 1993 and 1994', in Harris, D. R. (ed.) *Origins of Agriculture in Western Central Asia: An Environmental-Archaeological Study*, pp. 174–9. Philadelphia: University of Pennsylvania Press.

Ermolova, N. M. (1983) 'Skotovodstvo i okhota v tsentralnoj Azii v epokhu Eneolita i Bronzi', in *Arkheologiya Srednei Azii I Blizhnevo Vostoka: II Sevetsko-Amerikanskii symposium*, pp. 44–8. Tashkent: Fan.

Frachetti, M. and Benecke, N. (2009) 'From sheep to (some) horses: 4500 years of herd structure at the pastoralist settlement of Begash (south-eastern Kazakhstan)', *Antiquity*, 83, 1023–37.

Frachetti, M. D., Benecke, N., Mar'yashev, A. N., and Doumani, P. N. (2010) 'Eurasian pastoralists and their shifting regional interactions at the steppe margin: settlement history at Mukri, Kazakhstan', *World Archaeology*, 42(4), 622–46.

Harris, D. R. (ed.) (2010) *Origins of Agriculture in Western Central Asia: An Environmental-Archaeological Study*, Philadelphia: University of Pennsylvania Press.

Hiebert, F. T. (1994) *Origins of the Bronze Age Civilization of Central Asia*. American School of Prehistoric Research Bulletin 42. Philadelphia: The University Museum.

Kalieva, S. S. and Logvin, V. N. (1997) *Skotovody Turgaja v Tret'em Tysjachletii do Nashej Ery*, Almaty: Institut Arkheologii.

Kasparov, A. (1994) 'Environmental conditions and farming strategy of protohistoric inhabitants of south Central Asia', *Paléorient*, 20(2), 143–9.

Khazanov, A. M. (1994) *Nomads and the Outside World*, 2nd edn, Madison: University of Wisconsin Press.

Lasota-Moskalewska, A., Piątkowska-Małecka, J., Gręzak, A., and Szymczak, K. (2006) 'Animal bone remains from Ayakagytma 'The Site'', in Szmyczak, K. and Khudzhanazarov, M. (eds) *Exploring the Neolithic of the Kyzyl-Kums: Ayakagytma 'The Site' and Other Collections*, pp. 206–17. Warsaw: Institute of Archaeology, Warsaw University.

Levine, M. A. (2004) 'Exploring the criteria for early horse domestication', in Jones, M. (ed.) *Traces of Ancestry: Studies in Honour of Colin Renfrew*, pp. 115–26. Cambridge: McDonald Institute for Archaeological Research.

Ludwig, A., Pruvost, M., Reismann, M., Benecke, N., Brockmann, G. A., Castaños, P., Cieslak, M., Lippold, S., Llorente, L., Malaspinas, A.-S., Slatkin, M., and Hofreiter, M. (2009) 'Coat color variation at the beginning of horse domestication', *Science*, 324, 485.

Masson, V. M. and Sarianidi, V. I. (1972) *Central Asia: Turkmenia before the Achaemenids*, London: Thames and Hudson.

Matjushin, G. (2003) 'Problems of inhabiting central Eurasia: Mesolithic-Eneolithic exploitation of the central Eurasian steppes', in Levine, M. A., Renfrew, A. C., and Boyle, K. (eds) *Prehistoric Steppe Adaptation and the Horse*, pp. 367–93. Cambridge: McDonald Institute for Archaeological Research.

Moore, K. M., Ermolova, N. M., and Forsten, A. (2003) 'Animal herding, hunting, and the history of animal domestication at Anau depe', in Hiebert, F. T. (ed.) *A Central Asian Village at the Dawn of Civilization: Excavations at Anau, Turkmenistan*. University Museum Monograph 116, pp. 154–9. Philadelphia: University of Pennsylvania Museum of Archaeology and Anthropology.

Moore, K. M., Miller, N., Hiebert, F., and Meadow, R. (1994) 'Agriculture and herding in the early oasis settlements of the Oxus civilization', *Antiquity*, 68, 418–27.

O'Connell, T. C., Levine, M. A., and Hedges, R. E. M. (2003) 'The importance of fish in the diet of central Eurasian peoples from the Mesolithic to the Early Iron Age', in Levine, M. A., Renfrew, A. C., and Boyle, K. (eds) *Prehistoric Steppe Adaptation and the Horse*, pp. 253–68. Cambridge: McDonald Institute for Archaeological Research.

Olsen, S. L. (2003) 'The exploitation of horses at Botai, Kazakhstan', in Levine, M. A., Renfrew, A. C., and Boyle, K. (eds) *Prehistoric Steppe Adaptation and the Horse*, pp. 83–104. Cambridge: McDonald Institute for Archaeological Research.

Olsen, S. L. (2006) 'Early horse domestication: weighing the evidence', in Olsen, S. L., Grant, S., Choyke, A. M., and Bartosiewicz, L. (eds) *The Evolution of Human-Equine Relationships*. BAR International Series 1560, pp. 81–113. Oxford: Archaeopress.

Outram, A., Stear, N., Bendrey, R., Olsen, S., Kasparov, A., Zaibert, V., Thorpe, N., and Evershed, R. (2009) 'Earliest horse harnessing and milking in the Eneolithic of prehistoric Eurasia', *Science*, 323, 1332–5.

Peters, J. and von den Driesch, A. (1997) 'The two-humped camel (*Camelus bactrianus*): new light on its distribution, management and medical treatment in the past', *Journal of Zoology*, 242, 651–79.

Pollock, S. and Bernbeck, R. (2011) 'Excavations at Monjukli Depe, Meana-Čaača Region, Turkmenistan, 2010', *Archäologische Mitteilungen aus Iran und Turan* 43, 169–237.

Privat, K., Schneeweiß, J., Benecke, N., Vasil'ev, S. K., O'Connell, T., Hedges, R., and Craig, O. (2006) 'Economy and diet at the Late Bronze Age-Iron Age site of Čiča: artefactual, archaeozoological and biochemical analyses', *Eurasia Antiqua*, 11, 419–48.

Rolle, R. (1980) *Die Welt der Skythen*, Luzern: C. J. Bucher.

Rudenko, S. I. (1970) *Frozen Tombs of Siberia*, London: Dent & Sons.

Trinks, J., Burger, P., Benecke, N., and Burger, J. (2012) 'Ancient DNA reveals domestication process: the case of the two-humped camel', in Knoll, E.-M. and Burger, P. (eds) *Camels in Asia and North Africa: Interdisciplinary Perspectives on their Past and Present*

Significance. Österreichische Akademie der Wissenschaften, Philosophisch-Historische Klasse, Veröffentlichungen zur Sozialanthropologie 18, pp. 79–86. Wien: Verlag der Österreichischen Akademie der Wissenschaften.

Wechler, K.-P. (2001) *Studien zum Neolithikum der Osteuropäischen Steppe*. Archäologie in Eurasien 12. Mainz: Philipp von Zabern.

Zaibert, V. F. (1992) *Atbasarskaja Kul'tura*, Ekaterinburg: Rossijskaya Akademiya Nauk.

CHAPTER 21

INTRODUCTION OF DOMESTIC ANIMALS TO THE JAPANESE ARCHIPELAGO

HITOMI HONGO

INTRODUCTION

IT is generally considered that all domestic animals were brought to Japan from the Asian continent at different periods throughout history (see Fig. 21.1 for location of sites discussed in this chapter). Domestic dogs (*Canis familiaris*) arrived first, during the Jomon period (*c.*14,000–500 cal BC). Later, during the Yayoi period (*c.*500 cal BC–AD 300), pigs (*Sus domesticus*), and chickens (*Gallus gallus*) were introduced, together with rice cultivation and metallurgy, from the Chinese mainland. This introduction is thought to have occurred via the Korean peninsula. However, the timing of the appearance of domestic pigs in Japan, and the possibility of local domestication of wild boar (*Sus scrofa*) are debated. Horses (*Equus caballus*) and cattle (*Bos taurus*) arrived later, in the Kofun period (mid-third–end of sixth century AD), mostly beginning in the fifth century AD. Animal husbandry for meat and dairy products was not widespread until the twentieth century AD. However, due to poor preservation of animal bones at archaeological sites between the seventh and nineteenth century AD, estimation of dietary and economic importance of these animals based on zooarchaeological evidence is difficult. Other species of domestic animals such as sheep as well as more exotic animals, like camels, have occasionally been brought in small numbers, as gifts or attractions: they are depicted in historical texts or works of art but archaeological findings of their bones are only sporadic.

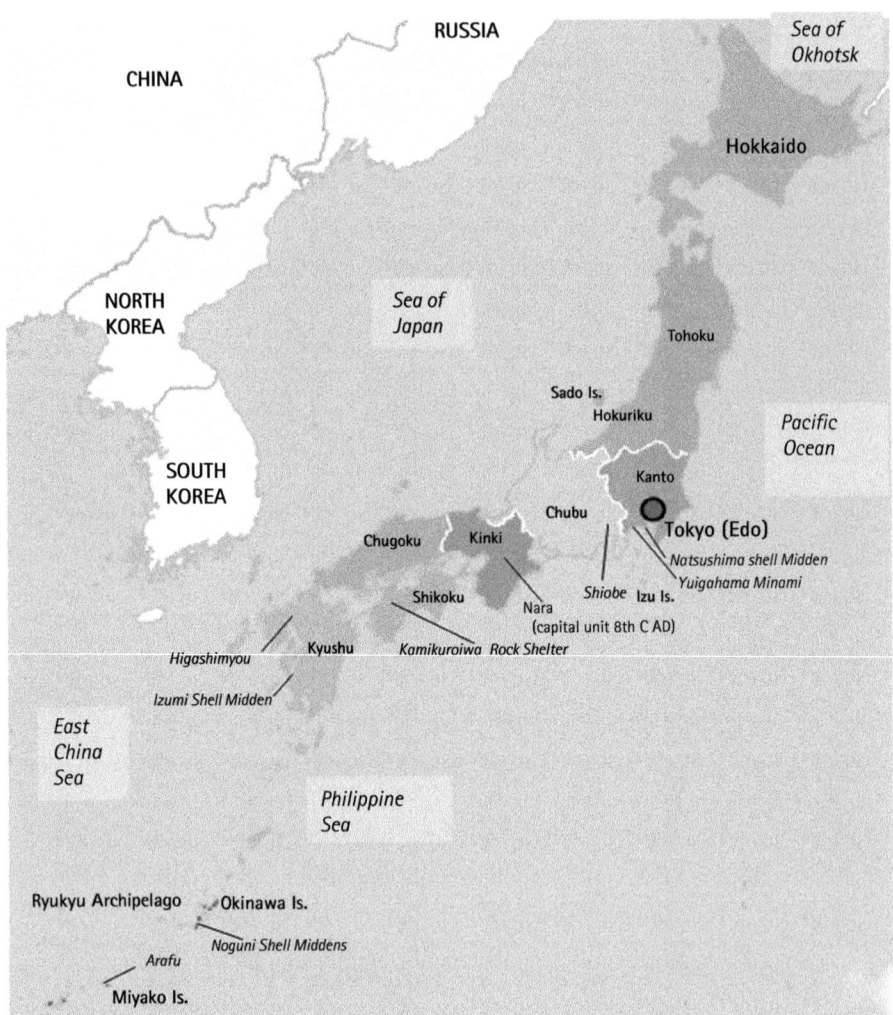

FIGURE 21.1 Map of Japan with regional divisions. Archaeological sites mentioned in the text are shown in italics. Author's own image.

Dog

Dogs were the only domestic species kept in Japan during the Jomon period. Two sub-species of grey wolf inhabited the Japanese Archipelago: the Hokkaido wolf (*Canis lupus hattai*) and the Japanese wolf (*C. l. hodophilax*), but they are not considered to be the wild progenitors of domestic Japanese dogs. The Hokkaido wolf, distributed on Hokkaido Island, northern Japan, was similar in size to the continental grey wolf. The Japanese wolf was smaller than the Hokkaido wolf and inhabited the three main islands of Honshu, Shikoku and Kyushu. Both the Japanese and the Hokkaido wolf were extinct by the beginning of the twentieth century (Walker, 2005).

Six Japanese indigenous dog breeds—Hokkaido, Akita, Kai, Shiba, Kishu, and Shikoku—are recognized today. These modern breeds are, however, indistinguishable by mtDNA analysis (Okumura et al., 1996).

Osteological studies on the limited number of specimens in museums and private collections indicate that the maximum skull length of the Japanese wolf ranges from 206 to 226 mm (Hasebe, 1924; Naora, 1965; Imaizumi, 1970a; 1970b; Miyamoto, 1991). The size of the Japanese wolf overlaps with that of the Akita (maximum cranial length *c*.220 mm) (Shigehara, 1986), the largest modern Japanese native dog breed, but this breed is said to have been cross-bred with western dogs in the nineteenth century, to achieve its large body size.

Analyses of mtDNA obtained from museum specimens of Hokkaido and Japanese wolf, including archaeological and recent specimens, revealed that haplotypes among Japanese wolf samples are grouped in a single lineage distinct from both the continental grey wolf and Japanese domestic dogs (Ishiguro et al., 2009; 2016). Hokkaido wolf specimens are genetically related to the Asian and Canadian grey wolves (Ishiguro et al., 2010; 2016). The results of these molecular phylogenetic analyses rule out the possibility of a local domestication of the Japanese wolf. Domestication of the Hokkaido wolf is also unlikely, as the earliest evidences of dogs have been found from sites outside Hokkaido (see below).

The earliest evidence for domestic dog in Japan is represented by a mandible fragment found during excavations at Natsushima Shell Mound, located on a small island in the western part of Tokyo Bay. This specimen is considered to belong to the Initial Jomon period (*c*.7300 cal BC), based on the dating of charcoal found in a contemporaneous layer at the site. The earliest examples of intentional burial of dog are represented by two individuals found at Kamikuroiwa Rock Shelter in Shikoku Island (Komiya et al., 2015) (Fig. 21.2). C14 dating of these dog bones attributes them to the end of the Initial or the beginning of the Early Jomon period (5400–5300 cal BC) (Gakuhari et al., 2015). The Kamikuroiwa dogs are the oldest directly dated dog bones in Japan. Dogs from the end of the Initial Jomon period (*c*.6000–5700 cal BC) were also reported from the shell midden at Higashimyou Site, northwestern Kyushu. MtDNA analysis revealed that the dogs from Higashimiyou Site have the same haplotypes found among modern Japanese dogs (Masuda and Sato, 2015).

Morphologically, dogs from Jomon sites are similar, or slightly smaller, in size than Shibas, the smallest modern Japanese breed. The cranial length of Jomon dogs is *c*.160 mm for males and *c*.150 mm for females (Shigehara and Hongo, 2000), with estimated shoulder height of *c*.38 cm. The postcranial bones are robust, and the skull has a relatively straight profile, intermediate in character between wolves and dogs. The relatively narrow zygomatic width, together with its relatively wide snout means that the skulls of Jomon dogs were narrow and elongated compared with modern Japanese breeds. Throughout the Jomon period, dogs were morphologically uniform, implying that breeding selection by humans did not normally occur. However, those from sites in northern Japan tended to be larger. Some genetic variation is found in Jomon dogs, including some haplotypes that no longer exist among modern Japanese dogs, but the

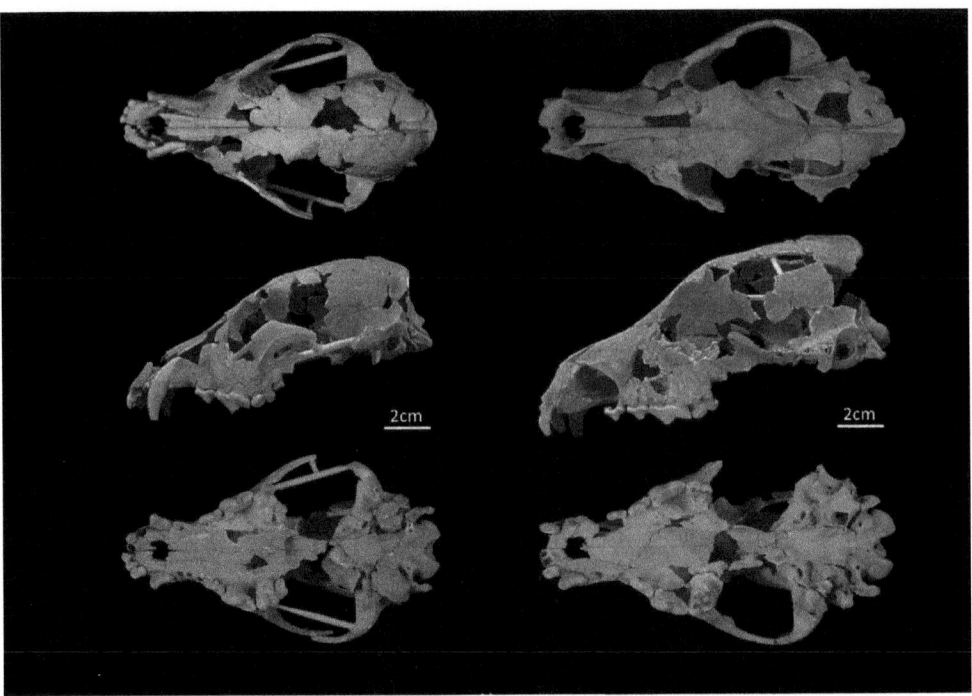

FIGURE 21.2 Skulls of two dogs excavated from Kamikuroiwa Rock Shelter (courtesy of Prof. Takao Sato, Keio University).

haplotypes of Jomon dogs all cluster into a single group, suggesting a relatively homogeneous but widely distributed population (Okumura et al., 1999).

Evidence for periodontal disease is rare in Jomon dogs. However, their teeth often display heavy attrition or breakage, and tooth loss in some cases. This has been used to suggest that dogs were primarily kept as hunting companions, as tooth stress could have occurred in the course of the hunting process (Shigehara and Hongo, 2000). Examples of dog burials increased in the Middle Jomon period (c.3500–2500 BC), indicating a close relationship between dogs and the Jomon people. There are also finds of scattered dog bones, suggesting that the treatment of dogs was not uniform, though rare occurrence of cut marks suggests that dogs were not commonly utilized as food sources during the Jomon period (Shigehara and Hongo, 2000). There was an influx of rice-farming populations from the Asian continent in the following Yayoi period (c.500 BC– AD 300), but morphological and genetic evidence (Okumura et al., 1999; Ishiguro et al., 2000) do not indicate that a large number of dogs accompanied the migrating population. The cranial length of dogs increased, but their overall morphology did not change significantly (Shigehara, 1994). The treatment of dogs changed during the Yayoi period. Incidences of intentional burial and broken teeth decrease, suggesting that dogs kept as hunting companions became less important in Yayoi society, which was based primarily on rice agriculture. Increased occurrence of cut-marks indicate that dogs were now more frequently eaten, probably a custom brought in from the Asian continent by the

Yayoi population (Shigehara and Hongo, 2000). Dogs continued to be used as sources of meat until early modern times. Although there is sporadic evidence of the importation of dogs after contact with European traders in the fifteenth century AD, intentional selection and breeding was not practised until the Edo period (seventeenth to nineteenth centuries AD).

Pig

The earliest historical mention of pigs is found in an eighth-century AD text, the *Harimanokuni Fudoki*, which states that pigs were brought on a boat from Kyushu to Harima (present Hyogo Prefecture in western Japan) and released into an open field. It is not clear from this document whether these pigs were domestic stock or captured wild boar, although the former is more likely. It is also known through historical sources that there was a vocational group specializing in pig management in Japan during this period, and that these pigs were probably kept free-range. The archaeological evidence, however, suggests that domestic pigs were introduced much earlier, from mainland China, together with rice farming, during the Yayoi period.

There are two sub-species of wild boar in Japan today; the Japanese wild boar (*Sus scrofa leucomystax*) and the Ryukyu wild boar (*S. s. riukiuanus*). They had been important sources of meat during the Jomon period. Today, the Japanese wild boar inhabits the Honshu, Shikoku, and Kyushu Islands, but not Hokkaido. The Ryukyu wild boar is found on several islands of the Ryukyu Archipelago. The Ryukyu wild boar, with a body length of *c.*90–110 cm and mandibular third molar length of *c.*26–28 mm, is much smaller than the Japanese wild boar, which is itself smaller than the wild boar on the Asian continent. In the past, some researchers have argued that the Ryukyu wild boar descends from feral domestic pigs introduced to the islands in prehistory (e.g. Naora, 1937; Hayashida, 1960). However, on the basis of cranial morphology, others consider them descendants of the Pleistocene wild boar that migrated from the Asian continent (e.g. Imaizumi, 1973) and became smaller according to Foster's rule (Foster, 1964). Fossil remains of *Sus* from Pleistocene layers in Okinawa Island, southern Japan, support this theory (Oshiro and Nohara, 2000).

Possible cases of local domestication or management of wild boar during the Jomon period have been discussed since the early 1900s (e.g. Naora, 1937; Kato, 1980; Ono, 1984; Nishimoto, 2003). Three types of evidence are considered (Nishimoto, 2003). First, *Sus* remains have been found from Jomon sites on islands that lie outside of the natural distribution of wild boar today, such as Izu, Sado, and Hokkaido. Secondly, figurines and pottery motifs depicting wild boar increase after the Middle Jomon period. Thirdly, burials of *Sus* associated with humans have been found (Sugaya and Toizumi, 1998; Sonan Bunkazai Center, 2003) (Fig. 21.3).

In the Izu islands southwest of Tokyo, there have been discoveries of *Sus* remains as early as the Initial Jomon Period (Kaneko, 1987; Yamazaki et al., 2005), and the specimens from Izu Island were found to be smaller than those from other sites. Although

FIGURE 21.3 Pig skeleton associated with human burial No. 105 (Middle Jomon period) at Simo-ota Shell Midden, Kanagawa Prefecture (Sonan Bunkazai Center, 2003: Fig. 56; reproduced with permission).

it is considered that wild boar never occurred north of the Blakiston's Line, a zooge-ographical boundary lying between Honshu and Hokkaido (Inukai, 1960; Nishimoto, 1985; Kawamura, 1991), *Sus* remains have been found from Late-Final Jomon period sites in Hokkaido.

Analysis of mtDNA from Jomon *Sus* remains suggests that Jomon people brought the *Sus* from Honshu to southern Hokkaido and the Izu Islands (Watanobe et al., 2004: 227). The archaeological *Sus* samples from both islands had haplotypes closely related to those of the Japanese wild boar in Honshu. Interestingly, archaeological *Sus* remains from Sado Island were found to be genetically unique, indicating that they may have belonged to an endemic population that became extinct after the Jomon period (Watanobe et al., 2004: 226).

Analysis of the mtDNA of archaeological *Sus* samples also suggests that domestic pigs may have been introduced from mainland Asia to western Japan as early as the Late-Final Jomon period (Morii et al., 2002). Hunting for wild boar continued at many sites in the following Yayoi period, while introduced domestic pigs were readily accepted at others.

Morphological examination of *Sus* remains found from Jomon and Yayoi period sites, however, found that the Jomon *Sus* were morphologically wild (Hongo et al., 2007; Anezaki et al., 2008). The Japanese Archipelago stretches over 3,000 km from north to south; the modern habitat of Japanese wild boar ranges approximately from latitudes 31 to 41N leading to considerable regional size cline from southwest to northeast following Bergman's rule (Bergmann, 1847; but see Endo et al., 2000) (Fig. 21.4a). Archaeological specimens from Initial (Earliest) and Early Jomon sites are generally larger than modern wild boar, but the southwest to northeast regional size cline is clearly observed in the size of teeth (Anezaki et al., 2008) (Fig. 21.4b) as well as in postcranial measurements (Hongo et al., 2007), indicating that the animals were probably wild. Overall, the size of *Sus* did not change during the Jomon period, but exhibited a sudden reduction in size in the Yayoi period, suggesting that domestic pigs were introduced, rather than being domesticated local stock (Fig. 21.4c). Analyses of kill-off patterns also do not indicate temporal changes during the Jomon period (Hongo et al., 2007). After the Yayoi period, pig size increased again, possibly due to interbreeding between introduced domestic pigs and local wild boar, as a result of the practice of free-ranging management (see above). Closer examination of *Sus* remains from the Yayoi period sites reveals a greater

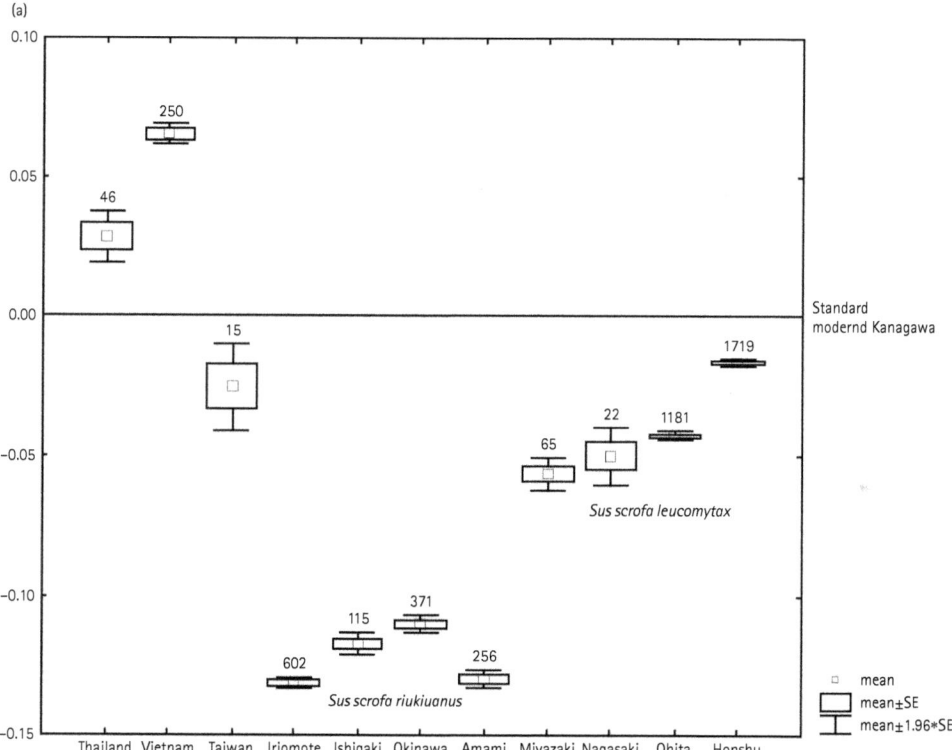

FIGURE 21.4 Geographical variation of the buccolingual measurements of lower P3, P4, M1, and M2 of *Sus*. Comparisons are made using the Log Size Index. The standard is the measurements of 26 modern wild pig specimens from Kanagawa Prefecture in southern Kanto. The numbers on each box plot indicate the number of measurements taken for the population (n). The geographical groups from southwestern Japan are plotted toward the left side of the chart, and those from northeastern Japan toward the right. a. Geographical variation observed in the size of modern *Sus* populations; b. Size variation of *Sus* in the Initial and Early Jomon periods; c. Temporal changes in the size of *Sus* from the Initial Jomon to the Kofun-Heian periods; d. Size variation of *Sus* in the Yayoi period. Names of the relevant sites are indicated on each box plot. Based on images by T. Anezaki, originally published in Anezaki et al. (2008); reproduced with permission.

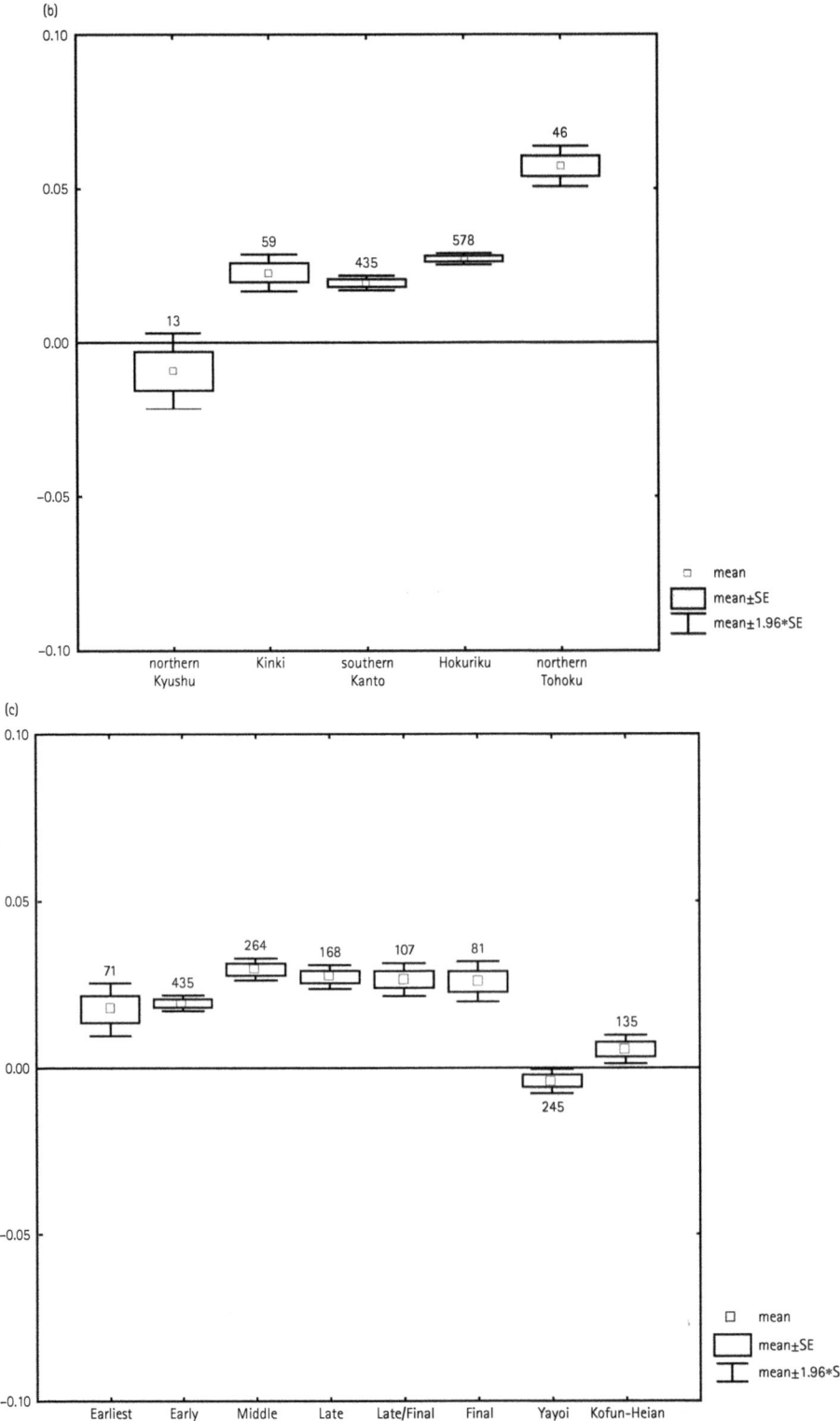

FIGURE 21.4 Continued

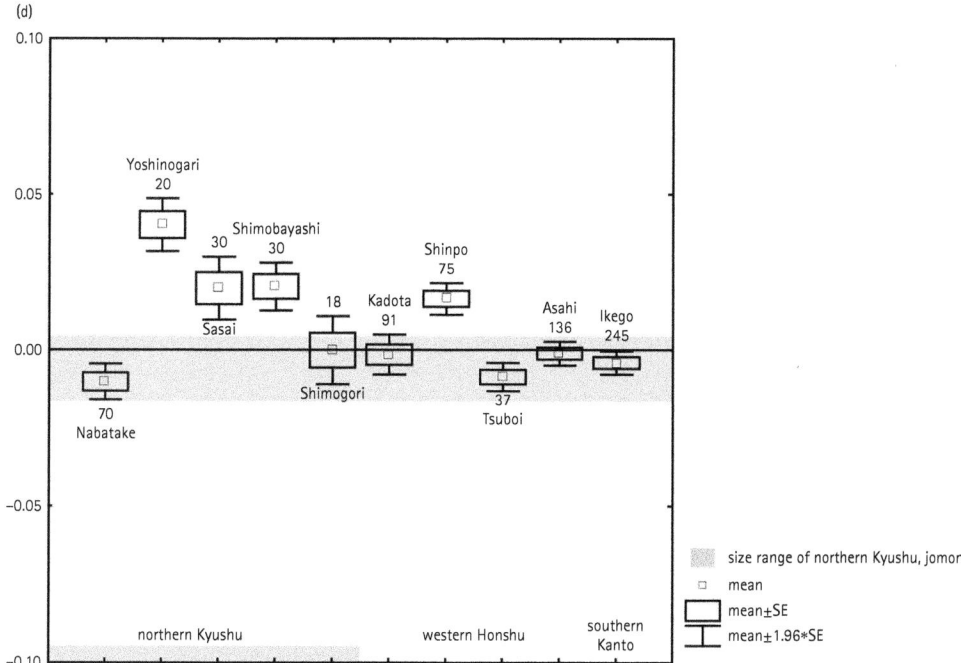

FIGURE 21.4 Continued

size variation than the Jomon *Sus* in both northern Kyushu and western Honshu sites. Notable increase of the size of *Sus* at some sites in Kyushu in the Yayoi period suggests that wild boar that were larger than native Kyushu population were also brought in, possibly from northeastern Japan (Anezaki et al., 2008) (Fig. 21.4d).

The analyses of the size and kill-off patterns of *Sus* from Jomon sites do not point toward local domestication of wild boar in the main islands of Japan, at least as concerns the emergence of a reproductively distinct population. Changes in the relationship between *Sus* and humans in the Middle Jomon and later can be inferred by the appearance of boar motifs and the burial of wild boars with humans (Fig. 21.3). The capturing and feeding of individual wild boar piglets is a common practice in rural areas today, and it is possible that such practices have a long tradition. This sporadic practice of keeping individual wild boar for only one generation, however, should be regarded as delayed consumption rather than management; it is unlikely that this practice led to full domestication during the Jomon period when wild boar were easily available.

A detailed examination of the introduction of *Sus* to the Izu Islands is a subject for future research. Both the teeth and postcranial bones of *Sus* from Jomon sites on the Izu Islands are much smaller than those from Honshu Island, implying that they lived on the islands over generations. They were not necessarily under management, because there are abundant ethnographic and historical examples of wild boar being brought and released on Japanese islands lacking in animal resources (Hongo et al., 2007).

The introduction of domestic pigs into the Ryukyu Islands might involve a different route and process from that of Kyushu and Honshu. The oldest historical accounts date to the fourteenth century AD and mention the introduction of pigs from China. Molecular phylogeny has determined that the Ryukyu wild boar did not genetically contribute to the local domestic pig population on the Ryukyu Islands (Watanobe et al., 1999). MtDNA haplotypes related to the East Asian domestic *Sus* lineage were found in archaeological specimens dating to *c.* AD 1 to 300 cal (Watanobe et al., 2002), suggesting that the introduction of pigs dates much earlier than the historical account states. At the shell middens of Noguni (*c.*6000–3000 cal BC), located on Okinawa Main Island, the *Sus* mandibular third molars are smaller than those of the modern Ryukyu wild boar or *Sus* remains from other archaeological sites (Takahashi et al., 2011). Ancient DNA analysis of *Sus* remains from Noguni B Shell Midden found haplotypes distinct from the modern Ryukyu wild boar, and instead related to East Asian domestic pigs (Takahashi et al., 2011). High variability in the nitrogen isotope ratios of *Sus* remains from the Noguni Shell Middens, and other prehistoric sites on Ie Island, near Okinawa Main Island, where the Ryukyu wild boar is not present today, suggest that some of these *Sus* had been imported from elsewhere, or were being fed meal scraps (Matsui et al., 2005; Minagawa et al., 2005). Ancient DNA analysis also ascertained that haplotypes distinct from those of the modern Ryukyu wild boar exist among the *Sus* remains from Arafu Site (*c.*2800–2200 BP), in the Miyako Islands of the southern Ryukyu Islands (Takahashi, 2012). Since the southern Ryukyu Islands had stronger connections with island Southeast Asia in prehistory than with the northern Ryukyu Islands, the possibility of a southern route of domestic pig introduction should be explored. The evidence at hand suggests that the Ryukyu wild boar were transported between islands in prehistory, that domestic pigs might have been introduced much earlier than previously thought, and that multiple routes of introduction should be considered.

Pig bones are commonly found from Okhotsk Culture archaeological sites of the fifth to ninth centuries AD in northern Hokkaido. MtDNA analysis of these remains revealed that they were related to the domestic pigs of the Amur River Basin (Watanobe et al., 2001), suggesting introduction of domestic pigs from the northeastern part of the Asian continent through Sakhalin. These pigs were, however, limited in distribution to Okhotsk Culture sites and were not introduced farther south to Honshu.

Horse

A historical account written in the Three Kingdom period of China in the third century AD states that there were no horses, cattle, or sheep in Japan. Domesticated horses were likely introduced into Japan during the Kofun period, via the Korean Peninsula. Claims of horse remains from Jomon period sites (e.g. Izumi Shell Midden in Kagoshima; Hayashida and Yamaguchi, 1956) were all found to be intrusions from later periods based on direct C14 dating and fluorine content analysis (Kondo et al., 1991; 1992). Amongst the earliest evidence for horses in Japan is the tooth remains found at the late fourth-century AD Shiobe Site in Yamanashi Prefecture, central Japan (Nishimoto, 1996; Muraishi, 1998). While it is possible that horses arrived earlier in northern Kyushu, which lies along the route by which they were transported from mainland Asia (Nishimoto and Niimi, 2010: 144), the majority of the archaeological evidence comes from contexts of the fifth century AD and later. An eighth-century AD historical source states that two horses were received from the King of Paekche in the fifth century AD, which corresponds in timing with a rise in the quantity of archaeological horse remains. Offerings of decorative horse tacks in tumuli and clay horse figurines, as well as findings of horse bones, increase in archaeological sites from the second half of the fifth century AD. It is likely that horses were accompanied by trainers and caretakers from the Korean Peninsula at the time of their introduction, and were initially prestigious animals used in transportation and to create an elite cavalry. Horses were sacrificed at their owners' funerals and their skeletons are often found buried around tumuli. A mid-seventh-century AD legal code banning the sacrifice of horses and humans at funerals suggests that the practice was not uncommon. The historical evidence implies that horses were exclusively controlled by the ruling class; a text written in the late seventh century AD refers to state farms in central and eastern Japan where horses and cattle were bred. Horses were sent from the state breeding farms to the capital city of Nara and were systematically allocated to messengers' posts along major road networks connecting the central government and local administration offices (according to *Yoro Ritsuryo*, AD 757, whose content succeeded the earlier legal codes from the early eighth century AD). Horses were also used as labour in the construction of the capital city in the late seventh and early eighth century AD. In the eleventh to twelfth century AD, a warrior class, who managed horse production for military purposes, emerged. Horses also began to be used for farming and as beasts of burden.

Measurements of excavated horse skeletons have determined that the horses from the fifth century AD sites are all small (Nishinakagawa and Matsumoto, 1991; Nishimoto and Niimi, 2010: 160). Medium-size horses with withers heights of over 130 cm appear by the eighth century AD (Nishimoto, 2008). A large number of horse bones were found at Yuigahama South site (fourteenth to fifteenth century AD) in Kamakura, the seat of the Shogunate in this period. While most of the horse remains were found scattered in the refuse area, there was also a burial of a relatively large horse, with a withers height of 140 cm. In addition to its large size, the individual had relatively longer legs compared with modern native Japanese breeds (Uzawa and Hongo, 2006). It is suggested that this individual is an example of a small number of fine horses that were kept by the elite class and received special treatment.

There are eight modern native breeds of horses in Japan today. Those in the islands in southwestern Japan—the Tokara, Miyako, and Yonaguni breeds—as well as Noma, Taishu, and Misaki breeds, all found in western Japan, are smaller, with a withers height of 108–120 cm, while the Kiso horses in central Honshu and Hokkaido breeds are larger, with withers heights of 130–135 cm. It is considered that horses in Japan had been rather isolated, aside from occasional imports from China and Korea, until crossing with Arab breeds was enforced for the production of cavalry horses in the late nineteenth century

AD. Based on the size of modern native horses, Hayashida (1956) speculated that small-sized horses are descendants of mountain ponies in southwestern China, introduced through the Ryukyu Islands, while middle-sized horses, similar to the Mongolian horse, were brought in through the Korean Peninsula. Nishinakagawa (1991), while supporting the multiple-origin theory of small- and medium-sized horses, argued that both types arrived in Japan by a single route through the Korean Peninsula, because no Kofun period horse remains have been found in the Ryukyu Islands. Both mtDNA and blood protein polymorphic analyses of modern native breeds indicate that they belong to a single group (Nozawa et al., 1998; Kawashima, 2004). There is considerable variability between some breeds; this could be the result of bottleneck and crossbreeding with western horses in the early twentieth century AD. Nozawa et al. (1975; 1998) and Nozawa (1992) proposed a single origin of Japanese native breeds from the Mongolian horses. They postulate that horses arrived in Japan through the Korean Peninsula and that the diminutive horses found in the southern islands were brought in from Honshu. They could not find a correlation between body-size and genetic distance (Nozawa et al., 1998: 67), and suggested that the reduced size is the result of geographic isolation and bottleneck effect.

Interestingly, donkeys (*Equus asinus*) were never adopted in Japan. There is a record of a donkey being gifted from Paekche at the end of the sixth century AD, but they were never accepted as a domestic animal, and no donkey remains have been unearthed from any archaeological sites. It is possible that the small strong horses with their relatively short legs filled the role of donkeys, even in mountainous terrain.

Domestic Cattle

It is thought that cattle were introduced into Japan with horses, and were likewise under control of state administration. Findings of cattle skeletons date back to the fifth century AD, increasing in number in the sixth century AD, and are more common on sites in western Japan. In a similar manner as horses, clay figurines of cattle were placed around tumuli in the Kofun period, suggesting that they were a symbol of prestige. It is apparent from the excavated remains that cattle were small, with estimated withers height of 100–130 cm (Nishinakagawa and Matsumoto, 1991; Nishimoto and Niimi, 2010: 162). Two modern local breeds of cattle are considered to have preserved the appearance of traditional cattle before their hybridization with western breeds in the last 150 years; the Mishima, kept on an island in western Honshu, and the feral cattle of Kuchinoshima Island in southern Kyushu. They are small, weighing about 250–320 kg, and are genetically close to the cattle of the Korean Peninsula and northern China (Sasazaki et al., 2006; Tsuda et al., 2013). Cattle were mainly used as draft animals and for the transportation of goods. Milk is mentioned in eighth-century AD texts, but its production was limited, and only used by the royal family for medicinal purposes. Decorated carts pulled by cattle became a common means of transportation in the capital city of Kyoto among the aristocratic class in the Heian period (AD 794–1185).

Chicken

The domestic chicken was introduced in the Mid-Late Yayoi period (first to third century AD), probably alongside pigs and rice farming (Nishimoto, 1993). These chickens were generally small, and distinctions between domestic chickens and Japanese pheasant (*Phasianus versicolor*) or copper pheasant (*Syrmaticus soemmerringii*) remain difficult (Eda and Inoue, 2011). Chicken bones are sporadic from sites in the subsequent Kofun period, but clay figurines of chickens are found in tumuli from the early fourth century AD, suggesting that they played an important role in funeral rites (Nishimoto and Niimi, 2010: 176). Most figurines depict cockerels, likely related to their symbolic importance as birds whose call at dawn signals the break of night. Cock-fighting was performed originally as a method of divination, but it became popular among the upper class, and later among the ordinary people, as a form of gambling. Consumption of chicken meat and eggs was negatively referred in the seventh-century AD texts (see below). Chickens were selected and bred for their appearance (feather colour and length of tails), as well as for their call. Chicken meat became a common part of the diet in the Edo period (e.g. recipes published in the seventeenth century AD).

ANIMAL HUSBANDRY IN JAPAN

Most of the domestic animal arrived in Japan during the Yayoi and Kofun periods, when preservation of animal bone remains at archaeological sites is poor. Information on animal husbandry comes mainly from historical sources after the seventh century AD and works of art. The supply of meat and dairy products was not the main purpose of animal husbandry until very recently, as Buddhism, the state religion which was first introduced in the seventh century AD, discouraged the consumption of meat, especially of domestic animals. Avoidance of meat continued until the beginning of the Meiji period (AD 1868) when the consumption of beef was encouraged as part of westernization policies.

Historical sources, however, suggest that consumption of the meat, mainly wild pigs and deer, have continued in spite of the general avoidance of meat. An imperial decree was issued in AD 675 that prohibited the consumption of cattle, horse, dog, monkey, and chickens, between April and September. It can be assumed, however, that these animals were eaten in autumn and winter. Wild animals other than monkey are not mentioned in the decree, suggesting that the decrees were meant to provide special treatment for domestic animals, rather than prohibition of meat consumption itself. Similar announcements were repeatedly issued in the eighth century AD, sometimes emphasizing that horses and cattle should not be slaughtered since they work to benefit people, suggesting that eating domestic animal meat continued in spite of the state's prohibition. Exemptions were applied if the animal died a natural death. The legal code in *Yoro Ritsuryo* (AD 757) stated that, if state-owned horses or cattle were to die, their hides, brains, horns, and livers should be delivered to the administration office. If they died on the road, their skin and meat could be sold; money from the sale was to go to the authorities. It is difficult to see to what extent the meat of horse and cattle was actually consumed, based on animal bone remains from this period. Animal brains, especially of horses, were used to tan leather. There are archaeological examples of horse and cattle skulls that have been opened in order to remove the contents (Matsui, 2002). Cattle liver was used as medicine, and bones were used for the production of bone tools such as combs (Matsui, 2002). Domestic pigs rarely appear in historical texts, but production of pork was probably limited because of the Buddhist belief that mammal meat was impure.

After contact with European traders began in the fifteenth century AD, various domestic animals were occasionally brought in, including western dogs, which were valued as exotic pets by the ruling class. Pigs as well as sheep (*Ovis aries*) and goats (*Capra hircus*) were kept at a Dutch trading post in Nagasaki. Although these domestic animals were raised and consumed mostly by the foreigners living there, their meat was sometimes presented as gifts to the local ruling class. Some local rulers in western Japan, those with close contact with the Dutch tradesmen, adopted pig husbandry and ate pork, as well as presenting gifts of meat to the Shogun. Large amounts of pig bones found at the Edo residence of Satsuma, a feudal lord from southern Kyushu, support the account (Yamane, 2013). Dogs continued to be a source of meat.

Zooarchaeological data from the Yayoi and Kofun period sites as well as the historical era have been gradually accumulating, helping the interpretation of textual records as well as supplementing them. Especially important is the fact that the zooarchaeological data often shed lights on the aspects of animal husbandry that tend not to appear in historical sources, such as meat consumption against prohibition by the authority.

References

Anezaki, T., Yamazaki, Y., Hongo, H., and Sugawara, H. (2008) 'Chronospatial variation of dental size of Holocene Japanese wild pigs (*Sus scrofa leucomystax*)', *The Quaternary Research*, 47(1), 29–38.

Bergmann, C. (1847) 'Über die Verhältniße der Wärmeökonomie der Tiere zu ihrer Größe', *Göttinger Studien*, 3(1), 595–798.

Eda, M. and Inoue, T. (2011) 'Identification of Phasianidae (chicken and indigenous fowls in Japan) bones using nonmetric characteristics and its application for reevaluation of chickens from the Yayoi period', *Zoo-archaeology*, 28, 23–33.*

Endo, H., Hayashi, Y., Sasaki, M., Kurosawa, Y., Tanaka, K., and Yamazaki, K. (2000) 'Geographical variation of mandible size and shape in the Japanese wild pig (*Sus scrofa leucomystax*)', *Journal of Veterinary Medical Science*, 62, 815–20.

Foster, J. B. (1964) 'Evolution of mammals on islands', *Nature*, 202, 234–5.

Gakuhari, T., Komiya, H., Sawada, J., Anezaki, T., Sato, T., Kobayashi, Ke., Itoh, S., Kobayashi, Ko., Matsuzaki, K., and Yoneda, M. (2015) 'Radiocarbon dating of one human and two dog burials from the Kamikuroiwa Rock Shelter site, Ehime Prefecture', *Anthropological Science*, 123(2), 87–94.

Hasebe, K. (1924) 'Über die Schadel und Unterkiefer von den steinzeitlich japanischen Hunderassen', *Sendai*, 10, 1–33.

Hayashida, S. (1956) 'Observations on the ancient horses of Japan', *Anthropological Science*, 64, 197–211.*

Hayashida, S. (1960) 'Studies on wild boar and dog found at shell-mounds in the Amami-Oshima archipelago', *Anthropological Science*, 68, 96–115.*

Hayashida, S. and Yamaguchi, C. (1956) 'On the horses found in the shell-mound of Izumi', *Bulletin of the Faculty of Agriculture, Kagoshima University*, 4, 70–7.*

Hongo, H., Anezaki, T., Yamazaki, K., Takahashi, O., and Sugawara, H. (2007) 'Hunting or management? Status of pigs in the Jomon period, Japan', in Albarella, U., Dobney, K., Ervynvk, A., and Rowley-Conwy, P. (eds) *Pigs and Humans: 10,000 Years of Interaction*, pp. 109–30. Oxford: Oxbow Books.

Imaizumi, Y. (1970a) 'Systematic status of the extinct Japanese wolf, *Canis hodophilax*. 1. Identification of specimens', *Journal of Mammal Society of Japan*, 5, 27–32.*

Imaizumi, Y. (1970b) 'Systematic status of the extinct Japanese wolf, *Canis hodophilax*. 2. Similarity relationship of *hodophilax* among species of the genus *Canis*', *Journal of Mammal Society of Japan*, 5, 62–6.*

Imaizumi, Y. (1973) 'Taxonomic study of the wild boar from the Ryukyu Islands, Japan', *Memoirs of the National Science Museum*, 6, 113–29.*

Inukai, T. (1960) 'Wild boar of Hokkaido from the ethnological view point', *Bulletin of the Institute for the Study of North Eurasian Cultures, Hokkaido University*, 15, 1–6.**

Ishiguro, M., Okumura, N., Matsui, A., and Shigehara, N. (2000) 'Molecular genetic analysis of ancient Japanese dog', in Crockford, S. J. (ed.) *Dogs through Time: An Archaeological Perspective*. BAR International Series 889, pp. 287–92. Oxford: Archaeopress.

Ishiguro, N., Inoshima, Y., and Shigehara, N. (2009) 'Mitochondrial DNA analysis of the Japanese wolf (*Canis lupus hodophilax* Temminick, 1839) and comparison with representative wolf and domestic dog haplotypes', *Zoological Science*, 26, 765–70.

Ishiguro, N., Inoshima, Y., Shigehara, N., Ichikawa, H., and Kato, M. (2010) 'Osteological and genetic analysis of the extinct Ezo wolf (*Canis lupus hattai*) from Hokkaido Island, Japan', *Zoological Science*, 27, 320–4.

Ishiguro, N., Inoshima, Y., Yanai, T., Sasaki, M., Matsui, A., Kikuchi, H., Maruyama, M., Hongo, H., Vostretsov, Y.E., Gasilin, V., Kosintsev, P.A., Quanjia, C., and Chunxue, W. (2016) 'Japanese wolves are genetically divided into two groups based on an 8-Nucleotide insertion/deletion within the mtDNA control region', *Zoological Science* 33: 44–9.

Kaneko, H. (1987) 'Vertebrate animal remains and artifacts made of bones, antlers, and canines found from Kurawa Site in Hachijyo Island', in Hachijyo Town Board of Education, Tokyo (ed.) *Tokyo-to Hachijyo-cho Kurawa Iseki* (*Kurawa Site in Hachijo Town, Tokyo*), pp. 87–103. Hachijo Town: Hachijo Town Board of Education.**

Kato, S. (1980) 'Animal keeping by Jomon people—in particular on the problem of wild pigs', *Rekishi Koron*, 54, 45–50.**

Kawamura, Y. (1991) 'Quaternary mammalian faunas in the Japanese Islands', *Quaternary Research*, 30, 213–20.

Kawashima, S. (2004) 'A Study on the Relationship between 8 Populations of Japanese Native Horse'. Unpublished PhD dissertation, Graduate School of Agricultural and Life Sciences, Tokyo University (Tokyo).**

Komiya, T., Sawada, J., Saeki, F., and Sato, T. (2015) 'Morphological characteristics of buried dog remains from Kamikuroiwa Rock Shelter site, Ehime Prefecture, Japan', *Anthropological Science*, 123(2), 73–85.

Kondo, M., Matsuura, S., Matsui, A., and Kanayama, Y. (1991) 'Fluorine dating of horse remains from Osaki shellmound in Noda City—Debatable existence of horses in Jomon period', *Anthropological Science*, 99, 93–9.*

Kondo, M., Matsuura, S., Nakai, N., Nakamura, T., and Matsui, A. (1992) 'Dating of horse remains recovered from Late Jomon layers of the Izumi shellmound', *Archaeology and Natural Science*, 26, 61–71.*

Masuda, R. and Sato, T. (2015) 'Mitochondrial DNA analysis of Jomon dogs from the Kamikuroiwa Rock Shelter site in Shikoku and the Higashimyo site in Kyushu, Japan', *Anthropological Science*, 123(2), 95–8.

Matsui, A., Ishiguro, N., Hongo, H., and Minagawa, M. (2005) 'Wild pig? Or domesticated boar? An archaeological view on the domestication of *Sus scrofa* in Japan', in Vigne, J.-D., Peters, J., and Helmer, D. (eds) *The First Steps of Animal Domestication: New Archaeological Approaches*, pp. 148–59. Oxford: Oxbow Books.

Matsui, A. (2002) 'Use of horse in ancient Japan', *Archaeology Journal*, 483, 12–16.**

Minagawa, M., Matsui, A., and Ishiguro, N. (2005) 'Patterns of prehistoric boar *Sus scrofa* domestication, and inter-islands pig trading across the East China Sea, as determined by carbon and nitrogen isotope analysis', *Chemical Geology*, 218, 91–102.

Miyamoto, F. (1991) 'On the skull of Japanese wolf (*Canis hodophilax* Temminck) taken out from the mounted specimen preserved in Wakayama University', *Bulletin of Faculty of Education, Wakayama University, Natural Science*, 39, 55–60.*

Morii, Y., Ishiguro, N., Watanobe, T., Nakano, M., Hongo, H., Matsui, A., and Nishimoto, T. (2002) 'Ancient DNA reveals genetic lineage of *Sus scrofa* among archaeological sites in Japan', *Anthropological Science*, 110, 313–28.

Muraishi, M. (1998) 'The origin of horse breeding in Kai Province', *Zoo-archaeology*, 10, 17–36.*

Naora, N. (1937) 'On the pig in the prehistoric age of Japan', *Journal of the Anthropological Society of Tokyo*, 52, 286–96.*

Naora, N. (1965) *On Wolves in Japan*, Tokyo: Azekura Shobo.**

Nishimoto, T. (1985) 'On wild pigs in Hokkaido in the Jomon period', *Kodai Tansou*, II, 137–52.**

Nishimoto, T. (1993) 'Chickens in the Yayoi period', *Zoo-archaeology*, 1, 45–8.

Nishimoto, T. (1996) 'Horse from Shiobe Site', in Archaeological Research Center of Yamanashi Prefecture (ed.) *Shiobe Iseki*. Reports of Archaeological Research Center of Yamanashi Prefecture 123, p. 30. Kofu: Archaeological Research Center of Yamanashi Prefecture.**

Nishimoto, T. (2003) 'Domestication of pigs in the Jomon period', *Bulletin of the National Museum of Japanese History*, 108, 1–16.*

Nishimoto, T. (2008) 'Animal bone remains', in Tokushima Cultural Property Center (ed.) *Kannonji Iseki IV, part 2*, pp. 203–19. Tokushima: Tokushima Cultural Property Center.**

Nishimoto, T. and Niimi, M. (eds) (2010) *Jiten: Hito to Dobutsu no Kokogaku (Encyclopedia of the Archaeology of Humans and Animals)*, Tokyo: Yoshikawa Kobunkan.**

Nishinakagawa, H. (ed.) (1991) 'A Study on the Time and the Route of the Introduction of Cattle and Horses into Japan, as Examined from Skeletal Remains from Archaeological Sites'. Report for JSPS Grant-in-Aid (B).**

Nishinakagawa, H. and Matsumoto, M. (1991) 'Morphometrical studies of cattle and horse bones from archaeological sites', in Nishinakagawa, H. (ed.) 'A Study on the Time and the Route of the Introduction of Cattle and Horses into Japan, as Examined from Skeletal Remains from Archaeological Sites', pp. 18–41. Report for JSPS Grant-in-Aid (B).**

Nozawa, K. (1992) 'Origin and ancestry of native horses in eastern Asia and Japan', *Japanese Journal of Equine Science*, 3(1), 1–18.*

Nozawa, K., Shootake, T., Ito, S., and Kawamoto, Y. (1998) 'Phylogenetic relationships among Japanese native and alien horses estimated by protein polymorphisms', *Journal of Equine Science*, 9, 53–69.

Nozawa, K., Shootake, T., and Namikawa, T. (1975) 'Gene constitution and phylogenetic inter-relationship among native livestock in Japan and its adjacent area, with special reference to native horses and cattle', *JIBP Synthesis*, 5, 130–7.

Okumura, N., Ishiguro, N., Nakano, M., Matsui, A., and Sahara, M. (1996) 'Intra-and inter-breed genetic variations of mitochondrial DNA major non-coding regions in Japanese native dog breeds (*Canis familiaris*)', *Animal Genetics*, 27, 397–405.

Okumura, N., Ishiguro, N., Nakano, M., Matsui, A., Shigehara, N., Nishimoto, T., and Sahara, M. (1999) 'Variations in mitochondrial DNA of dogs isolated from archaeological sites in Japan and neighbouring islands', *Anthropological Science*, 107, 213–28.

Ono, T. (1984) 'On the problem of boar keeping during the Jomon period', in Local History Research Association (ed.) *Kofu Bonchi: Sono Rekishi to Chiikisei* (*Kofu Basin, Its History and Regionality*), pp. 47–76. Tokyo: Yuzankaku.**

Oshiro, I. and Nohara, T. (2000) 'Distribution of Pleistocene terrestrial vertebrates and their migration to the Ryukyus', *Tropics*, 10, 41–50.

Sasazaki, S., Odahara, S., Hiura, C., Mukai, F., and Mannen, H. (2006) 'Mitochondrial DNA variation and genetic relationships in Japanese and Korean cattle', *Asian-Australasian Journal of Animal Sciences*, 19, 1394–8.

Shigehara, N. (1986) *Catalogue of the Ancient and Recent Canid Skeletons Collected by Dr. Kotondo Hasebe and Preserved in the University Museum*. University of Tokyo Museum Material Report 13. Tokyo: University of Tokyo Museum.

Shigehara, N. (1994) 'Morphological changes in Japanese ancient dogs', *Archaeozoologica*, IV, 78–94.

Shigehara, N. and Hongo, H. (2000) 'Ancient remains of Jomon dogs from Neolithic sites in Japan', in Crockford, S. J. (ed.) *Dogs through Time: An Archaeological Perspective*. BAR International Series 889, pp. 61–7. Oxford: Archaeopress.

Sonan Bunkazai Center (Sonan Cultural Property Center) (ed.) (2003) *Shimo-ota Kaizuka*, Mobara: Sonan Bunkazai Center.**

Sugaya, M. and Toizumi, T. (1998) 'An extensive Jomon cemetery with the human, a dog and pigs, Shimo-ota shell mound, Mobara City, Chiba Pre', *Zoo-archaeology*, 11, 69–74.**

Takahashi, R. (2012) 'Zooarchaeological Study of Introduction of *Sus scrofa* into the Prehistoric Ryukyu Islands Based on Ancient DNA Analysis'. Unpublished PhD dissertation, School of Advanced Science, Graduate University of Advanced Studies.*

Takahashi, R., Ishiguro, N., Matsui, A., Anezaki, T., and Hongo, H. (2011) 'Morphological and molecular phylogenetic characteristics of dwarf *Sus* specimens from the Noguni shell middens in the Ryukyu Islands', *Anthropological Science*, 120(1), 39–50.

Tsuda, K., Kawahara-Miki, R., Sano, S., Imai, M., Noguchi, T., Inayoshi, Y., and Kono, T. (2013) 'Abundant sequence divergence in the native Japanese cattle Mishima-Ushi (*Bos taurus*) detected using whole-genome sequencing', *Genomics*, 102(4), 372–8.

Uzawa, K. and Hongo, H. (2006) 'A morphological study of the Medieval horses from Yuigahama-minami site, Kamakura', *Archaeology and Natural Sciences*, 53, 57–67.

Walker, B. L. (2005) *The Lost Wolves of Japan*, Seattle: University of Washington Press.

Watanobe, T., Okumura, N., Ishiguro, N., Nakano, M., Matsui, A., Sahara, M., and Komatsu, M. (1999) 'Genetic relationship and distribution of the Japanese wild boar (*Sus scrofa leucomystax*) and Ryukyu wild boar (*Sus scrofa riukiuanus*) analyzed by mitochondrial DNA', *Molecular Ecology*, 8, 1509–12.

Watanobe, T., Ishiguro, N., Okumura, N., Nakano, M., Matsui, A., Hongo, H., and Ushiro, H. (2001) 'Ancient mitochondrial DNA reveals the origin of *Sus scrofa* from Rebun Island, Japan', *Journal of Molecular Evolution*, 52, 281–9.

Watanobe, T., Okumura, N., Ishiguro, N., Nakano, M., Takamiya, H., Matsui, A., and Hongo, H. (2002) 'Prehistoric introduction of domestic pigs on to the Okinawa Islands: ancient mitochondrial DNA evidence', *Journal of Molecular Evolution*, 55, 222–31.

Watanobe, T., Ishiguro, N., Nakano, M., Matsui, A., Hongo, H., Yamazaki, K., and Takahashi, O. (2004) 'Prehistoric Sado Island populations of *Sus scrofa* distinguished from contemporary Japanese wild boar by ancient mitochondrial DNA', *Zoological Science*, 21, 219–28.

Yamane, Y. (2013) 'Use of animals and birds in the Edo period—zooarchaeological evidence from residences of feudal lords', *Zoo-archaeology*, 30, 105–19.**

Yamazaki, K., Takahashi, O., Sugawara, H., Ishiguro, N., and Endo, H. (2005) 'Wild boar remains from the Neolithic (Jomon period) sites on the Izu Islands and in Hokkaido, Japan', in Vigne, J.-D., Peters, J. and Helmer, D. (eds) *The First Steps of Animal Domestication: New Archaeozoological Techniques, Proceedings of the 9th Conference of the International Council of Archaeozoology, Durham 2002*, pp. 160–76. Oxford: Oxbow Books.

* In Japanese with English summary ** In Japanese

CHAPTER 22

....

FARMING, SOCIAL CHANGE, AND STATE FORMATION IN SOUTHEAST ASIA

....

CHARLES F. W. HIGHAM

INTRODUCTION

FARMING in mainland Southeast Asia today centres on rice and domestic pigs (*Sus domesticus*), cattle (*Bos taurus*), chickens (*Gallus domesticus*), and water buffaloes (*Bubalus bubalis*). Millet (*Setaria italica*) is a secondary grain crop that is particularly well adapted to arid conditions and marginal land. There are many methods of cultivating rice (*Oryza sativa*), depending on the climate and terrain. Slash and burn is undertaken in the uplands. In northeast Thailand, rice is transplanted into bunded fields dependent on monsoon rains, yielding one crop per annum. Diversion of rivers in coastal Vietnam to irrigate the fields provides for at least two harvests in a year. Yields have greatly increased through breeding new varieties. Water buffaloes have been an integral part of rice cultivation for their tractive power, where ploughing and harrowing prepare the soil. However, their use is dwindling as mechanical ploughs become available. Cattle, pigs, and chickens are raised in rural Southeast Asia for the market and local consumption.

The present situation is founded on four millennia of evolution in crop and animal husbandry in which seminal innovations have dovetailed with social changes that include the rise of social elites and the formation and maintenance of states. During the pioneering research that began in the 1960s, identifying the origins of rice domestication was described as the $64,000 question in Southeast Asian prehistory, so much so that the possible role played by millet, the key domesticate in the Central Plains of China, was barely referenced. Following excavations at Spirit Cave, in northeast Thailand, and Non Nok Tha, on the Khorat Plateau (Fig. 22.1), Southeast Asia was reported to have witnessed a very early and indigenous development of plant and animal domestication

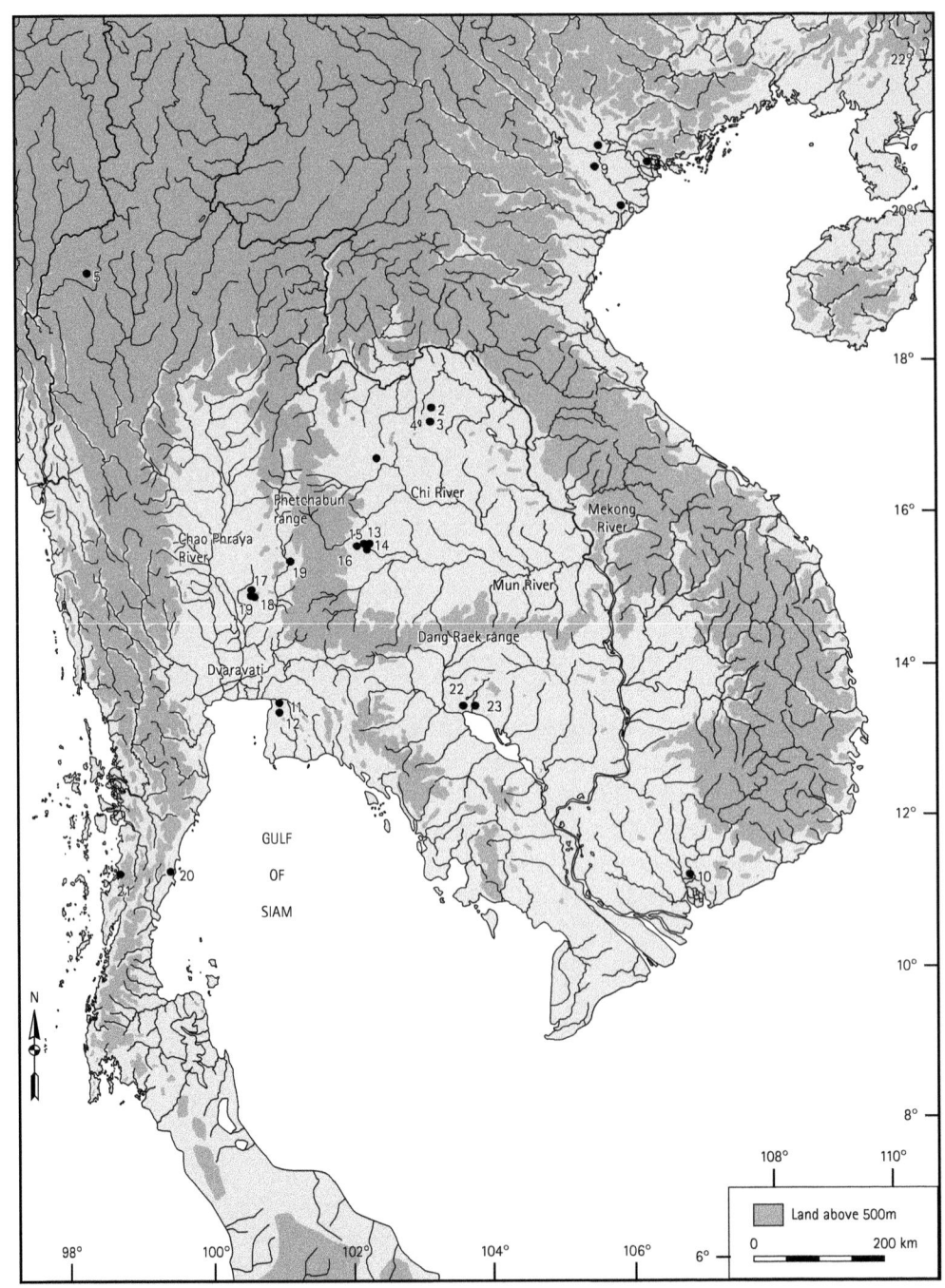

FIGURE 22.1 Map showing the location of sites mentioned in the text. 1. Non Nok Tha, 2. Ban Chiang, 3. Ban Na Di, 4. Lake Kumphawapi, 5. Spirit Cave, 6. Man Bac, 7. Co Loa, 8. Viet Khe, 9. Chau Can, 10. An Son, 11. Khok Phanom Di, 12. Nong Nor, 13. Ban Non Wat, 14. Ban Lum Khao, 15. Noen U-Loke, 16. Non Ban Jak, 17. Non Pa Wai, 18. Nil Kham Haeng, 19. Non Mak La, 20. Khao Sam Kaeo, 21. Phu Khao Thong, 22. Lovea, 23. Angkor.

(Solheim, 1972). Two developments in archaeological methodology have since provided better documented evidence. There is an increasing emphasis on flotation to recover plant remains and fine screening to sample microfauna, and a sound chronological framework based on the latest radiocarbon analytical techniques (Higham and Higham, 2009; Bellwood, 2011).

THE NEOLITHIC

In mainland Southeast Asia, the establishment of Neolithic communities that cultivated rice and/or millet, and maintained domestic pigs and cattle, began in the first half of the second millennium BC. Rapid advances in tracing the cultivation and ultimate domestication of rice in the Yangtze Valley play a basic role in illuminating the establishment of farming in Southeast Asia. The key point is that it took a very long time, measured in millennia, for the rice plant to evolve from a wild marsh grass, one of many wetland species exploited by Holocene hunter-gatherers, to a domesticated staple (Fuller et al., 2010; Zhao, 2010).

There is no matching developmental sequence in mainland Southeast Asia. Rather, the archaeological record reveals at least three distinct adaptations by hunter-gatherers. The first involved the occupation of rock shelters in forested, upland environments from southern China to Malaysia. Biological remains evidence broad-spectrum hunting and gathering strategies, just as is the case today with the Semang and Mani ethnic groups of Thailand and Malaysia. The second is represented between about 4000–2000 BC on raised beaches when the sea level stabilized at a level higher than at present. The third, represented best in Guangxi Province of China, but also hinted at in some Southeast Asian sites, involved large occupation and mortuary sites.

The two principal hypotheses for the timing and origins of farming in Southeast Asia are based on different sources of information. One comes from the biological remains recovered from cores bored into old lakebeds and other natural deposits. In this model, a rise in the frequencies of charcoal and open land indicators are interpreted as evidence for deliberate burning to clear land for crop production (Kealhofer, 2002; White et al., 2004). There is evidence for an increase in charcoal and vegetation disturbance dated to the mid-fourth millennium BC in a core from Lake Kumphawapi in northeast Thailand (Fig. 22.1). One of several interpretations for this is that there was forest clearance for the cultivation of domestic plants. White et al. (2004), incorporating radiocarbon determinations from basal levels at the nearby site of Ban Chiang, favour the establishment of farming communities in the region by about 3500 BC. There are two problems with this proposal. Forest burning is a well-known feature of hunter-gatherers to encourage game to the resulting clearings, while natural fires are endemic towards the end of the long dry season in this region. Secondly, there is no archaeological evidence for such early occupation: the new initiative to date the site of Ban Chiang on the basis of human bones has shown that the earliest evidence there for farming lies in the mid-second millennium BC (Higham et al., 2011a; 2015).

The archaeological evidence unanimously supports what is known as the 'two layer model'. Beginning in the Yangtze Valley, Fuller et al. (2010) have identified outward 'thrusts' of expansionary Neolithic rice farmers, one of which was directed to Southeast Asia. Rispoli (2008) and Zhang and Hung (2010) have traced this thrust into Southeast Asia on the basis of a 'Neolithic Package' centred on the presence of rice remains, related ceramic styles, and a distinctive mortuary ritual. A key site in documenting the impact of this expansionary movement is Man Bac, just south of the strategic Red River delta, where a cemetery has been shown to contain two groups of humans, distinguished biologically by the form of their heads and teeth and, most significantly, their DNA (Dodo, 2010; Matsumura, 2010; Oxenham et al., 2010). One group matches the local hunter-gatherers, the other is akin to the Neolithic farmers of the Yangtze. These latter individuals also share their DNA with the inhabitants of Weidun, a site in the Yangtze region (Shinoda, 2010). The demographic profile for these people, significantly, shows that the population was growing quickly (Bellwood and Oxenham, 2008). This is a widespread characteristic of early farming communities, since the sedentism necessary to safeguard crops is a spur to decreasing intervals between births. This very point has been pursued by Gignoux et al. (2011), who have applied Bayesian modelling to DNA haplogroup frequencies in Europe, Africa, and Southeast Asia, finding consistent evidence for a surge in population with early farming. That in East Asia is dated broadly as commencing within the period 5400–3000 BC.

Both archaeologically, and in terms of the survival of hunter-gatherer genes in modern inhabitants of Southeast Asia (Hill et al., 2006), the first farmer groups to penetrate and settle the area did not enter a *tabula rasa*. Even during periods of increasing cold in the world climate, Southeast Asia would still have offered abundant natural resources for hunter-gatherers to exploit. There are three indigenous species of cattle, deer, water buffalo, and wild pigs (Lekagul and McNeely, 1977). The lakes and rivers teemed with fish and shellfish. There would not seem to be any obvious reason why, amid such a range of resources, indigenous hunter-gatherers should be under any pressure to change. An emerging and widespread pattern rather suggests interaction between hunters and farmers that took place from about 2000 BC.

We have seen evidence for this at Man Bac. There, hunting and fishing continued to be a major part of the subsistence base, together with domestic pigs and rice. At An Son in southern Vietnam, there is an archaeological sequence that, in basal levels, might well involve hunter-gatherers during the late third millennium BC (Bellwood et al., 2013). This was followed from about 1800 BC by clear evidence for Neolithic settlement seen in the presence of rice identified by its DNA as *Oryza sativa japonica*, the sub-species that was domesticated in the Yangtze Valley (Bellwood et al., 2013; Castillo et al. 2015). The faunal remains are dominated by pigs and dogs (*Canis familiaris*) (Piper et al., 2014). The former were killed as they approached adult size, while dog bones were charred and bear cut-marks suggesting that they, too, were raised to eat. The dog is a particularly relevant species, because there is no native wolf in Southeast Asia from which to domesticate them. Like japonica rice, these are best sourced to the north, in China. As at Man Bac, the occupants turned to the river, swamps, and lakes to secure a range of resources, dominated numerically by the swamp eel (Synbranchidae), turtles, and shellfish. As for the inhabitants, the form of their skulls relates them to modern Southeast Asians, while there are aspects of the teeth that are more like the indigenous hunter-gatherers. Again, there may have been a mixed population.

Khok Phanom Di was located on or near the estuary of the Bang Pakong River as it entered the Gulf of Siam (Higham and Thosarat, 2004). Mangrove shores are not suitable locations for rice cultivation due to salinity. Occupation began in about 2000 BC by a community of marine hunter-gatherers. Fish and shellfish, crabs, and turtles dominated the fauna, although some rice remains were identified and thought to have been imported. During the third of seven mortuary phases, about 1800 BC, the sea level fell, leading to the formation of fresh-water swamps. At the same juncture, isotopes in the human teeth suggest that some female immigrants settled there (Bentley et al., 2007). It may well be that they brought with them knowledge of rice cultivation, for, at the same juncture, large granite hoes and shell-harvesting knives were fashioned. Domestic rice was now part of the diet, it has been found in human stomach contents and faeces, together with the beetle *Oryzaphilus*, which is attracted to rice stores. This part of the sequence also saw the first dog bones. The sea level soon rose again, and no more harvesting knives or hoes were found. It is quite possible that this sequence witnessed coalescence between indigenous marine hunter-gatherers and newly arrived rice farmers.

Ban Non Wat in northeast Thailand is the best-dated site in Southeast Asia. The initial Neolithic settlement took place in the seventeenth century BC. The middens contain the bones of domestic pigs, cattle, and dogs, but the water buffalo bones are the same size as those from wild animals (Kijngam, 2010). Fragments of rice chaff were also found in the middens. Wild deer, fish, shellfish, and turtles were also abundant: again, there was mixed subsistence. The Neolithic dead were laid out in a supine extended position, accompanied by finely decorated ceramic vessels, pig bones, bivalve shells, and stone adzes. Some distinctly different burials, however, have been dated within the Neolithic period. The corpse was interred in a flexed position, with a contrasting set of mortuary offerings. A flexed position was preferred by indigenous hunter-gatherers, and plausibly, we again encounter a mixed population. The isotopes in the human teeth indicate that while those interred in an extended position consumed rice, at least two of the flexed individuals did not (King et al., 2014).

Hitherto, all sites described have yielded the remains of rice. However, when we cross the Phetchabun range into the Lopburi region of central Thailand, the situation is quite different. Weber et al. (2010) have undertaken extensive flotation on samples from Non Pa Wai, Non Mak La, and Nil Kham Haeng. They found that foxtail millet dominated in the Neolithic, with rice becoming abundant only during the first millennium BC. Millet is better adapted to aridity, and the Lopburi region is one of the driest in Thailand. The presence of millet there strongly suggests that there was a series of expansionary movements from modern China into Southeast Asia by groups of farmers. Metaphorically speaking, this may well have been more in the form of related rivulets rather than a single deluge. However, one conclusion is inescapable: they set in place the foundations for subsequent social changes that led to increasing cultural complexity.

THE BRONZE AGE

The earliest secure dates for the establishment of copper base technology in Southeast Asia come from four sites in northeast Thailand: Ban Non Wat, Non Nok Tha, Ban Lum Khao, and Ban Chiang (Higham and Higham, 2009; Higham et al., 2011a; 2011b; 2014; 2015). They place the transition into the Bronze Age in the eleventh century BC. It might prove slightly earlier in northern Vietnam, since the Red River and the sea provided direct access to Bronze Age states in the Yangtze Valley. Excavations at Ban Non Wat fortuitously encountered exceptionally wealthy graves, set out in rows, which require a fresh appraisal of social changes that occurred at the same time as the adoption of metallurgy (Higham and Kijngam, 2012a). To a background of continuity in terms of ceramic forms from the Late Neolithic to the Early Bronze Age burials, there was a transformational rise in the range and quantity of mortuary offerings. This has been interpreted as the result of the rise of a social elite that maintained its dominance for at least six generations (Higham, 2011a). One of the most likely stimuli for this change lies in securing preferred access to exotic prestige valuables, and to judge from the mortuary offerings, these included marine shell, marble, and copper. In addition to a wide variety of ceramic vessels, the graves also include animal and fish bones that might well reflect the provision of mortuary feasts. Such rituals are a well-known means of projecting and maintaining elite status (Dietler and Hayden, 2001; Hayden, 2009).

The species represented in the Bronze Age burials should therefore reflect their ritual as well as their dietary significance. As with most other settlements of this period, pigs were the dominant mammal represented in human graves. One Early Bronze Age man was accompanied by a chicken skeleton. There are very few cattle bones, and no water buffalo remains. Numerically, fish dominated in the Early Bronze Age. This pattern continued though the succeeding phases of Bronze Age burials, but the quantity of animal and fish bones declined with the sharp drop in mortuary wealth that began with Bronze Age 3B. No contemporary midden deposits have been identified, so the domestic faunal spectrum is not known.

The Early Bronze Age cemetery at Ban Lum Khao, located 12 km east of Ban Non Wat, was contemporary with the elite graves at the latter site, but much poorer in terms of mortuary wealth. The occupation contexts reveal the presence of domestic pigs and cattle, together with a high proportion of deer. Water buffalo are rare, and quite possibly also hunted (Higham, 2005). Flotation and fine screening yielded a very large sample of fish bones (Thosarat, 2005). Young pig bones were, again, the most common food offering in the Early Bronze Age burials (Fig. 22.2).

The reliability of faunal assemblages as an accurate reflection of farming activities during the Bronze Age turns on the degree to which wet sieving and flotation were

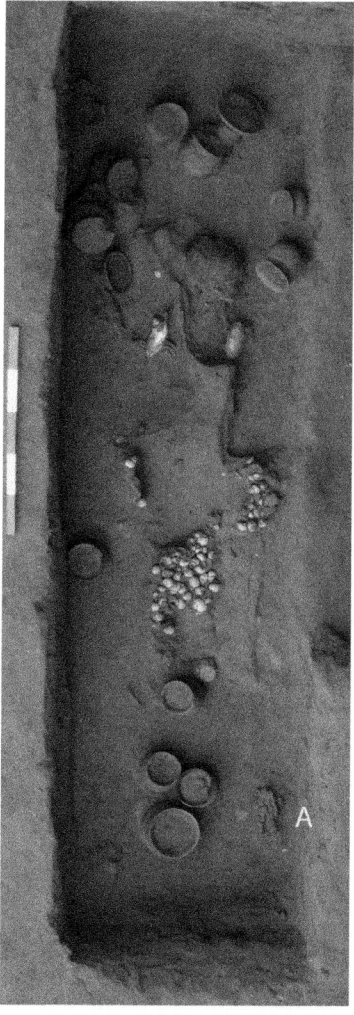

FIGURE 22.2 Burial 453, an infant Early Bronze Age burial from Ban Non Wat. Note the shell-fish placed over the body. A: articulating pig bones.

undertaken in the field. In the absence of sieving at Non Nok Tha, only large mammalian bones were recovered. These reveal that during the Bronze Age, complete pigs and articulating limbs of domestic cattle were placed with the dead. At Ban Chiang, cultural material was dry sieved but there was no wet sieving or flotation. Kijngam's analysis of the Bronze Age fauna from the 1975 excavation season concluded that the inhabitants exploited a range of wild species, particularly three species of deer (Kijngam, 1979). Pigs dominated among domestic animals, followed by cattle, dog, and chickens. No water buffaloes were identified until the Iron Age layers. Sieving did provide evidence for exploiting fish and frogs and a small number of turtles. Turning to the Bronze Age graves, we find that pigs were preferred as mortuary offerings.

At the nearby site of Ban Na Di, a sample of all cultural material was passed through a very fine-meshed wet sieving procedure. This resulted in the recovery of domestic rice grains (Chang and Loresto, 1984). The bones recovered reveal, from Bronze Age occupation contexts, a complete dominance of fish. Shellfish were also collected, and two pits were found filled with them. Once again pigs and cattle were the most abundant domesticates, followed by dog and chicken. The basal layers did not yield any water buffalo. Large, medium, and small deer figured prominently but all paled numerically when compared with the dominance of fish. There was a regular selection for the left fore articulating limb bones of cattle and pigs to place with the dead. One set of pig bones filled a ceramic vessel.

The Bronze Age graves at Nong Nor, located behind the shore of the Gulf of Siam, were cut into an earlier hunter-gatherer settlement (Higham and Thosarat, 1998). The cemetary probably dates within the period 700–500 BC. Only the animal bones placed with the dead evidence the Bronze Age economy. Again, there was a dominance of pigs' foot bones, but there were also domestic cattle remains, and dogs' crania.

The adequacy of the Bronze Age diet may be measured by the evidence for human health. Four sites are particularly relevant: Non Nok Tha (Douglas, 1996), Ban Lum Khao (Domett, 2004), Nong Nor (Tayles et al., 1998), and Ban Na Di (Houghton and Wiriyaromp, 1984). The inhabitants of Bronze Age settlements enjoyed a long local ancestry, and were therefore likely to have adapted to the potential health problems of their environment. Descent from the initial Neolithic settlers would have involved the passage of about 25–30 generations, but to judge from modern DNA patterns, the hunter-gatherer populations would also have been ancestral to the people of the Bronze Age (Hill et al., 2006).

The age structure of a population is one of the basic indicators of health. Age profiles for the Bronze Age in this region reveal far fewer infant deaths than at the coastal hunter-gatherer and Neolithic site of Khok Phanom Di, the figure for infants under one year of age ranging from 11% at Nong Nor to 19% at Ban Na Di and Ban Lum Khao. Most women from Ban Na Di showed evidence for pregnancy, but this figure fell markedly at Ban Lum Khao to about half the pelves examined. The average stature was very similar at each site. Douglas (1996) has divided the Non Nok Tha sample into early and late groups, and the men and women of the former were the shortest encountered. Later individuals were as tall as those at the other Bronze Age sites.

Generalized poor health in childhood may be seen in dental enamel hypoplasia, found in about a quarter of all adults at Khok Phanom Di, but between 11% and 15.6% for Ban Na Di, Ban Lum Khao, and Nong. Bone mass may be used as an indicator of the adequacy of the diet. Nordin's score, a measure of cortical bone thickness, shows that the men and women of Ban Na Di, Nong Nor, and Ban Lum Khao had similar values compatible with a good level of nutrition. Moreover, the diet was not cariogenic. Only 2.8% of the adult teeth at Non Nok Tha had caries, a figure which varied between 5.2% and 6.5% at the other three sites. Comparatively, the Bronze Age samples provide a generally similar picture of health, stature, and life expectancy. The people of Ban Lum Khao appear to have suffered more than those of the other sites, in terms of a higher death rate among younger women, while the men had the lowest stature of any of the Bronze Age sites. This group was also markedly poorer in terms of mortuary offerings, and it is possible that this was reflected in their diet and health.

THE IRON AGE

The millennium of the Iron Age witnessed profound changes in all aspects of society. Beginning in the fifth century BC, it coincided with the opening of an extensive maritime trade route that brought Southeast Asia into direct and regular contact with India and the Mediterranean world to the west, and with China to the east. Moreover, with the end of the Warring States period, the Qin and Han emperors showed increasing interest in imperial expansion to the south, which saw the incorporation of the Red River delta region into the sphere of Chinese hegemony.

This nodal region was then witnessing the establishment of Dong Son chiefdoms. This is best seen in the urban centre of Co Loa, where Kim et al. (2010) have sectioned the massive ramparts, the outer extending over a circumference of 8 km, the middle one still standing to a height of 10 m in places, with a breadth of 20 m. A fourth-century BC context for the defensive structures is suggested by ten radiocarbon dates derived from charcoal found in the moat fill. This was followed by a second and far larger rampart, constructed of layers of tamped earth, topped by a deposit of roof tiles that might well have come from a building of some sort. This wall has been radiocarbon dated to the third century BC, followed by a period of use that lasted at least into the first century AD.

Excavations within the walls led to the discovery of a large bronze drum. These drums were decorated with scenes presumably taken from the daily lives and rituals of elite leaders. While none illustrates agricultural activities, they do illuminate the importance of warfare, for we see large boats involved in conflict. Drums were evidently cast in fours, and played from a raised platform, while wooden houses were elaborately decorated. The Coa Loa drum contained a collection of bronzes that might have been destined for recasting. There are chisels, spearheads, arrows, daggers, and most significantly, socketed implements that have been interpreted as ploughshares (Higham, 2015). These are very similar in form to those used by rice farmers during the Han Dynasty, and it is virtually a given that this technique was introduced from China.

The Dong Son homeland has a less extreme climate than most of low-lying mainland Southeast Asia, largely because the dry season is tempered by moist winds which move across the gulf of Bac Bo. On reaching land, they form a low cloud cover often associated with drizzle. This moist climate permits two crops of rice per annum on favourable soil (Gourou, 1955). Given this supportive climate, the potential of harnessing the tractive power of domestic water buffalo to a plough is profound economically and socially. As Goody (1971) has emphasized, ploughing can open ten times more land to agriculture than a man with a hoe. By turning a furrow and then harrowing to break down the clods of earth, the plough aerates the soil, turns over dry season weeds to decompose, and furnishes a strong anchor to transplanted rice plants. The surpluses that can be generated are the fundamental requirement to sustain the social elites graphically depicted on their drums, and seen in the aristocratic boat burials of Viet Khe and Chau Can (Luu, 1977). Again, the craft specialists responsible for boatbuilding, casting bronzes, or forging or casting iron, would have been similarly freed from agricultural labour.

These innovations are echoed and confirmed in some of the Chinese texts. Wheatley, for example, has noted how in the late second century BC, the area was designated a Chinese protectorate in which 'the Lac chieftains, in whose persons were institutionalized customary rights to land, [are] confirmed in their traditional authority' (Wheatley, 1983: 368).

However, Chinese expansion to the south stalled at the Truong Son Cordillera, known to them as the 'Fortress of the Sky', and as we enter the coast and inland riverine plains, so there is a distinct prehistoric sequence devoid of Chinese control. The impact of the new maritime exchange network was most immediate in coastal settlements, and it is at Khao Sam Kaeo, on the eastern shore of peninsular Thailand, that excavations have uncovered a port city (Bellina-Pryce and Silapanth, 2006). The site incorporates four flat-topped hills flanking a river, separated by valley floors. Seventeen embankments or walls have been traced, and their pattern seems to demarcate the hills. They once stood to a height of about four metres, and may well have been defensive. Some parallels to these constructions have been identified in contemporary Indian sites. There is further occupation evidence on the terraces cut on the hillsides themselves, in the form of small walls, pathways, and post holes that would once have supported structures. In all, the site once covered 54 hectares. Compelling evidence for the presence of Indian craft workers is seen in the local manufacture of hard stone and glass beads, while unequivocal indications of Indian presence is seen on the opposite shore at Phu Khao Tong, where one potsherd bore a short inscription in the Tamil language (Bellina et al., 2014).

In her examination of the plant remains obtained by flotation, Castillo (2011) has identified the remains of mung beans and horsegram, both pointing to introductions from India. The principal crop was rice, which to judge from the associated weed seeds, was essentially a dry-land or rain-fed crop rather than grown in permanent irrigated fields. Foxtail millet (*Setaria italica*) was also recovered from Khao Sam Kaeo.

The tentacles of this new maritime trade soon penetrated inland. Perhaps the best-dated and documented impact is seen in the upper Mun Valley at Ban Non Wat. Here, the Late Bronze Age cemetery merged in an easterly direction with burials containing virtually identical pottery vessels, but also forged iron artefacts (Higham and Kijngam, 2012b). This transition is dated in the second half of the fifth century BC. Significantly, the first iron was contemporary with rare beads of carnelian and agate, as well as glass ear ornaments. At least in this part of Thailand, there are signals that knowledge of iron smelting originated in the context of Indian trade. The Iron Age in this region falls into four phases (Higham, 2011b). During the first and second phases, iron was employed for spears, knives, and awls, and also for heavy, socketed hoes that might well have been used in farming. Some of the Early Iron Age dead were also interred with the articulating limb bones of domestic water buffaloes.

There is now growing evidence for a major farming innovation that took hold during the third to the sixth centuries AD. A man interred in the Iron Age 3 cemetery at Noen U-Loke was accompanied by a socketed iron implement initially described as a spade (Connelly, 2007). When an identical tool was recovered from inside a kiln at Non Ban Jak, their forms were reviewed with the conclusion that they were actually ploughshares (Fig. 22.3). It is highly unlikely that many such tools will survive unless they were deliberately buried with the dead, for iron can so easily be recycled when an artefact is worn out. These two fortunate discoveries coincided with a set of other changes. This was the period in northeast Thailand when Iron Age settlements proliferated, and were enclosed by multiple banks that channelled river water into moats. There are five such moats at Noen U-Loke covering a linear distance of about 200 metres. These substantial and complex engineering works would have required the organization of a large workforce. Within the settlements themselves, some of the dead, seen as emerging elites, were now being interred with great wealth measured in exotic glass beads, carnelian, agate, silver and gold ornaments, and multiple bronzes, within graves filled with rice. Moreover, at Lovea and contemporary sites in northwest Cambodia, Hawken (2011) has traced the faint outline of Iron Age fields. By integrating these recent finds, it is possible to identify a veritable agricultural revolution. Against a background of a climate veering towards increasing aridity (Wohlfarth et al., 2016), Iron Age leaders created reservoirs round their sites that would have served more than defence: water could also have been reticulated into rice fields to ameliorate the effects of rainfall deficiency, and perhaps even to produce more than one crop

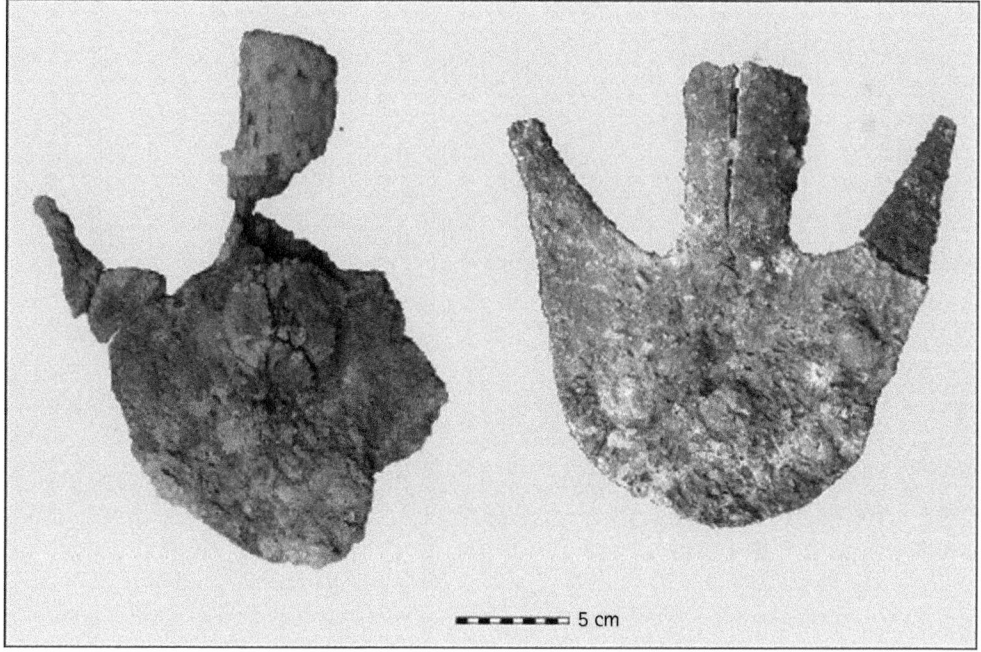

5 cm

FIGURE 22.3 Iron ploughshares from Non Ban Jak (left) and Noen U-Loke (right).

per annum. The importance of water buffaloes for their tractive power is also seen at Noen U-Loke, where a palimpsest of hoof prints suggests that animals were corralled within the settlement.

The improved land thus serviced, could have provided the basis for social changes based on private land ownership. As Rousseau stressed in the eighteenth century AD, 'the first man who, having enclosed a piece of ground, bethought himself of saying "this is mine", and found people simple enough to believe him, was the real founder of civil society' (Rousseau, 1913: 207). He went on to conclude that 'metallurgy and agriculture .... produced this great revolution ... it was iron and corn, which first civilized men ...' (Rousseau, 1913: 215).

The excavation at the moated site of Non Ban Jak in the upper Mun Valley has revealed substantial houses of this period, with thick clay walls and floors. One Late Iron Age house had been destroyed by fire, and the kitchen area survived, with hearths, pottery vessels still in place, and large quantities of carbonized rice grains (Fig. 22.4). The inference is clear: just as in northern Vietnam, so in the interior plains of Southeast Asia, a new and much more productive agricultural system was established, one which was capable of sustaining a large population divided socially into elites and less advantaged. This agricultural revolution was the prelude to the changes documented in early inscriptions. Sanskrit and old Khmer texts inscribed on stone stelae dating from the sixth century AD describe leaders, known as pon, who controlled access to land, organized the

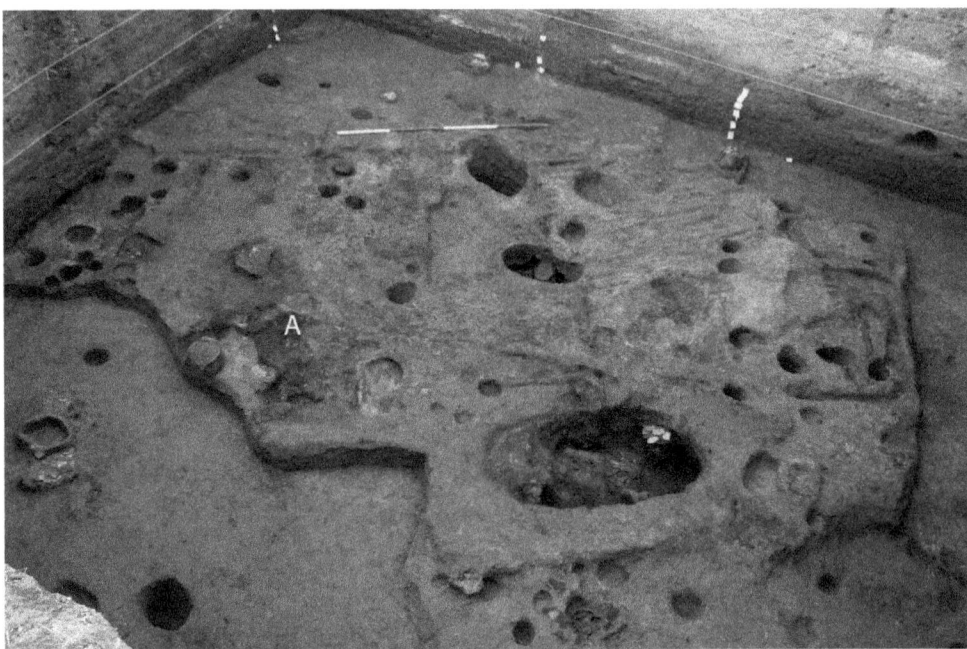

FIGURE 22.4 A Late Iron Age house foundation that was destroyed by fire. A: the kitchen area, with hearth, pots in place, and large quantities of carbonized rice.

provision and control of water, and led communities that incorporated a significant proportion of agricultural labourers (Vickery, 1998). The urban centres of this period contained large, brick temples dedicated to local and Indic gods, but also incorporated reservoirs and were surrounded by permanent rice fields. The emergence of early states in Southeast Asia best known at Angkor, but also in new Dvaravati centres in central Thailand, can thus be understood in the context of farming innovations of the Late Iron Age.

SUMMARY

Expansionary farmers originating in China brought domestic rice and millet, pigs, and cattle to Southeast Asia during the early second millennium BC. The rich natural resources they encountered led to a mixed subsistence in which fish, shellfish, and wild animals were a significant part of the diet. This may have been encouraged by growing evidence for integration with indigenous hunter-gatherers. Bronze Age graves, which date from the eleventh century BC, often incorporated animal bones as mortuary offerings, and these were dominated by pigs, with dogs, cattle, and chickens being relatively rare. Pig-rearing thus appears to have been widespread, and a potential source of wealth. There was an agricultural revolution during the course of the Iron Age, which began in the early centuries AD, seen in the construction of reservoirs, the increasing incidence of domestic water buffaloes together with iron ploughshares, and the creation of fixed, bunded rice fields. Iron knives, sickles, and hoes would have accelerated modifications to the landscape and efficiencies in agriculture. Improved productivity was not alone in contributing to the transition into early states that was underway by the sixth century AD, there was also increased competition over the ownership of exotic valuables, a rising population, and warfare. However, elite ownership of the best land and water resources, such a central feature of early Southeast Asian states, can be traced back into innovations that took place during the prehistoric Iron Age.

REFERENCES

Bellina-Pryce, B. and Silapanth, P. (2006) 'Weaving cultural identities on trans-Asiatic networks: Upper Thai-Malay Peninsula—an early socio-political landscape', *Bulletin de l'École Française d'Extrême-Orient*, 93, 257–94.

Bellina, B., Silapanth, P., Chaiswan, B., Thongcharoenchaikit, C., Allen, J., Bernard, V., Borrel, B., Bouvet, P., Castillo, C., Dussubieux, L., Malakie Laclair, J., Peronnet, S., and Pryce, T. O. (2014) 'The development of coastal polities in the upper Thai Malay Peninsula in the late first millennium BCE and the inception of long lasting economic and social exchange between polities on the east side of the Indian Ocean and the South China Sea', in Revire, N. and Murphy, S. A. (eds) *Before Siam Was Born: New Insights on the Art and Archaeology of Pre-Modern Thailand and Its Neighbouring Regions*, pp. 69–89. Bangkok: River Books.

Bellwood, P. (2011) 'The checkered Prehistory of rice movement southwards as a domesticated cereal—from the Yangzi to the equator', *Rice*, 4, 93–103.

Bellwood, P. and Oxenham, M. (2008) 'The expansions of farming societies and the role of the Neolithic Demographic Transition', in Bocquet-Appel, J. P. and Bar-Yosef, O. (eds) *The Neolithic Demographic Transition and Its Consequences*, pp. 13–34. Dordrecht: Springer.

Bellwood, P., Oxenham, M., Bui, C. H., Nguyen, T. K. D., Willis, A., Sarjeant, C., Piper, P. J., Matsumura, H., Tanaka, K., Beavan, N., Higham, T., Nguyen, Q. M., Dang, N. K., Nguyen, K. T. K., Vo, T. H., Van, N. B., Tran, T. K. Q., Nguyen, P. T., Campos, F., Sato, Y.-I., Nguyen L. C., and Amano, N. (2013) 'An Son and the Neolithic of southern Vietnam', *Asian Perspectives*, 50, 144–75.

Bentley, A., Tayles, N., Higham, C. F. W., Macpherson, C., and Atkinson, T. C. (2007) 'Shifting gender relations at Khok Phanom Di, Thailand: isotopic evidence from the skeletons', *Current Anthropology*, 48(2), 301–14.

Castillo, C. (2011) 'Rice in Thailand: the archaeobotanical contribution', *Rice*, 4, 114–20.

Castillo, C., Tanaka, K., Sato Y.-I., Ishikawa, R., Bellina, B., Higham, C., Chang, N., Mohanty, R., Kajale, M., and Fuller, D. (2015) Archaeogenetic study of prehistoric rice remains from Thailand and India: evidence of early japonica in South and Southeast Asia. *Archaeological and Anthropological Sciences* DOI 10.1007/s12520-015-0236-5.

Chang, T. and Loresto, E. (1984) 'The rice remains', in Higham, C. F. W. and Kijngam, A. (eds) *Prehistoric Investigations in Northeast Thailand*. BAR International Series 231, pp. 384–5. Oxford: Archaeopress.

Connelly, R. (2007) 'The iron and bimetallic artefacts', in Higham, C. F. W., Kijngam, A., and Talbot, S. (eds) *The Origins of Civilization of Angkor*, Vol. 2: *The Excavation of Noen U-Loke and Non Muang Kao*, pp. 431–46. Bangkok: The Thai Fine Arts Department.

Dietler, M. and Hayden, B. (2001) *Feasts: Archaeological and Ethnographic Perspectives on Food, Politics and Power*, Washington: Smithsonian Institution.

Dodo, Y. (2010) 'Qualitative cranio-morphology at Man Bac', in Oxenham, M., Matsumura, H., and Nguyen, D. K. (eds) *Man Bac: The Excavation of a Neolithic Site in Northern Vietnam. The Biology*, pp. 33–42. Canberra: Terra Australis.

Domett, K. (2004) 'The people of Ban Lum Khao', in Higham, C. F. W. and Thosarat, R. (eds) *The Origins of the Civilization of Angkor*, Vol. 1: *The Excavation of Ban Lum Khao*, pp. 113–58. Bangkok: Fine Arts Department.

Douglas, M. (1996) 'Paleopathology in Human Skeletal Remains from the Pre-Metal, Bronze and Iron Ages, Northeastern Thailand'. Unpublished PhD dissertation, University of Hawaii (Manoa).

Fuller, D., Sato, I., Castillo, C., Qin, L., Weisskopf, A. R., Kingwell-Banham, E. J., Song, J., Ahn, S.-M., and van Etten, J. (2010) 'Consilience of genetics and archaeobotany in the entangled history of rice', *Archaeological and Anthropological Science*, 2, 115–31.

Gignoux, C., Henn, B. M., and Mountain, J. L. (2011) 'Rapid, global demographic expansions after the origins of agriculture', *Proceedings of the National Academy of Sciences of the United States of America*, 108, 6044–9.

Goody, J. (1971) *Technology, Tradition and the State in Africa*, London: Hutchinson.

Gourou, P. (1955) *The Peasants of the Tonkin Delta*, New Haven: Human Relations Area Files.

Hawken, S. (2011) 'Metropolis of Ricefields: a Topographic Classification of a Dispersed Urban Complex'. Unpublished PhD dissertation, University of Sydney (Sydney).

Hayden, B. (2009) 'Funerals as feasts: why are they so important?', *Cambridge Archaeology Journal*, 19, 29–52.

Higham, C. F. W. (2005) 'The faunal remains', in Higham, C. F. W. and Thosarat, R. (eds) *The Origins of the Civilization of Angkor*, Vol. 1: *The Excavation of Ban Lum Khao*, pp. 159–70. Bangkok: The Fine Arts Department of Thailand.

Higham, C. F. W. (2011a) 'The Bronze Age of Southeast Asia: new insight on social change from Ban Non Wat', *Cambridge Archaeological Journal*, 21, 365–89.

Higham, C. F. W. (2011b) 'The Iron Age of the Mun Valley, Thailand', *The Antiquaries Journal*, 91, 101–44.

Higham, C.F.W. (2015) 'The Đông Sơn Chiefdom', in Reinecke, A. (ed.) *Perspectives on the Archaeology of Vietnam*, pp. 85–96. Berlin/Bonn: the German Archaeological Institute.

Higham, C. F. W. and Higham, T. F. G. (2009) 'A new chronological framework for prehistoric Southeast Asia, based on a Bayesian model from Ban Non Wat', *Antiquity*, 82, 125–44.

Higham, C. F. W., Higham, T. F. G., Douka, K., Ciarla, R., Kijngam, A., and Rispoli, F. (2011a) 'The origins of the Bronze Age of Southeast Asia', *Journal of World Prehistory*, 24, 227–74.

Higham, C. F. W., Higham, T. F. G., and Kijngam, A. (2011b) 'Cutting a Gordian Knot: the Bronze Age of Southeast Asia, timing, origins and impact', *Antiquity*, 85, 583–98.

Higham C. F. W., Higham T. F. G., and Douka, K. (2014) 'The chronology and status of Non Nok Tha, Northeast Thailand', *Journal of Indo-Pacific Archaeology*, 34, 61–75.

Higham C. F. W., Higham T. F. G., and Douka, K. (2015) 'A New Chronology for the Bronze Age of Northeastern Thailand and Its Implications for Southeast Asian Prehistory', http://journals.plos.org/plosone/article?id=10.1371/journal.pone.0137542.

Higham, C. F. W. and Kijngam, A. (eds) (2012a) *The Origins of the Civilization of Angkor*, Vol. 5: *The Excavation of Ban Non Wat: The Bronze Age*, Bangkok: The Fine Arts Department of Thailand.

Higham, C. F. W. and Kijngam, A. (eds) (2012b) *The Origins of the Civilization of Angkor*, Vol. 6: *The Excavation of Ban Non Wat: The Iron Age*, Bangkok: The Fine Arts Department of Thailand.

Higham, C. F. W. and Thosarat, R. (eds) (1998) *The Excavation of Nong Nor*. Studies in Prehistoric Anthopology 18. Dunedin: University of Otago.

Higham, C. F. W. and Thosarat, R. (2004) *The Excavation of Khok Phanom Di*, Vol. 7: *Summary and Conclusions*, London: The Society of Antiquaries of London.

Hill, C., Soares, P., Mormina, M., Macaulay, V., Meehan, W., Blackburn, J., Clarke, D., Maripa Raja, J., Ismail, P., Bulbeck, D., Oppenheimer, S., and Richards, M. (2006) 'Phylogeography and ethnogenesis of aboriginal Southeast Asians', *Molecular Biology and Evolution*, 12, 2480–91.

Houghton, P. and Wiriyaromp, W. (1984) 'The people of Ban Na Di', in Higham, C. F. W. and Kijngam, A. (eds) *Prehistoric Investigations in Northeast Thailand*. BAR International Series 231, pp. 391–412. Oxford: Archaeopress.

Kealhofer, L. (2002) 'Changing perceptions of risk: the development of agro-ecosystems in Southeast Asia', *American Anthropologist*, 104, 178–94.

Kijngam, A. (1979) 'The Faunal Remains from Ban Chiang and Its Implications for Thai Culture History'. Unpublished MA dissertation, University of Otago (Dunedin).

Kijngam, A. (2010) 'The mammalian fauna', in Higham, C. F. W. and Kijngam, A. (eds) *The Origins of the Civilization of Angkor*, Vol. 4: *The Excavation of Ban Non Wat: The Neolithic Occupation*, pp. 189–97. Bangkok: The Fine Arts Department of Thailand.

Kim, N., Toi, L. C., and Hiep, T. H. (2010) 'Co Loa: an investigation of Vietnam's ancient capital', *Antiquity*, 84, 1011–27.

King, C. L., Bentley, R. A., Higham, C. F. W., Tayles, N., Viðarsdóttir, U. S., Layton, R., Macpherson, C. G., and Nowell, G. (2014) 'Economic change after the agricultural revolution in Southeast Asia?', Submitted to *Antiquity*, 88, 112–25.

Lekagul, B. and McNeely, J. A. (1977) *Mammals of Thailand*, Bangkok: Kuruspha Ladprao Press.

Luu, T. T. (1977) *Khu Mo Co Chau Can*, Ha Noi: Archaeology Institute.

Matsumura, H. (2010) 'Quantitative cranio-morphology at Man Bac', Oxenham, M., Matsumura, H., and Nguyen, D. K. (eds) *Man Bac: The Excavation of a Neolithic Site in Northern Vietnam. The Biology*, pp. 43–63. Canberra: Terra Australis.

Oxenham, M., Matsumura, H., and Nguyen, D. K. (eds) (2010) *Man Bac: The Excavation of a Neolithic Site in Northern Vietnam. The Biology*, Canberra: Terra Australis.

Piper, P. J., Campos, F. Z., Ngoc Kinh, D., Amano, N., Oxenham, M., Chi Hoang, B., Bellwood, P., and Willis, A. (2014) 'Early evidence for pig and dog husbandry from the Neolithic site of An Son, southern Vietnam', *International Journal of Osteoarchaeology*, 24(1), 68–78.

Rispoli, F. (2008) 'The incised and impressed pottery style of mainland Southeast Asia: following the paths of Neolithisation', *East and West*, 57, 235–304.

Rousseau, J-J. (1913) *The Social Contract and Discourses by Jean Jacques Rousseau*, London: J. M. Dent.

Shinoda, K. (2010) 'Mitochondrial DNA of human remains at Man Bac', in Oxenham, M., Matsumura, H., and Nguyen, D. K. (eds) *Man Bac: The Excavation of a Neolithic Site in Northern Vietnam. The Biology*, pp. 95–116. Canberra: Terra Australis.

Solheim, W. G. II, (1972) 'An earlier agricultural revolution', *Scientific American*, 206, 34–41.

Tayles, N. G., Domett, K., and Hunt, V. (1998) 'The people of Nong Nor', in Higham, C. F. W. and Thosarat, R. (eds) *The Excavation of Nong Nor*. Studies in Prehistoric Anthopology 18, pp. 321–68. Dunedin: University of Otago.

Thosarat, R. (2005) 'The fish remains', in Higham, C: F. W. and Thosarat, R. (eds) *The Origins of the Civilization of Angkor*, Vol. 1: *The Excavation of Ban Lum Khao*, pp. 171–90. Bangkok: The Fine Arts Department of Thailand.

Vickery, M. (1998) *Society, Economics and Politics in Pre-Angkor Cambodia*, Tokyo: The Centre for East Asian Cultural Studies for Unesco.

Weber, S., Lehman, H., Barela, T., Hawks, S., and Harriman, D. (2010) 'Rice or millets: early farming strategies in prehistoric central Thailand', *Archaeological and Anthropological Sciences*, 72, 79–88.

Wheatley, P. (1983) *Nagara and Commandery*. Department of Geography Research Paper 207–8. Chicago: University of Chicago.

White, J. C., Penny, D., Kealhofer, L., and Maloney, B. (2004) 'Vegetation changes from the Late Pleistocene through the Holocene from three areas of archaeological significance in Thailand', *Quaternary International*, 113, 111–32.

Wohlfarth, B., Higham, C. F. W., Yamoah, K. A., Chabangborn, A., Chawchai, S., and Smittenberg, R. H. (2016) 'Human adaptation to mid- to late-Holocene climate change in Northeast Thailand', *The Holocene*, 26(4), 614–26.

Zhang, C. and Hung, H.-C. (2010) 'The emergence of agriculture in southern China', *Antiquity*, 84, 11–25.

Zhao, Z. (2010) 'New data and new issues for the study of the origin of rice agriculture in China', *Archaeological and Anthropological Science*, 2, 99–105.

CHAPTER 23

THE ZOOARCHAEOLOGY OF EARLY HISTORIC PERIODS IN THE SOUTHERN LEVANT

JUSTIN E. LEV-TOV
AND SARAH WHITCHER KANSA

INTRODUCTION

NUMEROUS zooarchaeologists (e.g. Hesse, 1990; Tchernov and Horwitz, 1990; Zeder, 1996), anthropologists (e.g. Douglas, 1972; Harris, 1985), and scholars from other disciplines have studied the intricate relationship between food and human cultures in the Levant. This region has received close attention not only because of its location between great powers to the north, east, and south, but also because the Levant's peoples produced two age-defying texts that both commented on food: the Old Testament codified a variety of dietary precepts, while the New Testament declared them invalid. The specific relationship between food and human cultures of the past is, of course, largely the realm of archaeologists, including archaeobotanists and zooarchaeologists. Here, we focus on foodways and complex societies in one part of the ancient Near East. Societies in that region were stratified, held one another in relations of power, were labeled by millennia-old texts with ethnic identities implying distinct origins and traditions, and shared many tenets of religious beliefs. All of these social factors affected the diets of the Levant's ancient populations, and indeed similar factors influenced the diets of complex societies the world over (cf. Crabtree, 1990; Gumerman, 1997; deFrance, 2009; Twiss, 2012).

The beginning of complex societies in the Levant, some 4,000 years ago, is defined by the separation of administrators, craftspeople, clergy, and other professions who did not spend the majority of their time in agriculture and pastoralism, from those who did. This intrinsically stratified situation created, as noted by Gumerman (1997: 106–7), conditions under which substantial differentiation in consumption patterns could flourish.

This chapter focuses on the zooarchaeology of the southern Levant, the region including modern-day Israel, the Palestinian territories, Jordan, and the southern part of Lebanon (cf. Golden, 2004: 15). We have chosen this geographic scope in order to provide a regional breadth great enough to address the diversity of human behaviours related to animals in the early historic periods of the southern Levant. This region encompasses diverse geographic and climatic zones, ranging from the well-watered, flat, and humid coastal Mediterranean zone of Israel and Gaza, eastward to a range of hills and low mountains forming a border zone between the coastal plain and the Jordan Rift Valley. The Jordan Rift Valley includes the Dead Sea, the lowest point on Earth at approximately 400 m below sea level. The northern portions of the southern Levant, northern Israel and southern Lebanon, are milder in terms of heat, receive more precipitation, and thus support extensive vegetation including oak forests (Danin, 1995). The southern parts of the region are desert, made up of the Negev Desert in Israel and the Wadi Rum of Jordan. Jordan makes up the eastern side of the rift valley. The extreme east of that country, the basalt and limestone desert Badia region, is an area today still populated by sheep and goat pastoralists (Lancaster and Lancaster, 1991). Various rivers and seasonal streams (Arabic: *wadi*) cross-cut the southern Levant, the most notable being the Jordan River. Westward-flowing perennial rivers water the coastal plain. Jordan's watercourses are mainly wadis that flow into the Jordan Valley or Dead Sea.

The temporal scope of this chapter is the Bronze and Iron Ages, that is, roughly the period 3500 to 550 BC. During that period, writing was developed, mainly in the neighbouring regions of Egypt and Mesopotamia, and spread to the Levant (Nissen, 1990). The Levant itself has comparatively little written evidence, but third- and second-millennium BC (Early and Middle Bronze Ages) texts have been found there, albeit few (e.g. Anbar and Na'aman, 1986–1987: 7–10). However, the region was written about by more powerful neighbouring states. Later, the Hebrew Bible was developed there by the mid-first millennium BC, from earlier written or oral sources. The lack of early writings and clay tablet archives indicates that the Levant was a region of secondary development to Mesopotamia and Egypt, involved with these regions in trade and politics. The end date of 550 BC is that used by archaeologists to demarcate the end of the Iron Age. In reality, it is an arbitrary date that marks the fall of the Neo-Babylonian Empire to the Persians, and the return to the Levant of various peoples deported by the Babylonians, including the Judahites (for overview of cultural chronology, see Mazar, 1990; Levy, 1995). Lev-Tov (2013) has recently summarized animal bone studies for the Hellenistic and Roman periods; therefore, we will not discuss those here.

Thematically, this chapter describes important themes addressed by scholars working in the discipline within the southern Levant, topics that are sometimes specific to the cultural milieu of a certain period, and others that transcend them. For example, one of the topics of interest to the Early Bronze Age is the effect of state administrations on the distribution of animal products to city residents not engaged in agriculture (e.g. Zeder, 1988). Other themes, such as how ethnic identity or religion affected diet (Hesse, 1990; London, 2011) and involved animals, span both the entire region and temporal periods. Another topic of concern among zooarchaeologists, empires, is an extension of the first topic mentioned, city-state administration of animal production. Discussion of empires and their effects on livestock management has been greatly influenced by Zeder's work in Iran (1988; 1991) as well as by Wapnish and Hesse (1988). These are the principal cultural questions that have preoccupied researchers in the area and time frame under discussion here, although other topics and methods, including the use of ethnographic analogy to understand ancient herding economies (Grigson, 1995; Sasson, 2006; 2008), the effects of environmental change on herders (Matthews, 2002), and taphonomy (Weissbrod and Bar-Oz, 2002) have also engaged scholars.

THE HISTORY OF ZOOARCHAEOLOGICAL ANALYSIS IN THE SOUTHERN LEVANT IN HISTORIC PERIODS—KEY AREAS OF RESEARCH

Zooarchaeology in the Levant and neighbouring regions started on prehistoric sites in the early twentieth century. Bate (1937) pioneered the study of animal bones from Levantine sites, with her studies of the Palaeolithic and Mesolithic fauna from the Mount Carmel caves in Israel. In the mid-twentieth century, members of Robert Braidwood's 'Hilly Flanks' project (commenced in 1947) included scholars from various disciplines, including Charles A. Reed, integrating them within field and laboratory research (cf. Braidwood and Howe, 1960). The idea of systematically saving and studying animal bones from historic period sites in the Near East took approximately another twenty years to catch on. Two projects, one each in Israel and Jordan, both influenced by the 'New Archaeology', began to save and study animal bones from excavations. One of these projects was the Heshbon Expedition in Jordan, where animal bones were saved from the first season in 1968 (LaBianca, 1995: 5–10; Younker, 1995: xxi). In Israel, the Gezer expedition employed an interdisciplinary approach (cf. Dever, 1992) and collected animal bones. In addition to the latter two projects, the excavators of Kamid el-Loz in southern Lebanon also saved faunal remains from the project's outset in 1964, later published by Bökönyi (1990).

The reason for the temporal gap between when prehistorians began collecting animal bones and those working in the historical periods started to do so was because, according to LaBianca (1995: 4),

> the concerns of prehistoric anthropologists ... failed to provide a compelling rationale for why the thousands of domestic animal bones routinely unearthed ... should be collected and analyzed ... [Biblical and classical archaeologists'] concerns were with seeking answers to historical questions ... In the minds of most of these scholars, there was little or nothing which the study of animal bones could yield which accorded with their research.

The assumption was that we already know about diet and economy through texts, or that the only problems animal bone studies could address were those of interest to pre-history (e.g. the origins of hunting or animal domestication). This is to our great detriment, as research shows that there is great benefit in comparing text-based evidence with zooarchaeological analyses (e.g. Wapnish, 1993; Lev-Tov and McGeough, 2007). Some such comparisons show great discrepancies between the written word and data-driven research, while others are more complementary but nonetheless lend insight to text interpretations.

It was still longer before zooarchaeology became standard practice, and Hesse (1995: 203) remarked 'in the eyes of some "regular" archaeologists of the historic periods, bone specialists are more like *talismans*. They are largely *symbols*, visible indications of a project's participation in scientific archaeology' (emphasis original). The first step toward normal relations between zooarchaeologists and excavation directors of historic-period sites came when, during the 1980s and 1990s, project directors regularly employed the interdisciplinary approach and incorporated zooarchaeological reports as un-integrated appendices in excavation volumes (cf. Garfinkel, 1993; Gopher, 1993; Lederman, 1993). Despite the late start and uncertain middle period of zooarchaeology in the Levantine historic periods, it has by now become *de rigueur* for excavations to save animal bones and employ a zooarchaeologist. Moreover, discussion of animal bones has become a key feature of ongoing debates concerning the ethnic make-up of many sites. The presence, absence, and relative abundance of pig bones on archaeological sites in Israel have become critical components in an ongoing debate about the ethnic identity of various Early Iron Age sites (e.g. Finkelstein, 1997; Sapir-Hen et al., 2013). Thus, although zooarchaeology has not reached the level of acceptance and integration

Table 23.1 Chronology of the early historic periods in the southern Levant (after Levy, 1995)

Period	Dates BC
Early Bronze Age IA/IB	3,500–3,000
Early Bronze Age II	3,000–2,700
Early Bronze Age III	2,700–2,200
Early Bronze Age IV/Intermediate Bronze Age	2,200–2,000
Middle Bronze Age IIA	2,000–1,750
Middle Bronze Age IIB	1,750–1,550
Late Bronze Age I	1,550–1,400
Late Bronze Age IIA	1,400–1,300
Late Bronze Age IIB	1,300–1,200
Iron Age IA	1,200–1,150
Iron Age IB	1,150–1,000
Iron Age IIA	1,000–900
Iron Age IIB	900–700
Iron Age IIC	700–586

it has in North America or Europe, the sub-discipline has made significant inroads in recent years. Indeed, of four hundred published faunal assemblages from 14,000 years of human settlement in the southern Levant (from the Natufian through the Ottoman periods), over half pertain to the 3,000-year span of early historic periods addressed by this chapter (Tsahar et al., 2009, Table 1). For ease of reference, traditional dates for the various historic periods discussed in this chapter are presented in Table 23.1. In addition to mapping the sites mentioned in this chapter (Fig. 23.1), we have also created a list and map of sites that report faunal analyses from southern Levantine excavations, along with references to where these can be found in the published literature. This resource is openly available in the web-based repository GitHub (http://dx.doi.org/10.6078/M7T151K3), where our colleagues may update the list and add references as new analyses are published.

EARLY BRONZE AGE

The Near East was an early centre of urbanism in the Early Bronze Age (EB). However, because the region's primary states developed in Mesopotamia and Egypt, the complex societies of the southern Levant are considered secondary states. Zooarchaeologists such as Redding (1985; 1992), Zeder (1988; 1991), and Zeder and Blackman (2003) have been active in assessing the role that the manipulation of herding economies had on developing increasingly complex societies. Zeder (1996: 307) examined this issue with reference to pigs at the relatively small but urban site of Tel Halif in southern Israel. There, pigs (*Sus domesticus*) were present in low numbers during the non-nucleated EB I and II periods, but absent during the earliest EB III phase. This fluctuation was interpreted to mean that, rather than being a self-sufficient rural settlement as before, the EB III city was incorporated into a regional network of animal production. Thus pigs—not being easily driven across arid parts of the Levant—were not utilized by the early urban dwellers. Hesse and Wapnish (2001) also examined early urban systems, from several sites in northern Israel. Their data demonstrated that, by the EB I, urban centres were recipients of animals raised at small outlying communities, as early cities had developed regional networks to supply their non-pastoralist workers.

Wapnish and Hesse (2000) examined religion and animal sacrifice when they published the animal bones from an EB I/EB III sacred compound at Megiddo. Of particular interest were mortality patterns in sheep and goats. Wapnish and Hesse (2000: 449) concluded that during the EB I the sacrificial system selected sheep and goats less than 18 months old for ritual slaughter. However, later debris within that compound demonstrated that by the EB III period the sacrificial system used mostly older sheep and goats greater than 42 months old. Similarly, London (2011) published a study of faunal remains from Tal al-'Umayri near Amman, Jordan. She examined evidence for calendrical feasting in the EB period, and argued that a pit filled with some 20,000 pieces

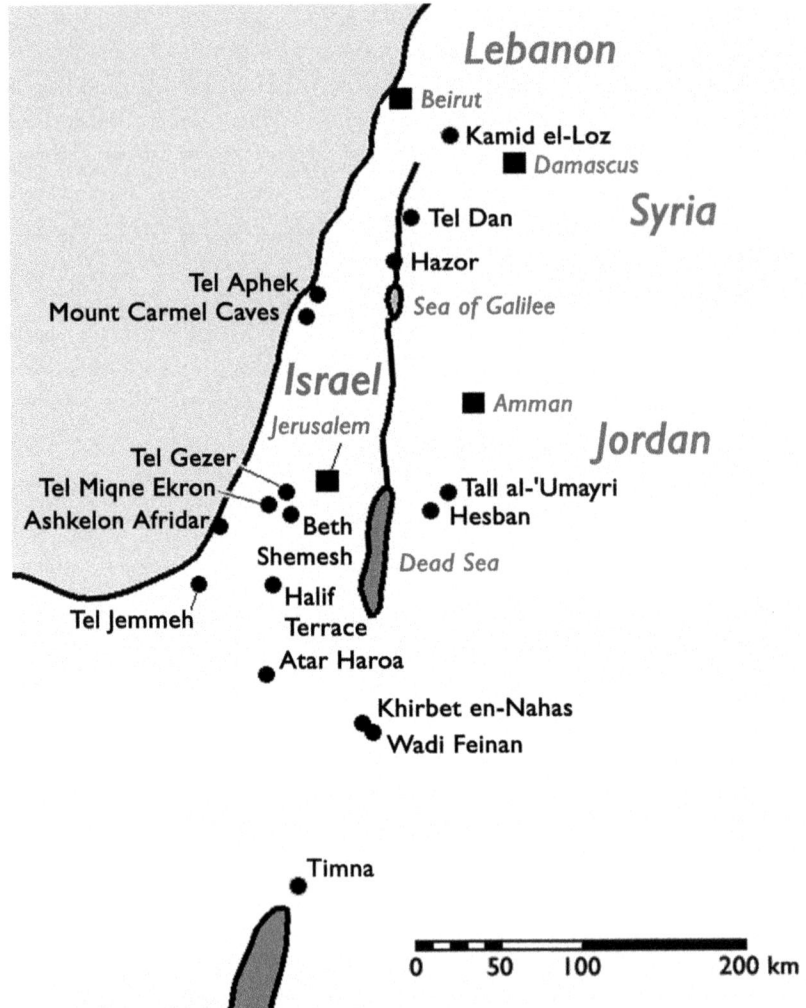

FIGURE 23.1 Map of the southern Levant, showing early historic period sites mentioned in the text.

of animal bone was the detritus of feasts held seasonally during pastoralist migrations through the area.

Kansa et al. (2006) explored the extent of Egyptian influence at the site of Nahal Tillah in the Negev Desert of Israel, since many scholars believe that Old Kingdom Egypt held the southern Levant as a colony during the EB IB. The authors first demonstrated that during the apparent period of Egyptian control, certain areas of the site had more Egyptian pottery than others that featured more local pottery, possibly indicating residence of Egyptians or an Egyptianizing local class. Zooarchaeological analysis suggested that there was no difference in diet/husbandry practices prior to the apparent Egyptian arrival/expansion. However, spatial analysis of ceramic and faunal remains showed only slight patterning between certain taxa, ages, or body parts with Egyptian-style and southern Levantine-style pottery (Kansa et al., 2006). This suggests that, though the inhabitants may have included some Egyptians, boundaries between locals and non-locals were indistinct. There was no clear pattern in the ceramics or the faunal remains to support the idea of a colony with distinct social boundaries. This evidence suggested that Egyptians were present as invited craft specialists or even family members, rather than invading administrators, such that the traditional understanding of imperial influence in the era may be mistaken.

The development of urban centres in the southern Levant during the Early Bronze Age was likely influenced by a key innovation during the early part of this period, the domestication of the donkey (*Equus asinus*) (Rossel et al., 2008). Though equid remains have been identified in small numbers in many sites dating to the Chalcolithic period, it is probable that donkeys became quickly widespread as a transport animal throughout the Near East only during the EB (Ovadia, 1992; Grigson, 1993). Several ceramic donkey figurines carrying loads exist from EB I sites (Epstein, 1985; Ovadia, 1992), which emphasize their value for transport. As discussed above, although it is clear that there was an Egyptian presence in the southern Levant during the EB I, the scale, organization, and motivations for the relationship between these areas are not well understood. The high numbers of equids during the EB IA at the Halif Terrace and in the same period at nearby sites such as Ashkelon Afridar (Kansa, 2004) attest to the donkey's economic importance, possibly for the movement of copper and other materials from far-away (e.g. Egypt) or difficult-to-access areas (e.g. the Wadi Feinan area in Jordan; Levy, 1995).

MIDDLE BRONZE AGE

Zooarchaeological research on Middle Bronze Age (MB) faunal assemblages has been limited; most studies are reports in excavation volumes of particular sites, and therefore do not address broader questions (e.g. Horwitz, 2002; Croft, 2004). Also, many bone collections are small, possibly because MB strata are frequently buried beneath later occupation layers, whereas many EB sites were not extensively reoccupied. Hellwing and Gophna (1984) studied a small MB IIA assemblage from the Israeli coastal site of Tel Aphek, and found that the population consumed higher amounts of pork than earlier, when the city was an Egyptian outpost. Zeder's (1998) hypothesis of low pig use being linked to empires is supported here. Wapnish and Hesse (1988) explored the process of urbanization and economic specialization at the MB site of Tel Jemmeh in southern Israel, generating three models to comprehend urbanization via faunal remains: the self-contained production/consumption economy, the consuming economy, and the producing economy. They concluded, on the basis of sheep/goat (*Ovis/Capra*) mortality patterning, that this city supported a self-contained economy, since the harvest profile included all mortality normally experienced by a herd, such that there was no evidence that the animals were either being exported from or imported to the settlement. The kill-off pattern was bimodal in profile, with peaks at 6–12 months and at 48–72 months. As well, carcass-part ratios showed even distributions of elements across the site, rather than differential allocation of certain portions to specific areas of the city.

Horwitz (2001) focused on religion and animal husbandry in her study of faunal remains from MB tombs. She concluded that, over time, more species were allowed as offerings for the dead. In addition Horwitz (2001: 89) offered two insights into the practice of leaving meat portions with the dead. First, since many such offerings were found in ceramic vessels, it indicates a practice 'reminiscent of a meal'. Also, although more-or-less contemporary texts describe mourners feasting at the tomb, faunal remains did not indicate that. Finally, body-part evidence from the tombs shows a preponderance of limb bones rather than heads and tails. Ancient texts assign both the latter to the gods as divine portions. Thus the meat was intended for the deceased rather than deities.

Late Bronze Age

The Late Bronze Age (LB) has been more intensively studied than preceding historical periods, possibly due to a greater number of ancient texts from the era. Some themes include effects of empire on herding decisions, and the intersection of religion and diet. Marom and Zuckerman (2012) studied LB I and LB II faunal remains from Hazor. The authors found significant and consistent differences in comparing collections from a royal/ritual precinct and a domestic area. Between the LB I and LB II and between the two areas there were declines in male sheep (*Ovis aries*) and goat (*Capra hircus*) kill-off rates, wild game abundance, and differences in the abundance of sideable limb bones. During the LB I, right sides were under-represented, while in the LB II left sides were over-represented. These trends were attributed to an increase in ritual slaughter activity over time in the royal precinct and appropriation by elites of wild game. Differences in the numbers of bones from the right *vs* left sides were suggested as an artifact of religious authorities reserving right sides of animals for sacrifice. Lev-Tov and McGeough (2007) studied the LB I royal/ritual precinct of the same city, interpreting a large bone assemblage to represent the remains of sacrifice and feasting. They employed contemporary religious texts from the Syrian city of Emar, examining the practice of animal sacrifice and body-part allotments to understand the rituals behind the remains at Hazor.

Horwitz and Milevski (2001) explored animal economics and empire demands, surveying published MB and LB faunal assemblages, finding significant changes in species abundances over time. Pigs, relatively frequent at many MB sites, decreased during the LB period. Further, sheep and goats in the MB had been herded in equivalent numbers, but in the LB sheep were more frequent throughout the region. The authors attributed the shift to an Egyptian demand for wool as tribute, as also stated in a contemporary Egyptian war booty list from Megiddo, wherein sheep outnumbered goats 10:1.

IRON AGE I

The Iron Age I (IA I) in the southern Levant has more zooarchaeological research devoted to it than most historical periods. It is the focus period for general archaeological research, because scholars have assigned the settlement of Canaan by the Israelites to this era. The Old Testament provides names of peoples and places, and records events that can be illuminated by archaeological findings. Given that background, it is not surprising that zooarchaeologists have focused on the question of ethnic identity (e.g. Hesse and Wapnish, 1997; 1998), as well as on religion (Wapnish and Hesse, 1991).

Ethnicity is often addressed via relative abundance of pigs, with the idea that Israelite settlements should have low percentages, whereas sites occupied by other peoples might have higher percentages. Hesse (1986), was the first to suggest that high pig percentages might be a Philistine ethnic marker, a people thought by archaeologists to have been immigrants from elsewhere in the Mediterranean world. More recently Sapir-Hen et al. (2013) examined the topic in detail, using a database of over seventy-five published assemblages. Their research supported the linkage between Philistines and pork consumption, but also noted other pig consumption dichotomies: pig consumption tended to be higher in large urban sites and in lowland areas, and not always strictly along ethnic lines.

Tamar et al. (2013) also discussed ethnic identity, based on animal bones from Beth Shemesh (central Israel). The authors showed that in the LB, caprines and cattle (*Bos taurus*) dominated the city's pastoral economy, while pigs were scarce. During the IA I, the same dietary pattern continued, despite the fact that the city, located just east of a major Philistine settlement, served as a border town. Some archaeologists had argued that, based on the presence of Philistine pottery at Beth Shemesh, it must have become a Philistine city. The faunal remains argue for population continuity through time.

IRON AGE II

Zooarchaeological research within Iron Age II (IA II) has addressed themes including religion and imperial effects on pastoral economies. Another avenue of research is the appearance of a new domestic mammal. The camel (*Camelus dromedarius*), the last domestic mammal to enter the region in antiquity, became common during the early first millennium BC. Wapnish (1981) identified over four hundred camel bones from Tell Jemmeh, a site in southern Israel located on a major trade route. Eight camel bones dated to the LB and IA I, while over two hundred dated to the IA II, with the remainder from Hellenistic levels. The marked increase in camel bones beginning in the IA II connects it to imperialism, as the IA II was when a succession of foreign empires conquered the Levant. These empires sought camels as war booty, and were enriched by their use in cross-desert camel caravan trade.

Other Levantine camel studies (Grigson, 2012; Sapir-Hen and Ben-Yosef, 2013) focused on the copper mining sites of Timna and Khirbet en-Nahas, in southern Israel and Jordan. Grigson (2012) interpreted the dating of camel bones from Timna as being LB/IA I and early IA II, thus pre-dating the incense trade and age of empires. She also suggested, on the basis of bone lesions, that camels were employed to bring copper to Levantine trade cities. Sapir-Hen and Ben-Yosef (2013) reinterpreted the camels from the two sites. Based on new high-precision radiocarbon dates, they placed nearly all the camel bones in the early IA II. The later dates match up with camel remains from Tel Jemmeh (Wapnish, 1981) but the connection to copper transport remains unassailed. Rather than link that trade to King Solomon as Grigson (2012: 97) did, Sapir-Hen and Ben-Yosef (2013: 282) instead attribute the animal's appearance to the reorganization of copper trade under Pharaoh Sheshonq I.

Lev-Tov (2010) examined imperial expansion and animal production, focusing on the Assyrian Empire. He employed the concept of staple finance to show that the Assyrians took from their vassal city of Ekron (central Israel) agricultural rather than craft products. Sheep-to-goat ratios increased through time, and in the Assyrian period sheep outnumbered goats 3 : 1. As well, there was a higher incidence of cattle foot-bone lesions, suggesting intensified agriculture or trade. Two other changes hint at an imperial economy: a near-complete disappearance of pigs, which Zeder (1998) argued indicates regional market connections, and an increase in Nile perch (*Lates niloticus*) from Egypt.

Finally, the subject of religion and cult practice has been addressed in the IA II, comparing bones with Old Testament descriptions of Israelite cult. Wapnish and Hesse (1991) compared a residential to a cultic area at Tel Dan, concluding the cultic area showed sacrifice evidence given it contained more bones from young animals, jars filled with bone ashes, and a greater amount of slaughter debris. Another question was whether the sacrificial bones could be attributed either to Israelite worship or other cults. The authors analysed both the Bible and Ugaritic mythological texts to derive assemblage expectations, but could not decide which religion's adherents had conducted the sacrifices (Wapnish and Hesse, 1991: 46–7). More recently, Greer (2013) reached the conclusion, based on analysis of a larger sample from the same site, that the Israelite cult was responsible for the sacrifices.

CURRENT TRENDS AND FUTURE DIRECTIONS

The early historic periods in the southern Levant have a long tradition of zooarchaeological research, as discussed above, and as evidenced by the long and growing list of faunal analyses published over the past fifty years (see: http://dx.doi.org/10.6078/M7T151K3). Researchers see the aggregation and integration of primary zooarchaeological data as an important priority. The ability to integrate multiple datasets allows zooarchaeologists to address large-scale questions, such as the impact of humans on the environment over time, ethnicity and foodways, the emergence of new domesticates, and a host of other questions. Attempts at such meta-analyses include Sasson's (2010) use of published data from seventy sites in his study of the ancient economy of the southern Levant, and the work of Tsahar and colleagues (2009) aggregating data from nearly four hundred reported faunal assemblages in the region to explore changes in the ungulate population during the Holocene.

More integration of published data not only creates new synthetic works of broad scale, but also opens new opportunities for zooarchaeologists working in the southern Levant to build collaborations and draw on data from other disciplines, such as archaeobotany, residue analysis, physical anthropology, and ethnography. Archaeobotany, zooarchaeology's sister discipline, is a natural partner, yet few joint foodways studies have been attempted (Smith and Miller, 2009; Smith and Munro, 2009; VanDerwarker, 2010), largely due to a lack of available data. Indeed, Smith and Munro (2009) lamented the paucity of sites with plant and animal data of sufficient size and quality to inform their large-scale regional survey of plant and animal exploitation in the Bronze and Iron Ages. Since ceramics and animal bones form the most populous artefact classes on historic-period sites, it is natural to attempt combined studies: animal products were stored in, cooked in, and served on ceramic vessels. Additionally, lipid and other residue extraction can now provide direct evidence of vessel use, tying animal remains directly to cuisine and documenting changing animal exploitation over time (Evershed et al., 2008). Ancient texts form a major source of information about human motivations. Lev-Tov and Maher (2001), Lev-Tov and McGeough (2007), and Wapnish (1993; 1995) have made efforts to integrate the two disparate sources of data, but otherwise very little integrative work has been done. Finally, the employment of a variety of techniques can provide a much more nuanced picture of ancient economies. Recent work at Iron Age Atar Haroa in Israel employed a holistic approach to exploring ancient use of plants and animals, incorporating micromorphology, dung and phytolith analysis, and ethnography, among others (Shahack-Gross and Finkelstein, 2008). These types of collaborative studies not only have better data and are more informative, but they also bring zooarchaeology out of the appendix and into the mainstream of archaeological reporting.

References

Anbar, M. and Na'aman, N. (1986–1987) 'An account tablet of sheep from ancient Hebron', *Tel Aviv*, 13–14, 3–12.

Bate, D. M. A. (1937) 'Palaeontology: the fossil fauna of the Wady el-Mughara caves', in Garrod, D. A. B. and Bate, D. M. A. (eds) *The Stone Age of Mount Carmel*, pp. 137–240. Oxford: Oxford University Press.

Bökönyi, S. (1990) *Kamid el-Loz 12: Tierhaltung und Jagd*, Bonn: Habelt.

Braidwood, R. J. and Howe, B. (1960) *Prehistoric Investigations in Iraqi Kurdistan*, Chicago: University Press of Chicago.

Crabtree, P. J. (1990) 'Zooarchaeology and complex societies: some uses of faunal analysis for the study of trade, social status, and ethnicity', in Schiffer, M. B. (ed.) *Archaeological Method and Theory*, pp. 122–205. Tucson: University of Arizona Press.

Croft, P. (2004) 'Archaeozoological studies section A: the osteological remains (mammalian and avian)', in Ussishkin, D. (ed.) *The Renewed Archaeological Excavations at Lachish (1973–1994).* Monograph Series of the Institute of Archaeology of Tel Aviv University 22, pp. 2254–348. Tel Aviv: Institute of Archaeology of Tel Aviv University.

Danin, A. (1995) 'Man and the natural environment', in Levy, T. E. (ed.) *The Archaeology of Society in the Holy Land*, pp. 24–39. New York: Facts on File.

deFrance, S. D. (2009) 'Zooarchaeology in complex societies', *Journal of Archaeological Research*, 17, 105–68.

Dever, W. G. (1992) 'Archaeology, Syro-Palestinian and biblical', in Freedman, D. N. (ed.) *The Anchor Bible Dictionary*, Vol. 1, pp. 354–67. New York: Doubleday.

Douglas, M. (1972) 'Deciphering a meal', *Daedalus*, 101, 61–81.

Epstein, C. (1985) 'Laden animal figurines from the Chalcolithic period in Palestine', *Bulletin of the American Schools of Oriental Research*, 258, 53–62.

Evershed, R. M., Payne, S., Sherratt, A. G., Copley, M. S., Urem-Kotsu, D., Kotsakis, K., Özdogan, M., Özdogan, A., Bailey, D., Nieuwenhuyse, O. P., Akkermans, P. M. M. G., Campbell, S., Farid, S., Hodder, I., Yalman, N., Özbarasan, M., Biçakci, E., Garfinkel, Y., Levy, T., and Burton, M. M. (2008) 'Earliest date for milk use in the Near East and southeastern Europe linked to cattle herding', *Nature*, 455, 528–31.

Finkelstein, I. (1997) 'Pots and people revisited: ethnic boundaries in the Iron Age I', in Silberman, N. A. and Small, D. (eds) *The Archaeology of Israel: Constructing the Past, Interpreting the Present.* Journal for the Study of the Old Testament Supplement Series 237, pp. 216–37. Sheffield: Sheffield Academic Press.

Garfinkel, Y. (1993) 'Respondents: Joseph Garfinkel', in Biran, A. and Aviram, J. (eds) *Biblical Archaeology Today 1990: Proceedings of the Second International Congress of Biblical Archaeology, Jerusalem, June–July 1990*, p. 482. Jerusalem: Israel Exploration Society.

Golden, J. (2004) *Ancient Canaan and Israel: An Introduction*, Oxford: Oxford University Press.

Gopher, A. (1993) 'Respondents: Avi Gopher', in Biran, A. and Aviram, J. (eds) *Biblical Archaeology Today 1990: Proceedings of the Second International Congress of Biblical Archaeology, Jerusalem, June–July 1990*, p. 483. Jerusalem: Israel Exploration Society.

Greer, J. S. (2013) *Dinner at Dan: Biblical and Archaeological Evidence for Sacred Feasts at Iron Age II Tel Dan and Their Significance*, Leiden: E. J. Brill.

Grigson, C. (1993) 'The earliest domestic horses in the Levant? New finds from the fourth millennium of the Negev', *Journal of Archaeological Science*, 20, 645–55.

Grigson, C. (1995) 'Plough and pasture in the early economy of the Southern Levant', in Levy, T. E. (ed.) *The Archaeology of Society in the Holy Land*, pp. 245–68. New York: Facts on File.

Grigson, C. (2012) 'Camels, copper and donkeys in the Early Iron Age of the southern Levant: Timna revisited', *Levant*, 44, 82–100.

Gumerman, G. I. IV, (1997) 'Food and complex societies', *Journal of Archaeological Method and Theory*, 4, 105–39.

Harris, M. (1985) *Good to Eat: Riddles of Food and Culture*, New York: Simon and Schuster.

Hellwing, S. and Gophna, R. (1984) 'The animal remains from the Early and Middle Bronze Age at Tel Aphek and Tel Dalit: a comparative study', *Tel Aviv*, 11, 48–59.

Hesse, B. (1986) 'Animal use at Tel Miqne-Ekron in the Bronze Age and Iron Age', *Bulletin of the American Schools of Oriental Research*, 264, 17–27.

Hesse, B. (1990) 'Pig lovers and pig haters: patterns of Palestinian pork production', *Journal of Ethnobiology*, 10, 195–225.

Hesse, B. (1995) 'Husbandry, dietary taboos and the bones of the ancient Near East: zooarchaeology in a post-processual world', in Small, D. B. (ed.) *Methods in the Mediterranean: Historical and Archaeological Views on Texts and Archaeology*, pp. 197–232. Leiden: E. J. Brill.

Hesse, B. and Wapnish, P. (1997) 'Can pig remains be used for ethnic diagnosis in the ancient Near East?', in Silberman, N. A. and Small, D. (eds) *The Archaeology of Israel: Constructing the Past, Interpreting the Present*. Journal for the Study of the Old Testament Supplement Series 237, pp. 238–70. Sheffield: Sheffield Academic Press.

Hesse, B. and Wapnish, P. (1998) 'Pig use and abuse in the ancient Levant: ethnoreligious boundary building with swine', in Nelson, S. (ed.) *Ancestors for the Pigs: Pigs in Prehistory*. MASCA Research Papers in Science and Archaeology 15, pp. 123–36. Philadelphia: University of Pennsylvania Museum of Archaeology and Anthropology.

Hesse, B. and Wapnish, P. (2001) 'Commodities and cuisine: animals in the Early Bronze Age of northern Palestine', in Wolff, S. R. (ed.) *Studies in the Archaeology of Israel and Neighboring Lands in Memory of Douglas L. Esse*, pp. 251–82. Atlanta: The American Schools of Oriental Research.

Horwitz, L. K. (2001) 'Animal offerings in the Middle Bronze Age: food for the gods, food for thought', *Palestine Exploration Journal*, 133, 78–90.

Horwitz, L. K. (2002) 'Archaeozoological remains', in Scheftelowitz, N. and Oren, R. (eds) *Tel Kabri: the 1986–1993 Excavation Seasons*, pp. 395–401. Tel Aviv: Emery and Claire Yass Publications in Archaeology, Institute of Archaeology, Tel Aviv University.

Horwitz, L. K. and Milevski, I. (2001) 'The faunal evidence for socioeconomic change between the Middle and Late Bronze Age in the southern Levant', in Wolff, S. R. (ed.) *Studies in the Archaeology of Israel and Neighboring Lands in Memory of Douglas L. Esse*. Studies in Ancient Oriental Civilization 50, pp. 283–306. Chicago: The Oriental Institute of the University of Chicago.

Kansa, E. C., Kansa, S. W., and Levy, T. E. (2006) 'Eat like an Egyptian?—A contextual approach to an Early Bronze I 'Egyptian colony' in the southern Levant', in Maltby, M. (ed.) *Integrating Zooarchaeology*, pp. 76–97. Oxford: Oxbow Books.

Kansa, S. W. (2004) 'Animal exploitation at Early Bronze Age Ashqelon, Afridar: what the bones tell us—initial analysis of the animal bones from Areas E, F and G', *Atiqot*, 45, 279–97.

LaBianca, Ø. S. (1995) 'The nature of the zooarchaeological record at [el] Hesban: taphonomical and zooarchaeological studies of the animal remains from Tell Hesban and vicinity', in LaBianca, Ø. S. and von den Driesch, A. (eds) *Hesban 13*, pp. 33–44. Berrien Springs: Andrews University Press.

Lancaster, W. and Lancaster, F. (1991) 'Limitations on sheep and goat herding in the eastern Badia of Jordan: an ethno-archaeological enquiry', *Levant*, 28, 125–38.

Lederman, Z. (1993) 'Respondents: Zvi Lederman', in Biran, A. and Aviram, J. (eds) *Biblical Archaeology Today 1990: Proceedings of the Second International Congress of Biblical Archaeology, Jerusalem, June-July 1990*, pp. 483–4. Jerusalem: Israel Exploration Society.

Lev-Tov, J. E. (2010) 'A plebeian perspective on empire economies: faunal remains from Tel Miqne-Ekron, Israel', in Campana, D., Crabtree, P., deFrance, S., Lev-Tov, J. E., and Choyke, A. (eds) *Anthropological Approaches to Zooarchaeology, Colonialism, Complexity, and Animal Transformations*, pp. 90–104. Oxford: Oxbow Books.

Lev-Tov, J. E. (2013) 'Diet, Hellenistic and Roman period', in Master, D. M. (ed.) *The Oxford Encyclopedia of the Bible and Archaeology*, pp. 296–302. Oxford: Oxford University Press.

Lev-Tov, J. E. and Maher, E. F. (2001) 'Food in Late Bronze Age funerary offerings: faunal evidence from Tomb 1 at Tell Dothan', *Palestine Exploration Quarterly*, 133, 91–110.

Lev-Tov, J. E. and McGeough, K. (2007) 'Examining feasting in Late Bronze Age Syro-Palestine through ancient texts', in Twiss, K. C. (ed.) *The Archaeology of Food and Identity*, pp. 85–111. Carbondale: Center for Archaeological Investigations, Southern Illinois University Carbondale.

Levy, T. E. (1995) *The Archaeology of Society in the Holy Land*, New York: Facts on File.

London, G. (2011) 'Late 2nd millennium BC feasting at an ancient ceremonial centre in Jordan', *Levant*, 43, 15–37.

Marom, N. and Zuckerman, S. (2012) 'The zooarchaeology of exclusion and expropriation: looking up from the Lower City in Late Bronze age Hazor', *Journal of Anthropological Archaeology*, 31, 573–85.

Matthews, R. (2002) 'Zebu: harbingers of doom in Bronze Age western Asia?', *Antiquity*, 76, 438–46.

Mazar, A. (1990) *Archaeology of the Land of the Bible 10,000–586 bce*, New York: Doubleday.

Nissen, H. J. (1990) *The Early History of the Ancient Near East, 9000–2000 bc*, Chicago: University of Chicago Press.

Ovadia, E. (1992) 'The domestication of the ass and pack transport by animals: a case of technological change', in Bar-Yosef, O. and Khazanov, A. (eds) *Pastoralism in the Levant: Archaeological Materials in Anthropological Perspectives*. Monographs in World Archaeology 10, pp. 19–28. Madison: Prehistory Press.

Redding, R. W. (1985) 'The role of faunal remains in the explanation of the development of complex societies in south-west Iran: potential, problems and the future', *Paléorient*, 11, 121–4.

Redding, R. W. (1992) 'Egyptian Old Kingdom patterns of animal use and the value of faunal data in modeling socioeconomic systems', *Paléorient*, 18, 99–107.

Rossel, S., Marshall, F., Peters, J., Pilgram, T., Adams, M. D., and O'Connor, D. (2008) 'Domestication of the donkey: timing, processes, and indicators', *Proceedings of the National Academy of Sciences of the United States of America*, 105, 3715–20.

Sapir-Hen, L., Bar-Oz, G., Gadot, Y., and Finkelstein, I. (2013) 'Pig husbandry in Iron Age Israel and Judah: new insights regarding the origin of the "Taboo"', *Zeitschrift des Deutschen Palastina-Vereins*, 129, 1–20.

Sapir-Hen, L. and Ben-Yosef, E. (2013) 'The introduction of domestic camels to the southern Levant: evidence from the Aravah Valley', *Tel Aviv*, 40, 277–85.

Sasson, A. (2006) 'Animal husbandry and diet in pre-modern villages in Mandatory Palestine, according to ethnographic data', in Maltby, J. M. (ed.) *Integrating Zooarchaeology*, pp. 33–40. Oxford: Oxbow Books.

Sasson, A. (2008) 'Reassessing the Bronze and Iron Age economy: sheep and goat husbandry in the southern Levant as a model case study', in Fantalkin, A. and Yasur-Landau, A. (eds) *Bene Israel: Studies in the Archaeology of Israel and the Levant during the Bronze and Iron Ages in Honour of Israel Finkelstein*, pp. 113–14. Boston: E. J. Brill.

Sasson, A. (2010) *Animal Husbandry in Ancient Israel: A Zooarchaeological Perspective on Livestock Exploitation, Herd Management and Economic Strategies*, London: Equinox.

Shahack-Gross, R. and Finkelstein, I. (2008) 'Subsistence practices in an arid environment: a geoarchaeological investigation in an Iron Age site, the Negev Highlands, Israel', *Journal of Archaeological Science*, 35, 965–82.

Smith A. and Miller, N. (2009) 'Integrating plant and animal data: delving deeper into subsistence. Introduction to the special section', *Current Anthropology*, 50, 883–4.

Smith, A. and Munro, N. (2009) 'A holistic approach to examining ancient agriculture: a case study from the Bronze and Iron Age Near East', *Current Anthropology*, 50, 925–36.

Tamar, K., Bar-Oz, G., Bunimovitz, S., Lederman, Z., and Dayan, T. (2013) 'Geography and economic preferences as cultural markers in a border town: the faunal remains from Tel Beth-Shemesh, Israel', *International Journal of Osteoarchaeology*, DOI: 10.1002/oa.2309.

Tchernov, E. and Horwitz, L. K. (1990) 'Herd management in the past and its impacts on the landscape of the southern Levant', in Bottema, S., Entjes-Nieborg, G., and Van Zeist, W. (eds) *Man's Role in the Shaping of the Eastern Mediterranean Landscape*, pp. 20–216. Rotterdam: Balkema.

Tsahar, E., Izhaki, I., Lev-Yadun, S., and Bar-Oz, G. (2009) 'Distribution and extinction of ungulates during the Holocene of the southern Levant', *PLoS ONE*, 4(4), e5316. DOI:10.1371/journal.pone.0005316.

Twiss, K. (2012) 'The archaeology of food and social diversity', *Journal of Archaeological Research*, 20, 357–95.

VanDerwarker, A. M. (2010) 'Simple measures for integrating plant and animal remains', in VanDerwarker, A. M. and Peres, T. M. (eds) *Integrating Zooarchaeology and Paleoethnobotany: A Consideration of Issues, Methods, and Cases*, pp. 65–74. New York: Springer.

Wapnish, P. (1981) 'Camel caravans and camel pastoralists at Tell Jemmeh', *Journal of the Ancient Near East Society*, 13, 101–21.

Wapnish, P. (1993) 'Archaeozoology: the integration of faunal data with biblical archaeology', in Biran, A. and Aviram, J. (eds) *Biblical Archaeology Today 1990: Proceedings of the Second International Congress of Biblical Archaeology, Jerusalem, June-July 1990*, pp. 426–42. Jerusalem: Israel Exploration Society.

Wapnish, P. (1995) 'Towards establishing a conceptual basis for animal categories in erchaeology', in Small, D. B. (ed.) *Methods in the Mediterranean: Historical and Archaeological Views on Texts and Archaeology*, pp. 233–73. New York: E. J. Brill.

Wapnish, P. and Hesse, B. (1988) 'Urbanization and organization of animal production at Tell Jemmeh in the Milddle Bronze Age Levant', *Journal of Near Eastern Studies*, 47, 81–94.

Wapnish, P. and Hesse, B. (1991) 'Faunal remains from Tel Dan: perspectives on animal production at a village, urban and ritual center', *Archaeozoologia*, 4, 9–36.

Wapnish, P. and Hesse, B. (2000) 'Mammal remains from the Early Bronze sacred compound', in Finkelstein, I., Ussishkin, D., and Halpern, B. (eds) *Megiddo III: the 1992–1996 Seasons*, Vol. 2, pp. 429–62. Tel Aviv: Emery and Claire Yass Publications in Archaeology, Institute of Archaeology, Tel Aviv University.

Weissbrod, L. and Bar-Oz, G. (2002) 'Caprines and toads: taphonomic patterning of animal offering practices in a Late Bronze Age burial assemblage', in O'Day, S., Van Neer, W., and Ervynk, A. (eds) *Behaviour behind Bones: The Zooarchaeology of Ritual, Religion, Status and Identity*, pp. 20–4. Oxford: Oxbow Books.

Younker, R. W. (1995) 'Foreword', in LaBianca, Ø. S. and von den Driesch, A. (eds) *Faunal Remains: Taphonomical and Zooarchaeological Studies of the Animal Remains from Tell Hesban and Vicinity*, pp. xxi–xxii. Berrien Springs: Andrews University Press.

Zeder, M. A. (1988) 'Understanding urban process through the study of specialized subsistence economy in the Near East', *Journal of Anthropological Archaeology*, 7, 1–55.

Zeder, M. A. (1991) *Feeding Cities: The Specialized Animal Economy in the Ancient Near East*, Washington: Smithsonian Institution Press.

Zeder, M. A. (1996) 'The role of pigs in Near Eastern subsistence: a view from the southern Levant', in Seger, J. D. (ed.) *Retrieving the Past: Essays on Archaeological Research and Methodology in Honor of Gus W. Van Beek*, pp. 297–312. Starkville: Cobb Institute of Archaeology, Mississippi State University.

Zeder, M. A. (1998) 'Pigs and emergent complexity in the ancient Near East', in Nelson, S. M. (ed.) *Ancestors for the Pigs: Pigs in Prehistory*. MASCA Research Papers in Science and Archaeology 15, pp. 109–22. Philadelphia: The University Museum of Archaeology and Anthropology, University of Pennsylvania.

Zeder, M. A. and Blackman, M. J. (2003) 'Economy and administration at Banesh Malyan: exploring the potential of faunal and chemical data for understanding state process', in Miller, N. F. and Abdi, K. (eds) *Yeki Bud, Yeki Nabud: Essays on the Archaeology of Iran in Honor of William M. Sumner*, pp. 121–39. Los Angeles: The Cotsen Institute of Archaeology, University of California.

PART IV

AFRICA

CHAPTER 24

MIDDLE AND LATER STONE AGE HUNTERS AND THEIR PREY IN SOUTHERN AFRICA

INA PLUG

INTRODUCTION

THE Stone Age of southern Africa dates back approximately two million years (Mitchell, 2002; Brain, 2004). This chapter deals with the faunal remains from some sites and surveys that cover the last *c.*70,000–90,000 years. A few sites are discussed in detail. I present mostly my experiences and perspectives and some of the challenges of faunal analysis in southern Africa.

The southern African Middle Stone Age (MSA) periods relevant to this chapter are MSA 2 (127–80 KA), Howieson's Poort (80–60 KA), and MSA 3 (60–21 KA) (Mitchell, 2002; Wadley, 2013). The MSA and the Later Stone Age (LSA) overlap between *c.*30–20 KA at the end of the Pleistocene. The association between the MSA and LSA is complex and not yet clear, as both periods have localized variants. The main LSA traditions recognized between 12 KA and *c.*2 KA are discussed by Mitchell (2002: 161). Stone Age technology still persists among hunter-gatherer communities in remote areas. Their modern tool kits contain LSA-type stone and bone tools, but also knives, arrowheads, and tools made from steel, porcelain insulators, and bottle glass (Peters et al., 2009; personal observations). Fig. 24.1 indicates the location of sites and areas mentioned in the text.

The faunal remains reflect aspects of human behaviour, past animal distributions, climate, and environments. The geography and climate zones of southern Africa are complex. From east to west, the landscape changes from semi-tropical to desert. From north to south it changes from dry savannah with summer rainfall in the north and summer rainfall in the semi-tropical northeast to winter rainfall in the far south and

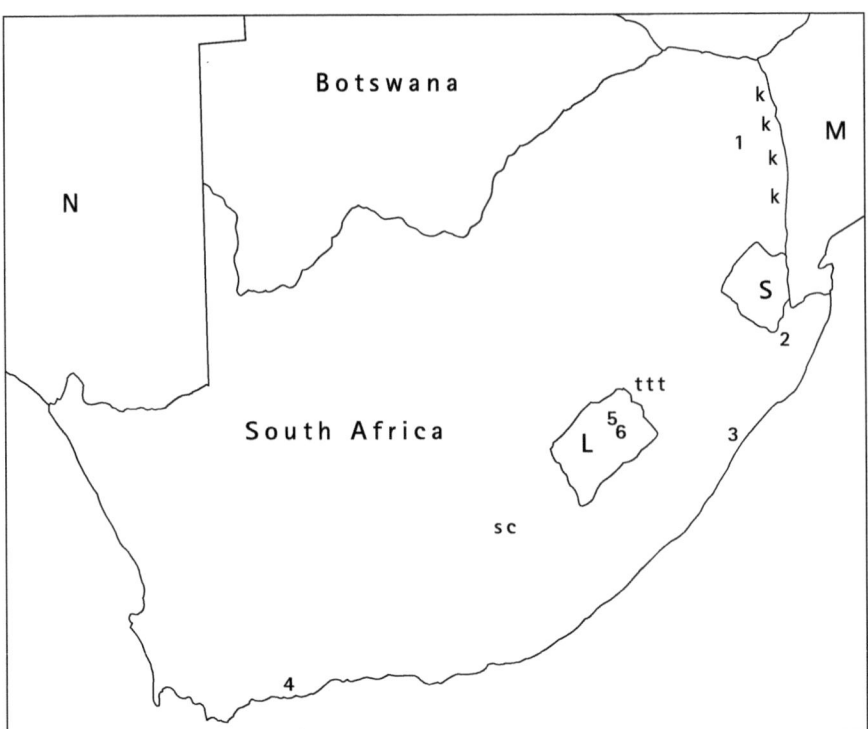

FIGURE 24.1 Southern Africa, location of sites and areas mentioned in the text. N: Namibia, M: Mozambique, S: Swaziland, L: Lesotho, k: Kruger Park, t: Thukela Basin, sc: Seacow Valley, 1. Bushman Rock Shelter, 2. Border Cave, 3. Sibudu Cave, 4. Blombos Cave, Lesotho Survey, 5: Likoaeng, and 6: Sehonghong. Author's own image.

in a narrow strip along the semi-arid to arid west coast. The interior plateau is between 1,000 and over 2,000 metres above sea level, with the northernmost parts within the Tropic of Capricorn. Because of its elevation, the climate is relatively cold, resulting in a variety of macro- and micro-habitats, each with its own endemic animal populations (Skinner and Chimimba, 2005). Knowledge of past animal distributions is essential for the archaeozoologist, but information is not easily come by. Publications on mammal distributions over the last 30,000 years, such as Skinner and Chimimba (2005) and Plug and Badenhorst (2001), represent a useful source of information.

IDENTIFICATION CHALLENGES

Excavation information is essential as the vast numbers of animal taxa and the variety of habitats compound the challenges of identification (Plug, 2014). These variables make it unreasonable to go into an analysis 'blind'. Sites can represent thousands of years of occupation and when no excavation information is supplied such assemblages should not be accepted for investigation. Heavy bone fragmentation also poses severe challenges. For example: the mean lengths of the Bushman Rock Shelter bone flakes varies between 25 mm and 40 mm with a combined mean of 29 mm (Plug, 1978). Identifications on such comminuted material require years of experience and a thorough knowledge of all morphological features of every bone and associated musculature of the different taxa. Access to a well-stocked comparative collection is also essential.

The large number of bovids (the dominant prey animals) and other taxa poses further problems. Publications on osteomorphology and osteometry provide some help (see Walker, 1985; Peters, 1989; Plug and Peters, 1991; Peters and Brink, 1992; Watson and Plug, 1995; Bezuidenhout and Seegers, 1996; Rossouw, 2000; Brink, 2005; Plug, 2005; De Klerk, 2007; Plug, 2014). More work is needed before a complete coverage of the African fauna skeletons is achieved. However, differences in morphological landmarks between geographically separated taxa may occur. In an unpublished study of four species of small bovids, de Wet and the author (Plug, 2014) observed that some of the features that distinguish all four species of bovids from southern Africa were not reliable when tested on the east African taxa.

With no comparative studies, identifications may be problematic. Researchers not thoroughly familiar with the bones of African taxa should base their identifications on comparing a specimen to more than one sample of the same species. Problems are encountered when poorly trained researchers conduct unsupervised analyses on collections that consist of only a few specimens, resulting in a superficial approach to identification.

Techniques that include DNA studies are useful but there are problems. Most old bones have no traceable collagen and DNA left; similarly, bones that have been in contact with other bones in storerooms and bones that were not excavated with DNA studies in mind can seldom be used, as contamination is a major problem. In addition, the costs involved are prohibitive. Most of the faunal assemblages have bone counts that number in the thousands if not a million or more. Attempting DNA analyses on such large samples is currently unaffordable, impractical, and ultimately unworthy. Apart from DNA studies, and as far as it was possible to assess at the time of writing this contribution, proteomics is not yet used on archaeological fauna in southern Africa.

SIZE MATTERS?

Bones that cannot be identified to taxon or genus can be listed according to size: micro, small, medium, large, and very large. These categories are useful, but are arbitrary and need to be properly defined. Classifying long bone flakes into animal size categories is also useful, but only if the flake can be identified to its skeletal element (Swanepoel et al., 2008; Reynard, 2011). Brain (1974) proposed four bovid size classes (I, II, III, IV). I added class V to accommodate extinct large bovids such as the giant buffalo, *Pelorovis*, and an antelope *Megalotragus*. Brain's (1974) size classes are based on the live animal and not on the individual bones. There is overlap in bone sizes with some of the largest animals in the Bov I class and the smallest in the Bov II class; for example, the metacarpals of a female Bov I, *Ourebia ourebi* (oribi) and of a female Bov II, *Pelea capreolus* (grey rhebok) measure 161.7 mm and 166.0 mm respectively, a difference of only 4.3 mm (Plug, 2014). Neither are the tallest nor the smallest of their kind. Overlaps also occur between other size classes: the metapodials of the Bov II *Aepyceros melampus* (impala), for example, are often longer than those of the Bov III *Connochaetes gnou* (black wildebeest) (Plug, 2014). Size ranges within a single species can vary depending on the region they inhabit (personal observation). Such differences may be genetic, but long-term geographic separation, chronic heat stress, and the quality of the grazing are factors that influence size. Size-range overlaps may confuse the unwary, but the morphology of the bones remains true to the taxon, even if size does not always conform.

Size matters also in the degree of preservation. The bones of smaller animals preserve better than those of large animals. Microfauna, fish, and bird bones are often found complete and can readily be identified. This leads to over-representation biases, which must be dealt with when the finds are interpreted. According to some theories, carnivore activities may account for the discrepancies between smaller and larger animal bone preservation (Klein et al., 1999), but in the sites I examined, there was seldom evidence for such activities and, rather, it seems that carcass size influences the survival rate and that differential transportation of larger animals accounts for the discrepancies (Klein et al., 1999; Plug, 2014).

Regardless of the state of preservation, the bones of some animals are more readily identified than others. Tortoise bones, even very small fragments of the carapace and plastron, are easily identified, leading to over-representation. High levels of identification are not achievable with the fragmented bones of most other animals. Such problems in representation must be taken into consideration when the finds are interpreted. Diagnostic zones are of use, but in the heavily fragmented tortoise material from Bushman Rock, for example, such zones were often not preserved.

SKELETAL COMPLEXITY AND UNEQUAL REPRESENTATION

Not all animals have the same number of bones. Animals with more bones than others have a better chance that more of their remains are represented. This leads to over-representation, giving a false impression of abundance. It can be compensated for by using the skeletal bone counts per taxon and using that bone count to work out a formula that equalizes the proportions of taxa recorded; this can be done by employing the QSP (Quantified Skeletal Parts) method proposed by Plug (1988), an adaptation of NISP counts. QSP can be applied to fragmented material, where each identified fragment counts as a specimen unit. The QSP method seems to be the best predictor of animal abundances. It does not estimate the number of individuals, but provides ratios of relative abundance (Plug, 1988; De Ruiter, 2004: 268–9). QSP is calculated by dividing the sum of the specimen units identified per taxon by the skeletal bone count of that taxon.

EXPECT THE UNEXPECTED

Like us, Stone Age people were attracted to the unusual. Out of context specimens may occur: secure identification is essential and possible explanations for unusual specimens must be provided. Animals out of context that I identified include a dinosaur claw, marine shells on inland sites, and bones from the antelope *Philantomba monticola* (blue duiker), a coastal forest dweller, from LSA sites in the highlands of Lesotho. In nature anomalies may also occur and some specimens may become incorporated in archaeological deposits. Such anomalies include horned female antelopes from taxa where horns are supposed to be carried by males only (Bredenkamp, 2008; De Man, 2008), the presence of the first premolar and of the lower second premolar in some bovid taxa, such as *Connochaetes taurinus* and *C. gnou* (blue and black wildebeest respectively), where these teeth are generally absent, and in *Antidorcas marsupialis* (springbok), where the lower and upper second premolars are usually absent but occasionally found. Marine birds are sometimes found in the deposits of inland sites such as Sibudu. Live marine birds have been recorded from the Kruger Park after a cyclone (Clark and Plug, 2008) and occur on mine dams and lakes more than 400 km from the nearest coastline (personal observations).

ARCHAEOLOGICAL SURVEYS: HABITATS, ANIMAL BEHAVIOUR, CLIMATE CHANGE AND HUMAN ACTIVITIES

Animal remains from archaeological surveys have contributed much to our understanding of animal distributions, human demography, and settlement patterns. These include surveys of the entire Kruger Park (Meyer, 1988; Plug, 1988), the central Seacow valley (Sampson, 1985; Plug, 1993; 1994; 1998; 1999; Plug et al., 1994; Plug and Sampson, 1996), the Lesotho highlands, the Drakensberg, and the Thukela Basin (Mazel, 1989; Plug, 1997; 1998; 2002; Mitchell and Charles, 2000; Plug et al., 2003; Plug and Mitchell, 2008a; 2008b).

The faunal analyst should be familiar with animal behaviour, as it largely dictates human hunting techniques. The faunal remains from Sibudu Cave and Bushman Rock, the more recent sites Likoaeng and Sehonghong (Lesotho), and sites in the Seacow valley provide evidence for meat-procuring techniques employed over the millennia. Various trapping and snaring methods, hunting with spears and bow and arrow were all used during the MSA and LSA. Bone-points in the Sibudu deposits (personal observation), as well as quartz arrowheads (Lombard, 2011), provide strong evidence that the hunters were well aware of the best method to obtain their targeted prey. Large and dangerous animals were also obtained. Scavenging may have played a part, but the compositions of the tool kits also suggest active hunting of these animals. No remains of traps and snares older than a few hundred years have been found, but their use on older sites can be inferred from the behaviour of the target prey and by analogies with modern and historical use.

Remains of migratory animals often indicate seasonal human occupation. Migrations can occur over large distances, for instance the springbok during historical times, elephants between South Africa, Mozambique, and Zimbabwe, and zebra and blue wildebeest in east Africa. Migrations also happen over more restricted areas. Eland (*Tragelaphus oryx*) migrations were, and still are, restricted to moving from high to lower elevations at the onset of winter and back again in spring. The remains of migrating eland and of spawning fishes demonstrate seasonal occupation in the highlands of Lesotho (for example at Likoaeng), as do the remains of springbok during the lambing season in the Seacow Valley (Plug, 1994; 1998; 2002; Plug and Mitchell, 2008a; 2008b; Mitchell et al., 2011).

Changes in the representation and frequency of taxa often indicate environmental and/or climate changes. Climatic fluctuations are clearly demonstrated through the movements of bovids in Lesotho and the change from a predominantly forest fauna during the Ice Age at 65–60 KA to a predominantly savannah fauna at c.39 KA at Sibudu (Plug et al., 2003; Plug, 2004; Clark and Plug, 2008; Plug and Mitchell, 2008a; 2008b). The Kruger Park survey identified numerous sites dating from the Early Stone Age to the historical period. Sites in each of the various climatic and vegetation zones were sampled. No deep shelters and caves were found in the Park, but many small shelters had shallow Later Stone Age deposits. Preservation was poor, and carnivores and burrowing animals often contributed bones to the deposits where no archaeological bone was preserved. Only one shelter near Skukuza, dating to between 3 KA and 6.8 KA, yielded bone material (Plug, 1988). The zooarchaeological work was challenging, not for the taxonomic identification in itself, but rather in terms of how the remains were to be interpreted. Hyrax (Procaviidae) remains were identified from areas where they cannot survive today, providing some evidence for possible climate change. The sudden appearance of tsetse fly between c.1750 and 1830 wiped out the cattle herds of the Iron Age people and forced many to move out of the Park and surrounds (Plug, 1988). Similarly, fluctuations in the tsetse absence/presence cycle could have influenced human activities during the Stone Ages. Springbok remains dominate the LSA mammal bone assemblage from Abbot's Cave in the Seacow Valley (Sampson, 1985; Plug, 1994). The unusually large number of foetal, neonatal, and adult female bones indicates that these antelopes were targeted during the lambing season (spring and early summer), therefore suggesting that LSA people were well aware of the migratory routes of these animals (Plug, 1994).

The Seacow Valley survey also provided evidence for the introduction of sheep in the pre-Colonial period. It is as yet unclear whether the hunter-gatherers (Bushman/San) of South Africa obtained sheep before the influx of Khoekhoe pastoralists during the pre-Colonial period, or whether they acquired sheep from these herders. Currently, the fierce Kalahari debate rages on the social status of the hunter-gatherer communities after contact with pastoralists, Iron Age farmers, and European colonists over the past 2,300 years, and how these contacts influenced Bushman communities (see Plug et al., 1994; Plug and Sampson, 1996; Sadr, 1997; Mitchell, 2002; Mitchell et al., 2008; Sampson, 2010).

SIBUDU CAVE AND BUSHMAN ROCK SHELTER

Sibudu and Bushman Rock are large shelters, both of which have been meticulously excavated. The work at these sites (still continuing at Sibudu, but discontinued at Bushman Rock in the 1970s), provides a rare chance to investigate relatively well-preserved faunal remains from a long sequence of Stone Age deposits. Sibudu lies in a cliff above the Tongaat river, northwest of Durban. Bushman Rock, what remains of a cave chamber, is situated near the edge of the escarpment in the northeastern interior, in rugged, mountainous terrain.

When Bushman Rock was excavated, it was one of the few inland sites with bone preservation for sequences older than ~30 KA. Apart from Border Cave, which lies further south on the Swaziland/South African border, Bushman Rock is to date the only site with a long sequence recorded in the escarpment region of northeast South Africa. The two top levels date to the last 200–300 years. Levels 3–17 date to the Early Holocene and Pleistocene/Holocene transition at ~9 KA to ~13 KA (Plug, 1978; Abell and Plug, 2000). Layer 21 is older than 45 KA (Plug, 1978). There are no dates for the rest of the deposits, but lithic and soil studies (Butzer, 1984) suggest that the lower levels could date from 50 KA below level 21 to older than 90 KA near the bottom of the excavation. Two rock-fall episodes represent the last glacial maximum and the cold period at ~60–65 KA (Butzer, 1984). My observations on the site and the fauna enabled us to identify MSA 1, MSA 2, and MSA 3 components (Plug, 1978; Badenhorst and Plug, 2012).

Sibudu has some Iron Age material at the top of the deposit, but no LSA deposits. The MSA deposits were accumulated during Marine Isotope Stages 2 and 4 (38–77 KA) (Wadley and Jacobs, 2006). Four OSL dating clusters are available, representing the final MSA at ~37 KA, late MSA, MSA, and post-Howieson's Poort at ~60 KA, and the Howieson's Poort at ~65 KA (Wadley and Jacobs, 2006; Lombard et al., 2012). The clustered dates suggest punctuated occupation, with long hiatuses in between. Below the Howieson's Poort follow the Still Bay and the pre-Still Bay (Wadley and Jacobs, 2006; Lombard et al., 2012). A preliminary OSL date for the Still Bay places it at > 70 KA. The pre-Still Bay should be older and may fall between ~72 to ~96 KA (Wadley, 2007). Date-wise Sibudu is somewhat similar to the coastal Blombos Cave, but where the fauna from the former is dominated by land animals, the latter is dominated by marine animals (Henshilwood et al., 2001). The pre-Still Bay and Still Bay periods fall within an inter-glacial period when the sea level was somewhat lower than it is today and the animal taxa indicate a predominantly forested environment. Similar taxa were represented during the Howieson's Poort period when the climate had cooled and the sea level was much lower (Wadley, 2013). During the MIS 3 at Sibudu the climate had ameliorated and became drier, forest species became less prominent and savannah species domi-nate the samples (Plug, 2004; Wadley, 2013). The animals of the post-Howieson's Poort and the later MSA indicate a drier climate with easy access to a savannah environment (Plug, 2004). Numerous bovid taxa of all sizes, many carnivores, and animals such as zebra (*Equus zebra*) and giraffe (*Giraffa camelopardis*) were identified (Plug, 2004).

A mosaic of habitats surrounds Sibudu and this was also the case in the past (Clark and Plug, 2008; Wadley, 2013), but the proportions of the mosaic components fluctu-ated over time. The fauna from the pre-Still Bay, Still Bay, and Howieson's Poort are dominated by a small antelope, the blue duiker, as well as by large numbers of bush-pig (*Potamochoerus larvatus*) remains. The blue duiker is a territorial forest dweller and lives either solitary or in small family groups. Bushpigs require bush or high grass cover and open water. These two species were less well represented during the drier phases.

The faunal profiles of Sibudu and Bushman Rock differ and reflect the different habitats wherein these two sites are located and the high-resolution climate data from Sibudu. The Bushman Rock fauna consists predominantly of woodland taxa with lim-ited riverine fauna throughout the LSA and MSA deposits. There is also little difference between the LSA and MSA faunal composition and no bushpigs were identified in any of the periods, nor were forest taxa. The faunal sample was more fragmented, less well preserved, and smaller than that of Sibudu. The taxa varied little over time and although there are indications of climate fluctuations in the MSA levels (Butzer, 1984), the faunal sample is too small to verify these (Plug, 1978; Badenhorst and Plug, 2012).

What of the Future?

Southern Africa has a limited number of archaeozoologists and needs more trained fau-nal specialists. At present training facilities and posts are few. More comparative collec-tions are needed and existing collections should be expanded. We need to convince the authorities that the study of old bones can contribute significantly to addressing con-temporary issues. With better education we may turn the situation around. In the mean-time, sterling work on sites such as Sibudu, Blombos, and many others continues.

References

Abell, P. and Plug, I. (2000) 'The Pleistocene/Holocene transition in South Africa: evidence for the Younger Dryas event', *Global and Planetary Change*, 26, 173–9.

Badenhorst, S. and Plug, I. (2012) 'The faunal remains from the Middle Stone Age levels of Bushman Rock Shelter in South Africa' *South African Archaeological Bulletin*, 67, 16–34.

Bezuidenhout, A. J. and Seegers, C. D. (1996) 'The osteology of the African elephant (*Loxodonta africana*): vertebral column, ribs and sternum', *Journal of Veterinary Research*, 63, 131–47.

Brain, C. K. (1974) 'Some suggested procedures in the analysis of bone accumulations from southern Africa', *Annals of the Transvaal Museum*, 29, 97–9.

Brain, C. K. (ed.) (2004) *Swartkrans: A Cave's Chronicle of Early Man*, Pretoria: Transvaal Museum.

Bredenkamp, J. (2008) 'Nog duiker-ooie met horings' *Wild & Jag; Game & Hunt*, 14(5), 33.

Brink, J. S. (2005) 'The Evolution of the Black Wildebeest, *Connochaetes gnou*, and Modern Large Mammal Faunas in Central Southern Africa'. Unpublished PhD Dissertation, University of Stellenbosch (Stellenbosch).

Butzer, K. W. (1984) 'Archeogeology and Quaternary environment in the interior of southern Africa', in Klein, R. D. (ed.) *Southern African Prehistory and Paleoenvironments*, pp. 1–64. Rotterdam: Balkema.

Clark, J. and Plug, I. (2008) 'Animal exploitation strategies during the South African Middle Stone Age: Howiesons Poort and Post-Howiesons Poort fauna from Sibudu Cave', *Journal of Human Evolution*, 54, 886–98.

De Klerk, B. (2007) 'An Osteological Documentation of Hybrid Wildebeest and Its Bearing on Lack Wildebeest Evolution'. Unpublished MSc Dissertation, University of the Witwatersrand (Johannesburg).

De Man, J. (2008) 'Nog duiker-ooie met horings', *Wild & Jag; Game & Hunt*, 14(5), 33.

De Ruiter, D. J. (2004) 'Relative abundance, skeletal part representation and accumulating agents of macromammals at Swartkrans', in Brain, C. K. (ed.) *Swartkrans: A Cave's Chronicle of Early Man*, pp. 265–78. Pretoria: Transvaal Museum.

Henshilwood, C., Yates, R., Cruz-Uribe, K., Goldberg, P., Grine, F. E., Poggenpoel, C., Van Niekerk, K., and Watts, I. (2001) 'Blombos Cave, Southern Cape, South Africa: preliminary report on the 1992–1999 excavations of the Middle Stone Age levels', *Journal of Archaeological Science*, 28, 421–48.

Klein, R. G., Cruz-Uribe, K., and Milo R. G. (1999) 'Skeletal part representation in archaeofaunas: comments on 'Explaining the 'Klasies Pattern': Kua ethnoarchaeology, the Die Kelders Middle Stone Age archaeofaunas, long bone fragmentation and carnivore ravaging' by Batram & Marean', *Journal of Archaeological Science*, 26, 1225–34.

Lombard, M. (2011) 'Quartz-tipped arrows older than 60 KA: further use-trace evidence from Sibudu, KwaZulu-Natal, South Africa', *Journal of Archaeological Science*, 38(8), 1918–30.

Lombard, M., Wadley, L., Deacon, J., Wurz, S., Parsons, I., Mohapi, M., Swart, J., and Mitchell, P. (2012) 'South African and Lesotho Stone Age sequence updated (I)', *South African Archaeological Bulletin*, 67(195), 120–44.

Mazel, A. (1989) 'People making history: the last ten thousand years of hunter-gatherer communities in the Thukela Basin', *Natal Museum Journal of Humanities*, 1, 1–168.

Meyer, A. (1988) ''n Kultuurhistoriese Interpretasie van die Ystertydperk in die Nasionale Kruger Wildtuin'. Unpublished PhD Dissertation, University of Pretoria (Pretoria).

Mitchell, P. (2002) *The Archaeology of Southern Africa*, Cambridge: Cambridge University Press.

Mitchell, P. J. and Charles, R. L. C. (2000) 'Later Stone Age hunter-gatherer adaptations in Lesotho', in Bailey, G. N., Charles, R. C. L., and Winder, N. (eds) *Human Ecodynamics: Proceedings of the Conference of the Association of Environmental Archaeology*, pp. 90–9. Oxford: Oxbow Press.

Mitchell, P., Plug, I., Bailey, G., Charles, R., Esterhuysen, A., Lee Thorp, J., Parker, A., and Woodburne, S. (2011) 'Beyond the drip-line: a high-resolution open-air Holocene hunter-gatherer sequence from highland Lesotho', *Antiquity*, 85, 1225–42.

Mitchell, P., Plug, I., Bailey, G., and Woodburne, S. (2008) 'Bringing the Kalahari debate to the mountains: late first millennium AD hunter-gatherer/farmer interaction in highland Lesotho', *Before Farming*, 2, 1–22.

Peters, J. (1989) 'Osteomorphological features of the appendicular skeleton of gazelles, genus *Gazella* Blainville 1816, Bohor reedbuck, *Redunca redunca* (Pallas, 1767) and bushbuck, *Tragelaphus scriptus* (Pallas, 1766)', *Anatomy, Histology, Embryology*, 18, 97–113.

Peters, J. and Brink, J. S. (1992) 'Comparative postcranial osteomorphology and osteometry of springbok, *Antidorcas marsupialis* (Zimmerman, 1780) and grey rhebok, *Pelea capreolus* (Forster, 1790) (Mammalia: Bovidae)', *Navorsinge van die Nasionale Museum Bloemfontein*, 8(4), 161–207.

Peters, J., Dieckmann, U., and Vogelsang, R. (2009) 'Losing the spoor: Haiǁom animal exploitation in the Etosha region', in Grupe, G., McGlynn, G., and Peters, J. (eds) *Tracking Down the Past: Ethnohistory Meets Archaeozoology*, pp. 103–85. Rahden: Verlag Marie Leidorf GmbH.

Plug, I. (1978) 'Die Latere Steentydperk van die Boesmanrotsskuiling in Oos-Transvaal'. Unpublished Masters Dissertation, University of Pretoria (Pretoria).

Plug, I. (1988) 'Hunters and Herders: An Archaeozoological Study of Some Prehistoric Communities in the Kruger National Park'. Unpublished PhD Dissertation, University of Pretoria (Pretoria).

Plug, I. (1993) 'The macrofaunal remains of wild animals from Abbot's Cave and Lame Sheep Shelter, Seacow Valley', *Koedoe*, 36, 15–26.

Plug, I. (1994) 'Springbok, *Antidorcas marsupialis* (Zimmerman, 1780) from the past', *Zeitschrift für Säugetierkunde*, 59, 246–51.

Plug, I. (1997) 'Late Pleistocene and Holocene hunter-gatherers in the eastern highlands of South Africa and Lesotho: a faunal interpretation', *Journal of Archaeological Science*, 24, 715–27.

Plug, I. (1998) 'Some evidence for seasonality amongst Later Stone Age hunter-gatherers in southern Africa', *Environmental Archaeology*, 3, 105–9.

Plug, I. (1999) 'The fauna from Later Stone Age and contact sites in the Karoo, South Africa', in Becker, C., Manhart, H., Peters, J., and Schibler, J. (eds) *Historia Animalium ex Ossibus*, pp. 343–53. Rahden: Verlag Marie Leidorf GmbH.

Plug, I. (2002) 'The exploitation of freshwater fish during the Later Stone Age of Lesotho: preliminary results', in Grier, G., Kim, J., and Uchiyama, J. (eds) *Beyond Affluent Foragers*, pp. 24–33. Oxford: Oxbow Books.

Plug, I. (2004) 'Resource exploitation: animal use during the Middle Stone Age at Sibudu Cave, KwaZulu-Natal', *South African Journal of Science*, 100, 151–8.

Plug, I. (2005) 'Osteomorphological differences between some skeletal elements of *Labeobarbus kimberleyensis, Labeobarbus aeneus* and *Labeo capensis* (Pisces: Cyprinidae)', *Annals of the Transvaal Museum*, 42, 5–17.

Plug, I. (2014) *What Bone Is That? A Guide to the Identification of Southern African Mammal Bones*, Pretoria: Rosslyn Press.

Plug, I. and Badenhorst, S. (2001) *The Distribution of Macromammals in Southern Africa over the Past 30,000 Years*, Pretoria: Transvaal Museum.

Plug, I., Bollong, C. A., Hart, T. J. G., and Sampson, C. G. (1994) 'Context and dating of pre-European livestock in the upper Seacow River valley', *Annals of the South African Museum*, 104, 31–48.

Plug, I. and Mitchell, P. (2008a) 'Sehonghong: hunter-gatherer utilization of animal resources in the highlands of Lesotho', *Annals of the Transvaal Museum*, 45, 31–5.

Plug, I. and Mitchell, P. (2008b) 'Fishing in the Lesotho highlands: 26,000 years of fish exploitation, with special reference to Sehonghong Shelter', *Journal of African Archaeology*, 6(1), 33–55.

Plug, I., Mitchell, P., and Bailey, G. (2003) 'Animal remains from Likoaeng, an open-air river site, and its place in the post-classic Wilton of Lesotho and eastern Free State, South Africa', *South African Journal of Science*, 99, 143–52.

Plug, I. and Peters, J. (1991) 'Osteomorphological differences in the appendicular skeleton of *Antidorcas marsupialis* (Zimmerman, 1789) and *Antidorcas bondi* (Cooke & Wells, 1951) (Mammalia: Bovidae) with notes on the osteometry of *Antidorcas bondi*', *Annals of the Transvaal Museum*, 35(17), 253–64.

Plug, I. and Sampson, C. G. (1996) 'European and Bushman impacts on Karoo fauna in the nineteenth century', *South African Archaeological Bulletin*, 51, 26–31.

Reynard, J. P. (2011) 'The Unidentified Long Bone Fragments from the Middle Stone Age Layers at Blombos Cave, Southern Cape, South Africa'. Unpublished MSc Dissertation, University of the Witwatersrand (Johannesburg).

Rossouw, L. (2000) 'The taxonomic status of fossil springbok, *Antidorcas australis* Hendey & Hendey 1968, as reflected by its postcranial osteomorphology'. Unpublished MSc Dissertation, University of the Witwatersrand (Johannesburg).

Sadr, K. (1997) 'Kalahari archaeology and the Bushman debate', *Current Anthropology*, 38, 104–12.

Sampson, C. G. (1985) *Atlas of Stone Age Settlements in the Central and Upper Seacow Valley*, Bloemfontein: National Museum.

Sampson, C. G. (2010) 'Chronology and dynamics of Later Stone Age herders in the upper Seacow valley, South Africa', *Journal of Arid Environments*, 74, 842–8.

Skinner, J. D. and Chimimba, C. (2005) *The Mammals of the Southern African Subregion*, Cambridge: Cambridge University Press.

Swanepoel, E., Steyn, M., and Scheepers, M. D. (2008) 'A preliminary report on animal bone classification through computerized methods', *Annals of the Transvaal Museum*, 45, 21–9.

Wadley, L. (2007) 'Announcing a Still Bay industry at Sibudu Cave', *Journal of Human Evolution*, 52, 681–9.

Wadley, L. (2013) 'MIS 4 and MIS 3 occupations in Sibudu, KwaZulu-Natal, South Africa', *South African Archaeological Bulletin*, 68, 41–51.

Wadley, L. and Jacobs, Z. (2006) 'Sibudu Cave: background to the excavations, stratigraphy and dating', *Southern African Humanities*, 18(1), 1–26.

Walker, N. (1985) *A Guide to Post-Cranial Bones of East African Mammals*, Norwich: Hylochoerus Press.

Watson, V. and Plug, I. (1995) '*Oreotragus major* Wells and *Oreotragus oreotragus* (Zimmerman) (Mammalia: Bovidae): two species?', *Annals of the Transvaal Museum*, 36(13), 183–91.

··

CHAPTER 25

PASTORALISM IN SUB-SAHARAN AFRICA

emergence and ramifications

··

DIANE GIFFORD-GONZALEZ

INTRODUCTION

···

THE phrase 'African pastoralism' may prompt iconic images to spring to readers' minds: Maasai warriors with long, plaited hair and spears, colourfully adorned Wodaabe men dancing before eligible girls at annual gatherings, Tuareg 'blue men' riding camels across the Sahara. Exotic males in their prime, captured by foreign cameras, appear timeless and free from modern political entanglements or even family ties.

On all counts, such images mislead. Herdsmen, warriors, and traders today live, as they have for centuries, in a web of social relations extending from the family and community to nation states that affect their lives. The picturesque garb of festival times is not just 'tradition' but an assertion of difference, of resistance to assimilative pressures. Today's African pastoralists use mobile phones, send family members to parliament, fight in civil wars, and suffer as the victims of them. They are neither timeless nor unchanging, nor they have they been in the past. This chapter traces the emergence, spread and ramifications of African pastoralism.

Pastoralists are here defined as people who depend mainly on the products of their hoofed domestic animals, and who organize their settlement and mobility strategies for the needs of their livestock. They may be integrated into market systems, selling herd products for currency with which to purchase other necessities. They may use barter to exchange products or animals with other specialized food producers or even foragers. They may logistically organize herding by some household members, while others farm (Hadjigeorgiou, 2011). So long as their animals' needs dictate their annual activities, they fit this definition.

This handbook's charge is to emphasize how *zooarchaeological* evidence elucidates key archaeological regions and research questions. As others, this chapter shows that zooarchaeological evidence now is not restricted to bones, teeth, hair, scales, and shells, as it was even a few decades ago. Our understanding of how past humans and animals interacted and coevolved has been enriched by human and animal genetics, stable isotope analyses of residues in ancient ceramics, and ethnographic studies of herders.

SAHARAN BEGINNINGS

Debates continue over details of animal domestication in Africa, but nearly all scholars agree that uniquely African forms of pastoralism emerged in what is now the Sahara, and that they did so well before any evidence for farming exists (Linseele, 2010). Palaeoenvironmental records testify that the span between 12,000 and 9,000 years ago was extraordinarily moist in African latitudes 0° to 30° N, with well-watered grasslands interspersed with lakes and rivers and forests in the Saharan highlands (e.g. Kuper and Kröpelin, 2006). Fluctuating cycles of desiccation and amelioration, but always trending in a drier direction, ensued over the next three millennia in the Sahara. This culminated with intense drought in what is now the Saharan core around 5,500 years ago, without return to appreciably moister conditions, as the Inter-tropical Convergence Zone (ICTZ) assumed annual movements approaching the modern.

During the Holocene Humid Phase, the Saharan region saw an influx and growth of foraging populations. Where the archaeological record is well enough known, it shows people immigrating from adjacent areas (Kuper and Kröpelin, 2006). Sites testify to hunter-gatherer-fishers settling around major lakes and rivers. Abundant grinding stones and storage pits reflect seasonal harvest of dense stands of wild African grains. From the mid-tenth millennium BP onwards, well-dated ceramics are widespread throughout the region (Haour, 2003).

African pastoralism emerged as a response to waning and less spatially predictable rainfall during the eighth to seventh millennia BP. Climatically conditioned declines in local land productivity likely inclined northeastern Sahara peoples to adopt sheep and goats from southwest Asia, along with or separate from cattle. Smith (1992, 2005) contends that the shift from forager to pastoralist ideology is so radical that few made such a transition, but Marshall and Hildebrand (2002) argue that Saharan foragers who stored wild grains would have already had ideas about ownership and delayed returns on invested effort. In the late ninth millennium BP, some foragers in the Fezzan (Libya) tried penning aoudads (*Ammotragus lervia*), the wild goat of North Africa, perhaps prompted by regional desiccation (di Lernia, 2001).

ORIGINS OF AFRICAN DOMESTIC LIVESTOCK

Although four of five animal *species* involved in African pastoralism are wholly or partly from southwest Asia, African pastoral *systems* are not derivative. Linseele (2010) notes the African pastoralist triad of sheep, goats, and cattle has no direct parallel in southwest Asia, where pastoralism is based on sheep, or on camels and goats in arid zones. Domestic cattle (*Bos taurus*), sheep (*Ovis aries*), and goats (*Capra hircus*) have been ubiquitous in Africa since the eighth millennium BP, augmented by the donkey (*Equus asinus*) in the sixth millennium BP (Rossel et al., 2008), and the camel (*Camelus dromedarius*) in northern Africa by the second millennium BP (Gifford-Gonzalez and Hanotte, 2013).

Goats and Sheep

Domestic goats, descendants of the wild bezoar goat (*Capra aegagrus*) of the Zagros Mountains, display at least six distinct mtDNA lineages (Luikart et al., 2001; see Peters et al., this volume), which display little geographic patterning in these haplogroups. Pereira et al. (2009) argue that modern goats' Y-chromosome haplotype uniformity from the Levant along the North African coast to the Maghreb reflects early maritime trading.

Sheep descend from the Asiatic mouflon, *Ovis orientalis*, with two common mtDNA haplogroups, A and B, and three less common lineages (Meadows et al., 2007), suggesting separate domestications. All contemporary eastern and southern African sheep studied so far bear haplogroup A, the widest-spread worldwide (Bruford and Townsend, 2006). However, archaeological sheep aDNA from the Western Cape displayed only haplogroup B (Horsburgh and Rhines, 2010). Y-chromosome research identifies two separate lineages in African sheep (Meadows et al., 2007). No clear phylogeographic pattern is apparent in African breeds. African sheep breeds may be thin-tailed, found today in western Africa to Sudan, or fat-tailed/fat-rumped, from eastern and southern Africa.

Caprines might be expected to have appeared in the Nile Valley, closest to the Sinai Peninsula, earlier than those in Sudan, but instead they occur there later, in deltaic Merimda Beni Salama by *c.*6100 BP and in the Fayûm Depression 7200–6500 BP (Gautier, 2002). Early Dynastic Egyptian tomb paintings (5100–4613 BP) depict thin-tailed sheep; fat-tailed types are first illustrated during the Middle Kingdom (3990–2630 BP) (Clutton-Brock, 1993).

Gautier (2002) reports caprine (sheep/goat) remains from early eighth-millennium BP sites in the eastern and western deserts of southern Egypt. They appear in the eighth-millennium BP Sodmein Cave-Tree Shelter site, east of the Red Sea in southern Egypt, with one goat definitively identified (Linseele and Van Neer, 2008). By the late eighth to early seventh millennia BP, caprines appear, again without cattle, in the Late Khartoum phase of the Nubian Nile, integrated with riverine resource exploitation and either intensive wild gathering or cultivation of wild-morphology sorghum (Wetterstrom, 1998). These data support Hassan's (2000) proposal that caprines were introduced to Africa about 7000 BP from the Mediterranean coast as well as via Egypt's Red Sea Hills coastal region. In both regions, they initially appear before cattle.

Were Wild African Cattle Domesticated or Southwestern Asian Cattle Introduced?

Morphologically domestic cattle are found in eastern Saharan sites and near the Nubian Nile from the mid-eighth millennium BP (Chenal-Vélardé, 1998; Linseele, 2013). Whether the African aurochs (*Bos primigenius africanus*), native to northern Africa to the Second Cataract of the Nile, contributed to African domestic cattle has been debated for four decades. On morphological grounds, southwest Asian cattle, domesticated by 10,000 BP, could be ancestral to early African domesticates, but strictly on logical grounds, so could African aurochs. In species with extensive geographic distributions, multiple domestications are possible, as has been shown for pigs (Larson et al., 2005).

In the 1980s–1990s two zooarchaeologists argued for an African domestication, Grigson (1991) basing her argument upon the morphology of ancient Egyptian cattle and Gautier (2001) on the early dates and ecological context of cattle specimens. In eleventh- to tenth-millennium BP archaeofaunas dominated by steppe-adapted gazelles and hares from Nabta and Bir Kiseiba (Egypt) (Fig. 25.1), Gautier (2001: 627–8) noted a few *Bos* specimens. He suggested that water-dependent cattle could only coincide with such species if under human management. Despite Gautier's cautious discussion of these rare specimens, Nabta-Kiseiba excavators Wendorf and Schild (1984, 2001) advocated an autochthonous North African cattle domestication. Others argued against this on multiple grounds (see Smith, 1992; Brass, 2013; Linseele, 2013; Stock and Gifford-Gonzalez, 2013). More recently, excavators of tenth-millennium BP Wadi al-Arab, east of the Nubian Nile, claimed that large bovid bones recovered there were domestic cattle (Honegger and Bastien, 2009). Linseele (2013) studied the specimens and reports that they are neither wild nor domestic *Bos*.

In sum, *archaeofaunal* evidence for domestic cattle in Africa older than the mid-eighth millennium BP is not compelling. Bovines of domestic morphology become abundant in archaeological sites over the next millennium, fitting a reasonable temporal window for their introduction from southwest Asia.

Genetics and African Cattle

Geneticists' original inferences about African cattle origins were based upon PCR analyses of modern African breeds' mitochondrial DNA control region. One mtDNA haplogroup, T1, has very high frequencies in African cattle, with low occurrences in the Iberian Peninsula and Levant, leading researchers to posit an independent domestication from wild North African aurochs (Bradley, 2003; Edwards et al., 2004).

FIGURE 25.1 Map of Africa, showing sites and regions mentioned in the text. Regions are represented in italics. Author's own image.

Later, whole-genome mtDNA analysis has produced different interpretations. Most domestic taurine cattle belong to the macro-haplogroup T, comprising six geographically distributed haplotypes. The southwest Asian T, T1, T2, and T3 are of particular interest. T3 predominates in Europe, reflecting dispersal there of Near Eastern stock. Achilli et al. (2009: R158) note that the dominant African T1 is distinguished from haplotype T by a single recurrent mutation in the control region, and that variants of the T1 haplotype, like those of European T3, radiate in the starburst pattern typical of rapid population expansion from a founding haplotype. They argue that T1 in Africa, like T3 in Europe, derives from southwest Asia, with founder's effect structuring haplogroup frequencies. Current data thus support a southwest Asian origin for African maternal lineages.

African Y-chromosomes may tell a different story. Pérez-Pardal et al. (2010) studied Y-specific microsatellites in 608 *Bos taurus* males from forty-five cattle populations in Europe and Africa. One Y2 'subfamily,' a group of nine exclusively African haplotypes among the thirty-two within the Y2 haplogroup, displays greater internal diversity than do all European Y2 haplotypes. They suggest that African aurochs bulls introduced multiple Y-chromosome lineages into southwest Asian derived domestic cattle populations but caution that study of southwest Asian Y2 haplotype diversity is needed before definitive pronouncement that this subfamily is exclusively African.

In sum, current genetic evidence testifies to a southwest Asian origin for maternal lines in African cattle, but allows for the possibility of admixture of African aurochs genes through the paternal line.

Pastoralism Emerges in Saharan Savannas

Radiocarbon dates indicate that domestic cattle spread west from northeastern Africa, across then-productive savannas to the central and western Sahara over several thousand years (Linseele, 2013), occupying the south central Sahara by 6000–5000 BP, and the far western grasslands by 4500–4000 BP (Manning, 2011). The final Saharan desiccation of the mid-sixth millennium BP transformed formerly productive grasslands to desert, and zooarchaeological evidence reflects movement of cattle southward into forest-margin West Africa and East African savannas (Linseele, 2010).

Such a smooth chronological narrative elides a pervasive problem in research on the emergence and nature of Saharan pastoralism: lack of thorough published analyses of whole archaeofaunas. Presence of domestic cattle is often interpreted as proof of pastoralism, rather than the first step in demonstrating its plausibility. Recent publications by di Lernia et al. (2013) on ritual practices in the Fezzan (Libya), by Manning and MacDonald (2005) on Tilemsi Valley (Mali) pastoralism, and by Gifford-Gonzalez and Parham (2008) on Adrar Bous (Niger) pastoral fauna go beyond species identification and simple osteometrics to construct mortality profiles, or to document butchery and cooking and ritualized animal use.

Dairying in Context: Genetics of Human Lactase Persistence and Lipid Analysis

Archaeologists have long speculated about when African pastoralism incorporated milking and the use of milk products, but until recently, evidence for dairying was equivocal. 'Bovidean Period' Saharan rock art attributed to the fifth to fourth millennia BP, depicts cows with full udders and scenes of milking (Muzzolini, 1986), but most lack actual dates. Comparative linguistics places words for milking early in the emergence of two Saharan language families, Afroasiatic and Nilo-Saharan, but precisely when is debated (Blench, 1993; Ehret, 1997).

New research on human genetics and lipid residue analysis has recently elucidated the antiquity of African dairying. Tishkoff et al. (2007) investigated genetics of lactose digestion among eastern Africans, discovering three mutations that enable continued lactase production in adults (a single Eurasian mutation for lactase persistence, C/T-13910, exists). The African C-14010 haplotype for lactase persistence is common among Nilo-Saharan speakers, with hints of high frequencies among Afroasiatic speakers. Assuming average mutation rates, Tishkoff et al. (2007: 198) estimated the age of the C-14010 mutation as 3000–7000 years BP, with a 95% confidence interval (1,200–23,200 years). This broad confidence interval makes genetic data suggestive rather than definitive.

Spectrographic analysis of lipid residues from the inner walls of pots offers greater temporal resolution. Such research had earlier established that southwest Asians were processing dairy products by the ninth millennium BP, shortly after the first appearance of domestic cattle (Evershed et al., 2008). Dunne et al. (2012) analyzed ceramics from the well-dated stratigraphy in Takarkori rock shelter (southwest Libya). Dairy fats appear in half of the vessels from Middle Pastoral levels (7200–5800 BP) but no earlier. Lipid signatures in ceramics indicate that milk was heated, suggesting technological tactics to break down lactose.

Together, these findings indicate that Africa's cattle-based dairy pastoralism appeared by the late eighth millennium BP, not long after the first appearance of domestic cattle, with selection for lactase persistence soon thereafter.

Pastoral Mobility Strategies

When pastoralists moved into what is now sub-Saharan Africa, they had donkeys as well as cattle to help carry household gear. Rossel et al. (2008) argued that the wild African ass (*Equus africanus*) came under domestication 6000–5000 BP, as an aid to pastoral mobility, when increasing aridity made residential moves longer and more arduous. Osteometric study of sacrificed equids in the Early Dynastic (c.5000 BP) tomb complex at Abydos (Egypt) established that these animals fall within the size ranges of wild *E. africanus*, but their domestic status is indicated by skeletal markers of weight-bearing (Rossel et al., 2008). mtDNA research indicates two domestication trajectories (Kimura et al., 2013), one from Nubian wild asses, another from an extinct lineage more closely related to Somali wild asses.

Tafuri et al. (2006) used stable isotope analysis of human bone to document changes from Middle (6100–5000 BP) to Late Pastoral Phases (5000–3500 BP) residential movements, when desiccation would have made herding more arduous. Cattle were replaced by caprines, and cattle-focused rituals shifted to practices centred on burials of male humans.

Working farther west in the Tilemsi Valley (Mali), Manning (2008) argues that this tributary of the Niger, then a flowing river and surrounding grasslands, linked the Sahara to the north with the Inland Niger Delta (IND). She interprets the 7–15 ha habitation mounds dating to 4600–4000 BP as semi-permanent residential bases of pastoralists who had no need to pursue a highly nomadic annual cycle of movement, given the richness of forage for cattle. Tilemsi cattle osteometrics show a cluster of animals the size of modern N'dama and other small West African shorthorn breeds, with some larger individuals. Manning notes similar size contrasts in other, slightly later, IND sites, including Jenné Jeno and Dia (MacDonald, 1995; Manning and MacDonald, 2005). These might represent either sexual dimorphism, with cows predominating in the samples, or existence of larger animals from more mobile herders.

Saharan Pastoral Ritual Practices

Cattle figure centrally in Saharan ritual practices, as sacrifices, as ritually placed remains of meals, or as common objects of representation in the region's stunning rock art. Even among late Pleistocene foragers along the Nubian Nile, aurochs played a significant symbolic role: Wendorf and Schild (2013: 667) report that aurochs skulls used to mark graves at the Qadan site of Tushka (14,500–12,000 cal BP two sigma), a practice continued with domestic cattle on the Nubian Nile at sixth-millennium BP El Barga (Honegger, 2005).

'Bovidean Phase' rock art of the Aïr, Tassili (Fig. 25.1), and elsewhere attests to the centrality of cattle. Representations of domestic life including houses with gourds or pots full of food and cows returning for milking portray an ideal of prosperity, while others depict dances and cattle sacrifices.

Di Lernia (2006) summarizes the Saharan record of domestic cattle sacrifices and burials. Burials of sacrificed heifers under tumuli or structures are common from the mid-seventh-millennium BP Late Neolithic of Nabta-Kiseiba, Egypt, to the mid-fifth-millennium BP of the Tilemsi Valley (Manning, 2008) (Fig. 25.2). Interment of cattle parts, often with evidence of intense burning, is common through the central Sahara (Paris, 2000; Gifford-Gonzalez and Parham, 2008), as is burial of cattle parts with humans.

Di Lernia et al. (2013) document cattle 'burnt offerings' in monumental complexes of the Middle Pastoral (9200–7800 BP) phase in the Fezzan, notable for their careful preparation of a basal sedimentary deposit, usually circular, and erection of 'standing stones'. Cattle, mostly adult males, were decapitated, their heads systematically deposited, and postcranial bones disarticulated, incinerated, and deposited.

Some Saharan pastoralist stone monuments are circular, being interpreted as sites of summer solstice rituals (Wendorf and Schild, 2013; di Lernia, 2006). The ICTZ shifts farthest north in summer, and during the millennia of increasingly unpredictable precipitation these may have been situated to ritually invoke the rains.

FIGURE 25.2 A heifer burial in a pit dug into the occupation mound at Karkarichinkat Nord, lower Tilemsi Valley, Mali. Excavators recovered evidence of a roofed wooden structure over the burial (Manning, 2008). Courtesy of K. M. Manning.

Saharan Social Arrangements

Tafuri et al. (2006) suggested that changes in $^{87}Sr/^{86}Sr$ isotopic signatures in burials from the Middle to Late Acacus sequence reflect the emergence of patrilocality and a male-centred social order. Holl (1995) made the case that a panel of paintings from Tikadourine (Tassili) represents a diachronic narrative of a male 'pathway to elderhood', from boyhood through to initiation to social maturity, marriage, and family, each stage accompanied by specific activities with domestic and wild animals.

While pastoral peoples worldwide espouse an egalitarian ethos, citing the leveling effects of herd losses due to drought and disease, studies of wealth transmission in Eurasian and African pastoralists reveal that affluent pastoralists actually do transmit differentially more wealth to their offspring (Bradburd, 1982; Borgerhoff Mulder et al., 2010). MacDonald (1998) and Brass (2007) argue that, at the apogee of ancient Saharan pastoralism, elites arose, who would have motivated the monument construction and sacrifices described above and been key agents in trade.

MacDonald (1998) suggested that the relatively standardized, small *hâches néolithiques*, found from the Nile to the west-central Sahara, served as circulating currency. Though these may not be currency in the strictest definition, they may have been markers for deferred exchanges of goods, including livestock, as were clay forms in Neolithic southwest Asia (Schmandt-Besserat, 1986).

PASTORAL MIGRATION
INTO SUB-SAHARAN AFRICA

By the fourth millennium BP, the inner Sahara had reverted to desert, and areas to its south underwent the recession of vegetation zones from their earlier northern extensions (Kröpelin et al., 2008). Pastoral groups moved southward into what is now the central and west African Sahel (Smith, 2005) and into East Africa. Breunig and Neumann (2002) traced pastoralists moving into the Lake Chad basin (Nigeria) in the third millennium BP. These Chad herders added low-level pearl millet cultivation (*Pennisetum glaucum*) to wild grain gathering, as did people in Mali (Manning, 2008) and Mauritania (Amblard, 1996). Jousse (2006), surveying diachronic changes in herding, hunting, and fishing, found that generalized hunting continued in tandem with herding and that specialized fishing emerged during the driest phases of regional history.

The time-honoured system of pastoralism with hunting and wild gathering appears finally to have been extinguished during an 800-year, pan-Sahelian drought in the first millennium BP (Breunig and Neumann, 2002). Later African pastoralists specialized in livestock rearing in complementary economic relations with ethnically distinct farmers and fishers.

The southward movement of cattle would have been stalled where pastoralists encountered bush and forest infested with tsetse fly, the vector for trypanosomiasis or sleeping sickness (Smith, 1992). Over time, pastoralists learned to manage this disease by avoiding such habitats in some areas and using fire and goats to clear brush in others (Gifford-Gonzalez, 2000). Several breeds of trypanotolerant cattle evolved in West African closed habitats (Gifford-Gonzalez and Hanotte, 2013).

EASTERN AFRICA

Political circumstances have impeded tracing a complementary trajectory of migration from the Sahara into East Africa. Chad, southwestern Sudan, and northern Uganda, logical zones to survey for such evidence, have been embroiled in civil wars for decades. We do know that general technological similarities link pre-pastoral, Holocene Humid Phase groups on the Nile with groups around Lakes Victoria and Turkana. Kansyore ceramics from Kenya, Uganda, Tanzania, and southeastern Sudan, dating from the ninth through mid-second millennia BP (Dale and Ashley, 2010), display special affinities with Sudanese pottery.

Pastoralists immigrated into the Lake Turkana basin by 5000–4500 BP (Hildebrand and Grillo, 2012). The most compelling evidence that the Lake Turkana immigrants were of Saharan tradition is their construction of standing-stone monuments, with basal platform preparation, human inhumations, and sacrifice of animals plus destruction of finely wrought ceramic vessels (Hildebrand and Grillo, 2012). Lake Turkana's Nderit ceramic tradition is found in south-central Kenya and northern Tanzania by the early fourth millennium BP, but thus far associated only with wild fauna and rare caprines, instead of cattle or monumental architecture (Gifford-Gonzalez, 2000).

By 3000 BP, cattle-based pastoralism was widespread in East Africa. Gifford-Gonzalez (2000) proposed that the 600- to 1,000-year gap between the first evidence of small stock and that of bovine-based pastoralism south of Lake Turkana may have been caused by novel endemic disease challenges from wild savanna bovids. Wildebeest-borne malignant catarrhal fever (WD-MCF), 99% fatal to exposed cattle, would have been encountered for the first time south of the Lake Turkana basin.

By 3000–2000 BP, two pastoralist groups with sufficient knowledge of disease vectors and etiology to allow cattle to thrive in the savannas existed in central Kenya to northern Tanzania. These display complementary geographic ranges, divergent lithic and ceramic *chaînes opératoires*, and obsidian distribution systems. The Elmenteitan occupied southwest Kenya to east of Lake Victoria. The Savanna Pastoral Neolithic is found in the Rift Valley and Athi-Kapiti Plains (Kenya), the Serengeti Plains south to Lake Eyasi, and the Mbulu Plateau (Tanzania) (Prendergast et al., 2013), where it may have been limited by tsetse-infested brush and forest habitats.

Relationships between indigenous foragers and immigrant pastoralists have recently begun to be explored systematically and with theoretical sophistication (e.g. Ashley, 2010; Prendergast and Mutundu, 2010).

Southern Africa

The routes and the nature of introductions of domesticate livestock into southern Africa remain obscure. The region does display another, later gap between the first appearance of caprines and that of cattle, at least in the Western Cape region. Sheep appear in southern Africa in the late third millennium BP (Sealy and Yates, 1996). Around 1200 BP, Bantu-speaking, iron-producing agropastoralists with cattle and caprines arrived in the Eastern Cape. By the fifteenth century AD, when Dutch and Portuguese encountered Khoisan speakers (Khoekhoen) of the Western Cape, these groups were practising an indigenous form of pastoralism with sheep, goats, and cattle (Smith, 2005).

Archaeologists and linguists have proposed several routes of entry for sheep into southern Africa, many citing transfers from eastern Africa across what is now Zimbabwe into northern Khoisan-speaking groups (Smith, 2008). However, archaeological evidence is very scarce (for cautions on using linguistics, see Heine and Konig, 2008). Muigai and Hanotte (2013) note a close genetic relationship between the modern Newala, a Tanzanian fat-tailed sheep breed, and southern African fat-tailed sheep. This link accords with the migration route noted above, as well as with one proposed by human geneticists. Based on high frequencies of a unique Y-chromosome haplogroup in both regions, Henn et al. (2008) argue for a pre-Bantu, pastoralist migration, at least by males from Tanzanian populations, whose genes ultimately entered Khoisan-speaking populations in Angola and Namibia. More genetic sampling of human Y-chromosome and mtDNA, and those of sheep populations, along this 3,700 km transect is needed, as are ecologically and anthropologically realistic models for such a moving front of pastoralists.

The earliest archaeological sites with sheep remains lie near the Namibian and Western Cape shores, such as Leopard Cave and Kasteelberg (Fig. 25.1). Smith (1998) and Sadr (1998) have argued that contemporaneous sites with early ceramics and/or domestic caprines in the Western Cape fall into two groups, reflecting coexistence of immigrant sheep-herders and indigenous foragers, with the foragers taking up the use of ceramics and some consumption of sheep through their contacts with the herders. One yields Smithfield or Wilton lithics, both with antecedents in the region, some pottery, and a preponderance of wild fauna. The second largely lacks formal stone tools, has more ceramics, and higher numbers of domestic sheep bones and/or substantial dung deposits. The latter, though as yet few in number, appear to have different geographical locations than the former.

SUBSEQUENT EXCHANGES AND MIXTURES

Genetic research on modern African cattle populations has clarified the later history of *Bos* on the continent, in the process revealing early trading contacts along the Indian Ocean coast. mtDNA and Y-chromosome analyses indicate that South Asian indicine (zebu) bulls crossbred with African taurine cows, producing the so-called sanga cattle breeds, occurring earliest in eastern Africa, then spreading to other regions (Hanotte et al., 2002). Analyses suggest at least two zebu introductions, the first occurring about 750 years ago, placing it in the florescence of the Indian Ocean trade. Whole-genome analysis indicates that this admixture has persisted long enough to produce an equilibrium of indicine-taurine traits at a 16: 84 proportion among a large sample of sanga cattle (Mbole-Kariuki et al., 2014).

The earliest archaeological evidence for zebu cattle are rock paintings of cattle and camels in the Horn of Africa, roughly dated around the mid-first millennium AD, a bit earlier than the genetically derived estimate. Sanga cattle reached southern Africa in the late first millennium AD, presumably traveling overland with Bantu-speaking agro-pastoralists. A direct introduction of breeding stock via the Indian Ocean trade into southern Africa remains a possibility.

A more recent introgression of male-mediated zebu-genes cattle likely results from British introduction of south Asian cattle into Sudan in the late 1800s, which also caused the devastating late nineteenth- to early twentieth-century rinderpest epidemics (Rossiter, 1994). Zebus are more heat-tolerant and resistant to rinderpest, and cross-breeds continue to spread in many areas (Linseele, 2010).

CONCLUSION

The emergence and spread of pastoralism in Africa was an ecological process, in which humans shifted from preying on hoofed animals to entering into close mutualist relations with a few such species. People and their domestic animals coevolved, humans developed genetic and technological means of digesting milk, and their livestock evolved to sustain milking and bleeding while remaining reproductively fit. Moving into sub-Saharan Africa, both humans and animals interacted with new species, including new diseases, hosts, and vectors. People responded to these challenges to livestock survival by cultural means and by developing new veterinary and ecological knowledge. Livestock often evolved to resist or tolerate novel infections.

Pastoralism was the foundational domesticate-based economy in much of sub-Saharan Africa, and its expansion introduced radically different approaches to social and property relations to foragers throughout the savanna regions of the continent.

REFERENCES

Achilli, A., Bonfiglio, S., Olivieri, A., Malusa, A., Pala, M., Kashani, B. H., Perego, U. A., Ajmone-Marsan, P., Liotta, L., Semoni, O., Bandelt, H.-J., Ferretti, L., and Torroni, A. (2009) 'The multifaceted origin of taurine cattle reflected by the mitochondrial genome', PLoS One, 4(6), DOI: 10.1371/journal.pone.0005753.

Amblard, S. (1996) 'Agricultural evidence and its interpretation on the Dhars Tichitt and Oualata, south-eastern Mauritania', in Pwiti, G. and Soper, R. (eds) Aspects of African Archaeology: Papers from the 10th Congress of the Panafrican Association for Prehistory and Related Studies, pp. 421–7. Harare: University of Zimbabwe Publications.

Ashley, C. Z. (2010) 'Towards a socialised archaeology of ceramics in Great Lakes Africa', African Archaeological Review, 27(2), 135–63.

Blench, R. M. (1993), 'Recent developments in African language classification and their implications for Prehistory', in Shaw, T., Sinclair, P., Andah, B., and Okpoko, A. (eds) *The Archaeology of Africa: Foods, Metals, and Towns*, pp. 126–38. London: Routledge.

Borgerhoff Mulder, M., Fazzio, I., Irons, W., McElreath, R. L., Bowles, S., Bell, A., Hertz, T., and Hazzah, L. (2010) 'Pastoralism and wealth inequality: revisiting an old question', *Current Anthropology*, 51(1), 35–48.

Bradburd, D. (1982) 'Volatility of animal wealth among southwest Asian pastoralists', *Human Ecology*, 10(1), 85–106.

Bradley, D. G. (2003) 'The DNA trail leading back to the origins of today's cattle has taken some surprising turns along the way', *Natural History*, 112(1), 36–42.

Brass, M. (2007) 'Reconsidering the emergence of social complexity in early Saharan pastoral societies, 5000–2500 BC', *Sahara*, 18, 7–22.

Brass, M. (2013) 'Revisiting a hoary chestnut: the nature of early cattle domestication in northeast Africa', *Sahara*, 24, 65–70.

Breunig, P. and Neumann, K. (2002) 'From hunters and gatherers to food producers: new archaeological and archaeobotanical evidence from the West African Sahel', in Hassan, F. A. (ed.) *Droughts, Food and Culture: Ecological Change and Food Security in Africa's Later Prehistory*, pp. 123–55. New York: Kluwer/Plenum.

Bruford, M. W. and Townsend, S. J. (2006) 'Mitochondridal DNA diversity in modern sheep', in Zeder, M. A., Bradley, D. G., Emshwiller, E., and Smith, B. D. (eds), *Documenting Domestication: New Genetic and Archaeological Paradigms*, pp. 306–16. Berkeley: University of California Press.

Chenal-Vélardé, I. (1998) 'Les premières traces de bœuf domestique en Afrique du Nord: état de la recherche centré sur les données archéozoologiques', *ArchaeoZoologia*, 9, 11–40.

Clutton-Brock, J. (1993) 'The spread of domestic animals in Africa', in Shaw, T., Sinclair, P., Andah, B., and Okpoko, A. (eds) *The Archaeology of Africa: Foods, Metals, and Towns*, pp. 61–70. London: Routledge.

Dale, D. D. and Ashley, C. Z. (2010) 'Holocene hunter-fisher-gatherer communities: new perspectives on Kansyore using communities of western Kenya', *Azania: Archaeological Research in Africa*, 45(1), 24–48.

di Lernia, S. (2001) 'Dismantling dung: delayed use of food resources among Early Holocene foragers of the Libyan Sahara', *Journal of Anthropological Archaeology*, 20, 408–41.

di Lernia, S. (2006) 'Building monuments, creating identity: cattle cult as a social response to rapid environmental changes in the Holocene Sahara', *Quaternary International*, 151, 50–62.

di Lernia, S., Tafuri, M. A., Gallinaro, M., Alhaique, F., Balasse, M., Cavorsi, L., Fullagar, P. D., Mercuri, A. M., Monaco, A., Perego, A., and Zerboni, A. (2013) 'Inside the "African Cattle Complex": animal burials in the Holocene central Sahara', *PLoS One*, 8(2), DOI: 10.1371/journal.pone.0056879.

Dunne, J., Evershed, R. P., Salque, M., Cramp, L., Bruni, S., Ryan, K., Biagetti, S., and di Lernia, S. (2012) 'First dairying in green Saharan Africa in the fifth millennium BC', *Nature*, 486, 390–4.

Edwards, C. J., MacHugh, D. E., Dobney, K. M., Martin, L., Russell, N., Horwitz, L. K., McIntosh, S. K., MacDonald, K. C., Helmer, D., Tresset, A., Vigne, J.-D., and Bradley, D. G. (2004) 'Ancient DNA analysis of 101 cattle remains: limits and prospects', *Journal of Archaeological Science*, 31(6), 695–710.

Ehret, C. (1997) 'African languages: a historical survey', in Vogel, J. O. (ed) *Encyclopedia of Precolonial Africa: Archaeology, History, Languages, Cultures, and Environments*, pp. 159–66. Walnut Creek: AltaMira Press.

Evershed, R. P., Payne, S., Sherratt, A. G., Copley, M. S., Coolidge, J., Urem-Kotsu, D., Kotsakis, K., Özdoğan, M., Özdoğan, A. E., Nieuwenhuyse, O., Akkermans, P. M. M. G., Bailey, D., Andeescu, R.-R., Campbell, S., Farid, S., Hodder, I., Yalman, N., Özbaşaran, M., Bıçakcı, E., Garfinkel, Y., Levy, T., and Burton, M. M. (2008) 'Earliest date for milk use in the Near East and southeastern Europe linked to cattle herding', *Nature*, 455, 528–31.

Gautier, A. (2001) 'The Early to Late Neolithic archaeofaunas from Nabta and Bir Kiseiba', in Wendorf, F., Schild, R., and Nelson, K. (eds) *Holocene Settlement of the Egyptian Sahara*, pp. 609–35. New York: Kluwer.

Gautier, A. (2002) 'The evidence of the earliest livestock in North Africa: or adventures with large bovids, ovicaprids, dogs and pigs', in Hassan, F. A. (ed.), *Droughts, Food and Culture: Ecological Change and Food Security in Africa's Late Prehistory*, pp. 195–207. New York: Kluwer.

Gifford-Gonzalez, D. (2000) 'Animal disease challenges to the emergence of pastoralism in sub-Saharan Africa', *African Archaeological Review*, 18, 95–139.

Gifford-Gonzalez, D. and Hanotte, O. (2013) 'Domesticating animals in Africa', in Lane, P., and Mitchell, P. J. (eds) *Oxford Handbook of African Archaeology*, pp. 491–505. New York: Oxford University Press.

Gifford-Gonzalez, D. and Parham, J. (2008) 'The fauna from Adrar Bous and surrounding areas', in Clark, J. D., Garcea, E. A. A., Gifford-Gonzalez, D., Smith, A. B., and Williams, M. A. J. (eds) *Adrar Bous: Archaeology of a Volcanic Ring Complex in Niger*, pp. 313–53. Tervuren: Royal Africa Museum.

Grigson, C. (1991) 'African origin for African cattle? Some archaeological evidence', *African Archaeological Review*, 9, 119–44.

Hadjigeorgiou, I. (2011) 'Past, present and future of pastoralism in Greece', *Pastoralism*, 1(1), 1–22.

Hanotte, O., Bradley, D. G., Ochieng, J. W., Verjee, Y., Hill, E. W., and Rege, J. E. O. (2002) 'African pastoralism: genetic imprints of origins and migrations', *Science*, 296, 336–9.

Haour, A. C. (2003) 'One hundred years of archaeology in Niger', *Journal of World Prehistory*, 17(2), 181–234.

Hassan, F. A. (2000) 'Climate and cattle in North Africa: a first approximation', in MacDonald, K. C. and Blench, R. M. (eds) *The Origins and Development of African Livestock: Archaeology, Genetics, Linguistics and Ethnography*, pp. 61–86. London: UCL Press.

Heine, B. and Konig, C. (2008) 'What can linguistics tell us about early Khoekhoe history?', *Southern African Humanities*, 20(1), 235–48.

Henn, B. M., Gignoux, C., Lin, A. A., Oefner, P. J., Shen, P., Scozzari, R., Cruciani, F., Tishkoff, S. A., Mountain, J. L., and Underhill, P. A. (2008) 'Y-chromosomal evidence of a pastoralist migration through Tanzania to southern Africa', *Proceedings of the National Academy of Sciences of the United States of America*, 105(31), 10693–8.

Hildebrand, E. A. and Grillo, K. M. (2012) 'Early herders and monumental sites in Eastern Africa: dating and interpretation', *Antiquity*, 86(332), 338–52.

Holl, A. F. C. (1995) 'Pathways to elderhood. Research on past pastoral iconography: the paintings from Tikadourine (Tassili-N-Ajjer)', *Origini. Preistoria e Protostoria delle civiltà Antiche*, 18, 69–113.

Honegger, M. (2005) 'Kerma et les débuts du Néolithique africain', *Genava N.S.*, 53, 239–49.

Honegger, M. and Bastien, J. (2009) 'The Early Holocene sequence of Wadi el-Arab', *Kerma*, 2009, 3–6.

Horsburgh, K. A. and Rhines, A. (2010) 'Genetic characterization of an archaeological sheep assemblage from South Africa's Western Cape', *Journal of Archaeological Science*, 37(11), 2906–10.

Jousse, H. (2006) 'What is the impact of Holocene climatic changes on human societies? Analysis of West African Neolithic populations dietary customs', *Quaternary International*, 151(1), 63–73.

Kimura, B., Marshall, F., Beja-Pereira, A., and Mulligan, C. (2013) 'Donkey domestication' *African Archaeological Review*, 30(1), 83–95.

Kröpelin, S., Verschuren, D., Lézine, A.-M., Eggermont, H., Cocquyt, C., Francus, P., Cazet, J.-P., Fagot, M., Rumes, B., Russell, J. M., Darius, F., Conley, D. J., Schuster, M., von Suchodoletz, H., and Engstrom, D. R. (2008) 'Climate-driven ecosystem succession in the Sahara: the past 6000 years', *Science*, 320, 765–8.

Kuper, R. and Kröpelin, S. (2006) 'Climate-controlled Holocene occupation in the Sahara: motor of Africa's evolution', *Science*, 313 (5877), 803–7.

Larson, G., Dobney, K., Albarella, U., Fang, M., Matisoo-Smith, E., Robins, E., Lowden, S., Finlayson, H., Brand, T., Willerslev, E., Rowley-Conwy, P., Andersson, L., and Cooper, A. (2005) 'Worldwide phylogeography of wild boar reveals multiple centers of pig domestication', *Science*, 307(5715), 1618–21.

Linseele, V. (2010) 'Did specialized pastoralism develop differently in Africa than in the Near East? An example from the West African Sahel', *Journal of World Prehistory*, 23, 43–77.

Linseele, V. (2013) 'Early stock keeping in northeastern Africa: Near Eastern influences and local developments', in Shirai, N. (ed.) *Neolithisation of Northeastern Africa*, pp. 97–108. Berlin: Ex Oriente.

Linseele, V. and Van Neer, W. (2008) 'Faunal remains from the Tree Shelter Site', in Vermeersch, P. (ed.) *A Holocene Prehistoric Sequence in the Egyptian Red Sea Area: The Tree Shelter*, pp. 79–84. Leuven: Leuven University Press.

Luikart, G., Gielly, L., Excoffier, L., Vigne, J. D., Bouvet, J., and Taberlet, P. (2001) 'Multiple maternal origins and weak phylogeographic structure in domestic goats', *Proceedings of the National Academy of Sciences of the United States of America*, 98(10), 5927–32.

MacDonald, K. C. (1995) 'Analysis of the mammalian, avian and reptilian remains', in McIntosh, S. K. (ed.) *Excavations at Jenné Jeno, Hambarketolo, and Kaniana (Inland Niger Delta, Mali): the 1981 Season*, pp. 291–318. Berkeley: University of California Press.

MacDonald, K. C. (1998) 'Before the Empire of Ghana: pastoralism and the origins of cultural complexity in the Sahel', in Connah, G. (ed.) *Transformations in Africa: Essays on Africa's Later Past*, pp. 71–103. London: Leicester University Press.

Manning, K. M. (2008) 'Mobility strategies and their social and economic implications for Late Stone Age Sahelian pastoral groups: a view from the lower Tilemsi Valley, eastern Mali', *Archaeological Review from Cambridge*, 23(2), 125–45.

Manning, K. M. (2011) 'The first herders of the West African Sahel: inter-site comparative analysis of zooarchaeological data from the lower Tilemsi Valley', in Jousse, H. and Lesur, J. (eds) *People and Animals in Holocene Africa: Recent Advances in Archaeozoology*, pp. 75–85. Frankfurt: Africa Magna.

Manning, K. M. and MacDonald, K. C. (2005) 'Analyse des restes d'animaux collectés à Dia', in Bedaux, R., Polet, J., Sanogo, K., and Schmidt, K. (eds) *Recherches archéologiques à Dia dans le Delta Intérieur du Niger (Mali): bilan des saisons de fouilles 1998–2003*, pp. 363–85. Leiden: CNWS Publications.

Marshall, F. B. and Hildebrand, E. (2002) 'Cattle before crops: the beginnings of food production in Africa', *Journal of World Prehistory*, 16(2), 99–143.

Mbole-Kariuki, M. N., Sonstegard, T. S., Orth, A., Thumbi, S. M., Bronsvoort, B. M. D. C., Kiara, H., Toye, P., Conradie, I., Jennings, A., Coetzer, K., Woolhouse, M. E. J., Hanotte, O., and Tapio, M. (2014) 'Genome-wide analysis reveals the ancient and recent admixture history of East African shorthorn Zebu from Western Kenya', *Heredity*, 113, 297–305.

Meadows, J. R. S., Cemal, I., Karaca, O., Gootwine, E., and Kijas, J. W. (2007) 'Five ovine mitochondrial lineages identified from sheep breeds of the Near East', *Genetics*, 175, 1371–9.

Muigai, A. W. T. and Hanotte, O. (2013) 'The origin of African sheep: archaeological and genetic perspectives', *African Archaeological Review*, 30(1), 39–50.

Muzzolini, A. (1986) 'Tassili-Sud, Aïr Oriental, Téneré, Djado: la frontière orientale des écoles rupestres du Sahara central', *Ars Praehistorica*, 5–6, 103–30.

Paris, F. (2000) 'African livestock remains from Saharan mortuary contexts', in Blench, R. M. and MacDonald, K. C. (eds) *The Origins and Development of African Livestock: Archaeology, Genetics, Linguistics and Ethnography*, pp. 111–26. London: UCL Press.

Pereira, F., Queirós, S., Gusmão, L., Nijman, I. J., Cuppen, E., Lenstra, J. A., Econogene Consortium, Davis, S. J. M., Nejmeddine, F., and Amorim, A. (2009) 'Tracing the history of goat pastoralism: new clues from mitochondrial and Y chromosome DNA in North Africa', *Molecular Biology and Evolution*, 26(12), 2765–73.

Pérez-Pardal, L., Royo, L. J., Beja-Pereira, A., Chen, S., Cantet, R. J. C., Traoré, A., Curik, I., Sölkner, J., Bozzi, R., Fernández, I., Álvarez, I., Gutiérrez, J. P., Gómez, E., Ponce de León, F. A., and Goyache, F. (2010) 'Multiple paternal origins of domestic cattle revealed by Y-specific interspersed multilocus microsatellites', *Heredity*, 105(6), 511–19.

Prendergast, M. E. and Mutundu, K. K. (2010) 'Late Holocene zooarchaeology in East Africa: ethnographic analogues and interpretive challenges', *Documenta Archaeobiologiae*, 7, 203–232.

Prendergast, M. E., Mabulla, A. Z. P., Grillo, K. M., Broderick, L. G., Seitsonen, O., Gidna, A. O., and Gifford-Gonzalez, D. (2013) 'Pastoral Neolithic sites on the southern Mbulu Plateau, Tanzania', *Azania: Archaeological Research in Africa*, 36(4), 1498–1520.

Rossel, S., Marshall, F., Peters, J., Pilgram, T., Adams, M. D., and O'Connor, D. (2008) 'Domestication of the donkey: timing, processes, and indicators', *Proceedings of the National Academy of Sciences of the United States of America*, 105(10), 3715–20.

Rossiter. (1994) 'Rinderpest', in *Infectious Diseases of Livestock with Special Reference to Southern Africa*. Vol. 2, Coetzer, J. A. W., Thomson, G. R., Tustun, R. C., and N. P. J. Kriek (eds), pp. 735–57. New York: Oxford University Press.

Sadr, K. (1998) 'The first herders at the Cape of Good Hope', *African Archaeological Review*, 15(2), 101–32.

Schmandt-Besserat, D. (1986) 'An ancient token system: the precursor to numerals and writing', *Archaeology*, 39, 32–9.

Sealy, J. and Yates, R. (1996) 'Direct radiocarbon dating of early sheep bones: two further results', *South African Archaeological Bulletin*, 51(164), 109–10.

Smith, A. B. (1992) 'Origins and spread of pastoralism in Africa', *Annual Review of Anthropology*, 21, 125–41.

Smith, A. B. (1998) 'Early domestic stock in southern Africa: a commentary', *African Archaeological Review*, 15(2), 151–6.

Smith, A. B. (2005) *African Herders: Emergence of Pastoral Traditions*, Walnut Creek: AltaMira Press.

Smith, A. B. (2008) 'Pastoral origins at the Cape, South Africa: influences and arguments', *Southern African Humanities*, 20(1), 49–60.

Stock, F. and Gifford-Gonzalez, D. (2013) 'Genetics and African cattle domestication', *African Archaeological Review*, 30(1), 51–72.

Tafuri, M. A., Bentley, R. A., Manzi, G., and di Lernia, S. (2006) 'Mobility and kinship in the prehistoric Sahara: strontium isotope analysis of Holocene human skeletons from the Acacus Mts. (southwestern Libya)', *Journal of Anthropological Archaeology*, 25(3), 390–402.

Tishkoff, S. A., Reed, F. A., Ranciaro, A., Voight, B. F., Babbitt, C. C., Silverman, J. S., Powell, K., Mortensen, H. M., Hirbo, J. B., Osman, M., Ibrahim, M., Omar, S. A., Lema, G., Nyambo, T. B., Ghori, J., Bumpstead, S., Pritchard, J. K., Wray, G. A., and Deloukas, P. (2007) 'Convergent adaptation of human lactase persistence in Africa and Europe', *Nature Genetics*, 39(1), 31–40.

Wendorf, F., Schild, R., and Close, A. E. (1984) *Cattle-Keepers of the Eastern Sahara: The Neolithic of Bir Kiseiba*. Dallas: Institute for the Study of Earth and Man, Southern Methodist University.

Wendorf, F. and Schild, R. (eds) (2013) *Holocene Settlement of the Egyptian Sahara: Volume 1: The Archaeology of Nabta Playa*. New York: Springer Science & Business Media.

Wetterstrom, W. (1998) 'The origins of agriculture in Africa: with particular reference to sorghum and pearl millet', *The Review of Archaeology*, 19(2), 30–46.

....

CHAPTER 26

CATTLE, A MAJOR COMPONENT OF THE KERMA CULTURE (SUDAN)

....

LOUIS CHAIX

INTRODUCTION

THE kingdom of Kerma, in the north of Sudan, covers a vast area, stretching over some 800 km from the area of the First Cataract in the north to the area of al Dabbah in the south; its western and eastern boundaries remain unknown (Gratien, 1978; Bonnet, 1986; 2004a) (Fig. 26.1). In terms of chronology, it is divided into three periods: Early Kerma (2600–2050 BC), Middle Kerma (2050–1750 BC), and Classic Kerma (1750–1500 BC). This chronology is based on the typological development of pottery as well as on the morphology of burials and changes in burial rite. A major archaeological project was conducted by the University of Geneva at Kerma between 1978 and 2001, excavating the ancient capital city of the kingdom and its accompanying cemetery (Bonnet, 2000; 2004b; 2004c).

Many archaeologists have studied Kerma and its culture since the nineteenth century (Ferrero, 1990a), largely because of the imposing mud-brick monument known locally as the *Deffufa*, which dominates the centre of the ancient town. This building, first interpreted as an Egyptian trading post by the American archaeologist G. Reisner, is now thought to be a temple, as documented by the recent excavations of the University of Geneva (Bonnet, 2000). Kerma was the capital of a vast independent kingdom which, from the third millennium BC onwards, became a military power threatening the empire of the Pharaohs. Relations between the two neighbours alternated between commercial exchanges and conflict ever since. At the end of the New Kingdom, around 1500 BC, the Pharaoh Thutmosis I brought the kingdom of Kerma to an end. Nevertheless that original Nubian culture was to continue influencing the subsequent Nabatean and Meroitic cultures.

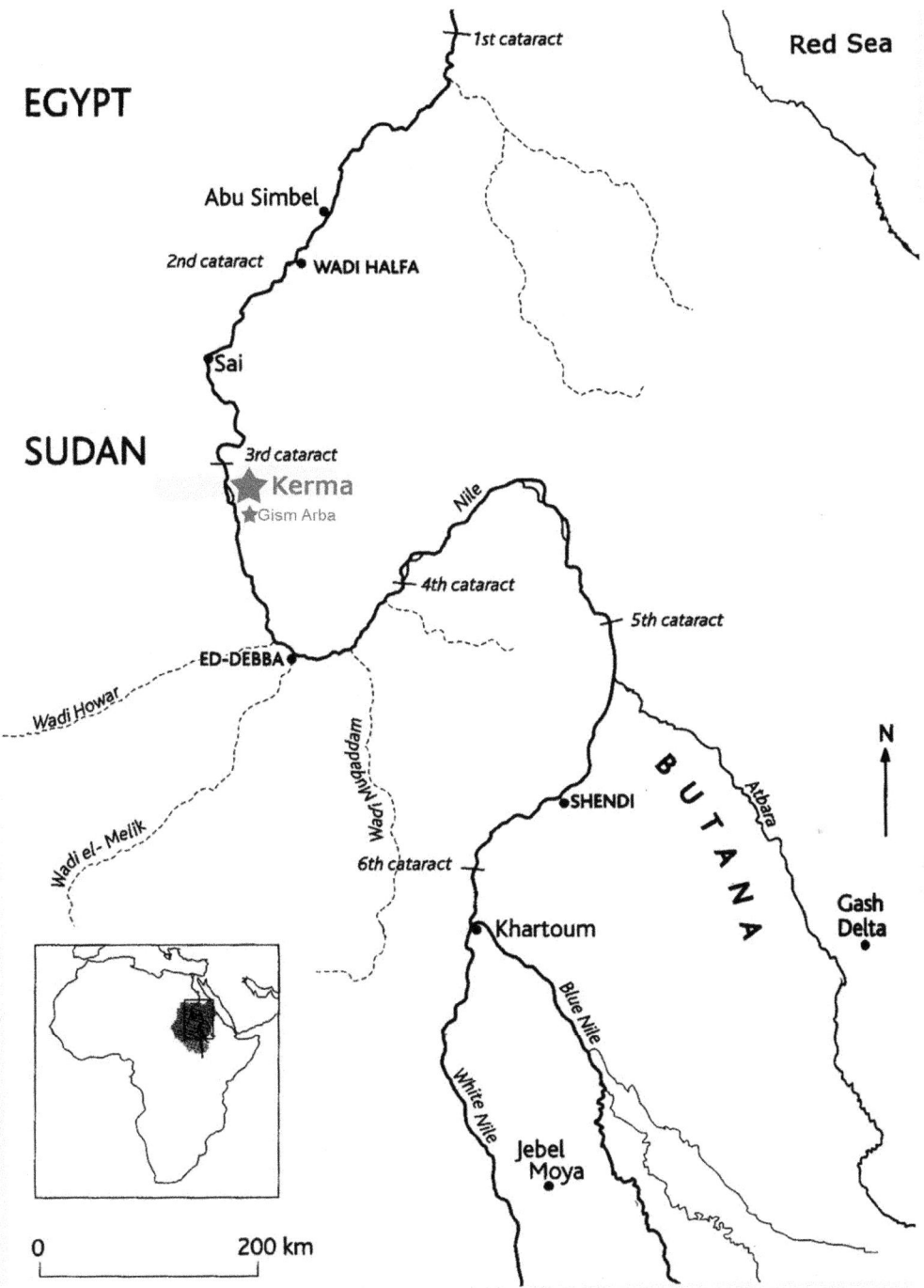

FIGURE 26.1 Map showing the position of Kerma and Gism-el-Arba. Author's own image.

The dwellings, pits, lanes, and ditches of the city have yielded large quantities of animal bones (*c*.36,000 NISP), which provide a reasonable idea of the inhabitants' diet (Chaix, 1990; 1993; Chaix and Grant, 1993). In general, the meat derived from domesticated animals, especially cattle and caprines; a few donkeys were also eaten. Wild species are barely represented and, apart from fish, they seem to have played a merely incidental role in the diet. The part that fish played in the diet, on the other hand, was substantial. Although fish remains have very rarely been found in the town, because of poor preservation of their fragile bones, stable isotope analysis of human skeletons indicates that an important percentage of the diet derived from fish, almost entirely freshwater fish (Iacumin et al., 2001).

The precise phasing of the various contexts at Kerma, especially the pits and floor levels of buildings, allows for a diachronic approach to the study of meat consumption based on a calculation of the number of bones. Among the dominant domesticates, i.e. cattle and caprines, a marked drop in cattle numbers between Early Kerma and Classic Kerma can be observed, and this seems to have taken place at the end of the Middle Kerma period, around 1800 BC. This trend is shown by the number of remains as well as by the estimate of meat weights, calculated on the basis of the weight of the living animal, minus a factor called 'gross butchery yield' (Vigne, 1988: 206–7). Meanwhile, the quantities of sheep and goat appear to increase evenly (Chaix, 1994; 2014).

Such phenomena are no doubt related to the development of the post-Neolithic arid phase. Indeed, a number of phenomena (end of lake sedimentation, drop in groundwater levels, retreat of savannah-type conditions towards the south in favour of desert steppes) are indicative of worsening climatic conditions between 3000 and 1500 BC. At the same time, an increase in the number of sites, corresponding to a growth in population, is documented over the entire region (Chaix, 2014). Worsening environmental conditions and an increase in population may explain why cattle became a rare and precious commodity. At the same time, the phenomenon of the 'accompanying dead' emerges. Human bodies (men, women, and children) replace the role once played by sheep in graves to the south and west of the burial ground (Bonnet, 2000; Testart, 2004). These deposits can contain several hundred human individuals, as in the great tumuli of the Classic Kerma phase excavated by Reisner (1923). In Testart's opinion (2004), this concept of the 'accompanying dead' is quite distinct from that of sacrifice. Indeed, in the latter case there is someone making the sacrifice, a sacrificial victim, and especially a person to whom the sacrifice is dedicated, whereas the 'accompanying dead' are merely following and serving the deceased in his or her life in the afterworld.

THE TOWN'S CATTLE

It is possible to give a reasonably accurate image of the cattle at Kerma, thanks to the teeth and bones excavated in settlement levels, to the bucrania recovered from the cemetery, and to the leather shrouds preserved in many graves.

The cattle were of reasonable size, measuring on average 1.3 m at the withers (the range being between 1.05 and 1.41 m (Matolcsi, 1970; Chaix, 2007). They were long-horned and very similar to the mummified oxen of ancient Egypt (Lortet and Gaillard, 1903; Pia, 1941; Boessneck, 1987). The examination of the zygomatic bone and the thoracic vertebrae makes it clear that all of them belong to the species *Bos taurus* L. (Grigson, 1980; Marshall, 1989). 'Humped cattle' or zebu emerge later. Their morphological similarity to cattle makes it difficult to identify the former with certainty, but remains recovered on Meroitic sites dated to around 300 BC appear to indicate that humped cattle were present there (Chaix, 2011a).

The few long bones that were preserved whole indicate that the animals had long and gracile limbs, especially the metapodials and phalanges. For the metapodials, the estimate of greater or lesser gracility is based on the robusticity index ((SD/GL)*100). The value for the metatarsals from Kerma is 11.9 whereas it is 14.8 for modern European Simmental cattle (Zalkin, 1960; Howard, 1963). The same goes for the first anterior and posterior phalanges, their index of robusticity being clearly different from that of the Simmental breed (for example the index for the first posterior phalanx is 3.86 for Kerma cattle and 4.23 for Simmental cattle). Similar observations have been made on the sheep from Kerma (Chaix and Grant, 1987; Guintard and Lallemand, 2003). It seems that the elongation of the feet (ankle and toes) is linked to both the soil type (sandy or rocky) and the temperature (tropical or temperate); indeed, animals living in more mountainous and cooler climates have shorter and more robust extremities (Tekkouk and Guintard, 2007).

The age determination of the town's cattle, using tooth eruption and wear (Grant, 1982), and epiphyseal fusion (Habermehl, 1961) shows that 30% of individuals were slaughtered at 3–4 years of age, and 67% between 4 and 5 years; a few older animals were also present (Grant, 1982). This age profile corresponds to a mixed exploitation, for meat on the one hand, but also for milk and traction, as documented by several skulls from the cemetery whose horns were deformed by a yoke (Bartosiewicz et al., 1997).

Determining the sex of the animals is done by measuring the articular part of the scapula and the transversal diameter of the distal tibia. The identifications made at Kerma show a proportion of 65–75% cows to 25–35% bulls and (probably) oxen. This unusually high representation of male individuals is best explained by the presence of castrated animals, and this aspect is currently being investigated through further research.

Numerous butchery marks indicate that cattle were exploited for beef and include evidence for slaughter, cutting into whole quarters, and dismemberment. Traces of use of heavy implements (cleavers) used to split long and robust bones, and of knives used for disarticulation and for defleshing, illustrate the different stages of butchering.

The bones themselves were exploited: the majority (99.9%) of the metapodials were split along the sagittal plane, probably for marrow extraction and to prepare bone objects. Indeed, cattle bones provided the raw material for making a number of tools. Cattle radii, ulnae, and metapodials were used to make robust awls and points for leather-working. Ribs (around 30% of the tools) could be used to make spatulae and smoothers or polishers used in ceramic manufacture and decoration (Choyke, 1990).

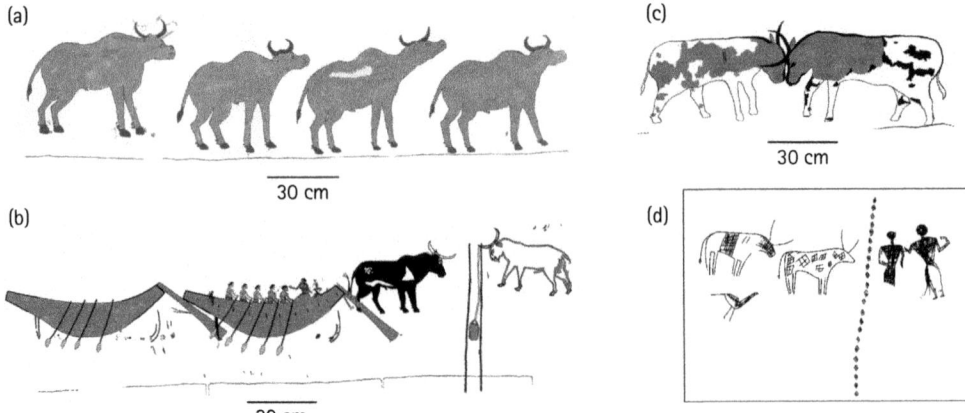

FIGURE 26.2 A, B, and C: images of cattle painted on the walls of the funerary chapel KXI in the eastern cemetery of Kerma (after Bonnet, 2000). D: ostrich egg with engraving of two cattle facing two people. This egg was found in the ancient city of Kerma (after Kantor, 1948). Courtesy of Charles Bonnet.

Several figurines representing bovines have been found in various contexts in the ancient city (religious area, settlement, defences). They have been identified as such because of the presence of four teats and the orientation of the base of the (often broken) horns. These crudely made clay miniatures were fired at low temperatures (Ferrero, 1984; 1990b). Similar pieces have been found in a rural settlement dating to the Middle Kerma phase, at Gism-el-Arba, a few kilometres south of the city (Chaix and Queyrat, 2003). They attest to the significance of cattle in the Kerma culture, whatever their interpretation: children's toys, elements of a counting system, or cult objects (Gratien et al., 2003).

Finally, two finds made in the town confirm the importance attached to cattle. The floor of a house dated to the Middle Kerma period (M 27) has yielded several fragments of an ostrich egg decorated with long-horned oxen featuring among human figures, giraffes, and a crocodile (Bonnet, 1993). Further, a fragment of an ostrich egg engraved with two human figures facing two cattle and a bird, separated by a garland or a barrier, has also been found in the town (Kantor, 1948) (Fig. 26.2 D).

THE CEMETERY'S CATTLE

A vast cemetery, at its greatest extent covering an area of over 70 ha and containing more than 40,000 graves, is located some 4 km east of the town. It grew from north to south and was used throughout the life of the kingdom. The study of its tombs, its ceramics, and its numerous radiocarbon dates provide a good understanding of its development and the evolution of funerary rites (Privati, 1999; Bonnet, 2000; Bourriau, 2004).

Cattle play an important part in these rituals, from as early as the end of the Early Kerma period and during the Middle Kerma period.

In the Early Kerma phase, the use of cattle is represented by leather sandals made from their hides. These sandals were worn by several of the deceased (Bonnet, 1982; Chaix, 2002). At the end of the Early Kerma phase and the beginning of Middle Kerma, cattle are represented by a number of elements detailed here. Still within the graves, the deceased were laid on a 'shroud' made of a tanned and cut ox hide (Ryder, 1984; Wills, 2002). The hide was shaved except for a border of a few centimetres where the hair was left in place. The colour of the coat appears to be a uniform dark brown in the examples that have been examined. At the tail attachment, part of which was preserved in these examples, there is a deformation caused by the 'shroud' having been hung up for a considerable time. This suggests that the hide had been used as a sleeping mat, and that it was an object used during the lifetime of the deceased (Chaix, 1982; 1990; 2002).

In a few graves of the Early Kerma (n: 5), Middle Kerma (n: 2), and Classic Kerma (n: 6) periods, deposits of cattle horns, often of large size and still retaining the outer horn sheath were found inside the grave and close to the body. Such horns accompanied adults as well as children, and occurred in both male and female graves.

Bucrania are another characteristic element found within the cemetery. These objects are the result of preparing a bovine skull so that only the horns and frontal face are preserved, with the lower part (premaxillar, maxillar, jugal, temporal, and basioccipital bones) removed. The occipital bone is cut transversally at the level of the nuchal line, above the *foramen magnum*. Such specific cutting requires a good mastery of technique and indicates that it was probably the work of a specialist (Fig. 26.3). The shape given to the bucrania evolved over time: in the early phases, frontal and nasal bones were kept, and the nasal bones, because of their tapered suture, were sometimes found separately and deposited next to the skull. At the beginning of the Middle Kerma phase, the bucrania were cut transversally at the suture between the nasal and frontal bones. During the full Middle Kerma period the cut was made even higher, in front of the horncores (Chaix, 2001; 2011b). This way of cutting the skull was employed for adult specimens as well as for newborn and very young animals.

The bucrania were deposited outside the burial mounds, to the south. When large numbers were present, they formed a semicircular arc whose extremities never went beyond the projected central line of the mound. They were arranged according to a precise order and packed close together, which makes their excavation difficult. In the larger assemblages, the bucrania of cows were arranged in the first row, and one or two young calves are often found behind them. The next row has the bucrania of bulls, followed by a row which we believe to be those of castrated animals (oxen). However, this is merely a hypothesis, which a study currently being carried out (using morphological, metrical, and statistical analyses) should help substantiate. This organization of bucrania may be repeated several times in the case of important tombs (Fig. 26.4).

During the Early Kerma phase, of the ninety-nine burials excavated, seven were accompanied by bucrania (i.e. 7%), ranging from one to nineteen bucrania per tomb; only one calf was recovered. There seems to be no link between the number of bucrania

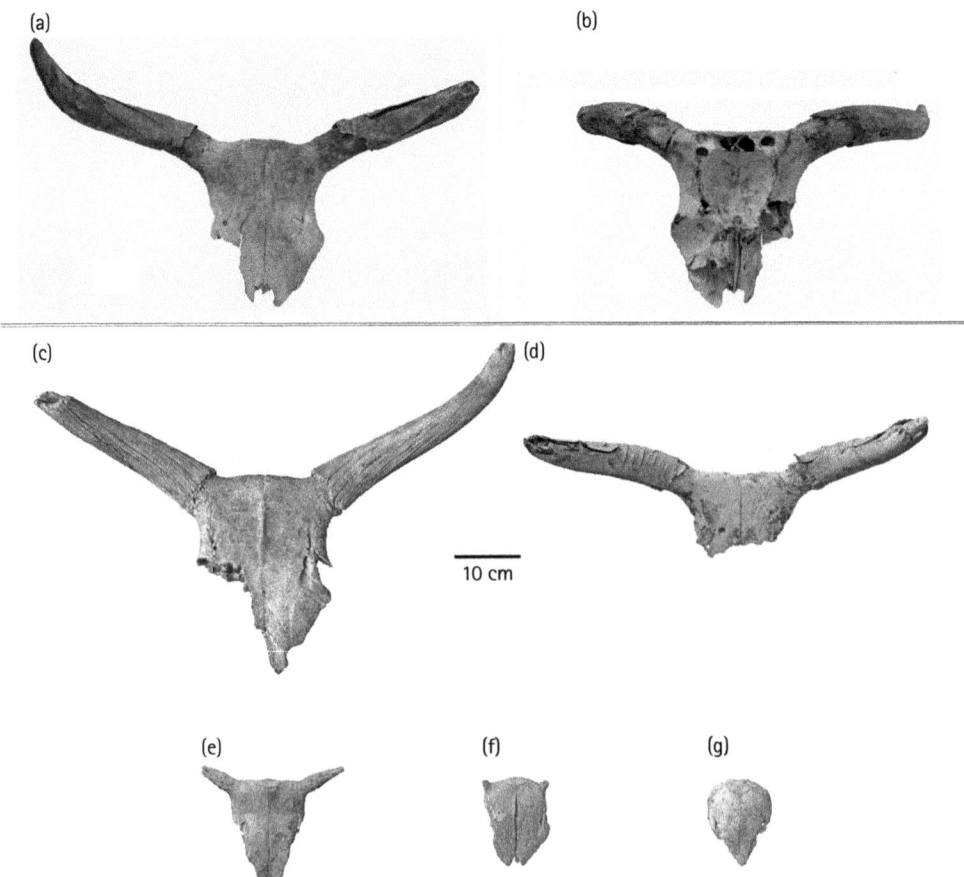

FIGURE 26.3 Cattle bucrania. A and B: frontal and basal views of an adult cow showing the typical carving. C and D: bucrania of a bull and a cow. E, F, and G: bucrania of calves of different ages. Author's own image.

and the age, sex, or status of the person buried. The burial of a small child aged 5–6 years was surrounded by nineteen bucrania, whereas a man of over 40 only had five, and a 40-year-old woman was only given one bucranium.

It is during the Middle Kerma period that the bucrania increase in importance. Out of a total of eighty-six excavated burials, seventeen yielded bucrania (19.3%) in groups ranging from 1 to 4,899! It seems that it is male burials that are most frequently provided with bucrania during this period, with ten men, one 17–18-year-old young man and a child aged 2–3 years (Simon et al., 1990). Five burials had no skeletal parts preserved, having been more or less completely robbed. Here again the quantities of bucrania in the graves where the human bodies are preserved do not seem to be linked to the age of the deceased. In the case of the very large assemblages—455 bucrania for Tomb 238, 4,899 for Tomb 253—the age and sex of the deceased are unknown because both burials were completely robbed out.

FIGURE 26.4 Typical organization of bucrania to the south of a large burial. Author's own image.

Several bucrania from the Middle Kerma phase have a deformation that was intended to make the two horns parallel (Chaix, 1996; 2001; 2004; Chaix and Hansen, 2003; Chaix et al., 2012). A number of observations show that this was achieved by placing a tightening device between the tips of the horns, fixed to them by means of a cut into the keratin to produce a knob at the horn's tip. This procedure was carried out on young calves and gradually the two horns were made to grow parallel to each other. In mechanical terms, the horn-cores exhibit an elliptical section and there is a significant swelling of the crest between the horns. In some extreme cases the two horn-sheaths become anastomosed and grow into a single one (Chaix and Hansen, 2003: 273). These 'parallel-horned' bucrania are similar to those depicted in numerous rock art representations in several areas of the Sahara and are named 'cattle with forward-pointing horns' (Chaix and Hansen, 2003; Chaix 2006).

The isotopic analysis of a number of cattle skulls in the major assemblages has shown that some individuals came from an area outside the region of Kerma, where they grazed on plants of Mediterranean type (C_3 plants) (Iacumin et al., 2001; Thompson et al., 2008). This would imply that at certain funerary ceremonies some cattle (or their bucrania) were brought to Kerma as offerings or as tribute.

The significance of cattle for both the economy and burial practices can be shown to have evolved in the following manner. In the Early Kerma phase cattle are present but in small quantities, and there are no bucrania in the graves. During the Middle Kerma phase cattle numbers greatly increase, in the town as well as in the cemetery. Bucrania become common and can reach considerable quantities. In the Classic Kerma phase cattle decrease in importance in favour of caprines. Even the large princely burials in the cemetery only yielded a few bucrania. On the other hand, the practice of burying the 'accompanying dead' becomes generalized. We believe that these phenomena are connected on the one hand to worsening climatic conditions, and on the other to demographic growth (see Introduction).

The cemetery has yielded iconographic evidence attesting to the importance attached to cattle in the Kerma culture. Temple K XI in particular is decorated with frescoes depicting various wild animals (giraffes, hippopotami, fishes) and several examples of cattle. There are no other domesticates depicted there, which seems to confirm that cattle had a special significance in the funerary sphere (Bonnet, 2000; Chaix, 2000) (Fig. 26.2).

CONCLUSION

As exposed here, cattle are a key constituent element of the civilization of Kerma. They played a major role in the diet of the population and as a source of secondary products such as milk, hides, and raw materials for making tools. They also contributed to agriculture as traction animals.

In the funerary and probably religious domain cattle represent, through the bucrania recovered, the power and wealth of the deceased. The deformation of the horns on certain specimens, the meaning of which is little understood, is rather reminiscent of some practices still alive today among many populations bordering the Nile (Chaix et al., 2012).

The presence of cattle iconography in the temples and the cemetery is further testimony to the role they played in the religious sphere, though at present it is not possible to go further in our interpretation.

As the title of this chapter indicates, cattle were an essential component of the economy of the communities living in the kingdom of Kerma. Besides caprines—which were also well represented—cattle were the main source of protein, as well as of a series of secondary products. They were of primary importance in funerary rites and probably also in the religious sphere. The Kerma cattle clearly belong to what several authors (Herskovits, 1926; Baroin and Boutrais, 2008) have called the 'African Cattle Complex', which characterizes most pastoralist communities in northeastern Africa and large parts of the Sahara from as early as the sixth millennium BC. This complex is distinguished by the preeminent role played by cattle, as documented by the analysis of animal bones, by the presence of cattle in burials dedicated to the dead (Paris, 2000; Di Lernia et al., 2013), and by the representations of these animals that are dominant in the rock art of the Sahara (Le Quellec et al., 2005).

REFERENCES

Baroin, C. and Boutrais, J. (eds) (2008) 'Le lien au bétail', Journal des Africanistes, 76(1/2), 7–217.

Bartosiewicz, L., Van Neer, W., and Lentacker, A. (1997) *Draught Cattle: Their Osteological Identification and History*. Annales du Musée Royal de l'Afrique Centrale 281. Tervuren: Koninklijk Museum voor Midden-Afrika.

Boessneck, J. (1987) 'Die Münchner Ochsenmumie', *Hildesheimer Ägyptologische Beiträge*, 25, 96.

Bonnet, C. (1982) 'Les fouilles archéologiques de Kerma (Soudan)', *Genava*, NS, 30, 1–25.

Bonnet, C. (1986) *Kerma: territoire et métropole. quatre leçons au Collège de France*, Paris: Institut Français d'Archéologie Orientale.

Bonnet, C. (1993) 'Les fouilles archéologiques de Kerma (Soudan)', *Genava*, NS, 41, 1–18.

Bonnet, C. (2000) *Edifices et rites funéraires à Kerma*, Paris: Éditions Errance.

Bonnet, C. (2004a) 'The Kerma culture', in Welsby, D. A. and Anderson, J. R. (eds) *Sudan: Ancient Treasures. An Exhibition of Recent Discoveries from the Sudan National Museum*, pp. 70–7. London: The British Museum Press.

Bonnet, C. (2004b) 'Kerma', in Welsby, D. A. and Anderson, J. R. (eds) *Sudan: Ancient Treasures. An Exhibition of Recent Discoveries from the Sudan National Museum*, pp. 78–89. London: The British Museum Press.

Bonnet, C. (2004c) *Le temple principal de la ville de Kerma et son quartier religieux*, Paris: Éditions Errance.

Bourriau, J. (2004) 'Egyptian pottery found in Kerma ancien, Kerma moyen and Kerma classique graves at Kerma', in Kendall, T. (ed.) *Proceedings of the Ninth International Conference of Nubian Studies (Boston, 21–26 August 1998)*, pp. 3–13. Boston: Northeastern University.

Chaix, L. (1982) 'Seconde note sur la faune de Kerma (Soudan). Campagnes 1981 et 1982', *Genava*, NS, 30, 39–42.

Chaix, L. (1990) 'Le monde animal', in Bonnet, C. (ed.) *Kerma, royaume de Nubie: l'antiquité africaine au temps des pharaons. Exposition organisée au Musée d'art et d'histoire, Genève, 14 Juin–25 Novembre 1990*, pp. 108–13. Geneva: Tribune de Genève.

Chaix, L. (1993) 'Archaeozoology of Kerma (Sudan)', in Davies, W. V. and Walker, R. (eds) *Biological Anthropology and the Study of Ancient Egypt*, pp. 175–85. London: The British Museum Press.

Chaix, L. (1994) 'Nouvelles données de l'archéozoologie au nord du Soudan. Hommages au Professeur J. Leclant. Vol. 2, Nubie, Soudan, Ethiopie', *IFAO* 106 (2), 105–10.

Chaix, L. (1996) 'Les boeufs à cornes parallèles: archéologie et ethnographie', *Sahara*, 8, 95–7.

Chaix, L. (2000) 'La faune des peintures murales du temple K XI', in Bonnet, C. (ed.) *Edifices et rites funéraires à Kerma*, pp. 163–74. Paris: Éditions Errance.

Chaix, L. (2001) 'Animals as symbols: the bucrania of the grave KN 24 (Kerma, Northern Sudan)', in Buitenhuis, H. and Prummel, W. (eds) *Animals and Man in the Past. Essays in Honour of Dr. A. T. Clason Emeritus Professor of Archaeozoology, Rijkuniversiteit Groningen, The Netherlands*. ARC Publicatie 41, pp. 364–70. Groningen: Archaeological Research and Consultancy.

Chaix, L. (2002) 'Omniprésence du cuir à Kerma (Soudan) au IIIe millénaire av. J.-C', in Audoin-Rouzeau, F. and Beyries, S. (eds) *Le travail du cuir de la préhistoire à nos jours. XXIIe rencontres internationales d'archéologie et d'histoire d'Antibes, Antibes*, pp. 31–40. Antibes: Éditions APDCA.

Chaix, L. (2004) 'Déformations anciennes et actuelles du cornage bovin en Afrique du Nord-Est', in Guintard, C. and Mazzoli-Guintard, C. (eds) *Elevage d'hier et d'aujourd'hui. Mélanges d'ethnozootechnie offerts à Bernard Denis*, pp. 21–32. Rennes: Presses Universitaires de Rennes.

Chaix, L. (2006) 'Bœufs à cornes déformées et béliers à sphéroïde: de l'art rupestre à l'archéozoologie', in Gauthier, Y., Le Quellec, J. L., and Simonis, R. (eds) *Hic sunt leones: mélanges sahariens en l'honneur d'Alfred Muzzolini*. Cahiers de l'AARS 10, pp. 49–54. St-Benoist-sur-Mer: Association des Amis de l'Art Rupestre Saharien.

Chaix, L. (2007) 'Contribution to the knowledge of domestic cattle in Africa: the osteometry of fossil *Bos taurus* L. from Kerma, Sudan (2050–1750 BC)', in Grupe, G. and Peters, J. (eds) *Skeletal Series and Their Socio-economic Context*. Documenta Archaeobiologiae 5, pp. 170–249. Rahden: Verlag Marie Leidorf GmbH.

Chaix, L. (2011a) 'A review of the history of cattle in the Sudan throughout the Holocene', in Jousse, H. and Lesur, J. (eds) *People and Animals in Holocene Africa. Recent Advances in Archaeozoology*. Reports in African Archaeology 2, pp. 13–26. Frankfurt: Africa Magna Verlag.

Chaix, L. (2011b) 'Cattle skulls (bucrania): an universal symbol all around the world. The case of Kerma (Sudan)', in Barta, M., Coppens, F., and Krejci, J. (eds) *Abusir and Saqqara in the Year 2010/11*, pp. 7–16. Prague: Czech Institute of Egyptology / Faculty of Arts, Charles University.

Chaix, L. (2014) 'Boeufs, moutons et chèvres à Kerma (Soudan) entre 2600 et 1500 av. J.-C. dans l'économie et les rites funéraires. Contraintes environnementales et démographiques', in Costamagno, S. (ed.) *Histoire de l'alimentation humaine: entre choix et contraintes*, pp. 26–40. Paris: Édition électronique du CTHS.

Chaix, L., Dubosson, J., and Honegger, M. (2012) 'Bucrania from the Eastern Cemetery at Kerma (Sudan) and the practice of cattle horn deformation', in Kabaciński, J., Chłodnicki, M. and Kobusiewicz, M. (eds) *Prehistory of Northeastern Africa: New Ideas and Discoveries*. Studies in African Archaeology 11, pp 185–208. Poznan: Poznan Archaeological Museum.

Chaix, L. and Grant, A. (1987) 'A study of a prehistoric population of sheep (*Ovis aries* L.) from Kerma (Sudan)—Archaeozoological and archaeological implications', *Archaeozoologia*, 1(1), 77–92.

Chaix, L. and Grant, A. (1993) 'Palaeoenvironment and economy at Kerma, Northern Sudan, during the third millennium BC: archaeozoological and botanical evidence', in Krzyzaniak, L., Kobusiewicz, M., and Alexander, J. (eds) *Environmental Change and Human Culture in the Nile Basin and Northern Africa until the Second Millennium BC*. Studies in African Archaeology 4, 399–404. Poznan: Poznan Archaeological Museum.

Chaix, L. and Hansen, J. W. (2003) 'Cattle with 'forward- pointing horns': archaeozoological and cultural aspects', in Krzyzaniak, L., Kroeper, K., and Kobusiewicz, M. (eds) *Cultural Markers in the Later Prehistory of Northeastern Africa and Recent Research*. Studies in African Archaeology 8, pp. 269–81. Poznan: Poznan Archaeological Museum.

Chaix, L. and Queyrat, I. (2003) 'Les figurines animales dans la culture de Kerma', *Anthropozoologica*, 38, 61–7.

Choyke, A. (1990) 'Travail de l'os et de l'ivoire à Kerma', in Bonnet, C. (ed.) *Kerma, royaume de Nubie: l'antiquité africaine au temps des pharaons. Exposition organisée au Musée d'art et d'histoire, Genève, 14 Juin–25 Novembre 1990*, pp. 140–1. Geneva: Tribune de Genève.

Di Lernia, S., Tafuri, M. A., Gallinaro, M., Alhaique, F., Balasse, M., Cavorsi, L., Fullagar, P. D., Mercuri, A. M., Monaco, A., Perego, A., and Zerboni, A. (2013) 'Inside "the African Cattle Complex": animal burials in the Holocene Central Sahara', *PLOS One*, 8(2), 1–28.

Ferrero, N. (1984) 'Figurines et modèles en terre mis au jour dans la ville de Kerma', *Genava*, NS, 32, 21–5.

Ferrero, N. (1990a) 'Des voyageurs du XIXe siècle aux campagnes de Nubie', in Bonnet, C. (ed.) *Kerma, royaume de Nubie: l'antiquité africaine au temps des pharaons. Exposition organisée au Musée d'art et d'histoire, Genève, 14 Juin–25 Novembre 1990*, pp. 24–7. Geneva: Tribune de Genève.

Ferrero, N. (1990b) 'Miniatures en terre', in Bonnet, C. (ed.) *Kerma, royaume de Nubie: l'antiquité africaine au temps des pharaons. Exposition organisée au Musée d'art et d'histoire, Genève, 14 Juin–25 Novembre 1990*, pp. 132–5. Geneva: Tribune de Genève.

Grant, A. (1982) 'The use of tooth wear as a guide to the age of domestic ungulates', in Wilson, B., Grigson, C., and Payne, S. (eds) *Ageing and Sexing Animal Bones from Archaeological Sites*. BAR British Series 109, pp. 91–108. Oxford: Archaeopress.

Gratien, B. (1978) *Les cultures Kerma: essai de classification*, Lille: Publications de l'Université de Lille III.

Gratien, B., Müller, A., and Parayre, D. (eds) (2003) *Figurines animales des mondes anciens: Actes de la journée d'étude de l'Institut des sciences de l'antiquité de l'Université Charles-de-Gaulle—Lille 3, Villeneuve d'Ascq, 8 Juin 2002*. Anthropozoologica 38. Paris: L'Homme et L'Animal.

Grigson, C. (1980) 'The craniology and relationships of four species of *Bos*. 5. *Bos indicus* L.', *Journal of Archaeological Science*, 7, 3–32.

Guintard, C. and Lallemand, M. (2003) 'Osteometric study of metapodial bones in sheep (*Ovis aries* L. 1758)', *Annals of Anatomy*, 185, 573–83.

Habermehl, K. H. (1961) *Die Altersbestimmung bei Haustieren, Pelztieren und beim jagdbaren Wild*, Berlin: Paul Parey.

Herskovits, M. J. (1926) 'The cattle complex in East Africa', *American Anthropologist*, 28(1), 230–72.

Howard, M. M. (1963) 'The metrical determination of the metapodials and skulls of cattle', in Mourant, A. E. and Zeuner, F. E. (eds) *Man and cattle*. Royal Anthropological Institute, Occasional Paper 18, pp. 91–100. London: Royal Anthropological Institute of Great Britain and Ireland.

Iacumin, P., Bocherens, H., and Chaix, L. (2001) 'Keratin C and N stable isotope ratios of fossil cattle horn from Kerma (Sudan): a record of dietary changes', *Il Quaternario, Italian Journal of Quaternary Sciences*, 14(1), 41–6.

Kantor, H. G. (1948) 'A predynastic ostrich egg with incised decoration', *Journal of Near Eastern Studies*, 7(1), 46–51.

Le Quellec, J. L., de Flers, P., and de Flers, P. (2005) *Peintures et gravures d'avant les pharaons du Sahara au Nil*, Paris: Fayard/Soleb.

Lortet, L. C. and Gaillard, C. (1903) 'La faune momifiée de l'ancienne Egypte', *Archives du Museum d'Histoire Naturelle de Lyon*, 8(2), 1–206.

Marshall, F. (1989) 'Rethinking the role of *Bos indicus* L. in sub-Saharan Africa', *Current Anthropology*, 30(2), 235–40.

Matolcsi, J. (1970) 'Historische Erforschung der Körpergrösse des Rindes auf Grund von ungarischem Knochenmaterial', *Zeitschrift für Tierzüchtung und Züchtungsbiologie*, 87(2), 89–137.

Paris, F. (2000) 'African livestock remains from Saharan mortuary contexts', in Blench, R. M. and MacDonald, K. (eds) *The Origins and Development of African Livestock: Archaeology, Genetics, Linguistics and Ethnography*, pp. 111–26. London: UCL Press.

Pia, J. (1941) 'Rassenkundliche Untersuchungen an Schädelresten des altägyptischen Hausrindes', *Zeitschrift für Tierzüchtung und Züchtungsbiologie*, 48(1), 17–55.

Privati, B. (1999) 'La céramique de la nécropole orientale de Kerma (Soudan): essai de classification', *Cahiers de Recherches de l'Institut de Papyrologie et d'Égyptologie de Lille*, 20, 41–69.

Reisner, G. A. (1923) *Excavations at Kerma. Parts I–III*. Harvard African Studies 5. Cambridge (MA): Peabody Museum of Harvard University.

Ryder, M. L. (1984) 'Skin, hair and cloth remains from the ancient Kerma civilization of Northern Sudan', *Journal of Archaeological Science*, 11, 477–82.

Simon, C., Kramar, C., and Susini, A. (1990) 'Étude des ossements humains', in Bonnet, C. (ed.) *Kerma, royaume de Nubie: l'antiquité africaine au temps des pharaons. Exposition organisée au Musée d'art et d'histoire, Genève, 14 Juin–25 Novembre 1990*, pp. 101–7. Geneva: Tribune de Genève.

Tekkouk, F. and Guintard, C. (2007) 'Approche ostéométrique de la variabilité des métacarpes de bovins et recherche de modèles applicables pour l'archéozoologie: cas de races rustiques françaises, algériennes et espagnole', *Revue de Médecine Vétérinaire*, 158(7), 388–96.

Testart, A. (2004) *Les morts d'accompagnement: la servitude volontaire*, Paris: Éditions Errance.

Thompson, A. H., Chaix, L., and Richards, M. P. (2008) 'Stables isotopes and diet at Ancient Kerma, Upper Nubia (Sudan)', *Journal of Archaeological Science*, 35, 376–87.

Vigne, J. D. (1988) *Les mammifères post-glaciaires de Corse, étude archéozoologique*. Gallia Préhistoire Supplement 26. Paris: CNRS Éditions.

Wills, B. (2002) 'Windows into ancient Nubian leatherwork', in Audoin-Rouzeau, F. and Beyries, S. (eds) *Le travail du cuir de la préhistoire à nos jours. XXIIe rencontres internationales d'archéologie et d'histoire d'Antibes*, pp. 41–64. Antibes: Éditions APDCA.

Zalkin, V. I. (1960) 'Die Veränderlichkeit der Metapodien und ihre Bedeutung für die Erforschung des grossen Hornviehs der Frühgeschichte', *Byulleten Moskovskogo Obshchestva Ispytatelei Prirody, Otdel Biologicheskii*, 66, 115–32.

CHAPTER 27

THE ZOOARCHAEOLOGY OF IRON AGE FARMERS FROM SOUTHERN AFRICA

SHAW BADENHORST

INTRODUCTION

DESPITE objections over the usage of the term 'Iron Age' (e.g. Maggs, 1992), the term refers to the period of the spread and settlement of Bantu-speaking farming communities from their original homeland in the Cameroon-Nigeria area of central-west Africa to central, eastern, and southern Africa (Mitchell, 2002). Farmers reached southern Africa in the first millennium AD (Huffman, 2007). The cultural continuity between first- and second-millennium AD farmers has been debated and the zooarchaeological evidence of cattle (*Bos taurus*) and caprines (sheep (*Ovis aries*) and goats (*Capra hircus*) collectively) has featured strongly in these discussions.

The Iron Age of southern Africa (Fig. 27.1), spanning at least the last 1,500 years, is divided into three periods, namely Early, Middle, and Late Iron Age. During the first millennium AD, known as the Early Iron Age (200–900 AD), the first farmers arrived in southern Africa. They settled and interacted with Later Stone Age hunter-gatherers (e.g. Jolly, 1996). The farmers cultivated plants such as millet, sorghum, pulses, cowpeas, and cucurbits, kept livestock, worked and produced metal objects, made ceramics of various kinds, and constructed wattle-and-daub huts, which they often positioned on valley floors and next to rivers. During the Middle Iron Age (900–1300 AD), socio-political developments in the Limpopo Valley brought farming communities into contact, through trade networks, with other societies on the coast of eastern Africa. The farmers from the Limpopo Valley and surrounding areas provided gold, ivory, skins, and other commodities in exchange for beads, porcelain, and cloth traded from as far away

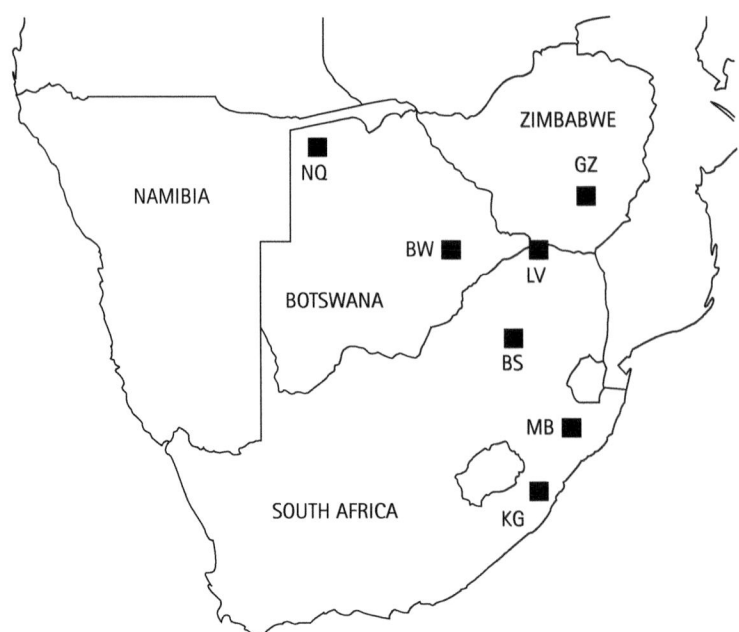

FIGURE 27.1 Location of sites and places mentioned in the text. BS: Broederstroom, BW: Bosutswe, GZ: Great Zimbabwe, KG: KwaGandaganda, LV: Limpopo Valley, MB: Mamba, NQ: Nqoma. Author's own image.

as Persia and China (Meyer, 1998). During the Late Iron Age (1300–1820s AD), ancestral Sotho-Tswana and Nguni-speakers arrived in southern Africa from East Africa. These farming groups are still dominant in South Africa today. The state of Great Zimbabwe, probably the best known Iron Age site in southern Africa, flourished during this time. Great Zimbabwe was built and used by Shona-speakers, who are related to the Venda. The Venda is a group of Bantu-speakers who still live in the northern parts of South Africa. Other states also developed during the Late Iron Age. Elaborate stone constructions appeared for the first time, and villages were often located on hilltops, probably for defensive reasons (overviews in Mitchell, 2002; Huffman, 2007).

Early traveller accounts, historical information, and particularly intensive ethnographic investigations during the twentieth century AD provided a wealth of information on farming communities in southern Africa (e.g. Schapera, 1953; Bruwer, 1956; Hammond-Tooke, 1974; Plug and Badenhorst, 2009). One aspect that is very apparent from these investigations is the pivotal role of cattle in these patrilineal farming societies. Great attachments were placed on cattle herds, and they were regarded as part of the household. Cattle played a major role in the social and ritual life of farmers, and these animals featured in all major events of daily life, including ancestral worship, marriage, and funerals. A man's wealth was measured according to the size of his cattle herd. Cattle were kept in a *kraal* (or byre) at night, and the *kraal* was an important place to perform ceremonies (e.g. Bruwer, 1956: 120–1).

ARCHAEOLOGICAL BACKGROUND

Archaeology as a science had a late beginning in the region. With the expansion of the economy of twentieth-century AD South Africa, archaeological research of Iron Age farmers only really gained momentum in the 1970s and beyond. Within this context, and inspired by historical and ethnographic information (e.g. Bruwer, 1956; Kuper, 1982), Huffman (1982) formulated a settlement pattern of farming villages, which he called the Central Cattle Pattern (CCP). The CCP is a settlement pattern of Iron Age farmers (Fig. 27.2), which is characterized by a male domain in the centre of a village, with cattle *kraals* where important individuals, usually men, are buried. In addition, this central male area also contains subterranean grain pits and elevated grain bins, a public smithy, and an assemblage area for men to settle disputes and make political decisions. The inner male zone is contrasted by an outer residential area of a village which is the domain of married women. This area includes the sleeping huts, kitchens, grain bins, storage pits, and graves. The households of the outer zone are arranged according to seniority, starting with the great hut built upslope of the court and *kraal*. The leading man of the community lives in the great hut. The space inside the village and huts are further divided into left and right sides (Mitchell, 2002: 279; Huffman, 2007: 25).

According to Huffman (2007: 25), the CCP is only found under eastern Bantu-speakers who follow a patrilineal form of descent, prefer cattle for bride-wealth, and have male hereditary leadership and a positive attitude towards the role and influence of ancestors in daily life. The presence of central *kraals* surrounded by huts led proponents of the CCP to argue that the model can be traced to the earliest farming settlements found in southern Africa. If the CCP does date to the Early Iron Age, patrilineal eastern Bantu-speakers with systems in which leadership is hereditary and cattle and ancestors have an important role, would have lived in southern Africa since the first millennium AD (Huffman, 2007).

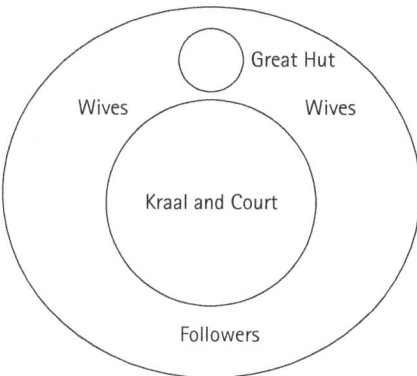

FIGURE 27.2 Simplified and ideal settlement arrangement of Iron Age CCP villages (redrawn from Huffman 2007: 25). Author's own image.

The CCP has dominated the interpretation of Iron Age archaeology since the early 1980s (Whitelaw, 2005) and many subsequent excavations identified the CCP settlement pattern at Early, Middle, and Later Iron Age sites in southern Africa. The few alternative explanations that have been offered over the years (Hall, 1987; Pearson, 1995; Lane, 1998) have had almost no impact whatsoever. More recently, it was shown that a range of human behaviours can mimic the CCP settlement. For example, the presence of lions that notoriously prey on livestock led to some groups, such as the herding Khoekhoen, placing *kraals* in the centre of villages (Badenhorst, 2009a).

ZOOARCHAEOLOGICAL EVIDENCE

Zooarchaeological evidence has played a decisive role in recent discussions about the antiquity of the CCP. Since the 1970s, an increasing number of faunal remains from farming sites have been analysed (Badenhorst, 2008a); it soon became apparent that the expectations of the CCP (what had previously been seen as the dominant cattle economy since the Early Iron Age) were at odds with the zooarchaeological evidence (Plug and Voigt, 1985). While Late Iron Age zooarchaeological assemblages are almost always dominated by cattle remains, this is not the case for the Early and Middle Iron Ages, where samples are typically dominated by caprines (Badenhorst, 2011). The CCP and zooarchaeological data were providing two different views. To provide support for the importance of cattle since the Early Iron Age, proponents of the CCP developed a number of alternative explanations over the years, which are discussed below (summary in Huffman, 2007). Zooarchaeological data have been used to oppose these ideas.

Livestock Dung

In one of the earliest attempts to counter the contradictory zooarchaeological evidence, proponents of the CCP argued that at Broederstroom, one of the earliest farming sites in southern Africa, at least four cattle *kraals* and three pits lined with cattle dung in two separate central zones were identified using phytolith analysis (Huffman, 1990). At Broederstroom, the faunal sample indicates that caprines outnumber cattle at a ratio of 42 : 1. Huffman (2001: 30) proposed that the zooarchaeological remains therefore did not accurately reflect the number of cattle in the settlement.

This argument has remained largely unchallenged for around two decades, but recently the proposition has been criticized. Cattle produce a much larger quantity of dung per day than caprines do. Although variations in environment, breed, health status, age, sex, body weight, lactation status, parasite infestation, and quantity of food intake varies, cattle usually produce at least 10 kg or more of wet dung per day (Badenhorst, 2009b); on the other hand, caprines produce less than 2 kg. Moreover, cattle dung preserves better compared to caprine dung (Castle and Macdaid, 1972). It is therefore not surprising to find cattle dung more commonly than caprine dung at Iron Age sites (Badenhorst, 2009b). While the location of the pits at Broederstroom is regarded as support for the CCP (Huffman, 2010), this may simply be a functional aspect as the preservative properties of animal dung are well known, a fact that is reflected in the wide distribution of these features in Africa, Europe, and Asia (Badenhorst, 2012).

Cattle Herd Sizes

Another argument was put forward by proponents of the CCP as an alternative explanation for the low number of cattle bones at Early and Middle Iron Age sites in southern Africa. This argument focused on the size of cattle herds. According to Huffman (2001), at least one hundred cattle individuals had to be present in a farming village for a herd to be sustainable. This then implies that even if one cattle individual is found in the archaeological record, at least one hundred individuals had to have been present for the herd to be sustainable. This argument was used to suggest that cattle numbers recorded at an Iron Age site are not a true reflection of herd sizes in the past (Huffman, 2007). Historical and ethnographic studies do not support Huffman's contention and the argument has been challenged. For example, Schapera (1962: 62) indicated that a herd of just ten cattle individuals are required for a sustainable herd. Elsewhere in Africa too, herds and flocks are often small, yet sustainable (Badenhorst, 2008b). As a result, this argument cannot be accepted.

Informants and Ethnography

Yet another approach to account for the despairingly low number of cattle during the Early and Middle Iron Ages in southern Africa was forwarded by proponents of the CCP. This time, personal observations and communications with descendants of Iron Age people were used, as well as an ethnographic example. A study of the Shona people in the former Rhodesia by Holleman (1952) showed that despite the devastating rinderpest epidemic of the late 1890s, when large numbers of cattle perished, they continued with their practice of using cattle for bride-wealth payments. This case study is used by proponents of the CCP to argue that low cattle numbers from Iron Age sites do not equate to limited social importance (Huffman, 2001).

However, bride-wealth was an essential part of marriage under the Shona and other Bantu-speaking people at the time. Bride-wealth influenced the rest of the marriage and cattle were of great social, political, and religious significance. In this case, the Shona had an existing practice of using cattle as bride-wealth, which continued after rinderpest struck the area. This existing importance explains the continuation of practice even in difficult times (Badenhorst, 2012). There is no evidence to suggest a similar practice of bride-wealth existed during the Early and Middle Iron Ages.

A new wave of arguments to prove that cattle were important since the Early Iron Age in southern Africa was recently presented (Huffman, 2010). Local informants from South Africa (presumably Nguni) and the Southern Shona from Zimbabwe indicated in 2010 that bones of cattle are ritually destroyed at farming villages and proponents of the CCP take this as evidence that cattle will be underrepresented at farming villages since the Early Iron Age (Huffman, 2010).

The evidence from informants that suggest cattle bones are ritually destroyed on farming sites is not supported by zooarchaeological evidence from Late Iron Age sites, which indicate an unambiguous dominance of cattle over caprines (Badenhorst, 2011). Moreover, events of the last two centuries had a great impact on farming practices of Bantu-speakers. These events include the Difaqane, a time of major social and political upheaval of Iron Age farmers in the 1820s and 1830s, the effect of which was felt over large parts of southern Africa. The rinderpest epidemic of the late 1890s led to huge losses of cattle herds in southern Africa. During conflicts such as the Anglo-Boer War of 1899–1902, the British 'scorched-earth' policy led to the mass killing of livestock in the former Boer Republics. The spread of British, Boer, Portuguese, and German influence included tax and stock control measures, and the formation and collapse of Bantu-speaking states also impacted on farming practices (Badenhorst, 2008b). It is possible that the ritual destruction of cattle bones by a few informants in recent times had been influenced by these and other events, and their views cannot be taken as evidence for the low number of cattle compared to caprine remains at Early and Middle Iron Age sites in southern Africa (Badenhorst, 2012).

Animal Weights

Huffman (2010) indicated that because live cattle weigh more than caprines, this suggests the importance of the former over the latter. This argument is untenable. One of the main problems with animal weight estimates is that larger species will always dominate smaller taxa. For example, live adult Nguni cattle from South Africa weigh between 400 and 800 kg, and Damara sheep from Namibia between 45 and 60 kg (Ramsay et al., 2000). Between seven and eighteen sheep therefore match the weight of one single cattle. Animal weights are therefore not a particularly useful indication of the importance of cattle during the Early and Middle Iron Ages (Badenhorst, 2012).

Ageing and Skeletal Parts

According to Huffman (2010), there are no changes in tooth-wear ageing data of cattle and caprines from the Early and Middle to Late Iron Ages in southern Africa. This lack of change in livestock ageing is considered to indicate that no change in herding practices (that would be from a dominant caprine to cattle economy) could have occurred. This argument ignores postcranial remains, which usually form the bulk of zooarchaeological specimens in the study area. In addition, due to fragmentation, often only a small proportion of all retrieved bones can be identified to taxa. Moreover, it is not necessary that the ageing data from cattle and caprines must show a change over time from a caprine-dominant to a cattle-dominant economy (Badenhorst, 2012).

Huffman (2010) also argued, as an example, that different skeletal part representation of cattle and caprines at K2 (an important capital farming site in the Limpopo Valley) provides support for the CCP. K2 was occupied during the Middle Iron Age (1,250 ± 40 BP) (Huffman, 2007). Huffman (2010) based his ideas on a study by Fatherley (2009), who assumed that all indeterminate large Bovidae from K2 were cattle. Similarly, it was assumed that all indeterminate medium Bovidae are caprines. The study suggested the court midden yielded more high-index meat (upper and lower fore and hind limbs) of cattle, whereas two residential middens had many more cattle feet. For caprines, high-index meat is high in the residential and court middens. Thus, high-index meat of cattle in the court middens is taken to provide support for cattle before 1300 AD (Huffman, 2010: 170). The latter study relied on already-analysed bone specimens from K2. However, the presence of wild Bovidae, Equidae, and Suidae in samples from K2 (Voigt, 1983; Hutten, 2005) makes this assumption unwarranted. Moreover, the study paid no attention to aspects such as different sample sizes and preservation between different parts of settlements. Without a consideration of these aspects, conclusions cannot be reached (Badenhorst, 2012).

Grazing Outposts

Although grazing outposts have never been mentioned as a reason for the low number of cattle at Early and Middle Iron Ages, it is worth considering them here. From various ethnographic sources (e.g. Schapera, 1953) it is clear that Bantu-speaking farmers made use of cattle and caprine outposts in recent times. Livestock would be moved out of a village to a camp further afield, perhaps to make better use of grazing further away from the village. However, this cannot account for the low number of cattle compared to caprines at Early and Middle Iron Age sites. Caprines were also moved to grazing outposts, and at terminal Late Iron Age sites where grazing outposts were utilized, the faunal samples are still dominated by cattle (Badenhorst, 2012).

RECENT DEVELOPMENTS

While the debate on the antiquity of the Central Cattle Pattern is by no means exhausted and will likely continue in the future, the possibility and opportunity now at least exists after several decades to investigate the archaeology of Iron Age farmers from a novel perspective outside the restrictions of the CCP. In a relatively short time, a few major changes in method and interpretation have already been established.

Methodologically, major changes have occurred. Since the early 1970s, successive faunal analysts, particularly those working at the then Transvaal Museum, developed, expanded, and applied a set of methods initiated by Brain (1974) with refinements by Voigt (1983) and Plug (1988). Essentially this method regards all vertebrae, ribs, indeterminate crania, and enamel fragments as 'unidentified specimens' and are therefore not included in species-lists and subsequent quantification. This method is still widely applied in zooarchaeological analyses of Iron Age farmers in southern Africa today.

Following developments in the theory of faunal identifications (e.g. Driver, 1982; 1991), a method developed by Driver (1999) and widely applied in the American Southwest (Badenhorst, 2008c) is now widely used to study Iron Age faunas (e.g. Brunton et al., 2013; Le Roux et al., 2013). According to the method of Driver (1999), any specimen that can be placed into an element category (e.g. humerus, rib, femur, crania, etc.) are regarded as 'identifiable specimens', even if only to order or class. This method firstly is very clear and unambiguous about which specimens can and cannot be identified. In addition, using a standardized method aims to allow more reliable comparisons of samples analysed by different zooarchaeologists.

Another methodological change in recent years has been the introduction of the Cattle Index, a ratio calculation based on the number of identified specimens (NISP) to measure the changes in cattle and caprine usage. The Cattle Index is calculated by the following formula:

$$\text{Cattle Index }(\text{CI}) = \frac{Cattle}{Cattle + Caprines}$$

The Cattle Index values can only range between 0 and 1. Values closer to 0 indicate samples with few cattle but many caprines, whereas a Cattle Index value nearer to 1 indicates samples where cattle significantly outnumber caprines. All firm (*Bos taurus, Ovis aries, Capra hircus, Ovis/Capra*) and possible identifications (denoted with *cf.*) are used in the Cattle Index calculations.

After the elimination of small samples, a few regional patterns were identified and confirmed through the application of the Cattle Index to Iron Age farming sites in southern Africa. These include: cattle outnumber caprines in Late Iron Age faunal assemblages, which supports the ethnographic and historical information on Bantu-speaking farmers that cattle were of great social importance; caprines outnumber cattle at most Early (n = 16 samples) and Middle (n = 13 samples) Iron Age sites; and only at three Early and Middle Iron Age sites do cattle outnumber caprines, namely Bosutswe in Botswana, and KwaGandaganda and Mamba in KwaZulu-Natal (eastern South Africa). These three sites share some important similarities: they all have large sample sizes and they were all occupied for several centuries (Badenhorst, 2011). A subsequent study revealed that the minimum number of individuals (MNI), despite its problematic nature, can be used to calculate the Cattle Index for large samples (Combrink, 2014). Another site, Nqoma, also in Botswana was identified as a pre-Late Iron Age site with a large sample size where cattle outnumber caprines (Fraser and Badenhorst, 2014). At Bosutswe, as high-ranked game (Equidae, Bovidae, and Suidae) declined compared to low-ranked game (small ground animals and wild birds), cattle were used in increasing numbers (Badenhorst, 2015). These changes were calculated using a Game Index (NISP), which is calculated with the following formula:

$$GameIndex = \frac{LowRanked}{Low + HighRanked}$$

Another major social change was suggested using the relative change in cattle and caprine numbers over time. In sub-Saharan Africa, ethnographic and historical studies of farmers found a strong correlation between dominant cattle-keeping and patrilineal descent (e.g. Holden et al., 2003). On the other hand, matrilineal descent patterns are associated with dominant caprine herding (e.g. Holden and Mace, 2003). The dominant caprine economies of the Early and Middle Iron Ages suggest matrilineal or perhaps other non-patrilineal communities, with the exception of Bosutswe, Nqoma, KwaGandaganda, and Mamba. The dominant cattle economy of the Late Iron Age, as expressed through the Cattle Index, supports ethnographic and historical records of the presence of patrilineal societies. In addition, this view is supported by the appearance of terraces during the second millennium AD. Terraces are indicative of intensive farming, and the latter associated with patrilineal descent (e.g. Ember, 1983). While these patterns were only established in a broad sense for southern Africa, individual areas as well as changing descent patterns must now be investigated (Badenhorst, 2010).

In southern Africa, changes in livestock usage are often regarded in relation to changes in environment and the presence of animal diseases. However, social changes should also be considered as a driving force. For example, the lack of cattle at many Iron Age sites in the Kruger National Park was seen as the result of the presence of animal diseases like nagana, which are fatal in cattle. Animal diseases such as nagana can appear and disappear within a very short period of time, making it difficult to identify from the archaeological record. However, viewed from a social perspective, the lack of cattle could rather indicate the presence of matrilineal and other non-patrilineal societies in the Kruger National Park in the past (Badenhorst, 2010).

Zooarchaeological data have been pivotal in disproving the Central Cattle Pattern as applied by archaeologists to the Early and Middle Iron Ages in southern Africa. These farming communities were clearly placing great economic and perhaps social importance on caprines before the Late Iron Age. There remain undoubtedly numerous issues to address using faunal assemblages in southern Africa, and a focus on social issues will be of great importance.

ACKNOWLEDGEMENTS

I wish to thank the National Research Foundation of South Africa and the Universities of South Africa and the Witwatersrand for financial support. Dr Ina Plug, formerly from the Transvaal Museum, read and commented on an earlier draft.

REFERENCES

Badenhorst, S. (2008a) 'Ina Plug: a tribute', in Badenhorst, S., Mitchell, P., and Driver, J. C. (eds) *Animals and People: Archaeozoological Papers in Honour of Ina Plug*. BAR International Series 1849, pp. 1–7. Oxford: Archaeopress.

Badenhorst, S. (2008b) 'Subsistence change among farming communities in southern Africa during the last two millennia: a search for potential causes', in Badenhorst, S., Mitchell, P., and Driver, J. C. (eds) *Animals and People: Archaeozoological Papers in Honour of Ina Plug*. BAR International Series 1849, pp. 215–28. Oxford: Archaeopress.

Badenhorst, S. (2008c) 'The Zooarchaeology of Great House Sites in the San Juan Basin of the American Southwest'. Unpublished PhD dissertation, Simon Fraser University (Burnaby).

Badenhorst, S. (2009a) 'The Central Cattle Pattern during the Iron Age of southern Africa: a critique of its spatial features', *The South African Archaeological Bulletin*, 190, 148–55.

Badenhorst, S. (2009b) 'Phytoliths and livestock dung at Early Iron Age sites in southern Africa', *The South African Archaeological Bulletin*, 189, 45–50.

Badenhorst, S. (2010) 'Descent of Iron Age farmers in southern Africa during the last 2000 years', *African Archaeological Review*, 27(2), 87–106.

Badenhorst, S. (2011) 'Measuring change: cattle and caprines from Iron Age farming sites in southern Africa', *The South African Archaeological Bulletin*, 194, 167–72.

Badenhorst, S. (2012) 'The significance of bone numbers in Iron Age faunal studies of southern Africa: a reply to Huffman', *The South African Archaeological Bulletin*, 196, 262–72.

Badenhorst, S. (2015) 'Intensive hunting during the Iron Age of southern Africa', *Environmental Archaeology*, 20(1), 41–51.

Brain, C. K. (1974) 'Some suggested procedures in the analysis of bone accumulations from southern African Quaternary sites', *Annals of the Transvaal Museum*, 29, 1–8.

Brunton, S., Badenhorst, S., and Schoeman, M. H. (2013) 'Ritual fauna from Ratho Kroonkop: a second millennium AD rain control site in the Shashe-Limpopo Confluence area of South Africa', *Azania: Archaeological Research in Africa*, 48(1), 111–32.

Bruwer, J. (1956) *Die Bantoe van Suid-Afrika*, Johannesburg: Afrikaanse Pers Boekhandel.

Castle, M. E. and Macdaid, E. (1972) 'The decomposition of cattle dung and its effect on pasture', *Journal of the British Grassland Society*, 27, 133–7.

Combrink, L. (2014) 'Changes in Livestock Utilisation during the Iron Age of Southern Africa'. Unpublished Honours thesis, University of Pretoria (Pretoria).

Driver, J. C. (1982) 'Minimum standards for reporting of animal bones in salvage archaeology: southern Alberta as a case study', in Francis, P. D. and Poplin, E. C. (eds) *Directions in Archaeology. A Question of Goals*, pp. 199–209. Calgary: The Archaeological Association of the University of Calgary.

Driver, J. C. (1991) 'Identification, classification and zooarchaeology', *Circaea*, 9(1), 35–47.

Driver, J. C. (1999) *Manual for Description of Vertebrae Remains*, 6th edn, Cortez: Crow Canyon Archaeological Centre.

Ember, C. R. (1983) 'The relative decline in women's contribution to agriculture with intensification', *American Anthropologist*, 85, 285–304.

Fatherley, K. (2009) 'Sociopolitical Status of Leokwe People in the Shashe-Limpopo Basin during the Middle Iron Age through Faunal Analysis'. Unpublished MSc thesis, University of the Witwatersrand (Johannesburg).

Fraser, L. and Badenhorst, S. (2014) 'Livestock use in the Limpopo Valley of southern Africa during the Iron Age', *The South African Archaeological Bulletin*, 69(200), 192–8.

Hall, M. (1987) *The Changing Past: Farmers, Kings and Traders in Southern Africa*, Cape Town: David Philip.

Hammond-Tooke, W. D. (ed.) (1974) *The Bantu-Speaking Peoples of Southern Africa*, London: Routledge & Kegan Paul.

Holden, C. J. and Mace, R. (2003) 'Spread of cattle led to the loss of matrilineal descent in Africa: a coevolutionary analysis', *Proceedings: Biological Sciences*, 1532, 2425–33.

Holden, C. J., Sear, R., and Mace, R. (2003) 'Matriliny as daughter-biased investment', *Evolution & Human Behavior*, 24, 99–112.

Holleman, J. F. (1952) *Shona Customary Law*, Cape Town: Oxford University Press.

Huffman, T. N. (1982) 'Archaeology and ethnohistory of the African Iron Age', *Annual Review of Anthropology*, 11, 133–50.

Huffman, T. N. (1990) 'Broederstroom and the origins of cattle-keeping in southern Africa', *African Studies*, 49, 1–12.

Huffman, T. N. (2001) 'The Central Cattle Pattern and interpreting the past', *Southern African Humanities*, 13, 19–35.

Huffman, T. N. (2007) *Handbook to the Iron Age: The Archaeology of Pre-Colonial Farming Societies in Southern Africa*, Scottsville: University of KwaZulu-Natal Press.

Huffman, T. N. (2010) 'Debating the Central Cattle Pattern: a reply to Badenhorst', *The South African Archaeological Bulletin*, 65, 164–74.

Hutten, L. (2005) 'K2 Revisited: an Archaeozoological Study of an Iron Age site in the Northern Province, South Africa'. Unpublished MSc thesis, University of Pretoria (Pretoria).

Jolly, P. (1996) 'Interaction between south-eastern San and southern Nguni and Sotho communities, c.1400–c.1880', *South African Historical Journal*, 35, 30–61.

Kuper, A. (1982) *Wives for Cattle. Bridewealth and Marriage in South Africa*, London: Routledge & Kegan Paul.

Lane, P. (1998) 'Engendered spaces and bodily practices in the Iron Age of southern Africa', in S. Kent (ed.) *Gender in African Prehistory*, pp. 179–203. Walnut Creek: AltaMira Press.

Le Roux, A., Badenhorst, S., Esterhuysen, A., and Cain, C. (2013) 'Fauna from the 1854 siege of Mugombane, Makapans Valley, South Africa', *Journal of African Archaeology*, 11(1), 97–110.

Maggs, T. M. O'C. (1992) 'Name calling in the Iron Age', *The South African Archaeological Bulletin*, 156, 131.

Meyer, A. (1998) *The Archaeological Sites of Greefswald: Stratigraphy and Chronology of the Sites and a History of Investigations*, Pretoria: University of Pretoria.

Mitchell, P. (2002) *The Archaeology of Southern Africa*, Cambridge: Cambridge University Press.

Pearson, N. (1995) 'Archaeological research at Modipe Hill, Kgatleng District: survey and excavation', *Botswana Notes and Records*, 27, 21–40.

Plug, I. (1988) 'Hunters and Herders: an Archaeozoological Study of some Prehistoric Communities in the Kruger National Park'. Unpublished DPhil dissertation, University of Pretoria (Pretoria).

Plug, I. and Badenhorst, S. (2009) 'Ethnography and southern African archaeozoology', in Grupe, G., McGlynn, G., and Peters, J. (eds) *Tracking Down the Past: Ethnohistory Meets Archaeozoology. Documenta Archaeobiologiae* Vol. 7, pp. 187–201. Rahden: Verlag Marie Leidorf.

Plug, I. and Voigt, E. A. (1985) 'Archaeozoological studies of Iron Age communities in southern Africa', in Wendorf, F. and Close, A. (eds) *Advances in World Archaeology*, Vol. 4, pp. 189–238. London: Academic Press.

Ramsay, K., Harris, L., and Kotzé, A. (2000) *Landrace Breeds: South Africa's Indigenous and Locally Developed Farm Animals*, Pretoria: Farm Animal Conservation Trust.

Schapera, I. (ed.) (1953) *The Bantu-Speaking Tribes of Southern Africa: An Ethnographic Survey*, London: Routledge & Kegan Paul.

Schapera, I. (1962) *The Tswana*, London: International African Institute.

Voigt, E. A. (1983) *Mapungubwe: An Archaeozoological Interpretation of an Iron Age Community*, Pretoria: Transvaal Museum.

Whitelaw, G. (2005) 'Comment on Greenfield and Van Schalkwyk's article on Ndondondwane, *Azania*, 2003', *Azania*, 60, 122–7.

CHAPTER 28

....

THE EXPLOITATION OF AQUATIC RESOURCES IN HOLOCENE WEST AFRICA

....

VEERLE LINSEELE

INTRODUCTION

....

Specialized Fishers of West Africa

Typical for Sahelian West Africa is the co-occurrence of different ethnic groups with their own specialized food procurement strategies (McIntosh, 1993). Settled farmers, pastoral nomads, and fishers, to name the three extremes, use different niches of the same geographical territory. Well-known fishermen from Sahelian West Africa include the Kotoko of the Lake Chad area (Bouquet, 1990) and the Bozo or Sorko of the Niger Valley (Sundström, 1972). A high degree of specialization can only exist because the different groups exchange food products and in that respect it is typical that all of them consume significant amounts of cultivated crops (cf. Bates and Lees, 1977).

In this chapter the focus will be on the exploitation of aquatic resources in Sahelian West Africa throughout the Holocene (Fig. 28.1). The fauna considered are fish and freshwater turtles, two groups which are caught in similar environments, with similar techniques. Emphasis will be on data from the Lake Chad area available in Linseele (2007), Linseele and Haour (2010), and Magnavita et al. (2009). At present, the Lake Chad area is one of West Africa's major fishing areas, together with the Niger and the Senegal flood-plain–river systems. Aspects discussed here are the relative importance of aquatic resources, the environments exploited, the seasons and techniques of fishing, and evidence for preparation, conservation, and trade of fish. Special attention will be paid to investigate how these different aspects reflect specialized fishing activities.

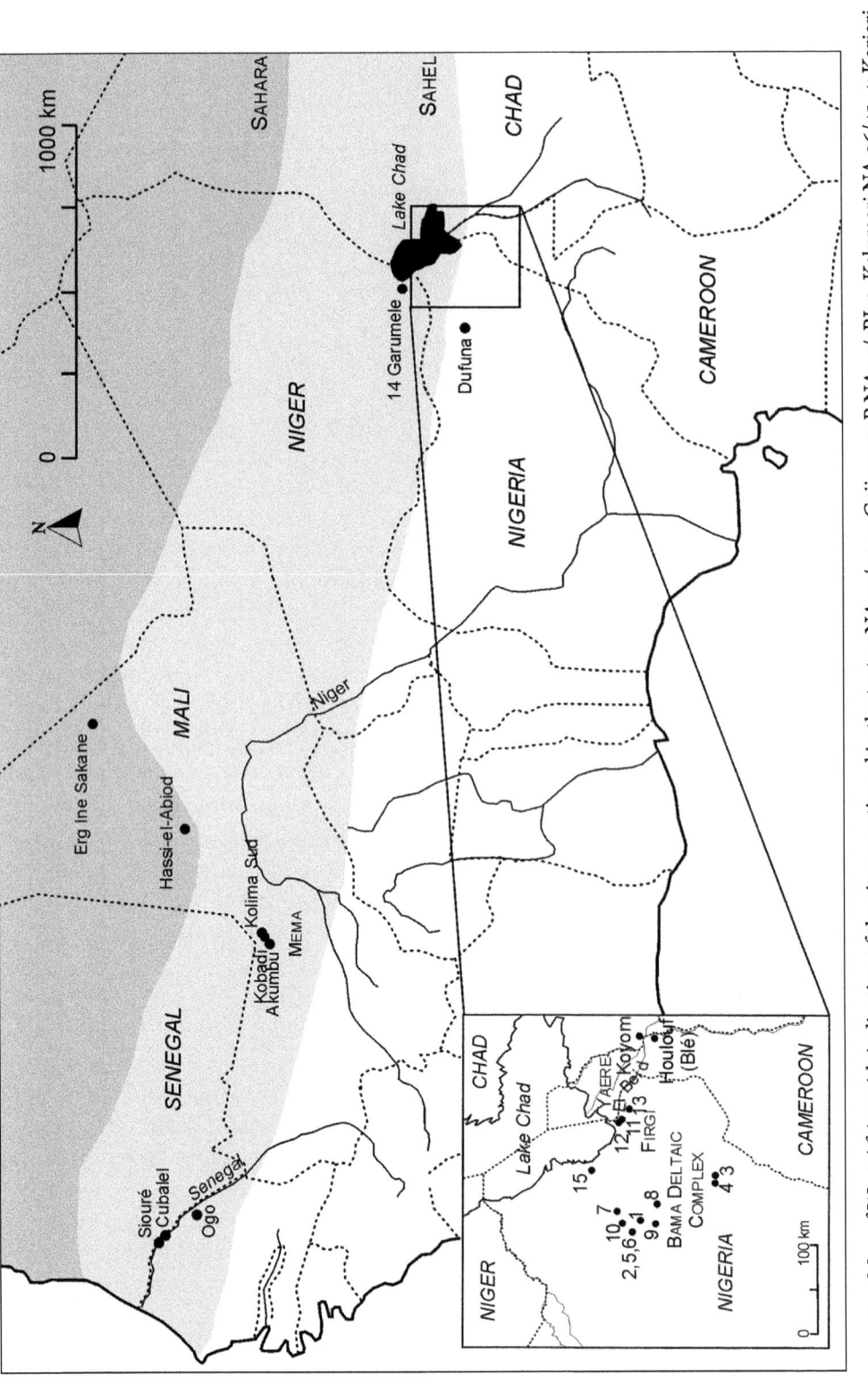

FIGURE 28.1 Map of West Africa with indication of the sites mentioned in the text. 1: NA 93/42, 2: Gajiganna B NA90/5BI, 3: Kelumeri NA 96/45, 4: Kariari C NA 95/1, 5: Gajiganna A NA 90/5A, 6: Gilgila NA 99/65, 7: Zilum NA 97/37, 8: Labe Kanuri NA97/26, 9: Elkido North NA 99/75, 10: Galaga NA 92/2C, 11: Kursakata NA 93/4, 12: Ngala NA 93/45, 13: Mege NA 94/7, 14: Garumele, situated north of the inset, 15: Garu Kime.

Archaeology of the Lake Chad Area

The Early and Middle Holocene were periods of a climatic optimum, with higher rainfall in the arid zones of West Africa, turning what is now the Saharan desert into a green savannah (Gasse, 2000). At Dufuna, a dugout canoe, about 8.5 m long, was found, dating to around 6000 BC (Breunig, 1996). It is the oldest recorded African boat and links have been made with the Mega Chad Lake, a lake the size of the Caspian Sea that existed during the Early and Middle Holocene humid phase. Gradual aridification, after the Middle Holocene, initiated large-scale population movements southwards from the Sahara. From that time onwards (about 2000 BC) substantial archaeological evidence is known from sub-Saharan West Africa. With the southward movements of people, domesticated livestock were also introduced (Breunig, 2013). From around 2500 BC, until the beginning of the current era, pearl millet (*Pennisetum americanum*) is the only known cultivated crop from Saharan and Sahelian West Africa (Kahlheber and Neumann, 2007; Manning et al., 2011). It is assumed that its cultivation did not require much expertise and could have been integrated into a mobile way of life (Kahlheber and Neumann, 2007). Full farming has been postulated from the early first millennium AD, with the appearance of new cultivated crops, most importantly sorghum (*Sorghum bicolor*), and diversified agricultural systems (Kahlheber and Neumann, 2007).

At about 1800 BC, archaeological evidence first appears along the southern shores of Lake Chad, in an area called the Bama Deltaic Complex (Breunig et al., 1993) that was previously under the waters of Lake Mega Chad. The central part of this complex consists of sand plains, interrupted by clay depressions, which fill up with rain water during the rainy season and sometimes contain water year-round (Breunig et al., 1993). The archaeological complex, representing the remains of the first colonists of the area, has been named the Gajiganna Complex (*c*.1800–400 BC) (Magnavita et al., 2004; Wendt, 2007). No archaeological traces are known from the Bama Deltaic Complex after the end of the Gajiganna Complex until the beginning of the Iron Age in the early first millennium AD. The period after the Early Iron Age remains largely unexplored, although one site dated to the mid-nineteenth century AD, Galaga, has been excavated.

Around 3,000 years ago, further shrinking of Lake Chad made also the *firgi* available for occupation, an area characterized by heavy lagoonal clay plains inundated for large parts of the year (Gronenborn, 1998). Several archaeological phases have been defined for the *firgi*, between its first occupation and the present day (Wiesmüller, 2001). On sand dunes, above inundation levels, large settlements were formed covering an area up to several hundred square metres and ten metres high (Breunig, 1995).

In addition to the sites in the Bama Deltaic Complex and the *firgi*, Garumele (*c*.1300–1800 AD) (Haour, 2008) and Garu Kime (seventeenth to eighteenth centuries AD) (Magnavita et al., 2009) will be discussed. Both have baked mud-brick structures and are associated with historical kingdoms in the Lake Chad area. Garumele is close to the northwestern shores of Lake Chad, in an area where flooding is infrequent, while Garu Kime is about 20 km away from the lake.

METHODOLOGY

Excavation, Sampling, and Preservation

The sites discussed were usually excavated in grids of 1 m² and by artificial spits of 10 cm, as structures or stratigraphy could not be recognized in the field. All archaeological sediment was dry-sieved on 5 mm meshes—1 mm at Garumele and Garu Kime. Often, the high clay content of the sediment caused it to stick together in large lumps, hampering find collection (see also Van Neer, 2008). For faunal remains, especially for fish, sampling methods were thus not ideal (Zohar and Belmaker, 2005), but in the logistically difficult circumstances of rural sub-Saharan Africa, with a lack of current and running, uncontaminated water, probably the best possible. All faunal material was shipped to Europe and studied there with the aid of appropriate reference collections, and identification keys where available (see Linseele, 2007). The material is poorly preserved, but this mainly had a negative impact on identification rates for mammals, while fish were usually still recognizable when fragmented.

Quantification

The main quantification technique used is number of identified specimens (NISP), which is most straightforward and most commonly used for Sahelian West Africa. Minimum numbers of individuals (MNI) were not calculated, for several reasons, one of them being that for the studied sites it is usually not clear how to define the archaeological units for which MNI calculations should be done, due to the lack of recognizable structures and stratigraphy (see similar arguments in Jousse et al., 2008, but contrary to MacDonald, 1995). To reconstruct the relative dietary/economic importance of aquatic animal resources, their NISPs are compared to those of other animal groups (including only taxa that were consumed). The different groups considered and connected economic activities are: fish and freshwater turtles (fishing), other reptiles, wild birds and wild mammals (hunting and fowling), and domestic birds and mammals (stock-keeping). NISPs are believed to reflect mainly the frequency in which animals are consumed. However, NISPs have poor connections with amounts of food: one fish bone for example obviously does not represent the same amount of food as one cattle bone. This can be resolved by additionally using weights of identified bones by taxon, because of the assumed correlation between bone weight and meat yield (see overview in Reitz and Wing, 1999: 225–8). Alternatively, for the sites under consideration, NISPs of consumed taxa have been multiplied by their estimated average live weight (Linseele, 2007). Still another technique that has been used in the study area is counting numbers of exploited taxa by strategy (Jousse, 2006). Similar considerations on quantification techniques can be made when comparing the importance of different aquatic taxa.

No quantification technique can be considered as a direct reflection of the proportion of different animal resources in the human diet due to differences in preservation, recovery, and identification, depending on the taxon and biases due to the quantification method itself. NISP tends to overemphasize the importance of small species, like fish, while weights will overemphasize larger species, such as cattle. All quantification methods have their drawbacks and their main value is as an instrument to compare sites and contexts, of which the fauna has preferably been sampled, studied, and quantified using the same methods.

Identified Taxa and Their Habitats

Typically, in arid West Africa, certain fish taxa are dominant in the archaeozoological samples. This is because they are usually large, resulting in higher preservation chances for their bones and smaller risks to be overlooked during sampling—for example Nile perch (*Lates niloticus*) and clariid catfish (Clariidae)—or because their bones remain recognizable when very fragmented (for example the neurocranium of clariid catfish). For our study, the species diversity has been calculated for all assemblages of aquatic fauna according to the formula in Cruz-Uribe (1988). The fish taxa were then subdivided into three groups according to their main habitat requirements (mainly based on Paugy et al., 2003; see similar subdivision in Jousse et al., 2008). The first group contains the taxa that are chiefly confined to shallow water and that can survive in adverse conditions with low oxygen contents, high temperatures, and high salinity, notably lungfish (*Protopterus annectens*), clariid catfish, and tilapia (tribe Tilapiini). Fish indicative of marshy, well-vegetated areas form the second group, including bichir (*Polypterus* sp.), African bonytongue (*Heterotis niloticus*), aba (*Gymnarchus niloticus*), Distichodontidae, Citharinidae, catfish *Auchenoglanis* sp., and snakehead (*Parachanna obscura*). Finally, the third group contains the fish that prefer open, deep, water: mormyrids (Mormyridae), tiger fish (*Hydrocynus* sp.), large cyprinids, catfishes *Bagrus* sp. and *Synodontis* sp., and Nile perch. Freshwater turtles have been counted with the shallow-water fish, as the identified taxa, terrapins (Pelomedusidae, mainly *Pelusios adansonii*) and soft-shell turtles (Trionychidae), can all survive periods of low water level by burying themselves in the mud, although the African softshell turtle (*Trionyx triunguis*) mainly occurs in large water bodies (Villiers, 1958: 197–219). Relative proportions of each of the three groups are again used for inter-site comparisons and are helpful in determining the preferred place, season, and technique of capture.

Fish Size and Skeletal Distribution

Fish sizes vary within one taxon depending on the place and season of capture, among other factors (Van Neer, 2004). Comparison of the studied fish remains with skeletons from modern fish of known size allowed a quick and simple subdivision of the fish specimens in size classes of 10 cm. Sizes are expressed as standard lengths (SL), i.e. the distance from the tip of the snout to the beginning of the tail. Skeletal elements of fish have been grouped by site and period according to the four major parts of the body: skull, pectoral girdle, body and tail, fins, as skeletal distribution can be informative for fish treatment.

Discussion

Sites of Specialized Fishers in West Africa

Fig. 28.2 summarizes the composition of the fauna from the sites considered in the Lake Chad area. The sites of the Bama Deltaic Complex generally yielded much lower proportions of fish bones than the other sites discussed and are poorer in taxa. In the central part of the Bama Deltaic Complex, no open-water fish have been recorded, presumably because its clay depressions were no longer connected with the main lake. The subrecent site, Galaga, did yield deep-water fish as well as more fish taxa compared to the older sites. Because of the incongruence of the Galaga fish fauna with the older assemblages in the same area, it has been assumed that fish were not locally obtained, but could have been traded in. Galaga also lacks turtle bones, contrary to the other sites where local fishing can be supposed. Gajiganna sites outside of the central area, Kelumeri and Kariari C, do have some open-water fish, probably obtained from the rivers running in their proximity.

In comparison with the sites from the Bama Deltaic Complex, the large numbers of fish bones for the *firgi* sites is very clear. With percentages usually around 90% of the identified faunal remains, fish are by far the most common animal group. The *firgi* sites usually also have a higher species diversity, including more fish from marshes, as well as from open, deep water. The site of Mege has the poorest species diversity and also the lowest proportion of open-water fish. It is probably no coincidence that Mege is the *firgi* site that is most distant from Lake Chad, and also the furthest from the river El Beid where floodwaters, seasonally inundating the *firgi* plains, were coming from. Exploitation of large water basins by people using the *firgi* sites is also confirmed through finds of African softshell turtle.

Despite the finer sampling techniques at Garumele and Garu Kime, the main characteristics of their aquatic fauna compared to the other sites can still be highlighted. At both sites, fish reach percentages comparable to the *firgi* area (i.e. around 90% of the identified fauna). At Garumele, they contain a high number of open water taxa, in proportions that much exceed those of the *firgi*. The lack of nearby flood-plains probably largely explains the different proportions. At Garumele, compared to the average for the other Lake Chad sites, both Nile perch and tilapia, the two most common taxa at the site, yielded much greater numbers of vertebrae (72% *vs* 25% of all skeletal elements for Nile perch; 75% *vs* 22% for tilapia). Since vertebrae are the skeletal elements of a fish to

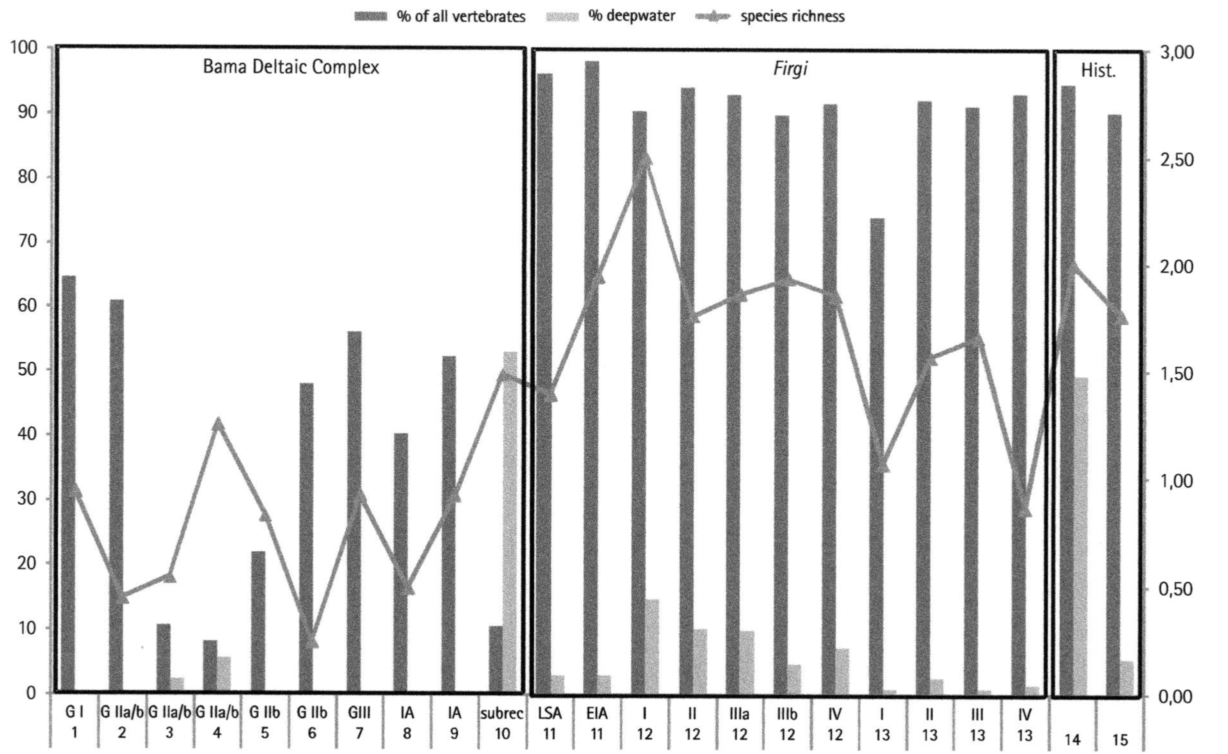

FIGURE 28.2 Lake Chad area. Visual representation of the proportion of fish in the total sample of identified vertebrates, the proportion of deep-water species in the total fish sample and the species richness of fish, by period and site. Phases and sites with less than 40 fish-bones have been excluded. For more details on the calculation methods, see Linseele (2007). Names of the 15 sites listed on the horizontal axis (sample size in brackets): 1: NA 93/42 (71), 2: Gajiganna B NA90/5BI (819), 3: Kelumeri NA 96/45 (40), 4: Kariari C NA 95/1 (54), 5: Gajiganna A NA 90/5A (415), 6: Gilgila NA 99/65 (59), 7: Zilum NA97/37 (226), 8: Labe Kanuri NA97/26 (58), 9: Elkido North NA 99/75 (77), 10: Galaga NA 92/2C (113), 11: Kursakata NA 93/ 4 (LSA: 647, EIA: 17,650), 12: Ngala NA 93/45 (I:1,824, II:603, IIIa:417, IIIb: 328, IV:406), 13: Mège NA 94/7 (I:920, II:368, III:287, IV:379), 14: Garumele (8,272), 15: Garu Kime (274). Absolute dates: Bama Deltaic Complex: G I: 1800–1400 BC, G IIa/b: 1500–1000 BC, G IIb: 1200–1000 BC, G III: 600–400 BC, IA: AD 1–600, sub-recent: mid-19th C AD. Firgi: Kursakata: LSA: 1300/1800–800/500 BC, EIA: 800/500 BC–AD 400/600, Ngala: I: Mixed/?–7th C AD, II: 8th–10th/11th C AD, IIIa: 10th–12th C AD, IIIb: 14th/16th–18th C AD, IV: 19th–20th C AD, Mège: I: 550 BC–AD 50, II: AD 50–700, III: AD 700–1150, IV: AD 1150–1983. Garumele: AD 1300–1800. Garu Kime: 17th–18th C AD. Graph by author.

which most food will adhere, their predominance has been taken as an indication that fish were processed prior to being taken to the site. Although the evidence is limited at present, this could imply trade or exchange between the inhabitants of Garumele and other groups. At Garu Kime, the faunal assemblage is diverse and also includes open-water fish. It has been assumed that the nearest source for open-water species was Lake Chad, at least 20 km away. Like at Galaga (and perhaps Garumele), they may therefore have been obtained through trade.

The sites in the Lake Chad area fall into two groups. One group with relatively few fish remains, poor species diversity, and usually a lack of open-water species, and another group with high percentages of fish, and more species, including open-water fish. It is among the second group that we need to look for specialized fishers. Other archaeological sites in arid West Africa yielded assemblages similar to the second group. In the Middle Senegal Valley, there is for example Cubalel and Siouré (1–1800 AD) (Van Neer, 2008). Multiple sites are known from the Niger Valley, where particularly Kobadi (1700–500 BC) in the Malian Méma (also known as the Dead Delta of the River Niger) has been connected with specialized fishers (Jousse et al., 2008). It has been suggested that the people from Kobadi originated from the area near Hassi-el-Abiod and Erg Ine Sakane, two sites in the Sahara with large numbers of fish remains but no domesticates (sixth to third millennium BC) (Van Neer and Gayet, 1988). At Akumbu (400–1400 AD), in the Méma, a contemporary habitation of both specialist and non-specialist fishers has been argued for by MacDonald and Van Neer (1994), based on a similar subdivision in parts with a rich fish fauna, and others poor in fish. In the *yaéré*, the Cameroonian equivalent of the *firgi*, the Blé sites (second to sixteenth century AD) yielded assemblages rich in fish as well (Linseele, 2007). Koyom (eighteenth to nineteenth century AD), in the Chadian part of the Lake Chad area, also yielded a lot of fish, including open-water species, but the main activity there appears to have been hunting of wild game, mainly kob (*Kobus kob*) and reedbuck (*Redunca redunca*) (Rivallain and Van Neer, 1983; 1984). This sets the site apart from other sites in arid West Africa where hunting is a minor component only (Linseele, 2007). All sites with a rich and varied fish fauna, including open-water fish, are in the vicinity of former large water bodies, either lakes or rivers. Exceptions are Galaga and Garu Kime where fish was presumably not obtained locally.

Seasons of Fishing

Connected to seasonal variations in water levels, fishing is particularly rewarding during two parts of the year (Van Neer, 2004). The first peak is when the water rises and spawning runs take place towards shallow, marginal areas. The spawning fish, mainly clariids and tilapia, are easy to catch, as are fish migrating between the main water body and the flood-plain through channels connecting the two. The predominance of larger, adult fish in the archaeological assemblages of the Lake Chad area can be connected to this peak. The second peak is at the end of the floods when residual pools are formed, where young, small fish can be found that did not make it back to the main water body in time. Small, young fish are not common at any of the studied sites, although this absence could also be due to a combination of differential destruction and coarse sieving. Fishing in the main channels is best at low water levels because of the reduction of turbulence. The height of the floods is probably least suitable for fishing as the fish tend to be much more dispersed. The peaks at the beginning and end of the flood make fish a predictable and often plentiful source of food. In addition, conservation of surpluses that allows storage for later use can help overcome shortages in periods of reduced resource abundance.

Fishing Techniques

Few recognizable remains of fishing gear have been found in archaeological contexts from Sahelian West Africa, probably because most were made from perishable materials. For the reconstruction of fishing techniques and equipment, reference is often made to studies describing traditional techniques that are still in use (e.g. Blache and Miton, 1962; Reed et al., 1967). Terra-cotta net weights were recorded at Kolima Sud in the Malian Méma (1700–500 BC) (MacDonald and Van Neer, 1994) and Koyom (eighteenth to nineteenth century AD) in southern Chad (Rivallain and Van Neer, 1983). At Kolima Sud also impressions of nets have been found on potsherds (MacDonald and Van Neer, 1994). A number of bone points and fragments of barbed hooks are recorded from Kolima-Sud (MacDonald and Van Neer, 1994). Bone harpoons were found at Kobadi (first half second millennium BC) (Raimbault cited in Jousse et al., 2008). Iron fish-hooks are known from Ogo (tenth to twelfth century AD) (Chavane, 1985), and other Iron Age locations in the Middle Senegal Valley (McIntosh, 1999).

Fish from marshes and flood-plains are usually easy to catch, using simple means (Van Neer, 2004). Spawning clariids can for example be captured with striking or wounding gears, using cover pots or simply by hand. The unspecialized nature of fishing in shallow water and marshes is also illustrated by the fact that agricultural groups of arid West Africa have events, usually in the dry season, whereby a complete village may take part in catching most of the fish from a single swamp or residual pool (Reed et al., 1967; Sundström, 1972: 18–19). For fishing in channels connecting the main water body and the inundation plains mainly fences are used. African tetras (Alestidae) are typically fished in this way (Sundström, 1972: 149). However, fish from this family are missing in archaeological contexts, probably due their small size, and as a consequence their poor preservation and poor chance of recovery.

Fishing in open, deep water is technically more advanced than flood-plain fishing, not in the least because it usually requires the use of boats (Van Neer, 2004). The dugout from Dufuna testifies that already around 6000 BC people from sub-Saharan West Africa had boats. Sites from the Lake Chad area with open-water fish generally yielded the most species. The use of boats should therefore probably be connected to the use of nets, a technique which only selects fish based on their size. As a result, a large number of taxa are captured. In the nets probably also freshwater turtles were caught.

Fish Processing and Conservation

Few butchery marks have been recorded on fish bones from archaeological contexts in the study area, suggesting that minimal processing of fish occurred, or that the techniques used did not frequently leave traces on the bones. Burning of the bones is mostly unrelated to fish preparation, but is rather due to post-depositional, accidental firing (see also Jousse et al., 2008). Traditionally in the Lake Chad area, fish are prepared for consumption by roasting or smoking, while cooking in pots is only rarely done (Blache and Miton, 1962; Reed et al., 1967). Curing allows fish to be consumed later in parts of the year when fishing is difficult and/or is not rewarding. Such fish can also be traded or exchanged more easily. The arid conditions in the Sahel are very favourable for sun-drying, and indeed this seems to be the most widely used fish conservation technique in the whole area (Sundström, 1972: 29, 153–4). However, contrary to other preservation techniques, it is unlikely to leave archaeological traces. Archaeological evidence for different techniques has been found in the Houlouf area (second to sixteenth century AD). Numerous fish-smoking features have been described there, as well as one site that was a special-purpose site for fish-smoking and fish processing (Holl, 2002: 120–30, 245–6). A set of small jugs from the same site was linked to the production of fish oil, while archaeological features elsewhere are thought to have been connected to the fermentation of fish (Holl, 2002: 245–7).

CONCLUSIONS

Comparison of the faunal composition of a set of sites in the Lake Chad area shows that these sites break down into one group with limited fish bones, which is usually also poor in species diversity and lacking open-water fish, and another group with a lot of fish remains, generally showing a higher species richness and containing the remains of open-water fish. Specialist fishers have to be looked for in the second group, which has also been attested in other areas of arid West Africa. The deep-water fish are indicative of an advanced technological level of fishing. The necessary know-how was apparently available from at least the Middle Holocene. Rather than being occupied by specialized fishers, a few sites were merely in contact with such groups, specifically Galaga, Garu Kime, and perhaps Garumele. The location of archaeological sites actually occupied by specialized fishers is in accordance with the pattern that fishers can at present be found near large water bodies. The availability of large and varied fishing grounds makes fishing a particularly rewarding economic activity. Fish must have been an abundant and predictable source of food during certain seasons of the year and various conservation techniques allowed surplus to be stored for other, leaner periods. In addition, surpluses in preserved fish could be traded or exchanged.

Sites considered as part of the specialized fishers group appear from the second millennium BC onwards, even earlier further north in the Sahara, and continue into the most recent periods. However, considering the relationship that recent fishers have with farmers in order to obtain crops (Sundström, 1972: 134–6), it is highly unlikely that specialized fishers in the past resembled those of today, at least not before the arrival of fully fledged farming at the beginning of our era. From that point other specialist economies, and more precisely pastoral nomadism, are also thought to have appeared (Linseele, 2010). The presence of pastoral nomads is indirectly shown in the study area. Due to their mobile lifestyles, sites of pastoral nomads themselves have poor archaeological visibility. Because fish are potentially year-round resources, like farming, fishing can be linked to sedentism and thus has a good archaeological visibility.

Unfortunately for our search for specialized fishers, there are no descriptions of what the waste of recent West African fishers looks like, and it can therefore not be compared to the archaeological data. To investigate whether large numbers of fish remains in faunal assemblages can indeed be connected to a high level of fish consumption, stable isotope studies on human remains are potentially useful. The few studies that have been conducted for arid West Africa do not include material expected to come from specialized fishers, and have all pointed to a high plant (or herbivore) component, C4 mainly, in the human diet (Breunig, 1995; Finucane et al., 2008). A recent survey in the Lake Chad area has shown that, regardless of their ethnographic background, rural communities all practice some mixture of farming, stock-keeping, and fishing (Béné et al., 2003), although one may wonder whether this is a very recent adaptation to high population densities and the modern arid climatic conditions (Morand et al., 2012). Regardless, this should probably make us reflect on the fact that usually we look for general subsistence patterns by period and by region, while in fact there may have been a lot of variation between and within communities.

References

Bates, D. and Lees, S. H. (1977) 'The role of exchange in productive specialization', *American Anthropologist*, 79, 824–41.

Béné, C., Neiland, A., Jolley, T., Ovie, S., Sule, O., Ladu, B., Mindjimba, K., Belal, E., Tiotsop, F., Baba, M., Dara, L., Zakara, A., and Quensiere, J. (2003) 'Inland fisheries, poverty, and rural livelihoods in the Lake Chad Basin', *Journal of Asian and African Studies*, 38, 17–51.

Blache, J. and Miton, F. (1962) *Première contribution à la connaissance de la pêche dans le bassin hydrographique Logone-Chari-Lac Tchad*, Paris: ORSTOM.

Bouquet, C. (1990) *Insulaires et riverains du Lac Tchad. Étude géographique*, Vol. 2, Paris: L'Harmattan.

Breunig, P. (1995) 'Gajiganna und Konduga. Zur frühen Besiedlung des Tschadbeckens in Nigeria', *Beiträge zur allgemeinen und vergleichenden Archäologie*, 15, 3–48.

Breunig, P. (1996) 'The 8000-year-old dugout canoe from Dufuna (NE Nigeria)', in Pwiti, G. and Soper, R. (eds) *Aspects of African Archaeology: Papers from the 10th Congress of the Panafrican Association for Prehistory and Related Studies*, pp. 461–8. Harare: University of Zimbabwe Publications.

Breunig, P. (2013) 'Pathways to food production in the Sahel', in Mitchell, P. and Lane, P. (eds) *The Oxford Handbook of African Archaeology*, pp. 555–70. Oxford: Oxford University Press.

Breunig, P., Garba, A., Gronenborn, D., Van Neer, W., and Wendt, P. (1993) 'Report on excavations at Gajiganna, Borno State, northeast Nigeria', *Nyame Akuma*, 40, 30–41.

Chavane, B. (1985) *Villages de l'ancien Tekrour: recherches archéologiques dans la Moyenne Vallée du Fleuve Sénégal*, Paris: Éditions Karthala.

Cruz-Uribe, K. (1988) 'The use and meaning of species diversity and richness in archaeological faunas', *Journal of Archaeological Science*, 15, 179–96.

Finucane, B., Manning, K., and Touré, M. (2008) 'Late Stone Age subsistence in the Tilemsi Valley, Mali: stable isotope analysis of human and animal remains from the site of Karkarichinkat Nord (KN05) and Karkarichinkat Sud (KS05)', *Journal of Anthropological Archaeology*, 27, 82–92.

Gasse, F. (2000) 'Hydrological changes in the African Tropics since the last Glacial Maximum', *Quaternary Science Reviews*, 19, 189–211.

Gronenborn, D. (1998) 'Archaeological and ethnohistorical investigations along the southern fringes of Lake Chad, 1993–1996', *African Archaeological Review*, 15, 225–57.

Haour, A. (2008) 'A pottery sequence from Garumele (Niger)—a former Kanem-Borno capital?', *Journal of African Archaeology*, 6, 3–20.

Holl, A. (2002) *The Land of Houlouf: The Genesis of a Chadic Polity, 1900 BC–AD 1800*, Ann Arbor: Museum of Anthropology, University of Michigan.

Jousse, H. (2006) 'What is the impact of Holocene climatic changes on human societies? Analysis of West African Neolithic populations dietary customs', *Quaternary International*, 151, 63–73.

Jousse, H., Obermaier, H., Raimbault, M., and Peters, J. (2008) 'Late Holocene economic specialisation through aquatic resource exploitation at Kobadi in the Méma, Mali', *International Journal of Osteoarchaeology*, 18, 549–72.

Kahlheber, S. and Neumann, K. (2007) 'The development of plant cultivation in semi-arid West Africa', in Denham, T., Iriarte, J., and Vrijdaghs, L. (eds) *Rethinking Agriculture: Archaeological and Ethnoarchaeological Perspectives*, pp. 320–46. Walnut Creek: Left Coast Press.

Linseele, V. (2007) *Archaeofaunal Remains from the Past 4000 Years in Sahelian West Africa. Domestic Livestock, Subsistence Strategies and Environmental Changes*, Oxford: Archaeopress.

Linseele, V. (2010) 'Did specialized pastoralism develop differently in Africa than in the Near East? An example from the West African Sahel', *Journal of World Prehistory*, 23, 43–77.

Linseele, V. and Haour, A. (2010) 'Animal remains from Medieval Garumele (Niger)', *Journal of African Archaeology*, 8, 167–84.

MacDonald, K. C. (1995) 'Analysis of the mammalian, avian and reptilian remains', in McIntosh, S. K. (ed.) *Excavations at Jenné-Jeno, Hambarketolo, and Kaniana (Inland Niger Delta), the 1981 Season*, pp. 291–318. Berkeley: University of California Press.

MacDonald, K. C. and Van Neer, W. (1994) 'Specialised fishing peoples in the Later Holocene of the Méma region (Mali)', in Van Neer, W. (ed.) *Fish Exploitation in the Past: Proceedings of the 7th Meeting of the Icaz Fish Remains Working Group*, pp. 243–51. Tervuren: Royal Museum of Central Africa.

Magnavita, C., Adebayo, O., Höhn, A., Ishaya, D., Kahlheber, S., Linseele, V., and Ogunseyin, S. (2009) 'Garu Kime: a Late Borno fired-brick site at Monguno, NE Nigeria', *African Archaeological Review*, 26, 219–46.

Magnavita, C., Kahlheber, S., and Eichhorn, B. (2004) 'The rise of organisational complexity in mid-first millennium BC Chad Basin', *Antiquity*, 78(301), URL: http://antiquity.ac.uk/ProjGall/magnavita/.

Manning, K., Pelling, R., Higham, T., Schwenniger, J.-L., and Fuller, D. Q. (2011) '4500-year old domesticated pearl millet (*Pennisetum glaucum*) from the Tilemsi Valley, Mali: new insights into an alternative cereal domestication pathway', *Journal of Archaeological Science*, 38, 312–22.

McIntosh, R. J. (1993) 'The pulse model: genesis and accommodation of specialization in the Middle-Niger', *Journal of African History*, 34, 181–220.

McIntosh, S. K. (1999) 'A tale from two floodplains: comparative perspectives on the emergence of complex societies and urbanism in the Middle Niger and Senegal Valleys', in Sinclair, P. (ed.) *Proceedings of the Second World Archaeological Congress Intercongress, Mombasa*, http://www.arkeologi.uu.se/afr/projects/BOOK/Mcintosh/mcintosh.htm.

Morand, P., Kodio, A., Andrew, N., Sinaba, F., Lemoalle, J., and Béné, C. (2012) 'Vulnerability and adaptation of African rural populations to hydro-climate change: experience from fishing communities in the inner Niger Delta (Mali)', *Climatic Change*, 115, 463–83.

Paugy, D., Lévêque, C. and Teugels, G. (2003) *Poissons d'Eaux Douces et Saumâtres de l'Afrique de l'Ouest. The Fresh and Brackish Water Fishes of West Africa (2 Vols)*, Paris: Éditions de l'IRD.

Reed, W., Burchard, J., Hopson, A. J., Jenness, J., and Yaro, I. (1967) *Fish and Fisheries of Northern Nigeria*, Zaria: Ministry of Agriculture Northern Nigeria, Gaskiya Corporation.

Reitz, E. J. and Wing, E. S. (1999) *Zooarchaeology, Cambridge Manuals in Archaeology*, Cambridge: Cambridge University Press.

Rivallain, J. and Van Neer, W. (1983) 'Les fouilles de Koyom (sud du Tchad). Étude du matériel archéologique et faunique', *L'Anthropologie*, 87, 221–39.

Rivallain, J. and Van Neer, W. (1984) 'Inventaire du matériel archéologique et faunique de Koyom, sud du Tchad', *L'Anthropologie*, 88, 441–8.

Sundström, L. (1972) *Ecology and Symbiosis: Niger Water Folk*, Uppsala: Almqvist & Wiksell.

Van Neer, W. (2004) 'Evolution of prehistoric fishing in the Egyptian Nile Valley', *Journal of African Archaeology*, 2, 251–69.

Van Neer, W. (2008) 'Fishing in the Senegal River during the Iron Age: the evidence from the habitation mounds of Cubalel and Siouré', in Badenhorst, S., Mitchell, P., and Driver, J. C. (eds) *Animals and People: Archaeozoological Papers in Honour of Ina Plug*, pp. 117–30. Oxford: Archaeopress.

Van Neer, W. and Gayet, M. (1988) 'Étude des poissons en provenance des sites holocènes du Bassin de Taoudenni-Araouane (Mali)', *Bulletin du Muséum National d'Histoire Naturelle Paris 4e sér.*, 10, 343–83.

Villiers, A. (1958) *Tortues et crocodiles de l'Afrique noire française: Initiations Africaines*, Dakar: Institut Français d'Afrique Noire.

Wendt, P. (2007) *Gajiganna: Analysis of Stratigraphies and Pottery of a Final Stone Age Culture of Northeast Nigeria*, Frankfurt: Africa Magna Verlag.

Wiesmüller, B. (2001) 'Die Entwicklung der Keramik von 3000 BP bis zu Gegenwart in den Tonebenen Südlich des Tschadsees'. Unpublished PhD dissertation, Johann Wolfgang Goethe-University (Frankfurt).

Zohar, I. and Belmaker, M. (2005) 'Size does matter: methodological comments on sieve size and species richness in fishbone assemblages', *Journal of Archaeological Science*, 32, 635–41.

....

CHAPTER 29

ANIMALS IN ANCIENT EGYPTIAN RELIGION

belief, identity, power, and economy

....

SALIMA IKRAM

INTRODUCTION

....

ANIMALS play an important part in religious rituals throughout the world. This was no more so than in ancient Egypt and Nubia, where animals were vital to religious practices. The most obvious and significant feature of Egyptian religion was that most divinities were theriomorphic, either completely or partially (Fig. 29.1), and that specific living animals, such as the Apis Bull, were, during their lifetime, revered as manifestations of particular deities on earth, with oracular powers (Kessler, 1986; Ikram, 2015a). Upon their death, these sacred animals were prepared for burial and interred with great pomp.

In addition to being potential manifestations of gods, animals also served deities. The flight of birds was used to celebrate certain festivals (Kessler and Nur el-Din, 2015: 128–9) and possibly for divination (Murnane, 1980: 37). Animals provided the raw materials, such as hides, shell, fur, guts, bones, ivory, and feathers, used to fabricate a variety of objects including amulets, regalia, furniture, and musical instruments for use in temple cults, as well as in private religious life. The foundation of temples and tombs was also sanctified through animal sacrifice, with bucrania and forelimbs of cattle deposited at the buildings' corners (Weinstein, 1973). Bucrania were also used in the external adornment of early Egyptian tombs (Emery, 1949; 1954; 1958; 1974), as well as playing a part in Nubian burials (Grant, 1991; Chaix and Grant, 1992; Davis, 2008; Ikram, 2012) as manifestations of wealth, offerings to the deceased, and for protection. Funerary offerings included meat and poultry, as attested both by texts and funerary

FIGURE 29.1 The raptor-headed god Horus, associated with solar worship and divine kingship, and the crocodile headed god Sobek, also a solar deity, as well as a fertility god, from Kom Ombo temple, near where mummified raptors and crocodiles were found. Author's own image.

remains, which sustained the deceased throughout eternity (Barta, 1963; Ikram, 1995; 2009; 2011; 2012).

Living animals played a crucial role in temple cult, and by extension, the Egyptian economy. There were hundreds of animals, primarily cattle (*Bos taurus*) and birds (for example *Anas* spp., *Anser* spp., and *Columba* spp.), although sheep (*Ovis ammon* f. *aries*) and goat (*Capra aegagrus* f. *hircus*) also feature in offering lists and were sacrificed on a daily basis as offerings in temples throughout Egypt. After consecration, much of this meat was redistributed as payment to temple personnel, who then either consumed it or used it for barter, thus making such offerings a significant component of the economic engine of Egypt (Posener-Kriéger, 1976; Ikram, 1995; Lehner, 2000; Warburton, 2000; Posener-Kriéger et al., 2006; Rossel, 2007). Huge herds of cattle, goats, and sheep had to be purpose-raised for this, with the temples having enormous holdings of livestock (Ghoneim, 1977).

Offerings did not only take the form of food; in later Pharaonic history, during the seventh century BC through the third century AD (Late and Graeco-Roman periods), a curious new type of animal offering came into prominence, associated with the formal sacred animal cults: votive animal mummies (Ikram and Iskander, 2002; Ikram, 2015a).

Studying animal use in religious contexts binds together many diverse strands of inquiry, allowing one to investigate the relationships between Ancient Egyptian society and culture and its fauna. These include:

- the human impact on the environment and ecology of an area;
- animal husbandry/management of animal resources;
- a study of society and social change;
- forms of ethnic identity and acculturation;
- cultural contacts;
- manifestation of political power;
- the economy and the role of trade within it;
- and religious constructs.

This brief essay focuses on the animal cults that reached their apogee in Late and Graeco-Roman Egypt, and the implications that these had for the relationship between humans and animals.

Animal Cults

The Romans derided the Egyptians for their reverence of animals (Smelik and Hemelrijk, 1984), with Juvenal writing in his *Satires* (XV) 'Who has not heard, Volusius, of the monstrous deities those crazy Egyptians worship? One lot adores crocodiles, another worships the snake-gorged ibis … you'll find whole cities devoted to cats, or to river-fish or dogs …' (Juvenal *Satires* 15.1, Rudd trans., 1991). For the Egyptians, however, animals seemed to be endowed with supernatural powers and gifts, with intimate access to the gods, and these attributes, together with their 'otherness', provided the basis for many Egyptian religious beliefs that linked specific animals to certain deities that shared their attributes and strengths (Dunand et al., 2005; Ikram, 2015a). Most gods had at least one totemic animal that exemplified his or her attributes, and often the heads of these animals were shown on human bodies as manifestations of the gods (Fig. 29.1). Thus, cats were identified with Bastet, goddess of love, beauty, and self-indulgence—all characteristics seen in living cats. Dogs and other canines were associated with Anubis, god of cemeteries, embalming, and travel, because these creatures frequented embalming houses (no doubt lured by the scent of flesh) and cemeteries, and were adept at navigating the desert. Raptors were associated with the sun god as they flew high into the sky, able to see even the smallest creature from their lofty height, as well as due to their coloration, and the way in which their eyes are evocative of the sun. The sacred ibis was an avatar of Thoth, the god of writing and wisdom, no doubt because the beak of the ibis took the form of a reed pen, and the bird was seen bent over, ever questing with its beak, in search of some truth—or at least a true lunch! However, despite their associations with divinities, the Egyptians did not worship all cats, dogs, and birds, although acknowledging their link with the gods.

Certainly, animals played a prominent role in religion in Egypt by the second century AD, when Juvenal knew the country, but the employment of living and dead creatures had not always been manifested as he saw them. Until about the seventh century BC, animal cults were limited in scope, with single sacred animals, recognized by special markings, being revered as the avatar of a particular deity at a particular location. It was believed that each god could manifest him- or herself in that specific creature during its lifetime, and after its death the spirit of the god would move to a different animal of its species, recognizable by identical markings, a concept not dissimilar to the manifestation of the Dalai and other Tibetan lamas. Upon their death, such animals were, at least from the middle of the New Kingdom (fourteenth century BC), elaborately mummified and buried in catacombs with considerable splendour (Kessler, 1986; Ray, 2001; Ikram, 2015a). Examples of sacred animal interments survive in the bull cemeteries at Saqqara, Heliopolis, and Armant, and of the rams at Elephantine. It was only from the seventh century AD onward that the cult of the living animals became widespread, as previously such cults seem to have been limited to only a few sites: Memphis/Saqqara, Heliopolis, and Thebes and its environs. It is at this time that large-scale votive mummies started to be offered to the animal manifestations of divinities, with cults appearing all over the country, from Alexandria to Aswan, as well as in the oases of the Western Desert (Fig. 29.2). Individual pilgrims apparently purchased mummies of animals associated with a particular deity, and dedicated it to its corresponding divinity so that the donor's prayers would reach the god through the medium of the deities' own animal. It may be that animal mummies, perhaps because they had once been alive, were deemed more effective intercessors than offerings of a statue or stela, and once transformed by mummification into a semi-divine state, they could interact eternally with the gods.

Thus far, there is no incontrovertible explanation as to why this era saw a massive expansion of animals employed in cults. No doubt a variety of factors fed into this phenomenon. Perhaps primarily, this was a manifestation of the uniqueness of Egyptian religion and a way of separating and defining Egyptians from other ethnic groups. The 26th Dynasty (c.664–525 BC) was a time when Egypt was recovering from foreign rule (first Nubian, then Assyrian) and struggling to re-establish its independence and reassert its former greatness. As a result, the rulers (probably in conjunction with the higher-ranking priesthood) consciously established a culture that hearkened back to times when Egypt was great. Archaism is apparent in art, literature, rhetoric, and modes of presenting the king and the gods. Religion was also key to uniting the country and reasserting the domination of kings and priests. It is probable that animal cults and votive mummies were a major part of this propagandistic programme, creating a unique way in which people could engage with the gods, recalling the earliest times of Egyptian history when the animal aspects of deities were emphasized. These cults, with their animal votive offerings, provided a more accessible route to the more important deities of the Egyptian pantheon, such as the sun god, who had hitherto been the preserve of rulers rather than the populace. Texts found inside catacombs suggest that a greater number of people had direct contact with these gods than with the deities who resided as statues within a temple (Kessler and Nur el-Din, 2015; Ray, 2011; 2013; Smith et al., 2011).

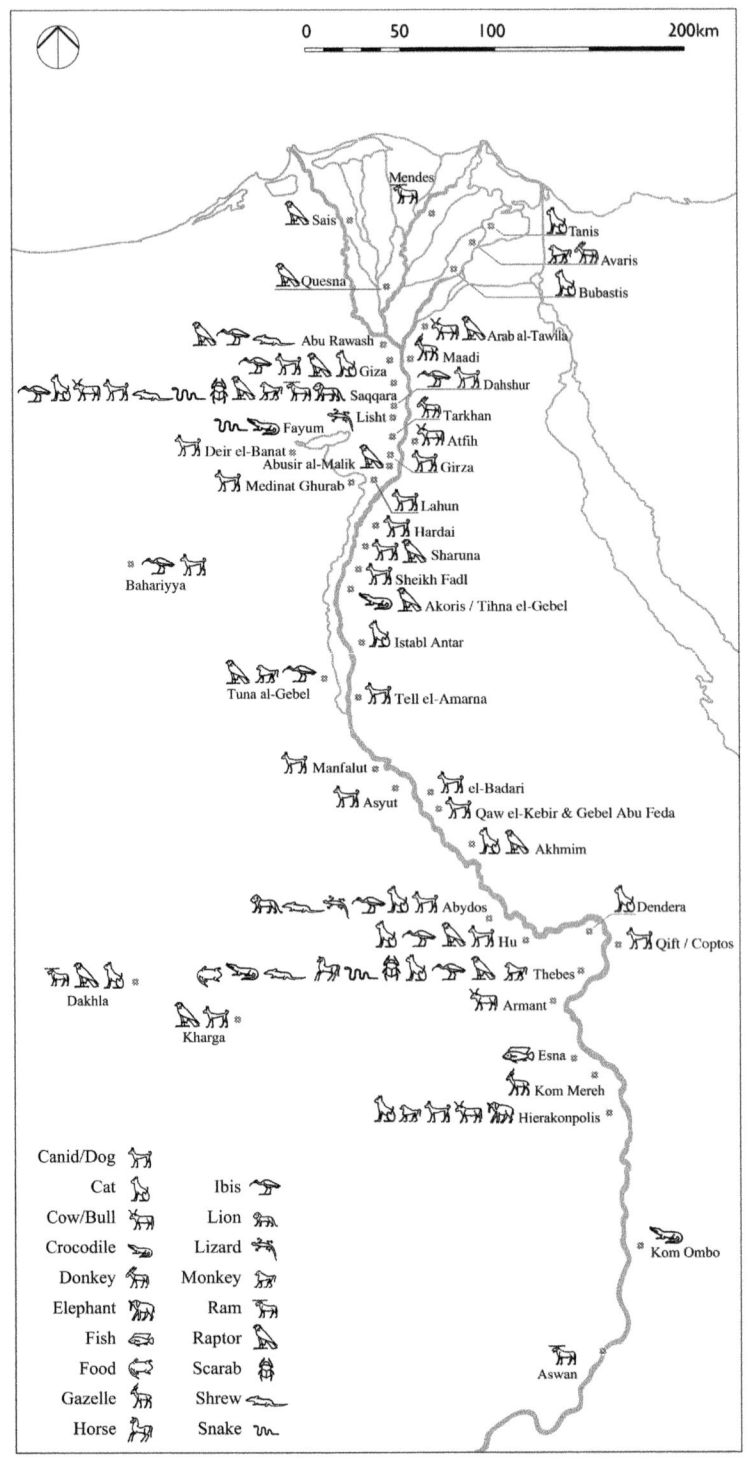

FIGURE 29.2 A map showing the major animal mummy cemeteries and sacred sites in Egypt. Drawing by N. Warner.

This 'democratization' of religion allowed for a more intimate relationship between people and gods, particularly as access to the living animal deities was more possible than access to the images of gods that were kept in close seclusion within a temple. No doubt this accessibility was significantly responsible for the success of these cults.

Subsequently, when the Persians seized Egypt in 525 BC, animal cults united the Egyptians and provided a very distinct way for them to define themselves religiously as separate from the Persians—an aspect of this cult that persisted through the Roman domination. The Macedonian dynasties (332–30 BC) seem to have embraced animal cults (Dodson, 2015; Kessler and Nur el-Din, 2015), which were extremely active during this time. The popularity of the cults also served to support the associated temples and local shrines, which burgeoned during this time, seemingly serving an increasing number of cities and towns throughout the country, as can be seen by the proliferation of sites that contain animal mummies (Fig. 29.2). As the temples and the rulers were allied (albeit in an uneasy alliance at times), it was in their interest to maintain and encourage popular cults that ultimately benefited the state, forming part of the economic (and social) web between state, temple, and the people. In the Roman period the practice continued, but with less state support; however, it maintained a way for Egyptians, and those who embraced their religion (Smelik and Hemelrijk, 1984), to forge and maintain a distinct and separate identity for themselves, and maybe even created a power-base from which to defy, in a small way, their conquerors periodically.

The Votive Mummies

Although sacred animals were buried throughout the course of Egyptian history, it is the votive mummies that make up the majority of animal mummies that are found in museums today, as well as excavated in catacombs and other tombs throughout the Nile Valley and the oases of the Western Desert. The number of species represented in these cults of the Late Period onward include almost all animals, with the notable exceptions of hippopotami and donkeys, both of which were associated with Seth, an inimical god in charge of deserts, among other things (Wilkinson, 2003: 197–9). Thus, cats, dogs, foxes, jackals, mongooses, sheep, goats, gazelles, shrews, monkeys, rodents, snakes, crocodiles, lizards, fish, raptors, ibises, other birds, scarab beetles, and even their dung balls are offered to various deities (Lortet and Gaillard, 1903–1909; Daressy and Gaillard, 1905; Kessler, 1986; Ikram and Iskander, 2002).

What is most striking about these mummies is their vast number. Unfortunately, it is only recently that a systematic approach to the study of animal mummies from excavations has been established, thus it is difficult to calculate the number of mummies that might have existed in each catacomb. However, some estimates are available. The Catacomb of Anubis at Saqqara contained some 7.8 million canine mummies (Ikram et al., 2013); five hundred canines were identified at el-Deir (Dunand et al., 2015); the Ibis Galleries at Saqqara yielded at least 4 million ibises (Ray, 1976); well over 1.8 million ibises were offered in Tuna el-Gebel (von den Driesch et al., 2005; Kessler and

Table 29.1 Table showing the MNI of votive
mummies from diverse cemeteries
dating from c.600 BC to c. AD 300

Species	Site	MNI
Canine	Saqqara	7,800,000
Canine	el-Deir	500
Ibis	diverse cemeteries	5,815,000
Baboons	Tuna el-Gebel	600+
Raptors	Thebes TT 12	2,000
Cats	Saqqara	1,000+
Crocodiles	Tebtunis	10,000+

Nur el-Din, 2015); more than five thousand ibises were discovered at Abydos (Loat, 1914; Ikram, personal observation); in a single chamber of a much larger burial complex associated with Theban Tomb (TT) 12, ten thousand ibises and two thousand raptors were identified (Ikram et al., in prep.); more than six hundred monkeys of different sorts were found at Tuna el-Gebel (Kessler and Nur el-Din, 2015); and at least one thousand cat mummies were recovered from the Bubasteion at Saqqara (A. Zivie, personal communication); at Tebtunis in the Fayum a deposit of ten thousand crocodiles is estimated (Bagnani, 1952). These figures represent a small proportion of the votive animal mummy deposits found throughout Egypt (Table 29.1), and do not, save for the estimates for the Catacomb of Anubis and the chamber in TT 12, provide an accurate estimate for the total number of animals in any single catacomb. If one were to be able to calculate the true number of votive animal mummies from all the known cemeteries in Egypt, the number for each species would be well into the millions.

Implications of the Production of Votive Mummies

Clearly, the huge amount of mummification materials needed to produce millions of mummies had a marked impact on the economy, both national and international (Ikram, 2015a: 16–43). Natron, the prime ingredient in mummification, used to desiccate and de-fat, came from two places in Egypt: the Wadi Natrun and el-Kab. This had to be processed and transported in vast quantities throughout the country in order to carry out basic mummification; at least 400 kg of natron are necessary to properly mummify a sheep (Ikram, 2015b); thus a single dog would need c.200 kg, while smaller creatures would require less. Oils (lettuce, castor, sesame) were locally produced, but resins for anointing the animals were imported from the Levant and East Africa as well as, possibly, Arabia. At least thousands of kilometres of linen had to be used for wrapping the animals, though these were locally available, and were often reused. More striking, though, is the sheer number of animals that were mummified. These had to play a major role in the economy, particularly providing revenue for temples and their dependents.

All the animals offered as ex-votos were indigenous to Egypt, save for baboons. Although once native, by the seventh century BC they were long extinct within Egypt, having retreated further south, and thus had to be imported for cult purposes. Quite possibly attempts at breeding them took place, but there is no evidence for a successful breeding programme. Osteological evidence from the extensively studied baboon mummies from Tuna el-Gebel (von den Driesch and Boessneck, 1985; von den Driesch, 1993; Nerlich et al., 1993) and from Saqqara (Goudsmit and Brandon-Jones, 2000), indicate that many of the animals showed pathologies indicative of ill health, some due to being kept in constrained spaces and poor diet, which most scholars think is due to the time the animals spent in the temple areas. Indeed, poor living conditions and care by people who were unversed in what these animals needed to survive and thrive might have been largely responsible for their condition, but responsibility for this did not lie solely with the care-takers associated with the temples. The long journey from sub-Saharan and northern East Africa, often taking months, even in the eighteenth and nineteenth centuries when camels were commonly used (Walz, 1978), necessitated that the animals be kept in confined spaces, most probably with a restricted and often insufficient diet. Thus, these creatures probably arrived, at great expense, in Egypt, already malnourished and prone to disease. Perhaps this is one reason why breeding groups could not be established successfully, thereby accounting for the limited number of baboon mummies in comparison with those of animals that were indigenous.

Given their number, the native animals must have been bred for the purpose as it was impractical to think of trapping so many animals, particularly those that could be easily bred, such as ibises, dogs, and cats. It is more likely that creatures such as raptors were trapped, although they too could be bred in captivity, albeit less effectively than the other animals. Indeed, if all these animals had been trapped and killed, rather than reared especially for their fate, it quite likely would have resulted in the extinction of these species (Ikram et al., 2015). The vast number of animals in the catacombs is not the only argument to support the idea of breeding programmes to supply the cult. Eggs (Lortet and Gaillard, 1903–1909; Ray, 1976; Bresciani, 2015; Ikram, personal observation from Sharuna, Thebes, and Abydos), particularly of ibises, feature amongst the votive offerings, with crocodile hatcheries being posited in association with temples dedicated to the crocodile god, Sobek (Bresciani, 2015).

Furthermore, the number of immature individuals found in animal cemeteries is remarkable, and argues for breeding programmes. In the Galleries of the Catacombs of Anubis at Saqqara 75% of the number of identified specimens (NISP) of 6,034 bones belonged to immature animals (Ikram et al., 2013) (Fig. 29.3). At el-Deir, out of five hundred individuals, 25% were puppies between 1 and 2 months of age, and 36% were juveniles between 6 and 14 months (Dunand et al., 2015).

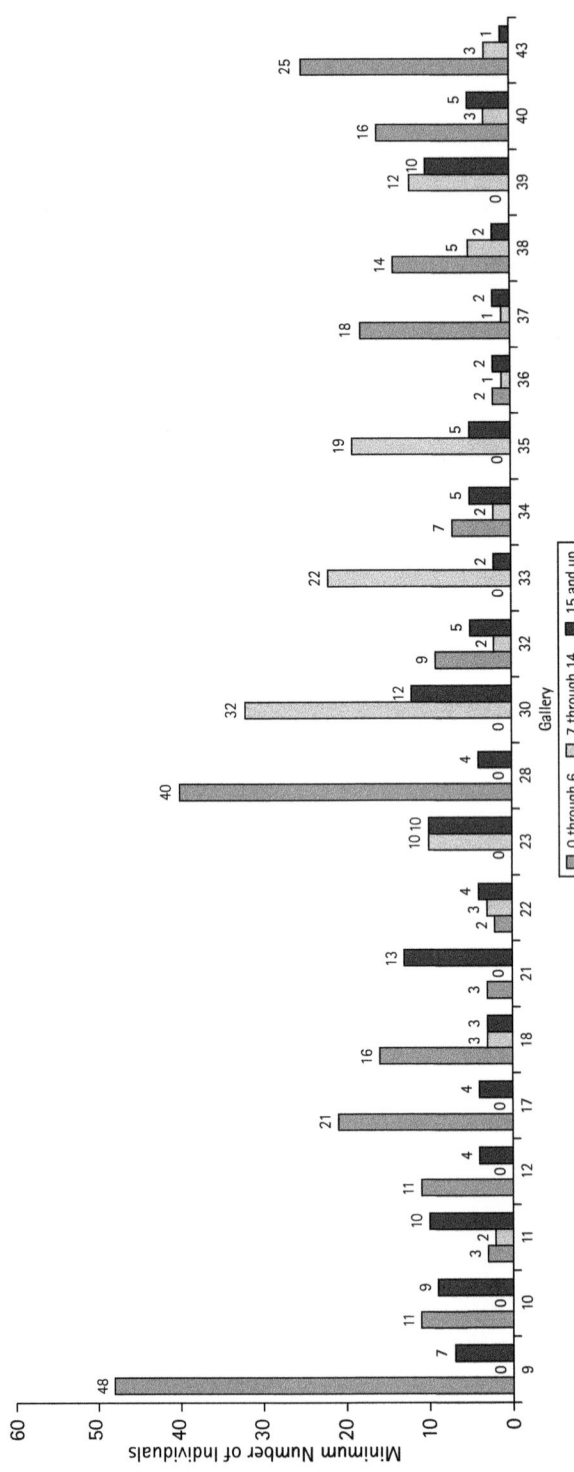

FIGURE 29.3 Distribution of the ages at death of the dogs from the Catacombs of Anubis at Saqqara. Prepared by L. Bertini.

At TT 12, a subterranean chamber measuring 9 m^2 was filled with the remains of dis-articulated and somewhat-burnt bird mummies. In order to sample the remains, a four-litre container was filled with bones chosen from three of the nine square metre contexts; the sample was taken at random, and was examined in detail. The samples were scooped up by hand and thus some anatomical elements escaped inclusion as we did not want to dig down too violently and grasp the bones too firmly lest they break. Each sample was analysed to obtain an overall idea of species represented, minimum number of birds placed in the room, their ages, and whether entire birds had been mummified or just specific portions. Information recorded also included pathologies, anatomical element and portion thereof, side, approximate age, and degree of burning. In addition, 'cherry-picked' bones (those not belonging to ibises) from the sieved remains of the chamber, extraneous to the four litre samples, were also analysed with a view to gaining a better perspective of the non-sacred ibis remains that were given as offerings. This yielded a total of 3,867 bones. Of these, a total of 756 immature/juvenile bones were noted, about 20% of the total number of bones. It is difficult to differentiate species in bones belong-ing to such young birds (including fledglings), but the general impression is that at least six hundred of these bones were from ibises, while the rest were of raptors (Ikram et al., in prep.). Although neither numbers nor percentages are currently available from the other ibis catacombs, it is likely that a large percentage of young or even eggs comprised their population.

The amount of disease and trauma found on the dog bones from the Anubis Catacomb also argues for their being kept in confined spaces (breeding pens) and not being well looked after. Out of the 3,867 bones examined, 266 of the canid bones (*c*.5%) showed evidence of pathology (Ikram et al., 2013). Another canine cemetery at Saqqara also yielded evidence for dogs being housed in a closely confined space, and poorly fed and looked after (Hartley et al., 2011).

It would seem that the puppies found in these cemeteries had been deliberately killed to supply the demand for animal mummies (Ikram et al., 2013; Dunand et al., 2015), as has been found in the case of kitten mummies recovered from similar catacombs at Saqqara and elsewhere (Armitage and Clutton-Brock, 1980; 1981; Charron, 1990; Zivie and Lichtenberg, 2015). Other votive animals probably also were deliberately killed in order to meet the demand of the thousands of pilgrims who required them. Quite pos-sibly birds such as ibises had their necks broken (strangled), although this is difficult to determine due to the twisted position of the necks necessitated by the way in which they were wrapped. Some young crocodiles with dented skulls appear to have been killed by blows to the head, while it has been posited that others were dispatched by the slitting of their nostrils (E. Bresciani, personal communication; Bresciani, 2015).

Clearly, animals such as dogs and cats were being farmed in dedicated spaces in order to support the temples and their pilgrims (Ikram et al., 2013). Such large-scale animal farming might well have extended beyond the immediate confines of temple personnel, and be part of the larger village/town/city economy, or even include a wider regional catchment area for people to supply the temples. Non-domesticates such as ibises and crocodiles were probably kept in environments such as pools or lakes where they were fed regularly and maintained by temple personnel, and thus were managed, and to some extent, one might say that they were 'farmed' (Strabo, Falconer trans., 1912; Preisigke and Spiegelberg, 1914; Ray, 1976; Herodotus, Gould trans., 1989; Bresciani, 2015). It is also possible that ibises were acquired by enterprising folk who lived near ibis breeding areas and thus could capture birds or raid nests and supply the temples with even more sacrificial victims. Thus, it is clear that the sourcing and maintenance, to whatever degree, of the animals was a major component of economic activity in animal cults. Further work on animal catacombs will elucidate exactly how significant a role they played in both the cult and the economy.

DISCUSSION

There is no doubt that animals played a major role in Egyptian religious and economic life, particularly as sacrificial victims. While one is accustomed to the idea of a single or even several animals being killed as food offerings to a deity (Posener-Kriéger, 1976; Posener-Kriéger et al., 2006), it is the vast scale of animal mummies that gives one pause. Unlike the meat from offerings, which was redistributed, animal mummies played no further role in the economy or the physical life of either the priests or the populace. Indeed, it is curious that creatures that were linked so closely with the gods were paradoxically bred (or imported) specifically to be killed, often brutally by strangulation, having their skulls bashed in, or their nostrils slit. Paradoxically, the continuations of these cults also guaranteed the local survival for many species, such as the Sacred Ibis, which is now extinct in Egypt.

One can take the cynical view that for the priests at least, the production of animal mummies as ex-votos was a way of wielding economic and social power: it provided employment for themselves and the villagers around them in terms of breeding, caring for, and feeding the animals and preparing the mummies; the sale of the mummies was a way of enriching temple coffers, as well as tying their donors and producers to the temples, the gods, and the state (Ikram, 2015c). However, the vast number of these votive mummies indicates a heartfelt belief in their efficacy on the part of their givers, and provided the donors with a mode of self-expression, identification, and interaction with the divine that was uniquely Egyptian. For the donors, probably, the animals not only conveyed prayers to the gods, but were themselves given a chance at an immortal existence in close proximity to their gods, and thus were achieving a state of grace and immortality that the donors themselves sought, and could only properly achieve with their own death.

REFERENCES

Armitage, P. L. and Clutton-Brock, J. (1980) 'Egyptian mummified cats held by the British Museum', *MASCA, Research Papers in Science and Archaeology*, 1, 185–8.

Armitage, P. L. and Clutton-Brock, J. (1981) 'A radiological and histological investigation into the mummification of cats from ancient Egypt', *Journal of Archaeological Science*, 8, 185–96.

Bagnani, G. (1952) 'The great Egyptian crocodile mystery', *Archaeology*, 5(2), 76–8.

Barta, W. (1963) *Die Altägyptische Opferliste von der Frühzeit bis zur Griechisch-Römischen Epoche*, Berlin: Hessling.

Bresciani, E. (2015) 'Sobek, Lord of the Land of the Lake', in Ikram, S. (ed.) *Divine Creatures: Animal Mummies from Ancient Egypt*, pp. 199–206. Cairo: American University in Cairo Press.

Chaix, L. and Grant, A. (1992) 'Cattle in ancient Nubia', in Grant, A. (ed.) *Les animaux et leurs produits dans le commerce et les échanges/Animals and Their Products in Trade and Exchange: actes du 3éme colloque internationale de l'homme et l'animal, Société de Recherche Interdisciplinaire (Oxford 8–11 Novembre 1990)*. Anthropozoologica 16, pp. 61–6. Clichy: L'Homme et L'Animal, Société de Recherche Interdisciplinaire.

Charron, A. (1990) 'Massacres d'animaux à la Basse Epoque', *Revue d'Égyptologie*, 41, 209–13.

Daressy, G. and Gaillard, C. (1905) *La faune momifiée de l'antique Égypte*, Cairo: Institut Français d'Archéologie Orientale.

Davis, S. J. M. (2008) '"Thou shalt take of the ram … the right thigh; for it is a ram of consecration . . .": some zoo-archaeological examples of body-part preferences', in D'Andria, F., De Grossi Mazzorin, J., and Fiorentino, G. (eds) *Uomini, piante e animali nella dimensione del sacro*, pp. 63–70. Lecce: Università degli Studi di Lecce/Consiglio Nazionale delle Ricerche.

Dodson, A. M. (2015) 'Bull cults', in Ikram, S. (ed.) *Divine Creatures: Animal Mummies from Ancient Egypt*, pp. 72–105. Cairo: American University in Cairo Press.

Driesch, von den, A. (1993) 'The keeping and worshipping of baboons during the Later Phase in Ancient Egypt', *Sartoniana*, 6, 15–36.

Driesch, von den, A. and Boessneck, J. (1985) 'Krankhaft veränderte Skelettreste von Pavianen aus altägyptischer Zeit', *Tierärztliche Praxis*, 13, 367–72.

Driesch, von den, A., Kessler, D., Steinmann, F., Berteaux, V., and Peters, J. (2005) 'Mummified, deified and buried at Hermopolis Magna—the sacred birds from Tuna El-Gebel, Middle Egypt', *Ägypten und Levante*, 15, 203–44.

Dunand, F., Lichtenberg, R. and Charron, A. (2005) *Des animaux et des hommes: une symbiose égyptienne*, Paris: Rocher.

Dunand, F., Lichtenberg, R., and Callou, C. (2015) 'Dogs at el-Deir', in Ikram, S., Kaiser, J., and Walker, R. (eds) *Egyptian Bioarchaeology: Humans, Animals, and the Environment*, pp. 169–76. Amsterdam: Sidestone.

Emery, W. B. (1949) *Great Tombs of the First Dynasty I*, Cairo: Service des Antiquités.

Emery, W. B. (1954) *Great Tombs of the First Dynasty II*, London: Egypt Exploration Society.

Emery, W. B. (1958) *Great Tombs of the First Dynasty III*, London: Egypt Exploration Society.

Emery, W. B. (1974) *Archaic Egypt*, Harmondsworth: Penguin.

Ghoneim, W. (1977) *Die ökonomische Bedeutung des Rindes im alten Ägypten*, Bonn: Habelt.

Goudsmit, J. and Brandon-Jones, D. (2000) 'Evidence from the Baboon Catacomb in North Saqqara for a west Mediterranean monkey trade route to Ptolemaic Alexandria', *Journal of Egyptian Archaeology*, 86, 111–19.

Grant, A. (1991) 'Economic or symbolic? Animals and ritual behaviour', in Garwood P., Jennings, D., and Toms, J. (eds) *Sacred and Profane: Archaeology, Ritual and Religion*, pp. 109–14. Oxford: Oxford University Committee for Archaeology.

Hartley, M., Buck, A., and Binder, S. (2011) 'Canine interments in the Teti Cemetery North at Saqqara during the Graeco-Roman period', in Coppens, F. and Krejsi, J. (eds) *Abusir and Saqqara in the Year 2010*, pp. 17–29. Prague: Czech Institute of Egyptology.

Herodotus, trans. by Gould, J. (1989) *The Histories II, III*, New York: St. Martin's Press.

Ikram, S. (1995) *Choice Cuts: Meat Production in Ancient Egypt*, Leuven: Peeters.

Ikram, S. (2009) 'Funerary food offerings', in Barta, M. (ed.) *Abusir XIII, Abusir South 2. Tomb Complex of the Vizier Qar, His Sons Qar Junior and Senedjemib, and Iykai*, pp. 294–8. Prague: Dryada/Czech Institute of Egyptology, Charles University.

Ikram, S. (2011) 'Food and funerals: sustaining the dead for eternity', *Polish Archaeology in the Mediterranean*, 20, 361–71.

Ikram, S. (2012) 'From food to furniture: animals in ancient Nubia', in Fisher, M., Lacovara, P., Ikram, S., and D'Auria, S. (eds) *Ancient Nubia: African Kingdoms on the Nile*, pp. 210–28. Cairo: American University in Cairo Press.

Ikram, S. (ed.) (2015a) *Divine Creatures: Animal Mummies in Ancient Egypt*, Cairo: American University in Cairo.

Ikram, S. (2015b) 'Experimental Archaeology: From Meadow to Em-baa-lming Table', in Graves-Brown, C. (ed.) *Egyptology in the Present: Experiential and Experimental Methods in Archaeology*, pp. 53–74. Swansea: The Classical Press of Wales.

Ikram, S. (2015c) 'Speculations on the Role of Animal Cults in the Economy of Ancient Egypt', in Massiera, M., Mathieu, B., and Rouffet, Fr. (eds) *Apprivoiser le sauvage/Taming the Wild* (CENiM 11), pp. 211–28. Montpellier: University Paul Valéry Montpellier 3.

Ikram, S., Bosch, C., and Spitzer, M. (in prep.) 'Offerings to Thoth and Horus: the avian deposit of Theban Tomb 12, the Chapel of Hery', *Journal of the American Research Center in Egypt*.

Ikram, S. and Iskander, N. (2002) *Catalogue Général of the Egyptian Museum: Non-Human Mummies*, Cairo: Supreme Council of Antiquities Press.

Ikram, S., Nicholson, P. T., Bertini, L., and Hurley, D. (2013) 'Killing man's best friend?', *Archaeological Review from Cambridge*, 28(2), 48–66.

Ikram, S., R. Slabbert, I. Cornelius, A. du Plessis, L. C. Swanepoel, and H. Weber. (2015) 'Fatal force-feeding or Gluttonous Gagging? The death of Kestrel SACHM 2575', *Journal of Archaeological Science*, 63, 72–7.

Juvenal, trans. by Rudd, N. (1991) *The Satires—Juvenal*, Oxford: Clarendon Press.

Kessler, D. (1986) 'Tierkult', in Helck, W. and Westendorf, W. (eds) *Lexikon der Ägyptologie VI*, pp. 571–87. Weisbaden: Otto Harrassowitz.

Kessler, D. and Nur el-Din, A. (2015) 'Tuna El-Gebel: millions of ibises and other animals', in Ikram, S. (ed.) *Divine Creatures: Animal Mummies from Ancient Egypt*, pp. 120–63. Cairo: American University in Cairo Press.

Lehner, M. (2000) 'The fractal house of pharaoh: ancient Egypt as a complex adaptive system, a trial formulation', in Kohler, T. A. and Gumerman, G. G. (eds) *Dynamics in Human and Primate Societies*, pp. 275–353. Oxford: Oxford University Press.

Loat, W. L. S. (1914) 'The ibis cemetery at Abydos', *Journal of Egyptian Archaeology*, 1, 40.

Lortet, L. C. and Gaillard, C. (1903–1909) *La faune momifiée de l'ancienne Egypte*, Lyon: Archives du Muséum Histoire Naturelle de Lyon VIII.

Murnane, W. J. (1980) *United with Eternity: A Concise Guide to the Monuments of Medinet Habu*, Chicago/Cairo: Oriental Institute, University of Chicago/American University in Cairo Press.

Nerlich, A. G., Parsche, F., Driesch, von den, A., and Löhrs, U. (1993) 'Osteopathological findings in mummified baboons from Ancient Egypt', *International Journal of Osteoarchaeology*, 3, 189–98.

Posener-Kriéger, P. (1976) *Les archives du temple funéraire de Neferirkare-Kakai, 1–2*, Cairo: Institut Français d'Archéologie Orientale.

Posener-Kriéger, P., Verner, M., and Vymazalova, H. (2006) *Abusir X: The Pyramid Complex of Raneferef, the Papyrus Archive*, Prague: Czech Institute of Egyptology.

Preisigke, F. and Spiegelberg, W. (1914) *Die Prinz-Joachim Ostraka*, Strasbourg: K. J. Trübner.

Ray, J. D. (1976) *The Archive of Hor*, London: Egypt Exploration Society.

Ray, J. D. (2001) 'Animal cults', in Redford, D. B. (ed.) *The Oxford Encyclopedia of Ancient Egypt*, pp. 345–8. Oxford: Oxford University Press.

Ray, J. D. (2011) *Texts from the Baboon and Falcon Galleries: Demotic, Hieroglyphic and Greek Inscriptions from the Sacred Animal Necropolis, North Saqqara*, London: Egypt Exploration Society.

Ray, J. D. (2013) *Demotic Ostraca and Other Inscriptions from the Sacred Animal Necropolis, North Saqqara*, London: Egypt Exploration Society.

Rossel, S. (2007) 'The Development of Productive Subsistence Economies in the Nile Valley: Zooarchaeological Analysis at El-Mahasna and South Abydos, Upper Egypt'. Unpublished PhD dissertation, Harvard University (Cambridge, MA).

Smelik, K. A. D. and Hemelrijk, E. A. (1984) "Who knows not what monsters demented Egypt worships?' Opinions on Egyptian animal worship in antiquity as part of the ancient conception of Egypt', in Haase, W. (ed.) *Aufstieg und Niedergang der Römischen Welt 17(4)*, pp. 1853–2000. Berlin: Walter de Gruyter.

Smith, H. S., Andrews, C. A. R., and Davies, S. (2011) *The Sacred Animal Necropolis at North Saqqara: The Mother of Apis Inscriptions 1–2*, London: Egypt Exploration Society.

Strabo, trans. by Falconer, W. (1912) *The Geography of Egypt*, Vol. 17, London: G. Bell and Sons.

Walz, T. (1978) *Trade Between Egypt and Bilad-as-Sudan, 1700–1820*, Cairo: Institut Français d'Archéologie Orientale.

Warburton, D. (2000) 'State and economy in ancient Egypt', in Denemark, R., Friedman, J., Gills, B. K., and Modelski, G. (eds) *World System History: The Social Science of Long Term Change*, pp. 169–84. London: Routledge.

Weinstein, J. M. (1973) 'Foundation Deposits in Ancient Egypt'. Unpublished PhD dissertation, University of Pennsylvania (Philadelphia).

Wilkinson, R. (2003) *The Complete Gods and Goddesses of Ancient Egypt*, London/Cairo: Thames and Hudson/American University in Cairo Press.

Zivie, A. and Lichtenberg, R. (2015) 'The cats of the goddess Bastet', in Ikram, S. (ed.) *Divine Creatures: Animal Mummies in Ancient Egypt*, pp. 106–19. Cairo: American University in Cairo Press.

CHAPTER 30

ANIMALS, ACCULTURATION, AND COLONIZATION IN ANCIENT AND ISLAMIC NORTH AFRICA

MICHAEL MACKINNON

INTRODUCTION

THREE great entities, the Roman, Vandal/Byzantine, and Islamic worlds, variously encompassed considerable geographic and cultural expanses. Building throughout the last few centuries BC and progressing into the first few centuries AD, the Roman Empire enveloped much of modern-day Europe, the Near East, and North Africa. Broad parts of this were incorporated into subsequent Vandal and Byzantine Empires, arising after Roman decline during Late Antiquity. The ensuing Islamic Empire developed chiefly thereafter. Politically, North Africa housed various Islamic dynasties from the ninth until the fifteenth century AD. The effect these domains had on North African cultural and economic life as regards animals, broadly spanning approximately 300 BC to approximately AD 1500, forms the focus of this chapter. How did these empires shape animal husbandry practices in North Africa? What dietary patterns surface? How were animals viewed and treated? How might social, religious, economic, and cultural conditions and values among the various peoples inhabiting, colonizing, or otherwise shaping life in North Africa affect animal use in the region, spatially and temporally? Attention focuses upon available zooarchaeological data, although it should be recognized that a rich and varied database of textual and iconographic information about animals from these periods under consideration also exists. Incorporation of such lines of evidence certainly can help enhance future exploration of animal use for ancient and Islamic North African contexts.

ZOOARCHAEOLOGICAL EVIDENCE—
TEMPORAL AND GEOGRAPHICAL BIASES

Acknowledging the wealth of information potentially available (including textual, iconographical, and zooarchaeological datasets) in assessing the role and use of animals in ancient North Africa contrasts, perhaps, with historic investigative development of the issue. Zooarchaeological analyses have been sporadic across North Africa over the course of excavation work for the periods under consideration. Much of this is a factor of traditional research foci within the disciplines of Classical Archaeology and Islamic Archaeology. Despite the long history of activity and scholarship in classical archaeology, attempts to recover and study animal bones from ancient sites in the Mediterranean world in a more systematic fashion, arguably, have only occurred over the last few decades (MacKinnon, 2007). Zooarchaeology, thus, is a relatively new participant to classical archaeology, admittedly facing steep tradition in a discipline founded upon philology, as well as stylistic and typological analyses of art and architecture. Similar concerns have historically affected zooarchaeological contributions to Islamic archaeology; however, debatably, the case here has lagged even more, with fewer Islamic sites excavated in North Africa overall and consequently an even smaller database than its Roman/Vandal/Byzantine equivalents. Insoll (1999: 96–9) reviews the potential of zooarchaeology to enhance our understanding of Muslim life, illustrating arguments with examples from sites in Jordan, Egypt, and Morocco, but the larger extent of Islamic North Africa, as regards available zooarchaeological materials, remains underexplored in this capacity.

Fig. 30.1 locates the principal North African sites with available Roman, Vandal/Byzantine, and Islamic period zooarchaeological data. References for the various sites listed are provided in Mattingly and Hitchner (1995), Van Neer (1997), King (1999), Hamilton-Dyer (2001; 2007), Leguilloux (2003), and MacKinnon (2010; 2012).

Samples are uneven, in terms of numbers, placement, reliability, recovery methods for the faunal materials provided, and time frames represented. Geographically, greater concentrations derive from work at Carthage, Tunisia, and in the eastern Egyptian desert zone. Both regions underwent considerable archaeological activity within recent times. Carthage was the centre of a large UNESCO-sponsored international archaeological campaign in the 1970s and 1980s, which itself spawned a series of subsequent projects. Broad, multidisciplinary frameworks underscored much of these ventures, with concerted efforts to retrieve faunal remains, alongside all categories of archaeological materials. Similar principles shaped research agendas affiliated with the intensive survey and excavation efforts invested in the eastern Egyptian desert and coastal area (which included work at a number of Roman military, settlement, and quarry sites in the region).

Temporally, Late Antique/Vandal/Byzantine contexts tend to dominate, in terms of sheer volume of available faunal material, making chronological comparisons difficult

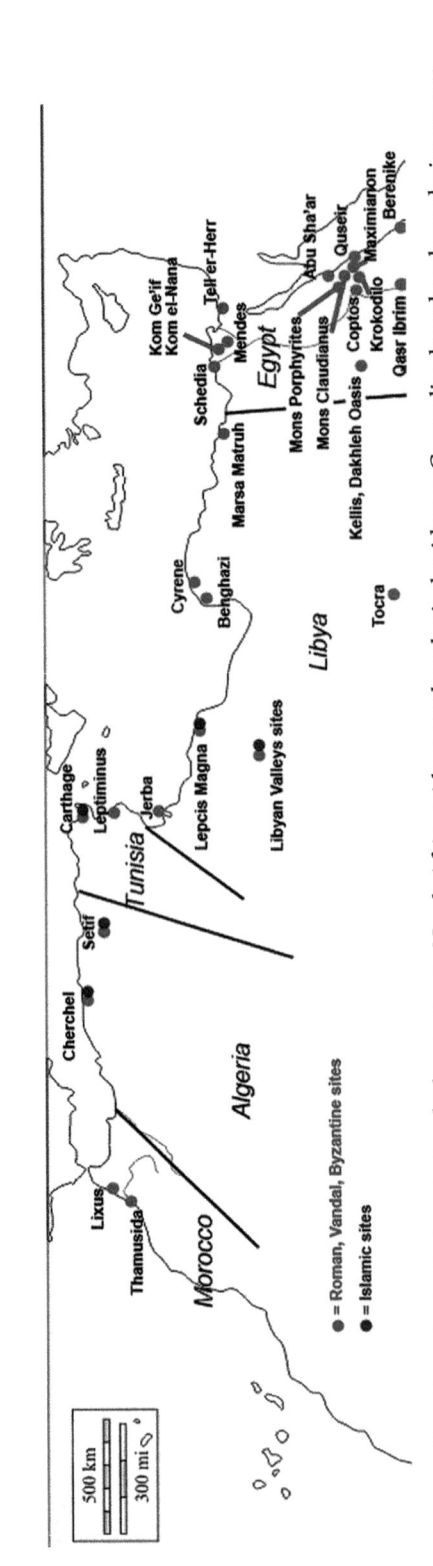

FIGURE 30.1 Location of principal ancient and Islamic sites in North Africa with zooarchaeological evidence. Generalized modern boundaries are provided for reference. Author's own image.

in many areas. This disparity is most apparent at Carthage (MacKinnon, 2010: 169–70). Presumably, rubbish was removed more regularly here during Roman times and deposited in outlying areas, probably beyond the immediate suburban region too, which was more heavily built-up. As Carthage declined through Late Antiquity, waste accumulated in abandoned buildings, disused cisterns, trash pits, and other convenient dumping groups both within and immediately outside the city (Leone, 1999: 122–3). Consequently, much of the archaeological materials (especially faunal remains) associated with Carthage derives from later contexts, even if structurally and architecturally more evidence exists from earlier phases in the city. Similar situations apply in various degrees among other North African sites.

The amount of detail provided across reports fluctuates greatly, as do aspects such as collection strategies, methodologies, etc. Biases certainly exist; however, broad patterns may be proposed to initiate an informational baseline for subsequent investigations of faunal deposits in ancient and Islamic North Africa—spatially, be these future studies of single or multiple deposits, or temporally, on a synchronic or diachronic scale.

ROMAN AND VANDAL/BYZANTINE PATTERNS

Great scholarly attention has focused on the impact of Roman expansion into North Africa, in whatever form such may have been felt, resisted, welcomed, displayed, controlled, manipulated, exploited, negotiated, and so forth (e.g. Mattingly and Hitchner, 1995; Crawley Quinn, 2009). Carthage fell first to the Romans in 146 BC, with further areas added in Tunisia, Libya, and Egypt during the first century BC (note: for ease of location, modern-day equivalents for ancient territories are used here). By the first century AD, Rome's conquered territories in Africa spanned the entire coast, from Egypt in the east to Morocco in the west. Archaeological and historical data outline changes or modifications that resulted from Roman cultural and military presence, conquest, and influence in North Africa, including a greater degree of urbanization, settlement, and wealth accumulation in some areas; increased agricultural demands due to taxation in kind for grain; and a push to develop, exploit, or enhance new areas and resources. Each of these aspects, in turn, affected the role and contribution of animals in Roman North Africa, observable, in some measure, through zooarchaeology.

Fig. 30.2 outlines changes in average NISP frequencies for three important mammalian meat taxa (cattle (*Bos taurus*), sheep/goat (*Ovis aries/Capra hircus*), and pig (*Sus scrofa* dom.)) for regions of North Africa. All site types are pooled (rural, urban, military, and so forth), so values reflect average dietary patterns. Areas are grouped by general geographic zone, as indicated, and by temporal phase: I = pre-Roman times (roughly sixth century BC to Roman conquest of the particular region); II = Roman times; III = post-Roman times (defined broadly as the fifth to seventh century AD, largely coinciding with major phases of Vandal/Byzantine occupation in sections of North Africa).

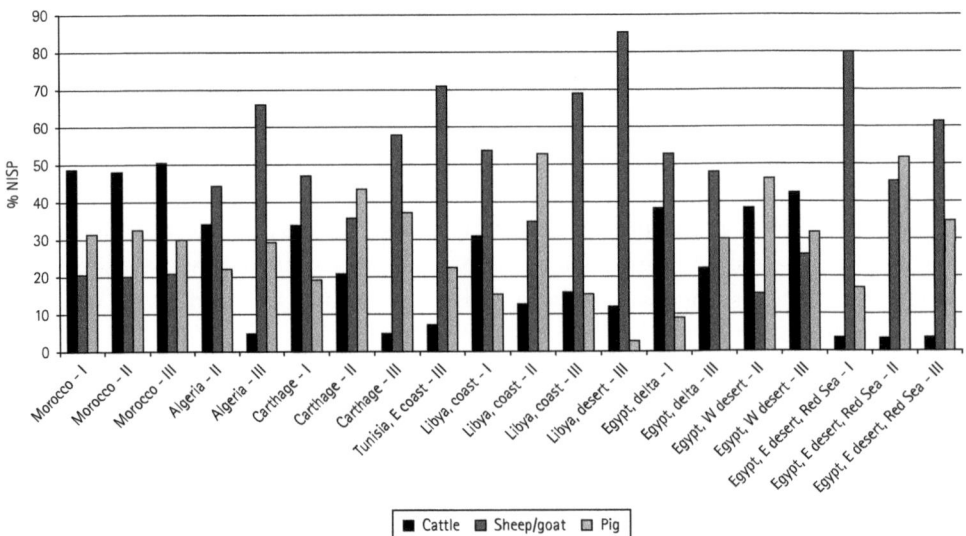

FIGURE 30.2 NISP relative frequencies for cattle, sheep/goat, and pig for North African sites by temporal period (codes: I = pre-Roman, II = Roman, III = post-Roman). Author's own image.

Increased pork consumption in many parts of the Empire often coincides with Roman cultural influence; however, variation does exist (King, 1999). Regional and local ecologies contribute in part. Sheep and goats are better suited to the dry scrublands in North Africa; hence they tend to dominate most assemblages here. Nevertheless, cultural choices, preferences, and dictates also affect patterns. For the Romans, such processes appear to take two forms: (1) a 'people-led' emulation of Rome; (2) a 'military-influenced' catalyst. Both factors are evident in Roman North Africa.

Roman conquest seems to have minimal effect in western North Africa, especially Morocco, where, as shown in Fig. 30.2, average values of cattle, sheep/goat, and pig remain fairly consistent over time. The high frequency for cattle here may be artificial to some degree, since deposits comprising this category (principally from the site of Lixus) contain a disproportionate percentage of cattle lower-leg bones, possibly connected with hide-processing and tanning waste, wherein such elements may be preferentially discarded. Still, this region houses some high-quality pastures, facilitating cattle-herding (Despois, 1940: 311–12). Among other taxa, pigs contribute about one-third of total NISP and are rather important to the diet, and from earlier times it seems. Morocco's ties with Iberia, where pigs also factor significantly, and from an earlier period, may help to partly explain this pattern.

For Algeria, the biggest change to NISP relative frequencies is a dramatic decline in cattle, with concomitant rise in sheep/goat between Roman and Late Antique/Vandal/ Byzantine phases. Tunisia registers a similar contemporaneous pattern. By Vandal and Byzantine times it seems that cattle are not required so much for intensive agriculture and transport, or become too expensive to maintain—probably because quality pasture lands had dried up, were shrinking, or were receiving less irrigation and attention;

administration that helped expand some of these lands in Roman times had dwindled; and nomadism increased. Hence, established practices of sheep- and goat-herding proliferate once more across ancient Algeria and Tunisia at this time.

Urban Tunisia, here dominated by Carthage and coastal Libya (the latter another region intensely settled or developed during Roman times), see comparable zooarchaeological patterns. NISP relative frequencies for pig between pre-Roman and Roman phases at Carthage more than double, a huge increase for Roman North Africa. Elevated levels of pork may signify some manner of 'Romanization', and it seems logical to argue that case here, presumably fueled by emulation of dietary patterns in Rome as brought by both military and elites populating or influencing Carthage (MacKinnon, 2010). Zooarchaeological evidence does not support the hypothesis that Carthage predominantly imported pigs or cuts of pork from overseas. There is no marked imbalance in skeletal part frequencies, nor any skewing of size or age parameters for this taxon to suggest export or import of vast quantities of meat cuts or of entire animals from great distances. Presumably, the city was supplied locally. As pigs cannot easily be herded vast distances, pig breeders would have displaced pastoral herders and grain farmers around Carthage as urban pork demands escalated. The suburban husbandry dynamic would have to change because of the Romans. Available age data reveal a predominance of adult cattle in pre-Roman examples at Carthage, but a mix of ages, including young calves in Roman contexts. Such a pattern suggests a change from a somewhat nomadic herding of cattle in pre-Roman times, where calves were raised to adulthood and maintained as status and wealth indicators to some degree, to market and agriculturally oriented systems in Roman times, where oxen were exploited as working beasts, superior brood stock were reared, usually in small, localized herds, and any surplus calves were sold to urban shops. Sheep and goat demographic patterns for Roman Carthage suggest slight age peaks separated by about twelve-month intervals, which might imply an annual cull. This could support the notion of some transhumant or seasonal rounds in herding, with flocks congregating near the coast probably in the spring/early winter, outside of harvesting seasons. Nevertheless, the presence of a wider range of age brackets overall in the sheep/goat data at Carthage confirms that some flocks were kept throughout the year in the immediate hinterland. These were not transhumant flocks, but localized smaller herds, presumably incorporated into mixed farming and herding operations around the city. It is important to add here that sheep predominate over goats in these Roman Tunisian and Libyan urban assemblages, a testament perhaps to city demand for wool. By contrast, goats outnumber sheep (and often largely so) in areas where pastoralism dominates, such as eastern Tunisia and the Libyan desert zone—drier, less populated regions where vaster grazing lands exist, suitable for herding flocks. It seems that where urban influence held less sway in North Africa, traditional husbandry patterns continued, perhaps ones geared around broader transhumant migration of flocks through Berber or Bedouin pastoralists.

Moving into Late Antique/Vandal/Byzantine times at Carthage, the following patterns emerge: cattle frequencies decline further from their already depleted Roman levels; sheep/goat values increase, a testament to augmented pastoralism at this time (as the economic schemes of the Romans declined and 'traditional' patterns returned); pig values drop only slightly, perhaps a reflection of Vandal and Byzantine rule in Carthage not dramatically changing earlier Roman dietary demands for pork, in other words keeping some uniformity in this regard. Overall, these patterns denote a continued, but perhaps less cohesive or networked, system of animal use in the area, compared to earlier Roman times.

The situation for Roman Egypt is mixed. Patterns for the Nile delta zone show some consistency with trends displayed for urban Tunisia and Libya, specifically an increase in pigs and a decrease in cattle, arguably a factor of Roman dietary influence, which favours pork, intensifying in the area. The importance of this region in the Roman grain trade seems to translate, in part, with relatively high frequencies for cattle (c.25%) at the sites of Schedia and Kom Ge'if. Cattle would fulfill greater roles here as plough and traction animals. Age and sex parameters at Schedia confirm the cattle there were chiefly older oxen, and of a relatively large stature.

Zooarchaeological results for the western desert area in Egypt, largely from the site of Kellis, a Roman settlement in the Dahkleh Oasis, display remarkably high frequencies of cattle and pigs during Roman times, but these decline into Late Antiquity as supply networks degrade. This desert area is not conducive for broad-scale localized herding of either taxon, although some cattle were presumably kept for traction and hauling purposes.

Available zooarchaeological evidence for the eastern Egyptian desert and Red Sea area warrants deeper investigation. Excavations here have been relatively extensive, and include military, settlement, port, and quarry sites. Fig. 30.3 provides NISP frequencies for a broader array of taxa for major sites in this zone.

Inspection of Fig. 30.3 shows the importance of marine resources across a variety of these sites, especially those along the coast (e.g. Abu Sha'ar and Berenike). Taxa inhabiting coral reefs such as groupers (Serranidae), emperors (Lethrinidae), and parrotfishes (Scaridae), as well as fish that inhabit inshore, sandy-bottom regions (e.g. sea breams, Sparidae) are most abundant. While fresh fish could be consumed more regularly at coastal sites, inland areas were likely more commonly provided with salted or other preserved fish along this active trade route (Van Neer, 1997: 146; Hamilton-Dyer, 2001: 285)

As regards consumed domesticates (Fig. 30.2), cattle play only a very minor role in these eastern regions. The area is simply not conducive for them. Sheep and goat register with some frequency, and dominate among the domestic mammalian taxa at sites along the Red Sea coast prior to, and after, Roman occupational influence. Presumably, such a pattern links with ecologically preferred traditional pastoral operations, unaffected by 'Romanized' lifestyles and economics. By contrast, pigs often comprise more that 80% of NISP totals for consumed domesticates at inland Roman sites of Krokodilo, Maximianon, Mons Claudianus, and Mons Porphyrites. For the most part, the army could operate a command economy and exercise dietary preferences without the constraints that affected those living closer to subsistence level. Pork was a prized meat, if it could be acquired, as well as a key meat for marking or displaying Roman identity, including a soldier's identity in Africa. Some pigs could have been kept at these

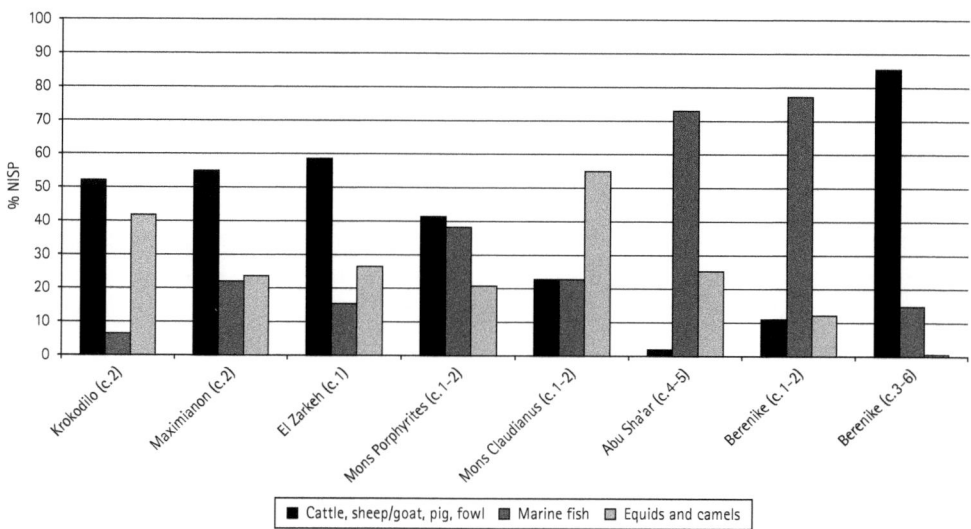

FIGURE 30.3 NISP relative frequencies for three taxonomic groups: (i) cattle, sheep/goat, pig, fowl, (ii) marine fish, (iii) equid and camel from Roman sites in the Eastern Egyptian desert and the Red Sea coast (dates in C AD). Author's own image.

eastern Egyptian sites, perhaps as urban garbage collectors in settlements or farms, but big herds were relatively impossible to maintain in these predominantly desert zones, without great expense. Presumably, a large part of the Roman demand for pork in these desert regions was met by imports of preserved meat from farms along the Nile, or perhaps further abroad from the Mediterranean. In support, skeletal part distribution for pigs at these sites is often skewed in favour of meatier cuts. Again, as in other areas of North Africa, as Roman presence and influence crumbled during Late Antiquity, values for pigs decline and more traditional sheep and goat (but probably more goat given the more elevated ratios of goat to sheep among these sites compared to elsewhere) nomadic herding amplified.

While the focus above has been on domestic mammalian food taxa, some comment is necessary about other animals consumed or otherwise utilized. First, equids (*Equus* sp.), dogs (*Canis familiaris*), and camels (*Camelus dromedarius*) are variously represented across many Roman North African sites. Data indicate great ranges in sizes, varieties, and use, with relatively weightier representation of donkeys (*Equus asinus*) among sites where pack animals were instrumental, especially among quarry and military sites investigated in the eastern Egyptian desert. This is apparent from inspection of Fig. 30.3, where equids (and camels) register rather significant NISP counts across sites that were linked with commercial traffic between the Nile and the Red Sea coast. Generally, equids were not eaten across the Roman world, but butchered donkey bones are noted at the Roman military and quarry sites of Mons Claudianus, Mons Porphyrites, Krokodilo, and Maximianon, implying some consumption, perhaps connected with limited availability or options in overall meat, and some measure then to augment local dietary resources. Equids and camels are very infrequent at Mediterranean North African sites, perhaps an indication of a diminished role as pack animals in these regions (duties possibly fulfilled more through cattle pulling carts), as well as a minimal dietary contribution (with cattle, sheep/goat, and pigs supplying ample mammalian meat in this respect).

A second point to note is that wild animals are very infrequent across assemblages in all areas and time frames across North Africa. Most sites never surpass 0.5% by total NISP. Presumably, wild animals were inconsequential in terms of North African diets, largely a factor of ecological limitations to promote them. Nevertheless, Roman rural villa sites, Roman urban sites, and some indigenous rural settlements (e.g. Libyan valley areas) tend to report higher frequencies of consumable wild animals (generally localized varieties of red deer (*Cervus elaphus*), hare (*Lepus* sp.), gazelle (*Gazella* sp.), and other taxa, all of which could be hunted) than do other site types or temporal phases. Incorporation of wild game on Roman sites probably links with dietary elitism, while traditional hunting practices might better explain higher percentages of gazelles and other wildlife among 'non-Romanized' sites. Also noteworthy is that Roman contexts across North Africa, and especially urban locales, often register the broadest diversification of bird and fish species compared to other temporal phases. This may support an increased segment of elites at this time, a social class cognisant of the importance of alimentary affluence. Not surprisingly, fish appear to make greater economic contributions among some coastal and riverine (especially the Nile) sites in North Africa, best exemplified zooarchaeologically in high fish NISP counts among Red Sea contexts (see Fig. 30.3), but also perhaps in archaeological evidence, notably tanks associated with fish sauce manufacturing (e.g. *garum, allec, liquamen*) discovered among various Mediterranean coastal sites (Mattingly and Hitchner, 1995: 200). Surprisingly, however, available faunal evidence suggests that fish comprised only a minor portion of the diet among Roman North African sites bordering the Mediterranean, even those for which faunal recovery was enhanced. Fishing was certainly undertaken along the Mediterranean coast, but dietary meat from domestic mammals and birds was probably more economical and abundant to meet demographic demands.

Finally, purported ancient textual and iconographic evidence recounting the importance of North Africa in supplying exotic beasts for Roman games, both locally and abroad, fails to translate into extensive numbers of such animals in the available faunal record (MacKinnon, 2006). Scattered gazelle, wild boar, deer, camel, bear, ostrich, wild goat, wild sheep, and hartebeest bones have been found at a number of sites in North Africa, especially at the ancient city of Carthage. No remains, however, from the traditionally 'exotic' fauna, such the lions, panthers, tigers, giraffes, elephants, and so forth are as yet noted in these deposits (MacKinnon, 2006: 152). While this zooarchaeological dearth need not imply vast exotic animal resources were not hunted and collected in North Africa for entertainment purposes, it does perhaps lend a sobering layer to the potentially exaggerated numbers outlined in ancient textual sources, as well as outlining the complications involved in capture and transport of such beasts.

ISLAMIC PATTERNS

Fig. 30.4 depicts NISP relative frequencies for principal domestic taxa from various Islamic sites in North Africa. Dates (century AD) for each sample are indicated in parentheses.

Although the geographic extent and political impact of different North African Islamic dynasties varied over time, it appears none had a major impact on the animal economy of the region. The faunal data show that pastoral herding of sheep and goats continued as the mainstay of the animal economy throughout North Africa during the ninth to fifteenth century AD. Cattle contribute more to the diet and economy in western regions of the Maghreb, such as Algeria and Morocco (represented here by sites of Setif and Cherchel), while sheep/goat largely predominate in the east (e.g. sites of Lepcis Magna and Libyan Valleys). These patterns show some consistency with what is displayed for Roman times, especially in the higher frequencies of cattle noted in western sections of North Africa. More pasture could have been available for cattle grazing in the western reaches of the Maghreb than in the relatively drier eastern sections. The protective nature of the Atlas Mountains, coupled with the moister Mediterranean climate that characterizes the western Maghreb, ensures that a much larger zone inland from the coast remains suitable for agriculture in this area.

It is also possible that enhanced NISP values for cattle among Islamic sites in the western Maghreb, as opposed to those in eastern North Africa, link to an urban demand for beef. Beef was permitted in Muslim faith, and its accepted dietary contribution is

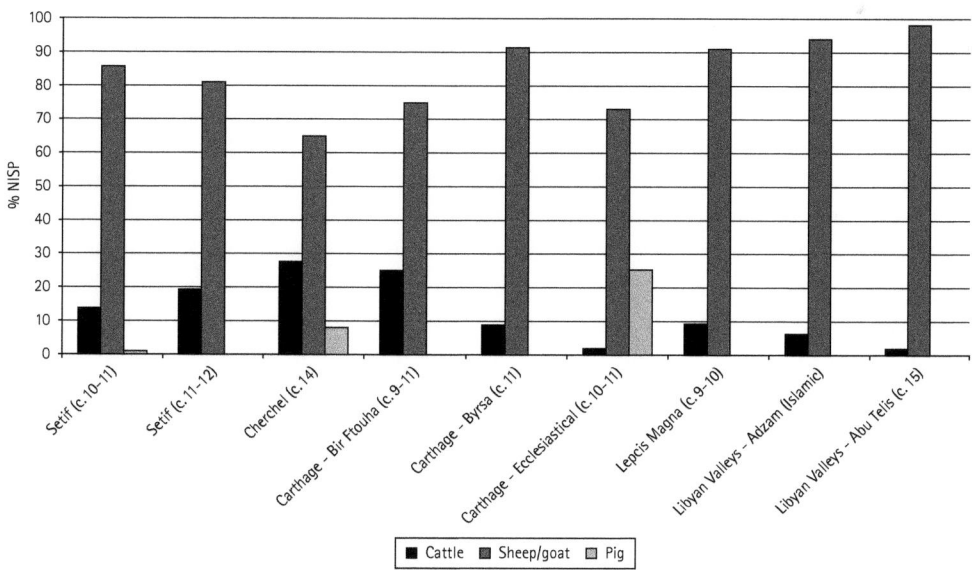

FIGURE 30.4 NISP relative frequencies for cattle, sheep/goat, and pig for North African Islamic sites (dates in C AD). Author's own image.

mentioned several times in the Qur'an (Qur'an 6.142, 16.5, 23.21, 36.71–3, 40.79). The aggregate demand for beef in Islamic cities could be substantial, and urban, as opposed to rural, locales may be predisposed to record higher frequency values for cattle bones. The argument for the import of beef cattle to Islamic urban areas is strengthened by the fact that the mortality profile for the cattle from these faunal samples accords with what is typical for meat consumption. Fusion and dental data indicate that many cattle were killed as immature animals, under three or four years of age. Additionally, some cattle bones might also have been transported to urban centres within hides destined for tanning. Hide processing and leather-working have traditionally been important industries in North Africa; the Qur'an specifically notes how Allah has created cattle, in particular, to provide people with leather for clothing (Qur'an 16.5) and tents (Qur'an 16.80).

Although available faunal data for Islamic sites seem to indicate some importance of beef at urban centres, essentially Islamic North Africa remained a pastoral region during the ninth to fifteenth centuries AD. Unlike other contemporary Islamic regions, such as Spain or Syria, North Africa had relatively few large towns; rather, it was thinly populated by pastoralists and marginal farming communities (Donner, 1999: 35). Berber tribes continued to exert great economic influence in the area, and perpetuated traditional pastoral schemes based on transhumant sheep- and goat-herding.

Patterns are variable, but available zooarchaeological evidence from Islamic North Africa show an overall decline in the number of sheep relative to goats over time. It is possible that changing environmental conditions, specifically increasing aridity, dictated a rise in the goat-to-sheep ratio. More goats were herded simply because they were better suited for the increasingly arid environment. Other reasons may include a heightened demand for milk and cheese among Islamic peoples, better satisfied by goats, which are more prolific milk producers than either cattle or sheep (White, 1970: 315). Additionally, with a religious ban on pork consumption, demand for other meats, such as kid and goat, may have grown. A review of the Muslim Sahih lends support to the argument of enhanced importance of goats over sheep in Islamic pastoral schemes of the past. Goats are mentioned far more often than sheep in these ancient recordings.

In light of data from modern pastoral operations in the Middle East and North Africa (Abu-Rabia, 1994), suggestions that Islamic herders practised a fairly unspecialized husbandry regime, and exploited their flocks for a variety of products including meat, milk, and wool, seem sensible. The flock is the livelihood for the pastoral herder, his supreme asset. Reproduction is a principal concern. More sheep and goats means greater wealth, but it also brings more responsibility, so a number of factors must be weighed in flock management. Nevertheless, the Bedouin herder of today views the flock as a resource that brings a higher 'interest' than any other investment. Meat, milk, wool, hides, brood stock, and other commodities from these animals can all be exploited as required by the shifting economic and social demands (Abu-Rabia, 1994: 110), so it is in the herders' best interest to maintain some flexibility in husbandry schemes.

The situation with pigs in Islamic North Africa is less complex than that for cattle, sheep, and goat. The Qur'an contains several references to pork abstinence (Qur'an

2.173, 5.4, 6.145, 16.115), and the absence of pig in the Carthage-Bir Ftouha, Carthage-Byrsa, Setif, Adzam, and Abu Telis faunal samples (Fig. 30.4) accords with these dietary restrictions. No part of the pig, including its skin and fat, can be used; the animal is strictly avoided by practitioners of Muslim faith.

The Qur'an also recognizes that transgressors exist, who may not abide by the dietary dictates and who do not bless the food they eat (Qur'an 6.119, 6.145). Zooarchaeological data might provide some support for this. The samples from Cherchel and the Ecclesiastical Complex at Carthage have a significant quantity of pig (c.10% and 25% respectively). While it is possible that these two particular sites did not follow Muslim dietary restrictions on pork consumption as strictly as elsewhere (be this due to Christian or other non-Muslim presence in the area), given the fact that both overlie substantial Roman deposits, it seems more probable that these pig bones represent residual material mixed in from earlier archaeological levels when pork was a favoured meat.

Lastly, osteometric data noted size diminution for cattle between Vandal/Byzantine and Islamic samples, and a general, although less pronounced (and in some measure contradictory), decrease in size for sheep and goats over the same period of time. This pattern occurs elsewhere in the Arab world; Toplyn (1994: 360–86) discusses a Near Eastern example involving sheep and goats. Essentially, this size decrease has been argued to reflect a breakdown in the selective breeding operations and improvements conducted by the ancient Romans and Byzantines.

CONCLUSION

In sum, it seems that regional identity within a loosely drawn Roman framework is the most appropriate way to characterize the contributions of animals to Roman North Africa. Some areas, such as urbanized Carthage and the Libyan coast, and 'militarized' parts of Egypt, saw great Roman dietary influence, predominantly displayed as increases in pork consumption. Others zones, such as Morocco and inland Tunisia and Libya, regions arguably less affected by, or exposed to, Roman customs or demands, as regards diets and husbandry practices, changed less. Patterns in Islamic times were less diversified, with sheep/goat pastoralism predominating, integrated husbandry schemes and animal breed manipulation generally diminishing, and cultural taboos against pork consumption registering in many areas.

REFERENCES

Abu-Rabia, A. (1994) *The Negev Bedouin and Livestock Rearing*, Oxford: BERG.

Crawley Quinn, J. (2009) 'North Africa', in Erskine, A. (ed.) *The Blackwell Companion to Ancient History*, pp. 260–72. Oxford: Blackwell.

Despois, J. (1940) *La Tunisie orientale: Sahel et basse steppe*, Paris: Les Belles Lettres.

Donner, F. (1999) 'Muhammad and the Caliphate', in Esposito, J. (ed.) *The Oxford History of Islam*, Oxford: Oxford University Press.

Hamilton-Dyer, S. (2001) 'The faunal remains', in Maxfield, V. and Peacock, D. (eds) *Mons Claudianus 1987–1993*. Vol. 2: *Excavations*, pp. 251–301. Cairo: IFAO.

Hamilton-Dyer, S. (2007) 'Faunal remains', in Maxfield, V. and Peacock, D. (eds) *The Roman Imperial Quarries: Survey and Excavation at Mons Porphyrites 1994–1998*. Vol. 2: *Excavations*, pp. 143–75. London: Egypt Exploration Society.

Insoll, T. (1999) *The Archaeology of Islam*, Oxford: Blackwell.

King, A. (1999) 'Diet in the Roman world: a regional inter-site comparison of the mammal bones', *Journal of Roman Archaeology*, 12, 168–202.

Leguilloux, M. (2003) 'Les animaux et l'alimentation d'après la faune: les restes de l'alimentation carnée des fortins de Krokodilô et Maximianon', in Cuvigny, H. (ed.) *La route de Myos Hormos. L'armée romaine dans le desert oriental d'Égypte*, pp. 549–88. Cairo: IFAO.

Leone, A. (1999) 'Change or no change? Revised perceptions of urban transformation in Late Antiquity', in Baker, P., Forcey, C., Jundi, S., and Witcher, R. (eds) *TRAC 98: Proceedings of the Eighth Annual Theoretical Roman Archaeology Conference*, pp. 121–30. Oxford: Oxbow.

MacKinnon, M. (2006) 'Supplying exotic animals for the Roman amphitheatre games: new reconstructions combining archaeological, ancient textual, historical and ethnographic data', *Mouseion*, 6, 137–61.

MacKinnon, M. (2007) 'State of the discipline: osteological research in classical archaeology', *American Journal of Archaeology*, 111, 473–504.

MacKinnon, M. (2010) 'Romanizing ancient Carthage: evidence from zooarchaeological remains', in Campana, D., Crabtree, P., deFrance, S., Lev-Tov, J. and Choyke, A. (eds) *Anthropological Approaches to Zooarchaeology: Complexity, Colonialism and Animal Transformations*, pp. 168–77. Oxford: Oxbow.

MacKinnon, M. (2012) 'Faunal remains', in Rossiter, J., Reynolds, P., and MacKinnon, M. (eds) A Roman Bath-House and a Group of Early Islamic Middens at Bir Ftouha, Carthage, pp. 274–80. *Archeologia Medievale*, 39, 245–80.

Mattingly, D. and Hitchner, R. B. (1995) 'Roman Africa: an archaeological review', *Journal of Roman Studies*, 85, 165–213.

Toplyn, M. R. (1994) 'Meat for Mars: Livestock, Limitanei and Pastoral Provisioning for the Roman Army on the Arabian Frontier (AD 284–551)'. Unpublished PhD dissertation, Harvard University (Cambridge, MA).

Neer, Van, W. (1997) 'Archaeozoological data on the food provisioning of Roman settlements in the Eastern Desert of Egypt', *Archaeozoologia*, 9, 137–54.

White, K. D. (1970) *Roman Farming*, London: Thames & Hudson.

CHAPTER 31

HISTORICAL ZOOARCHAEOLOGY OF COLONIALISM, MERCANTILISM, AND INDIGENOUS DISPOSSESSION

the Dutch East India Company's meat industry at the Cape of Good Hope, South Africa

ADAM R. HEINRICH

INTRODUCTION

SOME introductions to the study of historical archaeology simply define the field as the archaeology of the recent past since the fifteenth century, which heavily incorporates written documents. This does not sound terribly compelling. Historical archaeology can also be defined as the archaeology of the emerging and developing modern world. Beginning with the European diaspora around the globe, the discipline examines the modern world as it is characterized by an increasingly interconnected and interdependent mercantile system. Within this movement of Europeans across the globe, we can observe how they have had major impacts on the indigenous biology and cultures throughout the New Worlds (new in the European perspective) that they visited and often colonized. These foreign lands were rapidly established as nodes within the global mercantile network where the once distantly connected peoples would forever be changed.

Colonial contexts are terrific archaeological laboratories because the New Worlds were rarely barren and empty. New, unfamiliar people and ecosystems would mean that European colonists and the indigenous inhabitants had to adapt to succeed (Grove,

1997). On the trade route between Europe and southern and eastern Asia, the Cape of Good Hope at the southernmost tip of Africa was a strategic location for a colonial endeavour. Mercantile European powers such as the Portuguese and Dutch occasionally stopped to resupply trade ships with fresh water and meat acquired through trade from the Cape's indigenous Khoekhoen pastoralists. As the early, irregular European–Asian trade increased, high shipboard mortality and the challenge of supplying ships with fresh supplies pushed the Dutch under the authority of the joint-stock *Verenigde Oost-Indische Compagnie* (VOC), the Dutch East India Company, to establish a refreshment station at Table Bay at the Cape in 1652.

The primary aim of the VOC's Cape settlement was to provision sufficient fresh water, vegetables, fruits, and meat for their trade ships and the colonial contingent, and eventually profit by selling to ships of other nationalities (de Vries and van der Woude, 1997). The VOC's meat industry has left a rich archaeological record that can illustrate how the VOC adapted to their new environment, coped with increasing demands for meat, and ultimately penetrated and dominated the South African landscape.

DOCUMENTATION OF THE MEAT INDUSTRY

The strength of historical archaeology is through the ability to integrate the material cultural remains and a documentary record. Though the record is invaluable, historians are conscious about the inherent bias in documents as they were often recorded in a literate individual's perspective or simply incomplete as the mundane was often overlooked. The VOC is renowned for being the first international stock-offering megacorporation, and their success was partly maintained through meticulous recording of mercantile business that was able to coordinate hundreds of thousands of employees for more than a century over multiple continents. Curiously, the meat industry was relatively under-recorded as it became the mundane support operation within the Cape settlement and the seaborne trade route.

Due to the logistics of shipping livestock to the Cape, the Dutch hoped to trade with the Khoekhoen people for cattle (*Bos taurus*) and sheep (*Ovis aries*) in exchange for copper, beads, tobacco, and alcohol. The Dutch initially brought livestock on their first ships; the European cattle died off, and Bengali and Persian sheep were few and were quickly hybridized with the long-haired indigenous sheep (Kolben, 1731a; 1731b; Mentzel, 1921). Documents tell about how the trade was sporadic as the Khoekhoen were transhumant and they regarded cattle as wealth which structured status within their society. Therefore they traded sheep more frequently and the cattle they gave up were generally older or in poorer condition than those the Dutch desired (Elphick and Malherbe, 1988; Ross, 1988).

The Dutch holdings increased through trade, confiscation, and breeding, which prompted the need to contract burghers, free farmers, on farms along the Liesbeeck River near the Cape settlement in 1657 and further into the frontier at settlements such as Drakenstein and Stellenbosch by the later seventeenth century. With the greater exchange and breeding of sheep, they increasingly outnumbered cattle by ratios of 2.5:1 in 1720 and 4:1 by 1770, though individual farm ratios varied (Elphick, 1985; Ross, 1988). Jan van Riebeeck, the founding commander of the Cape settlement, intended to provision each calling ship with eight cattle and eight sheep but, due to the greater amount of sheep available, ship and settlement provisioning was increasingly reliant on mutton. By the 1660s, only a decade after the first landing, the number of sheep slaughtered for the ships was nine times higher than that proposed by Van Riebeeck, and ship provisioning became more heavily reliant on sheep into the later seventeenth century (Elphick, 1985). 'By 1683 deliveries of sheep had risen to the point where many ships were given a hundred each, and few got less than thirty: the quota was now close to one sheep for every four to five sailors on board' (Elphick, 1985: 153). By the eighteenth century, Otto Mentzel (1925) living at the Cape recorded that the VOC required a total of 390,000 pounds of slaughtered meat annually for the ships, with much of this being mutton. The increasing demand for meat by calling ships and colonists caused strain on the meat industry that struggled to keep up (Templin, 1984; Elphick, 1985).

Sites

In the historical period, European-derived meat industries are integrated systems with separate places of production and consumption. At the Cape, the settlement at Table Bay was the chief consumer, where an urbanizing population, a garrison, and the calling trade ships demanded meat. The settlement was surrounded by frontier outposts acting as defence and trading stations as well as the crucial farms where great herds of cattle and sheep were raised for the industry. Seven faunal samples were analysed that provide insights into change through time and across space to see how different contexts contributed to the development of the meat industry. Built on the shore of Table Bay, the Castle of Good Hope at the main settlement provided five samples: two from the rapidly infilled Moat (FG31 and DE25 c.1720–1725), two fills for ground-raising in the Granary (Phase 1 c.1666–1691 and Phase 7 c.mid-eighteenth century), and a fill for ground-raising in *Donkergat* (DKG), a prison (Layer 4 c.1790s). In the Cape interior, Elsenburg in Stellenbosch (c.1740s–1750s) was a farmstead and Oudepost I (occupied 1669–1673 and 1684/6–1732) was a military and trading outpost at Saldanha Bay. The sites and the taphonomically informed analyses are more fully described elsewhere (Heinrich, 2010).

Meat Provisioning

The fauna recovered from the sites within the Cape settlement was supplied from interior farms so they reflect the products of provisioning and husbandry strategies.

The Moat demonstrates evidence of an early stage in the provisioning process. The Moat was infilled to create an earthen path when the Castle's entranceway was relocated to the settlement side of the curtain wall. The deposits are only metres away from the dock where stock was slaughtered for the ships, and where the offal and waste were customarily dumped into the bay (Raven-Hart, 1970). The taxonomic representations are dominated by domesticates, where sheep outnumber cattle through minimum numbers of individuals (MNI) in ratios of 14: 1 (FG31 layer B) and 10: 1 (DE25 layer A-2), much higher than the 2.5: 1 ratio suggested by the documents for the 1720s (Elphick, 1985). The great proportion of sheep remains also suggests provisioning, through their skeletal parts representations. This is especially visible in FG31, the sample closest to the historical dock, which is heavily represented by cranial elements showing that most of the vertebral and appendicular elements were sent elsewhere (Fig. 31.1). DE25, the sample nearest the relocated Castle entrance is more evenly represented by the various skeletal elements.

Through the proportions of processing evidence, the Moat samples additionally demonstrate a provisioning signature where most of the bones derive from a pre-kitchen stage of preparation. Cooking evidence through burning or charring is observed minimally on 0–0.6% of the bones from the Moat contexts. Butchery marks show primary butchery. Chop marks, observed on 12–15% of the fragments in FG31 and 16–27% of those from DE25, are identified at joints and along the sagittal axis of the vertebral columns and the crania. Chopping was aimed at subdividing the carcasses into major units, rather than smaller pieces to fit in pots or on spits and roasting pans. Cut marks also show primary butchery, where in DE25 cut marks are only observed on 1–2% of the bone fragments. FG31 contrasts with DE25 with cut mark frequency at 7% of the fragments. FG31 is dominated by large-sized sheep cranial, hyoid, and mandibular fragments which contain 52% (n = 115 of 220 cut mark NISP) of the cut marks. This shows heavy skinning and tongue removal occurred within this sample as the sheep were processed before being sent to ships or other destinations.

The faunal samples from the inside of the Castle contrast with those recovered from the Moat. The bones from the fills in the Granary and DKG are the products of the kitchens and though the three samples are from moments across a century of time, they have comparable results providing a kitchen signature. Again, sheep dominate cattle by MNI but in more modest ratios between 3.5: 1 and 5: 1. Skeletal part representations for cattle and sheep from the late eighteenth-century DKG are dominated by postcranial elements with particularly high representations of forelimbs and hind quarters, while cranial elements are lacking (Fig. 31.1). For the Granary, the late seventeenth-century Phase 1 demonstrates a more even skeletal element representation, while the mid-eighteenth-century Phase 7 contains most elements, but meat-bearing limbs are more heavily represented.

Cooking and butchery show additional meat preparation that occurred after the initial processing demonstrated in the Moat, further showing that these deposits are kitchen residues. Chop marks are comparable to the Moat fauna as carcasses within the meat industry were initially processed in similar patterns before being sent to kitchens,

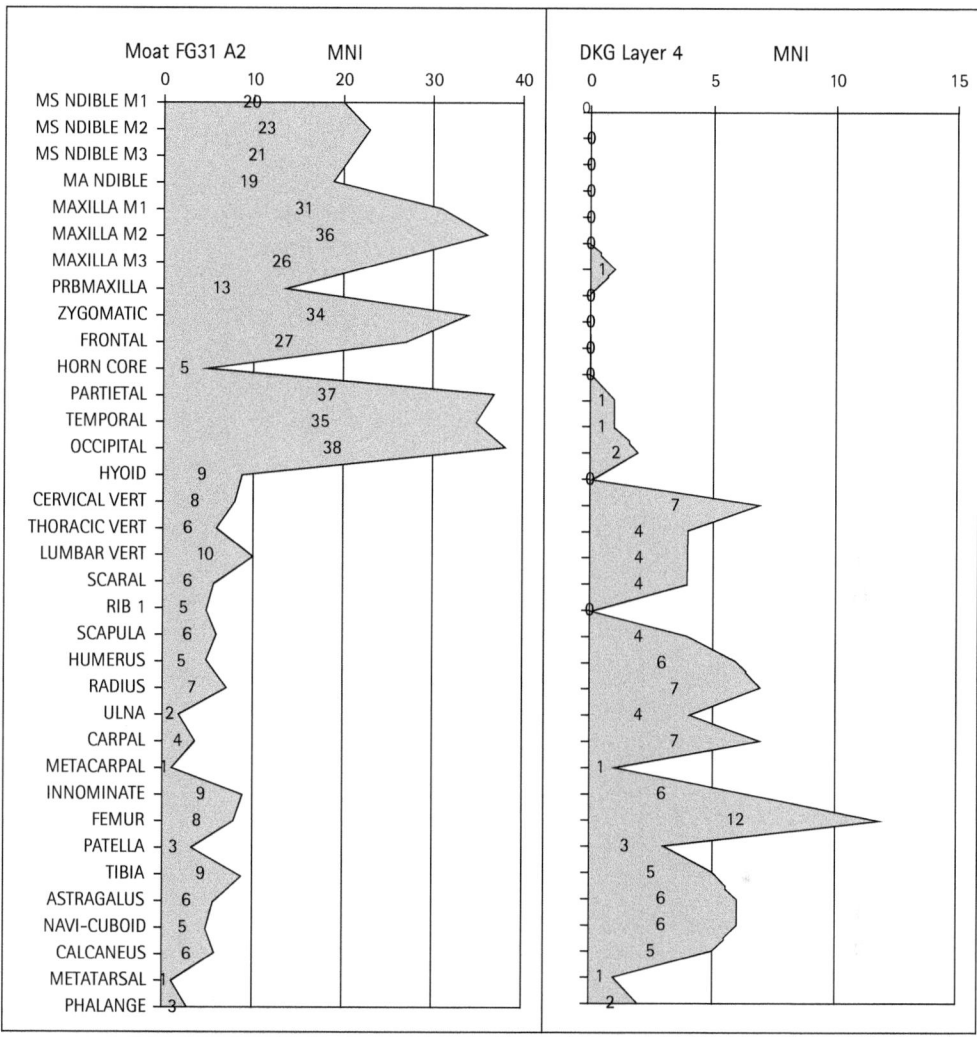

FIGURE 31.1 Sheep skeletal element frequencies for the jetty side end of the Moat, showing the high proportion of cranial elements contrasting with the DKG deposit, which is a kitchen dump dominated by meatier postcranial elements. Author's own image.

ships, or elsewhere. Cut marks are found in more moderate proportions in locales where meat masses were removed from the bones, such as on humeri and femora. DKG shows cut marks on 9% of the bone fragments, while in the Granary Phase 1 has cut marks on 6% of the bone fragments, and Phase 7 on 2%. Burning and charring is also more moderate than that observed from the Moat where the Granary's Phase 1 has 1% and Phase 7 has 6%, while DKG has 7% of the fragments affected. In DKG, cooking is observed as characteristic grayish-purple discolouration at bone and isolated spots on bone surfaces suggesting that they were discoloured through the burning of greasy, meat-laden bones.

MEAT PRODUCTION

Husbandry concerns in meat production systems include animal age and sex. Concern is made for the age when animals would be slaughtered. In European societies, the perception is that younger meat is seen as more tender or desirable as expression of status (Albarella, 1997; Ashby, 2002). Rearing animals to prime ages is ideal in a provisioning scenario to maximize meat yields. Regarding sex, stock-keepers desiring to increase their herds negotiate the ratios of breeding males and females with large proportions of castrated males to fulfil meat demands (Allison, 1958; Landon, 1997).

The sheep provide ages through dentition eruption and wear as well as epiphyseal development. When the late seventeenth-/early eighteenth-century contexts are contrasted with those dating from or after the first quarter of the eighteenth century, a difference can be noted. In the Granary's Phase 1, dentition and postcranial epiphyseal fusion shows that about half of the sheep were slaughtered younger than two years of age. Complementing Phase 1, the greater representation of dentition from the DP context at Oudepost I also shows that sheep were more likely to be slaughtered younger in the late seventeenth/early eighteenth century. Due to taphonomic processes, sheep teeth were not preserved in tooth rows, but tooth wear suggests that roughly a quarter of the sheep were slaughtered younger than two years of age while unworn second molars show that some of the sheep were slaughtered as young as nine months to one year of age. The large faunal collection from the terrestrial context at Oudepost I shows a similar pattern with that identified in DP with a significant proportion of younger sheep between six months and two years old (Cruz-Uribe and Schrire, 1991).

The post-1720 contexts contrast with those from the late seventeenth/early eighteenth century. Sheep were predominately slaughtered in prime ages between three to six years of age. This pattern of prime-age slaughter is seen in the Moat, Granary Phase 7, DKG, and in published data from the eighteenth-century sites at the ditch in the Grand Parade near the Castle (Abrahams, 1996) and the outpost Paradise (Avery, 1989). Elsenburg diverges from this pattern where most sheep were slaughtered at prime ages, but a greater proportion were older demonstrating the greatest frequency slaughtered between four to six years of age with some between six and eight and others even older than eight years of age.

Sex data also provides some revealing patterns that illustrate how the meat industry was established. Looking at sexual dimorphism of postcranial elements, the Oudepost I sheep from the terrestrial sample were slaughtered in equal proportions of males and females (Cruz-Uribe and Schrire, 1991; Heinrich, 2010). The later sites show that different sexes were more likely to be slaughtered at different points in the provisioning system. The Moat provides a large number of acetabulae that show males were predominately slaughtered at the urban settlement. Interior Castle sites of the Granary and DKG support the Moat findings though sample sizes are small in those collections. Strongly

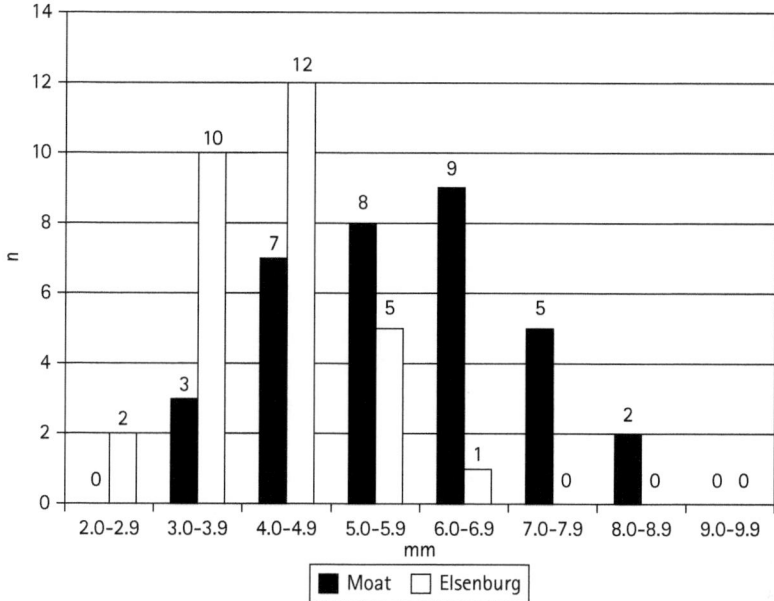

FIGURE 31.2 Histogram illustrating the distribution of medial acetabular measurements for sheep, showing that the Moat contained greater amounts of males, while Elsenburg contained more females. Author's own image.

contrasting with the Moat, the interior farm Elsenburg contains a greater proportion of females ($t = 3.58$, $P = 0.0008$) (Fig. 31.2).

THE CHARACTERISTICS
OF THE MEAT INDUSTRY

Evidence from the faunal remains clarify and add to the historical record of the meat industry. Documents tell us that the meat industry involved inland and frontier burghers who were contracted to supply specified amounts of animals to the official abattoirs at the main settlement as well as provisions for outposts further into the frontier (Cruz-Uribe and Schrire, 1991; Schrire, 1995). Tax and inventory rolls tell us how many sheep and cattle each farm managed, as well as pigs (*Sus domesticus*) and horses (*Equus caballus*) which were not contracted for the main meat supply.

Historical accounts give the impression that the colonial Cape diet was mundanely based on mutton. Otto Mentzel (1925: 101), a visitor to the Cape from 1733–1741, records 'mutton forms the staple diet; excellent mutton, undoubtedly, and made more appetizing by being prepared in a variety of ways, but—still mutton.' The historical accounts are challenged by the archaeology where sites containing major post-kitchen deposits such as the Granary, DKG, and Elsenburg show that beef was a much greater contribu-tor to the diet considering meat yield. Other contributions include pork, as pig bones are present in all samples and the evidence from Elsenburg shows that high-status burghers ate some pigs that were only a few months old. Some wild fauna was also consumed though it was restricted to the Governor's table within the Cape settlement as dictated by limiting laws emplaced in the face of wildlife depletion from overhunting (Mentzel, 1925; Cruz-Uribe and Schrire, 1991). Small amounts of wild fauna finds in the Moat and the Granary at the Castle suggest that residues from table preparations for the Governor's events are present. Wild fauna was a greater contribution to the diet of those in the interior, regardless of official laws. Oudepost I's terrestrial sample shows that wild fauna makes up over 70% of the MNI showing how the outpost's provisions were supplemented through hunting (Cruz-Uribe and Schrire, 1991). Elsenburg in the frontier demonstrates a narrower range of wild fauna than Oudepost I, but this fauna was probably consumed to add variety or exhibit wild game as domestic meat was plen-tiful and raised on site. Fish remains are also numerous from the Moat and interior Castle sites adding to dietary variety particularly for the garrison and other labourers (Poggenpoel, 1996).

The archaeological evidence suggests that provisioning was predominately based on mutton, where the Moat deposits have sheep-to-cattle ratios markedly higher than the documented farm holdings or post-kitchen deposits. Van Riebeeck's early estimate of eight sheep and eight cattle per ship was quickly determined to be unrealistic. By the late 1660s, ships calling at the port were consuming about two thousand sheep and over three hundred cattle in a year, a ratio of almost 7:1. Again, by the 1680s, most ships received from thirty to over one hundred sheep each. During Van Riebeeck's command (1652–1662) calling ships usually numbered less than forty a year. By the end of the cen-tury, often more than sixty-five came each year and by the time of the Moat's infilling in the 1720s, over eighty-five ships needed to be provisioned annually (Elphick, 1985). The Moat was very rapidly infilled with faunal remains that did not linger on a ground surface for a lengthy period of time showing a relatively discrete episode of provisioning slaughter. The Moat shows that provisioning increased from the documented 6:1 ratio in the 1660s to at least ten to fourteen sheep for each cow by the 1720s.

The faunal remains reveal other strategies the VOC had to undertake in order to keep up with the increasing demands for meat through increasingly numerous ships ply-ing between Europe and Asia. Initially, the VOC was slaughtering stock at age and sex ratios that were not ideal. By the 1720s, sheep were regularly slaughtered at prime ages, three to six years, when the animals were able to be reared to their maximum, efficient meat yields. The earlier contexts at Oudepost I and Phase 1 of the Granary show that a large proportion, about a quarter, of the sheep was slaughtered younger than two years of age with some as young as nine months. In addition to the ageing data, Oudepost I shows that males and females were slaughtered in equal proportions at a consumption site while the other sites show a greater demographic separation where mainly males, likely wethers, were sent to slaughter for provisioning and distribution to the Cape set-tlement's kitchens. The age and sex data emphasizes the VOC's struggles to keep up with meat demands where they had to diverge from their ideal husbandry practices, being forced to slaughter animals they would rather grow larger and breed. This struggle during the settlement's initial fifty to seventy-five years required continual trading and raiding expeditions into the interior to obtain even more stock from the Khoekhoen which would have major consequences for both sides (Elphick, 1985; Penn, 2005).

The early challenges of the VOC to provide enough meat to maintain their developing refreshment station mirrors patterns seen elsewhere, when Europeans attempted to gain colonial footholds. In the Chesapeake, the English colonists also experienced initial difficulties in establishing their herds. Cattle and pigs were slaughtered younger in the 1620–1660 period than during the 1660–1700s when better-established stock was slaughtered at prime ages (Bowen, 1999). Likewise, Spanish colonists in Florida and Dutch colonists in New Amsterdam initially slaughtered their cattle younger than what was typical in later periods (Reitz and Scarry, 1985; Reitz, 1991; Janowitz, 1993; Cantwell and DiZerega Wall, 2001).

WIDER IMPLICATIONS OF THE VOC'S MEAT INDUSTRY

The meat industry set in motion the dispossession of the indigenous Khoekhoen people (more fully discussed in Heinrich and Schrire, 2011). Since landing on the shores of Table Bay in 1652, the VOC engaged the Khoekhoen groups in an uneven trade where the Khoekhoen exchanged their economically significant stock for consumable and generally low-utility items such as beads, tobacco, brandy, and some metal (Schrire, 1995). The loss of stock removed the capital that maintained social status and leadership in the indigenous societies. The acquired trade items glutted the Khoekhoe economy, where the more widely used adornments along with uncontrolled trade relationships with the VOC blurred and subverted boundaries between the higher- and lower-status strata of the societies, diminishing the authority of the traditional leaders. In 1700, Commander Willem Adriaan van der Stel (1699–1707) promoted frontier settlement and opened free trade with the Khoekhoen. By 1708, the open trade devastated the Khoekhoe herds taking a probably under-recorded 35,562 sheep and 8,871 cattle (Elphick and Malherbe, 1988).

With greater numbers of colonists and the ever-growing demand for provisioned meat, the Khoekhoe herds were increasingly sought by traders and preyed upon by raiders removing their economic base faster than it was able to be replenished. The Khoekhoen, suffering as their societies fell into turmoil, sometimes resorted to retaliatory raids to steal back animals and occasionally resorted to violence with the killing of a colonist or the burning of structures throughout the frontier. Khoekhoe guerilla retaliation outraged the Dutch who meted out retributive justice with uneven violence through organized commando raids which included killing and additional confiscations of animals (Penn, 1986; 2005; Guelke, 1988; Legassick, 1988; van der Merwe, 1995). It was disjointed tribal groups with poisoned spears and arrows versus horsemen with guns.

The documents and archaeology show how the Dutch were consuming animals faster than their herds could reproduce (Templin, 1984; Elphick, 1985; Penn, 2005). Through these early difficulties, the Dutch were able to steadily grow their stock-holdings through breeding, trading, and stealing. Eventually, as they were able to acquire and produce more animals, the Dutch penetrated into the frontiers to occupy grazing lands and to claim the crucial watering holes in the dryer pastures. This further dispossessed the Khoekhoen, whose seasonally vacated lands were incorporated into the expanding colony. Losing water sources visited through millennia, the Khoekhoen were pushed further into the more arid frontier where their herds were not as well supported, forcing various Khoekhoe groups to occupy smaller, more crowded tracts, causing conflict and further impoverishing their societies. Dispossession ultimately forced many Khoekhoen to enter servitude with the colonists where their herding skills were now used for the success of the colonial endeavour, while scratching out the only living they could in their ancestral lands (Penn, 1986; Elphick and Malherbe, 1988; Ross 1998).

The European colonists themselves were not unchanged through the maturation of the meat industry. By the later seventeenth century, the colonial stock-holders penetrated further and further into the frontier. The region around Drakenstein and Stellenbosch was well watered, but quickly these lands were occupied and set for grazing. As their stock increased, needing more grazing land, burghers moved further inland into the increasingly empty frontier, which becomes dryer and sparser away from the more temperate coastal areas. To avoid overgrazing, these interior herders adopted a Khoekhoe-like lifestyle of seasonally transhumant pastoralism as they spread inland into increasingly arid areas where the stock required large tracts of land around suitable water sources. These occupations which emptied the lands of the traditional, indigenous pastoralists refilled them with European-descendent *trekboeren* pastoralists who would come to settle and claim much of the remaining South African landscape through the eighteenth and nineteenth centuries (Penn, 2005).

Trekboeren, of mixed Dutch, German, and French Huguenot ancestry with some indigenous African admixture, initially migrated further into the frontier to graze their stock. Within the more arid lands, their semi-nomadic lifeways pulled the *trekboeren* further and further from the cultural influence of the Cape settlement and the political authority of the VOC and, after 1795, the English governments. With distance, these frontier inhabitants and particularly their Cape-born descendents transformed their Dutch culture into a distinct form that would come to be called 'Afrikaner' (Idenburg, 1963; Guelke, 1988; Ross, 1988; 1998; 1999; Schutte 1988; Giliomee 2003). The Afrikaners disdained governmental regulation over their movements and acquisition of stock from the indigenous groups they encountered. In the frontiers, the Afrikaners were minimalistic, self-sufficient, and isolated. Their connections to the urban areas were limited to stock sales for the capital needed to purchase material goods such as the firearms they could not acquire elsewhere (Guelke, 1988; Schutte, 1988; Ross, 1998).

CONCLUSION

The VOC were master recorders who insisted on meticulous accounts in order to maintain an efficient global mercantile company. In spite of the impressive documentary record of the Cape settlement, details of their meat industry were not fully recorded for posterity, possibly because it was a mundane aspect of their overall scheme, maintained increasingly in the private and distant sectors. Faunal remains flesh out the details of the industry that are usually only hinted at in the documentary record. The remains speak of early struggles, identify the various components of the system of production and consumption, and reveal the general consumption of meat in a diasporic and eventually creolized cuisine.

The faunal remains at the Cape are one of the most valuable classes of material culture. Through the lens of the colonial VOC meat industry that governed the face-to-face relationships between the indigenous and European groups, one sees the process that dramatically and permanently changed the South African Cape. Through the endless appetite of a massive meat supply system in the unique Cape environment 'the disappearance of livestock must be counted the prime feature of the erosion of traditional [Khoekhoe] society' (Elphick, 1977: 164), while also catalysing the development of a new European-descendent ethnicity. One finds the Khoekhoe sheep that were traded and stolen, hybridized with Bengali and Persian rams, and paraded to the Cape settlement for slaughter. The Khoekhoe cattle, their wealth, remained completely recognizable to the pastoralists as the Europeans did not hybridize them until much later. The bones show us how a European system of husbandry would come to dominate the Khoekhoe grazing lands and watering holes, while also causing the loss of their wild prey. The Europeans who now controlled these sheep and cattle pushed further and further into the frontier. While some of the burghers would settle to establish permanent towns, a number would ultimately become the Afrikaner *trekboeren*, who, while ironically following a migratory Khoekhoe-like lifestyle, would ultimately create the disconnected rural, conservative foundation that would come to influence South Africa's apartheid policies over the following centuries.

ACKNOWLEDGEMENTS

Foremost, this work was only possible with the guidance and contributions of Carmel Schrire. Additional appreciation is extended to David Landon, Robert Blumenschine, and Robert Scott who provided support through the dissertation process. Thanks also to Royden Yates, Graham Avery, Margaret Avery, and Lalou Meltzer for access and help within the IZIKO: South African Museum, and the University of Cape Town for access to the faunal collections.

REFERENCES

Abrahams, G. (1996) 'Foodways of the Mid-eighteenth Century Cape: Archaeological Ceramics from the Grand Parade in Central Cape Town'. Unpublished PhD dissertation, University of Cape Town (Cape Town).

Albarella, U. (1997) 'Size, power, wool and veal: zooarchaeological evidence for Late Medieval innovations', in De Bow, G. and Verhaeghe, F. (eds) *Environment and Subsistence in Medieval Europe—Papers of the 'Medieval Europe Brugge 1997' Conference*, pp. 19–30. Bruges: Instituut voor het Archeologisch Patrimonium.

Allison, K. J. (1958) 'Flock management in the sixteenth and seventeenth centuries', *The Economic History Review*, 11, 98–112.

Ashby, S. P. (2002) 'The role of zooarchaeology in the interpretation of socioeconomic status: a discussion with reference to Medieval Europe', *Archaeological Review from Cambridge*, 18, 37–59.

Avery, D. M. (1989) 'Remarks concerning vertebrate faunal remains from the main house at Paradise', *South African Archaeological Bulletin*, 44, 114–16.

Bowen, J. (1999) 'The Chesapeake landscape and the ecology of animal husbandry', in Egan, G. and Michael, R. L. (eds) *Old and New Worlds: Historical/Post-Medieval Archaeology Papers from the Societies' Joint Conferences at Williamsburg and London 1997 to Mark Thirty Years of Work and Achievement*, pp. 358–67. Oxford: Oxbow Books.

Cantwell, A. and DiZerega Wall, D. (2001) *Unearthing Gotham: The Archaeology of New York City*, New Haven: Yale University Press.

Cruz-Uribe, K. and Schrire, C. (1991) 'Analysis of faunal remains from Oudepost I, an early outpost of the Dutch East India Company, Cape Province', *South African Archaeological Bulletin*, 46, 92–106.

Elphick, R. (1977) *Kraal and Castle*, New Haven: Yale University Press.

Elphick, R. (1985) *Khoikhoi and the Founding of White South Africa*, Johannesburg: Raven Press.

Elphick, R. and Malherbe, V. C. (1988) 'The Khoisan to 1828', in Elphick, R. and Giliomee, H. (eds) *The Shaping of South African Society, 1652–1840*, pp. 3–65. Middletown: Wesleyan University Press.

Giliomee, H. (2003) *The Afrikaners: Biography of a People*, London: Hurst and Company.

Grove, R. (1997) *Green Imperialism: Colonial Expansion, Tropical Island Edens and the Origin of Environmentalism, 1600-1860*, Cambridge: Cambridge University Press.

Guelke, L. (1988) 'Freehold farmers and frontier settlers, 1657–1780', in Elphick, R. and Giliomee, H. (eds) *The Shaping of South African Society, 1652–1840*, pp. 66–108. Middletown: Wesleyan University Press.

Heinrich, A. R. (2010) 'A Zooarchaeological Investigation into the Meat Industry Established at the Cape of Good Hope by the Dutch East India Company in the Seventeenth and Eighteenth Centuries'. Unpublished PhD dissertation, Rutgers University (Newark).

Heinrich, A. R. and Schrire, C. (2011) 'Colonial fauna at the Cape of Good Hope: a proxy for colonial impact on indigenous people', in Schablitsky, J. and Leone, M. (eds) *The Importance of Material Things*, Vol. 2, pp. 121–41. Washington: The Society for Historical Archaeology.

Idenburg, P. J. (1963) *The Cape of Good Hope at the Turn of the Eighteenth Century*, Leiden: Universitaire Pers.

Janowitz, M. (1993) 'Indian corn and Dutch pots: seventeenth-century foodways in New Amsterdam/New York', *Historical Archaeology*, 27, 6–24.

Kolben, P. (1731a) *The Present State of the Cape of Good Hope*. Vol. 1, *Containing the Natural History of the Cape, trans. Mr. Medley*, New York: Johnson Reprint Corporation.

Kolben, P. (1731b) *The Present State of the Cape of Good Hope*. Vol. 2, *Containing the Natural History of the Cape, trans. Mr. Medley*, New York: Johnson Reprint Corporation.

Landon, D. (1997) 'Interpreting urban food supply and distribution systems from the faunal assemblages: an example from colonial Massachusetts', *International Journal of Osteoarchaeology*, 7, 51–64.

Legassick, M. (1988) 'The northern frontier to c.1840: the rise and decline of the Griqua people', in Elphick, R. and Giliomee, H. (eds) *The Shaping of South African Society, 1652–1840*, pp. 358–420. Middletown: Wesleyan University Press.

Mentzel, O. F. (1921) *A Geographical and Topographical Description of the Cape of Good Hope, I, trans. H. J. Mandelbrote*, Cape Town: Van Riebeeck Society.

Mentzel, O. F. (1925) *A Geographical and Topographical Description of the Cape of Good Hope, II, trans. H. J. Mandelbrote*, Cape Town: Van Riebeeck Society.

Merwe, van der, P. J. (1995) *The Migrant Farmer in the History of the Cape Colony, 1657–1842. Roger Beck, translator*, Athens (Ohio): Ohio University Press.

Penn, N. (1986) 'Pastoralists and pastoralism in the northern Cape frontier zone during the eighteenth century', *South African Archaeological Society Goodwin Series*, 5, 62–8.

Penn, N. (2005) *The Forgotten Frontier: Colonist and Khoisan on the Cape's Northern Frontier in the Eighteenth Century*, Athens (Ohio): Ohio University Press.

Poggenpoel, C. A. (1996) 'The Exploitation of Fish during the Holocene in the South-Western Cape, South Africa'. Unpublished MA dissertation, University of Cape Town (Cape Town).

Raven-Hart, R. (1970) *Cape Good Hope 1652–1702: The First Fifty Years of Dutch Colonization as Seen by Callers*, Vol. II, Cape Town: Balkema.

Reitz, E. (1991) 'Animal use and culture change in Spanish Florida', in Crabtree, P. and Ryan, K. (eds) *MASCA Research Papers in Science and Archaeology, Supplement to Volume 8, 1991: Animal Use and Culture Change*, pp. 63–77. Philadelphia: University of Pennsylvania.

Reitz, E. and Scarry, M. (1985) *Reconstructing Historic Subsistence with an Example from Sixteenth-Century Spanish Florida*, Pleasant Hill: The Society for Historical Archaeology.

Ross, R. (1988) 'The Cape of Good Hope and the world economy, 1652–1835', in Elphick, R. and Giliomee, H. (eds) *The Shaping of South African Society, 1652–1840*, pp. 243–80. Middletown: Wesleyan University Press.

Ross, R. (1998) 'The first two centuries of colonial agriculture in the Cape colony: a historiographical review', in Lorimer, J. (ed.) *Settlement Patterns in Early Modern Colonization, 16th–18th Centuries*, pp. 301–16. Brookfield: Ashgate.

Ross, R. (1999) *Status and Respectability in the Cape Colony, 1750–1870*, Cambridge: Cambridge University Press.

Schrire, C. (1995) *Digging through Darkness: Chronicles of an Archaeologist*, Charlottesville: University of Virginia Press.

Schutte, G. (1988) 'Company and colonists at the Cape, 1652–1795', in Elphick, R. and Giliomee, H. (eds) *The Shaping of South African Society, 1652–1840*, pp. 283–323. Middletown: Wesleyan University Press.

Templin, J. A. (1984) *Ideology on a Frontier: The Theological Foundation of Afrikaner Nationalism, 1652–1910*, Westport: Greenwood Press.

Vries, de, J. and van der Woude, A. (1997) *The First Modern Economy: Success, Failure, and Perseverance of the Dutch Economy, 1500–1815*, Cambridge: Cambridge University Press.

PART V

NORTH AMERICA

CHAPTER 32

ZOOARCHAEOLOGY OF THE PRE-CONTACT NORTHWEST COAST OF NORTH AMERICA

GREGORY G. MONKS

INTRODUCTION

THE Northwest Coast of North America (NWC) is a culture area that extends from the Klamath River in northern California to Yakutat Bay in southeastern Alaska (Fig. 32.1). Key features in its definition include a fully maritime adaptation, social inequality, and abundant salmon use (Kroeber, 1939: 28–31). In effect, the culture area encompasses the coast from the Columbia River in the south to the Nass River in the north. The area's topography varies from a relatively linear open Pacific shoreline in Oregon and Washington to a highly irregular shoreline of islands, archipelagos, and fjords with mountains often descending precipitously into the sea. Temperature and rainfall vary considerably from south to north, but the ameliorating effect of the Pacific Ocean produces a generally cool-temperate environment with considerable rainfall. Forest vegetation and rapidly flowing rivers characterize the area. Aboriginal occupation of the area tended to focus on the littoral ecotone, although the geographic position of that ecotone has changed through the course of the Holocene because of post-Pleistocene eustatic and isostatic processes. The present coastline was achieved about 5000 BP, and much of the current archaeological knowledge falls after this date.

HISTORY

A history of zooarchaeological research in British Columbia was published twenty years ago (Driver, 1993), and Moss (2004) has provided a useful historical sketch of

Northwest
Coast

FIGURE 32.1 Map of North America showing the Northwest Coast culture area. Redrawn by the author from open access online map (Yahoo free maps North America).

archaeological work in southeast Alaska. Volumes on the general archaeology and prehistory of the area and its regions have also appeared in the same period (Lyman, 1991; Matson and Coupland, 1995; Ames and Maschner, 1999). Most recently, topical syntheses of zooarchaeological research on sea mammals (Braje and Rick, 2011) and fish (Moss and Cannon, 2011) have been published. None of these sources specifically addresses the zooarchaeology of the whole NWC, although Moss (2011a: 137–42) presents some directions in NWC zooarchaeological research. The present chapter will attempt to synthesize and expand on these existing contributions to formulate a useful areal overview.

The earliest recorded 'scientific' archaeology was conducted in the late 1800s and early 1900s at the Eburne midden in the southern part of Vancouver (Smith, 1903; Hill-Tout,

1895). Both Hill-Tout and Smith briefly described the animal remains they observed. The first systematic description of zooarchaeological remains was undertaken by Freed and Lane (1964) as part of Frederica de Laguna's 1949–1953 fieldwork at Yakutat Bay, Alaska (de Laguna, 1964). The report identifies faunal remains to species level, and quantifies the remains (NISP). Not until the late 1970s and early 1980s were detailed analyses similar to those of Freed and Lane applied to faunal remains (Stewart, 1977), although Gustafson (1968) specifically addressed northern fur seals. Mostly, mention of faunal remains was confined to 'laundry lists' of example taxa (Carlson, 1960; Mitchell, 1971; Ames, 1979).

Subsequent decades have seen an expansion of multiple interests in zooarchaeology, mostly pursued within a positivist paradigm. Early analyses of whole zooarchaeological assemblages have given way to more intensive scrutiny of selected aspects of those assemblages. The early attention to salmon continues to thrive, but the list of species now studied closely by zooarchaeologists has expanded dramatically to include many species of fish, shellfish, marine and terrestrial mammals, and birds. The methods used to examine zooarchaeological materials have expanded to include issues pertaining to field and laboratory sampling, advanced quantification, taphonomic processes, application of advanced analytic technology, integration of ethnographic and archaeological records, and interpretation of analytical results within different theoretical frameworks.

PEOPLE

The short history of NWC zooarchaeology, combined with the small community of researchers, make possible a consideration of individuals and their contributions in this area. Three generations of scholars have grown the study of zooarchaeology on the NWC. The first generation fostered zooarchaeology through direct or indirect support of it; the second generation directly focused on it; and the third generation continues to refine and advance zooarchaeology beyond the accomplishments of the second generation.

The first generation consisted of ethnographers and archaeologists such as George MacDonald (Archaeological Survey of Canada), William Folan (University of Colorado), Donald Mitchell (University of Victoria), Roy Carlson (Simon Fraser University), R. G. Matson (University of British Columbia), Richard Daugherty (Washington State University), Donald Grayson (University of Washington), and Wayne Suttles (Portland State University). Except for Grayson, none were zooarchaeologists, but they all trained and supported many of the second generation who were inspired by such concepts as cultural ecology, adaptive strategies, and procurement systems and by such works as Clarke's (1952) economic approach to prehistory.

The second generation emerged in the late 1970s and the 1980s and included Madonna Moss, Virginia Butler, Lee Lyman, David Huelsbeck, Dale Croes, Aubrey

Cannon, Gay Frederick, Becky Wigen, Susan Crockford, Jon Driver, Gary Coupland, and myself. This generation took methodological and theoretical approaches that were being developed elsewhere and applied them to faunal assemblages on the NWC. They also began to focus on specific aspects of faunal assemblages and led the way for the third generation's topical and technological analyses of zooarchaeological materials.

These researchers have provided a foundational corpus of zooarchaeological knowledge. Detailed identification and quantification of whole assemblages, in theses, dissertations, and publications (e.g. Moss, 1989; Cannon, 1991; Huelsbeck, 1994a; 1994b) began to show that older culture historical interpretations based on artefacts could be expanded and articulated with the natural environment. These full-assemblage analyses also quickly showed that, in shell-midden contexts where bone preservation was excellent and where fish bones existed in the tens of thousands, sampling was essential. The articulation with the natural environment became framed in terms of cultural ecological theory, subsistence strategies, and adaptive systems, all of which followed the dominant processualist-positivist approach of the time. The obvious need to sample large assemblages led not only to the concern with sampling but also with a concern about sample representativeness and interpretive reliability. Field sampling strategies were addressed by screen-mesh size, column sampling (Casteel, 1976), and bucket auger sampling (Cannon, 2000). Laboratory sampling of large collected assemblages has also been addressed (Lyman and Ames, 2004; Monks, 2004). A related issue was the concern about representativeness of faunal assemblages due to taphonomic processes (Butler and Chatters, 1994).

Ethnographies of NWC indigenous groups focused heavily on the importance of salmon, and early culture historical archaeologists tended to ignore other important resources. Salmon was indeed important, and it continues to receive scholarly attention (Moss et al., 1990; Cannon, 1991; 1996; 2000; 2001; Cannon and Yang 2006; Butler, 1993; Butler and O'Connor, 2004; Coupland et al., 2010); however, this 'salmonopia' (Monks, 1987) has recently been counteracted by interest in an increasingly wide range of taxa (e.g. Wigen and Stucki, 1988; Lyman, 1991; Bowers and Moss, 2001; Crockford et al., 2002; Monks et al., 2001; Moss, 2011b; McKechnie et al., 2014). Along with this broadening taxonomic interest has come a more sophisticated understanding of the zooarchaeological record and its cultural implications through the application of recent theoretical frameworks. Behavioural ecology, particularly optimization models, has been applied with interesting results. Butler and Campbell (2004) evaluated the proposition that resource depression of salmon had occurred in the central NWC area and adjacent interior regions, and they found no evidence for such an occurrence within the past 10,000 years. Also, Monks (2004) showed that the abundance of whale bones in archaeological sites on the west coast of Vancouver Island was directly correlated to their oil content.

The third generation of NWC zooarchaeologists is a larger and growing group. These researchers, alone or with second-generation researchers, have brought advanced technological studies of specific taxa to the fore. Ancient DNA analyses of salmon (Speller et al., 2005; Cannon and Yang, 2006), whales (Losey and Yang, 2007; Arndt, 2011), northern fur seals (Gifford-Gonzalez et al., 2005; Moss et al., 2006; Szpak et al., 2012) and herring (McKechnie et al., 2014) have provided valuable insights into the taxonomic identity of these species and the cultural reliance on them. Oxygen isotope analysis of clam shells (Hallmann et al., 2009; 2013; Burchell et al., 2013) has also provided details of both seasonal resource procurement and temporal and spatial environmental variability that were heretofore unattainable.

Comparative studies of taxonomic variability have yielded valuable information. The relationship between salmon and rockfish was discovered to be reciprocal on the major outer coastal islands (Frederick and Crockford, 2005; Orchard and Clark, 2005; Monks, 2006), whereas continuous emphasis on salmon appears to characterize those parts of the mainland near major rivers (Cannon and Yang, 2006; Coupland et al., 2010). An exception to this statement is the Hoko River wet and dry site complex where halibut were emphasized before salmon became dominant (Croes and Hackenberger, 1988). Elaboration of the comparative study of taxa has also brought about the study of food webs and their cultural implications (Szpak et al., 2009; 2013).

The addition of a historical ecology theoretical framework has also characterized NWC zooarchaeology in the last decade (e.g. Lyman, 1988). This framework positions zooarchaeologists to contribute a new understanding to the human–animal relationships of the past (e.g. chapters in Braje and Rick, 2011). Gender theory (Moss, 1993) and agency (Moss and Erlandson, 2002) have also been applied in NWC zooarchaeology, although these examples have yet to be widely emulated. As well, Moss (2011a) has applied a 'deep history' approach to understanding the pre-contact archaeology, including zooarchaeology, of the NWC.

Most archaeological knowledge of the NWC post-dates sea level stabilization at *c.*5000 BP, but geomorphological and zooarchaeologial studies of Haida Gwaii and Hecate Strait during the late Pleistocene and throughout the Holocene highlight the 'shifting baseline' problem (Fedje et al., 2001; Hetherington and Reid, 2003; McLaren et al., 2005). Sea levels as much as 150 m below present characterized the present Hecate Strait and Haida Gwaii between 15,000 and 10,400 BP, and mainland fauna traversed this plain and were captured by humans. Zooarchaeological analysis of Gaadu Din cave (13,000–10,800 BP) has confirmed the changes in animal availability over time in this area. Subsequently, relative sea levels rose to *c.*15 m above present sea level by 9000 BP, isolating the fauna of Haida Gwaii. The intertidal Kilgii Gwaii fauna (9300 BP) revealed that site occupants possessed technology that enabled them to harvest inshore and offshore fish, spawning salmon, land and sea mammals, and birds. After *c.*6000 BP, sea levels slowly dropped to their present levels by *c.*5000 BP (Josenhans et al., 1997; Fedje and Mathewes, 2005). This research has complemented interest, first, in waterlogged and intertidal sites that have long-term, excellent preservation of faunal remains (e.g. Hoko River, Ozette, Kilgii Gwaii) and, second, in alpine sites well away from the shoreline (e.g. Nagorsen et al., 1996).

Themes

Zooarchaeology continues to play an important role in the study of central themes in NWC anthropological archaeology. The interrelated themes of origins, storage, sedentism, social inequality, and culture-environment relations have been pursued throughout the history of research and by people discussed above.

Borden (1951) pointed out a number of material culture traits shared between NWC and 'Eskimoid' cultures of coastal Alaska. Subsequently, Matson (1976) included zooarchaeological remains in arguing, instead, that initial populations arrived post-glacially from the interior. Finally, Fladmark (1979) voiced the possibility of a coastal migration route from the Bering Land Bridge to unglaciated North America. The recent research in Haida Gwaii, cited above, indicates that a late Pleistocene coastal route was followed by accomplished maritime hunter-fisher-gatherers.

Storage and sedentism have heretofore been discussed together (Matson, 2006), but Cannon and Yang (2006) have recently argued for a separation of the two issues. They argue that sedentism was possible at Namu over the past 6,000 years, based on the sheer abundance of salmon there and that storage was not necessary to support the site occupants. This position is interesting and bears further examination elsewhere in order to evaluate the times and places at which this de-coupling of storage and sedentism may have applied. By implication, those areas without abundant salmon availability may be more likely to exhibit a linking of the two issues, although other resources need to be considered, e.g. halibut and other flatfish at Hoko River. Storage involves the quest for skeletal indicators in salmon and in other taxa. The cranial bones of salmon are thought not to preserve well (Hoffman et al., 2000), although Frederick and Crockford (2005) assert otherwise. The presence, absence, and abundance of salmon cranial and postcranial bones in sites thus remain debatable indicators of storage. Other issues that keep this issue murky are the uneven distribution and abundance of the five salmon species along the coast, the cultural preference for certain species for certain purposes, the varying practices for preparing salmon for storage, and the transportation of both fresh and stored salmon.

Sedentism has traditionally been thought to rely heavily on stored salmon (Ames, 1996), although, as noted above, this linkage has recently been questioned (Cannon and Yang, 2006). The search for proxy measures (site size, site distribution) and direct evidence (house depression size, excavated living floors) of sedentism often do not involve zooarchaeology directly, but the distribution of faunal remains on living floors has received initial attention. In the Strait of Georgia, the presence of extra-local salmon on a house floor at Dionisio Point site indicates possible storage and definite transport to a semi-sedentary dwelling (Grier, 2003). Similarly, zooarchaeological analysis of the Shingle Point site house floor suggested that two different families with unequal access to resources may have occupied the two compartments within the house (Matson, 2003). On western Vancouver Island, Frederick's (2012) analysis of faunal remains from a house floor at Huu7ii suggested that a sea mammal–focused activity area may be identified but that social differentiation of occupants was unclear. At the McNichol Creek site near Prince Rupert, Coupland et al. (2003) found a predominance of fish in one house and a predominance of land and sea mammals in another, and, within the fish assemblage of each house, salmon predominated, followed by herring. The abundance of food remains and the large house sizes involved in these analyses suggest that ethnographically described semi-sedentary winter villages were present on the NWC for some time in the past.

These and other zooarchaeological analyses also suggest that individual and corporate social inequality have been a feature of past NWC societies (see Coupland et al. 2016). While much has already been learned, greater understanding of the times, places, and circumstances in which ranked society emerged are still required, and zooarchaeology has a central role to play. This theme also holds the greatest potential to link NWC archaeology to more global discussions about the evolution of social complexity, e.g. coastal Peru, Mesolithic northern Europe, early Holocene southwest Asia. Following Orme (1977), I pointed out (Monks, 1987) that food chains and temporally and spatially limited resource blooms may have been instrumental in advancing sociocultural complexity (see Fig. 32.2).

Human relations with the environment underpin all these themes. Founded on functionalist cultural ecology theory, with later additions of human behavioural ecology and historical ecology, the intimate relationship between NWC human groups and their environments is most clearly seen in the zooarchaeological record. Because archaeological preservation of plant remains is uncommon except in carbonized and water-saturated form, animal bones are the main way in which researchers can see the panorama of human–environmental relations. Once viewed monolithically, these relations are now recognized as highly variable at the regional and local scales within the NWC culture area. As well, Suttles' (1962) identification of local, seasonal, annual, and unpredictable variation in resource abundance is now being fruitfully applied to analysis and interpretation in NWC zooarchaeology (e.g. Fedje and Mathewes, 2005; McMillan et al., 2008; Moss and Losey, 2011; Orchard and Szpak, 2011; Szpak et al., 2013). To this list will have to be added decadal, centennial, and millennial variations in environment and human response (e.g. Finney et al., 2002). Historical ecological studies (e.g. Braje and Rick, 2011) are a path in this direction.

SUMMARY

NWC zooarchaeology has made rapid strides in its short history. It has gone from relative invisibility before the mid-1970s to a globally integrated community of scholars working with leading-edge technology, methods, and theoretical constructs. Basic identification and quantification has led to understandings about settlement patterns,

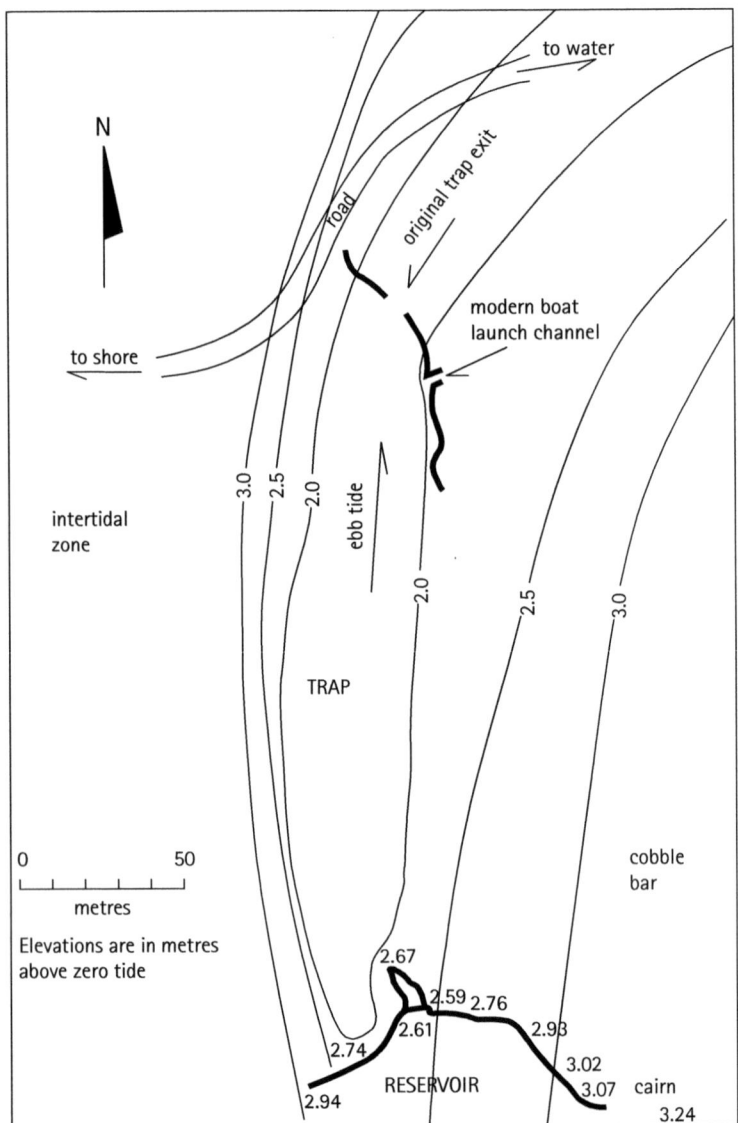

FIGURE 32.2 Plan of Deep Bay tidal herring trap, showing beach configuration and rock wall alignments forming the trap. Redrawn by the author after Monks 1987: Fig. 3. Used with permission of the Canadian Journal of Archaeology.

subsistence adaptations, seasonality and scheduling, storage, sedentism, the evolution of social inequality, and complex and variable interactions between human occupants of the area and their environments.

One encouraging and important trend is the attention paid by greater numbers of researchers to NWC zooarchaeology. Where once the number of researchers could be counted on one pair of hands, there are now dozens of scholars who occupy themselves with the subject. Hopefully, this trend will continue.

Another encouraging trend that should be emphasized more in the future is the rigorous application of theory in the study of zooarchaeological remains. These remains should be used to answer larger anthropological questions about past peoples and how they lived, organized themselves, and interacted with their environments.

Application of advanced analytic technologies to NWC zooarchaeological remains should be encouraged. Ongoing developments in isotopic analysis and molecular biology, for instance, should be constantly monitored and applied. Great strides in terms of oxygen isotope analysis and carbon/nitrogen analysis have already been made, as have studies of ancient DNA. Such studies should be expanded through application to a broader suite of culturally important resources. Another topic that should be expanded is the early geomorphology and human occupancy of ancient shorelines in areas outside Haida Gwaii. The islands of the Alaska Panhandle, Vancouver Island, and the coastal mainland have potential to reveal much about the zooarchaeology of the first half of the Holocene.

Archaeology in general provides what Moss (2011a) calls 'deep history', and zooarchaeology can provide a multi-faceted view of human subsistence and environmental relations over long expanses of time. Information of this type is valuable not only to archaeologists but also to such professionals as fisheries and wildlife managers, evolutionary ecologists, genomics researchers, and palaeoclimatologists. A new 'market' for zooarchaeology is thus opening up, and a reciprocal flow of benefits should be expected.

One area that needs to be addressed is the involvement of aboriginal peoples in zooarchaeology. It is their past that we study, and they have become actively involved in archaeology in the last four decades. We live in a post-colonial context, and Canadian provinces and American states on the NWC now require archaeologists to engage with local indigenous communities in the planning, conduct, interpretation, and reporting of archaeological work, but participation of aboriginal people in laboratory analysis, and particularly in zooarchaeology, remains limited.

REFERENCES

Ames, K. M. (1979) 'Report of excavations at GhSv2, Hagwilget Canyon', in MacDonald, G. F. and Inglis, R. I. (eds) *Skeena River Prehistory*. Mercury Series Paper 87, pp. 181–218. Ottawa: Archaeological Survey of Canada, National Museum of Man.

Ames, K. M. (1996) 'Life in the Big House: household labor and dwelling size on the Northwest Coast', in Coupland, G. and Banning, E. B. (eds) *People Who Lived in Big Houses: Archaeological Perspectives on Large Domestic Structures*. Monographs in World Archaeology 27, pp. 131–50. Madison: Prehistory Press.

Ames, K. M. and Maschner, H. D. G. (1999) *The Peoples of the Northwest Coast: Their Archaeology and Prehistory*, London: Thames and Hudson.

Arndt, U. M. (2011) 'Ancient DNA Analysis of Northeast Pacific Humpback Whale (*Megaptera novaeangliae*)'. Unpublished PhD dissertation, Simon Fraser University (Burnaby).

Borden, C. E. (1951) 'Facts and problems of Northwest Coast prehistory', *Anthropology in British Columbia*, 2, 35–52.

Bowers, P. M. and Moss, M. L. (2001) 'The North Point Wet Site and the subsistence importance of Pacific cod on the northern Northwest Coast', in Gerlach, S. C. and Murray, M. S. (eds) *People and Wildlife in Northern North America: Essays in Honor of R. Dale Guthrie*, pp. 159–77. Oxford: Archaeopress.

Braje, T. J. and Rick, T. (eds) (2011) *Human Impacts on Seals, Sea Lions and Sea Otters: Integrating Archaeology and Ecology in the Northeast Pacific*, Berkeley: University of California Press.

Burchell, M., Cannon, A., Hallmann, N., Schwarcz, H. P., and Schöne, B. R. (2013) 'Refining estimates for the season of shellfish collection on the Pacific Northwest Coast: applying high-resolution stable oxygen isotope analysis and sclerochronology', *Archaeometry*, 55(2), 258–76.

Butler, V. L. (1993) 'Natural versus cultural salmonid remains: origin of the Dalles Roadcut bones, Columbia River, Oregon, USA', *Journal of Archaeological Science*, 20, 1–24.

Butler, V. L. and Campbell, S. K. (2004) 'Resource intensification and resource depression in the Pacific Northwest of North America: a zooarchaeological review', *Journal of World Prehistory*, 18(4), 327–405.

Butler, V. L. and Chatters, J. C. (1994) 'The role of bone density in structuring prehistoric salmon bone assemblages', *Journal of Archaeological Science*, 21, 413–24.

Butler, V. L. and O'Connor, J. E. (2004) '9000 years of salmon fishing on the Columbia River, North America', *Quaternary Research*, 62, 1–8.

Cannon, A. (1991) *The Economic Prehistory of Namu: Patterns in Vertebrate Fauna*. Publication 19. Burnaby: SFU Archaeology Press.

Cannon, A. (1996) 'Scales of variability in Northwest Coast salmon fishing', in Plew, M. G. (ed.) *Prehistoric Hunter-Gatherer Fishing Strategies*, pp. 25–40. Boise: Department of Anthropology, Boise State University.

Cannon, A. (2000) 'Assessing variability in Northwest Coast salmon and herring fisheries: bucket-auger sampling of shell midden sites on the central coast of British Columbia', *Journal of Archaeological Science*, 27(8), 725–37.

Cannon, A. (2001) 'Was salmon important in Northwest Coast prehistory?', in Gerlach, S. C. and Murray, M. S. (eds) *People and Wildlife in Northern North America: Essays in Honor of R. Dale Guthrie*, pp. 178–87. Oxford: Archaeopress.

Cannon, A. and Yang, D. (2006) 'Early storage and sedentism on the Pacific Northwest Coast: ancient DNA analysis of salmon remains from Namu, British Columbia', *American Antiquity*, 71(1), 123–40.

Carlson, R. L. (1960) 'Chronology and culture change in the San Juan Islands, Washington', *American antiquity*, 25, 562–86.

Casteel, R. W. (1976) 'Comparison of column and whole unit samples for recovering fish remains', *World Archaeology*, 8(2), 192–6.

Clarke, J. G. D. (1952) *Prehistoric Europe: The Economic Basis*, London: Methuen.

Coupland, G., Colten, R. H., and Case, R. (2003) 'Preliminary analysis of socioeconomic organization at the McNichol Creek Site, British Columbia', in Matson, R. G., Coupland, G., and Mackie, Q. (eds) *Emerging from the Mist: Studies in Northwest Coast Culture History*, pp. 152–69. Vancouver: University of British Columbia Press.

Coupland, G., Stewart, K., and Patton, K. (2010) 'Do you never get tired of salmon? Evidence for extreme salmon specialization at Prince Rupert harbour, British Columbia', *Journal of Anthropological Archaeology*, 29(2), 189–207.

Coupland, G., Bilton, D., Clark, T., Cybulski, J., Frederick, G., Holland, A., Letham, B., and Williams, G. (2016) 'A wealth of beads: Evidence for material wealth-based inequality in the Salish Sea region, 4000–3500 cal BP'. *American Antiquity*, 81(2), 294–315.

Crockford, S. J., Frederick, S. G., and Wigen, R. J. (2002) 'The Cape Flattery fur seal: an extinct species of *Callorhinus* in the eastern North Pacific?', *Canadian Journal of Archaeology*, 26(2), 152–74.

Croes, D. R. and Hackenberger, S. (1988) 'Hoko River archaeological complex: modeling prehistoric Northwest Coast economic evolution', *Research in Economic Anthropology, Supplement*, 3, 19–85.

de Laguna, F. (1964) *The Archaeology of Yakutat Bay, Alaska*, Washington: Smithsonian Institution.

Driver, J. C. (1993) 'Zooarchaeology in British Columbia', *British Columbian Studies*, 99, 77–105.

Fedje, D. W. and Mathewes, R. (2005) *Haida Gwaii: Human History and Environment from the Time of Loon to the Time of the Iron People*, Vancouver: UBC Press.

Fedje, D. W., Wigen, R. J., Mackie, Q., Lake, C. R., and Sumpter, I. D. (2001) 'Preliminary results from investigations at Kilgii Gwaay: an Early Holocene archaeological site on Ellen Island, Haida Gwaii, British Columbia', *Canadian Journal of Archaeology*, 25(1–2), 98–120.

Finney, B. P., Gregory-Eaves, I., Douglas, M. S. V., and Smol, J. P. (2002) 'Fisheries productivity in the northeastern Pacific Ocean over the past 2,200 years', *Nature*, 416, 729–33.

Fladmark, K. (1979) 'Routes: alternative migration corridors for early man in North America', *American Antiquity*, 44(1), 55–69.

Frederick, S. G. (2012) 'Appendix A: vertebrate fauna from the Huu-ay-aht archaeology project. Results from the 2006 Huu7ii village excavations and summary of 2004 and 2006 data', in McMillan, A. D. and St. Claire, D. E. (eds) *Huu7ii: Household Archaeology at a Nuu-chah-nulth Village Site in Barkley Sound*, pp. 115–53. Burnaby: Archaeology Press, Simon Fraser University.

Frederick, S. G. and Crockford, S. J. (2005) 'Appendix D: analysis of the vertebrate fauna from Ts'ishaa village, DfSi-16, Benson Island', in McMillan, A. D. and St. Claire, D. E. (eds) *Ts'ishaa: Archaeology and Ethnography of a Nuu-chah-nulth Origin Site in Barkley Sound*, pp. 301–61. Victoria: Parks Canada and Tseshaht Nation.

Freed, J. A. and Lane, K. S. (1964) 'Analysis of faunal remains from Old Town, Knight Island', in de Laguna, F. (ed.) *The Archaeology of Yakutat Bay, Alaska*, pp. 77–84. Washington: Smithsonian Institution.

Gifford-Gonzalez, D. E., Newsome, S. D., Koch, P. L., Guilderson, T. P., Snodgrass, J. J., and Burton, R. K. (2005) 'Archaeofaunal insights on pinniped-human interactions in the northeastern Pacific', in Monks, G. G. (ed.) *The Exploitation and Cultural Importance of Sea Mammals*, pp. 19–38. Oxford: Oxbow Books.

Grier, C. (2003) 'Dimensions of regional interaction in the prehistoric Gulf of Georgia', in Matson, R. G., Coupland, G., and Mackie, Q. (eds) *Emerging from the Mist: Studies in Northwest Coast Culture History*, pp. 170–87. Vancouver: University of British Columbia Press.

Gustafson, C. E. (1968) 'Prehistoric use of fur seals: evidence from the Olympic Coast of Washington', *Science*, 161, 49–51.

Hallmann, N., Burchell, M., Brewster, N., Martindale, A., and Schöne, B. R. (2013) 'Holocene climate and seasonality of shell collection at the Dundas Islands Group, northern British Columbia, Canada: a bivalve sclerochronological approach', *Palaeogeography, Palaeoclimatology, Palaeoecology*, 373(0), 163–72.

Hallmann, N., Burchell, M., Schöne, B. R., Irvine, G. V., and Maxwell, D. (2009) 'High-resolution sclerochronological analysis of the bivalve mollusk *Saxidomus gigantea* from Alaska and British Columbia: techniques for revealing environmental archives and archaeological seasonality', *Journal of Archaeological Science*, 36(10), 2353–64.

Hetherington, R. and Reid, R. G. B. (2003) 'Malacological insights into the marine ecology and changing climate of the Late Pleistocene–Early Holocene Queen Charlotte Islands archipelago, western Canada, and implications for early peoples', *Canadian Journal of Zoology*, 81(4), 626–61.

Hill-Tout, C. (1895) 'Later prehistoric man in British Columbia', *Transactions of the Royal Society of Canada*, 1(2), 103–13.

Hoffman, B. W., Czederpiltz, J. M. C., and Partlow, M. A. (2000) 'Heads or tails: the zooarchaeology of Aleut salmon storage on Unimak Island, Alaska', *Journal of Archaeological Science*, 27, 699–708.

Huelsbeck, D. R. (1994a) 'The utilization of whales at Ozette, part V', in Samuels, S. R. (ed.) *Ozette Archaeological Project Research Reports*, Vol. II, *Fauna*, pp. 265–303: Pullman: Department of Anthropology, Washington State University.

Huelsbeck, D. R. (1994b) 'Mammals and fish in the subsistence economy of Ozette, part II', in Samuels, S. R. (ed.) *Ozette Archaeological Project Research Reports*, Volume II, *Fauna*, pp. 17–92. Pullman: Department of Anthropology, Washington State University.

Josenhans, H., Fedje, D., Plenitz, R., and Southon, J. (1997) 'Early humans and rapidly changing Holocene sea levels in the Queen Charlotte Islands—Hecate Strait, British Columbia, Canada', *Science*, 277, 71–4.

Kroeber, A. L. (1939) *Cultural and Natural Areas of North America*, Berkley: University of California Press.

Losey, R. and Yang, D. (2007) 'Opportunistic whale hunting on the southern Northwest Coast: ancient DNA, artifact, and ethnographic evidence', *American Antiquity*, 72(4), 657–76.

Lyman, R. L. (1988) 'Zoogeography of Oregon coast marine mammals: the last 3,000 years', *Marine Mammal Science*, 4(3), 247–64.

Lyman, R. L. (1991) *Prehistory of the Oregon Coast: The Effects of Excavation Strategies and Assemblage Size on Archaeological Inquiry*, San Diego: Academic Press.

Lyman, R. L. and Ames, K. M. (2004) 'Sampling to redundancy in zooarchaeology: lessons from the Portland Basin, northwestern Oregon and southwestern Washington', *Journal of Ethnobiology*, 24(2), 329–46.

Matson, R. G. (1976) *The Glenrose Cannery Site*, Ottawa: National Museum of Man.

Matson, R. G. (2003) 'The Coast Salish House: lessons from Shingle Point, Valdes Island, British Columbia', in Matson, R. G., Coupland, G., and Mackie, Q. (eds) *Emerging from the Mist: Studies in Northwest Coast Culture History*, pp. 76–104. Vancouver: University of British Columbia Press.

Matson, R. G. (2006) 'The Coming of the Stored Salmon Economy to Crescent Beach, B.C.' Paper presented at the 39th Annual Meeting of the Canadian Archaeological Association, Toronto.

Matson, R. G. and Coupland, G. (1995) *The Prehistory of the Northwest Coast*, San Diego: Academic Press.

McKechnie, I., Lepofsky, D., Moss, M. L., Butler, V. L., Orchard, T. J., Coupland, G., Foster, F., Caldwell, M., and Lertzman, K. (2014) 'Archaeological data provide alternative hypotheses on Pacific herring (*Clupea pallasii*) distribution, abundance, and variability', *Proceedings of the National Academy of Sciences of the United States of America*, 111(9), E807–E816.

McLaren, D., Wigen, R. J., Mackie, Q., and Fedje, D. W. (2005) 'Bear hunting at the Pleistocene/Holocene transition on the northern Northwest Coast of North America', *Canadian Zooarchaeology*, 22, 3–29.

McMillan, A. D., McKechnie, I., St. Claire, D. E., and Frederick, S. G. (2008) 'Exploring variability in maritime resource use on the Northwest Coast: a case study from Barkley Sound, Western Vancouver Island', *Canadian Journal of Archaeology*, 32(2), 214–38.

Mitchell, D. H. (1971) *Archaeology of the Gulf of Georgia Area: A Natural Region and Its Culture Types*, Victoria: University of Oregon.

Monks, G. G. (1987) 'Prey as bait: the Deep Bay example', *Canadian Journal of Archaeology*, 11, 119–42.

Monks, G. G., et al. (2001) 'Nuu-chah-nulth whaling: archaeological insights into antiquity, species preferences and cultural importance', *Arctic Anthropology*, 38(1), 60–81.

Monks, G. G. (2006) 'The fauna from Ma'acoah (DfSi-5), Vancouver Island, British Columbia: an interpretive summary', *Canadian Journal of Archaeology*, 30(2), 272–301.

Moss, M. L. (1989) 'Analysis of the vertebrate assemblage', in Davis, S. D. (ed.) *The Hidden Falls Site, Baranof Island, Alaska*, pp. 93–129. Anchorage: Alaska Anthropological Association Monograph Series.

Moss, M. L. (1993) 'Shellfish, gender and status on the Northwest Coast: reconciling archaeological, ethnographic, and ethnohistorical records of the Tlingit', *American Anthropologist*, 95(3), 631–52.

Moss, M. L. (2004) *Archaeological Investigations of Cape Addington Rockshelter: Human Occupation of the Rugged Seacoast on the Outer Prince of Wales Archipelago, Alaska*, Eugene: Department of Anthropology and Museum of Natural History, University of Oregon.

Moss, M. L. (2011a) *Northwest Coast: Archaeology as Deep History*, Washington: Society for American Archaeology.

Moss, M. L. (2011b) 'Pacific cod in southeast Alaska, the 'Cousin' of the fish that changed the world', in Moss, M. L. and Cannon, A. (eds) *The Archaeology of North Pacific Fisheries*, pp. 149–60. Fairbanks: University of Alaska Press.

Moss, M. L. and Cannon, A. (eds) (2011) *The Archaeology of North Pacific Fisheries*, Fairbanks: University of Alaska Press.

Moss, M. L. and Erlandson, J. M. (2002) 'Animal agency and coastal archaeology', *American Antiquity*, 67(2), 367–9.

Moss, M. L., Erlandson, J. M., and Stuckenrath, R. (1990) 'Wood stake wiers and salmon fishing on the Northwest Coast: evidence from southeast Alaska', *Canadian Journal of Archaeology*, 14, 143–58.

Moss, M. L. and Losey, R. (2011) 'Native American use of seals, sea lions, and sea otters in estuaries of northern Oregon and southern Washington', in Braje, T. J. and Rick, T. C. (eds) *Human Impacts on Seals, Sea Lions and Sea Otters: Integrating Archaeology and Ecology in the Northeast Pacific*, pp. 167–96. Berkeley: University of California Press.

Moss, M. L., Yang, D. Y., Newsome, S. D., Speller, C. F., McKechnie, I., McMillan, A. D., Losey, R., and Koch, P. L. (2006) 'Historical ecology and biogeography of North Pacific pinnipeds: isotopes and ancient DNA from three archaeological assemblages', *The Journal of Island and Coastal Archaeology*, 1(2), 165–90.

Nagorsen, D. W., Keddie, G., and Luszcz, T. (1996) *Vancouver Island Marmot Bones from Subalpine Caves: Archaeological and Biological Significance*, Victoria: B.C. Parks, Ministry of Environment, Lands, and Parks.

Orchard, T. and Clark, T. (2005) 'Multidimensional scaling of Northwest Coast faunal assemblages: a case study from southern Haida Gwaii, British Columbia', *Canadian Journal of Archaeology*, 29(1), 88–112.

Orchard, T. and Szpak, P. (2011) 'Identification of salmon species from archaeological remains on the Northwest Coast', in Moss, M. L. and Cannon, A. (eds) *The Archaeology of North Pacific Fisheries*, pp. 17–29. Anchorage: University of Alaska Press.

Orme, B. (1977) 'The advantages of agriculture', in Megaw, J. V. S. (ed.) *Hunters, Gatherers and First Farmers Beyond Europe: An Archaeological Survey*, pp. 41–9. Leicester: Leicester University Press.

Smith, H. I. (1903) 'Shell heaps of the Lower Fraser, British Columbia', *Memoirs of the American Museum of Natural History*, 3(4).

Speller, C. F., Yang, D. Y., and Hayden, B. (2005) 'Ancient DNA investigation of prehistoric salmon resource utilization at Keatley Creek, British Columbia, Canada', *Journal of Archaeological Science*, 32(9), 1378–89.

Stewart, F. (1977) 'Vertebrate Faunal Remains from the Boardwalk Site (GbTo-31) of Northern British Columbia'. Archaeological Survey of Canada Archives, Ottawa, Manuscript 1263.

Suttles, W. P. (1962) 'Variation in habitat and culture on the Northwest Coast', in Grunow, O. (ed.) *Proceedings of the 34th International Congress of Americanists*, pp. 522–37. Chicago: University of Chicago Press.

Szpak, P., Orchard, T. J., and Gröcke, D. R. (2009) 'A Late Holocene vertebrate food web from southern Haida Gwaii (Queen Charlotte Islands, British Columbia)', *Journal of Archaeological Science*, 36(12), 2734–41.

Szpak, P., Orchard, T. J., McKechnie, I., and Gröcke, D. R. (2012) 'Historical ecology of Late Holocene sea otters (*Enhydra lutris*) from northern British Columbia: isotopic and zooarchaeological perspectives', *Journal of Archaeological Science*, 39(5), 1553–71.

Szpak, P., Orchard, T. J., Salomon, A. K., and Gröcke, D. R. (2013) 'Regional ecological variability and impact of the maritime fur trade on nearshore ecosystems in southern Haida Gwaii (British Columbia, Canada): evidence from stable isotope analysis of rockfish (*Sebastes* spp.) bone collagen', *Archaeological and Anthropological Sciences*, 5(2), 159–82.

Wigen, R. J. and Stucki, B. R. (1988) 'Taphonomy and stratigraphy in the interpretation of economic patterns at Hoko River Rockshelter', *Research in Economic Anthropology, Supplement*, 3, 87–146.

CHAPTER 33

FAUNA AND THE EMERGENCE OF INTENSIVE AGRICULTURAL ECONOMIES IN THE UNITED STATES SOUTHWEST

REBECCA M. DEAN

INTRODUCTION

THE Hohokam of the Sonoran desert, southern Arizona, USA, created one of the most intensive agricultural economies in North America, based on the largest irrigation networks on the continent (Fig. 33.1; Table 33.1). The roots of Hohokam culture are found in the region's first agricultural societies, beginning in about 1,200 BC, but pottery and other artefacts with identifiably Hohokam style first developed around AD 650. At its peak, during the Classic period (AD 1150–1450) well over 200,000 people are estimated to have lived in the core Hohokam cultural region. With this high population density, significant complexity in the economic and social organization of Hohokam communities developed, as seen in evidence for social inequality and interdependent, specialized or semi-specialized trade networks (Bayman, 2001; Abbott, 2009).

ENVIRONMENTAL CONTEXT

Unlike similar agricultural societies in other regions of the world, the Hohokam had no domestic animals except dogs. Even turkeys, common in other parts of North America, were absent from the Sonoran desert; presumably they could not survive

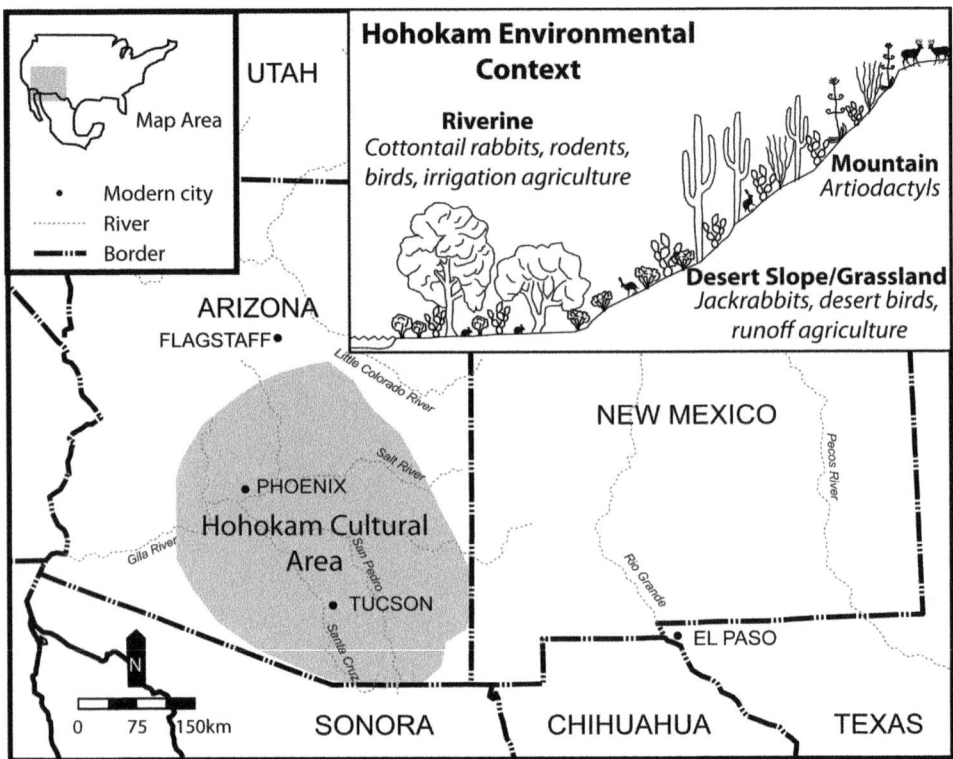

FIGURE 33.1 The extent of the Hohokam cultural area during the Classic period. Insert shows the environmental context of the Hohokam in the Basin and Range geographic zone of the United States southwest, where elevation is a key determinant of available resources. The most significant faunal and agricultural resources from each zone are indicated (with reference to Roth, 1996). Author's own image.

the heat, which can reach above 50°C in the summer. Without domestic animals, the Hohokam could not intensify their production of meat in the same way they could intensify their production of maize, beans, squash, and other plant staples. While agricultural yield per unit of land could be increased through the application of more time and energy (whether in the form of irrigation technology, terraces, domesticating local weeds, or simply spending more time caring for growing plants), the animal yield per unit of land could not be directly increased through the intensive breeding or care of domestic stock. The Hohokam were reliant on hunting for critical protein needs

The environmental context of the Sonoran desert posed unique challenges to Hohokam agriculture and hunting. Average rainfall in the Sonoran desert is only 300–400 mm each year in the eastern deserts, and less than 100 mm per year in the west (Dunbier, 1968: 6). This is well below the needs of the staple crops, so maize agriculture was only possible on floodplains (with or without irrigation canals),

Table 33.1 The cultural sequence of southern Arizona

Dates	Periods
AD 1450–1650	Protohistoric
AD 1150–1450	Classic
AD 950–1150	Sedentary
AD 750–950	Colonial
AD 650–750	Pioneer
AD 1–650	Early Ceramic
800 BC – AD 1	Early Agricultural (Cienega)
1,200–800 BC	Early Agricultural (San Pedro)
3,000–1,200 BC	Middle Archaic
8,000–3,000 BC	Early Archaic
10,000–8,000 BC	Palaeoindian

or in areas where rainfall runoff was channelled, such as along seasonal streams or talus slopes. As a result, Hohokam habitation was largely concentrated along the three permanent rivers of the region: the Santa Cruz, the Salt, and the Gila. By the Sedentary period, (AD 950–1150), an extensive network of irrigation canals along those rivers was absolutely essential for feeding the large and growing population (Bayman, 2001).

While the most critical and productive agricultural resources were located in the floodplains, river basins provided poor habitat for the four main artiodactyl species of the Sonoran desert, the only common large game found in the region: mule deer (*Odocoileus hemionus*), whitetail deer (*Odocoileus virginianus*), pronghorn antelope (*Antilocapra americana*), and bighorn sheep (*Ovis canadensis*). Although home to a diverse fauna, the river basins had mainly animals of small size, particularly jackrabbits (*Lepus* sp.), cottontail rabbits (*Sylvilagus* sp.), and a variety of rodents (Hoffmeister, 1986).

Habitats more favourable to large fauna are found away from rivers. The Sonoran desert is part of the Basin and Range Province of western North America, a series of north–south running mountain ranges crossing the western United States and northwestern Mexico. Significantly, different vegetation zones can be reached within a relatively short distance by moving away from the river and up in elevation (see insert in Fig. 33.1). Desert slopes and desert grasslands house deer, pronghorns, and bighorns, as well as valuable small-sized prey, such as jackrabbits and prairie dogs (*Cynomys gunnisoni*) (Hoffmeister, 1986; Roth 1996).

The first hunter-farmers in the Sonoran desert, then, faced a critical choice. They could focus on hunting, an activity that would take them away from the river basins and into the upland and off-river areas, but that would severely curtail their ability to grow crops. Or, they could focus on agriculture, which required their presence in the river basins, where larger preys were rare. Among the Hohokam, agricultural needs consistently won out, and this trade-off between agricultural and non-agricultural labour is critical to understanding hunting choices throughout the cultural sequence. Because of the constraints imposed by the landscape and agricultural labour demands, increasing meat yield occurred primarily in two ways: by embedding hunting activities in the agricultural exploitation of off-river landscapes; and by intentionally or unintentionally using irrigation to increase faunal yield per unit of land by creating agricultural habitats that supported populations of small game.

Socio-Economic Context

Table 33.1 presents a brief overview of the Hohokam cultural chronology. Agriculture was first introduced into the Sonoran desert in the Early Agricultural period. Despite the use of irrigation canals from the earliest agricultural periods, these first communities lack evidence for the socio-economic complexity and intensive cultivation techniques seen among the later Hohokam (Huckell, 1996). Three critical transitions are important for understanding Hohokam hunting patterns in their social and economic context: the first emergence of intensive agricultural systems in the Colonial period; the social and economic upheavals associated with the transition to the Classic period; the demographic decline (or 'collapse') of the Late Classic/post-Classic period.

The Colonial (AD 750–950) and Sedentary (AD 950–1150) Periods

In the Colonial period, an increase in habitation sites away from riverine environments, the logistic use of upland and desert habitats, the increasing size and formality of riverine sites, and overall population growth are all hallmarks of the first significant agricultural intensification in the Hohokam sequence (through investment in larger irrigation systems), as well as of agricultural diversification (increasing yield through the use of more extensive, far-flung, and often marginal agricultural land).

In the river valleys, especially along the Salt and Gila rivers, aggregated villages of 300–500 people were connected through complex canal systems (Cordell et al., 1994: 120; Bayman, 2001: 271). Sites became more formalized in their layout, which included true house clusters with multiple structures and outdoor work areas, more formal disposal patterns for trash, as well as central plazas and public architecture in the form of ball courts (Gumerman, 1991; Bayman, 2001: 271). Colonial period populations lengthened irrigation systems and brought more land into production to intensify agricultural adaptations (Gumerman, 1991: 8). Agricultural intensification also took the form of diversification of the plants exploited. Little barley (*Hordeum pusillum*), a wild, weedy plant that probably grew in or along fields, was increasingly encouraged and cultivated (Bohrer, 1991; Gasser and Kwiatkowski, 1991).

The Colonial period saw the expansion of Hohokam villages into off-river environments. Some sites in peripheral environmental zones represent seasonal logistic sites (Bayman, 2001: 273), but field-house and farmstead sites are seen in both riverine environments and in non-riverine areas, as either seasonal occupations of farmers who otherwise lived in larger villages, or small, year-round habitations (Henderson, 1989). Year-round or seasonal occupation of the fields would have been particularly important for run-off agricultural strategies, since diverting water to or from fields at the time of rainfall would have been critical (Fish and Nabhan, 1991).

The Sedentary period (AD 950–1150) continued the trends of agricultural expansion and population growth that had begun in the Colonial period. This furtherance of earlier economic and socio-political patterns came to an abrupt end during the transition from the Sedentary to the Classic periods.

Transition to the Classic Period (AD 1150–1450)

The Classic period transition was marked by significant social and economic reorganization, albeit a reorganization deeply rooted in economic trends already visible in the Colonial and Sedentary periods. House construction styles changed during the Classic period, from semi-subterranean pit-house architecture to mostly above-ground, adobe-walled constructions (Bayman, 2001: 281). Groups of adobe houses within villages formed compounds, usually enclosed by an adobe wall, and housing one or more extended families (Reid and Whittlesey, 1997: 101).

The large Classic period populations reorganized into aggregated communities with increasing evidence for social differentiation. Exotic materials were concentrated at larger sites, where platform mounds replaced ball courts as the monumental architecture of choice (Bayman, 2001: 285). These platform mounds were the focus of public ceremonies and, in the Late Classic period, rooms built on top of the platform mounds were used as residences, probably for elite families (Gumerman, 1991). Further evidence for social differentiation includes burials with variable amounts and qualities of grave goods (Fish and Fish, 1994: 26; Bayman, 2001: 287).

Changes in population organization had effects on subsistence practices. The integration of multiple sites into large communities led to the pooling of resources from multiple environmental zones, and the intensification of agricultural production in non-riverine settings. This pattern seems to be fully developed in the Classic period (e.g. Fish et al., 1992), but may have its roots in earlier Sedentary period settlement patterns. In later periods, these villages may have concentrated certain high-status food items, such as large game or cactus fruits, through trade and exchange networks (Gasser and Kwiatkowski, 1991; Bayman, 2001). This economic complexity allowed (and fuelled) specialization and semi-specialization in craft production (Abbott, 2009).

A wider variety of dry-land farming techniques were used to bring non-riverine areas into production, including check dams, rock piles, and reservoirs (Bayman, 2001: 275). A wider variety of domesticates were grown in the Sedentary and Classic periods as well, including cotton, barley grass, and tobacco. Cheno-ams, weedy annuals that thrive on disturbed, irrigated ground, had long been used by the Hohokam, but the first domestic forms appeared in the Middle Sedentary period (Gasser and Kwiatkowski, 1991). Wild plants, especially mesquite, agave, and cacti, continued to be an important component of Sedentary and Classic period economies (Gasser and Kwiatkowski, 1991; Bayman, 2001: 276).

The environmental impact of these Hohokam populations must have been great. Pollen and macro-botanical data suggest that weedy plants were thriving and possibly deliberately encouraged by the clearing and irrigation of land for agricultural fields (Fish, 1985; 2000; Gasser and Kwiatkowski, 1991). The extensive use of both riverine and non-riverine habitats would have also affected small-animal populations, including rodents and lagomorphs (Szuter, 1991; Dean, 2005a; 2007a; 2010).

The Late Classic Period and Hohokam 'Collapse'

Hohokam communities become archaeologically invisible after AD 1450. Palaeopathological studies suggest that the expansion of villages during the Classic period co-occurred with nutritional stress on the population as a whole (Sheridan, 2001; 2003). The large populations of the Classic period lived in a marginal environment for agricultural purposes, relying on extensive irrigation canals. Classic period Hohokam communities exploited multiple habitat zones and agricultural technologies within their territories to increase the odds of successful harvests (Fish and Fish, 1994). This complex social and agricultural system partially or completely collapsed after AD 1450. The details of this reduction in population are not clear, with debates continuing over the relative importance of catastrophic versus long-term demographic factors in explaining this phenomenon (Waters and Ravesloot, 2001; James, 2003; Hill et al., 2004; Dean, 2007b; Wasley and Doyel 2009).

Hohokam Hunting Intensification

Given the diversification and intensification of the agricultural system from the Colonial period through the post-Classic, it is not surprising that faunal remains reflect similar processes. This study uses a database of 118 faunal assemblages from across the agricultural sequence in southern Arizona to explore the nature and variability of hunting intensification.

The database was comprised of published analyses from Hohokam and pre-Hohokam sites, some of which were analysed by the author. All assemblages had a number of identified specimens (NISP) of at least one hundred, although logistic camp and farmstead/field-house sites were included if they had a NISP of at least fifty, in order to bring together a fully representative sample. Only those sites that were consistently screened were included, and where the majority of faunal remains was clearly the result of human introduction. The sites come from a variety of locations within the core area of the Hohokam culture, from large riverine villages to small desert hunting sites. Assemblages were assigned to the most specific time period possible, but where more than one phase was lumped by the original analyst, then all of the fauna was included in calculations relating to the last phase of site occupation.

Without considering the environmental and historical context of Hohokam hunting, simple diet breadth models would assume that hunting intensification (increasing meat yield through greater expenditure of energy by the hunter) would be seen in a growing proportion of small game in faunal assemblages. Diet breadth models are based on the premise that hunters seek the optimal investment in searching for and processing prey, such that their caloric returns are maximized relative to their caloric output. Since large animals have greater caloric returns than small animals, but require similar costs, large prey, such as deer, are normally considered higher ranked in hunters' prey preferences than small prey, such as rabbits. In other words, the model assumes that hunters seeking to maximize returns relative to the amount of energy invested will focus on large animals, only including small animals when higher-ranked animals are scarce or inadequate (Winterhalder, 1981; Bayham, 1982).

Intensification of the hunting economy is difficult to recognize in Hohokam assemblages because the fauna from the region is dominated by small animals, particularly rabbits, from the earliest agricultural periods, as a result of the landscape use choices that were imposed by the constraints of the agricultural system. Lagomorphs (cottontail rabbits and jackrabbits) made up more than 85% of NISP in three of the fifteen Early Agricultural period faunal assemblages in the database, and in five cases they reached over 90%. Hunting intensification among the Hohokam, therefore, required more than just increasing the diet breadth by including more small animals.

Assemblage diversity offers one measure of intensification in Hohokam hunting (Dean, 2007b). Faunal assemblages become more diverse when a wider variety of taxa are exploited. This is a reflection of diet breadth, because it is assumed that hunters who focus on a narrow range of prey tend to focus on the highest-return resource (for that specific socio-economic and environmental context), so exploiting a wider range of resources suggests that a larger number of lower-return taxa are included in the diet. Unlike measures of diet breadth that only consider the use of small *vs* large game, however, measures of faunal diversity do not assume that diets broaden through the inclusion of small game alone, and therefore they are more appropriate for the analysis of farming societies whose hunting decisions are constrained by factors unrelated by meat yield.

This study measures hunting diversification through the use of the Inverse of Simpson's Index of Evenness (Simpson, 1949; Levins, 1968). Evenness is a measure of prey dominance: was one prey species the focal target of hunting, or were a wide variety of different prey species used more or less equally? The Inverse of Simpson's Index of Evenness (B) is calculated as stated below, where NT = number of groups and p_i = proportion of the total number of individuals in group i:

$$B = 1 / \sum_{i=1}^{NT} p_i^2$$

The lowest possible value of the index is one, signifying that all fauna falls into only one category. Greater diversification in hunting strategies results in a more even distribution of faunal remains among the categories. Therefore, the index value will be higher, up to a maximum value equal to the number of categories used in the analysis.

The evenness index can be used simply by defining each group as discrete species. There are two problems with this approach using the current database. First, the published assemblages were analysed by a wide variety of researchers, many of whom used different levels of specificity in their analyses. For example, some analysts broke the artiodactyl category into four different species, where possible, while other analysts, perhaps because they lacked access to a sufficient comparative collection, lumped them all as 'artiodactyl'. Second, 343 different taxonomic groups were identified in assemblages in this database, but variations in sample size and specificity of analysis means that most of these taxa are absent from any specific site. An evenness analysis that defines groups simply as species will include a large number of empty cells.

An alternative to using species as the groups in calculating the index is to lump species into larger taxonomic or behavioural categories, allowing all assemblages to be used, and eliminating most empty categories. In this analysis, the identified faunal remains from each site were lumped into the following groups, based on similar habitats, size, and technologies required for successful exploitation: (1) lagomorphs (cottontail rabbits and jackrabbits); (2) artiodactyls (deer, pronghorn antelope, and bighorn sheep); (3) rodents (all rodents large enough to be likely prey species, including wood rats, *Neotoma* sp., cotton rats, *Sigmodon* sp., pocket gophers, *Thomomys bottae*, and squirrels from the genera *Spermophilus* and *Ammospermophilus*); (4) birds (all bird species large enough to be likely prey, including quail, *Callipepla* sp./*Cyrtonyx* sp., ducks and other members of the Anatidae, and pigeons, *Zenaidura* sp.); (5) riparian animals (all terrestrial or aquatic resources from riparian areas, including canal or river fish, muskrats, *Ondatra zibethicus*, beavers, *Castor canadensis*, and turtles—both mud turtles, Testudinidae and freshwater turtles, Emydidae). These categories reflect major groupings of hunted prey for the Hohokam, and, as will be discussed below, reflect changes that occurred in landscape use and labour organization across the Hohokam sequence.

Fig. 33.2 shows index values for all sites in the database, throughout the sequence of agricultural societies in the Sonoran desert (Table 33.2). As would be expected, evenness increases through time, and in particular increases with the Colonial period, the period of growing human populations and more intensive agricultural regimes. It is important to recognize, however, that while the *average* evenness of assemblages is increasing

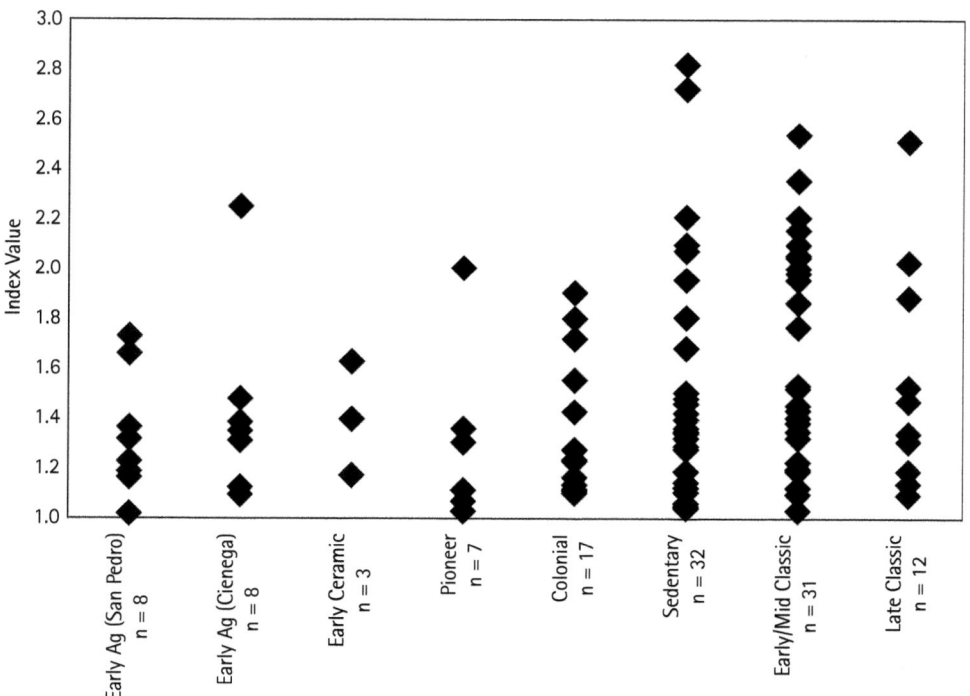

FIGURE 33.2 The Simpson's Inverse of Evenness Index values for the 118 faunal assemblages used in this analysis. At low values (close or equal to one), the index signifies that all fauna from an assemblage falls into one category. Greater evenness in the assemblage increases the index value, up to a maximum value of five. Author's own image.

Table 33.2 Average of Simpson's Inverse of Evenness Index values by time period

Time period	Average value	Number of assemblages
Late Classic	1.49	12
Early/Mid Classic	1.62	31
Sedentary	1.51	32
Colonial	1.33	17
Pioneer	1.27	7
Early Ceramic	1.40	3
Early Agricultural (Cienega)	1.39	8
Early Agricultural (San Pedro)	1.33	8

through time, a number of faunal assemblages from the later time periods continue to have highly uneven hunted assemblages. Not only is the evenness increasing, but so too is the variance. Some sites are diversifying their economies, while others are not. Hunting intensification, therefore, is not a universal correlate of agricultural intensification at Hohokam sites.

Furthermore, increasing evenness could be achieved through a variety of changes to hunting behaviours, from the relatively straightforward addition of small rodents and birds in the diet, to the increasing presence of artiodactyls due to changes in landscape use and labour organization in the later periods of the Hohokam chronology. Intensification through diversification of the diet was a complex process, one that was mitigated by the particular location of the site, by the use of the landscape dictated by the agricultural system, and by the availability and value of non-agricultural labour in a period of major social reorganization (Bayham, 1982; Speth and Scott, 1989; Dean, 2010).

Landscape Use and Variability in Hunting Intensification

A more detailed look at the type of hunting diversification (by taxonomic group and landscape location) reveals that much of the variability in index values that began in the Colonial period can be accounted for through changes in landscape use, including both the increased use of off-river environments for agricultural production, and the modifications to riverine environments for irrigation agriculture. Fig. 33.3 shows the average percentage out of total NISP for each of the categories of prey discussed above (birds and riparian animals were combined as 'other', since both were quite rare), divided into two sub-samples: riverine villages and off-river sites. The sub-samples highlight the importance of economic complexity and integration in understanding Hohokam hunting intensification, as well as the different paths toward intensification taken by communities in different environmental and economic contexts.

Large Game Use

The increased hunting of large game was a major component of diversification in later Hohokam communities. In riverine sites (Fig. 33.3a), artiodactyls are more common during Early Classic and Middle Classic times, corresponding to major reorganizations of the economy and social order. This increase is seen in a number of sites, including large, core platform-mound sites such as Muchas Casas and Las Casitas (Sparling, 1974; James, 1987). Since most sites in these core regions show no increase in artiodactyl remains, this is not a reflection of improved artiodactyl habitat. In fact, such an improvement would be unusual, given the rapidly growing human populations in the river valleys.

Instead, a variety of social and economic factors fuel this change in the Early Classic period. The most critical is the increasing use of upland environments, where large game was more abundant. Beginning in the Colonial period, but significantly

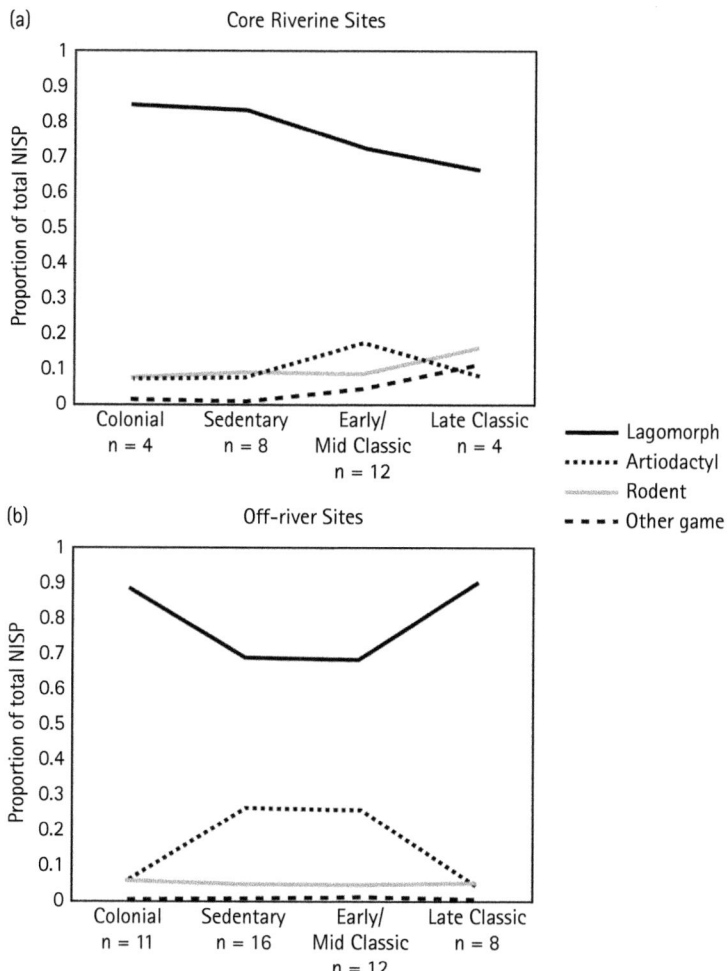

FIGURE 33.3 Change through time in the percentage of major prey taxa out of total NISP for faunal assemblages from A) core riverine sites and B) off-river sites, located along washes or on desert slopes. The time periods shown (Colonial through Classic) cover the period of agricultural and hunting intensification during the Hohokam cultural sequence. Author's own image.

increasing into the Early Classic period, off-river agricultural and habitation sites were economically tied to core villages on the floodplain. These integrated economies allowed large riverine populations to reap the fruits of the desert slopes, and promoted the survival of small communities in more marginal environments. The off-river villages, located on small, seasonal streams and washes, were largely absent from the sample prior to the Colonial period; indeed, only three of the fifty off-river assemblages date to before AD 750. Their very existence reflects the diversification of the economy overall.

Fig. 33.3b shows changes in the average proportions of major taxonomic groups in off-river villages and habitations from the Colonial through Classic periods. Rabbits dominate the off-river assemblages and, unlike riverine sites, these off-river sites show no change through time in the proportion of very small game (rodents and 'other', namely birds and riparian animals). Instead, large game is significantly more common in off-river sites than in river valley sites. Economic integration with riverine sites, then, likely explains much of the increase in artiodactyls seen in some river-valley platform-mound centres in the Early and Middle Classic period.

Landscape use alone is insufficient to understand the growing presence of artiodactyls in either off-river or riverine sites. There are no environmental indications of ameliorating artiodactyl habitat between the Colonial and Sedentary periods, nor of deteriorating artiodactyl habitat in the Late Classic period; yet, off-river sites contain fewer artiodactyl bones from those time periods. Increased demand for artiodactyls, despite the high search costs involved in their capture, was important in fuelling the shifting focus toward large game. Artiodactyls were a critical component of feasts, and their bones and skins were used for ritual paraphernalia. These products were particularly valuable in the Early Classic period, when major social reorganization led to competition among potential community leaders (Dean, 2005b; 2007b; Grimstead and Bayham, 2010).

Furthermore, the major social reorganization of the Classic period allowed increasing specialization (or semi-specialization) in hunting (Bayham, 1982; Speth and Scott, 1989). Higher populations and more intensive agricultural technology allowed some individuals freedom from agricultural pursuits, particularly those individuals whose farms were located on marginal land, like that around off-river villages. This level of specialization was possible during the peak of Sedentary and Classic period economic integration, but fell off in the Late Classic period, as community stress broke economic integration apart.

Hunting diversification through the use of artiodactyls, then, was mitigated by a variety of factors: the increasing use of upland environments, where larger game was more abundant, as agricultural and habitation zones tied into economic systems centred on core riverine sites; the reorganization of labour to allow for greater specialization in hunting; and the increasing demand for artiodactyls as important components in feasting and ritual at a time when social reorganization made potential leaders particularly willing to pay the high costs for such resources.

Small Game Use

In off-river sites, diversification took place largely through the increased hunting of artiodactyls. There is no evidence for the increasing use of small game (rodents, birds, and small riparian animals). This is not the case for sites in the core river valleys. Fig. 33.3a shows changes in the average proportions out of total NISP of major taxonomic groups in riverine villages and habitations from the Colonial through Classic periods (Table 33.3). Sites near and on major rivers were used throughout the cultural sequence.

Table 33.3 Average percentage out of total NISP (Number of Identified Specimens) for four categories of game

Site type	Time period	% lagomorph	% artiodactyl	% rodent	% other	Number of assemblages
Riverine sites	Late Classic	65.2	8.3	15.8	10.7	4
	Early/Mid Classic	71.4	17.1	8.7	4.8	12
	Sedentary	81.4	8.1	9.0	1.5	8
	Colonial	83.7	7.3	7.3	1.7	4
Off-river sites	Late Classic	89.5	4.4	5.4	0.8	8
	Early/Mid Classic	67.7	26.1	5.0	1.3	12
	Sedentary	68.3	26.1	5.0	0.6	16
	Colonial	87.9	5.9	5.9	0.3	11

River valleys supported the highest populations, saw the largest population increases, and were the centre of the major social and economic restructuring that took place in the Classic period. Although all sites are dominated by rabbits, the riverine village sites show an increase in even smaller game through time. These changes fit the classic expectations of a diet breadth model: the focus of the hunting economy (in this case rabbits) decreases in importance with the growth of human populations and the diet is supplemented with even smaller animals.

However, this increase in small game is also a reflection of human-made changes to the landscape, in particular the creation of vast, artificial riparian zones through the use of wide-spread irrigation (Dean, 2005a). Irrigation agriculture created habitats for small animals that otherwise would not have existed. Agricultural activities aimed at creating optimal environments for staple plant foods also created environments for certain animals, particularly cottontail rabbits and some small rodent species (Dean, 2005a; 2010). Increased access to small game as a result of irrigation helps to explain why riverine assemblages show an increase in small animals, while there is no concurrent increase in off-river sites, where different agricultural regimes were prevalent. It also explains why the increase in rodents does not include a significant component of larger desert rodents, like prairie dogs, that would have been prized due to their size, but which are normally found in desert grasslands. Only three prairie dog bones were identified during the Late Classic period, all from the site of Pueblo Salado (Stratton, 1995; Goodman, 1996).

Also for small game, landscape modifications alone cannot explain the different patterns of exploitation. This increase in very small game in river valley sites does not begin until the Classic period, despite the general trend toward diversification seen in faunal assemblages starting in the Colonial period, and despite the presence of irrigation agriculture since the beginning of the sequence. If the increased hunting of large game was a complex phenomenon, mediated by a number of social, economic, and environmental factors, then so too was the use of small game. As shown in Fig. 33.3a, rodents, riverine animals, and birds all increase notably over time in riverine sites. This increase is greatest in the Late Classic period, when, as discussed above, artiodactyl resources from off-river sites no longer made their way into riverine communities in the same frequency.

Conclusions

Hunting economies intensified along with agricultural economies in the desert southwest, but intensification (as measured through the diversification of hunting strategies) was mitigated and expressed through a variety of processes, not all of which are easily understood by more traditional methods of measuring intensification. The economic integration of core and periphery sites, which spanned riverine and off-river environments, allowed intensification through the inclusion of large game from areas marginally suited to agriculture, while the increasing use of this marginal land, at the same time as economic and social complexity increased both demand and market support for hunted products, fuelled the growth of semi-specialization in hunting. Even the use of very small animals, namely rodents and birds, as part of the intensification process, cannot be understood without reference to the significant changes that took place on the Hohokam landscape during the period of emerging intensive agricultural economies. Small game hunting was part of a larger economic strategy of irrigation agriculture, which effectively created thousands of acres of riparian habitat in what otherwise would have been desert. These habitats increased the availability of certain types of game, while limiting the presence of others.

In sum, the changes that occurred in the hunting economy during the emergence of intensive agricultural in the southwestern United States were complex, and critically tied to decisions that were made to increase agricultural yield, either through the intensification of irrigation agriculture or the diversification of landscape use. The hunting behaviour of the Hohokam cannot be understood solely in its own terms, as a product of optimal decision-making based on the availability of prey in the landscape at large. Rather, the decisions were highly contextualized within the particular constraints of the social and labour organization of the agricultural system, and they were contingent on the changes that had been made to that landscape as a result of agricultural demands.

References

Abbott, D. (2009) 'Extensive and long-term specialization: Hohokam ceramic production in the Phoenix Basin, Arizona', *American Antiquity*, 74, 531–7.

Bayham, F. (1982) 'A Diachronic Analysis of Prehistoric Animal Exploitation at Ventana Cave'. Unpublished PhD dissertation, Arizona State University (Tempe).

Bayman, J. (2001) 'The Hohokam of Southwest North America', *Journal of World Prehistory*, 15, 257–311.

Bohrer, V. (1991) 'Recently Recognized and Encouraged Plants among the Hohokam', *Kiva*, 56, 227–35.

Cordell, L., Doyel, D., and Kintigh, K. (1994) 'Processes of aggregation in the Prehistoric southwest', in Gumerman, G. (ed.) *Themes in Southwestern Prehistory*, pp. 109–34. Santa Fe: School for American Research Press.

Dean, R. (2005a) 'Site use intensity, cultural modification of the environment, and the development of agricultural communities in southern Arizona', *American Antiquity*, 70, 403–31.

Dean, R. (2005b) 'Old bones: the effects of curation and exchange on the interpretation of artiodactyl remains in Hohokam sites', *Kiva*, 70, 255–72.

Dean, R. (2007a) 'The lagomorph index: rethinking rabbit ratios in Hohokam sites', *Kiva*, 73, 7–30.

Dean, R. (2007b) 'Hunting intensification and the Hohokam "collapse"', *Journal of Anthropological Archaeology*, 26, 109–32.

Dean, R. (2010) 'The effect of cultivation techniques on small game populations: an archaeological example from the Hohokam region', in Dean, R. (ed.) *The Archaeology of Anthropogenic Environments*, pp. 250–65. Carbondale: Center for Archaeological Investigations Press.

Dunbier, R. (1968) *The Sonoran Desert: Its Geography, Economy, and People*, Tucson: University of Arizona Press.

Fish, S. (1985) 'Prehistoric disturbance flora of the lower Sonoran Desert and their implications', in Jacobs, B., Fall, P., and Davis, O. (eds) *Late Quaternary Vegetation and Climates of the American Southwest*, pp. 77–88. Dallas: American Association of Stratigraphic Palynologists Foundation.

Fish, S. (2000) 'Hohokam impacts on Sonoran Desert environment', in Lentz, D. (ed.) *Imperfect Balance: Landscape Transformations in the Precolumbian Americas*, pp. 251–80. New York: Columbia University Press.

Fish, S. and Fish, P. (1994) 'Prehistoric desert farmers of the Southwest', *Annual Review of Anthropology*, 23, 83–108.

Fish, S., Fish, P., and Madsen, J. (eds) (1992) *The Marana Community in the Hohokam World*, Tucson: University of Arizona Press.

Fish, S. and Nabhan, G. (1991) 'Desert as context: the Hohokam environment', in Gumerman, G. (ed.) *Exploring the Hohokam: Prehistoric Desert Peoples of the Southwest*, pp. 29–60. Albuquerque: University of New Mexico Press.

Gasser, R. and Kwiatkowski, S. (1991) 'Food for thought: recognizing patterns in Hohokam subsistence', in Gumerman, G. (ed.) *Exploring the Hohokam: Prehistoric Desert Peoples of the Southwest*, pp. 417–59. Albuquerque: University of New Mexico Press.

Goodman, J. (1996) 'Faunal remains from Areas 6 and 15, Pueblo Salado', in Greenwald, D., Ballagh, J., Mitchell, D., and Anduze, R. (eds) *Life on the Floodplain: Further Investigations at Pueblo Salado for Phoenix Sky Harbor International Airport*. Volume 2: *Data Recovery and Re-evaluation*, pp. 245–470. Phoenix: City of Phoenix Parks, Recreation and Library Department.

Grimstead, D. and Bayham, F. (2010) 'Evolutionary ecology, elite feasting, and the Hohokam: a case study from a southern Arizona platform mound', *American Antiquity*, 75, 841–64.

Gumerman, G. (1991) 'Understanding the Hohokam', in Gumerman, G. (ed.) *Exploring the Hohokam: Prehistoric Desert Peoples of the American Southwest*, pp. 1–27. Albuquerque: University of New Mexico Press.

Henderson, K. (1989) 'Farmsteads to fieldhouses: the evidence from La Cuenca del Sedimento', in Henderson, K. (ed.) *Prehistoric Agricultural Activities on the Lehi-Mesa Terrace: Excavations at La Cuenca del Sedimento*, pp. 334–57. Flagstaff: Northland Research.

Hill, B., Clark, J., Doelle, W., and Lyons, P. (2004) 'Prehistoric demography in the Southwest: migration, coalescence, and Hohokam population decline', *American Antiquity*, 69, 689–716.

Hoffmeister, D. (1986) *Mammals of Arizona*, Tucson: University of Arizona Press.

Huckell, B. (1996) 'The Archaic Prehistory of the North American Southwest', *Journal of World Prehistory*, 10, 305–373.

James, S. (1987) 'Hohokam patterns of faunal exploitation at Muchas Casas', in Rice, G. (ed.) *Studies in the Hohokam Community of Marana*, pp. 171–286. Tempe: Office of Cultural Resource Management, Department of Anthropology, Arizona State University.

James, S. (2003) 'Hunting and fishing patterns leading to resource depletion', in Abbott, D. (ed.) *Centuries of Decline during the Hohokam Classic Period at Pueblo Grande*, pp. 70–81. Tucson: University of Arizona Press.

Levins, R. (1968) *Evolution in Changing Environments*, Princeton: Princeton University Press.

Reid, J. and Whittlesey, S. (1997) *The Archaeology of Ancient Arizona*, Tucson: University of Arizona Press.

Roth, B. (1996) 'Regional land use in the Late Archaic of the Tucson Basin: a view from the Upper Bajada', in Roth, B. (ed.) *Early Formative Adaptations in the Southern Southwest*, pp. 37–48. Madison: Prehistory Press.

Sheridan, S. (2001) 'Morbidity and mortality in a Classic-period Hohokam community', in Mitchell, D. and Brunson-Hadley, J. (eds) *Ancient Burial Practices in the American Southwest: Archaeology, Physical Anthropology and Native American Perspectives*, pp. 191–222. Albuquerque: University of New Mexico.

Sheridan, S. (2003) 'Childhood health as an indicator of biological stress', in Abbott, D. (ed.) *Centuries of Decline during the Hohokam Classic Period at Pueblo Grande*, pp. 82–106. Tucson: University of Arizona Press.

Simpson, E. (1949) 'Measurement of diversity', *Nature*, 163, 688.

Sparling, J. (1974) 'Analysis of faunal remains from the Escalante Ruin Group', in Doyel, D. (ed.) *Excavations in the Escalante Ruin Group, Southern Arizona*, pp. 215–53. Tucson: Arizona State Museum, University of Arizona.

Speth, J. and Scott, S. (1989) 'Horticulture and large-mammal hunting: the role of resource depletion and the constraints of time and labor' in Kent, S. (ed.) *Farmers as Hunters: The Implications of Sedentism*, pp. 71–9. Cambridge: Cambridge University Press.

Stratton, S. (1995) 'Faunal analysis', in Greenwald, D. H., Chenault, M., and Greenwald, D. M. (eds) *The Sky Harbor Project: Early Desert Farming and Irrigation Settlements*. Vol. 3: *Pueblo Salado*, pp. 321–34. Flagstaff: SWCA.

Szuter, C. (1991) *Hunting by Prehistoric Horticulturists in the American Southwest*, New York: Garland.

Wasley, W. and Doyel, D. (2009) 'Classic period Hohokam', *Kiva*, 75, 193–208.

Waters, M. and Ravesloot, J. (2001) 'Landscape change and the cultural evolution of the Hohokam along the Middle Gila river and other river valleys in south-central Arizona', *American Antiquity*, 66, 285–99.

Winterhalder, B. (1981) 'Optimal foraging strategies and hunter-gatherer research in anthropology: theory and models', in Winterhalder, B. and Smith, E. (eds) *Hunter-Gatherer Foraging Strategies*, pp. 13–35. Chicago: University of Chicago Press.

CHAPTER 34

13,000 YEARS OF COMMUNAL BISON HUNTING IN WESTERN NORTH AMERICA

JOHN D. SPETH

INTRODUCTION

THE bison (or 'buffalo') is a familiar North American symbol, conjuring up images of explorers, frontier days, and our Native American heritage (Fig. 34.1). Bison appear on coins, stamps, flags, and the insignia of the US National Park Service. Countless geographical features incorporate the word 'buffalo' in their name, and at least twenty-four states have towns or cities named after the buffalo.

Although bison were abundant in the Canadian boreal forests, and the woodlands and prairies of the eastern United States, by far the greatest numbers were found in the Great Plains. This vast grassland encompasses over 1,800,000 km^2, extending about 2,575 km north–south and 645 km east–west (Fig. 34.2). Most sources estimate that bison once numbered between thirty and sixty million animals (Gates et al., 2010).

Modern bison (*Bison bison*) are the largest living native mammal in North America, with adult bulls weighing 544–907 kg and females 318–545 kg. Throughout most of the year females, calves, and young males are together in cow–calf groups which often contain dozens and at times hundreds of animals. Mature bulls remain solitary or in small 'bull groups' for most of the year. Only during the late summer rut (mid-July to late August/early September) do the bulls join the cow–calf groups and attempt to isolate cows in order to mate. After a gestation period of about 285 days, cows give birth in the spring (April/May in the Northern Plains, March/April in the Southern Plains). Calves are weaned in the fall, after nursing for seven to eight months. Bison have relatively poor eyesight, but a keen sense of smell, a fact which hunters took into account when approaching a herd (Lott,

FIGURE 34.1 American bison (*Bison bison*). Photo by Jack Dykinga, released in the public domain by the Agricultural Research Service, United States Department of Agriculture (Image ID K5680-1).

2002). Despite their size, bison are agile and fast, capable of running at speeds up to 56 km per hour, equivalent to running the 100 m dash in 6.4 seconds.

Many nineteenth-century Euroamerican observers believed that bison migrated annually over vast distances, summering in the Northern Plains and wintering in the Southern Plains (Roe, 1970). Until recently there was no effective way to test these reconstructions. However, stable isotope analyses ($^{87}Sr/^{86}Sr$, $\delta^{13}C$, $\delta^{18}O$) now offer objective ways to examine the migration hypothesis, revealing a far more complex reality (Chisholm et al., 1986; Graves, 2010; Widga et al., 2010; Carlson and Bement, 2013). Most bison did migrate annually, and some did so over considerable distances, but on the order of tens to hundreds of kilometres, not thousands. Moreover, migrations were not necessarily annual events, nor did every animal participate, some remaining resident in an area for extended periods. In addition, not all movements were north–south, and some involved annual moves to higher elevations. Many factors influenced migratory patterns, including forage quality, competition with other herbivores, snow conditions, insects, predator and human hunting pressure, and range fires. Lightning ignited many fires, but some were deliberately set by Native Americans. In certain seasons burning enhanced forage quality and attracted animals, but at other seasons it reduced forage quality, keeping herds away (Loscheider, 1977). Thus, Indians used fire as a political tool to keep bison away from competitors and enemies—both Native and Euroamerican.

FIGURE 34.2 Map of the extermination of the American bison to 1889. Intermediate gray: original range, dark gray: range in 1870; black: range in 1889; black numbers: estimated number of bison in 1889; white numbers: date of local extermination. Adapted from a drawing by William T. Hornaday in Scobel (1902). Digital map by Cephas.

Taxonomy

Bison (*Bison bison*) today are represented by two subspecies—*B. b. bison*, the plains bison, and *B. b. athabascae*, the wood bison (McDonald, 1981; Shapiro et al., 2004; Gates et al., 2010). Bison originated in the Old World, entering Beringia between 300,000 and 130,000 BP, long before humans were present in North America. During the last interglacial—130,000–75,000 BP—Beringian steppe bison (*B. priscus*) expanded south of the ice sheets into the North American grasslands, evolving into a large, long-horned variety, *B. latifrons*. During the Last Glacial Maximum (LGM ~20,000 BP), coalescence of the Laurentide and Cordilleran ice sheets separated Beringian and Plains bison populations, the latter evolving into *B. antiquus*, the first bison to be hunted by Native Americans ('Palaeoindians'). These early hunters had entered Beringia from Siberia sometime after the LGM and by 14,000 BP expanded southward, perhaps via the Pacific coast, into the mid-continent. As the ice sheets retreated, an 'ice-free corridor' opened between them that allowed *B. antiquus* to expand northward. Palaeoindian hunters apparently moved northward as well, exploiting these populations of Plains bison (Wilson et al., 2008). However, rapid colonization of the corridor by boreal forests precluded substantial mixing between *B. antiquus* and Beringian bison.

B. antiquus was 20–25% larger than modern *B. bison*. Interestingly, while other 'megafauna' (e.g. mammoths (*Mammuthus* sp.), mastodons (*Mammut americanum*), horses (Equidae), camels (*Camelops* sp.)) went extinct in North America by the end of the Pleistocene, bison did not. Instead, early in the Holocene they underwent a decline in body size and changes in horn size and morphology (Lewis et al., 2007; Hill et al., 2008; Widga, 2013). The body-size decline may reflect decreasing forage quality as the climate became warmer and drier. Changes in the horns, which play a key role in agonistic interactions between bulls, likely reflect a shift in bison social organization, from smaller, less cohesive aggregations of animals to larger, more hierarchically structured herds, a change that human hunters quickly learned to exploit.

EXTERMINATION

Indians relied heavily on bison for food and shelter. Nevertheless, while they undoubtedly killed large numbers of bison annually to meet their many and varied needs, it is doubtful their level of exploitation would ever have driven bison to extinction. Tragically, that 'honour' belongs to Euroamericans (see Fig. 34.2). By 1830 bison were gone east of the Mississippi River, and by 1874 the Southern Plains herd was exterminated. The Northern Plains herd suffered a similar fate a decade later (1880–1884). Of the millions of bison that inhabited North America in 1800, fewer than six hundred remained in 1890.

Once the slaughter was over the Plains were littered with rotting carcasses, and a business sprang up scavenging the bones for the fertilizer and bone char industry (Fig. 34.3). However, after about twenty years the bone supply was exhausted and the business declined, only to resurface briefly during the Depression of the 1930s, when local residents found they could earn supplemental income by mining bison kills (Davis, 1978). In the process, many kill sites were damaged or obliterated.

FIGURE 34.3 Pile of bison skulls awaiting processing into fertilizer and bone char used for refining sugar and bone black, Michigan Carbon Works, Detroit, Michigan, late 19th C. Image (ID DPA4901600jpg) courtesy of the Burton Historical Collection, Detroit Public Library.

COMMUNAL BISON HUNTING: THE ETHNOHISTORIC RECORD

For readers not familiar with North American prehistory, horses (*Equus caballus*) were not part of the picture until the late sixteenth century, when they were introduced by the Spaniards. Thus, prehistoric bison hunting was done on foot and without firearms. Early on, Native Americans hunted with spears and atlatls (spear-throwers), the bow and arrow appearing early in the first millennium AD (Blitz, 1988).

Much of what we know about prehistoric communal bison hunting is informed by the rich eighteenth- and nineteenth-century historic record (Roe, 1970; Wheat, 1972; Bamforth, 1988; Haines, 1995; Brink, 2008). These invaluable eyewitness accounts tell us in varying detail how Indians conducted large-scale drive operations. Unfortunately, few of these accounts pre-date the introduction of horses, creating uncertainty about the extent to which they can serve as models for kills in the pre-horse period. Horses made it easier for Indians to reach distant hunting grounds and increased the loads they could transport back. And horse-mounted hunters could more easily manoeuvre herds into traps and pursue fleeing animals. There were negative sides to the use of horses as well. The increased mobility of equestrian hunters altered the political landscape, bringing formerly distant groups into direct, often lethal, competition for access to dwindling herds. As a result, bison hunting became a far more dangerous enterprise.

Despite these pitfalls, the historic record is invaluable. For one thing, these accounts make it clear that the most common way to hunt bison was not in massive communal drives, but singly or in small numbers. But it is the spectacular mass kills that have attracted most attention by archaeologists. Judging from historic documents, and supported by archaeological evidence, we know that Indians used many ingenious methods to trap bison in substantial numbers (Kornfeld et al., 2010). Most people, when they think of communal bison hunts, conjure up the image of a jump, where animals were stampeded to their death over a cliff. But Indians also trapped bison in steep-sided arroyos (deep gullies in semi-arid and arid landscapes that have been downcut by ephemeral streams during occasional heavy downpours), driving them upstream from the mouth until the animals were cornered at the head of the arroyo. Bison were also driven into rivers and ponds, or on to thin ice, and occasionally into bogs, sand dunes, and even large snow drifts. Perhaps the most dangerous technique, though one commonly used, was the 'foot surround', where lines of pedestrian hunters encircled a herd, closing off the upwind side at the last moment, and then shooting as many animals as possible before the remainder escaped.

Operating a drive required skill and intimate familiarity with the behaviour of the animals (Brink, 2008). Herds seldom conveniently grazed near a cliff or arroyo entrance, and they were difficult to manoeuvre over a precipice or into a trap if they were aware of the danger. Hence, the animals had to be 'called' by a 'runner', and led to the kill. This was done by an individual who took advantage of the wariness and curiosity of the herd's leader—always an adult female. Bison do not instantly flee from a predator such as a wolf. Instead, they bunch together and watch the predator, even if it approaches, so long as it shows no sign of feeding. If the predator disappears from sight, the lead female may move forward to check on its whereabouts and, cattle-like, the rest of the herd follows behind her. So a skilled 'runner', disguised as a wolf, could use this peculiarity of bison behaviour to move them, sometimes many kilometres, from the gathering area to the cliff or arroyo. Wind direction was critical during this manoeuvre. If the animals caught the human scent, they would flee. They also could not be stampeded at this stage because the runner would lose control of the animals.

Beginning up to a kilometre or more from the jump, the route often was marked with two lines of rock cairns that formed a V-shaped 'drive lane' converging toward the jump. The cairns were neither close enough together nor tall enough to prevent the animals from escaping. According to historic descriptions the cairns sometimes supported branches whose rustling leaves would disturb but not stampede the animals, keeping them moving toward the kill. At other times hunters manned the cairns, holding up hides at the appropriate moment to keep the herd progressing. In this case, the cairns served as 'traffic markers' for the hunters. Drive lanes are still visible at many kills.

Not every cliff was suitable for a communal bison kill. One had to get the animals close to the jump point without their becoming aware of the danger until the very last moment. One way this was accomplished was to select a place where the final approach was uphill. At the last moment the hunters stampeded the animals at the rear. In their panic, these animals, not aware of the danger ahead, forced the lead animals over the precipice. If the drive ended in a pound or corral, the entry point was often equipped with a ramp, sometimes built of snow, so that the animals had to jump into the trap. That way, once inside, they could no longer escape.

THE OVERWINTERING MODEL

After many years of excavation and analysis archaeologists now have a 13,000-year-long record of prehistoric communal bison hunting. We also have an impressive set of tools and methods for reconstructing the way these kill facilities were operated, and determining the time(s) of year when kill events took place. While many archaeologists contributed to the development of this understanding, the work begun by George Frison (1978) more than four decades ago marks a watershed in the study of communal bison hunting (Kornfeld et al., 2010). The techniques that Frison pioneered for analysing kills, and the theoretical logic he articulated for interpreting them (the overwintering model), continue to provide the basic framework for understanding the evolution of communal bison hunting in North America.

The essence of Frison's overwintering model can be distilled down to the following. The majority of bison kills are in the Northern Plains, particularly in Wyoming, Montana, Alberta, and Saskatchewan. This region once supported vast herds of bison, which both ethnohistory and nineteenth-century ethnography clearly show to have been pivotal mainstays in the lives of the region's inhabitants. Based on bison ethology and the historic record, Frison and colleagues developed a compelling model for the way these communal events were organized and timed. In their view, because of the severity of Northern Plains winters, Indians timed their hunting activity so that it occurred in late fall or early winter, thereby providing them with stocks of meat (and hides) sufficient to assure their survival over the long period of hardship and deprivation that followed. Cows were the preferred targets because their physical condition was at its peak, having weaned their calves and taken advantage of lush summer forage. Moreover, by then the rut had ended and bulls were no longer disrupting the herd in their attempts to mate.

COMMUNAL BISON HUNTING: THE ARCHAEOLOGICAL RECORD

A great deal of archaeology has been done over the past century in the Plains, so it should come as no surprise that for such a huge and environmentally diverse region there is an equally diverse and complex terminology, both for the slices of time that archaeologists have delineated, and for the plethora of prehistoric cultures that are now recognized. Moreover, with new excavations, re-examination of old ones, and refinements in dating techniques, the boundaries of the various periods are constantly in flux. Fortunately, since our focus is specifically on communal bison hunting, we can sidestep much of this complexity and talk in very broad terms about the ways that Indians went about trapping these animals, and how these practices changed over the past 13,000 years, an undertaking that can be accomplished without becoming embroiled in all the details. Thus, in the interest of simplicity—and the readers' patience—I have reduced Plains culture history to its absolute minimum, using just four broad periods and three minor subdivisions—Palaeoindian (Clovis, Folsom, Late Palaeoindian, or Plano), Archaic, Late Prehistoric, and Historic.

As already noted, much of our knowledge about how Native Americans communally hunted bison in the past is drawn from the historic record, not from archaeology. The archaeological, and particularly the zooarchaeological, study of communal bison hunting was a 'Johnny-come-lately', not getting underway until the 1960s. To give the reader some appreciation of the number of communal kill sites that are now known, William Fawcett (1987) combed the literature, as well as federal, state, and provincial site files, and found information on 523 kill sites (all periods). More recently, Judith Cooper (2008) conducted a similar search, though her study focused exclusively on the Late Prehistoric period; she came up with a total of 386 kills.

Palaeoindian (13,150–8800 cal BP)

Native Americans were present in North America south of the ice sheets by at least 13,000 years ago, during the Clovis period (13,150–12,800 cal BP), although many archaeologists now believe that Palaeoindians were already present at least a millennium earlier (Pringle, 2011). In most Clovis sites where animal bones are preserved, the majority are from very large mammals, especially mammoths. Bison (*B. antiquus*) bones are sometimes present as well, but only two sites so far have yielded evidence of communal bison hunting—Murray Springs (Arizona), believed to have been a surround of at least eleven bison (Haynes and Huckell, 2007), and Jake Bluff (Oklahoma), an arroyo trap with twenty-two bison (Bement and Carter, 2010) (Fig. 34.4). These sites make it

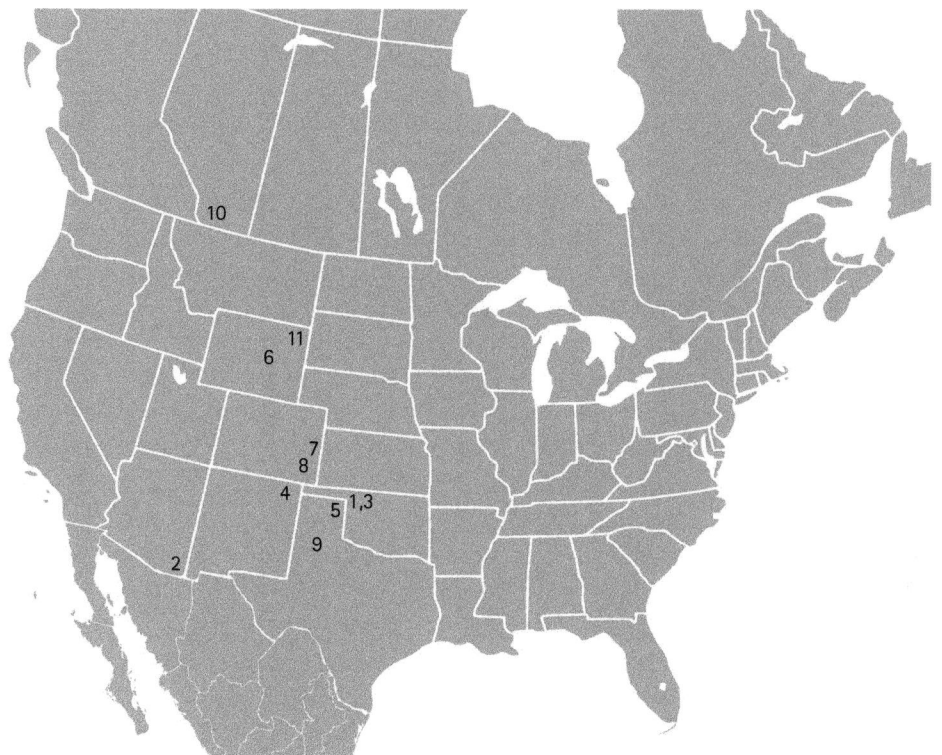

FIGURE 34.4 Location of bison kill sites discussed in the text. Palaeoindian (Clovis)—1: Jake Bluff (Oklahoma), 2: Murray Springs (Arizona). Palaeoindian (Folsom)—3: Cooper (Oklahoma), 4: Folsom (New Mexico), 5: Lipscomb (Texas). Palaeoindian (Plano)—6: Casper (Wyoming), 7: Jones-Miller (Colorado), 8: Olsen-Chubbuck (Colorado), 9: Plainview (Texas). Archaic— 10: Head-Smashed-In (Alberta). Late Prehistoric/Historic—10: Head-Smashed-In (Alberta), 11: Vore (Wyoming). Base map available from Wikimedia Commons (original SVG file: North America second level political division; 2253 x 1992 pixels; released into the public domain by the author, Alex Covarrubias).

clear that bison were being taken communally in small-scale operations almost from the start. No Clovis cliff jumps are known.

Lest the reader get the wrong impression, Clovis hunters were not the first to communally hunt bison. That distinction belongs to Neanderthals. The oldest example (> 130,000 BP) is the French Middle Palaeolithic site of Coudoulous I (level 4). At Coudoulous, Neanderthals, in an unknown number of late spring/early summer events, killed over 150 mostly female and young steppe bison (*B. priscus*) in what appear to have been communal efforts. Another example comes from Mauran, a French Middle Palaeolithic site dating to ~47,000 BP, where Neanderthals killed approximately 4,000 *B. priscus* in a series of late summer/early fall communal kills of mostly cow–calf groups (Jaubert and Delagnes, 2007; Rendu et al., 2012).

Palaeoindian bison kills become more common during the Folsom period (12,800–11,800 cal BP) (Fig. 34.4), but most again are small- to modest-sized operations. For example, in the Southern Plains thirty-two animals were taken at the Folsom site in New Mexico (Meltzer, 2006), twenty to thirty animals were killed in each of three kill events at the Cooper site in Oklahoma (Johnson and Bement, 2009), and at least fifty-five animals were trapped at Lipscomb in Texas (Todd et al., 1992). Most Folsom-age kills in the Northern Plains were also limited operations, ranging from a few animals to perhaps twenty or so. Again, most, if not all, were either foot surrounds or traps, usually in arroyos. There are no known Folsom cliff jumps.

It is in Late Palaeoindian (Plano) times (11,800–8800 cal BP) that we see the first significant increase in the size of the largest kills, as well as a greater diversity of trap types, but again cliff jumps are absent (Byerly et al., 2005). Among the larger kills, Casper (Wyoming) was a sand-dune trap with at least 74 bison (Frison, 1974), Jones-Miller (Colorado) may have been a snow-drift trap with 150 animals (Stanford, 1978), Olsen-Chubbuck (Colorado) was a variant of an arroyo trap with 143 animals (Wheat, 1972), and Plainview (Texas) with > 100 animals was a trap of some sort, perhaps utilizing a shallow pond (Sellards et al., 1947) (Fig. 34.4). Interestingly, the changes that gave rise to modern bison—decline in body size and changes in horn morphology—occurred quite rapidly during the Late Palaeoindian period (Widga, 2013). If these changes led to larger, more cohesive herds, as many suspect, Plano hunters very likely took advantage of these newly emerging behavioural patterns, allowing them to kill much larger numbers of animals in a single coordinated operation.

One particularly curious aspect of many Palaeoindian kill sites, regardless of time period or location, is the incomplete processing of many of the animals ('gourmet butchering'). Articulated skeletons are common, marrow bones are often ignored, brains are not extracted from skulls, and boiling pits and fire-cracked rock, the telltale signs of grease-rendering, are largely absent. Intensive processing of carcasses does not become a standard part of Plains bison kills until the Late Prehistoric period. I will return to this issue below.

Archaic (8800–1800 cal BP)

The number of bison kills increases in the Archaic, particularly after about 5000 BP (Kornfeld et al., 2010). And it is during the Archaic that we see the first use of cliff jumps. The earliest example is Head-Smashed-In, a spectacular jump complex in southern Alberta (Brink, 2008) (Fig. 34.4). Head-Smashed-In saw its first use about 5,700 years ago, though most activity at the site occurred during the Late Prehistoric period. No Archaic period jumps have as yet been identified in the Southern Plains; in this region, arroyo traps (and probably foot surrounds) remain the principal means for communally hunting bison (Buehler, 1997).

Late Prehistoric (1800–400 cal BP) and Historic (400 cal BP–Nineteenth Century)

The Late Prehistoric and Early Historic periods represent the zenith of communal bison hunting in the Northern Plains (Cooper, 2008). It is during these periods that we find the largest communal kills, including huge cliff jumps such as Head-Smashed-In (Brink, 2008) and Vore in Wyoming (Reher, 1978; Reher and Frison, 1980) (Fig. 34.4). At Head-Smashed-In thousands of bison were jumped off the 10–12 m high cliff, creating a massive deposit of butchered carcasses extending 260 m along the base of the cliff and attaining a depth of 4.5 m. At Vore, an estimated twenty thousand bison were driven into a large sinkhole between AD 1500 and 1800. Although Vore was used over and over again, finely varved (layered) sediments reflecting fluctuations in annual precipitation suggest that kill events only occurred, on average, once every twenty-five years (range 11–34 years), usually shortly after a period of elevated rainfall. These periodic wetter periods are thought to have improved forage conditions, fostering growth in bison numbers to some critical threshold when it became profitable to drive bison once again. This surprising finding suggests that large-scale drive operations were not necessarily annual events designed to stock the winter larder, as suggested by the 'overwintering' model, but instead were special events carried out periodically as a means of underwriting large regional or inter-regional social aggregations of otherwise small, widely dispersed bands. It is possible, of course, that the people who normally conducted kills at Vore relocated elsewhere during years when Vore could not be used. However, tree-ring records suggest that periods of low rainfall may have been synchronous over large areas of the Plains, so if herds were too small to drive at Vore, they may have been too small elsewhere as well.

Late Prehistoric bison hunting in the Southern Plains was quite different. While the Northern Plains remained the domain of mobile foraging bands who engaged at least periodically in large-scale drive operations, much of the Southern Plains was occupied by semi-sedentary village-farming communities whose inhabitants farmed part-time and hunted part-time. The faunal assemblages from these sites are usually quite diverse. While bison bones often occur in substantial quantities, particularly after AD 1200–1300 (Dillehay, 1974), communal kill sites are rare. Most of those that are known were small-scale operations, either foot surrounds or arroyo traps. Thus, it would seem that Southern Plains farmer-hunters, while doing a fair amount of bison hunting, took these animals in numbers small enough that the kill sites remain largely undetectable (Speth, 1983; 2004).

OVERWINTERING: AN UNRESOLVED ISSUE

Despite the rich historic and archaeological record of communal bison hunting, there still are many unresolved issues. Paramount among these is the validity of the overwintering model. Although more than four decades have passed since it was first formalized in the literature, it continues to provide the backbone for most explanations of large-scale communal bison hunting in the Northern Plains. However, as more kill sites are studied, it is becoming increasingly clear that bison were taken communally throughout the year, not just in late fall/early winter. This means that stockpiling of meat and hides for winter was not the sole reason for hunting bison in large numbers. In fact, many communal operations took place in late winter or during the spring, not the timing anticipated by the overwintering model (Fawcett, 1987; Cooper, 2008). These results do not necessarily disprove the model, but they suggest that more was going on than just winter provisioning.

Moreover, we really do not know how often these large drive operations were conducted, nor do we know how often they failed. One often gets the impression that such kills took place every year, and usually succeeded in generating the surpluses needed to make it through the winter. However, early accounts often mention drive efforts that failed, or that after repeated attempts only managed to take a few animals. Sometimes for days on end the winds blew from the wrong direction, making it impossible to approach the animals undetected. Often the herds simply did not materialize when and where they were anticipated. And in many cases competition with other hunting groups led to failure, either because another band had already hunted out an area, or because they had inadvertently or intentionally driven the herds elsewhere.

Additionally, if communal hunting in the Northern Plains was directed primarily at preparing for winter, why is there so little evidence of intensive carcass-processing prior to the Late Prehistoric period? Significant quantities of fat from brain, marrow, and cancellous tissue seem to have been wasted. These should have been a prime focus of processing efforts had overwintering been the principal goal of these drives (Speth, 2010).

As already noted, there was a significant upswing in the number and scale of communal bison kills in Late Palaeoindian times. Why? True, there may have been changes in herd structure that made bison easier to drive in large numbers, but why did Plano hunters bother? Since most archaeologists assume that Late Palaeoindian foragers, like their Clovis and Folsom predecessors, still lived in relatively small, mobile bands, what would they have done with such a substantial increase in the output of their hunting activities? Few archaeologists see this as an issue. The standard assumption is 'more is better'. But is it? The hunters would either have had to transport the booty of meat and hides—which in larger kills may have amounted to thousands of kilograms—or waste it; and, judging by the widespread evidence of 'gourmet butchering', it seems they often chose the latter course. Add in the risks of failure and bodily injury, and one cannot help but wonder whether the primary motivation for such large-scale endeavours extended beyond the provisioning of one's family or one's band. I suspect these massive communal operations were conducted first and foremost for social and political reasons, while food and hides were an added, though important, benefit. In the Late Prehistoric and Historic periods population growth and the economic benefits of exchange between hunters, village-dwelling farmers, and Euroamericans may have become increasingly important incentives.

While it is risky to build a general model on the basis of a single site, the Vore evidence raises the possibility that the kills were designed and timed to underwrite periodic aggregations of large numbers of people, drawn from a wide area, that were essential for maintaining the social, political, and reproductive viability of otherwise widely dispersed bands (Reher and Frison, 1980; Fawcett, 1987; Bamforth, 1988; Cooper, 2008).

There still is much we do not know about communal bison hunting in North America. And there is more at stake here than just filling in temporal or geographic gaps with additional examples of kills. Important questions remain unanswered and, in addressing these, the North American bison record can make a valuable contribution to more general issues in human evolution. Let me illustrate this with an example. Contrary to expectations, present-day hunter-gatherer societies are made up mostly of unrelated individuals (Hill et al., 2011). Meat from big game acquired by successful hunters is shared widely within the group, most of it going to non-kin, rather than the hunter or his family (Speth, 2010). In other words, meat from large animals is a public good that is manipulated by good hunters to form and maintain cooperative relationships among otherwise unrelated males and their families. In bands with good hunters there is less turnover in membership than in bands not so blessed. These bands also have longer 'half-lives' permitting their members to maintain rights to higher-quality resource areas over longer spans of time. Such groups also tend to be larger, their overall fertility and offspring survival higher, and their children when grown are more likely to remain resident in the group (Wiessner, 2002).

Communal bison hunting may have provided similar social and political cohesion for groups on the North American plains, by generating a public good in large quantities that could be widely shared both within and between participating bands. Obviously, these hunts fulfilled basic food and hide needs as well. However, periodic large-scale drive operations, especially those where animals were under-utilized, may have served a purpose beyond that of just satisfying subsistence needs; one that was situated squarely within the social and political domain. The sharing of meat from such kills provided the fabric that assured the long-term social, political, and reproductive viability of the groups that participated in these events.

Coming to understand why Plains bison hunters periodically engaged in what otherwise might seem like wasteful slaughters has implications well beyond North America and much further back in time. What, for example, motivated the large-scale communal bison hunts that were undertaken by Neanderthals at Coudoulous and Mauran? Since most archaeologists assume that Neanderthal groups were even smaller than those of Palaeoindians, were such massive events really needed to fulfil their calorie, protein, or hide needs? What did they do with the surplus? Or is it possible that these archaic humans, as much as 100,000 years ago or more, already possessed a mechanism—communal big-game hunting and meat-sharing—that assured the long-term cooperation of otherwise unrelated individuals?

There is so much we can yet learn about the human story from ongoing studies of the precious North American record of communal bison hunting ...

Acknowledgements

I am grateful to the editors for their invitation to participate in this volume. I am also grateful to them for being so gracious and understanding when I requested a delay in getting the final manuscript to them. *Mea culpa* …

References

Bamforth, D. (1988) *Ecology and Human Organization on the Great Plains*, New York: Plenum.

Bement, L. and Carter, B. (2010) 'Jake Bluff: Clovis bison hunting on the southern plains of North America', *American Antiquity*, 75, 907–33.

Blitz, J. (1988) 'Adoption of the bow in prehistoric North America', *North American Archaeologist*, 9, 123–45.

Brink, J. (2008) *Imagining Head-Smashed-In: Aboriginal Buffalo Hunting on the Northern Plains*, Edmonton: Athabasca University Press.

Buehler, K. (1997) 'Where's the cliff? Late Archaic bison kills in the southern plains', *Plains Anthropologist*, 42, 135–43.

Byerly, R., Cooper, J., Meltzer, D., Hill, M., and LaBelle, J. (2005) 'On Bonfire Shelter (Texas) as a Paleoindian bison jump: an assessment using GIS and zooarchaeology', *American Antiquity*, 70, 595–629.

Carlson, K. and Bement, L. (2013) 'Organization of bison hunting at the Pleistocene/Holocene transition on the plains of North America, *Quaternary International*, 297, 93–9.

Chisholm, B., Driver, J., Dube, S., and Schwarcz, H. (1986) 'Assessment of prehistoric bison foraging and movement patterns via stable-carbon isotopic analysis', *Plains Anthropologist*, 31, 193–205.

Cooper, J. (2008) 'Bison Hunting and Late Prehistoric Human Subsistence Economies in the Great Plains'. Unpublished PhD dissertation, Southern Methodist University (Dallas).

Davis, L. (1978) 'The 20th-century commercial mining of northern plains bison kills', *Plains Anthropologist*, 23, 254–86.

Dillehay, T. (1974) 'Late Quaternary bison population changes on the southern plains', *Plains Anthropologist*, 19, 180–96.

Fawcett, W. (1987) 'Communal Hunts, Human Aggregations, Social Variation, and Climatic Change: Bison Utilization by Prehistoric Inhabitants of the Great Plains'. Unpublished PhD dissertation, University of Massachusetts (Amherst).

Frison, G. (ed.) (1974) *The Casper Site: A Hell Gap Bison Kill on the High Plains*, New York: Academic Press.

Frison, G. (1978) *Prehistoric Hunters of the High Plains*, New York: Academic Press.

Gates, C., Freese, C., Gogan, P., and Kotzman, M. (eds) (2010) *American Bison: Status Survey and Conservation Guidelines 2010*, Gland: IUCN.

Graves, A. (2010) 'Investigating Resource Structure and Human Mobility: An Example from Folsom-Aged Bison Kill Sites on the U.S. Southern Great Plains'. Unpublished PhD dissertation, University of Oklahoma (Norman).

Haines, F. (1995) *The Buffalo: The Story of American Bison and Their Hunters from Prehistoric Times to the Present*, Norman: University of Oklahoma Press.

COMMUNAL BISON HUNTING IN WESTERN NORTH AMERICA   539

Haynes, C. and Huckell, B. (eds) (2007) *Murray Springs: A Clovis Site with Multiple Activity Areas in the San Pedro Valley, Arizona*. Anthropological Paper 71. Tucson: University of Arizona.

Hill, K., Walker, R., Božičević, M., Eder, J., Headland, T., Hewlett, B., Hurtado, A. M., Marlowe, F., Wiessner, P., and Wood, B. (2011) 'Co-residence patterns in hunter-gatherer societies show unique human social structure', *Science*, 331, 1286–9.

Hill, M., Hill, M., and Widga, C. (2008) 'Late Quaternary *Bison* diminution on the Great Plains of North America: evaluating the role of human hunting versus climate change', *Quaternary Science Reviews*, 27, 1752–71.

Jaubert, J. and Delagnes, A. (2007) 'De l'espace parcouru à l'espace habité au Paléolithique Moyen', in Vandermeersch, B. and Maureille, B. (eds) *Les Néanderthaliens: biologie et cultures*, pp. 263–81. Paris: Éditions du Comité des Travaux Historiques et Scientifiques.

Johnson, E. and Bement, L. (2009) Bison butchery at Cooper, a Folsom site on the Southern Plains', *Journal of Archaeological Science*, 36, 1430–46.

Kornfeld, M., Frison, G., and Larson, M. (eds) (2010) *Prehistoric Hunter-Gatherers of the High Plains and Rockies*, 3rd edn, Walnut Creek: Left Coast Press.

Lewis, P., Johnson, E., Buchanan, B., and Churchill, S. (2007) 'The evolution of *Bison bison*: a view from the Southern Plains', *Bulletin of the Texas Archeological Society*, 78, 197–204.

Loscheider, M. (1977) 'Use of fire in interethnic and intraethnic relations on the Northern Plains', *Western Canadian Journal of Anthropology*, 7, 82–96.

Lott, D. (2002) *American Bison: A Natural History*, Berkeley: University of California Press.

McDonald, J. (1981) *North American Bison: Their Classification and Evolution*, Berkeley: University of California Press.

Meltzer, D. (2006) *Folsom: New Archaeological Investigations of a Classic Paleoindian Bison Kill*, Berkeley: University of California Press.

Pringle, H. (2011) 'The first Americans', *Scientific American*, 305, 36–45.

Reher, C. (1978) 'Buffalo population and other deterministic factors in a model of adaptive process on the shortgrass plains', *Plains Anthropologist*, 23(82, Part 2: Memoir 14), 23–39.

Reher, C. and Frison, G. (1980) 'The Vore site, 48CK302, a stratified buffalo jump in the Wyoming Black Hills', *Plains Anthropologist*, 25 (88, Part 2: Memoir 16).

Rendu, W., Costamagno, S., Meignen, L., and Soulier, M.-C. (2012) 'Monospecific faunal spectra in Mousterian contexts: implications for social behavior', *Quaternary International*, 247, 50–8.

Roe, F. (1970) *The North American Buffalo: A Critical Study of the Species in Its Wild State*, 2nd edn, Toronto: University of Toronto Press.

Scobel, A. (ed.) (1902) *Geographisches Handbuch zu Andrees Handatlas*. 4th edn, Bielefeld und Leipzig: Velhagen und Klasing.

Sellards, E., Evans, G., and Meade, G. (1947) 'Fossil bison and associated artifacts from Plainview, Texas', *Bulletin of the Geological Society of America*, 58, 927–64.

Shapiro, B., Drummond, A. J., Rambaut, A., Wilson, M. C., Matheus, P. E., Sher, A. V., Pybus, O. G., Gilbert, M. T., Barnes, I., Binladen, J., Willerslev, E., Hansen, A. J., Baryshnikov, G. F., Burns, J. A., Davydov, S., Driver, J. C., Froese, D. G., Harington, C. R., Keddie, G., Kosintsev, P., Kunz, M. L., Martin, L. D., Stephenson R. O., Storer, J., Tedford, R., Zimov, S., and Cooper, A. (2004) 'Rise and fall of the Beringian steppe bison', *Science*, 306, 1561–5.

Speth, J. (1983) *Bison Kills and Bone Counts*, Chicago: University of Chicago Press.

Speth, J. (2004) *Life on the Periphery: Economic Change in Late Prehistoric Southeastern New Mexico*. Memoir 37. Ann Arbor: University of Michigan Museum of Anthropology.

Speth, J. (2010) *The Paleoanthropology and Archaeology of Big-Game Hunting: Protein, Fat or Politics?*, New York: Springer.

Stanford, D. (1978) 'The Jones-Miller site: an example of Hell Gap bison procurement strategy', *Plains Anthropologist*, 23(82, Part 2: Memoir 14), 90–7.

Todd, L., Hofman, J., and Schultz, C. (1992) 'Faunal analysis and Paleoindian studies: a reexamination of the Lipscomb bison bonebed', *Plains Anthropologist*, 37, 137–65.

Wheat, J. (1972) 'The Olsen-Chubbuck site: a Paleo-Indian bison kill', *American Antiquity* 37(1, Part 2: Memoir 26).

Widga, C. (2013) 'Evolution of the High Plains Paleoindian landscape: the paleoecology of Great Plains faunal assemblages', in Knell, E. and Muniz, M. (eds) *Paleoindian Lifeways of the Cody Complex*, pp. 69–92. Salt Lake City: University of Utah Press.

Widga, C., Walker, J., and Stockli, L. (2010) 'Middle Holocene bison diet and mobility in the eastern Great Plains (USA) based on δ^{13}C, δ^{18}O, and ^{87}Sr/^{86}Sr analyses of tooth enamel carbonate', *Quaternary Research*, 73, 449–63.

Wiessner, P. (2002) 'Hunting, healing, and hxaro exchange: a long-term perspective on !Kung (Ju/'hoansi) large-game hunting', *Evolution and Human Behavior*, 23, 407–36.

Wilson, M., Hills, L., and Shapiro, B. (2008) 'Late Pleistocene northward-dispersing *Bison antiquus* from the Bighill Creek Formation, Gallelli Gravel Pit, Alberta, Canada, and the fate of *Bison occidentalis*', *Canadian Journal of Earth Sciences*, 45, 827–59.

CHAPTER 35

....

ADVANCES IN HUNTER-GATHERER RESEARCH IN MEXICO

archaeozoological contributions

....

JOAQUÍN ARROYO-CABRALES
AND EDUARDO CORONA-M.

INTRODUCTION

....

GEOGRAPHICALLY, Mexico has an important role in regard to current discussions about the First Americans. The country has been considered as a corridor for the entrance of the first human groups coming from North to South, with a complex and rich faunal diversity that is largely due to the confluence of the two large biogeographical zones of the Americas: Nearctic and Neotropical (Corona-M., 2013; Ríos-Muñoz, 2013). There are many data on Late Pleistocene faunas but few of these concern the interactions of early people and large Pleistocene mammals in Mexico (e.g. Arroyo-Cabrales et al., 2006; Johnson et al., 2006). Data on the relationship between the extinction of large mammals and early people in Mexican localities are equally limited (Sanchez, 2001; González and Huddart, 2008).

The interest in early peopling of the study area has existed for over a century, mainly discussing chronology and the probable relationships with hominid species in Eurasia, and producing intense debates (e.g. Reyes, 1881; Mercer, 1975; Corona-M., 2008). However, scientists have not yet been able to define when and where the earliest people entered Mexico (e.g. González and Huddart, 2008; Mirambell, 2012).

In 1958, a new impetus was generated with the foundation of a department devoted to prehistoric research at the National Institute of Anthropology and History (INAH, by its acronym in Spanish), the federal agency devoted to research and protection of the historical, archaeological, and palaeontological heritages of Mexico. This department includes laboratories for palaeozoology, palaeobotany, soil chemistry, geology, and radiocarbon dating. Specialists interact to provide palaeoenvironmental hypotheses, thus providing a novel contribution to archaeological and palaeobiological research in Latin America (Lorenzo, 1991; Corona-M. et al., 2010).

The Palaeozoology (now 'Archaeozoology') Laboratory started in 1963 by studying a large number of skeletal remains from prehistoric contexts in Mexico (Álvarez, 1965; 1967). It was thought that these could provide evidence for the interaction between early human populations and extinct fauna, and the analysis was later expanded to incorporate the analysis of faunal remains from pre-hispanic localities.

Based on the technology utilized during lithic production, Lorenzo and Mirambell (1999) provided a relative chronological framework for the presence of the early humans in Mexico grouped under the term Lithic Age, with three cultural horizons known as Archaeolithic (35,000–14,000 years BP), Lower (14,000–9000 years BP) and Upper (9000–7000 years BP) Cenolithic, and Protoneolithic (7000–4500 years BP). In the Upper Cenolithic extinct faunas and small populations of hunter-gatherers predominated (García Bárcena, 2007), whereas the Protoneolithic saw the appearance of the first sedentary societies with an agriculture-based economy, supplemented by hunting. Territorial control is also characteristic of this period.

While the former Department of Prehistory existed, there were research endeavours in at least twenty locations, ten of them located in the Basin of Mexico. During those investigations there were findings of stone tools (i.e. arrow-heads, axes, etc.), anthropically modified bones, mainly those with butchery marks or artifacts made from bone. In the early 1980s, the department was transformed into the Archaeological Laboratory Section, with its research covering all cultural periods.

In the last fifteen years, there has been a steady increase in studies referring to the early peopling of America, including Mexico (e.g. Bonnichsen and Turnmire, 1999; Jiménez López et al., 2006). Beginning in 2002, and occurring every two years since then, there has been a symposium regarding the early peopling of the Americas, in which colleges from all over the continent and abroad have presented on-going research on the subjects. Originally, this event was promoted by scientists from INAH, but it has since been hosted by Argentina (2010) and Colombia (2012) and back in Mexico (2014). Other events on this theme were organized, such as Mexican Symposia in 2008, 2011, and 2014, where new localities were presented, and also one on the Americas organized by the Human Evolution, Adaptations, Dispersals, and Social Developments (HEADS) initiative of UNESCO World Heritage Thematic Programme (2013). Topics that have been dealt with include evolution, genetics, dating methods, migration, palaeoenvironments, megafauna, and geology.

Much of the controversy about the early peopling of the Americas deals with the value that is given to the indirect evidence of human presence when human skeletal remains are lacking at sites. Such evidence may include lithic, hearths, and culturally modified bone. Indirect evidence requires further detailed analyses that distinguish between natural processes the materials may have undergone and those cultural processes that are signatures of human intervention.

Animal bones are abundant in prehistoric archaeological sites, and are a valuable source for anthropological, archaeological, and generally scientific data as providing evidence for the reconstruction of palaeoenvironments, and the way-of-life of early hunter-gatherer populations. In the last few decades, several relatively new disciplines within the archaeological sciences have widened the kind of information that can be obtained from skeletal remains. Both palaeoclimates and ancient diets can be inferred through assays for learning bone chemical composition. Population genetic lines can be assessed utilizing ancient DNA whereas radiocarbon dating, now a well-recognized field, allows detailed and absolute chronologies to be built.

The following is a brief synthesis of current archaeozoological knowledge regarding early peopling in the Mexican Late Pleistocene.

LOCALITIES AND DATING

For the Quaternary in Mexico there are over seven hundred locations with mammal records known, some of them also including other groups such as birds and herptiles (Arroyo-Cabrales et al., 2002; Corona-M., 2014). Chronologically, most of these localities go back to the Rancholabrean North American Mammal Age (120,000–10,000 BP), and is considered that already in the Late Glacial Maximum (24,000–18,000 BP) human groups started colonizing the Americas. The localities of this period can be grouped into seven geographical regions (Fig. 35.1).

The Eastern region (Tamaulipas, Veracruz, Tabasco) has produced several sites, but it does not seem to have great faunal diversity and it does not offer any clues about the early population of Mexico. Further research in this area is required.

The Mexican Plateau region, with over 250 localities or sites, contains the San Josecito Cave, Nuevo Leon, which has the greatest diversity of vertebrate fauna known for the Mexican Late Pleistocene. Several localities in northern Zacatecas, for example Ojo de Agua (dating back to at least 20,000 BP) and Agua Dulce, promise to provide strong palaeoenvironmental data suggestive of early human presence, including the presence of megafauna, such as mammoths, gomphotheres, tapirs, and horses (Ardelean, 2013). The site of Cedral, San Luis Potosi, in addition to the great faunal diversity, has the purported oldest record of human inhabitants (c.33,000 BP), though there are serious doubts about its stratigraphic integrity. Valsequillo, Puebla, is another controversial site regarding early humans. Several archaeological excavations from the 1950s until the mid-2000s have provided contradictory evidence about the presence of the earliest people in the area. Occupation from the Sangamonian interglacial has been proposed (between 132,000 and 119,000 BP; Van Landingham, 2004), but questioned and eventually rejected (Feinberg et al., 2009; Mark et al., 2010). The most reliable dating of faunal

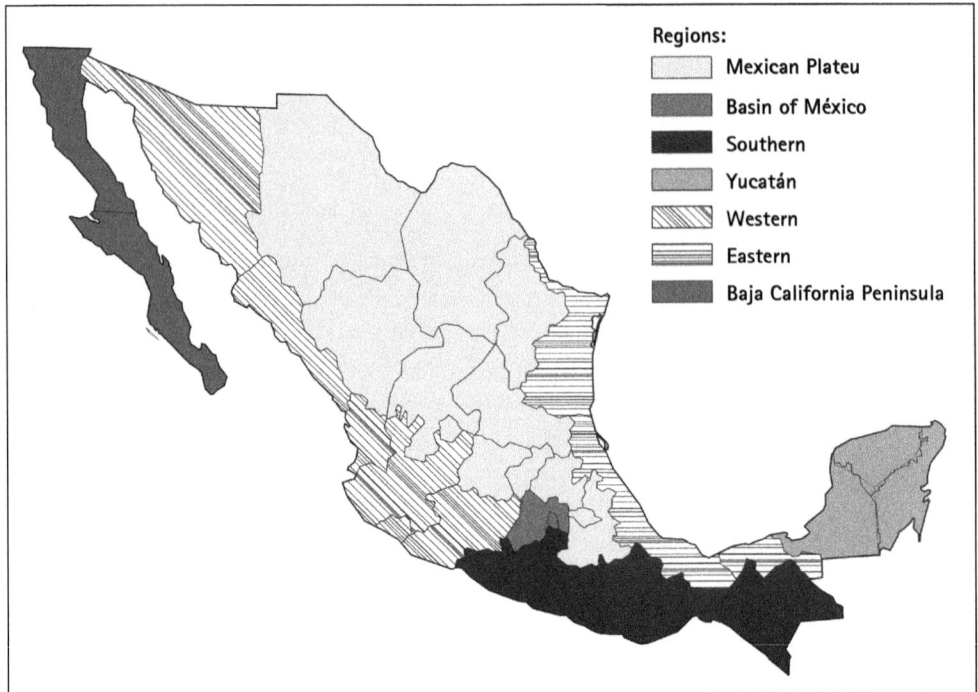

FIGURE 35.1 Map of Mexican regions discussed on the text. Author's own image.

remains is more conservative at around 20,780 BP (Arroyo-Cabrales et al., 2010). Sadly, further research at the site is precluded by urbanization and pollution associated with a nearby dam and reservoir (P. Ochoa-Castillo, pers. comm.).

The Basin of Mexico region, mainly covering the Federal District and the State of Mexico, is the best explored area in the country and has nearly 215 localities. It is the region with the highest density and best-known localities, with iconic sites such as Tequixquiac, Tlapacoya, and El Peñón de los Baños, among others. A large amount of construction continues to occur in this region resulting in the discovery of numerous palaeontological localities and archaeological sites. One of the earliest findings was that of a woman's bones at El Peñon de los Baños dated to 10,755 ± 75 BP (González et al., 2003; González and Huddart, 2008). This early age indicates that these are some of the oldest human skeletal remains to have been found in the Americas (Meltzer, 2009). At Tlapacoya researchers found hearths, extinct fauna, and human remains, not in a clear association but presumably contemporary (Lorenzo and Mirambell, 1986).

The Late Pleistocene Basin of Mexico was a rich environment that supported a large Columbian mammoth (*Mammuthus columbi*) population. Over one hundred mammoth localities are known for the area, yet very few show evidence of human interaction with the carcasses (Arroyo-Cabrales et al., 2006). Modified bone found in this area has mainly been identified as Columbian mammoth. Some authors identify up to fifteen sites in the Basin of Mexico with purported anthropic modification of mammoth bone, including Tequixquiac, Chimalhuacán, Los Reyes La Paz, Atenco, San Bartolo Atepehuacan, Santa Isabel Iztapan I and II, Los Reyes Acozac, Tlapacoya, Santa Lucía, Santa Marta Acatitla, Tepexpan, Tepexpan Hospital, Gertrudis Sánchez, Tocuila, and Villa de Guadalupe (Aveleyra, 1950; García Cook, pers. comm. in Corona-M., 2014), but a preliminary detailed analysis finds that only Santa Isabel Ixtapa, Villa de Guadalupe, and Tocuila should be considered (Arroyo-Cabrales et al., 2006). Further studies of bone modification are warranted to establish criteria for evaluating this kind of assemblage, which can then be applied to other prehistoric sites and localities from all over the country.

The Western region has fewer sites, but some are very important, such as the sites near the Chapala Lake (Jalisco), while in Sonora, in recent years, more than a dozen Clovis sites have been found, but few have stratigraphically controlled excavations such as the one carried out at El Fin del Mundo (Gaines and Sanchez, 2009; Sánchez-Miranda et al., 2009). At this site, probable interaction between people and gomphotheres (Gomphotheriidae, i.e. extinct proboscideans) was in the form of either hunting or scavenging activities (Sanchez et al., 2014). Clovis points are conventionally considered as a marker for the presence of an early technology among the first people arriving to the Americas. However, in Mexico evidence for the presence of Clovis people is minimal. Few Clovis points have been recovered from Baja California to Costa Rica, including some in the State of Hidalgo in the Mexican Plateau Region (Sanchez, 2001).

The southern region includes two sub-regions: one is represented by the states of Oaxaca and Chiapas and the other by the states of Morelos and Guerrero. The first sub-region has been actively explored in the last few years, including excavations at the site of Guila Naquitz, a small shelter in central Oaxaca. Flannery's 1986 excavation yielded both seeds and peduncles of squash (*Cucurbita pepo*), with indications of domestication as early as 9000 cal BP (Smith, 1997). This date supports the view that the earliest phase at Naquitz belongs to the early Archaic period (Flannery, 1986). More recently Pérez-Crespo et al. (2013) have inferred the environmental conditions from the Late Pleistocene to the present based on mammal faunas and palaeoclimatic modelling. At Mitla and Tehuantepec two sites have been discovered where the remains of gomphotheres were possibly associated with Clovis-like points elaborated with silex stone (Winters, 2014).

There is evidence that indicates the presence in the Americas of two early cultural traditions, North American Clovis from Oaxaca and Chiapas, and fish-tail fluted points from Central and South America (Santamaría Estévez and García-Bárcena, 1989). Recent studies from rock shelters nearby Ocozocuautla, Chiapas, have provided strong evidence of human presence in the state around 11,000 to 10,000 cal BP. These sites have yielded lithics reflecting expedient technology, as well as milling stones and botanical samples that may indicate incipient horticulture starting at the end of the Pleistocene and Early Holocene. Small- and medium-sized animals (such as deer, peccary, and rabbit) were the most hunted prey, while megafaunal remains were not found (Acosta-Ochoa, 2010).

The sub-region Morelos-Guerrero has been explored in recent years, and published reports include one of the most complete mammoth specimens, and seven reliable localities with herptiles, birds, and mammals, including extinct fauna (Corona-M., 2009; 2013). The findings show some changes in distributions as currently known, i.e. southern records of Nearctic mammals, in turn suggesting changes in vegetation distributions.

Research in the Baja California Peninsula has seen great advances recently with at least five sites with C^{14} dates from 11,000 to 9000 cal BP: Isla de Cedros, Abrigo Paredón-Laguna Chapala, Abrigo de los Escorpiones, Sierra de San Francisco, and Isla Espíritu Santo. Some hypothetical routes for human colonization are under debate, with the most widely accepted one being that people arrived from the north, either by terrestrial or maritime routes. Two kinds of subsistence strategies have been postulated: hunter-gathering focused on terrestrial resources and another, more specialized, system using marine resources (fishing, gathering edible molluscs, crustaceans, and urchins, and hunting sea lions, seals, marine turtles, and occasionally terrestrial mammals; Fujita and Porcayo Michelini, 2014).

Finally, the region of the Yucatan Peninsula also has important findings. Some decades ago Pleistocene records from the area were rare, and mainly consisted of equids and ground-sloths (*Paramylodon*) from dry caves. The most diverse was Loltún cave, which produced reptiles, birds, and mammals, including gomphotheres, carnivores, and meso- and microfauna (Arroyo-Cabrales and Alvarez, 2003). This assemblage suggests an arid steppe environment, very different to the recent tropical dry forest. Recently, molecular analysis undertaken on the remains of the tropical rodent *Ototylomys* cf. *phyllotis* found at the cave managed to extract ancient DNA and amplify six short overlapping fragments of the cytochrome b gene, totalling 666 bp, suggesting that the sister group to modern *O. phyllotis* arose during the Miocene–Pliocene, diversified during the Pleistocene and went extinct in the Holocene (Gutiérrez-García et al., 2014).

Current explorations in this region have increased the records with glyptodont (*Glypthotherium*) and camelids (*Hemiauchenia*). Some of those camelid specimens were found in a hearth with fire marks and are under investigation (González González et al., 2006).

Initially two early human specimens were found in Naharon Cave (11,670 ± 60 cal BP) and Las Palmas (8050 ± 130 cal BP), as well as evidence of hearths (8941 ± 39 cal BP) (González González et al., 2006), but currently remains pertaining to at least eight individuals are under study (González et al., 2013). Most recently, saber-tooth cat (*Smilodon*) and remains resembling spectacled bear (Tremarctinae) have been found in several *cenotes* (submerged caves) in the Yucatan Peninsula, along with detailed research on early human presence. In at least one of the *cenotes*, remains of these two animals were found near human remains. Further study and dating are required to be certain that the megacarnivore and human remains were contemporaneous (Chatters et al., 2014).

The analysis of stable isotopes provides data on the palaeodiet of herbivores in twenty-four Mexican localities. Mammoths and horses have resulted to be grazers that ate C4 plants, but with occasional inclusions of C3 plants, i.e. leaves and shrubs. Concerning gomphotheres, although those animals were traditionally considered as highly special-ized browsers, only the specimens of *Stegomastodon* showed a mixed C3/C4 plant diet, while *Cuvieronius* specimens resulted to be grazers (Pérez-Crespo et al., 2012). These analyses also support the palaeoenvironmental hypothesis that sites such as Cedral and Laguna de las Cruces, San Luis Potosi, and Valsequillo, Puebla were located in areas of mixed grassland and open forest.

SOME ZOOGEOGRAPHICAL CONSIDERATIONS

The confluence in Mexico of the two major biogeographic regions, Nearctic and Neotropical, led to a series of exchanges during the Pliocene-Pleistocene transi-tion, both northward and southward. In particular, the Great American Biotic Interchange (*c.*3.7 MYA) constituted an event caused by the emergence of the Isthmus of Panama, which facilitated the migration of terrestrial species, substantially chang-ing the composition of the faunas and biota, especially in the Mexican Plateau and Basin of Mexico regions. For example, a radical change occurred in the composi-tion of bird faunas, with an increase of Neotropical birds and a decrease of Nearctic birds (Corona-M., 2009) (Fig. 35.2). The ancient limit of mastodon distributions changed (Polaco et al., 2001), as did that of gomphotheres (Alberdi and Corona-M., 2005; Corona-M. and Alberdi, 2006), mammoths, and other mammals (Arroyo-Cabrales et al., 2010). Some other changes in the distributional ranges of individual species of birds, mammals, and herptiles were also recorded, suggesting changes in palaeoenvironments during the Late Pleistocene, but further research on this issue is necessary.

EXTINCTIONS

To evaluate the extent of the extinction process in the Late Pleistocene, extant species of vertebrates were compared with both the recorded taxa of Pleistocene vertebrates and the extinct vertebrates of the same period. This is a qualitative proxy due to the inequal-ity of recorded localities for each vertebrate group (Fig. 35.3). Mammals are the third largest group of extant vertebrates and have the highest losses (seventy-eight taxa going extinct), while birds and herptiles represent the largest groups but with fewer extinct taxa (twenty and four, respectively). It is possible that there is a bias produced by the large number of localities with mammal records, as well as the lack of other vertebrate remains, mainly due to general methodological issues (Arroyo-Cabrales and Polaco, 2008). However, this proxy suggests that the extinction process affected all of the ter-restrial vertebrates.

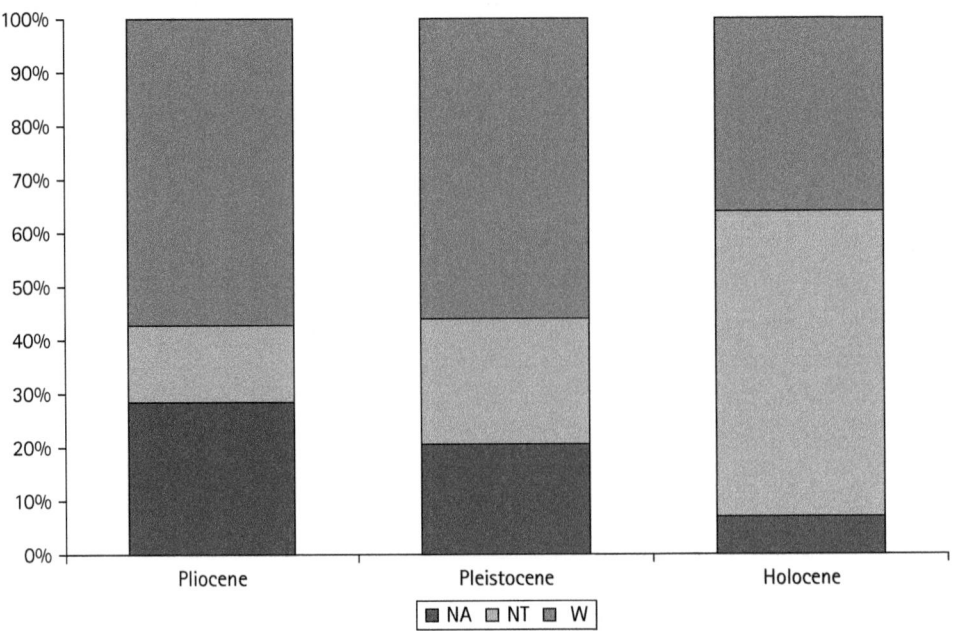

FIGURE 35.2 Changes in the biogeographic composition of birds in the Late Caenozoic. NA = Nearctic distribution, NT = Neotropical distribution, W = Wide distribution (Corona- M., 2009). Author's own image.

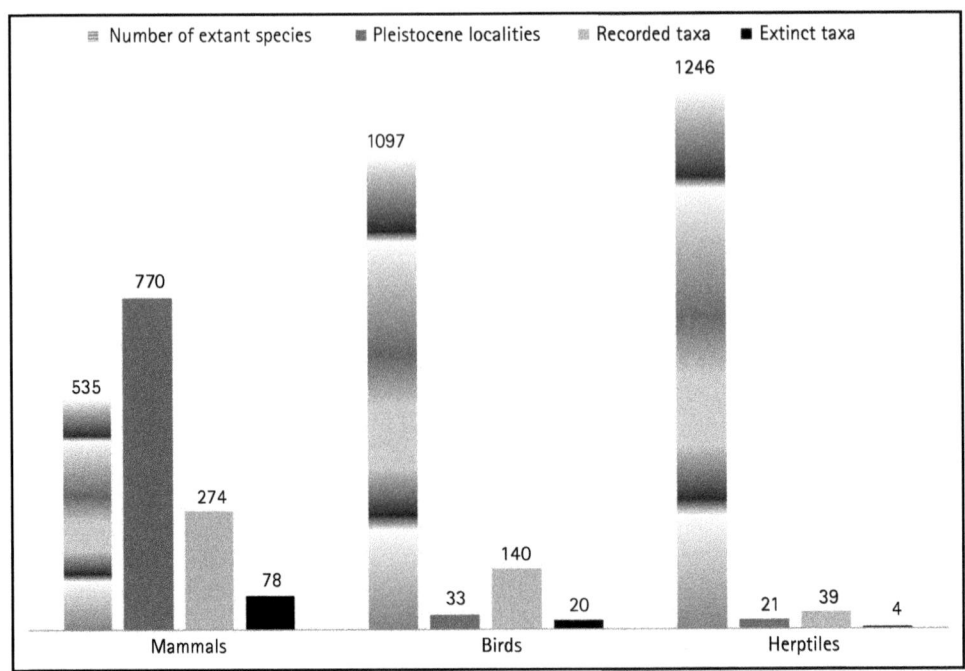

FIGURE 35.3 Comparison of records of terrestrial vertebrates in Mexico: extant species, total-records for the Late Pleistocene, extinct taxa and number of Pleistocene localities where each-group was recorded. Author's own image.

Another approach for analyzing the data is grouping the extinct mammals and birds by taxon-free categories related to their habitats (terrestrial and aquatic) and diets (herbivorous, carnivorous, etc.) (Fig. 35.4). The analysis shows that large mammalian herbivores were the most affected group, followed by aquatic birds (grouping swimmers and waders) and terrestrial predatory and scavenging birds (Corona-M., 2009; Arroyo-Cabrales et al., 2010).

Currently the four large species of extant mammals (those weighing over 100 kg) are: the jaguar (*Panthera onca*), tapir (*Tapirus bairdii*), white-tailed deer (*Odocoileus virginianus*), and bighorn sheep (*Ovis canadensis*), but for the Late Pleistocene there are sixty-two recorded species, most of which were herbivores that reached sizes ten times larger than extant ones (over a ton in some cases), such as mammoths (*Mammuthus* sp.), American mastodon (*Mammut americanum*), gomphotheres (*Cuvieronius* and *Stegomastodon*), and ground sloths (*Paramylodon* and *Eremotherium*). Therefore, some taxa unique to Mexico, namely the order Notoungulata (toxodons), and five families (Gomphotheriidae, Mammutidae, Glyptodontidae, Megatheriidae, and Mylodontidae) were completely lost. Other families lost their representatives in Mexico, but survived elsewhere (Camelidae, Herpestidae, Equidae, Elephantidae, Hydrochoeridae, and Megalonychidae). Other taxa only survived through a few representatives such as antilocaprids and bovids. Some taxa are extant in other parts of the world such as the red dog

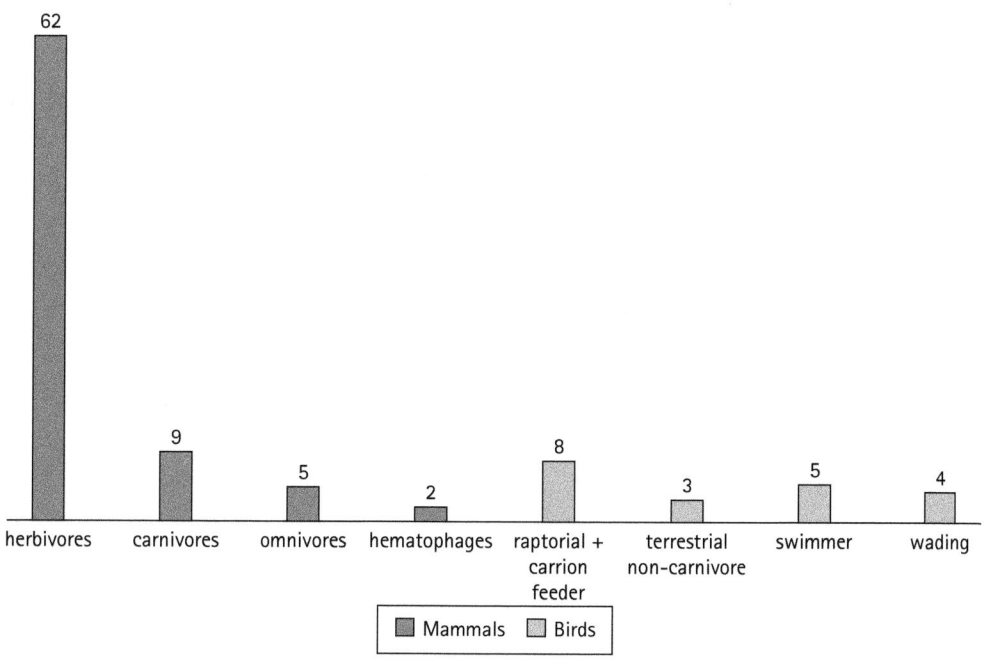

FIGURE 35.4 Comparison of Mexican extinct mammals and birds based on habitat and type of diet. Author's own image.

(*Cuon*), which survives today in Asia, the Andean spectacled bear (*Tremarctos*), and the South American giant anteater (*Myrmecophaga tridactyla*), which both inhabit South America. The yellow-bellied marmot (*Marmota flaviventris*) and the southern bog lemming (*Synaptomys cooperi*) today occur in North America north of Mexico (Arroyo-Cabrales et al., 2010).

In the case of birds, taxa affected by extinction processes include various water birds such as divers (*Plyolimbus baryosteus, Podiceps parvus*), ducks (*Oxyura zapatima*), cormorants (*Phalacrocorax goletensis* and *P. chapalensis*), and storks (*Mycteria wetmorei* and *Ciconia* cf. *maltha*). Terrestrial birds included a turkey (*Meleagris crassipes*) and a small parakeet (*Rhynchopsitta phillipsi*), along with several types of scavengers and predators (*Teratornis merriami, Breagyps clarki, Wetmoregyps daggetti, Neogyps errans, Neophrontops americanus, Buteogallus fragilis, Spizaetus grinelli*, and *Strix brea*) (Corona-M., 2009).

The overall pattern is consistent and similar to the extinction processes that have occurred in North America north of Mexico. However, the debate on the main causes for this mass extinction still continues. The strong dichotomy between environmental change and the role of humans has been slowly abandoned, and replaced with a multi-cause process in which environmental changes, specific histories of the species involved and the peopling of the Americas all contributed to the mass extinctions of the Late Pleistocene. The net effect was the loss of herbivores, carnivores, and scavengers, the simplification of food webs and a loss of the stability of ecosystems (Barnosky et al., 2004).

CONCLUSION

Based on data from many localities that lack a clear association between human and animal remains, archaeozoology provides data on palaeoenvironmental scenarios, important information about the palaeodiet of some herbivores, and evidence for biogeographical distributions. Some localities are quite diverse, with mammals, birds, and reptiles all represented, such as Tlapacoya, Cedral, Tequixquiac, Tepexpan, and Loltún Cave, among others.

For understanding the Late Pleistocene human occupation of Mexico an extensive programme that includes the study of fauna from central Mexico will be required. Those from primary contexts will be essential, as will be dating evidence for the association of tools and other elements of material culture. It would be advisable to look at the sites formally recognized as the earliest in the study area, and investigate them by the use of current geoarchaeological methods, finally undertaking a systematic programme to explore new sites based on predictive tools such as georadar and other techniques.

ACKNOWLEDGMENTS

We thank Umberto Albarella and other editors of the volume for their kind invitation to contribute to the Handbook, as well as for their helpful comments.

REFERENCES

Acosta-Ochoa, G. (2010) 'Late-Pleistocene/Early-Holocene tropical foragers of Chiapas, Mexico: recent studies', *Current Research in the Pleistocene*, 27, 1–4.

Alberdi, M. T. and Corona-M., E. (2005), 'Revisión de los Gonfoterios en el Cenozoico tardío de México', *Revista Mexicana de Ciencias Geológicas*, 21, 242–60.

Álvarez, T. (1965) *Catálogo paleomastozoológico mexicano*. Publicaciones 17. Mexico City: Instituto Nacional de Antropología e Historia.

Álvarez, T. (1967) 'El laboratorio de paleozoología', *Boletín del Instituto Nacional de Antropología e Historia*, 28, 43–7.

Ardelean, C. F. (2013) 'Archaeology of Early Human Occupations and the Pleistocene-Holocene Transition in the Zacatecas Desert, Northern Mexico', PhD Dissertation, Department of Archaeology, University of Exeter, United Kingdom.

Arroyo-Cabrales, J. and Alvarez, T. (2003) 'Chapter 10. A preliminary report of the late Quaternary mammal fauna from Loltún Cave, Yucatán, México', in Schubert, B. W., Mead, J. I. and Graham, R. W. (eds) *Ice Age Cave Faunas of North America*, pp. 262–72. Denver: Indiana University Press & Denver Museum of Nature & Science.

Arroyo-Cabrales, J. and Polaco, O. J. (2008) 'Fossil bats from Mesoamerica', *Arquivos do Museu Nacional, Rio de Janeiro*, 66, 155–60.

Arroyo-Cabrales, J., Polaco, O. J. and Johnson, E. (2002) 'La mastofauna del cuaternario tardío de México', in Montellano-Ballesteros, M. and Arroyo-Cabrales, J. (eds) *Avances en los estudios paleomastozoológicos en México*. Colección Científica 443, pp. 103–23. Mexico City: Instituto Nacional de Antropología e Historia.

Arroyo-Cabrales, J., Polaco, O. J., and Johnson, E. (2006) 'A preliminary view of the coexistence of mammoth and early peoples in México', *Quaternary International*, 142/143, 79–86.

Arroyo-Cabrales, J., Polaco, O. J., Johnson, E., and Ferrusquía-Villafranca, I. (2010) 'A perspective on mammal biodiversity and zoogeography in the Late Pleistocene of México', *Quaternary International*, 212, 187–97.

Aveleyra, L. (1950) *Prehistoria de México. Revisión de prehistoria mexicana: el hombre de Tepexpan y sus problemas*, Mexico City: Ediciones Mexicanas.

Barnosky, A. D., Koch, P. L., Feranec, R. S., Wing, S. L., and Shabel, A. B. (2004) 'Assessing the causes of Late Pleistocene extinctions on the continents', *Science*, 306, 70–5.

Bonnichsen, R. and Turnmire, K. L. (eds) (1999) *Ice Age Peoples of North America: Environments, Origins, and Adaptations of the First Americans*, Corvallis: Center for the Study of the First Americans, Oregon State University Press.

Chatters, J. C., Kennett, D. J., Asmerom, Y., Kemp, B. M., Polyak, V., Nava Blank, A., Beddows, P. A., Reinhardt, E., Arroyo-Cabrales, J., Bolnick, D. A., Malhi, R. S., Culleton, B. J., Luna Erreguerena, P., Rissolo, D., Morell-Hart, S., and Stafford, T. W. Jr (2014) 'Late Pleistocene human skeleton and mtDNA link Paleoamericans and modern Native Americans', *Science*, 344, 750–4.

Corona-M., E. (2008) 'An overview on the origin of Archaeozoology in México', *Quaternary International*, 185, 75–81.

Corona-M., E. (2009) '*Las aves del Cenozoico tardío de México: un análisis paleobiológico*, Madrid: Servicio de Publicaciones de la Universidad Autónoma de Madrid.

Corona-M., E. (2013) 'Localidades del Pleistoceno final en Morelos (México) y su importancia paleoambiental para el poblamiento temprano', *Archaeobios*, 7, 36–46.

Corona-M., E. (2014) 'Algunas consideraciones sobre las relaciones entre el hombre y la fauna en los estudios de prehistoria en México', in Corona-M., E. and Arroyo-Cabrales, J. (eds) *Perspectivas de los estudios de prehistoria en México: un homenaje a la trayectoria del Ing. Joaquín García-Bárcena*, pp. 199–220. Mexico: Instituto Nacional de Antropología e Historia.

Corona-M., E. and Alberdi, M. T. (2006) 'Two new records of Gomphotheriidae (Mammalia: Proboscidea) in southern México and some biogeographic implications', *Journal of Paleontology*, 80, 357–66.

Corona-M., E., Arroyo Cabrales, J., and Polaco, O. J. (2010) 'La arqueozoología en México, una reseña actual', in Mengoni Goñalons, G., Arroyo-Cabrales, J., Polaco, O. J., and Aguilar, F. J. (eds) *Estado actual de la arqueozoologia latinoamericana/Current Advances in Latin-American Archaeozoology*, pp. 165–72. Mexico: Instituto Nacional de Antropología e Historia, Consejo Nacional de Ciencia y Tecnología, International Council of Archaeozoology, Universidad de Buenos Aires.

Feinberg, J. M., Renne, P. R., Arroyo-Cabrales, J., Waters, M. R., Ochoa-Castillo, P., and Pérez-Campa, M. (2009) 'Age constraints on alleged 'footprints' preserved in the Xalnene Tuff near Puebla, Mexico', *Geology*, 37, 267–70.

Flannery, K. V. (ed.) (1986) *Guilá Naquitz: Archaic Foraging and Early Agriculture in Oaxaca, Mexico*, New York: Academic Press.

Fujita, H. and Porcayo Michelini, A. (2014) 'Poblamiento de la península de Baja California', in Corona-M., E. and Arroyo-Cabrales, J. (eds) *Perspectivas de los estudios de prehistoria en México: un homenaje a la trayectoria del Ing. Joaquín García-Bárcena*, pp. 95–121. Mexico: Instituto Nacional de Antropología e Historia.

Gaines, E. P. and Sánchez, G. (2009) 'Current Paleoindian research in Sonora, Mexico', *Archaeology Southwest*, 23, 4–5.

García-Bárcena, J. (2007) 'Etapa Lítica (30,000–2,000 a.C.)', *Arqueología Mexicana*, 86, 30–3.

Gutiérrez-García, T. A., Vázquez-Domínguez, E., Arroyo-Cabrales, J., Kuch, M., Enk, J., King, C., and Poinar, H. N. (2014) 'Ancient DNA and the tropics: a rodent's tale', *Biology Letters*, 10(6), 20140224.

González, S. and Huddart, D. (2008) 'The Late Pleistocene human occupation of Mexico', *FUMDHAMentos, Publicação da Fundação Museu do Homem Americano*, 7, 236–59.

González, S., Jiménez-López, J. C., Hedges, R., Huddart, D., Ohman, J. C., Turner, A., and Pompa y Padilla, J. A. (2003) 'Earliest humans in the Americas: new evidence from México', *Journal of Human Evolution*, 44(3), 379–87.

González González, A. H., Rojas Sandoval, C., Terrazas Mata, A., Benavente Sanvicente, M., and Stinnesbeck, W. (2006) 'Poblamiento temprano en la Península de Yucatán: evidencias localizadas en las cuevas sumergidas de Quintana Roo, México', in Jiménez López, J. C., Polaco, O. J., Martínez Sosa, G., and Hernández Flores, R. (eds) *2° Simposio Internacional el*

Hombre Temprano en América, pp. 73–92. Mexico City: Instituto Nacional de Antropología e Historia.

González, A. H., Terrazas, A., Stinnesbeck, W., Benavente, M. E., Avilés, J., Rojas, C., Padilla, J. M., Velásquez, A., Acevez, E., and Frey, E. (2013) 'Chapter 19. The first human settlers on the Yucatan Peninsula: evidence from drowned caves in the State of Quintana Roo (south Mexico)', in Graf, K. E., Ketron, C. V., and Waters, M. R. (eds) *Paleoamerican Odyssey*, pp. 323–38. College Station: Center for the Study of the First Americans, Texas A&M University.

Jiménez López, J. C., González, S., Pompa y Padilla, J. A., and Ortiz Pedraza, F. (coordinators), (2006) 'El hombre temprano y sus implicaciones en el poblamiento de la cuenca de México. Primer Simposio Internacional', *Colección Científica, Instituto Nacional de Antropología e Historia*, 500, 1–274.

Johnson, E., Arroyo-Cabrales, J. and Polaco, O. J. (2006) 'Climate, environment, and game animal resources of the Late Pleistocene Mexican grassland', in Jiménez López, J. C., González, S., Pompa, J. A., and Ortiz-Pedraza, F. (eds) *El hombre temprano en América y sus implicaciones en el poblamiento de la cuenca de México*, pp. 231–45. Mexico City: Instituto Nacional de Antropología e Historia.

Lorenzo, J. L. (1991) 'Las técnicas auxiliares de la arqueología moderna', in Lorenzo, J. L., and Mirambell Silva, L. (eds) *Prehistoria y arqueología*, pp. 72–131. Mexico City: Instituto Nacional de Antropología e Historia.

Lorenzo, J. L. and Mirambell, L. (1986) *Tlapacoya: 35.000 años de historia del Lago de Chalco*, Mexico City: Instituto Nacional de Antropología e Historia.

Lorenzo, J. L. and Mirambell, L. (1999) 'The inhabitants of México during the Upper Pleistocene', in Bonnichsen, R. and Turnmire, K. L. (eds) *Ice Age Peoples of North America: Environments, Origins, and Adaptations of the First Americans*, pp. 482–96. Corvallis: Center for the Study of the First Americans, Oregon State University Press.

Mark, D. F., González, S., Huddart, D., and Böhnel, H. (2010) 'Dating of the Valsequillo volcanic deposits: resolution of an ongoing archaeological controversy in Central Mexico', *Journal of Human Evolution*, 58, 441–5.

Meltzer, D. J. (2009) *First Peoples in a New World: Colonizing Ice Age America*, Berkeley: The University of California Press.

Mercer, H. C. (1975) *The Hill-Caves of Yucatan: A Search for Evidence of Man´s Antiquity in the Caverns of Central America*, Norman: University of Oklahoma Press.

Mirambell, L. E. (ed.) (2012) *Rancho 'La Amapola', Cedral, un sitio arqueológico-paleontológico Pleistocénico-Holocénico con restos de actividad humana*, Mexico City: Instituto Nacional de Antropología e Historia.

Pérez-Crespo, V. A., Arroyo-Cabrales, J., Alva-Valdivia, L. M., Morales Puente, P., Cienfuegos-Alvarado, E. E., and Otero, F. J. (2012) 'Estado actual de la aplicación de los marcadores biogeoquímicos en paleoecología de mamíferos del Pleistoceno tardío de México', *Archaeobios*, 6, 53–65.

Pérez-Crespo, V. A., Rodríguez, J., Arroyo-Cabrales, J., and Alva-Valdivia, L. M. (2013) 'Variación ambiental durante el Pleistoceno Tardío y Holoceno Temprano en Guilá Naquitz (Oaxaca, México)', *Revista Brasileira de Paleontologia*, 16(3), 487–94.

Polaco, O. J., Arroyo-Cabrales, J., Corona-M., E., and López Oliva, J. G. (2001) 'The American Mastodon *Mammut americanum* in México', in Cavaretta, G. P., Gioia, M., Mussi, M., and Palombo, M. R. (eds) *The World of Elephants*, pp. 237–42. Rome: Consiglio Nazionale delle Ricerche.

Reyes, J. M. (1881) 'Breve reseña de la emigración de los pueblos en el Continente Americano y especialmente en el territorio de la República Mexicana con la descripción de los monumentos de la Sierra Gorda del Estado de Querétaro, distritos de Cadereyta, San Pedro Toliman y Jalpan, y la extinción de la raza chichimeca', *Boletín de la Sociedad de Geografía y Estadística de la República Mexicana, Tercera Época,* 5, 385–490.

Ríos-Muñoz, C. (2013) '¿Es posible reconocer una unidad biótica entre América del Norte y del Sur?', *Revista Mexicana de Biodiversidad* 84(3), 1022–30.

Sanchez, M. G. (2001) 'A synopsis of Paleo-Indian archaeology in Mexico', *The Kiva,* 67, 119–36.

Sanchez, G., Holliday, V. T., Gaines, E. P., Arroyo-Cabrales, J., Martínez-Tagüeña, N., Kowler, A., Lange, T., Hodgins, G. W. L., Mentzer, S. M., and Sanchez-Morales, I. (2014) 'Human (Clovis)-gomphothere (*Cuvieronius* sp.) association ~13,390 calibrated YBP in Sonora, Mexico', *Proceedings of the National Academy of Sciences of the United States of America,* 111, 10972–7.

Sánchez-Miranda, G., Gaines, E. P., Holliday, V. T., and Arroyo-Cabrales, J. (2009) 'El fin del mundo', *Archaeology Southwest,* 23, 6–7.

Santamaría Estévez, D. and García-Bárcena, J. (1989) *Puntas de proyectil, cuchillos y otras herramientas de la Cueva de los Grifos, Chiapas.* Departamento de Prehistoria, Cuadernos de Trabajo 40. Mexico City: Instituto Nacional de Antropología e Historia.

Smith, B. D. (1997) 'The initial domestication of *Cucurbita pepo* in the Americas 10,000 years ago', *Science,* 276, 932–4.

Van Landingham, S. L. (2004) 'Corroboration of Sangamonian age of artefacts from the Valsequillo region, Puebla, Mexico by means of diatom biostratigraphy', *Micropalaeontology,* 50, 313–42.

Winters, M. (2014) 'La prehistoria en Oaxaca: avances recientes', in Corona-M., E. and Arroyo-Cabrales, J. (eds) *Perspectivas de los estudios de prehistoria en México: un homenaje a la trayectoria del Ing. Joaquín García-Bárcena,* pp. 123–41. Mexico City: Instituto Nacional de Antropología e Historia.

CHAPTER 36

THE EXPLOITATION OF AQUATIC ENVIRONMENTS BY THE OLMEC AND THE EPI-OLMEC

TANYA M. PERES

INTRODUCTION

HYDROGRAPHIC features dominate the Olmec heartland. Fishing, travel, transport, and trade via watercraft were essential parts of daily life. It has been well established that aquatic resources were important to the livelihood and sustenance of the Olmec and their successors. In this chapter I use previously published data to examine how the marine, estuarine, and fresh-water animal resources of the Gulf Coastal lowlands in Veracruz, Mexico were incorporated into Olmec and Epi-Olmec lifeways, belief systems, ideological expressions, and foodways.

ENVIRONMENTS OF THE OLMEC AND EPI-OLMEC

The area known as Olman was home to the cultural group we know today as the Olmec (Bernal, 1969). However, the Spanish gave them other lesser-known names, such as, Uixtotin ('people of the salt water') and olmecas uixtotin (Arnold, 2005: 2). Arnold (2005: 2) notes that, in the *Popol Vuh* narratives, the Quiche Maya identify a group of people that came from the east coast and were known as the 'fish-keepers'. Together these names underscore the importance of the aquatic environments to the Olmec.

The Olmec heartland lies in the modern states of Veracruz and Tabasco, Mexico and contains several major physiographic features that dominate the region and influenced Formative period peoples (Fig. 36.1). The Gulf of Mexico marks the northern border of the Olmec area and includes the continental shelf and slope, with waters ranging from < 200 m to 2,000 m in depth. The coast-line is a mix of salt lagoons/marshes, estuaries, mangroves, and sand-bars (CONABIO, 1998). The coastal plain is a narrow strip that fronts the Gulf of Mexico and is home to highly productive estuaries and marshes, especially the Alvarado mangrove community (Olson et al., 1996) (Fig. 36.2). Birds often stop in this area as they migrate to wintering grounds in South America (Ortíz-Pulido et al., 1995; Benitez et al., 1999).

Numerous fresh-water rivers, streams, lakes, and ponds also drain the Olmec heartland. Due to the high sediment load and lack of channelization, the rivers are navigable only by watercraft. Depending on the season, some areas in the Olmec lowlands are more aquatic than others. During the rainy season the fertile *terra firma* may be underwater, forming large lakes, creating seasonal habitats such as ponds and inundated flood-plains, which are home to fish, migratory birds, reptiles, and amphibians.

The highlands are dominated by the Sierra de los Tuxtlas that include several volcanoes. This area was a rich resource of basalt favoured by the Olmec for crafting into monuments and obsidian for tools. The fresh-water Lake Catemaco punctuates the centre of the Tuxtlas (Fig. 36.3). Additionally, two salt domes (Cerro El Mixe and Cerro El Manati) near the site of San Lorenzo are the highest points in the lower Coatzacoalcos River drainage (Cyphers, 1996). The soils in this area are rich for agriculture, and the numerous streams in the region allowed for large-scale harvests of aquatic resources that were reliable, predictable, and could produce a surplus for future consumption or trade, allowing the Olmec and Epi-Olmec to thrive for centuries.

FORMATIVE PERIOD SITES IN REGIONAL CONTEXT

The Olmec of the southern Mexican Gulf Coast thrived during the Formative period (1400 BC–AD 300). An in-depth literature review of previous archaeological work in this area is beyond the scope of this chapter, thus readers are referred to Arnold (2000), VanDerwarker (2006), Pool (2007), Diehl (2010), and Nichols and Pool (2012). Information included here is synthesized from published sources.

Multiple factors influenced the ultimate siting of a settlement, whether it was a resource-extraction station or a large civic-ceremonial centre. Permanent sites were situated for ease of access to flood-plain resources (Arnold, 2009: 401). Many of the sites dating to the Early Formative period were seasonally occupied and are suggestive of group control of flood-plain resource access (Symonds et al., 2002; Arnold, 2009).

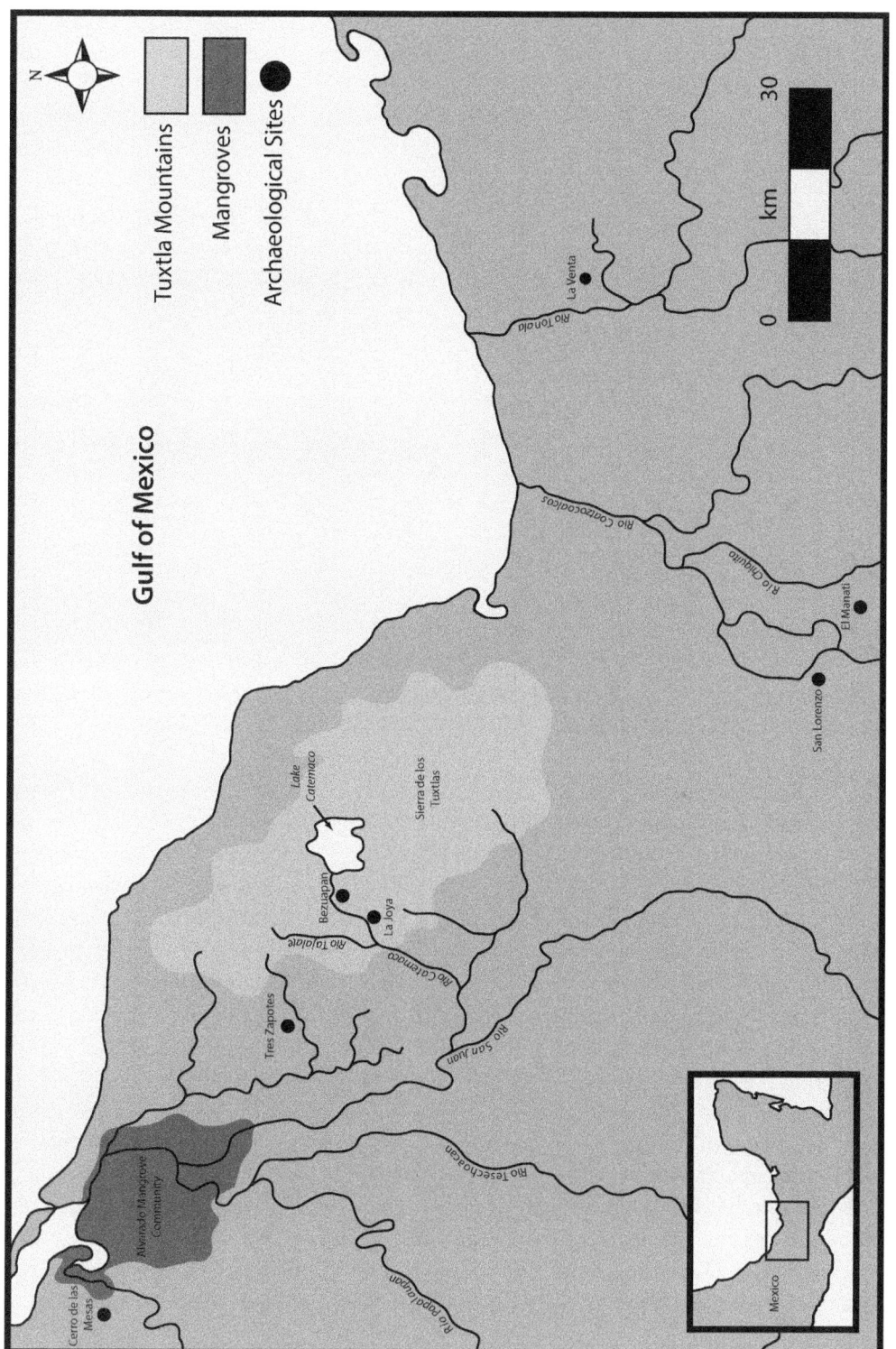

FIGURE 36.1 Map of Olmec heartland showing sites and environments discussed in the text. Created by Joseph Keasler.

FIGURE 36.2 View of estuarine mangroves along the Gulf Coast of Veracruz. Author's own image.

The earliest record of the Olmec dates to the Early Formative (1400–1000 BC) when mounds and monuments were constructed at the large political centre of San Lorenzo (Pool, 2007). Olmec culture was an outgrowth of the smaller farming communities in the region (Diehl, 2004). La Joya, a small farming village was situated along the Catemaco River within 10 km of Lake Catemaco and 20 km of the Gulf of Mexico (Peres et al., 2013). Thus, while the rich estuarine and ocean resources were a half-day's walk from La Joya, fresh-water resources were just minutes away. Farming villages such as La Joya persisted and new ones formed. One such village is Bezuapan, situated along the Bezuapan River, a lesser tributary of the Catemaco River, in the southwestern part of the Tuxtlas (Pool, 1997; VanDerwarker, 2006). While this site was occupied in the Early Formative (Pool, 1997; Santley et al., 1997), it was abandoned following a volcanic eruption at the close of this period. People resettled in this village during the Late Formative, only to be impacted by at least two more volcanic eruptions.

During the Middle Formative (1000–400 BC) San Lorenzo waned in importance and La Venta became the prominent civic-ceremonial centre. Tres Zapotes, on the western edge of the Olmec heartland, was occupied during the Early Formative but thrived during the Late (400 BC–AD 100) and Terminal Formative periods (AD 100–300) and beyond (Pool, 2003a; 2003b; 2008). It is the largest known Epi-Olmec regional centre with an occupational history that spans two thousand years (Pool, 2007). Ultimately

FIGURE 36.3 View of Lake Catemaco, with Sierra de los Tuxtals in the background. Author's own image.

populations decreased and the Olmec culture declined around 400 BC, probably due to major environmental changes (Santley et al., 1997; 2001; Pool, 2007). More recent settlements in the Alvarado are situated along the river, with houses constructed on raised earthen platforms—some of these platforms are previous archaeological occupations, which become islands during seasonal flooding. Modern houses in the area do not occur along interior swamps, though archaeological sites are known from those areas (Stark, 1974). Access and exploitation of aquatic resources helped to structure Olmec subsistence and settlement patterns, even though modern settlements may not entirely reflect these ancient patterns.

FOOD FROM LAND, SEA, ESTUARY, AND RIVER—ANIMALS FOR PHYSICAL SUSTENANCE

Across the Americas research into ancient diets primarily focuses on terrestrial animals, mostly due to poor bone preservation, and/or sampling strategies inadequate for the recovery of smaller and less robust aquatic resources. Evidence is mounting that for many ancient people, aquatic resources from fresh and saline environments were sustaining and sustainable. Previous research shows that the Olmec diet included terrestrial and aquatic animals (Wing, 1980a; 1980b; Rust and Sharer, 1988; VanDerwarker, 2006; Peres et al., 2010; 2013) as well as wild and domesticated foods (VanDerwarker, 2006; Peres et al., 2010). Here I summarize the results of previous studies of aquatic *vs* terrestrial animals at both lowland and Tuxtla Olmec sites.

Terrestrial Animal Resources

A total of twenty-seven unique terrestrial taxa are identified from four Olmec sites (Table 36.1). Wing (1980a) estimated that less than 40% of the edible protein from all contexts at San Lorenzo came from terrestrial resources. At Tres Zapotes, terrestrial resources increased through time (from 33% to 38%) and were largely restricted to elite and ceremonial contexts (Peres et al., 2010; 2013). At the Tuxtla sites of La Joya and Bezuapan, terrestrial animals comprised the majority of the assemblages (average of 85%), but no single taxon dominated (VanDerwarker, 2006; Peres et al., 2013). It appears that the farmers living at La Joya and Bezuapan diversified their intake of animal protein by garden-hunting a variety of terrestrial mammals, turkey, and turtles, and capturing aquatic turtles, birds, and fish (VanDerwarker, 2006; Peres et al., 2013).

Aquatic Animal Resources

The aquatic nature of the lowlands is markedly different from the volcanic Tuxtlas, though both are biologically productive. In the absence of weather and climatic extremes, the lowlands would have allowed for minimal subsistence risk. Indeed, more than eighty taxa combined are identified from San Lorenzo, Tres Zapotes, La Joya, and Bezuapan (Table 36.1).

Peres et al. (2013) conclude that aquatic animals are better represented at San Lorenzo and Tres Zapotes (urban political centres in the lowlands), though they are still present at Bezuapan and La Joya (rural outlying Tuxtla sites). Aquatic animals comprised over 60% of the MNI at San Lorenzo, 62–67% of the MNI at Tres Zapotes, 14–20% of MNI at La Joya, and 11–16% of MNI at Bezuapan (Peres et al., 2013: Table 4). The fact that fewer aquatic animals are present in the highland assemblages is not unexpected, given the sites' distance from the coast and seemingly greater focus on farming (VanDerwarker, 2006). Terrestrial mammals, especially dog and deer, were reserved for consumption by the elites at sites like Tres Zapotes, whereas rural non-elite sites show more even consumption of aquatic and terrestrial animals based on an opportunistic garden-hunting strategy (Peres et al., 2013).

One of the largest aquatic mammals available in the Olmec region, the manatee, was identified at only one site—Tres Zapotes. The twenty-two specimens are

Table 36.1 Identified taxa, their preferred environments and NISP from Bezuapan, La Joya, San Lorenzo and Tres Zapotes. A = Aquatic; T = Terrestrial; I = Indeterminate; B = Brackish; FW = Freshwater; M = Marine

Taxon	Common Name	Aquatic vs. Terrestrial	Preferred water salinity	Bezuapan NISP[1,2]	La Joya NISP[2]	San Lorenzo NISP[3]	Tres Zapotes NISP[4]
Cairina moschata	muscovy duck	A	FW	2	2		
Dermatemys mawii	Mesoamerican river turtle	A	FW			67	
Staurotypus triporcatus	giant Mexican musk turtle	A	FW	5	9	17	163
cf. *Staurotypus triporcatus*	giant Mexican musk turtle	A	FW				2
Chrysemys scripta	pond turtle	A	FW			12	
Trachemys scripta	slider	A	FW	23	49		
Claudius angustatus	mud turtle	A	FW			188	
Anatidae	ducks	A	FW		2	6	
Anas sp.	duck	A	FW		4		
Aythya affinis	lesser scaup	A	FW			1	
Kinosternidae	mud and musk turtles	A	FW				104
Emydidae	water and box turtles	A	FW	3	7		39
cf. *Ariopsis melanopus*	Aguadulce sea catfish	A	FW			2	
Anas carolinensis	green-winged teal	A	FW/B			1	
Chelydridae	snapping turtles	A	FW/B				1
Chelydra serpentina	snapping turtle	A	FW/B			5	
Atractosteus spatula	alligator gar	A	FW/B		2		
Ictaluridae	catfish	A	FW/B				11
Rhamdia guatemalensis	juil descolorido	A	FW/B			3	
Catostomidae	suckers	A	FW/B		2		
Cichlasoma sp.	mojarra	A	FW/B	5	28	2	
Anas clypeata	shoveler	A	FW/B/M			1	

(Continued)

Table 36.1 Continued

Taxon	Common Name	Aquatic vs. Terrestrial	Preferred water salinity	Bezuapan NISP[1,2]	La Joya NISP[2]	San Lorenzo NISP[3]	Tres Zapotes NISP[4]
Centropomus sp.	snook	A	FW/B/M	1	43	120	29
Megalops atlanticus	Atlantic tarpon	A	FW/B/M			1	
Caranx sp.	jack	A	FW/B/M		4	2	
Caranx hippos	jack crevalle	A	FW/B/M				2
Trichechus manatus	manatee	A	FW/B/M				4
cf. Trichechus manatus	manatee	A	FW/B/M				22
Ariidae	marine catfish	A	FW/B/M				9
Mugil sp.	mullet	A	FW/B/M				1
Pimelodidae	flat-nosed & long-whiskered catfish	A	M		6		1
Lutjanus sp.	snapper	A	M	3	7	2	1
Chondrichthyes	sharks, rays, skates	A	M				1
cf. Kinosternon leucostomum	mud turtle	A				7	
Rana sp.	frog	A		7	10		
Osteichthyes	bony fish	A		36	482	15	219
Testudines	turtles	I		9	38	6	1,892
Vertebrata	vertebrates	I		94	411	7	3,730
Mammalia	mammals	I		750	1,629	12	5,689
Aves	birds	I		12	26	4	33
Aves, small to medium	small - medium birds	I					3
Reptilia	reptiles	I		9			1
Serpentes	snakes	I		56	5		2
Amphibia	amphibians	I					
Bufo/Rana sp.	toad/frog	I		25	86		
Iguana iguana	green iguana	T	FW	3	160		

Taxon	Common name	T	FW/B			
Buteogallus anthracinus	black hawk	T			2	
cf. *Rhinoclemmys areolata*	furrowed wood turtle	T			1	
Dasypus novemcinctus	nine-banded armadillo	T	77	4		
Didelphis sp.	opossum	T	14	148		
Carnivora	carnivores	T				80
Canis lupus familiaris	domestic dog	T	80	79	45	37
Mustelidae	skunk/weasel family	T		1		
Procyon lotor	northern raccoon	T			3	
Leopardus pardalis	ocelot	T	2	2		
Artiodactyla	even-toed ungulates	T				2
Tayassuidae	peccary family	T		1		
Pecari spp.	peccary	T				1
cf. *Pecari tajacu*	peccary	T				1
Pecari tajacu	collared peccary	T	2	3	9	
Bos taurus	cattle	T				
Cervidae	deer, elk, wapiti	T	1	43		7
Mazama americana	red brocket	T	4	2		
Odocoileus virginianus	white-tailed deer	T	48	88	2	66
Rodentia	rodents	T				16
cf. *Orthogeomys hispidus*	pocket gopher	T				1
Orthogeomys hispidus	hispid pocket gopher	T	8	54	3	
Sylvilagus sp.	rabbit	T	5	26	1	
Muridae	mouse/rat family	T	6	22		
cf. *Neotoma mexicana*	Mexican wood rat	T	10	2		
Oryzomys sp.	rice rat	T	7	35		
Peromyscus sp.	mouse	T	23	94		
Sigmodon hispidus	hispid cotton rat	T	3	24		
Sciurus sp.	squirrel	T	20	3		

(Continued)

Table 36.1 Continued

Taxon	Common Name	Aquatic vs. Terrestrial	Preferred water salinity	Bezuapan NISP[1,2]	La Joya NISP[2]	San Lorenzo NISP[3]	Tres Zapotes NISP[4]
cf. *Dasyprocta punctata*	agouti	T					1
Soricidae	shrew family	T		1			
Primate	non-human primate	T					1
Homo sapiens	humans	T				7	1
Phasianidae	turkey & quail family	T			1		
Colinus virginianus	northern bobwhite	T			1		
Meleagris spp.	wild turkey	T		18	3		
Falconidae	falcons	T			2		
Buteo sp.	hawk	T		1	2		
Picidae	woodpecker family	T		1			
Sphyrapicus varius	yellow-bellied sapsucker	T			1		
Terrapene carolina	eastern box turtle	T					4
cf. *Terrapene carolina*	eastern box turtle	T					1
Boa constrictor	boa constrictor	T			262		
Bufo sp.	toad	T		200	345	85	
Bufo marinus	marine toad	T					

[1]Table modified after Peres et al., 2013. Data originally published in: [2]VanDerwarker, 2006; [3]Wing, 1980a; FLMNH Acc#0074; [4]Peres, 2008.

all rib fragments, some of which exhibit cut marks while others are burned. All are from an elite residential-administrative context (Peres et al., 2010; 2013). The inclusion of manatee in the prehistoric diet has been documented elsewhere in the Caribbean, with extensive use of manatee known from Moho Cay, Belize in the Maya area (McKillop, 1984; 1985). Non-food uses of manatee are in the form of carved bone figurines, musical instruments, model canoes, and possible fishing weights (McKillop, 1985).

The only identified shark in this sample of Olmec sites is from Tres Zapotes, and is a complete tooth (Peres et al., 2010; 2013). Certainly the use of shark teeth is known from other sites; at El Manati the blade of a ritual knife was tipped with shark's teeth (Oritz et al., 1997 in Arnold, 2005), and a shark's tooth was recovered along with other probable ritual blood-letting artefacts from a bundle on top of Tomb A at La Venta (Drucker, 1952). While the tooth from Tres Zapotes suggests a non-food use of this animal, we must bear in mind that other usable parts are not visible in the archaeological record due to lack of preservation of soft tissue and the cartilaginous nature of the skeleton. Indeed, shark fishing was recorded among the Yucatan Maya in the late nineteenth century by the American archaeologist Edward H. Thompson (de Borhegyi, 1961: 278–9). Thompson specifically recorded the killing of sharks for their livers by the Maya fishermen; these were then boiled to extract oil (Thompson, 1932 in de Borhegyi, 1961: 279–80).

Site-Specific Exploitation of Aquatic Environments

Published data based on number of identified specimens (NISP) for each site as well as the original data collection cards for San Lorenzo were consulted (Wing, 1980a; VanDerwarker, 2006; Peres, 2008; Peres et al., 2013). The NISP are listed for taxa that could be assigned to more specific aquatic environments (see Table 36.1). All of the sites yielded taxa assigned to all fresh-water categories. The two Tuxtla sites of La Joya and Bezuapan yielded the highest NISP derived from fresh-water environs. The lowland civic-ceremonial sites of San Lorenzo and Tres Zapotes are also comprised largely of taxa from the fresh-water category.

The biological literature was consulted to determine preferred habitats for each of the identified taxa from all four sites. Some taxa are better studied and have more detailed information. A number of taxa easily move between areas of ranging salinity, and some taxonomic identifications are too general (i.e. Osteichthyes) to be assigned to fresh-water, fresh-water/brackish/marsh, or marine habitats, and these are listed as semi-aquatic. Of the twenty-six taxa that spend all or a major part of their lifecycle in or around aquatic environments, seven taxa prefer fresh-water, eight taxa prefer fresh-water/brackish environments, six taxa can be found in any of the fresh-water/brackish/marine environments, and three taxa are fully marine. Using the NISP data from the four Olmec sites, it is clear that fresh-water taxa were the preferred resource (see Table 36.1).

IMPLICATIONS FOR HUMAN ACTIVITIES

Fishing Technologies

The fish identified in these assemblages would have necessitated several types of fishing techniques and technologies. Smaller schooling fish are generally caught with mass-capture techniques such as nets, baskets, or weirs, while larger fish may be caught with hook and line, leisters, and gigs. To better understand the types of technologies employed in the capture of different size-classes of fish, I provide a summary of maritime cultures with long-lived coastal fishing traditions in close geographical proximity to the Olmec.

Williams (2009: 613–14) describes the methods used by contemporary fisherfolk in the Lake Cuitzeo region of Mexico. He describes the technologies employed, such as *rudea*—fine mesh circular nets with a long handle—fishhooks, and different types of traps including *corrales* made of reeds, and the *tumbo*, a long narrow net with reed posts and buoys to keep it afloat and with weights at the bottom. The larger *chinchorro* is similar in concept to the *tumbo*, with buoys at the top, but requires more individuals to set and control. Williams (2009: 613) also comments on some techniques that are no longer in use today. These include the *nasa*, made from twigs woven together like a basket with bait inside, and the *tregua*, a line with several hooks along the length, typically baited with shrimp.

Walker (2000) outlines the case for fishing-tackle artefacts from the southwest coast of Florida. Her review of gorges, hooks in various configurations, sinkers, weights, and net-mesh gauges is correlated to size of fish captured. It is likely that equipment similar to that described for coastal fishing traditions from North and South America, including west-central Mexico, was used by the Olmec of the Gulf Coast. However, most of these pieces of fishing tackle were crafted from perishable materials and thus do not survive in the archaeological record. If they do survive they are likely not identified as fishing tackle as Walker (2000: 28) laments: 'the most prevalent [interpretation] maintains that the bone points were used as projectile points for terrestrial hunting or warfare'.

The hunting of the slow-moving, shallow-water manatee was a different occupation than capturing small schooling fish. Based on ethnohistoric and ethnographic accounts from the West Indies, harpoons were the preferred method for hunting manatee (see Romero et al., 2002: Table 1; McKillop, 1985: 339–40). The harpoons themselves may have been fashioned from bone, stone, reed, or obsidian, while the ropes attached to the harpoons would have been formed from plant resources.

The use of dugout canoes in the San Lorenzo region was common in historic times, and it is likely to be a tradition that stretched back for thousands of years. Canoes would have increased the diversity of environments and fish species that could be exploited. Coe and Diehl (1980a; 1980b) observed the use of dugouts in the late 1960s, while Wendt and Cyphers (2008) note the use of plank canoes in this region today. Bitumen is a waterproofing agent for canoes (Coe and Diehl, 1980a: 17; 1980b: 54; Wendt and Cyphers, 2008: 180), and the extraction, use, and trade/exchange of bitumen by the ancient Olmec has been the focus of recent archaeological research (Wendt and Cyphers, 2008; Wendt, 2009; Humphreys, 2012).

Preservation Methods

Contrary to claims that the pre-hispanic Mesoamerican diet 'lacked abundant animal protein, thus suffering from a deficit in sodium chloride' (Williams, 2002: 237), abundant animal protein was available, mainly in the form of turtles and fish. Salt was used for many things, including food preservation. In some wetland zones, where people depend on aquatic resources for a major part of their subsistence economy (see Essuman, 1992; Zohar and Cooke, 1997; Williams, 2009), it is typical for them to be preserved for longer-term consumption. We must therefore be careful to not rule out the likely widespread use of salting, drying, and fermenting as important preservation techniques used by the Olmec. Salt extraction and production by local Tuxtla populations 'was organized at the household level or by small groups of families' (Santley, 2004: 219). I would argue that salt extraction and production was an industry necessary for food preservation and thus actively engaged in by families to this end.

Sixteenth-century ethnohistoric documents from the Lake Cuitzeo area of Mesoamerica note preservation of the *charao* fish via sun-drying and sardine-like fishes (*curuenga*) as being salted (Acuña, 1987: 85 in Williams, 2009: 612). Based on ethnohistorical and ethnographic research (Zohar and Cooke, 1997; Carvajal-Contreras et al., 2008) salting and/or drying fish for future consumption, transport, and trade was a common practice in Panama and probably other areas of Central and Mesoamerica. Carvajal-Contreras and colleagues (2008) also postulate that turtle meat was cured at two rock shelters on the Azuero Peninsula of Panama. They outline markers for smoking and/or dehydration of fish that include: large numbers of burnt fish bones; frequent cut marks, some anatomically patterned; furniture for such activities (i.e. racks); many hearths and pits; and an unusually high proportion of fish branchial bones (Carvajal-Contreras et al., 2008: 103). At this time in the Olmec region there is a lack of available information concerning fish anatomical elements, though the presence of burned bone is often recorded or noted. More feature and contextual analysis at the site level must be undertaken to tease out processing areas with their associated features.

Trading and Sharing of Resources

When access to certain resources is limited to specific individuals, families, or groups, trade becomes one of the few ways to gain such resources. As Arnold (2009: 408) notes: 'Individuals or families controlled specific extraction points, access technologies, and natural holding tanks for live-storage.' McKillop (1985: 345) notes that the sharing of manatee meat was an important and essential activity for the Rama of Caribbean Nicaragua, which saw it as a way of promoting community unification. The provisioning of certain classes of people with preserved foodstuffs is likely, though not always visible archaeologically. Williams (2009: 624) states: 'in prehispanic times fish and other lake products may have been sent to other areas of the [Tarascan] empire (either smoked or salted for preservation) to support troops or colonists.' Peres and colleagues (2013) suggest that the elite groups at San Lorenzo and Tres Zapotes may have been provisioned by rural outlying settlements based on the greater quantities of deer in elite contexts.

Effigies, Teeth, and Images—Animals for Spiritual Sustenance

The roles of animals in past lifeways, as in modern times, go beyond the edible. Human–animal interactions structured people's worldviews, settlement patterns, and technological skills, all of which translated into social and political relationships and ideological expressions. Through artistic representations on artefacts, textual information, context, distribution of faunal materials, and other evidence, we can gain a better understanding of belief systems about animals (cf. Russell, 2012) and how animals feed the soul.

There is a vast body of literature describing animal-effigy and animal-motif artefacts ranging from figurines and ceramics to larger-than-life sculptures. Both terrestrial and aquatic animals are represented, though exotic animals such as jaguars and serpents have been examined more closely than mundane fish or turtle (Arnold, 2005). This overview highlights the aquatic-orientation of many of these objects. These faunal themes (Drucker, 1952: iv) are discussed in terms of iconographic images and objects crafted from animal bone, shell, or teeth that depict aquatic animals or the use of aquatic environments. Notably, a distinction within the aquatic icons suggest that marine creatures were the subjects of the iconographer's craft more often than fresh-water animals, which were the preferred food source.

Aquatica as Iconography and Ritual Paraphernalia

Sharks and sting rays have skeletons made of cartilage rather than bone, and thus very few parts are durable enough to survive in the archaeological record of Mesoamerica—the exceptions are shark teeth and sting-ray spines. The teeth and spines are notable as sting-ray spines are composed of vasodentin, a bone-like cartilage with serrated edges. The number of serrations on the spine is sex-dependent—males have nearly twice as many serrations as females—and habitat-dependent. Sting rays living in open marine waters have more serrations on their spines than do rays living in fresh-water rivers (Schwartz, 2007).

Joyce et al. (1991) note that Olmec perforators are either a shark's tooth or sting-ray spine. The only published record of sting-ray spines recovered from an Olmec or Epi-Olmec site is at La Venta. A bundle of six perforated spines was recovered from the same tomb as a great white shark's tooth and a jade sting-ray spine pendant (Drucker, 1952: 162, Plate 53d). De Borhegyi (1961: Table 1) interprets them as a necklace; however, given their association with the shark's tooth and jade spine, they probably served as ritual perforators. None of the identified sting-ray spines from Olmec or Maya sites have been sexed, but determining if one sex of sting ray was preferred over the other is an intriguing research avenue. Additionally, de Borhegyi (1961: 283) notes that more than half of the recorded shark's teeth from Mesoamerican sites were recovered in association with sting-ray spines (both of which have serrations). This suggests the teeth may have served a similar ritual function as the spines. I include here published occurrences of shark's teeth in ritual settings. A juvenile bundle burial from a tomb in Mound A-2 at La Venta yielded a great white shark's tooth on which a blue jade figurine was placed (de Borhegyi, 1961; Reilly, 1995; Arnold, 2005). A single shark's tooth was recovered from elite-associated deposits at Tres Zapotes, and Peres et al. (2010: 302) suggest the shark's tooth and other aquatic resources from elite contexts at this site are the archaeological correlates of an ocean-based iconography. Arnold (2005: 10) interprets the use of shark imagery and shark products (teeth) in association with ritual regalia or in ritualized contexts as indicative of its sacred importance, and describes various known monuments and artefacts related to the shark-monster. The most well-known jade figurine from the Mound 1 cache at Cerro de las Mesas depicts the shark-monster (Drucker, 1952: Figure 1, Plate 40c; Arnold, 2005). Arnold (2005) suggests the shark-monster was an integral part of the Olmec world-creation story in which it is defeated by a mythical hero figure. The body of the shark monster then becomes the surface of the Earth and the foundation for the *axis mundi* (Grove and Angulo, 1987; Joyce et al., 1991; Stross, 1994; Joralemon, 1996; Arnold, 2005: 2).

Of the bony fish (Osteichthyes), there are a number of families of catfish that are native to the Olmec heartland, both marine and fresh-water. Undoubtedly catfish were caught as a food source, but they are also notable for their serrated pectoral spines that are similar to sting-ray spines. Catfish use these spines as an envenomation defence mechanism and are known to cause painful puncture wounds in humans (Blomkalns and Otten, 1999). Interestingly, these spines also produce sound when the associated muscles are abducted or adducted (Gainer, 1967). A grating or cracking sound is produced by stridulation, the rubbing together of bony parts, in this case the friction between the joints of the pectoral girdle and pectoral fins (Gainer, 1967). This sound was reproduced with dissected pectoral spines and girdles in a laboratory setting (Gainer, 1967). Catfish spines recovered from archaeological contexts are often interpreted as needles or perforator tools. While this is a fair interpretation, the possibility that these elements were valued for their acoustic capacity must also be taken into consideration. A more thorough examination of catfish spines and related elements, including the portion of the elements represented (i.e. articulating ends) and wear patterns need to be conducted to explore this interpretation. Likewise, some turtles may have been used as drums (Miller and Taube, 1993; Emery, 2007).

Other Olmec figurines or effigies made of jade, serpentine, or ceramic depict aquatic animals such as ducks, fish, alligators, and molluscs, many from the site of La Venta (Bernal, 1969). Duck and human/duck ceramic effigies are known from the Olmec area, including the Tuxtla statuette (Bernal, 1969: Plate 47) and Monument 5 from Cerro de Las Mesas (Stirling, 1943: Figure 14b [which is mislabeled as Stela 4 in the caption], Plate 28). Additionally, Altar 7 at La Venta exhibits a human face with duck beak (Drucker, 1952: Plate 65). The clam-shell pendant from La Venta (Drucker, 1952: 163, Plate 53) is very similar to species of the short razor clam family (Solecurtidae). These clams are known from the Gulf of Mexico in the area of study. Additional jade pendants from La Venta (Drucker, 1952: Plate 57c) may be stylized crab claws.

Overall, animals portrayed in art, iconography, and ritual paraphernalia are extant to marine environments in the Olmec heartland. Some of these animals may also have been useful as food sources, but given the elements represented in the faunal assemblages it is unclear. Regardless, the marine motif in the Olmec social and spiritual ideology elevates *aquatica* from the mundane to the sublime.

SUMMARY AND CONCLUSIONS

Previously published faunal, material culture, iconographic, settlement, and environmental data were synthesized to better understand the relationships between the Formative period cultures and the neotropical environments of the southern Gulf Coast of Mexico. In doing this I have shown that different ecological zones sustained the ancient Olmec and Epi-Olmec both physically and spiritually. While the Olmec heartland has a variety of aquatic environments, they differ in regards to salinity, elevation, currents, depths, and representative taxa. The Sierra de los Tuxtlas are dominated by Lake Catemaco and several large fresh-water rivers and their lesser tributaries. The lowlands are dominated by the Gulf of Mexico and the estuarine environments that form the boundary between land and sea.

The two Tuxtla sites had the most diverse representation of aquatic environments, which is likely to be due to the environmental setting of La Joya and Bezuapan and the increasing diversity of the farming villager diets through time. Both sites are situated to easily access fresh-water, estuarine, and marine resources. If food insecurity was an issue, diversifying the animals included for daily subsistence practices helped to offset risk associated with changing subsistence strategies or environmental changes. An avenue for future research is the impact of the Formative period volcanic eruptions on the aquatic resources—how did the ashfall that blanketed the *terra firma* affect the streams, lagoons, rivers, and lakes in the area? Possible evidence in sediment cores of silting-in events might help explain the increased diversification of the diet to include all aquatic resources, not just those nearest the villages.

The use of fresh-water taxa over estuarine and marine taxa at San Lorenzo and Tres Zapotes is interesting, and geographically appropriate. The higher abundance of aquatic species in domestic contexts at Tres Zapotes is probably the result of ease of capture, local availability and abundance, and ease of preservation over terrestrial taxa. The same explanations may extend to the use of fish and other aquatic fauna at San Lorenzo as well.

Overall, the use of animals at Olmec and Epi-Olmec sites extends beyond physical sustenance. Aquatic animals appear early on in Formative period iconography, sculptures, ceramics, and effigies and continue after the Olmec declined. Using marine animals in this way elevates them from the natural worldly realm to the spiritual realm. In this way the aquatic environments of the Olmec and Epi-Olmec fed both body and soul.

ACKNOWLEDGMENTS

I am grateful to Elizabeth Wing, Curator Emeritus at the Florida Museum of Natural History, for granting permission to use the San Lorenzo faunal data she originally collected in 1969. Kitty F. Emery, Associate Curator of Environmental Archaeology, and Irv Quitmyer, Senior Biological Scientist, of the Florida Museum of Natural History assisted with access to the original San Lorenzo data cards (Accession #0074). Joseph Keasler created Fig. 36.1.

REFERENCES

Acuña, René (ed.) (1987) *Relaciones geográficas del siglo XVI: Michoacán.* Mexico, UNAM.

Arnold, P. J, III. (2000) 'Sociopolitical complexity and the Gulf Olmecs: a view from the Tuxtla mountains', in Clark, J. E. and Pye, M. E. (eds) *Olmec Art and Archaeology in Mesoamerica*, pp. 117–35. New Haven: Yale University Press.

Arnold, P. J, III. (2005) 'The shark-monster in Olmec iconography', *Mesoamerican Voices*, 2, 1–31.

Arnold, P. J, III. (2009) 'Settlement and subsistence among the Early Formative Gulf Olmec', *Journal of Anthropological Archaeology*, 28, 397–411.

Benitez, H., Arizmendi, C., and Marquez, L. (1999) *Base de Datos de las AICAS*, Mexico City: CIPAMEX, CONABIO, FMCN and CCA.

Bernal, I. (1969) *The Olmec World*, Berkeley: University of California Press.

Blomkalns, A. L. and Otten, E. J. (1999) 'Catfish spine enovenomation: a case report and literature review', *Wilderness & Environmental Medicine*, 10(4), 242–6.

Carvajal-Contreras, D. R., Cooke, R. G., and Jimenez, M. (2008) 'Taphonomy at two contiguous coastal rockshelters in Panama: preliminary observations focusing on fishing and curing fish', *Quaternary International*, 180, 90–106.

Coe, M. D. and Diehl, R. A. (1980a) *In the Land of the Olmec.* Vol. 1: *The Archaeology of San Lorenzo Tenochtitlan*, Austin: University of Texas Press.

Coe, M. D. and Diehl, R. A. (1980b) *In the Land of the Olmec.* Vol. 2: *The People of the River*, Austin: University of Texas Press.

CONABIO (1998) Regiones Hidrológicas Prioritarias: Fichas Técnicas y Mapa, Mexico City: CONABIO.

Cyphers, A. (1996) 'Reconstructing Olmec life at San Lorenzo', in Benson, E. P. and de la Fuente, B. (eds) *Olmec Art of Ancient Mexico*, pp. 61–72. Washington: National Gallery of Art.

de Borhegyi, S. F. (1961) 'Shark teeth, sting ray spines, and shark fishing in ancient Mexico and Central America', *Southwestern Journal of Anthropology*, 17(3), 273–96.

Diehl, R. A. (2004) *The Olmecs: America's First Civilization*, London: Thames and Hudson.

Diehl, R. A. (2010) *Death Gods, Smiling Faces and Colossal Heads: Archaeology of the Mexican Gulf Lowlands: Recent Publications in Gulf Lowlands Archaeology.* http://www.famsi.org/research/diehl/section06.html.

Drucker, P. (1952) *La Venta, Tabasco: A Study of Olmec Ceramics and Art.* Bureau of American Ethnology Bulletin 153. Washington: Smithsonian Institution.

Emery, K. F. (2007) 'Aprovechamiento de la fauna en Piedras Negras: dieta, ritual y artesanía del period Clásico Maya', *Mayab: Journal of the Sociedad Española de Estudios Mayas*, 19, 51–69.

Essuman, K. M. (1992) *Fermented Fish in Africa: A Study on Processing, Marketing, and Consumption.* FAO Fisheries Technical Paper 329. Rome: Food and Agriculture Organization of the United Nations.

Gainer, H. (1967) 'Neuromuscular mechanisms of sound production and pectoral spine locking in the banjo catfish, *Bunocephalus* species', *Physiological Zoology*, 40(3), 296–306.

Grove, D. C. and Angulo V., J. (1987) 'A catalog and description of Chalcatzingo's monuments', in Grove, D. C. (ed.) *Ancient Chalcatzingo*, pp. 114–31. Austin: University of Texas Press.

Humphreys, T. (2012) 'Bitumen Control and Exchange among the Olmec'. Unpublished MA dissertation, California State University (Chico).

Joralemon, P. D. (1996) 'Bottle with carved fish monster', in Benson, E. P. and de la Fuente, B. (eds) *Olmec Art of Ancient Mexico*, p. 199. Washington: National Gallery of Art.

Joyce, R. A., Edging, R., Lorenz, K., and Gillespie, S. D. (1991) 'Olmec bloodletting: an iconographic study', in Robertson, M. G. and Fields, V. M. (eds) *Sixth Palenque Round Table*, pp. 143–50. Norman: University of Oklahoma Press.

McKillop, H. I. (1984) 'Prehistoric Maya reliance on marine resources: analysis of a midden from Moho Cay, Belize', *Journal of Field Archaeology*, 11(1), 25–35.

McKillop, H. I. (1985) 'Prehistoric exploitation of the manatee in the Maya and circum-Caribbean areas', *World Archaeology*, 16(3), 337–53.

Miller, M. and Taube, K. (1993) *An Illustrated Dictionary of the Gods and Symbols of Ancient Mexico and the Maya*, London: Thames and Hudson.

Nichols, D. L. and Pool, C. A. (eds) (2012) *The Oxford Handbook of Mesoamerican Archaeology*, Cambridge: Oxford University Press.

Olson, D. M., Dinerstein, E., Cintrón, G., and Iolster, P. (1996) *A Conservation Assessment of Mangrove Ecosystems of Latin America and the Caribbean. Final Report for The Ford Foundation*, Washington: World Wildlife Fund.

Ortíz-Pulido, R., Gómez de Silva, H., González-García, F., and Alvarez, A. (1995) 'Avifauna del Centro de Investigaciones Costeras La Mancha, Veracruz, México', *Acta Zoologica Mexicana, Nueva Serie*, 66, 87–118.

Peres, T. M. (2008) Zooarchaeological Analysis of Animal Remains from Tres Zapotes, Veracruz, México. Report submitted to Christopher A. Pool, Department of Anthropology, University of Kentucky (Lexington).

Peres, T. M., VanDerwarker, A. M., and Pool, C. A. (2010) 'The farmed and the hunted: integrating floral and faunal data from Tres Zapotes, Veracruz', in VanDerwarker, A. M. and Peres, T. M. (eds) *Integrating Zooarchaeology and Paleoethnobotany: A Consideration of Issues, Methods, and Cases*, pp. 281–308. New York: Springer Press.

Peres, T. M., VanDerwarker, A. M., and Pool, C. A. (2013) 'The zooarchaeology of Olmec and Epi-Olmec foodways along Mexico's Gulf Coast', in Götz, C. M. and Emery, K. F. (eds) *The Archaeology of Mesoamerican Animals*, pp. 95–128. Georgia: Lockwood Press.

Pool, C. A. (1997) 'The spatial structure of Formative houselots at Bezuapan', in Stark, B. L. and Arnold, P. J, III. (eds) *Olmec to Aztec: Settlement Patterns in the Ancient Gulf Lowlands*, pp. 40–67. Tuscon: University of Arizona Press.

Pool, C. A. (2003a) 'Introduction', in Pool, C. A. (ed.) *Settlement Archaeology and Political Economy at Tres Zapotes, Veracruz, Mexico*, pp. 1–5. Los Angeles: Cotsen Institute of Archaeology, University of California.

Pool, C. A. (2003b) 'Centers and peripheries: urbanization and political economy at Tres Zapotes', in Pool, C. A. (ed.) *Settlement Archaeology and Political Economy at Tres Zapotes, Veracruz*, pp. 90–8. Los Angeles: Cotsen Institute of Archaeology, University of California.

Pool, C. A. (2007) *Olmec Archaeology and Early Mesoamerica*, Cambridge: Cambridge University Press.

Pool, C. A. (2008) 'Architectural plans, factionalism, and the Protoclassic-Classic at Tres Zapotes', in Arnold, P. J. and Pool, C. A. (eds) *Classic Veracruz: Cultural Currents in the Ancient Gulf Lowlands*, pp. 121–57. Washington: Dumbarton Oaks.

Reilly, F. K, III. (1995) 'Art, ritual, and rulership in the Olmec world', in Guthrie, J. and Benson, E. P. (eds) *The Olmec World: Ritual and Rulership*, pp. 27–45. Princeton: The Art Museum, Princeton University.

Romero, A., Baker, R., Cresswell, J. E., Singh, A., McKie, A., and Manna, M. (2002) 'Environmental history of marine mammal exploitation in Trinidad and Tobago, W. I., and its ecological impact', *Environment and History*, 8(3), 255–74.

Russell, N. (2012) *Social Zooarchaeology: Humans and Animals in Prehistory*, Cambridge: Cambridge University Press.

Rust, W. F. and Sharer, R. J. (1988) 'Olmec settlement data from La Venta, Tabasco, Mexico', *Science*, 242, 102–4.

Santley, R. S. (2004) 'Prehistoric salt production at El Salado, Veracruz, Mexico', *Latin American Antiquity*, 15(2), 199–221.

Santley, R. S., Arnold, P. J., III, and Barrett, T. P. (1997) 'Formative period settlement patterns in the Tuxtla Mountains', in Stark, B. L. and Arnold, P. J, III. (eds) *Olmec to Aztec: Settlement Patterns in the Ancient Gulf Lowlands*, pp. 174–205. Tuscon: University of Arizona Press.

Santley, R. S., Nelson, S., Reinhardt, B., Pool, C. A., and Arnold, P. J. III, (2001) 'When day turned to night: volcanism and the archaeological record from the Tuxtla Mountains, Southern Veracruz, Mexico', in Bawden, G. and R. Reycraft (eds) *Environmental Disaster and the Archaeology of Human Response*. Anthropological Papers of the Maxwell Museum of Anthropology 7, pp. 143–62. Albuquerque: University of New Mexico.

Schwartz, F. J. (2007) 'A survey of tail spine characteristics of stingrays frequenting African, Arabian to Chagos-Maldive archipelago waters', *Smithiana Bulletin*, 8, 41–52.

Stark, B. L. (1974) 'Geography and economic specialization in the lower Papaloapan, Veracruz, Mexico', *Ethnohistory*, 21(3), 199–221.

Stirling, M. W. (1943) *Stone Monuments of Southern Mexico*. Bureau of American Ethnology Bulletin 138. Washington: Government Printing Office.

Stross, B. (1994) 'The iconography of power in Late Formative Mesoamerica', *RES: Anthropology and Aesthetics*, 25, 10–35.

Symonds, S., Cyphers, A., and Lunagomez, R. (2002) *Asentamiento Prehispanico en San Lorenzo Tenochtitlan*, Mexico City: Universidad Nacional Autonoma de Mexico.

VanDerwarker, Amber M. (2006) *Farming, Hunting, and Fishing in the Olmec World*, Austin: Univeristy of Texas Press.

Walker, K. J. (2000) 'The material culture of Precolumbian fishing: artefacts and fish remains from coastal southwest Florida', *Southeastern Archaeology*, 19(1), 24–45.

Wendt, C. J. (2009) 'The scale and structure of bitumen processing in early Olmec households', in Hirth, K. G. (ed.) *Housework: Specialization, Risk, and Domestic Craft Production in Mesoamerica*. Archaeological Papers 19, pp. 33–44. Washington: American Anthropological Association.

Wendt, C. J. and Cyphers, A. (2008) 'How the Olmec used bitumen in ancient Mesoamerica', *Journal of Anthropological Archaeology*, 27(2), 175–91.

Williams, E. (2002) 'Salt production in the coastal area of Michoacan, Mexico: an ethnoarchaeological study', *Ancient Mesoamerica*, 13, 237–53.

Williams, E. (2009) 'The exploitation of aquatic resources at Lake Cuitzeo, Michoacan, Mexico: an ethnoarchaeological study', *American Antiquity*, 20(4), 607–27.

Wing, E. S. (1980a) 'Zooarchaeology', in Coe, M. D. and Diehl, R. A. (eds) *In the Land of the Olmec*. Vol. 1: *the Archaeology of San Lorenzo Tenochtitlan*, pp. 375–86. Austin: University of Texas Press.

Wing, E. S. (1980b) 'Human-animal relationships', in Coe, M. D. and Diehl, R. A. (eds) *In the Land of the Olmec*. Vol. 2: *People of the River*, pp. 97–123. Austin: University of Texas Press.

Zohar, I. and Cooke, R. G. (1997) 'The impact of salting and drying on fish bones: preliminary observations on four marine species from Parita Bay, Panama', *Archaeofauna*, 6, 59–66.

CHAPTER 37

TRACKING THE TRADE IN ANIMAL PELTS IN EARLY HISTORIC EASTERN NORTH AMERICA

HEATHER A. LAPHAM

INTRODUCTION

THIS chapter examines zooarchaeology's contribution to our understanding of the trade in animal pelts (furs, skins, and hides) and a wide array of other goods that developed between Native Americans and Europeans in early historic eastern North America. Hides from white-tailed deer (*Odocoileus virginianus*) dominated exchanges in the southern trades, whereas the northern trades focused on acquiring the furs of American beaver (*Castor canadensis*). Pelts from other fur-bearing animals, large and small, entered into the commerce as well, including black bear (*Ursus americanus*), red and gray fox (*Vulpes vulpes* and *Urocyon cinereoargenteus*), raccoon (*Procyon lotor*), river otter (*Lontra canadensis*), and mink (*Mustela vison*), among many others. The driving forces behind the acquisition of animal pelts in both regions was the exportation of these items overseas to Europe for use in the manufacture of fine leather goods, fashionable beaver felt hats, and other furry accoutrements. Declining deer populations and the near extinction of beavers in western Europe, due to over-harvesting and a reduction in their natural habitats, resulted in millions of pelts exported from colonial ports in eastern North America to Europe during the seventeenth and eighteenth centuries and beyond.

European merchants obtained these desired pelts through trade with Native Americans residing up and down the eastern seaboard to far inland locales, with middlemen of European, Native, and Mestizo descent negotiating exchanges at designated forts (especially in northern regions) and at indigenous towns and villages throughout the region (a more southernly practice). Many traders operated independently, selling their pelts directly to merchants, or to smaller, lesser-known firms, while others were employed by one of the larger companies that vied to monopolize the lucrative northern fur trade business at various times throughout history (e.g. the Company of New France, Hudson's Bay Company, Northwest Company, and American Fur Trade Company). As this trade in animal pelts and European-manufactured goods burgeoned, Native hunters were presented, rather suddenly, with new opportunities to use their keen hunting and trapping skills for different purposes. The exploitation of certain animals, especially deer and beaver, now had the potential to involve more than hunting for household consumption, and it could also include hunting for the production of surplus skins and furs for commercial trade.

These new business ventures (commonly called the deerskin trade in the south and the fur trade in the north) forever changed, in very complex ways, many aspects of Native American society, including subsistence economies and hunting practices. The literature on the pelt trades is expansive, and this chapter does not begin to attempt to summarize what is known about these commercial enterprises. The focus here is to examine how the pelt trades influenced Native American animal procurement and consumption patterns at two contemporaneous, yet geographically different locales, one in southwestern Virginia and the other in south-central Pennsylvania. There is ample historical and archaeological evidence that Native Americans participated in the pelt trades, but what evidence exists in the zooarchaeological record to indicate that certain seventeenth-century indigenous communities were hunting and trapping animals for commercial trade? Drawing on two case studies, this chapter explores the ways in which the deerskin and fur trades can be identified and tracked in the zooarchaeological record.

Hunting to Procure Hides for Trade: Some Zooarchaeological Signatures

The beginnings of the deerskin trade in the southern American colonies evolved over several decades before becoming a viable and profitable business. By the mid- to late seventeenth century, the deerskin trade was well established in Virginia (Crane, 1928; Martin, 1994; Stine, 1990; Rountree, 2002), with merchants and traders from the Carolina colony joining these lucrative business ventures in the 1670s following the founding of Charleston on the present-day South Carolina coast (Moore, 1973; Robinson, 1979). Over time, as the trade grew, commercial activities shifted southward and westward. Charleston, along with Savannah farther south on the Georgia coast, became key ports of export for deerskins shipped overseas from Great Britain's Atlantic colonies (Fig. 37.1). French traders joined the burgeoning commerce as well, export-ing their pelts from coastal ports along the Gulf of Mexico, such as New Orleans and Mobile in the Louisiana colony (Usner, 1992). Traders acquired hundreds of thousands of deerskins and other animal pelts annually from the Catawba, Cherokee, Creek, Chickasaw, and Choctaw Indians, among other southeastern Native groups (Braund, 1993; Perdue, 1998; Wesson, 2010). Trade paths and rivers linked Anglo merchants and traders to Indian settlements located hundreds of kilometres inland, allowing hides and furs to be transported from Native villages in the interior regions of eastern North America to colonial ports along the Atlantic and Gulf coasts.

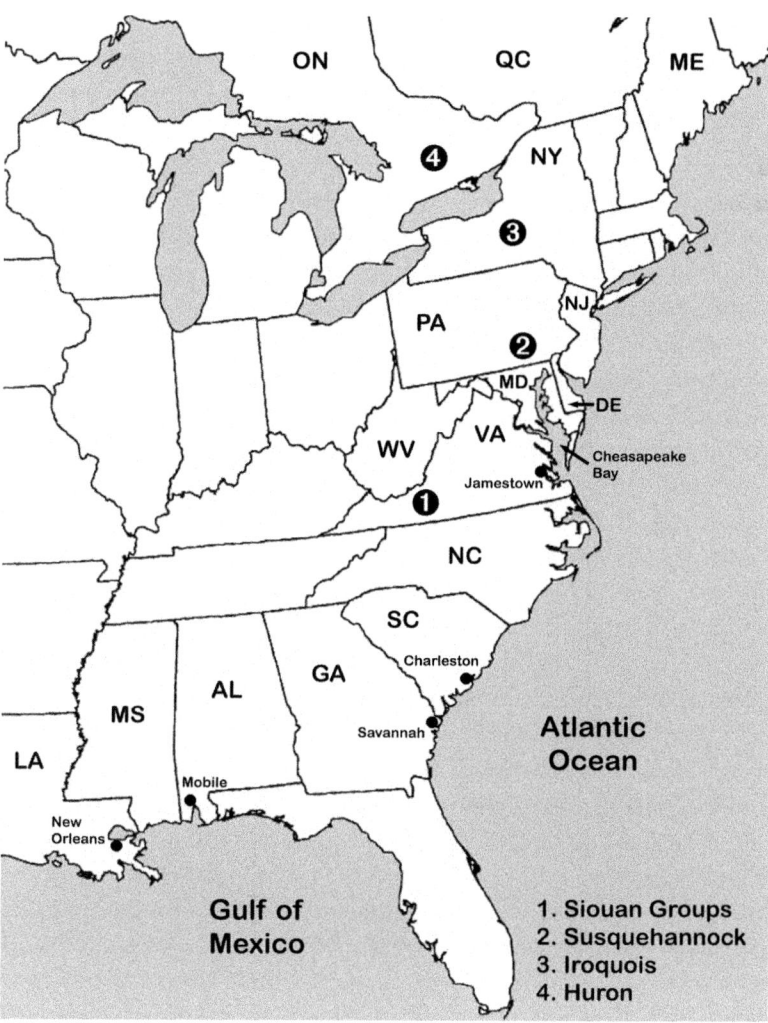

FIGURE 37.1 Map locating Native American groups and archaeological sites and colonial settle-ments mentioned in the text.

By the late seventeenth and eighteenth centuries, the deerskin trade had become a predominant transformative force in many southeastern Native American societies. Deerskins were now an important marketable resource, with procurement and production going beyond household consumption. On the commercial market, several factors influenced the value of a deerskin, with larger or heavier hides that had been well processed bringing the highest exchange rates, at least during the initial decades of the trade. Historic valuations of deerskins combined with a knowledge of white-tailed deer biology have been used to define relationships between the various types of deerskins mentioned in colonial documents and the sex and age of the deer hunted, which can be distinguished in the zooarchaeological record (Lapham, 2004; 2005). A comparative study by the author of animal remains from three Native American village sites located deep in the Virginia interior provided insights into how hunting to procure hides for commercial trade manifested in the zooarchaeological record with regard to seventeenth-century subsistence practices, deer-hunting strategies, and hide-production methods (Lapham, 2004; 2005). The results of this research are briefly summarized here to provide a foundation to examine zooarchaeological evidence of participation in the pelt trades at contemporaneous Susquehannock settlements in south-central Pennsylvania.

In the fifteenth and sixteenth centuries, prior to European colonization and the commencement of the commercial deerskin trade, deer harvests in southwestern Virginia and the surrounding region typically resembled the natural structure of a wild, unmanaged deer population in that fawns were the single most common age group killed, followed generally by progressively fewer animals in older age classes. Hunters killed male and female deer in similar numbers, or killed female deer more often. Hunting occurred most frequently during the late summer through early winter months, largely taking place as a seasonal activity after agricultural harvests had slowed, but occasionally year-round (Lapham, 2004; 2005).

By the early to mid-seventeenth century, deer-hunting strategies had changed at some probable Sioaun speaking Native villages in southwestern Virginia, and these changes were directly related to hunting deer to produce hides for commercial trade. At the seventeenth-century Trigg site, for example, hunters focused on procuring prime-age deer (deer in the 1.5 to 4.5 year age classes), killing bucks more often than does, and hunting deer year-round with the fewest numbers of deer taken during peak moulting season when hide quality was poor. These selective harvest strategies procured deer whose hides would bring the highest exchange rates on the commercial market. Deer butchery practices also showed an increase in skinning events intended to maximize the size of the hide removed from the animal. Likewise, beamers, hide-processing tools made from deer metatarsals, increased markedly in production and use. In addition, Native residents of the Trigg site exploited more deer than their Late Woodland-period predecessors, resulting in the increased consumption of venison, which replaced meat once provided by other large mammals, such as black bear and wapiti (*Cervus elaphus*). Other fur-bearing animals important to trade (specifically beaver, raccoon, and fox) were hunted and trapped slightly more often than they had been in earlier times (Lapham, 2004; 2005).

Drawing on this hunting-for-hides model, the following section examines seventeenth-century Susquehannock animal economies to see how, and if, similar trends are apparent in subsistence and deer hunting in south-central Pennsylvania. In this region, beaver was the most sought after fur in the commercial pelt trades, so the exploitation of this large rodent also must be considered to gain a more comprehensive perspective on animal-use practices.

THE NORTHERN FUR TRADES: A CASE STUDY FROM SOUTH-CENTRAL PENNSYLVANIA

The European beaver (*Castor fiber*), once plentiful in western Europe, began to decline in numbers as early as the tenth century in the British Isles due to over-harvesting and a reduction in its natural habitat as more and more lands were put to agricultural uses (Spriggs, 1998). With local beaver populations nearing extinction, the English sought new sources of beaver, reaching out to Scandinavia and Russia, until expeditions to North America in the sixteenth and early seventeenth centuries opened the doors to a vast, untapped supply of this large, highly valued rodent (Martin, 1892; Sleeper-Smith, 2009). In the decades that followed, sustained settlement by the French along the St. Lawrence River and the far northeastern seaboard, the Dutch along the Hudson River, and the English, who claimed lands along the Hudson Bay and the eastern seaboard from present-day Maine south to Maryland and Virginia, gave rise to sizeable commercial enterprises focused largely (from a European perspective) on the acquisition and exportation of beaver pelts (Lawson, 1943; Innis, 1956; Phillips, 1961; Ray, 1974).

Beaver was highly valued for its fur by the European fashion industry, particularly for use in hat-making. Hatters extracted the soft underfur from the pelt to produce a supple yet strong felt that when formed into a hat held its shape exceptionally well (Crean, 1962; Grant, 1989). The beaver felt hat soared in popularity in the seventeenth century, quickly becoming an essential element of high fashion attire, and spilling over into commoner dress as well. Hat styles and details changed rapidly, defining and identifying those who had wealth (and those who did not), and reflecting political and religious affiliations (Crean, 1962; Grant, 1989). The seemingly insatiable demands of the European market led to the export of many hundreds of thousands of beaver pelts from North America. Between 1700 and 1763, for example, more than 2.75 million beaver pelts were acquired by the English-owned Hudson's Bay Company alone, along with nearly as many obtained from Native Americans by French traders (Carlos and Lewis, 2010: 107). Spanning more than two centuries, these commercial ventures resulted in the depletion of beaver populations throughout regions where they were once plentiful (Martin, 1892; Spriggs, 1998). About 1840, European fashion trends in hats changed to favour a high sheen that could only be produced using silk, and the beaver felt hat fell out of style, essentially ending the profitable beaver fur trade in the Americas (Crean, 1962;

Lawson, 1972). The rise in popularity of silk hats was well timed, as beaver populations in North America were in rapid decline by the 1820s due to over-harvesting (Hanson, 2000; Carlos and Lewis, 2010).

The Hudson's Bay Company, founded in 1670 and still in existence today, was a corporate giant in the fur trade, controlling much of the trade on English-claimed lands in northern North America. Under the company, beaver became a standard of barter and trade, where European goods were valued based on their worth in beaver pelts (Martin, 1892: 104–5; Carlos and Lewis, 2010: 51–2). By the early eighteenth century, the list of trade goods supplied to Native Americans in exchange for beaver furs was extensive, representing upwards of seventy different items. These goods ranged from beads and bells, to needles and awls, broad cloth and flannel, tobacco and brandy, knives, hatches, and guns, just to name a few (Martin, 1892: 104–5; Carlos and Lewis, 2010: 80–3). Despite its previously overstated importance, alcohol was just one of many items exchanged for animal pelts, and one that played an insignificant role in the trade and in indigenous lifeways, at least in the northeast, prior to the 1740s (Carlos and Lewis, 2010: 11–12).

Supplying satisfactory goods was of upmost importance to a successful business transaction in the northeast and southeast alike (Ray, 1980; Anderson, 1994; Stern, 2012). An item had to meet the wants, needs, standards, and aesthetic preferences of Native American consumers. Supplying anything less than acceptable goods could result in furs being taken to the competitor for trade. Each item had a known value in beaver skins. One beaver pelt could be used to purchase twlece dozen buttons or one pair of breeches or shoes, two yards of gartering, one half pound of thread, two pounds of sugar, two hatchets, twenty gun flints, or one pound of lead, and the list goes on and on (Martin, 1892: 104–5). Although beaver reigned supreme in these northern commercial enterprises, historic documents (merchant and trader records, government export and import numbers, etc.) indicate the furs and hides of many other animals also entered into the trade in smaller numbers, including martens (*Martes americana*), otters, raccoons, foxes, wolves (*Canis lupus*), black bears, deer, and wapiti. Like the manufactured goods, each animal was assigned a standard value in beaver (Martin, 1892: 103). These values fluctuated as English and French traders vied for a bigger share of available pelts, attempting to establish a monopoly in this lucrative business.

Susquehannock Traders on the Chesapeake Bay and Beyond

The Susquehannock Indians played a leading role in the fur trade in the greater Chesapeake Bay area and beyond for a good part of the seventeenth century. They arrived in the lower Susquehanna River Valley in south-central Pennsylvania sometime in the late sixteenth century, having migrated south from the river's headwaters in southern New York, perhaps to better position themselves in the fur trade and gain better access to sources of furs and European goods (Kent, 1984: 19). When Captain John Smith, leader of the Jamestown colony in Virginia (the first permanent English settlement in North America), first encountered the Susquehannock in 1608 they were already in possession of European goods, apparently acquired from French traders to the north, and the Susquehannock had been trading these items to Native peoples residing at the north end of the Chesapeake Bay (Kent, 1984: 26).

By the initial decades of the seventeenth century, the Susquehannock controlled lands along the Susquehanna River and its tributaries in central Pennsylvania south to where the river enters the Chesapeake Bay, with satellite settlements identified archaeologically as far west as the upper Potomac River Valley in western Maryland and West Virginia (Brashler, 1987; Wall and Lapham, 2003; Wyatt, 2012). Their power and predominance on the colonial stage faded as quickly as it had appeared. By the mid-1670s, the Susquehannock, having been decimated by disease and ongoing warfare with the Iroquois to the north, were forcibly removed from their homelands in south-central Pennsylvania. Historical accounts identify the remaining populations as transient, living briefly in Maryland, Virginia, and then North Carolina, until they became obscured by history (Kent, 1984).

Early Historic Susquehannock Sites

The case study presented here considers zooarchaeological evidence of participation in the pelt trades at three Susquehannock villages (Schultz, Eschelman, and Strickler) located in Lancaster County, Pennsylvania. The earliest site in the sequence, the Schultz site (36La7), represents the Susquehannock's first settlement in the lower Susquehanna River Valley, following their migration south from the river's headwaters in southern New York. Population estimates suggest between 1,200 and 1,500 people resided at the Schultz village (Kent, 1984: 325), which was occupied c.1590s–1620 (Sempowski, 1994; for alternative dates at all three sites see Kent, 1984). Archaeological excavations that began in the late 1960s revealed multiple stockades, numerous multi-family domestic residences (known as longhouses), and hundreds of storage and refuse pits and other features (Kent, 1984: 321–6). Objects of European manufacture included several iron axes and knives and a variety of brass ornaments (Cadzow, 1936). Excavations at two adjacent cemeteries in the late 1970s yielded more than 3,000 glass beads (Smith and Graybill, 1977; Kent, 1983).

The Eschelman site (36La12) is an extensive midden area that represents a small portion of a large Susquehannock village located within the present-day town of Washington Boro, Pennsylvania (Guilday et al., 1962). An estimated 1,700 persons lived at this settlement c.1620–1630s (Kent, 1984: 364; Sempowski, 1994). Excavations in 1949 recovered tens of thousands of artefacts from this area alone, including more than 58,000 animal remains (Guilday et al., 1962). European goods acquired through trade (both directly with European merchants and indirectly via other indigenous groups) included iron tools and cutting implements, lead shot (although no guns), brass projectile points, several whole brass kettles, and other functional and decorative items (Cadzow, 1936; Kent, 1984). The Washington Boro settlement as a whole contained more than 7,000 glass beads (Kent, 1983). The glass bead assemblage most closely resembles those types associated with Dutch and English trade c.1624–1660s, although some beads indicative of earlier periods also occur (Lapham and Johnson, 2002). The quantity and diversity of these objects provide ample evidence of the Susquehannock's ability to acquire large numbers of European goods through new and existing trade networks.

The Susquehannock eventually abandoned the Washington Boro village and relocated to several other settlements, one closer to the Delaware River where Swedish and Dutch traders operated (Kent, 1984: 339), where they lived for a decade or so. During the mid-1640s, the Susquehannock returned to an area located several hundred yards south of the Schultz village and established a large settlement known as the Strickler site (36La3). An estimated 2,900 residents occupied the village until the mid-1660s. This period in Susquehannock history was characterized by hostile relationships with the Iroquois in New York and fluctuating friend-to-foe relationships with the English in Maryland (Kent, 1984: 22, 363). Excavations beginning in 1968 revealed three co-existing stockade lines surrounding numerous longhouses, with the settlement further protected by European-style bastions at two corners, along with more than five hundred storage and refuse pits and other features (Kent, 1984: 351–3). The sheer number and variety of European objects recovered during excavations at the Strickler site reflects the Susquehannock's successful economic status and military strength at this moment in history. European-manufactured military items, hunting implements, and edged tools were numerous along with a wide variety of imported decorative objects, functional household items, and more than 50,000 glass beads (Cadzow, 1936; Futer, 1959; Kent, 1983; 1984).

Susquehannock Animal Economies

All three Susquehannock sites had large, well-preserved zooarchaeological assemblages. The data presented in this chapter rely on the analyses conducted by Webster (1983) and Guilday and colleagues (1962), however the interpretations of the data are my own, unless otherwise noted. The number of identified specimens (NISP) is the one piece of primary data reported consistently for all three Susquehannock sites, therefore it is the comparative measure used here. In addition to wild game, numerous birds, turtles, and fishes contributed to Susquehannock meat diet (Guilday et al., 1962; Webster, 1983). This chapter examines zooarchaeological evidence for hunting and trapping animals to produce hides and furs for commercial trade, so the following summaries focus solely on mammalian taxa. Discussing other classes of animals, although important to Susquehannock animal economies, is beyond the scope of this chapter.

At the Schultz site, more than 2,300 specimens analyzed from a sample of domestic areas within the palisaded village were identified to the taxonomic level of genus and lower (Webster, 1983: 256, Table 62). White-tailed deer comprised nearly half of the mammalian assemblage. Other important species based on NISP included black bear, wapiti, gray fox, raccoon, and beaver. The Eschelman site midden at Washington Boro contained more than 23,500 identified specimens, with white-tailed deer being the most frequent taxon represented, followed at a distance by bear, wapiti, raccoon, and bea-ver (Guilday et al., 1962: Table 1). Domestic dog (*Canis familiaris*), although present, does not appear to have been part of the diet (Guilday et al., 1962: 64–5), though some ethnohistoric evidence suggests the Susquehannock occasionally consumed dog meat (Webster, 1983: 265). At the Strickler village, a sample of more than 1,400 identified spec-imens was studied (Webster, 1983: 257, Table 64). White-tailed deer comprised nearly three-quarters of the wild mammals. Other important animals included bear, beaver, cottontail rabbit (*Sylvilagus floridanus*), raccoon, and wapiti.

Changes over time in the proportion of deer and beaver remains within each assemblage provide an interesting window on the changing role of these animals in Susquehannock society and potentially the changing role of the Susquehannock in the regional fur trade. At the Schultz site, the earliest village in the Susquehannock sequence occupied *c.*1590s–1620, white-tailed deer comprise half (51% NISP) of the identified wild mammalian taxa, a category that includes specimens identified to the taxonomic level of genus and lower, excluding dog (because of its domestic status) and commensal taxa such as mice and rats (Table 37.1). Beaver are present, but they make up a minor percent-age (5% NISP) of small- to medium-sized wild mammals. By the time the Eschelman midden is in use at Washington Boro in the *c.*1620–1630s, deer show a substantial increase in number, comprising 85% NISP of wild mammals. Beaver are nearly five times more frequent than they had been in the past, rising to 24% NISP of small- to medium-sized wild mammals. Later in time, between the mid-1640s to mid-1660s, deer decline somewhat in proportion to three-quarters (75% NISP) of the mammal remains at the

Table 37.1 Summary of the comparative Susquehannock faunal samples

	Schultz[1]	Eschelman[2]	Strickler[3]
NISP of all vertebrate taxa[4]	2,314	23,605	1,421
NISP of wild mammals[5]	1,632	20,260	1,239
% NISP of wild mammals among all identified vertebrate taxa	71%	86%	87%
No. of wild mammal species	15	18	15
% NISP of white-tailed deer among wild mammals	51%	85%	75%
% NISP of beaver among small- to medium-sized wild mammals[6]	5%	24%	21%

[1]Data compiled from Webster (1983: Table 62). [2]Data compiled from Guilday et al. (1962: Table 1). [3]Data compiled from Webster (1983: Table 64). [4]Includes specimens identified to the taxonomic level of genus and lower. [5]Includes identified specimens; excludes domestic dog and commensal taxa. [6]Includes identified specimens; excludes wapiti, deer, bear, dog, and commensal taxa.

Strickler village, and beaver decreases in frequency slightly (to 21% NISP). Although the proportion of deer and beaver decline between the Eschelman and Strickler assemblages, both taxa remain substantially more important in Susquehannock animal economies than they had been at the turn of the century.

NISP data, as used above, consider the proportion of certain taxa relative to the sample as a whole. These same data can also be viewed in their numerical rank order, which provides another perspective on the importance of deer and beaver. Deer is the top ranked mammal at all three Susquehannock sites (Table 37.2). This pattern is not unexpected given the importance of venison in the Eastern Woodlands meat diet and the range of other products deer provided, from organ meat and fat to raw materials such as sinew, bones, and hides (Lapham, 2011). Beaver, in comparison, progresses steadily upward in ranking, becoming consistently more important in Susquehannock animal economies over time. This valued rodent is the third-ranked small- to medium-sized mammalian resource at the Schultz site, the second-ranked at the Eschelman site, and the first-ranked at the Strickler site (Table 37.3). Beaver increases both in rank and proportion over time, indicating that the Susquehannock intensified beaver hunting within local habitats. This, in turn, suggests that, despite their status as middlemen in the fur trade, the Susquehannock did

Table 37.2   Ranking of wild mammals (NISP)

Rank	Schultz[1]	Eschelman[2]	Strickler[3]
1st	*White-tailed deer*	*White-tailed deer*	*White-tailed deer*
2nd	Gray fox	Black bear	Black bear
3rd	Racoon	Wapiti	*Beaver*
4th	Black bear	Racoon	Cottontail rabbit
5th	Wapiti	*Beaver*	Raccoon

[1]Data compiled from Webster (1983: Table 62). [2]Data compiled from Guilday et al. (1962: Table 1). [3]Data compiled from Webster (1983: Table 64).

Table 37.3   Ranking of small- to medium-sized wild mammals (NISP)

Rank	Schultz[1]	Eschelman[2]	Strickler[3]
1st	Gray fox	Racoon	*Beaver*
2nd	Racoon	*Beaver*	Cottontail rabbit
3rd	*Beaver*	Gray squirrel	Raccoon
4th	Squirrel	Gray fox	Gray fox
5th	Bobcat	Gray wolf	Gray wolf

[1]Data compiled from Webster (1983: Table 62). [2]Data compiled from Guilday et al. (1962: Table 1). [3]Data compiled from Webster (1983: Table 64).

not, or could not, rely solely on pelts supplied from other Native communities to provide them with an ample surplus of furs to trade to colonial merchants.

Deer and Beaver Butchery

Guilday and colleagues (1962) conducted an extensive study of Susquehannock butchery techniques on the faunal remains recovered from the Eschelman midden at Washington Boro. Animals were processed using stone tools along with iron knives and axes. They identified skinning cut marks intended to remove the pelt from the carcass on a variety of mammals. Skinning scars included cuts to the symphysis region of the mandible (bear, fox, and racoon), cuts encircling the distal tibia (beaver), cuts encircling the distal metapodials (wapiti and deer), and occasionally the feet (deer). They noted that deer hide removal generally began several inches above the distal metapodials, often excluding the hoof-encased phalanges from the skinning process, and continued to the top of the skull. Care was taken to remove the skin up to the antler pedicle on male deer. Beavers were skinned and their carcasses disarticulated prior to consumption. Butchery marks indicate removal of the incisor teeth and the tail, which may have been a delicacy (Guilday et al., 1962).

The Schultz and Strickler assemblages also contained butchery scars on deer and beaver skeletal elements, but few interpretable data were provided other than the number of scars per taxon (Webster, 1983: 263–4, Table 66). With only qualitative data published, it is impossible to say whether Susquehannock butchery practices changed over time towards skinning techniques that maximized pelt size, or in other ways.

Early Historic Deer-Hunting Strategies

Susquehannock deer-hunting strategies are evaluated based on the same three criteria used in the southwestern Virginia case study: age of deer killed, season of death, and sex of deer hunted. Webster (1983: Table 28) and Guilday and colleagues (1962: 72) aged more than 360 deer mandibles from the three Susquehannock sites using a classification system developed by Severinghaus (1949), which allowed age profiles to be constructed based on an analysis of tooth eruption and occlusal wear patterns. Susquehannock deer harvests show a shift over time, with hunters targeting more prime-age animals and fewer juvenile deer as the seventeenth century progressed. At the Schultz site, the earliest village in the Susquehannock sequence, deer in the 2.5 year age class were the most frequent age group killed, followed by juvenile animals less than one year of age (0.5 year age class) (Fig. 37.2). Hunters at the Eschelman site targeted deer in the 3.5 year age class along with adjacent age groups (2.5 and 4.5 year age classes). By the occupation of the Strickler village, nearly three-quarters (71%) of aged deer fall into prime age classes (2.5 to 4.5 year age groups). The harvesting of juvenile animals declines steadily over time, from 18% to 12% to 6% at the Schultz, Eschelman, and Strickler sites respectively (Fig. 37.2).

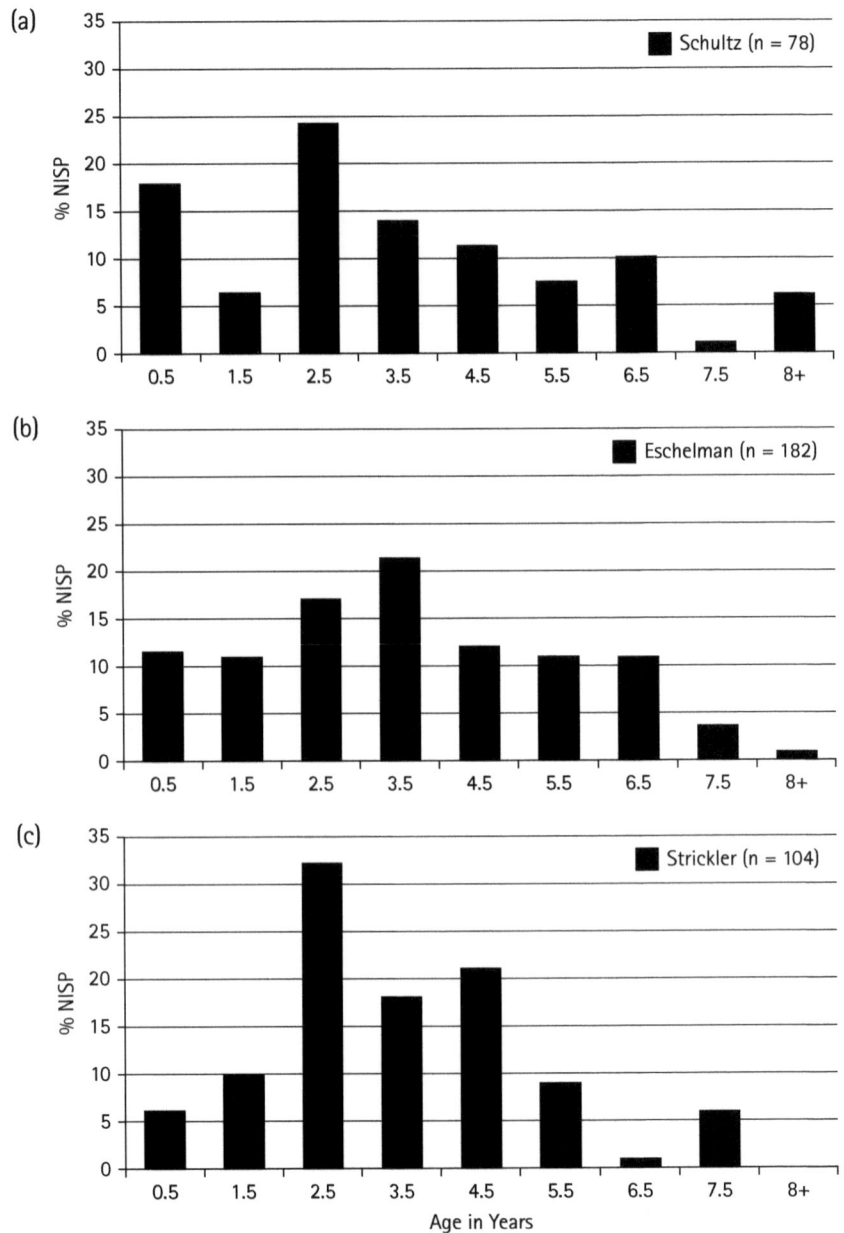

FIGURE 37.2 Distribution of white-tailed deer age at death for the (a) Schultz, (b) Eschelman, and (c) Strickler sites. Data from Webster (1983: Table 28) and Guilday et al. (1962: 72). Graph by author.

Juvenile deer are the most common age class in wild, unmanaged populations (Emerson, 1980; Klein and Cruz-Uribe, 1984), so their continued reduction in the zooarchaeological record suggests Susquehannock hunters made strategic choices to avoid killing juvenile deer—deer whose hides would have been too small or light to be marketable.

Season of death was evaluated using mandibles of deer aged younger than 20 months, which were placed by Webster (1983: 282–3) into 2–3 month age classes following Severinghaus (1949). This age range was then added to the average fawn drop date for the region, which was estimated at 1 June. This method assumes that season of kill for young animals reflects the season of death for hunted deer of all ages, which may or may not be an accurate assumption. During the Schultz and Strickler occupations, deer hunting occurred primarily as a seasonal activity during the autumn and winter months, although some hunting took place at other times of the year (Webster, 1983: Table 29) (Fig. 37.3). Seasonality data were not available for the Eschelman site, but Guilday and colleagues (1962: 72) estimated, based on male deer antler development and the presence of young fawns in the assemblage, that hunting took place primarily in the autumn and winter months, with some summer kills evident. Deer moult twice a year, shedding their existing coats over a several week period. In Pennsylvania, the first moult begins in late spring (May or June) and the second occurs in late summer (August or September) (Merritt, 1987). At the Schultz and Strickler sites, few deer for which season of death could be determined were killed during moulting season (11% and 17% of the respective assemblages), a time when hide quality would have been poor (Fig. 37.3).

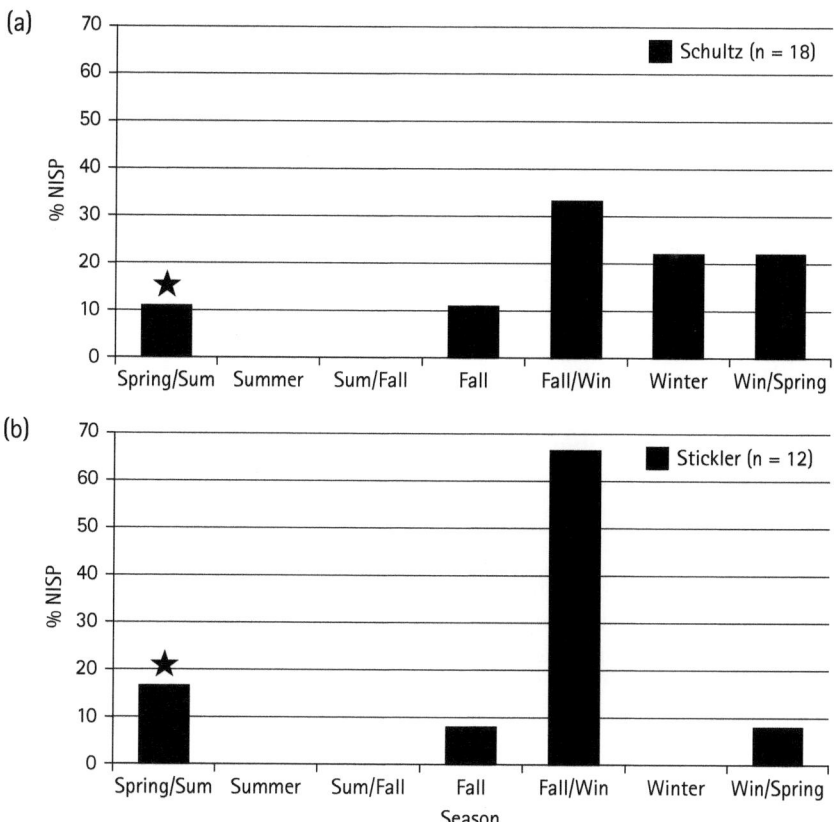

FIGURE 37.3 Distribution of white-tailed deer season at death for the (a) Schultz and (b) Strickler sites. Data from Webster (1983: Table 29). The solid stars represent young deer killed during moulting season. Graph by author.

The sex of the deer hunted cannot be assessed accurately for the Susquehannock sites. The original analyses determined sex based on the presence and absence of antlers on the frontal bones of the skull (Guilday et al., 1962: 72; Webster, 1983: Table 31). This method is biased because female deer frontal bones that lack antlers are more delicate and prone to breakage, and thus less frequently recovered and identified than the thicker, more durable, frontal bones of male deer. A more accurate technique developed by Edwards and colleagues (1982) uses the morphology of the pubis portion of the pelvic bone to determine deer sex, but this article appeared in print after the completion of the two Susquehannock zooarchaeological studies.

TRACKING THE PELT TRADES IN THE ZOOARCHAEOLOGICAL RECORD

In the seventeenth century in southwestern Virginia and south-central Pennsylvania, venison became a more important dietary resource as the importance of obtaining deerskins for trade became a goal of the deer hunt, and perhaps even a priority later in time. Hunting and trapping other fur-bearing animals important in the historic trades increased as well, just slightly so in southwestern Virginia and more markedly in south-central Pennsylvania. These differences can be explained by regional market preferences for deerskins in the south and beaver furs in the north. An increase over time in the importance of beaver in Susquehannock animal economies is clearly apparent, reflecting a change in the value assigned to this large rodent, whose luxurious pelt was held in high esteem by the European fashion industry and thus highly valued in the commercial pelt trades. Where data are available, deer skinning techniques intentionally maximized the size of the hide removed from the carcass—a practice that is not apparent in earlier faunal assemblages in the two study areas (Lapham, 2005: 90; Beisaw, 2012). Also, in both regions, deer-hunting strategies became more selective over time. Deer harvests increasingly focused on killing prime-age animals, whose hides would bring the highest exchange rate on the commercial market. In Pennsylvania, the Susquehannock were well armed with European muskets and rifles by the occupation of the Strickler settlement in the 1640s–1660s (Cadzow, 1936; Futer, 1959; Kent, 1984: 243–4). Hunting with guns may account, in part, for an increase in prime-age deer killed by Strickler hunters, since guns would have increased hunting efficiency (Webster, 1983: 267–8). Prey selection, however, was ultimately determined by the choices and decisions made by human hunters.

Indigenous groups throughout eastern North America hunted animals to procure pelts for the commercial deerskin and fur trades that developed between Native Americans and Europeans in the sixteenth and seventeenth centuries. Drawing on preexisting, well-honed animal procurement and processing skills, Native peoples produced surplus pelts specifically for European and Euro-American commercial markets.

The degree to which different families, kin groups, and communities participated in the trades differed over time and space, from minor engagement to intense involvement. These new economic ventures influenced subsistence practices and hunting strategies in different ways and to varying degrees, resulting in subtle and marked changes in historic-period Native American animal economies. This chapter highlights the importance of zooarchaeological studies to inform our understanding of the impacts that hunting and trapping animals for commercial trade had on Native American lifeways.

REFERENCES

Anderson, D. (1994) 'The flow of European trade goods into the western Great Lakes region, 1715–1760', in Brown, J., Eccles, W., and Heldman, D. (eds) *The Fur Trade Revisited: Selected Papers of the Sixth North American Fur Trade Conference*, pp. 93–115. East Lansing: Michigan State University Press.

Beisaw, A. (2012) 'Environmental history of the Susquehanna valley around the time of European contact', *Pennsylvania History: A Journal of Mid-Atlantic Studies*, 79(4), 366–76.

Brashler, J. (1987) 'A middle 16th century Susquehannock village in Hampshire county, West Virginia', *West Virginia Archeologist*, 39(2), 1–30.

Braund, K. (1993) *Deerskins and Duffels: The Creek Indian Trade with Anglo-America, 1685–1815*, Lincoln: The University of Nebraska Press.

Cadzow, D. (1936) *Archaeological Studies of the Susquehannock Indians of Pennsylvania*. Safe Harbor Report 2. Harrisburg: Pennsylvania Historical Commission.

Carlos, A. and Lewis, F. (2010) *Commerce by a Frozen Sea: Native Americans and the European Fur Trade*, Philadelphia: University of Pennsylvania Press.

Crane, V. (1928) *The Southern Frontier, 1670–1732*, Durham: Duke University Press.

Crean, J. (1962) 'Hats and the fur trade', *The Canadian Journal of Economics and Political Science*, 28(3), 373–86.

Edwards, J., Marchinton, R., and Smith, G. (1982) 'Pelvic girdle criteria for sex determination of white-tailed deer', *The Journal of Wildlife Management*, 46(2), 544–7.

Emerson, T. (1980) 'A stable white-tailed deer population model and its implications for interpreting prehistoric hunting patterns', *Mid-Continental Journal of Archaeology*, 5(1), 117–32.

Futer, A. (1959) 'The Strickler site', in Witthoft, J. and Kinsey, W. (eds) *Susquehannock Miscellany*, Harrisburg: The Pennsylvania Historical and Museum Commission.

Grant, H. (1989) 'Revenge of the Paris hat: the European craze for wearing headgear had a profound effect on Canadian history', *The Beaver*, Dec. 1988–Jan. 1989, 37–44.

Guilday, J., Parmalee, P., and Tanner, D. (1962) 'Aboriginal butchering techniques at the Eschelman site (36LA12), Lancaster county, Pennsylvania', *Pennsylvania Archaeologist*, 32, 59–83.

Hanson, J. (2000) 'The myth of the silk hat and the end of the rendezvous', *Museum of the Fur Trade Quarterly*, 36(1), 2–11.

Innis, H. (1956) *The Fur Trade in Canada: An Introduction to Canadian Economic History*, Toronto: University of Toronto Press.

Kent, B. (1983) 'The Susquehanna bead sequence', in Hayes, C. (ed.) *Proceedings of the 1982 Glass Bead Conference*. Research Records 16, pp. 75–81. Rochester: Rochester Museum and Science Division.

Kent, B. (1984) *Susquehanna's Indians*. Anthropological Series 6. Harrisburg: Commonwealth of Pennsylvania.

Klein, R. and Cruz-Uribe, K. (1984) *The Analysis of Animal Bones from Archaeological Sites*, Chicago: University of Chicago Press.

Lapham, H. (2004) "'Their complement of deer-skins and furs': changing patterns of white-tailed deer exploitation in the seventeenth-century southern Chesapeake and Virginia hinterlands', in Blanton, D. and King, J. (eds) *Indian and European Contact in Context: The Mid-Atlantic Region*, pp. 172–92. Gainesville: University Press of Florida.

Lapham, H. (2005) *Hunting for Hides: Deerskins, Status, and Cultural Change in the Protohistoric Appalachians*, Tuscaloosa: University of Alabama Press.

Lapham, H. (2011) 'Animals in southeastern Native American subsistence economies', in Smith, B. (ed.) *Subsistence Economies of Indigenous North American Societies: A Handbook*, pp. 401–29. Washington, DC: Smithsonian Institution Scholarly Press.

Lapham, H. and Johnson, W. (2002) 'Protohistoric Monongahela trade relations: evidence from the Foley Farm phase glass beads', *Archaeology of Eastern North America*, 30, 97–120.

Lawson, M. (1943) *Fur: A Study in English Mercantilism, 1700–1775*, Toronto: The University of Toronto Press.

Lawson, M. (1972) 'The beaver hat and the fur trade', in Bolus, M. (ed.) *People and Pelts: Selected Papers of the Second North American Fur Trade Conference*, pp. 27–38. Winnipeg: Peguis Publishers.

Martin, H. (1892) *Castorologia or the History and Traditions of the Canadian Beaver*, London: Edward Stanford.

Martin, J. (1994) 'Southeastern Indians and the English trade in skins and slaves', in Hudson, C. and Tesser, C. (eds) *The Forgotten Centuries: Indians and Europeans in the American South, 1521–1704*, pp. 304–26. Athens: The University of Georgia Press.

Merritt, J. (1987) *Guide to the Mammals of Pennsylvania*, Pittsburgh: University of Pittsburgh Press.

Moore, W. (1973) 'The largest exporters of deerskins from Charles Town, 1735–1775', *The South Carolina Historical Magazine*, 74(3), 144–50.

Perdue, T. (1998) *Cherokee Women: Gender and Culture Change, 1700–1835*, Lincoln: University of Nebraska Press.

Phillips, P. (1961) *The Fur Trade*, Norman: University of Oklahoma.

Ray, A. (1974) *Indians in the Fur Trade: Their Role as Trappers, Hunters, and Middlemen in the Lands Southwest of Hudson Bay, 1660–1870*, Toronto: University of Toronto Press.

Ray, A. (1980) 'Indians as consumers in the eighteenth century', in Judd, C. and Ray, A. (eds) *Old Trails and New Directions: Papers of the Third North American Fur Trade Conference*, pp. 255–68. Toronto: University of Toronto Press.

Robinson, W. (1979) *The Southern Colonial Frontier, 1607–1763*, Albuquerque: University of New Mexico Press.

Rountree, H. (2002) 'Trouble coming southward: emanations through and from Virginia, 1607–1675', in Ethridge, R. and Hudson, C. (eds) *The Transformation of the Southeastern Indians, 1540–1760*, pp. 65–78. Jackson: University Press of Mississippi.

Sempowski, M. (1994) 'Early historic exchange between the Seneca and the Susquehannock', in Hayes, C. (ed.) *Proceedings of the 1992 People to People Conference*. Research Records 23, pp. 194–218. Rochester: Rochester Museum and Science Center.

Severinghaus, C. (1949) 'Tooth development and wear as criteria of age in white-tailed deer', *Journal of Wildlife Management*, 13(2), 195–216.

Sleeper-Smith, S. (2009) 'Cultures of exchange in a North Atlantic world', in Sleeper-Smith, S. (ed.) *Rethinking the Fur Trade: Cultures of Exchange in an Atlantic World*, pp. xvii–lxii. Lincoln: University of Nebraska Press.

Smith, I. and Graybill, J. (1977) 'A report on the Shenks Ferry and Susquehannock components at the Funk site, Lancaster county, Pennsylvania', *Man in the Northeast*, 13, 45–65.

Spriggs, J. (1998) 'The British beaver—fur, fact and fantasy', in Cameron, E. (ed.) *Leather and Fur: Aspects of Early Medieval Trade and Technology*, pp. 91–101. London: Archetype Publications Ltd.

Stern, J. (2012) 'Native American taste: re-evaluating the gift-commodity debate in the British colonial Southeast', *Native South*, 5, 1–37.

Stine, L. (1990) *Mercantilism and Piedmont Peltry: Colonial Perceptions of the Southern Fur Trade, circa 1640–1740*. Volumes in Historical Archaeology 14. Columbia: University of South Carolina.

Usner, D. (1992) *Indians, Settlers, and Slaves in a Frontier Exchange Economy*, Chapel Hill: University of North Carolina Press.

Wall, R. and Lapham, H. (2003) 'Material culture of the Contact period in the upper Potomac valley: chronological and cultural implications', *Archaeology of Eastern North America*, 31, 151–77.

Webster, G. (1983) 'Northern Iroquoian Hunting: An Optimization Approach'. Unpublished PhD dissertation, Pennsylvania State University (State College).

Wesson, C. (2010) 'When moral economies and capitalism meet: Creek factionalism and the colonial southeastern frontier', in Scheiber, L. and Mitchell, M. (eds) *Across a Great Divide: Continuity and Change in Native North American Societies, 1400–1900*, pp. 61–78. Tucson: The University of Arizona Press.

Wyatt, A. (2012) 'Reconsidering early seventeenth century AD Susquehannock settlement patterns: excavation and analysis of the Lemoyne site, Cumberland county, Pennsylvania', *Archaeology of Eastern North America*, 40, 71–98.

CHAPTER 38

··

ANIMAL USE AT EARLY COLONIES ON THE SOUTHEASTERN COAST OF THE UNITED STATES

··

ELIZABETH J. REITZ

INTRODUCTION

MUCH of the research into culture contact in what became the southeastern United States focuses on historical events, social tensions, political interactions, and economic undertakings. Less frequently is the focus on plant and animal use, apparently under the assumption that use of these resources was little changed from practices in the nation sponsoring the colony. Studies of plant and animal remains recovered from colonial sites, however, provide evidence of highly varied responses at colonies in the Americas and elsewhere (e.g. Crabtree and Ryan, 1991; Deagan and Reitz, 1995; Gremillion, 1995; DeCorse, 1998; Stein, 1998; Pavao-Zuckerman, 2000; 2007; deFrance, 2003; Mondini et al., 2004; Twiss, 2007; deFrance and Hanson, 2008). Colonists did not necessarily retain the traditions of their 'homeland', nor did Native Americans abandon their traditions of animal use in favour of European ones. Instead, Native American, European, and African traditions merged to form novel patterns reflecting place, time, and cultural context (e.g. Pavao-Zuckerman and Reitz, 2011). This flexibility is particularly clear in the vertebrate record from the first permanent Spanish and English settlements on the southeastern Atlantic coast and French settlements on the northern coast of the Gulf of Mexico (Fig. 38.1; Table 38.1).

Zooarchaeological studies of sixteenth- through nineteenth-century Native American, Spanish, English, French, and American uses of animals show that colonial strategies on these two coasts combined a rich array of indigenous wild animals with a few introduced domestic ones, resulting in new cultural forms that shared some

FIGURE 38.1 Map of study area. Author's own image.

basic features but also were unique to each colony. Early settlers in these multi-ethnic colonies modified traditional husbandry, economic, and dietary practices to include resources better suited to these coasts. Many of the wild animal species in early assemblages are represented by turtles and fishes. White-tailed deer (*Odocoileus virginianus*) are prominent among the wild, terrestrial mammals, accompanied by such mammals as opossums (*Didelphis virginianus*), rabbits (*Sylvilagus* spp.), squirrels (*Sciurus* spp.), raccoons (*Procyon lotor*), and black bears (*Ursus americanus*). Introduced domestic animals were primarily pigs (*Sus scrofa*), cattle (*Bos taurus*), and chickens (*Gallus gallus*), but also included some cats (*Felis catus*), goats (*Capra hircus*), sheep (*Ovis aries*), and pigeons (*Columba livia*). Dogs (*Canis familiaris*) may be introduced breeds or indigenous ones.

This pattern is characteristic of assemblages from coastal Native American communities in Florida, Georgia, and Alabama; the Spanish towns of St. Augustine (Florida) and Santa Elena (South Carolina); the Spanish missions in Florida and Georgia; the English

Table 38.1 Characteristics of the faunal assemblages

	Time Period	MNI	NISP	Biomass, kg[1]	Location
Fountain of Youth Park (FOY, at St. Augustine)	1565	142	5709	7.781	St. Johns Co., Florida[2]
St. Augustine	1783–1821	152	10735	54.322	St. Johns Co., Florida[3]
St. Giles Kussoe (near Charleston)	1674–1684	12	250	2.49	Dorchester Co., South Carolina[4]
Miller Site (Charles Towne, near Charleston)	1670–1680	16	332	1.521	Charleston Co., South Carolina[5]
Charleston	1750s–1820s	606	50990	970.897	Charleston Co., South Carolina[6]
Old Mobile (near Mobile)	1702–1711	26	47384	2.2102	Mobile Co., Alabama[7]
La Pointe-Krebs Plantation (near Mobile)	1780s–1850s	57	4204	19.196	Jackson Co., Mississippi[8]

[1] Biomass as reported here is estimated only for taxa for which the Minimum Number of Individuals (MNI) also is estimated. NISP refers to the Number of Identified Specimens. [2] Orr and Colaninno, 2008. [3] Reitz et al., 2010. [4] Agha, 2012; Agha and Philips, 2010; Reitz and Bergh, 2012. [5] Jones and Beeby, 2010; Reitz and Bergh, 2012. [6] Zierden and Reitz, 2009. [7] Clute and Waselkov, 2002; Waselkov, 2002. [8] Reitz et al., 2013.

colonies in Florida, Georgia, and South Carolina; Charles Towne (modern Charleston, South Carolina); the French colonies on the Gulf coast; and American plantations in South Carolina and Georgia (e.g. Reitz and Honerkamp, 1983; Reitz et al., 1985; Reitz, 1986; 1994; Scott, 2001; Orr and Lucas, 2007; Zierden and Reitz, 2009; Reitz et al., 2010; Hardy, 2011; Scott and Dawdy, 2011; Reitz et al., 2013).

Methods

Each site in this study is referred to in terms of its most prominent European claimant. The major players were the Spanish Empire, or entities that had been or were to become part of that empire (e.g. the Philippines, the Netherlands, Germany, northern Africa, New Spain, the Canary and Caribbean islands, South America), England (becoming Great Britain in 1707), and France. Each site is referred to in terms of its present geopolitical affiliation for sake of brevity.

Differences among faunal collections often are attributed to ethnicity; but it is difficult to directly associate these assemblages with a single ethnic affiliation in a region where political dominance changed from Spanish, British, French, and American authority within less than three hundred years. Colonial populations were much more diverse than each colony's political affiliation implies (e.g. Deagan, 1973; Joseph and Zierden, 2002; Waselkov, 2009). Some colonists originated in Spanish, English, and French colonies elsewhere in the Americas or in the emerging global network of these and other European nations. Africans were part of the colonial mix from the beginning, often as slaves, but also as free people of colour. Some Africans were directly from Africa, others came to this region from other colonies, especially from Bermuda and elsewhere in the Americas. Native Americans from within the southeastern region and beyond were present at all of these colonies. The presence of people from many different parts of the world with very different perspectives on animal use is likely one of the principle explanations for the rich and eclectic use of animals seen in the colonial faunal record in this region.

The southeastern Atlantic and Gulf coast is a narrow band skirting the North American seaboard, extending inland c.30–40 km. The region is characterized by high humidity, numerous swamps, streams, flatwoods, maritime forests, pinelands, and daily tides. Colonists and Native Americans interacted over a much larger area, but all of the sites in this study lie within this coastal strand. From the 1500s into the 1700s, Spanish Florida included peninsular Florida, the Georgia coast, and part of the Carolina coast. British and French authorities claimed much the same area. Political boundaries were fluid but characterized by a steady retreat of Spanish interests in the face of advancing British ones from the mid-1600s into the late 1700s.

Vertebrate remains were studied following methods described in Reitz and Wing (2008) and the references in Table 1. Most attributions were made using the comparative skeletal collection of the Zooarchaeology Laboratory, Georgia Museum of Natural History, Athens (USA). Biomass is estimated using the allometric equation: $Y = aX^b$. The species identified are summarized into faunal categories based on vertebrate class and economic status. These categories are wild animals (sharks, rays, bony fishes, alligators, turtles, wild birds, deer, other wild mammals), domestic animals (domestic birds and mammals), and commensal taxa. Frogs and toads (Anura), snakes (Serpentes), moles (*Scalopus aquaticus*), mice and rats (Muridae), dogs, cats, horses or donkeys (*Equus* spp.), and similar animals, are considered commensal in this study. Although commensal animals might be consumed, they are commonly associated with people and their built environment as vermin, pets, or working animals. Similarly, some animals classified as consumed might have been commensal.

Spanish Florida

Spanish St. Augustine was founded in 1565. The first location of the town is known as Fountain of Youth Park (FOY, 8SJ31). The FOY materials indicate that Spanish efforts to colonize this setting were marked by rapid adoption of the suite of resources used by local Native Americans in this same area (Figs 38.2 and 38.3).

Initially, attempts were made to introduce domestic livestock in proportions that maintained traditions of consuming mutton and pork prevailing in some parts of Iberia (Reitz and Scarry, 1985: 96–7). When this failed, the gap was filled by wild species, especially marine fishes. After a few decades, a small, hardy cattle breed, known as criollos, emerged from the original Spanish stock. Criollos flourished in the coastal plain environment under a free-range regime. They supported a flourishing Spanish cattle industry in the 1600s, which persisted despite devastating British raids in the early 1700s (Arnade, 1961; Rouse, 1977; Bushnell, 1978). Although domestic meats never completely replaced fish and other wild resources, by the time Spain ceded Florida to the United States in 1821, meat from domestic mammals, primarily beef, contributed over half of the biomass (Reitz et al., 2010: 82–3) (Fig. 38.4).

THE ENGLISH CAROLINAS

A combination of indigenous and introduced animals is also found in vertebrate collections from two seventeenth-century English sites in South Carolina. English Carolina data are from two sites associated with Charles Towne, founded in 1670. One of these sites is the St. Giles Kussoe House/Lord Ashley settlement and trading post (38DR83A), owned by Lord Anthony Ashley Cooper. The primary economic activities at St. Giles Kussoe were trade with Native Americans and cattle ranching. By 1682, there were nearly six hundred head of cattle at St. Giles. The second site is known as the Miller Site (38CH1-MS), which was just outside Charles Towne's palisade and may have been a tavern operating between 1670 and 1680. Wild indigenous animals contribute less of the biomass at these early Carolina sites compared to FOY (Figs 38.2 and 38.3). The

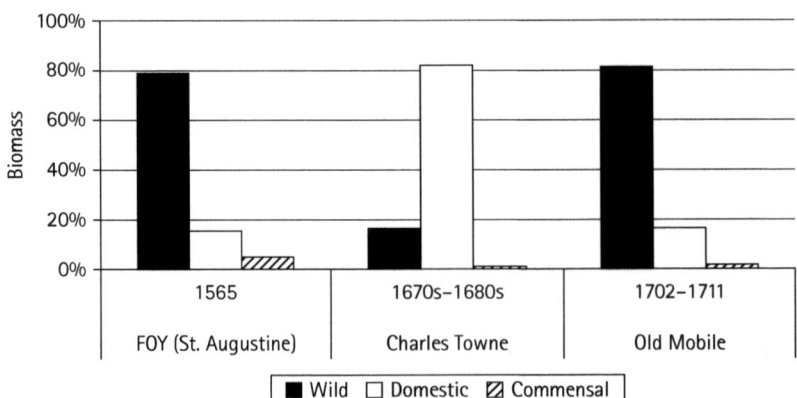

FIGURE 38.2 Summary of wild, domestic, and commensal vertebrate biomass. See Table 38.1 for sources. Author's own image.

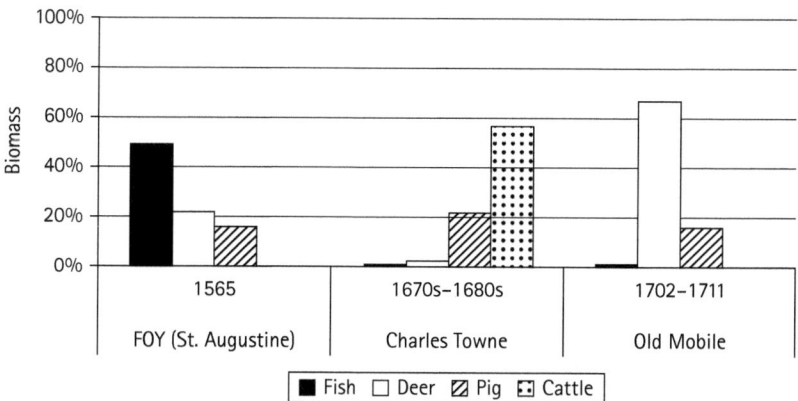

FIGURE 38.3 Summary of fish, deer, pig, and cattle biomass. See Table 38.1 for sources. Author's own image.

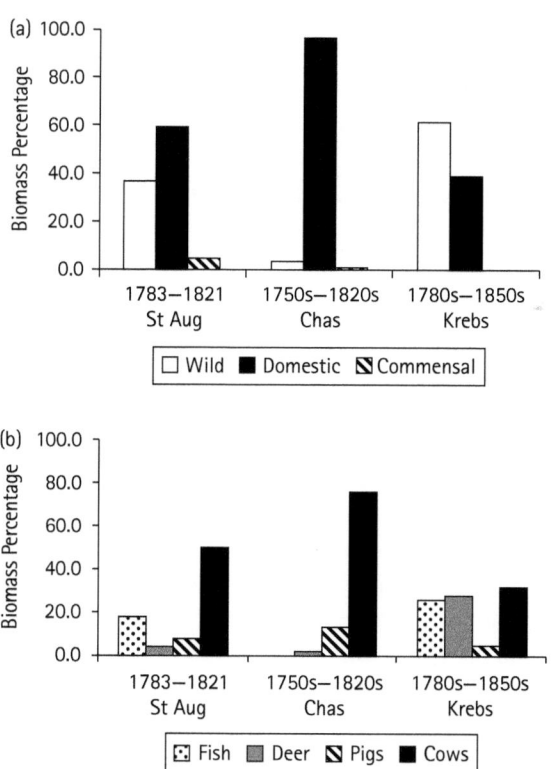

FIGURE 38.4 (A) Summary of wild, domestic, and commensal vertebrate biomass at St. Augustine, Charleston, and La Pointe-Krebs Plantation (French Gulf Coast). (B) Summary of fish, deer, pig, and cattle biomass. See Table 38.1 for sources. Author's own image.

primary wild sources of biomass were turtles and small mammals. The low quantity of biomass from wild animals reflects the dominance of beef, a characteristic that persisted throughout the English Carolina's colonial and subsequent history (Zierden and Reitz, 2009) (Fig. 38.4).

THE FRENCH GULF COAST

A similar combination of indigenous and introduced animals is found at French Old Mobile, Alabama (1MB94). Old Mobile was established in 1702 and abandoned in 1711. This French-dominated site had a major Native American presence and a small African one, as well as close ties with Spanish trading partners in the circum-Caribbean basin. As at FOY, local wild animals were the primary source of biomass (Fig. 38.2); however, the dominant source of wild meat was venison, not fish (Fig. 38.3). This early deposit contains no remains of either cattle or caprines, though pigs and chickens are present. Sheep failed to flourish in the French Gulf coast colonies in the eighteenth century, just as they did for Spanish colonists in the sixteenth century and English colonists in the seventeenth century.

A NEW CULTURAL FORM

Changes in these assemblages reflect site formation processes, the specific functions of the deposits used in this study, analytical biases, and broader social, economic, political, and environmental contexts influencing the availability of indigenous wild resources and enabling beef production. People at all three locations made use of both wild and domestic resources from the earliest colonial occupations into the 1800s (Figs 38.2–38.4). Regardless of the European affiliation of the colony or colonists, at least 80% of the biomass is from fish, venison, pork, and beef, supplemented by a wide range of turtles and other wild animals. Explanations for the similarities and differences among these collections include: national origins and ethnicity, exchanges between Native Americans and colonists, adaptations, mimicry or invention on the part of colonists, African influences, and broader economic patterns related to the cattle industry.

Differences among these collections could be evidence that the national origins or ethnic identities of settlers at each location informed animal use; however, such ties may have been ephemeral or largely for social convenience in these multi-ethnic settlements. The similar lists of taxa used at all three locations are aspects of these data that should be explored further. It is likely that ethnic identity informed public displays and cuisine, but that choosing the ingredients of that cuisine was more pragmatic.

Exchanges between Native Americans and colonists were probably significant in forming new Spanish, English, and French economies. Most of the indigenous resources are ones that Native Americans in these same locations had used for millennia. Indians provided commodities to the colonists as trade goods. In particular, trade in deer hides and venison was extensive in all three colonies (e.g. Braund, 1993; Lapham, 2005). Some Native Americans were slaves and others were allied with colonists of European or African descent through marriage or other social arrangements (Deagan, 1973; Waselkov and Gums, 2000; Silvia, 2002). Reciprocity within kin groups is a particularly likely source of wild foods given that Native American women were present in some households. Some colonists obtained resources in the form of tithes and tribute, commandeered what they wanted, or relied upon enslaved native peoples to provide them. Many of the goods obtained from Native Americans were the traditional local fare, primarily fish and venison, and these continued to be used into the nineteenth century at former colonies. Although some Native Americans did raise European livestock, this was limited (e.g. Reitz, 1993; Pavao-Zuckerman, 2000; 2007; Reitz et al., 2010; Pavao-Zuckerman and Reitz, 2011).

Some aspects of these new strategies might be inventive adaptations that would have developed even in the absence of native contributions to colonial economies. Evidence for this is seen in the similarities in animal use by early settlers at these three different locations in three different centuries. Some elements of these early colonial strategies persisted for centuries after native populations were extinct or dispersed (Reitz and Cumbaa, 1983; Reitz and Honerkamp, 1983; Reitz et al., 1985; Reitz, 1986; 1994; Zierden and Reitz, 2009; Reitz et al., 2010). Bökönyi (1975) argues that when people with an animal husbandry tradition immigrate into a region where animal husbandry is unknown, settlers attempt to maintain their original husbandry system. People do this even when the original system is unproductive (e.g. sheep), compensating for shortfalls initially by increasing their use of wild animals and, subsequently, incorporating a different suite of domestic resources as colonial economies matured. Bökönyi's prediction broadly characterizes the rapidity with which traditional indigenous resources were incorporated into these economies and the subsequent increase in beef at all three colonies.

It might be that the characteristic coastal economies of these colonies reflect African influences instead of Native American ones, although Africans were also strangers in a strange land and had to learn productive techniques just as other colonists had to do, including Native American colonists from other parts of the Americas.

Cattle raising was an important commercial activity in all three colonies, either initially or eventually (Arnade, 1961; Bushnell, 1978; Otto, 1986; 1987; Stewart, 1991). In the early days of both the Spanish and English colonies, cattle thrived in the pinewoods and lowlands of the Atlantic coastal plain. Winters are generally mild and native forage is available throughout the year. Much of the coastal plain is prone to natural fires and favourable pastures sometimes were enlarged though the intentional use of fire (Dunbar, 1961). Cattle generally were provided minimal or no supplemental feed or shelter, though in some cases calves and nursing females were penned to keep them from running wild (Otto, 1986; Stewart, 1991; Groover and Brooks, 2003). The term 'cow hunter', used to refer to people who worked with these animals, suggests the wild nature of the cattle and the prevailing herd management strategy. Such herds flourished among colonists and, to a lesser extent, among some Native Americans (Otto, 1986; 1987; Stewart, 1991; Braund, 1993; Orr and Lucas, 2007).

The cattle industry subsequently was replaced by other commodities (Dunbar, 1961; Otto, 1986). The decline observed in the archaeological record could be due to better commercial opportunities offered by rice, tobacco, and cotton and to a shift in marketing patterns such that refuse, including bones and teeth, were discarded elsewhere. It also may reflect the rapid and devastating spread of what was called 'Spanish staggers' in the mid-1700s. 'Spanish staggers' may have been babesiosis (i.e. Texas or Southern fever), which is caused by a tick-borne red blood cell parasite and results in massive organ damage (Haygood 1986; Stewart 1991). Babesiosis is believed to be the cause of disease outbreaks in the Carolinas in 1760s–1770s and perhaps earlier, which disrupted trade in cattle and led to laws restricting their movements (Bierer, 1939). The decline in cattle biomass in St. Augustine (later assemblages) and Charles Towne during the mid-1700s may be evidence of this disease's impact (see Reitz and Waselkov, 2015). The French colonization of the northern Gulf coast had the misfortune of occurring just as this disease emerged, which could explain the low levels of beef in assemblages from Old Mobile and La Pointe-Krebs Plantation (this latter located on the coast just southwest of Old Mobile).

Another explanation for attributes observed in the 1700s may be climate-related landscape changes. Multiple proxies associate a North American 'megadrought' with the inability of the Spanish and English colonies to thrive (e.g. Stahle et al., 1998; Blanton and Thomas, 2008: 801). Oscillating periods of wetter-drier/colder-warmer conditions are documented for both the 1600s and 1700s in this area. Climate variability and other phenomena related to the Little Ice Age (1300–1870) could have influenced colonial enterprises on both coasts.

MESTIZAJE, TRANSCULTURATION, DIETARY ACCULTURATION, OR CREOLIZATION?

On the southeastern coast, the role of human agency in culture contact and change is referred to as *mestizaje* (Deagan, 1973), transculturation and ethnogenesis (Deagan, 1998), dietary acculturation (Gremillion, 2002), and creolization (Hardy, 2011), among other terms (see Majewski and Gaimster, 2009). A fundamental distinction among these concepts is whether the outcomes are mixtures of several cultural strains, with roots that can be traced in a more or less linear fashion back to an original ancestry, or new cultural forms with multiple origins and agents (Deagan, 1998). These early colonial data suggest that patterns of animal use in each colony were new cultural forms with multiple origins produced by multiple agents, and cannot be traced back to a single ancestral tradition.

The early colonial record on the southeastern coasts of North America indicates that transformations in animal use occurred almost immediately, with diverse outcomes likely reflecting the dynamics of the specific setting. Regardless of ethnicity or national affiliation, colonists at Spanish, English, and French colonies followed strategies that merged fish and venison, long used by Native Americans in these areas, with beef or pork. Use of wild game, particularly venison, declined in later centuries while use of domestic meats increased as southeastern coastal colonies became nineteenth-century American states, but the cultural forms that emerged in the colonial past persist in regional cuisines today.

Conclusion

Early colonists at all three early colonies merged Native American and European animal resources regardless of the national affiliation of the colony, the ethnicity of the colonists, or the century in which colonization occurred. The economies that emerged on the southeastern Atlantic and Gulf coasts represent eclectic fusions of indigenous and introduced animals. Colonists continued to use the Eurasian suite of domestic animals, but the primary sources of meat might have been fish, venison, or beef. The distinctive aspect of the colonial strategy, however, was the richness of the faunal assemblages and reliance on indigenous wild animals. In many respects the colonial strategy was largely a local one with the addition of domestic taxa that could survive and prosper in these coastal settings with minimal attention, primarily free-range cattle. Although many aspects of animal use at each colony had an indigenous flavour, they retained European ingredients, indicating that a new cultural form had emerged that cannot be traced back to single Native American, European, or African sources. These data suggest that it is important to explore evidence for changes in the use of animals by both colonized and colonizing populations in studies of culture contact.

Acknowledgements

I am grateful to Brockington and Associates, The Center for Archaeological Studies at the University of South Alabama, The South Carolina State Park Service, the Historic Charleston Foundation, The Charleston Museum, the Donnelly Foundation, Magnolia Plantation Foundation, Mississippi Department of Archives and History, Andrew Agha, Sarah G. Bergh, Carol E. Colaninno-Meeks, Kathleen A. Deagan, Kevin S. Gibbons, Bonnie L. Gums, David Jones, Maran E. Little, Gregory S. Lucas, Kelly L. Orr, Elizabeth M. Scott, Gregory A. Waselkov, and Martha A. Zierden for their contributions to this study. Fig. 38.1 was drafted by Susan Duser. An earlier version of this paper was presented at the 69th Annual Meeting of the Southeastern Archaeological Conference.

REFERENCES

Agha, A. (ed.) (2012) 'St. Giles Kussoe and "The Character of a Loyal States-Man": Historical Archaeology at Lord Anthony Ashley Cooper's Carolina Plantation'. Report on file, Historic Charleston Foundation, Charleston, South Carolina.

Agha, A. and Philips, C. F., Jr (2010) 'Archaeological Investigations at 38DR83A, St. Giles Kussoe House/Lord Ashley Settlement, Dorchester County, South Carolina'. Final report submitted by Brockington and Associates, Atlanta, Georgia, to the Historic Charleston Foundation, Charleston, South Carolina.

Arnade, C. W. (1961) 'Cattle raising in Spanish Florida', *Agricultural History*, 35, 3–11.

Bierer, B. W. (1939) *History of Animal Plagues of North America*, 1974 edn, Washington, DC: United States Department of Agriculture.

Blanton, D. B. and Thomas, D. H. (2008) 'Paleoclimates and human responses along the central Georgia coast: a tree-ring perspective', in Thomas, D. H. (ed.) *Native American Landscapes of St. Catherines Island, Georgia*, pp. 778–806. New York: Anthropological Papers of the American Museum of Natural History 88.

Bökönyi, S. (1975) 'Effects of environmental and cultural changes on prehistoric fauna assemblages', in Arnott, M. C. (ed.) *Gastronomy. The Anthropology of Food and Food Habits*, pp. 3–12. The Hague: Mouton.

Braund, K. E. H. (1993) *Deerskins & Duffles: Creek Indian Trade with Anglo-America, 1685–1815*, Lincoln: University of Nebraska Press.

Bushnell, A. (1978) 'The Menendez Marquez cattle barony at La Chua and the determinants of economic expansion in seventeenth-century Florida', *Florida Historical Quarterly*, 56, 407–31.

Clute, J. R. and Waselkov, G. A. (2002) 'Faunal remains from Old Mobile', in Waselkov, G. A. (ed.) *French Colonial Archaeology at Old Mobile: Selected Studies*, pp. 129–34. *Historical Archaeology*, 36 (1), 1–148.

Crabtree, P. J. and Ryan, R. (eds) (1991) *Animal Use and Culture Change*, Philadelphia: University of Pennsylvania, Museum of Archaeology and Anthropology Research Papers in Science and Archaeology, MASCA 8, Supplement.

Deagan, K. A. (1973) '*Mestizaje* in colonial St. Augustine', *Ethnohistory*, 20, 55–65.

Deagan, K. A. (1998) 'Transculturation and Spanish American ethnogenesis: the archaeological legacy of the Quincentenary', in Cusick, J. G. (ed.) *Studies in Culture Contact: Interaction, Culture Change, and Archaeology*, pp. 23–43. Carbondale: Southern Illinois University Center for Archaeological Investigations, Occasional Paper 25.

Deagan, K. A. and Reitz, E. J. (1995) 'Merchants and cattlemen: the archaeology of a commercial structure at Puerto Real', in Deagan, K. A. (ed.) *Puerto Real: The Archaeology of a Sixteenth-Century Spanish Town in Hispaniola*, pp. 231–84. Gainesville: University Press of Florida.

DeCorse, C. R. (1998) 'Culture contact and change in West Africa', in Cusick, J. G. (ed.) *Studies in Culture Contact: Interaction, Culture Change, and Archaeology*, pp. 358–77. Carbondale: Southern Illinois University Center for Archaeological Investigations, Occasional Paper 25.

deFrance, S. D. (2003) 'Diet and provisioning in the high Andes: a Spanish colonial settlement on the outskirts of Potosí, Bolivia', *International Journal of Historical Archaeology*, 7, 99–126.

deFrance, S. D. and Hanson, C. A. (2008) 'Labor, population movement, and food in sixteenth-century Ek Balam, Yucatán', *Latin American Antiquity*, 19, 299–316.

Dunbar, G. S. (1961) 'Colonial Carolina cowpens', *Agricultural History*, 35, 125–30.

Gremillion, K. J. (1995) 'Comparative paleoethnobotany of three native southeastern communities of the historic period', *Southeastern Archaeology*, 14, 1–16.

Gremillion, K. J. (2002) 'Archaeobotany at Old Mobile', in Waselkov, G. A. (ed.) *French Colonial Archaeology at Old Mobile: Selected Studies*, pp. 117–28. *Historical Archaeology*, 36 (1), 1–148.

Groover, M. D. and Brooks, R. D. (2003) 'The Catherine Brown Cowpen and Thomas Howell Site: Material characteristics of cattle raisers in the South Carolina backcountry', *Southeastern Archaeology*, 22, 92–111.

Hardy, M. D. (2011) 'Living on the edge: Foodways and early expressions of creole culture on the French colonial Gulf Coast frontier', in Kelly, K. G. and Hardy, M. D. (ed.) *French Colonial Archaeology in the Southeast and Caribbean*, pp. 152–88. Gainesville: University Press of Florida.

Haygood, T. M. (1986) 'Cows, ticks, and disease: a medical interpretation of the southern cattle industry', *The Journal of Southern History*, 52, 551–64.

Jones, D. and Beeby, C. (2010) 'Miller Site Excavations: Fall Field Season 2009, 38CH1-MS, Charles Towne Landing State Historic Site'. Manuscript on file, The Charleston Museum, Charleston, South Carolina.

Joseph, J. W. and Zierden, M. (2002) 'Cultural diversity in the southern colonies', in Joseph, J. W. and Zierden, M. (ed.) *Another's Country: Archaeological and Historical Perspectives on Cultural Interactions in the Southern Colonies*, pp. 1–12. Tuscaloosa: University of Alabama Press.

Lapham, H. (2005) *Hunting for Hides: Deerskins, Status, and Cultural Change in the Protohistoric Appalachians*, Tuscaloosa: University of Alabama Press.

Majewski, T. and Gaimster, D. (ed.) (2009) *International Handbook of Historical Archaeology*, New York: Springer Science+Business Media.

Mondini, M., Muñoz, S., and Wickler, S. (ed.) (2004) *Colonisation, Migration and Marginal Areas: A Zooarchaeological Approach*, Oxford: Oxbow Books.

Orr, K. L. and Colaninno, C. (2008) 'Native American and Spanish Subsistence in Sixteenth-Century St. Augustine: Vertebrate Faunal Remains from Fountain of Youth (8SJ31), St. Johns Co., Florida'. Manuscript on file, Zooarchaeology Laboratory, Georgia Museum of Natural History, University of Georgia, Athens.

Orr, K. L. and Lucas, G. S. (2007) 'Rural-urban connections in the southern colonial market economy: zooarchaeological evidence from the Grange Plantation (9CH137) trading post and cowpens', *South Carolina Antiquities*, 39(1 and 2), 1–17.

Otto, J. S. (1986) 'The origins of cattle-ranching in colonial South Carolina, 1670–1715', *South Carolina Historical Magazine*, 87, 117–24.

Otto, J. S. (1987) 'Livestock-raising in early South Carolina, 1670–1700: prelude to the rice plantation economy', *Agricultural History*, 61, 13–24.

Pavao-Zuckerman, B. (2000) 'Vertebrate subsistence in the Mississippian-Historic transition', *Southeastern Archaeology*, 19, 135–44.

Pavao-Zuckerman, B. (2007) 'Deerskins and domesticates: Creek subsistence and economic strategies in the Historic period', *American Antiquity*, 72, 5–33.

Pavao-Zuckerman, B. and Reitz, E. J. (2011) 'Eurasian domestic livestock in Native American economies', in Smith, B. D. (ed.) *The Subsistence Economies of Indigenous North American Societies*, pp. 577–91. Washington, DC: Smithsonian Institution Scholarly Press.

Reitz, E. J. (1986) 'Urban/rural contrasts in vertebrate fauna from the southern coastal plain', *Historical Archaeology*, 20, 47–58.

Reitz, E. J. (1993) 'Evidence for animal use at the missions of Spanish Florida', in McEwan, B. G. (ed.) *The Spanish Missions of La Florida*, pp. 376–98. Gainesville: University Press of Florida.

Reitz, E. J. (1994) 'Zooarchaeological analysis of a free African community: Gracia Real de Santa Teresa de Mose', *Historical Archaeology*, 28, 23–40.

Reitz, E. J. and Bergh, S. G. (2012) 'Animal remains from two early South Carolina sites', in Agha, A. (ed.) *St. Giles Kussoe and 'The Character of a Loyal States-Man': Historical Archaeology at Lord Anthony Ashley Cooper's Carolina Plantation*, pp. 119–77. Charleston: Historic Charleston Foundation.

Reitz, E. J. and Cumbaa, S. L. (1983) 'Diet and foodways of eighteenth century Spanish St. Augustine', in Deagan, K. A. (ed.) *Spanish St. Augustine: The Archaeology of a Colonial Creole Community*, pp. 147–81. New York: Academic Press.

Reitz, E. J., Gibbons, K. S., and Little, M. E. (2013) 'Animal remains from La Pointe-Krebs House (22JA526), Mississippi', in Gums, B. L., and Waselkov, G. A. (ed.) *Archaeology at La Pointe-Krebs Plantation in Old Spanish Fort Park (22JA526), Pascagoula, Jackson County, Mississippi*, pp. 166–91. Jackson: Mississippi Department of Archives and History Monograph.

Reitz, E. J., Gibbs, T., and Rathbun, T. A. (1985) 'Archaeological evidence for subsistence on coastal plantations', in Singleton, T. A. (ed.) *The Archaeology of Slavery and Plantation Life*, pp. 163–91. New York: Academic Press.

Reitz, E. J. and Honerkamp, N. (1983) 'British colonial subsistence strategy on the southeastern coastal plain', *Historical Archaeology*, 17, 4–26.

Reitz, E. J., Pavao-Zuckerman. B., Weinand, D. C., Duncan, G. A., and Thomas, D. H. (2010) *Mission and Pueblo of Santa Catalina de Guale, St. Catherines Island, Georgia: A Comparative Zooarchaeological Analysis*, New York: Anthropological Papers of the American Museum of Natural History 91.

Reitz, E. J. and Scarry, C. M. (1985) *Reconstructing Historic Subsistence with an Example from Sixteenth-Century Spanish Florida*, Pleasant Hill: The Society for Historical Archaeology Special Publication 3.

Reitz, E. J. and Waselkov, G. A. (2015) 'Vertebrate use at early colonies on the southeastern coasts of Eastern North America'. *International Journal of Historical Archaeology*, 19, 21–45.

Reitz, E. J. and Wing, E. S. (2008) *Zooarchaeology*, 2nd edn, Cambridge: Cambridge University Press.

Rouse, J. E. (1977) *The Criollo*, Norman: University of Oklahoma Press.

Scott, E. M. (2001) 'Food and social relations at Nina Plantation', *American Anthropologist*, 103, 671–91.

Scott, E. M. and Dawdy, S. L. (2011) Colonial and creole diets in eighteenth-century New Orleans', in Kelly, K. G. and Hardy, M. D. (ed.) *French Colonial Archaeology in the Southeast and Caribbean*, pp. 97–116. Gainesville: University Press of Florida.

Silvia, D. E. (2002) 'Native American and French cultural dynamics on the Gulf coast', *Historical Archaeology*, 36, 26–35.

Stahle, D. W., Cleaveland, M. K., Blanton, D. B., Therrell, M. D., and Gay, D. A. (1998) 'The Lost Colony and Jamestown droughts', *Science*, 280, 564–7.

Stein, G. J. (1998) 'World system theory and alternative modes of interaction in the archaeology of culture contact', in Cusick, J. G. (ed.) *Studies in Culture Contact: Interaction, Culture Change, and Archaeology*, pp. 220–55. Carbondale: Southern Illinois University Center for Archaeological Investigations, Occasional Paper 25.

Stewart, M. A. (1991) ' "Whether wast, deodand, or stray": cattle, culture, and the environment in early Georgia', *Agricultural History*, 65, 1–28.

Twiss, K. C. (ed.) (2007) *The Archaeology of Food and Identity*, Carbondale: Southern Illinois University Center for Archaeological Investigations Occasional Paper 34.

Waselkov, G. A. (2002) 'French colonial archaeology at Old Mobile: an introduction', in Waselkov, G. A. (ed.) *French Colonial Archaeology at Old Mobile: Selected Studies*, pp. 3–12. *Historical Archaeology*, 36 (1), 1–148.

Waselkov, G. A. (2009) 'French colonial archaeology', in Majewski, T. and Gaimster, D. (ed.) *International Handbook of Historical Archaeology*, pp. 613–28. New York: Springer Science+Business Media.

Waselkov, G. A. and Gums, B. L. (2000) *Plantation Archaeology at Rivière aux Chiens, ca. 1725–1848*, Mobile: University of South Alabama, Center for Archaeological Studies, Archaeological Monograph 7.

Zierden, M. A. and Reitz, E. J. (2009) 'Animal use and the urban landscape in colonial Charleston, South Carolina, USA', *International Journal of Historical Archaeology*, 13, 327–65.

CHAPTER 39

ZOOARCHAEOLOGY OF THE MAYA

KITTY F. EMERY

INTRODUCTION

MAYA zooarchaeology is a rapidly growing field with more researchers and research studies each year. Several reviews of Maya zooarchaeology have presented the basic parameters of history, methods, and animal distribution (Emery, 2004a; 2004d; Corona et al., 2010; Götz, 2013a; 2014a). The examples I present in this chapter instead focus on the contributions of zooarchaeology to the major questions of current Maya archaeology and indeed of archaeology in general (Kintigh et al., 2014). This review, based primarily on my own research, is not exhaustive, but merely intended to provide a sample of current studies in this fascinating world area.

DESCRIPTION OF THE MAYA WORLD

The ancient Maya world (1500 BC–1500 AD), encompassed the modern political regions of southeastern Mexico, Guatemala, Belize and western Honduras, and El Salvador (Fig. 39.1; for excellent texts on the ancient Maya, see Henderson, 1997; Sharer and Traxler, 2005). The first residents of the region probably arrived in the Archaic (~8000–900 BC), though we have very little archaeological evidence of their presence. The classic Maya florescence, defined by peak settlement size and density, was in the Late Classic (AD ~500–900). Settlement and political activity diminished in the Terminal Classic and post-Classic (AD ~900–1200) in the southern lowlands, but continued well into the contact period (AD ~1500) in the northern Yucatan. The lowland Maya were among the final holdouts against the Spanish and the Maya world is still inhabited by the modern Maya whose oral traditions often inform our research on their ancestors.

FIGURE 39.1 Map of the Maya area. Sites mentioned in the text and other sites of interest are indicated. The Petexbatun region includes the sites of Dos Pilas, Aguateca, Cancuen, and Las Pacayas mentioned in the text. The Motul region includes the site of Motul de San Jose, Trinidad, and other sites of the Motul polity. Map drawn by the author, all site locations are approximate.

Extant Maya fauna are a diverse mix of Nearctic and Neotropical taxa and vary only slightly across these regions although biodiversity generally increases at the lower elevations and southward through the Yucatan Peninsula. Animal distributions are more specifically tied to the availability of surface water, both marine and inland fresh water, and to natural and anthropogenic variations in forest cover and settled lands, as they doubtless also were in the past (see Emery, 2001).

The animal remains most often recovered in Maya deposits are from deer, dog, peccary, river and sea turtles, smaller agoutis, pacas, rabbits, armadillos, tuzas, iguanas, wild game birds, local fish and molluscs, the very large tapir and manatee, spider and howler monkeys, rare wild cats, and other high forest species (see Table 39.1). These proportions vary spatially, reflecting the availability of different animals and other resources, and temporally, reflecting cultural shifts and perhaps natural and anthropogenic impacts on animals and habitats. Animals were used as food, but also as construction material, for tools, for their hides, as medicine, for musical instruments, and for adornment. Maya animals were (and are) agents in the mythic cosmology, representing and acting as liaisons with gods, ancestors, and spirit alter-egos. They were emblematic of genders, lineages, occupations, rulership roles, and status groups. Our research on this complex relationship between the Maya and their animals is therefore also multi-faceted and difficult to summarize. In this review, I will touch on three primary zooarchaeological and archaeological themes—environments, community organization, and ceremony—and will focus on my own research although this will only represent a few of the many Maya zooarchaeological studies.

Did Climate Variability Cause the Rise and Fall of Maya Culture?

We begin this review of Maya zooarchaeology with studies of animals as proxy for habitats. Keeping in mind that such proxy models cannot effectively take into account the human agent in choosing specific animals or the effects of discard behaviour and taphonomy, they can provide insight to some long-standing questions in Maya archaeology.

Palaeoenvironmental reconstructions of Holocene climate change, primarily detected through palaeolimnological and speleological research, give precipitation a primal role in causality for cultural change in the Maya world (for a review see Iannone, 2013). These studies reveal that precipitation in early occupation periods was higher (pre-Classic) and more stable (Late Classic) than in later Terminal Classic and post-Classic periods, encouraging increased settlement, agriculture, and predictability of harvest and economic growth. Associated sea-level rise during the Classic period resulted in shoreline and inland hydrologic changes. These patterns are hypothesized to have been causal to the pre-Classic and Late Classic florescences as well as the Early and Terminal Classic cultural hiatuses, sometimes termed 'collapses'.

However, these palaeolimnological proxies are not easily correlated with archaeological deposits and relative cultural chronologies, and often occurred quite distantly from sites where cultural changes are most visible (as an extreme example, see Haug et al., 2003). By comparing animal remains from dated archaeological deposits, we can instead track the specific impact of climatological droughts on the hydrology of large and small, temporary and permanent, local water resources. In one such study we found that the proportions of animals that prefer to inhabit wet environments correlated well with the general patterns of wet and dry periods shown by other palaeoenvironmental studies during the 'collapse period' droughts (Emery and Thornton, 2013). However, the correlation was primarily

Table 39.1 Taxa of importance to the Maya peoples and mentioned in the text. This is not a complete list of all taxa found at Maya archaeological sites

Scientific Name: Higher Taxonomy	Scientific Name: Taxon	Common English Name	Common Spanish Name in Maya Area	Common Nahuatl and Maya Names
Architaenioglossa, Ampullariidae	*Pomacea flagellata* (Say, 1929)	Apple Snail	Caracol Manzana Maya	
Gastropoda, Sorbeoconcha, Pachychilidae	*Pachychilus* sp (Morelet, 1849)	Fresh-water Jute	Jute	Tutu, Jute (Maya)
Bivalvia, Unionoida, Unionidae	*Nephronaias* sp (Fischer & Crosse, 1893)	River Clam	Concha de Río	Xkin (Maya: K'iche')
Bivalvia, Unionoida, Unionidae	*Psoronaias* sp (Crosse & Fischer, 1894)	River Clam	Concha de Río	Xkin (Maya: K'iche')
Arthropoda, Malacostraca, Decapoda	*Brachyura* sp (Latreille, 1802)	Short-Tailed Crabs, True Crabs	Cangrejos	Atecuicitli, Tecuicitli (Nahuatl)
Chondrichthyes, Myliobatiformes, Dasyatidae	*Dasyatis americana* (Hildebrand and Schroeder, 1928)	Southern Sting Ray	Raya Redonda, Raya Americana	
Chondrichthyes, Carcharhiniformes, Carcharhinidae	Carcharhinidae (Jordan and Evermann, 1896)	Sharks	Tiburones	
Osteichthyes, Actinopterygii, Perciformes, Cichlidae	*Petenia splendida* (Günther, 1862)	Bay Snook, Giant Cichlid	Blanco	Tenguajagua, Tenguayaca (Nahuatl)
Osteichthyes, Actinopterygii, Perciformes, Cichlidae	Cichlidae	small fresh-water cichlids		
Osteichthyes, Actinopterygii, Siluriformes, Pimelodidae	*Rhamdia guatemalensis* (Meek, 1907)	Catfish, Guatemalan Chulín	Barbudo, Juil Descolorido	
Osteichthyes, Actinopterygii, Semionotiformes, Lepisosteidae	*Atractosteus spatula* (Lacepède, 1803)	Alligator Gar, Garpike	Gaspar Baba, Pejelagarto	
Osteichthyes, Actinopterygii, Semionotiformes, Lepisosteidae	*Atractosteus tropicus* Gill, 1863	Tropical Gar	Catán, Gaspar, Pejelagarto	
Reptilia, Testudines, Dermatemydidae	*Dermatemys mawii* (Gray, 1847)	Central American River Turtle, Giant River Turtle	Tortuga Blanca	

(Continued)

Table 39.1 Continued

Scientific Name: Higher Taxonomy	Scientific Name: Taxon	Common English Name	Common Spanish Name in Maya Area	Common Nahuatl and Maya Names
Reptilia, Testudines, Kinosternidae	*Kinosternon* sp (Spix, 1824)	Mud and Musk Turtles		
Reptilia, Testudines, Kinosternidae	*Staurotypus triporcatus* (Wiegmann, 1828)	Mexican Giant Musk Turtle	Tortuga Almizclera Mexicana Gigante	Jolom kok (Maya)
Reptilia, Testudines, Emydidae	*Trachemys scripta* (Schoepff, 1792)	Common Slider, Elegant Terrapin	Jicotea, Tortuga Pintada	Ka'a nix (Maya)
Reptilia, Testudines, Emydidae	*Trachemys venusta* (Gray, 1856)	Mesoamerican slider		
Reptilia, Testudines, Cheloniidae	Cheloniidae (Oppel, 1811)	Sea Turtles	Tortuga Marina	áak (Nahuatl)
Reptilia, Squamata, Iguanidae	*Ctenosaura similis* (Gray, 1831)	Black Spiny-Tailed Iguana	Iguana Espinosa Rayada	Juuj (Maya)
Reptilia, Squamata, Iguanidae	*Iguana iguana* (Linnaeus, 1758)	Common Green Iguana	Iguana Verde, Iguana de Mar	Topitl, Topitzin (Nahuatl), T'ool huuh (Maya Yucateco)
Reptilia, Crocodilia, Crocodylidae	*Crocodylus acutus* (Cuvier, 1807)	American Crocodile	Cocodrilo, Cocodrilo Americano, Lagarto	Ain, áayin (Maya)
Reptilia, Crocodilia, Crocodylidae	*Crocodylus moreletti* (Duméril and Bibron, 1851)	Central American Crocodile, Morelet's Crocodile	Cocodrilo de Morelet	Ain, áayin (Maya)
Aves, Psittaciformes, Psittacidae	*Ara macao* (Linnaeus, 1758)	Scarlet Macaw	Guacamaya Roja	Kiaq k'ix (Maya: K'iche')
Aves, Craciformes, Cracidae	*Crax rubra* (Linnaeus, 1758)	Great Curassow	Hocofaisán, Faisan Real	
Aves, Galliformes, Phasianidae	*Meleagris gallopavo* (Linnaeus, 1758)	Domestic Turkey, Wild Turkey, Northern Turkey	Chompipe, Guajolote Norteño, Pavo, Pavo Salvaje	Uexolotl (Nahuatl), tso' (Maya)
Aves, Galliformes, Phasianidae	*Meleagris ocellata* (Cuvier, 1820)	Ocellated Turkey, Wild Turkey	Guajolote Ocelado, Pavo de Monte, Pavo Ocelado	Kuuts (Maya)
Aves, Trogoniformes, Trogonidae	*Pharomachrus mocinno* (De la Llave, 1832)	Northern Quetzal, Resplendent Quetzal	Quetzal	Quetzalin, Quetzalli (Nahuatl)
Aves, Passeriformes, Cardinalidae	*Cardinalis cardinalis* (Linnaeus, 1758)	Northern Cardinal	Cardenal Común, Cardenal Norteño	Chak ts'íits'ib (Maya)

Scientific Name: Higher Taxonomy	Scientific Name: Taxon	Common English Name	Common Spanish Name in Maya Area	Common Nahuatl and Maya Names
Aves, Passeriformes, Icteridae	*Icterus* sp (Brisson, 1760)	Orioles	Turpial	Toch'ich' (Maya; K'iche')
Mammalia, Cingulata, Dasypodidae	*Dasypus novemcinctus* (Linnaeus, 1758)	Armadillo, Nine-Banded Armadillo	Armadillo de Nueve Bandas	Ayotl-tochtli (Nahuatl), Weech (Maya Yucateco), Ip (Maya Tzeltales)
Mammalia, Chiroptera	Chiroptera (Blumenbach, 1779)	Bats	Murciélago	Sootz', Zotz (Maya)
Mammalia, Primates, Cebidae	*Ateles geoffroyi* Kuhl, 1820	Central American Spider Monkey	Mono Araña	Ozomatli, Oçomatli (Nahuatl), Tucháaj, C'oy, Chouen, Chuen (Maya)
Mammalia, Primates, Cebidae	*Alouatta pigra* Lawrence, 1933	Black Howler Monkey	Saraguato Yucateco	Baats' (Maya)
Mammalia, Carnivora, Canidae	*Canis lupus familiaris* (Linnaeus, 1758)	Domestic Dog	Perro domesticado	Chichi, Itzcuintli (Nahuatl), Peek', Xibil peek' (Maya)
Mammalia, Carnivora, Felidae	*Panthera onca* (Linnaeus, 1758)	Jaguar	Jaguar	Ocelotl (Nahuatl), Báalam, Chak mo'ol (Maya)
Mammalia, Perissodactyla, Tapiridae	*Tapirella bairdii* (Gill, 1865)	Baird's Tapir, Central American Tapir	Tapir	Tlalpizotl (Nahuatl), Tzimin (Maya Yucateco)
Mammalia, Sirenia, Trichechidae	*Trichechus manatus* (Linnaeus, 1758)	American Manatee, West Indian Manatee	Manatí del Caribe	Teek, Baklam (Maya)
Mammalia, Artiodactyla, Cervidae	*Odocoileus virginianus* (Zimmermann, 1780)	White-Tailed Deer	Venado Cola Blanca	Mazatl (Nahuatl), Kéej (Maya)
Mammalia, Artiodactyla, Cervidae	*Mazama americana* (Erxleben, 1777)	Red Brocket Deer	Cabro de Monte	Yuk (Maya)
Mammalia, Artiodactyla, Cervidae	*Mazama pandora* (Merriam, 1901)	Yucatan Brown Brocket Deer	Temazate Yucateco	
Mammalia, Artiodactyla, Tayassuidae	*Pecari tajacu* (Linnaeus, 1758)	Collared Peccary	Coche de Monte, Pecarí de Collar	Coyámel (Nahuatl), K'itam (Maya)
Mammalia, Artiodactyla, Tayassuidae	*Tayassu pecari* (Link, 1795)	White-Lipped Peccary	Pecarí Labios Blancos, Jabali	Coyámel (Nahuatl), K'itam (Maya)
Mammalia, Rodentia, Agoutidae	*Cuniculus paca* (Linnaeus, 1766)	Agouti, Spotted Paca	Tepezcuintle	Tepeitzcuintli (Nahuatl), Jaaleb, Jaale' (Maya)

(Continued)

Table 39.1 Continued

Scientific Name: Higher Taxonomy	Scientific Name: Taxon	Common English Name	Common Spanish Name in Maya Area	Common Nahuatl and Maya Names
Mammalia, Rodentia, Dasyproctidae	*Dasyprocta punctata* (Gray, 1842)	Central American Agouti	Aguti Centroamericano, Sereque	
Mammalia, Rodentia, Geomyidae	*Orthogeomys hispidus* (LeConte, 1852)	Hispid Pocket Gopher	Tuza Crespa	Ba (Maya; K'iche')
Mammalia, Lagomorpha, Leporidae	*Sylvilagus brasiliensis* (Linnaeus, 1758)	Forest Rabbit	Conejo Tropical	Tapeti (Nahuatl)
Mammalia, Lagomorpha, Leporidae	*Sylvilagus floridanus* (J. A. Allen, 1890)	Eastern Cottontail, Florida Cottontail	Conejo Serrano	Tapeti (Nahuatl)

with small swamp-dwelling species while animals from larger water systems were continuously available at all sites. Thus, we propose that the impact of precipitation decline on people was dependent on the proportional availability of large to small water bodies. As well we hypothesize that either the Maya went far afield to continue harvesting the large-water source animals depleted locally by disappearing water, or the Maya area did not experience the megadroughts that would have removed or significantly changed the water systems. The second hypothesis is supported by studies tracing the ratios of oxygen isotopes in the bones and teeth of white-tail deer (Repussard et al., 2013), which found that the water supply for these animals was unaffected by the precipitation variability.

McKillop and Winemiller's (2004: 71) study of marine mollusc distributions at Frenchman's Caye, Belize revealed increases in shallow water and mangrove species that correlated with sea-level rise which eventually submerged the island community. However, further north in the Yucatan, Götz (2012: 434) reports no significant changes in the types of coastal resources recovered from Early through Terminal Classic periods, suggesting that, in this area, sea-level changes had little impact on resource availability or choices.

DID THE ANCIENT MAYA DEFOREST THEIR LANDSCAPE?

The density of Late Classic Maya settlements and size of their populations has also led many to suggest significant decimation of forest cover and soils (for a review, see

Turner and Sabloff, 2012). Certainly the Maya lowlands were much more densely settled during the Late Classic than they are today and sedimentary and palynological
research from palaeolimnological cores reveal increases in corn pollen, decreases in
forest tree pollen, and a buildup of clay deposits. Chronological assessment suggests,
however, that soil erosion, if associated at all with human activity, was an early phenomenon resulting from initial land clearance before population growth (Anselmetti
et al., 2007). In addition, palynological markers are ineffective for understanding
forest cover since the only wind-pollinated tree in the area is the ramon (*Brosimum
alicastrum*), itself a cultivated tree; the botanical makeup of the modern forests alongside the traditional agroforestry systems are instead evidence that the Maya forest
was managed, not decimated (Ford and Nigh, 2009). In local and regional analyses at
southern lowland sites, terrestrial animal habitats again correlate broadly with other
palaeoenvironmental studies showing deforestation following regional settlement
growth. But once more the zooarchaeological data provide greater detail. Regional
assessment shows that mature forest animal species such as the wild cats, curassow,
and shy brocket deer do decline in frequency through time, but species associated
with secondary forest such as the crop-raiding white-tail deer, agouti, and collared
peccary are ubiquitous throughout and often rise in frequency during periods of
high settlement (Emery and Thornton, 2008b). This habitat preference research correlates well with chemical studies that use carbon isotopes in white-tail deer bone,
reflective of corn (*Zea mays*) browsing (the primary C4 plant in the Maya region), as
a proxy for agricultural expansion. In the Pasion region of the Peten, isotopic analyses
revealed a stable and low rate of corn browsing among deer throughout the occupation of the region and no correlation between site expansion or abandonment and
deer corn consumption at any of six sites (Emery, 2004c). In a broader regional study,
carbon isotope variation was high between sites, even when those with similar population histories and sizes were compared, and there was no evidence of correlation
between land clearance and expansion of settlement or population density (Emery
and Thornton, 2008c).

DID THE ANCIENT MAYA OVERHUNT
THEIR ANIMALS?

Also related to the questions of Maya environmental resource management is the question of whether they decimated their animal populations through over-hunting. Unlike
in many other world areas, there is no evidence for extinction or even local extirpation of animals in the Maya region. Some zooarchaeological studies have looked for
subtle clues of resource depression caused by unsustainable harvest of single favoured
(often keystone) or large-bodied game such as white-tail deer, peccary, jaguar, or tapir.
My regional study in the southern lowlands found proportional reductions in large- to

small-bodied prey and also in proportions of the culturally favourite species, white-tail deer. But those reductions were variable and most closely correlated to political centres and periods of local political stress (Emery, 2007b). For example, at the six sites of the Petexbatun polity, resource depression was found in two distinct periods: quite early and quite late in each site history, both times when political activity and competition for status and power was high (Emery, 2008). It seems possible that hunting intensity was not a predetermined result of population growth but was instead linked to political activity and an associated demand for status goods (high-value animals) during these competitive periods. In the northern lowlands, Götz (2013b) noted a reduction in the proportion of large- to small-game fauna at the end of the Classic period at two sites, but very high proportions of large game (especially white-tail deer) in later Terminal Classic to post-Classic elite middens and domestic contexts, suggesting that patterns of resource depression were also variable across this northern region, but possibly also linked to political pressure.

Evidence supporting an alternate model of ancient hunting sustainability under conditions of low political stress can be found in the various studies of locally specific adaptations through resource scheduling, the most commonly cited being the garden hunting model (Linares, 1976) that seems to apply well to reconstructions of Maya animal procurement. Maya milpas, a complex agricultural and resource management system, are a predictable, high-yield resource that reliably attracts forest-fringe game species, which are encouraged to feed in the milpas where they are then available for hunters to harvest (Atran, 1993). Götz's comparative analysis of inland and coastal site assemblages supports this model of local adaptations for efficient harvest. He found that, although inland site assemblages focused on species known to be resilient edge-dwellers and common milpa visitors, on the coast not only were coastal species such as sea turtle, manatee, and fish more prevalent, but different site assemblages reflected the local marine taxa very closely (Götz, 2013b). In a comparison of a range of ecological statistics used by recent studies of bush meat harvest between modern hunter assemblages from the Guatemalan Atitlan villages with those from the lowland archaeological sites, by all measures past hunting activities were more sustainable than are traditional modern practices (Emery and Brown, 2012).

Did the Maya Domesticate Animals for 'Meat on the Foot'?

Yet another way to consider Maya resource management and its links to other social trends is through studies of animal domestication (Valadez Azúa, 2003). The only two ancient Maya vertebrate domesticates were the northern turkey and the dog. Dogs accompanied humans into Mesoamerica and genetic and zooarchaeological evidence shows that the species was intentionally bred starting in the Early Classic to create different varieties for eating, hunting, and perhaps even sacrifice (for review, see Valadez Azúa et al., 2013). Dogs were most abundant during the earliest (pre-Classic) and latest (Terminal Classic and post-Classic) periods of Maya prehistory (based on a regional review from thirty sites, Emery, 2004b). One interpretation of pre-Classic frequencies and butchery patterns is that these reflect a response to dietary pressure during these early formative periods. However, more recent studies of early correlations between dog abundances and elite structures, particularly in political centres, suggests that dog breeding was linked to politics rather than simple dietary necessity (for review, Emery et al., 2013). In the Maya area the use of dogs as feast food, sacrifice, ritual offering, and tribute is well documented in ethnohistoric chronicles and archaeological distributions. Carbon and nitrogen isotopic studies of dogs from ritual deposits indicate that some individuals were intentionally fattened on corn for elite feasts and sacrifices (White et al., 2001; White, 2004).

Although ocellated turkeys are local to the Maya region, they were never domesticated. The northern turkey, later our domestic turkey, was thought to have been domesticated in Northern Mexico in the Classic period and introduced to the Maya area only in the post-Classic (c. AD 1000–1500) (for reviews see Steadman, 1980; Valadez Azúa, 2003). The recent identification of northern turkeys in elite, pre-Classic ritual caches at El Mirador, far outside their range, reveals a much earlier introduction (Thornton et al., 2012). Although it is not clear if the pre-Classic El Mirador birds were domesticated, that they were at least husbanded (hand-raised) is indicated by the mix of males, females, adults, and juveniles in the group. The context suggests that this early turkey husbandry was linked to elite ritual use, much as is documented by the post-Classic codices and described by de Landa (Tozzer, 1941). Regardless of the turkeys' early initial arrival, its appearance in large numbers in post-Classic deposits from the northern Yucatan coast (for example, Götz, 2008a: 114–22) to the southern inlands (Emery, 1999) suggests it was introduced or reintroduced at that time as a domestic species. This was likely to be part of the increased post-Classic economic interactions along the Maya coasts. As with the dogs, these birds are most common in elite deposits, and ethnohistoric literature again points to a significant role for the turkey in ritual feasting (Pohl and Feldman, 1982). These studies suggest that animal domestication or husbandry in the Maya area was also more clearly linked to struggles for power and status rather than a simple need for supplemental meat.

DID THE MAYA ELITE HAVE PREFERENTIAL RESOURCE ACCESS?

Maya zooarchaeology must consider the overlapping roles of social, economic, and political organization in any evaluation of environmental relationships or the process of resource management. Archaeological reconstructions of ancient Maya systems emphasize organizational complexity and diversity at community, polity, and inter-polity scales (for review see Foias, 2013).

Again, zooarchaeology is well equipped to answer this question by reconstructing correlations between animal distributions and archaeological markers of social inequality. Our reconstructions often generalize a separation between elite and non-elite status groups, but a more complex multi-tiered hierarchy of status, lineage, and occupational groups likely existed (discussion in Chase and Chase, 1992). Unfortunately many of these groups are zooarchaeologically invisible except in special circumstances (for example, see Collins, 2002, a zooarchaeological study of slave residents in elite households at Copan). The variable distribution of commodities and luxury goods among social classes provides one avenue for exploration. Zooarchaeological analyses agree that the ancient Maya elite had preferential access to highly valued (e.g. deer, tapir) and exotic or ritual (marine shell, wild cats) species. That this relationship between the elite and certain prized species has great antiquity is shown by the elaborate pre-Classic San Bartolo murals illustrating a number of ancient myths nearly all of which incorporate animal sacrifice, symbolic otherworld links, and royal accession (Sharpe et al., 2014).

Some zooarchaeological analyses support a model of elite preferential access to meat by recording higher diversity in elite deposits of food animals, a larger quantity of animal products, and elements from the meatier cuts of the game animals (Pohl, 1994; 1995; Masson, 1999; Masson and Peraza Lope, 2008; Thornton and Emery, in press). But other studies do not agree. For example, at post-Classic Isla Civiltuk, Alexander et al. (2013: 292) find no evidence that elites consumed different animal resources, greater proportions of species yielding high-quality meat, or a different diversity of taxa than non-elites (see also, Pohl, 1985: 141; Emery, 2007a: 67). More detailed status reconstructions reveal more complexity. In the Petexbatun polity (Emery, 2006), although elite deposits generally had more remains from favourite species, in fact, only the most highly favoured species, the white-tail deer, was more common in deposits of the ruling family (Rank 1a). Upper middle-class (Rank 2) households had relatively less deer among their assemblages than either the non-noble highest elite (Rank 1b) or Rank 3, the lower middle class. Some generally favoured species were excluded from Rank 1a households and some were found only in Rank 1b or 2 households. At post-Classic Mayapan, Masson and Peraza Lope (2013: 253, 272) reveal outlier proportions of white-tail deer and dog in centrally located elite ritual structures, while outlying elite houses had very high proportions of rabbit and turkey, indicating specialized and spatially separated activities.

Inter-community differences of rank and political centrality within a polity may also impact the differential distribution and access to animal resources. In a study of Late Classic period Maya sites in the Yaxchilan, Piedras Negras, and Petexbatun polities, Sharpe's (2015) results show diminishing access to diverse taxa, exotics, and ritual species between capital, secondary, and tertiary subordinate sites. In addition, highest-ranking elites in capitals had highest proportions of ritual and exotic species and favoured white-tail deer portions, but those of intermediate status living at the capitals had access to the greatest diversity of species.

WAS ANCIENT MAYA TRADE RESTRICTED TO ELITE STATUS-REINFORCING PRODUCTS?

These findings bring us to the role of animals in economics. Clearly animal products were moving through Maya communities. How did that function? To what extent were animal remains being exchanged solely as elite status-reinforcing commodities or also as domestic products? Maya zooarchaeology easily distinguishes long-distance exchange between elites of exotic species from the highlands (e.g. quetzals; Sharpe, 2014), the lowlands (e.g. macaws; Somerville et al., 2010), and marine products from coastal to inland sites (e.g. sting rays; Cunningham-Smith et al., 2014: 48). More complicated is recognizing the movement of locally available and ubiquitous taxa and those used in the domestic sphere.

Elite trade of exotic products such as marine shell, ubiquitous at Classic period Maya sites large and small, has been recognized from the earliest Maya archaeology. But coastal trade products also included lesser-recognized sea turtle, bony and cartilaginous marine fish, corals, and other invertebrates also found in elite and special deposits. Marine fish are in particular surprisingly ubiquitous in elite inland deposits, to the extent that in an early contribution to the subject Lange (1971) proposed that these may have provided dietary supplements. Carr (1986) hypothesized a specialized marine fish industry based on predominance of cranial remains at the coastal site of Cerros. Marine-shell debitage at some coastal sites (Caye Coco, Rosenswig and Masson, 2002) in combination with cranial fish element predominance at coastal salt-production sites also suggest specialized production of marine shell-fish ornaments and preserved (salted/dried) fish for exchange with inland sites—Northern River Lagoon sites (Mock, 1997; Masson, 2004), Mayapan (Masson and Peraza Lope, 2008), and Wild Cane Cay (McKillop, 1996).

Arguments against this model are the ubiquity of whole marine shells and marine shell debitage at many inland sites suggesting *in situ* artefact crafting (for example, Emery, 2014), and the predominance of postcranial fish remains at some coastal sites indicates local consumption rather than trade (see Götz, 2012: 426). Also, where marine products are found at inland sites, they are restricted to elite deposits, and human biological isotopic studies show only limited elite marine-product consumption at inland sites indicating that the diet did not rely substantially on marine resources. Inland products such as fresh-water turtles, ocellated turkey, white-tail deer (see Götz, 2012: 429, 432), jaguars (Marco Gonzalez; Pendergast and Graham, 1990), tapirs (Ek Luum; Shaw, 1995), and riverine molluscs (Frenchman's Caye; McKillop and Winemiller, 2004: 71) have also been found at Atlantic island sites. Most were recovered in elite deposits or as worked artefacts indicating these also represent exotic, status-reinforcing imports on the islands rather than dietary products.

In contrast to the exchange of exotic species, direct acquisition is generally assumed for locally available animals such as white-tail deer, peccary, and dog, which are ubiquitous at all inland Maya sites (see McAnany, 1993 for a review). Strontium isotopes used to source animal remains (Thornton, 2011b) have revealed the trade of locally available subsistence species such as deer and peccary from the Early (at the site of Copan) through post-Classic periods (at Lamanai and Tipu). Evidence for such trade was found in all ritual deposits tested and is consistent with ethnographic accounts of trade in deer and other local species in tribute and gift exchanges (Tozzer, 1941). Evidence again points to elite trading spheres since all the non-local individuals recovered by Thornton were found in elite deposits, often in areas with possible special functions such as specialist crafting (at the site of Motul de San Jose) and caching (at Lamanai).

Recent zooarchaeological studies also hint at household and site-level exchange of subsistence animal resources. For example, within the small polity of Motul de San Jose each of five satellite centres had greater proportions of locally available animal species and also small numbers of animals local to other polity sites. Polity-capital elite residents (Rank 2) and particularly the royalty (Rank 1) had higher proportions of species from all habitats. High-ranked polity capital households had high-value species and body portions but so too did the lowest-rank residents (Rank 4) of the low-status satellite site of Chäkokot. The residents of Chäkokot, however, did not have the most highly valued portions of these valued animals. This distribution of species and body portions suggests specialized hunting at the satellite site of Chäkokot and provisioning from Chäkokot to the capital site of Motul de San Jose (Emery, 2012). Together such data suggest a combination of local trade, possibly even via markets, and capital provisioning through tribute and taxation.

WHO WERE THE MAYA CRAFT SPECIALISTS?

We also find tight economic interconnectedness and heterogeneity linked to craft specialization. Bone and shell production debris (debitage) is overall scarce and broadly distributed across Maya settlements, and few formal bone or shell workshops or crafting areas have been identified. Maya crafting has traditionally been interpreted as a combination of unspecialized household production for personal consumption and attached production by artisans producing prestige items for elite use and gifting (for a review, see McAnany, 1993). Recent zooarchaeological studies add to a growing body of evidence that crafting also included occasional large-scale specialized production, elite and even royal family crafting, and possibly the production of goods for market-type exchange (Moholy-Nagy, 1997; Isaza-Aizpurúa, 2004; Emery and Aoyama, 2007). Large deposits of marine shell debitage and shell-working implements at various pre-Classic sites, such as Pacbitun (Hohmann, 2014), Kaxob (Isaza-Aizpurúa and McAnany, 1999), and Chan (Keller, 2013) indicate a great antiquity for inland shell-working while Late Classic shell- and bone-working is less clear. In elite deposits at Late Classic Caracol (Pope, 1994: 150) and Early Classic Copan (Aoyama, 2011), concentrated numbers of whole shells, shell pieces, and shell debitage suggest shell-crafting workshops. In a Terminal Classic household deposit outside the abandoned royal central plaza at the site of Dos Pilas, massive collections of debitage from the production of utilitarian perforators likely intended for trade rather than household consumption are one of the only examples of large-scale bone-crafting in the Maya world (Emery, 2010: 187–267).

Who were the Maya crafters? At Late Classic Aguateca, residents of high-status households manufactured luxury goods from bone and shell that were likely intended for consumption by the sites' royal families, a pattern seen at other sites such as Copan (Aoyama, 2011). We compared lithic use-wear and bone and shell artefact distributions in neighbouring Aguateca households of equivalent rank. Overlaps between lithic use-wear and bone/shell artefacts at Aguateca revealed household-level specialization in everything from bone blank production to marine shell-working, suggesting that occupational specialization may have rested, as it does today in many areas of the highlands, at the family/household/lineage level (Emery and Aoyama, 2007). Studies both at Motul de San Jose and Aguateca indicate that even the ruling family was involved in a palace economy that included specialized crafting of animal products such as animal hides and fresh-water clam nacre textile decorations (Emery, 2014). This practice has also been noted in the palaces of Motul de San Jose, where Rank 2 (non-noble elites) crafted marine shell adornments then used by the ruling family (Emery, 2012).

DID THE MAYA RULERS COMBINE POLITICS AND THE POWER OF RITUAL FOR COMMUNITY INTEGRATION?

Clearly, the economic and social structures of the ancient Maya were complex and heterarchical in terms of which local and exotic resources were used, and who acquired, prepared, and consumed them. How did the political leaders at the core of this integrated organization, imbued with divine authority and participants in politics at the regional, local, and spiritual level, obtain and retain their power in a system that has shown no archaeological evidence of the use of forceful political coercion? It is likely that ceremony was the glue that allowed the rulers to coordinate the integration of community members within the bounds of polities, regions, and a cosmos that included animate and ever-present ancestor spirits.

As in other world areas, we struggle with how we recognize rituals because the uses of animals and their products, and even the role of animals themselves as actors and agents, was combined in a rich symbolic and animistic tapestry that is not easily divorced from a view of secular life. The most recognizable 'ritual' find is that of a complete carcass, or significant portions thereof, in a contextually significant place, suggesting a sacrifice without later dismemberment. Excellent examples of these range from the famous wild cats recovered from beneath the accession altar at Copan (Sugiyama and Fash, 2010), to the lesser-known complement of sub-adult dogs recovered from a ritual water source in front of the palace of Cancuen in Guatemala (Thornton, 2011a: 150). But these clear examples are quite rare and it seems likely that sacrifice victims were subsequently incorporated into feasting, sectioned for tribute to various practitioners or supplicants, and their products (hides, bones, teeth) used in other sacred ways. Thus, the study of Maya ritual animal use is based on characters such as the presence of animals considered sacred or reflective of specific ritual activities in ethnographic, ethnohistoric, and icono-graphic texts, but also includes over-representation of managed or husbanded species (Emery, 2002), young individuals (see, for example, Carr, 1996) and unequal distributions of specific elements or body sides (e.g. Pohl, 1983: 89–90; 1985). As an example, at Chinikiha, Montero-Lopez (2009; 2013) combines these characters with a detailed study of depositional history to classify a palace-midden deposit as accumulated waste from repetitive ritual events rather than secular food debitage. A similar study of the large ritual deposit at the site of Lagartero suggested that the deposit resulted from feasting on dog, deer, and rabbit, and might have been related to the role of women (Kozelsky, 2005).

How were such rituals used in maintaining the fabric of ancient Maya society? Starting at the level of community cohesion, many authors have proposed that 'feasting' (e.g. LeCount, 2001), was an essential part of community solidarity and manifestation of power by the Maya rulers. Hayden's (2001) classic definitions of feast signatures include large quantities of animal remains, rare, exotic, or labour-intensive animal species, articulated body portions, and unprocessed bone. Where Maya deposits are proposed as feasting-related, the animal remains have been helpful in understanding the specific purpose of the festive ceremony that created the deposit (see, for example, Stanchly and Ianonne, 2011)

At the next level, reconstructing maintenance of political ties between polity leaders, animal remains again are pivotal. As shown earlier, some animal species were apparently restricted to the ruling nobility, and some were clearly emblematic of rulership (the prime example being the jaguar). But reconstructing the details of ceremonies among rulers is more complicated. One study compared surface site deposits with those from a cave deposit within the palace grounds of the site of Las Pacayas and suggested, on the basis of characters related to status (adornments, emblems), animal management, and private ritual activities (musical instruments), that these particular rites were public exclusionary rituals associated with elite power and landscape control and conducted by the ruling elite of the site (Emery, 2002).

And what of the negotiations between a ruler on behalf of his people, and the ancestors, gods, and alter-egos that, for the Maya, controlled interactions between people and their climate and environments? Here we might consider ritual caching, a practice quite common across the Maya world in the past and still today (Brown and Emery, 2008). The most common archaeological caches are found beneath structure floors and are likely linked to the modern practice of caching associated with house construction or regeneration to placate the earth gods for the use of wilderness space and resources (Taube, 2003). The caches placed by ancient rulers often appear to represent cosmograms of the Maya universe (e.g. Bell et al., 2000). As an example, a well-preserved cache from the site of Joyanca included complete individuals of taxa that inhabit water and those that fly. The chosen species were colourful and very small, and the otherwise very large Central American river turtle and crocodile were tiny juveniles suggesting that small size and colour were of more importance than age and species. Together these may have represented the heavens and underworld, placing the ruler at the centre of the universe (Emery and Thornton, 2008a).

Another type of caching is marked by the overabundance of crania, often of deer or other large game, in natural landscape features and also on ritual platforms such as those at pre-Classic Cuello (Wing and Scudder, 1991), and post-Classic Mayapan, (Masson, 1999: 54). Correspondingly, in most residential deposits the proportion of cranial remains to other body portions is very low (Götz, 2008b: 319–33). These cranial caches may be linked to hunting rites and negotiation with T'zip, the guardian of the animals depicted iconographically from the pre-Classic onward, and described by Landa (Tozzer, 1941) and revealed as a continuing force in animal management decisions by hunters today (Brown, 2009).

CONCLUSIONS: WHAT QUESTIONS ARE STILL LEFT TO EXPLORE?

This review is broad, but it emphasizes the interpretive power of animal remains at a local scale in understanding Maya life. Combined habitat proxy and isotopic studies, for example, argue that the local impact of broad climatological patterns was less extreme than predicted by Maya megadrought models based on other proxies. They also provide evidence countering current models of deforestation indicating regional reductions of high canopy forest and reveal instead increases in secondary, often managed, tree species. Many models of human–environment relationships fail to recognize the diversity of local resource management strategies developed over millennia by the Maya. As a case in point, despite high population densities in the lowland forests, the only evidence for resource depression caused by animal harvest seems to correlate with time periods in which political pressure may have superseded wise local decision-making. The examples presented here of Maya husbandry and domestication indicate that many management decisions had much to do with politics and religion, and less with the challenges of basic subsistence. Maya zooarchaeology suggests a complexity of provisioning, tribute, trade, and gifting that, while distributing animal products broadly, must also have factored in animal management decision-making. The maintenance of animal-product distribution links, particularly when considering species and elements used in crafting and as emblems, was vital to the success of economic and political control at the site, polity, and broader regional levels. Undoubtedly, negotiations with the spirit world also defined some of the primary considerations for management decisions.

Needless to say though, none of the topics that I have mentioned here are fully researched in the Maya zooarchaeological record, and there remain other topics that have been even less explored and will undoubtedly be central to much future research. Foremost are the methodological questions that are so vital in ensuring the representativeness of our samples such as the relationship between activity areas and animal remains (Alexander et al., 2013: 308), how the ancient Maya conceptualized and dealt with animal-product waste (Alexander et al., 2013: 307; Montero López, 2013: 319; for discussion, Pendergast, 2004), the effects of taphonomy (see for review, Stanchly, 2004; Götz, 2014b), or the impact of our research methods on our interpretations (Thornton, 2012). Other areas of rising interest are those of cuisine (Götz, 2014a), pre-Classic development of the Classic Maya social system, and the effects of Spanish (e.g. Emery, 1999; deFrance and Hanson, 2008) and, later, British contact (Thornton and Ng Cackler, 2013). All this is pivotal to our understanding of the evolution of Maya–animal relationships.

Although Maya archaeologists are still sometimes reluctant to include zooarchaeologists on their teams, particularly at the outset of research planning, this review should indicate the value of our studies to reconstructions of ancient Maya life in all spheres. A decade ago the science was plagued by a lack of analysts and studies, but this is changing rapidly as more researchers recognize the multi-faceted nature of Maya zooarchaeology and the remarkable range of topics that the study can approach.

REFERENCES

Alexander, R. T., Hunter, J. A., Arata, S., Martínez Cervantes, R., and Scudder, K. (2013) 'Archaeofauna at Isla Cilvituk, Campeche, Mexico: residential site structure and taphonomy in Postclassic Mesoamerica', in Götz, C. M. and Emery, K. F. (eds) *The Archaeology of Mesoamerican Animals*, pp. 281–313. Atlanta: Lockwood Press.

Anselmetti, F. S., Hodell, D. A., Ariztegui, D., Brenner, M., and Rosenmeier, M. F. (2007) 'Quantification of soil erosion rates related to ancient Maya deforestation', *Geology*, 35(10), 915–18.

Aoyama, K. (2011) 'Socioeconomic and political implications of regional studies of Maya lithic artifacts: two case studies of the Copan Region, Honduras and the Aguateca Region, Guatemala', in Hruby, Z. X., Braswell, G. E., and Chinchilla, O. (eds) *The Technology of Maya Civilization: Political Economy and Beyond in Lithic Studies*, pp. 37–53. Oakville: Equinox Publishing Ltd.

Atran, S. (1993) 'Itza Maya tropical agro-forestry', *Current Anthropology*, 34(5), 633–700.

Bell, E. E., Sharer, R. J., Sedat, D. W., Canuto, M. A., and Grant, L. A. (2000) 'The Margarita Tomb at Copan, Honduras: a research update', *Expedition*, 42(3), 21–5.

Brown, L. A. (2009) 'Communal and personal hunting shrines around Lake Atitlan, Guatemala', *Maya Archaeology*, 1, 36–59.

Brown, L. A. and Emery, K. F. (2008) 'Negotiations with the animate forest: hunting shrines and houses in the Maya highlands', *Journal of Archaeological Method and Theory*, 15 (4), 300–37.

Carr, H. S. (1986) 'Faunal Utilization in a Late Preclassic Maya Community at Cerros, Belize'. Unpublished PhD dissertation, Tulane University (New Orleans).

Carr, H. S. (1996) 'Precolumbian Maya exploitation and management of deer populations', in Fedick, S. L. (ed.) *The Managed Mosaic: Ancient Maya Agriculture and Resource Use*, pp. 251–61. Salt Lake City: University of Utah Press.

Chase, A. F. and Chase, D. Z. (1992) 'Mesoamerican elites: assumptions, definitions and models', in Chase, D. Z. and Chase, A. F. (eds) *Mesoamerican Elites: An Archaeological Assessment*, pp. 3–17. Norman: University of Oklahoma Press.

Collins, L. M. (2002) 'The Zooarchaeology of the Copan Valley: Social Status and the Search for a Maya Slave Class'. Unpublished PhD dissertation, Harvard University (Cambridge, MA).

Corona-M., E., Arroyo Cabrales, J., and Polaco, O. J. (2010) 'Arqueozoología en México, una reseña actual', in Goñalons Mengoni, G., Arroyo Cabrales, J., Polaco, O. J., and Aguilar, F. (eds) *Estado actual de la arqueozoología latinoamericana/Current Advances for the Latin American Archaeozoology*, pp. 165–72. México City: INAH, CONACyT, CONACULTA.

Cunningham-Smith, P., Chase, A. F., and Chase, D. Z. (2014) 'Fish from afar: marine resource use at Caracol Belize', *Research Reports in Belizean Archaeology*, 11, 43–53.

deFrance, S. D. and Hanson, C. A. (2008) 'Labor, population movement and food in sixteenth-century Ek Balam, Yucatan', *Latin American Antiquity*, 19(3), 299–316.

Emery, K. F. (1999) 'Continuity and variability in Postclassic and colonial animal use at Lamanai and Tipu, Belize', in White, C. D. (ed.) *Reconstructing Ancient Maya Diet*, pp. 61–82. Salt Lake City: University of Utah Press.

Emery, K. F. (2001) 'Fauna', in Evans, S. T. and Webster, D. (eds) *Archaeology of Ancient Mexico and Central America—an Encyclopedia*, pp. 255–65. New York: Garland Publishing, Inc.

Emery, K. F. (2002) 'Animals from the Maya underworld: reconstructing elite Maya ritual at the Cueva de los Quetzales, Guatemala', in Jones O'Day, S., Van Neer, W., and Ervynck, A. (eds) *Behaviour behind Bones: The Zooarchaeology of Ritual, Religion, Status and Identity*, pp. 101–13. Oxford: Oxbow.

Emery, K. F. (2004a) 'Maya zooarchaeology: in pursuit of cultural variability and environmental heterogeneity', in Golden, C. and Borgstede, G. (eds) *Continuities and Changes in Maya Archaeology: Perspectives at the Millennium*, pp. 217–41. New York: Routledge Press.

Emery, K. F. (2004b) 'In search of the "Maya Diet": is regional comparison possible in the Maya area?', *Archaeofauna*, 13, 37–56.

Emery, K. F. (2004c) 'Environments of the Maya collapse: a zooarchaeological perspective from the Petexbatun, Guatemala', in Emery, K. F. (ed.) *Maya Zooarchaeology: New Directions in Method and Theory*, pp. 81–96. Los Angeles: Cotsen Institute of Archaeology, UCLA.

Emery, K. F. (ed.) (2004d) *Maya Zooarchaeology: New Directions in Method and Theory*, Los Angeles: Cotsen Institute of Archaeology, UCLA.

Emery, K. F. (2006) 'Definiendo el aprovechamiento de la fauna por la elite: evidencia en Aguateca y otros sitios en Petexbatún, Guatemala', *Ut'zib*, 4(1), 1–16.

Emery, K. F. (2007a) 'Aprovechamiento de la fauna en Piedras Negras: dieta, ritual y artesanía del periodo clásico Maya', *Mayab*, 19, 51–69.

Emery, K. F. (2007b) 'Assessing the impact of ancient Maya animal use', *Journal of Nature Conservation*, 15(3), 184–95.

Emery, K. F. (2008) 'A zooarchaeological test for dietary resource depression at the end of the Classic period in the Petexbatun, Guatemala', *Human Ecology*, 36(5), 617–34.

Emery, K. F. (2010) *Dietary, Environmental, and Societal Implications of Ancient Maya Animal Use in the Petexbatun: A Zooarchaeological Perspective on the Collapse*, Nashville: Vanderbilt University Press.

Emery, K. F. (2012) 'Zooarchaeology of Motul de San Jose, an economic perspective', in Foias, A. and Emery, K. F. (eds) *Politics, History, and Economy at the Classic Maya Polity of Motul de San Jose, Guatemala*, pp. 291–325. Gainesville: University of Florida Presses.

Emery, K. F. (2014) 'Aguateca animal remains', in Inomata, T. and Triadan, D. (eds) *Life and Politics at the Royal Court of Aguateca: Artifacts, Analytical Data and Synthesis*, pp. 158–200. Salt Lake City: University of Utah Press.

Emery, K. F. and Aoyama, K. (2007) 'Bone tool manufacturing in elite Maya households at Aguateca, Guatemala', *Ancient Mesoamerica*, 18(2), 69–89.

Emery, K. F. and Brown, L. A. (2012) 'Maya hunting sustainability: perspectives from past and present', in Chacon, R. J. and Mendoza, R. G. (eds) *The Ethics of Anthropology and Amerindian Research: Reporting on Environmental Degradation and Warfare*, pp. 79–116. New York: Springer Press.

Emery, K. F. and Thornton, E. K. (2008a) 'Reporte preliminar de los restos faunísticos recuperados en el sitio de La Joyanca, Peten, Guatemala'. Unpublished report, Instituto de Antropologia e Historia de Guatemala.

Emery, K. F. and Thornton, E. K. (2008b) 'Zooarchaeological habitat analysis of ancient Maya landscape changes', *Journal of Ethnobiology*, 28(2), 154–79.

Emery, K. F. and Thornton, E. K. (2008c) 'Zooarchaeological isotopic chemistry: a regional perspective on biotic change during the Classic Maya occupation of the Guatemalan Peten', *Quaternary International*, 191, 131–43.

Emery, K. F. and Thornton, E. K. (2013) 'Tracking climate change in the ancient Maya world through zooarchaeological habitat analysis', in Iannone, G. (ed.) *The Great Maya Droughts in Cultural Context*, pp. 301–32. Boulder: University Press of Colorado.

Emery, K. F., Thornton, E. K., Cannarozzi, N. R., Houston, S., and Escobedo, H. (2013) 'Archaeological animals of the southern Maya highlands: zooarchaeology of Kaminaljuyu', in Götz, C. M. and Emery, K. F. (eds) *The Archaeology of Mesoamerican Animals*, pp. 381–415. Atlanta: Lockwood Press.

Foias, A. (2013) *Ancient Maya Political Dynamics*, Gainesville: University of Florida Press.

Ford, A. and Nigh, R. B. (2009) 'Origins of the Maya forest garden: a resource management system', *Journal of Ethnobiology*, 29(2), 213–36.

Götz, C. M. (2008a) 'Coastal and inland patterns of faunal exploitation in the Prehispanic northern Maya lowlands', *Quaternary International*, 191, 154–69.

Götz, C. M. (2008b) *Die Verwendung von Wirbeltieren durch die Maya des nördlichen Tieflandes während der Klassik und Post Klassik (600–1500 n. Chr). Internationale Archäologie 106*, Westfalen: Verlag Marie Leidorf GmbH.

Götz, C. M. (2012) 'Caza y pesca prehispánicas en la costa norte peninsular Yucateca', *Ancient Mesoamerica*, 23(2), 421–39.

Götz, C. M. (2013a) 'Introduction', in Götz, C. M. and Emery, K. F. (eds) *The Archaeology of Mesoamerican Animals*, pp. 1–21. Atlanta: Lockwood Press.

Götz, C. M. (2013b) 'The sustainability of Prehispanic Maya agroecosystems: implications of hunting and animal domestication in the northern Maya lowlands', in Stanton, T. W. (ed.) *The Archaeology of Yucatan: New Directions and Data*. BAR International Series 108, pp. 477–86. Oxford: Archaeopress.

Götz, C. M. (2014a) 'La alimentación de los Mayas prehispánicos: vista desde la zooarqueología', *Anales de Antropología*, 48(1), 167–99.

Götz, C. M. (2014b) 'Solamente contextos culturales? Evaluación del papel de la tafonomía en la zooarqueología Maya de las tierras bajas del norte de la península de Yucatán, México', *Etnobiología*, 12(2), 20–38.

Haug, G. H., Gunther, D., Peterson, L. C., Sigman, D. M., Hughen, K. A., and Aeschlimann, B. (2003) 'Climate and the collapse of Maya civilization', *Science*, 299, 1731–5.

Hayden, B. (2001) 'Fabulous feasts: a prolegomenon to the importance of feasting', in Dietler, M. and Hayden, B. (eds) *Feasts: Archaeological and Ethnographic Perspectives on Food, Politics, and Power*, pp. 23–64. Washington, DC: Smithsonian Institution Press.

Henderson, J. S. (1997) *The World of the Ancient Maya*, 2nd edn, Ithaca: Cornell University Press.

Hohmann, B., (2014) 'Middle Preclassic shell working at Pacbitun, Belize', in Healy, P. F. and Emery, K. F. (eds) *Zooarchaeology of the Ancient Maya Centre of Pacbitun, (Belize)*, pp. 56–78. Trent University, Occasional Papers in Anthropology No. 16, Trent University, Peterborough, Ontario.

Iannone, G. (ed.) (2013) *The Great Maya Droughts in Cultural Context*, Boulder: University Press of Colorado.

Isaza-Aizpurúa, I. I. (2004) 'The art of shell working and the social uses of shell ornaments', in McAnany, P. A. (ed.) *K'axob: Ritual, Work and Family in an Ancient Maya Village*, pp. 335–51. Los Angeles: The Cotsen Institute of Archaeology Press.

Isaza-Aizpurúa, I. I. and McAnany, P. A. (1999) 'Adornment and identity: shell ornaments from formative K'axob', *Ancient Mesoamerica*, 10, 117–27.

Keller, A. (2013) 'Creating community with shell', in Robin, C. (ed.) *Chan: An Ancient Maya Farming Community*, pp. 253–70. Gainesville: University Press of Florida.

Kintigh, K. W., Altschul, J. H., Beaudry, M. C., Drennan, R. D., Kinzig, A. P., Kohler, T. A., Limp, W. F., Maschner, H. D. G., Michener, W. K., Pauketat, T. R., Peregrine, P., Sabloff, J. A., Wilkinson, T. J., Wright, H. T., and Zeder, M. A. (2014) 'Grand challenges for archaeology', *American Antiquity*, 79(1), 5–24.

Kozelsky, K. (2005) 'Identifying Social Drama in the Maya Region: Fauna from the Lagartero Basurero, Chiapas, Mexico'. Unpublished MA thesis, Florida State University (Tallahassee).

Lange, F. W. (1971) 'Marine resources: a viable subsistence alternative for the prehistoric lowland Maya', *American Anthropologist*, 73(3), 619–39.

LeCount, L. J. (2001) 'Like water for chocolate: feasting and political ritual among the Late Classic Maya at Xunantunich, Belize', *American Anthropologist*, 103(4), 935–53.

Linares, O. F. (1976) '"Garden hunting" in the American tropics', *Human Ecology*, 4(4), 331–49.

Masson, M. A. (1999) 'Animal resource manipulation in ritual and domestic contexts at Postclassic Maya communities', *World Archaeology*, 31(1), 93–120.

Masson, M. A. (2004) 'Fauna exploitation from the Preclassic to the Postclassic periods at four Maya settlements in northern Belize', in Emery, K. F. (ed.) *Maya Zooarchaeology: New Directions in Method and Theory*, pp. 97–122. Los Angeles: Cotsen Institute of Archaeology, UCLA.

Masson, M. A. and Peraza Lope, C. (2008) 'Animal use at the Postclassic Maya center of Mayapan', *Quaternary International*, 191, 170–83.

Masson, M. A. and Peraza Lope, C. (2013) 'Animal consumption at the monumental center of Mayapan', in Götz, C. and Emery, K. F. (eds) *The Archaeology of Mesoamerican Animals*, pp. 233–80. Atlanta: Lockwood Press.

McAnany, P. A. (1993) 'The economics of social power and wealth among eighth-century Maya households', in Sabloff, J. A. and Henderson, J. S. (eds) *Lowland Maya Civilization in the Eighth Century A.D.*, pp. 65–89. Washington, DC: Dumbarton Oaks.

McKillop, H. I. (1996) 'Ancient Maya trading ports and the integration of long-distance and regional economies: Wild Cane Cay in south-coastal Belize', *Ancient Mesoamerica*, 7, 49–62.

McKillop, H. I. and Winemiller, T. (2004) 'Ancient Maya environment, settlement, and diet: quantitative and GIS analyses of mollusca from Frenchman's Cay, Belize', in Emery, K. F. (ed.) *Maya Zooarchaeology: New Directions in Method and Theory*, pp. 57–80. Los Angeles: Institute of Archaeology, UCLA Press.

Mock, S. B. (1997) 'Monkey business at Northern River Lagoon: a coastal-inland interaction sphere in northern Belize', *Ancient Mesoamerica*, 8, 165–83.

Moholy-Nagy, H. (1997) 'Middens, construction fill, and offerings: evidence for the organization of Classic period craft production at Tikal, Guatemala', *Journal of Field Archaeology*, 24(3), 293–313.

Montero López, C. (2009) 'Sacrifice and feasting among the Classic Maya elite and the importance of the white-tailed deer: is there a regional pattern?', *Journal of Historical and European Studies*, 2, 53–68.

Montero López, C. (2013) 'Inferring archaeological context through taphonomy: the use of the white-tailed deer (*Odocoileus virginianus*) in Chinikihá, Chiapas', in Götz, C. M. and Emery, K. F. (eds) *The Archaeology of Mesoamerican Animals*, pp. 315–50. Atlanta: Lockwood Press.

Pendergast, D. M. (2004) 'Where's the meat? Maya zooarchaeology from an archaeological perspective', in Emery, K. F. (ed.) *Maya Zooarchaeology: New Directions in Method and Theory*, pp. 239–48. Los Angeles: Institute of Archaeology, UCLA Press.

Pendergast, D. M. and Graham, E. (1990) 'An island paradise (??): Marco Gonzalez 1990', *Royal Ontario Museum Newsletter*, 2(41), 1–4.

Pohl, M. D. (1983) 'Maya ritual faunas: vertebrate remains from burials, caches, caves, and cenotes in the Maya lowlands', in Leventhal, R. M. and Kolata, A. L. (eds) *Civilization in the Ancient Americas: Essays in Honor of Gordon R. Willey*, pp. 55–103. Albuquerque: University of New Mexico Press.

Pohl, M. D. (1985) 'The privileges of Maya elites: prehistoric vertebrate fauna from Seibal', in Pohl, M. D. (ed.) *Prehistoric Lowland Maya Environment and Subsistence Economy*. Papers of the Peabody Museum of Archaeology and Ethnology 77, pp. 133–45. Cambridge (MA): Harvard University.

Pohl, M. D. (1994) 'The economics and politics of Maya meat eating', in Brumfiel, E. M. (ed.) *The Economic Anthropology of the State*. Monographs in Economic Anthropology 11, pp. 119–47. New York: University Press of America.

Pohl, M. D. (1995) 'Late Classic Maya fauna from settlement in the Copán Valley, Honduras: assertation of social status through animal consumption', in Willey, G. R., Leventhal, R., Demarest, A. A., and Fash, W. (eds) *Excavations at Copán Honduras*, pp. 459–76. Cambridge (MA): Peabody Museum, Harvard University.

Pohl, M. D. and Feldman, L. H. (1982) 'The traditional role of women and animals in lowland Maya economy', in Flannery, K. (ed.) *Maya Subsistence*, pp. 295–311. New York: Academic Press.

Pope, C. L. (1994) 'Preliminary analysis of small chert tools and related debitage at Caracol, Belize', in Chase, D. Z. and Chase, A. F. (eds) *Studies in the Archaeology of Caracol, Belize*, pp. 148–56. San Francisco: Pre-Colombian Art Research Institute.

Repussard, A., Schwarcz, H. P., Emery, K. F., and Thornton, E. K. (2013) 'Oxygen isotopes from Maya archaeological deer remains: experiments in tracing droughts using bones', in Iannone, G. (ed.) *The Great Maya Droughts in Cultural Context*, pp. 231–54. Boulder: University Press of Colorado.

Rosenswig, R. M. and Masson, M. A. (2002) 'Transformation of the terminal Classic to Postclassic architectural landscape at Caye Coco, Belize', *Ancient Mesoamerica*, 13, 213–35.

Sharer, R. J. and Traxler, L.P. (2005) *The Ancient Maya*, 6th edn, Stanford: Stanford University Press.

Sharpe, A. (2014) 'A reexamination of the birds in the central Mexican codices', *Ancient Mesoamerica*, 25(2), 317–36.

Sharpe, A. E. and Emery, K. F. (2015) 'Differences in animal use between three late Classic Maya states: implications for politics and economics', *Journal of Anthropological Archaeology*, 40, 280–301.

Sharpe, A., Saturno, W. A., and Emery, K. F. (2014) 'Shifting patterns of Maya social complexity through time: preliminary zooarchaeological results from San Bartolo, Guatemala', in Arbuckle, B. S. and McCarty, S. A. (eds) *Animals and Inequality in the Ancient World*, pp. 85–106. Boulder: University Press of Colorado.

Shaw, L. C. (1995) 'Analysis of faunal materials from Ek Luum', in Guderjan, T. H. and Garber, J. F. (eds) *Maya Maritime Trade, Settlement, and Population on Ambergris Caye, Belize*, pp. 175–81. San Antonio: Maya Research Programs and Labyrinthos.

Somerville, A. D., Nelson, B. A., and Knudson, K. J. (2010) 'Isotopic investigation of prehistoric macaw breeding in northwest Mexico', *Journal of Anthropological Archaeology*, 29, 125–35.

Stanchly, N. (2004) 'Picks and stones may break my bones: taphonomy and Maya zooarchaeology', in Emery, K. F. (ed.) *Maya Zooarchaeology: New Directions in Method and Theory*, pp. 35–43. Los Angeles: Cotsen Institute of Archaeology, UCLA.

Stanchly, N. and Ianonne, G. (2011) 'Examining variability in ancient Maya feasting: insights from Belize'. Paper presented at the 40th annual Chacmool Conference, Department of Archaeology, University of Calgary (Calgary).

Steadman, D. W. (1980) 'A Review of the Osteology and Paleontology of Turkeys (Aves: Meleagridinae)'. Contributions in Science 330. Los Angeles: Natural History Museum of Los Angeles County.

Sugiyama, N. and Fash, W. (2010) 'Reinterpreting the Copan felines: reconstructing the "jaguar stew" assemblage by Altar Q'. Annual Meetings of the Society for American Archaeology, Atlanta, Georgia.

Taube, K. A. (2003) 'Ancient and contemporary Maya conceptions about forest and field', in Gomez-Pompa, A., Allen, M. F., Fedick, S. L., and Jiménez-Osornio, J. J. (eds) *The Lowlands Maya Area: Three Millennia at the Human-Wildland Interface*, pp. 461–92. New York: Food Products Press.

Thornton, E. K. (2011a) 'Zooarchaeological and Isotopic Perspectives on Ancient Maya Economy and Exchange'. Unpublished PhD dissertation, University of Florida (Gainesville).

Thornton, E. K. (2011b) 'Reconstructing ancient Maya animal trade through strontium isotope (87Sr/86Sr) analysis', *Journal of Archaeological Science*, 38, 3254–63.

Thornton, E. K. (2012) 'Animal resource use and exchange at an inland Maya port: zooarchaeological investigations at Trinidad de Nosotros', in Foias, A. and Emery, K. F. (eds) *Politics, History, and Economy at the Classic Maya Polity of Motul de San Jose, Guatemala*, pp. 326–56. Gainesville: University of Florida Presses.

Thornton, E. K. and Ng Cackler, O. (2013) 'Late-nineteenth and early twentieth-century animal use by San Pedro Maya and British populations at Holotunich, Belize', in Götz, C. M. and Emery, K. F. (eds) *The Archaeology of Mesoamerican Animals*, pp. 351–80. Atlanta: Lockwood Press.

Thornton, E. K. and Emery, K. F. (in press) 'Preliminary analysis of zooarchaeological remains from El Mirador', *Archaeofauna*.

Thornton, E. K., Emery, K. F., Speller, C., Steadman, D., Matheny, R., and Yang, D. (2012) 'Earliest Mexican turkeys (*Meleagris gallopavo*) in the Maya region: implications for pre-Hispanic animal trade and the timing of turkey domestication', *PLoS ONE*, 7(8), DOI: 10.1371/journal.pone.0042630.

Tozzer, A. M. (1941) *Landa's Relación de las Cosas de Yucatán*. Papers of the Peabody Museum of American Archaeology and Ethnology 18. Cambridge (MA): Harvard University.

Turner, B. L. I. and Sabloff, J. A. (2012) 'Classic period collapse of the central Maya lowlands: insights about human-environment relationships for sustainability', *Proceedings of the National Academy of Sciences of the United States of America*, 109(35), 13908–14.

Valadez Azúa, R. (2003) *La domesticación animal*, Mexico City: Universidad Nacional Autónoma de México, Instituto de Investigaciones Antropológicas.

Valadez Azúa, R., Blanco Padilla, A., Rodríguez Galicia, B., and Pérez Roldán, G. (2013) 'The dog in the Mexican archaeozoological record', in Götz, C. M. and Emery, K. F. (eds) *The Archaeology of Mesoamerican Animals*, pp. 557–82. Atlanta: Lockwood Press.

White, C. D. (2004) 'Stable isotopes and the human-animal interface in Maya biosocial and environmental systems', *Archaeofauna*, 13, 183–98.

White, C. D., Pohl, M. D., Schwarcz, H. P., and Longstaffe, F. J. (2001) 'Isotopic evidence for Maya patterns of deer and dog use at Preclassic Colha', *Journal of Archaeological Science*, 28, 89–107.

Wing, E. S. and Scudder, S. J. (1991) 'The exploitation of animals', in Hammond, N. (ed.) *Cuello: An Early Maya Community*, pp. 84–97. Cambridge: Cambridge University Press.

PART VI

SOUTH AMERICA

ZOOARCHAEOLOGICAL APPROACHES TO PRE-COLUMBIAN ARCHAEOLOGY IN THE NEOTROPICS OF NORTHWESTERN SOUTH AMERICA

PETER W. STAHL

INTRODUCTION

THE environments of northwestern South America are renowned for harbouring elevated species richness and high rates of endemism (Gentry, 1992). This astonishing biodiversity is due in great part to the area's remarkable environmental heterogeneity which, compared to many other continental areas, supports a proportionately more complex and often unique biota. The modern nations of Ecuador, Colombia, and Panama encompass an unprecedented range of contiguous ecosystems associated with marine environments of two oceanic systems, coastal mangroves, and beaches, biogeographically distinct eastern and western lowland areas, savannahs, montañas, mountainous Andean valleys, páramo, and snow-capped volcanoes (Fig. 40.1).

Despite its current biological renown, the area was historically denigrated in the archaeological literature as 'intermediate' in both geographical and cultural developmental perspectives. Unlike neighbouring centres of 'high culture' in Mexico to the north and the Central Andes to the south, it was assumed that this intermediate area did not possess the appropriate environmental potential for developing autochthonous complex civilization. For example, even though early evidence of highly precocious ceramic assemblages was identified here in the mid-twentiethth century, it was still

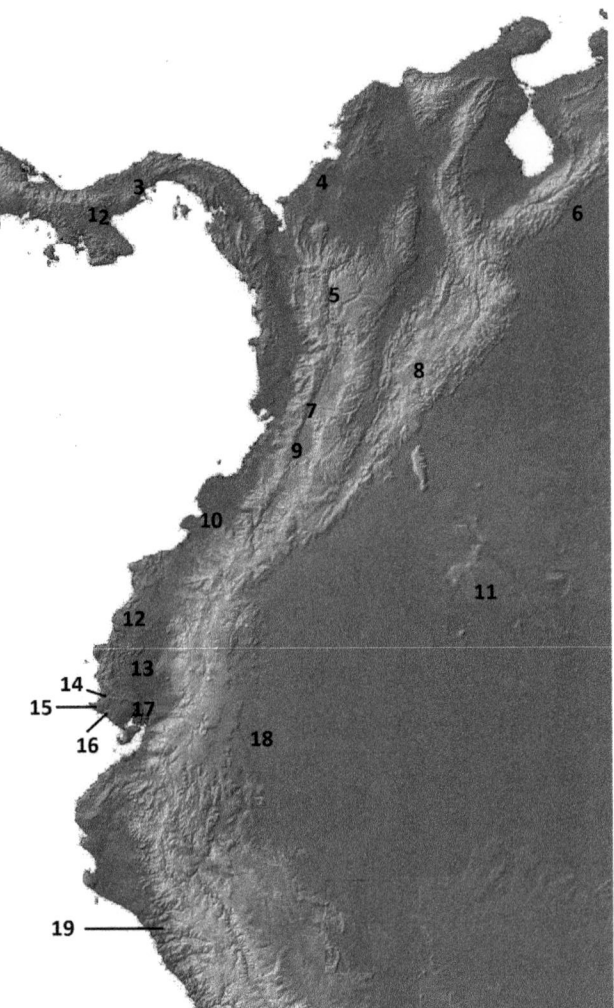

FIGURE 40.1 Archaeological sites, cultural and geographical names mentioned in text. 1. Lake Yeguada, Carabalí, 2. Aguadulce, Cerro Mangote, Cueva de los Ladrones, Vampiros-1, 3. Monte Oscuro, 4. Sinú and San Jorge River Basins, San Jacinto-1, 5. Middle Porce and Cauca Valleys, El Jazmín, 6. Barinas Region, 7. Calima, 8. Muisca, 9. San Isidro, 10. Tumaco, 11. Peña Roja, 12. Jama Valley, 13. Gran Cacao, 14. Loma Alta, 15. Santa Elena Peninsula, 16. Real Alto, 17. Lower Guayas Basin, 18. Lake AyauchI, 19. Zaña and Nanchoc Valleys. Base map modified from the United States National Oceanic and Atmospheric Administration (NOAA) National Satellites Data and Information Services (NESDIS).

considered credible to seek their ultimate origins in the accidental voyage of lost fisher-men from Japan (Meggers, 2011).

Archaeological and historical research has dramatically altered our depiction of the area's pre-Columbian past. At the time of Spanish conquest, much of northwestern South America was densely occupied by complex indigenous cultures many of which were subsequently subjected to the most devastating depopulation rates encountered anywhere during the colonial experience. Archaeology has uncovered long developmental sequences of indigenous culture with extensive and complex histories supported in some areas by evidence that extends back into the terminal Pleistocene. The area offers the prospect of an elaborate human history which rivals its renowned biodiversity.

Northwestern South America is important for many issues involving indigenous occupation of the western hemisphere. At a time when standard orthodoxy denigrated the area's culturally 'intermediate' development, historical geographer Carl Sauer (1952) considered it as one of two global hearths for early plant domestication. Indeed, throughout the Holocene, large areas of northwestern South America were likely heavily managed and complexly interrelated through a variety of contiguous humanized landscapes. Today, the area transcends the interests of anthropology as it offers potential answers to broader questions about the very nature and origin of regional biodiversity. These issues are applicable to contemporary conservation and management, particularly within a context of rapid transformation and disappearance of unique local ecosystems. A critical approach to zooarchaeology in the northern neotropics of South America that combines taphonomy and allied methodologies in archaeobiology can be used in different ways to engage related research questions within a framework of Historical Ecology. In particular, zooarchaeological and corroborating archaeological data can be integrated to understand various aspects of the area's early and continuously humanized landscapes that supported human agro-ecological systems, domesticated plants and animals, inter-regional trade, and exchange and cultural complexity.

The Neotropical Zooarchaeological Record

Neotropical zooarchaeology is often confronted by a range of difficulties that are not unique to the neotropics but frequently worse in matters of degree. Temporal and spatial patchiness in the archaeological record is often the result of differential preservation and recovery, but also reflects a variety of related issues ranging from the different research priorities of archaeologists to problems associated with site discovery. Throughout much of the area, prevailing research agendas have produced an archaeological record that is often markedly biased toward either early or late temporal contexts. Early contexts often include: pre-ceramic sites associated with the earliest human populations of the area; Early Holocene sites (beginning around 12,000 BP), particularly those yielding evidence for plant manipulation; and village sites with assemblages characterized by examples of early and aesthetically precocious pottery. Later contexts usually reflect a focus on the material legacy of complex polities, such as highland Incan sites which appeared as far north as southern Colombia in the final centuries before the arrival of Europeans during the early sixteenth century. In these cases, archaeological activity can be guided by highly conspicuous examples of preserved architecture and the pursuit of sumptuous elite burials, whose discovery is often aided and abetted by construction activities and omnipresent *huaquerismo* (looting). Although of great interest, and often serving as important sources for heritage tourism and national pride, these contexts may not necessarily have served as depositional locations from which archaeofaunal materials are recovered. In many areas, site discovery is generally hindered by heavy vegetation cover, archaeological activities may be logistically difficult to undertake, or investigations can become politically problematic. The archaeological record is generally far more robust in accessible areas, and typically archaeological activities are today increasingly guided by agendas stressing cultural remediation over research.

A particularly important issue in neotropical archaeology is the differential preservation of organic residues. The varied environments of the northern neotropics offer equally varied contexts for faunal preservation. Differential preservation is particularly important in open air sites and certainly one reason why rock shelters in mountainous areas are preferred excavation locales, especially for early contexts. Humidity and acidic soils hinder preservation of all but the most durable associations (Stahl, 1995); however, excellent preservation of bone material can be found in regularly inundated areas (Stahl, 1988), while highly fragmented bone specimens are often recovered from time-averaged contexts in arid environments (Stahl, 1991). Even with conditions of poor organic preservation, a minimal understanding of available faunal taxa can be otherwise accessed in some areas through vivid and durable artistic representations (e.g. Cadena and Bouchard, 1980; Echeverría, 1988; Rodríguez, 1992; Cooke et al., 2008).

The differential preservation and fragmentation of preserved bone specimens require neotropical zooarchaeologists to adjust their techniques and methods when interpreting recovered faunal assemblages. Issues of preservation can require redoubling intensive recovery of organic specimens through fine screening or water flotation which affects the quantitative and qualitative structure of the studied assemblage (Stahl, 1992; 1995; Pearsall, 1995). Basic identification of archaeofaunal specimens can be seriously encumbered by the state of zoological knowledge and lack of access to comprehensive comparative collections. Earlier research in the area was often disinterested in recovering archaeofaunal specimens or lacked the expertise to analyse them; in some fortunate cases zooarchaeological data may be retrieved through the study of curated assemblages.

Reported data can be highly variable and difficult to compare. At times, essential statistics are missing, whereas many sites may have nothing to report. In some cases site records may include only partial assemblages or lists of identified taxa with little or no accompanying detail. Evidence for understanding assemblage accumulation and differential preservation is often lacking, which has important implications for understanding assemblage origin, the validity of research questions, the relevance of numbers, and the reliability of results. Generally, the only way to proceed is through the judicious use of presence/absence data which can be corroborated through examining ecological communities in association with botanical data (Grayson, 1981).

The nature of contextual associations is essential for understanding the limitations of data and for asking appropriate questions. Although data are often retrieved in different ways from relatively small-scale excavation units, unique contexts, when encountered, should be recovered opportunistically. These include various kinds of depositional contexts with fine-grained stratigraphy (e.g. Stahl and Oyuela-Caycedo, 2007), spatially extensive living floors (e.g. Stahl and Zeidler, 1990), special burial features (e.g. Zeidler et al., 1998; Cooke and Jiménez, 2010; Stahl, 2012), elite precincts (e.g. Stahl, 2005; Cooke and Jiménez, 2010), tool workshops (e.g. Stahl and Athens, 2001), and high-resolution pit features (e.g. Stahl, 2000), which can be significant for asking relevant questions and providing potentially reliable answers.

HUMANIZED LANDSCAPES IN THE PRE-COLUMBIAN NEOTROPICS

Despite its biological prominence, and perhaps in part due to its complex environmental heterogeneity, our understanding of northwestern South America's palaeoecology, especially in its extensive lowlands, is relatively poor. The area was neither heavily impacted by extensive glacial cover nor affected by extreme aridity during the Pleistocene; even during the glacial maximum, temperatures in the Andean slopes were on average eight degrees Celsius cooler, and in the Amazonian lowlands four to six degrees cooler, while both areas were covered in forest (Bush et al., 2007a; 2007b). With the onset of warmer and wetter Holocene conditions and forest expansion by 11,000–10,000 BP, evidence for human landscape alteration and plant manipulation becomes increasingly apparent in the archaeological record throughout northwestern South America.

In Panama, preserved phytoliths and particulate carbon in core sediments recovered at Lake Yeguada and Monte Oscuro suggest burning and forest clearance after at least 11,000 BP. Later evidence for arrowroot (*Maranta* sp.), llerén (*Calathea allouia*), squash (*Cucurbita* sp.), and bottle gourd (*Lagenaria siceraria*) was recovered in rock-shelter sites Vampiros-1 and Aguadulce, with the eventual appearance of exotic maize (*Zea mays*) and manioc (*Manihot esculenta*) by 7000 BP at Aguadulce and Cueva de los Ladrones (Cooke et al., 2013). Stone tools and different forms of archaeobotanical data corroborate the claim for anthropogenic vegetation clearance, burning, and plant manipulation in Early Holocene Colombia. Evidence for tree crops including avocado (*Persea americana*), along with arrowroot, sweet potato (*Ipomea batatas*), and manioc, is associated with 10,000 BP contexts at San Isidro in the Popayan Plateau (Aceituno et al., 2013). Beginning as early as 7500 BP, fruits, palms, roots, and grasses co-evolve with anthropogenic forest disturbance, culminating in horticultural experimentation and the appearance of maize, amaranth (*Amaranthus* sp.), manioc, and squash in the middle Río Porce valley (Castillo and Aceituno, 2006). Multiple lines of evidence from various sites at 1,600 m asl in the middle Cauca valley also indicate the early exploitation of maize, beans (*Phaseolus* sp.), manioc, and sweet potato, particularly at El Jazmín which yields an early date of 10,100 BP (Aceituno et al., 2013). Excavations at Peña Roja in Colombia's southern Amazon recovered charred seeds of many different palms along with phytolith evidence suggesting the cultivation of squash, bottle gourd, and llerén by 8000 BP (Mora, 2003).

Further south, on Ecuador's Santa Elena peninsula, phytolith and starch grain analysis identifies bottle gourd, squash, and llerén in Early Las Vegas contexts before 9000 BP, with subsequent additions of maize, peanut (*Arachis hypogaea*), and possibly bean and achira (*Canna edulis*). Llerén and achira are also identified in roughly coeval deposits in the Guayas Basin at Gran Cacao (Stothert and Sánchez, 2011). Phytolith and pollen evidence in core records from Lake AyauchI implicate forest clearance and maize cultivation in the southern Ecuadorian Amazon by at least 5000 BP (Piperno and Pearsall, 1998: 258). Squash seeds directly dated to 10,300 BP were recovered in El Palto Phase (11,500–9800 BP) contexts in the upper Zaña and Nanchoc valleys of northern Peru, and evidence for house gardens, squash, peanut, manioc, bean, cotton (*Gossypium barbadense*), medicinal plants including coca (*Erythroxylum* sp.), and tree crops is added in later Las Pircas Phase (9800–7800 BP) contexts (Rossen, 2011).

Preserved animal bone assemblages from these earliest contexts are relatively rare, but when available, tend to corroborate the palaeobotanical evidence which indicates an early and precocious human manipulation of neotropical ecosystems. It is important to emphasize that the majority of identifiable faunal specimens associated with early contexts in the middle Río Porce valley consist of smaller mammalian generalists like armadillos (*Dasypus novemcinctus*), agoutis (*Dasyprocta fulginosa*), pacas (*Cuniculus paca*), and spiny rats (*Proechimys semispinosus*) (Castillo and Aceituno, 2006). Considered together with other identified animals, especially porcupines (*Coendou prehensilis*), howler monkeys (*Alouatta seniculus*), and red brocket deer (*Mazama americana*), the recovered assemblage suggests the presence of ecological generalists often associated with disturbed habitats and cultivated clearings. Although scarce, faunal associations in the most ancient Panamanian contexts hint at similar observations. The earliest horticultural period contexts at Carabalí, Ladrones, Cerro Mangote, and Aguadulce beginning at 7000 BP tend to be dominated by mammalian generalists like opossum (*Didelphis marsupialis*), tamandua (*Tamandia mexicana*), armadillos, rabbits (*Sylvilagus brasiliensis*), agoutis, pacas, raccoons (Procyonidae), and deer (Cooke and Ranere, 1992). Similarly, Vegas mammalian faunas include numerous specimens of opossum, tamandua, rabbit, squirrel (*Sciurus stramineus*), spiny rat, and deer (Stothert and Sánchez, 2011; Stahl, unpublished data), while the early pre-ceramic faunal inventory in northern Peru contains an assortment of deer and smaller rodents (Stackelbeck, 2011).

Neotropical archaeofaunal assemblages are often comprised of eurytopic taxa, or generalists with broad niche requirements, which have a tendency to thrive under conditions of disturbed vegetation. This can include anthropogenic clearance, particularly in the form of temporally and spatially intermediate disturbance of forest cover. As universal footprints of human activity, vegetative disturbance and clearing can create or minimally increase favourable conditions for eurytopic generalists who thrive in, and exploit, culturally modified habitats (Stahl, 2006). This can be especially pronounced where plant management or cultivation substantially increases local resource supply.

These sites can attract and concentrate foraging animals, which can be incidentally or intentionally pursued by humans in a form of garden hunting, which provides important dietary supplements, particularly animal-derived protein, for neotropical gardeners (Linares, 1976; Stahl, 2014).

It is interesting, but not coincidental, that all of the earliest Holocene archaeofaunal assemblages in northwestern South America also include preserved specimens of gray (*Lycalopex griseus*), culpeo (*L. culpaeus*), and/or Sechuran fox (*L. sechurae*). The latter species is particularly evident in Las Vegas contexts where its preserved bone specimens overwhelmingly dominate the faunal assemblage. Foxes, particularly the endemic South American varieties, share a number of important behavioural characteristics. They are all opportunistic habitat generalists inclined to increasing omnivory, adjusting their dietary preferences and timing of activity to local and seasonal conditions, and incorporating variable amounts of plant food into their diet. Comfortable in a range of habitats but preferring open settings, they tolerate anthropogenic disturbance particularly where it increases local resource supply. Throughout the continent, wherever preserved animal bones are recovered in Early Holocene contexts, fox specimens are included. Their appearance also coincides with the relatively late arrival of the exotic domestic dog in South America (Stahl, 2012). It is conceivable that the earliest canids described by European explorers in northern South America and the Caribbean islands included some form of tamed native fox or forest dog (Stahl, 2013).

Historic interest in the archaeology of northwestern South America has been dominated in great part by investigations of the area's early and precocious pottery-producing cultures. Following European scholarship, attention was focused on the origin, nature and diffusion of a bundled Neolithic (Formative in Americanist terminology) 'package' of agriculture, pottery, and sedentary village life. The Formative parcel was subsequently unpacked to reveal that sedentism and agriculture preceded the appearance of pottery and that plant cultivation was not obligatorily associated with pottery production and consumption.

To date, the earliest pottery in the area was recovered from the buried riverside site of San Jacinto 1 in the seasonal coastal savannas of northern Colombia. High-resolution stratified contexts at San Jacinto 1 revealed the association of fibre-tempered pottery with uncalibrated dates in the early fourth millennium BC. Recurring associations of a restricted range of smaller vertebrate and invertebrate prey items dominate the faunal assemblages through time. These assemblages may represent human occupation of a preferred logistic camp, where specific game items were procured at the height of the dry season. Although only wild plants were identified in the recovered assemblage, it may be erroneous to assume that the focused seasonal foraging at this encampment was representative of the entire prehistoric diet (Stahl and Oyuela-Caycedo, 2007).

Intensive study of Early Formative Valdivia contexts in the western lowlands of Ecuador have uncovered two millennia of village development beginning as early as 6250 cal BP in association with pottery and an expanding range of plant domesticates including manioc, arrowroot, chili pepper (*Capsicum* sp.), jack bean (*Canivalia* sp.), and llerén at the sites of Loma Alta and Real Alto (Zarillo et al., 2008). Within this context of early and later Formative developments, which lasts to around 300 BC, we see augmentation of tropical botanical domesticates (Pearsall, 2003) and an accompanying list of faunal resources that is both strikingly rich and comprehensive in its range of represented habitats from deep offshore pelagic waters to the frigid high-elevation páramos (Stahl, 2003a). The Formative archaeofaunas include certain taxa that are consistently present throughout the entire sequence, all of which have been identified in earlier Holocene contexts. Some were undoubtedly important food sources, yet the archaeofaunal record also suggests that others were used as tools, adornments, and ritual adjuncts (Zeidler et al., 1998; Stahl and Athens, 2001; 2004; Stahl, 2003a: 182–4). Throughout the larger region, later contextual associations and increasingly better preservation augment the list of utilized animals and cultural functions through time.

Indigenous agroecological landscape management throughout the region was based upon agroforestry, specifically forms of agri-silviculture combining crops, pasture, and trees (Nair, 1985), which in some areas were later elaborated into intensive systems of food production. Poly-cultural horticulture appears to have been practised by ceramic-producing middle Holocene cultures along alluvial floodplains in Colombia, Ecuador's western lowlands, and northern Peru, with subsequent relocation to adjacent hill slopes in some areas. Early interfluvial swiddening in Panama eventually relocated to alluvial bottomlands only after a protracted period of time around 2000 BP (Piperno and Pearsall, 1998).

The persistence and evolution of indigenous pre-Columbian agroforestry, and its relationship to cataclysmic disturbance have been well documented in Ecuador's lowland coastal Jama Valley, where stratified probabilistic survey located some 239 archaeological sites spanning over at least 3,500 calendar years of human occupation. A total of twenty-one archaeological sites, representing forty-two discrete archaeological assemblages with associated archaeobiological evidence beginning around 2000 BC, provide data for inferring the local evolution of indigenous agroforestry up to the arrival of Europeans in the early sixteenth century. Stratigraphic associations with tephra fallout throughout the river system also suggest how local pre-Columbian populations responded to intermittent cataclysmic volcanic events emanating from the Andean highlands. Phytoliths from a deep sediment profile corroborate the archaeobotanical data which indicate the early appearance of particulate carbon and open-air taxa along with a highly consistent suite of cultivated and utilized plants. The continued association of a restricted range of ubiquitous eurytopic generalists that frequent anthropogenic clearings corroborates the archaeobotanical data. In certain high-resolution contexts, these faunas were associated with isolated specimens of primarily human dietary taxa that frequent forest gaps. The entire data set suggests that the preserved faunas may have originally accumulated through both natural and cultural mechanisms within a mosaic environment of anthropogenic forest fragments (Stahl, 2000; 2006; 2011; Stahl and Pearsall, 2012). The spatial and temporal persistence of these patterns, even after cataclysmic devastation by intermittent volcanism, suggests that the inclusion of diversified upland agroforestry was at least partially responsible for cultural resilience in the face of natural disaster (Zeidler and Isaacson, 2003; Pearsall, 2004; Stahl, 2011).

Later intensive agro-ecological systems appeared, particularly where increased productivity could be secured in agriculturally challenging habitats prone to seasonal flooding, through the construction of raised planting beds. Raised field complexes appear throughout the area during the first centuries AD in the Sinú and San Jorge flood plains (Oyuela-Caycedo, 2008), the Barinas region of western Venezuela (Spencer and Redmond, 1992), Tumaco, Calima and Muisca (Drennan, 2008), and by at least 500 BC in the Guayas Basin (Muse, 1991). Archaeofaunal assemblages from two sites associated with extensive raised-field agriculture in the seasonally inundated savannas of the lower Guayas Basin have been studied. The system reached its maximum extension in the area during the Milagro-Quevedo culture which is identified with the historic Chono nation from AD 900 to the arrival of Spanish invaders in the early sixteenth century. In addition to a wide array of marine, fresh-water, and terrestrial gastropods, crustaceans, marine fish, reptiles, birds, and mammals, the full complement of domesticated vertebrates in South America, including muscovy duck (*Cairina moschata*), guinea pig (*Cavia procellus*), dog (*Canis lupus familiaris*), and camelids (*Lama* spp.) were identified (Stahl, 1988; Stahl et al., 2006).

The appearance of domesticated animals in the neotropics of northwestern South America is of special interest because, with the possible exception of the unknown cultural origins of domesticated Muscovy ducks, all are exotic introductions (Stahl, 2003b). In addition, the archaeological identification of possibly introduced allochthonous faunas can be suggestive of longer distance trade connections throughout the neotropics and beyond (Cooke, 1984; Cooke et al., 2008; Stahl, 2009). This is a critical consideration for the entire region which was interconnected through both local and long-distance terrestrial and marine trade networks. Indeed, the secure subsistence base provided through agricultural introduction, coupled with the acquisition of exotics through long-distance trade likely underscored the rise of political-ideological leadership throughout the entire region (Helms, 1993).

CONCLUSION

Despite a host of potential problems associated with the practice of zooarchaeology in the neotropics, archaeologists have managed to recover impressive evidence from caves and open-air sites for early landscape management and food production in northwestern South America. The trajectory for all subsequent pre-Columbian cultural developments in the area was established very early through the precocious achievements of its earliest Holocene human occupations. The study of preserved archaeobiological evidence outlines the subsequent development and elaboration of indigenous agricultural systems and trade networks up to their cataclysmic encounter with invading European populations in the early sixteenth century.

REFERENCES

Aceituno, F. J., Loaiza, N., Delgado-Burbano, M. E., and Barrientos, G. (2013) 'The initial human settlement of northwest South America during the Pleistocene/Holocene transition: synthesis and perspectives', *Quaternary International*, 301, 23–33.

Bush, M. B., Gosling, W. D., and Colinvaux, P. A. (2007a) 'Climate change in the lowlands of the Amazon basin', in Bush, M. B. and Flenley, J. R. (eds) *Tropical Rainforest Responses to Climate Change*, pp. 55–76. Chichester: Springer-Praxis.

Bush, M. B., Hanselman, J. A., and Hooghiemstra H. (2007b) 'Andean montane forests and climate change', in Bush, M. B. and Flenley, J. R. (eds) *Tropical Rainforest Responses to Climate Change*, pp. 33–54. Chichester: Springer-Praxis.

Cadena, A. and Bouchard, J.-F. (1980) 'Las figurillas zoomorfas de cerámica del litoral Pacífico ecuatorial', *Bulletin de l'Institut Française d'Ètudes Andines*, 9, 49–86.

Castillo, N. and Aceituno, F. J. (2006) 'El bosque domesticado, el bosque cultivado: un proceso milenario en el valle medio del Río Porce en el noroccidente Colombiano', *Latin American Antiquity*, 17, 561–78.

Cooke, R. G. (1984) 'Birds and men in prehistoric Central Panama', in Lange, F. W. (ed.) *Recent Developments in Isthmian Archaeology: Advances in the Prehistory of Lower Central America*. BAR International Series 212, pp. 243–81. Oxford: Archaeopress.

Cooke, R. and Jiménez, M. (2010) 'Animal-derived artefacts at two pre-Columbian sites in the ancient savannas of central Panama: an update on their relevance to studies of social hierarchy and cultural attitudes towards animals', in Campana, D., Crabtree, P., deFrance, S. D., Lev-Tov, J., and Choyke, A. (eds) *Anthropological Approaches to Zooarchaeology*, pp. 30–55. Oxford: Oxbow Books.

Cooke, R., Jiménez, M., and Ranere, A. J. (2008) 'Archaeozoology, art, documents, and the life assemblage', in Reitz, E. J., Scarry, C. M., and Scudder, E. J. (eds) *Case Studies in Environmental Archaeology*, pp. 95–121. New York: Springer.

Cooke, R. and Ranere, A. J. (1992) 'Precolumbian influences on the zoogeography of Panama: an update based on archaeofaunal and documentary data', *Tulane Studies in Zoology and Botany, Supplementary Publication*, 1, 21–58.

Cooke, R., Ranere, A. J., Pearson, G., and Dickau, R. (2013) 'Radiocarbon chronology of early human settlement on the Isthmus of Panama (13,000–7,000 BP) with comments on cultural affinities, environments, subsistence, and technological change', *Quaternary International*, 301, 3–22.

Drennan, R. D. (2008) 'Chiefdoms of southwestern Colombia', in Silverman, H. and Isbell, W. H. (eds) *Handbook of South American Archaeology*, pp. 381–403. New York: Springer.

Echeverría, J. (1988) *El lenguaje simbólico en los Andes Septentrionales*, Otavalo: Instituto Otavaleño de Antropología.

Gentry, A. H. (1992) 'Tropical forest biodiversity: distributional patterns and their conservational significance', *Oikos*, 63, 9–28.

Grayson, D. K. (1981) 'A critical view of the use of archaeological vertebrates in paleoenvironmental reconstruction', *Journal of Ethnobiology*, 1, 2–38.

Helms, M. W. (1993) *Craft and the Kingly Ideal: Art, Trade, and Power*, Austin: University of Texas.

Linares, O. (1976) 'Garden hunting in the American tropics', *Human Ecology*, 4, 331–49.

Meggers, B. J. (2011) 'Review of Handbook of South American Archaeology', *Chungara, Revista de Antropología Chilena*, 43, 147–57.

Mora, S. (2003) *Early Inhabitants of the Amazonian Tropical Rain Forest: A Study of Humans and Environmental Dynamics*. Latin American Archaeology Reports 3. Pittsburgh: University of Pittsburgh.

Muse, M. (1991) 'Products and politics of a Milagro entrepôt: Peñón del Río, Guayas basin, Ecuador', *Research in Economic Anthropology*, 13, 269–323.

Nair, P. K. R. (1985) 'Classification of agroforestry systems', *Agroforestry Systems*, 3, 97–128.

Oyuela-Caycedo, A. (2008) 'Late pre-hispanic chiefdoms of northern Colombia and the formation of anthropogenic landscapes', in Silverman, H. and Isbell, W. H. (eds) *Handbook of South American Archaeology*, pp. 405–28. New York: Springer.

Pearsall, D. M. (1995) 'Doing paleoethnobotany in the tropical lowlands: adaptation and innovation in methodology', in Stahl, P. W. (ed.) *Archaeology in the Lowland American Tropics: Current Analytical Methods and Applications*, pp. 113–29. Cambridge: Cambridge University Press.

Pearsall, D. M. (2003) 'Plant food resources of the Ecuadorian Formative: an overview and comparison to the central Andes', in Raymond, J. S. and Burger, R. L. (eds) *Archaeology of Formative Ecuador*, pp. 213–57. Washington: Dumbarton Oaks Research Library and Collection.

Pearsall, D. M. (2004) *Plants and People in Ancient Ecuador: The Ethnobotany of the Jama River Valley*, Belmont: Wadsworth.

Piperno, D. R. and Pearsall, D. M. (1998) *The Origins of Agriculture in the Lowland Neotropics*, San Diego: Academic Press.

Rodríguez, E. E. (1992) *Fauna Precolombina de Nariño*, Bogotá: Fundación de Investigaciones Arqueológicas Nacionales, Instituto Colombiano de Antropología.

Rossen, J. (2011) 'Preceramic plant gathering, gardening, and farming', in Dillehay, T. D. (ed.) *From Foraging to Farming in the Andes: New Perspectives on Food Production and Social Organization*, pp. 177–92. Cambridge: Cambridge University Press.

Sauer, C. O. (1952) *Agricultural Origins and Dispersals*, New York: The American Geographical Society.

Spencer, C. S. and Redmond, E. M. (1992) 'Prehispanic chiefdoms of the western Venezuelan llanos', *World Archaeology*, 24, 134–57.

Stackelbeck, K. (2011) 'Faunal remains', in Dillehay, T. D. (ed.) *From Foraging to Farming in the Andes: New Perspectives on Food Production and Social Organization*, pp. 192–204. Cambridge: Cambridge University Press.

Stahl, P. W. (1988) 'Prehistoric camelids in the lowlands of western Ecuador', *Journal of Archaeological Science*, 15, 355–65.

Stahl, P. W. (1991) 'Arid landscapes and environmental transformations in ancient southwestern Ecuador', *World Archaeology*, 22, 346–59.

Stahl, P. W. (1992) 'Diversity, body size, and the archaeological recovery of mammalian faunas in the neotropical forests', *Journal of the Steward Anthropological Society*, 20, 209–33.

Stahl, P. W. (1995) 'Differential preservation histories affecting the mammalian zooarchaeological record from the forested neotropical lowlands', in Stahl, P. W. (ed.) *Archaeology in the Lowland American Tropics: Current Analytical Methods and Applications*, pp. 154–80. Cambridge: Cambridge University Press.

Stahl, P. W. (2000) 'Archaeofaunal accumulation, fragmented forests, and anthropogenic landscape mosaics in the tropical lowlands of prehispanic Ecuador', *Latin American Antiquity*, 11, 241–57.

Stahl, P. W. (2003a) 'The zooarchaeological record from Formative Ecuador', in Raymond, J. S. and Burger, R. L. (eds) *Archaeology of Formative Ecuador*, pp. 175–212. Washington: Dumbarton Oaks Research Library and Collection.

Stahl, P. W. (2003b) 'Pre-Columbian Andean animal domesticates at the edge of empire', *World Archaeology*, 34, 470–83.

Stahl, P. W. (2005) 'Selective faunal provisioning in the southern highlands of Formative Ecuador', *Latin American Antiquity*, 16, 313–28.

Stahl, P. W. (2006) 'Microvertebrate synecology and anthropogenic footprints in the forested neotropics', in Balée, W. and Erickson, C. L. (eds) *Time and Complexity in Historical Ecology*, pp. 127–49. New York: Columbia University Press.

Stahl, P. W. (2009) 'Adventive vertebrates and historical ecology in the pre-Columbian neotropics', *Diversity*, 1, 151–65.

Stahl, P. W. (2011) 'Periodic volcanism, persistent landscapes, and the archaeofaunal record in the Jama Valley of western Ecuador', in Miller, N., Moore, K., and Ryan, K. (eds) *Sustainable Lifeways: Cultural Persistence in an Ever-Changing Environment*, pp. 273–309. Philadelphia: University of Pennsylvania Museum of Archaeology and Anthropology.

Stahl, P. W. (2012) 'Interactions between humans and endemic canids in Holocene South America', *Journal of Ethnobiology*, 32, 108–27.

Stahl, P. W. (2013) 'Early dogs and endemic South American canids of the Spanish main', *Journal of Anthropological Research*, 69, 515–33.

Stahl, P. W. (2014) 'Garden hunting', in Smith, C. (ed.) *Encyclopedia of Global Archaeology*, Vol. 5, pp. 2945–52. New York: Springer Reference.

Stahl, P. W. and Athens, J. S. (2001) 'A high elevation zooarchaeological assemblage from the northern Andes of Ecuador', *Journal of Field Archaeology*, 28, 161–76.

Stahl, P. W. and Athens, J. S. (2004) 'Aprovechamiento prehistórico de animales y manufactura de utensilios de hueso en la parte alta de los Andes, al norte del Ecuador', *Cuadernos de Historia y Arqueología*, 54–6, 115–65.

Stahl, P. W., Muse, M. C., and Delgado-Espinoza, F. (2006) 'New evidence for pre-Columbian Muscovy duck *Cairina moschata* from Ecuador', *Ibis*, 148, 657–63.

Stahl, P. W. and Oyuela-Caycedo, A. (2007) 'Early prehistoric sedentism and seasonal animal exploitation in the Caribbean lowlands of Colombia', *Journal of Anthropological Archaeology*, 26, 329–49.

Stahl, P. W. and Pearsall, D. M. (2012) 'Late pre-Columbian agroforestry in the tropical lowlands of western Ecuador', *Quaternary International*, 249, 43–52.

Stahl, P. W. and Zeidler, J. A. (1990) 'Differential bone refuse accumulation in food preparation and traffic areas on an Early Ecuadorian house floor', *Latin American Antiquity*, 1, 150–69.

Stothert, K. E. and Sánchez, A. (2011) 'Culturas del Pleistoceno final y el Holoceno temprano en el Ecuador', *Boletín de Arqueología PUCP*, 15, 1–39.

Zarillo, S., Pearsall, D. M., Raymond, J. S., Tisdale, M. A., and Quon, D. J. (2008) 'Directly dated starch residues document Early Formative maize (*Zea mays* L.) in tropical Ecuador', *Proceedings of the National Academy of Science*, 105, 5006–11.

Zeidler, J. A. and Isaacson, J. S. (2003) 'Settlement process and historical contingency in the western Ecuadorian Formative', in Raymond, J. S. and Burger, R. L. (eds) *Archaeology of Formative Ecuador*, pp. 69–123. Washington: Dumbarton Oaks Research Library and Collection.

Zeidler, J. A., Stahl, P. W., and Sutliff, M. J. (1998) 'Shamanistic elements in a Terminal Valdivia burial, northern Manabí, Ecuador: implications for mortuary symbolism and social ranking', in Oyuelo-Caycedo, A. and Raymond, J. S. (eds) *Recent Advances in the Archaeology of the Northern Andes: In Memory of Gerardo Reichel-Dolmatoff*. Institute of Archaeology Monograph 39, pp. 109–20. Los Angeles: University of California.

CHAPTER 41

ZOOARCHAEOLOGY OF BRAZILIAN SHELL-MOUNDS

DANIELA KLOKLER

INTRODUCTION

SHELL-MOUNDS were among the first archaeological sites to be identified in Brazil due to their visibility, aided by their size, prominent locations, and proximity to the coast. Since the seventeenth century, religious missionaries, travellers, naturalists, and researchers proposed explanations for the origins and significance of these shell concentrations. Shell sites have many distinct forms and characteristics, ranging from large mounds that reached up to 50 metres in height, to small accumulations of shells barely discernible on the modern surface. The former sites, however, were (and still are) more alluring to researchers due to their physical characteristics and the lack of ethno-historical information regarding their function and significance. For the latter, researchers rather rely on descriptions of their use as temporary camps.

Brazilian zooarchaeology originated and primarily developed through the study of these coastal sites. The element that unifies these sites is the presence of faunal remains as their main component, so a focus on these assemblages often spurred development of archaeological research. Faunal components of shell-mounds, as a consequence, have been used to categorize sites, define phases, reconstruct palaeoenvironmental conditions, and infer subsistence practices, among other issues. More recently, shell site analyses have contributed to our understanding of social organization and interpretations of ritual activities (Klokler, 2001; 2012; 2013; 2014a; Nishida, 2007; Gaspar et al., 2008; 2013; 2014; Klokler and Gaspar, 2013). Here the trajectory of Brazilian studies of shell-mounds is explored.

AQUATIC RESOURCES AND THEIR MULTIPLE ROLES

Levi-Strauss (1962) highlights that animal species are selected not only for their nutritional or caloric values but also because of their importance within a determined cultural system. Little by little, this perspective also became salient in zooarchaeological studies exploring coastal sites, besides being good to eat and good to think with, we can add that some animals are also good for building; they provide useful construction materials.

Most Brazilian archaeologists, especially those influenced by cultural ecology, explained the presence of shells in shell-mounds as pure reflections of the diet of their creators and the surrounding environment (Lima, 1991; Bandeira, 1992). Shell-fish were considered a main staple, while fishing was argued to become an important activity only after 2000 BP. The massive amounts of shell valves found at these sites were directly attributed to the estimated quantity needed to fulfil the dietary needs of human groups.

Figuti demonstrated that coastal populations always relied heavily on procurement of fish for their subsistence (1989; 1993; 1995; 1998); his primary innovation was to treat the site's matrix as an artefact. Through the analysis of bulk samples, inspired mainly by the work of Casteel (1970), Bailey (1975), and Botkin (1980), and by transforming bones and shells into proxy measures of edible meat, Figuti was able to prove that previous interpretations of shell-mound groups' diet were based on an impressionistic rather than quantitative approach. Through the use of this new methodology, Figuti affirms that the high quantity of shell-fish found in most sites (approximately 80% of the matrix) in fact corresponds to only 15% of available meat, whereas vertebrates (mostly fish) account for approximately 80% of edible meat. The remaining 5% included mammals, birds, reptiles, and amphibians. This redirected questions regarding the importance of shell-fish.

Gaspar and DeBlasis (1992) and Afonso and DeBlasis (1994) proposed that large shell sites are mounded structures resulting from the organization of sizeable work groups following specific social rules, thus changing the perception that we had of shell-mound societies. Rather than small, mobile bands of fisher-hunter-gatherers, these populations are now understood as larger, more sedentary groups that not only dominated the coastal landscape but also transformed it through the construction of large features built using shells and fish bones.

The transformation in the study of coastal populations in Brazil is met with the broadening of research topics that use faunal remains to interpret the behaviours of shell-mound groups. Some authors focus on advancing the discussion about the connection between site formation and ritual uses of animal resources (Klokler, 2001; 2008; 2012; 2014a; Nishida, 2007; Plens, 2010; Gaspar et al., 2013; Klokler and Gaspar, 2013). This new focus explores the possibility that shell-fish found in shell-mounds could result from uses that did not directly involve consumption as food. As seen elsewhere (Claassen, 1998; 2010; De Masi, 1999; Klokler, 2001), molluscs and their shells have been used extensively around the world as bait, dye, medicine, containers, raw material for tools and jewellry, construction material, and pottery temper, amongst many other purposes. Claassen (1998) highlights the symbolism associated with shells in several societies and demonstrates that bivalve and gastropod shells are frequently linked to fertility, death, and abundance.

SHELL MOUND CONSTRUCTION: WHY AND HOW?

According to ethnographic and archaeological research, shell-mound sites can have diverse functions, being used for habitation (both short and long term), as workshops, as processing locations, as seasonal camps, and for burial, amongst other uses. If we can identify the human activities that established, created, and developed a site, we can also determine its function(s).

My research has focused on identifying the distinct forms of deposits composing these sites, and the activities responsible for creating them, in order to achieve an interpretation of the function of shell-mounds and their faunal contents (Figuti and Klokler, 1996; Klokler, 2001; 2008; 2012; 2013; 2014a; Klokler and Gaspar, 2013). The selection of shells as building materials demonstrates the central role of aquatic resources, especially shell-fish for these coastal groups. Faunal analysis of assemblages from shell-mounds at Espinheiros II and Jabuticabeira II shows that the accumulations of shell valves forming mounds was not simply fortuitous, but was arranged according to specific sets of rules.

Initial Research: Espinheiros II

Changing the focus from subsistence to formation processes, Figuti and Klokler (1996) analysed materials from Espinheiros II, a site located in Santa Catarina state (Fig. 41.1). The layers from the site's lower levels primarily consist of clam shells (Veneridae), with a high percentage (approximately 70% of the weight) of just one species of mollusc: the Carib or West Indian pointed venus, or *Anomalocardia brasiliana* (Fig. 41.2a). Mytilidae specimens are the second most common shell-fish at this site, accounting for just over 5% of the sample. In other words, intensive exploitation of *Anomalocardia brasiliana* banks occurred. Fish bones account for only 0.11% of the matrix components. Little evidence of activities related to subsistence was found in the lower levels of the shell-mound. For example, no artefacts were recovered from the samples analysed for faunal studies. However, remains of basketry were found in the two lowest layers. The base of the site, whose initial construction is dated at 2970 ± 60 years BP (Afonso and DeBlasis, 1994), with layers composed almost exclusively of shell-fish remains (considered part of a 'clean site' according to the 1950's typology; Gaspar, 1998), was a product of the construction of a platform above the mangrove, since there were no features or other signs of activities.

The upper portions of Espinheiros II site exhibited evidence of several features similar to what would be expected for habitation areas such as hearths, activity areas

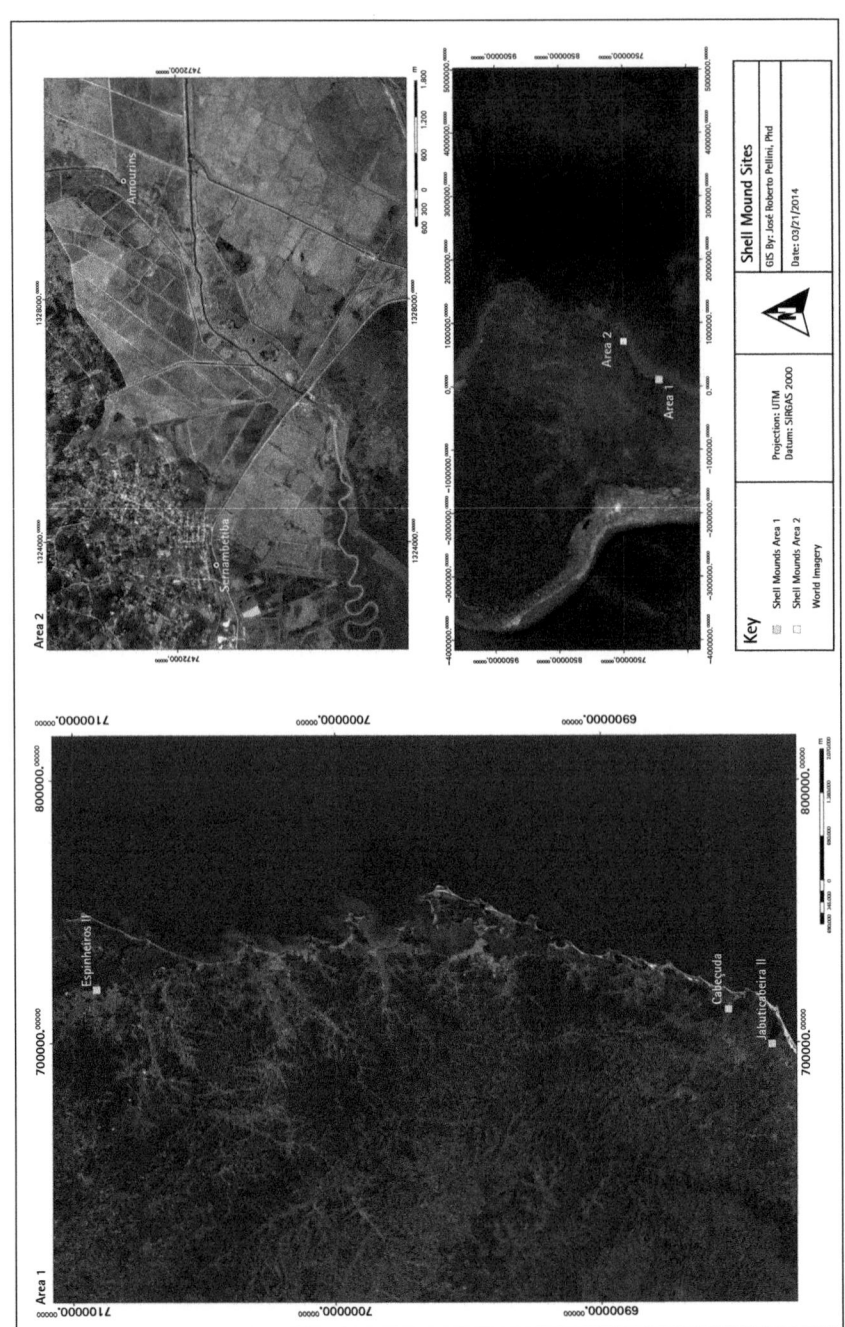

FIGURE 41.1 Location of sites mentioned in the text (original by J. R. Pellini, reproduced with permission).

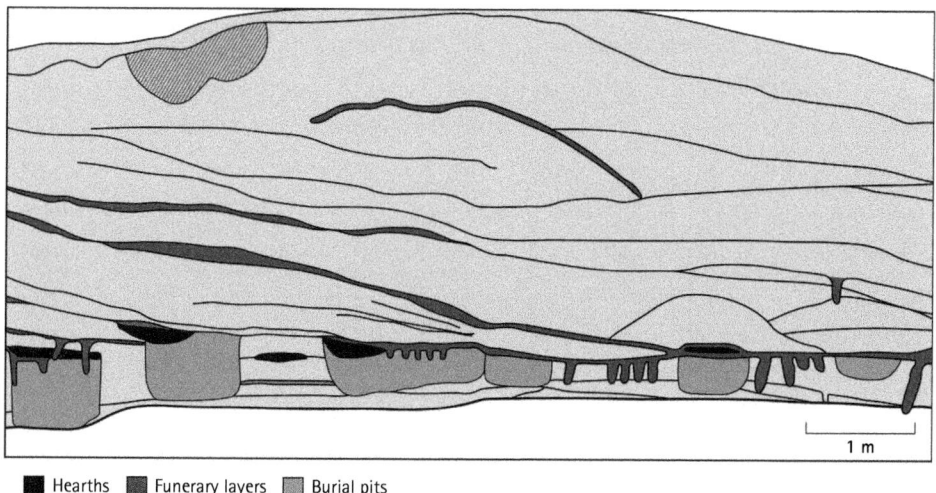

■ Hearths   ■ Funerary layers   ▨ Burial pits

FIGURE 41.2   Shell layers covering burials in Locus 2.

and artefacts. Also, there is an important change in the composition of the sediment: Veneridae (*A. brasiliana*), Mytilidae (*Mytella* sp. and *Brachidontes* sp.), and Ostreidae (*Crassostrea* sp.) (Table 41.1) specimens appear in relatively similar weight percentages (15%, 25%, and 32% respectively). At this point, collection is not solely focused on *A. brasiliana* in the mud banks, but is more diversified also exploiting mussels and oysters in the mangroves (Figuti and Klokler, 1996). In the upper layers, there is a significant presence of fish bones and otoliths (8%), confirming repeated fishing activities.

For the first time, research identified a shell-mound with two distinct sets of activities with specific purposes: initially, rapid construction of an elevated area, then a slower accumulation of materials from diversified day-to-day activities. Instead of random aggregates of refuse, large shell sites began to be interpreted as intentionally built structures.

Mounding for the Dead: Jabuticabeira II

Jabuticabeira II is a shell-mound site located in southern Brazil (Fig. 41.1) in a region that was dominated by a large palaeo-lagoon at the time of its construction (around 3,000 years ago). The area, conflating the coastal plain with a rich estuary, rivers, and easy access to a mountainous area, was very attractive to prehistoric populations and approximately seventy sites have been identified in the region. Jabuticabeira II is a medium-sized shell-mound, with a maximum height of nine metres and a footprint of approximately 100,000 m². Archaeological research began in 1997 and included the sampling, recording, and analyses of more than 350 m of profiles, eighteen trenches, and three excavation areas. The site has been extensively dated with over thirty radiocarbon

Table 41.1 Species mentioned in the text recovered from Espinheiros II and Jabuticabeira II

Common name	Scientific name
Carib or West Indian pointed venus	*Anomalocardia brasiliana*
Mangrove mussel	*Mytella* sp.
Native mussel	*Brachidontes* sp.
Oyster	*Ostrea* sp., *Crassostrea* sp.
Thick lucine	*Phacoides pectinatus*
Southern oyster drill	*Stramonita haemastoma*
Catfish	*Genidens barbus, Genidens genidens*
Sheepshead seabream	*Archosargus probatocephalus*
Grunt	Haemulidae
Smooth puffer	*Lagocephalus laevigatus*
Atlantic spadefish	*Chaetodipterus faber*
Mullet	*Mugil* sp.
Leatherjacket	*Oligoplites* sp.
Weakfish	*Cynoscion acoupa, Cynoscion leiarchus*
Whitemouth croaker	*Micropogonias furnieri*
Black drum	*Pogonias cromis*
Ground croaker	*Bairdiella ronchus*
Snook	*Centropomus* sp.
Shorthead drum	*Larimus breviceps*
Cownose ray	*Rhinoptera bonasus*
Shark	Selachimorpha

dates on different materials such as charcoal, shell, and human bones. The oldest date for Jabuticabeira II is 2890 ± 55 (BETA A10633 on shell) while the most recent is 1400 ± 40 (BETA 234201 on bone collagen).

After the results achieved at Espinheiros II, Klokler (2001) developed a mound-sampling strategy seeking to explore formation processes using faunal data. Using volume-controlled bulk column and 'strategic' sampling (samples from features such as postholes, hearths, and graves from profiles and excavation areas) allowed evaluation of the differences in proportions of animal remains in several areas of the site, possible changes through time, and distinction between and among features (both horizontally and vertically) (Fig. 41.3). Columns from three different loci, from the periphery and core of sites, were collected for an initial assessment of the site's construction process.

Jabuticabeira II's complex stratigraphy exhibits a pervasive combination of thick, mostly clean, light-coloured shell layers interspersed with thin, dark layers (Fig. 41.2b), topped by a dark-coloured fish-dominant deposit. The thick, shell-dominated layers enclose very thin dark lenses composed mostly of charcoal fragments and fish bones and

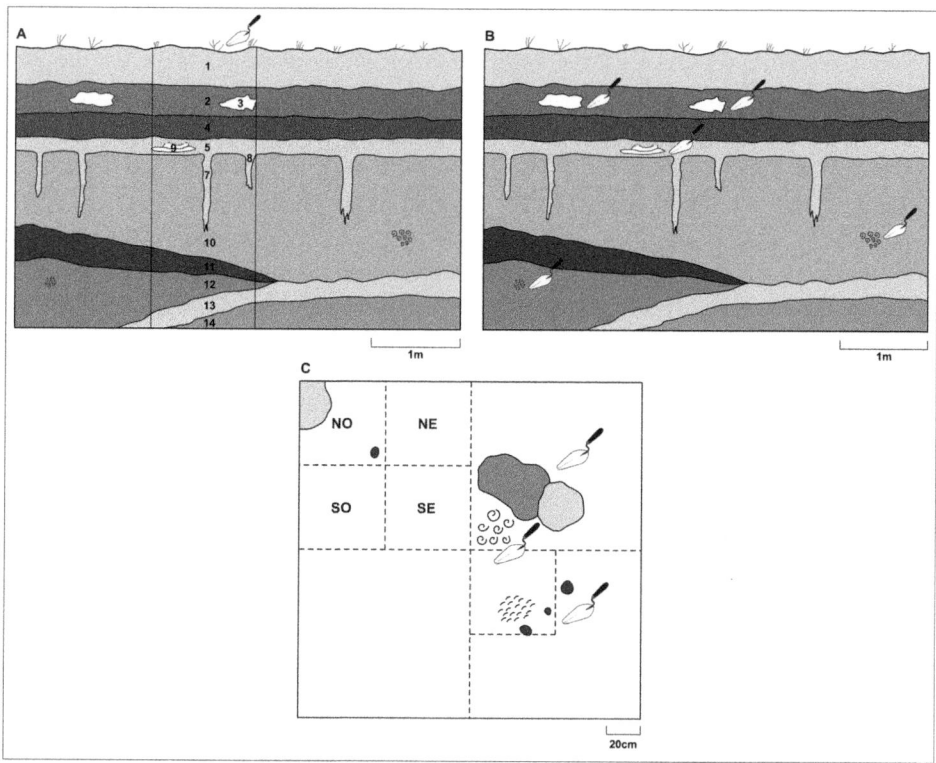

FIGURE 41.3 Examples of sampling strategies utilized in shell mounds: (a) columns, samples are recovered in each identified deposit (including any features) in a selected profile portion, (b) features on profiles, samples that contemplate any distinct deposit identified in profiles, (c) samples made during excavation (in bulk from quads, and from features) (modified from Klokler 2013).

are generally devoid of features. The thin, dark layers occur repeatedly along the profiles, both horizontally and vertically. They have variable thicknesses (rarely over 20 cm), are shorter horizontally than the shell layers, and include features such as hearths, post-holes, and burials. Almost all burials originate from the dark compacted deposits, and these deposits have become known as funerary areas (DeBlasis et al., 1998). The frequencies of mollusc remains decrease sharply, from an average of 80% of the total components in shell layers, to 15% in the dark ones.

Analyses demonstrated that both types of layers had other significant differences in composition. The light-coloured, thick, and loose layers include clam shells, mostly whole *A. Brasiliana* (many still articulated). By weight, this clam comprises around 60% of the bulk layers on average. Other molluscs, such as thick lucine (*Phacoides pectinatus*), native mussel (*Brachidontes* sp.), southern oyster drill (*Stramonita haemastoma*), oysters (*Ostrea* sp. and *Crassostrea* sp.), along with several species of gastropods (Table

41.1) account for 15% of the sample. As samples were collected closer to the site's surface, frequencies of mussels increased slightly, while basal layers have the lowest amounts. However, it is unlikely that these differences were related to shifts in subsistence or collection strategies, since the same pattern occurs within some of the site's internal mounds (Klokler, 2001) suggesting that patterns of mound building in the smaller mounds are reproduced in the large mound. The mollusc valves did not show signs of intentional breakage, exhibit low fragmentation rates, and just a small percentage (2%) show signs of burning. The fish remains recovered from shell layers include mostly estuarine species such as whitemouth croaker, catfish, black drum, and weakfish (*Micropogonias furnieri, Genidens barbus* and *G. genidens, Pogonias cromis*, and *Cynoscion acoupa* and *C. leiarchus* respectively) (Table 41.1).

These shell layers appear to 'close' the burial clusters located in the dark layers (Fig. 41.2c), and instead of being almost horizontal, they exhibit a mounded arrangement. Study of these shell layers indicates that the deposits were made through the collection, transport, and deposition of massive quantities of mollusc remains. The repetition of depositional episodes forming the layers and lenses of the site demonstrates that they were part of a long set of activities involved in the interment, mourning, and celebration rites for the deceased.

Analyses demonstrated that burials were made among large accumulations of fish bones, suggesting that the matrix composition of the dark layers corresponds to the remains of feasts (Klokler, 2001; 2008; 2012; 2014a). The shell layers represent repeated episodes of collection and transportation of massive amounts of molluscs used to cover the funerary areas or particular graves within funerary layers, while also adding volume to the mound. The collection of one particular species forming large shell deposits is similar to the lower levels of Espinheiros II.

Fish bone quantities greatly surpass amounts recovered from shell layers by weight. In contrast, few mollusc remains are recovered in these lenses; *A. brasiliana* and mussels account on average for 15% of the samples by weight. Weathering affects the surface of most of the shell valves (a brownish-reddish pellicle), as well as causing some loss of their calcium carbonate and other materials; this process seems to have been associated with higher organic matter content in the dark layers (Klokler, 2008). Most of the components in the funerary areas show varying degrees of burning, indicating differential processing of the remains.

Funerary areas include an average of 275 kg of available fish meat per cubic metre, enough meat to feed large gatherings of people, especially considering reports about fish consumption estimates in recent Brazilian fishing communities estimate daily consumption of approximately 130 to 400 grams per person (Mazzilli, 1975; Garcez and Sanchez-Botero, 2005). A detailed study of these lenses suggests that funerary areas were formed through the deposition of large amounts of food (mostly fish) in few occasions, corresponding to the interment of individuals. To illustrate the quantities of food available in these dark layers, within layer 2.25.11, which has an estimated size of 27 m x 5 m x 10 m and the lowest biomass value per cubic metre (according to Klokler, 2008), we calculated/estimated the total amount of food refuse deposited, and the corresponding fish meat available for guests would be more than one metric ton.

The excavation of 32 m³ of a dark layer revealed 28 hearths, 12 graves (including 21 individuals), and 384 postholes. Their spatial and temporal relationships confirmed the sequence of mortuary activities inferred from profiles. However, multiple test pits, extensive excavations, and profile analyses throughout the site were unable to identify evidence of daily domestic activities. Jabuticabeira II lacks domestic artefact assemblages and manufacturing debris, habitation floors, processing areas, and other features that would indicate its use as a habitation (Klokler, 2014a). The few artefacts recovered appear exclusively in association with burials, either as grave goods or in the immediate surroundings (Klokler, 2008). Grave goods were found in most burials and include personal adornments made of faunal and stone materials. Moreover, 73% of faunal artefacts are associated with burials, 68% of which are adornments recovered from graves (Hering, 2005). Shell beads comprise most of the adornments; gastropods (*Olivella* sp.) were largely used to manufacture the beads, while bivalves were chosen mostly to produce pendants. Of the total of artefacts recovered from Jabuticabeira II, 8% correspond to adornments made of mammal teeth (Hering, 2005). Bird long bones were chosen to make bone points, used as fishing implements. Almost 75% of lithic artefacts were recovered from funerary contexts at the site.

Conical features, believed to have been postholes, originate from the dark layers. Almost four hundred were recorded across a space of approximately 150 metres of profiles in just one area of the site. The depths vary between 6 and 82 cm. Postholes are found in association with graves and hearths (a characteristic also noticed in the profiles) and they often surround or mark these features (Klokler, 2001).

In general, the composition of the materials in features such as hearths, burial pits, and postholes closely resembled those of the funerary area matrix with minor—and expected—differences. Small and shallow burial pits were usually dug into the preceding shell layer, over which a subsequent funerary area developed. Similarities in the fish contents from burial pits and surrounding areas (the thin dark layers), and the slightly higher frequency of shells in burial pits, indicate that graves were filled with a mix of materials from the shell layer removed to create the pit, plus the faunal materials from the funerary surface that include the remains of feasts. Some graves contained large amounts of fish remains, particularly otoliths, and during excavation researchers noticed whole fish skeletons in anatomical connection (Table 41.2), suggesting that they might also have been used as grave goods deposited to surround the dead (Klokler, 2001; 2008; 2014a; 2014b). There was a higher presence of mammal and bird remains in graves compared to other site features and deposits (Klokler, 2014a).

Hearths were lit on top of, or near burial pits. Most hearths were 40 to 60 cm in diameter, and were sometimes superimposed, showing the importance of fires for the funerary process. Some hearths contained large amounts of ash, indicating not only that they were lit for an extended period of time, but also that they were most probably covered

Table 41.2 MNI of fish recovered from selected burial pits at Jabuticabeira II. The taxa included in the various families are: Ariidae (*G. genidens* and *G. barbus*), Scianidae (*Cynoscion acoupa, C. leiarchus, Micropogonias furnieri, Pogonias chromis, Larimus breviceps,* and *Bairdiella ronchus*), Haemulidae, Mugilidae (*Mugil* sp.), Carangidae (*Oligoplites* sp.), Sparidae (*Archosargus probatocephalus*), Centropomidae (*Centropomus* sp.), Tetraodontidae (*Lagocephalus laevigatus*), Myliobatidae (*Rhinoptera bonasus*). Selected burials include pits fully excavated in which bulk samples were collected

Burial pits	Ariidae	Scianidae	Haemulidae	Mugilidae	Carangidae	Sparidae	Centropomidae	Tetraodontidae	Ephippidae	Myliobatidae	Selachimorpha	TOTAL
32	53	107		2	1	2	1	1		1	1	168
34	122	125		3	2	2		1		1		256
36	34	37			1	2				1		75
37	18	14				2				1		35
40	32	36		1	1	3	1			1		75
41	58	54	1	1	1	4				1		121
101	26	6	2	1	1	4	2		4			46

after use. Their small size and their location suggest that they were used for ritual fires to increase the visibility of the ceremonies performed at the funerary areas and not for cooking large amounts of food.

Previously, the presence of hearths, the great quantity of faunal remains, and the dark compacted layers were taken as evidence that shell-mounds functioned as habitation areas. The small amount of artefacts in relation to the sites' sizes was attributed to the shell accumulation, whose large volume masked the importance of lithic and bone technology. However, the recurrence of areas reserved for interments throughout Jabuticabeira II and the association of hearths and deposition of artefacts in these areas changed this perception.

Fishing activities that contributed to the shell layers seem to be the same as those described for funerary feasts. Fish size and species composition are very similar in both deposits, but meat quantities are lower in shell layers. Very low degrees of burning indicate that food refuse was not burned after consumption, indicating differences regarding refuse management.

The location of Jabuticabeira II at the margins of a large lagoon greatly facilitated the exploitation of molluscs and the transport of valves to the site. The use of *A. brasiliana* shells to build the mound can be explained by their ease of collection, and some of their properties. The shells are thick and light-coloured making them good materials for mounding, since they can add volume and retain some of the shape of the structure. The shells also provide good drainage and reflect light, making the mound even more visually distinctive in the landscape.

New Research Focus: Micro Scale Analysis and Ritual

Besides discussing the ritual aspect of fauna for the construction of shell-mounds, we should also include a smaller scale in which we could identify the ritual use of animal remains in shell-mounds. Otoliths are ubiquitous in all samples recovered from Jabuticabeira II and are commonly recovered in other shell sites. The most common otoliths come from whitemouth croaker (*Micropogonias furnieri*), catfish (*Genidens genidens*), black drum (*Pogonias cromis*), weakfish (*Cynoscion* sp.), and mullet (*Mugil* sp). At Jabuticabeira II, a total of 9,258 otoliths was recovered.

Otoliths are the second most common fish element at the site, following vertebrae. A greater amount of elements that occur in high numbers within individual fish, such as vertebrae, is expected. However, otoliths are only present in fish skeletons as pairs. If processing was a factor affecting the presence of otoliths, it would be expected that other cranial elements would also have high frequencies, since otoliths are located inside the neurocrania. Otoliths are concretions of aragonite (calcium carbonate) a composition similar to shells, which helps explain their preservation within shell-mound deposits.

Some interesting patterns emerge when considering deposition of these elements. On average, hearths have eighty-six otoliths per 8-litre sample, while funerary areas have sixty. Although otoliths are more common in hearths, these features do not contain an increased amount of fish bones. Their deposition in higher numbers in these features must have been intentional even though at this point we cannot identify clear explanations as to why this might be. Two burial pits (32 and 34) at the site contain over three hundred otoliths. Both graves are small and could not fit 150 whole fish, fish skeletons, or even fish-heads, suggesting that the elements were removed from the neurocrania and intentionally deposited within both contexts.

While the use of these fish elements has greatly helped researchers study fishing economies, less attention has been given to other possible roles that otoliths may have played in the past lives of prehistoric communities (Klokler, 2014b). Today, otoliths are reportedly used as a medicine and are added to little pouches or used in necklaces due to their purported magical powers recognized by fishing communities around the world (Klokler, 2014b). Historical and ethnographic studies in Brazil identify the use of otoliths for both magical and medicinal purposes. During colonial times, healers used these ear bones in divinatory sessions and to cure certain diseases (Souza, 1984). Fishing communities in northeastern Brazil still use otoliths in teas as a remedy for renal failure (Alves et al., 2007).

Due to their physical characteristics, such as smoothness and colour, these elements may have been perceived differently and assigned with powers by shell-mound groups. It is possible to suggest that part of the otoliths recovered from Jabuticabeira II (and other shell-mound sites) were collected, used, and deposited due to associations with healing, or other powers. Further studies, with inclusion of more sites and ethnographic reviews, could offer more information regarding the use of otoliths by coastal peoples.

A New Perspective, or Maybe Not: Ritual in Shell Sites

Ritual in shell sites has been gaining increased attention from researchers over the last decade, and published research, especially in the United States and Brazil, demonstrates the performance of feasts and/or funerary rituals as the main impetus for site construction (Klokler, 2001; 2008; 2012; 2014a; Luby, 2004; Claassen, 2010; Russo, 2014; Saunders, 2014). The ritual roles of mounds are now commonly discussed, if not widely accepted. However, the opening of more research topics is welcomed in the previously subsistence-dominated field. While the advancement of interest in the ritual aspect of mounds gained strength during the late 1990s, some researchers had already began to debate this topic in the early nineteenth century.

Wiener (1876) identified the large shell-mounds in Santa Catarina State as mortuary monuments erected to honour the deceased members of society. He also first noted the association between dark lenses and the presence of human burials. Wiener's close attention to the shell-mounds' stratigraphy allowed him to identify sites that contained evidence of both daily activities and burials from shell-mound sites that were solely erected for the deceased (1876: 18). Lacerda (1885) and Ihering (1903) conceded that shell-mounds could have served as graveyards but Lacerda refused to identify evidence of their exclusive use as funerary loci while Ihering asserted that only large mounds were associated with burials (interestingly he never accepted the anthropic origin of the sites). Unfortunately, archaeologists from the twentieth century mostly ignored these early statements. The only exception was Paulo Duarte who put forth an interpretation of sites as graveyards in the 1950s (Duarte, 1967).

Tenório (1995) proposed that shell-mound clusters are associated with areas rich in resources, and therefore strategic for settlement. Mound building would then be a strong indicator of territoriality. In fact, recent research confirms that shell-mounds are not only powerful indicators of territoriality, reaffirming a groups' privileged access to local resources, but their funerary character also reinforced the ancestral aspect of landscape dominance.

The selection of locations away from depressions, generally on terraces that provide good views of the surrounding landscape, is another important characteristic of shell sites. Many authors (Claassen, 1998: 231; DeBlasis et al., 1998; Klokler, 2001; 2008; 2014a; Villagrán, 2010; Saunders, 2014) identify the mounds as sites associated with the construction of a cultural landscape, with the mounds being territorial markers, probably associated with a group's identity (Klokler, 2001).

Studies in Santa Catarina (Klokler and Gaspar, 2013) and Rio de Janeiro (Gaspar et al., 2013), reviewing previous analyses, demonstrate the presence of large quantities of fish bones associated with burials in some of the mounds recently identified as cemeteries, repeating a similar pattern found at Jabuticabeira II, notably in Cabeçuda (Santa Catarina state), Amourins, and Sernambetiba (located in Rio de Janeiro state) sites (Fig. 41.1). The recurrent presence of features with funerary characteristics and the lack of clear-cut evidence of habitation areas within or near the sites, allied with the rarity or absence of artefacts or features dissociated from the funerary layers, demonstrate the ritual nature of these sites.

At Amourins research indicates that large communal meals with a menu composed basically of catfish and whitemouth croaker were already held approximately 3,800 years ago (Gaspar et al., 2013), more than 1,000 years earlier than the feasts held at Jabuticabeira II. Preliminary analyses from Sernambetiba, built between 2,600 and 1,800 years ago (Gaspar et al., 2013) and Cabeçuda, a large site with a date of 4120 BP (Rodrigues-Carvalho et al., 2011) indicate increased deposition of fish close to burials, though there is no clear evidence for feasting so far (Klokler, 2010; Klokler and Gaspar, 2013).

In these sites we can identify an inter-relationship between ritual, diet, and construction uses of faunal remains. The superposition of all these uses and activities shows us how they are not mutually exclusive and must be interpreted as part of the mortuary ritual of shell-mound groups.

The perspective that identifies shell-mounds as ceremonial structures suggests that their builders had a complex social organization and a sedentary lifestyle, as seen in the shell-mounds from the Archaic period in the Americas (Gaspar, 1998; Luby, 2004; Claassen, 2010; Russo, 2014; Saunders, 2014). Shell-mounds were locales for large gatherings, integrating people from different groups that inhabited the region during funerary events. These events served as a means to socially connect different groups. The feasts provided the perfect moment to form or reinforce alliances, exchange information, and reassure friendships. Since the mounds were composed of the remains of centuries of such events, they were public reminders of the importance of these relationships and a testament to their endurance.

Most importantly, the mounds also enclosed generations of groups' members, augmenting their importance as sacred places. Shell-mounds modified and sacralized the landscape, also establishing the territory of these fisher-gatherer groups. Shell-mounds are links to ancestors and symbols of group aggregation and solidarity, as well as indicators of their relationships with the environment.

The results put forth the perspective that mollusc shells were used as construction material, and caution the dangers of simply equating a site's faunal remains with diet, thereby expanding our understanding of the relationships between human groups and the animal world. This perspective forces archaeologists to see faunal remains as indications of a much richer relationship in which molluscs and fish were captured not only for their nutritional value but also because of their utility as building materials and their merit as sacred or meaningful elements. Aside from the recognition that animals symbolize ideals and beliefs, thereby expanding their importance to coastal groups well beyond their value as food, this perspective forces zooarchaeologists to face avenues of research that have not yet been fully explored.

Acknowledgements

The author would like to thank the editors for the invitation to participate in the volume and their comments on this chapter, as well as the financial support of CAPES (process 1501-02-0), NSF doctoral dissertation grant (SBR-0652177), and CNPq (processes 151457/2009-3 and 409428/2013-2). I am also grateful to David Mehalic, JR Pellini, and Gabriela Farias for their valuable help with the images.

References

Afonso, M. C. and DeBlasis, P. (1994) 'Aspectos da formação de um grande sambaqui: alguns indicadores em Espinheiros II, Joinville', *Revista do Museu de Arqueologia e Etnologia*, 4, 21–30.

Alves, R., Rosa, I., and Santana, G. (2007) 'The role of animal-derived remedies as complementary medicine in Brazil', *Bioscience*, 57(11), 949–55.

Bailey, G. (1975) 'The role of molluscs in coastal economies: the results of midden analysis in Australia', *Journal of Archaeological Science*, 2, 45–62.

Bandeira, D. (1992) 'Mudança na Estratégia de Subsistência—o Sítio Arqueológico Enseada I— um Estudo de Caso'. Unpublished MA dissertation, Universidade Federal de Santa Catarina (Florianópolis).

Botkin, S. (1980) 'Effects of human exploitation on shellfish populations at Malibu Creek, California', in Earle, T. and Christenson, A. (eds) *Modeling Changes in Prehistoric Subsistence Economies*, pp. 121–39. New York: Academic Press.

Casteel, C. (1970) 'Core and column sampling', *American Antiquity*, 35(4), 465–7.

Claassen, C. (1998) *Shells*, Cambridge: Cambridge University Press.

Claassen, C. (2010) *Feasting with Shellfish in the Southern Ohio Valley: Archaic Sacred Sites and Rituals*, Knoxville: University of Tennessee Press.

DeBlasis, P., Fish, S. K., Gaspar, M., and Fish, P. R. (1998) 'Some references for the discussion of complexity among the sambaqui moundbuilders from the southern shores of Brazil', *Revista de Arqueologia Americana*, 15, 75–105.

De Masi, M. (1999) 'Prehistoric Hunter-Gatherer Mobility on the Southern Brazilian Coast: Santa Catarina Island'. Unpublished PhD Dissertation, Stanford University (Stanford).

Duarte, P. (1967) 'O sambaqui visto através de alguns sambaquis', *Ciência e Cultura*, 19, 643–5.

Figuti, L. (1989) 'Estudos dos vestígios faunísticos do sambaqui Cosipa-3, Cubatão, São Paulo', *Revista de Pré-História*, 7, 112–26.

Figuti, L. (1993) 'O homem pré-histórico, o molusco e o sambaqui: considerações sobre a subsistência dos povos sambaquianos', *Revista do Museu de Arqueologia e Etnologia*, 3, 67–80.

Figuti, L. (1995) 'Os sambaquis Cosipa (4200 a 1200 anos AP): estudo da subsistência dos povos pescadores coletores pré-históricos da Baixada Santista', *Anais da Revista de Arqueologia*, 8, 267–83.

Figuti, L. (1998) 'Estórias de arqueopescador', *Revista de Arqueologia*, 11, 57–70.

Figuti, L. and Klokler, D. (1996) 'Resultados preliminares dos vestígios zooarqueológicos do sambaqui Espinheiros II (Joinville, SC)', *Revista do Museu de Arqueologia e Etnologia*, 6, 169–87.

Garcez, D. and Sanchez-Botero, J. (2005) 'Comunidades de pescadores artesanais no Estado do Rio Grande do Sul, Brasil', *Atlantica*, 27, 17–29.

Gaspar, M. D. (1998) 'Considerations of the sambaquis of the Brazilian Coast', *Antiquity*, 75, 592–615.

Gaspar, M. D. and DeBlasis, P. (1992) 'Construção de sambaqui', *Anais da Reunião Científica da Sociedade de Arqueologia Brasileira*, 6(2), 811–20.

Gaspar, M. D., DeBlasis, P., Fish, S., and Fish, P. (2008) 'Sambaqui (shell mound) societies of coastal Brazil', in Silverman, H. and Isbell, W. (eds) *Handbook of South American Archaeology*, pp. 319–35. New York: Springer-Verlage LLC.

Gaspar, M. D., Klokler, D., and DeBlasis, P. (2014) 'Were sambaqui people buried in the trash? Archaeology, physical anthropology, and the evolution of the interpretation of Brazilian shell mounds', in Roksandic, M., Souza, S., Eggers, S., Burcell, M., and Klokler, D. (eds) *The Cultural Dynamics of Shell Middens and Shell Mounds: A Worldwide Perspective*, pp. 91–100. Albuquerque: University of New Mexico Press.

Gaspar, M. D., Klokler, D., Scheel-Ybert, R., and Bianchini, G. (2013) 'Sambaqui de Amourins: mesmo sítio, perspectivas diferentes. Arqueologia de um sambaqui 30 anos depois', *Revista del Museo de Antropología*, 6, 7–20.

Hering, A. (2005) 'Estudo da Indústria Osteodontomalacológica do Sambaqui Jabuticabeira II, Jaguaruna, SC'. Report on file, São Paulo: Museu de Arqueologia e Etnologia, Universidade de São Paulo, Brazil.

Ihering, H. Von (1903) 'El hombre prehistórico del Brasil', *Historia*, 1(I), 161–70

Klokler, D. (2001) 'Construindo ou Deixando um Sambaqui? Análise de Sedimentos. Região de Laguna—SC'. Unpublished MA dissertation, Universidade de São Paulo (São Paulo).

Klokler, D. (2008) 'Food for Body and Soul: Mortuary Ritual in Shell Mounds (Laguna—Brazil)'. Unpublished PhD dissertation, University of Arizona (Tucson).

Klokler, D. (2010) 'Alimento, sacrifício ou oferenda: possíveis usos da fauna do Sambaqui de Cabeçuda'. Report on file, Conselho Nacional de Desenvolvimento Científico e Tecnológico.

Klokler, D. (2012) 'Consumo ritual, consumo no ritual: festins funerários e sambaquis', *Revista Habitus*, 10(1), 83–104.

Klokler, D. (2013) 'Em um mar de conchas, por onde começar? Amostragem zooarqueológica em sambaquis', in Gaspar, M. and Souza, S. (eds) *Abordagens Estratégicas em sambaquis*, pp. 177–91. Erechin: Editora Habilis.

Klokler, D. (2014a) 'A ritually constructed shell mound: feasting at the Jabuticabeira II site', in Roksandic, M., Souza, S., Eggers, S., Burcell, M., and Klokler, D. (eds) *The Cultural Dynamics of Shell Middens and Shell Mounds: A Worldwide Perspective*, pp. 151–62. Albuquerque: University of New Mexico Press.

Klokler, D. (2014b) 'Fishing for "Lucky Stones": Presence of Otoliths in Brazilian Shell Mound Sites'. Paper presented at the 79th Society for American Archaeology Annual Meeting, 23rd–27th April 2014, Austin.

Klokler, D. and Gaspar, M. (2013) 'Há uma estrutura funerária em meu sambaqui … , esse sambaqui é uma estrutura funerária!', in Gaspar, M. and Souza, S. (eds) *Abordagens em Sambaquis*, pp. 109–26. Erechin: Editora Habilis.

Lacerda, J. (1885) 'O homem dos sambaquis, contribuição para a antropologia brasileira', *Arquivos do Museu Nacional*, 6, 175–204.

Levi-Strauss, C. (1962) *The Savage Mind*, Chicago: University of Chicago Press.

Lima, T. (1991) 'Dos mariscos aos peixes: um estudo zooarqueológico da mudança de subsistência na pré-história do Rio de Janeiro'. Unpublished PhD dissertation, Universidade de São Paulo (São Paulo).

Luby, E. (2004) 'Shell mounds and mortuary behavior in the San Francisco Bay area', *North American Archaeologist*, 25, 1–33.

Mazzilli, R. (1975) 'Algumas considerações sobre o consumo de alimentos em Icapara e Pontal do Ribeira, São Paulo, Brasil', *Revista de Saúde Pública*, 9, 49–55.

Nishida, P. (2007) 'A coisa ficou preta: estudo do processo de formação da Terra Preta do sítio arqueológico Jabuticabeira II'. Unpublished PhD dissertation, Universidade de São Paulo (São Paulo).

Plens, C. (2010) 'Animals for humans in life and death', *Revista do Museu de Arqueologia e Etnologia*, 20, 31–51.

Rodrigues-Carvalho, C., Scheel-Ybert, R., Gaspar, M. D., Bianchini, G., Klokler, D., Andrade, M., and Borges, D. (2011) 'Cabeçuda-II: um conjunto de amoladores—polidores evidenciado em Laguna, SC', *Revista do Museu de Arqueologia e Etnologia*, 21, 402–5.

Russo, M. (2014) 'Ringed shell features of the Southeast US: architecture and midden', in Roksandic, M., Souza, S., Eggers, S., Burcell, M., and Klokler, D. (eds) *The Cultural Dynamics of Shell Middens and Shell Mounds: A Worldwide Perspective*, pp. 21–39. Albuquerque: University of New Mexico Press.

Saunders, R. (2014) 'Shell rings along the lower Atlantic coast of the United States: Defining function by contrasting details, with reference to Ecuador, Columbia, and Japan', in Roksandic, M., Souza, S., Eggers, S., Burcell, M., and Klokler, D. (eds) *The Cultural*

Dynamics of Shell Middens and Shell Mounds: A Worldwide Perspective, pp. 41–55. Albuquerque: University of New Mexico Press.

Souza, L. (1984) *O diabo e a Terra de Santa Cruz: feitiçaria e religiosidade popular no Brasil colonial*, São Paulo: Companhia das Letras.

Tenório, M. (1995) 'Estabilidade dos grupos litorâneos pré-históricos: uma questão para ser discutida', in Beltrão, M. (ed.) *Arqueologia do estado do Rio de Janeiro*, pp. 43–50. Rio de Janeiro: Arquivo Público do Estado do Rio de Janeiro.

Villagrán, X. (2010) *Estratigrafias que falam: geoarqueologia de um sambaqui monumental*, São Paulo: Editora Annablume-FAPESP.

Wiener, C. (1876) 'Estudos sobre os sambaquis do sul do Brasil', *Archivos do Museu Nacional do Rio de Janeiro*, 1, 3–20.

....

CHAPTER 42

CAMELID HUNTING AND HERDING IN INCA TIMES

a view from the south of the empire

....

GUILLERMO L. MENGONI GOÑALONS

INTRODUCTION

....

SOUTH American Camelids (SAC) occupied a central role in the development of Andean societies and were integral to their cultural landscape. It is important to highlight that camelids are the only large herd mammals that were domesticated in all the Americas, a co-evolutive process that gave origin to two domesticated forms: the alpaca (*Vicugna pacos*) and the llama (*Lama glama*), and their different breeds. The present wild forms—vicuña (*Vicugna vicugna*) and guanaco (*Lama guanicoe*)—were also exploited since the Early Holocene and hunting was maintained during the Inca period (AD 1450–1536 for northwestern Argentina), and persisted until Colonial (AD 1536–1816) and Republican (AD 1853 onwards) periods, but with a different impact on the sustainability of wild camelid populations.

During the Inca period camelids had a major significance, integrating economy, social, political, and ritual life. Camelids were a key instrument for the expansion and establishment of the empire. Llamas were used as beasts of burden for transporting goods along extensive redistribution networks that connected the highlands, valleys, and Pacific coast. From a utilitarian perspective camelids provided different products (e.g. meat, wool). During Inca times, as well as in previous periods, camelids were used for domestic consumption but also sacrificed in public festivities during the religious annual cycle (Dedenbach-Salazar Sáenz, 1990). They were widely depicted in rock art and appear represented as figurines made of raw materials that had symbolic value, such as certain kinds of stones (*illas* and *conopas*), *mullu* seashells (*Spondylus* sp.), and metals (gold and silver), and are found in offerings in different religious and ritual contexts.

In colonial times the general number of wild and domesticated camelids declined drastically due to over-slaughtering, competition with introduced species (e.g. sheep and goat), and illnesses (e.g. mercury pollution due to the amalgamation process used in silver mines). As a consequence, their geographic distribution was significantly reduced. Nevertheless, they still remain a core element of many rural communities in the Puna region (above *c*.3,200 m). Camelids, both wild and domesticated, are emblematic animals of the Andes and Patagonia.

They have been exported all over the world, and their use as wool producers and pets has expanded due to economic globalization. Several sustainable programs are currently under development, either to improve their living conditions and that of the local traditional herder communities that rely on them, or as an alternative commercial project for present-day farmers.

RECENT GENETIC AND ARCHAEOLOGICAL STUDIES ON CAMELIDS

A review of background information about the biology of these ungulates and the criteria used for their classification, both from a Linnaean and ethnographical perspective, is necessary to understand the use and management practices developed in the past. At present, SAC are represented by four different species: two wild, the vicuña and the guanaco, and two domesticated, the llama and the alpaca.

Recent genetic studies have confirmed the validity of the two main genera (*Vicugna* and *Lama*) and the existence of two monophyletic groups (Marín et al., 2007b). In central and south-central Andes there are two subspecies of vicuña: a northern form (*V. vicugna mensalis*) and a southern form (*V. vicugna vicugna*) that comprise separate mitochondrial lineages (Marín et al., 2007a). There is also a clear genetic difference between the two existing guanaco (*L. guanicoe*) populations. In Peru and northern Chile (Ayacucho and Putre) there is one subspecies (*L. g. cacsilensis*), while another one (*L. g. guanicoe*) is present in the rest of the known geographical distribution all the way down to Patagonia and Tierra del Fuego (Marín et al., 2013). However, genetic studies are still lacking for guanaco populations from northwestern Argentina.

Combined analyses of chromosomal and molecular markers have shown close genetic similarity between the vicuña (*V. vicugna*) and the alpaca (*V. pacos*) and between the guanaco (*L. guanicoe*) and the llama (*L. glama*). Although these same studies also show the existence of a hybridization process among the domesticated forms, these results support the idea that the alpaca is derived from the northern vicuña and the llama from the northern guanaco (Wheeler et al., 2006; Marín et al., 2007b).

Certain external features (e.g. fibre shape and colour, ear and tail form) and body proportions can be used for differentiating between the four present species. In addition, size variation is very significant among camelids. There is a body-size gradient among present camelids starting with vicuñas as the smallest (35–50 kg), followed by the alpaca (55–65 kg), then the guanaco (80–130 kg), and finally the llama (80–150 kg) as the largest. This means that vicuñas and alpacas overlap in size, as do the guanaco and llamas. Guanaco and llamas also have a wide variation that needs to be accounted for when undertaking osteometric studies (Mengoni Goñalons and Yacobaccio, 2006; Mengoni Goñalons, 2008). The remarkable variability in the size of domestic camelids across the Andean region is particularly pronounced in llamas (e.g. Miller and Gill, 1990; Yacobaccio, 2010).

The two subspecies of vicuña differ in live weight and body size (Yacobaccio, 2006), with the northern form (*V. v. mensalis*) smaller than the southern one (*V. v. vicugna*). Also the north-Andean guanaco (*L. g. cacsilensis*) is smaller than some of the representatives of the Patagonian forms (*L. g. guanicoe*), which have a broader distribution (Mengoni Goñalons, 2008). This variation in size emphasizes the importance of choosing the correct standard when measuring and comparing bones. The osteometric standard should be based on contemporary wild camelids from the same or a neighbouring region from which the archaeological material is derived (Mengoni Goñalons and Yacobaccio, 2006).

The study of pre-Hispanic camelid mummies has discovered alpaca and llama breeds that have no present-day counterpart in Peru (Wheeler et al., 1995). This evidence reveals a greater diversity of morphotypes in the past that probably lasted until the Inca period, and that was reduced in the aftermath of the Spanish conquest.

The Spanish terms for the four camelid species derive from the *Quechua* and *Aymara* languages but the Spanish chroniclers from the sixteenth and seventeenth centuries also recorded ethno-categories based on different criteria for classifying them. The domesticated forms (llama and alpaca) were described with categories based on utilitarian purposes (e.g. cargo animals, wool producers) or certain physical traits (e.g. wool colour and fibre characteristics). Wild camelids (guanaco and vicuña) were clearly differentiated from the domesticated ones.

CRITERIA FOR TAXONOMIC IDENTIFICATION: INDICATORS OR MARKERS

Different criteria have been used to study the variability of archaeological camelid remains (Mengoni Goñalons and Yacobaccio, 2006). Both direct and indirect indicators can be used, either based on characteristics of the materials involved (teeth, bone, or fibre) or on assemblage properties (Mengoni Goñalons, 2008). For example, dental morphology and bone size are used for classifying camelid remains and assigning them to a known morphotype category. Mortality profiles and stable isotope analyses can be used to assess management practices, either for wild species or domesticates. All this is usually complemented with contextual information in order to understand the processes involved in the formation of bone assemblages and the significance of faunal remains within a broader social context.

The Incas at Their Southeastern Quarter

A series of coordinated policies were implemented by the Incas during their occupation of northern Chile and Argentina. This vast and diverse region formed part of the Qollasuyu, which was only incorporated into the empire during the first half of the fifteenth century AD. These strategies involved the construction of a great number of administrative and state installations along the road network (*qhapaq ñan*), the placing of several shrines at high elevations, and the development of several state farms.

Other state policies included the intensification of craft, mine, agricultural, and pastoral production. In many cases these activities involved corvée labour (*mit´a*). Power was exercised through a system based on reciprocity and hospitality relationships in which ceremonial feasting acted as a way for reproducing social order. These practices varied regionally as well as in their timing, and in some cases the Inca presence is only revealed by certain architectural features and cultural material (Williams et al., 2009).

The intensification of an economy usually implies an increase in the control of several aspects of production, distribution, and consumption of products and goods (e.g. D'Altroy and Hastorf, 2001). In an agro-pastoral economy this may involve herd specialization. This means keeping the domestic animals segregated due to the specificity of their products (meat, fibre, or transportation), or for particular purposes (e.g. religious ceremonies) or for maintaining certain segments of society (e.g. feeding those in charge of the production of certain goods).

It is reasonable to assume that slaughter concentrated on individuals whose killing did not compromise the stability of the herds (e.g. unproductive adults), as has been highlighted by several authors (e.g. Flannery et al., 1989). Whenever prime-age animals were culled it was because there was a surplus of individuals which otherwise would be kept alive, especially in herds that are used for mixed production (e.g. meat and fibre). Management of vicuñas and guanacos that belonged to the Inca state could imply restrictions for their slaughter which may have been limited to ceremonial purposes or other special occasions.

It is also expected that access to and, therefore, redistribution of some products (e.g. meat for cooking and fibre for crafts) could have been regulated or even centralized, if the provisioning system was indirect. An outcome of this increase in control would be a relative standardization in the production of primary products (meat) and goods (textiles) and the regulation of their circulation.

Certain aspects of the products consumed, the techniques used for food preparation, the culinary equipment, and the social context of their use, either private or public, can contribute to the interpretation of faunal bone materials. For example, different segments of the society (elite, commoners) could have preferential access to certain products (e.g. animals with a special diet). Additionally, the context of use (daily meals or feasting) might determine the kinds and amounts of products to be used or consumed.

SOME CASE STUDIES FROM NORTHWESTERN ARGENTINA (NWA)

This section will focus on zooarchaeological markers used to analyse production, distribution, and consumption patterns. The zooarchaeological indicators are represented by the size variation of the camelids, information pertaining to their diet derived from stable isotope data (δ^{13}C and δ^{15}N), and age profiles derived from each assemblage. These criteria were applied to a selection of sites located in the northern section of northwestern Argentina (NWA), whose faunal assemblages were studied by the writer. This data will be complemented with information from sites studied by other colleagues.

Three major environmental units can be identified in the NWA region, each with its particular topography, water supply, and vegetation. Major changes in altitude and vegetation occur within a relative short distance (*c.*100 km). The *puna* is a high-altitude plateau (in general, above 3,000 m) traversed by mountains and valleys (*quebradas*), covered by different kinds of herbaceous and shrub steppes. The sierras that run at a lower altitude are covered by a tall shrub steppe, which may include either trees or large cacti, depending on the local topographical conditions of the *bolsones* (basins placed between mountains) and more open valleys. The eastern flank of the sierras is called *yungas* and captures the humid winds coming from the Atlantic and allows the growth of a very diverse and dense rainforest.

The last two zones have been heavily transformed by deforestation and overgrazing, but also by the installation of different kinds of urban and agricultural architectures, which can often be traced back to pre-Hispanic times. The *puna* and high *quebradas* provide the ideal pastures for rearing camelids and other herbivores. Literature about current populations shows that camelids living above 4,000 m have a diet dominated by C3 plants and their average δ^{13}C values are lower (guanaco: 19,4‰; vicuña: 18,3‰; llama: 19,3‰) than those which live below that altitude with a C3 and C4 mixed diet (vicuña: 15,3‰; llama: 17,0‰). These data indicate that altitude correlates with the quality of pastures available in these different environments and, therefore, suggest the probable provenance of the camelids exploited at the sites analysed in this chapter. This present set of values will be used as a baseline for interpreting isotope variability within the archaeological samples.

The site Pucará de Volcán is placed in the southern portion of the Quebrada de Humahuaca (Jujuy), at *c.*2,100 m (Fig. 42.1). It is a large and complex site with a main residential area, a cemetery, agro-pastoral structures, and other installations. The faunal materials come from a trash feature (Tum1B2) associated with a series of habitation structures and open spaces surrounding an artificial mound located to the west of the main residential area. The ceramics in the deposit show it as culturally homogenous and are dated to 1390–1640 cal AD. The site occupied a strategic position within the cultural landscape by connecting different localities placed to the east (*yungas*) and west (*prepuna*) of the Quebrada de Humahuaca (Cremonte and Scaro, 2010).

Esquina de Huajra is also located at the Quebrada de Humahuaca, a few kilometres north of Pucará de Volcán at *c*.2,000 m (Fig. 42.1). There are a series of architectural installations, mainly residential structures, burial features, and terracing structures. The bones, together with other elements of material culture, come from a domestic context accumulated in a patio (Terrace 1—Floor) adjacent to a residence. Several indicators

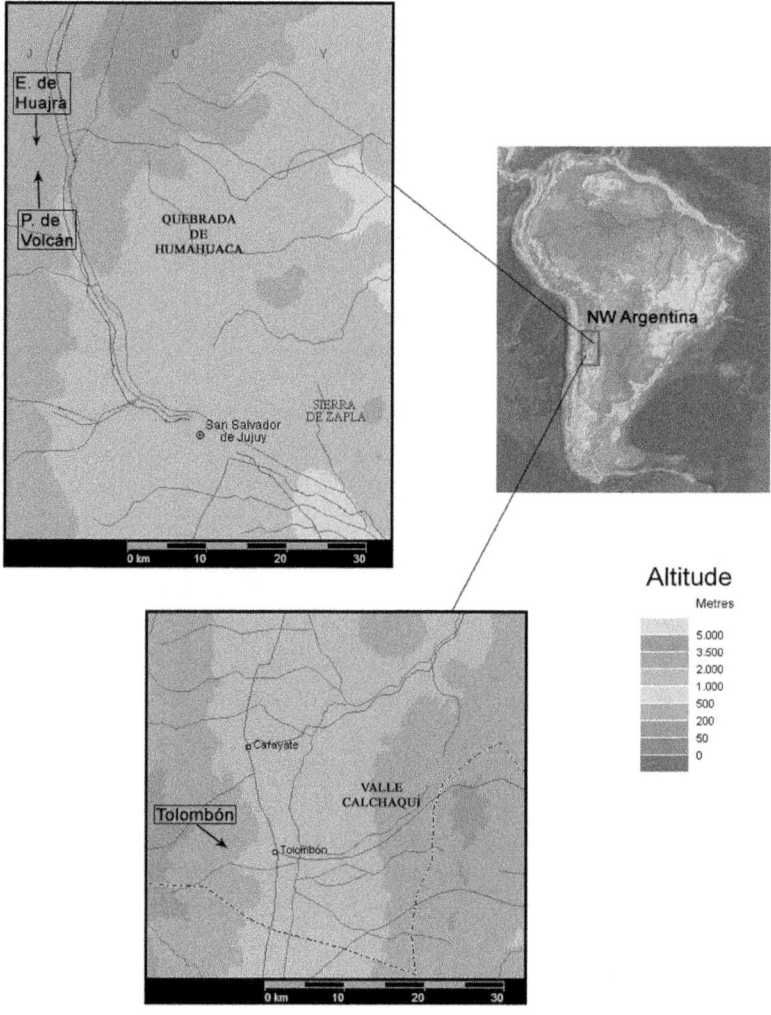

FIGURE 42.1 Location of sites mentioned in the text. Author's own image.

(pottery and other elements) suggest a high-status context dating to 1520–1620 cal AD. The locality probably had a strategic role connecting the *quebrada* with the eastern *yungas* (Cremonte et al., 2006–2007; Scaro and Cremonte, 2012).

The locality of Tolombón is situated close to the western bank of Santa María river (Valle Yocavil-Calchaquí, Salta), at *c.*1,700 m (Fig. 42.1). The architecture includes residential areas, public spaces, burial grounds, and defence and agricultural structures. The faunal assemblage comes from a residential space (Structure 6, Architectural Division A) where excavations revealed several layers containing ceramics, stone artefacts, and shell ornaments, among other elements, and also plant and animal bone remains. It was occupied during the Late period (1291–1628 cal AD) and is the only residential structure from which Inca pottery was retrieved. This site played a very important role during native Indian resistance against the Spaniards (Williams, 2002–2005; Williams, 2010).

CAMELID IMPORTANCE AND THEIR SIZE VARIABILITY

Camelids dominate most of the faunal assemblages of this time period while other species are also present but in low frequencies. Other ungulates, such as deer, are present at some sites. For example, the taruca (*Hippocamelus antisensis*) was identified at Esquina de Huajra and Tolombón and has also been found in other NWA contexts (e.g. Mercolli, 2010). Nevertheless, the camelid index (proportion of NISP camelids in relation to total NISP artiodactyls) is extremely high: 0.94 for Esquina de Huajra, 0.97 for Pucará de Volcán, and 0.93 for Tolombón. Medium- and small-size mammals (rodents, edentates, and carnivores) are occasionally present but in low frequencies. The presence of the muscovy duck (*Cairina moschata*), a domesticated bird, at Esquina de Huajra is a highlight. Ungulates (basically camelids) appear to be the most abundant taxa in most assemblages.

Most of the Late period (1000–1536 AD) NWA sites show a great variability in the size of camelids present (e.g. Mercolli, 2010). Using the present NWA guanaco as a modern standard and the log-difference technique (following Meadow, 1999) it was possible to classify the bone specimens from these three sites within the two main size groups: small and large camelids. But it was also possible to further visually distinguish three subgroups: one overlaps in size with the present vicuñas, another which probably corresponds to large llamas and a third that falls around the standard of the guanaco that corresponds to value 0 (Fig. 42.2). At present no alpacas have been found in northwestern Argentina, probably due to the unsuitable environment (Mengoni Goñalons and Yacobaccio, 2006). This last size sub-group could be composed of either small llamas, guanacos, or some kind of domestic (?) hybrid. All bones shown in the figure belong to osteologically mature animals. It is important to have in mind that camelids lack substantial sexual dimorphism (Mengoni Goñalons and Yacobaccio, 2006) that could obscure this interpretation.

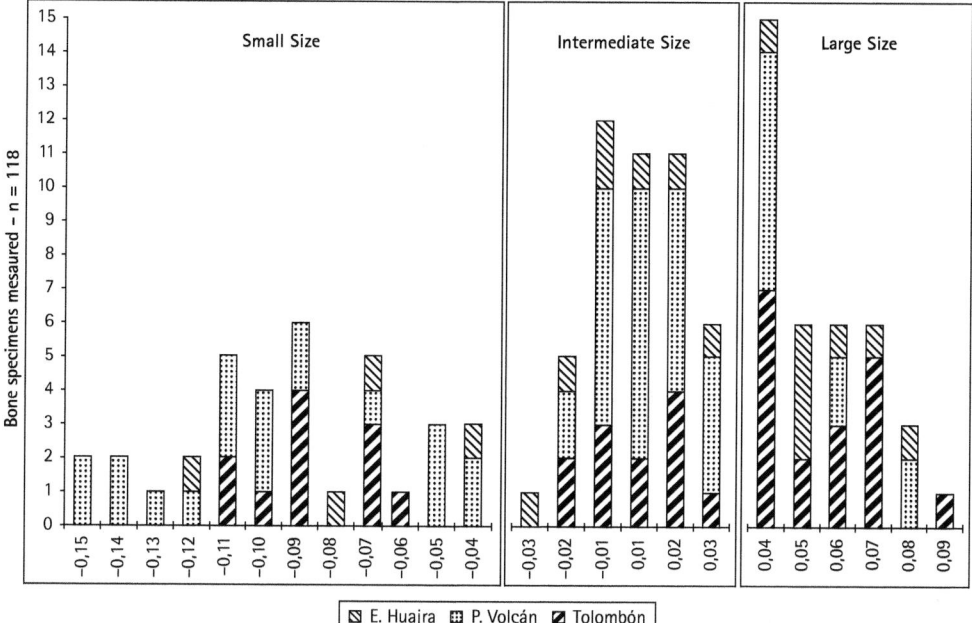

FIGURE 42.2 Log-difference between the archaeological specimens of the three case studies discussed and the modern standard. Measurements used as a standard were taken from two modern guanacos of northwestern Argentina from Cumbres Calchaquíes and Nevado de Aconquija. Author's own image.

At these three sites large camelids are represented in similar proportions (50–60%) to small ones (40–50%). But, if we consider the whole range of size variability some interesting patterns emerge. At Esquina de Huajra and Pucará de Volcán the three size sub-groups are represented, the very small (vicuñas) as well as the very large (llamas) individuals, but additionally a notable number of intermediate-size animals. However, at Tolombón there is a clear bimodal size distribution, with very small camelids (vicuñas) and very large ones (cargo llamas?) dominant. Vicuñas seem to have been used continuously and llamas probably varied in size and management conditions, as we will see in the next section.

DIET VARIABILITY AND MANAGEMENT PRACTICES

We were able to expand the importance of the size variability issue by using other indicators. Herd management involves several aspects that include the kinds of animals, the nature and distance of the feeding grounds, and the slaughter patterns.

The first of these indicators is represented by the stable isotopes $\delta^{13}C$ and $\delta^{15}N$, which can help in investigating the variability in management practices of each size sub-group. A selected sample of the measured bones that was used for isotope analyses had acceptable values of bone collagen carbon and nitrogen concentrations (Mengoni Goñalons, 2007). None of the isotopic values obtained from the sampled bones differ from those observed in modern counterparts (*puna* vicuñas, guanacos, and llamas) and fall within their expected range.

At Esquina de Huajra (n = 5) camelid diet is relatively homogeneous with $\delta^{13}C$ values that range between -16.3‰ and -13.9‰ (with a coefficient of variation (CV) = 6.7%). These values (Fig. 42.3) are similar to those from camelids that graze below 4,000 m, as we have seen above. This suggests that vicuñas (very small-sized individuals) and domesticates (larger individuals) fed on a mixture of C3 and C4 plants.

At Pucará de Volcán the animals (n = 12) vary notably in carbon isotope (CV = 16.2%), ranging from -17.9‰ to -9.0‰ (Fig. 42.3), though most signatures overlap with those of camelids (llamas and vicuñas) living below 4,000 m. Additionally, the intermediate and large sub-group includes individuals that have carbon enriched diets (> -13.0‰). One individual especially has a very positive carbon and high nitrogen signature, suggesting the use of fertilizers in some pasturelands (e.g. Szpak et al., 2012). These values also indicate that llama-sized animals had access to different pastures. This is to be expected when herds are kept segregated and the locality is provisioned from different altitudinal zones, a scenario compatible with the existence of complex redistribution networks of products and goods.

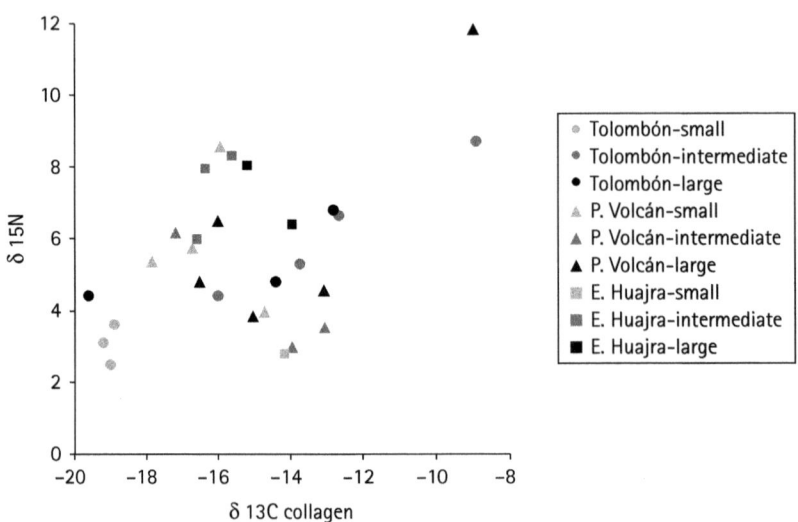

FIGURE 42.3 Carbon and nitrogen values of a selected sample of measured bones and camelid group size. Author's own image.

The Tolombón samples (n = 11) also present a high variability (CV = 21.8%) with carbon isotopic values between -19.6‰ and -8.9‰ (Fig. 42.3). This is also indicative of camelids being fed on different pastures. Small-sized individuals (vicuñas) had a similar diet to animals that live above 4,000 m. The intermediate sized sub-group also shows great variability but it is compatible with that of camelids living below 4,000 m on a C3 and C4 diet. Nonetheless, there is an individual within this size sub-group with a very enriched carbon diet (-8.9‰), suggesting a diet centred on C4. Again these values show that people had access to a wide range of camelids coming from different as well as distant sources, thanks to an articulated redistribution system.

This stable isotope data uncovered aspects of variability that could not be appreciated by using size variation alone. Animals of the same size may feed on different plants and, consequently, have different diets. The use of multiple zooarchaeological markers helps to understand the complexity of the production and redistribution system.

AGE PROFILES: FROM PRODUCTION TO CONSUMPTION

Certain aspects of faunal assemblages need to be considered when analysing slaughter patterns. One is their aggregative character and the nature of its corresponding archaeological context, aspects that have been considered when presenting the above case studies. Another issue is that certain practices take place at a particular age within the life cycle of an animal. In camelids 2 to 3 years is a critical age at which their future role is defined, either as producers of secondary products (wool or transportation) or for reproduction. Beyond the age of 7 years animals are, generally, not considered productive anymore, although they may be kept alive. These aspects determine certain developments in the life of an animal, which influence cultural preferences and decisions. The three age classes used in this study were based on fusion stages: early (< 12–18 months), intermediate (< 18–36 m), and late fusion (< 36–48 m). These categories were applied to the individuals of the main large-size group (most probably llamas).

At Esquina de Huajra consumption was focused on young (< 18 months) and young-adult (24–36 months) individuals. Nearly 50% died between 18 and 48 months. Only 25% survived above 36–48 months (Fig. 42.4). This indicates a culinary preference and access to prime cuts probably associated with the high status of those occupying this sector of the site. The pottery associated with the bone assemblage includes prestige goods (Scaro and Cremonte, 2012) and probably consumption of *hauté cuisine* dishes prepared with tender meat belonging to young animals besides more common dishes served as one-pot meals.

At Pucará de Volcán slaughter was concentrated on very young animals (newborns or *teques* and weaned individuals or *tuis*) and young adults of prime age (< 4 years old). A little more than half of the animals were slaughtered before 4 years of age (Fig. 42.4).

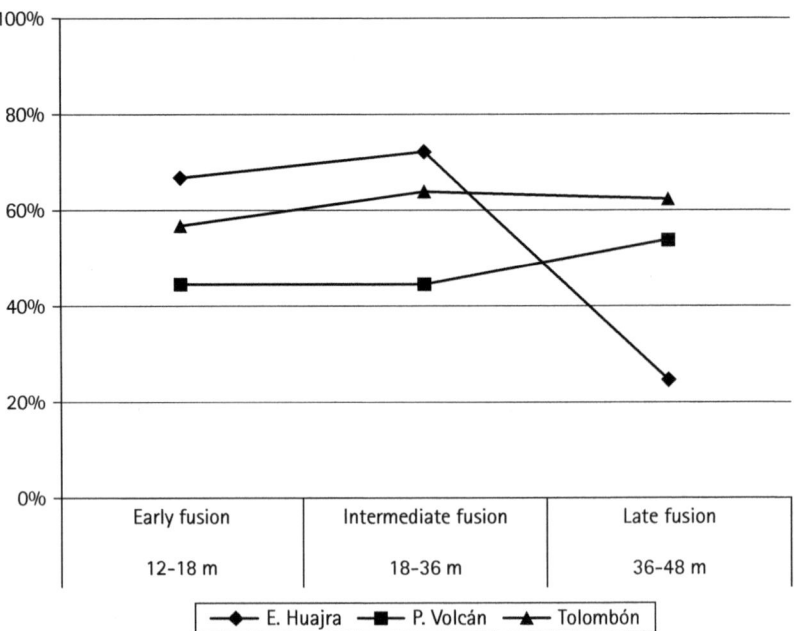

FIGURE 42.4 Survivorship curves based on fused elements (innominate, scapula, long bones, and first phalanx) relative counts (%). Author's own image.

The age category < 36–48 months surprisingly shows a slight increase in the surviving fraction, which in many contexts has been referred to as 'resurrection'. This signals the introduction of body parts of animals belonging to that particular age category and, therefore, implies access to a surplus stock of prime-age animals that could be sacrificed at will. In this case we think that quality was the choice, implying the access to body parts of high-value animals.

At Tolombón nearly 40% of the animals died before 48 months (Fig. 42.4). The survival of a high percentage of animals between 2 and 4 years old suggests an interest in keeping a stock of animals of an age typical of producers of secondary products, either wool or transportation. In this particular case, sacrifice was concentrated on those animals that were considered to be surplus and did not affect the reproduction and maintenance of the herd population.

SOME COMPARISONS AND THOUGHTS
FOR FUTURE RESEARCH

The three cases briefly presented here allow discussion of how Inca strategies of political dominance and resource administration developed in the southeastern section of their empire. The Incas always adapted their policies to local conditions so we cannot generalize the patterns observed at these sites but it is possible to dwell upon some aspects that have a wider significance.

Contrary to what is traditionally assumed, wild camelids (vicuñas and guanacos) were hunted and consumed, both in domestic and public contexts. In this sense, the exploitation of these animals shows a similar pattern to that recorded in some pre-Inca Late period sites of both Quebrada de Humahuaca and Calchaquí valley (e.g. Mercolli, 2010). The persistence of hunting is even seen in some isolated areas such as the highlands of central-west and northwest Argentina above 3,000 m where vicuñas and guanacos were hunted and/or captured for shearing, probably associated with communal practices (*chaku*) during Inca times (Bárcena et al., 2008; Ratto and Orgaz, 2002–2004).

Domesticated llama herds were bred at different altitudes and their meat travelled relatively long distances through a redistribution network that connected the camelid productive zones (probably located at the *puna* or close by) with those in the lower valley areas that concentrated most of the urban and agricultural activities. Herds were kept segregated as is evidenced by the isotope information and were used for different purposes.

Consumption occurred at the domestic and public level. Culling patterns show certain meat preferences for young individuals and the existence of a surplus of animals of prime age that was either used in elite households or during feasting performances. As camelids were readily accessible they were also probably used in common daily domestic contexts, but as an ingredient of stews and soups, with its associated specific culinary equipment.

In sum, the Qollasuyu region offered abundant land, animals, and crops allowing the Incas to intensively manage locally available resources and provision more distant products that were obtained at the very southern end of their empire. All of this was accomplished by means of strategic political control on all aspects of the productive-consumption system based on the material evidence already mentioned above.

Acknowledgments

Several colleagues collaborated generously with information and I sincerely appreciate the help of Félix Acuto, Pablo Cahiza, Isabel Cartajena, Beatriz Cremonte, Andrés Izeta, Pablo Mercolli, Norma Ratto, Beatriz Ventura, and Verónica Williams. Isotope analyses were carried out by Randy Culp at CAIS-University and by Robert Tykot at the Department of Anthropology—University of Florida. Carolina Mengoni Goñalons and Mercedes Rocco kindly helped with the preparation of the figures.

References

Bárcena, J. R., Cahiza, P. A., García Llorca, J., and Martín, S. E. (2008) *Arqueología del sitio inka de La Alcaparrosa: Parque Nacional San Guillermo*, Mendoza: CONICET.

Cremonte, M. B., Peralta, S. M., and Scaro, A. (2006–2007) 'Esquina de Huajra (Tum 10, Dpto. Tumbaya, Jujuy) y el poblamiento prehispánico tardío en el sur de la Quebrada de Humahuaca', *Cuadernos del INAPL*, 21, 27–38.

Cremonte, M. B. and Scaro, A. (2010) 'Consumo de vasijas cerámicas en un contexto público tardío del Pucara de Volcán (Dto. Tumabaya, Jujuy, Argentina)', *Revista do Museu Arqueologia e Etnologia*, 20, 147–61.

D'Altroy, T. N. and Hastorf, C. A. (2001) *Empire and Domestic Economy*, New York: Kluwer Academic/Plenum.

Dedenbach-Salazar Sáenz, S. (1990) *Inka Pachaq Llamampa Willaynin: uso y crianza de los camélidos en la época incaica*, Bonn: Universitat Bonn.

Flannery, K., Marcus, J., and Reynolds, R. G. (1989) *The Flocks of the Wamani*, San Diego: Academic Press.

Marín, J. C., Casey, C. S., Kadwell, M., Yaya, K., Hoces, D., Olazabal, J., Rosadio, R., Rodriguez, J., Spotorno, A., Bruford, M. W., and Wheeler, J. C. (2007a) 'Mitochondrial phylogeography and demographic history of the Vicuña: implications for conservation', *Heredity*, 99, 70–80.

Marín, J. C., Zapata, B., Gonzalez, B. A., Bonacic, C., Wheeler, J. C., Casey, C., Bruford, M., Palma, R. E., Poulin, E., Angelica Alliende, M., and Spotorno, A. E. (2007b) 'Sistemática, taxonomía y domesticación de alpacas y llamas: nueva evidencia cromosómica y molecular', *Revista Chilena de Historia Natural*, 80, 121–40.

Marín, J. C., González, B. A., Poulin, E., Casey, C. S., and Johnson, W.E. (2013) 'The influence of the arid Andean high plateau on the phylogeography and population genetics of guanaco (*Lama guanicoe*) in South America', *Molecular Ecology*, 22, 463–82.

Meadow, R. H. (1999) 'The use of size index scaling techniques for research on archaeozoological collections from the Middle East', in Becker, C., Manhart, H., Peters, J., and Schibler, J. (eds) *Historia Animalium ex Ossibus, Festschrift für Angela von den Driesch*, pp. 285–300. Rahden: Verlag Maire Leidorf GmbH.

Mengoni Goñalons, G. L, (2007) 'Camelid management during Inca times in NW Argentina: models and archaeological indicators', *Anthropozoologica*, 42(2), 129–41.

Mengoni Goñalons, G. L. (2008) 'Camelids in ancient Andean societies: a review of the zooarchaeological evidence', *Quaternary International*, 185, 59–68.

Mengoni Goñalons, G. L. and Yacobaccio, H. D. (2006) 'The domestication of South American camelids: a view from the South-Central Andes', in Zeder, M. A., Bradley, D., Emshwiller, E., and Smith, B. D. (eds) *Documenting domestication: new genetic and archaeological paradigms*, pp. 228–44. Berkeley: University of California Press.

Mercolli, P. (2010) 'Estrategias de subsistencia en la Quebrada de Humahuaca, provincia de Jujuy. Dos casos de estudio relacionados al manejo ganadero y la trascendencia de la caza a través del tiempo en las sociedades humanas', in Gutiérrez, M., De Nigris, M., Fernández, P., Giardina, M., Gil, A., Izeta, A., Neme, G., and Yacobaccio, H. (eds) *Zooarqueología a principios del siglo XXI*, pp: 273–84. Buenos Aires: Ediciones Libros del Espinillo.

Miller, G. R. and Gill, A. R. (1990) 'Zooarchaeology at Pirincay: a Formative period site in highland Ecuador', *Journal of Field Archaeology*, 17, 49–68.

Ratto, N. and Orgaz, M. (2002–2004) 'La cacería en los Andes: registro material del *chaku* en la Puna meridional catamarqueña (Cazadero Grande, Dpto Tinogasta, Catamarca)', *Arqueología*, 12, 72–102.

Scaro, A. and Cremonte, M. B. (2012) 'La vajilla de servicio de Esquina de Huajra (Dpto. Tumbaya, Jujuy, Argentina). Alternativas teóricas para interpretar su significado', *Revista del Museo de Antropología*, 5, 31–44.

Szpak, P., Millaire, J. F., White, C. D., and Longstaffe, F. J. (2012) 'Influence of seabird guano and camelid dung fertilization on the nitrogen isotopic composition of field-grown maize (*Zea mays*)', *Journal of Archaeological Science*, 39, 3721–40.

Wheeler, J. C., Chikhi, L., and Bruford, M. W. (2006) 'Genetic analysis of the origins of domestic South American camelids', in Zeder, M. A. (ed.) *Documenting Domestication: New Genetic and Archaeological Paradigm*, pp. 329–41. Berkeley: University of California Press.

Wheeler, J. C., Russel, A. J. F., and Redden, H. (1995) 'Llamas and alpacas: pre-conquest breeds and post-conquest hybrids', *Journal of Archaeological Science*, 22, 833–40.

Williams, V. I. (2002–2005) 'Provincias y capitales. Una visita a Tolombón, Salta, Argentina', *Xama*, 15–18, 177–98.

Williams, V. I. (2010) 'El uso del espacio a nivel estatal en el sur del Tawantisuyu', in Albeck, M. E., Scattolin, M. C., and Korstanje, M. A. (eds) *El habitat prehispánico*, pp. 77–114. Jujuy: Universidad Nacional de Jujuy.

Williams, V. I., Santoro, C. M., Romero, A. L, Gordillo, J., Valenzuela, D., and Standen, V. G. (2009) 'Dominación inca en los Valles Occidentales (Sur del Perú y Norte de Chile) y el Noroeste Argentino', in Ziółkowski, M. S, Jennings, J., Belan Franco, L. A., and Drusini, A. (eds) *Arqueología del area centro sur Andina, ANDES 7*, pp. 615–54. Warsaw: Warsaw University Press.

Yacobaccio, H. D. (2006) 'Variables morfométricas de vicuñas (*Vicugna vicugna vicugna*) en Cieneguillas, Jujuy', in Vilá B. (ed.), *Investigación, conservación y manejo de vicuñas*, pp. 101–12. Buenos Aires: Proyecto MACS—Argentina.

Yacobaccio, H. D. (2010) 'Osteometría de llamas (*Lama glama* L.) y sus consecuencias arqueológicas', in Gutiérrez, M., De Nigris, M., Fernández, P., Giardina, M., Gil, A., Izeta, A., Neme, G., and Yacobaccio, H. D. (eds), *Zooarqueología a principios del siglo XXI*, pp. 65–75. Buenos Aires: Ediciones Libros del Espinillo.

....

CHAPTER 43

FORESTS, STEPPES, AND COASTLINES

zooarchaeology and the prehistoric exploitation of Patagonian habitats

....

LUIS A. BORRERO

INTRODUCTION

....

THIS chapter will discuss the archaeology of southern Patagonia, from the south of the Santa Cruz river to the Strait of Magellan (Fig. 43.1). The geography of that area can be summarized as including the Andean range, a narrow belt of forests, and a wide eastern steppe. Coastal habitats in the maze of southwestern Pacific archipelagos are forested, indented, and steep, while they are long and with low terraces on the Atlantic side. The plateaux of the hinterland are dominated by a steppe environment, with shrub patches. The Andean range, with altitudes of *c.*3,000 m asl, constitutes the western limit. The precipitation gradient varies from >5,000 mm/year in the western Pacific coast to *c.*200 mm/year in the eastern steppes (Mazzoni and Vázquez, 2010). This extraordinary range is basic to the existence of two extreme cultural configurations recognized at the end of the nineteenth century, the so-called canoe peoples in the west and the terrestrial hunters in the east. These extremes were usually treated as complete opposites. However, variation in material culture is not always so marked and cultural formations observed during the nineteenth century at the Strait of Magellan included both canoe and terrestrial people (Borrero et al., 2011). Accordingly, prehistoric hunter-gatherers were rarely limited in their distribution to any one of these zones.

In prehistoric times people adapted to highly variable conditions. The available palaeoclimatic information points to the existence of arid environments since the end of the Pleistocene in most of extra-Andean Patagonia, while the forests began to expand after the retreat of the Pleistocene glaciers *c.*14,000 BP. Before the end of the Pleistocene there

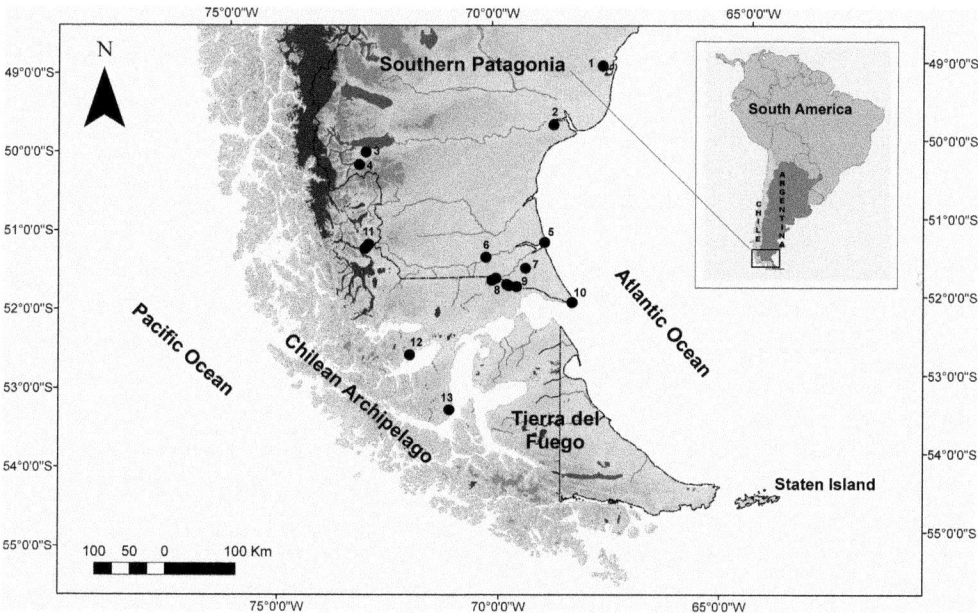

FIGURE 43.1 Location map. Sites code: 1: Floridablanca, 2: Monte Entrance, Monte León, 3: Alice 2, 4: Chorrillo Malo 2, 5: Punta Bustamante, 6: Las Buitreras, 7: Cóndor, 8: Orejas de Burro 1, Zurdo valley, 9: Fell's Cave, Cueva del Puma, Cueva de los Chingues, Pali Aike, 10: Nombre de Jesús, 11: Cueva del Medio, Cueva Lago Sofía 1, 12: Pizzulic 2, Bahía Colorada, 13: Punta Santa Ana 1, Bahía Buena. Author's own image.

was another cold period in synchrony with the Antarctic Cold Reversal (McCulloch et al., 2005). A warmer trend started at the beginning of the Holocene and was only interrupted by short cold periods (Mancini et al., 2008).

First Human Inhabitants

The human occupation of Patagonia began in Late Pleistocene times, from about 13,000–12,000 BP, during the Antarctic Cold Reversal (Borrero, 2012). The oldest archaeological evidence only dates from around 11,500 BP. The zooarchaeological remains of extinct mammals physically associated with artefacts of the first human explorers are difficult to recognize, since they are often mixed with naturally deposited faunas, especially accumulated by carnivores (Martin, 2013). In the early 1980s a taphonomic study focused on the preservation and burial potential of guanaco (*Lama guanicoe*) bones (Borrero, 1990a) showed that many of the conditions under which bones become deposited by different agents were indistinguishable. Taphonomic studies of Late Pleistocene bone assemblages are increasingly abundant, offering some clarification about which bones were truly associated with humans and which were the result of transport by carnivores, natural deaths, and other causes (Borrero et al., 1997; Martin, 2008; 2013; Borrero and Martin, 2012). The use of fossil DNA on ground sloth (*Mylodon darwinii*) bones initiated an important research avenue, now extended to extinct horses. This research revised aspects of equid phylogenetic relationships, particularly in relation with *Hippidion* (Orlando et al., 2009).

The dominant ungulate in Patagonian bone assemblages is the guanaco, which experienced changes through time. A morphometric approach was used to study the variation of guanaco appendicular bones during the last 12,000 radiocarbon years. Particularly the first and second phalanges, distal humeri, and distal metacarpals were selected. The results obtained show no significant difference in size between Late Pleistocene samples of second phalanges from southern Patagonia and Tierra del Fuego. The study concluded that, during the following period, reproductive isolation and vicariance resulted in guanaco populations from Tierra del Fuego retaining their Pleistocene size, while they were diminishing in size on the mainland (L'Heureux, 2005). This diminution is clearly indicated in the Late Holocene distal humeri and distal metacarpals of the continent, which are significantly smaller than those from the island.

Only a handful of the known early bone assemblages of southern Patagonia were thoroughly studied. For example, early excavations at Fell's Cave demonstrated an association between megamammals, particularly horse (*Hippidion saldiasi*) and ground sloth, with humans (Bird, 1938). However, most animal bone assemblages dating to 11,000–10,000 years were, for decades, only known through mere faunal lists (Poulain Josien, 1963; Saxon, 1979; Bird, 1988). Only recently there have been more detailed studies of the bone collections from Fell's Cave (Martin, 2013). Martin's study suggested that the *Mylodon* remains probably resulted from humans scavenging carnivore kills. Few cutmarks on selected ground sloth bones were found which were interpreted as part of the scavenging process (Martin, 2013). Carnivore activity was identified at nearby Cueva del Puma and Cueva de los Chingues (Martin, 2008). Several other sites, such as Pali Aike, Las Buitreras, and Cóndor also contain ground sloths bone assemblages, but in none is there any indication of human association (Martin, 2013).

Horse and camelid bones from Fell's Cave and other sites display both cut- and breakage-marks attesting to their regular use in human diet (Martin, 2013). At the Cueva del Medio, Ultima Esperanza, a Late Pleistocene human occupation dated to 11,120–9600 BP was recorded (Nami and Menegaz, 1991), and butchered bones of horse and camelids are associated with hearths and abundant lithics. Extinct faunas associated with artefacts dated to 11,500–10,100 BP were also found at nearby Cueva Lago Sofía (Prieto, 1991). Cutmarks were recorded on horse bones (Alberdi and Prieto, 2000; Martin, 2013). Camelid bones were dominant at all these Late Pleistocene bone assemblages.

HOLOCENE BONE ASSEMBLAGES

After the end of the Pleistocene, when several genera of mammals went extinct, the guanacos remained the main human prey. Early Holocene bone assemblages are dominated by bones of this ungulate (Mengoni and Silveira, 1976; Borrero, 1990b; Miotti, 1998). Most of these occupations were short term, and the sites were rarely revisited. The analysis of guanaco remains from Middle Holocene sites such as Fell's Cave, Chorrillo Malo 2, and Las Buitreras indicated selection of bones with high marrow fat content (Caviglia and Figuerero Torres, 1976; Bird, 1988; Otaola and Franco, 2008). During the Middle to Late Holocene, archaeological sites increase in number and variety, and can be found in virtually all Patagonian environments including the oceanic coasts (see below). Around 4000–3500 BP intensive use of animals, particularly guanacos, was recorded at Orejas de Burro, Cóndor, and other sites in Pali Aike (Bird, 1988; Barberena et al., 2004; L'Heureux, 2008). This is a time during which there is repetitive use of rock shelters (De Nigris, 2004; Rindel, 2004), but open-air camps such as Punta Bustamante (Mansur et al., 2004) or large sites with intensive occupations are also recorded in different parts of Patagonia (Mengoni, 1999).

The study of several bone assemblages showed that butchery marks are abundant, which is consistent with a thorough exploitation of guanacos (Muñoz, 1997: 217). Also, numerous studies indicate selective transport of guanaco anatomical parts (Mena and Jackson, 1991; Muñoz, 1997; Mengoni, 1999; De Nigris, 2004; Rindel, 2004; Mengoni and De Nigris, 2005) as demonstrated by the greater abundance of the appendicular than the axial skeleton. Ethnographic observations carried out in the nineteenth century lend support to this pattern by showing that the axial skeleton was usually abandoned at kill sites (Claraz, 2008: 77). Nevertheless, there are a few Late Historic archaeological examples of bone assemblages where appendicular and axial parts are equally represented (Mena and Jackson, 1991: 181). The vicinity of the kill site may explain such a pattern, but differential availability of transport facilities, such as horses, may also explain this pattern.

Studies of the guanaco anatomy and its economic exploitation have led to the development of meat and marrow indices which are used to quantify the economic value attached to different bones (Borrero, 1990b; Mengoni, 1999), and drying indices, used to measure the drying potential of different body parts (Mengoni and De Nigris, 2005). Together with densitometric studies, used to evaluate the survival of different bones, the use of these indices has contributed to the interpretation of the economic value of several bone assemblages. Along with the distribution of cut-marks (Muñoz, 1997), these indices were fundamental in the study of the exploitation of marrow and fat contained in different bones. In this context, it must be noted that Muñoz observed the abundance of cranial elements in several bone assemblages (Muñoz, 1997: 217). Beyond that, the study of cut-marks indicates that the final consumption of guanaco is mostly represented at rock shelters, while other activities characterize open-air sites (Borrero, 1987; Rindel et al., 2011). More specifically, it was observed that most archaeological sites present a high frequency of transversal fractures in guanaco long bones. This fracture pattern is known as 'perimetral marking', and is characterized by its regular edges (Fig. 43.2). These fractures were variously explained as a way to access the marrow cavity, to facilitate the transport of guanaco parts, to process frozen carcasses, or to make tools (Mengoni and Silveira, 1976; Caviglia and Borrero, 1978; Muñoz and Belardi, 1998;

FIGURE 43.2 Proximal humerus of *Lama guanicoe* with 'perimetral marking', Cóndor site. Author's own image.

Bourlot et al., 2008). One widely accepted explanation is that they were bone instruments used to process meat or other products like plants (Hajduk and Lezcano, 2005).

Thorough studies of guanaco sex differences, epiphyseal fusion, and tooth eruption were published by Kaufmann (2009). On that basis L'Heureux and Kaufmann (2012a) developed a model to estimate the age of immature guanacos by measuring the length of their long bones. Using bones from a modern population of guanacos they found that the correlation between the greatest length and age was very high for all unfused long bones and were therefore able to establish the age-at-death of juvenile guanacos based on biometrical analysis. An application of this technique to the guanaco bone sample from the Orejas de Burro 1 site shows the usefulness of the technique in refining the age-at-death of archaeological bone specimens (L'Heureux and Kaufmann, 2012b). Also, the application of tooth microwear and mesowear (i.e. observations of tooth cusps relief and shape) for the analysis of dietary traits of ungulates was recently initiated. Mesowear analysis evaluated the attrition and abrasion in teeth cusps, indicating that the guanacos from the Cardiel Lake area are mixed feeders, and thus have a seasonally shifting diet (Rivals et al., 2013).

Several studies indicated that even when guanacos were the dominant mammals, there were also other animals contributing to the subsistence (Caviglia and Figuerero

Torres, 1976; Mengoni and Silveira, 1976). Since these remains were usually not abundant, it was not always clear if they were related to anthropic activities or not. This was at least one reason to engage in taphonomic studies, a field slowly expanding in Patagonia. Very few studies deal with the huemul (*Hippocamelus bisulcus*), an endemic deer of Patagonia, mainly because it is not abundant in bone assemblages (Mena, 1992; De Nigris, 2004). Work by Belardi and Gómez Otero (1998) on the economic value of different body parts provided a framework to understand exploitation strategies. The few sites where huemul is abundant are located in the forest, and are characterized by intensive utilization of all parts of the carcass (De Nigris, 2004).

Not all the evidence can be explained in purely economic or subsistence terms. One example is provided by the Alice 2 site, where a pile of guanaco skulls and mandibles of at least eight individuals dated to *c*.800 BP was found. It is not clear what purpose those bones served, but on the basis of ethnographic observations showing the placement of guanaco skulls at hunting stands (Moreno, 1979: 114–15), the possibility that they were associated with hunting rituals has been suggested.

COASTAL BONE ASSEMBLAGES

At the time when hinterland sites indicate intensive utilization, coastal habitats were also exploited. Human occupation of the western side of the Strait of Magellan and along the Pacific coast is dated to *c*.6500 BP onwards and indicates that a fully maritime economy was in place. However, the study of coastal assemblages is not well developed in southern Patagonia. Early studies rarely went beyond taxonomic and anatomical lists (e.g. Ortiz Troncoso, 1975), but the study of marine mammal bones and teeth is gradually developing. Studies of the economic utility of the South American sea lion (*Otaria flavescens*) (San Román, 2009) and of the sex and age (Legoupil, 1989–1990), volume density (Borella et al., 2007), and morphometry of other pinnipeds (L'Heureux and Borella, 2012) have been developed.

Initial applications of these studies recorded trends in the exploitation and consumption of marine faunas (Morello et al., 2012). Particularly, the bone assemblages of the western Strait of Magellan were studied using economic utility and bone mineral density indices. On the basis of the intensity of use of pinnipeds, San Román (2010) proposed a differentiation between residential sites intensively used, such as Bahía Buena, and those with more ephemeral occupations, such as Punta Santa Ana 1 and Bahía Colorada. A predominance of the South American fur seal (*Arctocephalus australis*) was recorded at Bahía Buena, where age and sex studies showed the predominance of adult males. Also, cut-marks are abundant, and their distribution on the skull is indicative of hide removal in the South American fur seal (San Román, 2007). At Punta Santa Ana 1 and Bahía Colorada reverse utility curves were recorded. This indicates that the bones that are present are not those with a higher economic importance and therefore that a secondary use of pinnipeds took place, in comparison to the larger sites (Legoupil, 1997; San Román, 2010). Terrestrial mammals, particularly guanaco, were also exploited.

There are also indications of the utilization of resources on the Atlantic coast, a process that started slightly before 5000 BP (Moreno, 2008). The exploitation of this marine ecosystem appears to be complementary to that of the interior, with populations relying both on terrestrial and coastal resources (Gómez Otero, 1995; Miotti, 1998; Mansur et al., 2004; Borrero and Barberena, 2006).

In Tierra del Fuego birds constituted an important component of the human diet in marine economies, though this was rarely as intensive as on the main continent (Lefèvre, 1989; Tivoli, 2010), where they are mainly restricted to the coasts (Cruz, 2007; Tivoli, 2010). The majority of the species identified in Patagonia and Tierra del Fuego are coastal or pelagic, and are often gregarious (Lefèvre, 2010). Flightless birds (Rheidae) are never abundant in archaeological sites and are mostly represented by their lower legs (Fernández, 2000; Cruz and Elkin, 2003). Although taphonomic programs have been only recently systematically implemented (Cruz, 2003; 2007), it has been recognized for decades that many of the bird bones were introduced by non-human agents (Lefèvre, 1989).

The remains of fish are not abundant at continental archaeological sites. Fish bones occasionally occur on sites on the Atlantic coast, but on the Beagle Channel and the western side of the Strait of Magellan the evidence is more substantial. Studies of two Middle Holocene bone assemblages, Punta Santa Ana 1 and Pizzulic 2 showed a focus on the tadpole codling (*Salilota australis*), which is a species that inhabits the greater depths of the tidal plain (Torres and Ruz, 2011). Rays (Rajidae) were also recovered at both of these sites at which archaeozoological remains are dominated by pinnipeds (San Román, 2010; Morello et al., 2012). The evidence from the Beagle Channel shows that fish, together with birds, were intensively consumed resources during the Late Holocene (Zangrando, 2003; Tivoli, 2010). The intensive study of otoliths recently started and will be of great help in clarifying the seasonality of human occupation of coastal habitats (Torres and Ruz, 2011; Scartascini and Volpedo, 2012) (Fig. 43.3).

Several sites attest to the exploitation of malacological resources on the Atlantic coast at least since the Middle Holocene. One peculiar site (Orejas de Burro 1) is located some 20 km off the coast of the Strait of Magellan. Molluscs were recorded in large numbers there, and these are likely to have been collected during excursions to the sea coast (L'Heureux, 2008). Elsewhere, seasonality of occupation at Monte León was determined from the shells of *Mytilus* spp. through sclerochronological analysis, a technique that identifies growth increments in mollusc shells. It was found that they were mostly collected in summer, but with about 30% indicating winter foraging (Lobbia, 2012), thus confirming previous suggestions (based on ethnographic observations) of the coasts being used at different seasons of the year. Shell middens are relatively abundant north of the mouth of the Santa Cruz river (Hammond et al., 2013), but are less frequent south of that river. Late Holocene shell middens at Monte Entrance display evidence of

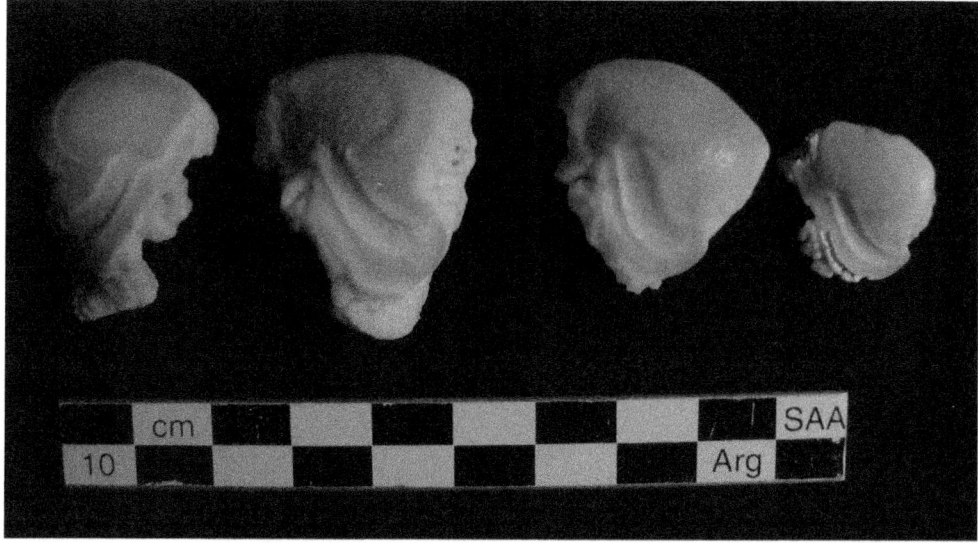

FIGURE 43.3 Archaeological specimens of *Micropogonias furnieri* otoliths, San Matías Gulf, Patagonia. Photo by Federico Scartascini.

utilization of both marine and terrestrial resources. Limpets (*Nacella magellanica*) are the most abundant molluscs, suggesting their preferential selection by human populations. The most recent occupation at this site indicates less reliance on molluscs (Franco et al., 2010), which is probably associated with an overall greater emphasis on terrestrial resources in southern Patagonia, to the point that marine resources went out of use in historical times (Moreno and Videla, 2008).

Studies of historical zooarchaeology in Southern Patagonia are, however, few. An exception is the analysis of faunal remains from the early Spanish occupation at Nombre de Jesús (sixteenth century), on the Strait of Magellan, which indicated a diet dominated by local faunas. Historical sources showed that the inhabitants were living under severe stress due to hard living conditions and scarcity of food (De Nigris et al., 2010). This condition was confirmed by the study of five human skeletons recovered at Nombre de Jesús (Suby et al., 2009). Zooarchaeological reports from the Spanish occupation at Floridablanca, a site dated from the end of the eighteenth century, produced abundant evidence for the use of both imported and local faunas (Marschoff, 2007; Senatore et al., 2009). Historical occupation in interior Patagonia dating to the end of the nineteenth and the beginning of the twentieth century produced evidence for the sedentarization of the tribe of the *Aónikenk* chief Mulato in the río Zurdo valley (Martinic et al., 1995). The *Aónikenk* of the nineteenth century were sheep ranchers, exploiting both sheep and horses, and hunting guanacos for their skins for commerce (Martinic, 1995). The consumption of horse meat was common, including also for ceremonial reasons (Martinic, 1995; Nuevo Delaunay, 2012).

CONCLUSION

Archaeozoological studies display relatively monotonous trends in Patagonia. After the disappearance of the Pleistocene megamammals, the intense exploitation of guanacos in the interior and marine mammals on the coasts followed. The main observed variations were in the intensity of utilization, a fact probably related with increasing human demography. Huemul and Rheidae were discontinuously exploited in the interior, particularly in the forests. On the coasts, molluscs, fish, and birds complemented the human diet, especially during the Late Holocene. The main subsistence changes introduced by the European contact were the sheep and the horse, which was intensively used for hunting.

ACKNOWLEDGMENTS

I am grateful to Cecilia Pallo and Federico Scartascini for their help with the figures.

REFERENCES

Alberdi, M. T. and Prieto, A. (2000) 'Hippidion (Mammalia, Perissodactyla) de las cuevas de Magallanes y Tierra del Fuego', Anales del Instituto de la Patagonia, 28, 147–71.

Barberena, R., L'Heureux, G. L., and Borrero, L. A. (2004) 'Expandiendo el alcance de las reconstrucciones de subsistencia. Isótopos estables y conjuntos arqueofaunísticos', in Civalero, M. T., Fernández, P. M., and Guráieb, A. G. (eds) Contra viento y marea: arqueología de Patagonia, pp. 417–33. Buenos Aires: INAPL-SAA.

Belardi, J. B., Gómez Otero, J. (1998) 'Anatomía económica del huemul (Hippocamelus bisulcus): una contribución a la interpretación de las evidencias arqueológicas de su aprovechamiento en Patagonia', Anales del Instituto de la Patagonia, 26, 195–207.

Bird, J. (1938) 'Antiquity and migrations of the early inhabitants of Patagonia', Geographical Review, 28(2), 250–75.

Bird, J. (1988) Travels and Archaeology in South Chile, Iowa City: Iowa University Press.

Borella, F., Gutiérrez, M., Fodere, H., and Merlo, J. (2007) 'Estudio de densidad mineral ósea para dos especies de otáridos frecuentes en el registro arqueofaunístico patagónico (Otaria flavescens y Arctocephalus australis)', in Morello, F., Prieto, A., Martinic, M., and Bahamonde, G. (eds) Desenterrando huesos, recolectando piedras y develando arcanos ... arqueología de Fuego-Patagonia, pp. 421–6. Punta Arenas: CEQUA.

Borrero, L. A. (1987) 'Variabilidad de sitios arqueológicos en la Patagonia meridional', in Comunicaciones, primeras jornadas de arqueología de la Patagonia, pp. 41–9. Rawson: Dirección de Impresiones Oficiales.

Borrero, L. A. (1990a) 'Taphonomy of guanaco bones in Tierra del Fuego', Quaternary Research, 34, 361–71.

Borrero, L. A. (1990b) 'Fuego-Patagonian bone assemblages and the problem of communal guanaco hunting', in Davis, L. B. and Reeves, B. O. K (eds) *Hunters of the Recent Past*, pp. 373–99. London: Unwin Hyman.

Borrero, L. A. (2012) 'The human colonization of the High Andes and southern South America during the Cold Pulse of the Late Pleistocene', in Eren, M. (ed.) *Hunter-Gatherer Behavior: Human Response during the Younger Dryas*, pp. 57–78. Walnut Creek: Left Coast Press.

Borrero, L. A. and Barberena, R. (2006) 'Hunter-gatherer home ranges and marine resources', *Current Anthropology*, 47(5), 855–67.

Borrero, L. A. and Martin, F. M. (2012) 'Ground sloths and humans in southern Fuego-Patagonia: taphonomy and archaeology', *World Archaeology*, 44(1), 102–17.

Borrero, L. A., Martin, F. M., and Barberena, R. (2011) 'Visits, 'Fuegians', and Information Networks', in Whallon, R., Lovis, W. A., and Hitchcock, R. K. (eds) *Information and its Role in Hunter-Gatherer Bands*, pp. 249–65. Los Angeles: The Cotsen Institute of Archaeology at UCLA.

Borrero, L. A., Martin, F. M., and Prieto, A. (1997) 'La Cueva Lago Sofía 4, Ultima Esperanza: una madriguera de felino del Pleistoceno tardío', *Anales del Instituto de la Patagonia*, 25, 103–22.

Bourlot, T., Rindel, D., and Aragone, A. (2008) 'La fractura transversa/marcado perimetral en sitios a cielo abierto durante el Holoceno tardío en el noroeste de Santa Cruz', in Salemme, M., Santiago, F., Álvarez, M., Piana, E., Vázquez, M., and Mansur, E. (eds) *Arqueología de Patagonia: una mirada desde el último confín*, pp. 693–705. Ushuaia: Utopías, II.

Caviglia, S. and Borrero, L. A. (1978) 'Bahía Solano: su interpretación paleoetnozoológica en un marco regional'. Paper presented at the V Congreso Nacional de Arqueología Argentina, San Juan.

Caviglia, S. and Figuerero Torres, M. J. (1976) 'Material faunístico de la cueva Las Buitreras', *Relaciones*, 10, 315–19.

Claraz, G. (2008) *Viaje al río Chubut: aspectos naturalísticos y etnológicos*, Buenos Aires: Ediciones Continente.

Cruz, I. (2003) 'Paisajes tafonómicos de restos de aves en el Sur de Patagonia continental: aportes para la interpretación de conjuntos avifaunísticos en registros arqueológicos del Holoceno'. Unpublished PhD Dissertation, Universidad de Buenos Aires (Buenos Aires).

Cruz, I. (2007) 'Avian taphonomy: observations at two Magellanic penguin (*Spheniscus magellanicus*) breeding colonies and their implications for the fossil record', *Journal of Archaeological Science*, 34, 1252–61.

Cruz, I. and Elkin, D. (2003) Structural bone density of the Lesser Rhea (*Pterocnemia pennata*) (Aves: Rheidae). Taphonomic and archaeological implications', *Journal of Archaeological Science*, 30, 37–44.

De Nigris, M. (2004) *El consumo en grupos cazadores recolectores: un ejemplo zooarqueológico de Patagonia meridional*, Buenos Aires: Sociedad Argentina de Antropología.

De Nigris, M., Palombo, P. S., and Senatore, X. (2010) 'Craving for hunger: a zooarchaeological study at the edge of the Spanish Empire', in Campana, D., Crabtree, P., deFrance, S. D., Lev-Tov, J., and Choyke, A. (eds) *Anthropological Approaches to Zooarchaeology*, Oxford: Oxford Books, pp. 131–7.

Fernández, P. M. (2000) 'Rendido a tus pies: acerca de la composición anatómica de los conjuntos arqueofaunisticos con restos de Rheiformes de Pampa y Patagonia', in *Desde el país de los gigantes*, pp. 573–86. Río Gallegos: Universidad Nacional de la Patagonia Austral.

Franco, N. V., Zubimendi, M. A., Cardillo, M., and Guarido, A. L. (2010) 'Relevamiento arqueológico en Cañadón de los Mejillones: primeros resultados', *Magallania*, 38(1), 269–80.

Gómez Otero, J. (1995) 'Bases para una arqueología de la costa patagónica central (entre Golfo San José y Cabo Blanco)', *Arqueología*, 5, 61–103.

Hajduk, A. and Lezcano, M. J. (2005) 'Un "nuevo-viejo" integrante del elenco de instrumentos óseos de Patagonia: los machacadores óseos', *Magallania*, 33(1), 63–80.

Hammond, H., Zubimendi, M. A., and Zilio, L. (2013) 'Composición de concheros y uso del espacio: aproximaciones al paisaje arqueológico costero en Punta Medanosa', *Anuario de Arqueología*, 5, 67–84.

Kaufmann, C. A. (2009) *Estructura de edad y sexo en guanaco. Estudios actualísticos y arqueológicos en Pampa y Patagonia*, Buenos Aires: Sociedad Argentina de Antropología.

Lefèvre, C. (1989) 'L'avifaune de Patagonie austral et ses relations avec l'homme au cours des six derniers millenaires'. Unpublished PhD dissertation, Université Paris I Panthéon-Sorbonne (Paris).

Lefèvre, C. (2010) 'Birds in maritime hunter-gatherers subsistence: case studies from southern Patagonia and the Aleutian Islands', in Prummel, W., Zeiler, J. T., and Brinkhuizen, D. C. (eds) *Birds in Archaeology*, pp. 117–30. Groningen: Barkhuis.

Legoupil, D. (1989–1990) 'La identificación de los mamíferos marinos en los sitios canoeros de Patagonia: problemas y constataciones', *Anales del Instituto de la Patagonia*, 19, 101–15.

Legoupil, D. (1997) *Bahía Colorada (île Englefield). Les premiers chasseurs de mammifères marins de Patagonie australe*, Paris: Recherche sur les Civilisations.

L'Heureux, L. G. (2005) 'Variación morfométrica en restos óseos de guanaco de sitios arqueológicos de Patagonia austral continental y de la Isla Grande de Tierra del Fuego', *Magallania*, 33(1), 81–94.

L'Heureux, L. G. (2008) 'La arqueofauna del campo volcánico Pali Aike. El sitio Orejas de Burro 1', *Magallania*, 36(1), 65–78.

L'Heureux, L. G. and Borella, F. (2012) *Guía osteométrica para el estudio de elementos óseos de Otaria flavescens*, Olavarría: Universidad Nacional del Centro.

L'Heureux, L. G. and Kaufmann, C. (2012a) 'Age estimation of juvenile guanaco (*Lama guanicoe*) individuals using diaphyseal long bone length', in Lefèvre, C. (ed.) *Proceedings of the General Session of the 11th International Council for Archaeozoology Conference*, pp. 33–9. Oxford: BAR International Series 2354.

L'Heureux, L. G. and Kaufmann, C. (2012b) 'Estimación de la edad de muerte de guanacos juveniles a partir de las dimensiones de los huesos largos no fusionados. Estructura de edad y estacionalidad en el campo volcánico de Pali Aike (Orejas de Burro 1)', *Magallania*, 40(2), 151–220.

Lobbia, P. A. (2012) 'Esclerocronología en valvas de *Mytilus* spp.: análisis del sitio CCH4 (Parque Nacional Monte León, Santa Cruz, Argentina) e implicaciones para la arqueología de Patagonia', *Magallania*, 40(2), 221–31.

Mancini, M. V., Prieto, A., Páez, M. M., and Schäbitz, F. (2008) 'Late Quaternary vegetation and climate of Patagonia', in Rabassa J. (ed.) *The Late Cenozoic of Patagonia and Tierra del Fuego*, pp. 351–68. Amsterdam: Elsevier.

Mansur, E., Lasa, A., and Vázquez, M. (2004) 'Investigaciones arqueológicas en Punta Bustamante, Santa Cruz: el sitio RUD01BK', in Civalero, T., Fernández, P., and Guráieb, A. G. (eds) *Contra viento y marea: arqueología de Patagonia*, pp. 755–74. Buenos Aires: INAPL.

Marschoff, M. (2007) *Gato por liebre. Prácticas alimenticias en Floridablanca*, Buenos Aires: Teseo.

Martin, F. M. (2008) 'Bone crunching felids at the end of the Pleistocene in Fuego-Patagonia, Chile', *Journal of Taphonomy*, 6(3–4), 337–72.

Martin, F. M. (2013) *Tafonomía y paleoecología de la transición Pleistoceno-Holoceno en Fuego-Patagonia: interacción entre humanos y carnívoros y su importancia como agentes en la formación del registro fósil*, Punta Arenas: Universidad de Magallanes.

Martinic, M. (1995) *Los aónikenk: historia y cultura*, Punta Arenas: Ediciones de la Universidad de Magallanes.

Martinic, M., Prieto, A., and Cárdenas, P. (1995) 'Hallazgo del asentamiento del Jefe Aónikenk Mulato en el valle del Zurdo. Una prueba de sedentarización indígena en el período histórico tardío', *Anales del Instituto de la Patagonia*, 23, 87–94.

Mazzoni, E. and Vázquez, M. (2010) 'Desertification in Patagonia', *Developments in Earth Surface Processes*, 13, 351–77.

McCulloch, R. D., Bentley, M. J., Tipping, R. M., and Clapperton, C. M. (2005) 'Evidence for late-glacial ice dammed lakes in the central Strait of Magellan and Bahía Inútil, southernmost South America', *Geografiska Annaler*, 87, 335–62.

Mena, F. (1992) 'Mandíbulas y maxilares: un primer acercamiento a los conjuntos arqueofaunísticos del alero Fontana', *Boletín del Museo Nacional de Historia Natural*, 43, 179–91.

Mena, F. and Jackson, D. (1991) 'Tecnología y subsistencia en Alero Entrada Baker, Región de Aisén, Chile', *Anales del Instituto de la Patagonia*, 20, 169–203.

Mengoni, G. (1999) *Cazadores de guanacos de la estepa Patagónica*, Buenos Aires: Sociedad Argentina de Antropología.

Mengoni, G. and De Nigris, M. (2005) 'The guanaco as a source of meat and fat in the southern Andes', in Mulville, J. and Outram, A. K. (eds) *The Zooarchaeology of Fats, Oils, Milk and Dairying*, pp. 160–6. Oxford: Oxbow Books.

Mengoni, G. and Silveira, M. J. (1976) 'Análisis e interpretación de los restos faunísticos de la Cueva de las Manos, Alto Río Pinturas', *Relaciones*, 10, 261–70.

Miotti, L. (1998) *Zooarqueologia de la meseta central y costa de Santa Cruz: un enfoque de las estrategias adaptativas aborigenes y los paleoambientes*, San Rafael: Museo de Historia Natural.

Morello, F., Torres, J., Martínez, I., Rodriguez, K., Arroyo-Kalin, M., French, C., Sierpe, V., and San Román, M. (2012) 'Arqueología de la Punta Santa Ana: reconstrucción de secuencias de ocupación de cazadores-recolectores marinos del Estrecho de Magallanes, Chile', *Magallania*, 40(2), 129–49.

Moreno, F. P. (1979) *Reminiscencias*, Buenos Aires: Eudeba.

Moreno, E. J. and Videla, B. A. (2008) 'Rastreando ausencias: la hipótesis del abandono del uso de los recursos marinos en el momento ecuestre en la Patagonia continental', *Magallania*, 36(2), 91–104.

Moreno, J. E. (2008) *Arqueología y etnohistoria de la costa patagónica central en el Holoceno tardío*, Rawson: Secretaría de Cultura del Chubut.

Muñoz, A. S. (1997) 'Explotación y procesamiento de ungulado en Patagonia meridional y Tierra del Fuego', *Anales del Instituto de la Patagonia*, 25, 201–22.

Muñoz, A. S. and Belardi, J. B. (1998) 'El marcado perimetral en los huesos largos de guanaco de Cañadón Leona (colección Junius Bird): implicaciones arqueofaunísticas para Patagonia meridional', *Anales del Instituto de la Patagonia*, 26, 107–18.

Nami, H. and Menegaz, A. N. (1991) 'Cueva del Medio: aportes para el conocimiento de la diversidad faunística hacia el Pleistoceno-Holoceno en Patagonia austral', *Anales del Instituto de la Patagonia*, 20, 117–32.

Nuevo Delaunay, A. (2012) 'Disarticulation of Aónikenk hunter-gatherer lifeways during the late nineteenth and early twentieth centuries: two case studies from Argentinean Patagonia', *Historical Archaeology*, 46(3), 149–64.

Orlando, L., Metcalf, J. L., Alberdi, M. T., Telles-Antunes, M., Bonjeane, D., Otte, M., Martin, F., Eisenmann, V., Mashkouri, M., Morello, F., Prado, J. L., Salas-Gismondil, R., Shockey, B. J., Wrinn, P. J., Vasil'ev, S. K., Ovodov, N. D., Cherry, M. I., Hopwood, B., Male, D., Austin, J. J., Hänni, C., and Cooper, A. (2009) 'Revising the recent evolutionary history of equids using ancient DNA', *Proceedings of the National Academy of Sciences*, 106(51), 21754–9.

Ortiz Troncoso, O. (1975) 'Los yacimientos de Punta Santa Ana y Bahía Buena. Excavaciones y fechados radiocarbónicos', *Anales del Instituto de la Patagonia*, 7, 93–122.

Otaola, C. and Franco, N. V. (2008) 'Procesamiento y consumo de guanaco en el sitio Chorrillo Malo 2', *Magallania*, 36(2), 205–19.

Poulain-Josien, T. (1963) 'La Grotte Fell: étude de la faune', *Journal de la Société des Américanistes*, 52, 230–54.

Prieto, A, (1991) 'Cazadores tempranos y tardíos en Cueva Lago Sofía 1', *Anales del Instituto de la Patagonia*, 20, 75–99.

Rindel, D. (2004) 'Patrones de procesamiento faunístico en el sitio Alero Destacamento Guardaparque durante el Holoceno tardío', in Civalero, M. T., Fernández, P. M., and Guráieb, A. G. (eds) *Contra viento y marea: arqueología de Patagonia*, pp. 263–76. Buenos Aires: INAPL-SAA.

Rindel, D., Martínez, C., and Dellepiane, J. M. (2011) 'Evidencias de procesamiento de guanaco en sitios a cielo abierto y aleros estratificados del noroeste de la provincia de Santa Cruz', in Acosta, A., Loponte, D., and Mucciolo, L. (eds) *Estudios tafonómicos y zooarqueológicos (II)*, pp. 107–36. Buenos Aires: Instituto Nacional de Antropología y Pensamiento Latinoamericano.

Rivals, F., Rindel, D., and Belardi, J. B. (2013) 'Dietary ecology of extant guanaco (*Lama guanicoe*) from southern Patagonia: seasonal leaf browsing and its archaeological implications', *Journal of Archaeological Science*, 40, 2971–80.

San Román, M. (2007) 'La explotación de mamíferos en el sitio de Bahía Buena: economía de canoeros tempranos de Patagonia (Estrecho de Magallanes, Chile)', in Morello, F., Prieto, A., Martinic, M., and Bahamonde, G. (eds) *Desenterrando huesos, recolectando piedras y develando arcanos ... arqueología de Fuego-Patagonia*, pp. 295–310. Punta Arenas: Ediciones CEQUA.

San Román, M. (2009) 'Anatomía económica de *Otaria flavescens* (Shaw, 1800)', in López, P., Cartajena, I., García, C., and Mena, F. (eds) *Zooarqueología y tafonomía en el confín del mundo*, pp. 169–79. Santiago de Chile: Universidad Internacional Sek-Chile.

San Román, M. (2010) 'La explotación de recursos faunísticos en el sitio Punta Santa Ana 1: estrategias de subsistencia de grupos de cazadores marinos tempranos de Patagonia meridional', *Magallania*, 38(1), 183–98.

Saxon, E. C. (1979) 'Natural prehistory: the archaeology of Fuego-Patagonian ecology', *Quaternaria*, 21, 329–56.

Scartascini, F. L. and Volpedo, A. V. (2012) 'White croaker (*Micropogonias furnieri*) paleodistribution in the southwestern Atlantic Ocean. An archaeological perspective', *Journal of Archaeological Science*, 40(2), 1059–66.

Senatore, M. X., Bianchi Villelli, M., Buscaglia, S., Marschoff, M., Nuviala, V., Bosoni, C., and Starópoli, L. (2009) 'Arqueología en Floridablanca: historias de una colonia española en la costa patagónica a fines del siglo XVIII', in Herbst, R. (ed.) *Estado actual de las investigaciones realizadas sobre patrimonio cultural en Santa Cruz*, pp. 207–12. Río Gallegos: Dirección de Patrimonio Cultural, Subsecretaría de Cultura de Santa Cruz.

Suby, J., Guichón, R. A., and Senatore, M. X. (2009) 'Los restos oseos humanos de Nombre de Jesús: evidencias de la salud en el primer asentamiento europeo en Patagonia austral' *Magallania*, 37(2), 23–40.

Tivoli, A. M. (2010) 'Las aves en la organización socioeconómica de cazadores-recolectores-pescadores del extremo sur sudamericano'. Unpublished PhD dissertation, Universidad de Buenos Aires (Buenos Aires).

Torres, J. and Ruz, J. (2011) 'Datos preliminares de la modalidad de pesca en la tradición cultural Englefield (ca. 6000 años) en la zona del Estrecho de Magallanes y mar de Otway, Chile', *Magallania*, 39(2), 165–76.

Zangrando, A. F. J. (2003) *Ictioarqueología del Canal Beagle: explotación de peces y su implicación en la subsistencia humana*, Buenos Aires: Sociedad Argentina de Antropología.

PART VII

OCEANIA

CHAPTER 44

··

THEMES IN THE ZOOARCHAEOLOGY OF PLEISTOCENE MELANESIA

··

MATTHEW LEAVESLEY

INTRODUCTION

··

NOTIONS of human 'adaptation' are key to any consideration of Pleistocene human dispersals. They allow archaeologists to conceive of how hunter-gatherers were able to occupy new environments. Adaptation is usually thought of in terms of land-use and mobility. In regions such as the Bismarck Archipelago, where the archaeological record includes large faunal assemblages, it is possible to investigate human adaptation by analysing the human selection of prey within the context of local biodiversity and cultural ecology. One form of adaptation is animal translocation. An influential paper by Grayson (2001) brought together evidence from across the globe indicating the fundamental importance of animal translocation to human dispersals. This chapter describes the impact of animal translocations on hunter-gatherer life ways in terms of resource-use, land-use, and mobility.

Melanesia consists of an area ranging from New Guinea in the west to Vanuatu in the east. New Guinea, or the northern end of Sahul (a former continent stretching from Tasmania to New Guinea), was first colonized by anatomically modern human hunter-gatherers around 50,000 years ago (Summerhayes et al., 2010); further dispersals followed both southwards (see Cosgrove and Garvey, p. 000 this volume) and eastwards into the Bismarck Archipelago by c.40,000 BP (Pavlides and Gosden, 1994; Leavesley et al., 2002; Leavesley and Chappell, 2004; Torrence et al., 2004), Bougainville by c.30,000 BP (Wickler and Spriggs, 1988), and Manus perhaps by 25,000 BP (Fredericksen et al., 1993; Minol, 2000). The case study presented in this chapter is from the Buang Merabak cave site, in central New Ireland (Bismarck Archipelago) (Fig. 44.1).

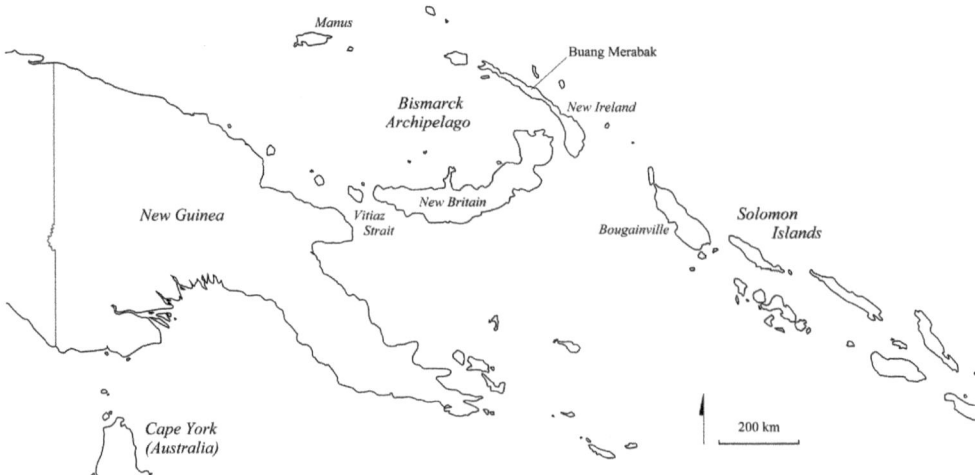

FIGURE 44.1 Map of northern Melanesia with location of Buang Merabak (New Ireland). Editors' own image.

By the time the first humans arrived in Sahul, the Bismarck Archipelago had already separated from mainland New Guinea (Vitiaz Strait), and New Ireland from New Britain (St Georges Channel). Global cycles of glaciation are thought to have had a local impact of lowering sea levels; during the last glacial maximum, this was in the order of 125 m (Thom and Chappell, 1977; Lambeck and Chappell, 2001). The Bismarck Archipelago submarine topography is characterized by relatively steep coasts and 1,000+ m ocean troughs (Marlow et al., 1988), the result being that sea-crossings remained relatively constant for the duration of human occupation.

In times past, anthropologists conceived of islands as relatively insulated and isolated microcosms of the world. While this model no longer strictly holds true for humans, it is still partially true for animals: indeed, island environments have relatively depauperated faunas compared to continental landscapes. In this case, when humans first crossed from New Guinea into the Bismarck Archipelago, New Ireland had (and still has) a relatively similar flora, but a dramatically reduced suite of fauna. New Guinea is known to have two anteaters, four tree kangaroos, seven bandicoots, nine wallabies, twenty-seven phalangers, and 265 bird species, while New Ireland is known to have one bandicoot, one wallaby, two phalangers, and eighty bird species (Allen, 2000). This provides a very useful opportunity to investigate how human hunter-gatherers adapted to reduced protein availability in terms of landscape use and mobility.

BUANG MERABAK—THE SITE

The Buang Merabak cave site is located in central New Ireland on land traditionally owned and occupied by Mandak-speaking people. The site is *c.*200 m inland of the current coastal zone. The submarine topography in front of the site is consistent with much of New Ireland in that it is relatively steep; on this basis it is highly likely that the site was never far from the coast, even during the last glacial maximum. The cave itself is situated at the bottom of a series of fourteen uplifted limestone terraces that rise $c.1,480$ m asl to the Lelet Plateau (Hohnen, 1978). The cave is a collapsed doline with three chambers, the deepest of which is periodically occupied by a colony of cave-dwelling bats. The second (or middle) chamber was used for ritual activities associated with rain-making and also contains some rock art of the stencil variety. The first chamber has a large mouth measuring $c.15$ m wide and 8 m high. Evidence still remains of its use during WWII as a hide-out. The excavations that provided the data presented and discussed here were undertaken in the first chamber within the light zone of the cave mouth.

There were two series of excavations undertaken at the site of which some data has been reported elsewhere (Balean, 1989; Rosenfeld, 1997; Leavesley and Allen, 1998). The cave floor area has been calculated at $c.450$ m^2, of which 1.5 m^2 was excavated in 1985–1986 and an additional 2 m^2 in 2000–2001. Only data from the latter excavations are reported here. The site was excavated in 5 cm spits following the stratigraphy as determined from the earlier excavations. The sediment was sieved through 3 mm mesh and all material was collected for analysis in the laboratory.

The assemblage is presented and interpreted within four chrono-stratigraphic units (Table 44.1). The calibrations presented in Table 44.1 are derived from Calib 4.3 (Stuiver and Reimer, 1993).

FISH AS THE ARCHETYPICAL ISLAND PROTEIN RESOURCE

When exploring notions of island adaptation, the potential exploitation of fish often comes to mind. Presumably the invocation of fish as a protein source comes from ethnography and contemporary observation by archaeologists themselves. While it is true to say that fish are an important part of today's island economy, there is (perhaps surprisingly) relatively little evidence of fish in the archaeological record until Holocene times. Indeed, only one pelagic fish element was recovered from Buang Merabak (Schmidt, 2004), in addition to a reef shark tooth (Kemp, 2004). While Pleistocene shell-fish midden remains are highly visible in the New Ireland archaeological record (Balean, 1989; Gosden and Robertson, 1991; Rosenfeld, 1997), it was proved that in terms of calorific value their contribution is relatively small. Reef fish-bones are also present in relatively small amounts, while pelagic fish are present in even smaller quantities (Schmidt, 2004). The low levels of pelagic fish in the New Ireland Pleistocene assemblages is even more intriguing in light of a Pleistocene fish-hook reported from East Timor (O'Connor et al., 2011), suggesting that fish-hook technology was not beyond the means of the modern humans who colonized and inhabited New Ireland. In summary, while we do have evidence of coastal occupation

Table 44.1 Chrono-stratigraphic units from Buang Merabak (New Ireland).
Calibrations are derived from Calib 4.3 (Stuiver and Reimer, 1993)

	Spits	cm below the surface	Uncalibrated dates bp	Calibrated dates bp
Unit 1	1–8	0–40	3500–1800	3300–1300
Unit 2	9–17	40–95	12,000–7000	13,150–8200
Unit 3	18–27	85–135	20,000–17,000	23,050–19,650
Unit 4	28–40	135–200	39,590–27,000	—

and exploitation (at least of the littoral zone), we see little evidence of a systematic exploitation of deep water/pelagic fish during the Pleistocene.

ANIMAL TRANSLOCATION

The movement by people of wild animals into environments that the animals themselves would not naturally occupy is a relatively common practice across the globe (Grayson, 2001). While this is not uncommon amongst later horticulturalists, it is relatively rare for Pleistocene hunter-gatherers to translocate wild animals. The Bismarck Archipelago has seen a number of animal translocations; those for New Ireland include species such as the northern common cuscus (*Phalanger orientalis*) (*c.*20,000 BP), one wallaby species—the Brown's pademelon (*Thylogale browni*) during the early Holocene, the Polynesian rat (*Rattus exulans*) with the Lapita migration (3300 cal BP) and the common spotted cuscus (*Spilocuscus maculatus*, a slightly larger cuscus species) in more recent times (Flannery and White, 1991; Heinsohn, 2001). The introduction of the northern common cuscus is here hypothesized to have had a major impact on Pleistocene hunter-gatherer lifeways, for the reasons outlined later.

PREY SELECTION (RESOURCE USE) AT BUANG MERABAK

Given the relatively narrow range of potential preys available in New Ireland 40,000 years ago, those selected by humans for consumption may be predictable. The earliest human occupation Unit (4) at Buang Merabak predominantly contains evidence of the consumption of the Andersen's bare-backed fruit bat (*Dobsonia anderseni*), a cave-dwelling bat, and of the mangrove monitor (Varanidae). In addition to abundant shell-fish throughout the entire assemblage, this unit also contains minor contributions from various microchiroptera and from the eastern rat (*Rattus mordax*). This unit also contains a single perforated tiger shark (*Galeocerdo cuvier*) tooth (Leavesley, 2007) and a single pelagic fish-bone belonging to the family of Thunnidae (Schmidt, 2004).

As with the previous period, Unit 3 is dominated by the Andersen's bare-backed fruit bat and by Varanids. For the first time at this site, evidence of the consumption of the northern common cuscus occurs. Other minor contributions to the assemblage are represented by microchiroptera and rats such as the large New Guinea spiny rat (*Rattus praetor*), while the eastern rat present in Unit 4 is thought to have gone extinct by this time.

Unit 2 continues to exhibit evidence of the consumption of the Andersen's bare-backed fruit bat and of Varanids, but it is dominated by the consumption of the northern common cuscus. The wallaby (*Thylogale browni*) appears in the assemblage for the first time and, like the cuscus, is also thought to have arrived in New Ireland by human translocation.

The most recent period of human use of Buang Merabak, Unit 1, post-dates the introduction of horticulture with Lapita migrants (Kirch, 1997; Summerhayes, 2001), and is therefore less relevant to this discussion. However, it continues to be dominated by cuscus bones.

Essentially, when the first hunter-gatherers entered New Ireland, they focused on the locally available cave-roosting bats and lizards. After its introduction, the cuscus became the dominant protein resource in the human diet.

These taxa have a range of specific ecological and behavioural traits to which human hunters have had to adapt in order to exploit them for food. These traits, in turn, inform notions of Pleistocene hunter-gatherers land use and mobility.

PREY SPECIES ECOLOGY AND HUNTING ETHNOGRAPHY

The Andersen's bare-backed fruit bat roosts in relatively large colonies within caves by day and forages throughout the forest during the night (Flannery, 1995; Bonaccorso, 1998). The northern common cuscus is a herbivore that spends the day either in tree hollows or thick foliage high up in the canopy and forages during the night. Its range is limited only by the extent of the rainforest trees themselves that, prior to human occupation, will have covered the entire length of New Ireland (Flannery, 1995). Varanids are known to inhabit the entire lowland zone including notional inland forests, but are more abundant in the coastal mangroves that stretch along much of the New Ireland coast (although these are more patchy on the west coast, owing in part to the steeper gradient of the beach and higher energy wave action). The Brown's pademelon occupies the forest floor throughout New Ireland and is therefore only limited by the land area of New Ireland (Flannery, 1995).

Hunting ethnography indicates that hunters can track and catch the dominant preys represented at Buang Merabak with relatively low technology and allows reconstructing models of protein acquisition during the Pleistocene.

Cuscus tend to be creatures of habit in that they will often use the same routes through their respective territories, leaving tell-tale signs such as subtle pathways marked by scats and scratch marks on trees and branches. Hunting cuscus is basically a two-step activity. First, the hunters will undertake reconnaissance of the bush by day, to identify the signs of recent cuscus activity. They will then return in the evening and lay in wait with bow and arrow or spear for the cuscus to return. Latinis (1996) described a situation whereby the hunters took him to a tree hole which had previously been identified as a cuscus nest. The hunter essentially put his arm in the hole retrieving a cuscus by the tail and upon extracting it, simply dashed its head against the tree.

LAND USE

Prey selection can very usefully inform us about land use. The guiding principle is that hunter-gatherers must have sought out their prey in the specific ecological niche inhabited by the prey itself.

In terms of protein acquisition and due to the general lack of alternative prey sources in New Ireland, the evidence from Buang Merabak suggests a specific emphasis in terms of land use during the earliest unit (40,000 BP to 27,000 BP), when people focused on the consumption of cave-dwelling bats with a relatively minor contribution to the diet from Varanids. The selective predation of the latter suggests periodic focus on the lowlands if not the mangroves themselves, while the former suggests a focus on caves.

Tropical rainforest fauna, more commonly than not, forages for food as isolated individuals, as opposed to the grazing herds of grassland plain taxa. However, on some occasions, tropical rainforest fauna does exhibit degrees of gregarious behaviour. For example, a female cuscus will live with her juvenile offspring in a tree hollow. The rest of the time it will remain alone. Some species of bats are an important exception to this mode of behaviour: while the Andersen's bare-backed fruit bat and other larger Pteropodidae forage as individuals, conventionally they roost in colonies. The significance of this roosting behaviour lies in the fact that, as a colony, they might be conceived of as the largest food resource in a given locality.

The apparent emphasis on cave bats is interesting given that larger bats (*Pteropus* sp.) were also extant in New Ireland during this time. Unlike *Pteropus* bats, which roost in large colonies in tree tops, the Andersen's bare-backed fruit bat roosts in large colonies in caves. The different skills required to capture cave-dwelling bats as opposed to tree-roosting bats may be instructive in conceiving of why the former were the preferred prey at this time and this, in turn, reveals pattern of land use and mobility.

Bat-hunting ethnography can usefully inform notions of Pleistocene bat hunting by providing a model of how this activity might be achieved with minimal technology

(Bulmer, 1968). An example from the Kalam country illustrates the process quite nicely. Once a colony of cave-roosting bats is identified, the hunters, often in small groups, will reconnoitre the cave in order to identify all the exit points. All but one of the exit points are covered with barriers of sticks and leaves with the express purpose of trapping the entire colony inside. The single exit point which is left open may be partially covered providing an opportunity for the bats to escape. Once all the barriers are set, an individual will be nominated to carry a torch made of burning/smoking leaves into the back of the cave. As the cave fills with smoke the bats become distressed and fly around in the dark looking for an exit point. Once they identify the partially closed exit they seek to depart the cave only to be struck by a man waving a stick with the intention of maiming them. This results in a pile of screeching bats in various states of decline flapping around on the ground, which are finally dispatched by the hunters once they have accumulated a viable quantity of bats for a meal. An important side effect of this strategy is that those bats that have somehow escaped are less likely to return to the roost and therefore each roost is unlikely to sustain more than one hunting event at a time. Ultimately, the law of diminishing returns requires hunters to move on from the cave fairly rapidly and not return until a given colony has had the chance to recover its population numbers.

In contrast to this large return, although tree-dwelling bats tend to be larger on average than cave-dwelling bats, they are much harder to corral, and therefore the ultimate returns are lower.

There are two important things to note from the bat hunting ethnography. First, the species found at Buang Merabak and other Pleistocene sites (described below) can be captured in slightly different ways using minimal technology. Secondly, of the medium- to large-size bats, those that roost in colonies in caves are the easiest to catch in large numbers and, therefore, are the most economical protein resource for human consumption.

The distribution of caves across New Ireland can be inferred by reference to both topography and geology. New Ireland is long and narrow. At the north is a large bay (Bagail Bay) and associated swamp. Moving to the south is a spine of hills that rise to mountains in central New Ireland, where the Lelet Plateau is situated at 1,480 m asl. Further south the mountain spine dips before peaking above the Weitin River (which follows the Weitin fault-line as it cuts across the southern end of New Ireland) at 2,000+ m asl. Additionally, and in general, the southern half of New Ireland consists of upraised volcanic rocks, while the northern half is made of uplifted limestone (Hohnen, 1978). The geochemical nature of these rock types and the nature of the topography suggest that central New Ireland is highly likely to have a greater density of caves conducive to cave-dwelling bat occupancy. Indeed, many thousands of caves are likely to be present in central New Ireland (Bourke and Gallasch, 1974: 202; Wilde, 1975), the region within which Buang Merabak is located.

Unit 3 (23,050–19,650 cal BP) and following units exhibit an increase in the predation of the northern common cuscus. This species signals a significant alteration in land use. The cuscus currently occupies the forests which cover the length and breadth of New Ireland. Although cuscus generally live as isolated individuals and therefore in low densities across a given landscape (Montague, 2000), they represent a protein source that is much more widely and evenly dispersed across New Ireland than the cave-roosting bats. Essentially, roosting bat colonies represented high-value nodes of protein across the landscape, while cuscus represented a low-density but more evenly distributed food resource across the same landscape.

During the time period represented by Unit 3, hunters were no longer primarily dependent on caves with roosting bat colonies. With the introduction of the cuscus, they had an alternative protein resource that was available throughout the island.

Landscape use changed as protein availability and preferences changed. In Unit 4, the caves represented high-density protein resource patches within vast forests with relatively little accessible protein. In Unit 3, cave-roosting bat colonies still represented a high-density resource patch, but now the rainforests also contained a relatively even distribution of protein sources in the form of cuscus populations. At least in terms of protein acquisition, landscape use altered from Unit 4, during which caves were paramount for protein acquisition, to a situation where protein could be acquired across the entire rainforest. Hunting patterns changed from the exploitation of these high-density (protein) nodes (caves) to an increased exploitation of a low-density protein source (cuscus within the forests). Dependence on the high-density nodes was reduced and replaced with exploitation of a lower-density resource, thereby allowing landscape use to become more dispersed. The trend towards selection for the cuscus only gets stronger into the second period represented at Buang Merabak (Unit 2, 13,150–8200 cal BP).

The data presented here suggest that the introduction (through animal translocation) of the northern common cuscus, and the subsequent preferential selection of it by hunters, had a significant impact on Pleistocene and Early Holocene land use in New Ireland.

Implications for Mobility

Prey selection can very usefully inform hunter-gatherer mobility by taking into consideration the ecology of the prey themselves.

In the context of this study, mobility is conceived as the relative rate of movement across the landscape. Mobility can be inferred as a product of the dispersal of the respective prey species. It can be interpreted as changing over time if the preferred prey species (as indicated by the relative proportion of each species in a given assemblage) also changes. In the case of New Ireland, as based on the faunal assemblage from Buang Merabak, there is a trend from greater mobility to lesser mobility. Equally at the level of interpretation specific to the site the trend is reversed.

Considering the Buang Merabak assemblage, in the early phase it can be hypothesized that, in relation to protein acquisition, hunters moved relatively quickly through the forests from cave system to cave system in search of a viable bat colony to exploit. Once a satisfactorily large colony was identified the hunters would stay until the law of diminishing returns meant that pursuing bats at a given cave was no longer viable, with the consequence that they would necessarily have to move. Given the nature of hunting cave-dwelling bat colonies, any given colony might only be viable for a few days as those individuals that were not caught during the first hunt were increasingly less likely to return in large numbers. Relatively little emphasis was placed on searching the forests, because they contained relatively few consistently viable prey species. After the introduction into New Ireland of the cuscus, in tandem with the human preference for them, mobility through the landscape slowed down as the forests themselves contained a viable protein source of its own, namely the cuscus itself. In effect, the presence of low-density protein resources did not require such rapid movement to the next protein resource node (i.e. a cave site).

At a site-specific level, and in Buang Merabak in particular, mobility patterns also changed. As the faunal assemblage indicates, the site continued to be used for protein consumption after the introduction of the cuscus. It appears that, having added cuscus to the menu, the human hunters now had two separate sources of protein to exploit in the proximity of Buang Merabak. Cave-roosting bats were still on the menu and their consumption remains important throughout the use of the site into the Holocene. The appearance of cuscus indicates that they too were exploited in the adjacent forests and at least some were brought back to Buang Merabak for consumption. This site-use changed from initially focusing on the hunting and consumption of bats, to a situation in which the consumption of cuscus was also important. As described above, the hunting of cuscus requires some knowledge of local conditions and, therefore, at least some time to reconnoitre the local forests. In terms of site-use itself, it changed from relatively shorter stays (bat hunting and consumption) with little exploration of the adjacent rain-forests, to relatively longer stays at the site to facilitate more intensive exploitation of the adjacent forests (i.e. hunting cuscus). In essence, it is highly likely that, with the addition of the cuscus to the diet, mobility through the landscape reduced, occupations of cave sites such as Buang Merabak became longer, and the exploitation of adjacent forests intensified.

A Model for Site-Use at Buang Merabak

From the beginning of the human colonization of New Ireland, the mouth of the Buang Merabak site was used for consumption of protein in addition to other activities such as tool-making. During the Early period (40,000 BP to 27,000 BP), the back of the cave was also used periodically for hunting the cave-roosting bats that formed a colony there. Other protein, such as shell-fish, was brought up from the nearby littoral zone, while Varanids were hunted in the nearby lowland forest and mangroves.

After the introduction of the cuscus, the use of the site changed to some extent. Hunting cave-roosting bats remained important, but their contribution to the total protein intake diminished. Hunting the cuscus that resided in the surrounding forest and bringing the catch back to the site for consumption became customary, and in the early Holocene the remains from this species dominate the assemblage. Buang Merabak came to resemble something much more akin to a classic base-camp site, from where hunters departed to exploit the local forests and bring their preys back for consumption.

Conclusion

While islands are no longer conceived of as insulated laboratories, they are still extremely useful to test notions of adaptation amongst Pleistocene hunter-gatherers. When Pleistocene hunter-gatherers crossed into the Bismarck Archipelago and arrived in New Ireland, they adapted to an island that was relatively depauperated in terms of terrestrial protein resources compared to where they came from. In time, they sought to bolster the available protein resources by introducing the northern common cuscus into the region, with important implications for land use and mobility.

Acknowledgements

I wish to acknowledge Glenn Summerhayes (Otago), Matthew Spriggs (ANU), Peter Hiscock (Uni of Syd), and Herman Mandui (NMAG) for their support over the life of this project. The late Michael Boxos (Chief at Konogusngus) and Tuvu Telexas (the owner of Buang Merabak) gave their permission to undertake research at Buang Merabak, while Jim Robins (NRI) and the New Ireland Provincial Government granted permission to enter the Province. Sean Ulm (JCU), and various editors and reviewers read versions of this chapter prior to publication.

References

Allen, J. (2000) 'From beach to beach: the development of maritime economies in pre-historic Melanesia', in O'Connor, S. and Veth, P. (eds) *East of Wallace's Line: Studies of Past and Present Maritime Cultures of the Indo-Pacific Region*, pp. 139–76. Rotterdam: A. A. Balkema.

Balean, C. (1989) 'Caves as Refuge Sites: An Analysis of the Shell Material from Buang Merabak, New Ireland'. Unpublished BA dissertation, Australian National University (Canberra).

Bonaccorso, F. J. (1998) *Bats of Papua New Guinea*, Washington: Conservation International.

Bourke, R. M. and Gallasch, H. (1974) 'Caves of the New Ireland District', *Nuigini Caver*, 2(3), 193–204.

Bulmer, R. (1968) 'The strategies of hunting in New Guinea', *Oceania*, 38, 302–18.

Flannery, T. F. (1995) *Mammals of the South-West Pacific and Moluccan Islands*, Chatswood: Australian Museum/Reed Books.

Flannery, T. F. and White, J. P. (1991) 'Animal translocation', *National Geographic Research and Exploration*, 7(1), 96–113.

Fredericksen, C. F. K., Spriggs, M. J. T., and Ambrose, W. (1993) 'Pamwak Rockshelter: a Pleistocene site on Manus Island, Papua New Guinea', in Smith, M. A., Spriggs, M. J. T., and Fankhauser, B. (eds) *Sahul in Review: Pleistocene Archaeology in Australia, New Guinea and Island Melanesia*. Occasional Papers in Prehistory 24, pp. 131–52. Canberra: Department of Prehistory, RSPacS, Australian National University.

Gosden, C. and Robertson, N. (1991) 'Models for Matenkupkum: interpreting a Late Pleistocene site from southern New Ireland, Papua New Guinea', in Allen, J. and Gosden, C. (eds) *Report of the Lapita Homeland Project*. Occasional Papers in Prehistory 20, pp. 20–45. Canberra: Department of Prehistory, Australian National University.

Grayson, D. K. (2001) 'The archaeological record of human impacts on animal populations', *Journal of World Prehistory*, 15(1), 1–68.

Heinsohn, T. E. (2001) 'Human influences on vertebrate zoogeography: animal translocation and biological invasions across and to the east of Wallace's Line', in Metcalfe, I., Smith, J. M. B., Morwood, M., and Davidson, I. (eds) *Faunal and Floral Migrations and Evolution in SE Asia-Australasia*, pp. 154–70. Lisse: A. A. Balkema.

Hohnen, P. D. (1978) *Geology of New Ireland, Papua New Guinea*. Bulletin 194. Canberra: Bureau of Mineral Resources, Geology and Geophysics.

Kemp, N. R. (2004) 'Appendix 7: Buang Merabak Shark Teeth', in Leavesley, M. (ed.) 'Trees to the Sky: Prehistoric Hunting in New Ireland, Papua New Guinea', pp. 240–2. Unpublished PhD dissertation, Australian National University (Canberra).

Kirch, P. V. (1997) *The Lapita Peoples: Ancestors of the Oceanic World*, Oxford: Blackwell.

Lambeck, K. and Chappell, J. (2001) 'Sea level change through the Last Glacial cycle', *Science*, 292, 679–86.

Latinis, K. (1996) 'Hunting the cuscus in Seram: the role of the phalanger in subsistence economies in central Maluku', *Cakalele*, 7, 17–32.

Leavesley, M. (2007) 'A shark-tooth ornament from Pleistocene Sahul', *Antiquity*, 81, 308–15.

Leavesley, M. and Allen, J. (1998) 'Dates, disturbance and artefact distributions: another analysis of Buang Merabak, a Pleistocene site on New Ireland, Papua New Guinea', *Archaeology in Oceania*, 33, 68–82.

Leavesley, M. and Chappell, J. (2004) 'Buang Merabak: additional early radiocarbon evidence of the colonisation of the Bismarck Archipelago, Papua New Guinea', *Antiquity*, 78(301), http://www.antiquity.ac.uk/projgall/leavesley/index.html.

Leavesley, M., Bird, M. I., Fifield, L. K., Hausladen, P. A., Santos, G. M., and di Tada, M. I. (2002) 'Buang Merabak: early evidence of human occupation in the Bismarck Archipelago, Papua New Guinea', *Australian Archaeology*, 54, 55–7.

Marlow, M. S., Dadisman, S. V., and Exon, N. F. (eds) (1988) *Geology and Offshore Resources of Pacific Island Arcs—New Ireland and Manus Region*. Earth Science Series 9. Houston: Circum-Pacific Council for Energy and Mineral Resources.

Minol, B. (2000) *Manus from the Legends to Year 2000*, Port Moresby: University of Papua New Guinea Press.

Montague, T. L. (ed.) (2000) *The Brushtail Possum: Biology, Impact and Management of an Introduced Marsupial*, Lincoln: Manaaki Whenua Press.

O'Connor, S., Ono, R., and Clarkson, C. (2011) 'Pelagic fishing at 42,000 years before present and the maritime skills of modern humans', *Science*, 334, 1117–21.

Pavlides, C. and Gosden, C. (1994) '35,000 year old sites in the rainforests of west New Britain, Papua New Guinea', *Antiquity*, 68, 604–10.

Rosenfeld, A. (1997) 'Excavations at Buang Merabak, central New Ireland', *Bulletin of the Indo-Pacific Prehistory Association*, 16, 213–24.

Schmidt, L. (2004) 'Appendix 6: Buang Merabak fish', in Leavesley, M. (ed.) 'Trees to the Sky: Prehistoric Hunting in New Ireland, Papua New Guinea', pp. 237–9. Unpublished PhD dissertation, Australian National University (Canberra).

Stuiver, M. and Reimer, P. J. (1993) 'Extended 14C database and revised CALIB 3.0 14C age calibration program', in Stuiver, M., Long, A., and Kra, R. S. (eds) Calibration 1993. *Radiocarbon*, 35(1), 215–30.

Summerhayes, G. R. (2001) *Lapita Interactions*. Terra Australis 15. Canberra: Archaeology and Natural History, RSPAS, Australian National University.

Summerhayes, G. R., Leavesley, M., Fairbairn, A., Mandui, H., Field, J., Ford, A., and Fullagar, R. (2010) 'Human adaptation and plant use in highland New Guinea 49,000 to 44,000 years ago', *Science*, 330, 78–81.

Thom, B. G. and Chappell, J. (1977) 'Sea levels and coasts', in Allen, J., Golson, J., and Jones, R. (eds) *Sunda and Sahul: Prehistoric Studies in Southeast Asia, Melanesia and Australia*, pp. 275–92. London: Academic Press.

Torrence, R., Neall, V., Doelman, T., Rhodes, E., McKee, C., Davies, H., Bonetti, R., Guglielmetti, A., Manzoni, A., Oddone, M., Parr, J., and Wallace, C. (2004) 'Pleistocene colonization of the Bismarck Archipelago: new evidence from west New Britain', *Archaeology in Oceania*, 39, 101–30.

Wickler, S. and Spriggs, M. J. T. (1988) 'Pleistocene human occupation of the Solomon Islands, Melanesia', *Antiquity*, 62, 703–6.

Wilde, K. A. (1975) 'More caves from the Lelet Plateau—New Ireland', *Nuigini Caver*, 3(1), 6–12.

CHAPTER 45

....

BEHAVIOURAL INFERENCES FROM LATE PLEISTOCENE ABORIGINAL AUSTRALIA

seasonality, butchery, and nutrition in southwest Tasmania

....

RICHARD COSGROVE AND JILLIAN GARVEY

INTRODUCTION

....

FEW studies have been made of the seasonal and nutritional aspects of terrestrial Australian Aboriginal prey species, or the butchery patterns and economic utility patterns compared to the work of European and North American archaeologists and zoo-archaeologists. Those Australian studies that have tackled this issue predominately focused on marine and fresh-water taxa such as shell-fish and fish (Bird et al., 2002; Faulkner, 2013; Garvey, 2016).

Here we focus on the use of marsupials in past Aboriginal diets, although it is recognized that animals such as reptiles, birds, and fish have also played a significant role in the subsistence of Indigenous Australians. The marsupial fauna of Australia have quite different behaviours from the placental, seasonal breeding mammals of the Northern Hemisphere. They are often solitary or live in small groups; they do not migrate over large distances and their breeding patterns are strongly mediated by the availability of biomass containing essential moisture. One significant problem for Australian zooarchaeologists is establishing predictable seasonal skeleto-chronological responses, such as development of teeth annuli, age-related bone fusion, and tooth eruption in prey species that live in arid and semi-arid zones that do not have such regular seasonal cycles.

Macropod populations of the semi-arid and arid zones of Australia increase when unpredictable and highly variable moisture is available or when local or regional biomass increases after rains (Caughley et al., 1987). However, the population recovery is rarely predictable since macropod population numbers are strongly influenced by previous grazing pressure on biomass, which is the result itself of either historically low or high kangaroo populations (Caughley et al., 1987: 170). It is difficult to distinguish in their archaeological bones to what extent resulting physiological changes are due to seasonal human hunting, moisture variability, or biomass availability. In these cases, establishing modern comparative seasonal data is very difficult, and a major limitation to extending these sorts of studies beyond the temperate climatic zones of Australia, which make up only c.20% of the continent. Identifying suitable skeleto-chronological methods for studying modern and fossil bone assemblages to reconstruct Aboriginal subsistence behaviour will be a challenge, and some studies may be limited in the questions they can ask, as correlations between bone physiology and seasonal change are not as straightforward as those developed for the Northern Hemisphere.

Previous Work

There has been detailed investigation into the role that fat and protein play in human selection of prey body parts. Such studies focus predominantly on placental species, specifically ungulates, in North America, Europe, and Africa (Binford, 1978; Jones and Metcalfe, 1988; Metcalfe and Jones, 1988; Brink, 1997; Morin, 2007). These analyses relied on modern studies of physiological changes in bone marrow of animals in response to seasonal fluctuations (Irving et al., 1957; Meng et al., 1969; West and Shaw, 1975; Turner, 1979). This research has provided significant contemporary baseline data for comparison with archaeological bone. Economic utility or butchery experiments also provide estimates of the relative food values represented by different skeletal elements. This enables archaeologists to predict the likelihood of specific body parts being selected and transported, and such analyses have also been used to infer site use (Binford, 1978; Kooyman, 1984; Blumenschine and Caro, 1986; Metcalfe and Jones, 1988; Lyman, 1992; Diab, 1998; Outram and Rowley-Conwy, 1998).

In Australia there has been very little research on the butchery, nutrition, and seasonality of human prey animals. This is probably due to the apparent lack of well-preserved faunal assemblages of sufficient sample size; at the same time, very few zooarchaeologists specialize in endemic terrestrial Australian fauna. In addition, only a handful of modern zoological studies include ecological and physiological investigations. This contrasts with analyses of Northern Hemisphere placental ungulates to detect variability in human subsistence and mobility during the Middle and Upper Palaeolithic (Pike-Tay, 1991; 1993; 2000; Burke, 1993; 1995; Burke and Pike-Tay, 1997).

The majority of nutritional research on Australian mammals has focused on their potential role in modern diets (O'Dea and Spargo, 1982; O'Dea, 1984; 1988; 1991; White, 1985; 1989; 1990; 2001; Naughton et al., 1986; Sinclair et al., 1987; O'Dea et al., 1988; Sinclair, 1988). There has also been a focus on the marketability of native animals for the commercial game meat industry (Hopwood et al., 1976; Hopwood, 1981; 1988; 1999; Hopwood and Griffiths, 1984), and the medicinal and cosmetic properties of Australian emu oil (Whitehouse et al., 1998; Yoganathan et al., 2003; Bennett et al., 2008; Howarth et al., 2008). With the exception of recent studies focused primarily on the Bennett's wallaby (*Macropus rufogriseus*) (Garvey, 2011), Common wombat (*Vombatus ursinus*) (Garvey et al. 2016) and on the Australian emu (*Dromaius novaehollandiae*) (Garvey et al., 2011), there have been no studies combining modern nutritional analyses with patterns identified in the Australian zooarchaeological record.

Steps to rectifying this were undertaken by Pike-Tay and Cosgrove (2002). Their pilot analysis on the skeleto-chronology of Bennett's wallaby teeth annuli was the first to compare modern and archaeological bones. Its application to the Late Pleistocene record of southwest Tasmania threw new light on Ice Age human seasonal occupation of this area (Cosgrove and Pike-Tay, 2004; Pike-Tay et al., 2008).

Tasmania

Tasmania is one of the few places in Australia that has unambiguous evidence of systematic terrestrial human prey selection and processing during the Late Pleistocene. Research has focused on the question of differential animal butchery between sites, based on climatic, chronostratigraphic, and seasonal criteria (Cosgrove et al., 1990; Cosgrove and Allen, 2001; Pike-Tay and Cosgrove, 2002; Garvey, 2006; Pike-Tay et al., 2008).

The Late Pleistocene archaeology of southwest Tasmania is renowned for its faunal and lithic richness, deposited in limestone caves by Aboriginal people between *c.*40,000 BP and *c.*13,000 BP (Cosgrove and Allen, 2001). All sites indicate a preferential selection of two prey species, the Bennett's wallaby (*Macropus rufogriseus)* and Common or Bare-nosed wombat (*Vombatus ursinus*). Other minor but important species, such as Tasmanian pademelon (*Thylogale billardierii*), platypus (*Ornithorhynchus anatinus*), Tasmanian emu (*Dromaius novaehollandiae diemenensis*), Tasmanian native hen (*Gallinula mortierii*), and Brush-tail possum (*Trichosurus vulpecula*) were exploited, but no extinct (megafauna) fauna above 60 kg have been recorded (Cosgrove et al. 2010).

Species Composition

The total Number of Identified Specimens (NISP) for all fauna in the seven cave sites is shown in Table 45.1.

At least twenty-nine different taxa have been identified from all the cave sites. They comprise small- and medium-sized carnivores (dasyurids), possums, flightless birds (emu, native hen), macropods, native rats, bandicoots, wombat, and the two extant

Table 45.1 Total NISP for all sites. BC = Bone Cave; KC = Kutikina Cave; MC = Mackintosh Cave; NC = Nunamira Cave; PT = Pallawa Trountra; SC = Stone Cave; WC = Warreen Cave

Taxon	Common name	Site						
		BC	KC	MC	NC	PT	SC	WC
Antechinus minimus	Swamp antechinus	0	4	0	0	0	0	0
Antechinus sp.		17	1	0	39	0	0	125
Antechinus swainsonii	Dusky antechinus	13	6	0	0	0	0	0
Antechinus/Sminthopsis		0	0	181	0	0	12	0
Aves	Bird	65	1	32	0	0	6	86
Caryodes dufresnii	Endemic land mollusc	0	0	0	6	0	0	0
Cercartetus lepidus	Little pygmy possum	1	1	0	0	0	0	61
Cercartetus nanus	Eastern pygmy possum	2	0	0	0	0	0	71
Cercartetus sp.		0	0	11	14	1	0	0
Dasyuridae		10	6	0	13	59	0	0
Dasyurus maculatus	Spotted-tailed quoll	3	2	0	4	0	0	0
Dasyurus viverrinus	Eastern quoll	83	8	5	6	4	66	243
Dromaius novaehollandi	Emu	4	3	0	4	0	0	0
Dromaius egg-shell		0	0	0	5	0	0	0
Pisces	Fish	0	1	0	0	0	0	0
Gallinula mortierii	Native hen	21	3	0	4	0	0	0
Hydromys chrysogaster	Water rat	1	0	5	0	0	0	0
Indeterminate		729	0	352	120	170	95	439
Isoodon obesulus	Southern brown bandicoot	3	1	0	8	2	0	0
Large Macropodidae	Large macropod	1,486	3	258	0	1	164	1,752
Large Mammalia	Large mammal	6,695	902	521	254	234	283	1,608
Large–medium Macropodidae	Large–medium macropod	0	0	0	0	15	0	0
Large–medium Mammalia	Large–medium mammal	0	1,443	0	0	32	0	0
Macropus giganteus	Eastern grey kangaroo	0	1	0	3	0	0	0
Macropus rufogriseus	Bennett's wallaby	4,665	5,483	507	266	618	379	3,235

Mastacomys fuscus	Broad-tooth rat	201	35	166	61	67	140	360
Medium Macropodidae	Medium macropod	0	10	0	0	2	0	0
Medium Mammalia	Medium mammal	621	157	479	100	152	66	411
Medium–small Mammalia	Medium–small mammal	0	25	0	0	3	0	0
Muridae		283	45	363	173	6	50	1,546
Ornithorhynchus anatinus	Platypus	51	0	1	17	2	14	22
Strigiformes	Owl	1	0	0	0	0	0	0
Perameles gunnii	Barred bandicoot	0	0	0	0	0	0	1
Petaurus breviceps	Sugar glider	1	0	0	0	0	0	0
Potorous tridactylus	Long-nosed potoroo	4	2	161	10	0	2	5
Pseudocheirus peregrinus	Ring-tailed possum	34	1	260	13	1	0	19
Pseudomys higginsi	Long-tailed mouse	11	6	23	7	0	1	12
Pseudomys sp.		3	1	0	1	0	0	0
Rattus lutreolus	Swamp rat	12	3	89	32	52	0	0
Reptilia	Reptile	2	0	13	0	0	0	17
Sarcophilus harrisii	Tasmanian devil	10	6	1	0	1	0	3
Small Macropodidae	Small macropod	71	0	58	0	0	33	148
Small Mammalia	Small mammal	303	96	514	202	641	77	3,703
Small Marsupialia	Small marsupial	2	0	7	0	0	4	276
Sminthopsis leucopus	White-footed dunnart	2	1	0	0	0	0	0
Tachyglossus aculeatus	Echidna	0	0	2	0	0	0	7
Thylogale billardierii	Tasmanian pademelon	34	14	20	13	0	0	18
Trichosurus vulpecula	Brush-tailed possum	13	1	0	3	0	17	27
Vombatus ursinus	Wombat	1,208	586	142	39	121	169	961

Australian monotremes—platypus and echidna. One of the striking aspects of the prey assemblage however, is the sheer dominance of the wallaby in both NISP and Minimum Number of Individuals (MNI), in addition to the patterns of selected butchery and targeted nutritional extraction.

The origin of the fauna has been discussed elsewhere (Cosgrove et al., 1990; Cosgrove, 1995: 98–100; Cosgrove, 1999; Cosgrove and Allen, 2001; Garvey, 2006; Allen et al. 2016) and will not be reiterated here in detail. It has been observed, however, that all sites contain small fauna that has a wide range of complete elements preserved, including crania, long bone, vertebrae, and ribs. All the very small animals are likely to be the product of owl pellets (Cosgrove and Allen, 2001), occur in relatively high numbers, and range from 50–60 g for *Antechinus* sp. and *Cercartetus* sp. to 50–100 g for *Rattus* sp., *Mastacomys fuscus*, and *Sminthopsis* sp. The number of the medium-sized mammals is, conversely, relatively low with a wide distribution of body parts. These animals are between approximately 1 to 14 kg in weight (Table 45.2).

Humans could easily transport these medium-sized animals to the caves and thus would have the greatest chance of being represented if hunted. However, their low numbers suggests that these animals were generally overlooked by people in favour of the larger wallaby and wombat for reasons explained below. The minor presence of Tasmanian devil bone in the seven sites suggests the possibility of introduction by this carnivore, although these scavengers do not accumulate bone in the same way as Old World carnivores, so this seems unlikely. Devils do not den, nor have latrine areas or carry bones back to living sites (Marshall and Cosgrove, 1990). They transport much of the bone around in their stomachs and eat most of their prey close to the capture or scavenge site. Strigiformes (owls), dasyurids, including devils and thylacines (marsupial carnivores), or both, probably introduced these medium-sized mammals to the sites. At present we do not precisely know which of these was most significant for their introduction, but it seems humans did not play a dominant role.

The presence of two flightless birds, the Tasmanian emu (*Dromaius Novaehollandiae diemenensis*) and the Tasmanian native hen (*Gallinula mortierii*), occurs in Bone, Kutikina, and Nunamira caves (Fig. 45.1). This suggests that, at the time of human occupation, around these sites there were open environmental conditions, favourable to the bird's behavioural requirements. Emus occur in groups and have a home range of 5–10 km^2, while the native hen is sedentary and lives in family groups requiring open grassy habitats close to water, food, and vegetation cover.

A total of 333 emu egg-shell fragments were identified from all the excavated squares at Nunamira Cave, suggesting Aboriginal people were targeting eggs (Cosgrove, 1995). The very small number of emu bones (n = 12) in all sites, comprising vertebrae, ribs and pelvis, however, suggests that people rarely hunted the birds themselves. The Tasmanian native hen is represented by a relatively larger number of bones (n = 28), mostly derived from long bones (73%) and cranium (14%). These birds are elusive, easily alarmed, and are difficult to approach and capture without some form of collaborative hunting and/ or facilities such as nets and fire drives. Although their long bones have been fractured systematically, we cannot identify their mode of capture at present.

The presence of the platypus is interesting, since they are relatively small (weighing between 800 g to 3 kg) and are solitary. They live in burrows along river banks, occupying stretches of river and pools for over a decade, are sedentary, and contain significant amounts of fat in summer/autumn and meat relative to their size (Marshall, 1993). These figures may also vary due to platypus prey productivity in different Australian regions. Marshall (1993) further suggests that the platypus was also exploited for its fur. However, the presence of a relatively large macropod, the Bennett's wallaby, in the southwest assemblages is thought to have been a primary resource of skins and fur for clothing (Gilligan, 2014), though the smaller platypus pelts may also have been used to make garments.

To calculate each animal's relative importance to human subsistence, the NISP has been corrected to indicate number of bone specimens per cubic metre of deposit (Table 45.2), as different amounts of sediment were excavated from each site. Nevertheless, macropod, wombat, and small mammals make up the bulk of the identified specimens recovered from the seven southwest sites. The large mammal category probably includes a high proportion of macropod remains. Where there are several species within the same genus, they have been combined to allow for comparative clarity between the sites. This obviates the problem of underestimation of species due to the similarity of postcranial elements among the microfauna. In addition, the high number of unidentified fragments is attributed to larger mammals, almost certainly the Bennett's wallaby and wombat. Where microfauna could not be assigned to species, the category 'small mammal' was used.

Cosgrove and Allen (2001: 404) ranked nine potential human prey animals found in the Late Pleistocene southwest Tasmanian archaeological deposits. They based their conclusions on a number of factors: prey size, search time, pursuit/ease of capture, field processing, and transport. They concluded that, although the Tasmanian emu was an attractive prey based on its weight and fat potential, its capture was uncertain, once the bird had been alarmed, due to its speed and endurance. A further reason for its low representation in the southwest assemblages is that these formed in areas outside the emu's ecological range, as the sites occur in steeper, less fertile valleys. Although not discounting the other medium-sized animals and birds as prey, Cosgrove and Allen (2001) identified the wombat and wallaby as the most preferred prey species.

PREVIOUS ANALYSES

The detailed analysis of wallaby bones from Bone and Warreen caves indicates that butchery was systematic, with the hindquarter elements, particularly the two metatarsals IV/V, tibia and femur, the preferred bones for processing (Cosgrove and Allen, 2001) (Fig. 45.2A).

These contain the largest marrow cavities and all were methodically smashed to extract the marrow. These elements were the best represented in the assemblage, with

Table 45.2 NISP per cubic metre of each species. The species weight is according to Van Dyck et al. (2013)

Taxon	Common name	Weight (Kg)	Site						
			BC	KC	MC	NC	PT	SC	WC
Antechinus sp.	Antechinus	0.05–0.12	34	6	197	137	0	12	125
Aves	Bird		83	1	32	0	0	6	96
Caryodes dufresnii	Land snail		0	0	0	38	0	0	0
Cercartetus sp.	Pygmy-possum	0.06–0.10	3	1	11	28	1	0	133
Dasyurus sp.	Quoll	1.5–5.0	97	18	5	29	65	66	246
Dromaius novaehollandi	Emu	18.0–60.0	4	3	0	8	0	0	0
Dromaius egg shell			0	0	0	43	0	0	0
Pisces	Fish		0	1	0	0	0	0	0
Gallinula mortierii	Native hen	0.4	21	3	0	6	0	0	0
Hydromys chrysogaster	Water rat	0.4–1.2	1	0	5	0	0	0	0
Indeterminate			54,942	0	7,090	31,278	13,607	9,427	124,486
Isoodon obesulus	Brown bandicoot	0.4–1.2	3	1	0	12	4	0	0
Macropus rufogriseus	Bennett's wallaby	15.0–24.0	25,895	27,030	1,309	1,015	757	379	3,469
Macropus giganteus	Eastern grey kangaroo	19.0–85.0	0	1	0	4	0	0	0
Mastacomys fuscus	Broad tooth rat	0.14	215	35	173	424	73	140	367
Medium Macropodia	Medium macropod		1,196	195	507	823	412	66	417
Muridae	Native rat	0.15	421	45	388	1,724	6	50	1,551
Ornithorhynchus anatinus	Platypus	0.8–3.0	55	0	1	47	2	14	22
Strigiformes	Owl		1	0	0	0	0	0	0

Perameles gunnii	Barred bandicoot	0.5–1.4	0	0	0	0	0	0	1
Petaurus breviceps	Sugar glider	0.14	1	0	0	0	0	0	0
Potorous tridactylus	Potoroo	0.7–1.6	5	2	164	14	0	2	5
Pseudocheirus peregrinus	Ring-tail possum	1.0	45	1	267	74	1	0	19
Pseudomys sp.	Long-tail mouse	0.70	15	7	23	46	0	1	12
Rattus lutreolus	Swamp rat	0.05–0.2	13	3	89	100	55	0	0
Reptilia	Reptile		2	0	14	0	0	0	50
Sarcophilus harrisii	Tasmanian devil	8.0–14.0	10	9	1	0	1	0	3
Small Macropodia	Small macropod		84	0	66	0	0	37	430
Small Mammalia	Small mammal		582	103	521	3,893	1,100	77	3,737
Sminthopsis leucopus	White-footed dunnart	0.01–0.03	2	1	0	0	0	0	0
Tachyglossus aculeatus	Echidna	2.0–7.0	0	0	2	0	0	0	7
Thylogale billardierii	Pademelon	4.0–12.0	34	14	20	34	0	0	18
Trichosurus vulpecula	Brush tail possum	1.3–4.0	13	4	0	3	0	17	28
Vombatus ursinus	Wombat	22.0–32.0	1,491	721	145	187	153	169	1,157

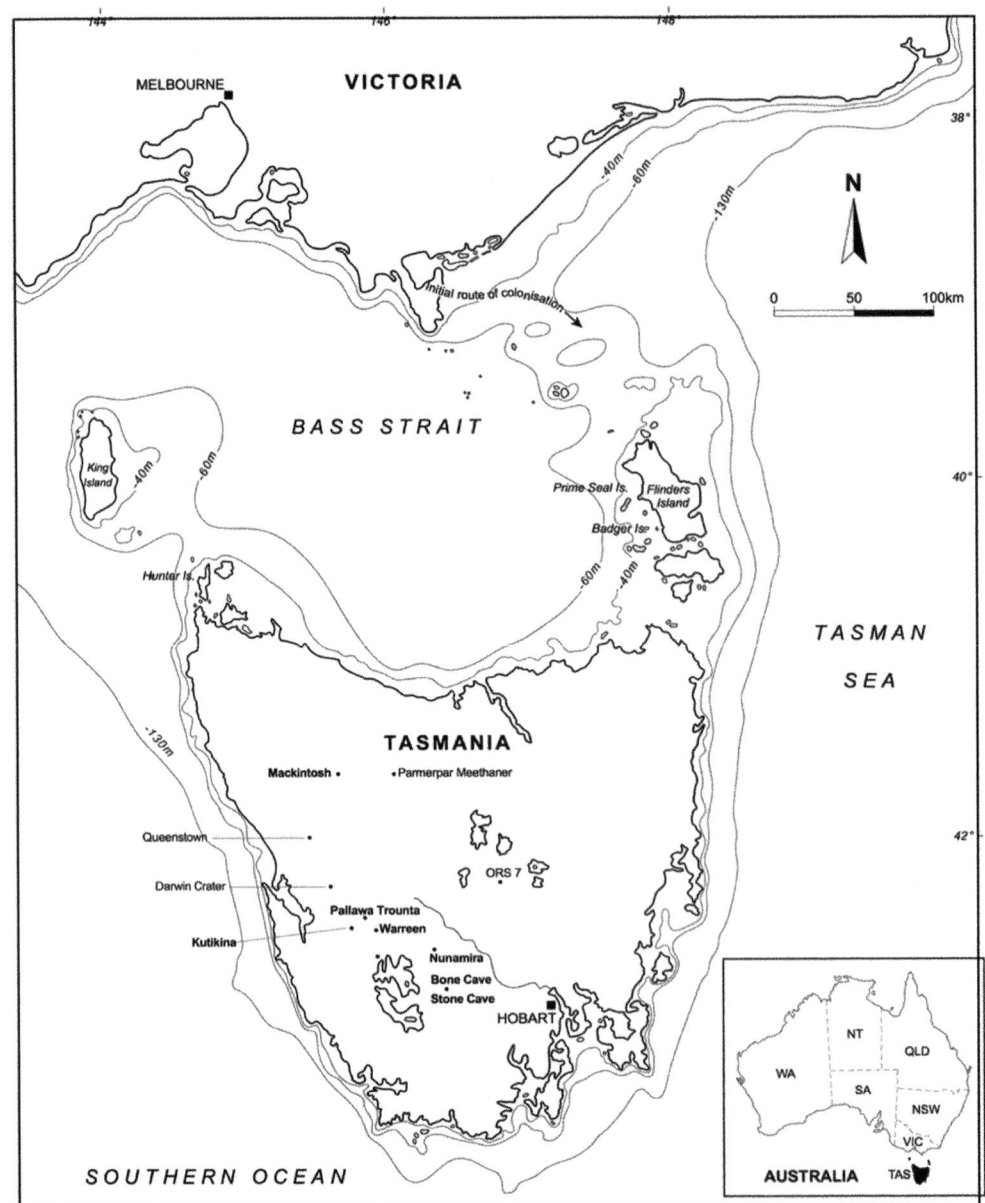

FIGURE 45.1 Tasmanian Late Pleistocene sites mentioned in the text. The modern capital cities of Melbourne and Hobart are included for reference. Authors' own image.

other body parts, such as the pelvic and pectoral girdle, under-represented (Cosgrove and Allen, 2001: 420; Garvey, 2010) (Fig. 45.2A). The anterior, medial, and posterior axial elements only made up 1.6% of the wallaby skeleton in the Kutikina assemblages, probably indicating field butchery away from the sites. Taphonomic influences were ruled out as an explanation for this pattern (Garvey, 2010) since there was excellent

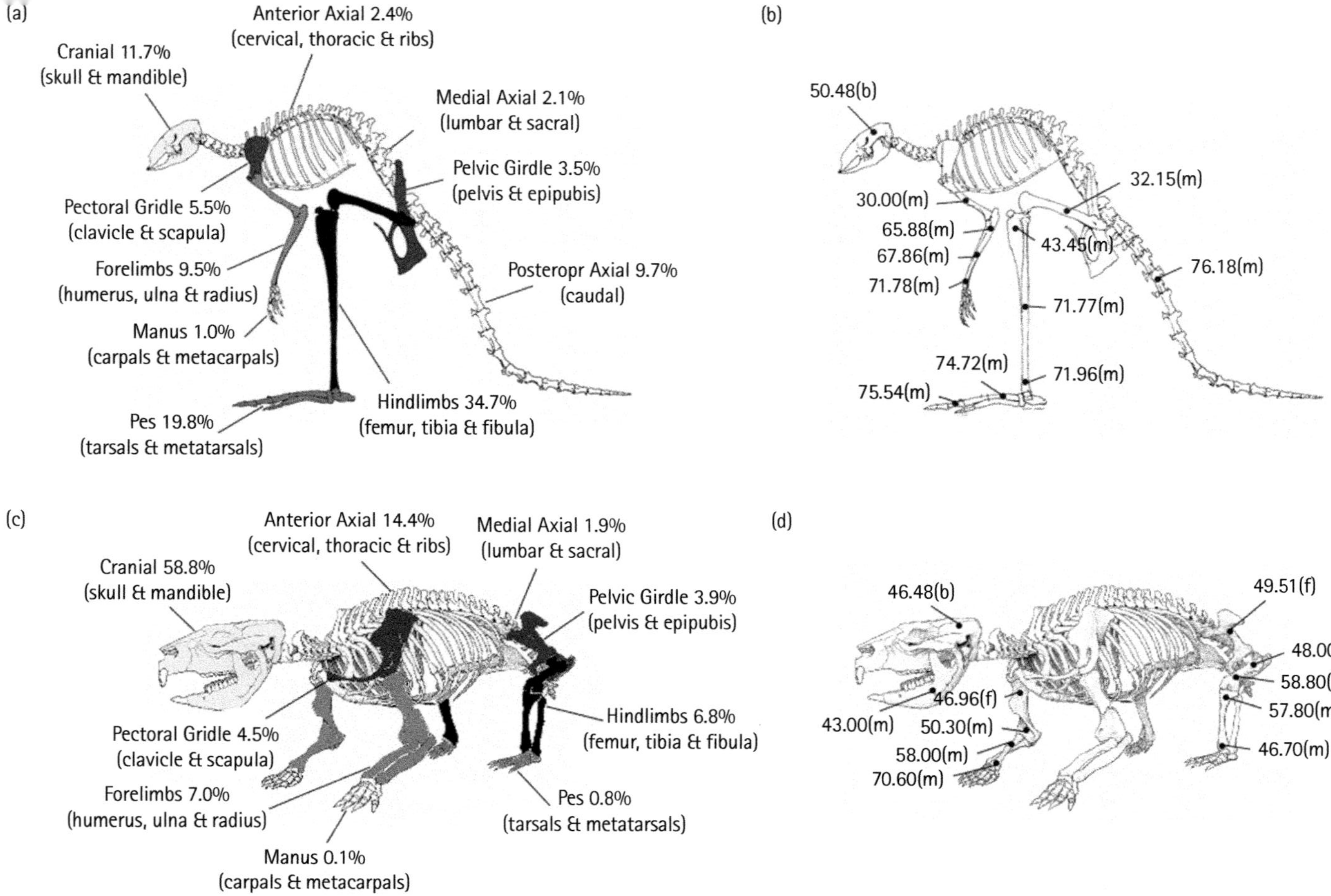

(a)

Anterior Axial 2.4%
(cervical, thoracic & ribs)

Cranial 11.7%
(skull & mandible)

Medial Axial 2.1%
(lumbar & sacral)

Pelvic Girdle 3.5%
(pelvis & epipubis)

Posteropr Axial 9.7%
(caudal)

Pectoral Gridle 5.5%
(clavicle & scapula)

Forelimbs 9.5%
(humerus, ulna & radius)

Hindlimbs 34.7%
(femur, tibia & fibula)

Manus 1.0%
(carpals & metacarpals)

Pes 19.8%
(tarsals & metatarsals)

(b)

50.48(b)

32.15(m)

43.45(m)

76.18(m)

30.00(m)
65.88(m)
67.86(m)
71.78(m)

71.77(m)

71.96(m)

74.72(m)

75.54(m)

(c)

Anterior Axial 14.4%
(cervical, thoracic & ribs)

Cranial 58.8%
(skull & mandible)

Medial Axial 1.9%
(lumbar & sacral)

Pelvic Girdle 3.9%
(pelvis & epipubis)

Hindlimbs 6.8%
(femur, tibia & fibula)

Pectoral Gridle 4.5%
(clavicle & scapula)

Forelimbs 7.0%
(humerus, ulna & radius)

Pes 0.8%
(tarsals & metatarsals)

Manus 0.1%
(carpals & metacarpals)

(d)

49.51(f)

46.48(b)

48.00(m)
58.80(m)
57.80(m)
46.70(m)

43.00(m)
46.96(f)
50.30(m)
58.00(m)
70.60(m)

FIGURE 45.2 A: The distribution of skeletal elements of the Bennett's wallaby (*Macropus rufogriseus*) from the seven Late Pleistocene south-west caves. B: The percentage of the total unsaturated fatty acids in the bone marrow (m) and brain tissue (b) of the Bennett's wallaby (after Garvey, 2011: Fig. 2A). C: The distribution of skeletal elements of the common wombat (*Vombatus ursinus*) from the seven Late Pleistocene southwest caves. D: The percentage of the total unsaturated fatty acids in the bone marrow (m), brain tissue (b), and flesh/muscle (f) of the common wombat (after Garvey, 2011: Fig. 2B). The macropod skeleton is adapted from Hume et al. (1987: Fig. 29.4), and the wombat skeleton from Wells (1989: Fig. 32.6). Authors' own image.

preservation of murid (e.g. *Pseudomys* sp.) scapulae in deposits with a pH of 10 in the 30,000-year levels. Similar excellent preservation was recorded for the other cave sites (Cosgrove and Allen, 2001). In comparison, the second most common prey species, the common or bare-nosed wombat, was mainly represented by cranial, anterior axial, and forelimb elements (Fig. 45.2C).

Garvey (2011) has investigated the reasons for the preoccupation with the wallaby hind limb bones and found that they contain the larger proportion of marrow that is high in unsaturated fatty acids (Fig. 45.2B). Similar patterns are also seen in the Common or Bare-nosed wombat (Fig. 45.2D). The wallabies also remain nutritionally stable between seasons, irrespective of sex, and are less variable in body condition than the human placental prey animals in the Northern Hemisphere, such as the reindeer or horse. In particular, wallabies do not appear to suffer the depletion of fat stores and marrow like bison (Speth and Spielman, 1983) or reindeer (Morin, 2007; Prior, 2008; Chan-McLeod et al., 2013). This stability of seasonal fat quality would also likely preclude the need for meat and bone storage. This is further supported by Tasmania's cold glacial climate and absence of competition from major carnivores. It suggests that wallabies were a predictable and reliable source of essential nutrients, with slightly elevated concentrations of fatty acids in the distal tibia during winter and spring (Garvey, 2011).

The other major advantage to hunters was the wallaby's relative sedentism. Modern studies show that they have home ranges of 15–20 ha and do not migrate over the seasonal cycle. Therefore, hunting was probably via encounter within restricted patches of grassland (Cosgrove et al., 1990; Allen et al. 2016). These fertile grasslands were attractive to grazing wallabies and concentrated them around the limestone geology that also contained caves, used by people for shelter and as observation points to target game. Wallabies were ecologically tethered to these nutrient-rich patches (Cosgrove and Allen, 2001); such areas were predictable and it is hypothesized that people moved from the patch after wallaby numbers fell below return rates (Cosgrove and Allen, 2001). Although this model best explained the rich archaeological record of southwest Tasmanian caves, little was known until recently about seasonal use of the region, or contact with other eastern areas outside the western river valleys.

Seasonality Studies

One of the limiting factors in understanding past seasonal Aboriginal occupation was the lack of data on Bennett's wallaby and wombat skeletal responses to seasonal variability. Analysing macropod teeth based on techniques developed for Northern Hemisphere ungulates had been attempted using transverse thin sections of the root apices (Innes, 1982). The results were limited, as this method did not record the full growth history of the wallaby. Cutting the teeth in longitudinal sections attained a better resolution of the seasonal responses (Pike-Tay and Cosgrove, 2002). Using a large set of modern comparative samples and additional archaeological teeth sourced from four limestone cave sites was crucial in identifying the systematic seasonal exploitation of the region by humans during the last Glacial Maximum (Pike-Tay et al., 2008). The results indicated that the lowlands (Warreen and Kutikina caves) were mostly used in the autumn and winter, while the upland cave sites (Nunamira and Bone caves) were mainly occupied during spring and summer.

Given that there seemed to be little nutritional seasonal variation in the wallaby, it begged the question whether these butchery strategies differed in winter due to the increased need for thermoregulation for human populations in these glacial environments. The archaeological attributes of such an approach would be increased processing of the lower limb bones, the cranium, and a higher frequency of small fragments and forequarter bones associated with visceral fat (Outram, 2001; Garvey, 2011).

To test whether there were systematic changes to butchery between seasons, time periods or at site locations, seven sites are analysed here; Kutikina, Pallawa Trountra, Warreen, and Mackintosh caves in the lowland representing winter/autumn occupations, and the upland sites Bone, Stone, and Numamira caves representing spring/summer occupations (Fig. 45.1).

Bennett's Wallaby Butchery Patterns

To understand whether there were different approaches to seasonal butchery, an examination of the treatment of the macropod remains is undertaken. There are several limitations to the analysis based on the sampling of wallaby jaws and the allocation of a seasonal signal to each site (MNI = 130). Taken together, the results are assumed here to reflect the seasonal presence of people at least at the times of year indicated by the last annuli deposited in each wallaby tooth. This is not to say that people were not there at other times, but it is the time of the year at which each wallaby was hunted, butchered, and marrow extraction occurred. In addition, the selection of wallaby jaws from different layers and chronologies also conflates thousands of years of Aboriginal occupation. The individual hunting events carried out over yearly cycles cannot be identified due to the low chronological resolution in each site (Cosgrove and Allen, 1996: 21–30). Nevertheless it is extraordinary that such robust seasonal patterns survive within the archaeological record over a 25,000-year period. Put simply, if Aboriginal people were highly variable in their subsistence behaviour without any systematic approach, the patterns would be far less clear. Thus, we assume that the caves in the uplands were usually inhabited in summer, while those at lower altitudes were used in winter (Pike-Tay et al., 2008). The following analysis is based on this premise.

For this comparative analysis *Macropus rufogriseus*, large macropod, and large mammal categories have been combined, in the belief that the large majority of these bones are likely to derive from Bennett's wallabies.

In an analysis of fragment weight *vs* body part, no difference has been identified between seasons, with long bones of similar fragment size consistently dominating the assemblage (Cosgrove and Allen, 2001). It indicates a preoccupation with the macropod hindquarter lower limbs. Fragmentation patterns based on fragment weight from all seven sites were compared. An ANOVA test shows no significant difference between the mean weights of each bone fragment of the main marrow bearing bones (n = 7242; F Ratio = 2.25; Prob > F = 0.133) as a measure of the intensity of bone processing between seasons. There is no significant difference between tibia (n = 2212; F Ratio = 1.69; Prob > F = 0.193) and metatarsal weights (n = 1263; F Ratio = 0.48; Prob > F = 0.484). There are significant differences in femur weights (n = 1085; F Ratio = 15.07; Prob > F = 0.0001) and vertebra weights (n = 2837; F Ratio = 38.31; Prob > F = 0.001), in particular the caudal vertebra (n = 1207; F Ratio = 19.06; Prob > F = 0.001). We expected to find smaller fragments in the winter period, reflecting more intense marrow processing, but in fact this occurred in summer.

Conclusions

The results do not suggest a more intensive marrow-extraction strategy as measured by bone fragmentation in winter. Considered together, it appears that people using the caves produced similar amounts of fragmented bone, and over the long term this pattern seems to suggest that there were very few differences between the body part selections, the intensity of bone processing and function at different sites. This may be in large part due to the relatively stable amount and quality of fat in the Bennett's wallaby and wombat throughout the seasonal cycle. Having a predictable and reliable supply of essential fatty acids in both lowland and upland sites at crucial times throughout the last glacial would have lowered economic risk and uncertainty. The behavioural ecology of both these major prey animals would also have been advantageous to hunters, given that they have high tolerance for variable climatic conditions. This is especially true for the wallaby, an animal that would have been ecologically tethered to grassland patches within the limestone geology containing the caves that were suitable for human habitation.

Despite the wallaby's relatively small size when compared to larger Northern Hemisphere prey taxa like horse, red deer, and reindeer, they remained the preferred prey animals during the entire Late Pleistocene period in Tasmania. If extinct megafauna were encountered and hunted, there is no evidence for this. Claims have been made for the involvement of humans in the extinction of Tasmania's megafauna (Turney et al., 2010), but no bones from c.900,000 fragments have been found in any of the archaeological deposits to suggest this (Cosgrove et al., 2010). If people and larger animals did overlap, it is curious that these taxa were ignored in favour of the smaller, but more predictable and possibly more nutritious wallabies and wombats. It is of some interest that in experimental studies, some larger macropods (the 25 to 50 kg Eastern grey kangaroo, and the 40 to 85 kg Red kangaroo) contain lower amounts of unsaturated fatty acids, and perhaps this is one reason why they were generally ignored as major prey animals (Garvey, 2011). Garvey suggested that if there was a correlation between large size and lowered unsaturated fatty acids, this may be one reason why extinct Australian megafauna are not recorded in archaeological sites as human prey. Further work is needed to clarify this interaction, but it seems that wallabies and wombats were the glacial equivalent of a low-key dependable resource, probably serving a similar dietary and secondary produce role as the reindeer did in Upper Palaeolithic Europe over the same time periods (Cosgrove et al., 2013).

References

Allen, J., R. Cosgrove, and J. Garvey (2016) 'Optimality models and the food quest in Pleistocene Tasmania'. *Journal of Anthropological Archaeology*, http://dx.doi.org/10.1016/j.jaa.2016.07.009.

Bennett, D. C., Code, W. E., Godin, D. V., and Cheng, K. M. (2008) 'Comparison of the antioxidant properties of emu oil with other avian oils', *Australian Journal of Experimental Agriculture*, 48, 1345–50.

Binford, L. R. (1978) *Nunamiut Ethnoarchaeology*, New York: Academic Press.

Bird, B., Richardson, J., Veth, P., and Barham, A. (2002) 'Explaining shellfish variability in middens on the Meriam Islands, Torres Strait, Australia', *Journal of Archaeological Science*, 29, 457–69.

Blumenschine, R. J. and Caro, T. M. (1986) 'Unit flesh weights of some East African bovids', *African Journal of Ecology*, 24, 273–86.

Brink, J. W. (1997) 'Fat content in leg bones of *Bison bison*, and applications to archaeology', *Journal of Archaeological Science*, 24, 259–74.

Burke, A. M. (1993) 'Observation of incremental growth structures in dental cementum using the scanning electron microscope', *Archaeozoologia*, 5(2), 41–54.

Burke, A. M. (1995) *Prey Movements and Settlement Patterns during the Upper Palaeolithic in Southwestern France*. BAR International Series 619. Oxford: Tempus Reparatum.

Burke, A. M. and Pike-Tay, A. (1997) 'Reconstructing "l'Age du Renne"', in Jackson, L. and Thacker, P. (eds) *Caribou/Reindeer Hunters of the Northern Hemisphere*. Worldwide Archaeology Series 6, pp. 69–81. Farnham: Ashgate Publishing.

Caughley, G., Shepherd, N., and Short, J. (1987) *Kangaroos: Their Ecology and Management in the Sheep Rangelands of Australia*, Cambridge: Cambridge University Press.

Chan-McLeod, A. C. A., White, R. G., and Russell, D. E. (2013) 'Comparative body composition strategies of breeding and non-breeding female caribou', *Canadian Journal of Zoology*, 77(12), 1901–7.

Cosgrove, R. (1995) *The Illusion of Riches: Scale, Resolution and Explanation in Tasmanian Pleistocene Human Behaviour*. BAR International Series 608. Oxford: Tempus Reparatum.

Cosgrove, R. (1999) 'Forty-two degrees south: the archaeology of Late Pleistocene Tasmania', *Journal of World Prehistory*, 13(4), 357–402.

Cosgrove, R. and Allen, J. (1996) 'Research strategies and theoretical perspectives', in Allen, J. (ed.) *Report of the Southern Forests Archaeological Project*, Vol. 1, pp. 21–30. Bundoora: School of Archaeology, La Trobe University.

Cosgrove, R. and Allen, J. (2001) 'Prey choice and hunting strategies in the Late Pleistocene: evidence from southwest Tasmania', in Anderson, A., O'Connor, S., and Lilley, I. (eds) *Histories of Old Ages: Essays in Honour of Rhys Jones*, pp. 397–430. Canberra: Coombs Academic Publishing, Australian National University.

Cosgrove, R., Allen, J., and Marshall, B. (1990) 'Palaeoecology and Pleistocene human occupation in south central Tasmania', *Antiquity*, 64(242), 59–78.

Cosgrove, R., Field, J., Garvey, J., Brenner-Coltrain, J., Goede, A., Charles, B., Wroe, S., Pike-Tay, A., Grün, R., Aubert, M., Lees, W., and O'Connell, J. (2010) 'Overdone overkill—the archaeological perspective on Tasmanian megafaunal extinctions', *Journal of Archaeological Science*, 37(10), 2486–503.

Cosgrove, R., Geneste, J.-M., Chadelle, J.-P., and Castel, J.-C. (2013) 'Perspectives on global comparative hunter-gatherer archaeology', in Frankel, D., Webb, J., and Lawrence, S. (eds) *Intersections and Transformations: Archaeological Studies in Environment and Technology*, pp. 13–30. Oxford: Routledge.

Cosgrove, R. and Pike-Tay, A. (2004) 'The Middle Palaeolithic and Late Pleistocene Tasmania hunting behaviour: a reconsideration of the attributes of modern human behaviour', *International Journal of Osteoarchaeology*, 14, 321–32.

Diab, M. C. (1998) 'Economic utility of the ringed seal (*Phoca hispida*): implications for Arctic archaeology', *Journal of Archaeological Science*, 25, 1–26.

Faulkner, P. (2013) *Life on the Margins: An Archaeological Investigation of Late Holocene Economic Variability, Blue Mud Bay, Northern Australia*. Terra Australis 38. Canberra: ANU Press.

Garvey, J. (2006) 'Preliminary zooarchaeological interpretations from Kutikina Cave, southwest Tasmania', *Australian Aboriginal Studies*, 1, 58–63.

Garvey, J. (2010) 'Economic anatomy of the Bennett's wallaby (*Macropus rufogriseus*): implications for understanding human hunting strategies in Late Pleistocene Tasmania', *Quaternary International*, 211, 144–256.

Garvey, J. (2011) 'Bennett's wallaby (*Macropus rufogriseus*) bone marrow quality *vs* quantity: evaluating human decision making and seasonal occupation in Late Pleistocene Tasmania', *Journal of Archaeological Science*, 38, 763–83.

Garvey, J. (2016) 'Australian Aboriginal freshwater middens from late Quaternary northwest Victoria: prey choice, economic variability and exploitation'. Quaternary International. http://dx.doi.org.ez.library.latrobe.edu.au/10.1016/j.quaint.2015.11.065.

Garvey, J., G. Roberts, and R. Cosgrove (2016) 'Economic utility and nutritional value of the Common wombat (*Vombatus ursinus*): Evaluating Australian Aboriginal hunting and butchery patterns'. *Journal of Archaeological Science: Reports*, 7, 751–63.

Garvey, J., Cochrane, B., Field, J., and Boney, C. (2011) 'Emu butchery and economic utility: implications for understanding Australian zooarchaeology and megafauna extinctions', *Environmental Archaeology*, 16(2), 97–112.

Gilligan, I. (2014) 'Clothing and modern human behaviour: the challenge from Tasmania', in Dennell, R. and Porr, M. (eds) *Southern Asia, Australia and the Search for Human Origins*, pp. 289–300. New York: Cambridge University Press.

Hopwood, P. R. (1981) 'Carcass muscle weight distribution and yield: a comparison between Grey Kangaroos, *Macropus giganteus*, and Red Kangaroos, *M. rufus*', *Australian Wildlife Research*, 8, 263–8.

Hopwood, P. R. (1988) 'Kangaroos as game meat animals: carcass yields and meat inspection', *Australian Zoologist*, 24, 169–77.

Hopwood, P. (1999) 'Kangaroo meat', *Leatherwood Food RA Food Industry Journal*, 2, 304–9.

Hopwood, P. R. and Griffiths, D. A. (1984) 'Carcass muscle-weight distribution and yield: a comparison between male and female Grey Kangaroos (*Macropus giganteus*)', *Australian Wildlife Research*, 11, 299–302.

Hopwood, P. R., Hilmi, M., and Butterfield, R. M. (1976) 'A comparative study of the carcass composition of kangaroos and sheep', *Australian Journal of Zoology*, 24, 1–6.

Howarth, G. S., Lindsay, R. J., Butler, R. N., and Geier, M. S. (2008) 'Can emu oil ameliorate inflammatory disorders affecting the gastrointestinal system?', *Australian Journal of Experimental Agriculture*, 48, 1276–9.

Hume, I. D., Jarman, P. J., Renfree, M. B., and Temple-Smith, P. D. (1987) 'Macropodidae', in Walton, D. W. and Richardson, B. J. (eds) *Fauna of Australia*, Vol. 1B *Mammalia*, pp. 679–715. Canberra: Australian Government Publishing Service.

Innes, R. W. (1982) 'Age determination in the Kangaroo Island wallaby, *Macropus eugenii*', *Australian Wildlife Research*, 9, 213–20.

Irving, L., Schmidt-Nielsen, K., and Abrahamsen, N. S. B. (1957) 'On the melting points of animal fats in cold climates', *Physiological Zoology*, 30, 93–105.

Jones, K. T. and Metcalfe, D. (1988) 'Bare bones archaeology: bone marrow indices and efficiency', *Journal of Archaeological Science*, 15, 415–23.

Kooyman, B. (1984) 'Moa utilisation at Owens Ferry, Otago, New Zealand', *New Zealand Journal of Archaeology*, 6, 47–57.

Lyman, R. L. (1992) 'Anatomical considerations of utility curves in zooarchaeology', *Journal of Archaeological Science*, 19, 7–22.

Marshall, B. (1993) 'Late Pleistocene human exploitation of the Platypus in southern Tasmania', in Augee, M. L. (ed.) *Platypus and Echidnas*, pp. 268–76. Sydney: Royal Zoological Society of New South Wales.

Marshall, B. and Cosgrove, R. (1990) 'Tasmanian Devil (*Sarcophilus harrisii*) scat bone: signature criteria and archaeological implications', *Archaeology in Oceania*, 25, 102–13.

Meng, M. S., West, G. C., and Irving, L. (1969) 'Fatty acid composition of caribou bone marrow', *Comparative Biochemistry and Physiology*, 30, 187–91.

Metcalfe, D. and Jones, K. T. (1988) 'A reconsideration of animal body-part utility indices', *American Antiquity*, 53, 486–504.

Morin, E. (2007) 'Fat composition and Nunamiut decision-making: a new look at the marrow and bone grease indices', *Journal of Archaeological Science*, 34, 69–82.

Naughton, J. M., O'Dea, K., and Sinclair, A. J. (1986) 'Animal foods in traditional Aboriginal diets: polyunsaturated and low in fat', *LIPIDS*, 21, 684–90.

O'Dea, K. (1984) 'Marked improvement in carbohydrate and lipid metabolism in diabetic Australian aborigines after temporary reversion to traditional lifestyle', *Diabetes*, 33, 596–603.

O'Dea, K. (1988) 'Kangaroo meat—polyunsaturated and low in fat: ideal for cholesterol-lowering diets', *Australian Zoologist*, 24, 140–3.

O'Dea, K. (1991) 'Traditional diet and food preferences of Australian Aboriginal hunter-gatherers', *Proceedings of the Royal Society of London, Series B Biological Sciences*, 334, 233–40.

O'Dea, K. and Spargo, R. M. (1982) 'Metabolic adaptation to a low carbohydrate-high protein ("traditional") diet in Australian aborigines', *Diabetologica*, 23, 494–8.

O'Dea, K., White, N. G., and Sinclair, A. J. (1988) 'An investigation of nutritional-related risk factors in an isolated Aboriginal community in Northern Australia: advances of a traditionally-orientated life-style', *The Medical Journal of Australia*, 148, 177–80.

Outram, A. (2001) 'A new approach to identifying bone marrow and grease exploitation: why the "indeterminate" fragments should not be ignored', *Journal of Archaeological Science*, 28, 401–10.

Outram, A. and Rowley-Conwy, P. (1998) 'Meat and marrow utility indices for horse (*Equus*)', *Journal of Archaeological Science*, 25, 839–49.

Pike-Tay, A. (1991) *Red Deer Hunting in the Upper Paleolithic of Southwest France: A Study in Seasonality*. BAR International Series 569. Oxford: Tempus Reparatum.

Pike-Tay, A. (1993) 'Hunting in the Upper Périgordian: a matter of strategy or expedience?', in Knecht, H., Pike-Tay, A., and White, R. (eds) *Before Lascaux: The Complex Record of the Early Upper Paleolithic*, pp. 85–100. Boca Raton: CRC Press.

Pike-Tay, A. (2000) 'Seasonality studies of archaeofaunas within a multiscalar framework: a case study from Cantabrian Spain', in Rowley-Conwy, P. (ed.) *Animal Bones, Human Societies*, pp. 1–11. Oxford: Oxbow Books.

Pike-Tay, A. and Cosgrove, R. (2002) 'From reindeer to wallaby: recovering patterns of seasonality, mobility, and prey selection in the Palaeolithic Old World', *Journal of Archaeological Method and Theory. Special Issue, Paleolithic Zooarchaeology*, 9(2), 101–46.

Pike-Tay, A., Cosgrove, R., and Garvey, J. (2008) 'Archaeological evidence for seasonal land use patterns by Late Pleistocene Tasmanian aborigines', *Journal of Archaeological Science*, 35, 2532–44.

Prior, A. J. E. (2008) Following the fat: food and mobility in the European Upper Palaeolithic 45,000 to 18,000 BP', *Archaeological Review from Cambridge*, 23(2), 161–79.

Sinclair, A. J. (1988) 'Nutritional properties of kangaroo meat', *Australian Zoologist*, 24, 146–8.

Sinclair, A. J., O'Dea, K., Dunstan, G., Ireland, P. D., and Niall, M. (1987) 'Effects on plasma lipids and fatty acid composition of very low fat diets enriched with fish or kangaroo meat', *LIPIDS*, 22, 523–9.

Speth, J. D. and Spielman, K. A. (1983) 'Energy source, protein metabolism, and hunter-gatherer subsistence strategies', *Journal of Anthropological Archaeology*, 2, 1–31.

Turner, J. C. (1979) 'Adaptive strategies of selective fatty acids deposition in the bone marrow of desert bighorn sheep', *Comparative Biochemistry and Physiology*, 62A, 599–604.

Turney, C. S. M., Flannery, T. F., Roberts, R. G., Reid, C., Fifield, L. K., Higham, T. F. G., Jacobs, Z., Kemp, N., Colhoun, E. A., Kalin, R. M., and Ogle, N. (2010) 'Late-surviving megafauna in Tasmania, Australia, implicate human involvement in their extinction', *Proceedings of the National Academy of Science of the United States of America*, 105(34), 12150–3.

Dyck, Van, S., Gynther, I., and Baker, A. (eds) (2013) *Field Companion to the Mammals of Australia*, London: New Holland Publishers.

Wells, R. T. (1989) 'Vombatidae', in Walton, D. W. and Richardson, B. J. (eds) *Fauna of Australia*, Vol. 1B, pp. 755–68. Canberra: Australian Government Publishing Service.

West, G. and Shaw, D. (1975) 'Fatty acid composition of Dall sheep bone marrow', *Comparative Biochemistry and Physiology*, 50B(4), 599–601.

White, N. G. (1985) 'Sex differences in Australian Aboriginal subsistence: possible implications for the biology of hunter-gatherers', in Ghesquiere, J., Martin, R. D., and Newcombe, F. (eds) *Human Sexual Dimorphism*, pp. 323–61. London: Taylor and Francis.

White, N. G. (1989) 'Cultural influences on the biology of Aboriginal people: examples from Arnhem Land', in Schmitt, L. H., Freeman, L., and Bruce, N. M. (eds) *Growing Scope of Human Biology*. Proceedings of the Australasian Society for Human Biology 2, pp. 171–8. Nedlands: The Centre for Human Biology, The University of Western Australia.

White, N. G. (1990) 'Food intake patterns in a traditionally-orientated Aboriginal community: dietary fat as an example', *Proceedings of the Nutrition Society of Australia*, 15, 225.

White, N. G. (2001) 'In search of the traditional Australian Aboriginal diet—then and now', in Anderson, A., O'Connor, S., and Lilley, I. (eds) *Histories of Old Ages: Essays in Honour of Rhys Jones*, pp. 343–59. Canberra: Coombs Academic Publishing, Australian National University.

Whitehouse, M. W., Turner, A. G., Davis, C. K. C., and Roberts, M. S. (1998) 'Emu oils(s): a source of non-toxic transdermal anti-inflammatory agents in Aboriginal medicine', *Inflammopharmacology*, 6, 1–8.

Yoganathan, S., Nicolosi, R., Wilson, T., Handelman, G., Scollin, P., Tao, R., Binford, P., and Orthoefer, F. (2003) 'Antagonism of croton oil inflammation by topical Emu oil in CD-1 mice', *Lipids*, 38, 603–7.

CHAPTER 46

....

REGIONAL AND CHRONOLOGICAL VARIATIONS IN ENERGY HARVESTS FROM PREHISTORIC FAUNA IN NEW ZEALAND

IAN W. G. SMITH

INTRODUCTION

....

REGIONAL variations in the subsistence practices of New Zealand's indigenous Maori were recognized by the first Europeans who studied them closely in the late eighteenth century, with the restriction of domesticated plant production to the northern and central parts of the country being the most obvious (Salmond, 1991). Soon after archaeological investigations commenced in the mid-nineteenth century, chronological changes in the composition of faunal assemblages were recognized, initially through Haast's (1870) stratigraphically defined distinction between earlier deposits containing bones of moa (ten species of now-extinct large birds of the order Dinornithiformes) and later layers dominated by shell-fish. The advent of radiocarbon dating in the 1950s provided a time-scale by which such changes could be measured, and archaeological research throughout the following decades provided finer-grained understanding of variation through both space and time. Although population replacement was initially posited as a driver of change, by the late 1980s there was general acceptance that explanations for these patterns could be found in the gradual adaptations made by Neolithic settlers from tropical

Polynesia to the varied temperate environments of New Zealand and their impacts on those environments over a period of about one thousand years since first human settlement (Davidson, 1985).

The last decade of the twentieth century undid much of that certainty. With the application of rigorous criteria for assessing the suitability of radiocarbon determinations (Anderson, 1991; Higham and Hogg, 1997; Petchey, 1999; Schmidt, 2000a), a large proportion of the archaeological assemblages on which the accepted models had been based were left without secure dates. For example, application of these criteria reduced the seventy-three reliably dated moa-hunting sites identified by Anderson (1989) to just fifteen (Schmidt, 2000b). Furthermore, the 'chronometric hygiene revolution' revised the date for first human arrival in New Zealand to c. AD 1230–1280 (Wilmshurst et al., 2011), shortening the period of human occupancy by about 25%. This has implications for understanding the nature and timing of human impacts on the environment and the processes of cultural adaptation in response to them. There is now a critical need to reassess the evidence for both regional and chronological variations in evidence for the types and relative importance of the foods that prehistoric Maori ate to establish when, where, and how changes took place.

The analysis of prehistoric diet has, until recently, relied upon animal remains excavated from archaeological sites. Since the 1990s the development of isotopic approaches has enabled estimation of dietary composition directly from human remains (Ambrose, 1993; Katzenberg, 2008), with the distinct advantage that it can incorporate estimation of the contributions of plant foods. Only two such studies have been undertaken in New Zealand to date (Leach et al., 2003; Kinaston et al., 2013), due in part to the infrequency with which human skeletal remains are recovered from archaeological contexts. For the present study, attention is focused on the more widely available faunal remains, and what they reveal about the meat component of past human diets. The method used to investigate this is based on the approach first proposed by White (1953) who calculated the meat yields of exploited species, which Clark (1954) adapted by converting meat weights to energy yields. This procedure has been refined significantly through improved procedures for determining the parts of animals actually consumed, the weights of meat that they represent and the energy yields for particular taxa, both internationally (Reitz and Wing, 2008) and in New Zealand (Leach et al., 1996; 2001; Smith, 2004; 2011b).

Throughout New Zealand's prehistory, settlement patterns were mobile (with communities hypothesized to have made regular intra-annual shifts of residence to facilitate the exploitation of dispersed, seasonally available resources) and communities made occasional territorial shifts over time (Anderson and Smith, 1996; Walter et al., 2006). Thus no single site can be expected to provide a complete picture of resource exploitation, making it essential to aggregate data at a regional level from a judiciously selected range of sites.

Study Areas, Sample Selection, and Chronological Ordering

This study scrutinizes existing zooarchaeological data in two study areas on the northern and southern coasts of New Zealand (Fig. 46.1). These were selected as part of a wider study examining the long-term effects of climate variation and human impacts on New Zealand's marine shelf ecosystems (Smith, 2011a; 2013), hence their coastal location. This mirrors the very strong coastal emphasis in Maori settlement; 53% of all known prehistoric sites are within one kilometre of the shore, and less than 0.5% beyond ten kilometres (McFadgen, 2007: 103–7). More than ten thousand sites of presumed prehistoric age have been recorded in the Greater Hauraki study area, and some eight hundred in Otago-Catlins (CINZAS, 2008), the difference reflecting the marked concentration of pre-European Maori population in the northern third of the country.

Only a small proportion of recorded sites in each area have been excavated, and information from these was assessed through a review of published literature, unpublished reports, and data archives. A sample for each study area was selected for detailed analysis on the basis of two criteria: the availability of data on faunal remains suitable for the methodology described below; and the availability of reliable chronological information enabling the sites, or specific assemblages from them to be placed securely in time.

The requirements for data on faunal remains were (1) that taxonomic identifications had been reported for all classes of animal remains in the excavated sample, and (2) the number of animals assigned to each taxon was reported. In this context identifications include determinations to species level along with assignments to genus, family, or higher-level designations as necessitated by the nature of archaeozoological material. Identification data were accepted as reported, except where assemblages or components of them had been re-examined and any revisions of identification were thus incorporated. Where necessary, identifications were updated to accommodate revisions of nomenclature, based on the following sources: for shell-fish, Spencer et al. (2009); for fin-fish, Froese and Pauly (2010); for birds, Checklist Committee (OSNZ) (2010); for mammals, King (1995), and Baker et al. (2010).

Quantification of identified taxa was in terms of the minimum number of individuals (MNI): the smallest number of individual animals necessary to account for all of the remains of a taxon in an archaeological assemblage (Reitz and Wing, 2008). While NISP (Number of Identified Specimens) is sometimes preferred for inter-assemblage comparisons (Lyman, 2008), this measure was not reported for the majority of assemblages under consideration here. The primary dataset used in this analysis included only assemblages for which MNI were reported for all classes of fauna represented at the site. Because this did not provide sufficient information for some time periods in each study area, a secondary dataset was compiled of assemblages for which only some of the taxa noted as present were quantified by MNI. All the taxonomic identifications, MNI values, and presence data utilized in this study, along with the data sources that they were drawn from, are reported in detail by Smith and James-Lee (2010).

With regard to chronology, the primary requirement was for radiocarbon determinations that closely dated formation of the occupation deposit from which the faunal sample under analysis derived. All dates reported in publications were treated according to protocols that are set out in detail elsewhere (Smith, 2010). The 155 admissible dates employed in this study are reported in Smith and James-Lee (2010).

Together the selection criteria admitted a total of 109 assemblages from sixty-eight sites for analysis. For the Greater Hauraki area a total of 77 assemblages from forty-nine sites were included, and 32 assemblages from nineteen sites in the Otago-Catlins area. These

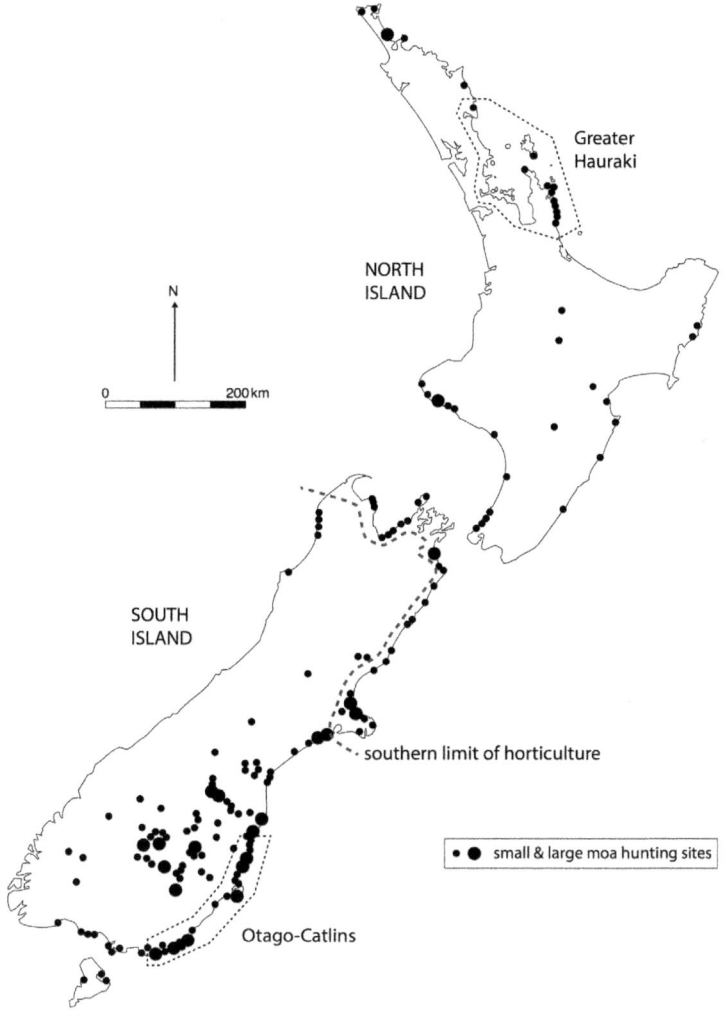

FIGURE 46.1 Map of New Zealand showing the study areas in relation to moa hunting sites and limits of horticulture. Author's own image.

Table 46.1 Frequency of faunal assemblages employed by study area, period, and size. Assemblage size (total MNI): S: small (< 100); M: medium (100–999); L: large (1,000–9,999); VL: very large (≥ 10,000)

Area/Period	Primary assemblages					Secondary	Total
	S	M	L	VL	Total	assemblages	
Greater Hauraki							
Early	1	2	3	1	7	1	8
Early/Middle	2	4	5	–	11	–	11
Middle	1	9	8	–	18	7	25
Middle/Late	–	6	3	2	11	7	28
Late	–	7	3	1	11	4	15
subtotal	4	28	22	4	58	19	77
Otago-Catlins							
Early	–	3	4	1	8	2	10
Early/Middle	–	–	3	2	5	4	9
Middle	–	–	–	–	0	2	0
Middle/Late	–	–	1	1	2	–	2
Late	2	2	1	–	5	4	9
subtotal	2	5	9	4	20	12	32
TOTAL	**6**	**33**	**31**	**8**	**78**	**31**	**109**

were grouped according to three broad period designations: Early (*c.* AD 1250–1450), Middle (AD 1450–1650), and Late (AD 1650–1800). Assemblages were allocated to time periods using a protocol that uses both 1σ and 2σ calibrated age ranges to distinguish those that can be assigned with confidence to a discrete period from those that overlap the period boundaries (Smith, 2010). On this basis almost two-thirds of the assemblages were assigned to one of the target periods, and the remainder to one of the two overlap zones (Table 46.1). Although the latter do not represent discrete time spans, they usefully group assemblages that cluster in age around the arbitrary period boundaries and were used here to provide a finer-grained assessment of trends in faunal assemblage composition over time.

METHODS

Preliminary analysis incorporated data from both primary and secondary datasets to maximize information about the range of taxa exploited. A total of 320 species was identified (Table 46.2). The proportion of assemblages in which each of these was represented for each area/study period was calculated to provide the broadest level analysis of changes in harvest patterns over time. This disclosed taxa likely to be under- or over-represented in the fully quantified samples (Smith, 2013).

Table 46.2 Number of species represented in the faunal assemblages from the two study areas. Identified species are listed in Smith and James–Lee (2010)

Faunal class		Greater Hauraki	Otago-Catlins	Total
moa		3	6	9
small birds	– marine/coastal	22	30	39
	– terrestrial/wetland	33	43	54
mammals	– terrestrial	5	7	7
	– marine	2	2	2
fish		35	32	39
shell-fish		147	90	170
Total		**247**	**210**	**320**

Calculation of the energy harvested from animal foods was undertaken with the primary dataset. The procedures used to do this are summarized in Fig. 46.2 and described in detail elsewhere (Smith, 2011b). In brief, this involved converting the frequency of each species in an assemblage into the weight of meat that they represent and then to the calorific value of energy that this would produce. While algorithms for estimating body size from bone or shell dimensions are available for several New Zealand fish and shell-fish (Leach, 2006), the measurements necessary to apply them have been taken from only a small number of assemblages, and thus could not be employed with the consistency required for the present analysis. Instead estimates of adult body weight, derived from the most precise data available, were used as a starting point, except in the case of the two major seal species, for which age- and sex-related size differences had already been determined for the majority of archaeological assemblages (Smith, 1985). For smaller-sized classes of fauna it was presumed that all usable meat on each animal would have been available for consumption. Usable meat was calculated as 70% of body weight for fish, small birds, and the Pacific rat (*Rattus exulans*), and 60% for the Polynesian dog (*Canis lupus familiaris*). Mean wet meat weights were used for shell-fish (Smith, 2011b). For larger-sized fauna the archaeological evidence frequently disclosed only partial carcass representation, and these taxa were quantified in terms of the minimum number of butchery units (MNBU): the smallest number of butchery units necessary to account for all of the skeletal elements in an assemblage. For the most common marine mammal species, New Zealand fur seal (*Arctocephalus forsteri*), New Zealand sea lion (*Phocarctus hookeri*), Southern elephant seal (*Mirounga leonina*) and Leopard seal (*Hydrurga leptonyx*), meat weights for butchery units were derived from data on the distribution of muscle mass (Smith, 1985: 120–7). An arbitrary value of 10% of meat weight was assumed for butchery units in the small numbers of cetaceans represented. For the extinct moa species body weights were based on the body mass values calculated from femur lengths of Holocene specimens (Worthy and Holdaway, 2002: 144–5, 154). Usable meat was assumed to be 60% of body weight, and meat yield for the major butchery unit, the leg, was calculated at 14.5% of body weight, based on the leg muscle mass of ostriches (Smith, 2011b).

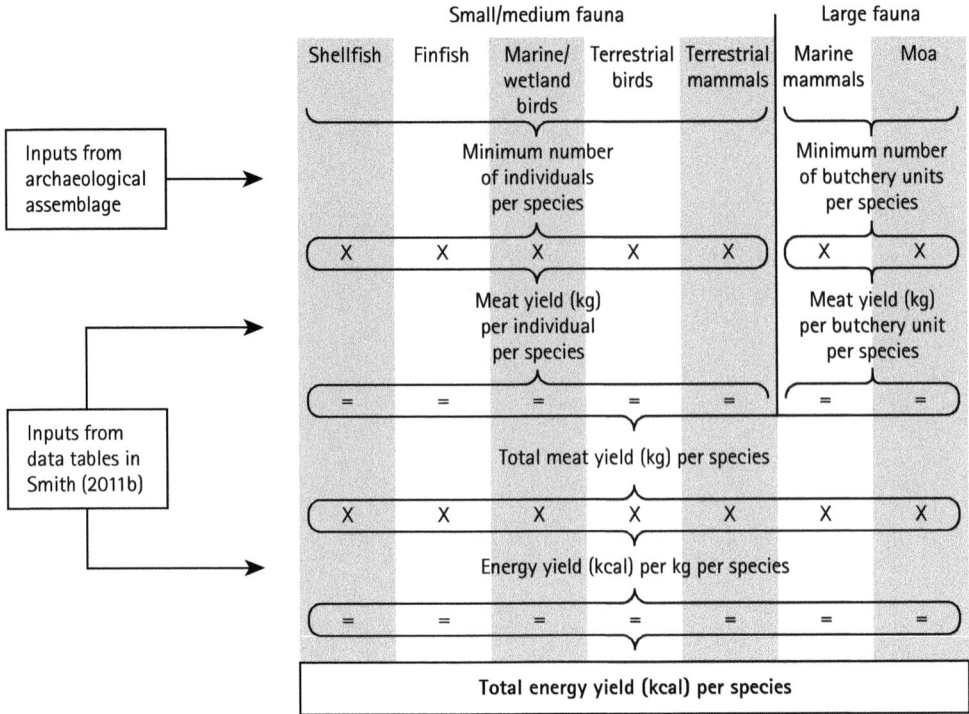

FIGURE 46.2 Procedures used in determining the energy yield per species. Author's own image.

Energy yields (kcal per kg) for each species were derived from a range of sources. Data from detailed proximate composition analysis was available for most of the fish and a small number of the shell-fish species. For all other taxa it was necessary to adopt values from comparable species or derive them from the best available sources (Smith, 2011b).

For the present analysis species were grouped into seven classes of fauna reflecting the manner in which they were harvested and the environment from which they derived. The proportion of total energy harvest derived from each faunal class was calculated for the seventy-six assemblages with complete quantified data. Data for each study area/period was summarized in two steps. First the mean energy contribution of each faunal class was calculated from the suite of assemblages from each area/period. Second, a series of adjustments were made to account for known or suspected biases within the study sample.

Results

Greater Hauraki

The seven Early period assemblages from this region exhibited a very consistent pattern. Marine mammals, almost exclusively fur seals, provided more than half of total kcal in

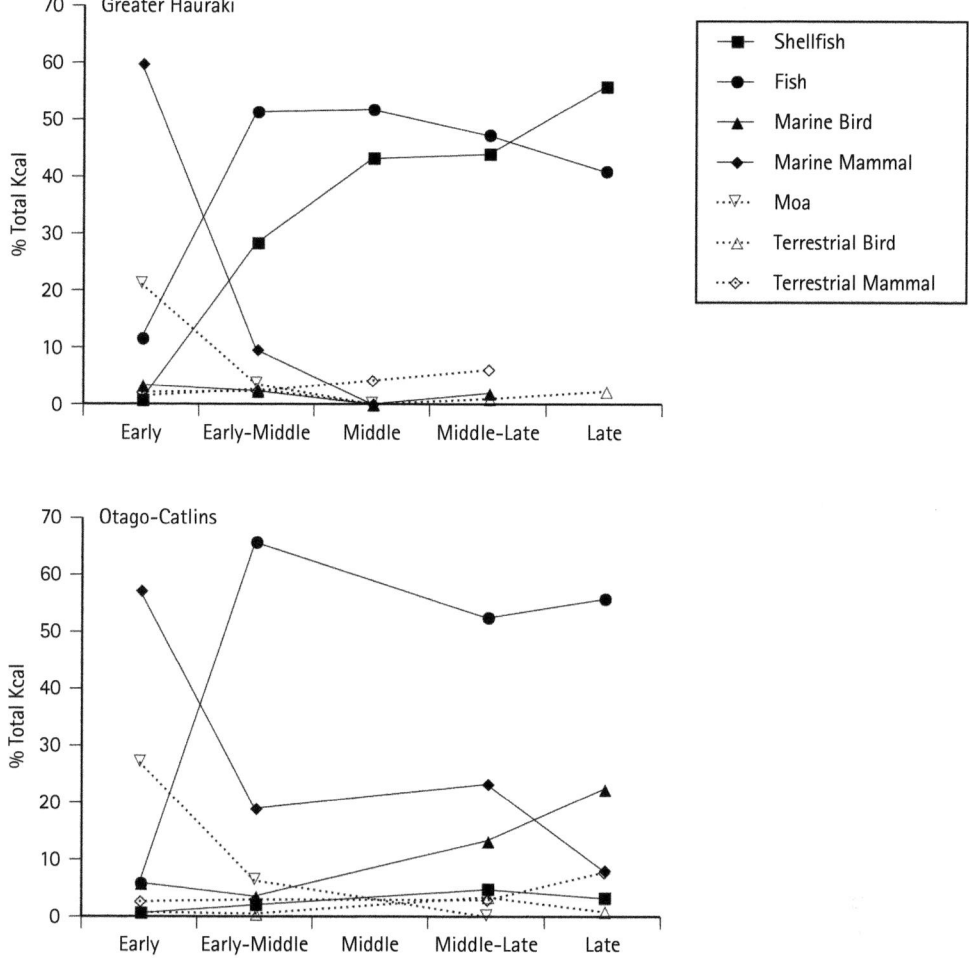

FIGURE 46.3 Mean proportions of total energy from different faunal class per period for the two study areas. Author's own image.

the six assemblages in which they occurred, contributing on average 60% of the energy harvest (Fig. 46.3). Moas dominated the one assemblage in which seals were absent, and were the second major contributor in four others. Fish were the only other faunal class to provide more than 10% of total energy. One notable feature of these assemblages is the breadth of the energy harvest with fish, shell-fish, and marine/coastal birds present in every assemblage and the remaining faunal classes each absent from only one.

Assemblages dated to the Early–Middle overlap zone disclose a narrowing of the energy harvest. The full range of faunal classes is represented in only one of the eleven assemblages and just two more have one missing component. In part this reflects declining reliance on marine mammals and moas; the former occur in only three assemblages and are the leading source of energy in just one, while the latter are the main source in one and occur in only one other. Alongside this is the emergence of specialized harvesting sites; two assemblages are comprised solely of shell-fish, and three of fish and shell-fish. Overall, these two resources dominate the energy harvest, with a mean of 51% of kcal from fish and 28% from shell-fish.

Specialization in energy procurement is even more strongly evident in the Middle period sample. Ten of the eighteen assemblages are comprised solely of shell-fish, six have fish and shell-fish with the former the dominant energy source in all but one, and there are two with fish, shell-fish, and terrestrial mammals. In these last two assemblages, dogs yielded 32% and 61% of total energy. Marine mammals, moas, and small birds were absent from all of the fully quantified assemblages.

Fish and shell-fish provided nearly identical proportions of the mean energy yield from the eleven assemblages dated to the Middle–Late overlap zone. Shell-fish are the sole animal food represented in three of these, and occurred alongside fish in another five. The remaining three assemblages exhibited a more generalized energy harvest, comprising terrestrial mammals, either or both marine/coastal and terrestrial/wetland birds, alongside the ubiquitous fish and shell-fish.

The Late period assemblages indicate that the energy harvest was very restricted in scope. Four of the eleven assemblages were shell-fish only and six contained both fish and shell-fish. In the remaining sample these two were joined by marine/coastal and terrestrial/wetland birds. On average shell-fish provided 56% of the total energy and fish 41%.

Otago-Catlins

The Early period samples comprised eight assemblages from five sites. These showed a broadly consistent pattern with marine mammals (57%) and moas (27%) dominating the energy harvest (Fig. 46.3). The former were the major energy source in all but two assemblages, where they were surpassed by the latter. Smaller animals were widely represented (fish were absent from one assemblage and terrestrial birds from another) but their contributions to the energy harvest were modest, with values greater than 10% from fish in two assemblages and marine/coastal birds in another two.

Assemblages assigned to the Early–Middle overlap zone show a substantial rise in the importance of fish to 66% of total kcal. While moas continue to be represented in every assemblage and marine mammals in all but one, both decline significantly in overall importance. Small birds, terrestrial mammals, and shell-fish make modest contributions at all sites.

No fully quantified samples were available for the Middle period, but the secondary dataset demonstrated that all faunal classes continued to be exploited.

Only two fully quantified assemblages were assigned to the Middle–Late overlap. Fish continued to provide the largest proportion of total energy, with substantial contributions from marine mammals and marine/coastal birds. Moas were absent from both assemblages.

The five Late period assemblages present a similar picture. Fish again dominate, marine/coastal birds rise further in importance, while marine mammals fall but continue to be exploited. Other than moas, all faunal classes are represented in virtually every site.

Adjustments

The energy harvest estimates were refined to address issues concerning sample representativeness, taphonomy, and ethnohistoric data that inform on late prehistoric subsistence practices (Smith, 2011a). The secondary dataset showed that small birds continued to be exploited through the Middle period in Greater Hauraki (Smith,

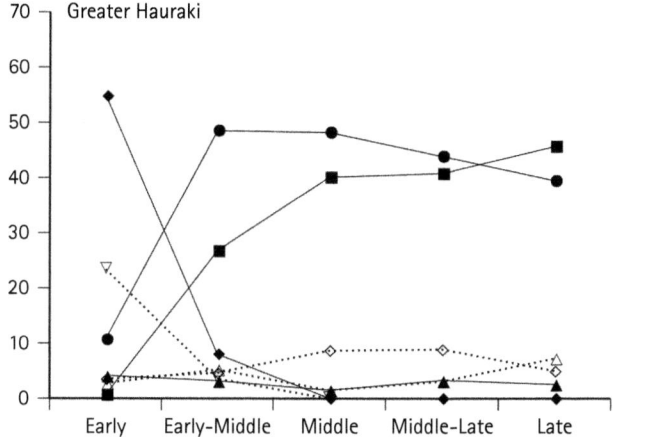
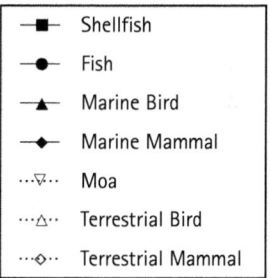
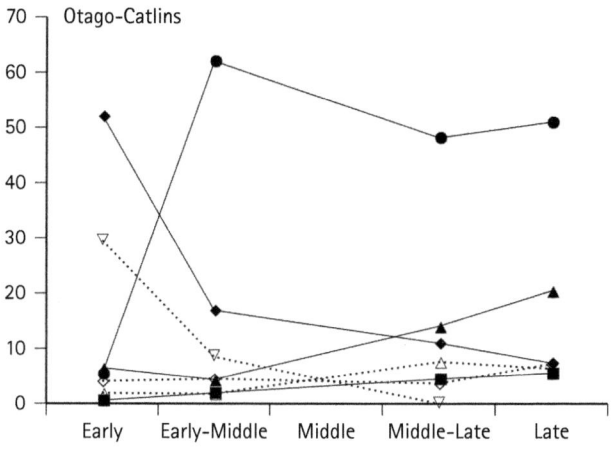

FIGURE 46.4 Proportional energy yields per period for the two study areas after adjustments. Author's own image.

2013), and allowance for that needs to be added to the model. Likewise the presence of dolphin bones and bone harpoon heads thought to have been used in their hunting (Smith, 1989) attest to continued exploitation of marine mammals through to the Late period in Greater Hauraki. Broader assessment of the archaeological and ethnohistorical records suggests that both classes of small birds are likely to be under-represented in all study area/periods (Smith, 2013), due to either or both the manner in which their carcasses were processed for preservation and transport and the taphonomic impact of scavenging by dogs. Similarly, sharks and rays are likely to be under-represented archaeologically, due to their predominantly cartilaginous skeletons, and ethnohistoric information indicates that they are likely to have been a significant food resource during the Late period in Greater Hauraki (Smith, 2013). Finally, due to the coastal location of nearly all study sites it is likely that the terrestrial animals exploited by the Maori communities in each study area are under-represented. A semi-quantitative analysis of relative frequencies of the main faunal classes in Early period sites in the southern South Island (Anderson, 1982: Table 5) shows that when inland sites are added to those of the coast the weighted abundance of terrestrial taxa increases by about 7%. With generally lower-energy yields from terrestrial animals compared to those of the sea, this suggests under-representation of total energy harvested from the former may be about 5%.

To accommodate these issues a series of adjustments, described in detail elsewhere (Smith, 2011a: 11–12), were made to the mean values calculated from the complete dataset (Fig. 46.4). This makes only minor changes to the proportionality of each faunal class, and does not modify the trends already identified.

Discussion

Using only data that meet strict chronometric protocols, this analysis has confirmed the long-recognized transition from a diet dominated by large game to one based around smaller animals, and is broadly consistent with recent attempts to date the extinction of moa and extirpation of seals (Holdaway and Jacomb, 2000; Smith, 2005). However it shows that moa persisted longer in the southern study area, suggesting that there were regional differences in the balance between human and moa population sizes. This was also the case with seals which continued to figure in Otago-Catlins diets throughout the prehistoric sequence. Earlier studies have tended to emphasize the importance of moa hunting in early economies, but this study has shown that seal hunting provided around twice as much dietary energy even in Otago-Catlins, where there are numerous large moa-hunting sites (Fig. 46.1). While it is possible that moas were of greater importance in other parts of New Zealand, it is noteworthy that isotopic analysis of bone from eighteen human burials at Wairau Bar, a large site with considerable evidence of moa hunting in northeastern South Island, indicates that high trophic level marine foods formed a significant part of their diet (Kinaston et al., 2013).

Both study areas show that availability of large game did not limit the use of smaller animals, with all Early period sites indicating a broad spectrum food quest. While these include some examples of smaller temporary camps (Smith, 1999), most have been interpreted as permanently occupied villages, even in Otago-Catlins, which is beyond the southern limit of prehistoric crop production (Anderson and Smith, 1996). This similarity in subsistence and settlement patterns between the horticultural and non-horticultural zones has implications for a broader understanding of the relationships between economic and social organization in segmentary societies (Walter et al., 2006; Marcus, 2008).

Differences between the two study regions are apparent after the Early period. In Otago-Catlins fish were quickly established as the primary source of dietary energy and remained in that role to the end of the sequence. Nonetheless the food quest remained broadly based with substantial inputs from marine mammals and birds, along with smaller contributions from terrestrial fauna and shell-fish. In Greater Hauraki fishing also rose to prominence alongside shell-fish gathering which became the primary source of energy by the Late period, while small birds and terrestrial mammals made only minor and irregular contributions. Unless there are serious deficiencies in the archaeological record, the animal component of Maori diets in the northern study was narrowly based from the Middle period onward.

One of the most striking differences between the two areas is the emergence in Greater Hauraki of specialized middens derived solely from shell-fish gathering. While the majority of these appear to be from small temporary camps close to beaches or estuaries, they also occur in activity areas adjacent to gardens and at larger more permanent settlements, sometimes several kilometres inland from the sea. Although the sample sizes recovered from excavations at some of these larger sites may have contributed to the apparent absence of scarce taxa, it is difficult to escape the conclusion that, wherever they were within their settlement pattern, northern Maori were highly dependent on shell-fish as a source of dietary energy.

Both the chronological changes and regional differences identified in this study can be explained through human impacts on the pristine environment colonized by early Maori. Extinction of moas and localized extirpation of seals have been attributed unequivocally to hunting pressure (Anderson, 1989; Smith, 2005; Oskam et al., 2012), and the more rapid and wide-ranging depletion of other resources in the northern study area reflects sustained growth in the human population there, enabled by horticultural production (Smith, 2011a; 2013). At a broader level the New Zealand case study emphasizes just how quickly the human quest for food could leave measureable impacts on both terrestrial and marine faunas.

References

Ambrose, S. (1993) 'Isotopic analysis of paleodiets: methodological and interpretive considerations', in Sandford, M. K. (ed.) *Investigations of Ancient Human Tissue: Chemical Analysis in Archaeology*, pp. 59–130. Philadelphia: Gordon and Breach Science Publishers.

Anderson, A. (1982) 'A review of economic patterns during the Archaic phase in southern New Zealand', *New Zealand Journal of Archaeology*, 4, 45–75.

Anderson, A. (1989) *Prodigious Birds: Moas and Moa-Hunting in Prehistoric New Zealand*, Cambridge: Cambridge University Press.

Anderson, A. (1991) 'The chronology of colonization in New Zealand', *Antiquity*, 65, 767–95.

Anderson, A. and Smith, I. (1996) 'The transient village in southern New Zealand', *World Archaeology*, 27, 359–71.

Baker, C. S., Chilvers, B., Constantine, R., Du Fresne, S., Mattlin, R., van Helden, A., and Hitchmough, R. (2010) 'Conservation status of New Zealand marine mammals (suborders Cetacea and Pinnipedia), 2009', *New Zealand Journal of Marine and Freshwater Research*, 44, 101–15.

Checklist Committee (OSNZ) (2010) *Checklist of the Birds of New Zealand, Norfolk and Macquarrie Islands, and the Ross Dependency of Antarctica*, 4th edn, Wellington: Ornithological Society of New Zealand and Te Papa Press.

CINZAS (2008) *Central Index of New Zealand Archaeological Sites*, Wellington: New Zealand Archaeological Association & Department of Conservation.

Clark, J. G. D. (1954) *Excavations at Star Carr*, London: Cambridge University Press.

Davidson, J. (1985) 'New Zealand prehistory', *Advances in World Archaeology*, 4, 239–91.

Froese, R. and Pauly, D. (2010) 'FishBase: World Wide Web Electronic Publication'. Version (09, 2010). http://www.fishbase.org.

Haast, J. (1870) 'On certain prehistoric remains discovered in New Zealand and on the nature of the deposits in which they occurred', *Journal of the Ethnological Society of London*, 2, 110–20.

Higham, T. and Hogg, A. (1997) 'Evidence for late Polynesian colonisation of New Zealand: University of Waikato radiocarbon measurements', *Radiocarbon*, 39, 149–92.

Holdaway, R. and Jacomb, C. (2000) 'Rapid extinction of the moas (Aves: Dinornithiformes): model, test and implications', *Science*, 287, 2250–4.

Katzenberg, M. (2008) 'Stable isotope analysis: a tool for studying past diet, demography and life history', in Katzenberg, M. and Saunders, S. (ed.) *Biological Anthropology of the Human Skeleton*, pp. 413–41. Hoboken: Wiley-Liss.

Kinaston, R., Walter, R., Jacomb, C., Brooks, E., Tayles, N., Halcrow, S., Stirling, C., Reid, M., Gray, A., Spinks, J., Shaw, B., Fyfe, R., and Buckley, H. (2013) 'The first New Zealanders: patterns of diet and mobility revealed through isotope analysis', *PLOS ONE*, 8(5), DOI: 10.1371/journal.pone.0064580.

King, C. (1995) *The Handbook of New Zealand Mammals*, Auckland: Oxford University Press.

Leach, B. F. (2006) *Fishing in Pre-European New Zealand*. Archaeofauna Series 15. Wellington: New Zealand Journal of Archaeology.

Leach, B. F., Davidson, J., Horwood, L. M., and Anderson, A. (1996) 'The estimation of live fish size from archaeological cranial bones of the New Zealand barracouta (*Thyrsites atun*)', *Tuhinga: Records of the Museum of New Zealand Te Papa Tongarewa*, 6, 1–25.

Leach, B. F., Davidson, J., Robertshawe, M., and Leach, P. (2001) 'Identification, nutritional yield, and economic role of tuatua shell-fish, Paphies spp., in New Zealand archaeological sites', *People and Culture in Oceania*, 17, 1–26.

Leach, B. F., Quinn, C., Mosrrsion, J., and Lyon, G. (2003) 'The use of multiple isotope signatures in reconstructing prehistoric human diet from archaeological bone from the Pacific and New Zealand', *New Zealand Journal of Archaeology*, 23, 31–98.

Lyman, R. L. (2008) *Quantitative Paleozoology*, Cambridge: Cambridge University Press.

Marcus, J. (2008) 'The archaeological evidence for social evolution', *Annual Review of Anthropology*, 37, 251–66.

McFadgen, B. (2007) *Hostile Shores: Catastrophic Events in Prehistoric New Zealand and Their Impact on Maori Coastal Communities*, Auckland: Auckland University Press.

Oskam, C., Allentoft, M., Walter, R., Scofield, R. P., Haile, J., Holdaway, R., Bunce, M., and Jacomb, C. (2012) 'Ancient DNA analyses of early archaeological sites in New Zealand reveal extreme exploitation of moa (Aves: Dinornithiformes) at all life stages', *Quaternary Science Reviews*, 52, 41–8.

Petchey, F. (1999) 'New Zealand bone dating revisited: a radiocarbon discard protocol for bone', *New Zealand Journal of Archaeology*, 19, 81–124.

Reitz, E. and Wing, E. (2008) *Zooarchaeology*, 2nd edn, Cambridge: Cambridge University Press.

Salmond, A. (1991) *Two Worlds: First Meetings between Maori and Europeans 1642–1772*, Auckland: Viking.

Schmidt, M. (2000a) *Radiocarbon Dating New Zealand Prehistory Using Marine Shell*. BAR International Series 842. Oxford: Archaeopress.

Schmidt, M. (2000b) 'Radiocarbon dating the end of moa-hunting in New Zealand prehistory', *Archaeology in New Zealand*, 43, 314–29.

Smith, I. (1985) 'Sea Mammal Hunting and Prehistoric Subsistence in New Zealand'. Unpublished PhD dissertation, University of Otago (Dunedin).

Smith, I. (1989) 'Maori impact on the marine megafauna: pre-European distributions of New Zealand sea mammals', in Sutton, D. (ed.) *Saying So Doesn't Make It So: Papers in Honour of B. Foss Leach*, pp. 76–108. Dunedin: New Zealand Archaeological Association.

Smith, I. (1999) 'Settlement permanence and function at Pleasant River Mouth, east Otago, New Zealand', *New Zealand Journal of Archaeology*, 19, 27–79.

Smith, I. (2004) 'Nutritional perspectives on prehistoric marine fishing in New Zealand', *New Zealand Journal of Archaeology*, 24, 5–31.

Smith, I. (2005) 'Retreat and resilience: fur seals and human settlement in New Zealand', in Monks, G. (ed.) *The Exploitation and Cultural Importance of Marine Mammals*, pp. 6–18. Oxford: Oxbow Books.

Smith, I. (2010) 'Protocols for organising radiocarbon dated assemblages from New Zealand archaeological sites for comparative analysis', *Journal of Pacific Archaeology*, 1, 184–7.

Smith, I. (2011a) *Estimating the Magnitude of Pre-European Maori Marine Harvest in Two New Zealand Study Areas*. New Zealand Aquatic Environment and Biodiversity Report 82. Wellington: Ministry of Fisheries.

Smith, I. (2011b) *Meat Weights and Nutritional Yield Values for New Zealand Archaeofauna*. Otago Archaeological Laboratory Report 8. Dunedin: Anthropology Department, University of Otago.

Smith, I. (2013) 'Pre-European Maori exploitation of marine resources in two New Zealand case study areas: species range and temporal change', *Journal of the Royal Society of New Zealand*, 43, 1–37.

Smith, I. and James-Lee, T. (2010) *Data for an Archaeozoological Analysis of Marine Resource Use in Two New Zealand Study Areas*, revised edn. Otago Archaeological Laboratory Report 7. Dunedin: Anthropology Department, University of Otago.

Spencer, H., Willan, R., Marshall, B., and Murray, T. (2009) 'Checklist of the Recent Mollusca Recorded from the New Zealand Exclusive Economic Zone'. http://www.molluscs.otago.ac.nz/index.html.

Walter, R., Smith, I., and Jacomb, C. (2006) 'Sedentism, subsistence and socio-political organization in prehistoric New Zealand', *World Archaeology*, 38, 274–90.

White, T. (1953) 'A method of calculating the dietary percantages of various animal foods utilised by aboriginal peoples', *American Antiquity*, 18, 396–7.

Wilmshurst, J., Hunt, T., Lipo, C., and Anderson, A. (2011) 'High precision radiocarbon dating shows recent rapid initial human colonisation of east Polynesia', *Proceedings of the National Academy of Sciences of the United States of America*, 108, 1815–20.

Worthy, T. and Holdaway, R. (2002) *The Lost World of the Moa*, Christchurch: Canterbury University Press.

CHAPTER 47

SPATIAL VARIABILITY AND HUMAN ECO-DYNAMICS IN CENTRAL EAST POLYNESIAN FISHERIES

MELINDA S. ALLEN

INTRODUCTION

IN the oceanic world of Polynesia, marine fish are typically the most common vertebrate remains in coastal archaeological sites, and fish continue to be an important food resource today. Nonetheless, indigenous fisheries have not been static but have evolved in dynamic response to changing cultural and natural influences. East Polynesia, the last Pacific area to be colonized, was settled by biologically and culturally closely related peoples around the end of the first millennium AD (Kirch, 2000; Allen, 2014a). They entered a region with familiar fishes but varied marine environments. This combination of factors renders East Polynesia an ideal setting for investigating the interplay between cultural practices, island geography, and long-term natural and socio-cultural processes. The deep time perspectives afforded by zooarchaeology are of value to not only social scientists interested in human eco-dynamics, but also contemporary island communities and conservation managers (e.g. Dalzell, 1998; Pitcher, 2001; Pinca et al., 2009).

THE STUDY AREA

This paper focuses on the indigenous marine fisheries of central East Polynesia (CEP), namely of the islands east of Samoa and between the Equator and 30° S. Economically, CEP colonists were foraging agriculturalists, who translocated several useful plants and three domesticated animals (Kirch, 2000). Additionally, they harvested locally available fish and shell-fish, land and sea birds, turtles, and occasionally sea mammals. Colonists arriving in CEP would have found a poorer but nonetheless familiar fish fauna, derived largely from the Indo-Pacific biogeographic province, where there are an estimated 4,000 shore-fish species (Springer, 1982: 127). The non-marginal Pacific Plate in particular supports around 111 fish families, 461 genera, and over 1,300 species, with biodiversity declining from west to east: Samoa (102 families), Society Islands (83), Rapa (63), Rapa Nui (52) (Springer, 1982). Given the close cultural affinities to and biogeographic continuities with western Polynesia, we might expect CEP fisheries to be fairly uniform, especially early in time.

Over the millennium or so of human settlement, CEP coastlines evolved, shaped by both natural and anthropogenic processes (Kirch, 2000). Sea-level fluctuations, tsunami, tropical storms, and tectonic processes figure among the former, while human colonists effected erosion, near-shore siltation, and shoreline progradation as they cleared lowland forests to make way for cultivated crops. Land and sea birds, abundant on human arrival, often went extinct or were extirpated (through hunting, habitat disturbance, and/or predation by introduced mammals), changes which disrupted ecosystem functioning and reduced dietary choices (e.g. Steadman, 2006). Late prehistory also saw the occasional extirpation of domesticated pig (*Sus scrofa*) and dog (*Canis familiaris*), probably the outcome of inter-specific competition (Walter, 1998; Allen and Craig, 2009). In light of these processes, a second expectation is that marine fisheries would have become increasingly important over time. These two hypotheses, the uniformity of early CEP fisheries and the increasing contribution of marine fisheries through time, are considered here using archaeofish assemblages from sites across the CEP region (Fig. 47.1; Table 47.1).

ASSEMBLAGES AND ANALYTICAL ISSUES

Twenty relatively large (NISP ≥ 200 and/or MNI ≥ 50) fish-bone assemblages from twelve dated and well-described sites are examined (Table 47.1). Smaller assemblages from Rapa Nui (Martinsson-Wallin and Crockford, 2002) and the late Peva occupation (Weisler et al., 2010), are considered in some comparisons. Stratigraphic and chronological evidence allows the assemblages to be assigned to either an Early East Polynesian (EEP) or a Late East Polynesian (LEP) period. The former typically are penecontemporaneous with initial island settlement (*c.* AD 1100 to 1300), but occasionally the associated ^{14}C age ranges extend to *c.* AD 1450. LEP assemblages date from *c.* AD 1450 to 1800, the latter date representing the beginning of sustained European contact. In cases where ^{14}C dates indicate individual strata are close in age, layer assemblages were combined to increase sample sizes. Overall, the fifteenth century is a useful divide, as important demographic, social, and economic changes often date from around this time (see Kirch, 2000).

Table 47.1 Geographical features of assemblage localities, arranged by latitude, and references for archaeological fish–bone analyses (excavation references, excluded due to space limitations, can be sourced from latter)

Archipelago	Island	Geology	Latitude	Land area (km²)	Lagoon	Reef	Site	Specialist references
Marquesas Is.	Ua Huka	volcanic	8°54'S	83	no	minor	Hane Dune (Areas A, B)	Davidson et al., 2000
Marquesas Is.	Ua Pou	volcanic	9°24'S	106	no	minor	Anapua Rockshelter	Leach et al., 1997
Marquesas Is.	Tahuata	volcanic	9°56'S	50	no	minor	Hanamiai (TH1)	Rolett, 1998
Society Is.	Huahine	volcanic	16°44'S	74.8	yes	extensive	Fa'ahia (ScH1-2) (Sinoto excavation)	Leach et al., 1984; Fraser, 1998
Cook Is.	Aitutaki	almost-atoll	18°85'S	18.3	yes	extensive	Ureia (AIT-10)	Allen, 2002
Cook Is.	Aitutaki	almost-atoll	18°85'S	18.3	yes	extensive	Moturakau Rockshelter (MR-1)	Allen, 2002; Allen and Morrison, 2013
Cook Is.	Ma'uke	makatea	20°16'S	18.4	no	minor	Anai'o	Walter, 1998
Cook Is.	Mangaia	makatea	21°55'S	51.8	no	minor	Tangatatau Rockshelter (MAN-44)	Kirch et al., 1995; Butler, 2001
Austral Is.	Rurutu	makatea/ volcanic	22°27'S	38.5	no	minor	Peva (ON1)	Weisler et al., 2010
Gambier Is.	Mangareva	volcanic	23°06'S	24.4	yes	extensive	Kamaka Is. Rockshelter (GK-1)	Weisler and Green, 2013
Pitcairn Group	Henderson	makatea	24°24'S	37.3	no	minor	HEN-5	Weisler, 1994; Weisler and Green, 2013
Austral Is.	Rapa	volcanic	27°36'S	40	no	none	Tangarutu Cave (R2002-29)	Vogel and Anderson, 2012

To assess the impact of differing environmental conditions on assemblage composition, islands are categorized on the basis of geological history into volcanic (high) islands, makatea, or atolls (Table 47.1). Volcanic islands are the products of mid-plate volcanism and typically are well-watered, agriculturally productive, and have complex marine environments with barrier reefs, deep or large lagoons, and/or extensive fringing reefs. Makatea are uplifted reef limestone islands, sometimes with exposed and weathered volcanic centres and fresh or brackish water features at the limestone-volcanic interface (e.g. Mangaia). Makatea are typified by narrow fringing reefs and usually lack lagoons or bays. Lastly, atolls are remnant barrier reefs of eroded or subsided

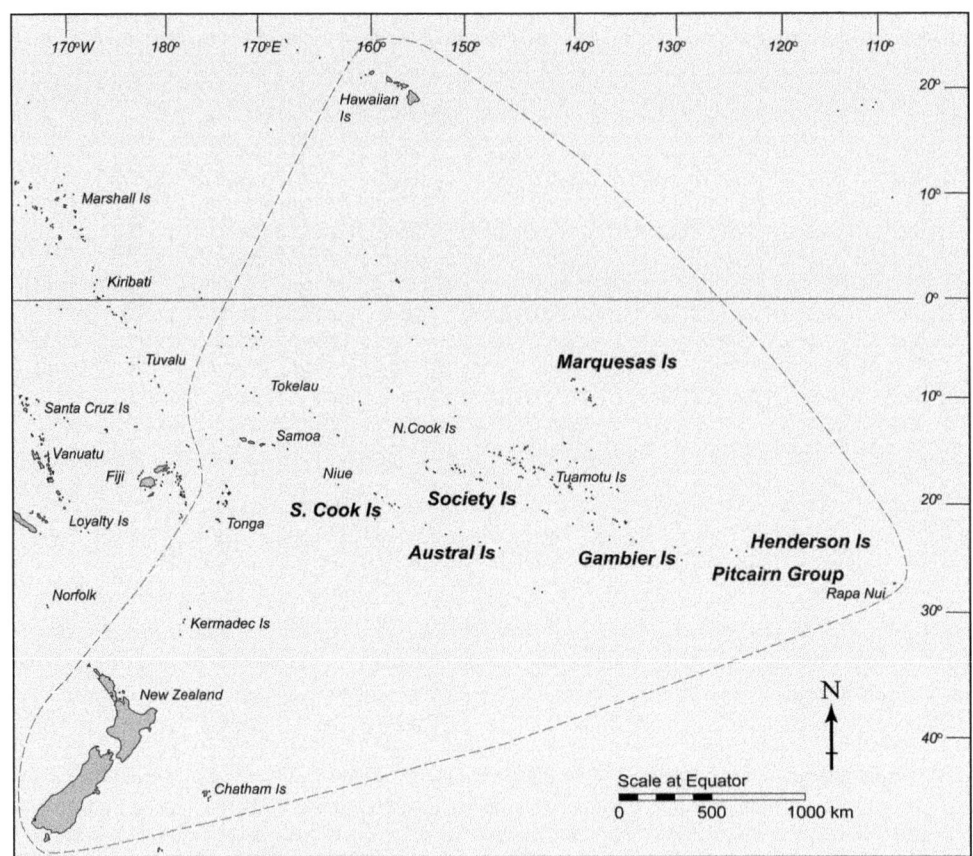

FIGURE 47.1 The Polynesian triangle, with locations of the analysed assemblages in bold. Author's own image.

volcanic islands, usually encircling a lagoon. Atolls are typically rich in marine life, but fresh-water resources are limited and the soils usually of poor quality.

Protocols used in Pacific fish-bone analyses have varied, making comparative studies a challenge and potentially introducing biases. Four analytical factors warrant consideration here: (1) mesh size of screens used in processing; (2) quality of the reference collections; (3) elements used in identifications; and (4) choice of quantitative measure(s). All of the assemblages considered here were recovered with at least 6.4 mm mesh sieves, but in several cases 3.2 mm or smaller meshes were also used (Table 47.2). Problematically, screen size can have a marked effect on how much and what kind of fish-bone is recovered. Gordon (1993), for example, illustrates how screen size alone can alter interpretations of dietary strategies (e.g. generalists *vs* specialists). She found that fine sieves (3.2 mm) improved bone recovery, better represented small-bodied and small-mouthed fishes, and altered assemblage evenness (see also Nagaoka, 1994; Leach et al., 1997; Allen, 2014b).

Reference collections also are crucial in obtaining accurate and precise results. Biodiversity in the tropical Pacific is high and no extant zooarchaeological reference collection represents the full range of natural biodiversity. Additionally, the skeletal morphology of many fish groups is little studied and identifications are often limited to family (but see Butler, 2001; Weisler et al., 2010). Consequently, this analysis examines regional variation in bony fishes (superclass Osteichthyes) at the level of family. Sharks and rays are included in NISP and MNI totals (Table 47.2) but, given that identifications are usually to subclass (Elasmobranchii), these are not further assessed (see also Weisler et al., 2010).

Decisions regarding which bone elements to identify also affect taxonomic results. Bone elements vary in abundance and morphology across fish taxa and a given element may not be uniformly identifiable across all families. Pacific analysts have tradition- ally focused on five elements that readily preserve and maximally differentiate families (dentary, premaxilla, maxilla, articular, quadrate), along with a smaller set of elements that are distinctive for some taxa (e.g. pharyngeals, distinctive spines) (Leach, 2006). Butler (1988: 109; 2001) has long advocated the use of additional elements to increase sample sizes and improve taxonomic representation. As part of an increasingly com- mon practice, six additional paired bones have proven especially useful: opercular, hyomandibula, cleithrum, preopercular, ceratohyal, and palatine (Walter, 1998; Butler, 2001; Weisler et al., 2010; Vogel and Anderson, 2012). The assemblages considered here vary with respect to the array of elements analysed but, in all cases, at least the five paired jaw bones were identified.

Finally, the choice of quantitative measure(s) affects results, especially in archaeo- fish studies (Grayson, 1984; Nagaoka, 1994; Lyman, 2008). Pacific analysts have used both number of identified specimens (NISP) and minimum number of individu- als (MNI), increasingly in combination, to maximize cross-assemblage comparisons. Problematically, NISP can be inflated by fragmentation and fish present additional problems as distinctive elements can vary considerably by taxon. Diondontidae (por- cupinefishes), for example, have around three hundred distinctive dermal spines, while Scaridae[1] (parrotfishes) and Labridae (wrasses) have three durable and easily recognized toothed pharyngeals, elements that can be non-descript or fragile for other taxa. MNI, a derived measure, also is problematic. MNI values can radically change as archaeological proveniences (areas, layers, features) are aggregated or subdivided—procedures which have little to do with actual bone abundances. Further, MNI can be calculated in varied ways which are not always specified. NISP and MNI are often correlated and, when not, regression analysis can help identify sources of bias (Grayson, 1984; Lyman, 2008). As a compromise, Allen (2002) proposed limiting NISP to paired jaw elements that are rela- tively distinctive for all taxa (NISP jaw).

MNI is used for most of the comparisons here because five economically impor- tant families have distinctive specialized elements that make them susceptible to over- representation by NISP: Scaridae and Labridae (pharyngeals), Carangidae or jacks/ trevallies (scutes), Scombridae or tunas and mackerels (vertebrae), and Diodontidae (body spines). MNI is a less than ideal measure and use of a consistent set of elements for all families should be a future goal (e.g. NISP of a large set of paired elements). In the assemblages reviewed here, where MNI and NISP could be compared, the most com- mon families were usually consistent, although rank orders varied.

Table 47.2 Details of archaeological contexts and fish-bone samples, arranged by latitude. Site values (columns 10–12) refer to the entire fish-bone assemblage, sometimes including materials from chronological units not analysed here. NISP values (column 11) refer to all teleost specimens identified within the bone sample area (column 7). Number of taxa (column 12) refers to families and supra-family identifications within a site; to simplify comparisons, sharks and rays are counted as one taxon (Elasmobranchii). Chronological unit values (columns 13–15) refer to the fish-bone derived from a specific phase, level, zone, or layer, excluding unspecified teleost (see text for explanations). The HEN–5 NISP value excludes Diodontidae spines.

Archipelago	Island	Site	Chronological unit	Time period	Site type	Bone sample area	Screen size (mm)	Activity	Site values			Chronological unit values		
									Total fish-bone(all phases)	Fish NISP	No. taxa	NISP	NISPjaw	MNI
Marquesas Is.	Ua Huka	Hane Dune	Phases I, II	early	dune	35.5 m²	6.35	residential	1,246	1,199	23	na	na	321
			Phases III,IV	late	dune	35.5 m²	6.35	residential	1,246	1,199	23	na	na	80
	Ua Pou	Anapua	Level II	early	rockshelter	uncertain	2	habitation	na	1,430	24	na	na	199
			Level IV	late	rockshelter	uncertain	2	habitation	na	1,430	24	na	na	213
	Tahuata	Hanamiai (TH1)	Phases I, II	early	open	20 m²	3.2	residential	11,411	497	17	205	na	na
			Phase IV	late	open	20 m²	3.2	residential	11,411	497	17	205	na	na
Society Is.	Huahine	Fa'ahia (ScH1-2); lower layer	Section 3, Zone A	early	open	uncertain	6.35	residential	na	na	27	1,693	na	545
Cook Is.	Aitutaki	Ureia (AIT-10)	Zones E, G	early	open	8 m²	6.4 to 1.5	residential	na	2,930	22	889	542	na
			Zone C	late	open	8 m²	6.4 to 1.5	residential	na	2,930	22	839	81	na

Archipelago	Island	Site	Layer	Period	Site type	Area	Levels	Function						
	Aitutaki	Moturakau (MR–1)	Zone F, H	early	rockshelter	2 m²	3.2, 6.4	campsite	na	11,183	29	4,032	2,414	na
			Zone D	late	rockshelter	2 m²	3.2, 6.4	campsite	na	11,183	29	2,368	1,402	na
	Ma'uke	Anai'o	Layer 4	early	open	uncertain	5, 3	residential	1,816	346	16	287	na	61
	Mangaia	Tangatatau (MAN-44)	Zones 2–4	early	rockshelter	25 m²	3.2	habitation	na	1,475	30	603	na	na
			Zone 8	late	rockshelter	25 m²	3.2	residential	na	1,475	30	375	na	na
Austral Is.	Rururtu	Peva (ON1)	Phase I (Layer D)	early	dune	33 m²	3.2	residential	5,021	1,484	21	1,081	na	141
			Phase II (Layer A)	late	dune	33 m²	3.2	ritual	5,021	1,484	21	403	na	24
Gambier Is.	Mangareva	Kamaka Is (GK-1)	Layers H–J	early	rockshelter	~22 m²	6.4, some 3.2	campsite	5,769	824	13	311	na	73
			Layers C–G	late	rockshelter	~22 m²	6.4, some 3.2	campsite	5,769	824	13	446	na	128
Pitcairn Group	Henderson	HEN-5	not specified	early	open	na	6, 3	habitation	33,194	1,725	18	1,725	na	na
Austral Is.	Rapa	Tangarutu (R2002-29; E1/E2)	Level I	early	rockshelter	1.5 m²	3, some 2	habitation	11,417	1,554	21	505	na	82
			Level III	late	rockshelter	1.5 m²	3, some 2	habitation	11,417	1,554	21	712	na	115

Table 47.3 Fish–bone assemblages (arranged by latitude within island types) and rank order Assemblages with MNI < 50 or NISP < 200 are excluded. Legend for major trophic (column 23) is the number of identified families based on all identified elements;

Island	Site: Provenience	Phase	rank measure	Nets, weirs, traps							Angling
				Scaridae (H)	Acanthuridae (H)	Mullidae (IN)	Balistidae (IN)	Diodontidae (IN)	Chaetodontidae (IN)	Eleotridae (IN)	Serranidae (P)
Makatea Islands											
Ma'uke	Anai'o: Layer 4	early	MNI	5	3						1
Mangaia	Tangatatau: Zones 2–4	early	NISP		3					1	5
	Tangatatau: Zone 8	late	NISP		4					2.5	5
Rurutu	Peva: Layer D	early	MNI	1	3			4			2
Henderson	HEN-5	early	NISP		2						1
Volcanic Islands											
Ua Huka	Hane: Phases I–II	early	MNI								2
	Hane: Phases III–IV	late	MNI					4			1
Ua Pou	Anapua: Level II	early	MNI				3.5				2
	Anapua: Level IV	late	MNI	6			4				3
Tahuata	Hanamiai: Phases I–II	early	NISP		4.5						3
	Hanamiai: Phase IV	late	NISP		3						
Huahine	Fa'ahia: lower layer	early	MNI	3							4
Rapa	Tangarutu: Level I	early	MNI	1					3.5		5
	Tangarutu: Level III	late	MNI	2					4		3
Mangareva	Kamaka GK-1: Layers H–J	early	MNI	3	5						1.5
	Kamaka GK-1: Layers C–G	late	MNI	3	1						2
Atolls											
Aitutaki	Ureia: Zones E–G	early	NISPjaw	1		3					2
	Ureia: Zone C	late	NISPjaw	2		4.5					1
Aitutaki	Moturakau: Zones F–H	early	NISPjaw	2		5					1
	Moturakau: Zone D	late	NISPjaw	2		4					1
No. of occurrences (Ubiquity)				12	9	4	2	2	2	2	19

abundances of five most common families (or up to seven in instances of tied ranks). guilds: H = herbivore; IN = invertebrate feeder; P = piscivore; PL = planktivore. Ntaxa Elasmobranchii (sharks and rays) are excluded from these counts.

| | | | | | | | Trolling | | Other | |
Labridae (IN)	Lutjanidae (P)	Lethrinidae (IN)	Holocentridae (PL)	Carangidae (P)	Cirrhitidae (IN)	Muraenidae (P)	Scombridae (P)	Sphyraenidae (P)	Belonidae (P)	Ntaxa
2		5	5							15
4					2					24
1					2.5					18
5										20
4				3	5					17
	4	5		3			1			17
		5		3			2			18
	3.5		5				1			18
	6	6	2				1			23
	2			4.5			1			14
			5	5			5	1	2	16
		5		2			1			26
3.5						2				16
5						1				16
1.5		4								11
4		5								12
5.5	5.5		4							20
	4.5			3						15
4	3									26
3	5									27
12	8	7	5	7	3	2	7	1	1	

These foregoing analytical factors require better control if unbiased intra- and inter-site comparisons are a goal. Table 47.2 takes some steps in this direction, making explicit differences in site type, sampling and recovery procedures, and quantitative measures. Further, recognizing that the biases outlined above could affect comparisons, an ordinal quantitative measure (rank order) is used to assess taxonomic abundances and the focus is on the five most common taxa where sample sizes are comparatively robust (Table 47.3). Table 47.3 also provides information on associated trophic guilds and common capture methods (after Dye, 1983; Butler, 1994; Allen, 2002).

Spatial Patterning

General Observations

The combined assemblages include an NISP of 16,679 fish remains and an additional MNI of 813 (which lack NISP values). The number of fish families varies from eleven to twenty-seven, with an average of nineteen (Table 47.3). Eighteen families figure in the five highest-ranked families (up to seven in case of tied ranks) which account for the bulk ($c. \geq 60\%$) of any given assemblage. In most assemblages, both high- and low-trophic species, and both targeted angling technologies and mass harvesting techniques (netting, weirs, traps) are represented. Ubiquity (% present) values for the main families are as follows: Serranidae (rockcods/groupers, 95%), Scaridae (60%), Labridae (60%), Acanthuridae (surgeonfishes, 45%), Lutjanidae (snappers, 40%), and 35% each for Lethrinidae (emperors), Carangidae, and Scombridae (Table 47.3). Overall, the diversity of top families is remarkable for what are presumed to be closely related peoples deriving from a geographically delimited area (i.e. West Polynesia). However, to a large extent, variation in favoured families is consistent with natural abundances and taxonomic diversity. Acanthuridae and Scaridae are major contributors to natural biomass today, while Serranidae and Labridae are among the five most speciose fish families in the region, comprising 36% of the Pacific Plate non-marginal inshore species (Springer, 1982). Further, both makatea island and sub-tropical adaptations have been previously proposed and are considered below.

Most assemblages are dominated by one of three families: Serranidae, Scaridae, or Scombridae. Excluding Tangatatau Rockshelter (Mangaia), these families combined comprise 35–57% of any given EEP assemblage (Fig. 47.2). Serranidae are a diverse group of carnivores that are not necessarily well represented in terms of natural biomass or density (Pinca et al., 2009). Scaridae are high biomass herbivores that are strongly associated with coral reefs, whereas the piscivorous Scombridae are another high biomass but pelagic family. The EEP assemblages (Fig. 47.2) are of particular interest as they are most likely to reflect pristine marine environments. At Tangatatau, an interior site, the focus was on a local fresh-water Eleotridae (sleeper gobies, 24%), along with Cirrhitidae

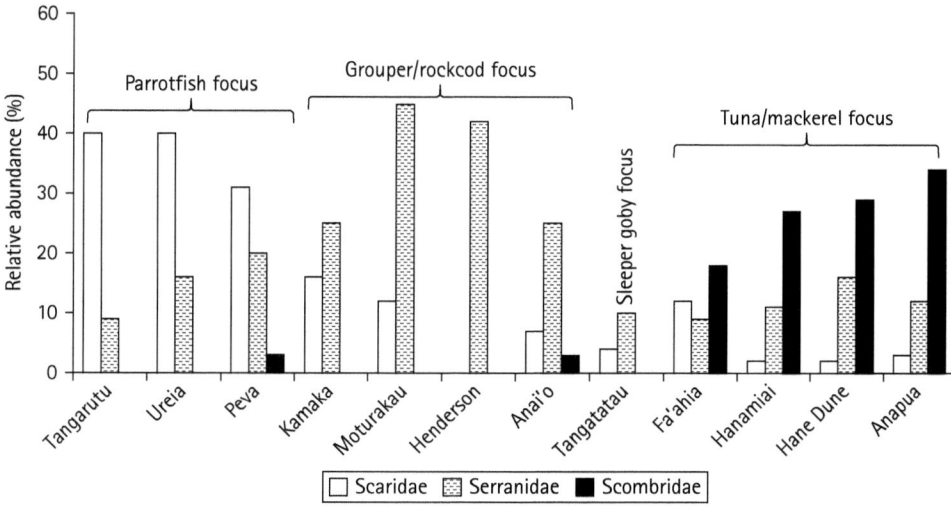

FIGURE 47.2 Regional patterning in the three top-ranking fish families within early East Polynesian assemblages. Author's own image.

(hawkfishes) (15%), imported from the coast. In LEP assemblages, discussed below, these foci hold for most localities, although another four families ascend to top-ranking positions in single assemblages (Table 47.3).

Makatea Island Patterns

The possibility of a makatea-specific pattern has been raised by Weisler et al. (2010), who compared assemblages from Rurutu with those from Mangaia, Ma'uke, Henderson, and Paluki (Niue Island, outside CEP). They draw attention to the poor showing of Scaridae (a coral-grazing family) and the prevalence of serranids. As Table 47.3 illustrates, along with Serranidae (4/4 assemblages), Labridae (4/4) and Acanthuridae (4/4) are major contributors to makatea assemblages. On these islands the lack of lagoons, narrow reefs, poor coral growth, and strong wave action limit the abundance of and access to Scaridae, and perhaps promoted exploitation of taxa that frequent rocky shores and tolerate turbulent surge zones (Walter, 1998; Weisler et al., 2010). However, the larger set of CEP assemblages examined here indicates that Serranidae were important throughout the region.

Notably, Serranidae is a speciose family with considerable variation in habitats, dietary preferences, body size, and life history traits. Sub-familial identifications are critically needed to better understand unique aspects of makatea island fisheries and more generally, Polynesian use of this large and diverse family. Molecular analyses have proven useful in securing sub-familial fish identifications from archaeological remains and warrant further exploration. For example, Nicholls et al. (2003) identified five serranid species on Aitutaki (Motutakau) and determined that *Epinephelus merra* (dwarf spotted rockcod) was an assemblage dominant (66%).

Sub-Tropical Patterns

Vogel and Anderson (2012) propose a distinct sub-tropical (23.5° to 38° S) adaptation. Their Rapa Island assemblages were dominated by Scaridae, Chaetodontidae (butter-flyfishes), Muraenidae (moray eels), and Labridae. They attribute the identification of chaetodontids, an unusual finding, to their use of an extended range of elements. On sub-tropical Rapa Nui (28° S), labrids and eels also were common, especially in late sites where they contribute up to 60%, and chaetodontids are present but rare (Ayres, 1985; Martinsson-Wallin and Crockford, 2002). Tables 47.1 and 47.3 show labrids are important in the sub-tropical assemblages of Henderson and Mangareva, but also at five tropical localities. Moray eels, in contrast, are poorly represented everywhere and chaetodontids are altogether lacking. As with serranids, Labridae is a very diverse family and sub-familial identifications (e.g. Butler, 2001: Table 1) would greatly aid interpretations.

Offshore Fishing

Three Marquesan sites and one Society Island site are unusual in their emphasis on pelagic fishing, with Scombridae contributing between 18% and 34% (Leach et al., 1984; 1997; Fraser, 1998; Davidson et al., 2000). Potential targets were *Katsuwonus pelamis* (skipjack tuna) and *Thunnus albacares* (yellowfin tuna), both common and economically important today. Pelagic fishing also is indicated by the associated artefact assemblages that include pearl-shell trolling lures (e.g. Rolett, 1998: 154). Given that pelagic fishing requires specialized knowledge and fishing gear, as well as seaworthy canoes, it is unlikely that the patterning relates to natural availability alone. In her Pacific-wide survey of tuna fishing, Fraser (2001) argued that the evidence points to targeted capture strategies, rather than opportunistic catches, which she suggests were motivated by cultural choice rather than environmental dictates. Further, there are indications that early CEP tuna fishing was part of a more comprehensive offshore orientation. The foregoing assemblages also contain other carnivorous taxa that are common in offshore waters and typically caught by angling: Serranidae (4/4 assemblages), Lethrinidae (3/4), Carangidae (3/4), and Lutjanidae (3/4). Carangids are the highest-ranked family in an early Rapa Nui assemblage as well, while Scombridae rank third, but samples sizes are small (total NISP 56) (Martinsson-Wallin and Crockford, 2002). These same sites evidence sea mammal (dolphin and porpoise) hunting, probably involving harpooning from canoes (Leach et al., 1984: 189; Steadman et al., 1994: 91).

A pelagic orientation is consistent with people who were comfortable on the open ocean. Early CEP colonists were not only well versed in blue-water sailing, but engaged in long-distance inter-archipelago travel following colonization, especially the inhabitants of the Marquesas (Allen, 2014a). It may be that colonists of the Society and Marquesas Islands (where the region's earliest ^{14}C dates are found), and possibly those of Rapa Nui, had specific cultural links, distinct from those of other CEP settlers.

TEMPORAL TRENDS

General Observations

Eight assemblages allow for assessment of temporal trends (Table 47.3). Although rank orders shift within the main families, the overwhelming pattern is one of stability, suggesting that site patterns are in large part tied to natural abundances. There are no changes in the top five taxa at Moturakau, Tangatatau, Tangarutu, or Kamaka. At another two sites, the highest-ranked families persist, but a greater number of tied ranks in the corresponding LEP assemblages intimate that fishing became more generalized (Hane: two families tie for Rank 5; Anapua: three families tie for Rank 5). At Ureia, Holocentridae (squirrelfishes and soldierfishes) and labrids fall below Rank 5 in the LEP, being replaced by carangids. Only at Hanamiai is there a major change in family composition, with Scombridae and Lutjanidae giving way to Sphyraenidae (barracudas) and Belonidae (longtoms) in late prehistory (Rolett, 1998: 132–45).

Declines in Fishing

Given early losses of native birds on many islands (Steadman, 2006) and, occasionally, animal domesticates (Kirch, 2000), one might expect that marine resources became more important over time. A surprising trend, therefore, is the marked decline in fishbone deposition in some localities. Changes in fish-bone abundance can be systematically evaluated at four sites: Moturakau and Ureia (Allen, 2002), Hanamiai (Rolett, 1998: Table 5.4), and Peva (Weisler et al., 2010: 140). Information on bone abundance (NISP or weight) and sediment volume, allows $NISP/m^3$ to be calculated for strata or analytic zones and variation in the rate of fish-bone deposition to be assessed. At these four sites, fish-bone declines through time, markedly so in three cases (Fig. 47.3). Additionally, casual observations at Anapua (Leach et al., 1997: 62) and Kamaka (Weisler and Green, 2013: 82) hint at reductions in fishing, as does quantitative evidence from Rapa Nui (Ayres, 1985). Changing bone deposition could arise from several processes, including new site functions, over-harvesting, adverse environmental conditions, or new economic priorities, but the regional scale patterning is notable.

On Aitutaki, stable isotopic analyses were undertaken to obtain direct information on human, pig, and dog diets and test the idea that fish remains might have been redirected to animal fodder in late prehistory (Allen and Craig, 2009). Results ($n = 47$) show that, on the whole, late prehistoric individuals were more $\delta^{13}C$ depleted relative to those of

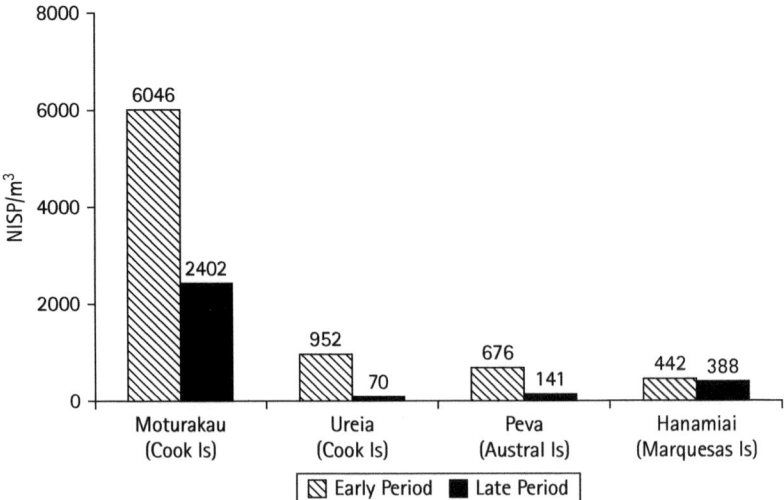

FIGURE 47.3 Changes in fish-bone density (NISP/m³) at four East Polynesian localities. Author's own image.

earlier times, indicating increased consumption of terrestrial C₃ plants and declines in marine protein. Isotopic analyses at Hanamiai also reveal that early colonists and their pigs consumed more marine protein relative to individuals of late prehistory (Richards et al., 2009). Stable isotope analyses from longer-occupied islands to the west hint at comparable trends (Field et al., 2009; Valentin et al., 2011). These findings, over different temporal scales and from diverse geographic settings, suggest that declines in marine fishing may accompany increasing commitments to agricultural activities.

Anthropogenic Impacts

Could these changes derive from resource depression, specifically predator-driven declines in prey capture rates? Foraging theory and the prey choice model in particular, offer formal predictions about predator-prey interactions and long-term outcomes (Butler, 2001). In archaeological applications, prey body size is used as a proxy for return rates and taxa are ranked on this basis. Differentially ranked prey are then examined within 'patches' or habitat types (e.g. inshore fishes) and variability in the ratio of high- to low-ranked prey assessed using abundance indices. Declines in the proportion of high-ranked (large-bodied) taxa, reductions in body size within high-ranked taxa, and increases in diet breadth are expected signals of declines in foraging efficiency and resource depression. Importantly, however, other factors that can affect fish size, such as technological or environmental change, also need to be evaluated (Butler, 2001).

Butler (2001) used the prey choice model and abundance indices to evaluate temporal changes at Tangatatau Rockshelter. She established that there were significant increases in small-bodied marine and fresh-water fishes (those with lengths < 30 cm). To assess whether this trend related to changes in prey demography, she measured the dentaries of two high-ranked prey: *Anguilla* (fresh-water eels; $n = 23$) and Serranidae ($n = 13$). The results for *Anguilla* were inconclusive, but serranids became progressively smaller over time, suggesting resource depression. Butler (2001) was able to discount capture technologies, but the possibility of intra-familial prey-switching remains a possibility.

Allen (2002) also applied the prey choice model on Aitutaki, examining changes in the proportion of large-bodied inshore fish (lengths typically > 35 cm) at two sites. At the permanently occupied mainland site (Ureia), the Large Inshore Fish index declined from 0.51 to 0.31 over the prehistoric sequence, while at the offshore campsite (Moturakau) there was an increase, from 0.20 to 0.35. These results suggest that resource depression probably was not a major factor in the fish-bone declines observed on Aitutaki.

Elsewhere variation in fish size has been considered independent of the prey choice model. At Hanamiai, vertebral measurements (early: $n = 329$; late: $n = 1,360$) indicated marked changes in fish size. However, these were attributed to changes in catch composition, based on the associated families and fishing gear (Rolett, 1998: 142–5). At Peva, the reverse was found (Fig. 47.4), with the LEP assemblage being more variable in terms of fish size and containing a greater proportion of large individuals (Weisler et al., 2010). The authors link these differences to changes in site function, from habitation to religious, and propose the late assemblage represents offerings, illustrating the necessity of considering functional context(s) in inter-assemblage comparisons. Finally, measurements elsewhere indicate stability (Weisler and Green, 2013).

These studies demonstrate the difficulties of unambiguously identifying harvesting pressures with archaeological materials. Even in New Zealand, where sub-familial diversity is reduced, separating anthropogenic impacts from those of climate, technology or prey-switching is challenging (Leach, 2006). More analyses combining formal models, direct measures of body size, and palaeoenvironmental evidence (e.g. Butler, 2001) are needed to better assess harvesting effects and differentiate them from regional-scale processes such as climate change.

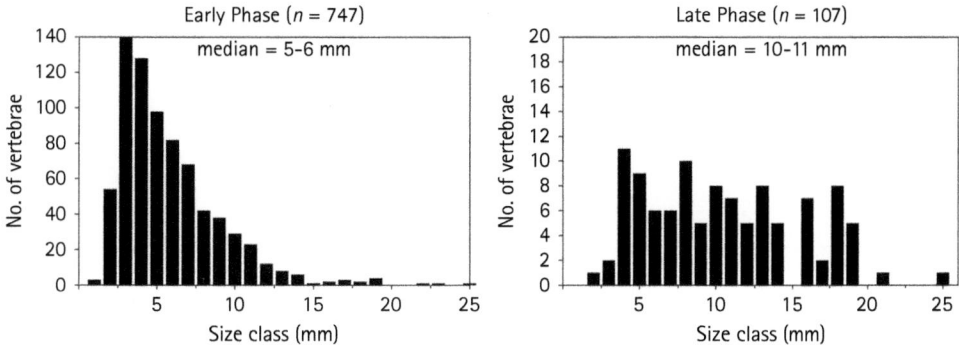

FIGURE 47.4 Temporal patterning in bony fish vertebrae width from Peva site, Austral Islands (data from Weisler et al., 2010). Author's own image.

Intensification of Inshore Fisheries and Mass Harvesting

Several island sequences register compositional changes, sometimes with parallel technological shifts, suggesting intensified use of inshore habitats in late prehistory. Faunal remains from Aitutaki's stratified and well-dated sites (Allen, 2002; Allen and Morrison, 2013) along with stable isotope evidence (Allen and Craig, 2009) and material culture studies (Hiroa, 1927) illustrate this process. Inshore fish are well represented at all times and most assemblages are co-dominated by Scaridae and Serranidae. Carnivorous fish, including Lutjanidae, Carangidae, and Belonidae, however, are best represented in the EEP assemblages. These taxa, along with shell fish-hooks, suggest that early in time fishermen regularly worked the productive outer reef. By the fifteenth to sixteenth century AD, however, both carnivorous fish and fish-hooks are on the decline. In late prehistory, fish-hooks are rare and mid- to low-trophic families, such as Scaridae (herbivore) and Mullidae (goatfishes) (invertebrate feeder), are more abundant (Allen, 2002).

These changes are loosely supported by stable isotope analyses (Allen and Craig, 2009). Although there is variability in human and domesticate δ^{15}N values throughout the sequence, by late prehistory (after c. AD 1650) most individuals ($n = 14$) are more δ^{15}N depleted than those of earlier times ($n = 33$) (Allen and Craig, 2009). The combination of decreases in fish-bones from carnivorous taxa, the loss of shell fish-hooks, and depleted δ^{15}N values suggest changes not only in the kinds of fishes consumed, but also where and how they were caught.

When ethnographer Te Rangi Hiroa (1927) undertook his early twentieth-century research, a diversity of nets, trapping devices, and weirs were widely used on Aitutaki. Contemporary time allocation studies demonstrate that mass harvesting technologies often are more productive than angling or spear-fishing in terms of catch-kilograms per person-hour. They also involve less travel time and fewer risks relative to pelagic fishing. Although netting and trapping probably were used on Aitutaki from initial settlement, given local conditions, their importance apparently increased over time. Other late prehistoric CEP fish-bone assemblages show similar shifts to inshore fisheries (Table 47.4). Ureia, an apparent exception, may signal use of more pristine but distant fishing grounds following local declines in large-bodied fish (Allen, 2002).

CONCLUSIONS

Over the last decade the pace of Polynesian archaeofish studies has accelerated and methodologies are improving (cf. Butler, 1988; Allen, 2003), allowing for this multiscale assessment of local and regional patterns. This review of twelve localities shows that CEP fisheries were regionally diverse, with eighteen families represented in the top five ranks of the twenty assemblages investigated and between eleven and twenty-seven families targeted at any given site. This diversity of fishing patterns is remarkable for

Table 47.4 Temporal variation in rank order abundances across two trophic guilds at sites with EEP and LEP assemblages; — indicates a decline, + an increase, NC = no change (only families in Table 47.3 are considered)

Island	Site	Piscivores			Herbivores		
		Lutjanidae	Carangidae	Scombridae	Acanthuridae	Scaridae	Mullidae
Ua Huka	Hane	—	NC	—			
Ua Pou	Anapua	—		NC		+	
Tahuata	Hanamiai	—	—	—	+		
Aitutaki	Ureia	+	+			—	—
Aitutaki	Moturakau	—				NC	+
Mangareva	Kamaka				+	NC	
Rapa	Tangarutu					—	

a relatively homogenous set of colonists from a circumscribed geographic homeland. However, in many cases cultural patterns appear to be strongly tied to natural abundances and/or speciose families. In most cases three families (serranids, scarids, and scombrids) comprise the bulk of any given assemblage (≥ 60%). The specialized focus on scombrids and other offshore taxa is especially apparent early in time for a subset of islands along a north to far-southeast arc (Societies-Marquesas-Rapa Nui). Given that pelagic fishing requires both specialized gear and knowledge, and that orientation cross-cuts island geographies and latitude, a particularly close cultural historical relationship is suggested. Early assemblages of central and southern archipelagos, in contrast, are dominated by inshore taxa such as scarids, serranids, acanthurids, and labrids.

Over time, serranids, labrids, and scarids emerge as some of the region's most important, enduring, and resilient fisheries. Nevertheless, general declines in fish are prominent at several sites, including Aitutaki (Allen, 2002), Rurutu (Weisler et al., 2010) Tahuata (Rolett, 1998), and Rapa Nui (Ayres, 1985). Harvesting pressures could have played a role, but the evidence to date is not particularly convincing and variability in fish size, where demonstrated, is sometimes related to other processes (e.g. changing site functions).

On Aitutaki, where multiple lines of evidence are available, intensification of inshore fisheries is evidenced, involving a diversity of nets, traps, and more permanent facilities (weirs). These mass harvesting technologies can be both efficient and productive, although perhaps less prestigious with respect to catch. Aitutaki's combined faunal, artefact, and bone chemistry records suggest that these developments followed from, or co-occurred with, diminishing use of carnivorous fish and offshore habitats and increasing commitments to agricultural resources (Allen, 2002; Allen and Craig, 2009). Other regional records also hint that this is a common trajectory (e.g. Field et al., 2009; Valentin et al., 2011). Within the larger set of LEP assemblages there is a shift away from high-trophic families (piscivores) and an increase in inshore families, including mid- to low-trophic taxa (herbivores and invertebrate feeders).

This analysis highlights the considerable methodological challenges of archaeofish studies in tropical environments. Improved regional comparisons will require standardized and explicit analytical protocols and more comparability in the quantitative representation of taxa. Sub-family identifications also are needed to make interpretive headway. Particularly useful in this regard will be in-depth morphological studies of key taxa, continued development of digital resources (e.g. Colley, 2002–2014; Dye and Longenecker, 2004), and additional molecular studies (e.g. Nicholls et al., 2003).

ACKNOWLEDGEMENTS

This research was supported by the Marsden Fund Council, Royal Society of New Zealand Grant 11-UOA-027 (2011–2014) to M. S. Allen, A. M. Lorrey, and M. N. Evans. I thank the volume editors for the opportunity to be involved in this effort and for their helpful comments.

NOTE

1. Scaridae is treated as a distinct family, following Parenti and Randall (2011).

REFERENCES

Allen, M. S. (2002) 'Resolving long-term change in Polynesian marine fisheries', *Asian Perspectives*, 41, 195–212.

Allen, M. S. (2003) 'Human impact on Pacific nearshore marine ecosystems', in Sand, C. (ed.) *Pacific Archaeology: Assessments and Prospects*. Le Cahiers de l'Archéologie en Nouvelle-Calédonie 15, pp. 317–25, Noumea: Le Cahiers de l'Archéologie en Nouvelle-Calédonie.

Allen, M. S. (2014a) 'Marquesan colonisation chronologies and post-colonisation interaction: implications for Hawaiian origins and the "Marquesan Homeland" hypothesis' *Journal of Pacific Archaeology*, 5, 1–17.

Allen, M. S. (2014b) 'Variability is in the mesh-size of the sorter: Harataonga Beach and spatiotemporal patterning in northern Māori fisheries', *Journal of Pacific Archaeology*, 5, 21–38.

Allen, M. S. and Craig, J. A. (2009) 'Dynamics of Polynesian subsistence: insights from archaeofauna and stable isotope studies, Aitutaki, southern Cook Islands', *Pacific Science*, 63, 477–506.

Allen, M. S. and Morrison, A. E. (2013) 'Modelling site formation dynamics: geoarchaeological, chronometric and statistical approaches to a stratified rockshelter sequence, Polynesia', *Journal of Archaeological Science*, 40, 4560–75.

Ayres, B. (1985) 'Easter Island subsistence', *Journal de la Société des océanistes*, 80, 103–24.

Butler, V. L. (1988) 'Lapita fishing strategies: the faunal evidence', in Kirch, P. V. and Hunt, T. L. (eds) *Archaeology of the Lapita Cultural Complex: A Critical Review*. Thomas Burke

Memorial Washington State Museum Research Report 5, pp. 99–115. Seattle: Thomas Burke Memorial Washington State Museum.

Butler, V. L. (1994) 'Fish feeding behaviour and fish capture: the case for variation in Lapita fishing strategies', *Archaeology in Oceania*, 29, 81–90.

Butler, V. L. (2001) 'Changing fish use on Mangaia, southern Cook Islands: resource depression and the prey choice model', *International Journal of Osteoarchaeology*, 11, 88–100.

Colley, S. (2002–2014) 'Archaeological fish bone images, partnership initiative of Dr. S. Colley and University of Sydney Library'. Available at: http://fish.library.usyd.edu.au/.

Dalzell, P. (1998) 'The role of archaeological and cultural-historical records in long-range coastal fisheries resources management strategies and policies in the Pacific Islands', *Ocean & Coastal Management*, 40, 237–52.

Davidson, J. M., Fraser, K., Leach, B. F., and Sinoto, Y. H. (2000) 'Prehistoric fishing at Hane, Ua Huka, Marquesas Islands, French Polynesia', *New Zealand Journal of Archaeology*, 21, 5–28.

Dye, T. (1983) 'Fish and fishing on Niuatoputapu', *Oceania*, 53, 242–71.

Dye, T. S. and Longenecker, K. S. (2004) *Manual of Hawaiian Fish Remains Identification Based on the Skeletal Reference Collection of Alan C. Ziegler and including Otoliths*, Honolulu: Society for Hawaiian Archaeology Special Publication 1.

Field, J. S., Cochrane, E. E., and Greenlee, D. M. (2009) 'Dietary change in Fijian prehistory: isotopic analyses of human and animal skeletal material', *Journal of Archaeological Science*, 36, 1547–56.

Fraser, K. (1998) 'Fishing for Tuna in Pacific Prehistory'. Unpublished MA dissertation, University of Otago (Dunedin).

Fraser, K. (2001) 'Variation in tuna fish catches in Pacific prehistory', *International Journal of Osteoarchaeology*, 11, 127–35.

Gordon, E. A. (1993) 'Screen size and differential faunal recovery: a Hawaiian example', *Journal of Field Archaeology*, 20, 453–60.

Grayson, D. K. (1984) *Quantitative Zooarchaeology: Topics in the Analysis of Archaeological Faunas*, San Diego: Academic Press.

Hiroa, Te Rangi (1927) *The Material Culture of the Cook Islands (Aitutaki)*. Memoirs of the Board of Ethnological Research 1. New Plymouth, New Zealand: Board of Ethnological Research.

Kirch, P. V. (2000) *On the Road of the Winds: An Archaeological History of the Pacific Islands before European Contact*, Berkeley: University of California Press.

Kirch, P. V., Steadman, D. W., Butler, V. L., Hather, J., and Weisler, M. I. (1995) 'Prehistory and human ecology in Eastern Polynesia: excavations at Tangatatau Rockshelter, Mangaia, Cook Islands', *Archaeology in Oceania*, 30, 47–65.

Leach, F. (2006) *Fishing in pre-European New Zealand*. New Zealand Journal of Archaeology Special Publication and Archaeofauna vol. 15. Wellington: New Zealand.

Leach, F., Davidson, J., Horwood, M., and Ottino, P. (1997) 'The fishermen of Anapua Rock Shelter, Ua Pou, Marquesas Islands', *Asian Perspectives*, 36, 51–66.

Leach, F., Intoh, M., and Smith, I. W. G. (1984) 'Fishing, turtle hunting, and mammal exploitation at Fa'ahia, Huahine, French Polynesia', *Journal de la Société des océanistes*, 79, 183–97.

Lyman, R. L. (2008) *Quantitative Paleozoology*, Cambridge: Cambridge University Press.

Martinsson-Wallin, H. and Crockford, S. J. (2002) 'Early settlement of Rapa Nui (Easter Island)', *Asian Perspectives*, 40, 244–78.

Nagaoka, L. A. (1994) 'Differential recovery of Pacific Island fish remains: evidence from the Moturakau Rockshelter, Aitutaki, Cook Islands', *Asian Perspectives*, 33, 1–17.

Nicholls, A., Matisoo-Smith, E., and Allen, M. S. (2003) 'A novel application of molecular techniques to Pacific archaeofish remains', *Archaeometry*, 45, 133–47.

Parenti, P. and Randall, J. E. (2011) 'Checklist of the species of the families Labridae and Scaridae: an update', *Smithiana Bulletin*, 13, 29–44.

Pinca, S., Awira, R., Kronen, M., Chapman, L., Lasi, F., Pakoa, K., Boblin, P., Friedman, K., Magron, F., and Tardy, E. (2009) Cook Islands Country Report: Profiles and Results from Survey Work at Aitutaki, Palmerston, Mangaia and Rarotonga (February and October 2007), Noumea: Secretariat of the Pacific Community, Coastal Fisheries Program.

Pitcher, T. J. (2001) 'Fisheries managed to rebuild ecosystems? Reconstructing the past to salvage the future', *Ecological Applications*, 11, 601–17.

Richards, M. P., West, E., Rolett, B., and Dobney, K. (2009) 'Isotope analysis of human and animal diets from the Hanamiai archaeological site (French Polynesia)', *Archaeology in Oceania*, 44, 29–37.

Rolett, B. V. (1998) *Hanamiai: Prehistoric Colonization and Cultural Change in the Marquesas Islands (East Polynesia)*. Yale University Publications in Anthropology No. 81. New Haven: Department of Anthropology and The Peabody Museum, Yale University.

Springer, V. G. (1982) *Pacific Plate Biogeography, with Special Reference to Shorefishes*. Smithsonian Contributions to Zoology No. 367. Washington, DC: Smithsonian Institution Press.

Steadman, D. W. (2006) *Extinction and Biogeography of Tropical Pacific Birds*, Chicago: University of Chicago Press.

Steadman, D. W., Casanova, P. V., and Ferrando, C. C. (1994) 'Stratigraphy, chronology, and cultural context of an early faunal assemblage from Easter Island', *Asian Perspectives*, 33, 79–96.

Valentin, F., Herrscher, E., Petchey, F., and Addison, D. J. (2011) 'An analysis of the last 1000 years human diet on Tutuila (American Samoa) using carbon and nitrogen stable isotope data', *American Antiquity*, 76, 473–86.

Vogel, Y. and Anderson, A. (2012) 'Prehistoric fishing on Rapa Island', in Anderson, A. and Kennett, D. J. (eds) *Taking the High Ground: The Archaeology of Rapa, a Fortified Island in Remote East Polynesia*. Terra Australis 37, pp. 115–33. Canberra: Australian National University E Press.

Walter, R. (1998) *Anai'o: The Archaeology of a Fourteenth Century Polynesian Community in the Cook Islands*. New Zealand Archaeological Association Monograph 22. Auckland: New Zealand Archaeological Association.

Weisler, M. I. (1994) 'The settlement of marginal Polynesia: new evidence from Henderson Island', *Journal of Field Archaeology*, 21, 83–102.

Weisler, M. I., Bollt, R., and Findlater, A. (2010) 'Prehistoric fishing strategies on the *makatea* island of Rurutu', *Archaeology in Oceania*, 45, 130–43.

Weisler, M. I. and Green, R. C. (2013) 'Mangarevan fishing strategies in regional context: an analysis of fish bones from five sites excavated in 1959', *Journal of Pacific Archaeology*, 4, 73–89.

A GLOSSARY OF ZOOARCHAEOLOGICAL METHODS

MAURO RIZZETTO AND UMBERTO ALBARELLA

Note

This methodological glossary presents brief explanations of the main analytical methods employed by zooarchaeologists and makes reference to those chapters in the Handbook that provide examples of their applications. The aim is to provide non-expert readers with a basic understanding of how the evidence presented in this volume has been obtained. The limitations and biases of each type of analysis are also briefly outlined, as their acknowledgement represents an essential prerequisite to research on archaeological materials. The definitions always refer to zooarchaeological applications of the term, although many of them may be employed in other disciplines with similar or different meanings. Each term makes reference to the chapters where it is mentioned, with the exception of approaches that are invariably adopted (e.g. taxonomic identification, quantification).

Ageing (see Chapters 2, 5–21, 23, 26–31, 37, 39, 42, 43, 45)

The determination of the age-at-death of animals from their archaeological remains can provide insights into culling, hunting, and foraging strategies. Kill-off patterns can be reconstructed when a statistically reliable group of aged remains from the same taxon is considered. For domestic animals, such patterns can be interpreted as the intentional attempt of herders to focus animal exploitation on one or more services/outputs, since each of them requires the animal to be killed at more or less specific age-stages in order to optimize production. Hunters could also prefer (or be forced to choose) specific age-stages, for optimizing the results of hunting activities, maintaining a stable animal population, or for other non-economic reasons. Kill-off patterns are usually represented as bar charts or mortality curves, often in combination with an equivalent survival curve.

Most of the ageing methods devised by zooarchaeologists are based on tooth eruption and wear, as well as the epiphyseal fusion of postcranial bones. The former method provides more detailed and reliable reconstructions of culling strategies, since the gradual eruption and wear of teeth allows the identification of specific age-stages (e.g. Ewbank et al., 1964; Payne, 1973; Grant, 1982; O'Connor, 1988). The different timing at which diaphyses and epiphyses of long bones fuse allows grouping skeletal elements into broader fusion age-stages (e.g. Silver, 1969), providing alternative or complementary ageing information.

Other more sophisticated (though expensive and time-consuming) techniques are also used. Growth rings on teeth and periosteal surface of bones, and the relative frequency or area of osteons observed on bone thin sections can provide ageing information with a variable degree of accuracy (Dammers, 2006). Counts of annual layers of cementum from tooth sections can also be used for determining the age-at-death of an animal and, in some cases, the season in which it died (Stallibrass, 1982).

Information on the overall age-at-death of an archaeological animal population can be provided by those postcranial skeletal measurements that are more susceptible to growth with age. Joints or parts of the bone that are less constricted by other bones or ligaments tend to continue growing even after skeletal maturity; examples include the smallest breadth of the diaphysis of long bones (SD) and the smallest length of the scapular neck (SLC). Variability in such measurements may help in identifying age groups, particularly for very young animals (Rowley-Conwy, 2001).

Attempts have been made to age cattle using size, shape, and texture of horn-cores (Armitage, 1982); the size and complexity of cervid antlers are also correlated to age (Brown, 1983), although other variables are at play (Clutton-Brock, 1981). The incremental structures observed in many mollusc shells and in otoliths (fish ear bones) represent specific growth periods and can be used for ageing (Jones, 1983; Secor et al., 1995).

Allometry This is the study of morphometrical relations between body parts. See **Biometry**.

Analytical error (see Chapters 12, 38)

Any error resulting from the mistaken use of primary data can be defined as such. Analytical errors can be determined by the misuse of analytical methods in relation to the research question or to the nature of the material, by errors made by the researcher when processing data, or by overlooking methodological pitfalls. Analytical errors are distinguished from 'input errors', which are instead the result of recovery biases and other taphonomic processes.

Archaeoentomology (see Chapter 15)

This is the study of insect remains from archaeological sites. Insects may represent reliable palaeoenvironmental indicators and can provide information about the occurrence of animal parasitic diseases (see **Palaeopathology**).

Archaeomalacology (see Chapters 17, 22, 32, 39, 41, 43–7)

This is the study of mollusc remains from archaeological sites. Molluscs (mainly marine) have been extensively exploited as a source of food and raw materials, and their shells (mainly terrestrial) can also be used in palaeoenvironmental reconstructions and seasonality studies (see **Seasonality**).

Artistic sources. See **Documentary and artistic sources.**

Biometry (see Chapters 2, 5–9, 11–13, 15, 16, 18–21, 25, 26, 28, 30, 34, 42, 43)

The size and shape of past animals can be studied through biometrical analysis of their remains. Measurements can be taken in many different ways and this is useful in order to address different research questions, but some form of standardization is valuable to allow direct comparison of different datasets. Many different systems of taking measurements for specific taxa or across different taxa have been proposed, but the most universally used manual for mammals and birds is that of von den Driesch (1976).

Biometry is regularly employed to distinguish between taxa that cannot be easily separated on morphological grounds, such as sheep and goat (Payne, 1969), pig and wild boar (Payne and Bull, 1988), equids (Davis, 1982; Johnstone, 2005), camelids (Wapnish, 1984), felids (O'Connor, 2007), and fish (Libois and Hallet-Libois, 1988). It can be particularly useful for investigating the appearance of domesticates, as these are usually smaller than their wild ancestors (e.g. Albarella et al., 2006; Zeder, 2006; Dobney et al., 2007). Biometrical differences can also be used to separate breeds of the same species (e.g. Harcourt, 1974; De Cupere et al., 2005; MacKinnon, 2010), or identify successful attempts of 'improvement' of domesticates (e.g. Albarella et al., 2008; Davis, 2008), as well as reverse processes of size decrease (e.g. Teichert, 1984; Forest and Rodet-Belarbi, 1998). Biometrical differences can also be the consequence of changing environmental conditions (e.g. Davis, 1981; Hill et al., 2008; Albarella et al., 2009) (see **Palaeoenvironmental reconstruction**).

Genuine size and shape differences between two animal populations, however, can be masked or over-emphasized by the existence of different sex ratios, age patterns, or pathological incidences. Sexual dimorphism affects certain elements more than others (e.g. distal metapodials, distal humerus); these can be ruled out to avoid the interference of sexual dimorphism, or purposely selected when investigating changes in the sex ratio or the relative proportions of male, female, and castrate individuals. The same procedure can be applied for those measurements that are more affected by ageing (e.g. the smallest breadth of long bone diaphyses) (see **Ageing**). Teeth are relatively unaffected by sexual dimorphism and age and tend to react slower to size changes, unless new genetic stock is introduced into a population; this latter property of tooth biometry can be employed, for example, to distinguish between local improvements and introduction of new breeds (e.g. Albarella et al., 2008).

Biometrical analyses are usually visualized in histograms or scatter plots, and results are often supported by statistical tests. The analyses can be applied to anatomical elements individually or by merging the available data through a scaling index technique (see **Scaling index technique**). Single measurements can also be used to estimate the withers height of an animal.

Butchery (see Chapters 2, 3, 8, 10, 11, 13–15, 21, 26, 28, 30, 31, 34, 36–9, 43, 45)

Butchery marks generally refer to any traces left by humans when processing animal carcasses; such activities include culling, skinning, splitting of the carcass, and removal or extraction of flesh, fat, ligaments, and marrow. A basic classification of the types of butchery marks may distinguish between cuts, chops, saw marks, and percussion marks, although much more detailed classifications have been produced on the basis of size, morphology, and location of marks (e.g. Lauwerier, 1988). Attempts are sometimes also made to distinguish between fractures occurring on fresh and dry bones (e.g. Binford, 1981), which has obvious consequences for their interpretation.

Butchery evidence tends to be recorded more intensely on large species (e.g. cattle), due to the multiple steps required by the processing of larger carcasses and the reduction of larger muscle masses into sizeable portions. For these animals, it is usually possible to detect and distinguish between primary and secondary butchery; the former consists of the initial removal of inedible or less-meaty appendices (e.g. lower limb bones, head) and a first dismemberment of the carcass, while the latter is much more variable and includes further sub-division into smaller units of meat, careful cutting or more energetic chopping through the joints, filleting, marrow extraction, and evidence for curing. In addition, the inedible wastes of carcass-processing can represent an abundant and valuable source of raw materials, and have been variably exploited for the production of functional, ritual, and ornamental items (Choyke and O'Connor, 2013). A number of manufacture traditions even developed into specialized industries, as in the case of leather production, horn-working, or comb-making (e.g. Albarella, 2003; Rijkelijkhuizen, 2011).

A better understanding of the nature and purpose of carcass-processing marks can be achieved through experimental studies (e.g. Jones, 1980; Seetah, 2008; Lloveras et al., 2009).

The analysis of butchery marks, along with the distribution of body parts, the material culture recovered from the site, and documentary sources, can provide information on the types of tools employed for carcass-processing and their purpose (e.g. Seetah, 2006; Maltby, 2007).

Comparative collection. See **Reference collection.**

Coprolites (fossilized dung). See **Dung analysis.**

Dating (most chapters)

Faunal assemblages are organized into chronological phases according to the dating of archaeological contexts. Often this is achieved through a typological analysis of associated artefacts (relative dating), but direct dating of archaeological materials, including fauna, is also sometimes possible (absolute dating). A number of dating techniques based on the radioactive decay of naturally occurring elements (e.g. ^{14}C, Uranium-series, Potassium-Argon, and thermoluminescence) or the magnetic properties of some minerals can be used.

Because of their organic component and abundance, faunal remains are among the best archaeological materials for ^{14}C (radiocarbon) dating, when they are less than approximately fifty thousand years old. For this to be effective it is important that the association between faunal material and stratigraphic context is secure; articulated bones in primary deposit are therefore ideal for radiocarbon dating.

The identification of allochthonous species can be used to establish a *terminus post quem* for the formation of the deposit under consideration, should the date of arrival of such species be known. This is the case, for example, of the deliberate introduction of domesticates or wild animals in some areas, or of the accidental spread of commensal species (see examples in O'Connor and Sykes, 2010). For more ancient periods and larger timespans, the rapid speciation undergone by some micromammals can be used in relative dating; the evolution of three sub-species of water vole with different dental conformations, for example, allowed the reconstruction of a chronological sequence for the European Pleistocene (Wilkinson and Stevens, 2008; though see Martin, 2014 for a critique of the method). The succession of distinctive combinations of mammal species (biostratigraphy) during the Pliocene and Pleistocene, as a result of consistent climatic changes, allowed identification of specific biochrones, which can be used for dating archaeological and palaeontological contexts (e.g. Currant and Jacobi, 2001; Kahlke et al., 2011).

Density (see Chapters 5, 6, 41)

Density represents the quantification of animal remains (either wholly or for specific taxa) per volume unit. Density analysis can provide information on taphonomic processes, differential use of archaeological contexts, duration, and nature of site occupation, intensity of animal exploitation, and formation of archaeological contexts.

Diagenesis This term describes the physical, chemical, and biological processes of decay undergone by organic remains after burial. It refers to all taphonomic processes occurring after burial and before excavation (see **Taphonomy**).

Diagnostic zone method This is a recording method that identifies a pre-determined set of anatomic zones as those requiring consistent recording and counting. These 'zones' are known as 'diagnostic' because they are generally chosen on the basis of their degree of identifiability at the anatomic and taxonomic level. Diagnostic zones are also selected on the basis of their likely high chance of preservation in the archaeological context and the amount of information that they can provide. The method was first introduced by Watson (1979), but has then been adopted and developed by many other researchers (e.g. Dobney and Rielly, 1988; Serjeantson, 1991; Davis, 1992; Albarella and Davis, 1994).

The diagnostic zone method allows researchers to overcome many of the biases related to the nature of archaeological faunal assemblages and the different conditions under which research is conducted. The degree of preservation of faunal assemblages is highly variable, depending on a wide range of largely unpredictable taphonomic processes (see **Taphonomy**); in addition, researchers with different experience and access to different facilities will tend to produce results of variable detail. All these variables must be taken into account, with the result that, in most cases, different studies conducted in different research environments cannot be reliably compared. The selection of diagnostic zones, however, mitigates problems deriving from unpredictable taphonomic variables, subjectivity, and other research biases; at the same time, the method counteracts over-representation of larger species, over-counting of fragments from the same element, and, to some extent, differences in skeletal complexity between taxa.

In order to avoid potential loss of information, a diagnostic zone method can be integrated with the recording of 'non-countable' specimens, which will provide qualitative rather than quantitative information. The number of chosen diagnostic zones can vary substantially, but, in case of highly fragmented assemblages, the number of chosen zones will need to be large enough to capture sufficient information.

Distribution of anatomical elements (see Chapters 2, 8, 10, 13–15, 22–4, 27–32, 36, 39, 41, 45, 46)

The distribution of anatomical elements is the result of human choices of carcass processing and disposal as well as post-depositional taphonomic processes. These latter may generate biases such as the preferential destruction of less robust elements (e.g. Brain, 1967), the partial recovery of smaller bones and teeth when sieving is not undertaken, and the further selection imposed by the recording process. Once the extent of such biases has been acknowledged, information on carcass-processing and disposal can be extrapolated. For food mammals, a distinction is commonly made between meat-bearing (e.g. upper limb bones) and non-meat-bearing elements (e.g. lower limb bones, cranial elements). A higher incidence of the former tends to suggest the prevalence of waste from consumption and secondary butchery, whereas primary butchery will generally result in a predominance of non-meat-bearing elements (see **Butchery**). Alternatively, an even distribution of body parts can be interpreted as the result of unspecialized butchery and/or waste disposal practices. Butchery practices can, however, be determined by the need to maximize energy optimization, as well as by cultural and religious traditions (e.g. Milne and Crabtree, 2001).

Ritual practices, bone-working, and the presence of specific animal products (e.g. hides, furs) can produce distinctive body part distributions (e.g. Albarella, 2003).

The analysis of body part distribution can be undertaken using a variety of different methods, but these will need to consider taphonomic biases as well potentially uneven distribution of anatomical elements in different taxa.

DNA. See **Genetics**.

Documentary and artistic sources (see Chapters 3, 8–16, 21, 26, 29, 31, 36, 40, 41)

This is the study of textual evidence, and iconographic and other artistic reproductions for the investigation of the economic, religious, and symbolic role of animals in past societies. These studies can complement the results of zooarchaeological analysis (e.g. MacKinnon, 2001; Soderberg, 2004).

Dung analysis (see Chapters 10, 16)

This is the analysis of animal palaeofeces recovered from archaeological contexts. The morphology and content of dung is distinctive for each species and can signal an animal's presence even when its physical remains are not recovered. The content of animal dung can provide information on diet, health, and palaeoenvironmental conditions (e.g. Maldre, 2006); dung could have also been used as fuel or building material (Lancelotti and Madella, 2012).

Epiphyseal fusion. See Ageing.

Ethnozooarchaeology (see Chapters 10–12, 20, 21, 27, 36, 41, 44, 46)

This is a zooarchaeological research strategy (David and Kramer, 2001) that employs ethnographic evidence as an aid to the interpretation of faunal assemblages from archaeological sites (e.g. Binford, 1978; Bartosiewicz and Greenfield, 1999; Albarella and Trentacoste, 2011; Halstead, 2014).

Experimental zooarchaeology Zooarchaeologists can learn from the performance of experiments involving animals, animal remains, or animal products in order to formulate or verify theories and interpretations of past animal exploitation practices.

Experiments have been classified as imitative (when reproducing activities performed in the past), reconstructive (when aiming at conserving or recreating rare or extinct species and breeds), or as reproductions and observations of taphonomic processes (van Wijngaarden-Bakker, 1987). This latter approach represents a prominent line of research in zooarchaeology, due to the important impact that taphonomic biases have on the nature and preservation of faunal assemblages. Experiments focusing on butchery and cooking practices, as well as diagenetic processes and recovery biases, have improved our ability to evaluate taphonomic biases (e.g. Payne, 1975; Nagaoka, 1994; Nicholson, 1996; Fernández-Jalvo and Andrews, 2003). In addition, they have also helped the identification of potential cooking patterns (e.g. Lupo and Schmitt, 1997; Vacca, 2008).

Most ageing and sexing methods used by zooarchaeologists are also based on observations of modern animal populations (e.g. Payne, 1973; Munro et al., 2011; Popkin et al., 2012).

Fragmentation (see Chapters 8, 24, 37, 45)

This is the degree of breakage of animal remains. It can provide information on the nature and impact of taphonomic processes, including specific butchery activities and recovery biases (e.g. Outram, 2001; Maltby, 2007) (see **Taphonomy**).

Frequency of species (all chapters)

This is the relative abundance of identified taxa within a faunal assemblage. It relies on what has been recorded and the adopted quantification methods, and can be visualized in charts, ternary plots, or tables. The changing frequency of one or more taxa in relation to others can be investigated through the use of indices (e.g. the cattle index used by Badenhorst, this volume).

Genetics (see Chapters 7, 12–14, 16, 18–21, 25, 32, 42, 43, 47)

This is the study of genetic information from modern and ancient animal specimens to aid zooarchaeological research; it is mainly based on nuclear and mitochondrial DNA (mtDNA).

Indirect genetic evidence can also be obtained from the amino-acid sequences of proteins, which are encoded in the DNA (Weiner, 2010).

DNA analysis on archaeological specimens can provide a range of data, including taxonomic identification, sex, and inherited pathological conditions, as well as genetic affinity to a known haplotype. This latter line of research has contributed to the identification of the wild ancestors of many domestic species. In particular, the combination of studies on modern and ancient DNA (aDNA) is employed to reveal the geographical area of origin of domestication and patterns of diffusion of domesticates into other areas (e.g. Larson et al., 2005; Beja-Pereira et al., 2006; Cai et al., 2011). Comparisons between mtDNA (maternally inherited) and Y-chromosome (only present in male individuals) studies allow investigation of differences in genetic contributions on the maternal and paternal lines (e.g. Götherström et al., 2005).

Preservation biases and high costs restrict the use of DNA to specific case studies, which can nonetheless complement and verify the results from other types of analysis (e.g. Davis et al., 2012; Telldahl et al., 2012).

Geometric morphometrics This is the study of the shape of animal remains based on the location of a determined set of features or landmarks. Variations on the position of landmarks in relation to each other on different specimens are analysed using statistics; the morphological differences highlighted by such methods can be used in taxonomic identification or in the analysis of morphometrical changes within an animal species or other taxonomic group (e.g. Bignon et al., 2005; Ottoni et al., 2013).

Iconography. See **Documentary and artistic sources.**

Identification. See **Taxonomic identification.**

Input error. See **Analytical error.**

Isotope analysis. See **Stable isotope analysis.**

Kill-off pattern. See **Ageing.**

MAU (Minimum number of Animal Units) A quantification method calculated from the MNE, whereby this latter is adjusted according to the frequency of each element within the skeleton of a species (Binford, 1984; Lyman, 2008). In order to allow comparisons between different samples, values can be expressed as a percentage of the highest MAU value.

Microwear. See **Tooth microwear.**

MNE (Minimum Number of anatomical Elements) (see Chapter 5)
This is a quantification method that identifies the minimum number of anatomical elements necessary to account for the remains observed for each taxon.

The nature of the basic unit count for this method—the 'element'—has been widely debated. According to different researchers, it can refer to complete anatomical elements, incomplete bones (whose integrity can be estimated as a percentage), anatomical parts 'reconstructed' by overlapping landmark features, or skeletal portions (e.g. lower fore limb, thoracic girdle); in addition, the MNE can be calculated by considering or disregarding differences in age, sex, and size (Grayson, 1984; Reitz and Wing, 1999; for an overview of different approaches, see Lyman, 2008). Paired elements can be distinguished between left and right.

The MNE provides the basis for calculations of the MAU.

MNI (Minimum Number of Individuals) (see Chapters 13, 27, 28, 31, 36, 45–7)

This is a quantification method that identifies the minimum number of animals necessary to account for the number of remains observed for each taxon. The MNI can be regarded to be either the highest MAU or MNE for each taxon.

Since the unit of analysis (and interpretation) is the entire animal, differences in skeletal complexity between taxa do not bias this method, which is also less affected than others by recovery bias, as it relies on the most common element, regardless of its size. At the same time, however, it may lead to an over-estimation of rarer species within an assemblage (especially when samples are small) and may vary substantially with different aggregations of archaeological contexts (Grayson, 1984; Lyman, 2008). In addition, adopting single individuals as the unit of analysis may be misleading due to the fact that meat exploitation may have focused on carcass portions rather than entire animals, which would have been processed and disposed of accordingly (Binford, 1978).

It is important to consider that the main aim of the MNI is not to calculate, or even approximate, the actual number of animals present on site at a given time—very rarely possible for archaeological assemblages—but rather provide another measure of relative taxonomic frequency.

Morphometry. See **Biometry.**

NISP (Number of Identified Specimens) (most chapters)

This is a quantification method that considers every countable element for each taxon, as defined by the counting method employed. Although it overcomes the problem of aggregation and over-estimation of rarer species (see **MNI**), it does not take into account the different size and skeletal composition of different taxa, leading to an over-estimation of larger species. For this reason, some authors provide adjusted measures of NISP (e.g. Albarella and Davis, 1996). In addition, the NISP introduces the problem of interdependence, whereby more fragments from the same animal may contribute to the relative frequency of a given species (Grayson, 1984; Lyman, 2008).

The basic unit count for this method—the 'specimen'—can be variably interpreted, either explicitly (see **Diagnostic zone method**) or implicitly.

Organic residue analysis (see Chapters 6, 8, 14, 17, 20)

Ancient organic materials can sometimes be detected on tools and containers recovered from archaeological sites. Organic materials that are less susceptible to diagenetic processes (e.g. fats and resins) can be identified through chemical analysis, mass spectrometry, or gas liquid chromatography (Weiner, 2010). Organic residue analysis has been contributing to many important archaeological questions, such as food preferences, cuisine and the origins of domestication and dairying (e.g. Spangenberg et al., 2006; Evershed et al., 2008; Mukherjee et al., 2008; Salque, 2012).

Osteometry. See **Biometry.**

Palaeoenvironmental reconstruction (see Chapters 6, 16, 19, 39–41, 45)

The reconstruction of past environmental conditions represents an important aspect of zooarchaeological research, though some taxa are better than others for this purpose (Evans and O'Connor, 1999; Wilkinson and Stevens, 2008). Micromammals, land molluscs, and insects tend to be sensitive to environmental conditions, and can provide valuable information about vegetation cover, climate, moisture level, and many other palaeoenvironmental features (e.g. Montuire, 1999; Davies, 2008; Vickers et al., 2011).

Larger mammals are usually more flexible in terms of environmental adaptability and often move or migrate over large distances; however, their presence or absence can still provide a broad indication of regional environmental conditions (e.g. Jousse and Escarguel, 2006; Kahlke et al., 2011). In addition, biometrical changes can be the result of important climatic developments, since body size and temperature tend to be inversely related in large mammal species (e.g. Davis, 1981; Bell and Walker, 2005). Stable isotope analysis can also be relevant to environmental analysis (see **Stable isotope analysis**).

Palaeogenetics. See **Genetics.**

Palaeopathology (see Chapters 6–8, 19, 20, 23, 29, 38)

In zooarchaeology, palaeopathological studies try to understand how animal disease may be related to, or can affect, the interaction between humans and animals.

As in humans, pathological manifestations in animals can be the result of developmental deficiencies or anomalies, ageing (including arthropathies), traumatic lesions, inflammatory diseases (infections), parasitic activities, and other environmental stresses, as well as inherited (genetic) disorders. The pathological evidence detected on faunal remains is often non-specific and its interpretation is hindered by equifinality, whereby different pathological conditions would result in similar reactions and manifestations (Bartosiewicz with Gál, 2013). Since palaeopathological analyses on archaeological materials are limited to observations of the normally preserved hard tissues (e.g. bones, teeth, and shells) those diseases leaving no traces on such tissues will not be detected. In the absence of soft tissues, indirect evidence for these pathological conditions can be provided by, for example, the recovery of parasites or carriers of parasites, the occurrence of animal mass graves suggestive of epidemics, documentary sources, and studies combining environmental conditions and animal biology (Baker and Brothwell, 1980).

Particular attention has been paid by zooarchaeologists to those pathological insurgences resulting from human exploitation of animals. Animal domestication led to a number of anatomical, biological, and behavioural changes that resulted in an increase in developmental anomalies and pathological conditions. Overcrowding and intensive husbandry practices could also facilitate the insurgence and spread of diseases, while protracted inbreeding would have increased the incidence of genetic anomalies. A number of bone and dental pathologies are the direct result of the use of animals as a workforce. Overworking in draught cattle, for example, could lead to arthritic conditions in the shoulder and pelvic girdle as well as on the lower limb bones, while horseback riding often resulted in spondylotic fusion and ankyloses of vertebral elements (Bartosiewicz et al., 1997).

Quantification (all chapters)

Through quantitative methods zooarchaeologists make estimates of the relative frequencies of different categories, such as age, sex, and size groups, but, most typically, the term 'quantification' is associated with an evaluation of body part and taxonomic frequencies. Quantitative methods can rely on raw fragment counts (e.g. NISP), adjustments of raw fragment counts based on animal anatomy (e.g. MNI, MNE, MAU), or meat weight estimations. More than one counting method can be included in zooarchaeological analyses, in order to enhance comparability with other studies and evaluate the biases generated by different methods.

Recording (all chapters)

This is the initial process of selection and data collection performed by zooarchaeologists on faunal assemblages. The material is recorded in order to provide quantitative and qualitative information to be used in data analysis and interpretation.

The set of recorded data varies according to the type and degree of preservation of each specimen, as well as to research questions, experience of the researcher, and time and facilities available.

Recovery bias (see Chapters 5, 6, 10–12, 14, 15, 17, 20, 22, 32, 33, 40, 46)

The retrieval of faunal remains from archaeological excavations is almost invariably incomplete, and this will generate a bias. Typically, the bias consists of an under-representation of smaller species and smaller anatomical elements. The extent of recovery bias is determined by excavation and collection strategies, as well as by a range of case-specific variables.

When only hand collecting is implemented, the quality of recovery relies on variable circumstances such as the physical characteristics of the soil, work and climatic conditions, excavation techniques (e.g. shovelling, trowelling), and the excavators' experience and tiredness.

The impact of such variables is considerably reduced by sieving (especially water-sieving) and flotation. The size of the sieve mesh plays an important role in rectifying the recovery bias produced by hand collection; small meshes are obviously more efficient but require longer time to process samples, and choices must be made according to research questions and time/money constraints. An alternative solution is to devise a systematic strategy for the sieving of sub-samples. Due to the qualitative and quantitative differences of sieved and unsieved samples, it is important to keep sieved material separate from the rest; mixing them would simply blur the recovery bias of hand-collected remains and compromise the information added by sieved materials. There will be loss of faunal material even through the sieving mesh, but at least sieving will allow monitoring of the degree of loss, as indicated by the mesh size.

The importance of sieving (or acknowledging the stronger recovery bias of unsieved assemblages) has been the focus of a number of experimental studies employing modern and archaeological materials, which highlighted how partial recovery can result in a biased frequency of species, lower taxonomic abundance, and misleading distribution of anatomical elements (e.g. Payne, 1975; Nagaoka, 1994; Zohar and Belmaker, 2005).

The extent of the recovery bias can be estimated in order to verify the reliability of zooarchaeological analyses. Methods of bias estimation may rely on comparisons of the frequency of a small element in relation to a large one it articulates with (e.g. astragali and distal tibiae, first phalanges and distal metapodials, loose teeth and jaws), with the assumption that adjoining elements are likely to be deposited in roughly equal quantities; such frequencies are then compared between species of different size.

Reference collection This is a collection of known animal specimens used for taxonomic and anatomic identification of unknown faunal remains. Reference collections often specialize in specific geographical areas and/or tend to focus on some classes of animals, unless owned by large museums, and therefore not primarily focused on their zooarchaeological potential. They can be organized by individual or anatomical element and mainly consist of modern material, although ancient specimens can also be included (e.g. extinct breeds or species). For ease of use bones are generally kept loose rather than as mounted skeletons, as the latter are more appropriate for anatomical studies and public displays.

Scaling index technique (see Chapters 12, 42)

This is a biometrical method that compares different measurements taken on archaeological material to a standard set of the same measurements, obtained from either modern or ancient material. The absolute values of different measurements are thus converted into relative values, which can be plotted on the same scale. This can result in a substantial amplification of the size of the sample being analysed; however, some measurements are affected to different extents by different biological variables, compromising the readability of the results (i.e. potential patterns or changes are either masked or over-emphasized) (Meadow, 1999; Albarella, 2002). For this reason, measurements should be selected according to the research question. In addition, length, breadth, and depth measurements should ideally be considered separately, as dimensions lying on different axes can react differently and their analysis may reveal changes in shape (Davis, 1996; Meadow, 1999). Even more importantly, tooth and bone measurements, which respond differently to environmental and genetic factors, need to be analysed separately. A very important added value of this method is that, even when measurements are plotted separately, they can be compared directly as they will be using the same scale. This can allow inferences regarding shape and not just size.

Different scaling techniques have been used by zooarchaeologists since the introduction of the size index (SI) in the late 1960s (Ducos, 1968); Uerpmann (1979) re-elaborated the same concept by calculating the relative size index (RSI) and the variability size index (VSI); Meadow (1981) sought to simplify Uerpmann's indices by adopting Simpson et al.'s (1960) log size index (LSI) (Meadow, 1999). The LSI calculates the decimal logarithm of the ratio between each measurement and the standard (i.e. of the SI); this latter method is currently the most widely used in zooarchaeology, since it is easy to calculate and provides a better visual representation of results (Albarella, 2002; Albarella and Payne, 2005).

Sclerochronological analysis (see Chapter 43)

This is the study of incremental growth structures in the hard tissues of some invertebrates (e.g. mollusc shells). It provides information on ageing, palaeoenvironmental conditions, and on seasonality of human exploitation of these resources (see **Ageing, Palaeoenvironmental reconstruction, Seasonality**).

Seasonality (see Chapters 10, 13, 32, 34, 37, 40, 43, 45, 46)

Zooarchaeologists are interested in the season when an animal died as this can provide important information about past human behaviour. Seasonality is here intended in terms of warm/dry and cold/wet seasons, whose alternation affects population dynamics of wild species and decision-making of herders.

Patterns of seasonality are determined by a number of variables, including the biological characteristics of animal species, environmental conditions, residence patterns, technology, and the level of success of agricultural and husbandry practices, as well as the economic and socio-political choices of a community (Reitz and Wing, 1999). Seasonal periodicity can be investigated through the presence and incidence of some animal species (e.g. migratory birds, anadromous fish from inland sites), stable isotope analysis (e.g. the $^{18}O/^{16}O$ ratio from the last growth ring on mollusc shells, reflecting a warm or cold season), observation of incremental growth structures (on mammal teeth, mollusc shells, bird medullary bone, on some fish bones and otoliths) or seasonal anatomic features (e.g. antlers in cervids, when the pedicle is present), ageing (e.g. evidence for seasonal culling of juvenile mammals), biometry, and disposal patterns (for some examples, see Lefèvre, 1997; Cerón-Carrasco, 1998; Uchiyama, 1999; Higham and Horn, 2000; Pike-Tay and Cosgrove, 2002; Balasse et al., 2012; Colonese et al., 2012).

Zooarchaeologists may in some cases detect the seasonality of a certain activity (e.g. hunting or foraging patterns, culling of domesticates) but this does not necessarily coincide with the seasonal occupation of a site.

Sexing (see Chapters 8, 12, 13, 16, 20, 23, 26, 30, 31, 37, 43)

The sexual identification of animal remains relies on morphological, physiological, and biometrical characters, as well as genetics. Most sexing methods based on morphological and physiological characters are conceived for specific taxa. A study of the shape and size of cattle horn-cores from modern animals has been used as an aid to sex remains from archaeological sites (Armitage, 1982). Attempts have been made using other morphometric traits of the skull and a number of postcranial bones (Grigson, 1982a; 1982b); in particular, the conformation of the pelvis proved to be useful in distinguishing sexes in bovids (Grigson, 1982a; Prummel and Frisch, 1986), and the same method has been extended to other ungulates (Greenfield, 2006). In other species, some specific anatomical elements are distinctive of male or female individuals.

Antlers, for example, are only grown by male deer, with the exception of reindeer/caribou (*Rangifer tarandus*) (Schmid, 1972). Similarly, a number of animal species (including some primates, rodents, insectivores, and carnivores) have a penis bone (*baculum*) in males and a clitoral bone (*baubellum*) in females.

Male birds grow a spur near the distal end of the tarsometatarsus; its presence or absence is often used to estimate sex-ratios in faunal assemblages, especially within large samples of domestic fowl remains (Sadler, 1990). Limitations to this method include the timing of appearance of the spur (immature specimens cannot be sexed) and the effects of caponization (i.e. castration), which results in a more or less visible scar on the bone (West, 1982). Birds can also be sexed by verifying the presence or absence of the medullary bone, which is deposited by laying birds inside their long bones (Driver, 1982).

Male suids develop clearly distinctive canines, which can be easily distinguished from female teeth.

Biometrical analysis of sexually dimorphic elements rarely allows sex identification on individual specimens but can provide estimates of the sex-ratio at the population level (see **Biometry**).

The analysis of aDNA allows sexing even small fragments with a certain degree of confidence, although it remains bounded to preservation biases, methodological limitations, and financial constraints (Weiner, 2010) (see **Genetics**). For this reason, it is performed only occasionally and in specific circumstances; DNA sex identification on sub-samples, for example, can verify the validity of more widely applicable sexing methods, such as biometry (e.g. Davis et al., 2012; Telldahl et al., 2012).

Sieving. See **Recovery bias.**

Site catchment analysis (see Chapters 10, 11, 14, 16, 32, 33, 40, 41, 44–6)
This is the study of the area of procurement of natural resources ordinarily used by the members of a community for their basic needs (Higgs and Jarman, 1975). Procurement strategies within the catchment area evaluate costs, risks, and advantages; these, in turn, depend on a number of variables, such as resource availability, technology, residence pattern, and a range of economic and socio-political choices (Reitz and Wing, 1999).

Information on a site's catchment area can be provided by frequency of animal species, taxonomic abundance, density of remains, and biometry. Zooarchaeological analyses can highlight evidence for over-exploitation of faunal resources, extent and nature of the territory exploited, environmental changes, and changes in procurement strategies (i.e. choices and technology). Evaluation of the roles of all these variables, however, must be supported by the analysis of complementary contextual information.

Spatial distribution (see Chapters 8, 14, 23, 32, 40, 41)

Patterns of spatial distribution can provide information on the nature and scale of animal exploitation, site function, area use, and waste disposal practices (e.g. Enloe et al., 1994; Navas et al., 2008). Information from faunal remains should be complemented by other lines of archaeological evidence (e.g. structures and features, other materials), and can be biased by the partial excavation of a site.

Species frequency. See **Frequency of species**.

Stable isotope analysis (see Chapters 4, 10, 16, 17, 19, 21, 25, 26, 28, 32, 34, 39, 42, 47)

The analysis of the relative frequencies of stable isotopes of specific elements can be useful for the investigation of animal diet, management, and provenance, as well as the reconstruction of palaeoenvironmental conditions.

Each chemical element is represented in nature by a number of isotopes, which have the same number of protons (i.e. the same chemical properties) but a different number of neutrons (i.e. atomic weight, determined by the sum of protons and neutrons). Isotopes can be unstable (or radioactive), when the excess of neutrons causes instability and gradual release of energy (e.g. ^{14}C, which decays into ^{12}C), or stable. The proportion of stable isotopes in the biosphere varies according to various chemical processes and can be analysed to investigate a number of issues (Price et al., 1985). In zooarchaeology, samples for isotopic analysis can be extracted from both the organic (collagen) and inorganic (hydroxyapatite) components of bones, from tooth enamel and from mollusc shells. The different isotopic compositions of carbon ($^{14}C/^{13}C$), nitrogen ($^{15}N/^{14}N$), oxygen ($^{18}O/^{16}O$), sulphur ($^{34}S/^{32}S$), and strontium ($^{87}Sr/^{86}Sr$) measured on samples of animal tissues are a reflection of the food intake, climatic conditions, and local isotope geochemistry, and therefore can provide important dietary and environmental information (e.g. Bocherens and Drucker, 2003; Stevens et al., 2008; Rawlings and Driver, 2010). Since tooth enamel is not remodelled once formed, intra-tooth variability in isotopic composition can be analysed to detect changes in an animal's diet and residence (Balasse, 2002), providing useful insights in animal management and movement (e.g. Viner et al., 2010; Balasse et al., 2012).

Taphonomy (see Chapters 2, 4–6, 11, 14, 24, 29, 31, 32, 36, 37, 40, 41, 43, 46)

Taphonomy was defined by Efremov (1940, 85) as 'the study of the transition (in all of its details) of animal remains from the biosphere to the lithosphere'. Post-mortem taphonomic processes in particular are the focus of zooarchaeological analyses, as they usually include most of the modifications undergone by animal remains and, more importantly, most of those resulting from human animal exploitation, such as butchery, selection of body parts, and burning. Non-human modifications include gnawing and partial digestion by other animals, as well as exposure to weathering agents (e.g. water, wind). The incorporation of animal remains into archaeological deposits (either as deliberate burial or through sedimentation) results in further modifications and selection of the remains. The composition, acidity, and humidity of sediments surrounding animal remains can affect their preservation by initiating, accelerating, or arresting physical and/or chemical processes; the type and extent of microbial attack also depends on such variables. Finally, excavation, recovery, analysis, and storage can also affect the preservation of animal remains (Lyman, 1994).

The nature and extent of taphonomic alterations can be estimated by analysing the taxa and anatomical elements present, the degree of fragmentation, and the type and degree of macroscopic and microscopic diagenetic (i.e. after burial) surface modifications (Lyman, 1994; Child, 1995; Weiner, 2010). Such analyses and the ability to recognize taphonomic biases have been improved by a number of experimental studies (e.g. Shipman et al., 1984; Nicholson, 1996;

Fernández-Jalvo and Andrews, 2003). Aside from estimating effects on preservation, the analysis of taphonomic processes provides useful information on practices of animal exploitation, as well as on palaeoenvironmental and sedimentological conditions (e.g. Tappen, 1994; Jans et al., 2004; Turner-Walker and Jans, 2008; Vacca, 2008).

Taxonomic abundance (see Chapters 3, 11, 19, 22, 32, 33, 38, 45, 46)
This method aims to measure the degree of taxonomic variability within an assemblage. Low levels of taxonomic abundance are typical of specialized practices of animal exploitation (e.g. agricultural communities focusing on few domesticates); on the contrary, high taxonomic variability characterizes broad-spectrum economies, as in the case of the Epipalaeolithic 'broad-spectrum revolution' (Flannery, 1969).

Taxonomic abundance, however, is usually biased (sometimes to a considerable extent) by taphonomic biases; differential preservation and incomplete recovery can exponentially reduce the number of taxa recorded within an assemblage, especially when sieving is not implemented (e.g. Payne, 1975; Zohar and Belmaker, 2005) (see **Recovery bias**).

Taxonomic identification (all chapters)
The attribution of an animal specimen to a taxon relies on morphological and biometrical observations of taxon-specific traits, based on the researcher's experience and comparisons with anatomical atlases and reference collections (see **Reference collection**).

Biometrical analyses can be employed to distinguish between taxa that are difficult to separate on morphological grounds (see **Biometry**). Species-level identification using aDNA can be fairly precise, although it is subjected to a number of practical limitations (see **Genetics**). Similarly, the analysis of peptide mass fingerprinting in bone collagen can provide reliable taxonomic identification (Buckley, 2009).

Textual evidence. See **Documentary and artistic sources**.

Tooth microwear (see Chapters 21, 43)
Microwear and mesowear patterns on teeth are analysed for reconstructing animal diet, which can provide information on past animal management and palaeoenvironmental conditions (e.g. Hogue, 2006; Vanpoucke et al., 2009; Kahlke and Kaiser, 2011). The results from faunal remains can be compared to experimental studies on modern animals (e.g. Mainland, 1998; 2003).

Management issues addressed by microwear studies include the distinction between grazing, free-ranging, and stall-fed animals (e.g. Mainland, 1998; Vanpoucke et al., 2009); differential feeding practices observed on samples from the same domestic population can be the result of the different use of animals deposited in distinctive archaeological contexts (e.g. Mainland and Halstead, 2005). Changing microwear patterns can also reflect the seasonality of hunting activities (e.g. Rivals and Deniaux, 2005).

Tooth wear. See **Ageing**.

Weight method (see Chapters 27, 28, 38, 41, 46)
This is a quantification method that measures relative abundance on the basis of weight rather than counts. It relies on raw weight itself or estimations of the potential edible output represented by faunal remains. For this latter version of the method, calculations are made by multiplying a factor, empirically derived from modern animals, by either the bone weight or the MNI of a specific taxon. Meat weight quantifications aim to interpret the relative

importance of food animals on the basis of their potential meat output rather than numerical frequency (Reitz and Wing, 1999).

The reliability of establishing a constant relationship between bone weight/MNI and meat weight has been widely debated (e.g. Casteel, 1978; Jackson, 1989; Vigne, 1991). Meat quantification implies a series of methodological biases and subjective assumptions about past practices of animal exploitation. Multiplying factors are obtained from modern animals: the considerable improvement and fattening of modern breeds do not make them an ideal proxy for past domestic types; at the same time, weight is not a fixed parameter and fluctuates according to age, sex, and seasonality (Stewart and Stahl, 1977). In addition, species-level identification is not always possible, leading to further increased error margins when dealing with higher taxonomic groups (e.g. Caprinae—sheep/goat, *Sus* sp.—pig/wild boar, *Bos* sp.—cattle/aurochs, Cervidae, etc.) (Vigne, 1991).

Meat weight quantification methods based on bone weight also overlook patterns of body part distribution: different elements are associated with different amounts of meat and, therefore, distributions skewed towards meat-bearing or non-meat-bearing elements imply sharply different meat weights. In addition, immature bones are lighter (not only smaller) than those of adult and elderly individuals, potentially leading to an underestimation of meat weight from young animals.

The meat and offal weight (MOW) method proposed by Vigne (1988) tries to overcome these methodological drawbacks by including age, sex, and biometrical information into the calculations. However, this also results in an accumulation of analytical biases and observer errors typical of such analyses; in addition, since it is based on the MNI, it carries the assumption that complete carcasses were introduced and consumed at the site (Vigne, 1991), which is not necessarily the case.

It is important to consider that the meat weight method relies on the assumption that carcasses were fully exploited for food, and therefore this cannot be applied to non-food animals or cases where consumption was incomplete.

Withers height (see Chapters 2, 26)

This is the height of an animal calculated from the withers (highest point of the shoulders) to the ground. In zooarchaeology, it is calculated by multiplying bone lengths by a factor derived from studies of modern animals (Reitz and Wing, 1999). Similarly to the scaling index technique (see **Scaling index technique**), it allows comparing different measurements, thus potentially amplifying the sample size of biometrical data (O'Connor, 2000); however, since complete lengths are rarely measurable on fragmented archaeological material, this potential is rarely fulfilled. It has been argued that the multiplying factors devised from modern animals cannot reliably account for the variety of morphometrical differences characterizing ancient faunal populations, especially when considering domestic breeds or geographical types (Forest, 1998). For these reasons, more recent research has placed less emphasis on the use of withers heights.

Zoogeography (see Chapters 2, 4, 9–11, 16, 18, 40, 44, 45)

The study of the geographic distribution of animal species is very important in zooarchaeology as this can change over time, as a consequence of the modification of environmental and/or cultural contexts. Zooarchaeology can contribute to reconstruct past ranges of certain species, both in terms of human-mediated movement of animals and natural expansion/shrinkage of animal populations (e.g. O'Connor and Sykes, 2010; Forcina et al., 2015).

Acknowledgements

We are grateful to Jane Ford for comments on an earlier draft.

References

Albarella, U. (2002) ' "Size matters": how and why biometry is still important in zooarchaeology', in Dobney, K. and O'Connor, T. (eds) *Bones and the Man: Studies in Honour of Don Brothwell*, pp. 51–62. Oxford: Oxbow Books.

Albarella, U. (2003) 'Tanners, tawyers, horn working and the mystery of the missing goat', in Murphy, P. and Wiltshire, P. (eds) *The Environmental Archaeology of Industry*, pp. 71–86. Oxford: Oxbow Books.

Albarella, U. and Davis, S. J. M. (1994) 'The Saxon and Medieval Animal Bones Excavated 1985–1989 from West Cotton, Northamptonshire'. Ancient Monuments Laboratory Report 17/94.

Albarella, U. and Davis, S. J. M. (1996) 'Mammals and birds from Launceston Castle, Cornwall: decline in status and the rise of agriculture', *Circaea*, 12(1), 1–156.

Albarella, U. and Payne, S. (2005) 'Neolithic pigs from Durrington Walls, Wiltshire, England: a biometrical database', *Journal of Archaeological Science*, 32(4), 589–99.

Albarella, U. and Trentacoste, A. (eds) (2011) *Ethnozooarchaeology: The Present and Past of Human-Animal Relationships*, Oxford: Oxbow Books.

Albarella, U., Dobney, K., and Rowley-Conwy, P. (2006) 'The domestication of the pig (*Sus scrofa*): new challenges and approaches', in Zeder, M. A., Bradley, D. G., Emshwiller, E., and Smith, B. D. (eds) *Documenting Domestication: New Genetic and Archaeological Paradigms*, pp. 209–27. Berkeley: University of California Press.

Albarella, U., Dobney, K., and Rowley-Conwy, P. (2009) 'Size and shape of the Eurasian wild boar (*Sus scrofa*), with a view to the reconstruction of its Holocene history', *Environmental Archaeology*, 14(2), 103–36.

Albarella, U., Johnstone, C., and Vickers, K. (2008) 'The development of animal husbandry from the late Iron Age to the end of the Roman period: a case study from south-east Britain', *Journal of Archaeological Science*, 35, 1828–48.

Armitage, P. (1982) 'A system for ageing and sexing the horn cores of cattle from British Post-Medieval sites (17th to early 18th century) with special reference to unimproved British longhorn cattle', in Wilson, B., Grigson, C., and Payne, S. (eds) *Ageing and Sexing Animal Bones from Archaeological Sites*. BAR British Series 109, pp. 37–54. Oxford: Archaeopress.

Baker, J. and Brothwell, D. (1980) *Animal Diseases in Archaeology*, London: Academic Press.

Balasse, M. (2002) 'Reconstructing dietary and environmental history from enamel isotopic analysis: time resolution of intra-tooth sequential sampling', *International Journal of Osteoarchaeology*, 12, 155–65.

Balasse, M., Boury, L., Ughetto-Monfrin, J., and Tresset, A. (2012) 'Stable isotope insights (δ^{18}O, δ^{13}C) into cattle and sheep husbandry at Bercy (Paris, France, 4th millennium BC): birth seasonality and winter leaf foddering', *Environmental Archaeology*, 17(1), 29–44.

Bartosiewicz, L. and Greenfield, H. J. (eds) (1999) *Transhumant Pastoralism in Southern Europe: Recent Perspectives from Archaeology, History and Ethnology*, Budapest: Archaeolingua.

Bartosiewicz, L. with Gál, E. (2013) *Shuffling Nags, Lame Ducks: The Archaeology of Animal Disease*, Oxford: Oxbow Books.

Bartosiewicz, L., Van Neer, W., Lentacker, A., and Fabiš, M. (1997) *Draught Cattle: Their Osteological Identification and History*. Annalen Zoölogische Wetenschappen 281. Tervuren: Koninklijk Museum voor Midden-Afrika.

Beja-Pereira, A., Caramelli, D., Lalueza-Fox, C., Vernesi, C., Ferrand, N., Casoli, A., Goyache, F., Royo, L. J., Conti, S., Lari, M., Martini, A., Ouragh, L., Magid, A., Atash, A., Zsolnai, A., Boscato, P., Triantaphylidis, C., Ploumi, K., Sineo, L., Mallegni, F., Taberlet, P., Erhardt, G., Sampietro, L., Bertranpetit, J., Barbujani, G., Luikart, G., and Bertorelle, G. (2006) 'The origin of European cattle: evidence from modern and ancient DNA', *Proceedings of the National Academy of Sciences of the United States of America*, 103(21), 8113–18.

Bell, M. and Walker, M. J. C. (2005) *Late Quaternary Environmental Change: Physical and Human Perspectives*, 2nd edn, Harlow: Pearson-Prentice Hall.

Bignon, O., Baylac, M., Vigne, J.-D., and Eisenmann, V. (2005) 'Geometric morphometrics and the population diversity of Late Glacial horses in Western Europe (Equus caballus arcelini): phylogeographic and anthropological implications', *Journal of Archaeological Science*, 32, 375–91.

Binford, L. R. (1978) *Nunamiut Ethnoarchaeology*, New York: Academic Press.

Binford, L. R. (1981) *Bones: Ancient Men and Modern Myths*, Orlando: Academic Press.

Binford, L. R. (1984) *Faunal Remains from Klasies River Mouth*, New York: Academic Press.

Bocherens, H. and Drucker, D. (2003) 'Trophic level isotopic enrichment of carbon and nitrogen in bone collagen: case studies from recent and ancient terrestrial ecosystems', *International Journal of Osteoarchaeology*, 13, 46–53.

Brain, C. (1967) 'Hottentot food remains and their bearing on the interpretation of fossil bone assemblages', *Scientific Papers of the Namib Desert Research Station*, 32, 1–11.

Brown, R. D. (ed.) (1983) *Antler Development in Cervidae*, Kingsville: Caesar Klebery Wildlife Research Institute.

Buckley, M., Collins, M., Thomas-Oates, J., and Wilson, J. C. (2009) 'Species identification by analysis of bone collagen using matrix-assisted laser desorption/ionisation time-of-flight mass spectrometry', *Rapid Communications in Mass Spectrometry*, 23(23), 3843–54.

Cai, D., Tang, Z., Yu, H., Han, L., Ren, X., Zhao, X., Zhu, H., and Zhou, H. (2011) 'Early history of Chinese domestic sheep indicated by ancient DNA analysis of Bronze Age individuals', *Journal of Archaeological Science*, 38, 896–902.

Casteel, R. W. (1978) 'Faunal assemblages and the "Wiegemethode" or weight method', *Journal of Field Archaeology*, 5, 71–7.

Cerón-Carrasco, R. (1998) 'Fishing: evidence for seasonality and processing of fish for preservation in the northern isles of Scotland during the Iron Age and Norse times', *Environmental Archaeology*, 3, 73–80.

Child, A. M. (1995) 'Towards an understanding of the microbial decomposition of archaeological bone in the burial environment', *Journal of Archaeological Science*, 22, 165–74.

Choyke, A. and O'Connor, S. (2013) *From These Bare Bones: Raw Materials and the Study of Worked Osseous Objects*, Oxford: Oxbow Books.

Clutton-Brock, T. H. (1981) 'The functions of antlers', *Behaviour*, 79(2/4), 108–25.

Colonese, A. C., Verdún-Castelló, E., Álvarez, M., Briz i Godino, I., Zurro, D., and Salvatelli, L. (2012) 'Oxygen isotopic composition of limpet shells from the Beagle Channel: implications for seasonal studies in shell middens of Tierra del Fuego', *Journal of Archaeological Science*, 39, 1738–48.

Currant, A. and Jacobi, R. (2001) 'A formal mammalian biostratigraphy for the late Pleistocene of Britain', *Quaternary Science Reviews*, 20, 1707–16.

Dammers, K. (2006) 'Using osteohistology for ageing and sexing', in Ruscillo, D. (ed.) *Recent Advances in Ageing and Sexing Animal Bones: Proceedings of the 9th Conference of the International Council of Archaeozoology, Durham, August 2002*, pp. 9–39. Oxford: Oxbow Books.

David, N. and Kramer, C. (2001) *Ethnoarchaeology in Action*, Cambridge: Cambridge University Press.

Davies, P. (2008) *Snails: Archaeology and Landscape Change*, Oxford: Oxbow Books.

Davis, S. J. M. (1981) 'The effects of temperature change and domestication on the body size of Late Pleistocene to Holocene mammals of Israel', *Paleobiology*, 7(1), 101–14.

Davis, S. J. M. (1982) 'A Trivariate Morphometric Method to Discriminate between First Phalanges of *Equus hydruntinus, asinus/hemionus*, and *caballus*', Tubingen meet 1982.

Davis, S. J. M. (1992) 'A Rapid Method for Recording Information about Mammal Bones from Archaeological Sites'. Ancient Monuments Laboratory Report 19/92.

Davis, S. J. M. (1996) 'Measurements of a group of adult female Shetland sheep skeletons from a single flock: a baseline for zooarchaeologists', *Journal of Archaeological Science*, 23, 593–612.

Davis, S. J. M. (2008) 'Zooarchaeological evidence for Moslem and Christian improvements of sheep and cattle in Portugal', *Journal of Archaeological Science*, 35, 991–1010.

Davis, S., Svensson, E., Albarella, U., Detry, C., Götherström, A., Pires, A. E., and Ginja, C. (2012) 'Molecular and osteometric sexing of cattle metacarpals: a case study from 15th century AD Beja, Portugal', *Journal of Archaeological Science*, 39, 1445–54.

De Cupere, B., Van Neer, W., Monchot, H., Rijmenants, E., Udrescu, M., and Waelkens, M. (2005) 'Ancient breeds of domestic fowl (*Gallus gallus* f. domestica) distinguished on the basis of traditional observations combined with mixture analysis', *Journal of Archaeological Science*, 32, 1587–97.

Dobney, K. and Rielly, K. (1988) 'A method for recording archaeological animal bones: the use of diagnostic zones', *Circaea*, 5(2), 79–96.

Dobney, K., Ervynck, A., Albarella, U., and Rowley-Conwy, P. (2007) 'The transition from wild boar to domestic pig in Eurasia, illustrated by a tooth developmental defect and biometrical data', in Albarella, U., Dobney, K., Ervynck, A., and Rowley-Conwy, P. (eds) *Pigs and Humans: 10,000 Years of Interaction*, pp. 57–82. Oxford: Oxford University Press.

Driesch, von den, A. (1976) *A Guide to the Measurement of Animal Bones from Archaeological Sites*, Harvard: Peabody Museum.

Driver, J. C. (1982) 'Medullary bone as an indicator of sex in bird remains from archaeological sites', in Wilson, B., Grigson, C., and Payne, S. (eds) *Ageing and Sexing Animal Bones from Archaeological Sites*. BAR British Series 109, pp. 251–4. Oxford: Archaeopress.

Ducos, P. (1968) *L'origine des animaux domestiques en Palestine*, Bordeux: Delmas.

Efremov, I. A. (1940) 'Taphonomy: a new branch of paleontology', *Pan-American Geologist*, 74, 81–93.

Enloe, J. G., David, F., and Hare, T. S. (1994) 'Patterns of faunal processing at Section 27 of Pincevent: the use of spatial analysis and ethnoarchaeological data in the interpretation of archaeological site structure', *Journal of Anthropological Archaeology*, 13, 105–24.

Evans, J. and O'Connor, T. (1999) *Environmental Archaeology: Principles and Methods*, Stroud: Sutton Publishing.

Evershed, R. P., Payne, S., Sherratt, A. G., Copley, M. S., Coolidge, J., Urem-Kotsu, D., Kotsakis, K., Özdogan, M., Özdogan, A. E., Nieuwenhuyse, O., Akkermans, P. M. M. G., Bailey, D., Andeescu, R.-R., Campbell, S., Farid, S., Hodder, I., Yalman, N., Özbasaran, M., Bıçakcı, E., Garfinkel, Y., Levy, T., and Burton, M. M. (2008) 'Earliest date for milk use in the Near East and southeastern Europe linked to cattle herding', *Nature*, 455, 528–31.

Ewbank, J. M., Phillipson, D. W., Whitehouse, R. D., and Higgs, E. S. (1964) 'Sheep in the Iron Age: a method of study', *Proceedings of the Prehistoric Society*, 30, 423–6.

Fernández-Jalvo, Y. and Andrews, P. (2003) 'Experimental effects of water abrasion on bone fragments', *Journal of Taphonomy*, 1(3), 147–63.

Flannery, K. V. (1969) 'Origins and ecological effects of early domestication in Iran and the Near East', in Ucko, P. J. and Dimbleby, G. W. (eds) *The Domestication and Exploitation of Plants and Animals*, pp. 73–100. London: Duckworth.

Forcina, G., Guerrini, M., van Grouw, H., Gupta, B. K., Panayides, P., Hadjigerou, P., Al-Sheikhly, O. F., Awan, M. N., Khan, A. A., Zeder, M. A., and Barbanera, F. (2015) 'Impacts of biological globalization in the Mediterranean: unveiling the deep history of human-mediated gamebird dispersal', *Proceedings of the National Academy of Sciences of the United States of America*, 112(11), 3296–301.

Forest, V. (1998) 'De la hauteur au garrot des espèces domestiques en archéozoologie', *Revue de Médecine Vétérinaire*, 149(1), 55–60.

Forest, V. and Rodet-Belarbi, I. (1998) 'Ostéométrie du métatarse des bovins en Gaule de la conquête romaine à l'Antiquité Tardive', *Revue de Médecine Vétérinaire*, 149(11), 1033–56.

Götherström, A., Anderung, C., Hellborg, L., Elburg, R., Smith, C., Bradley, D. J., and Ellegren, H. (2005) 'Cattle domestication in the Near East was followed by hybridisation with aurochs bulls in Europe', *Proceedings: Biological Sciences*, 272(1579), 2345–50.

Grant, A. (1982) 'The use of tooth wear as a guide to the age of domestic ungulates', in Wilson, B., Grigson, C., and Payne, S. (eds) *Ageing and Sexing Animal Bones from Archaeological Sites*. BAR British Series 109, pp. 91–108. Oxford: Archaeopress.

Grayson, D. K. (1984) *Quantitative Zooarchaeology: Topics in the Analysis of Archaeological Faunas*, Orlando: Academic Press.

Greenfield, H. J. (2006) 'Sexing fragmentary ungulate acetabulae', in Ruscillo, D. (ed.) *Recent Advances in Ageing and Sexing Animal Bones: Proceedings of the 9th Conference of the International Council of Archaeozoology, Durham, August 2002*, pp. 68–86. Oxford: Oxbow Books.

Grigson, C. (1982a) 'Sex and age determination of some bones and teeth of domestic cattle: a review of the literature', in Wilson, B., Grigson, C., and Payne, S. (eds) *Ageing and Sexing Animal Bones from Archaeological Sites*. BAR British Series 109, pp. 7–23. Oxford: Archaeopress.

Grigson, C. (1982b) 'Sexing Neolithic domestic cattle skulls and horncores', in Wilson, B., Grigson, C., and Payne, S. (eds) *Ageing and Sexing Animal Bones from Archaeological Sites*. BAR British Series 109, pp. 25–35. Oxford: Archaeopress.

Halstead, P. (2014) *Two Oxen Ahead: Pre-Mechanized Farming in the Mediterranean*, Chichester: Wiley-Blackwell.

Harcourt, R. A. (1974) 'The dog in prehistoric and early historic Britain', *Journal of Archaeological Science*, 1, 151–75.

Higgs, E. S. and Jarman, M. R. (1975) 'Palaeoeconomy', in Higgs, E. S. (ed.) *Palaeoeconomy*, pp. 1–7. Cambridge: Cambridge University Press.

Higham, T. F. G. and Horn, P. L. (2000) 'Seasonal dating using fish otoliths: results from the Shag River Mouth site, New Zealand', *Journal of Archaeological Science*, 27, 439–48.

Hill, M. E., Hill, M. G., and Widga, C. C. (2008) 'Late Quaternary *Bison* diminution on the Great Plains of North America: evaluating the role of human hunting versus climate change', *Quaternary Science Reviews*, 27, 1752–71.

Hogue, H. S. (2006) 'Carbon isotope and microwear analysis of dog burials: evidence for maize agriculture at a small Mississippian site', in Maltby, M. (ed.) *Integrating Zooarchaeology*, pp. 125–32. Oxford: Oxbow Books.

Jackson, H. E. (1989) 'The trouble with transformations: effects of sample size and sample composition on meat weight estimates based on skeletal mass allometry', *Journal of Archaeological Science*, 16, 601–10.

Jans, M. M. E., Nielsen-Marsh, C. M., Smith, C. I., Collins, M. J., and Kars, H. (2004) 'Characterisation of microbial attack on archaeological bone', *Journal of Archaeological Science*, 31, 87–95.

Johnstone, C. (2005) 'Those elusive mules: investigating osteometric methods for their identification', in Mashkour, M. (ed.) *Equids in Time and Space: Proceedings of the 9th Conference of the International Council of Archaeozoology, Durham, August 2002*, pp. 183–91. Oxford: Oxbow Books.

Jones, D. S. (1983) 'Sclerochronology: reading the record of the molluscan shell', *American Scientist*, 71(4), 384–91.

Jones, P. R. (1980) 'Experimental butchery with modern stone tools and its relevance for Palaeolithic archaeology', *World Archaeology*, 12(2), 153–65.

Jousse, H. and Escarguel, G. (2006) 'The use of Holocene bovid fossils to infer palaeoenvironment in Africa', *Quaternary Science Reviews*, 25, 763–83.

Kahlke, R.-D. and Kaiser, T. M. (2011) 'Generalism as a subsistence strategy: advantages and limitations of the highly flexible feeding traits of Pleistocene *Stephanorhinus hundsheimensis* (Rhinocerotidae, Mammalia)', *Quaternary Science Reviews*, 30(17–18), 2250–61.

Kahlke, R.-D., García, N., Kostopoulos, D. S., Lacombat, F., Lister, A. M., Mazza, P. P. A., Spassov, N., and Titov, V. V. (2011) 'Western Palaearctic palaeoenvironmental conditions during the Early and early Middle Pleistocene inferred from large mammal communities, and implications for hominin dispersal in Europe', *Quaternary Science Reviews*, 30 (11–12), 1368–95.

Lancelotti, C. and Madella, M. (2012) 'The 'invisible' product: developing markers for identifying dung in archaeological contexts', *Journal of Archaeological Science*, 39, 953–63.

Larson, G., Dobney, K., Albarella, U., Fang, M., Matisoo-Smith, E., Robins, J., Lowden, S., Finlayson, H., Brand, T., Willerslev, E., Rowley-Conwy, P., Andersson, L., and Cooper, A. (2005) 'Worldwide phylogeography of wild boar reveals multiple centres of pig domestication', *Science*, 307, 1618–21.

Lauwerier, R. C. G. M. (1988) *Animals in Roman Times in the Dutch Eastern River Area*, Amersfoort: Rijksdienst voor het Oudheidkundig Bodemonderzoek.

Lefèvre, C. (1997) 'Sea bird fowling in southern Patagonia: a contribution to understanding the nomadic round of the Canoeros Indians', *International Journal of Osteoarchaeology*, 7, 260–70.

Libois, R. M. and Hallet-Libois, C. (1988) *Éléments pour l'identification des restes crâniens des poissons dulçaquicoles de Belgique et du nord de la France. 2—Cypriniformes*. Fiches d'ostéologie animale pour l'archéologie, Série A: Poissons. Juan-les-Pins: APDCA.

Lloveras, Ll., Moreno-García, M., and Nadal, J. (2009) 'Butchery, cooking and human consumption marks on rabbit (*Oryctolagus cuniculus*) bones: an experimental study', *Journal of Taphonomy*, 7(2/3), 179–201.

Lupo, K. D. and Schmitt, D. N. (1997) 'Experiments in bone boiling: nutritional returns and archaeological reflections', *Anthropozoologica*, 25/26, 137–44.

Lyman, R. L. (1994) *Vertebrate Taphonomy*, Cambridge: Cambridge University Press.

Lyman, R. L. (2008) *Quantitative Palaeozoology*, Cambridge: Cambridge University Press.

MacKinnon, M. (2010) 'Cattle 'breed' variation and improvement in Roman Italy: connecting the zooarchaeological and ancient textual evidence', *World Archaeology*, 42(1), 55–73.

Mainland, I. and Halstead, P. (2005) 'The diet and management of domestic sheep and goat at Neolithic Makriyalos', in Davies, J., Fabiš, M., Mainland, I., Richards, M., and Thomas, R. (eds) *Diet and Health in Past Animal Populations: Current Research and Future Directions. Proceedings of the 9th ICAZ Conference, Durham 2002*, pp. 104–12. Oxford: Oxbow Books.

Mainland, I. L. (1998) 'The lamb's last supper: the role of dental microwear analysis in reconstructing livestock diet in the past', *Environmental Archaeology*, 1, 55–62.

Mainland, I. L. (2003) 'Dental microwear in grazing and browsing Gotland sheep (*Ovis aries*) and its implications for dietary reconstruction', *Journal of Archaeological Science*, 30, 1513–27.

Maldre, L. (2006) 'What did the Bronze Age dogs eat? Coprolithic analyses', *Dogs and People in Social, Working, Economic or Symbolic Interaction: Proceedings of the 9th Conference of the International Council of Archaeozoology, Durham, August 2002*, pp. 44–8. Oxford: Oxbow Books.

Maltby, M. (2007) 'Chop and change: specialist cattle carcass processing in Roman Britain', in Croxford, B., Ray, N., and Roth, R. (eds) *TRAC 2006: Proceedings of the 16th Annual Theoretical Roman Archaeology Conference*, pp. 59–76. Oxford: Oxbow Books.

Martin, R. A. (2014) 'A critique of vole clocks', *Quaternary Science Reviews*, 94, 1–6.

Meadow, R. H. (1981) 'Early animal domestication in South Asia: a first report of the faunal remains from Mehrgarh, Pakistan', in Härtel, H. (ed.) *South Asian Archaeology 1979*, pp. 143–79. Berlin: Dietrich Reimer Verlag.

Meadow, R. H. (1999) 'The use of size index scaling techniques for research on archaeozoological collections from the Middle East', in Becker, C., Manhart, H., Peters, J., and Schibler, J. (eds) *Historia Animalium ex Ossibus: Festschrift für Angela von den Driesch*, pp. 285–300. Rahden: Verlag Marie Leidorf GmbH.

Milne, C. and Crabtree, P. J. (2001) 'Prostitutes, a rabbi, and a carpenter: dinner at the Five Points in the 1830s', *Historical Archaeology*, 35(3), 31–48.

Montuire, S. (1999) 'Mammalian faunas as indicators of environmental and climatic changes in Spain during the Pliocene-Quaternary transition', *Quaternary Research*, 52, 129–37.

Mukherjee, A. J., Gibson, A. M., and Evershed, R. P. (2008) 'Trends in pig product processing at British Neolithic Grooved Ware sites traced through organic residues in potsherds', *Journal of Archaeological Science*, 35, 2059–73.

Munro, N. D., Bar-Oz, G., and Hill, A. C. (2011) 'An exploration of character traits and linear measurements for sexing mountain gazelle (*Gazella gazella*) skeletons', *Journal of Archaeological Science*, 38, 1253–65.

Nagaoka, L. (1994) 'Differential recovery of Pacific island fish remains: evidence from the Moturakau rockshelter, Aitutaki, Cook Islands', *Asian Perspectives*, 33(1), 1–17.

Navas, E., Esquivel, J. A., and Molina, F. (2008) 'Butchering patterns and spatial distribution of faunal animal remains consumed at the Los Millares Chalcolithic settlement (Santa Fe de Mondújar, Almería, Spain)', *Oxford Journal of Archaeology*, 27(4), 325–39.

Nicholson, R. A. (1996) 'Bone degradation, burial medium and species representation: debunking the myths, an experiment-based approach', *Journal of Archaeological Science*, 23, 513–33.

O'Connor, T. (1988) *Bones from the General Accident Site, Tanner Row*, York: Council for British Archaeology.

O'Connor, T. (2000) *The Archaeology of Animal Bones*, Stroud: Sutton Publishing.

O'Connor, T. and Sykes, N. (eds) (2010) *Extinctions and Invasions: A Social History of British Fauna*, Oxford: Oxbow Books.

Ottoni, C., Flink, L. G., Evin, A., Geörg, C., De Cupere, B., Van Neer, W., Bartosiewicz, L., Linderholm, A., Barnett, R., Peters, J., Decorte, R., Waelkens, M., Vanderheyden, N., Ricaut, F.-X., Hoelzel, A. R., Mashkour, M., Karimlu, A. F. M., Sheikhi Seno, S., Daujat, J., Brock, F., Pinhasi, R., Hongo, H., Perez-Enciso, M., Rasmussen, M., Frantz, L., Megens, H.-J., Crooijmans, R., Groenen, M., Arbuckle, B., Benecke, N., Viðarsdóttir, U. S., Burger, J., Cucchi, T., Dobney, K., and Larson, G. (2013) 'Pig domestication and human-mediated dispersal in western Eurasia revealed through ancient DNA and geometric morphometrics', *Molecular Biology and Evolution*, 30(4), 824–32.

Outram, A. K. (2001) 'A new approach to identifying bone marrow and grease exploitation: why the "indeterminate" fragments should not be ignored', *Journal of Archaeological Science*, 28, 401–10.

Payne, S. (1969) 'A metrical distinction between sheep and goat metacarpals', in Ucko, P. J. and Dimbleby, G. W. (eds) *The Domestication and Exploitation of Plants and Animals*, pp. 296–305. London: Gerald Duckworth & Co.

Payne, S. (1973) 'Kill-off patterns in sheep and goats: the mandibles from Aşvan Kale', *Anatolian Studies*, 23, 281–303.

Payne, S. (1975) 'Partial recovery and sample bias', in Clason, A. T. (ed.) *Archaeozoological Studies*, pp. 7–17. New York: North Holland.

Payne, S. and Bull, G. (1988) 'Components of variation in measurements of pig bones and teeth, and the use of measurements to distinguish wild from domestic pig remains', *Archaeozoologia*, 2, 27–65.

Pike-Tay, A. and Cosgrove, R. (2002) 'From reindeer to wallaby: recovering patterns of seasonality, mobility, and prey selection in the Palaeolithic Old World', *Journal of Archaeological Method and Theory*, 9(2), 101–46.

Popkin, P. R. W., Baker, P., Worley, F., Payne, S., and Hammon, A. (2012) 'The Sheep Project (1): determining skeletal growth, timing of epiphyseal fusion and morphometric variation in unimproved Shetland sheep of known age, sex, castration status and nutrition', *Journal of Archaeological Science*, 39, 1775–92.

Price, T. D., Schoeninger, M. J., and Armelagos, G. J. (1985) 'Bone chemistry and past behaviour: an overview', *Journal of Human Evolution*, 14(5), 419–48.

Prummel, W. and Frisch, H.-J. (1986) 'A guide for the distinction of species, sex and body side in bones of sheep and goat', *Journal of Archaeological Science*, 13, 567–77.

Rawlings, T. A. and Driver, J. C. (2010) 'Paleodiet of domestic turkey, Shields Pueblo (5MT3807), Colorado: isotopic analysis and its implications for care of a household domesticate', *Journal of Archaeological Science*, 37, 2433–41.

Reitz, E. J. and Wing, E. S. (1999) *Zooarchaeology*, Cambridge: Cambridge University Press.

Rijkelijkhuizen, M. (2011) 'Dutch Medieval bone and antler combs', in Baron, J. and Kufel-Diakowska (eds) *Written in Bones: Studies on Technological and Social Contexts of Past Faunal Skeletal Remains*, pp. 197–206. Wrocław: Institute of Archaeology, University of Wrocław.

Rivals, F. and Deniaux, B. (2005) 'Investigation of human hunting seasonality through dental microwear analysis of two Caprinae in late Pleistocene localities in southern France', *Journal of Archaeological Science*, 32, 1603–12.

Rowley-Conwy, P. (2001) 'Determination of the season of death in European Wild Boar (*Sus scrofa ferus*): a preliminary study', in Millard, A. (ed.) *Archaeological Sciences '97: Proceedings of the Conference Held at the University of Durham 2–4 September 1997*. BAR International Series 939. Oxford: Archaeopress.

Sadler, P. (1990) 'The use of tarsometatarsi in sexing and ageing domestic fowl (*Gallus gallus* L.), and recognising five toed breeds in archaeological material', *Circaea*, 8(1), 41–8.

Salque, M. (2012) 'Was milk processed in these ceramic pots? Organic residue analyses of European prehistoric cooking vessels', in Feulner, F., Doorn, N. L., and Leonardi, M. (eds) *May Contain Traces of Milk: Investigating the Role of Dairy Farming and Milk Consumption in the European Neolithic*, pp. 127–41. York: The University of York.

Schmid, E. (1972) *Atlas of Animal Bones for Prehistorians, Archaeologists and Quaternary Geologists*, Amsterdam: Elsevier.

Secor, D. H., Dean, J. M., and Campana, S. E. (eds) (1995) *Recent Developments in Fish Otolith Research*, Columbia: University of South Carolina Press.

Seetah, K. (2006) 'Multidisciplinary approach to Romano-British cattle butchery', in Maltby, M. (ed.) *Integrating Zooarchaeology*, pp. 111–18. Oxford: Oxbow Books.

Seetah, K. (2008) 'Modern analogy, cultural theory and experimental replication: a merging point at the cutting edge of archaeology', *World Archaeology*, 40(1), 135–50.

Serjeantson, D. (1991) ' "Rid grasse of bones": a taphonomic study of the bones from midden deposits at the Neolithic and Bronze Age site of Runnymede, Surrey, England', *International Journal of Osteoarchaeology*, 1(2), 73–89.

Shipman, P., Foster, G., and Schoeninger, M. (1984) 'Burnt bones and teeth: an experimental study of color, morphology, crystal structure and shrinkage', *Journal of Archaeological Science*, 11, 307–25.

Silver, I. A. (1969) 'The ageing of domestic animals', in Brothwell, D. and Higgs, E. (eds) *Science in Archaeology*, pp. 283–302. London: Thames and Hudson.

Simpson, G. G., Roe, A., and Lewontin, R. C. (1960) *Quantitative Zoology*, revised edn, New York: Harcourt, Brace, and World.

Soderberg, J. (2004) 'Wild cattle: red deer in the religious texts, iconography, and archaeology of Early Medieval Ireland', *International Journal of Historical Archaeology*, 8(3), 167–83.

Spangenberg, J. E., Jacomet, S., and Schibler, J. (2006) 'Chemical analyses of organic residues in archaeological pottery from Arbon Bleiche 3, Switzerland—evidence for dairying in the late Neolithic', *Journal of Archaeological Science*, 33, 1–13.

Stallibrass, S. (1982) 'The use of cement layers for absolute ageing of mammalian teeth: a selective review of the literature, with suggestions for further studies and alternative applications', in Wilson, B., Grigson, C., and Payne, S. (eds) *Ageing and Sexing Animal Bones from Archaeological Sites*. BAR British Series 109, pp. 109–26. Oxford: Archaeopress.

Stevens, R. E., Jacobi, R., Street, M., Germonpré, M., Conard, N. J., Münzel, S. C., and Hedges, R. E. M. (2008) 'Nitrogen isotope analyses of reindeer (*Rangifer tarandus*), 45,000 BP to 9,000 BP: palaeoenvironmental reconstructions', *Palaeogeography, Palaeoclimatology, Palaeoecology*, 262, 32–45.

Stewart, F. L. and Stahl, P. W. (1977) 'Cautionary note on edible meat poundage figures', *American Antiquity*, 42(2), 267–70.

Tappen, M. (1994) 'Bone weathering in the tropical rain forest', *Journal of Archaeological Science*, 21, 667–73.

Teichert, M. (1984) 'Size variation in cattle from Germania Romana & Germania Libera', in Grigson, C. and Clutton-Brock, J. (eds) *Animals and Archaeology: 4. Husbandry in Europe*. BAR International Series 227, pp. 93–103. Oxford: Archaeopress.

Telldahl, Y., Svensson, E. M., Götherström, A., and Storå, J. (2012) 'Osteometric and molecular sexing of cattle metapodia', *Journal of Archaeological Science*, 39, 121–7.

Turner-Walker, G. and Jans, M. (2008) 'Reconstructing taphonomic histories using histological analysis', *Palaeogeography, Palaeoclimatology, Palaeoecology*, 266, 227–35.

Uchiyama, J. (1999) 'Seasonality and age structure in an archaeological assemblage of Sika deer (*Cervus nippon*)', *International Journal of Osteoarchaeology*, 9, 209–18.

Uerpmann, H.-P. (1979) *Probleme der Neolithisierung des Mittelmeerraums*, Wiesbaden: Dr. Ludwig Reicher Verlag.

Vacca, B. (2008) 'Ambrosetti 1 (Sesto Fiorentino, Florence): an experimental study of burning damage of the faunal remains', in Baioni, M., Leonini, V., Lo Vetro, D., Martini, F., Poggiani Keller, R., and Sarti, L. (eds) *Bell Beaker in Everyday Life: Proceedings of the 10th Meeting 'Archéologie et Gobelets' (Florence—Siena—Villanuova sul Clisi, May 12–15, 2006)*. Millenni, Studi di Archeologia Preistorica 6, pp. 353–6. Firenze: Museo Fiorentino di Preistoria «Paolo Graziosi».

Vanpoucke, S., Mainland, I., De Cupere, B., and Waelkens, M. (2009) 'Dental microwear study of pigs from the classical site of Sagalassos (SW Turkey) as an aid for the reconstruction of husbandry practices in ancient times', *Environmental Archaeology*, 14(2), 137–54.

Vickers, K., Erlendsson, E., Church, M. J., Edwards, K. J., and Bending, J. (2011) '1000 years of environmental change and human impact at Stóra-Mörk, southern Iceland: a multiproxy study of a dynamic and vulnerable landscape', *The Holocene*, 21(6), 979–95.

Vigne, J.-D. (1988) *Les mammifères post-glaciaires de Corse: étude archéozoologique*. «Gallia Préhistoire» supplement 26. Paris: Éditions du CNRS.

Vigne, J.-D. (1991) 'The meat and offal weight (MOW) method and the relative proportion of ovicaprines in some ancient meat diets of the north-western Mediterranean', *Rivista di Studi Liguri*, 57(1–4), 21–47.

Viner, S., Evans, J., Albarella, U., and Parker Pearson, M. (2010) 'Cattle mobility in prehistoric Britain: strontium isotope analysis of cattle teeth from Durrington Walls (Wiltshire, Britain)', *Journal of Archaeological Science*, 37, 2812–20.

Wapnish, P. (1984) 'The dromedary and Bactrian camel in Levantine historical settings: the evidence from Tell Jemmeh', in Clutton-Brock, J. and Grigson, C. (eds) *Animals and Archaeology: 3. Early Herders and Their Flocks*. BAR International Series 202, pp. 171–200. Oxford: Archaeopress.

Watson, J. P. N. (1979) 'The estimation of the relative frequencies of mammalian species: Khirokitia 1972', *Journal of Archaeological Science*, 6, 127–37.

Weiner, S. (2010) *Microarchaeology: Beyond the Visible Archaeological Record*, Cambridge: Cambridge University Press.

West, N. (1982) 'Spur development: recognizing caponized fowl in archaeological material', in Wilson, B., Grigson, C., and Payne, S. (eds) *Ageing and Sexing Animal Bones from Archaeological Sites*. BAR British Series 109, pp. 255–61. Oxford: Archaeopress.

Wijngaarden-Bakker, van, L. H. (1987) 'Experimental zooarchaeology', in Groenman-van Waateringe, W. and van Wijngaarden-Bakker, L. H. (eds) *Farm Life in a Carolingian Village: A Model Based on Botanical and Zoological Data from an Excavated Site*. Studies in Prae- en Protohistorie 1, pp. 107–17. Assen: Van Gorcum.

Wilkinson, K. and Stevens, C. (2008) *Environmental Archaeology: Approaches, Techniques and Applications*, 2nd edn, Stroud: Tempus Publishing.

Zeder, M. A. (2006) 'A critical assessment of markers of initial domestication in goats (*Capra hircus*)', in Zeder, M. A., Bradley, D. G., Emshwiller, E., and Smith, B. D. (eds) *Documenting Domestication: New Genetic and Archaeological Paradigms*, pp. 181–208. Berkeley: University of California Press.

Zohar, I. and Belmaker, M. (2005) 'Size does matter: methodological comments on sieve size and species richness in fishbone assemblages', *Journal of Archaeological Science*, 32, 635–41.

Notes on Contributors

Umberto Albarella is a Professor in Zooarchaeology at the University of Sheffield (UK). He studied Natural Sciences at the University of Naples (Italy) and obtained his PhD from the University of Durham (UK). He has also worked at the Universities of Lecce (Italy), Birmingham (UK), and Durham (UK), as well as for English Heritage. His main areas of research include domestication, pastoralism, ethnography, husbandry innovations, and the integration of different strands of archaeological research. His work is predominantly based in Britain and Italy, but he has also worked in Armenia, Greece, the Netherlands, Germany, Switzerland, France, and Portugal. Within archaeology, he has been an advocate for global and social justice.

Melinda S. Allen is an Associate Professor at the University of Auckland, New Zealand. She also is Editor of the *Journal of the Polynesian Society*, and a former staff member and current Research Affiliate of the Bernice Pauahi Bishop Museum in Hawai'i. She has worked throughout the Pacific but has particular interests in East Polynesia, with major research projects in the southern Cook and Marquesas Islands. Her research focuses on Polynesian colonization and chronometrics, human ecodynamics, sustainability and resilience, and human–climate interactions. She is currently leading an interdisciplinary project on palaeoclimate change and effects on Polynesian marine fisheries with funding from the Royal Society of New Zealand (Marsden Fund).

Benjamin S. Arbuckle is an Associate Professor in the Department of Anthropology at the University of North Carolina at Chapel Hill. He is an archaeologist whose research focuses on the changing relationship between humans and animals in the prehistory and early history of the ancient Near East. He has published articles on Neolithic animal exploitation in journals including *PLoS ONE, The Levant, International Journal of Osteoarchaeology*, and the *Journal of Field Archaeology* and is co-editor of the book *Animals and Inequality in the Ancient World* (University Press of Colorado).

Joaquín Arroyo-Cabrales is a Senior Scientist at the Archaeozoology Laboratory in the Instituto Nacional de Antropología e Historia, the Mexican federal agency that takes care of the historical, archaeological, and palaeontological heritage. Here, he is currently in charge of the Paleontological Collection. He holds a BSc in Biology from the National Polytechnic Institute, and an MA in Museum Science and a PhD in Zoology focused on Quaternary vertebrate palaeontology, both from Texas Tech University. He is a member of the National Researchers System (Level II), has published over one hundred and twenty peer-reviewed papers (including fifty-three Web of Science indexed), and co-authored more than two hundred oral or poster presentations.

Levent Atici is an Associate Professor of Archaeology in the Department of Anthropology at the University of Nevada, Las Vegas (UNLV). After receiving his PhD from Harvard University, Atici joined UNLV in 2007 and founded the UNLV Zooarchaeology Laboratory in 2009. His research covers the full spectrum of human–animal interactions from hunting to specialized pastoralism in Southwest Asia. His current research program in Turkey comprises active field work at sites ranging from the Epipalaeolithic to the Bronze Age. Atici has published on Epipalaeolithic forager adaptations, food provisioning systems in early complex societies, and the relationships between food and ethnicity. Atici is committed to advancing zooarchaeology through fostering international collaboration, data sharing, and best practices.

Shaw Badenhorst is Senior Curator of the Archaeozoology and Large Mammal Section at the Ditsong National Museum of Natural History (former Transvaal Museum) in Pretoria, South Africa. He also serves as Head of the Vertebrate Department at the Museum, and is an Honorary Lecturer at the Department of Anthropology and Archaeology of the University of South Africa as well as Honorary Senior Research Fellow at the Evolutionary Studies Institute of the University of Witwatersrand. He obtained his PhD from Simon Fraser University in Canada on faunal remains from the American Southwest. His current research focuses on Pleistocene-Holocene archaeofaunas from southern Africa and is funded by the National Research Foundation of South Africa.

László Bartosiewicz obtained his degrees in animal science from the University of Gödöllő, Hungary. He has worked for over thirty-five years as an archaeozoologist mostly studying large mammals and fish, including research in Switzerland, Belgium, and the UK. His interests encompass animal–human relations in all post-Palaeolithic periods in Europe, the Near East, and South America. His work was published in four books and over three hundred scholarly articles. He has taught zooarchaeology at the Universities of Budapest and Edinburgh and served two consecutive terms (2006–2014) as president of the International Council for Archaeozoology.

Norbert Benecke is head of the Department of Natural Sciences at the German Archaeological Institute, Berlin, and a Honorary Professor at the Humboldt University. He has worked on faunal remains from many archaeological sites throughout Europe, the Near East, and Central and Southeast Asia, covering all periods from the Mesolithic to Medieval times. His main research interests include the exploitation of animal resources by prehistoric man in the different types of landscape of Eurasia, the history of animal domestication, and the environmental history of Europe (Late Glacial period and Holocene) with special emphasis on fauna.

Luis A. Borrero obtained his PhD from the Universidad de Buenos Aires, Argentina (1986) and a Honorary PhD from the Universidad de Magallanes, Chile (2005). He is currently a Professor at the Universidad de Buenos Aires and a Research Fellow for the National Research Council of Argentina (CONICET). He received the *Bernardo Houssay's Award to Scientific & Technological Research* (Argentina, 2004). He works in southern Patagonia and Tierra del Fuego, especially on the early exploration and colonization of South America. He recently published in *World Archaeology, Comechingonia, Quaternary International, Magallania*, and several book chapters.

Katherine Boyle is an archaeozoologist at the University of Cambridge and a Fellow of Homerton College. Her primary interests include the archaeozoological record for hunting during the Palaeolithic through Neolithic of Western Europe. Major publications on the Palaeolithic include *Upper Palaeolithic Faunas of South West France: A Zoogeographic Perspective* (1990), *The Middle Palaeolithic Geography of Southern France* (1998), and 'Rethinking the "ecological basis of social complexity"', in *The Upper Palaeolithic Revolution in Global Perspective: Essays in Honour of Paul Mellars* (2010). Current projects include analysis of pan-European evidence for hunting through the Neolithic, MIS3/2 biogeography, and landscape investigations (Palaeolithic through Medieval) along the banks of the River Granta (Cambridgeshire).

Seth Brewington is a Research Associate with the Department of Anthropology at Hunter College in New York. His research focuses primarily on past human–environment dynamics in the islands of the North Atlantic. In particular, he is interested in themes such as human impacts on biota and landscapes, island and coastal archaeology, human adaptation to climate change, and indigenous management of natural resources. Seth's current research includes ongoing collaboration in a multidisciplinary comparative study of social–ecological resilience in the Norse North Atlantic and prehistoric US Southwest.

Canan Çakirlar is an Assistant Professor at the Groningen University Institute of Archaeology in the Netherlands. She received her PhD from Tübingen University (2007) and held research positions at the Smithsonian Institution and the Belgian Institute of Natural Sciences (2007–2012). Canan's research interests include dispersals of animal husbandry, coastal adaptations, cultural resilience, and human impact on ecosystems. She has recently published widely on early herding and aquatic foraging in western Turkey. Canan is involved with fieldwork projects in Turkey, Iraq, Lebanon, Bulgaria, and the Netherlands. She is also an active promoter of the international standards of zooarchaeological research and teaching both in her native Turkey and her home, the Netherlands.

Louis Chaix is Emeritus Professor of Archaeozoology at the University of Geneva (Switzerland) and Honorary Curator of the Archaeozoological Department of the Natural History Museum of Geneva. He is also a member of the International Council of Archaeozoology (ICAZ) Committee of Honour. His significant publications include: Chaix, L. (1976) *La faune néolithique du Valais (Suisse): ses caractères et ses relations avec les faunes néolithiques des régions proches*, Document du Département d'Anthropologie et d'Écologie de l'Université de Genève 3, Geneva: Université de Genève, Département d'Anthropologie; Chaix, L. (1993) 'The archaeozoology of Kerma (Sudan)', in Davies, W. V. and Walker, R. (eds) *Biological Anthropology and the Study of Ancient Egypt*, pp. 175–85, London: British Museum Press; and Chaix, L. and Méniel, P. (2001) *Archéozoologie: les animaux et l'archéologie*, Paris: Éditions Errance. His research interests mainly concern the archaeozoology of Holocene Africa, the European Mesolithic, ritual activities, and malacology.

Eduardo Corona-M. is a Researcher at the Instituto Nacional de Antropología e Historia. His main research field is change and continuity in the human and animal relationship during the Quaternary, developing works on vertebrate palaeontology, archaeozoology, and ethnozoology. His most recent books are: *Clásicos de la etnobiología en México* (2012); *Saberes colectivos y diálogo de saberes en México* (2011); *Las aves del Cenozoico tardío de México: un análisis paleobiológico* (2009). He authored over one hundred articles, including contributions in *Arqueobios, International Journal of Osteoarchaeology, Current Research in the Pleistocene*, and *Journal of Paleontology*. Webpages: Seminario Relaciones Hombre-Fauna (http://sites.google.com/site/shofaun1/home) and http://inah.academia.edu/EduardoCoronaM.

Richard Cosgrove is a Reader in the Department of Archaeology and History at La Trobe University, Melbourne. Over the last thirty years, he has undertaken fieldwork in most Australian states and, internationally, in France, China, Jordan, and England. His research and teaching experience has been in Late Pleistocene human behavioural ecology, rock art studies, palaeoecology, zooarchaeology, and hunter-gatherer archaeology. His recent faunal research has focused on human prey animal exploitation during the period 40,000–10,000 BP in southwest Tasmania and southwest France. He has published widely on zooarchaeology, lithic analysis, human colonization and settlement of temperate and tropical rainforests, toxic food exploitation, and Aboriginal use of fire.

Simon J. M. Davis works in the archaeological science laboratory of the DGPC (Portuguese Ministry of Culture) in Lisbon. After studying zoology at University College, London, he did his PhD in Jerusalem on 'Late Pleistocene mammals' directed by the late Eitan Tchernov. His interests include developing ways to distinguish between wild and domesticated animals via their archaeological remains, the origin of domesticated animals and why they were domesticated, and using zooarchaeology to reconstruct the palaeoenvironment and to aid an understanding of economic history. Places worked in include England, Persia, Israel, Cyprus, Italy, Greece, and Portugal. He spends much time building up reference collections of modern vertebrates. Other interests include carpentry, cooking, and the music of the Hellenic and Muslim worlds.

Rebecca M. Dean (University of Minnesota, Morris) has a research focus on early agricultural societies, their impacts on local environments, and the integration of hunting in domestic economies. She has published on her work in the southwestern deserts of the United States, as well as on the earliest farming societies in the coastal areas of southern Portugal and Southwest Asia.

Jacopo De Grossi Mazzorin is Lecturer in Archaeozoology at the University of Salento (Lecce, Italy), Department of Cultural Heritage. His work has resulted in more than 180 scientific papers. He is involved in numerous Italian and international archaeological excavations in the Mediterranean area. He specializes in the study of faunal remains from archaeological sites and their cultural interpretation, mainly focusing on the Bronze Age, the Iron Age, and the Roman period. His research also includes the integration of different archaeological disciplines related to the relationship between humans and animals (economy, rituals, etc.). He is currently the President of the Italian Association of Archaeozoology (AIAZ).

Céline Dupont-Hébert is a PhD candidate in archaeology at Université Laval (Québec). Her thesis aims to reconstruct the landscape economy of the Icelandic estates from the Landnám to the Early Modern period through the study of zooarchaeological assemblages, settlement patterns, and social and climatic contexts. Since 2009, she is responsible for midden excavations in the *Archaeology of Settlement and Abandonment of Svalbarð* project in northeast Iceland.

Kitty F. Emery is an environmental archaeologist (Associate Curator, Florida Museum of Natural History/Associate Professor, Department of Anthropology, University of Florida), with a regional focus on Mesoamerica. She uses archaeological animal remains to evaluate climate, land use, and hunting practices as well as the role of animal resources in ancient communities and households. In broader studies she coordinates combined zooarchaeological, archaeobotanical, and geoarchaeological research to provide an ecological perspective on questions of Late Classic Maya life. These two types of study aim at a greater understanding of decision-making and sustainable living within the environmental context.

Frank J. Feeley is a graduate student at the CUNY Graduate Center in New York City. He is a zooarchaeologist who is interested in early commercial fishing in fifteenth-century Iceland. His thesis work focuses on the site of Gufuskálar in western Iceland, one of the few large commercial fishing stations dotting the coast of Iceland in the fifteenth century.

Jillian Garvey is an Australian Research Council DECRA Fellow specializing in Late Quaternary Indigenous archaeology. With a background in zoology and archaeology, her main research focus is on the role of native animals in Australian archaeology. She has integrated her zoological background on modern Australian vertebrates and invertebrates by conducting fatty acid nutritional analyses, economic utility or anatomy experiments, and butchery and cooking experiments. These modern experiments are combined with the zooarchaeology and ethnographic record to provide an interpretation of patterns in the archaeological record. With a PhD in palaeontology, Jillian is also interested in studying natural faunal assemblages, and what these can reveal about past palaeoenvironments and palaeoecology.

Mietje Germonpré has worked since 1989 at the Royal Belgian Institute of Natural Sciences. She visited Siberia for the first time in 1992, when she worked at the Denisova

Cave in the Altai and studied the Pleistocene mammal remains excavated during that field season. She returned to Siberia to do fieldwork in 1993 and 1994, this time in Buriatia; she visited Siberia again in 2014 (Yakutia) to look at Pleistocene canids stored at the Mammoth Museum in Yakutsk. Beside the visits to Siberia, she has had the opportunity to go many times to the European part of Russia (Saint-Petersburg, Bryansk area, Voronesh area). Over the last seven years she has studied the question of the early domestication of the wolf.

Diane Gifford-Gonzalez is Distinguished Professor of Anthropology at the University of California, Santa Cruz. She has published on the emergence of pastoralism in Africa, human–animal interactions in coastal California, and archaeological method and theory. Past President of the Society of Africanist Archaeologists, she was elected President of the Society for American Archaeology for 2015–2017. She likes animals and plants.

Paul Halstead has a BA and PhD in archaeology from Cambridge University and has taught zooarchaeology and later European prehistory since 1984 in the Department of Archaeology at the University of Sheffield. His research has focused primarily on Greek Neolithic and Bronze Age society and economy, including animal husbandry and consumption, and integrates faunal and textual evidence in the light of ethnographic studies of 'traditional' animal exploitation.

George Hambrecht is an Assistant Professor in the Anthropology Department at the University of Maryland, College Park. He is a zooarchaeologist interested in the intersection between human action and environmental change. He is also interested in the impacts that contemporary climate change is having on cultural heritage, especially given the potential for cultural heritage to produce data relevant to our current climate situation. Dr Hambrecht is currently working with the US National Parks Service and the United States Department of Agriculture, as well as an agriculture bioscience company, Acceligen, investigating these issues.

Ramona Harrison is Associate Professor of Archaeological Methods with specialization in Zooarchaeology at the Department of Archaeology, History, Cultural Studies and Religion, University of Bergen. She is an active member of NABO and has worked in the field in Iceland, Greenland, the Faroes, Austria, Hungary, Antigua, and the US. Her doctoral thesis was a study of the effects of international Medieval exchange on Northern Icelandic society. She is co-director of the Gásir Hinterlands Project (GHP), which most recently includes extensive research of the Viking Age site at Skuggi in northern Iceland. She also co-directs the Siglunes rescue and research excavation project as part of an investigation of the long-term human-ecodynamics of the Eyjafjörður Region (www.nabohome.org).

Adam R. Heinrich is currently an Adjunct Instructor in Anthropology at Monmouth University (New Jersey). His research interests focus on faunal and botanical foodways as well as material culture related to the development of mercantile networks. A recent contribution identified the reason for the appearance of cherub and related iconography on eighteenth-century grave markers found across the American colonies and in the British Isles (*International Journal of Historical Archaeology*, DOI: 10.1007/ s10761-013-0246-x).

Megan Hicks is a zooarchaeologist and a PhD candidate at the City University of New York. She is currently working toward her dissertation, which investigates long-term human and animal relationships and animal economies in the Mývatn region. Her work focuses on household-level economies, changing trade relationships, and local and traditional ecological knowledge relating to farming and wild resource use from the Viking Age through the Early Modern period. She has been involved in collaborative work—with the NABO and GHEA networks—in the Mývatn region of Iceland since 2008, including archaeological excavations, laboratory analysis, archival research, and education and outreach activities.

Charles F. W. Higham is a Research Professor in the Department of Anthropology and Archaeology, University of Otago. After graduating in prehistory at Cambridge University, he was appointed the foundation Professor at Otago in 1968, and began his fieldwork in Southeast Asia a year later. He has directed excavations at several key sites, with a particular emphasis on the millennia from the initial Neolithic settlement to the origins of the civilization of Angkor. He is a Fellow of the British Academy, the Royal Society of New Zealand, and an Honorary Fellow of St. Catharine's College, Cambridge.

Hitomi Hongo is an Associate Professor at the Department of Evolutionary Studies of Biosystems, Graduate University for Advanced Sciences (SOKENDAI). Her research interests include animal domestication, the Neolithic in Southwest Asia and East Asia, and the introduction of pigs and horses to Japan. She has been working on a number of excavations in Turkey, Jordan, Mongolia, and Japan. Her publications include: Hongo, H. and Auetrakulvit, P. (2011) 'Ethnozooarchaeology of the Mani (Orang Asli) of Trang Province, southern Thailand: a preliminary result of faunal analysis at Sakai Cave', in Albarella, U. and Trentacoste, A. (eds) *Ethnozooarchaeology: The Present and Past of Human–Animal Relationships*, pp. 82–9, Oxford: Oxbow Books; and Hongo, H., Omar, L., Nasu, H., and Fujii, S. (2014) 'Faunal remains from Wadi Abu Tulayha: a PPNB outpost in the steppe-desert of southern Jordan', in De Cupere, B., Linseele, V., and Hamilton-Dyer, S. (eds) *Archaeozoology of the Near East X*, pp. 1–25, Leuven: Peeters Publishers.

Salima Ikram is Distinguished University Professor of Egyptology at the American University in Cairo and has worked throughout Egypt on faunal assemblages as well as animal mummies, in addition to working as a mortuary archaeologist. Her research interests focus on the changing role that animals played in the diet and economy of ancient Egypt. She has published extensively on these and other topics.

Valasia Isaakidou studied for an MSc in Environmental Archaeology and Palaeoeconomy at the University of Sheffield between a BA in Classical Archaeology and a PhD on the zooarchaeology of Knossos at University College, London. She has explored human–animal relations in the Neolithic and Bronze Age Aegean through analyses of iconography, faunal remains, and worked bone and ethnographic study of 'traditional' animal husbandry and consumption.

Daniela Klokler received her Master's degree from the Universidade de São Paulo in 2001 and PhD from the University of Arizona in 2008. She is currently an Associate Professor at the Department of Archaeology of the Universidade Federal de Sergipe in Brazil. Her research interests focus on the relationships between humans and animals, shell-mound archaeology, formation processes, and ritual practices. Her publications include co-editing the volume *The Cultural Dynamics of Shell Middens and Shell Mounds: A Worldwide Perspective* (2014).

Heather A. Lapham (Research Archaeologist, Research Laboratories of Achaeology, University of North Carolina at Chapel Hill) has been conducting zooarchaeological research in eastern North America for nearly twenty years. Her current research examines differential animal resource use and social aspects of Early Historic-period Native American and Spanish economies in the American Southeast, along with urban animal economies among the Classic-period Zapotec in Oaxaca, Mexico. She is the author of the book *Hunting for Hides*, which explores cultural change at a seventeenth century Native American settlement in southwestern Virginia, and more than a dozen other book chapters and journal articles.

Matthew Leavesley is a Senior Lecturer in Archaeology at the University of Papua New Guinea (PNG) and an Adjunct Senior Research Fellow at James Cook University (Australia). His research interests revolve around notions of prehistoric human adaptation(s) to depauperate/marginal environments with particular reference to case studies in PNG.

Justin E. Lev-Tov was trained in zooarchaeology at the Smithsonian Institution under Melinda Zeder, and at the University of Tennessee under Walter Klippel and Paul Parmalee, where he received his PhD (2000). His research focuses on the intersection between complex societies and the social implications of dietary choices, mainly concerning the eastern Mediterranean region. Temporally, his focus is on the Bronze and Iron Ages, but sometimes ranges as recent as the Roman era. When not physically in the Mediterranean region, or researching the above themes, Justin is an archaeologist at Cogstone Resource Management, Inc., an archaeology and heritage firm located in California.

Veerle Linseele is a post-doctoral Research Fellow of the FWO-Flanders, affiliated with the Center for Archaeological Sciences at the KU Leuven (Belgium) and with the Royal Belgian Institute of Natural Sciences. She is also part-time Lecturer in Environmental Archaeology at Ghent University (Belgium). Her research focuses on early food production and early complex societies in western and northeastern Africa. She analyses bone remains of all groups of vertebrates, but mainly fish and mammals.

Li Liu (PhD in Anthropology from Harvard University) is the Sir Robert Ho Tung Professor in Chinese Archaeology in the Department of East Asian Languages and

Cultures at Stanford University, USA. Previously she taught archaeology at La Trobe University in Melbourne, Australia, for fourteen years and was elected as Fellow of the Academy of Humanities in Australia. Her research interests include archaeology of early China (Neolithic and Bronze Age), domestication of plants and animals in China, development of complex societies and state formation, settlement archaeology, use-wear analysis of stone tools, and starch analysis.

Lembi Lõugas, educated in biology-zoology at Tartu University, Estonia, has worked in zooarchaeology since 1991. From the year 1991, she has held a research position at the Institute of History, Tallinn University, and from 1997 she has led the Department of Archaeobiology and Ancient Technology (the structure unit was renamed the Tallinn University Archaeological Research Collection in 2015). Lembi has contributed to collaborative projects in Russia, Latvia, Sweden, Finland, Germany, Belgium, France, and the UK. Her main interests include the history of aquatic fauna (fish and seals) and their environments, as well as past relationships between animals, as investigated though palaeogenetics. She has been a member of the International Council of Archaeozoology (ICAZ) since 1994 and has actively participated to the activities of several ICAZ working groups. In 2014 she was also elected to the ICAZ International Committee.

Xiaolin Ma is the deputy director of the Henan Provincial Administration of Cultural Heritage, China. He obtained a PhD at the Department of Archaeology of La Trobe University, Australia, in 2004. He has established a large zooarchaeological laboratory including both ancient faunal remains and modern comparative materials. He has directed archaeological excavations of several Neolithic sites, which have provided important insights in Chinese Neolithic archaeology. He has launched dozens of archaeological programs and published over fifty papers, as well as authored or co-authored several books related to Neolithic archaeology and zooarchaeology. Xiaolin is a member of the International Committee (IC) of the International Council of Archaeozoology (ICAZ), and a visiting scholar at the Universities of Michigan and Harvard (USA). He organized two international conferences of zooarchaeology in the Henan province in 2007 and 2013.

Michael MacKinnon is Professor of Classics at the University of Winnipeg. His research interests encompass interdisciplinary investigation of animals in Greek and Roman contexts, principally throughout the Mediterranean, where he has worked at more than fifty different archaeological projects. He is the author of 'State of the Discipline: Osteological Research in Classical Archaeology' (*American Journal of Archaeology*, 2007) and *Production and Consumption of Animals in Roman Italy: Integrating the Zooarchaeological and Textual Evidence* (2004).

Mark Maltby is Professor of Archaeology at Bournemouth University, UK. He has been studying animal bones since 1974 and has worked on a large range of assemblages from prehistoric and historic periods in England. He has also been studying material from Novgorod and other sites in northeastern Europe since 1994. He is particularly interested in the role of towns in animal exploitation systems.

Finbar McCormick is a Senior Lecturer in Queen's University Belfast, with a special interest in early settlement and livestock history. Much of his work is concerned with the study of faunal assemblages from early sites in Scotland and Ireland. He is co-author of *Early Medieval Ireland 400–1200: The Evidence from Archaeological Excavations* (2014).

Thomas H. McGovern directs the Hunter College CUNY Zooarchaeology Laboratory and has served as coordinator for the North Atlantic Biocultural Organization (NABO, www.nabohome.org) since 1992. His doctoral thesis (Columbia U. 1979) was on the zooarchaeology of Norse Greenland and he has done fieldwork in the US, Greenland, Iceland, Scotland, Faroes, Norway, France, England, and Barbuda in the Caribbean. He is on the science advisory panel of the IHOPE group (Integrated History and Future of People on Earth, http://ihopenet.org/) and participates in the Circumpolar Networks, Islands, and Threats to Heritage and the Distributed Observing Network of the Past IHOPE teams.

Richard H. Meadow has been undertaking zooarchaeological research for more than forty years. He has analysed collections from sites in Iran, Pakistan, Thailand, Syria, China, and the USA. In northwestern South Asia he has studied animal remains from and carried out archaeological fieldwork at Balakot and Mehrgarh in Balochistan and Harappa in Punjab, as well as in Iran at Tepe Yahya. He directs the Zooarchaeology Laboratory of the Peabody Museum, Harvard University, where he has overseen the establishment of a large comparative collection including specimens of domestic river buffalo and zebu cattle. His academic interests include the development of agriculture and animal husbandry and the Indus civilization.

Guillermo L. Mengoni Goñalons graduated from the Universidad de Buenos Aires, UBA (Argentina). He is full Professor at the Facultad de Filosofía y Letras (FFyL-UBA) and Scientific Researcher from the Consejo Nacional de Investigaciones Científicas y Tecnológicas (Argentina). Since 2008 he is Director of the Instituto de Arqueología (FFyL-UBA). His research interests include camelid zooarchaeology, archaeology and history of the indigenous peoples of Patagonia, and archaeological practice and community.

Claudia Minniti is currently a Research Associate at the University of Salento (Lecce, Italy), Dipartimento di Beni Culturali. She was until recently staff member at the University of Sheffield (UK) as a Marie Curie Fellow. Her research interests include the study of animal remains from archaeological sites with integration of various disciplines (e.g. history, ethnography, zoology, and ethology). Her work has resulted in more than ninety papers in national and international journals and in conference proceedings and books, including her recent book *Hunting, Animal Exploitation and Social Complexity in Central Italy between the Middle Bronze Age and Iron Age*, published by Archaeopress in the British Archaeological Reports (BAR) International Series.

Gregory G. Monks received his PhD in 1977 from the University of British Columbia and is now Professor and Head of the Anthropology Department at the University of Manitoba in Winnipeg, Canada. His research time is divided unequally between the archaeology of the northwest coast of North America and fur trade archaeology of western Canada. His research interest focuses primarily on zooarchaeology in each of these areas. Within that research focus, the topics of seasonality, fish, whales, and climate change continue to be major themes. Application of theory and advanced analytic methods also underlie his work.

Emily Murray is a Research Fellow in Queen's University Belfast. She has a particular interest in the Medieval period and recent excavations she has directed include the site of a Medieval monastery in Derry city centre and the church and well complex at Struell Wells in Co. Down. Her zooarchaeological research has also focused on the Medieval period, with a particular interest in the exploitation of marine species. With Finbar McCormick she co-wrote the principal reference work on Irish zooarchaeology, *Knowth and the Zooarchaeology of Early Christian Ireland* (2007).

Terry O'Connor studied zooarchaeology in London with I. W. Cornwall and D. R. Brothwell, with a particular focus on biometric variation, and went on to work closely with Lincoln and York Archaeological Trusts. Terry has held research and lecturing positions at the Universities of York and Bradford, UK. He has published extensively on Medieval zooarchaeology and late Upper Palaeolithic zooarchaeology of northern England. He is particularly interested in the archaeology of human–animal relations such as domestication and in the taphonomy of bone assemblages.

Ajita K. Patel has been carrying out zooarchaeological research on South Asia and the Middle East for more than twenty-five years. She has worked extensively at archaeological sites or on collections in India, France, Turkey, Syria, and China. In northwestern South Asia she has excavated at and analysed faunal material from Nagwada, Loteshwar, and Santhli in North Gujarat, and Dholavira in Kutch, among others. She is based at Harvard University, where she teaches and carries out research in the Zooarchaeology Laboratory of the Peabody Museum. She is interested in the changing effects of human–animal relations in the past, including domestication, pastoralism, and urban economies. Her current projects include the comparative osteology of water buffalo and zebu cattle, the study of the ancient water buffalo of China, and the analysis of faunal remains from Tell Leilan, Syria.

Tanya M. Peres received a Doctorate in Anthropology from the University of Florida in 2001. She is a member of the Anthropology faculty at Florida State University. Her research interests include ethnozooarchaeology, the use of aquatic environments, shell-bearing sites of the Americas, and non-food uses of animals. Publications include *Integrating Zooarchaeology and Paleoethnobotany: A Consideration of Issues, Methods, and Cases*, co-edited with Amber VanDerwarker, the edited volume *Trends and Traditions in Southeastern Zooarchaeology*, and numerous journal articles, book chapters, and technical reports.

Joris Peters studied Biology/Zoology at the University of Ghent, Belgium. He obtained his PhD in Natural Sciences (Archaeozoology) from the University of Ghent, and a

Habilitation in Palaeoanatomy, Domestication Research, and History of Veterinary Medicine from the LMU Munich (Veterinary Faculty). Since 2000, he holds the Chair of Palaeoanatomy, Domestication Research, and History of Veterinary Medicine at the LMU Munich and is the director of the Palaeoanatomy Department of the State Collection of Anthropology and Palaeoanatomy, Munich. He is also laureate of the Royal Academy of Sciences, Literature and Arts, Belgium, and Corresponding Member of the German Archaeological Institute. His main research interests include human–animal–environment relationships in prehistoric times, animal domestication, animal husbandry and breeding in antiquity, and Medieval hippiatry.

Ina Plug was born in the Netherlands but emigrated to South Africa after WWII. She graduated from the University of South Africa and continued her postgraduate studies at the University of Pretoria, where she obtained her doctorate in Archaeology in 1988. She worked at the Transvaal Museum (the Ditsong National Museum of Natural History) from 1976 and retired in 1999 as Deputy Director. She has over 130 publications, including books, book chapters, and journal articles. She works mainly on faunas from southern African sites and has collaborated with colleagues from Africa, Europe, and the USA.

Nadja Pöllath studied Prehistoric Archaeology, Archaeology of the Roman Provinces, and Zooarchaeology in Munich and Kiel, Germany, where she obtained her PhD in Humanities (Prehistoric Archaeology). Since 2007 she is scientific assistant at the Institute of Palaeoanatomy and History of Veterinary History, LMU Munich, and since 2010 she is also in charge of zooarchaeological analyses as part of the long-term project 'The prehistoric societies of Upper Mesopotamia and their subsistence', funded by the German Science Foundation (DFG). Her main research interests include the archaeozoology of Bavaria, northeastern Africa, and the Near East, animal domestication, and ichthyoarchaeology.

Brenda Prehal is an archaeologist and PhD student at the City University of New York. She is currently researching for her dissertation, which focuses on pagan Icelandic burial practices. Her work involves an interdisciplinary approach to understanding mortuary ritual and identity by combining archaeology, Medieval literature studies, hard sciences such as aDNA, and environmental sciences. Since 2010, she has been doing fieldwork in Iceland and collaborating on projects of the NABO and GHEA networks.

Elizabeth J. Reitz received her PhD in 1979 from the University of Florida and is Professor of Anthropology at the University of Georgia, where she manages the zooarchaeology comparative collection and research laboratory. Her research focuses on interpreting animal remains from coastal archaeological sites dating from the Late Pleistocene into the twentieth century throughout the Americas. She has studied collections from Peru, Ecuador, Nicaragua, Jamaica, Montserrat, Barbuda, Haiti, the Dominican Republic, and the southeastern United States. She has authored or co-authored over one hundred books and articles on topics ranging from climate change to symbolism in the archaeofaunal record.

Mauro Rizzetto is a PhD student at the University of Sheffield. His research concerns the development of animal husbandry during the Late Roman–Early Medieval transition in Britain and in the lower Rhine region, with particular regards to biometrical changes. He obtained his undergraduate degree in Archaeological Science (2013) and a Master's degree in Osteoarchaeology (2015) at the same university. He has been working at a number of archaeological sites in Italy, Britain, France, Greece, and Spain, dating from the Neolithic to the post-Medieval period.

Hannah Russ is a freelance zooarchaeologist working with archaeology.biz, based in Barnard Castle, and Honorary Research Fellow at the Universities of Sheffield and Wales, Trinity Saint David. She is a zooarchaeologist specializing in the study of aquatic animals including fish, molluscs, and crustaceans. She has worked on remains from five UNESCO World Heritage sites, as well as other sites in Western Europe and the Middle East dating from the Upper Palaeolithic through to the Post-Medieval period. Hannah completed her undergraduate degree in Bioarchaeology (2004), a Master's degree in Biological Archaeology (2006), and her PhD in Archaeological Sciences (2011) at the University of Bradford. After completing her PhD Hannah held positions at the University of Sheffield and Oxford Brookes University before working as a post-excavation manager at Northern Archaeological Associates.

Mikhail V. Sablin works at the Zoological Institute of the Russian Academy of Sciences in Saint-Petersburg. He has a special interest in the domestication of the wolf and has worked on the fauna of the Epigravettian mammoth sites in Russian, as well as on the archaeozoology of the Black Sea Greek settlements.

Jörg Schibler studied Prehistory and Biology at the University of Basel and in 1981 completed his PhD thesis about the archaeozoological, typological, and chronological contents of more than 8,000 Neolithic bone artefacts from the lake shore site of Twann, at the lake of Bienne. After working as an archaeologist in the archaeological services of Berne and Solothurn he returned to the University of Basel where he worked on Neolithic animal bone assemblages from several sites. At the university of Basel he built up a research group in archaeozoology and got a permanent teaching position in 1988. In 1995 he was awarded a Professorship in Archaeozoology and Prehistory. Beside Basel he has also been a Lecturer at the Universities of Berne, Heidelberg and Frankfurt/M.

Dale Serjeantson is an Honorary Research Fellow in archaeology at the University of Southampton. She has researched and taught zooarchaeology since 1980, first at Birkbeck College, University of London, and then at the University of Southampton, where she was the English Heritage Research Fellow in Zooarchaeology from 1991 to 2000. Her research interests are varied: they include the social and economic roles of birds in archaeology; mammals, fish and birds from coastal sites in Scotland; husbandry and the social importance of animals in prehistoric Britain; and food as a marker of rank and status in the Middle Ages.

Konrad Smiarowski is a zooarchaeologist and a doctoral candidate at the City University of New York Graduate Center. His research interests focus on past human ecodynamics in the northern landscapes, especially where challenging climatic conditions influenced significant social response and societal reorganization. His main research area is in Norse Greenland, where he conducted multidisciplinary NABO field projects in 2006–2016, and studied the complex use of wild terrestrial and marine resources, as well as climate effects on animal husbandry practices. As an active member of NABO and GHEA Konrad works with other researchers on projects in the North Atlantic, Iceland, Norway, Poland, the United States, and the Caribbean.

Ian W. G. Smith is an Associate Professor in the Department of Anthropology and Archaeology at the University of Otago (Dunedin), where he undertakes research in both prehistoric and historical archaeology of New Zealand. His primary research interests are in the impacts of human colonization on the environment, and in the interactions of cultures during colonial contact. Significant publications include *Shag River Mouth: The Archaeology of an Early Southern Maori Village*, and *Archaeology of the Hohi Mission Station*.

John D. Speth is Arthur F. Thurnau Professor (Emeritus) of Anthropology in the Department of Anthropology, University of Michigan (Ann Arbor). He studies hunter-gatherers, past and present, New World and Old World. He is interested in the evolution of forager diet and subsistence strategies, and the ways that hunter-gatherers cope with seasonal and inter-annual resource unpredictability. His most recent books include *The Paleoanthropology and Archaeology of Big-Game Hunting: Protein, Fat, or Politics?* (2010, Springer); and J. L. Clark and J. D. Speth (eds) *Zooarchaeology and Modern Human Origins: Human Hunting Behavior During the Later Pleistocene* (2013, Springer).

Peter W. Stahl is an archaeologist interested in South American archaeology, zooarchaeology, vertebrate taphonomy, historical ecology, tropical forest ecology, and lowland South American ethnography. His research focus is on the lowland neotropics, principally in western Ecuador. He has studied zooarchaeological assemblages from Ecuador, Colombia, Mexico, USA, Ghana, and Iraq at major research collections in Europe and North America. He received his PhD in 1984 from the University of Illinois at Urbana-Champaign, and is currently Professor (Limited Term) of Anthropology, University of Victoria (Canada), and Emeritus Professor of Anthropology, Binghamton University (State University of New York, USA).

Kim Vickers is a Honorary Research Associate in the Department of Archaeology at the University of Sheffield and specializes in the reconstruction of ancient environments and zooarchaeology. Her research interests include the Iron Age–Roman transition in Britain and the effects of the Roman invasion of Britain on farming practices and animal husbandry in the early first millennium AD. She has also worked extensively in Iceland,

Greenland, and the Faroe Islands. After completing her PhD on the palaeoentomology of the North Atlantic islands in 2007, her research has focused on the environmental impact of Medieval human settlement and activity in Iceland, Greenland, and the Faroe islands, and on the nature of resource use and contact between Norse and Inuit cultures in Greenland. She has also undertaken an extensive study of the biogeography of the North Atlantic beetle fauna.

Jean-Denis Vigne is an archaeozoologist at the CNRS, working in the National Museum of Natural History, in the laboratory 'Archaeozoology, archaeobotany: societies, practices and environments'. He directed this lab from 2002 to 2013 and is now leading the French network of bioarchaeology, and the Laboratory of Excellence entitled 'Natural and cultural diversities: origins, evolution, interactions, future'. He has led archaeological excavations in Corsica and Cyprus, including now the PPNA site of Klimonas. He has analysed numerous Mesolithic, Neolithic, and Bronze Age faunal assemblages, and developed numerous research projects about the impact of anthropization, the origins of commensalism, and the beginning of animal domestication and husbandry in the Mediterranean and Central Asia.

Sarah Viner-Daniels studied as an undergraduate at the University of Liverpool (UK) and completed her Masters and PhD at the University of Sheffield (UK). She was then appointed as a Research Associate to the Feeding Stonehenge project, still at the University of Sheffield. Sarah's main areas of interest include animal exploitation in the Mesolithic and Neolithic of Britain and the application of isotopic analysis (Strontium and Oxygen) to the understanding of prehistoric livestock mobility.

Sarah Whitcher Kansa directs the non-profit Alexandria Archive Institute, working with researchers to publish data sets in the open access data sharing platform, Open Context. Sarah received her PhD in archaeology from the University of Edinburgh in 2000. She has conducted zooarchaeological research for fifteen years at prehistoric and early historic sites in the Near East. She is currently studying fifty years of faunal remains in continuing excavations at the Etruscan site of Poggio Civitate (Italy).

James Woollett specializes in areas including the archaeology of the Arctic, the North Atlantic, and prehistoric northern North America. In zooarchaeology, his expertises include reconstructions of seasonality, palaeodemography, and life histories through analyses of periodic growth structures. His recent work is focused on regional-scale interdisciplinary studies of landscape history, settlement, and subsistence practices of Inuit seal hunters in northern Labrador, Canada, and of communities practising mixed sheep herding, fishing, and hunting economies in northeast Iceland. He is a faculty member of the Département des sciences historiques, Université Laval (Québec), and a regular member of the *Centre d'études Nordiques*.

Index

....

Tables and figures are indicated by an italic *t* and *f* following the page number.